# Lecture Notes in Computer Science 3992

*Commenced Publication in 1973*
Founding and Former Series Editors:
Gerhard Goos, Juris Hartmanis, and Jan van Leeuwen

Vassil N. Alexandrov
Geert Dick van Albada  Peter M.A. Sloot
Jack Dongarra (Eds.)

# Computational Science – ICCS 2006

6th International Conference
Reading, UK, May 28-31, 2006
Proceedings, Part II



Springer

Volume Editors

Vassil N. Alexandrov
University of Reading
Centre for Advanced Computing and Emerging Technologies
Reading RG6 6AY, UK
E-mail: v.n.alexandrov@rdg.ac.uk

Geert Dick van Albada
Peter M.A. Sloot
University of Amsterdam
Department of Mathematics and Computer Science
Kruislaan 403, 1098 SJ Amsterdam, The Netherlands
E-mail: {dick,sloot}@science.uva.nl

Jack Dongarra
University of Tennessee
Computer Science Department
1122 Volunteer Blvd., Knoxville, TN 37996-3450, USA
E-mail: dongarra@cs.utk.edu

Library of Congress Control Number: 2006926429

CR Subject Classification (1998): F, D, G, H, I, J, C.2-3

LNCS Sublibrary: SL 1 – Theoretical Computer Science and General Issues

ISSN 0302-9743
ISBN-10 3-540-34381-4 Springer Berlin Heidelberg New York
ISBN-13 978-3-540-34381-3 Springer Berlin Heidelberg New York



Springer is a part of Springer Science+Business Media

springer.com


Printed in Germany

Typesetting: Camera-ready by author, data conversion by Scientific Publishing Services, Chennai, India
Printed on acid-free paper SPIN: 11758525 06/3142 5 4 3 2 1 0

# Preface

The Sixth International Conference on Computational Science (ICCS 2006) was held in Reading, United Kingdom, May 28-31 and continued the traditions of previous conferences in the series: ICCS 2005 in Atlanta, Georgia, USA; ICCS 2004 in Krakow, Poland; ICCS 2003 held simultaneously at two locations in, Melbourne, Australia and St. Petersburg, Russia; ICCS 2002 in Amsterdam, The Netherlands; and ICCS 2001 in San Francisco, California, USA.

Since the first conference in San Francisco, rapid developments in Computational Science as a mainstream area facilitating multi-disciplinary research essential for the advancement of science have been observed. The theme of ICCS 2006 was "Advancing Science through Computation", marking several decades of progress in Computational Science theory and practice, leading to greatly improved applications science. The conference focused on the following major themes: tackling Grand Challenges Problems; modelling and simulations of complex systems; scalable algorithms and tools and environments for Computational Science. Of particular interest were the following major recent developments in novel methods and modelling of complex systems for diverse areas of science, scalable scientific algorithms, advanced software tools, computational grids, advanced numerical methods, and novel application areas where the above novel models, algorithms and tools can be efficiently applied such as physical systems, computational and systems biology, environmental systems, finance, and others.

Keynote lectures were delivered by Mateo Valero (Director, Barcelona Supercomputing Centre) - "Tackling Grand Challenges Problems"; Chris Johnson (Distinguished Professor, University of Utah) - "Visualizing the Future"; José Moreira (IBM, Chief Architect, Commercial Scale Out) - "Achieving Breakthrough Science with the Blue Gene/L Supercomputer"; Martin Curley (INTEL, Global Director of Innovation and IT Research) - "IT Innovation: A New Era"; Vaidy Sunderam (Samuel Candler Dobbs Professor of Computer Science, Emory University, USA) - "Metacomputing Revisited: Alternative Paradigms for Distributed Resource Sharing"; and Ron Bell (AWE plc.) - "The AWE HPC Benchmark".

In addition, two special sessions were held - one by industry and one by the funding bodies. Three tutorials preceded the main technical program of the conference: "Tools for Program Analysis in Computational Science" by Dieter Kranzlmüller; "P-GRADE Portal" by P. Kascuk, T. Kiss and G. Sipos; and "Scientific Computing on Graphics Hardware" by Dominik Göddeke. We would like to thank all the keynote, the invited, and the tutorial speakers for their inspiring talks.

Apart from the plenary sessions and tutorials the conference included twelve parallel oral sessions and two poster sessions. Since the first ICCS in San

Francisco the conference has grown steadily attracting increasing numbers of researchers in the field of Computational Science. For ICCS 2006 we received over 1,400 submissions, around 300 for the main track and over 1,100 for the originally proposed workshops. Of these submissions, 98 were accepted as a full papers and 29 as posters for the main track; and 500 were accepted as full papers, short papers or posters for the 32 workshops. This selection was possible due to the tremendous work done by the Program Committee and the 720 reviewers. The author index contains over 1,000 names and over 600 participants from all the major continents. The papers cover a wide variety of topics in Computational Science, ranging from Grand Challenges problems and modelling of complex systems in various areas to advanced numerical algorithms and new scalable algorithms in diverse application areas and software environments for Computational Science. The ICCS 2006 Proceedings consist of four volumes, 3991 to 3994, where the first volume contains the papers from the main track and all the posters; the remaining three volumes contain the papers from the workshops. ICCS this year is primary published on a CD and we would like to thank Springer for their cooperation and partnership. We hope that the ICCS 2006 Proceedings will be a major intellectual resource for many computational scientists and researchers for years ahead. During the conference the best papers from the main track and workshops as well as the best posters were nominated and commended on ICCS 2006 website. A number of selected papers will also be published in special issues of relevant mainstream journals.

We would like to thank all workshop organisers and the program committee for the excellent work, which further enhanced the conference's standing and led to very high quality event with excellent papers. We would like to express our gratitude to Advanced Computing and Emerging Technologies Centre staff, postgraduates and students for their wholehearted support of ICCS 2006. We would like to thank the School of Systems Engineering, Conference Office, Finance Department and various units at the University of Reading for different aspects of the organization and for their constant support in making ICCS 2006 a success. We would like to thank the Local Organizing Committee for their persistent and enthusiastic work towards the success of ICCS 2006. We owe special thanks to our sponsors: Intel, IBM, SGI, Microsoft Research, EPSRC and Springer; and to ACET Centre and the University of Reading for their generous support. We would like to thank SIAM, IMACS, and UK e-Science programme for endorsing ICCS 2006.

ICCS 2006 was organized by the Advanced Computing and Emerging Technologies Centre, University of Reading, with support from the Section Computational Science at the Universiteit van Amsterdam and Innovative Computing Laboratory at the University of Tennessee, in cooperation with the Society for Industrial and Applied Mathematics (SIAM), the International Association for Mathematics and Computers in Simulation (IMACS), and the UK Engineering and Physical Sciences Research Council (EPSRC). We invite you to visit the ICCS 2006 website (http://www.iccs-meeting.org/iccs2006/) and ACET Centre website (http://www.acet.reading.ac.uk/) to recount the events leading up to the conference, to view the technical programme, and to recall memories of three and a half days of engagement in the interest of fostering and advancing Computational Science.

June 2006

Vassil N. Alexandrov
G. Dick van Albada
Peter M.A. Sloot
Jack J. Dongarra

# Organisation

ICCS 2006 was organised by the Centre for Advanced Computing and Emerging Technologies (ACET), University of Reading, UK, in cooperation with the University of Reading (UK), the Universiteit van Amsterdam (The Netherlands), the University of Tennessee (USA), Society for Industrial and Applied Mathematics (SIAM), International Association for Mathematics and Computers in Simulation (IMACS) and Engineering and Physical Sciences Research Council (EPSRC). The conference took place on the Whiteknights Campus of the University of Reading.

## Conference Chairs

Scientific Chair - Vassil N. Alexandrov (ACET, University of Reading, UK)
Workshops Chair - G. Dick van Albada (Universiteit van Amsterdam, The Netherlands)
ICCS Series Overall Chair - Peter M.A. Sloot (Universiteit van Amsterdam, The Netherlands)
ICCS Series Overall Co-Chair - Jack J. Dongarra (University of Tennessee, USA)

## Local Organising Committee

Vassil N. Alexandrov
Linda Mogort-Valls
Nia Alexandrov
Ashish Thandavan
Christian Weihrauch
Simon Branford
Adrian Haffegee
David Monk
Janki Dodiya
Priscilla Ramsamy
Ronan Jamieson
Ali Al-Khalifah
David Johnson
Eve-Marie Larsen
Gareth Lewis
Ismail Bhana
S. Mehmood Hasan
Sokratis Antoniou

## Sponsoring Institutions

Intel Corporation
IBM
SGI
Microsoft Research
EPSRC
Springer
ACET Centre
University of Reading

## Endorsed by

SIAM
IMACS
UK e-Science Programme

## Program Committee

D. Abramson - Monash University, Australia
V. Alexandrov - University of Reading, UK
D.A. Bader - Georgia Tech, USA
M. Baker - University of Portsmouth, UK
S. Belkasim - Georgia State University, USA
A. Benoit - Ecole Normale Superieure de Lyon, France
I. Bhana - University of Reading, UK
R. Blais - University of Calgary, Canada
A. Bogdanov - Institute for High Performance Computing and Information Systems, Russia
G. Bosilca - University of Tennessee, USA
S. Branford - University of Reading, UK
M. Bubak - Institute of Computer Science and ACC Cyfronet - AGH, Poland
R. Buyya - University of Melbourne, Australia
F. Cappello - Laboratoire de Recherche en Informatique, Paris Sud, France
T. Cortes - Universitat Politecnica de Catalunya, Spain
J.C. Cunha - New University of Lisbon, Portugal
F. Desprez - INRIA, France
T. Dhaene - University of Antwerp, Belgium
I.T. Dimov - University of Reading, UK
J. Dongarra - University of Tennessee, USA
C. Douglas - University of Kentucky, USA
G.E. Fagg, University of Tennessee, USA
M. Gerndt - Technical University of Munich, Germany

Y. Gorbachev - Institute for High Performance Computing and Information Systems, Russia
A. Goscinski - Deakin University, Australia
A. Haffegee - University of Reading, UK
L. Hluchy - Slovak Academy of Science, Slovakia
A. Hoekstra - Universiteit van Amsterdam, The Netherlands
A. Iglesias - University of Cantabria, Spain
R. Jamieson - University of Reading, UK
D. Johnson - University of Reading, UK
J. Kitowski - AGH University of Science and Technology, Poland
D. Kranzlmüller - Johannes Kepler University Linz, Austria
A. Lagana - Universita di Perugia, Italy
G. Lewis - University of Reading, UK
E. Luque - University Autonoma of Barcelona, Spain
M. Malawski - Institute of Computer Science AGH, Poland
M. Mascagni - Florida State University, USA
E. Moreno - Euripides Foundation of Marilia, Brazil
J. Ni The - University of Iowa, Iowa City, IA, USA
G. Norman - Russian Academy of Sciences, Russia
S. Orlando - University of Venice, Italy
B. Ó Nulláin - UUniversiteit van Amsterdam, The Netherlands
M. Paprzycki - Computer Science Institute, SWSP, Warsaw, Poland
R. Perrott - Queen's University of Belfast, UK
R. Renaut - Arizona State University, USA
A. Rendell - Australian National University, Australia
D. Rodriguez-García - University of Reading, UK
P. Roe Queensland - University of Technology, Australia
S.L. Scott - Oak Ridge National Laboratory, USA
D. Shires - U.S. Army Research Laboratory, USA
P.M.A. Sloot - Universiteit van Amsterdam, The Netherlands
G. Stuer - University of Antwerp, Belgium
R. Tadeusiewicz - AGH University of Science and Technology, Poland
A. Thandavan - University of Reading, UK
P. Tvrdik - Czech Technical University, Czech Republic
P. Uthayopas - Kasetsart University, Thailand
G.D. van Albada - Universiteit van Amsterdam, The Netherlands
J. Vigo-Aguiar - University of Salamanca, Spain
J.A. Vrugt - Los Alamos National Laboratory, USA
J. Wasniewski - Technical University of Denmark, Denmark
G. Watson - Los Alamos National Laboratory, USA
C. Weihrauch - University of Reading, UK
Y. Xue - Chinese Academy of Sciences, China
E. Zudilova-Seinstra - Universiteit van Amsterdam, The Netherlands

## Reviewers

A. Adamatzky
A. Arenas
A. Belloum
A. Benoit
A. Bielecki
A. Bode
A. Cepulkauskas
A. Chkrebtii
A. Drummond
A. Erzan
A. Fedaravicius
A. Galvez
A. Gerbessiotis
A. Goscinski
A. Griewank
A. Grösslinger
A. Grzech
A. Haffegee
A. Hoekstra
A. Iglesias
A. Jakulin
A. Janicki
A. Javor
A. Karpfen
A. Kertész
A. Knuepfer
A. Koukam
A. Lagana
A. Lawniczak
A. Lewis
A. Li
A. Ligeza
A. Mamat
A. Martin del Rey
A. McGough
A. Menezes
A. Motter
A. Nasri
A. Neumann
A. Noel
A. Obuchowicz
A. Papini
A. Paventhan
A. Pieczynska
A. Rackauskas
A. Rendell
A. Sánchez
A. Sánchez-Campos
A. Sayyed-Ahmad
A. Shafarenko
A. Skowron
A. Sosnov
A. Sourin
A. Stuempel
A. Thandavan
A. Tiskin
A. Turan
A. Walther
A. Wei
A. Wibisono
A. Wong
A. Yacizi
A. Zelikovsky
A. Zhmakin
A. Zhou
A.N. Karaivanova
A.S. Rodinov
A.S. Tosun
A.V. Bogdanov
B. Ó Nualláin
B. Autin
B. Balis
B. Boghosian
B. Chopard
B. Christianson
B. Cogan
B. Dasgupta
B. Di Martino
B. Gabrys
B. Javadi
B. Kahng
B. Kovalerchuk
B. Lesyng
B. Paternoster
B. Payne
B. Saunders
B. Shan
B. Sniezynski
B. Song
B. Strug
B. Tadic
B. Xiao
B.M. Rode
B.S. Shin
C. Anthes
C. Bannert
C. Biely
C. Bischof
C. Cotta
C. Douglas
C. Faure
C. Glasner
C. Grelck
C. Herrmann
C. Imielinska
C. Lursinsap
C. Mastroianni
C. Miyaji
C. Nelson
C. Otero
C. Rodriguez Leon
C. Schaubschläger
C. Wang
C. Weihrauch
C. Woolley
C. Wu
C. Xu
C. Yang
C.-H. Huang
C.-S. Jeong
C.G.H. Diks
C.H. Goya
C.H. Kim
C.H. Wu
C.K. Chen
C.N. Lee
C.R. Kleijn
C.S. Hong
D. Abramson

D. Brinza
D. Brown
D. Che
D. Déry
D. Donnelly
D. Evers
D. Göddeke
D. Johnson
D. Kim
D. Kranzlmüller
D. Laforenza
D. Li
D. Luebke
D. Maringer
D. Pfahl
D. Plemenos
D. Rodriguez-García
D. Shires
D. Stoffer
D. Stokic
D. Szczerba
D. Taniar
D. Thalmann
D. Vasuinin
D. Wang
D. Xu
D.A. Bader
D.B. Davies
D.B.D. Birkbeck
D.C. Ghosh
D.C. Lee
D.J. Roberts
D.M. Chiu
D.M. Tartakovsky
D.R. Green
D.S. Kim
D.S. Perry
E. Atanasov
E. Grabska
E. Huedo Cuesta
E. Jaeger-Frank
E. Lee
E. Luque
E. Macias
E. Moreno
E. Nawarecki
E. Puppo
E. Roanes-Lozano
E. Valakevicius
E. Zeng
E. Zotenko
E. Zudilova-Seinstra
E.A. Castro
E.N. Huh
E.S. Quintana-Orti
F. Capkovic
F. Cappello
F. Desprez
F. Gava
F. Hirata
F. Iavernaro
F. Kiss
F. Lamantia
F. Lee
F. Loulergue
F. Markowetz
F. Melendez
F. Perales
F. Rogier
F. Terpstra
F. Zuccarello
F.-X. Roux
F.J. Keil
G. Alexe
G. Allen
G. Bosilca
G. Chen
G. Cheng
G. Dobrowolski
G. Dong
G. Erlebacher
G. Farin
G. Felici
G. Frenking
G. Gheri
G. Jeon
G. Kolaczek
G. Kou
G. Lewis
G. Lin
G. Mauri
G. Messina
G. Mounié
G. Narasimhan
G. Norman
G. Pavesi
G. Rojek
G. Slusarczyk
G. Stuer
G. Szabó
G. Tempesti
G. Volkert
G. Watson
G. Zheng
G.-L. Park
G.D. van Albada
G.D. Vedova
G.E. Fagg
G.J. Rodgers
H. Bungartz
H. Choo
H. Diab
H. Fangohr
H. Jin
H. Kaltenbach
H. Kosina
H. Labiod
H. Lee
H. Moradkhani
H. Müller
H. Munakata
H. Oh
H. Sarafian
H. Stockinger
H. Suzuki
H. Umeo
H. Wang
H. Yanami
H.-K. Choi
H.-K. Lee
H.C. Chojnacki
H.F. Schaefer III
H.K. Kim
H.P. Luehi
H.S. Nguyen

H.Y. Lee
I. Bhana
I. Boada
I. Kolingerova
I. Lee
I. Mandoiu
I. Moret
I. Navas-Delgado
I. Podolak
I. Schagaev
I. Suehiro
I. Tabakow
I. Taylor
I.T. Dimov
J. Abawajy
J. Aroba
J. Blower
J. Cabero
J. Cai
J. Cao
J. Chen
J. Cho
J. Choi
J. Davila
J. Dolado
J. Dongarra
J. Guo
J. Gutierrez
J. Han
J. He
J. Heo
J. Hong
J. Humble
J. Hwang
J. Jeong
J. Jurek
J. Kalcher
J. Kang
J. Kim
J. King
J. Kitowski
J. Koller
J. Kommineni
J. Koo
J. Kozlak
J. Kroc
J. Krueger
J. Laws
J. Lee
J. Li
J. Liu
J. Michopoulos
J. Nabrzyski
J. Nenortaite
J. Ni
J. Owen
J. Owens
J. Pang
J. Pjesivac-Grbovic
J. Quinqueton
J. Sanchez-Reyes
J. Shin
J. Stefanowski
J. Stoye
J. Tao
J. Utke
J. Vigo-Aguiar
J. Volkert
J. Wang
J. Wasniewski
J. Weidendorfer
J. Wu
J. Yu
J. Zara
J. Zhang
J. Zhao
J. Zivkovic
J.-H. Nam
J.-L. Koning
J.-W. Lee
J.A. Vrugt
J.C. Cunha
J.C. Liu
J.C. Teixeira
J.C.S. Lui
J.F. San Juan
J.H. Hrusak
J.H. Lee
J.J. Alvarez
J.J. Cuadrado
J.J. Korczak
J.J. Zhang
J.K. Choi
J.L. Leszczynski
J.M. Bradshaw
J.M. Gilp
J.P. Crutchfield
J.P. Suarez Rivero
J.V. Alvarez
J.Y. Chen
K. Akkaya
K. Anjyo
K. Banas
K. Bolton
K. Boryczko
K. Chae
K. Ebihara
K. Ellrott
K. Fisher
K. Fuerlinger
K. Gaaloul
K. Han
K. Hsu
K. Jinsuk
K. Juszczyszyn
K. Kubota
K. Li
K. Meridg
K. Najarian
K. Ouazzane
K. Sarac
K. Sycara
K. Tai-hoon Kim
K. Trojahner
K. Tuncay
K. Westbrooks
K. Xu
K. Yang
K. Zhang
K.-J. Jeong
K.B. Lipkowitz
K.D. Nguyen
K.V. Mikkelsen
K.X.S. Souza
K.Y. Huang

L. Borzemski
L. Brugnano
L. Cai
L. Czekierda
L. Fernandez
L. Gao
L. Gonzalez-Vega
L. Hascoet
L. Hluchy
L. Jia
L. Kotulski
L. Liu
L. Lopez
L. Marchal
L. Neumann
L. Parida
L. Taher
L. Xiao
L. Xin
L. Yang
L. Yu
L. Zheng
L. Zhigilei
L.H. Figueiredo
L.J. Song
L.T. Yang
M. Aldinucci
M. Baker
M. Bamha
M. Baumgartner
M. Bhuruth
M. Borodovsky
M. Bubak
M. Caliari
M. Chover
M. Classen
M. Comin
M. Deris
M. Drew
M. Fagan
M. Fras
M. Fujimoto
M. Gerndt
M. Guo
M. Hardman
M. Hobbs
M. Houston
M. Iwami
M. Jankowski
M. Khater
M. Kim
M. Kirby
M. Kisiel-Dorochinicki
M. Li
M. Malawski
M. Mascagni
M. Morshed
M. Mou
M. Omar
M. Pérez-Hernández
M. Palakal
M. Paprzycki
M. Paszynski
M. Polak
M. Rajkovic
M. Ronsse
M. Rosvall
M. Ruiz
M. Sarfraz
M. Sbert
M. Smolka
M. Suvakov
M. Tomassini
M. Verleysen
M. Vianello
M. Zhang
M.A. Sicilia
M.H. Zhu
M.J. Brunger
M.J. Harris
M.Y. Chung
N. Bauernfeind
N. Hu
N. Ishizawa
N. Jayaram
N. Masayuki
N. Murray
N. Navarro
N. Navet
N. Sastry
N. Sundaraganesan
N.T. Nguyen
O. Beckmann
O. Belmonte
O. Habala
O. Maruyama
O. Otto
O. Yasar
P. Alper
P. Amodio
P. Balbuena
P. Bekaert
P. Berman
P. Blowers
P. Bonizzoni
P. Buendia
P. Czarnul
P. Damaschke
P. Diaz Gutierrez
P. Dyshlovenko
P. Geerlings
P. Gruer
P. Heimbach
P. Heinzlreiter
P. Herrero
P. Hovland
P. Kacsuk
P. Li
P. Lingras
P. Martineau
P. Pan
P. Praxmarer
P. Rice
P. Roe
P. Sloot
P. Tvrdik
P. Uthayopas
P. van Hooft
P. Venuvanalingam
P. Whitlock
P. Wolschann
P.H. Lin
P.K. Chattaraj
P.R. Ramasami
Q. Deng

R. Aspin
R. Blais
R. Buyya
R. Dondi
R. Drezewski
R. Duran Diaz
R. Jamieson
R. Jothi
R. Kakkar
R. Katarzyniak
R. Kobler
R. Lambiotte
R. Liu
R. Marcjan
R. Mikusauskas
R. Nock
R. Perrott
R. Ramaroson
R. Rejas
R. Renaut
R. Rizzi
R. Ruiz
R. Sander
R. Schaefer
R. Simutis
R. Strzodka
R. Tadeusiewicz
R. Walentynski
R. Westermann
R. Wismüller
R. Wolff
R.G. Giering
R.Q. Wu
S. Abe
S. Aluru
S. Ambroszkiewicz
S. Balla
S. Bandini
S. Belkasim
S. Bhowmick
S. Böcker
S. Branford
S. Chen
S. Chiu
S. Cho
S. Dong
S. El Yacoubi
S. Forth
S. Gilmore
S. Gimelshein
S. Gorlatch
S. Green
S. Gremalschi
S. Han
S. Jhang
S. Kawano
S. Kim
S. Lee
S. Lightstone
S. Maniccam
S. Olariu
S. Orlando
S. Pal
S. Rahmann
S. Rajasekaran
S. Sanchez
S. Thurner
S. Tsunekawa
S. Turek
S. Valverde
S. Yi
S. Yoon
S.-B. Scholz
S.-R. Kim
S.-Y. Han
S.C. Lo
S.H. Cho
S.J. Han
S.K. Ghosh
S.L. Gargh
S.L. Scott
S.S. Manna
T. Angskun
T. Atoguchi
T. Cortes
T. Dhaene
T. Dokken
T. Ezaki
T. Fahringer
T. Hu
T. Ida
T. Korkmaz
T. McKenzie
T. Milledge
T. Politi
T. Przytycka
T. Recio
T. Strothotte
T. Suzudo
T. Takahashi
T. Tsuji
T. Wang
T. Ward
T. Worsch
T.-J. Lee
T.B. Ho
T.C. Lu
T.L. Zhang
T.N. Troung
T.V. Gurov
T.W. Kim
U. Ruede
U. Ufuktepe
U. Vaccaro
U.N. Naumann
V. Alexandrov
V. Aquilanti
V. Debelov
V. Hargy
V. Korkhov
V. Parasuk
V. Rafe
V. Robles
V. Srovnal
V. Weispfenning
V.A. Emanuele II
V.C. Chinh
V.V. Krzhizhanovskaya
V.V. Shakhov
W. Alda
W. Bronsvoort
W. Choi
W. Dou
W. Funika
W. Lee

## Workshop Organisers

### Third International Workshop on Simulation of Multiphysics Multiscale Systems

V.V. Krzhizhanovskaya - Universiteit van Amsterdam, The Netherlands and St. Petersburg State Polytechnical University, Russia
Y.E. Gorbachev - St. Petersburg State Polytechnic University, Russia
B. Chopard - University of Geneva, Switzerland

### Innovations in Computational Science Education

D. Donnelly - Department of Physics, Siena College, USA

### Fifth International Workshop on Computer Graphics and Geometric Modeling (CGGM 2006)

A. Iglesias - University of Cantabria, Spain

### Fourth International Workshop on Computer Algebra Systems and Applications (CASA 2006)

A. Iglesias - University of Cantabria, Spain
A. Galvez - University of Cantabria, Spain

**Tools for Program Development and Analysis in Computational Science**

D. Kranzlmüller - GUP, Joh. Kepler University, Linz, Austria
R. Wismüller - University of Siegen, Germany
A. Bode - Technische Universität München, Germany
J. Volkert - GUP, Joh. Kepler University, Linz, Austria

**Collaborative and Cooperative Environments**

C. Anthes - GUP, Joh. Kepler University, Linz, Austria
V.N. Alexandrov - ACET, University of Reading, UK
D.J. Roberts - NICVE, University of Salford, UK
J. Volkert - GUP, Joh. Kepler University, Linz, Austria
D. Kranzlmüller - GUP, Joh. Kepler University, Linz, Austria

**Second International Workshop on Bioinformatics Research and Applications (IWBRA'06)**

A. Zelikovsky - Georgia State University, USA
Y. Pan - Georgia State University, USA
I.I. Mandoiu - University of Connecticut, USA

**Third International Workshop on Practical Aspects of High-Level Parallel Programming (PAPP 2006)**

A. Benoît - Laboratoire d'Informatique du Parallélisme, Ecole Normale Supérieure de Lyon, France
F. Loulergue - LIFO, Université d'Orléans, France

**Wireless and Mobile Systems**

H. Choo - Networking Laboratory, Sungkyunkwan University, Suwon, KOREA

**GeoComputation**

Y. Xue - Department of Computing, Communications Technology and Mathematics, London Metropolitan University, UK

**Computational Chemistry and Its Applications**

P. Ramasami - Department of Chemistry, University of Mauritius

**Knowledge and Information Management in Computer Communication Systems (KIMCCS 2006)**

N.T. Nguyen - Institute of Control and Systems Engineering, Wroclaw University of Technology, Poland

A. Grzech - Institute of Information Science and Engineering, Wroclaw University of Technology, Poland
R. Katarzyniak - Institute of Information Science and Engineering, Wroclaw University of Technology, Poland

### Modelling of Complex Systems by Cellular Automata (MCSCA 2006)

J. Kroc - University of West Bohemia, Czech Republic
T. Suzudo - Japan Atomic Energy Agency, Japan
S. Bandini - University of Milano - Bicocca, Italy

### Dynamic Data Driven Application Systems (DDDAS 2006)

F. Darema - National Science Foundation, USA

### Parallel Monte Carlo Algorithms for Diverse Applications in a Distributed Setting

I.T. Dimov - ACET, University of Reading, UK
V.N. Alexandrov - ACET, University of Reading, UK

### International Workshop on Intelligent Storage Technology (IST06)

J. Shu - Department of Computer Science and Technology, Tsinghua University, Beijing, P.R. China

### Intelligent Agents in Computing Systems

R. Schaefer - Department of Computer Science, Stanislaw Staszic University of Science and Technology in Kraków
K. Cetnarowicz - Department of Computer Science, Stanislaw Staszic University of Science and Technology in Kraków

### First International Workshop on Workflow Systems in e-Science (WSES06)

Z. Zhao - Informatics Institute, University of Amsterdam, The Netherlands
A. Belloum - University of Amsterdam, The Netherlands

### Networks: Structure and Dynamics

B. Tadic - Theoretical Physics Department, J. Stefan Institute, Ljubljana, Slovenia
S. Thurner - Complex Systems Research Group, Medical University Vienna, Austria

**Evolution Toward Next Generation Internet (ENGI)**

Y. Cui - Tsinghua University, P.R. China
T. Korkmaz - University of Texas at San Antonio, USA

**General Purpose Computation on Graphics Hardware (GPGPU): Methods, Algorithms and Applications**

D. Göddeke - Universität Dortmund, Institut für Angewandte Mathematik und Numerik, Germany
S. Turek - Universität Dortmund, Institut für Angewandte Mathematik und Numerik, Germany

**Intelligent and Collaborative System Integration Technology (ICSIT)**

J.-W. Lee - Center for Advanced e-System Integration Technology, Konkuk University, Seoul, Korea

**Computational Methods for Financial Markets**

R. Simutis - Department of Informatics, Kaunas Faculty, Vilnius University, Lithuania
V. Sakalauskas - Department of Informatics, Kaunas Faculty, Vilnius University, Lithuania
D. Kriksciuniene - Department of Informatics, Kaunas Faculty, Vilnius University, Lithuania

**2006 International Workshop on P2P for High Performance Computational Sciences (P2P-HPCS06)**

H. Jin - School of Computer Science and Technology, Huazhong University of Science and Technology, Wuhan, China
X. Liao - Huazhong University of Science and Technology, Wuhan, China

**Computational Finance and Business Intelligence**

Y. Shi - Graduate School of the Chinese Academy of Sciences, Beijing, China

**Third International Workshop on Automatic Differentiation Tools and Applications**

C. Bischof - Inst. for Scientific Computing, RWTH Aachen University, Germany
S.A. Forth - Engineering Systems Department, Cranfield University, RMCS Shrivenham, UK
U. Naumann - Software and Tools for Computational Engineering, RWTH Aachen University, Germany
J. Utke - Mathematics and Computer Science Division, Argonne National Laboratory, IL, USA

**2006 Workshop on Scientific Computing in Electronics Engineering**

Y. Li - National Chiao Tung Univeristy, Hsinchu City, Taiwan

**New Trends in the Numerical Solution of Structured Systems with Applications**

T. Politi - Dipartimento di Matematica, Politecnico di Bari, Itali
L. Lopez - Dipartimento di Matematica, Universita' di Bari, Itali

**Workshop on Computational Science in Software Engineering (CSSE'06)**

D. Rodríguez García - University of Reading, UK
J.J. Cuadrado - University of Alcalá, Spain
M.A. Sicilia - University of Alcalá, Spain
M. Ruiz - University of Cádiz, Spain

**Digital Human Modeling (DHM-06)**

Y. Cai - Carnegie Mellon University, USA
C. Imielinska - Columbia University

**Real Time Systems and Adaptive Applications (RTSAA 06)**

T. Kuo - National Taiwan University, Taiwan
J. Hong - School of Computer Science and Engineering, Kwangwoon University, Seoul, Korea
G. Jeon - Korea Polytechnic University, Korea

**International Workshop on Grid Computing Security and Resource Management (GSRM'06)**

J.H. Abawajy - School of Information Technology, Deakin University, Geelong, Australia

**Fourth International Workshop on Autonomic Distributed Data and Storage Systems Management Workshop (ADSM 2006)**

J.H. Abawajy - School of Information Technology, Deakin University, Geelong, Australia

# Table of Contents – Part II

## Third International Workshop on Simulation of Multiphysics Multiscale Systems


Numerical Modeling of Plasma - Flow Interaction
*Jean-Charles Matéo-Vélez, Francois Rogier, Frédéric Thivet, Pierre Degond* ........ 1

Numerical Methods for Reacting Gas Flow Simulations
*S. van Veldhuizen, C. Vuik, C.R. Kleijn* ........ 10

Reduced Flame Kinetics Via Rate-Controlled Constrained Equilibrium
*Stelios Rigopoulos* ........ 18

Flow Patterns in the Vicinity of Triple Line Dynamics Arising from a Local Surface Tension Model
*J. Monnier, I. Cotoi* ........ 26

A Multilevel-Multigrid Approach to Multiscale Electromagnetic Simulation
*Peter Chow, Tetsuyuki Kubota, Takefumi Namiki* ........ 34

Scalable Simulation of Electromagnetic Hybrid Codes
*Kalyan Perumalla, Richard Fujimoto, Homa Karimabadi* ........ 41

Numerical Modelling of Poroviscoelastic Grounds in the Time Domain Using a Parallel Approach
*Arnaud Mesgouez, Gaëlle Lefeuve-Mesgouez, André Chambarel, Dominique Fougère* ........ 50

Numerical Modeling of Tidal Effects and Hydrodynamics in the Po River Estuary
*Célestin Leupi, Michel Deville, Mustafa Siddik Altinakar* ........ 58

Adaptive Mesh Refinement and Domain Decomposition: A Framework to Study Multi-physical and Multi-scale Phenomena. First Application to Reacting Gas Flows
*J. Ryan* ........ 66

Time Splitting and Grid Refinement Methods in the Lattice Boltzmann Framework for Solving a Reaction-Diffusion Process
*Davide Alemani, Bastien Chopard, Josep Galceran, Jacques Buffle* ........ 70

Mesoscopic Simulations of Unsteady Shear-Thinning Flows
*Abdel Monim Artoli, Adélia Sequeira* ........ 78

A Multiphysics Model of Capillary Growth and Remodeling
*Dominik Szczerba, Gábor Székely, Haymo Kurz* ........ 86

Liquid Computations and Large Simulations of the Mammalian Visual Cortex
*Grzegorz M. Wojcik, Wieslaw A. Kaminski* ........ 94

Which Meshes Are Better Conditioned: Adaptive, Uniform, Locally Refined or Locally Adjusted?
*Sanjay Kumar Khattri, Gunnar Fladmark* ........ 102

Parallel Simulation of Three–Dimensional Bursting with MPI and OpenMP
*S. Tabik, L.F. Romero, E.M. Garzón, J.I. Ramos* ........ 106

Numerical Simulation of Phase Transformations in Shape Memory Alloy Thin Films
*Debiprosad Roy Mahapatra, Roderick V.N. Melnik* ........ 114

A Virtual Test Facility for Simulating Detonation-Induced Fracture of Thin Flexible Shells
*Ralf Deiterding, Fehmi Cirak, Sean P. Mauch, Daniel I. Meiron* ........ 122

Data-Driven Inverse Modelling of Ionic Polymer Conductive Composite Plates
*John G. Michopoulos, Moshen Shahinpoor* ........ 131

## Innovations in Computational Science Education

Exploiting Real-Time 3d Visualisation to Enthuse Students: A Case Study of Using Visual Python in Engineering
*Hans Fangohr* ........ 139

Involving Undergraduates in Computational Science and Engineering Research: Successes and Challenges
*R.M. Kirby, C.R. Johnson, M. Berzins* ........ 147

A Project Based Approach to Teaching Parallel Systems
*Alistair P. Rendell* ........................................ 155

Learning by Doing: Software Projects in CSE Education
*Martin Bernreuther, Hans-Joachim Bungartz* ........................ 161

Computational Math, Science, and Technology (CMST): A Strategy to Improve STEM Workforce and Pedagogy to Improve Math and Science Education
*O. Yaşar, L. Little, R. Tuzun, K. Rajasethupathy, J. Maliekal, M. Tahar* ........................................ 169

The School of Computational Science at Florida State University
*Gordon Erlebacher, Janet Peterson* ........................ 177

Teaching the Foundations of Computational Science on the Undergraduate Level
*C. Freundl, H. Köstler, U. Rüde* ........................ 185

Computational Science: An Intermingling of Science, Mathematics, and Computer Science
*Frederick R.W. McCourt* ........................ 193

A Framework for Conceptually Modelling the Domain Knowledge of an Instructional System
*Emilia Pecheanu, Luminita Dumitriu, Diana Stefanescu, Cristina Segal* ........................ 199

Platyhelminthes Are [Re]constructed Recursively
*Alberto de la Encina, Mercedes Hidalgo-Herrero, Olga Marroquín-Alonso* ........................ 207

## Fifth International Workshop on Computer Graphics and Geometric Modeling (CGGM 2006)

Extensions for 3D Graphics Rendering Engine Used for Direct Tessellation of Spline Surfaces
*Adrian Sfarti, Brian A. Barsky, Todd J. Kosloff, Egon Pasztor, Alex Kozlowski, Eric Roman, Alex Perelman* ........................ 215

An Evolution Computation Based Approach to Synthesize Video Texture
*Yu Meng, Wen-hui Li, Yan Wang, Wu Guo, Wei Pang* ........................ 223

Deformation of Dynamic Surfaces
*L.H. You, Jian J. Zhang* ........ 231

A New Smoothing Algorithm for Quadrilateral and Hexahedral Meshes
*Sanjay Kumar Khattri* ........ 239

The Calculation of Parametric NURBS Surface Interval Values Using Neural Networks
*Erkan Ülker, Ahmet Arslan* ........ 247

Characterizing and Covering Some Subclasses of Orthogonal Polygons
*Ana Mafalda Martins, António Leslie Bajuelos* ........ 255

Techniques for Computing Viewpoint Entropy of a 3D Scene
*Pascual Castelló, Mateu Sbert, Miguel Chover, Miquel Feixas* ........ 263

3D Object Repair Using 2D Algorithms
*Pavlos Stavrou, Pavlos Mavridis, Georgios Papaioannou, Georgios Passalis, Theoharis Theoharis* ........ 271

Extraction of Ridges-Valleys for Feature-Preserving Simplification of Polygonal Models
*Soo-Kyun Kim, Sun-Jeong Kim, Chang-Hun Kim* ........ 279

Multiresolution 3D Rendering on Mobile Devices
*Javier Lluch, Rafa Gaitán, Miguel Escrivá, Emilio Camahort* ........ 287

Multiresolution Remeshing Using Weighted Centroidal Voronoi Diagram
*Chao-Hung Lin, Chung-Ren Yan, Ji-Hsen Hsu, Tong-Yee Lee* ........ 295

Metric 3D Surface Mesh Generation Using Delaunay Criteria
*Tomasz Jurczyk, Barbara Głut* ........ 302

A Multiresolution Model for Non-photorealistic Rendering of Trees
*Celso Campos, Ricardo Quirós, Joaquin Huerta, Emilio Camahort, Roberto Vivó, Javier Lluch* ........ 310

Model Creation by Velocity Controlled Surface Deformation
*Risto Rangel-Kuoppa, David Mould* ........ 318

Rendering of Unorganized Points with Octagonal Splats
*Sun-Jeong Kim, Chang-Geun Song* ........ 326

Statistical Based Vectorization for Standard Vector Graphics
*Sebastiano Battiato, Giovanni Maria Farinella, Giovanni Puglisi* ........ 334

Robustly Computing Intersection Curves of Two Canal Surfaces with Quadric Decomposition
*Jinyuan Jia, Ajay Joneja, Kai Tang* ............................ 342

Triangle Strip Multiresolution Modelling Using Sorted Edges
*Ó. Belmonte Fernández, S. Aguado González, S. Sancho Chust* ...... 350

Improvement Construction for Planar G2 Transition Curve Between Two Separated Circles
*Zhong Li, Lizhuang Ma, Mingxi Zhao, Zhihong Mao* ................ 358

B-Spline Curve Fitting Using Dominant Points
*Hyungjun Park, Joo-Haeng Lee* ................................ 362

Quality and Performance Evaluation of Ray-Space Interpolation for Free Viewpoint Video Systems
*Fan Liangzhong, Yu Mei, Yu Zhou, Jiang Gangyi* .................. 367

Framework for Adaptive Sampling of Point-Based Surfaces Using Geometry and Color Attributes
*Duck Bong Kim, Eui Chul Kang, Kwan H. Lee, Renato B. Pajarola* ........................................ 371

## Fourth International Workshop on Computer Algebra Systems and Applications (CASA 2006)

Normalizing Relational Database Schemas Using Mathematica
*Ali Yazici, Ziya Karakaya* ...................................... 375

Extending Maple Capabilities for Solving and Displaying Inequalities
*A. Iglesias, R. Ipanaqué* ...................................... 383

Phase Response Curves, Delays and Synchronization in MATLAB
*W. Govaerts, B. Sautois* ....................................... 391

A Hybrid Approach for Normal Factorization of Polynomials
*Nicos Karcanias, Marilena Mitrouli, Dimitrios Triantafyllou* ........ 399

Computer Algebra for the Formation of Structural Matrices of Piezoceramic Finite Elements
*Algimantas Čepulkauskas, Regina Kulvietienė, Genadijus Kulvietis* ... 407

Symbolic Analysis of Economical Models with Mathematica
*A. Gálvez, A. Iglesias* ........................................ 414

Polarizable Theta-Stable Parabolic Subalgebras and $K_{\mathbb{C}}$-Saturation in the Non-compact Real Forms of $G_2$ and $F_4$
*Steven Glenn Jackson, Alfred G. Noël* ... 422

Dynamic Load Balancing with MatlabMPI
*Ricolindo L. Cariño, Ioana Banicescu, Wenzhong Gao* ... 430

Maple Implementation of the Chor-Rivest Cryptosystem
*L. Hernández Encinas, J. Muñoz Masqué, A. Queiruga Dios* ... 438

Development of TRIP: Fast Sparse Multivariate Polynomial Multiplication Using Burst Tries
*Mickaël Gastineau, Jacques Laskar* ... 446

A Symbolic Approach to Quantum Computation Simulation
*António Pereira, Rosália Rodrigues* ... 454

Development of SyNRAC
*Hitoshi Yanami, Hirokazu Anai* ... 462

Automated Discovery in Elementary Extrema Problems
*Francisco Botana, José L. Valcarce* ... 470

Stabilizing Second-Order Linear Dynamic Systems Via Hybrid Output Feedback Controls
*Liguo Zhang, Yangzhou Chen, Pingyuan Cui* ... 478

Computation of the Adjoint Matrix
*Alkiviadis Akritas, Gennadi Malaschonok* ... 486

MathBlackBoard as Effective Tool in Classroom
*Deguchi Hiroaki, Hashiba Hirokazu* ... 490

## Tools for Program Development and Analysis in Computational Science

Finding Inefficiencies in OpenMP Applications Automatically with Periscope
*Karl Fürlinger, Michael Gerndt* ... 494

Analysis of the Spatial and Temporal Locality in Data Accesses
*Jie Tao, Siegfried Schloissnig, Wolfgang Karl* ... 502

A Performance Profile and Test Tool for Development of Embedded Software Using Various Report Views
*Yongyun Cho, Chae-Woo Yoo* ..... 510

SCE Toolboxes for the Development of High-Level Parallel Applications
*J. Fernández, M. Anguita, E. Ros, J.L. Bernier* ..... 518

Introducing the Open Trace Format (OTF)
*Andreas Knüpfer, Ronny Brendel, Holger Brunst, Hartmut Mix, Wolfgang E. Nagel* ..... 526

Dynamic Instrumentation of Distributed Java Applications Using Bytecode Modifications
*Włodzimierz Funika, Pawel Swierszcz* ..... 534

Fine-Grained Instrumentation and Monitoring of Legacy Applications in a Service-Oriented Environment
*Bartosz Baliś, Marian Bubak, Krzysztof Guzy* ..... 542

Monitoring of WS-Based Applications
*Lechoslaw Trebacz, Piotr Handzlik, Wlodzimierz Funika, Marcin Smetek* ..... 549

Using Sequential Debugging Techniques with Massively Parallel Programs
*Christian Schaubschläger, Dieter Kranzlmüller, Jens Volkert* ..... 557

## Collaborative and Cooperative Environments

Workflow for Integrated Object Detection in Collaborative Video Annotation Environments
*Lars Grunewaldt, Kim Möller, Karsten Morisse* ..... 565

RMIX: A Dynamic, Heterogeneous, Reconfigurable Communication Framework
*Christian Engelmann, Al Geist* ..... 573

Developing Collaborative Social Software
*Ismail Bhana, David Johnson* ..... 581

An Efficient and Reflective Event Filtering for Context-Awareness in Ubiquitous Computing
*Kyu Bong Cho, Sung Keun Song, Hee Yong Youn, Gyung Leen Park* ..... 587

Creation and Control of Interactive Virtual Environments
*Adrian Haffegee, Priscilla Ramsamy, Ronan Jamieson, Vassil Alexandrov* ........ 595

Using Haptics to Improve Immersion in Virtual Environments
*Priscilla Ramsamy, Adrian Haffegee, Ronan Jamieson, Vassil Alexandrov* ........ 603

A Novel Navigation Algorithm for Locomotion Interfaces with Programmable Platforms
*Jungwon Yoon, Jeha Ryu* ........ 610

## Second International Workshop on Bioinformatics Research and Applications (IWBRA06)

Efficient and Practical Algorithms for Deducing the History of Recombination in Populations
*Dan Gusfield* ........ 618

Chordal Graphs in Computational Biology - New Insights and Applications
*Teresa M. Przytycka* ........ 620

Exemplar Longest Common Subsequence
*Paola Bonizzoni, Gianluca Della Vedova, Riccardo Dondi, Guillaume Fertin, Stéphane Vialette* ........ 622

Synonymous Codon Substitution Matrices
*Adrian Schneider, Gaston H. Gonnet, Gina M. Cannarozzi* ........ 630

SEPA: Approximate Non-subjective Empirical *p*-Value Estimation for Nucleotide Sequence Alignment
*Ofer Gill, Bud Mishra* ........ 638

Multiple Sequence Alignment by Ant Colony Optimization and Divide-and-Conquer
*Yixin Chen, Yi Pan, Juan Chen, Wei Liu, Ling Chen* ........ 646

COMBAT: Search Rapidly for Highly Similar Protein-Coding Sequences Using Bipartite Graph Matching
*Bing Sun, Jacob T. Schwartz, Ofer H. Gill, Bud Mishra* ........ 654

Missing Values Estimation in Microarray Data with Partial Least Squares Regression
*Kun Yang, Jianzhong Li, Chaokun Wang* ........ 662

Boost Feature Subset Selection: A New Gene Selection Algorithm for Microarray Dataset
*Xian Xu, Aidong Zhang* .................................... 670

A Hybrid Feature Selection Approach for Microarray Gene Expression Data
*Feng Tan, Xuezheng Fu, Hao Wang, Yanqing Zhang, Anu Bourgeois* .................................... 678

A Self-supervised Learning Framework for Classifying Microarray Gene Expression Data
*Yijuan Lu, Qi Tian, Feng Liu, Maribel Sanchez, Yufeng Wang* ...... 686

Pooling Evidence to Identify Cell Cycle–Regulated Genes
*Gaolin Zheng, Tom Milledge, E. Olusegun George, Giri Narasimhan* .................................... 694

Discovering Sequence-Structure Patterns in Proteins with Variable Secondary Structure
*Tom Milledge, Gaolin Zheng, Giri Narasimhan* .................... 702

Clustering Support Vector Machines and Its Application to Local Protein Tertiary Structure Prediction
*Jieyue He, Wei Zhong, Robert Harrison, Phang C. Tai, Yi Pan* ...... 710

Extracting Protein-Protein Interactions from the Literature Using the Hidden Vector State Model
*Deyu Zhou, Yulan He, Chee Keong Kwoh* ........................ 718

A Multilevel Approach to Identify Functional Modules in a Yeast Protein-Protein Interaction Network
*S. Oliveira, S.C. Seok* .................................... 726

Towards Detecting Protein Complexes from Protein Interaction Data
*Pengjun Pei, Aidong Zhang* .................................... 734

High-Throughput SNP Genotyping by SBE/SBH
*Ion I. Măndoiu, Claudia Prăjescu* .................................... 742

Tag SNP Selection Based on Multivariate Linear Regression
*Jingwu He, Alex Zelikovsky* .................................... 750

Minimum Multicolored Subgraph Problem in Multiplex PCR Primer Set Selection and Population Haplotyping
*M.T. Hajiaghayi, K. Jain, L.C. Lau, I.I. Măndoiu, A. Russell, V.V. Vazirani* .................................... 758

Phasing of 2-SNP Genotypes Based on Non-random Mating Model
*Dumitru Brinza, Alexander Zelikovsky* .......... 767

Event Models for Tumor Classification with SAGE Gene Expression Data
*Xin Jin, Anbang Xu, Guoxing Zhao, Jixin Ma, Rongfang Bie* .......... 775

Genomes Containing Duplicates Are Hard to Compare
*Cedric Chauve, Guillaume Fertin, Romeo Rizzi, Stéphane Vialette* ... 783

Rearrangement of Noisy Genomes
*Chunfang Zheng, David Sankoff* .......... 791

Simple Reconstruction of Binary Near-Perfect Phylogenetic Trees
*Srinath Sridhar, Kedar Dhamdhere, Guy E. Blelloch, Eran Halperin, R. Ravi, Russell Schwartz* .......... 799

Reconstructing Ancestor-Descendant Lineages from Serially-Sampled Data: A Comparison Study
*Patricia Buendia, Timothy M. Collins, Giri Narasimhan* .......... 807

Robustness of Greedy Type Minimum Evolution Algorithms
*Takeya Shigezumi* .......... 815

Space and Time Efficient Algorithms for Planted Motif Search
*Jaime Davila, Sudha Balla, Sanguthevar Rajasekaran* .......... 822

Predictability of Rules in HIV-1 Protease Cleavage Site Analysis
*Hyeoncheol Kim, Tae-Sun Yoon, Yiying Zhang, Anupam Dikshit, Su-Shing Chen* .......... 830

Statistical Feature Selection from Chaos Game Representation for Promoter Recognition
*Orawan Tinnungwattana, Chidchanok Lursinsap* .......... 838

Blue Matter: Strong Scaling of Molecular Dynamics on Blue Gene/L
*Blake G. Fitch, Aleksandr Rayshubskiy, Maria Eleftheriou, T.J. Christopher Ward, Mark Giampapa, Yuri Zhestkov, Michael C. Pitman, Frank Suits, Alan Grossfield, Jed Pitera, William Swope, Ruhong Zhou, Scott Feller, Robert S. Germain* ...... 846

DigitalTree: A Tool for Displaying Biological Data in Tree Structure
*Robin Kramer, Victor Olman, Ying Xu, Dong Xu* .......... 855

HiSP: A Probabilistic Data Mining Technique for Protein Classification
*Luiz Merschmann, Alexandre Plastino* .......... 863

Cross-Ontological Analytics: Combining Associative and Hierarchical Relations in the Gene Ontologies to Assess Gene Product Similarity
*C. Posse, A. Sanfilippo, B. Gopalan, R. Riensche, N. Beagley, B. Baddeley* .......... 871

A GO-Based Method for Assessing the Biological Plausibility of Regulatory Hypotheses
*Jonas Gamalielsson, Patric Nilsson, Björn Olsson* .......... 879

Delays in Biological Regulatory Networks (BRN)
*Jamil Ahmad, Adrien Richard, Gilles Bernot, Jean-Paul Comet, Olivier Roux* .......... 887

Phase Transitions in Gene Knockdown Networks of Transitive RNAi
*Shibin Qiu, Terran Lane* .......... 895

## Third International Workshop on Practical Aspects of High-Level Parallel Programming (PAPP 2006)

Compile-Time Energy Optimization for Parallel Applications in On-Chip Multiprocessors
*Juan Chen, Huizhan Yi, Xuejun Yang, Liang Qian* .......... 904

Using SBASCO to Solve Reaction-Diffusion Equations in Two-Dimensional Irregular Domains
*Manuel Díaz, Sergio Romero, Bartolomé Rubio, Enrique Soler, José M. Troya* .......... 912

Higher Order Flattening
*Roman Leshchinskiy, Manuel M.T. Chakravarty, Gabriele Keller* .......... 920

Combining Measurement and Stochastic Modelling to Enhance Scheduling Decisions for a Parallel Mean Value Analysis Algorithm
*Gagarine Yaikhom, Murray Cole, Stephen Gilmore* .......... 929

Joint Structured/Unstructured Parallelism Exploitation in `muskel`
*M. Danelutto, P. Dazzi* .......... 937

Co-Array Collectives: Refined Semantics for Co-Array Fortran
*Matthew J. Sottile, Craig E Rasmussen, Richard L. Graham* .......... 945

An Approach to Buffer Management in Java HPC Messaging
*Mark Baker, Bryan Carpenter, Aamir Shafi* .......... 953

## Wireless and Mobile Systems

A Low Complexity and Robust Frequency Offset Estimation Algorithm for OFDM-Based WLAN Systems
*Sanghun Kim, Seokho Yoon, Hyoung-Kee Choi, Sun Yong Kim* ...... 961

Simplified Signal Detection for BLAST Architecture with ML and DFE Detectors
*Myung-Sun Baek, Byung-Su Kang, So-Young Yeo, Young-Hwan You, Hyoung-Kyu Song* ...... 969

Scenario Decomposition Based Analysis of Next Generation Mobile Services
*Dongchun Shin, Jinbae Kim, Seungwan Ryu, Donsung Oh, Joowan Lee, Minhyung Kang* ...... 977

A Power Saving Scheme for Integrated WLAN and Cellular Networks
*SuKyoung Lee* ...... 985

The Optimal Subchannel and Bit Allocation Problem for OFDM
*Taehyung Park, Sungbin Im* ...... 992

Bluetooth Broadcasting Performance: Reliability and Throughput
*Kaan Dogan, Guray Gurel, A. Kerim Kamci, Ibrahim Korpeoglu* ...... 996

An Optimized Two Factor Authenticated Key Exchange Protocol in PWLANs
*Eun-Jun Yoon, Kee-Young Yoo* ...... 1000

Adaptive Clustering with Virtual Subnets Support in Ad Hoc Networks
*Tzu-Chiang Chiang, Ming-Hui Tsai, Yueh-Min Huang* ...... 1008

A Service Management Architecture for NEMO in IPv4 and IPv6 Networks
*Jin Ho Kim, Choong Seon Hong, Dae Sun Kim* ...... 1016

Overlapped Detection Via Approximate Entropy Estimation Against Flooding Attack in Mobile Sensor Networks
*Mihui Kim, Kijoon Chae* ...... 1024

Implementation of Next Generation Mobile Service: The Context-Aware Follow-Me Service
*Jungsook Bae, Seungwan Ryu, JaeYong Lee, ByungChul Kim* ...... 1033

Multi-piconet Formation to Increase Channel Utilization in IEEE 802.15.3 High-Rate WPAN
*Ssang-Bong Jung, Soon-Bin Yim, Tae-Jin Lee, Sun-Do June, Hyeon-Seok Lee, Tai-Gil Kwon, Jin-Woong Cho* .................. 1041

Unencapsulated Mobile Multicast Routing for Next Generation Video Networks
*Thomas C. Schmidt, Matthias Wählisch, Hans L. Cycon, Mark Palkow, Henrik Regensburg* .................................. 1050

Channel Estimation of High Rate WPAN System with Diversity Technique
*Byung-Su Kang, Myung-Sun Baek, Dong-Jun Cho, Young-Hwan You, Hyoung-Kyu Song* ............................ 1058

A Timestamp Tree-Based Cache Invalidation Report Scheme in Mobile Computing Environments
*Hakjoo Lee, Jonghyun Suh, Sungwon Jung, Sooyoung Lee, Junguck Lee* ................................................ 1065

Clustering Versus Evenly Distributing Energy Dissipation in Wireless Sensor Routing for Prolonging Network Lifetime
*Guangyan Huang, Xiaowei Li, Jing He* .......................... 1069

On the Effect of Heterogeneous Traffic Sources on the Network Availability for Wireless Sensor Grids
*Ali Hammad Akbar, Ki-Hyung Kim, Shaokai Yu, Won-Sik Yoon* ..... 1073

Selective Handover Technique on Multihomed Mobile Network Environment
*Kiyong Park, Sunyong Han, Jungwook Song* ....................... 1081

Collaborative Trust-Based Shortest Secure Path Discovery in Mobile Ad Hoc Networks
*Seungtak Oh, Chilgee Lee, Hyunseung Choo* ...................... 1089

An Efficient Neighbor Knowledge Based Broadcasting for Mobile Ad Hoc Networks
*Sung-Hee Lee, Young-Bae Ko* ................................... 1097

Maximum Lifetime Paths for the High Packet Delivery Ratio Using Fast Recovery in a Mobile Ad Hoc Network
*HyoJin Kim, SeungJae Han, JooSeok Song* ........................ 1101

**Author Index** ................................................... 1105

# Numerical Modeling of Plasma - Flow Interaction

Jean-Charles Matéo-Vélez[1], Francois Rogier[1],
Frédéric Thivet[1], and Pierre Degond[2]

[1] ONERA, Département Modèles pour l'Aérodynamique et l'Énergétique,
2 avenue Edouard Belin, 31055 Toulouse, France
mateo@onera.fr
http://www.onera.fr

[2] UMR 5640, Laboratoire Mathématiques pour l'Industrie et la Physique,
118 route de Narbonne, 31 062 Toulouse, France

**Abstract.** In the frame of the internal project PUMA (Plasma Used to Master Aerodynamics), ONERA is conducting fundamental studies of plasma-flow interactions. In this paper, the ionic wind created by corona discharges is studied in the case of a subsonic flow over a flat plate. The proposed mechanism of the ionic wind proposed is the addition of momentum by collisions between charged and neutral particles. In order to evaluate the effect of plasma on aerodynamics, a kinetic modeling of the discharge is coupled with a Fluid Dynamics code.

## 1 Introduction

A large number of studies has been carried out on the ionic wind since the 60's [1]. This phenomenon was first discovered by Hauksbee in 1709. It is a way to provide energy to flows thanks to an electric power supply. The strong advantages of such an energy supply are that there is no need for external mass addition and it can act over a wide range of frequencies (0-10 kHz). Numerous experiments [2]-[6] have pointed out the possibility of adding significant momentum in the boundary layer of subsonic flows.

Boeuf and Pitchford [7] give a detailed description of the origin of ionic wind created by dielectric barrier discharges. When charged particles are submitted to an intense electric field $E$, they collide with the neutral molecules of the gas. By this way they can transfer an amount of their momentum. The force $f^i_{k\to n}$ that exerts a specie $k$ on a neutral $n$ in the direction $i$ is:

$$f^i_{k\to n} = m_k N_k \nu_{k-n} \left(U^i_k - U^i_n\right) \tag{1}$$

where $m_k$ is the mass of specie $k$, $N_k$ its density, $U^i_k$ the $i^{th}$ component of its mean velocity vector; $\nu_{k-m}$ is the momentum transfer frequency for collisions between $k$ and $n$. This expression simplifies because the velocity of charged species are three or four orders of magnitude higher than the neutral velocity.

V.N. Alexandrov et al. (Eds.): ICCS 2006, Part II, LNCS 3992, pp. 1–9, 2006.

Let us consider the density current $j_k^i$ and the mobility $\mu_k$ of specie $k$:

$$j_k^i = qN_kU_k^i \tag{2}$$

$$\mu_k = \frac{q}{m_k\nu_{k-n}} \tag{3}$$

where $q$ is the electric charge of specie $k$. Neglecting the gas velocity with respect to ions and electrons velocities, the volumetric force $f$ exerted by all the charged particles reads:

$$f^i = \sum_k \frac{j_k^i}{\mu_k} \tag{4}$$

It is important to note that all these contributions to the global force are algebraic. So, the total force can be equal to zero even if the single contributions are not equal to zero. The conductive density current $j_c^i$ is the sum of the particles current:

$$|j_c^i| = \sum_k |j_k^i| \tag{5}$$

And finally, the total density current $j_t^i$ is the sum of the conductive and displacement density currents and reads:

$$|j_t^i| = \sum_k |j_k^i| + \varepsilon_0\frac{\partial E}{\partial t} \tag{6}$$

where $\varepsilon_0$ is the dielectric permittivity of air.

Moreau's team [4] used a wire-to-wire discharge on a flat plate (Figure 1) and observed an ionic wind around 5 m/s in the absence of uncoming flow. Two thin electrodes are flushed mounted in a dielectric plate. The first electrode is a 0.7 mm diameter anode set to a +22 kV potential. The second one has a 2 mm diameter and is set to -10 kV. They are placed at a distance of 4 cm. The regime of the discharge depends on many parameters such as the potential difference, the space between the electrodes, air humidity degree, flow velocity, etc. From the aerodynamic point of view, the most efficient regime is the corona regime in which a luminescent discharge appears in the vicinity of both electrodes. The purpose of this paper is to propose a model of this regime.

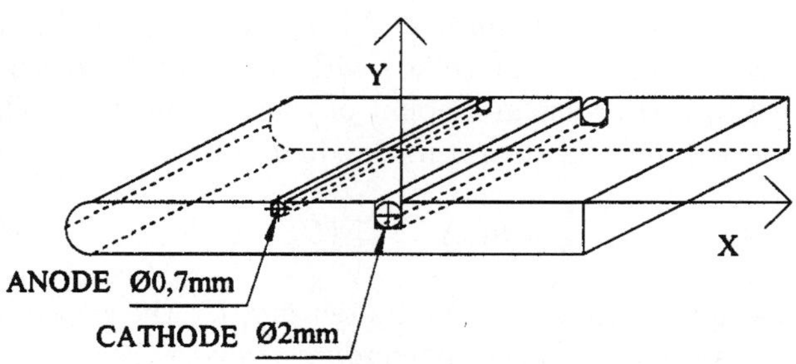


**Fig. 1.** Experimental setup of [4]

## 2 Modeling Corona Discharges - Flow Interaction

A modeling of the experimental device proposed in [4] and described in section 1 is proposed here. This modeling follows previous work [8].

### 2.1 Space and Time Scales

In order to simulate numerically the effect of corona discharges on a gas flow, two different time and space scales issues must be solved. First, to be correctly described, the discharge must be calculated with a maximal time increment of about $10^{-11}$ s, which is representative of the kinetic time scale and the characteristic time of electron transport. In the same time, the characteristic response time of the flow to the discharge effect is about $10^{-3}$ s. As a consequence, a full coupling of both parts of the problem is hardly realizable. The next three sections present the set of equations governing the problem, the numerical way of coupling and the results obtained.

### 2.2 Equations

The equations are written here in cartesian coordinates with the Einstein notation. The aerodynamic part of the problem follows the continuum and momentum equations (7) and (8):

$$\frac{\partial \rho U^i}{\partial x^i} = 0 \tag{7}$$

$$\rho\left(\frac{\partial U^i}{\partial t} + U^j \frac{\partial U^i}{\partial x^j}\right) = -\frac{\partial P}{\partial x^i} + \mu \frac{\partial^2 U^i}{\partial x^j \partial x^i} + f^i \tag{8}$$

where $U^i$ is the $i^{th}$ component of the flow velocity vector, $\rho$ the volumetric mass of air, $P$ the pressure and $\mu$ the dynamic viscosity of air. The total electric force $f$ is calculated solving the density transport equations (9) for a set of $N$ particles species coupled with the Poisson equation (10) for the electric potential $V$:

$$\frac{\partial N_k}{\partial t} + \frac{\partial N_k U_k^i}{\partial x^i} = \omega_k \tag{9}$$

$$\Delta V = -\frac{\rho_c}{\varepsilon_0} \tag{10}$$

$$\rho_c = \sum_k Z_k N_k \tag{11}$$

In these equations, $\omega_k$ is the rate of production of specie $k$. It is the balance between chemical production and decomposition. The $\Delta$ symbol stands for the divergence operator. $\rho_c$ is the space charge density, namely the difference between positive and negative charges densities. $Z_k$ is the charge of specie $k$.

The neutral particles velocity is the flow velocity $U^i$. The charged particles velocities are given by a drift-diffusion model. The ions velocity $U^i_{ion}$ is the sum

**Table 1.** Air plasma kinetics of [9]

| | |
|---|---|
| Ionization | $e, N_2 \rightarrow e, e, N_2^+$<br>$e, O_2 \rightarrow e, e, O_2^+$ |
| Attachment | $e, O_2 \rightarrow O_2^-$ |
| Recombination | $e, N_2^+ \rightarrow N_2$<br>$e, O_2^+ \rightarrow O_2$<br>$O_2^-, O_2^+ \rightarrow O_2, O_2$<br>$O_2^-, N_2^+ \rightarrow O_2, N_2$ |
| Excitation | $e, O_2 \rightarrow O_2^*$ |
| Detachment | $O_2^-, O_2^* \rightarrow e, O_2, O_2$ |
| Quenching | $O_2, O_2^* \rightarrow O_2, O_2$ |

of the flow velocity and the drift velocity due to the electric field. The electrons velocity reads the same way, adding the diffusion velocity due to heterogeneity of their densities. Ions are heavy particles in comparison to electrons. As a consequence, their diffusion velocity is neglected :

$$U^i_{ion} = U^i \pm \mu_{ion} E^i \tag{12}$$

$$U^i_e = U^i - \mu_e E^i - \frac{D_e}{N_e}\frac{\partial N_e}{\partial x^i} \tag{13}$$

Following [9], the discharge is described by the kinetics of air in high electric field which includes electrons, positive ions $O_2^+$ and $N_2^+$, negative ions $O_2^-$ and metastables $O_2^*$. Table 1 gives the reactions taken into account in the present work. The kinetics coefficients corresponding to these reactions are given by [9] and [10].

The cathode is considered as a source of electrons thanks to ionic bombardment. The number of electrons $N_e^{bbt}$ emitted by this way is supposed proportional to the number of positive ions colliding with the cathode.

$$N_e^{bbt} = \gamma \frac{N_+ |U_+|}{|U_e|} \tag{14}$$

where $\gamma$ is the ionic bombardment coefficient.

Finally, a resistance $R$ is connected in series between the anode and the power supply which voltage is $V_0$. The cathode is directly connected to the power supply mass. Thanks to eq. 6, the potential difference $\Delta V$ applied between the electrodes is:

$$\Delta V = V_0 - RI \tag{15}$$

where $I$ is the total electric current in amperes. Its expression is given in the next section.

### 2.3 1D Discharge Model with Plasma Thickness

Figure 2 presents the way the two parts of the problem (eq. (7)-(8) and eq. (9)-(14)) will be coupled. The aerodynamic part is solved thanks to a Fluid

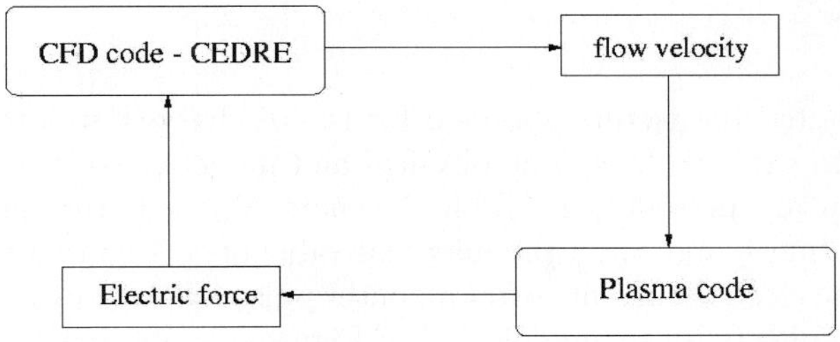


**Fig. 2.** Plasma - Aerodynamics Coupling

Dynamics code developed at ONERA, namely CEDRE. The space integration scheme follows a generalized MUSCL approach and the time integration scheme can be explicit ot implicit [12]. For the flow calculation, the volumetric force exerted by the discharge is needed. The input data for plasma calculation is the global flow velocity. In this work, the discharge development in one space dimension without any effect of the external flow is first calculated. It consists in writing equations (9)-(14) in 1D and assuming $U = 0$. This system is discretized by a finite volume method. The integration scheme is a second order scheme in time and space with a minmod slope limitor for the fluxes computation. The electric field $E_{sp}$ created by space charges is obtained thanks to a LU decomposition. The electric field $E_{ext}$ created by the electrodes with no space charge is calculated thanks to a 2D analytical formula [11], which permits to take into account the singularity of the two thin electrodes.

We suppose that all the variables (densities, velocities, electric potential, etc.) are constant over a thickness $S(x)$. It allows the total electric current conservation by ensuring the consistancy of the 2D electric field $E_{ext}$ with a 1D approximation. As a consequence, the equations (9)-(14) become:

$$\frac{\partial N_k}{\partial t} + \frac{1}{S}\frac{\partial S N_k U_k}{\partial x} = \omega_k \tag{16}$$

$$\frac{1}{S}\frac{\partial}{\partial x}\left(S\frac{\partial V}{\partial x}\right) = -\frac{\rho_c}{\varepsilon_0} \tag{17}$$

$$\rho_c = \sum_k Z_k N_k \tag{18}$$

$$U_{ion} = U \pm \mu_{ion} E \tag{19}$$

$$U_e = U - \mu_e E - \frac{D_e}{S N_e}\frac{\partial S N_e}{\partial x} \tag{20}$$

According to (17), $S$ is obtained by solving $S(x)E_{ext}(x)$ equal to a constant. The intoduction of $S$ allows to consider a non trivial electric field respecting a zero divergence in 1D. The total current $I$ is the same in every cell of the meshing and is calculated with respect to eq. 21, where $L$ is the electrodes length:

$$I = j_t L S \tag{21}$$

## 3 Results

Here are presented the results obtained for the discharge development without any effect of an external flow. The physical and numerical parameters used for this simulation are presented in Table 2, where $N_{mesh}$ is the number of cells of the uniform mesh and $S_{max}$ the maximal value of $S$. Figure 3 represents the evolution of the electric current. After an onset pulse followed by a long transient regime, the regular pulse regime plotted in Figure 3 is obtained. The frequency of these pulses is about 100 kHz. This description seems qualitatively close to the experimental results of [4] obtained for the same configuration (Figure 4). The minimal computed and experimental currents are about 0.8 mA and 0.5 mA respectively. Yet, the maximal computed current is one order of magnitude higher than the experimental one. This can be explained by two facts. First, the experimental potential difference is about 30 kV instead of 40 kV here. Then, the value of $S$ might be over estimated. This represents a limit of the 1D approach. 2D simulations would determine the thickness $S$ more accurately.

The fact that the pulsed discharge does not numerically appear for potential smaller than 35 or 40 kV indicates that the kinetics model is under evaluated. We can argue for the absence of an additional electrons provider: the photoionization. Yet, this regime seems to be similar to the Trichel pulses regime of negative corona discharges. So the simulation of the cathode region is quite in good agreement with previous works.

Figure 5 represents the evolution of electron density during a single pulse. On Figures 5, 6 and 7, the time $t = 0$ s corresponds to the ignition of a single pulse. The pulses observed in Figure 3 are due to the periodic creation of electrons thanks to ionization. When the space charge is high enough to diminish the electric field at the electrodes (Figure 6), the discharge is stopped and the species are evacuated. When the electric field reaches its initial value, a new pulse initiates.

**Table 2.** Parameters

| $V_0$ | $R$ | $\gamma$ | $N_{mesh}$ | $S_{max}$ |
|---|---|---|---|---|
| 40 kV | 10 k$\Omega$ | $10^{-4}$ | 400 | 5 mm |

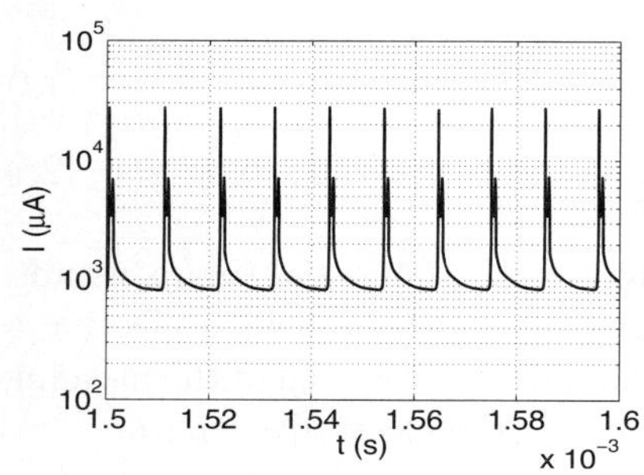


**Fig. 3.** Calculated electric current

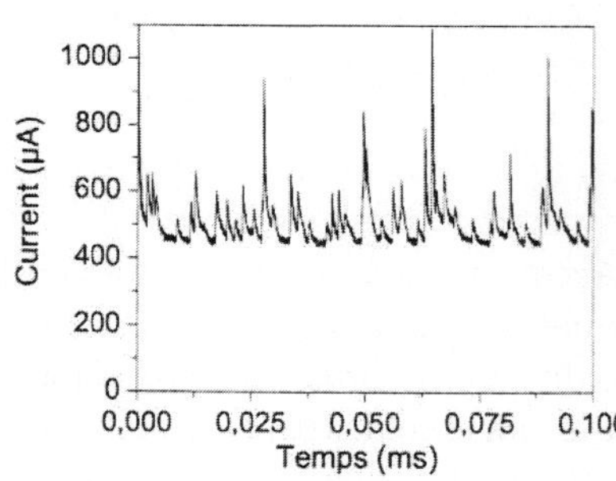


**Fig. 4.** Exp. electric current [4]

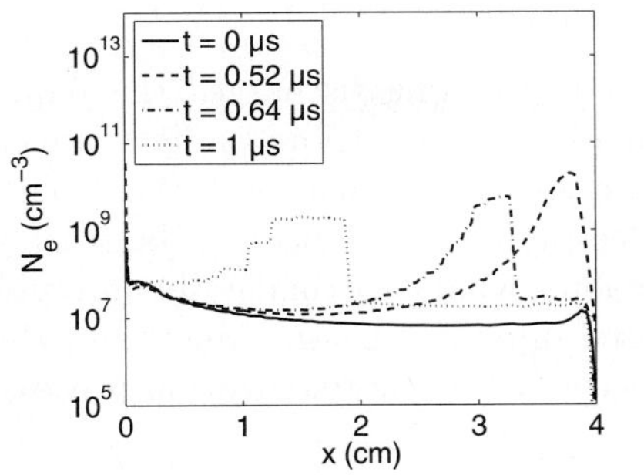


**Fig. 5.** Repartition of electron number density. $t = 0$ s corresponds to the beginning of a pulse.

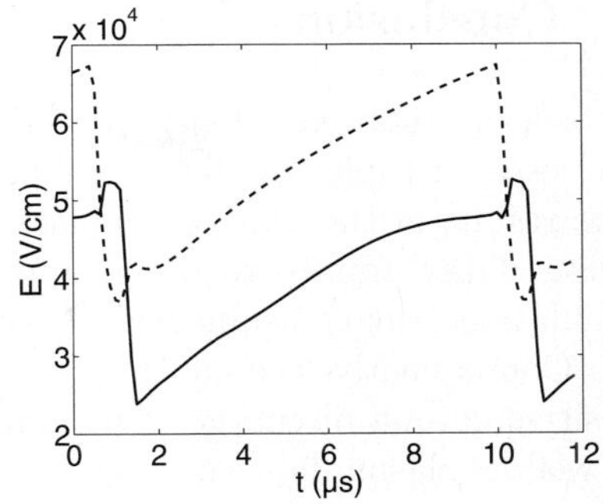


**Fig. 6.** Evolution of the electric field at the anode (—) and at the cathode (- -)

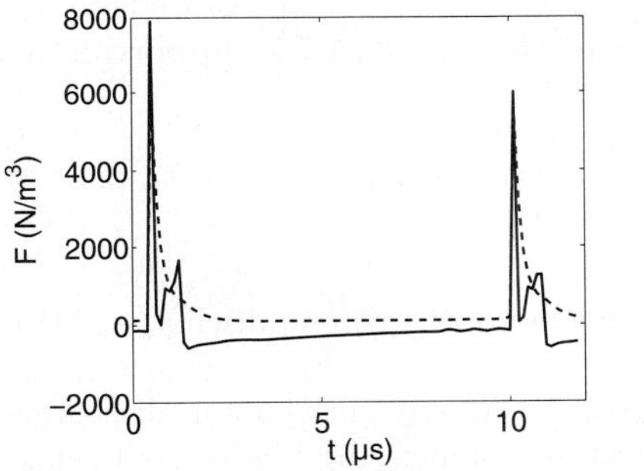


**Fig. 7.** Force at the anode (—) and at the cathode (- -)

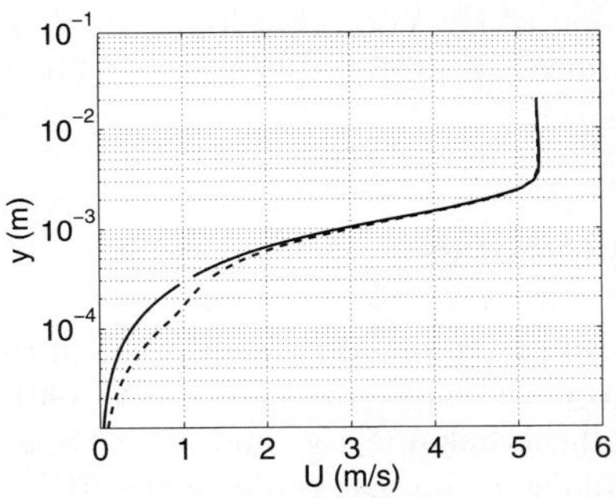


**Fig. 8.** flow velocity profile without plasma force (—) and with plasma force (- -)

From the aerodynamic point of view, the forces exerted by the discharges are also periodic. One of the most prominent features is that the forces are concentrated in the cathode and anode regions. The force levels are very high there, about 1000 to 5000 N/m$^3$. There is almost no force in the space between the electrodes because the density currents of positive and negative ions $j_+$ and $j_-$ are almost equal, but with opposite signs, see eq. (4). Figure 7 shows the evolution of the force near the anode and the cathode. The forces are locally important near electrodes but weak on average. The CFD code CEDRE has been used in order to simulate the effect of the calculated forces on a subsonic laminar flow over a flat plate. The external flow velocity is 5 m/s. Figure 8 presents the flow velocity profiles obtained at location $X = 1$ mm on Figure 1 when applying or not applying the force. The gain of velocity is obtained near the electrodes and is about 0.5 m/s. This influence is smaller than observed in experiments [4], where the gain of velocity can reach 1 m/s on a thickness of 5 mm, for an uncoming flow of 5 m/s. The force is insufficient to explain the experimental velocity gains because it is located just around both electrodes. To act more efficently on flows, this force should apply on a larger region.

## 4 Conclusion

In order to represent the creation of ionic wind by corona discharges, this paper proposes a simple coupling between a plasma kinetics model and a flow model. The results of the plasma simulations show a good agreement with the Trichel pulses. Once applied to the gas flow, the electric force obtained by the plasma simulations seems insufficient to explain the ionic wind experimentally observed [4]. Consequently, the efforts are now focused on enriching the kinetics of the positive corona discharges by taking into account the photoionization process, as well as plasma-flow coupling.

## Aknowledgements

The authors would like to thank G. Quinio from ONERA who realized the first version of the code developed in this work [13], and J. P. Boeuf from CPAT for a number of usefull discussions. The first author thanks DGA for supporting his research.

## References

1. Robinson M.: Movement of air in the electric wind of the corona discharge. AIEE transactions, New York, USA (1961).
2. Shcherbakov Y. V. and coll.: Drag reduction by AC streamer corona discharges along a wing-like profile plate. $31^{st}$ AIAA Plasmadynamics and Lasers Conference, Denver, USA, AIAA paper 2000-2670 (2000).
3. Roth J. R., Sin H., Madhan R., Wilkinson S.: Flow re-attachment and acceleration by paraelectric and peristaltic electrohydrodynamic effects. $41^{st}$ Aerospace Sciences Meeting and Exhibit, Reno, USA, AIAA paper 2003-531 (2003).
4. Pons J., Moreau E., Touchard G.: Surface DC corona discharges and AC barrier discharges in ambient air at atmospheric pressure: measurements of the induced ionic wind velocity. $15^{th}$ International Conference on Gas Discharges and their Applications, Toulouse, France (2004).
5. Post M. L., Corke T. C.: Plasma actuators for separation control on airfoils. $11^{th}$ International Symposium on Flow Visualization, Notre-Dame, USA (2004).
6. Zavialov I. N., Roupassov D. V., Starikovskii A. Yu., Saddoughi S. G.: Boundary layer control by gas discharge plasma. European Conference for Aerospace Sciences (EUCASS), Moscow, Russia (2005).
7. Boeuf J. P., Pitchford L. C.: Electrohydrodynamic force and aerodynamic flow acceleration in surface barrier discharge. J. Appl. Phys. **97**, 103307 (2005).
8. Mateo-Velez J. C., Thivet F., Rogier F., Degond P., Quinio G.: Numerical modeling of corona discharges and their interaction with aerodynamics. European Conference for Aerospace Sciences (EUCASS), Moscow, Russia (2005).
9. Morrow R.: The theory of positive glow corona. J. Phys. D: Appl. Phys. **30**, 3099-3114 (1997).
10. Bolsig, CPAT & Kinema Software, http://www.siglo-kinema.com/bolsig.htm.
11. Li S. Z., Uhm H. S.: Investigation of electrical breakdown characteristics in the electrodes of cylindrical geometry. Phys. Plasmas **11**, 3088-3095 (2004).

12. Chevalier P., Courbet B., Dutoya D., Klotz P., Ruiz E., Troyes J., Villedieu P.: CEDRE, Development and Validation of a Multiphysic Computational Software. European Conference for Aerospace Sciences (EUCASS), Moscow, Russia (2005).
13. Quinio G.: Modélisation numérique de la génération d'un plasma d'air dans un écoulement. PhD thesis, INSA Toulouse, France (december 2005).

# Numerical Methods for Reacting Gas Flow Simulations

S. van Veldhuizen[1], C. Vuik[1], and C.R. Kleijn[2]

[1] Delft University of Technology, Delft Institute of Applied Mathematics and J.M. Burgers Center, Mekelweg 4, 2628 CD Delft, The Netherlands
{s.vanveldhuizen, c.vuik}@tudelft.nl
[2] Delft University of Technology, Department of Multi Scale Physics and J.M. Burgers Center, Prins Bernardlaan 6, 2628 BW Delft, The Netherlands
c.r.kleijn@tudelft.nl

**Abstract.** In this study various numerical schemes for simulating 2D laminar reacting gas flows, as typically found in Chemical Vapor Deposition (CVD) reactors, are proposed and compared. These systems are generally modeled by means of many stiffly coupled elementary gas phase reactions between a large number of reactants and intermediate species. The purpose of this study is to develop robust and efficient solvers for the stiff heat-reaction system. The velocities are assumed to be given. For non-stationary CVD simulation, an optimal combination in terms of efficiency and robustness between time integration, nonlinear solvers and linear solvers has to be found. Besides stability, which is important due to the stiffness of the problem, the preservation of non-negativity of the species is crucial. It appears that this extra condition on time integration methods is much more restrictive towards the time-step than stability.

## 1 Introduction

In Chemical Vapor Deposition (CVD) literature, and also other reactive flow literature, one is usually looking for the steady state solution of the species equations (1). The usual procedure to find this steady state solution is to perform a (damped/relaxed) Newton iteration with an (arbitrary) initial solution. Hopefully, the Newton iteration converges to the steady state. If this is not the case one performs some (artificial) time stepping in order to find a better initial solution for the next Newton iteration. In this paper we present suitable time integration methods for stiff problems. Furthermore, we compare these integration methods by their performance, in terms of efficiency.

In our research we are not looking for the steady state solution only, but we also want the transient solution. In order to simulate this transient, we have to use a time integration method that can handle stiff problems.

## 2 Model for CVD Simulation

The mathematical model describing the CVD process consists of a set of PDEs with appropriate boundary and initial conditions, which describe the gas flow, the transport of energy, the transport of species and reactions in the reactor.

V.N. Alexandrov et al. (Eds.): ICCS 2006, Part II, LNCS 3992, pp. 10–17, 2006.

The gas mixture in the reactor is assumed to behave as a continuum. The gas flow in the reactor is assumed to be laminar. Since no large velocity gradients appear in CVD gas flows, viscous heating due to dissipation will be neglected. We also neglect the effects of pressure variations in the energy equation. The composition of the $N$ component gas mixture is described in terms of the dimensionless mass fractions $\omega_i = \frac{\rho_i}{\rho}$, $i = 1, \ldots, N$, having the property $\sum_{i=1}^{N} \omega_i = 1$. The transport of mass, momentum and heat are described respectively by the continuity equation , the Navier-Stokes equations and the transport equation for thermal energy expressed in terms of temperature $T$. See for instance [4, 6].

We assume that in the gas-phase $K$ reversible reactions take place. For the $k^{\text{th}}$ reaction the *net* molar reaction rate is denoted as $R_k^g$ $\left(\frac{\text{mole}}{m^3 \cdot s}\right)$. For an explicit description of the net molar reaction rate, we refer to [4, 6]. The mass diffusion flux is decomposed into concentration diffusion and thermal diffusion. In this study we describe ordinary diffusion in terms of effective diffusion coefficients $\mathbb{D}_i'$, such that we obtain

$$\frac{\partial(\rho\omega_i)}{\partial t} = -\nabla \cdot (\rho \mathbf{v} \omega_i) + \nabla \cdot (\rho \mathbb{D}_i' \nabla \omega_i) + \nabla \cdot (\mathbb{D}_i^T \nabla (\ln T)) + m_i \sum_{k=1}^{K} \nu_{ik} R_k^g, \quad (1)$$

where $\mathbb{D}_i^T$ the multi-component thermal diffusion coefficient for species $i$.

The main focus of our research is on efficient solvers for the above species equation(s) (1). Typically the time scales of the slow and fast reaction terms differ orders of magnitude from each other, and from the time scales of the diffusion and advection terms, leading to extremely stiff systems.

**Simplified System.** Since our research focuses on solving the species equations (1), we will only solve the coupled system of $N$ species equations, where $N$ denotes the number of gas-species in the reactor. The velocity field, temperature field, pressure field and density field are computed via another simulation package developed by Kleijn [5]. Other simplifications are the omission of surface reactions and thermal diffusion.

We consider a CVD process,which is a simplification of the CVD process considered in [5], that deposits silicon Si from silane $SiH_4$. The gas-mixture consists of 7 species and the reaction mechanism

$$\begin{array}{lccc}
G1: & SiH_4 & \rightleftarrows & SiH_2 + H_2 \\
G2: & Si_2H_6 & \rightleftarrows & SiH_4 + SiH_2 \\
G3: & Si_2H_6 & \rightleftarrows & H_2SiSiH_2 + H_2 \\
G4: & SiH_2 + Si_2H_6 & \rightleftarrows & Si_3H_8 \\
G5: & 2SiH_2 & \rightleftarrows & H_2SiSiH_2.
\end{array}$$

The studied reactor configuration is illustrated in Figure 1. As computational domain we take, because of axisymmetry, one half of the $r-z$ plane. The pressure in the reactor is 1 atm. From the top a gas-mixture, consisting of silane, with mass fraction $f_{\text{in},SiH_4} = 0.001$, and helium (the rest), enters the reactor with a uniform temperature $T_{\text{in}} = 300$ K and a uniform velocity $u_{\text{in}}$. At a distance of

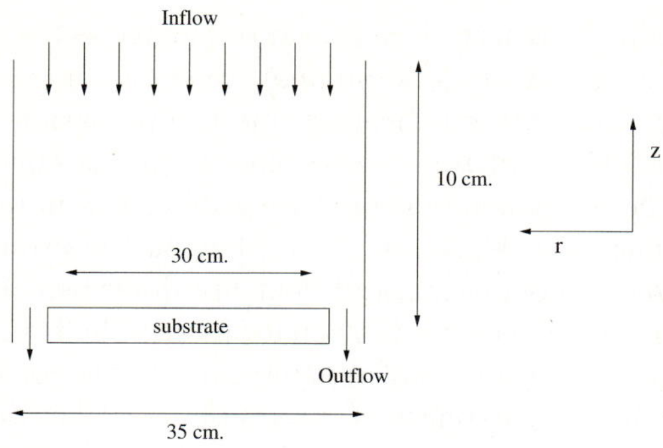


**Fig. 1.** Reactor geometry

10 cm. below the inlet a non rotating susceptor with temperature $T = 1000$ K and a diameter 30 cm. is placed.

We emphasize that this test-problem is not representing a practical process, but representing its computational problems. Further details on the test-problem can be found in [6].

## 3 Properties of Numerical Methods for Solving the Species Equations

As seen in the previous section the species eqns. (1) are PDEs of the advection-diffusion-reaction type. In order to have a unique solution appropriate boundary conditions and initial values have to be chosen.

To approximate the solution we use the Method of Lines (MOL), i.e., we first discretize in space, resulting into the ODE system

$$w'(t) = F(t, w(t)), \quad 0 < t \leq T, \quad w(0) \text{ given.} \tag{2}$$

The next step is to integrate the ODE system (2) with an appropriate time integration method. We remark that the stiff reaction terms in CVD motivates to integrate parts of $F(t, w(t))$ implicitly. In general, due to the nonlinearities in the reaction term, (huge) nonlinear systems have to be solved.

The topic of this research is to find the best combination of time integration, nonlinear and linear solvers in terms of efficiency. Note that if the computational cost of one time step is (very) expensive, then a time integration method that needs more, but computational cheaper, time steps is better in terms of efficiency.

Besides the efficiency criteria, also some other properties are desired for the numerical methods. As already mentioned in Section 2 the system of species equations is stiff. Following [2], we say that stiffness indicates a class of problems for which implicit methods perform (much) better than explicit methods. The eigenvalues of the Jacobian $\frac{\delta f}{\delta y}$ play certainly a role in this decision, but quantities such as the dimension of the system and the smoothness of the solution are also important.

**Positivity.** A natural property for mass fractions is that they are non-negative. As a consequence, it should also hold for the mathematical model, spatial discretization and time integration of the process. While the first one is obvious, the latter two should not introduce any (small) negative components causing blow up of the solution. It appears that this extra condition on time integration methods is much more restrictive towards the time step than stability. We remark that positivity for spatial discretization can be assured by locally first order upwinding.

An ODE system $w'(t) = F(t, w(t))$, $t \geq 0$, is called positive if $w(0) \geq 0$ implies $w(t) \geq 0$ for all $t > 0$. It is easy to prove that linear systems $w'(t) = Aw(t)$ are positive if and only if $a_{ij} \geq 0$ for $i \neq j$. See [7]. For general nonlinear semi-discretizations $w'(t) = F(t, w(t))$, it appears that unconditional positivity is a very restrictive requirement. Suppose that $F(t, w(t))$ satisfies the condition:

**Condition 1.** *There is an $\alpha > 0$, with $\alpha$ as large as possible, such that $\alpha\tau \leq 1$ and*

$$v + \tau F(t, v) \geq 0 \quad \textit{for all} \quad t \geq 0, v \geq 0. \tag{3}$$

Application of Euler Forward to the nonlinear system $w'(t) = F(t, w(t))$ gives

$$w_{n+1} = w_n + \tau F(t_n, w_n). \tag{4}$$

Provided that $w_n \geq 0$, Condition 1 guarantees positivity for $w_{n+1}$ computed via Euler Forward (4). Furthermore, assume that $F(t, w(t))$ also satisfies:

**Condition 2.** *$F(t, v)$ is continuously differentiable and*

$$\|I - \tau J_F(t, v)\| \leq C, \quad \textit{for any } v \in \mathbb{R}^n, \ t \geq 0 \textit{ and } \tau > 0, \tag{5}$$

*whereby $C$ is a positive constant, $I$ the identity matrix and $J_F(t, v)$ the Jacobian matrix of derivatives of $F(t, v)$ with respect to $v$.*

In [2] it has been proven that Condition 1 and 2 imply positivity for Euler Backward for any step-size $\tau$. However, in practice the solutions of the resulting nonlinear systems have to be approximated, which might introduce small negative components. In the case the negative components of the solution are the result of rounding errors, then it is justified to set them to zero. In the case one has negative components in the solution as consequence of the nonlinear (Newton) solver, then the most common method to avoid negative concentrations is *clipping*. Clipping has the disadvantage that mass is not longer preserved. In practice it is desired not to use clipping. In our experience, the implementation of variable time step size algorithm, see [6], avoids negative concentrations without the use of clipping.

We conclude this section with the claim that Euler Backward is the only time integration that is unconditionally positive. For a proof we refer to [1]. This means that for any higher order (implicit) time integration method a time step criterion is needed to ensure preservation of non-negativity.

# 4 Suitable Time Integration Methods (TIM)

In this section we briefly present integration methods that are suitable, from a theoretic point of view, for the time integration of the species equations. More comprehensive descriptions are given in [2, 6]. At the end of this section we will also make some remarks on the linear and nonlinear solvers.

From the previous section it is clear that the Euler Backward method is a suitable method to perform time integration. It has the advantage of being unconditionally positive. Disadvantages are the first order consistency and the probably high computational costs for one time step. The latter is due to the fact that the succeeding approximations are computed in an implicit manner.

## 4.1 Time Integration Methods

We will discuss a selection of time integration methods that have good properties in both stability and positivity, or TVD.

**Rosenbrock Methods.** The two stage Rosenbrock method

$$\begin{aligned} w_{n+1} &= w_n b_1 k_1 + b_2 k_2 \\ k_1 &= \tau F(w_n) + \gamma\tau \mathbf{A} k_1 \\ k_2 &= \tau F(w_n + \alpha_{21} k_1) + \gamma_{21}\tau \mathbf{A} k_1 + \gamma\tau \mathbf{A} k_2, \end{aligned} \tag{6}$$

with $\mathbf{A} = F'(w_n)$ is the Jacobian matrix of $F$, and $b_1 = 1 - b_2$, $\alpha_{21} = \frac{1}{2b_2}$ and $\gamma_{21} = -\frac{\gamma}{b_2}$, is interesting. The method is of order two for arbitrary $\gamma$ as long as $b_2 \neq 0$ . The stability function is given as

$$R(z) = \frac{1 + (1-2\gamma)z + (\gamma^2 - 2\gamma + \frac{1}{2})z^2}{(1-\gamma z)^2}. \tag{7}$$

For $\gamma \geq \frac{1}{4}$ the method is unconditionally stable. For $\gamma_+ = 1 + \frac{1}{2}\sqrt{2}$, we have the property that $R(z) \geq 0$, for all negative real $z$. For diffusion-reaction problems, which have negative real eigenvalues, this property ensures positivity of the solution. It appears that the second order Rosenbrock method performs quite well with respect to the positivity property, as has been shown in [8]. In [8] it is conjectured that the property $R(z) \geq 0$ for all negative real $z$ plays a role.

**Backward Differentiation Formulas (BDF).** The $k$-step BDF methods are implicit, of order $k$ and defined as

$$\sum_{j=0}^{k} \alpha_j w_{n+j} = \tau F(t_{n+k}, w_{n+k}), \quad n = 0, 1, \ldots, \tag{8}$$

which uses the $k$ past values $w_n, \ldots, w_{n+k-1}$ to compute $w_{n+k}$. Remark that the most advanced level is $t_{n+k}$ instead of $t_{n+1}$. The 1-step BDF method is Backward Euler. The 2-step method is

$$\frac{3}{2} w_{n+2} - 2w_{n+1} + \frac{1}{2} w_n = \tau F(t_{n+2}, w_{n+2}), \tag{9}$$

and the three step BDF is given by

$$\frac{11}{6}w_{n+3} - 3w_{n+2} + \frac{3}{2}w_{n+1} - \frac{1}{3}w_n = \tau F(t_{n+3}, w_{n+3}). \tag{10}$$

*Remark 1.* A disadvantage of linear multi-step methods is that the first $k-1$ approximations cannot be computed with the linear $k$-step scheme. To compute the first $(k-1)$ approximations, one could use for the first step a BDF 1-step method, for the second approximation a BDF 2-step method, ... and for the $(k-1)^{\mathrm{st}}$ approximation a BDF $(k-1)$-step scheme.

For the 2-step BDF method we obtain positivity, under Conditions 1 and 2, of $w'(t) = F(t, w(t))$ whenever $\alpha\tau \le \frac{1}{2}$, provided that $w_1$ is computed from $w_0$ by a suitable starting procedure, i.e., $w_1$ has been computed such that $w_1 \ge 0$ holds. For a derivation we refer to [2, 6].

**IMEX Runge-Kutta Chebyshev Methods.** The second order Runge-Kutta Chebyshev method is given as

$$\begin{aligned} w_{n0} &= w_n, \\ w_{n1} &= w_n + \tilde{\mu}_1 \tau F(t_n + c_0\tau, w_{n0}), \\ w_{nj} &= (1 - \mu_j - \nu_j) w_n + \mu_j w_{n,j-1} + \nu_j w_{n,j-2} + \qquad j = 1, \ldots, s \\ &\quad + \tilde{\mu}_1 \tau F(t_n + c_{j-1}\tau, w_{n,j-1}) + \tilde{\gamma}_j \tau F(t_n + c_0\tau, w_{n0}), \\ w_{n+1} &= w_{ns}. \end{aligned} \tag{11}$$

The coefficients $\omega_0, \omega_1, b_j, c_j, \tilde{\mu}_j, \ldots$ can be found in [6]. In Figure 2 its stability region is given. The parameter $\beta(s)$ moves to $-\infty$ when the number of stages $s$ increases.

The IMEX extension of the above scheme is as follows. Suppose we have an ODE system $w'(t) = F(t, w(t))$, where $F(t, w(t))$ can be split as

$$F(t, w(t)) = F_E(t, w(t)) + F_I(t, w(t)). \tag{12}$$

In (12) the term $F_I(t, w(t))$ is the part of $F$ which is (supposed to be) too stiff to be integrated by an explicit Runge-Kutta Chebyshev method. Obviously, the term $F_E(t, w(t))$ is the moderate stiff part of $F$ that can be integrated in an explicit manner using RKC methods. The first stage of (11) becomes in the IMEX-RKC scheme

$$w_{n1} = w_n + \tilde{\mu}_1 \tau F_E(t_n + c_0\tau, w_{n0}) + \tilde{\mu}_1 \tau F_I(t_n + c_1\tau, w_{n1}), \tag{13}$$

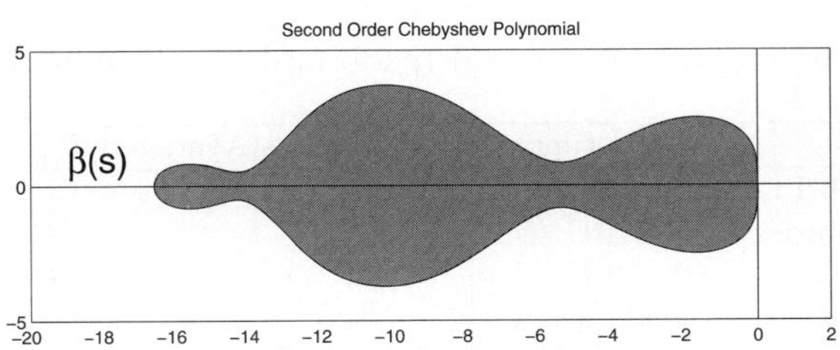


**Fig. 2.** Stability region of (11) with $s = 5$

with $\tilde{\mu}_1$ as defined before. Note that the highly stiff part of $F$ is treated implicitly. The other $(s-1)$ subsequent stages of (11) will be modified in a similar way.

With respect to stability of this IMEX extension of (11) we remark that the implicit part is unconditionally stable, whereas the stability condition for the explicit part remains unchanged.

### 4.2 Nonlinear and Linear Solvers

In all suitable TIM nonlinear systems $F(x) = 0$, $x \in \mathbb{R}^n$ have to be solved. The Newton iteration is, with its second order convergence, an obvious choice. The disadvantage of having local convergence will disappear if one uses a line-search algorithm, such that

$$\|F(x^{k+1})\| \leq \|F(x^k)\| \quad k = 0, 1, 2, \ldots, \tag{14}$$

for some norm in $\mathbb{R}^n$. More background information can be found in [3, 6].

In the Newton iteration linear systems have to be solved. In most 2D applications direct solvers like LU factorization are still applicable. To reduce the amount of work one usually re-arranges the unknowns, in order to reduce the bandwidth of the matrix. Also in our case it is possible to reduce the bandwidth of the Jacobian considerably. The way to do this is described in [7].

For 3D problems direct solvers (LU factorization) are no longer applicable. To approximate the solution of the linear systems one has to switch to iterative linear solvers like, for instance, Krylov Subspace methods.

## 5 Numerical Results

In this section we compare the performance of the TIM of Section 4.1 for solving the species equations. If necessary, nonlinear systems will be solved by the full Newton iteration. Linear systems will be solved using the LU factorization with rearranging, as mentioned in Section 4.2. Since the solutions differ not much, we are only interested in the performance of the TIM.

For a comparison between the workloads of the various TIM, we look to the amount of CPU time, the number of time steps and the workload per time step it takes to reach steady state. We say that steady state is reached when the norm of the right hand side of (1) is below a threshold, in our case $10^{-4}$, and the fluxes at the boundaries are constant. The threshold is equal to the accuracy

**Table 1.** Workloads of various TIM (performed on a Pentium IV 3.2 GHz)

| TIM | CPU time | # time steps | Average CPU time / time step |
|---|---|---|---|
| Euler Backward | 600 CPU sec | 163 | 3 CPU sec |
| $2^{\text{nd}}$ order Rosenbrock | 375 CPU sec | 100 | 3 CPU sec |
| BDF-2 | 600 CPU sec | 63 | 8 CPU sec |
| IMEX-RKC(5) | > 5000 CPU sec | > 500 | 25 CPU sec |

of the solution of the system of Navier Stokes eqns., pressure eqn., etc. For the different TIM mentioned in Section 4.1 the workloads are given in Table 1.

## 6 Conclusions

Based on Table 1, we conclude that for this 2D test-problem Rosenbrock is the cheapest TIM to solve (1). The 'bad' performance of the IMEX RKC scheme is due to that per time step more than one nonlinear system has to be solved. Although these nonlinear systems are cheaper to solve, it did not pay off in this 2D test case. However, this property can become interesting in 3D simulations.

## References


1. C. Bolley and M. Crouzeix, *Conservation de la Positivité Lors de la Discrétisation des Problèmes d'Évolution Paraboliques*, RAIRO Anal. Numer. 12, pp. 237-245, (1973)
2. W. Hundsdorfer and J.G. Verwer, *Numerical Solution of Time-Dependent Advection-Diffusion-Reaction Equations*, Springer Series in Computational Mathematics, 33, Springer, Berlin, (2003)
3. C.T. Kelley, *Solving Nonlinear Equations with Newton's Method*, Fundamentals of Algorithms, SIAM, Philadelphia, (2003)
4. C.R. Kleijn, *Transport Phenomena in Chemical Vapor Deposition Reactors*, PhD thesis, Delft University of Technology, (1991)
5. C.R. Kleijn, *Computational Modeling of Transport Phenomena and Detailed Chemistry in Chemical Vapor Deposition- A Benchmark Solution*, Thin Solid Films, 365, pp. 294-306, (2000)
6. S. van Veldhuizen, *Efficient Solution Methods for Stiff Systems of Advection-Diffusion-Reaction Equations*, Literature Study, Technical Report at the Delft University of Technology, TWA-05-05 , (2005)
7. S. van Veldhuizen, *Time Integration for Reacting Gas Flow Simulations*, Technical Report at the Delft University of Technology, to appear, (2006)
8. J.G. Verwer, E.J. Spee, J.G. Blom and W. Hundsdorfer, *A Second-Order Rosenbrock Method Applied to Photochemical Dipersion Problems*, SIAM Journal on Sci. Comp., 20, pp.1456-1480, (1999)

# Reduced Flame Kinetics Via Rate-Controlled Constrained Equilibrium

Stelios Rigopoulos

The University of Manchester, Manchester M60 1QD, UK

**Abstract.** The dynamical evolution of a chemically reacting system is governed by the equations of chemical kinetics, which exhibit a wide range of time scales thus giving rise to stiff equations. In the rate-controlled constrained equilibrium method (RCCE), the dynamical evolution of the system is governed by the kinetics of the species associated with the slower timescales (kinetically controlled), while the remaining species are calculated via a constrained minimisation of the Gibbs free energy of the mixture. This permits the derivation of a general set of differential-algebraic equations, which apply to any reduced system given a particular selection of kinetically controlled species. In this paper, it is shown how the differential-algebraic formulation of RCCE can be derived from first principles, in the form of an extension of the computation of chemical equilibrium via miminisation of the free energy. Subsequently, RCCE is employed to reduce a comprehensive combustion mechanism and to calculate the burning velocity of premixed $H_2$-$O_2$ and $CH_4$-air flames under a range of pressures and equivalence ratios.

## 1 Introduction

Chemical kinetics describes the dynamical evolution of a chemical system via a set of rate equations for all the species and radicals involved, and all possible elementary reactions. Such comprehensive mechanisms often comprise very large numbers of species and reactions, and their integration is additionally hampered by excessive stiffness, due to the presence of a vast range of chemical time scales. Combustion is a classic example, where hundreds of reactions are often included in detailed mechanisms, and in addition they must be coupled with fluid flow. Naturally ways have been sought to describe kinetics via a reduced set of variables, taking advantage of the separation of time scales. In principle, accounting for the kinetics of the species associated with the slow time scales should suffice, if one had a way of calculating the equilibrated ones. Conventional derivation of reduced mechanisms proceeds by examining the detailed mechanism and carefully applying steady-state and partial equilibrium assumptions to individual species and reactions. This laborious procedure must be carried out individually for every reaction mechanism and physical problem. Recently certain mathematical methods have appeared that attempt to identify the time scales separation in a methodical way [9], [8].

V.N. Alexandrov et al. (Eds.): ICCS 2006, Part II, LNCS 3992, pp. 18–25, 2006.

Rate-controlled Constrained Equilibrium (RCCE), first introduced in [7], also discussed in [6], provides a different viewpoint of the problem. As with all methods, the system is guided by the kinetics of the kinetically-controlled species. The remaining ones are then calculated via a minimisation of the Gibbs free energy of the mixture subject to the constraint that the kinetically-controlled species retain their present values. Thus the dynamical evolution is still governed by kinetics alone. One can draw an analogy with the computation of chemical equilibrium via equilibrium constants and via free energy minimisation. Previous work by the author [3] has addressed the formulation of RCCE as a differential-algebraic (DAE) problem, and shown how it can be applied to non-premixed flames. In this work the derivation of the DAE formulation from first principles is discussed, and subsequently RCCE is employed to predict the burning velocity of premixed $H_2$-$O_2$ and $CH_4$-air flames of varying composition and pressure.

## 2 Derivation of the Constrained Equilibrium Equations

*Minimisation of Gibbs free energy*
To derive the RCCE equations, we start from the concept of chemical equilibrium via minimisation of the Gibbs free energy. The mixture Gibbs free energy is:

$$g = \sum_{j=1}^{N} \mu_j n_j \tag{1}$$

where $n_j$ is the composition in kmol/kg of mixture, while the chemical potential, $\mu_j$, can be written in terms of the chemical potential in the standard state [2]:

$$\mu_j = \mu_j^o + RTln\frac{n_j}{n} + RTln\frac{P}{P_o} \tag{2}$$

The condition for the minimisation of Gibbs free energy is:

$$\frac{\partial g}{\partial n_j} = 0 \tag{3}$$

subject to the constraints expressing conservation of elements, enthalpy and pressure:

$$E_i = \sum_{j=1}^{N} a_{ij}^e n_j \quad (i = 1, ..., M_e) \tag{4}$$

$$\sum_{i=1}^{N} n_j h_j = h \tag{5}$$

$$\sum_{i=1}^{N} n_j \rho RT = P \tag{6}$$

*Full Equilibrium*
The computation of the species' concentrations at the equilibrium state is thus reduced to the solution of a constrained optimisation problem that can be solved with the method of Lagrange multipliers. We first introduce introducing the parameters $\lambda_i^e$ which we call element potentials and form the following function:

$$g' = g + \sum_{i=1}^{M_e} \lambda_i^e \left(E_i - a_{ij}^e n_j\right) \tag{7}$$

According to the method of Lagrange multipliers, the original operation is equivalent to computing the extrema of this function, i.e.:

$$\frac{\partial g'}{\partial n_j} = 0 \tag{8}$$

Substituting eq. 7 and eq. 1 in eq. 8 yields:

$$\mu_j + \sum_{i=1}^{M_e} \lambda_i^e a_{ij}^e = 0 \tag{9}$$

Substituting eq. 2 we obtain:

$$\mu_j^o + RTln\frac{n_j}{n} + RTln\frac{P}{P_o} + \sum_{i=1}^{M_e} \lambda_i^e a_{ij}^e = 0 \tag{10}$$

This system of equations, together with eqs. 4, 5, 6 can be solved to yield the composition, $n_j^*$, temperature, density and Lagrange multipliers at the equilibrium state - an efficient method is that of Gordon and McBride [2].

*Constrained Equilibrium*
We now extend the minimisation of free energy to the constrained equilibrium state by adding the following constraints to those defined by eqs. 4, 5, 6:

$$C_i = \sum_{j=1}^{N} a_{ij}^c n_j \quad (i = 1, ..., M_c) \tag{11}$$

In essence, this states that certain species (or linear combinations of) must retain a certain fixed value. Thus, if for instance a system contains C,H,O and N atoms, the full equilibrium may contain certain $NO_x$ species. If, however, we impose the additional constraint that $N_2$ retains a value such that all of the N atoms must occur in the form of $N_2$, we would not observe any $NO_x$ species in the constrained equilibrium state. To perform the mimimisation with the method of Lagrange multipliers, we introduce the additional parameters $\lambda_i^c$, which we call constraint potentials, and the function to be minimised is now formed as:

$$g' = g + \sum_{i=1}^{M_e} \lambda_i^e \left(E_i - a_{ij}^e n_j\right) + \sum_{i=1}^{M_c} \lambda_i^c \left(C_i - a_{ij}^c n_j\right) \tag{12}$$

Substituting eq. 12 and eq. 11 in eq. 1 yields:

$$\mu_j + \sum_{i=1}^{M_e} \lambda_i^e a_{ij}^e + \sum_{i=1}^{M_c} \lambda_i^c a_{ij}^c = 0 \tag{13}$$

Using eq. 2, we now obtain:

$$\mu_j^o + RTln\frac{n_j}{n} + RTln\frac{P}{P_o} + \sum_{i=1}^{M_e} \lambda_i^e a_{ij}^e + \sum_{i=1}^{M_c} \lambda_i^c a_{ij}^c = 0 \tag{14}$$

This system of equations must be solved together with the constraints, i.e. eqs. 4, 5, 6 and 11. The method of Gordon and McBride [2] can be straightforwardly extended to carry out this calculation.

*Rate-Controlled Constrained Equilibrium*
The extension to RCCE is now straightforward. Chemical kinetics yields the production rate for each species in terms of the reaction rates of individual reactions, $r_k$, and the stoichiometric factors, $\nu_{jk}$, as follows:

$$w_j = \sum_{k=1}^{NR} (\nu_{jk} r_k) \quad (j = 1, ..., N) \tag{15}$$

The main idea of RCCE is to allow certain species (or linear combinations of) to evolve according to kinetics, while computing the remaining species via a constrained equilibrium calculation where the kinetically controlled species are the constraints. The selection of kinetically controlled and equilibrated species can be based on criteria for analysing the separation of time scales, e.g. Peters' criteria [10] or Computational Singular Perturbation (CSP) [8]. RCCE is not a criterion itself; instead it provides a system of equations that describe the evolution of the reduced system, and which has the same general form whatever the selection.

The dynamical evolution of the constraints can be computed as follows:

$$\frac{dC_i}{dt} = \sum_{j=1}^{N} \left( a_{ij}^c \sum_{k=1}^{NR} (\nu_{jk} r_k) \right) \quad (i = 1, ..., M_c) \tag{16}$$

These ODEs must be computed together with the algebraic equations of constrained equilibrium, i.e. eqs. 14, 11, 4, 5, 6. A more convenient form of eq. 14 can be obtained by taking exponentials, so that the equation is directly solved in terms of the composition at the constrained equilibrium state ($n_j^*$):

$$n_j^* = n\,\frac{P}{P_o}\,exp\left(\frac{-\mu_j^o}{RT}\right)\,exp\left[\sum_{i=1}^{M_e} (a_{ij}^e \lambda_i^e)\right]\,exp\left[\sum_{i=1}^{M_c} (a_{ij}^c \lambda_i^c)\right] \quad (j = 1, ..., N) \tag{17}$$

Therefore the basic formulation of RCCE is a differential-algebraic system comprising the ODEs (16) for the constraints and the algebraic equations (17),

(11), (4), (5), (6). Solution of this system determines the constrained equilibrium composition, potentials, temperature and density at each time step. Differential-algebraic systems can be solved using programs such as DASSL [1]. An alternative, pure ODE form can be derived using a process called index reduction; this has been presented in [3] and will not be shown here.

# 3 Results and Discussion

We shall consider a freely propagating, 1-D premixed laminar flame of constant pressure, with equal diffusivities and Lewis numbers equal to unity. Under these conditions the governing equations assume a well-known form:

$$\frac{\partial \rho u}{\partial x} = 0 \tag{18}$$

$$\rho u \frac{\partial Y_i}{\partial x} = \frac{\partial}{\partial x}\left(\rho D \frac{\partial Y_i}{\partial x}\right) + w_i \tag{19}$$

$$\rho u \frac{\partial h}{\partial y} = \frac{\partial}{\partial y}\left(\frac{\lambda}{Cp}\frac{\partial h}{\partial y}\right) \tag{20}$$

where $Y_i$ are the mass fractions of the constrained species; at the same time, the algebraic equations of RCCE must be satisfied.

We will apply RCCE to reduce a comprehensive mechanism of $CH_4$ combustion due to Lindstedt and co-workers [4] [5] comprising 63 species and 415 reactions. We will calculate a $H_2$-air flame using the hydrogen part of the mechanism, comprising only 10 species - in which case, the reduction in CPU is not notable - and a $CH_4$-air flame using the entire mechanism - resulting in a typical

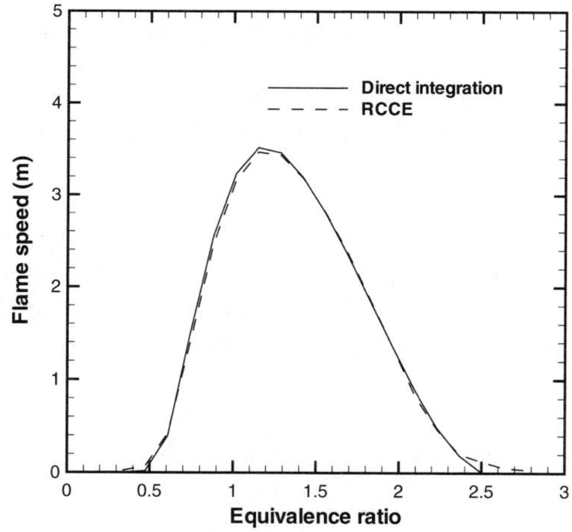


**Fig. 1.** Burning velocity of stoichiometric $H_2$ - air flame vs. equivalence ratio, comparison between direct integration and RCCE

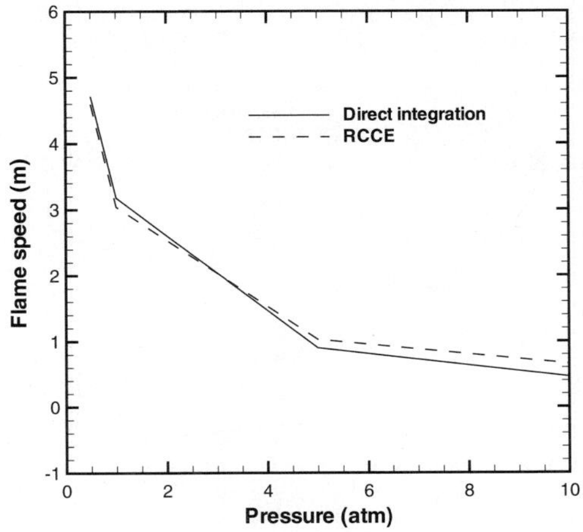


**Fig. 2.** Burning velocity of stoichiometric $H_2$ - air flame vs. equivalence ratio, comparison between direct integration and RCCE

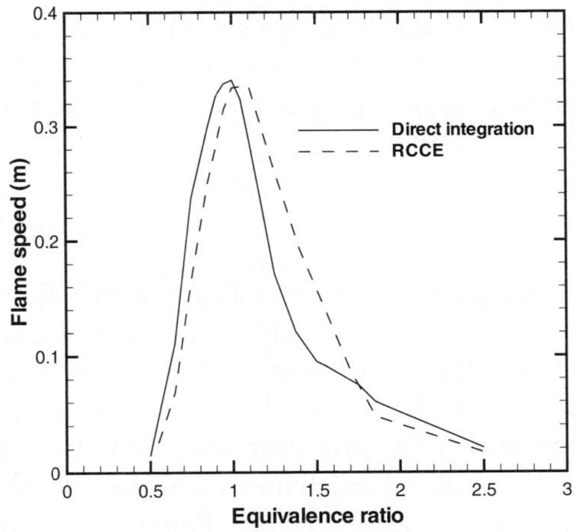


**Fig. 3.** Burning velocity of stoichiometric $CH_4$ - air flame vs. equivalence ratio, comparison between direct integration and RCCE

reduction of CPU time from 15 hours to 4. For the $H_2$ part we constrain $H_2$, $O_2$, $H_2O$, H and OH. For the $CH_4$ flame we constrain the above plus $CH_4$, $CO_2$, CO, O and $N_2$. These are the most common main species appearing in reduced mechanisms.

Figure 1 shows the predictions of the burning velocity as a function of equivalence ratio for the hydrogen flame. Results are excellent for the entire range of equivalence ratios. Results for different pressures are also shown in fig. 2; these also exhibit good agreement. Figure 3 shows results for the $CH_4$-air flame over a wide range of equivalence ratios, and good agreement is exhibited in the lean and stoichiometric mixture, with some discrepancy in the rich side. Results for

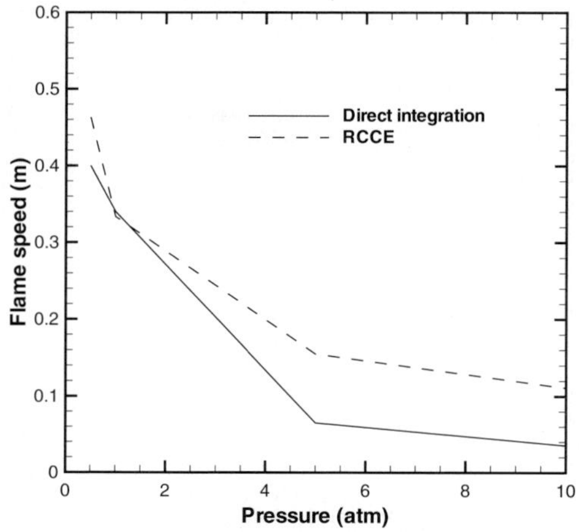


**Fig. 4.** Burning velocity of stoichiometric $CH_4$ - air flame vs. pressure, comparison between direct integration and RCCE

different pressures for the $CH_4$-air flame are also shown in fig. 4, and at elevated pressure the discrepancy is more significant, indicating that more constraints may be necessary to obtain quantitative agreement in that range.

## 4 Conclusions

In this paper we have presented the derivation of the RCCE equations, starting from the basic equations of chemical equilibrium and chemical kinetics, and shown results for the burning velocity of a laminar premixed flame. RCCE provides a rigorous framework for reduced chemistry computations; the main advantage is that a general differential-algebraic system of equations can be derived that describes the reduced mechanism, irrespective of the choice of kinetically controlled and equilibrated species. Thus once a computer code for solving this system has been set up, investigation of reduced mechanisms is greatly facilitated.

## References

1. Brenan, K.E., Campbell, S.L., Petzold, L.R.: Numerical Solution of Initial Value Problems in Differential-Algebraic Equations, North-Holland (1989).
2. Gordon, S., McBride, B.J.: Computer Program for Calculation of Complex Chemical Equilibrium Compositions and Applications I. Analysis, NASA Ref. Pub. **1311**, (1994).
3. Jones, W. P., Rigopoulos, S.: Rate-controlled constrained equilibrium: Formulation and application to nonpremixed laminar flames. Combustion and Flame, **142** (2005) 223–234.

4. Juchmann, W., Latzel, H., Shin, D.I., Peiter, G., Dreier, T., Volpp, H.R., Wolfrum, J., Lindstedt, R.P. and Leung, K.M.: Absolute radical concentration measurements and modeling of low-pressure $CH_4/O_2/NO$ flames, Proc. Combust. Inst. 27 (1998) 469-476.
5. Sick, V., Hildenbrand, F., Lindstedt, R.P.: Quantitative laser-based measurements and detailed chemical kinetic modelling of nitric oxide concentrations in methane-air counterflow diffusion flames, Proc. Combust. Inst., 27 (1998) 1401-1409.
6. Keck, J.C.: Rate-controlled constrained-equilibrium theory of chemical reactions in complex systems, Prog. Energy Combust. Sci. 16 (1990) 125-154.
7. Keck, J.C., Gillespie, D.: Rate- controlled partial- equilibrium method for treating reacting gas mixtures, Combust. Flame 17 (1971) 237-241.
8. Lam, S.H., Goussis, D.A.: Understanding Complex Chemical Kinetics with Computational Singular Perturbation, *24th Symp. (Int.) Comb.*, 931:941 (1988)
9. Maas, U., Pope, S.B.: Simplifying Chemical Kinetics: Intrinsic Low-Dimensional Manifolds in Composition Space, Combust. Flame 88 (3-4) (1992) 239-264.
10. Peters, N., Kee, R. J.: Computations of laminar stratched methane-air diffusion flames using a reduced four-step mechanism, Combust. Flame, 68, 17-29, (1987).

# Flow Patterns in the Vicinity of Triple Line Dynamics Arising from a Local Surface Tension Model

J. Monnier and I. Cotoi

Laboratoire de Modelisation et Calcul (LMC-IMAG),
Project-Team IDOPT, BP 53, F-38041 Grenoble Cedex 9, France
{Jerome.Monnier, Cotoi}@imag.fr

**Abstract.** We model and simulate numerically a droplet impact onto a solid substrate. The triple line dynamics modelling is implicit (as opposed to classical explicit mobility relations), it is based on the Shikhmurzaev equations. These equations include generalized Navier slip type boundary conditions with extra local surface tension gradient terms. Numerical results when spreading and recoiling are presented. A particular attention is paid to flow patterns near the contact line.

## 1 Introduction

One of the main difficulty to simulate present droplet flows is the contact line dynamics modelling. This problem has been widely studied and still remains an open problem. Let us cite macroscopic models with slip boundary conditions, see [4], [5], [3], mesoscopic models with diffuse interface, see [6] and also [9], and connection with molecular dynamics, see [11]. The most frequent contact line model is an explicit mobility relation giving the contact line velocity in function of the wetting angle value at any time (Tanner type laws). Nevertheless, it has been showed, [2], that the dynamic wetting angle cannot be determined inherently by such a mobility relation.

In the present paper, we consider an "implicit" model based on the Shikhmurzaev theory, [10], [2]. This model lead to generalized Navier slip type boundary conditions with local surface tension gradient terms. It does not impose the contact line velocity nor the wetting angle since they are a response of the full model. In Section 2, we present the equations: 2D axi-symmetric Navier-Stokes equations with an ALE formulation. In Section 3, we discretize the equations using a finite element scheme. We introduce a contact line algorithm which imitates caterpillar motion observed in experiments. Curvature is computed using a local Bezier least-square approximation. In Section 4, first we consider a Tanner type law and we test the robustness of our algorithm. Then, we consider both spreading phase and recoiling phase, using the present "implicit" model based on the Shikhmurzaev theory. The extra local terms appearing in this model are setted by preliminary results presented in [7] and [8]. We show different flow patterns generated by these terms near the triple point. Finally, we obtain encouraging results for both phases.

V.N. Alexandrov et al. (Eds.): ICCS 2006, Part II, LNCS 3992, pp. 26–33, 2006.

# 2 Mathematical Model

## 2.1 Equations

The droplet dynamics is modelled by the unsteady incompressible Navier-Stokes equations 2D axi-symmetric. We denote by $\boldsymbol{u} = (u_r, u_z)^T$ the fluid velocity, $p$ its pressure, $\Sigma$ the stress tensor, $D$ the deformation tensor and $Re$ the Reynolds number. We denote by $(\boldsymbol{\tau}, \boldsymbol{n})$ the unit tangential and external normal vectors such that it is direct. We set: $\boldsymbol{\Sigma_n} = \Sigma.\boldsymbol{n}$; $\boldsymbol{\Sigma_n} = \Sigma_n \boldsymbol{n} + \Sigma_\tau \boldsymbol{\tau}$. The r-momentum equations with ALE formulation is:

$$Re\left(\frac{\partial u_r}{\partial t} + (\boldsymbol{u} - \gamma) \cdot \nabla u_r\right) = -\frac{\partial p}{\partial r} + 2\left(\frac{1}{r}\frac{\partial(r\, D_{rr}(\boldsymbol{u}))}{\partial r} + \frac{\partial(D_{rz}(\boldsymbol{u}))}{\partial z} - \frac{u_r}{r^2}\right) \quad (1)$$

where $\boldsymbol{\gamma}$ is the equivalent velocity field. The z-momentum equation is similar. The continuity equation is: $\frac{1}{r}\frac{\partial(ru_r)}{\partial r} + \frac{\partial u_z}{\partial z} = 0$, with initial conditions. Boundary conditions on the free surface (liquid-gas) are:

$$\boldsymbol{\Sigma}_n = (-p_{ext} + \frac{\kappa}{Ca})\boldsymbol{n} + h\boldsymbol{\tau} \qquad \text{in } (0,T) \times \Gamma_f \quad (2)$$

where $Ca$ is the Capillary number, $\kappa$ is the mean curvature and $p_{ext}$ is the external pressure. The extra term $h$ is given (see Shikhmurzaev's theory below). We have classical boundary conditions on the symmetry axis. We decompose the liquid-solid interface into two parts: $\Gamma_{ad}$ and $\Gamma_{sl}$. $\Gamma_{sl}$ denotes a "small" part of the liquid-solid interface near the triple point. We consider a generalized Navier slip type boundary condition (local slipping):

$$\begin{cases} \boldsymbol{u}.\boldsymbol{n} = 0 & \text{in } (0,T) \times \Gamma_{sl} \\ \Sigma_\tau = -[\beta \boldsymbol{u} + \boldsymbol{g}].\boldsymbol{\tau} & \text{in } (0,T) \times \Gamma_{sl} \end{cases} \quad (3)$$

where $\beta$ is a given sliding coefficient. The extra term $\boldsymbol{g}$ is given, it models either a surface tension gradient in the Shikhmurzaev's theory or a uncompensated Young stress in the Qian-Wang-Sheng theory, see below. On $\Gamma_{ad}$, we impose adherence boundary conditions: $\boldsymbol{u} = 0$. The free surface $\Gamma_f$ is transported by $\boldsymbol{u}$, it is described by a function $\phi(t, r(t), z(t))$. We have: $\frac{d\phi}{dt} = \frac{\partial \phi}{\partial t} + \boldsymbol{u} \cdot \nabla \phi = 0$ in $(0,T) \times \Omega$, with initial conditions.

## 2.2 Contact Line Dynamics Modelling

We consider two different types of model for the contact line dynamics: an explicit model (Tanner type law) and an "implicit" one deriving from Shikhmurzaev theory.

*Mobility relation (Tanner type Law).* The contact line velocity and the wetting angle are related by:

$$U_{CL}(t) = k\, \frac{(\theta(t) - \theta^{eq})^\gamma}{(\theta^{eq})^\gamma} \quad \text{for } t \in ]0,T[ \quad (4)$$

where $U_{CL}$ is the contact line velocity, $\theta^{eq}$ is the wetting angle at equilibrium (Young's law), $\theta$ is the (dynamic) wetting angle. $k$ and $\gamma$ are parameters usually determined using experimental results.

*Local flow modelling and Shikhmurzaev's theory.* The "implicit" model does not impose the contact line velocity and the wetting angle but consider them as a response of the model. This model is based on the Shikhmurzaev's theory, [10], [2], which makes introduce the generalized Navier slip condition (3) and the condition (2). These conditions are local since the extra terms $\boldsymbol{g}$ and $h$ vanish except in a vicinity of the triple point. The basic idea of this theory is to consider that the rolling motion observed in experiments, [4], implies that particles of the liquid-gas interface become an element of the solid-liquid in a finite time. Then, the surface tension value associated to this particle must change to its new equilibrium value relative to the solid-liquid interface. This process would gives rise to a surface tension gradient in a small vicinity of the advancing contact line (hence a local Marangoni effect). In other respect, the Young equation would remain valid at any time. In this theory,

$$\boldsymbol{g} \;=\; -\frac{1}{2Ca}\,\boldsymbol{\nabla}\sigma \text{ and } h \;=\; \frac{1}{Ca}\,\boldsymbol{\nabla}\sigma_{LG}.\boldsymbol{\tau} \tag{5}$$

where $\sigma$ and $\sigma_{LG}$ are the liquid-solid and the liquid-gas surface tension coefficient respectively. In [7] and [8], a mathematical and numerical study presents some qualitative behaviors of $g$ and $h$ arising from Shikhmurzaev's theory.

*A connection with Qian-Wang-Sheng theory.* From molecular dynamics simulations on immiscible fluids, [11] shows that the relative slipping between the fluid and the solid wall follows a generalized Navier slip b.c. similar to (3) (if one phase fluid only is considered). In this theory, the extra term $\boldsymbol{g}$ in (3) would model an interfacial uncompensated Young stress. The extra tangential stress $\boldsymbol{g}$ is defined as follows: $\int_{\Gamma_{int}} g\, dy \;=\; \sigma(cos\theta - cos\theta^{eq})$, where $\int_{\Gamma_{int}} dy$ denotes the integral across the interface $\Gamma_f$.

## 3 Discretization and Algorithms

### 3.1 Finite Element Scheme

We discretize the full model in time using the Euler implicit scheme and a first order characteristic method. The space discretization is done using the second order Taylor-Hood finite element. The scheme is implemented using a public C++ finite element library. To compute the curvature , first we consider a second order local Bezier least-square approximation of points defining the interface, second we evaluate the curvature of the Bezier curve. We performed numerical tests which showed that this method filters noise quite reasonably while it allows to detect local and rapid variation of curvature.

### 3.2 The ALE Free Surface Algorithm

The free surface problem is solved using an ALE formulation and a characteristic method. The algorithm is the following. At time step $n$, given $\Omega^n$ and $(\boldsymbol{u}^n, p^n)$ in $\Omega^n$.

*Step 1.* Compute the new free surface position $\Gamma_f^n$ following the lagrangian characteristics lines (each point of the boundary is translated by $\boldsymbol{u}^n \Delta t$). Then, compute an equivalent deformation field $\gamma^n$ preserving the mesh by solving a linear elasticity system.

*Step 1bis.* Compute the new solution $(\boldsymbol{u}_\tau^{n+1}, p_\tau^{n+1})$ of Navier-Stokes in $\Omega^n$.

*Step 2.* Update the domain $\Omega^{n+1} = \gamma^n(\Omega^n)$ (mesh transport by $\gamma^n$).

*Step 3.* Obtain the new solution $(\boldsymbol{u}^{n+1}, p^{n+1})$ in $\Omega^{n+1}$ by setting: $\boldsymbol{u}^{n+1}|_{\Omega^{n+1}} = \boldsymbol{u}_\tau^{n+1}|_{\Omega^n}$ and $p^{n+1}|_{\Omega^{n+1}} = p_\tau^{n+1}|_{\Omega^n}$.

### 3.3 Spreading and Caterpillar Motion

When spreading, our algorithm imitates the caterpillar motion observed in experiments, [4]. Fig. 5 (Left) represents typical successive time steps. If at time step $n$, the first point (or several points) on the free surface are projected on the solid substrate, all the points on the solid wall but the last ones become no-slip points. The last two points remain slip point.

### 3.4 Recoiling Motion

When de-wetting (recoiling), the triple point and the slip points are allowed to move horizontally, also vertically if their vertical velocity is positive, see Fig. 5 (Right). If the triple point moves away from the solid wall, the first point to the left becomes the triple point, hence becoming a free point. The second point to the left becomes a slip point.

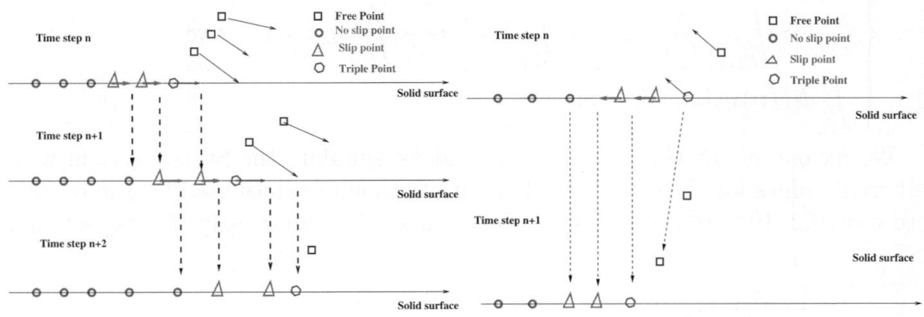


**Fig. 1.** *Left.* Circles represent no-slip nodes, triangles represent slip nodes $u_z = 0$, squares represent "free" nodes, diamond is the triple point. *Right.* Circles represent no-slip points $\boldsymbol{u} = 0$, triangles represent slip points $u_z = 0$, squares represent "free" points, diamond is the triple point.

# 4 Numerical Results

We consider a water droplet. The reference length and velocity are: $L_{ref} = 2.3\,mm$, $U_{ref} = 0.98\,m/s$. Then, $Re = \dfrac{\rho U_{ref} L_{ref}}{\mu} \approx 46$, Weber number $We = \dfrac{\rho U_{ref}^2 L_{ref}}{\sigma} \approx 68$ and $C_a = \dfrac{\sigma}{\mu U_{ref}} \approx 1.5$.

## 4.1 Spreading Phase Using a Tanner Type Law

We start by testing the efficiency of the algorithm. To this end, we consider the Tanner type law as triple line dynamic modelling, and we make fit this law with available experimental data, [1], related to a spreading phase. The triple point position is imposed at each time. We focus on the volume conservation, the height and diameter of the spreading (splat radius) and the deformation of the mesh (number of re-meshing necessary), see Fig. 2. With a time step $dt = 5.10^{-5}s$ and a coarse mesh ($\approx 600$ elements), volume lost after 650 iterations is roughly 3%. The loss occurs mainly at the very beginning of spreading. Mesh transport is efficient hence re-meshing occurs only when a projection occurs (see previous section). From a qualitative point of view, numerical results are similar to experimental data, [1].

## 4.2 Influence of the Local Parameters ($\boldsymbol{\beta, g, h}$)

We present some influence of terms specific to the Shikmurzaev's model by considering the following steady-state Stokes system: Find $(\boldsymbol{u}, p)$ such that:

$$\begin{cases} 2\displaystyle\int_\Omega D(\mathbf{u}) : D(\mathbf{v})dx - \int_\Omega p\, div(\mathbf{v})dx + 2\int_\Omega \mathbf{u}_a \mathbf{v} dx + \beta \int_{\Gamma_{slip}} u_r\, v_r\, ds \\ \quad = \displaystyle\int_{\Gamma_{free}} \frac{\kappa}{Ca}\, \mathbf{v}\, \mathbf{n} ds + \int_{\Gamma_f} h\, \mathbf{v}\, \tau\, ds - \int_{\Gamma_{sl}} g_r\, v_r ds, \quad \forall \mathbf{v} \\ \displaystyle\int_\Omega div(\mathbf{u})\, q\, dx = 0, \quad \forall q \end{cases}$$

We fix one of the three parameters and we simulate the Stokes system with different values for the other two. The set of absolute values used for the parameters is $\{0, 1, 10^2, 10^4\}$ with signs taken as described previously. We expect that

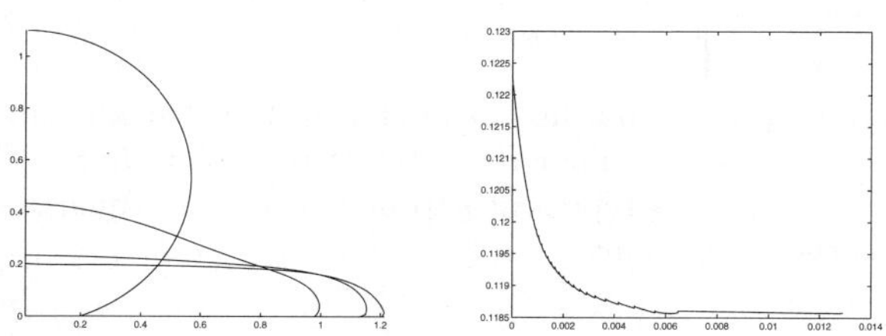

**Fig. 2.** Spreading: droplet profiles and volume in function of time steps

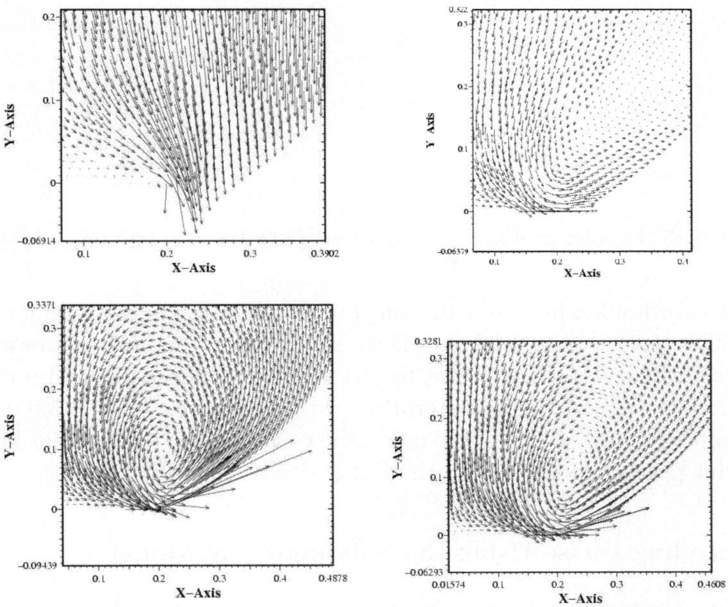

**Fig. 3.** Velocity in the vicinity of the triple point for $\beta = 10^2$ and $(g, h) = \{(0, 0), (-10^2, 0), (0, -10^2), (-10^2, -2 \times 10^2)\}$. Scale factors are $10, 0.5.0.1, 0.1$ respectively.

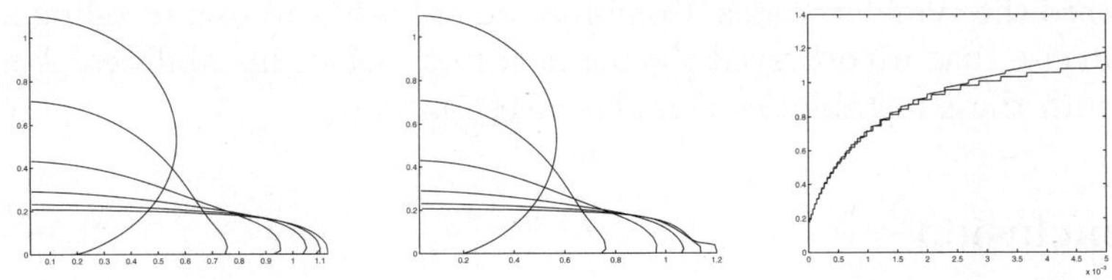

**Fig. 4.** Droplet profiles and triple point position (right) for $(\beta, g) = \{(10^3, -10), (10^3, -10^2)\}$

for a high $\beta$ value, the effect of this term will be similar to a no-slip/slip boundary condition on $\Gamma_{sl}$. For the $g$-term, it is clear that a high $|g|$ will be equivalent to accelerating/decelarating the spreading. The $h$-term creates an upward force which should be responsible for recoiling, see Fig. 3, for $\beta = 100$ and different values of $g$ and $h$.

### 4.3 Spreading Phase Using the Shikhmurzaev Model

We consider the spreading phase using the Shikhmurzaev's model and the algorithm presented previously. As mentioned above, $g$ acts like a control on the position of the triple point. A decrease for $g < 0$ translates in a faster spreading. If $g$ is small, this influence is likely not to be seen in the first part of the spreading, when the inertial forces dominate all other forces in presence. By increasing

**Fig. 5.** Droplet profiles for $(\beta, g, h) = \{(500, 100, 0), (100, 100, -10)\}$

$|g|$ over a threshold, which depends on $\beta$, it accelerates the spreading and modifies the flow nature. For $|g|$ large, large interface distortions occurs and mesh generator fails to re-mesh. In order to observe the influence of the $g$-term, we fix $\beta = 1000$, and we perform 1000 iterations with a step size of $dt = 5.0 \times 10^{-6}$ for $g \in \{-100, -10, -1\}$, Fig. 4. One can observe that after the inertial phase, the triple point position is farther to the right with a decreasing $g$.

### 4.4 Recoiling Phase Using the Shikhmurzaev Model

We consider the recoiling phase using the Shikhmurzaev model and the algorithm presented previously. Parameters are as follows: $(\beta, g, h) = (100, 100, -20)$ and $(500, 100, -20)$. As expected, the $h$-term controls the separation of the triple point from the solid surface, see Fig. 5, while the $g$-term helps the droplet to slide toward the symmetry axis. Combination of both lead to a recoiling motion. Let us precise that we observed the onset of free surface instabilities. A method to deal with these instabilities is under investigation.

## 5 Conclusion

We have discretized and implemented macroscopic part of the Shikhmurzaev model, [10], [2], using an ALE formulation and a finite element scheme. Also, a connection with Qian-Wang-Sheng's results, [11], is presented. The present model comprises generalized Navier slip boundary conditions with local surface tension gradient terms in the vicinity of the contact line. The local term values have been chosen upon a previous mathematical and numerical study done in [8]. To test the efficiency of the present algorithm, first we considered a classical mobility relation and we obtained spreading droplet profiles qualitatively comparable to experimental results. Then, we showed how the introduction of these new local terms in the vicinity of the contact line allow to model on one hand the spreading phase, on the other hand the recoiling phase. A full coupling between the macroscopic part and the microscopic part of the Shikhmurzaev model is under progress.

**Acknowledgements.** The authors would like to thank A. Soucemarianadin and JL. Achard from Laboratory of Geophysical and Industrial Fluid Flows -LEGI- Grenoble, France, for their numerous and fruitful discussions and remarks.

## References


[1] Allaman S., Desie G., Vadillo D., Soucemarianadin A. : Impact and spreading of micro-drops onto solid substrates. Mcaniques & Industries **4** (2003) 443-455.

[2] Blake T.D., Bracke M., Shikhmurzaev Y.D. : Experimental evidence of nonlocal hydrodynamic influence on the dynamic contact angle. Physics of fluids **11** (1999) 1995-2007.

[3] de Gennes P.G.: Wettings: statics and dynamics. Reviews Modern Phys. **57** (1985) 827-863.

[4] Dussan V.E.B., Davis S.H. : On the motion of a fluid-fluid interface along a solid surface. J. Fluid Mech. **65** (1974) 71-95 .

[5] Hocking L.M. : A moving fluid interface. Part II. The removal of the force singularity by a slip flow. J. Fluid Mech. **79** (1977) 209-229 .

[6] Jacqmin D. : Contact-line dynamics of a diffuse fluid interface. J. Fluid. Mech., **402** (2000) 57-88.

[7] Monnier J. : Modélisation Numérique de la Ligne Triple. Internal Report LMC-IMAG, Summer school CEA Grenoble. September (2003).

[8] Monnier J., Witomski P. : Analysis of a Local Hydrodynamic Model with Marangoni Effect. J. Sc. Comp. **21** (2004) 369-403.

[9] Pomeau Y. : Recent progress in the moving contact line problem: a review. C.R. Mécanique. **330** (2002) 207-222.

[10] Shikhmurzaev Y.D. : The moving contact line problem on a smooth solid surface. Int. J. Multiphase Flow. **19** (1993) 589-610.

[11] Qian T., Wang X-P., Sheng P. : Generalized Navier boundary condition for the moving contact line. Comm. Math. Sci. **1** (2003) 333-341.

# A Multilevel-Multigrid Approach to Multiscale Electromagnetic Simulation

Peter Chow[1], Tetsuyuki Kubota[2], and Takefumi Namiki[3]

[1] Fujitsu Laboratories of Europe,
Hayes Park Central, Hayes, UB4 8FE, United Kingdom
peter.chow@uk.fujitsu.com

[2] Corporate Component Engineering Center, Fujitsu Limited,
Nakahara-ku, Kawasaki 211-8588, Japan
t.kubota@jp.fujitsu.com

[3] Computational Science and Engineering Solutions Center, Fujitsu Limited,
Chiba 261-8588, Japan
t.namiki@jp.fujitsu.com

**Abstract.** The time-dependent Maxwell's equations are solved for mobile device applications using a multilevel-multigrid finite-difference time-domain (FDTD) method. For three-dimensional models that simulate system level details of mobile devices, the smallest features are in the nanometre ($10^{-9}$ m) range, leading to a time-step size in the attosecond ($10^{-18}$ s) range. The feature sizes of mobile devices are in the centimetre ($10^{-2}$ m) range, while for health and safety studies that include human models features are in the metre range.

## 1 Introduction

The drive towards compact, lightweight and fast mobile electronic devices has fuelled the demand for electromagnetic simulations/analyses that employ complete models in fine detail but at the same time show shortened simulation times. In general, detailed models and faster simulations are not compatible - detailed models by nature are associated with long computational times. The most popular method for computational electromagnetic simulation, the finite-difference time-domain (FDTD) method [1,2], is no different. The FDTD method solves the time-dependent Maxwell's equations using the explicit leapfrog time-stepping scheme. The Courant-Friedrichs-Lewy (CFL) stability condition restricts the size of the time-step directly to the smallest feature in the domain. For three-dimensional models that simulate system level details of mobile devices, the smallest features are in the nanometre ($10^{-9}$ m) range, leading to a time-step size in the attosecond ($10^{-18}$ s) range. The feature sizes of mobile devices are in the centimetre ($10^{-2}$ m) range, while for health and safety studies that include human models features are in the metre range. Direct analysis of such models with the FDTD method is extremely demanding in both memory and computer time. Here, we apply the multilevel-multigrid grid embedding approach to address such multiscale simulations in both spatial and temporal spaces. We also

V.N. Alexandrov et al. (Eds.): ICCS 2006, Part II, LNCS 3992, pp. 34–40, 2006.

introduce the block-multigrid concept which makes the multigrid FDTD method even more efficient.

## 2 Multilevel-multigrid FDTD Method

In this section, the multilevel-multigrid FDTD method is introduced one element at a time for clarity, with references being given to more detailed explanations. First the multigrid FDTD method is introduced, followed by the block-solve enhancement, and finally the multilevel element is discussed.

### 2.1 Multigrid FDTD

The multigrid FDTD method [3-9] solves Maxwell's equations based on the finite-difference time-domain (FDTD) method [1,2] for the calculation of electromagnetic wave propagation by strategically embedding finer grids in regions that require high resolution and not everywhere. Fig. 1 illustrates the grids of normal FDTD and multigrid FDTD methods. The multigrid approach greatly reduces memory and computational cost compared with an FDTD method of similar accuracy. Furthermore, the Cartesian grid used in the finite-difference scheme is considered the most efficient in terms of memory requirement and computational efficiency.

### 2.2 Block-Solve Multigrid FDTD

For most calculations the multigrid FDTD method is fast and efficient. However, there are situations where the embedded Cartesian grid is very inefficient, such as elongated irregular signal-line type structures. Owing to the rectangular nature of Cartesian grids, they can cover large empty spaces. Fig. 2 shows such a model where the embedded grid (the rectangle covering the signal-line structure) is covering some 50 per cent of empty signal-line space as illustrated by the two dotted-line ovals. Naturally, removing such empty space will reduce memory

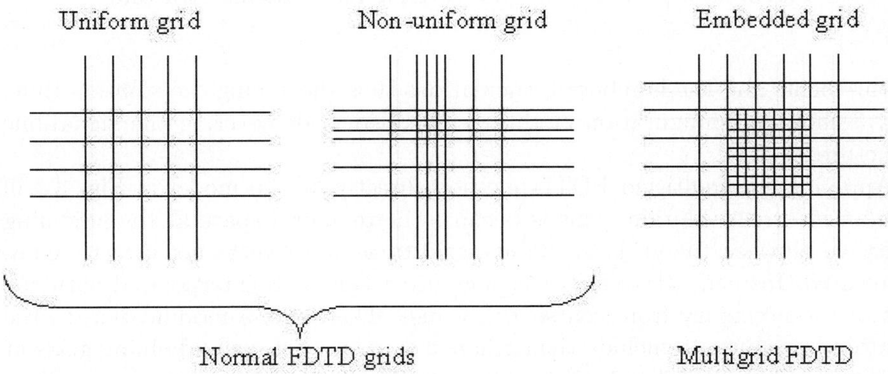


**Fig. 1.** Comparison of grid types used in FDTD and multigrid FDTD methods

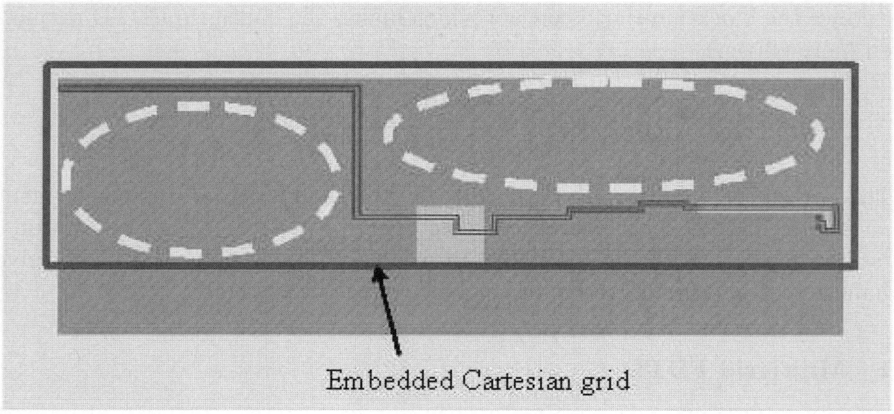


**Fig. 2.** Embedded grid covering an irregular signal-line structure

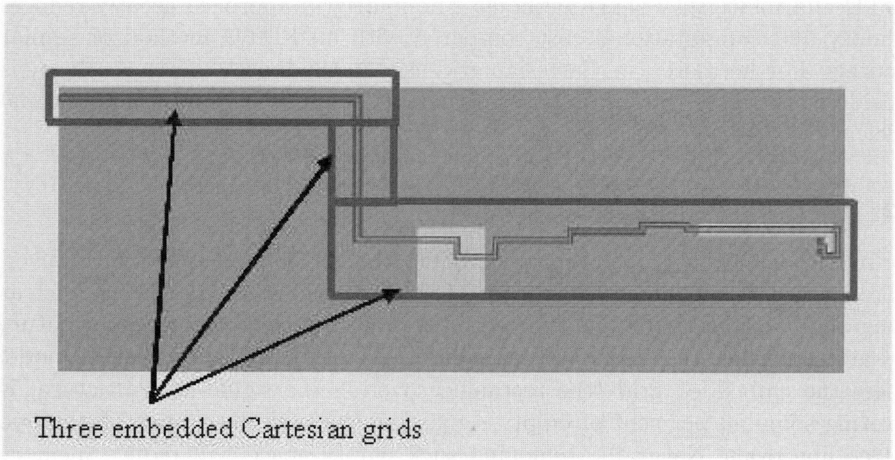


**Fig. 3.** Multiple Cartesian grids covering the signal-line structure

requirements and the number of calculations thus shortening the solution time. Fig. 3 shows a configuration of three embedded grids covering the signal-line structure.

Applying the multigrid FDTD method directly to the model in Fig. 3 will give an incorrect solution. This is because the temporal space at the adjoining interface between the grids in the fine-grid time-step level is not directly communicated. Instead, it is via the coarse grid where it is interpolated with the rest of the boundary from coarse-grid values. Therefore, a modification to the solution procedure to include the exchange of values between adjoining grids at each embedded time-step level is required. The block-solve solution procedure for multigrid FDTD is detailed in [10] and will not be repeated here.

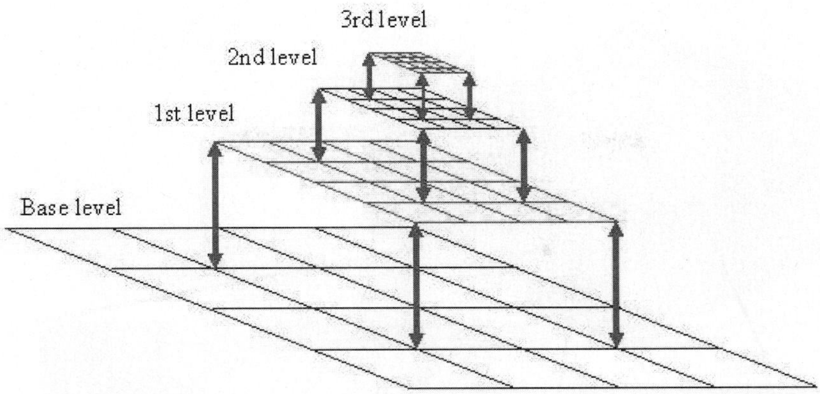


**Fig. 4.** Multi-level grid configuration

### 2.3 Multilevel-multigrid FDTD

Fig. 4 shows a multilevel grid configuration (in two dimensions) of three levels of grid embedding, with each embedded grid finer than the one before. The multilevel-multigrid FDTD method [11] provides a refinement process where the grid cell aspect scale changes gradually as opposed to a severe change. This refinement applies to both spatial and temporal spaces. Thus, a gradual refinement is commonly a more stable approach than a severe change all at once.

## 3 Example Model and Results

Fig. 5 shows a simple example model with geometric scale aspect ratio of about 1,000 between the smallest feature in the model and the largest. The smallest feature is the coil wire width of 1 micrometre ($10^{-6}$ m) and the largest component is the strip-line with a length in the millimetre range.

Table 1 shows five grid models and their respective results. The first two grid models, coarse and fine grids, are for the standard FDTD method. The fine grid model is created by refining the coarse grid with a refinement factor of 2. The coarse grid model has one cell for the width of the coil wire, and the fine grid has four (2x2) cells due to the refinement of 2. The respective time-step sizes calculated from the CFL stability condition are 2.07 and 1.035 picoseconds ($10^{-15}$ s).

The other grid models are for the multilevel-multigrid FDTD method and a refinement of 2 is applied at each grid level. The third grid model (MG 1-level) has one level of embedded grid, so the base grid has the same time-step size as the coarse grid model. The fourth grid model (MG 4-levels) has four levels of grid refinement at the coil region (giving one cell for the width of the coil wire) and one level at the port regions. The final grid model (MG 5-levels) has five levels of grid refinement at the coil region, giving four (2x2) cells

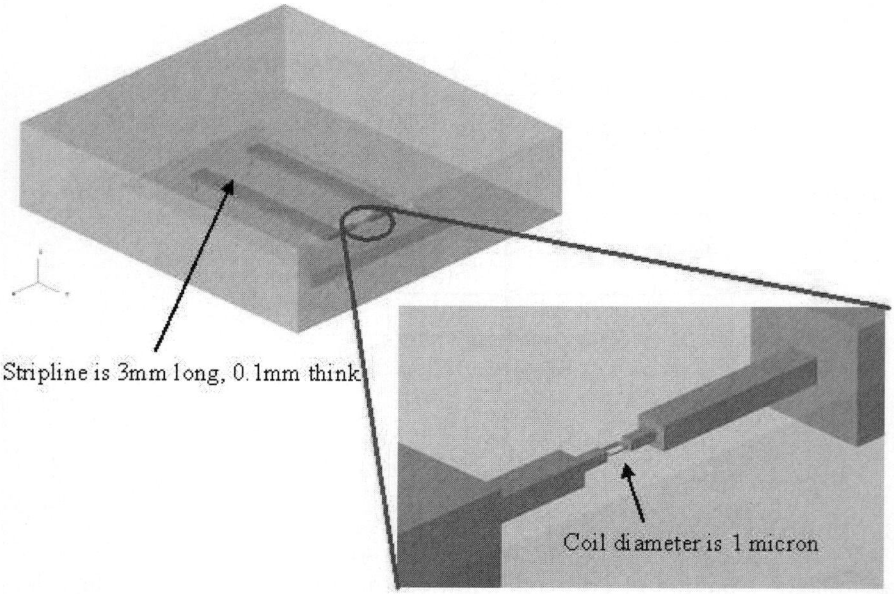


**Fig. 5.** Example model

**Table 1.** Computing time, grid levels, and size og grids

| Model | CPU seconds | Grid size | Speedup | Speedup |
|---|---|---|---|---|
| Coarse grid | 3,402 | 63 x 60 x 50 | 1x1 cell | |
| Fine grid | 47,954 | 125 x 119 x 99 | | 2x2 cells |
| MG 1-level<br>(2x2 cells width of coil) | 11,655 | 63 x 60 x 50 base<br>50 x 52 x 56 | | 4.1 |
| MG 4-levels<br>(1x1 cell coil centre) | 1,577 | 46 x 44 x 28 base<br>32 x 32 x 26 level-1<br>47 x 35 x 35 level-2<br>43 x 27 x 31 level-3<br>35 x 35 x 35 level-4 | 2.16 | |
| MG 1-level at each port | | 19 x 17 x 31 port-1<br>19 x 17 x 31 port-2 | | |
| MG 5-levels<br>(2x2 cells coil centre) | 10,028 | 46 x 44 x 28 base<br>32 x 32 x 26 level-1<br>47 x 35 x 35 level-2<br>51 x 43 x 43 level-3<br>63 x 59 x 57 level-4<br>39 x 55 x 55 level-5 | | 4.78 |
| MG 1-level at each port | | same size as before | | |

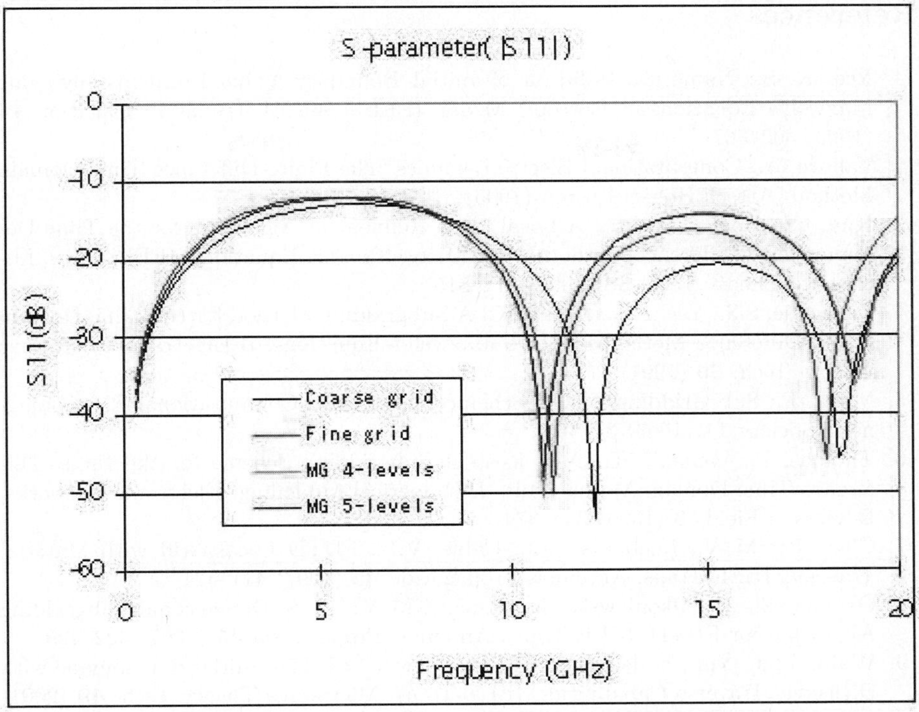


**Fig. 6.** Plot of S-parameters

for the width of the coil wire and, as for the fourth model, one level at the port regions. The last two grid models both have a base-grid time-step of 33.56 picoseconds.

The 1-level embedded grid model attains a speedup of 4.1 compared with the fine grid model, whereas the 5-levels model attains a 4.78 speedup. The 4-level embedded grid model attains a 2.16 speedup compared with the coarse grid model.

Fig. 6 shows the S-parameter for all the grid models. There is good agreement between the fine grid and multilevel-multigrid FDTD methods up to 15 GHz, with slight deviations starting to appear above 15 GHz. Also, the plot shows that the coarse grid result has the worst result due to insufficient grid resolution.

## 4 Conclusions

We have used a multilevel-multigrid approach to address multiscale problems for three-dimensional electromagnetic wave propagation. The result obtained for features in the micrometre ($10^{-6}$) range is encouraging. The challenge now is to advance the technique to the nanometre ($10^{-9}$) range.

## References


1. Yee, K. S.: Numerical Solution of Initial Boundary Value Problems Involving Maxwell's Equations in Isotropic Media. IEEE Trans. Antennas Propagation **14** (1996) 302–307
2. Taflove, A.: Computational Electrodynamics: The Finite-Difference Time-Domain Method. Artech House, Boston (1995)
3. Kim, I.S., Hoefer, W.J.R.: A Local Mesh Refinement Algorithm for the Time Domain Finite Difference Method Using Maxwell's Curl Equations. IEEE Trans. Microwave Theory Tech. **38** (1990) 812–815
4. Zivanovic, S.S., Yee, K.S., Mei, K.K.: A Subgridding Method for the Time-Domain Finite-Difference Method to Solve Maxwell's Equations. IEEE Trans. Microwave Theory Tech. **39** (1991) 471–479
5. Monk, P.: Sub-Gridding FDTD Schemes. J. Applied Computational Electromagnetic Society **11** (1996) 37–46
6. Thomas, P., Weiland, T.: A Consistent Subgridding Scheme for the Finite Difference Time Domain Method. Int. J. Numerical Modelling: Electronic Networks, Devices & Fields **9** (1996) 359–374
7. Chevalier, M.W., Luebbers, R.J., Cable, V.P.: FDTD Local Grid with Material Traverse. IEEE Trans. Antennas Propagation **45** (1997) 411–421
8. Okoniewski, M., Okoniewska, E., Stuchly, M.A.: Three-Dimensional Subgridding Algorithm for FDTD. IEEE Trans. Antennas Propagation **45** (1997) 422–429
9. White, M.J., Yun, Z., Iskander, M.F.: A New 3D FDTD Multigrid Technique with Dielectric Traverse Capabilities. IEEE Trans. Microwave Theory Tech. **49** (2001) 422–430
10. Chow, P., Kubota, T., Namiki, T.: A Block-Solve Multigrid-FDTD Method. To appear in the 22nd International Review of Progress in Applied Computational Electromagnetics (ACES 2006) conference in March 2006, Miami, Florida, USA
11. Chaillou, S., Wiart, J., Tabbara, W.: A Subgridding Scheme Based on Mesh Nesting for FDTD Method. Microwave and Optical Technology Letters **22** (1999) 211–214

# Scalable Simulation of Electromagnetic Hybrid Codes

Kalyan Perumalla[1], Richard Fujimoto[2], and Homa Karimabadi[3]

[1] Oak Ridge National Laboratory, Oak Ridge, TN, USA
[2] Georgia Institute of Technology, Atlanta, GA, USA
[3] SciberQuest Inc, Solana Beach, CA, USA

**Abstract.** New discrete-event formulations of physics simulation models are emerging that can outperform models based on traditional time-stepped techniques. Detailed simulation of the Earth's magnetosphere, for example, requires execution of sub-models that are at widely differing timescales. In contrast to time-stepped simulation which requires tightly coupled updates to entire system state at regular time intervals, the new discrete event simulation (DES) approaches help evolve the states of sub-models on relatively independent timescales. However, parallel execution of DES-based models raises challenges with respect to their scalability and performance. One of the key challenges is to improve the computation granularity to offset synchronization and communication overheads within and across processors. Our previous work was limited in scalability and runtime performance due to the parallelization challenges. Here we report on optimizations we performed on DES-based plasma simulation models to improve parallel performance. The mapping of model to simulation processes is optimized via aggregation techniques, and the parallel runtime engine is optimized for communication and memory efficiency. The net result is the capability to simulate hybrid particle-in-cell (PIC) models with over 2 billion ion particles using 512 processors on supercomputing platforms.

## 1 Introduction

New discrete-event approaches are being developed to speed up simulations of inhomogeneous physical systems in order to efficiently accommodate the variety of spatial and temporal scales in such systems. The new discrete-event approaches are being proposed in place of traditional time-stepped approaches in order to overcome the worst-case limitations imposed by the fastest processes in the system. Parallel execution of these discrete-event models is challenging due to a combination of their characteristics, including fine-grained event computation and dynamic inter-entity event communication patterns. In this paper, we document our optimizations to a discrete-event model of a one-dimensional hybrid shock simulation that uses a particle-in-cell method to simulate electromagnetic fields in a plasma environment. In our earlier work, we reported results from a preliminary parallel implementation on a cluster of workstations. The previous implementation uncovered several avenues for improvement, including computation granularity issues, memory usage requirements and inter-processor communication overheads. Our new implementation incorporates

V.N. Alexandrov et al. (Eds.): ICCS 2006, Part II, LNCS 3992, pp. 41–49, 2006.

optimizations to the discrete event model implementation to enable efficient parallel/distributed execution, and enabled scaling it to supercomputing platforms.

The rest of the document is organized as follows. Section 2 provides the motivation and background to this work. The one-dimensional hybrid shock application is outlined in Section 3. The optimizations to the parallel implementation are described in Section 4, followed by a parallel execution performance study in Section 5. Finally, Section 6 outlines the status and future work.

## 2 Background and Related Work

The conventional approach to realizing PIC models with spatial grid elements is to use time-stepped execution where the state of the model, e.g., particle position, velocity, charge, etc., is updated at fixed time increments. Discrete event simulation offers an alternative approach where particle and field updates are instead only carried out on an "as needed" basis, e.g., when field values cross certain thresholds, resulting in state updates at irregular (and less frequent) time points. The time interval between updates is therefore dictated by the predicted rate of change. Particle and field update "events" are used to denote when state updates occur. These events are queued and continuously processed over time to complete the simulation. Event-driven PIC simulations automatically guarantee that the progression of the system captures important state changes while reducing computation of less interesting, "idle" information. Further details of this approach are presented in [1, 2], where performance measurements were presented showing as much as two orders of magnitude speedup for certain PIC simulations.

Further increases in speed and scalability can be accomplished by applying parallel discrete event simulation (PDES) techniques. Here, the computation is divided into a collection of simulation processes that communicate by exchanging time stamped messages (events). A central question that must be addressed in PDES concerns ensuring proper synchronization of the computation. Unlike time-stepped simulations, PDES techniques allow some simulation processes to progress ahead of others in simulation time. This introduces the possibility of synchronization errors where a simulation process receives a message (event) with time stamp smaller than its current simulation time. Several approaches have been proposed to address this problem [3]. One class, termed conservative synchronization, blocks simulation processes to ensure no such synchronization errors occur [4, 5]. By contrast, optimistic synchronization techniques allow such errors to occur, but recover using a rollback mechanism [6].

PDES systems are typically composed of a simulation engine that handles issues such as synchronization, and invoking the simulation model entities at appropriate times. The μsik system [7] used here is one example of a PDES simulation engine that can be configured to handle either conservative or optimistic synchronization methods. μsik is based on a micro-kernel approach to parallel simulation engine design where fundamental mechanisms necessary for synchronization are implemented within the micro-kernel, and the rest of the kernel is built over the micro-kernel. Both conservative and optimistic parallel PIC simulations have been realized utilizing the μsik system [1, 2, 8].

A limited amount of work has examined the application of PDES techniques to physical system simulation. Perhaps the earliest was the "colliding pucks" application developed for the Time Warp Operating System (TWOS) [9]. Lubachevsky discusses the use of conservative simulation protocols to create cellular automata models of Ising spin [10] and other physical system problems [11]. A formal approach to both discrete event and continuous simulation modeling based on DEVS (Discrete EVent System Specification), was proposed by Zeigler et al. [12] and some numerical solutions have been examined based on the DEVS formalism [13].

## 3 One-Dimensional Hybrid Shock Discrete Event Model

Here we provide a brief description of our DES model. Additional information can be found in [2]. Electromagnetic hybrid algorithms with fluid electrons and kinetic ions are ideally suited for physical phenomena that occur on ion time and spatial scales. Maxwell's equations are solved by neglecting the displacement current in Ampere's law (Darwin approximation), and by explicitly assuming charge neutrality. There are several variations of electromagnetic hybrid algorithms with fluid electrons and kinetic ions [14]. Here we use the one-dimensional (1-D) resistive formulation which casts field equations in terms of vector potential.

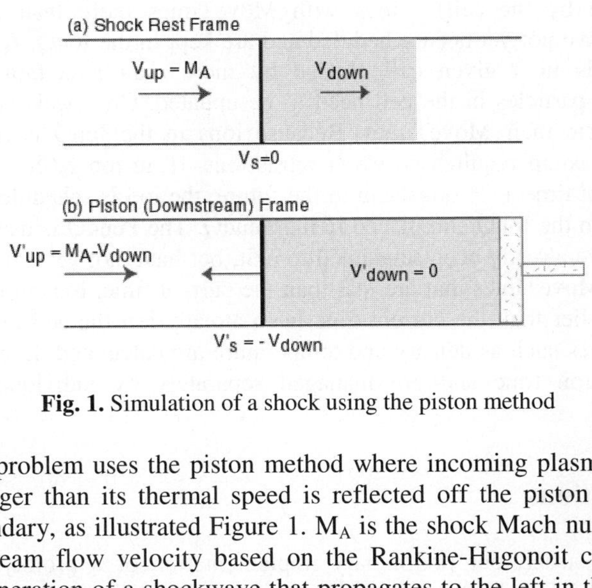


**Fig. 1.** Simulation of a shock using the piston method

The model problem uses the piston method where incoming plasma moving with flow speed larger than its thermal speed is reflected off the piston located on the rightmost boundary, as illustrated Figure 1. $M_A$ is the shock Mach number and $V_{down}$ is the downstream flow velocity based on the Rankine-Hugonoit condition. This leads to the generation of a shockwave that propagates to the left in the piston frame of reference. In this example, we use a flow speed large enough to form a fast magnetosonic shock. In all the runs shown here, the plasma is injected with a velocity of 1.0 (normalized to upstream Alfven speed), the background magnetic field is tilted at an angle of 30°, and the ion and electron betas are set to 0.1. The simulation domain is divided into cells [1], and the ions are uniformly loaded into each cell. Each cell is modeled as a Logical Process (LP) in µsik and the state of each LP includes the cell's field variables. The main tasks in the simulation are to (a) initialize fields,

(b) initialize particles, (c) calculate the exit time of each particle, (d) sort IonQ, (e) push particle, (f) update fields, (g) recalculate exit time, and (h) reschedule. This is accomplished through a combination of priority queues and three main classes of events.

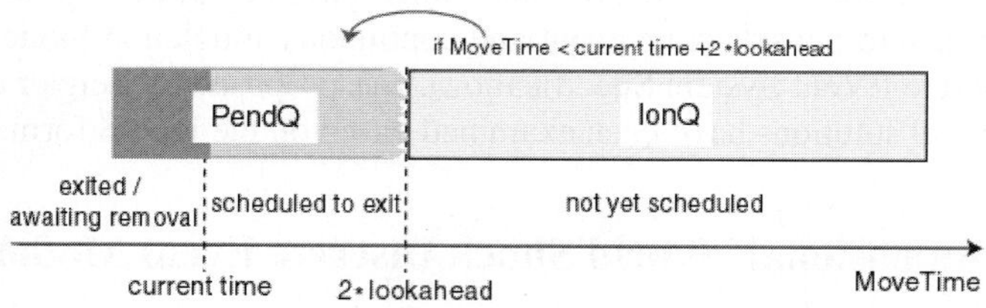


**Fig. 2.** Organization of particle events in PendQ and IonQ priority queues

The ions are stored in either one of two priority queues[2], as shown in Fig. 2. Ions are initialized within cells in an IonQ. As ions move out of the leftmost cell, new ions are injected into that cell in order to keep the flux of incoming ions fixed at the left boundary. MoveTime is the time at which an ion moves to an adjacent cell. The placement and removal of ions in IonQ and PendQ is controlled by comparing their MoveTimes to the current time and lookahead (lookahead is the shortest delay between the current simulation time of the cell and the time of any event scheduled into the future by the cell). Ions with MoveTimes more than current time + 2*lookahead have not yet been scheduled and are kept in the IonQ. A wakeup occurs when the fields in a given cell change by more than a certain threshold and MoveTimes of particles in the cell need to be updated. On a wakeup, ions in IonQ queue recalculate their MoveTimes. Because ions in the IonQ have not yet been scheduled, a wakeup requires no event retractions. If an ion's MoveTime becomes less than current time + 2*lookahead in the future, the ion is scheduled to move, and is removed from the IonQ and placed in the PendQ. The PendQ is used to keep track of ions that have already been scheduled to exit, but have not yet left the cell. These particles have MoveTimes that are less than the current time. Ions in the PendQ with MoveTimes earlier than the current time have already left the cell and are removed before cell values such as density and temperature are calculated. Events can happen at any simulation time and are managed separately by individual cells of the simulation.

# 4 Optimizations

As mentioned earlier, our preliminary implementation of a prototype for parallel execution of the 1-D hybrid shock model was limited in different ways. First, the per-event overhead incurred due to discrete event processing was found to be large due to the low granularity of event computation. Secondly, our parallel execution was constrained by sockets-based communication, which suffered from inefficiencies. Finally, the discrete event simulation engine itself was in an evolutionary state, and was consequently not optimized for memory usage. Our optimizations were aimed along these lines: The mapping from cells to simulation processes is changed to an aggregate scheme in order to minimize overheads. With communication subsystem optimizations, runtime performance has been significantly improved. Additionally, porting to a supercomputer enabled the simulation to scale up to 512 processors. By specializing the data structure to conservative synchronization (at runtime), the memory requirements to represent the cells and particles have been reduced. The largest configurations that can be simulated have been pushed significantly, to include over 2 billion ion particles overall. Some of these optimizations are described in detail next.

### 4.1 Mapping Cells to DES Logical Processes

One way to realize PIC models, shown in Fig. 3, is to map each cell to a logical process (LP). This provides maximum flexibility for load balancing, but makes every particle-transfer event to go through the (micro-kernel) PDES simulator, making it inefficient due to lack of optimization for locality of communication. Also, shared state is disallowed in this scheme, which makes it impossible for neighboring cells to exchange data via direct access to data structures.

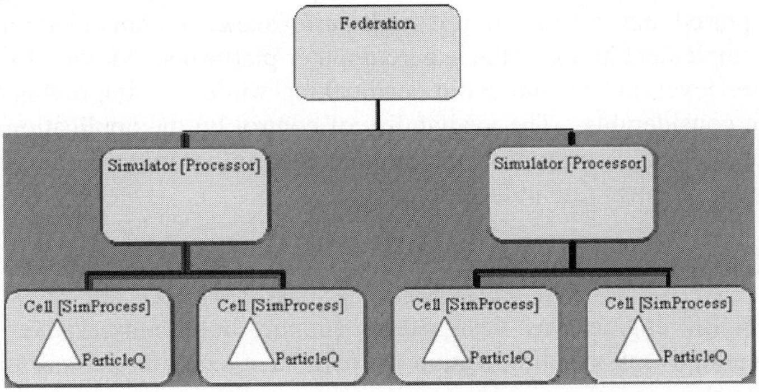


**Fig. 3.** A suboptimal way of realizing a particle-in-cell DES model

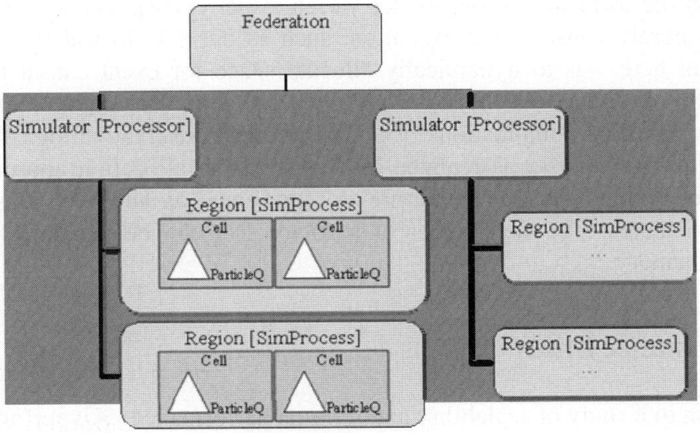


**Fig. 4.** An efficient way of realizing a particle-in-cell DES model

A more efficient alternative approach is shown in Fig. 4. The concept of a "region" is introduced, which is an aggregate that contains multiple cells. Instead of mapping one *cell* per LP, each *region* is mapped to an LP. It results in memory savings, because the memory overheads of an LP are not incurred for every cell. Also, it is more natural to model: each region can be viewed as a sequential engine that simulates multiple cells. Particles crossing regions (i.e., across sequential engines) are sent as μsik events across simulation processes (and, by natural implication, across processors). In our earlier work, we used the one-LP-per-Cell mapping scheme, which incurred overheads. We re-implemented the model with the new scheme based on multiple-cells-per-region, which significantly cut down event scheduling and event processing overheads.

### 4.2 Communication Subsystem

Our earlier system used Berkeley sockets-based inter-processor communication. However, sockets have limited buffering capacities, which led to deadlocks on large-scale configurations due to the fact that large number of events (particle transfers, field updates) needed to be transferred across processors simultaneously. We have since then ported our engine to use high-performance communications based on native MPI implementations of the supercomputer platforms. Moving to MPI helped use large user-level buffers and avoid deadlocking while also improving the runtime performance considerably. The availability of control by the application on the size of the buffers helped us customize the communication based on the largest expected event message exchange rate in the application.

### 4.3 μsik Engine Enhancements

Since the PDES engine was designed to support both conservative as well as optimistic methods of synchronization in parallel execution, it was organized to accommodate the general case. However, the generality in the initial versions of the engine resulted in overheads of optimistic synchronization encroaching into conservative execution as well (e.g., the causal list maintenance among events, required for rollbacks, in the form of several pointer variables per event). This overhead is unnecessary in purely conservative execution, such as our 1-D hybrid simulation. The improvement here was to dynamically allocate space for event causal list pointers only upon first reference for the same within each event. This automatically ensures resilience to arbitrary combinations of optimistic and conservative logical processes. A 40% memory savings was realized by this dynamic allocation approach. Since every particle (ion) arrival or departure is represented as an event, this translated directly into increase in the number of particles that can be simulated in a given amount of memory.

## 5 Performance Study

We now turn to a study of scalability and runtime performance. All performance data reported here are collected on the San Diego Supercomputing Center's IBM DataStar supercomputer (www.sdsc.edu/user_services/datastar). The DataStar is a cluster of

IBM P655 nodes, each node with 8 Power4 1.5GHz processors and 16GB memory (shared by the 8 processors). The nodes are connected by an IBM Federation Switch providing low latency and high bandwidth communication. The performance on up to 512 processors is shown in Fig. 5. The observed performance is significantly better than previously reported, as a cumulative result of all the optimizations. Since the amount of concurrency is dependent on the simulated number of cells, we experimented with three configurations: small (150 cells/CPU), medium (1,500 cells/CPU) and large (40,000 cells/CPU). The total number of cells is scaled with the number of processors.

It is observed that the speedup with small configuration is less than that with medium-sized configuration. This is due to lack of enough concurrency with the smaller number of cells, making parallel synchronization overheads dominate. On the other hand, we observe lower speedup with large-sized configuration than that with the medium-sized. This turns out to be due to the large amount of inter-processor event communication inherent in the larger run, imposing greater messaging overhead in the parallel run. To confirm this, we instrumented the code to obtain measures of inter-processor event types and their counts. It was observed that the number of "notify" events increases with the number of cells, which contributes significantly to the messaging overheads.

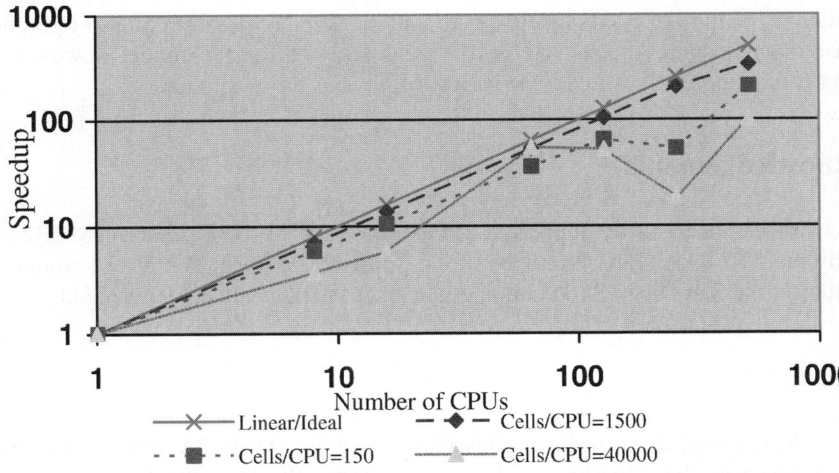


**Fig. 5.** Runtime speedup of region-based Hybrid Shock code on varying no. of CPUs

The next observation is on memory requirements. Each Ion takes approximately 150 bytes to be represented, and each cell has 100 ions. For 40,000 cells per CPU, the memory consumed to represent all the ions is 150bytes/ion*100ions/ cell*40,000cells/ CPU=600MB per CPU. Also, MPI buffers at each CPU have to be allocated sufficiently large to prevent deadlocks. In a 2 billion ion simulation on 512 CPUs, with a conservative estimate of one million incoming ions into a CPU between synchronization steps, an MPI buffer of size 2bloat/byte*400bytes/ message*$10^6$ messages =800MB is required at each CPU to avoid full buffers. The extra "bloat" factor of 2 on byte size is used to portably accommodate potential memory cost due to MPI pack/unpack data type conversions & representations. With these metrics, the number of particles has been increased linearly with the number of processors, reaching 20 million cells and over 2 billion ions in the largest execution using 40,000 cells per CPU on 512 CPUs.

Although this configuration is a bit large for the 1-D case, we are interested in observing the scaling properties of our system, with the goal of achieving efficient parallel execution for two-dimensional (2-D) and three-dimensional (3-D) versions as well. In a 2-D shock simulation, the number of cells and particles would be reasonable for some of the bigger runs. We in fact verified our expectation of similar performance on 2-D by observing a speedup of 194 on 256 processors and 248 on 512 processors on a configuration with 400 cells/CPU, and 10,000 ions/cell.

## 6 Status and Future Work

To our knowledge, the performance results reported here represent some of the largest executions of parallel discrete event-based physics simulation models. The techniques used here are fully extendable to multiple dimensions and non-uniform meshes. We are currently developing a uni-dimensional infrastructure with adaptive logical mapping capabilities. Our immediate application areas include global kinetic simulations of the Earth's magnetosphere and particle acceleration due to turbulence at fast magnetosonic shocks. Given the generality of the technique, however, we expect future applications to a wide variety of physics based simulations.

## Acknowledgements

This work has been partly supported at Georgia Tech by NSF grant ATM-0326431 and by the NSF ITR Grant No. 0539106 at SciberQuest, Inc. The use of computing facilities at the San Diego Supercomputing Center is gratefully acknowledged.

## References

[1] H. Karimabadi, J. Driscoll, Y. Omelchenko and N. Omidi, "A New Asynchronous Methodology for Modeling of Physical Systems: Breaking the Curse of Courant Condition," *Journal of Computational Physics*, vol. 205(2), 2005.

[2] H. Karimabadi, J. Driscoll, Y. Omelchenko, K. S. Perumalla, R. M. Fujimoto, and N. Omidi, "Parallel Discrete Event Simulation of Grid-based Models: Asynchronous Electromagnetic Hybrid Code," Springer LNCS Proceedings, pp. 580-588, 2005.

[3] R. M. Fujimoto, *Parallel and Distributed Simulation Systems*: Wiley Interscience, 2000.

[4] K. Chandy and J. Misra, "Asynchronous distributed simulation via a sequence of parallel computations," in *Communications of the ACM*, vol. 24, 1981.

[5] K. M. Chandy and J. Misra, "Distributed Simulation: A Case Study in Design and Verification of Distributed Programs," *IEEE Transactions on Software Engineering*, vol. SE-5(5), pp. 440-452, 1978.

[6] D. Jefferson, "Virtual Time," *ACM Transactions on Programming Languages and Systems*, vol. 7(3), pp. 404-425, 1985.
[7] K. S. Perumalla, "μsik - A Micro-Kernel for Parallel/Distributed Simulation Systems," Workshop on Principles of Advanced and Distributed Simulation, 2005.
[8] Y. Tang, K. S. Perumalla, R. M. Fujimoto, H. Karimabadi, J. Driscoll, & Y. Omelchenko, "Optimistic Parallel Discrete Event Simulations of Physical Systems using Reverse Computation," Workshop on Principles of Advanced and Distributed Simulation, 2005.
[9] P. Hontalas, et al., "Performance of the Colliding Pucks Simulation on the Time Warp Operating System," Distributed Simulation, 1989.
[10] B. D. Lubachevsky, "Efficient Distributed Event-Driven Simulations of Multiple-Loop Networks," *Communications of the ACM*, vol. 32(1), pp. 111-123, 1989.
[11] B. Lubachevsky, "Several Unsolved Problems in Large-Scale Discrete Event Simulations," Workshop on Parallel and Distributed Simulation, 1993.
[12] B. P. Zeigler, et al, *Theory of Modeling & Simulation*, 2nd ed: Academic Press, 2000.
[13] J. Nutaro, "Parallel Discrete Event Simulation with Application to Continuous Systems," in *Department of Electrical and Computer Engineering*, vol. Ph.D. Tucson, AZ: University of Arizona, 2003, pp. 182.
[14] H. Karimabadi, H. D. Krauss-Varban, J. Huba and H. X. Vu, "On Magnetic Reconnection Regimes and Associated Three-Dimensional Asymmetries: Hybrid, Hall-less Hybrid and Hall-MHD Simulations," *Journal of Geophysical Research*, vol. 109, pp. 1-21, 2004.

# Numerical Modelling of Poroviscoelastic Grounds in the Time Domain Using a Parallel Approach

Arnaud Mesgouez[1], Gaëlle Lefeuve-Mesgouez[1], André Chambarel[1], and Dominique Fougère[2]

[1] UMR A Climate, Soil and Environment, Université d'Avignon, Faculté des Sciences, 33 rue Louis Pasteur, F-84000 Avignon, France
{arnaud.mesgouez, gaelle.mesgouez, andre.chambarel}@univ-avignon.fr
[2] UMR 6181 Modélisation et Simulation Numérique en Mécanique et Génie des, Procédés, 38 rue F. J. Curie, Tech. de Château-Gombert, F-13451 Marseille, France
fougere@l3m.univ-mrs.fr

**Abstract.** In this paper, we present a parallelized finite element code developed to study wave propagation phenomena, specifically in porous soils problems which usually require millions of degrees of freedom. The parallelization technique uses an algebraic grid partitioning managed by a Single Program Multiple Data (SPMD) programming model. Message Passing Interface (MPI) library specification is the standard used to exchange data between processors. The architecture of the code is explained and numerical results show its performance.

## 1 Introduction

The study of the mechanical wave propagation in porous media is a subject of great interest in diverse scientific fields ranging from environmental engineering or vibration isolation to geomechanics. At the macroscopic scale, the medium is considered as a two-phase continuum. The Biot theory is known as the reference theory to deal with the macroscopic mechanical wave propagation phenomenon, see Biot [1] or Coussy [2] for instance.

Theoretical works are restricted to simple geometries. Consequently, they have to be completed by numerical approaches such as Finite Element or Boundary Element Methods, allowing the study of more complex problems to better represent the ground. The difficult study of transient regimes in geomechanics has been treated numerically by several authors but only for specific cases, Zienkiewicz and Shiomi [3], Simon et al. [4] and Gajo et al. [5] for example. In particular, in many cases, the tortuosity and the physical damping parameters are not taken into account.

Moreover, even with an efficient and optimized finite element code, only a restricted range of problems can be treated. As a matter of fact, solution of practical problems (for instance, realistic 3D geometries, and problems with short pulse load needing fine meshes for representing well the high frequencies)

V.N. Alexandrov et al. (Eds.): ICCS 2006, Part II, LNCS 3992, pp. 50–57, 2006.

usually requires millions of degrees of freedom. This is often virtually out of capabilities of contemporary sequential computers either because of lack of memory or abundantly long computation time. In all these cases, parallel programming techniques may be a good solution to overcome the computational complexity.

Nevertheless, to our knowledge, no parallelization of numerical approach on the complete Biot theory in the transient domain exists in the literature. In fact, papers presenting parallel computing of the finite element method often deal with elastodynamic problems using the domain decomposition method, Papadrakakis and Bitzarakis [6], Lirkov [7] and Bohlen [8].

In this paper, the authors propose a parallelized version of a finite element C++ code specifically developed at the CSE Laboratory to study transient wave propagation. This approach includes the whole Biot theory with all the couplings which represent the interactions between the solid and fluid phases. The sequential version has previously been presented at ICCS 2005, Mesgouez et al. [9].

The study of the different time-consuming parts yields a parallelization technique using an algebraic grid partitioning managed by a SPMD programming technique. MPI standard library is used to exchange data between processors. For this, numerical results, obtained for a two-dimensional problem, include the analysis of speed-up and efficiency on a SGI Origin 3800. Complementary results compare the performance obtained by another supercomputer (AMD-ATHLON cluster), on which MPICH and SCI-MPI implementations of MPI standard is used. First results with a three-dimensional geometry solving a problem with more than 2,000,000 unknowns are obtained.

## 2 Mechanical and Numerical Works

### 2.1 Spatial Scales and Macroscopic Approach

When we focus our attention on the description of a porous medium, the first question to be put is that of the spatial scale of analysis: indeed, two approaches are conceivable. The first one is situated at the microscopic scale. The characteristic length size is the dimension of the pore. In this configuration, the solid matrix is partially or completely filled with one or several viscous fluids. One geometric point is thus located in one of the different identifiable solid or fluid phases. Mechanical equations of each phase and mixture with compatible interface conditions are written. They correspond to those of linear elasticity in the solid and those of Stokes in the fluid. This approach deals with problems like interface modelling or description of microscopic geological structures. Homogenization is then obtained through asymptotic developments or averaging procedures and leads to a macroscopic description of the porous medium, see Terada and al. [10] or Coussy et al. [11] for instance. We obtain thus the famous set of macroscopic mechanical equations for a representative elementary volume. In this macroscopic spatial description, the porous medium is seen as a two-phase continuum. This scale, we study here, is well adapted to most of practical geomechanical problems.

Writing $u_i$ and $U_i$ respectively the macroscopic solid and fluid displacements components, Biot's equations can be written with usual notations as follows:

$$\sigma_{ij,j} = (1-\phi)\rho_s \ddot{u}_i + \phi\rho_f \ddot{U}_i \tag{1}$$

$$p_{,i} = -\frac{\phi}{K}(\dot{U}_i - \dot{u}_i) + \rho_f(a-1)\ddot{u}_i - a\rho_f\ddot{U}_i \tag{2}$$

$$\sigma_{ij} = \lambda_{0v}\varepsilon_{kk}\delta_{ij} + 2\mu_v\varepsilon_{ij} - \beta p\delta_{ij} \tag{3}$$

$$-\phi\left(U_{k,k} - u_{k,k}\right) = \beta u_{k,k} + \frac{1}{M}p \tag{4}$$

$\sigma_{ij}$ are the total Cauchy stress tensor components and $p$ is the pore pressure. The soil's characteristics are: $\lambda_{0v}$ and $\mu_v$ (drained viscoelastic equivalent Lamé constants), $\rho_s$ and $\rho_f$ (solid grains and fluid densities), $\phi$ (porosity), $K$ (hydraulic permeability representing the viscous coupling), $a$ (tortuosity standing for the mass coupling), $M$ and $\beta$ (Biot coefficients including the elastic coupling).

### 2.2 Finite Element Formulation and Numerical Resolution

To determine the solid and fluid displacements in the ground, we develop a numerical code based on the finite element method for the space integration, coupled to a finite difference method for the time integration. The main steps are:

- some boundary and initial conditions are associated to the previous partial differential system. Some modifications on the field equations are done in order to lead to a Cauchy's problem.
- integral forms are obtained using the weighted residual method. They are then spatially and analytically discretized and lead to a time differential system. The global differential system to be solved can be written as

$$[M]\frac{d}{dt}\{W^{(G)}\} + [K]\{W^{(G)}\} = \{F^{(G)}\} \tag{5}$$

  $[M]$ and $[K]$ are respectively the global mass and stiffness matrixes. $\{W^{(G)}\}$ and $\{F^{(G)}\}$ are the global vectors of unknowns and solicitation. With the developed technique, the mass matrix is diagonal and can be easily inverted.
- the backward finite difference method modified with an upward time parameter is used to obtain an approximate solution of the problem.

### 2.3 Structure of the Code and Parallelization

The sequential code called FAFEMO (Fast Adaptive Finite Element Modular Object), developed to solve the previous problem, constitutes an efficient code to deal with transient 2D problems and small 3D ones. The use of a matrix free technique, not necessary for small cases, becomes interesting for huge ones. An expert multigrid system is also used to optimize the problem size and yields a modification of the global matrixes at each time step. The two previous techniques lead to a high performance level both for the storage and the CPU costs.

**Table 1.** Time profile of the 2D sequential code

| reading of the data files and *element* class | *elementary matrixes* class | *building-resolution* class |
|---|---|---|
| 7.45% | 90.60% | 1.95% |

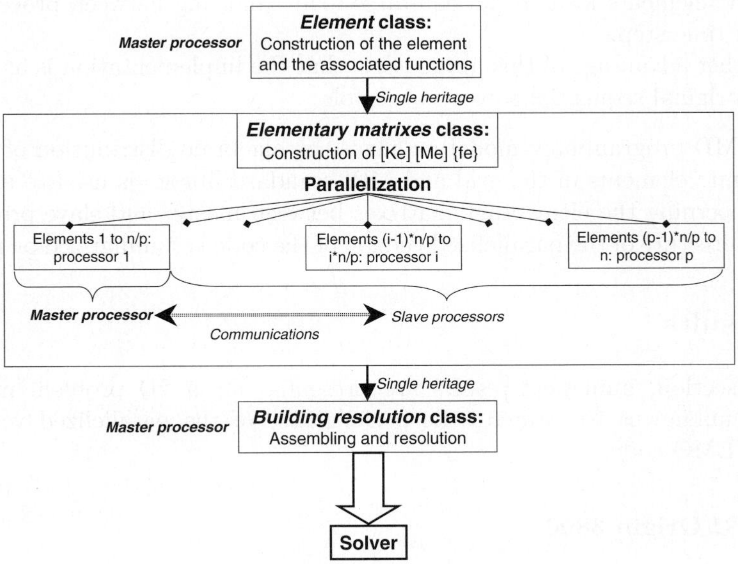


**Fig. 1.** Structure of the parallelized version of the C++ finite element code

The C++ code is organized in three classes connected by a single heritage: *element*, *elementary matrixes* and *building-resolution* classes.

For huge problems, the elementary vectors have to be calculated and assembled for each time step since they are too expensive in terms of Input/Output cost to be stored. In order to treat 3D problems and to perform intensive 2D parametric studies, we propose a parallelization of the code to reduce the time calculation.

The Unix/Linux *gprof* tool draws a time profile of the sequential code. For a two-dimensional application, the elapsed time is divided as presented in Table 1, for each of the three classes.

The part which is the largest consumer of elapsed time clearly appears to be the *elementary matrixes* class. This can be explained as the elementary matrixes have to be calculated for each time step. Besides, as we use a matrix free technique with a diagonal mass matrix, the resolution part is more efficient and needs little computational time. Moreover, the process of construction of $[K_e]$, $[M_e]$ and $\{f_e\}$ is iterative and independent element by element. This independent and time-consuming loop can thus be divided into several processors by distributing the $n$ elements between $p$ equitable parts.

We use a grid partitioning based on algebraic decomposition which is performed randomly without any geometric factors. Several advantages are:

- unlike the domain decomposition method, this technique does not need any particular interface management. This is particularly important when the expert multigrid system is activated, or when the geometry is changed.
- moreover, when the size of the grid is modified, the algebraic distribution of the elements leads to an equitable load balancing between processors at each time step.
- another advantage of this approach is that the implementation is as close to the original sequential solver as possible.

A SPMD programming model manages the algebraic distribution of the different finite elements of the grid and MPI standard library is used to exchange data concerning the elementary matrixes between master and slave processors. The architecture of the parallelized version of the code is summarized on figure 1.

## 3 Results

In this section, numerical results are presented for a 2D problem involving 200,000 unknowns to estimate the performance of the parallelized version of the FAFEMO code.

### 3.1 SGI Origin 3800

We have worked on SGI Origin 3800 installed at the National Computer Center of Higher Education (CINES, Montpellier, France). SGI ORIGIN is a shared global memory based on cc-NUMA architecture (cache-coherent Non-Uniform Access Memory) composed of 768 processors MIPS R14000 (500 MHz, 512 Mo RAM), distributed among 80 nodes. The internal network (Bristled hypercube) give a speed of 1,6 GB/s.

Figures 2 and 3 present the evolution of the elapsed times and the speed-up depending on the number of processors (up to 64). The main conclusions are:

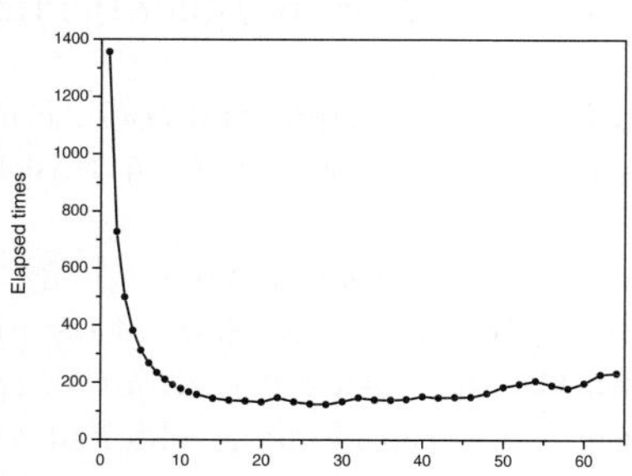


**Fig. 2.** Elapsed times in minutes up to 64 processors on SGI Origin 3800

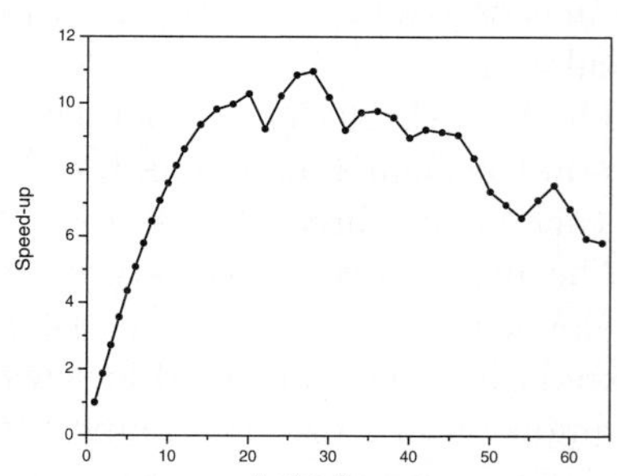


**Fig. 3.** Speed-up up to 64 processors on SGI Origin 3800

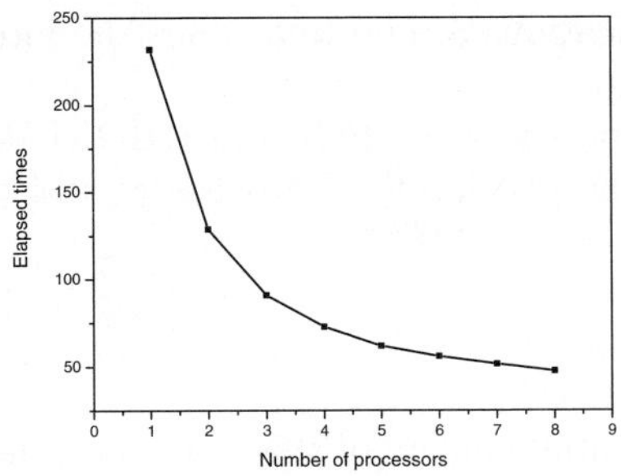


**Fig. 4.** Elapsed times in minutes up to 8 processors on cluster CHOEUR

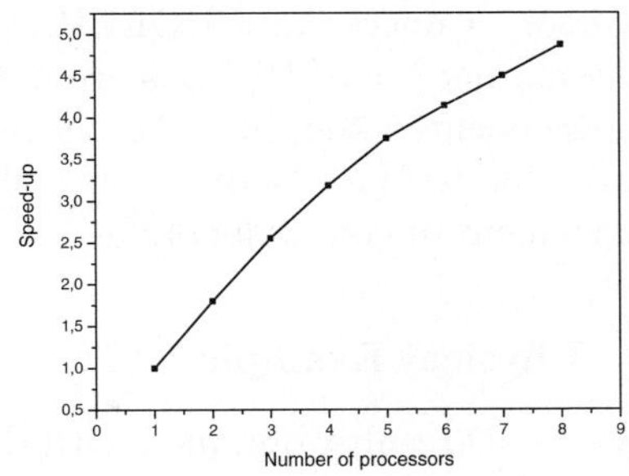


**Fig. 5.** Speed-up up to 8 processors on cluster CHOEUR

– the elapsed time is reduced considerably for a few number of processors and the communications do not affect the performance of the parallelization. For instance for 8 processors, the speed-up value equals 6.45. Compared to literature results on elastodynamic problems, these results are pretty good.
– for a larger number of processors, the performance of the code is not so interesting. Actually, the maximum of the speed-up curve corresponds to an optimal value of the number of processors situated between 20 and 30.

### 3.2 Cluster and Comparison

We have also worked on cluster CHOEUR, installed at the Laboratoire de Modélisation et Simulation Numérique en Mécanique et Génie des Procédés (MSNM-GP/L3M) of the CNRS and Aix-Marseille University. It is an AMD-ATHLON cluster, built upon 20 AMD-760MP 1800 bi-processors nodes interconnected with gigabit Ethernet, and 6 AMD-760MPX 1800+ bi-processors nodes (1530 Mhz, 1 GB RAM) interconnected with SCI-Dolphin High Speed 2,5 Gb network. The cluster is managed by an extra master node (2KMhz, 2 GB RAM).

Figures 4 and 5 present the elapsed times and the speed-up. Results are limited to 8 processors. As previously, the results show a real gain for the time calculation. Nevertheless, we can not visualize the maximum of speed-up for cluster CHOEUR which does not provide enough processors to reach it.

The elapsed times on SGI Origin 3800 are much longer than the ones obtained on cluster CHOEUR: for a single processor, more than 1300 min (i.e. 22 h) on SGI Origin 3800 and only 230 min (i.e. less than 4 h) on CHOEUR, because processors on CHOEUR are more recent and efficient.

The speed-up and the efficiency are better on SGI than on CHOEUR: for instance for 5 processors, the speed-up obtained on SGI equals 4.3 and 3.7 on CHOEUR. Calculation time is longer dominating compared to communications time on SGI due to less efficient processors.

Moreover, the inter-processor communication process is a very important parameter to take into account. Particularly, due to the hierarchical architecture of the memory of SGI (NUMA, non uniform memory access), the communication times decrease with the number of processors, but only over a limited range of processors. Concerning CHOEUR, communications are optimal since the limits of the mother board PCI bus are not reached.

Consequently, the difference in performance between SGI and CHOEUR is mainly due to the difference of quality of the two kinds of processors and the management of communications.

### 3.3 Physical Example

Previous 2D results, obtained with the sequential version of the code, have been presented for instance in [9] and [12]. Parametric studies on the influence of the different couplings and an approach of heterogeneous soil have been carried out.

The interest of the parallelized version is to be able to lead parametric study more rapidly and to tackle problem much more expensive in terms of storage and CPU costs, like 3D geometries. The 3D problem presented here is a 2,300,000 unknowns problem and concerns an homogeneous poroviscoelastic semi-infinite half-space ground subjected to an impulsional load applied on the center of the surface. Figure 6 presents for instance the solid displacement contour levels for a given time on the surface of the ground. The propagation of the Rayleigh wave is clearly perceptible and corresponds to the dark area. The first compressional wave has almost reached the boundary. Figure 7 shows similar results for the fluid phase for a high hydraulic permeability value: note that the two behaviors are strongly uncoupled. These results allow great perspectives for the study of 3D heterogeneous soils.

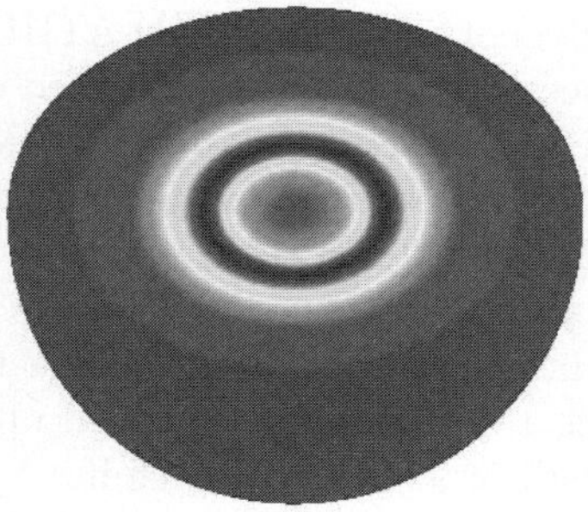

**Fig. 6.** Contour levels of the solid displacements (dimensionless time = 0.6)

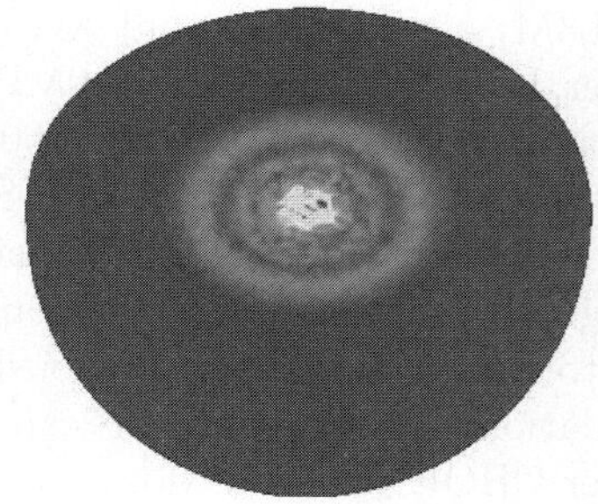

**Fig. 7.** Contour levels of the fluid displacements (dimensionless time = 0.6)

## 4 Conclusion and Further Works

A parallelized finite element code has been presented to study wave propagation phenomena in poroviscoelastic grounds. In fact, the applications are wider and can concern for instance porous bones or foams. Besides, the code can treat all propagation wave phenomena: a version studying electromagnetic wave propagation have been developed in the same way.

With the parallelized code, one can consider the work to be continued: *i)* a study of more complex ground : an approach of partially saturated ground is under progress *ii)* an analysis of 2D heterogeneous random media mixing two kinds of soils has shown the existence of different thresholds for fluid and solid phases: we would like to do similar analysis for a 3D geometry in order to know if conclusions are identical or not *iii)* a study of the coupling of electromagnetic and seismic waves, which represents a hard problem because of the two different time scales.

Moreover, concerning the numerical point of view, some improvements can still be done: for instance a compressed message passing technique for inter-node communication could allow a significant reduction of the communications time.

To achieve these aims, an access of more computational resources at the CINES has been asked for 2006. Besides, SGI will be soon replaced by a massive parallel supercomputer which will allow more efficient abilities.

## References

1. Biot, M.A.: Theory of propagation of elastic waves in a fluid-saturated porous solid. I- Low-frequency range. J. Acoust. Soc. Am. **28(2)** (1956) 168–178
2. Coussy, O.: Mécanique des milieux poreux. (1991) Paris: Ed. Technip
3. Zienkiewicz, O.C., Shiomi, T.: Dynamic behaviour of saturated porous media: the generalized Biot formulation and its numerical solution. Int. J. Numer. Anal. Methods Geomech. **8** (1984) 71–96
4. Simon, B.R., Wu, J.S.S., Zienkiewicz, O.C., Paul, D.K.: Evaluation of u-w and u-$\pi$ finite element methods for the dynamic response of saturated porous media using one-dimensional models. Int. J. Numer. Anal. Methods Geomech. **10** (1986) 461–482
5. Gajo, A., Saetta, A., Vitaliani, R.: Evaluation of three and two field finite element methods for the dynamic response of saturated soil. Int. J. Numer. Anal. Methods Geomech. **37** (1994) 1231–1247
6. Papadrakakis, M., Bitzarakis, S.: Domain decomposition PCG methods for serial and parallel processing. Adv. Eng. Softw. **25** (1996) 291–307
7. Lirkov, I.: MPI solver for 3D elasticity problems. Math. Comput. Simul. **61** (2003) 509–516
8. Bohlen, T.: Parallel 3-D viscoelastic finite difference seismic modelling. Comput. Geosci. **28** (2002) 887–899
9. Mesgouez, A., Lefeuve-Mesgouez, G., Chambarel, A.: Simulation of transient mechanical wave propagation in heterogeneous soils. Lect. Notes Comput. Sc. **3514** (2005) 647–654
10. Terada, K., Ito, T., Kikuchi, N.: Characterization of the mechanical behaviors of solid-fluid mixture by the homogenization method. Comput. Methods Appl. Mech. Eng. **153** (1998) 223–257
11. Coussy, O., Dormieux, L., Detournay, E.: From Mixture theory to Biot's approach for porous media. Int. J. Solids Struct. **35** (1998) 4619–4635
12. Mesgouez, A., Lefeuve-Mesgouez, G., Chambarel, A.: Transient mechanical wave propagation in semi-infinite porous media using a finite element approach. Soil Dyn. Earthq. Eng. **25** (2005) 421–430

# Numerical Modeling of Tidal Effects and Hydrodynamics in the Po River Estuary

Célestin Leupi[1], Michel Deville[1], and Mustafa Siddik Altinakar[2]

[1]ISE-STI-LIN, Ecole Polytechnique Fédérale,
Station 9, Lausanne 1015, Switzerland
Tel.: +41.21.693.25.07; Fax: +41.21.693.36.46
celestin.leupi@epfl.ch
Tel.: +41.21.693.53.18; Fax: + 41.21.693.36.46
michel.deville@epfl.ch
[2] NCCHE, The University of Mississipi,
Carrier Hall Room 102 University,
MS 38677 USA

**Abstract.** The present work contributes to the numerical simulation of complex turbulent multiphasic fluid flows encountered in estuarine channels. A numerical solution is based on Reynolds averaged Navier-Stokes equations using the mass preserving model based on the so-called Raviart-Thomas finite element on the unstructured mesh in the horizontal plane. In the vertical, the computational domain is divided into number of layers at predefined heights and the method uses a conventional conforming finite element scheme, with the advantage that the lowermost and uppermost layers variable height allow a faithful representation of the time-varying bed and free surface, respectively. A robust up-to-date algorithm is used for computing the eddy viscosity from the efficient $k-\varepsilon$ turbulence model for variable density fluid flows. Finally, the capability and the predicting performance of the model are successfully evaluated by applying it to the simulation of the Po River Estuary (PRE) in Italy.

**Keywords:** Finite Element model; multiphasic fluid flows; $k-\varepsilon$ turbulence model; multi-layers system; unstructured grid; Estuary.

## 1 Introduction

Estuaries, where are situated many famous ports of the world, are the meeting point between the freshwater charged with sediment from river and the saline sea cold water. The present research finds its motivation in the increasing need for efficient management of estuaries by taking into account the various conflicting environmental and socio-economical aspects.

Shallow water equations have been successfully applied to many engineering problems and their use has become common practice in environmental impact studies in estuarial and coastal regions.

V.N. Alexandrov et al. (Eds.): ICCS 2006, Part II, LNCS 3992, pp. 58–65, 2006.

Chen *et al.* [3] have successfully applied a three-dimensional finite difference model with hydrostatic pressure distribution for simulating the Pearl River Estuary. Casulli and Chen [1] have developed a semi-implicit finite difference scheme for the the $3D$ shallow water equations (SWE). Lie and Mellor [8], [9] have applied a $3D$ finite difference model for simulation the Hudson-Raritan Estuary with a second order turbulence-closure sub-model. Lu and Wai [10] developed an efficient operator splitting scheme for 3D hybrid hydrodynamic model using the so-called *sigma transformation* in the vertical direction. Chau and Jiang [2] have developed a model based on POM (Princeton Ocean Model) with an orthogonal curvilinear coordinate in the horizontal direction and sigma coordinate in the vertical direction for simulating the Pearl River Estuary.

Most of the existing models use a finite difference or finite volume approach based on a topographically conform, sigma-transformed grid which, regardless of the depth, forces the same number of divisions in the water column. However, these models cannot fit to complicated geometry and this may induce large large errors if applied to complex boundary configurations.

In this study, the hydrodynamics solutions are obtained by solving the shallow water equations using implicit time marching scheme (see Leupi et al. [7], [6]), while discretization is conducted using Euler or Runge-Kutta scheme to obtain a set of algebraic equations. The model uses hydrostatic pressure distribution, and an efficient fractional time-step scheme of Mohammadi and Pironneau [12] for the $k - \varepsilon$ turbulence closure. This paper deals with the simulation of the 3D turbulent flow in the Po River Estuary in Italy.

## 2 Mathematical Models

A brief description of the model is provided in [7], [5]. Let us consider an incompressible fluid body in a three-dimensional time varying domain $\widehat{\Omega}$. Let $\Omega$ the projection of $\widehat{\Omega}$ on the $xy$ horizontal plane, bounded by the free-surface $\Gamma_s$ given by $z = \eta(x, y, t)$, the bottom topography $\Gamma_b$ given by $z = -\mathfrak{h}(x, y)$, the open boundary denoted by $\Gamma_o$. Where $\mathfrak{h}(x, y)$ is the distance between the bottom and the reference plane $xy$ and $\eta(x, y, t)$ the elevation of the free-surface with the respect to the reference plane $xy$.

The RANS description of the turbulent motion relies on the Boussinesq approximation. For $t > 0$ and $(x, y, z) \in \widehat{\Omega}(t)$, and using the Boussinesq assumption for the hydrostatic approximation we have that

$$\frac{1}{\rho}\frac{\partial p}{\partial z} = -g \Rightarrow p = p_0 + \rho g\,(\eta - z) \tag{1}$$

where $p_0$ is the atmospheric pressure.

For a variable density fluid under the hydrostatic assumption the system of the shallow water equations for long waves (vertical scale is negligible compared to the horizontal one) reads

$$\begin{cases} \nabla_{xy} \cdot \mathbf{U} + \dfrac{\partial w}{\partial z} = 0, \\ \dfrac{D\mathbf{U}}{Dt} - \dfrac{\partial}{\partial z}\left(\nu_T \dfrac{\partial \mathbf{U}}{\partial z}\right) + g\nabla\eta + g\nabla\left(\displaystyle\int_z^\eta \dfrac{\Delta\rho}{\rho_0}\mathrm{d}z\right) = \mathbf{F}_{xy}, \\ \dfrac{\partial \eta}{\partial t} + \nabla \cdot \displaystyle\int_{-h}^{\eta} \mathbf{U}\mathrm{d}z = \tilde{F}_b, \\ \dfrac{D\zeta}{Dt} - \nabla \cdot \left[c_\zeta \dfrac{k^2}{\varepsilon}\nabla\zeta\right] = F_\zeta \quad \text{for} \quad \forall\zeta = (k, \varepsilon) \\ \dfrac{D\Phi}{Dt} - \nabla_{xy} \cdot (\nu_T \nabla\Phi) - \dfrac{\partial}{\partial z}\left(\nu_T \dfrac{\partial \Phi}{\partial z}\right) = F_\Phi, \quad \Phi = (T, S), \\ \rho = \rho(T, S), \end{cases} \tag{2}$$

where the unknowns are $\boldsymbol{v}, w, \eta, k, \varepsilon, T, S$. $\mathbf{U} = (u, v)^T$ is the horizontal velocity vector , $\mathbf{F}_{xy} = (\mathrm{f}v, -\mathrm{f}u)^T$ is vector of body forces with $f$ the Coriolis parameter, $g$ is the gravitational acceleration, $\nu_T$ is the eddy viscosity. $\nabla\cdot$ is the $3D$ divergence operator, $\frac{D}{Dt}$ represents the material derivative, and $\nabla_{xy}\cdot$ is the surface divergence operator. $\rho$, $\rho_0$ are respectively the fluid density and the basic water density. $\mathbf{n}_{xy}$, $n_z$ are respectively the horizontal and vertical components of the normal vector $\mathbf{n}$. $\boldsymbol{\Phi}(S, T)$ stands for the passive scalars: salinity (S), temperature (T) (for which $S_0$ and $T_0$ stand for the initial salinity and temperature).

The motion of the free surface is described by an alternative form of the equation, third equation of system (2), which is obtained by integrating the first equation [i.e. (local) mass balance] along the vertical coordinate by applying the Leibniz rule with the suitable kinematic boundaries conditions at the free-surface and bottom.

In the present work the state-of-the-art modeling of the $k-\varepsilon$ turbulence (fifth equation of system (2) is introduced to account for boththe stratified flows. The turbulence equations read, (see [12]).

$$\frac{Dk}{Dt} - \nabla \cdot \left[c_\mu \frac{k^2}{\varepsilon}\nabla k\right] = c_\mu \frac{k^2}{\varepsilon}G - \varepsilon - \lambda_T N^2 \tag{3}$$

$$\frac{D\varepsilon}{Dt} - \nabla \cdot \left[c_\varepsilon \frac{k^2}{\varepsilon}\nabla\varepsilon\right] = c_1 kG - \frac{\varepsilon}{k}\left[c_3\lambda_T N^2 + c_2\varepsilon\right] \tag{4}$$

The squared shear frequency or production term $G$ is (see [5]):

$$G = \frac{1}{2}\left(\|\nabla\mathbf{U}\| + \|\nabla\mathbf{U}\|^T\right)^2 \tag{5}$$

where $\|.\|$ represents the Euclidian norm, $\mathbf{U} = \mathbf{U}(u, v, w)$ is the 3D velocity vector.

The model constants are given: $c_1 = 0.126$, $c_2 = 1.92$, $c_\mu = 0.09$, $c_\varepsilon = 0.07$. $N$ is the Brunt-Väisälä frequency, with the related following squared buoyancy $N^2$ expression,

$$N^2 = -\frac{\nu_T}{\sigma_t} \frac{g}{\rho_0} \frac{\partial \rho}{\partial z} \tag{6}$$

Here $\sigma_t = 1$ is a constant, and the following expressions of eddy coefficients $\nu_T, \lambda_T$ should include

$$\nu_T = S_u \frac{k^2}{\varepsilon} + \nu \quad , \quad \lambda_T = S_b \frac{k^2}{\varepsilon} + \lambda_b \tag{7}$$

where $\sigma_t = 1$ is a constant. $S_u$ and $S_b$ are the stability parameters to account for the turbulence damping in the stratified fluid flows (see [5], Luyten et al. [11]).

In the system 2, the state equation for the variable density can be written as follows,

$$\rho = \alpha (S - S_0) - \beta (T - T_0) \tag{8}$$

where $\alpha$ and $\beta$ are the known coefficients, $\rho$ is the pore water density.

The equilibrium assumption is adopted at the bed (i.e local balance between production of turbulent kinetic energy and the rate of dissipation) and the wall functions are applied to relate the velocity to the bed shear velocity (see [7], [5]). The physical $3D$ domain is embedded in a parallelepiped composed of $\mathcal{N}$ layers and the refine description can be found in [6]. The horizontal components of the velocity vector are defined at the middle of edges of the triangular mesh elements, while the vertical component is associated with the lower horizontal faces of the element. The horizontal velocity is approximated combining the lowest order Raviart-Thomas element ($\mathbb{RT}_0$) in $xy$ plane with the $\mathbb{P}_1$ elements along the vertical direction.

The Lagrange-Galerkin (or characteristics Galerkin) approach is used to discretize the convection terms, while the Euler scheme or more accurate Runge-Kutta is used to obtain the algebraic system. The fractional time-step scheme from Mohammadi and Pironneau ([7], [12]), is adopted to solve turbulence equations. Consequently, the positivity of $k$ and $\varepsilon$ is preserved as well as the stability of the scheme (see [7]).

## 3 Application to the Po River Estuary in Italy

An efficient decoupling scheme is employed to solve hydrodynamic, turbulence, heat and salinity. The decoupling method partitions a time step into fractional steps according to physical phenomena. In the first step hydrodynamics is solved, followed by turbulence closure, and finally the advection-diffusion equations for passive scalars (temperature, salinity) are solved. PRE is located in the east of the Adriatic Sea along the direction North West-North East (NW-NE). The axis of PRE is east longitude 12.05° and north latitude 44°96667. The $M2$ main tidal component is taken into account in this study and the mean wind speed is 30 − 50[km/h]. The average discharge of the PRE is $1.5 \times 10^3$ $[m^3/\text{s}]$. For simulating the tidal currents and sediment transport in PRE, a large area was selected and the 2D horizontal plane mesh. The north upstream boundary has been set far upstream the five open boundaries. The available field data for the model calibration were mostly provided by the *MOX (Modeling and Scientific Computing), Department of Mathematics, Politecnico of Milano* and the *Istituto di ricerca Sulle Acque, CNR, Milan, Italy.*

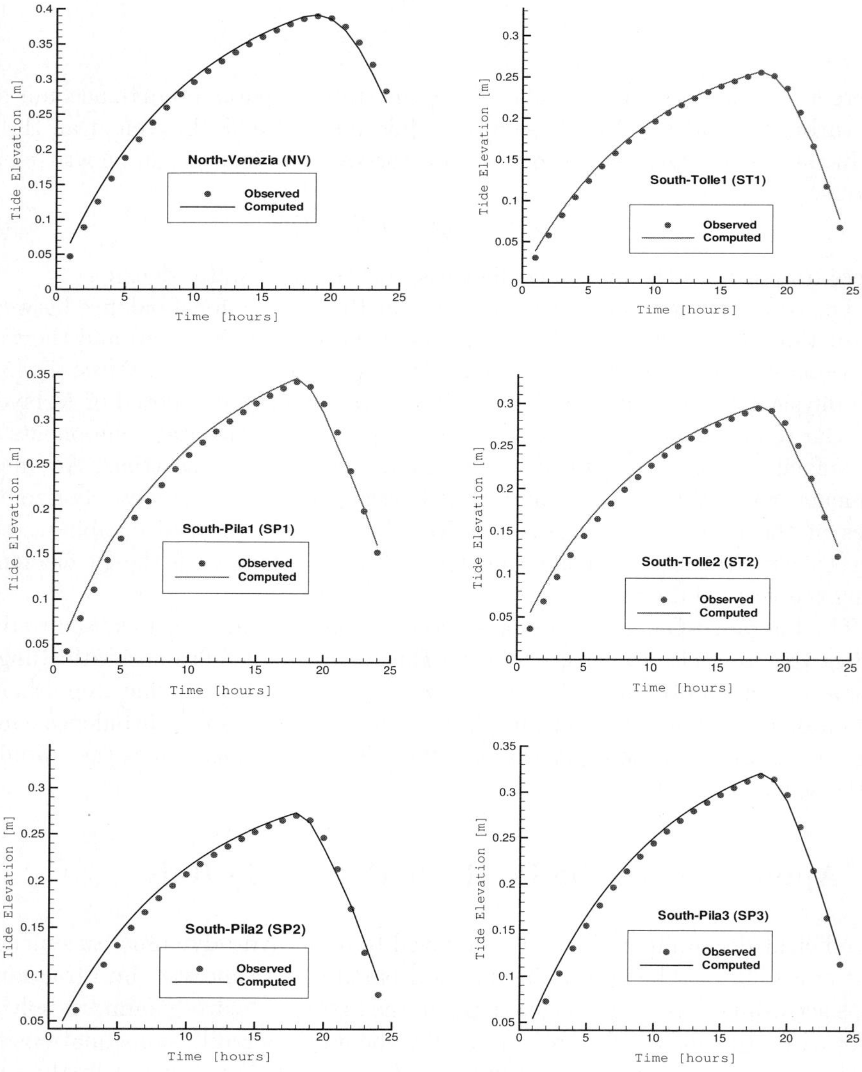


**Fig. 1.** Comparison of the computed and observed tidal elevations at river upstream and open sea boundaries in PRE

In this simulation, the horizontal unstructured mesh of simulated area has been divided into 3185 triangles in each horizontal layer, and 2172 nodes with the mean space step size ranging from $d^T = 6.2$ to 46.08 [m] (where $d^T$ denotes the diameter size of the circumscribed triangle). The vertical is divided into 30 layers. To account of the boundary layer effects the bottom is divided into several thin layers of thickness $\delta z_b = 0.5$[m]. The simulation period is from 14.00 of 1st of May to 14:00 of 2nd May 1999 for hydrodynamic calibration. The forcing is neap tide in wet season. There are seven tide gauges and seven tide stations and

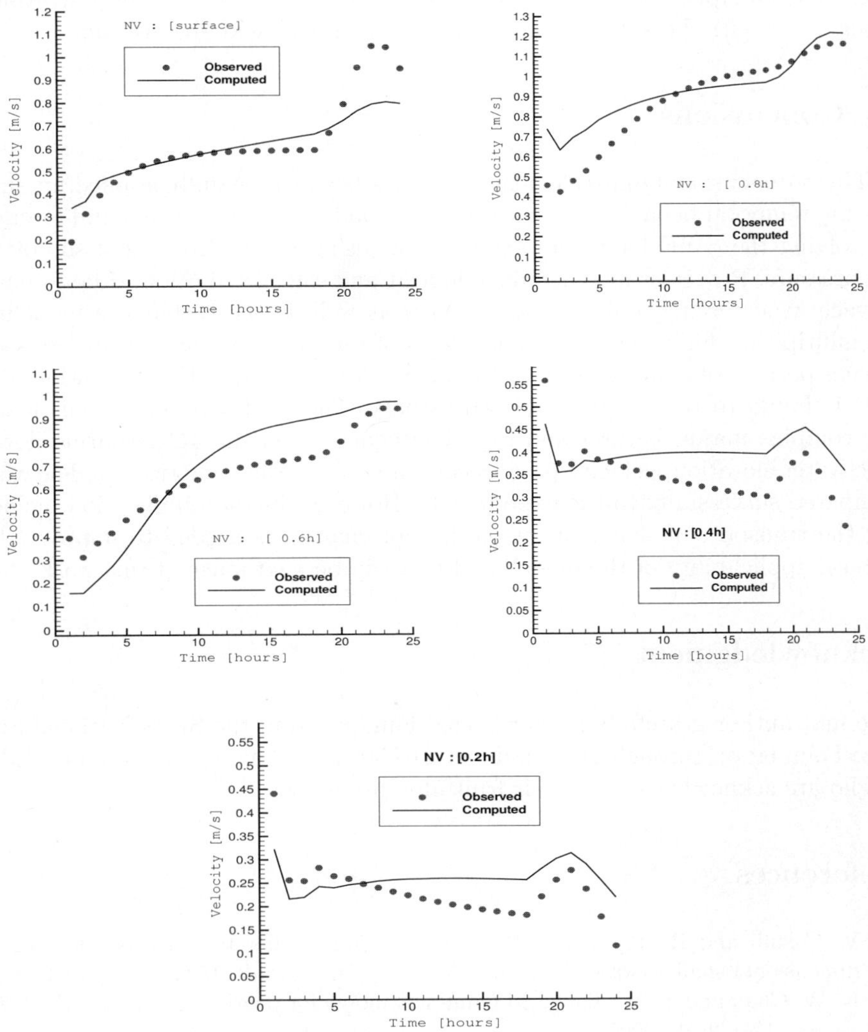


**Fig. 2.** Comparison of the computed and observed current velocities at North-Venezia station (NV) in PRE

water open boundaries were controlled by tidal levels located at $NV$, $NT$, $ST1$, $ST2$, $SP1$, $SP2$, $SP3$.

Figure 1 shows the comparisons of the computed and observed tidal elevations during the flood at different tide stations. As indicated, the model predicts the tidal elevations with good accuracy. Figure 2 shows the comparisons of the computed and observed flow velocity magnitude at the river upstream tide stations and different vertical layers position at respectively free-surface, $0.8H$, $0.6H$, $0.4H$ and $0.2H$. Good agreement is found between computed and observed velocity distributions. The hydrodynamics computations have shown the ability of the model. A wet spring current was chosen for one-year simulation period from May 1999 to April 2000. The PRE is found to be a stable and fairly stratified estuary (see [4]). This is in good accordance with the field observation.

## 4 Conclusions

A Three-dimensional finite element solver has been successfully applied for simulating temporal behavior of stratified water bodies with environmental forcing, by solving the SWE for incompressible flow using the hydrostatic assumption for pressure. Processes modeled include heat and salt distribution, tidal forcing, surface wind forcing, inflows and outflows as well as the turbulence modeling for multiphasic fluid-solid flow. The effects of salinity and temperature are seen during periods of simulation, and the micro-tidal PRE (i.e. tidal elevation $< 2$ [m]) is found to be a fairly stratified estuary. The model predicts realistically the complex major features and their consequences on the $3D$ estuarine flows. The water elevation, velocity profiles computed from the numerical models were compared successfully to the experiments. However, further application including the transport of sediment on the harbor channel is needed to improve the general applicability of the model, and this will be next stage of this work.

## Acknowledgment

The first author gratefully acknowledged Funding from the Swiss National Science Foundation through grant number 21-65095.01. Alfio Quarteroni and Edie Miglio are acknowledged for their fruitfully discussions.

## References

1. V. Casulli and R. T. Cheng. Semi-implicit finite difference methods for three-dimensional shallow-water flow. *Int. Numer. Meth. Fluids*, 15:629–648, 1992.
2. K. W. Chau and Y. W. Jiang. 3d numerical model for pearl river estuary. *J. Hydr. Engrg.*, 127:72–82, 2001.
3. Y. Chen, O. W. H. Wai, Y. S. Li, and Q. Lu. Three-dimensional numerical modeling of cohesive sediment transport by tidal current in Pearl River Estuary. *Int. J. Sediment Res.*, 14:107–123, 1999.

4. W. H. Graf and M. S. Altinakar. *Hydraulique Fluviale, Tome II.* Presses Polytechniques et Universitaires Romandes, CH-1015 Lausanne, Switzerland, 1996.
5. C. Leupi. *Numerical Modeling of Cohesive Sediment Transport and Bed Morphology in Estuaries.* PhD thesis, Ecole Polytechnique Fédérale de Lausanne-EPFL, No. 3266, 2005.
6. C. Leupi and M. S. Altinakar. Finite element modeling of free-surface flows with non-hydrostatic pressure and $k-\varepsilon$ turbulence model. *Int. J. Numer. Meth. Fluids*, page In press, 2005.
7. C. Leupi, E. Miglio, M. Altinakar, A. Quarteroni, and M. Deville. Quasi-3D finite element shallow-water flow with $k-\epsilon$ turbulence model. In Altinakar M. S., Wang S. S. Y., Holz K. P., and Kawahara M., editors, *Proc. of 6th Int. Conf. Hydro-Science and Engrg, 6, 400-402 & on CD-Rom*, Brisbane, Australia, May 31, June 03 2004. ICHE, SWang S. Y., NCCHE, University of Mississippi, Carrier Hall, University, MS38677, USA.
8. O. Lie-Yauw and L. G. Mellor. A three-dimensional simulation of the hudson-raritan estuary. part i: Comparison with observation. *J. Phys. Ocean.*, 15:1693–1709, 1985.
9. O. Lie-Yauw and L. G. Mellor. A three-dimensional simulation of the hudson-raritan estuary. part i: Description of the model and model simulations. *J. Phys. Ocean.*, 15:1676–1692, 1985.
10. Q. M. Lu and W. H. O. Wai. An efficient operator splitting scheme for three-dimensional hydrodynamics computations. *Int. J. Numer. Methods Fluids*, 26:771–789, 1998.
11. P. J. Luyten, E. Deleesnijder, J. Ozer, and K. G. Ruddick. Presentation of a family of turbulence closure models for stratified shallow water flows and preliminary application to the Rhine outflow region. *Continental Shelf Res.*, 16:101–130, 1996.
12. B. Mohammadi and O. Pironneau. *Analysis of $k-\epsilon$ Turbulence Model.* Research in Applied Mathematics. John Wiley & Sons, Chichester, 1994.

# Adaptive Mesh Refinement and Domain Decomposition: A Framework to Study Multi-physical and Multi-scale Phenomena. First Application to Reacting Gas Flows

J. Ryan

ONERA,B.P. 72, 92322 Châtillon cedex, France
ryan@onera.fr

**Abstract.** The aim of this work is to extend the Adaptive Mesh Refinement (AMR) technique . AMR is an ideal platform for coupling different discretisations and models, as each grid level can use a different solver. Compact conditions have to be defined and these are problems which are at the core of Domain Decomposition methods. An application to a planar shock interacting with a circular diffusion H2-air flame is presented.

## 1 Introduction

The complexity of combustion and the subsequent difficulty of describing the intricate coupling between the many physical phenomena due to the presence of multiple length scales and the large number of species involved still limit the impact of numerical studies. One of the themes which is presented here is to extend the Adaptive Mesh Refinement (AMR) technique initially written for a finite volume algorithm to finite differences in order to integrate a 3D combustion solver. One of the main problems with AMR is the connection between coarse and fine patches. Domain decomposition methods can help to improve these. The final aim of our work is the coupling of various combustion solvers from simplified combustion model on a coarse grid to high precision DNS Finite Difference reacting gas model.

## 2 Adaptive Mesh Refinement

The locally adaptive mesh refinement method written at Onera by M. Borrel et al [2], originally developed by Berger and Oliger [1] and many other authors, uses a sequence of nested levels of refined structured grids (patches), on which different solvers on each level can be applied. These patches are built around cells that have been flagged with a problem dependant sensor, with a grouping/clustering technique.

The system of overlaid grids with successively finer spacing in time and in space needs the following operators to couple the different levels Prolongation operator

V.N. Alexandrov et al. (Eds.): ICCS 2006, Part II, LNCS 3992, pp. 66–69, 2006.

**(P)**, Restriction operator **(R)**, Fine coarse boundary treatment **(FC)**, Fine fine boundary treatment(**FF**).

The first operator **(P)** usually consists in retrieving values from former patches and interpolating the missing values. The second operator ( **(R)** )replaces coarse values by the overlying newly computed fine values. The last operator **(FF)** , as the scheme is entirely time explicit, is only coding work as it consists in retrieving computed values from neighbouring patches. Instabilities arise from the choice of the third operator **(FC)** . In the original Finite Volume method , fine coarse boundary treatments are dealt with by creating around the fine patch a set of fictitious points with values interpolated from the coarse grid. Fluxes between fine and coarse interfaces are readjusted after computation. Applied to high order finite difference discretisation, replacing interpolated values by values such that fine and coarse gradients were equal proved to be stable in all considered cases.

## 3 AMR and Combustion

Thévenin's two-dimensional studies [3] have shown the necessity for a high precision solver (6th order in space and 4th order in time) with a large number of points. For three dimensional cases, this becomes quite prohibitive. Only locally adaptive and refined grids can resolve this impediment, especially as cases to be studied show locality of structures. To reduce even more computation costs , and as complex chemistry modeling is not necessary every where and all the time, a lower order algorithm MAJIC has been tested and integrated within the AMR platform. This method can run on coarse grids and detect areas that need to be refined in terms of geometry and model, and if necessary, a higher order solver can be used locally .

### 3.1 MAJIC

MAJIC is a 3D Finite Difference discretization code with a simplified combustion model developed by G. Billet [4]. MAJIC uses a time-splitting method . The 3-D finite difference operator is split into a product of simpler explicit operators. The convective terms are solved by a $AUSM^{+}-MUSCL-triad$ approach ( [6]). The "double flux model" presented in [6] is used to preserve pressure and velocity across the contact discontinuities . All the diffusion and dissipation terms are solved with a centered second-order scheme.

### 3.2 Planar Shock Interacting with a Circular Diffusion H2-Air Flame

A circular diffusion flame moving in a supersonic air flow ($M = 2.$) interacts with a steady planar shock. The bubble contains a H2-N2 mixture ($Y_{H_2} = 0.233, Y_{N_2} = 0.767$) at temperature $T = 300K$Upstream, the temperature of

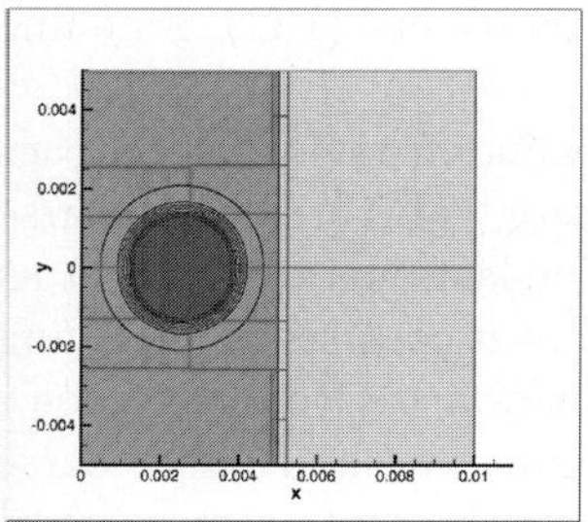


Fig. 1. Before the shock

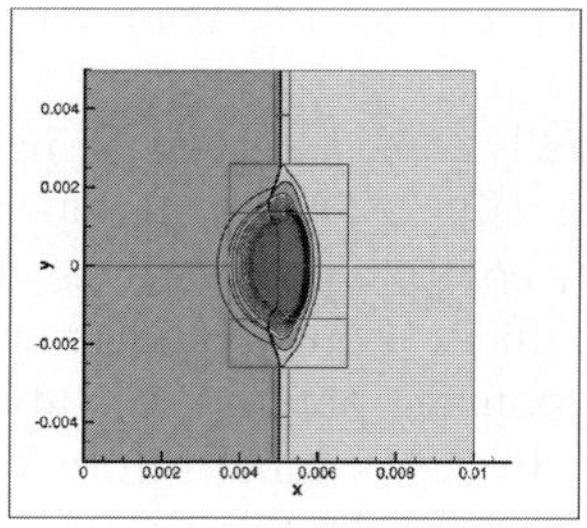


Fig. 2. Through the shock

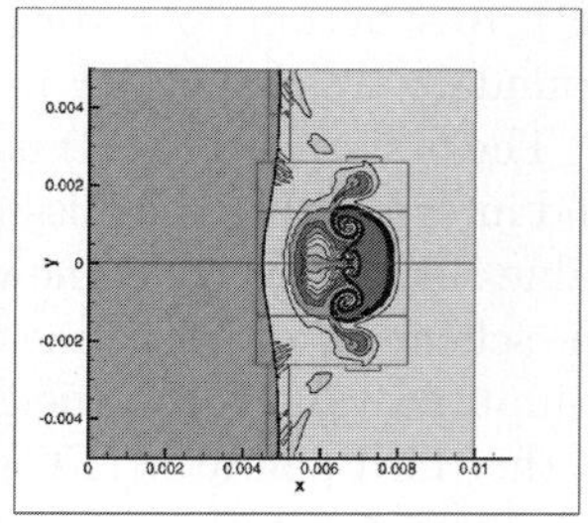


Fig. 3. After the shock

the ambient air is $1500K$ and the pressure is uniform ($p = 1bar$). This interaction sets up two counter-rotating vortices that are going to stretch the flame and enhance the total burning velocity. A detailed chemistry (9 species and 38 reactions) and detailed transport properties are used.

On figures (1,2,3), are shown temperature evolution through the shock and the attached patches.

## 4 AMR and Coupling

AMR is an ideal platform for coupling different discretizations and models, as each grid level can use a different solver. In the classical AMR technique, fictitious points and interpolated values are created to allow the fine grid to compute the needed derivatives. To extend this to high order schemes , compact conditions have to be defined and these are problems which are at the core of Domain Decomposition methods.

### 4.1 Artificial Boundary Conditions

J. Gander, L. Halpern and F. Nataf in [5] developed optimal transmission conditions for space - time domain decompositions for wave propagation problems. When defining a finer patch, and using explicit time schemes, the time step to preserve the CFL number is reduced by the refinement factor. Fictitious points have thus to be defined at all the intermediate time steps. The space-time domain boundary conditions (Robin type conditions in time and space) mentioned above simplifies computation of these boundary values. From the underlying coarse domain, space and time derivatives can be computed at the boundaries of the fine patch, and combined to provide a boundary condition for the finer grid. Computation can then proceed in the fine patch. Fine values are then injected in the coarse grid.

First comparisons between the fictitious points technique and the time domain method ( Artificial boundary conditions, ABC) were performed in the case of the one dimensional unsteady convection diffusion of a Gaussian The ABC method gives better results for low order schemes, and are as good for a high order scheme.

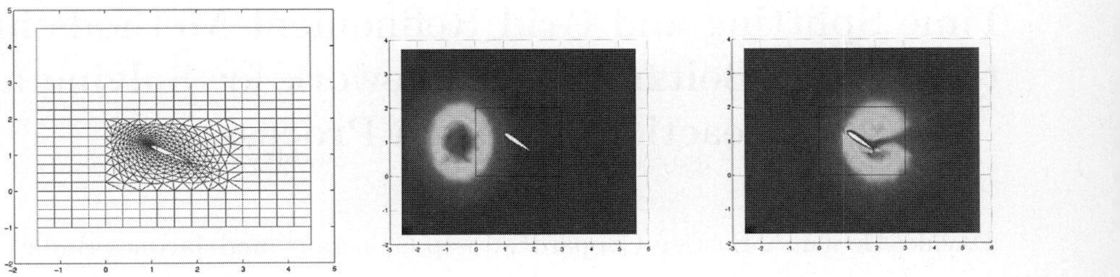

a) General mesh b) T=0 $\delta t$ d) T=50$\delta t$

**Fig. 4.** Time evolution of a temperature blob round a NACA airfoil

Another test to validate the artificial boundary conditions was the 2D convection diffusion equation round a NACA airfoil. Two domains were used with a different solver :P1 finite elements, on the triangular mesh close to the airfoil, second order finite differences on the far-away field.

## 5 Conclusion

The AMR technique has shown to be valid in the case of different discretisations of the fluid equations. Future work will concern coupling different (simple to complex) modeling of these equations. The AMR-DDM technique can be extended to other fields of physics (solid mechanics, electro-magnetism, ...). The use of common tools cannot but be fruitful to develop multi-physics coupling.

## References


1. Berger M. J. , Oliger J. Adaptive Mesh Refinement for Hyperbolic Partial Differential Equations. *J. Comput. Phys.*, **53**, pp. 484-512, (1984).
2. Borrel M., Huré N. A MUSCL/DG AMR Method in 3D: A Multiblock Multigrid AMR Method in 3D using a MUSCL or a Discontinuous Galerkin Approach. *Finite Volume for Complex Applications*, Porquerolles, France, June 2002
3. Thévenin D., Gicquel O., Hilka M. and Darabiha N., *Direct Numerical Simulation of turbulent premixed flames using intrinsic low-dimensional manifolds, Combustion Science and Technology* ,**149**, pp.297-337 (1999)
4. Billet G. Vers l'extension de l'approche MUSCL-AUSM-triade à la LES et la DNS des écoulements compressibles visqueux. Application à la combustion instationnaire des flammes de diffusion avec chimie simplifiée. Rapport ONERA (2001).
5. Gander M.J. ,Halpern L. ,Nataf. F. Optimal convergence for Overlapping and Non-Overlapping Schwartz Wave form, relaxation. *11th International Conference on Domain Decomposition Methods, (1999).*
6. Billet G., Abgrall R. An adaptive shock capturing algorithm for solvingunsteady reactive flows. *Computers and Fluids*, **32**, pp 1473-1495, (2003).

# Time Splitting and Grid Refinement Methods in the Lattice Boltzmann Framework for Solving a Reaction-Diffusion Process

Davide Alemani[1], Bastien Chopard[2], Josep Galceran[3], and Jacques Buffle[1]

[1] CABE - Analytical and Biophysical Environmental Chemistry
Sciences II, 30 quai E. Ansermet, CH-1211 Geneva 4, Switzerland
{Davide.Alemani, Jacques.Buffle}@cabe.unige.ch
[2] CUI - Computer Science Department
24 rue General-Dufour CH-1211 Geneva 4, Switzerland
Bastien.Chopard@cui.unige.ch
[3] ETSEA - Departament de Quimica, UdL
Av. Rovira Roure 191, 25198 Lleida, Spain
galceran@quimica.udl.es

**Abstract.** The paper shows how to combine together the Lattice Boltzmann Methods with the time splitting and the grid refinement techniques, in order to solve reaction-diffusion processes including very fast reaction dynamics, i.e. with time and length scales that vary in a wide range of values. The method is applied to the reaction prototype problem: $\mathrm{M}^0 \leftarrow \mathrm{M} + \mathrm{L} \rightleftharpoons \mathrm{ML}$ with semi-infinite diffusion conditions and in presence of an electrode where Nernst + flux balance conditions are considered. Two important geometries are considered, planar and spherical, and off-lattice boundary conditions are set up, for general irregular and curved boundaries. We discuss the need, for some cases, of applying the time splitting and the grid refinement approach to have a numerical scheme more easily handled and to substantially reduce the computational time. Furthermore, we point out the physico-chemical conditions to apply the time splitting and the grid refinement to optimise accuracy and performance. In particular, we stress: a) the range of values of the relaxation BGK parameter to have the best performance in solving the pure diffusive scheme and b) the best values of the grid refinement factor to preserve a good accuracy and drastically reduce the time of computation and the memory usage.

## 1 Introduction

Environmental physicochemical problems usually involve a large number of processes that take place over a wide range of space and time scales. We will consider, in this paper, the typical reaction-diffusion process as:

$$\mathrm{M}^0 \leftarrow \mathrm{M} + \mathrm{L} \rightleftharpoons \mathrm{ML} \tag{1}$$

V.N. Alexandrov et al. (Eds.): ICCS 2006, Part II, LNCS 3992, pp. 70–77, 2006.

wherein a metal ion M can react in solution with a ligand L forming a complex ML with given association and dissociation rate constants, $k_a$ and $k_d$ respectively. Furthermore, the metal ion is reduced at an electrode surface into $M^0$. All the species diffuse with diffusion coefficients, $D_M$, $D_L$, $D_{ML}$ and $D_{M^0}$ respectively. Process (1) is a typical electrochemical problem and in environmental systems is a typical multi-scale problem. In fact, the reaction rate constants and the diffusion coefficients of the species can vary over a large range of values. Typically, the reaction constants $k_a$ can take values between $10^{-6}\text{m}^3\text{mol}^{-1}\text{s}^{-1}$ and $10^8\text{m}^3\text{mol}^{-1}\ \text{s}^{-1}$, ranging from a complete inert to a fully dynamic case. On the other hand, the diffusion coefficients usually take values from $10^{-12}\text{m}^2\text{s}^{-1}$ to $10^{-9}\text{m}^2\text{s}^{-1}$. As a consequence, the reaction and the diffusion may occur at a different time and space scales. We shall focus the attention to the pure reaction-diffusion process, by applying a potential $E$ at the electrode surface. We take $E$ sufficiently negative relatively to the reduction-oxidation standard potential $E_0$. For instance, a value such that $E - E_0 = -0.3\text{V}$ is large enough in order to consider the electrode surface acting as a perfect sink for M.

The space-time scales of the system are described by four key parameters. The space scales of diffusion and reaction are described by the diffusion layer thickness $\delta$ and the reaction layer thickness $\mu$, respectively (see [1] for a complete explanation). In many cases, $\delta$, is expressed as [1]:

$$\delta = \sqrt{\pi D_M t} \tag{2}$$

where $t$ is the total time in which diffusion occurs.

The thickness $\mu$ depends on the ratio of the diffusion rate of M over its recombination rate with L [1]:

$$\mu = \sqrt{\frac{D_M}{k_a c_L^*}} \tag{3}$$

where $c_L^*$ is the bulk concentration of L.

The time scales of reaction and diffusion are described, respectively, by the recombination rate of M with L:

$$\frac{1}{k_a c_L^*} \tag{4}$$

and the diffusion rate of M

$$\frac{\delta^2}{D_M} \tag{5}$$

Relevant cases are those where the concentration of the complex ML is relatively large with respect to M. In fact, let us consider the typical set of values wherein the bulk concentration of L, $c_L^*$ is in excess compared with the bulk concentration of M, $c_M^* : c_M^* = 10^{-3}\text{mol m}^{-3}$, $c_L^* = 1\text{mol m}^{-3}$, $D_M = 10^{-9}\text{m}^2\text{s}^{-1}$, $k_a c_L^* = 10^8\text{s}^{-1}$. By applying the potential $E$ at the electrode surface, the reaction is forced to be far from the chemical equilibrium and a diffusion gradient is established close to the electrode surface. After one second, the above mentioned parameters take the following values: $\mu = 3.16 \cdot 10^{-9}\text{m}$, $\delta = 5.6 \cdot 10^{-5}\text{m}$,

$(k_a c_L^*)^{-1} = 10^{-8}$s and $\delta^2/D_M = 3.14$s. Clearly, the reaction and the diffusion processes take place at very different scales. In order to numerically solve this kind of multi-scale process we use two numerical methods, the time splitting which allows us to discriminate fast and slow processes and the grid refinement which allows us to localise a fine grid only within regions wherein large variations of the concentration profiles are expected.

### 1.1 Definition of the Problem

The semi-infinite diffusion-reaction problem (1) is described by the following system of partial differential equations defined in an open set $\Omega$ ($\forall t > 0$):

$$\frac{\partial c_M(x,t)}{\partial t} = D_M \nabla^2 c_M(x,t) + R(x,t) \tag{6}$$

$$\frac{\partial c_{ML}(x,t)}{\partial t} = D_{ML} \nabla^2 c_{ML}(x,t) - R(x,t) \tag{7}$$

$$\frac{\partial c_L(x,t)}{\partial t} = D_L \nabla^2 c_L(x,t) + R(x,t) \tag{8}$$

where

$$R(x,t) = k_d c_{ML}(x,t) - k_a c_M(x,t) c_L(x,t) \tag{9}$$

For the planar case, $\Omega = (0, +\infty)$, the electrode is placed at $x = 0$. For spherical case, $\Omega = \mathbb{R}^3 - S$, the electrode is placed at $S = r \in \mathbb{R}^3 : |r| < r_0$ where $r_0$ is the radius of the electrode, usually 5-10$\mu m$. We shall refer to the electrode surface as $\partial\Omega$ for both planar and spherical geometries. In order to compact the notation, we introduce the functions $c_X = c_X(x,t)$, with X=M, L, ML to represent the values of the concentrations of the species involved in the processes.
The equilibrium constant between the species of the reaction (1), $K = \frac{k_a}{k_d}$ is defined by:

$$K = \frac{c_{ML}^*}{c_M^* c_L^*} \tag{10}$$

where $c_X^*$ are the bulk concentrations of the species.

Initial conditions are the bulk values of $c_X$:

$$c_X = c_X^* \qquad t = 0 \tag{11}$$

The boundary conditions for semi-infinite diffusion are the following:

$$c_X \to c_X^* \qquad x \to \infty \tag{12}$$

ML and L are electroinactive, therefore no flux of ML and L is crossing the electrode-solution interface ($x = 0$, $\forall t > 0$):

$$\left(\frac{\partial c_{ML}(x,t)}{\partial x}\right)_{x=0} = 0 \qquad \left(\frac{\partial c_L(x,t)}{\partial x}\right)_{x=0} = 0 \tag{13}$$

Finally, the Nernst and flux balance conditions at the electrode surface $\partial\Omega$, in the limit $E - E_0 << -0.3$V, give:

$$c_M(x,t) = 0 \qquad x = 0 \tag{14}$$

## 2 The Numerical LBGK Scheme

The problem stated in equations (6) - (8) is numerically solved by using the Lattice Boltzmann (LB) method in the LBGK approximation [2, 3]. The LB approach is based on a mesoscopic description of the physical system. The space is discretized by a grid or a lattice of spacing $\Delta x$ and a discrete time step of duration $\Delta t$. The lattice is identified by its spatial dimension $d$ and its coordination number $z$ indicating the number of neighbours of each lattice. Traditionally, the lattice is then referred to as a D$d$Q$z$ lattice (D stands for Dimension and Q for Quantities). Important quantities of a LB model are the mesoscopic velocities $v_i$, $i = 0, ..., z$, which denote the admissible velocities of the particles. The admissible velocities reflect the chosen lattice topology so that, in one time step, the particles can reach any of the $z$ neighbouring lattice site. Following the model proposed by [4], the LBGK dynamics of the system is described by the following evolution equation:

$$f_{i,\mathrm{X}}(x + v_i \Delta t, t + \Delta t) = f_{i,\mathrm{X}}(x,t) + \omega_{\mathrm{X}}(f^{(\mathrm{eq})}_{i,\mathrm{X}}(x,t) - f_{i,\mathrm{X}}(x,t)) + \Delta t \frac{R(x,t)}{2d} \quad (15)$$

where $f_{i,\mathrm{X}}(x,t)$ is the density distribution function, $\omega_{\mathrm{X}}$ is the relaxation parameter, $f^{(\mathrm{eq})}_{i,\mathrm{X}}(x,t)$ is the equilibrium density distribution function and $R(x,t)$ is defined in (9). The second term in the rhs of equation (15) is the non-reactive term, $\Omega^{\mathrm{NR}}_{i,\mathrm{X}}$. The third term in the rhs of equation (15) is the reactive term, $\Omega^{\mathrm{R}}_{i,\mathrm{X}}$. The equilibrium density distribution and the relaxation parameter in equation (15) are [4]:

$$f^{(\mathrm{eq})}_{i,\mathrm{X}}(x,t) = \frac{c_{\mathrm{X}}}{2d} \qquad \omega_{\mathrm{X}} = \frac{2}{1 + \frac{2dD_{\mathrm{X}}\Delta t}{\Delta x^2}} \quad (16)$$

### 2.1 The Time Splitting Method

In order to perform simulations when the time scale of diffusion and reaction are very different, we will use the time splitting method, following the work presented in [1] wherein the time splitting method has been applied to the LBGK numerical scheme. The complete numerical scheme (15) is now split into its pure diffusive and reactive parts in the following way [1]:

$$\bar{f}_{i,\mathrm{X}}(x + v_i \Delta t, t + \Delta t) = \tilde{f}_{i,\mathrm{X}}(x, t + \Delta t) + \Omega^{\mathrm{NR}}_{i,\mathrm{X}}(\tilde{f}(x, t + \Delta t)) \quad (17)$$

and

$$\tilde{f}_{i,\mathrm{X}}(x, t' + \Delta t_{\mathrm{K}}) = \tilde{f}_{i,\mathrm{X}}(x, t') + \Omega^{\mathrm{R}}_{i,\mathrm{X}}(\tilde{f}(x, t')) \quad (18)$$

Here we have assumed that the time variable $t'$ for the reaction equation runs faster than $t$, the time variable of the diffusion equation. This means that $\Delta t_{\mathrm{K}} << \Delta t$. In addition, the initial condition of equation (18) is $\tilde{f}(x,t') = \bar{f}(x,t)$. For a detailed description of the technique and of the symbols introduced, see [1]. However, the time splitting method alone is not sufficient to completely solve processes with very large reaction rates. In the next subsection we will explain with a simple example, that applying the time splitting, also requires to adjust the grid size according to the reaction rate.

### 2.2 The Reason to Refine the Grid

Let us consider a simple reaction diffusion problem with only one species, say M, and let us vary the rate $k$. For a one dimensional problem, considering the initial and the boundary conditions (11) and (12) and the perfect sink condition on $\partial\Omega$, the time splitting exact solution reads [5] in the time interval $[0, t]$:

$$c_{\mathrm{M}}(x,t) = c_{\mathrm{M}}^{*} e^{kt} \mathrm{erf}\Big(\frac{x}{2\sqrt{D_{\mathrm{M}} t}}\Big) \tag{19}$$

If we consider the exact solution in two points, $x$ and $x + \Delta x$, we can write:

$$c_{\mathrm{M}}(x+\Delta x,t) - c_{\mathrm{M}}(x,t) = c_{\mathrm{M}}^{*} e^{kt} \Big[\mathrm{erf}\Big(\frac{x+\Delta x}{2\sqrt{D_{\mathrm{M}} t}}\Big) - \mathrm{erf}\Big(\frac{x}{2\sqrt{D_{\mathrm{M}} t}}\Big)\Big] \tag{20}$$

The first order Taylor series expansion around $x$ of an amount $\Delta x$, with $\Delta x \to 0$, allows us to write:

$$c_{\mathrm{M}}(x+\Delta x,t) - c_{\mathrm{M}}(x,t) = c_{\mathrm{M}}^{*} e^{kt} \Delta x \frac{e^{-\frac{x^2}{2\sqrt{D_{\mathrm{M}} t}}}}{\sqrt{\pi D_{\mathrm{M}} t}} \tag{21}$$

In general, we expect that $c_{\mathrm{M}}(x,t)$ will vary smoothly in space to ensure numerical accuracy. So, if we require that $|c_{\mathrm{M}}(x+\Delta x,t) - c_{\mathrm{M}}(x,t)| < \theta$ , then we obtain the condition:

$$\Delta x < e^{-kt} \frac{\sqrt{\pi D_{\mathrm{M}} t}}{c_{\mathrm{M}}^{*}} \exp\Big(\frac{x^2}{2\sqrt{D_{\mathrm{M}} t}}\Big) \cdot \theta \tag{22}$$

Inequality (22) says that: a) $\Delta x$ has to be bounded and b) if we consider two rates $k_1$ and $k_2$ such that $k_1 > k_2$, then $\Delta x_{k_1} < \Delta x_{k_2}$. Therefore, for very fast reaction dynamics, in order to get a variation between $x$ and $x + \Delta x$ less than $\theta$, the grid size $\Delta x$ has to be chosen small enough, accordingly to inequality (22). Furthermore, it is not necessary to use the restriction (22) in all the domain. It is advisable to restrict its usage only to region where this is really needed. In fact, the initial conditions on the concentration profiles $c_{\mathrm{X}}$ are chosen such that equation (10) is fulfilled. This fact suggests to use a grid size that satisfy condition (22) only close to the electrode surface and more precisely within the region comparable to the reaction layer thickness $\mu$ ($\sim 2 - 3\mu$).

### 2.3 The Time Splitting - Grid Refinement - LBGK Scheme

To simplify the description, we will refer only to the simpler refinement made by just two sub-grids. Let us divide the domain $\Omega$ into two sub-domains $\Omega_1$ and $\Omega_2$. Let us build two grids, $G_1$ , with discretization parameters $\Delta x_1$ and $\Delta t_1$, and $G_2$, with discretization parameters $\Delta x_2 = g\Delta x_1$ and $\Delta t_2 = g\Delta t_1$, where $g$ is the grid refinement factor. The grid $G_1$ covers the set $\Omega_1$, while the grid $G_2$ covers the set $\Omega_2$. The strategy suggested is based on the conservation of the function $c_{\mathrm{X}}$ and of its first derivative $\frac{\partial c_{\mathrm{X}}}{\partial x}$ at the subgrids $\Gamma = G_1 \cap G_2$. If we define $c_{\mathrm{X}}^{(1)}$

and $c_{\mathrm{X}}^{(2)}$ as the concentrations defined in the subgrids $G_1$ and $G_2$ respectively, the conservation conditions are:

$$c_{\mathrm{X}}^{(1)} = c_{\mathrm{X}}^{(2)} \qquad \frac{\partial c_{\mathrm{X}}^{(1)}}{\partial x} = \frac{\partial c_{\mathrm{X}}^{(2)}}{\partial x} \qquad x \in \Gamma \tag{23}$$

which ensures the continuity of the concentration profiles and of the fluxes of the species X. Following [1], the coupling conditions at $x \in \Gamma$ are obtained by solving the following system in the unknowns $f_{1,\mathrm{X}}^{(2)}$ and $f_{2,\mathrm{X}}^{(1)}$:

$$\begin{pmatrix} 1 & -1 \\ -\Delta x_1 & -\Delta x_2 \end{pmatrix} \begin{pmatrix} f_{1,\mathrm{X}}^{(2)} \\ f_{2,\mathrm{X}}^{(1)} \end{pmatrix} = \begin{pmatrix} f_{1,\mathrm{X}}^{(1)} - f_{2,\mathrm{X}}^{(2)} \\ \Delta x_2 [f_{1,\mathrm{X}}^{(1)} - c_{\mathrm{X}}^{(1)}(x_1)] + \Delta x_1 [f_{2,\mathrm{X}}^{(2)} - c_{\mathrm{X}}^{(2)}(x_2)] \end{pmatrix} \tag{24}$$

where $x_1 = x - \Delta x_1$ and $x_2 = x + \Delta x_2$. The final numerical scheme in the time interval $[t, t + \Delta t_2]$ is the following:

1. Solve the pure diffusive problem by applying equation (17) in the coarser sub-grid $G_2$ and then $g$-times in the finer sub-grid $G_1$. Apply the coupling conditions (23) on $\Gamma$ by solving the system (24)
2. Solve the pure reactive problem by applying equation (18) in the coarser sub-grid $G_2$ and then in the finer sub-grid $G_1$.

## 2.4 The Off-Lattice Numerical Boundary Conditions

In order to preserve the accuracy of the numerical scheme also for irregular boundaries, we propose a numerical discretization of the boundary conditions where the boundary is not on a grid point. For simplicity only discuss the 1D problem. Let us consider the grid $G_1$ and two points, $x_1$ and $x_2$, where $x_1$ is inside the boundary, $x_2$ outside, and $x_1 < x_b < x_2$. The $f_{i,\mathrm{X}}(x,t)$ are computed at $x_1$ only to ensure the appropriate boundary condition at $x_b$, assuming that interpolation is considered. The Taylor expansion of $c_{\mathrm{X}}$ at $x_b$ reads:

$$c_{\mathrm{X}}(x_2) = c_{\mathrm{X}}(x_b) + (\Delta x - x_b) \left.\frac{\partial c_{\mathrm{X}}}{\partial x}\right|_{(x=x_b)} + o(\Delta x - x_b)^2 \tag{25}$$

Furthermore, we can write the spatial derivative in the rhs of equation (25) as:

$$\left.\frac{\partial c_{\mathrm{X}}}{\partial x}\right|_{(x=x_b)} = \frac{c_{\mathrm{X}}(x_2) - c_{\mathrm{X}}(x_1)}{\Delta x} + o(\Delta x^2) \tag{26}$$

By combining expressions (25) and (26) we obtain the numerical Dirichlet and Neumann boundary conditions:

$$c_{\mathrm{X}}(x_1) = c_{\mathrm{X}}(x_2) - \frac{\Delta x}{\Delta x - x_b} (c_{\mathrm{X}}(x_2) - c_{\mathrm{X}}(x_b)) \tag{27}$$

$$c_{\mathrm{X}}(x_1) = c_{\mathrm{X}}(x_2) - \Delta x \left.\frac{\partial c_{\mathrm{X}}}{\partial x}\right|_{(x=x_b)} \tag{28}$$

All the values of the rhs in equations (27) and (28) are known from the boundary condition or already computed from the numerical scheme. Note that imposing $c_{\mathrm{X}}$ in the LB scheme can be done as explained in [1].

## 3 Numerical Validation

We show some numerical validations of the method proposed, for a planar electrode case wherein $D_{\mathrm{M}} = D_{\mathrm{L}} = D_{\mathrm{ML}} = 10^{-9}\mathrm{m}^2\mathrm{s}^{-1}$, $k_{\mathrm{a}}c_{\mathrm{L}}^* = 10^5\mathrm{s}^{-1}$, $k_{\mathrm{d}} = 10^3\mathrm{s}^{-1}$. Initial conditions are $c_{\mathrm{M}}^* = 9.9 \cdot 10^{-6}$ mol m$^{-3}$, $c_{\mathrm{ML}}^* = 9.9 \cdot 10^{-4}$ mol m$^{-3}$ and $c_{\mathrm{L}}^* = 10^{-2}$ mol m$^{-3}$. For this case, the scales are: $\mu = 10^{-7}$m, $\delta = 5.6 \cdot 10^{-5}\sqrt{t}$ m, $\frac{1}{k_{\mathrm{a}}c_{\mathrm{L}}^*} = 10^{-5}$s, $\frac{\delta^2}{D_{\mathrm{M}}} = 3.14 \cdot t$ s. The simulations are run up to $t = 10^{-2}$s. We use the time splitting method by applying the discrete equations (17) and (18) and the reaction is solved with an implicit scheme. We use the grid $G_1$ from the electrode surface ($x = 0$) to the distance $x_b = 3\mu$ and the the grid $G_2$ for $x > x_b$. The domain size is $\sim 5 \cdot \delta$ in order to fulfil the semi-infinite diffusion conditions (12).

Tables 1 and 2 show the error on the flux computation computed for several values of the discretization parameters. The relative error $\epsilon$ with respect to an analytical result is computed as in [1].

Table 1 shows that the error decreases if the grid size close to the electrode surface becomes smaller, until $\Delta x_1 = \mu/20$. For smaller grid sizes, the error increases again, due to the small values of the relaxation parameter. In fact, when $\Delta x_1 = \mu/50$, the corresponding values of $\omega_1$ and $\omega_2$ are too small and the pure diffusive scheme loses accuracy. It is reasonable to fix the relaxation parameters in the range $0.8 < \omega < 1.5$ to get the best performance. A way to circumvent this problem would be to have a grid refinement technique where $\frac{\Delta x^2}{\Delta t}$ is constant across the grid instead of $\frac{\Delta x}{\Delta t}$. This topic is under investigation.

Table 2 shows the effect of different grid refinement factors, $g$, for $\Delta x_1 = \mu/20$ and $\Delta x_2 = g\Delta x_1$, on accuracy and CPU time. The case where only $G_1$ and $G_2$ are used in all the domain, is shown for comparison. If $g$ increases, then the error increases as well, but the computation time is shorter. However, for $g = 8$ the error is more than 5 percent, a value no longer acceptable. It is preferable to use $g = 2 - 4$ which allows us to gain time of computation without losing too much in precision. It is interesting to compare the errors and the computation time shown for several $g$ with the results obtained by applying only the grid $G_1$ everywhere. The results show that the grid refinement allows us to reduce the computation time while keeping a good precision. With $g = 2$, the error still remains less than 1% and the computation time is reduced 20% in comparison to the time required using only the grid $G_1$. Furthermore, in the last line of table 2,

**Table 1.** Error for different space grid sizes. Parameters of the simulation: $D_{\mathrm{M}} = D_{\mathrm{L}} = D_{\mathrm{ML}} = 10^{-9}\mathrm{m}^2\mathrm{s}^{-1}$, $k_{\mathrm{a}}c_{\mathrm{L}}^* = 10^5\mathrm{s}^{-1}$, $k_{\mathrm{d}} = 10^3\mathrm{s}^{-1}$. $g = 2$, $\Delta x_2 = g\Delta x_1$, $x_b = 3\mu$, $t = 10^{-2}$ s. (1000 time steps).

| $\Delta x_1$ | $\omega_1$ | $\omega_2$ | $\epsilon$ |
|---|---|---|---|
| $\mu/5$ | 1.90 | 1.95 | 8.71% |
| $\mu/10$ | 1.67 | 1.82 | 3.80% |
| $\mu/20$ | 1.11 | 1.43 | 0.50% |
| $\mu/50$ | 0.33 | 0.57 | 3.90% |

**Table 2.** Error for different grid refinement factors $g$. Same parameters as in table 1, but $\Delta x_1 = \mu/20$ and $\Delta x_2 = g\Delta x_1$.

| $g$ | Computation time | $\epsilon$ |
|---|---|---|
| 2 | 0.80 | 0.50% |
| 4 | 0.77 | 2.03% |
| 6 | 0.72 | 4.79% |
| 8 | 0.70 | 7.52% |
| only grid $G_1$ everywhere | 1 | 0.02% |
| only grid $G_2$ ($g = 2$) everywhere | 0.50 | 0.71% |

we show the error and the computation time obtained with $\Delta x_2 = 2\Delta x_1$, by applying only the grid $G_2$. The computation time is still reduced, but the error obtained by using only the coarser grid everywhere is larger.

## 4 Conclusions

We have seen that for processes involving fast reactions the grid size should be small enough to follow the strong variations of the concentration profiles close to the electrode surface. However, the grid size should be chosen carefully in order to avoid values of the relaxation parameters too small. It is advisable to choose grid size values, such that the corresponding relaxation parameter values range between 0.8 and 1.5. The grid refinement is very useful because it allows us to keep a good precision and to reduce the time of computation.

## Acknowledgements

The authors gratefully acknowledge support of this work by the Swiss National Foundation. Josep Galceran also thanks the support of this research by the Spanish Ministry of Education and Science (DGICYT: Project BQU2003-07587).

## References

1. Alemani, D., Chopard, B., Galceran, J., Buffle, J.: Phys. Chem. Chem. Phys., 7 (2005) 3331-3341.
2. Wolf-Gladrow, D.A.: Lattice Gas Cellular Automata and Lattice Boltzmann Models: An introduction, Printed in Berlin, Springer-Verlag (2000).
3. Chopard, B., Droz, M.: Cellular Automata Modelling of Physical Systems, Cambridge Univ. Press. UK (1988).
4. Dawson, S., Chen, S., Doolen, G.D.: Journal of Chemical Physics 98 (1992) 1514.
5. Hundsdorfer, W., Verwer, J.G.: Numerical Solution of Time-Dependent Advection-Diffusion-Reaction Equations, Printed in Berlin, Springer-Verlag (2003).

# Mesoscopic Simulations of Unsteady Shear-Thinning Flows

Abdel Monim Artoli and Adélia Sequeira

Centre for Mathematics and its Applications (CEMAT)
Departamento de Matemática , Instituto Superior Técnico
Av. Rovisco Pais, 1049-001 LISBOA
Portugal
Tel.: +351 21 8417073; Fax: +351 218417048
{artoli, adelia.sequeira}@math.ist.utl.pt
http://www.math.ist.utl.pt/cemat/

**Abstract.** The capability of the lattice Boltzmann method as an accurate mesoscopic solver for unsteady non-Newtonian flows is shown by investigating pulsatile shear-thinning blood flow in a three-dimensional idealised vessel. The non-Newtonian behaviour of blood flow is modelled by the Carreau-Yasuda viscosity model. Higher velocity and shear stress magnitudes, relative to Newtonian cases, are observed for the shear-thinning simulations in response to changes in the shear-rate dependent Womersley parameter. We also investigate the flexibility of the method through the shear-thinning behaviour of the lattice Boltzmann relaxation parameter at different Deborah numbers.

## 1 Introduction

The role of hemodynamics on the development of cardiovascular diseases and the high rate of failure in vascular implantation of artificial devices is still under research focus, as these diseases are a major cause of mortality in the world [1]. Since long, analytical, numerical and experimental investigations of steady and pulsatile blood flow of Newtonian and non-Newtonian models have been carried out (see e.g. [2] for a review). It is now well established that localised low and oscillatory shear stress and high shear-rate gradients promote, among other factors, most of the diseases of the cardiovascular system.

With recent advances in mathematical modelling and computer technology, research in this field is now very active, targeting a better understanding of the complex nature of the fluid and the transporting vessels. Blood is a heterogeneous suspension of small deformable red blood cells, white blood cells, platelets and other matter in plasma. Red blood cells characterise the main rheological behaviour of the fluid. These cells aggregate and deform in response to the applied shear rate, change the fluid viscosity and relax back within about 100 msec. The non-Newtonian behaviour of many fluids, including blood, may be well described by the shear-thinning Carreau-Yasuda viscosity model

$$\eta = \eta_\infty + (\eta_0 - \eta_\infty)(1 + (\lambda\dot{\gamma})^a)^b \tag{1}$$

V.N. Alexandrov et al. (Eds.): ICCS 2006, Part II, LNCS 3992, pp. 78–85, 2006.

where $\eta_0$ and $\eta_\infty$ are the asymptotic low and high shear-rate viscosities, respectively, $\lambda$ is a characteristic viscoelastic time of the fluid, $\dot{\gamma}$ is the shear rate and the parameters $a$ and $b$ are determined from experimental data, with $b < 0$ for the shear thinning fluids. For blood, $\eta_0 = 0.022$ Pa S, $\eta_\infty = 0.0022$ Pa S, $\lambda = 0.11$ S, $a = 0.644$ and $b = -0.944099$ (see e.g. [3]). The fluid behaves as Newtonian for small $\dot{\gamma}$ or negligible values of $\lambda$ and also when $a = 0$ or $b = 0$.

In addition to hemodynamics, many industrial activities involve shear-thinning fluid processing and numerical simulations are found viable. In most numerical studies, accurate and robust conventional Navier-Stokes solvers are required to simulate flow of shear-thinning fluids [4]. However, these solvers yield three major difficulties: the non-trivial and time consuming grid generation, the necessity to solve the Poisson equation for the pressure field with additional corrective terms, and the approximations involved in computing the shear rate from computed velocity field. All these shortcomings are eliminated with the newly established lattice Boltzmann method (LBM). Firstly, LBM works with a Cartesian grid, making it highly suitable for high performance computing. Secondly, the pressure is directly computed from the linear equation of state without a need to solve the Poisson equation. Last and more important in computational rheology is that the shear rate is directly obtained from the non-equilibrium part of the distribution functions, independent of the velocity field. This adds a great flexibility to the LBM in modelling and simulation of most non-Newtonian fluids, as it is easy to establish constitutive equations with the use of simple equations. This will be explained in the next section.

In a previous article [5], it has been demonstrated that LBM is an accurate and robust hemodynamic numerical solver for unsteady Newtonian flows in realistic geometries. Localised shear stress and velocity field have well compared with values in the available literature. Following arguments that the non-Newtonian influence should not be ignored, the main objective of this study is to explore in general the capability of the lattice Boltzmann method in simulating shear-thinning flows. A number of lattice Boltzmann simulations of shear-thinning flows using the Casson viscosity model have been reported in the literature[6, 7]. Results on LBM simulations of 3D Carreau-Yasuda flows are lacking. Due to the limitations of Casson model for blood in particular, the Carreau-Yasuda model is adopted in this study.

In the rest of this paper, we will briefly describe the adaptation of LBM to non-Newtonian shear-thinning fluids and present our simulation results for a benchmark test of three-dimensional pulsatile blood flow in a straight tube. Gijsen et al.[3] have compared shear thinning results with analytical Newtonian solutions, finite elements simulations and experimental results and demonstrated the deviation of shear-thinning flow from Newtonian solutions. Taking this procedure into account, here we compare the obtained results with analytical and simulated results. Comparison with other CFD methods is a subject of ongoing research.

## 2 The Lattice Boltzmann Method for Shear-Thinning Flows

The lattice Boltzmann method is a special finite difference discretisation of the simplified Boltzmann equation [8, 9, 10] which describes transport phenomena at the mesoscale level. The dynamics of the fluid is modelled by the transport of simple fictitious particles on the nodes of a Cartesian grid. Simulations with this method involve two simple steps; streaming to the neighbouring nodes and colliding with local node populations represented by the probability $f_i$ of a particle moving with a velocity $\mathbf{e}_i$ per unit time step $\delta t$. Populations are relaxed towards their equilibrium states during a collision process. The equilibrium distribution function

$$f_i^{(eq)} = w_i \rho \left( 1 + \frac{3}{v^2} \mathbf{e}_i \cdot \mathbf{u} + \frac{9}{2v^4} (\mathbf{e}_i \cdot \mathbf{u})^2 - \frac{3}{2v^2} \mathbf{u} \cdot \mathbf{u} \right) \tag{2}$$

is a low Mach number approximation to the Maxwellian distribution. Here, $w_i$ is a weighting factor, $v = \delta x / \delta t$ is the lattice speed, and $\delta x$ is the lattice spacing.

The lattice Boltzmann equation

$$f_i(\mathbf{x} + \mathbf{e}_i \delta t, \mathbf{e}_i, t + \delta t) - f_i(\mathbf{x}, \mathbf{e}_i, t) = -\frac{1}{\tau} [f_i(\mathbf{x}, \mathbf{e}_i, t) - f_i^{(0)}(\mathbf{x}, \mathbf{e}_i, t)] \tag{3}$$

can be obtained by discretizing the evolution equation of the distribution functions in the velocity space using a finite set of velocities $\mathbf{e}_i$. In this equation, $\tau$ is the dimensionless relaxation time. By Taylor expansion of the lattice Boltzmann equation up to $O(\delta t^2)$ and application of the multiscale Chapman-Enskog technique [10], the Navier-Stokes equations and the momentum flux tensor up to second order in the Knudsen number are obtained. The hydrodynamic density, $\rho$, and the macroscopic velocity, $\mathbf{u}$, are determined in terms of the particle distribution functions from the laws of conservation of mass and momentum: $\rho = \sum_i f_i = \sum_i f_i^{(eq)}$ and $\rho \mathbf{u} = \sum_i \mathbf{e}_i f_i = \sum_i \mathbf{e}_i f_i^{(eq)}$. The pressure is given by $p = \rho c_s^2$ and the kinematic viscosity is $\nu = c_s^2 \delta t (\tau - \frac{1}{2})$, where $c_s$ is the lattice speed of sound. A number of lattice Boltzmann models have been introduced, being characterised by the choice distribution functions, the number of moving particles and the lattice speed of sound. Furthermore, by expanding $f_i$ about its equilibrium distribution $f_i^0$

$$f_i = f_i^0 + \varepsilon f_i^{(1)} + \varepsilon^2 f_i^{(2)} + \ldots \tag{4}$$

where $\varepsilon$ is of the order of the Knudsen number and in the limit of small $\varepsilon$, the momentum flux tensor is obtained by [10]

$$\Pi_{\alpha\beta}^{(1)} = \sum_i f_i^{(1)} \mathbf{e}_{i\alpha} e_{i\beta} = -2\rho \delta_t \tau_c C(S_{\alpha\beta}), \tag{5}$$

where $C$ is a lattice-dependent constant (for the three-dimensional model with 19 particles (D3Q19), $C = 1/3$). The momentum flux is therefore directly computed from the non-equilibrium part of the distribution functions and therefore, the strain rate tensor is

$$S_{\alpha\beta} = -\frac{1}{2C\delta_t \tau_c \rho} \sum_i f_i^{(1)} \mathbf{e}_{i\alpha} e_{i\beta} \tag{6}$$

and the stress tensor from

$$\sigma_{\alpha\beta} = -\rho c_s^2 \delta_{\alpha\beta} - \left( 1 - \frac{1}{2\tau_c} \right) \sum_{i=0} f_i^{(1)} e_{i\alpha} e_{i\beta}. \tag{7}$$

In constitutive equations of shear-thinning generalised Newtonian fluids the viscosity depends on the magnitude or the second invariant of the strain rate tensor which can be computed from the double inner product of $S_{\alpha\beta}$ by itself

$$|S| \equiv \dot{\gamma} = \sqrt{2S_{\alpha\beta} : S_{\alpha\beta}}. \tag{8}$$

More specifically, for the $D3Q19$ model, after making use of the symmetry of the strain rate tensor, this is locally computed during collision from the simple relation

$$\dot{\gamma} = \dot{\gamma}_c \sqrt{0.5(S_{xx}^2 + S_{yy}^2 + S_{zz}^2) + (S_{xy}^2 + S_{xz}^2 + S_{yz}^2)} \tag{9}$$

where $\dot{\gamma}_c = \frac{3}{2\rho\tau_c}$ could be used as a characteristic shear rate. In this study we propose that $\tau_c = 1$ to benefit from the simplicity and the accuracy of the scheme at $\tau = 1$ since the relaxation towards equilibrium in this case (for Newtonian fluids) is direct. This will allow us to avoid rescaling dimensionless numbers when Newtonian solutions are simulated for comparison. In this article the Carreau-Yasuda model is used (see Eq. 1) to model non-Newtonian blood viscosity. Other non-Newtonian characteristics of blood may be ignored [3]. In LBM simulations, we can implement the Carreau-Yasuda model in terms of the dimensionless relaxation time. Making use of the fact that $\eta = \rho\nu = \rho(2\tau - 1)/6$, the Carreau-Yasuda model in its dimensionless form is

$$\tau = \tau_\infty + (\tau_0 - \tau_\infty)(1 + (\lambda\dot{\gamma})^a)^b \tag{10}$$

where $\tau_0$ and $\tau_\infty$ correspond to $\eta_0$ and $\eta_\infty$, respectively. The stability of the method is controlled by the difference $(\tau_0 - \tau_\infty)$ which is normally large for shear-thinning fluids. This slightly enhances the stability of the LBM scheme as one more free parameter is introduced. However, $\tau_\infty$ shall be in the working stability region if $\tau_0$ is small. Another way to avoid instability is to tune the material relaxation time (in lattice units) by grid-refinement (or coarsening) in time.

The variety of numerous dimensionless numbers defined for the shear-thinning fluids implies great restrictions on the choice of working free parameters and has motivated research towards specialised LBM schemes for viscoelasticity [11]. However, for shear-thinning fluids, the standard BGK [8] algorithm allows us to introduce two additional parameters, one for the high shear-rate viscosity and the other for the difference between high-shear rate and low-shear rate viscosities. This adds a great flexibility to the lattice BGK method and it would not be necessary to go beyond the BGK for shear-thinning fluids. This might not be the case for other viscosity models.

## 3 Simulations

We have conducted a number of benchmark simulations for pulsatile shear-thinning blood flow in a three-dimensional rigid straight tube with circular cross-section of constant radius. The main objective of this simulation is to investigate the influence of non-Newtonian behaviour on the velocity field and the shear stress. The shear-thinning behaviour is captured by the Carreau-Yasuda model explained in the previous section. The diameter $D$ of the tube is represented by 20 lattice points and the

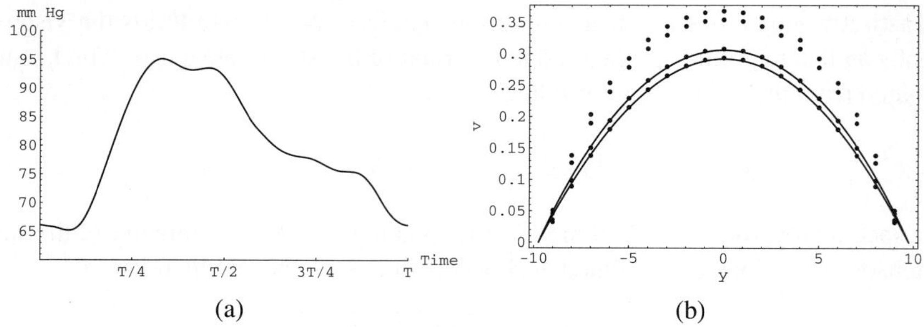


**Fig. 1.** (a) Pressure pulse and (b) LBM unsteady shear-thinning velocity profiles at 0.2 T (upper dots, lower values) and 0.3 T (upper dots, higher values) compared to analytical Newtonian solutions (lines). Shown are also Newtonian simulations (lower dots attached to the lines).

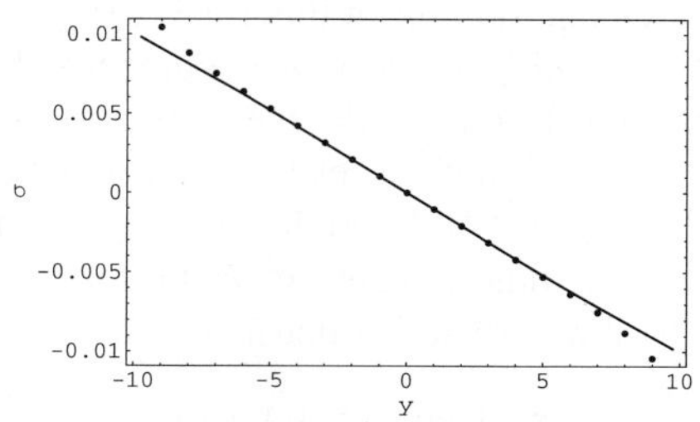


**Fig. 2.** LBM shear stress results (dots) for pulsatile shear-thinning blood flow at the mid-diastole compared to the analytical Newtonian solution (line)

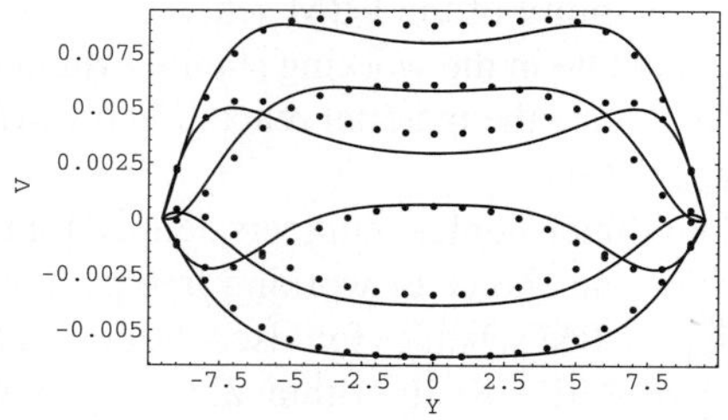


**Fig. 3.** Simulated (dots) harmonic velocity components compared to the analytical Newtonian solutions (lines)

period $T$ of the pressure pulse by 100 time steps. The dimensionless numbers are $De = \frac{\lambda}{T} = 0.1164, Cr = 0.2, Re = UD/\nu = 42$ and $\alpha = D/2\sqrt{\frac{2\pi}{T\nu}} = 6.2$ at their maximum values, where $\lambda$ is the material parameter, $De$ is the Deborah number, $Cr$ is the Carreau number, $Re$ is the Reynolds number and $\alpha$ is the Womersley number. The boundary conditions are applied as follows. The bounce-back boundary condition is applied on the walls. A pressure gradient computed from a pulse (see Fig. 1 (a)) measured at the entrance of the human carotid artery is Fourier transformed into a steady and 16 harmonic terms and applied at the entrance of the tube as an inlet condition [5]. At the outlet, a constant pressure is implemented. The velocity and unknown distributions at the

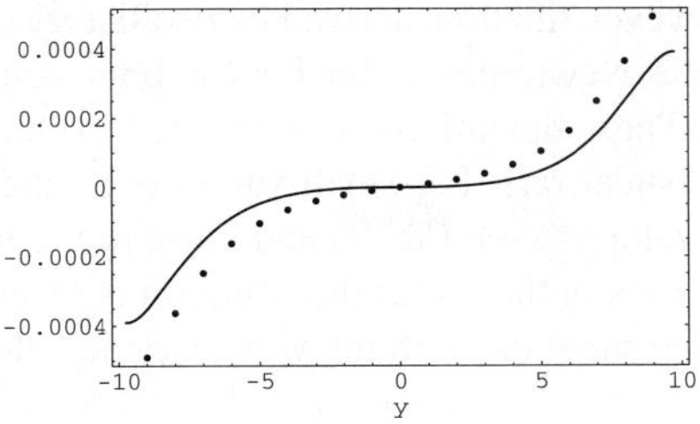


(a)

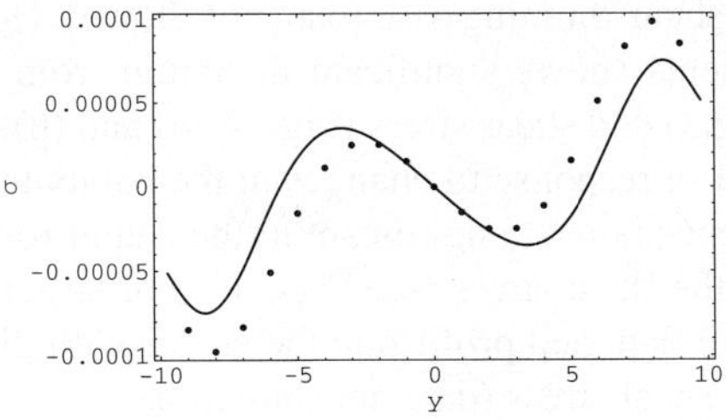


(b)

**Fig. 4.** Deviation from Newtonian profile for simulated (dots) harmonic shear-stress at (a) the begining of systole and (b) at the mid-diastole

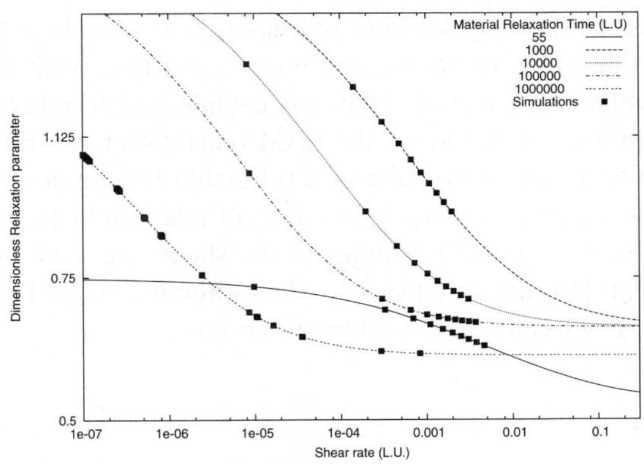


**Fig. 5.** Shear-thinning behavior of the LBM relaxation time for different material relaxation times

boundaries are computed from the implemented pressure boundary conditions. Figure 1 (b) shows simulation results for the shear-thinning (upper dots) flow case compared to the analytical Newtonians (lines). Newtonian simulation results are also shown as dots on the lines demonstrating an excellent agreement with the theory. The unsteady non-Newtonian profile is almost parabolic with slightly varying amplitudes throughout the cardiac cycle. The difference between Newtonian and non-Newtonian results varies in time and across the tube with a maximum of 20% at the centerline. We have observed significant deviations from the non-Newtonian in the shear stress close to the walls through the whole cardiac cycle, with a maximum difference at the mid-diastole (see Fig. 2), in qualitative agreement with shear-thinning results compared to Newtonian solutions and experiments in Ref. [3]. The comparison with Newtonian cases is not meant for direct validation but by transitivity as the study by Gijsen et al. [3] have compared numerical and experimental values for velocity and shear stress deviation of Carreau-Yasuda fluids with Newtonian solutions.

The pressure pulse can be splitted into a steady mean component and oscillatory components. For better understanding of the problem, we have performed a number of steady and oscillatory simulations. The steady component produced a simple parabolic profile for the Newtonian case and a semi-parabolic profile with higher amplitude for the shear-thinning case (data not shown). However, the flow derived by oscillatory components shows significant deviation from the Newtonian behavior for both velocity (Fig.3) and shear stress (Figs. 4 (a) and (b)). These deviations are attributed to changes in $\alpha$ in response to changes in the computed shear rate. For small values of $\alpha$, the difference is more significant in the center for velocity (see Fig. 3) and close to the walls for the shear stress (see Figs. 4 ). For larger $\alpha$ (as in the aorta) the situation is reversed; more flattened profiles in the center with differences close to the walls for the velocity and shear stress (data not shown).

In general, for this specific benchmark, the non-Newtonian shear-thinning flow gives higher velocities and higher shear stress values than when the fluid is considered Newtonian. The oscillatory profile of the shear stress, controlled by the Womersley number behaves almost similarly, with peaks more distant from the walls than for the Newtonian cases, as is clearly shown in Fig. 4.

To investigate the flexibility of LBM in simulating flow of a shear-thinning fluid, we have explored possible stable solutions at different shear rates with different values of $\lambda_L$ (different *De* numbers) and recorded the dimensionless LBM relaxation time. Figure 5 shows the shear-thinning behavior of the LBM relaxation time. From this figure we observe that the dimensionless relaxation time is flexible enough in simulations of shear thinning flows as the shear rate changes six orders of magnitude. For a fluid like blood, at most three orders of magnitude changes in the shear rate have been reported and therefore, the scheme is quite suitable as a shear-thinning hemodynamic solver. For other polymeric fluids we can argue similar conclusions.

## 4 Conclusions

Unsteady non-Newtonian shear-thinning blood flow in a straight rigid tube is investigated using the Carreau-Yasuda model and simulated with the lattice Boltzmann method. A characteristic relaxation time has been added to the standard BGK scheme to avoid implicit dependency of the strain rate and the relaxation parameter. The capabilities of the method as an efficient and accurate numerical solver for this type of problems have been successfully demonstrated and benefits over the commonly used Navier-Stokes solvers have been highlighted. The most relevant advantage in using this numerical solver lies in the fact that the shear rate is directly computed from the non-equilibrium part of the distribution functions without a need for computing the velocity gradients. This enables more sophisticated shear-rate dependent fluid viscosities to be modeled accurately and efficiently. Comparison with other numerical methods is underway. Future work would involve simulating shear-thinning fluid flows in realistic geometries from hemodynamics and industrial applications.

## Acknowledgments

This work has been partially supported by the grant SFRH/BPD/20823/2004 of Fundação para a Ciência e a Tecnologia (A. Artoli), by the Center for Mathematics and its Applications - CEMAT through FCTs funding program and the projects HPRN-CT-2002-00270 and POCTI/MAT/41898/ 2001.

## References


1. The World Health Report 2005, WHO publications (2005).
2. Wooton, D.M., Ku, D.N.: Fluid mechanics of vascular systems, diseases, and thrombosis. Annual Review of Biomedical Engineering **1** (1999) 299-329.
3. Gijsen, F.J.H., de Vosse, F.N., Janssen F.N.: The influence of the non-Newtonian properties of blood on the flow in large arteries: steady flow in a carotid bifurcation model. Journal of Biomechanics **32** (1999) 601-608.
4. Arada, N., Pires, M., Sequeira, A.: Numerical simulations of shear-thinning Oldroyd-B fluids in curved pipes. IASME Transactions 6 **2** (2005) 948-959.
5. Artoli, A.M., Hoekstra A.G., Sloot P.M.A.: Simulation of a Systolic Cycle in a Realistic Artery with the Lattice Boltzmann BGK Method, International Journal of Modern Physics B, **17** (2003) 95-98.
6. Ouared, R.,Chopard B.; Lattice Boltzmann Simulations of Blood Flow: Non-Newtonian Rheology and Clotting Processes. Journal of Statistical Physics **121** 1-2 (2005) 209- 221.
7. Aharonov, E., Rothman, D.H.:Non-newtonian flow (through porous media): a lattice boltzmann method. Geophys. Research Letters **20** (1993) 679-682.
8. Sauro Succi: The Lattice Boltzmann Equation for Fluid Dynamics and Beyond. Oxford University Press (2001).
9. Benzi R., Succi S., Vergassola M.: The Lattice Boltzmann Equation - Theory and Applications. Physics Reports **222** (1992) 145-197.
10. Chopard B. and Droz M.: Cellular Automata Modeling of Physical Systems. Cambridge University Press (1998).
11. Giraud, L., d'Humires, D., Lallemand, P.: A lattice Boltzmann model for visco-elasticity. International Journal of Modern Physics C **8** (1997) 805-815.

# A Multiphysics Model of Capillary Growth and Remodeling

Dominik Szczerba[1], Gábor Székely[1], and Haymo Kurz[2]

[1] Computer Vision Lab, ETH, CH-8092 Zürich, Switzerland
[2] Institute of Anatomy and Cell Biology, University of Freiburg, D-79104 Freiburg, Germany

**Abstract.** We report on an enhanced computational framework for simulating flow-tissue interactions that significantly expands the capabilities of our previous model [1]. We adhere to the basic structural concept of the so-called *intussusceptive* growth and remodeling which does not only generate capillaries and terminal vessels but also rebuilds them into a highly perfused system [2]. Present enhancements comprise calculation and visualization in three dimensions, refined tissue and fluid mechanics, and the transport of molecules that act as biochemical growth or signaling factors. Our present model explains formation of capillary meshes and bifurcations, and the emergence of feeding and draining microvessels in an interdigitating pattern that avoids arterio-venous shunts. In addition, it predicts detailed hydrodynamic properties and transport characteristics for oxygen, metabolites or signaling molecules. In comparison to the previous work, the complexity of our approach is dramatically increased by using a *multiphysics* modeling environment, where many independent computational components are combined and the data structure is unified.

## 1 Introduction

The long-standing interest of biology and medicine in a thorough quantitative understanding of the vascular system and its development has gained new impetus due to increased efficiency of computational resources during the past two decades. Different aspects of normal and tumor angiogenesis have been subject to *in silico* modeling, and such different approaches have been chosen as fractal self-similar constructs [3], cellular automata [4], functional macroscopic pipelines ([5], [6], [7], [8]), optimal design principles [9] and various mathematical formulations ([10], [11], [12]). A more comprehensive review of available literature can be found in our previous publications ([6], [7], [1]) and will not be repeated here. Although many of those models are successful in predicting structurally realistic systems with relevant biophysical properties, they do not demonstrate fundamental understanding of how such structures actually come into existence from the microscopic point of view. Such understanding is, however, mandatory if realistic simulations of, e.g., anti-angiogenic cancer treatments or the effects of irradiation are considered.

In our previous approach, formation of blood vessels was treated as flow-driven remodeling of a planar capillary network or *plexus* demonstrated to be present at many anatomical sites. The tissue was represented as a set of discrete entities (pixels) displacing under shear stress determined by the local flow conditions. The model successfully

V.N. Alexandrov et al. (Eds.): ICCS 2006, Part II, LNCS 3992, pp. 86–93, 2006.

predicted bifurcation formation and micro-vessel separation from a capillary bed, but failed to ensure correct topology of the resulting arterial and venous branches, thus producing shunt vessels.

We therefore attempted to implement cellular response mechanisms that would result in self-avoidance of arterial and venous branches. We found that the interplay of production and transport of molecular compounds controlling the development, refined tissue and fluid dynamics (including non-linear material responses) and remodeling into the third dimension result in self-avoidance and interdigitating patterns of the terminal branches. This allowed us to achieve better similarity to real patters and hence improved predictive power.

Fundamental differences between other models that we are aware of and our current approach are: 1) addressing each and every single capillary loop and terminal arterial or venous branch - not just vector fields or concentration gradients, or only ramified trees; 2) using exclusively local interactions - without explicitly enforcing global optimality or *a priori* prescribing bifurcation parameters. We thus hope to demonstrate that the emergence of gross functional optimality in macroscopically tree-like networks can be realized by molecular mechanisms that act on the microscopic scale. Major innovations in this mathematically and computationally challenging project are presented here.

## 2 Numerical Background

Most generally spoken, we base our modeling on in vivo observations (cf. cited references) that strongly justify mechanistic assumptions about vascular tissues that are able to respond to hydrodynamic conditions and signaling molecules by shape changes. Numerically, the core of our model underlying all computations is a differential form of the general transport equation for property $\phi$, usefully written in the following form:

$$\frac{\partial \rho\phi}{\partial t} + div(\rho\phi\boldsymbol{u}) = div(\Gamma grad\phi) + S_\phi \,, \tag{1}$$

with $\phi$ denoting the transported property, $\Gamma$ diffusivity, $\rho$ density, $\boldsymbol{u}$ convective vector field and $S_\phi$ being the sources (or sinks) of the transported property.

Invoking the Gauss theorem and integrating over a three dimensional control volume $V$ yields an integral (conservative) equation for the transported property:

$$\frac{\partial}{\partial t}\int_V \rho\phi dV + \int_A (\rho\phi\boldsymbol{u})\boldsymbol{da} = \int_A (\Gamma grad\phi)\boldsymbol{da} + \int_V S_\phi dV \,, \tag{2}$$

The above *finite volume* discretization is further linearized and written as:

$$A_P^\phi \phi_P = A_i^\phi \phi_i + S_P^\phi \,, \tag{3}$$

with summation over explicit cell neighbours and compass notation as in [13]. The above equation has to hold true for every point in the grid. This can be formulated as a linear problem $Ax = b$, where $x$ is a vector of unknown scalar values, $A$ is a domain-dependent (and property-specific) coefficient matrix and $b$ is given by the boundary conditions and the solution from the previous time steps.

The above sketched general computational technique allows to address a wide range of transport phenomena by setting $\phi$ to an implicit geometry function, velocity components, concentrations, internal energy, temperature, etc. and identifying the diffusive coefficient accordingly (mass diffusivity, fluid viscosity, etc.).

## 3 Tissue Modeling

Simulation of biological phenomena most often requires dealing with complex geometries. Unstructured body-fitted meshes, while offering obvious advantages like delivering accurate solutions at the boundaries, are computationally too expensive in case of complex anatomical domains. Therefore, in the presented model we rely on Cartesian grids, eventually using adaptive refining, as seen in Fig. 1.

We represent the tissue formally as a mathematical *continuum*, separated from external fluid (blood) by an interface defined in a *level set* sense. An introduction to this family of methods is beyond the scope of this report and can be found elsewhere [14]. Here we only briefly discuss the most important issues relevant to our modeling. The key ingredient is to define an *implicit* or *embedding* function to represent the fluid-structure interface to be tracked. In our case it clearly will be the deforming blood-tissue interface. Deforming such an interface is computationally equivalent to solving a transport problem with $\phi = \phi_{ls}$ and $\boldsymbol{u}$ being an *interface advancing* vector field. Such a convenient formulation offers a natural treatment of topological changes like break-up and fusion of the boundaries, however, does not provide *explicit* geometrical information about the domain. In particular, this means that we will not be able to (directly) describe local geometry or topology as no geometric primitives as pipes or tripods will be available. On the other hand, a big advantage of this technique is that the level set function is continuous and the associated partial differential equations can be handled with a common, previously introduced numerical method. Such a treatment allows to drive the domain remodeling in an elegant manner using deformation fields (available near the boundaries) instead of arbitrary displacements introduced in our previous work [1].

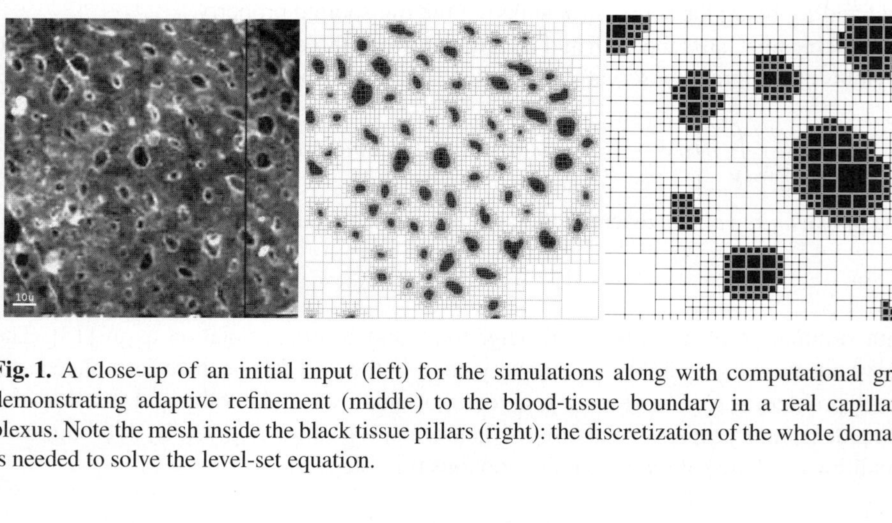


**Fig. 1.** A close-up of an initial input (left) for the simulations along with computational grid demonstrating adaptive refinement (middle) to the blood-tissue boundary in a real capillary plexus. Note the mesh inside the black tissue pillars (right): the discretization of the whole domain is needed to solve the level-set equation.

Whereas previously we only considered the tangential shear stress as the remodeling factor, using the current formulation we are able to derive the deformation fields generated by any sources, including boundary viscous and pressure forces, surface tension and gradients of growth factors. In addition, we propose a non-linear tissue response to the shear stress of the form $f(\tau) \sim 1/(\tau + \tau_0)$. This means that due to decreased cellular sensitivity the tissue will deform less to increasingly stronger stimulus exerted by the flowing fluid.

## 4 Flow Conditions

In order to provide hemodynamic conditions in the domain of interest, the Navier-Stokes equations governing the fluid flow have to be solved. Generally speaking, this can be done by using the general transport equation (taking $\phi = u, v, w$), relating the source terms to the pressure gradient and enforcing the mass-conservation equation (taking $\phi = 1$) - and additionally neglecting the rate of change of $\rho$ for an incompressible fluid. Unfortunately, such treatment will not guarantee smooth pressure solution, as pressure will not be present in the resulting mass-conservation equation. Solving such a system will result in oscillatory pressure solutions, therefore this case needs special treatment. We chose a recent *smoothing pressure correction* approach [15], which unlike most other established techniques, allows calculations to be performed using a colocated variable arrangement. The whole solution procedure results in discrete distributions of velocity and pressure in the domain of interest, which can be used to derive any further variables like, e.g., shear stress, experimentally known to modulate endothelial cell remodeling via realignment and elongation [2]. To control the remodeling process using flow derived forces we first calculate the viscous stress tensor $\tau$:

$$\tau = \eta(\nabla \boldsymbol{u} + (\nabla \boldsymbol{u})^T) \, , \tag{4}$$

In addition to the stress due to viscous forces, there can also be internal stresses due to the fluid pressure:

$$\sigma = -p\boldsymbol{I} \, , \tag{5}$$

where $I$ is the identity matrix. From these stress tensors, the viscous and pressure boundary forces (normal, tangential) per unit area can be derived and directly added to the *interface advancing* vector field displacing the tissue boundary. Another component of this field can be surface tension:

$$\sigma_s = \sigma_0 \kappa \boldsymbol{n} \delta \, , \tag{6}$$

where $\sigma_0$ is a material coefficient, $\kappa$ denotes the *curvature*, $\boldsymbol{n}$ is an interface normal vector and $\delta$ is a delta function of the implicit level-set function (non-zero only at the zero level-set). We also attempt to include some approximations of the known non-linear properties of blood. As a first order approximation to the experimentally known *shear-thinning* behaviour of blood we pick a *power law fluid*, where the apparent or *effective* viscosity is a function of the shear rate as given by:

$$\eta^* = KD^{n-1} \, , \tag{7}$$

where $D$ is the shear rate (independent of fluid orientation) and $K, n$ are the model parameters to be found experimentally. More sophisticated versions of such power law rheology are also available in our model to simulate the non-newtonian regime of the flow, however, they are all only valid for homogeneous fluids. In capillaries with a few microns diameter, however, blood does not fulfill this assumption. In particular, the finite sizes of the red blood cells could lead to explicit momentum or molecular transfer to the walls during collisions or cause mechanical obstructions in thinner capillaries. We do not currently address these effects in our model.

## 5 Transport Properties

We regard the mechanical response of a tissue exposed to physical influence of external flow to be a major factor in capillary plexus growth. It is, however, not yet understood, how molecular signaling interacts with this process and how these two effects can be integrated within a single comprehensive model of vascular tissue morphogenesis. Once the computational method is available to solve the general transport equation it becomes straight-forward to model mass transport of any chemical molecules like oxygen, carbon dioxide or growth factors. Technically speaking, nothing restricts us from introducing a few such molecular partners and allowing them to react with each other. First order chemical reactions can easily be modeled by modifications of the corresponding source terms, e.g.:

$$S_{O_2} = -\theta_{O_2}\phi_{O_2}, \quad \text{(burning oxygen)} \tag{8}$$

$$S_{CO_2} = -S_{O_2}, \quad \text{(into carbon dioxide)} \tag{9}$$

$$S_{AGF} = \theta_{AGF}\phi_{CO_2}, \quad \text{(producing growth factors in ischemic areas)} \tag{10}$$

$$S_{ls} = \theta_{ls}\phi_{AGF}, \quad \text{(enforcing/inhibiting tissue remodeling)} \tag{11}$$

etc. with $\phi_x$ relevant properties, $S_x$ production/consumption rates (positive/negative), and $\theta_x$ local tissue activities. Modeling possibilities here are limited solely by experimental knowledge.

## 6 Results

The simulation workflow is composed of all the components described so far. First, initial structural conditions are derived experimentally from confocal microscopy images of an early developmental stage of chicken embryo capillary plexus. The resulting images are enhanced and segmented using standard, semi-automatic image processing techniques. Binarized blood-tissue boundaries become the input for the flow simulation, thereby determining hemodynamic conditions in the domain. Currently, the input is two-dimensional, which is a sufficient approximation of the initial stage of a real capillary plexus. The tissue remodeling during the simulation, however, is permitted into a thin three-dimensional layer (see [2] for background information about the real remodeling process). The next version of our framework, however, is planned to use full 3-D segmentation for structural input.

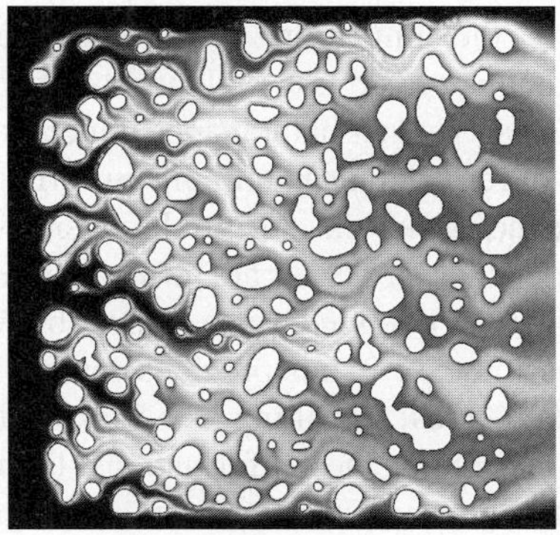

**Fig. 2.** Simulation of transport from the arterial (left) to the venous (right) side of a respiratory capillary plexus. Oxygen is diffused into the domain through the top face of the square image area and then convected with the flowing blood. In response to its time-averaged exposure, a heavy growth factor is produced and also convected with the flowing fluid. Color intensity values correspond to its concentration levels and the white areas correspond to the tissue pillars (black in Fig. 1). Note the inhomogeneous concentration that results from greatly varying passage times through the capillary plexus, which can in itself represent a growth or remodeling signal.

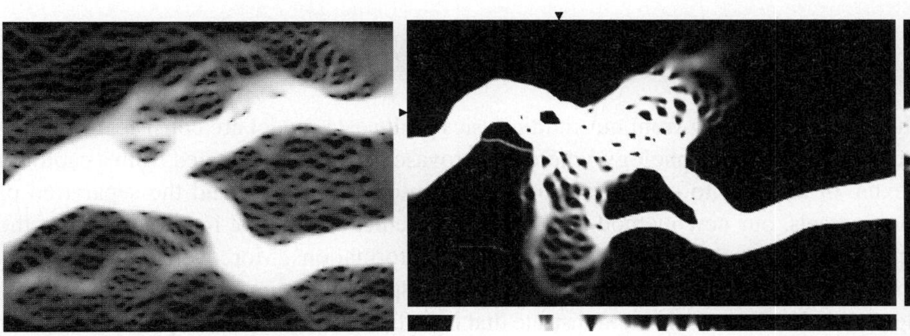

**Fig. 3.** Final stage of a virtual vascular development. **Left:** a 2-D network. Due to linear tissue response and lack of chemical signalling a shunt is formed. **Right:** a 3-D network. Formation of an arterio-venous shunt is avoided and a primitive capillary exchange plexus is formed in between the feeding artery (left) and draining vein (right). At the right and bottom sides, cross-sections of the structure are shown. Both results originate from the same input geometry. Size of the 3-D grid: 600x400x20.

Second, the global deformation field is computed, and molecular components ($O_2$, $CO_2$, growth factor) are dissolved and eventually allowed to react with the tissue surface (Fig. 2). The resulting vector field is transferred to the tissue simulation module, where shape changes are calculated as physical deformations (including surface tensions). Thus, the solution determined for a given input domain is used for structural remodeling of the entire system. This procedure is iterated in subsequent steps for a preset time period or until steady-state is reached. Note, that opposed to our previous 2-D approach, we do not anymore attempt to *a priori* separate between the experimentally known sub-modes of the intussusceptive remodeling. Instead, we presume they all should emerge as a result of the simulations once a sufficient level of modeling complexity is reached. In particular, we do not explicitly insert new pillars into low shear stress regions; instead, we observe their natural appearance from adjacent layers in 3-D.

A possible stable result of remodeling an initially uniform capillary plexus into a differentiated microvascular network via our multiphysics simulation is shown in Fig. 3 (right). Obviously, higher order vessels are formed that avoid direct fusion (arterio-venous shunt) by interdigitation: one feeder artery aims between two draining venous branches. This will not be the case if our previous remodeling procedure is applied Fig. 3 (left). We identified chemical signalling in 3-D as well as a non-linear tissue response to strain as the keys to avoiding the malformations. The non-linear nature of blood, at least in the homogeneous regime we address, is not mandatory to achieve these results and can be accounted for in the tissue response function. This will noticeably increase the performance without seriously compromising the modeling efficiency. It could also be noted that the simulation process itself has reached a considerable level of complexity. Even if it is still far away from inherent complexity of real vascular networks, however, it excels most other models that we are aware of in this aspect, hopefully leading to superior predictive power. While we are currently evaluating the performance of our multiphysics framework with respect to robustness and similarity to real microvessels, this work is beyond the scope of this article.

## 7 Conclusions

The findings resulting from our multiphysics multiscale model are consistent with observations on real morphology, where a microvascular tree is generated from a capillary plexus. In addition to explaining the formation of bifurcations and the separation of microvessels, our new method is also able to predict a primitive interdigitation of the terminal branches and avoids arterio-venous malformations. Moreover, the current version allows computational studies about the transport of oxygen, metabolites or signaling molecules. Our results demonstrate that interdisciplinary multicomponent computer models of blood vessel networks can integrate experimental data on the cellular level to simulate supracellular morphogenesis with unprecedented detail.

## References

1. Szczerba, D., Székely, G.: Computational model of flow-tissue interactions in intussusceptive angiogenesis. Journal of Theoretical Biology 234(1) (2005) 87–97
2. Kurz, H., Burri, P., Djonov, V.: Angiogenesis and vascular remodeling by intussusception: From form to function. News in Physiological Sciences 18 (2003) 65–70
3. Lindenmayer, A.: Mathematical models for cellular interaction in development. Journal of Theoretical Biology 18 (1968) 280–315
4. Alarcon, T., Byrne, H., Maini, P.: A cellular automaton model for tumour growth in inhomogeneous environment. Journal of Theoretical Biology 225 (2003) 257–274
5. Sandau, K., Kurz, H.: Modelling of vascular growth processes: a stochastic biophysical approach to embryonic angiogenesis. Journal of Microscopy 175 (1994) 205–213
6. Gödde, R., Kurz, H.: Structural and biophysical simulation of angiogenesis and vascular remodeling. Developmental Dynamics 220 (2001) 387–401

7. Szczerba, D., Székely, G.: Macroscopic modeling of vascular systems. In: Medical Image Computing and Computer-Assisted Intervention - MICCAI 2002. Volume 2489 of Lecture Notes in Computer Science., LNCS Springer (2002) 284–292
8. Szczerba, D., Székely, G.: Simulating vascular systems in arbitrary anatomies. In: Medical Image Computing and Computer-Assisted Intervention - MICCAI 2005: 8th International Conference, Palm Springs, CA, USA,. Volume 3750 of Lecture Notes in Computer Science., Springer-Verlag GmbH (2005) 641–648
9. Schreiner, W., Buxbaum, P.F.: Computer optimization of vascular trees. IEEE Transactions on Biomedical Engineering 40 (1993) 482–491
10. Baish, J.W., Gazit, Y., Berk, D.A., Nozue, M., Baxter, L.T., Jain, R.K.: Role of tumor vascular architecture in nutrient and drug delivery: An invasion percolation-based network model. Microvascular Research 51 (1996) 327–346
11. Anderson, A., Chaplain, M.: Continuous and discrete mathematical models of tumor-induced angiogenesis. Bulletin of Mathematical Biology 60 (1998) 857–899
12. McDougall, S.R., Anderson, A.R.A., Chaplain, M.A.J., Sherratt, J.A.: Mathematical modelling of flow through vascular networks: Implications for tumour-induced angiogenesis and chemotherapy strategies. Bulletin of Mathematical Biology 64 (2002) 673–702
13. Ferziger, J., Perić, M.: Computational Methods for Fluid Dynamics. Springer-Verlag (2002)
14. Sethian, J.A.: Level Set Methods and Fast Marching Methods. Cambridge University Press (1999)
15. Date, A.W.: Solution of transport equations on unstructured meshes with cell-centered colocated variables. International Journal of Heat and Mass Transfer 48 (2005) 1117–1127

# Liquid Computations and Large Simulations of the Mammalian Visual Cortex

Grzegorz M. Wojcik and Wieslaw A. Kaminski

Institute of Computer Science
Department of Complex Systems and Neurodynamics
Maria Curie-Sklodowska University
Plac Marii Curie-Sklodowskiej 5, 20-031 Lublin, Poland
gmwojcik@gmail.com

**Abstract.** Large artificial Hodgkin-Huxley neural networks are examined. The structures discussed in this article simulate the cortex of the mammalian visual system. We use a modular architecture of the cortex divided into sub-regions. Results of parallel simulations based on the liquid computing theory are presented in some detail. Separation ability of groups of neural microcircuits is observed. We show that such property may be useful for explaining some edge or contrast detection phenomena.

## 1 Introduction

The human brain and its cortex are probably the most complex systems known. A structure built of about $10^{11}$ interacting neural cells is always a hard object for simulation, even for the fastest super-computers. A new idea of brain modelling was suggested by Maass [1] and since then it has been called Liquid State Machine (LSM) [2]. In general, the brain (or its fragment) is treated as a liquid. The cortex is built of neurons organised in microcircuits [3] which form columns and the function of each column depends on its location in the brain. Cortical microcircuits turn out to be very good "liquids" for computing on perturbations. They are characterised by the large diversity of their elements, neurons, synapses, the large variety of mechanisms and time constants characterising their interactions, involving recurrent connections on multiple spatial scales. Like the Turing machine, the model of LSM is based on a strict mathematical framework that guarantees, under ideal conditions, universal computational power [1]. Applying ideas of liquid computing [1] allows to decrease the number of neurons in the constructed model. In addition, the simulation time can be dramatically shortened using cluster-based parallelised simulations of groups of microcircuits.

It has been discovered that simple cells in cat's primary visual cortex (V1) are specialised for the orientation of light and dark borders [4]. The orientation selectivity of simple cells in V1 comes from an oriented arrangement of the input from the Lateral Geniculate Nuclei (LGN). Namely, ON-centre LGN inputs have receptive fields centres aligned over simple cell's ON sub-regions, and similarly for OFF-centre inputs [5]. Some observations of monkey cortex suggest that the microstructure of V1 is spatially periodic [6]. Thanks such architecture many

V.N. Alexandrov et al. (Eds.): ICCS 2006, Part II, LNCS 3992, pp. 94–101, 2006.

phenomena of signal processing occurring in visual systems can be explained. Visual pathway and detailed description of ON-OFF centres can be found elsewhere [7].

In this paper we present some results of mammalian visual cortex simulations. We prove that our model can help to understand some edge or contrast detection phenomena.

## 2 The Model of Mammalian Visual System

Discussed model of mammalian visual system consists of two main modules (Fig. 1). Because the idea of LSM calls for such an architecture, our model includes "Input" (Retina) and "Liquid" (Cortex) [1]. All simulations discussed in this paper are conducted in parallel version of GENESIS for MPI environment (for parallelisation effectiveness, time of typical runs and other detail see Appendix A). Neurons used in the simulations are built according to the Hodgkin-Huxley model [8] and are relatively simple (for detail see Appendix B).

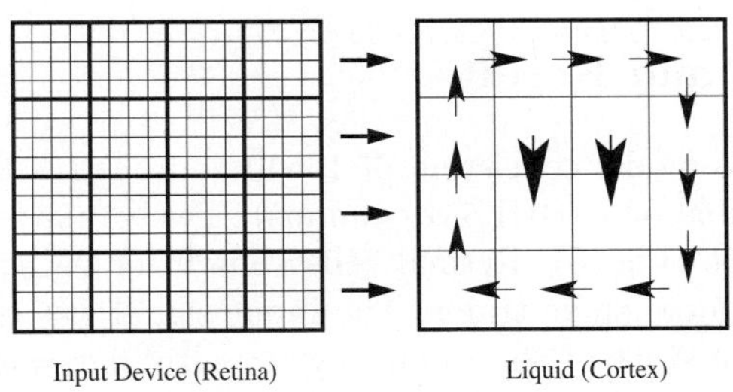


**Fig. 1.** The model of simulated visual system. The arrows represent connections of HHLSM columns (for detail see text).

The Retina is built on $16 \times 16$ square-shaped grid and divided into 16 patches ($4 \times 4$). Each patch is connected with one of 16 HHLSM (Hodgkin- Huxley Liquid State Machine) columns which simulate LGN and the ensemble of cortical microcircuits (retinal cells are connected only with LGN). HHLSM consists of 1024 cells put on a $8 \times 8 \times 16$ grid. There are layers arranged in each column (Fig. 2) and the set of columns simulates the Liquid. There are 80% of excitatory connections established among layers and neurons of each layer and 20% of inhibitory connections. In addition, layers L6 of some columns are connected with LGNs of other HHLSMs in the same way, simulating the corticothalamic feedback. Intercolumn connections are presented in Fig. 1. Each connection in the model is characterised with some "delay" parameter and random weight. The thick arrows in Fig. 1 represent connections with short delays ($d_1 = 10^{-4}$ s) and the thin arrows correspond to connections with long delays ($d_2 = 10^{-3}$ s). We choose such topology to model ON and OFF-centre subregions of V1. We can treat the "Liquid" as a hypercolumn in some part of periodic structure of the cortex.

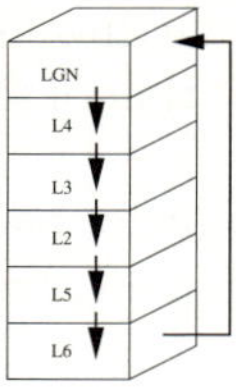


**Fig. 2.** Structure of HHLSM column as the fundamental microcircuit

Such a model can be easily scaled into multiprocessor simulation. In discussed research each column and its corresponding retinal patches should be simulated on one node. It should be noted that 16 processors were required for the best realisation of the model and additional one for the simulation control. However, the Retina may be easily divided into 4 ($2 \times 2$), 64 ($8 \times 8$) or 256 ($16 \times 16$) patches, depending on the number of processors available. Thus, if each patch is connected with corresponding HHLSM column - it should be possible to conduct a simulation of about 256 thousands Hodgkin-Huxley neural cells.

## 3 Simulation and Results

We investigated the model consisting of 16640 neurons (as the Liquid is simulated by the ensemble of 16 HHLSM columns). Twelve patterns were arranged on the Input Device (Fig. 3). Retinal cells chosen for each pattern were then stimulated with random spike trains. The input signal was encoded in Liquid's state. We define the state of the Liquid by a multidimensional vector with binary coordinates 0 for a "sleeping neuron" and 1 for an "active" neuron. The "dead-state" corresponds to the Liquid with all neurons "sleeping".

We simulated 500 ms of biological work of our system. The main objective of current research was to check the Euclidean distance of states of the liquid for different couples of input patterns.

The results confirm liquid computing abilities of neural microcircuits. In each case a meaningful difference in states of the liquid was observed for different

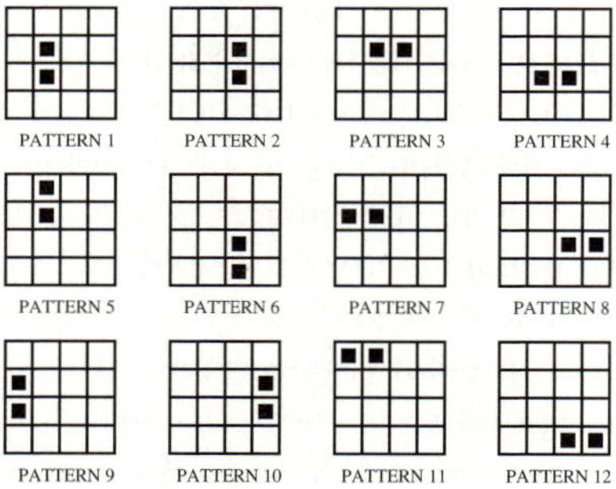


**Fig. 3.** Set of input patterns stimulating the retina

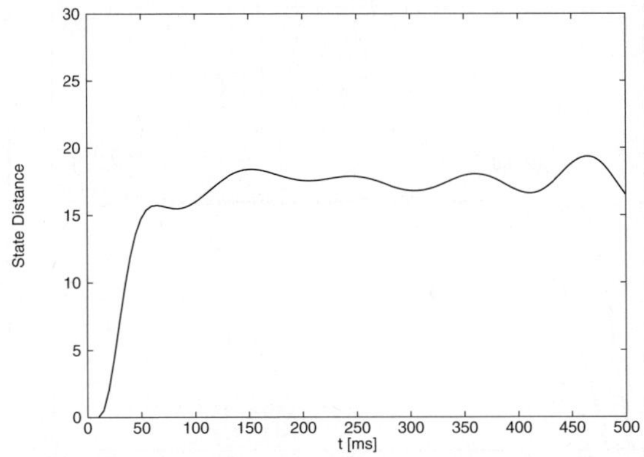


**Fig. 4.** The liquid state distance for two different spike trains given as an input to retinal pattern 1

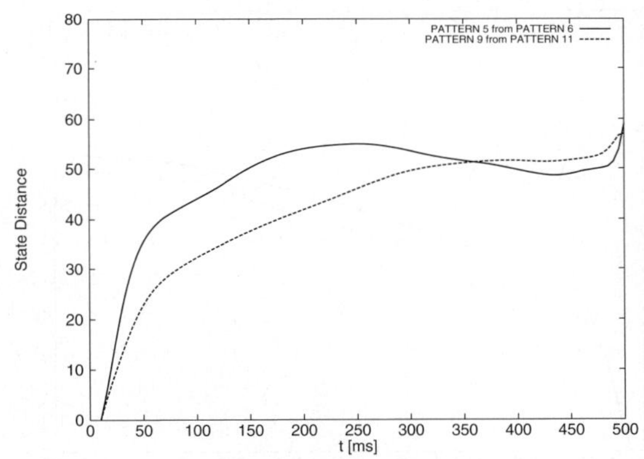


**Fig. 5.** Liquid state distances for different input spike trains given as input to retinal patterns 5,6 (solid line) and 9,11 (dashed line)

spike trains stimulating even the same retinal pattern (Fig. 4). It is usually a very typical behaviour for LSM. However, Maass' LSM has been built of integrate and fire neurons, while our structure consists of much more biologically realistic Hodgkin- Huxley neural cells. Additionally, for the states of two geometrically different input patterns even larger "distance of liquids" was observed (Fig. 5).

In general, we can divide the set of input patterns into three groups: the Retina for patterns 1 – 4 (group 1) has ON-centre receptive fields stimulated, patterns 9 – 12 (group 2) stimulate the OFF-centre sub-regions and patterns 5 – 8 (group 3) send impulses both to ON and OFF-centre patches of the Input Device. Figs. 6-8 present the state of liquids measured as the distance from the so-called "dead state" (liquid is in the "dead-state" when nothing is stimulating its cells). One can note different tendencies for each of three pattern groups. In Figs. 6-7 we present typical plots from pattern group 1 and 2 respectively. Network's architecture implies such behaviour: large delays in OFF-centre V1 microcircuits

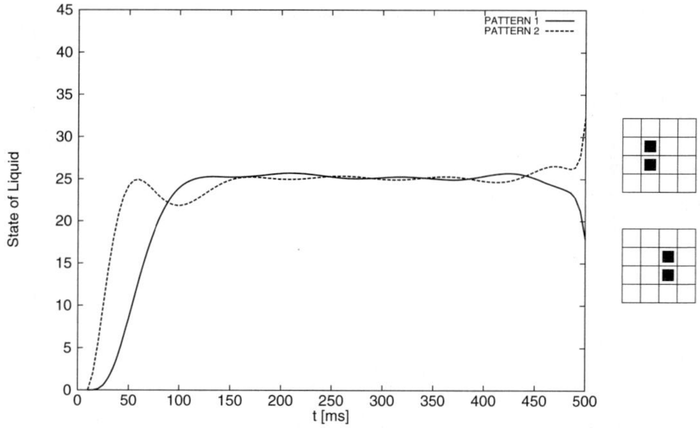


**Fig. 6.** Liquid distance from the dead-state for two patterns from group 1

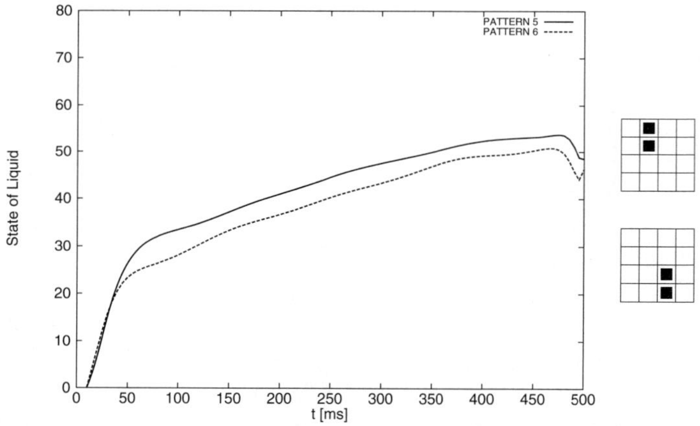


**Fig. 7.** Liquid distance from the dead-state for two patterns from group 2

cause more action spike potentials in the late phases of simulations. In addition, for patterns from the 2nd group we obtain activity of both ON and OFF-central HHLSMs. Fig. 8 shows characteristics for patterns 7 and 12. It should be noted that 12th pattern belongs to the 3rd group. In this case only the OFF-centre area of the cortex is stimulated and whole liquid activity is relatively smaller.

The distance curve in Figs. 6-8 goes down after 475 ms as result of intended and decreased retinal activity in the end of simulation. In the case of simulations longer than 500 ms the typical behaviour (like from 100 ms - 400 ms parts of the plots) was observed.

Following Maass' [1] ideas and applying a readout for liquid state analysis we can imagine some expert-devices able to classify ON and OFF-central patterns. Implementing different delays for connections and arranging a proper architecture of the simulated cortex can then lead to a better understanding of, for example, contrast or edge detection phenomena.

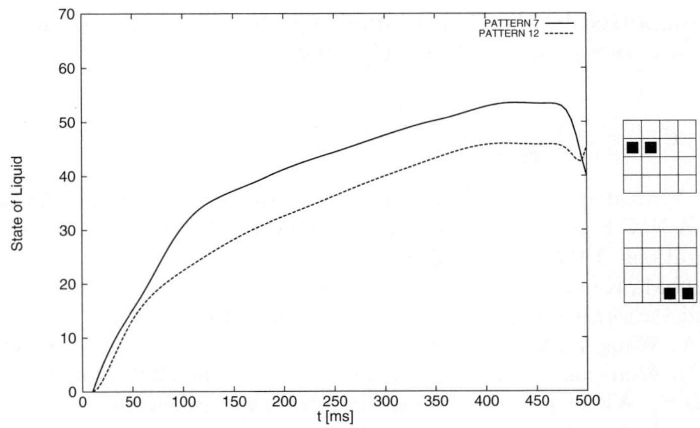


**Fig. 8.** Liquid distance from the dead-state for two patterns from group 2 and 3

## 4 Conclusions

In this paper we report results of the mammalian visual cortex' simulations. We simulated about 16 thousands Hodgkin-Huxley neurons organised in layers and cortical microcircuits. Some biologically-inspired topology was arranged. The results prove that such organisation of the cortex has good separation ability characteristic for LSM and model used for this article can explain some natural pattern recognition phenomena.

The modular structure of visual cortex makes possible the application of good parallelisation as particular microcircuits can be simulated on separate nodes. Our model is scalable and we can easily increase the number of neurons in each cortical column which will let us run simulations consisting of more than 256 thousands Hodgkin-Huxley neurons. This will help us build more realistic model of visual cortex. Most of the discussed simulations were conducted on the local cluster. Our machine is part of the CLUSTERIX grid project [10]. With access to 800 processors and by increasing the number of simulated microcircuits a structure consisting of several millions of neural cells simulated in a similar way can be imagined. This could lead to the creation of very sophisticated models and such possibility can open for us quite new field of computational complex systems' research.

## Acknowledgements

The initial stage of this work has been performed under the HPC-EUROPA Project (RII3-CT-2003-506079), with the support of the European Community - Research Infrastructure Action under the FP6 Structuring the European Research Area Programme. The author (GMW) would like to thank the EPCC and HPC-Europa organisation, especially to my hosts: Prof. L. Smith, Dr. B. P. Graham and Dr. M. Hennig of Stirling University, Scotland. This work has been also supported by the Polish Ministry of Science and Information Society Technologies under grant 6T11 2003C/06098.

## References

1. Maass W., Natschlager T., Markram H.: Real-time Computing Without Stable States: A New Framework for Neural Computation Based on perturbations. Neural Computations. **14(11)** (2002) 2531–2560
2. Wojcik G.M., Kaminski W.A.: Liquid State Machine Built of Hodgkin-Huxley Neurons and Pattern Recognition. Neurocomputing **239** (2004) 245–251
3. Gupta A., Wang Y., Markram H.: Organizing principles for a diversity of GABAergic interneurons and synapses in the neocortex. Science **287** (2000) 273–278
4. Hubel D.H., Wiesel T.N.: Functional Architecture of Macaque Monkey Visual Cortex. Proc. R. Soc. Lond. B. **198** (2002) 1–59
5. Kolesnik M.: Iterative Orientation Tuning for Contrast Detection in Images. ERCIM News. **5** (2003)
6. Bresloff P.C., Cowan J.D.: A Spherical Model for Orientation and Spatial Frequency Tuning in a Cortical Hypercolumn. Phil. Trans. R. Soc. Lond. B. **357** (2002) 1643–1667
7. Remington L. A.: Clinical Anatomy of the Visual System. Butterworth-Heinemann. (2004)
8. Hodgkin A. L., Huxley A. F.: A Quantitative Description of Membrane Current and its Application to Conduction and Excitation in nerve. J. Physiol. **117** (1952) 500–544
9. Bower J. M., Beeman D.: The Book of GENESIS - Exploring Realistic Neural Models with the GEneral NEural SImulation System. Telos, New York (1995)
10. CLUSTERIX - National Cluster of Linux Systems: http://www.clusterix.pcz.pl

## Appendix A: Details of Simulations' Hardware and Software Environment

The local cluster used for all simulations and discussed in this contribution was built of 12 machines and 1 additional machine - the so-called "access node". Each SMP machine had two 64-bit 1.4 GHz Itanium2 IA64 processors with 4 GB of RAM memory. The cluster works under control of Debian Linux Sarge (v. 3.1) and 2.6.8-1 kernel version. The model is simulated in GEneral NEural SImulation System GENESIS v.2.2.1 with its MPI extension. A gcc compiler was used for the general system configuration.

The length of a typical run was about 3000 s. The problem was parallelised for 17 nodes. Some benchmarking was done for the parallelisation. The speedup of $11-12$ if compared to $1-2$ processor runs was obtained.

## Appendix B: Properties of Hodgkin-Huxley Neurons

Our HHLSMs consist of multicompartmental neurons with two dendrite compartments, a soma, and an axon. The dendrites contain synaptically activated channel and the soma has voltage activated Hodgkin-Huxley sodium and potassium channels. The behaviour of each compartment is equivalent to the behaviour of some electrical circuit [9]. Thus, each circuit is characterised by a typical for GENESIS group of parameters set as follows: resistances $R_a = 0.3\ \Omega$, $R_m = 0.33\ \Omega$, capacity $C_m = 0.01$ F, and potential $E_m = 0.07$ V. For the soma compartment $E_k = 0.0594$ V whilst for the dendrite $E_k = 0.07$ V. Conductance for each type of ionic channels is chosen to be: $G_K = 360\ \Omega^{-1}$ and $G_{Na} = 1200\ \Omega^{-1}$. These parameters originate from neurophysiological experiments [9] and are chosen to make the model biologically more realistic. The soma has a circular shape with the diameter of 30 $\mu$m, while dendrites and axon are cable like with the length of 100 $\mu$m. All the other parameters are chosen as suggested by GENESIS authors to simulate the behaviour of the biological-like neurons [9]. More details concerning Hodgkin-Huxley model can be found elsewhere [8].

# Which Meshes Are Better Conditioned: Adaptive, Uniform, Locally Refined or Locally Adjusted?

Sanjay Kumar Khattri and Gunnar Fladmark

Department of Mathematics, University of Bergen, Norway
{sanjay, Gunnar.Fladmark}@mi.uib.no
http://www.mi.uib.no/∼sanjay

**Abstract.** Adaptive, locally refined and locally adjusted meshes are preferred over uniform meshes for capturing singular or localised solutions. Roughly speaking, for a given degree of freedom a solution associated with adaptive, locally refined and locally adjusted meshes is more accurate than the solution given by uniform meshes. In this work, we answer the question which meshes are better conditioned. We found, for approximately same degree of freedom (same size of matrix), it is easier to solve a system of equations associated with an adaptive mesh.

## 1 Introduction

Uniform, locally adjusted, adaptive and locally refined meshes are shown in Figures 3, 4, 5 and 6, respectively. Here, for each mesh the number of cells (or degree of freedom) are approximately 1024 ($2^5 \times 2^5$). Let us consider the steady state pressure equation of a single phase flow in a porous medium $\Omega$ [1]

$$-\operatorname{div}(\mathrm{K}\operatorname{grad} p) = f \quad \text{in} \quad \Omega \quad \text{and} \quad p(x,y) = p^{\mathrm{D}} \quad \text{on} \quad \partial\Omega_{\mathrm{D}} \ . \tag{1}$$

Here, $\Omega$ is a polyhedral domain in $\mathbb{R}^2$, the source function $f$ is assumed to be in $L^2(\Omega)$ and the diagonal tensor coefficient $\mathrm{K}(x,y)$ is positive definite and piecewise constant. K (permeability) is allowed to be discontinuous in space. We are discretizating the equation (1) on the meshes (see Figures 3, 4, 5 and 6) by the method of Finite Volumes [1, 3, 7, 8]. For discretization of the problem (1) on uniform and localised meshes (see the Figures 3 and 4), we refer to the References [1, 7, 8]. Discretization of the equation (1) on adaptive and locally refined meshes is given in the following References [3, 5]. Finite Volume discretization of the problem (1) on a mesh results in a matrix system $\boldsymbol{A}\,\mathbf{p}_h = \mathbf{b}$. Here, $\boldsymbol{A}$ is symmetric positive definite matrix associated with a mesh.

Let us define a problem to be solved on the four meshes. Let the domain be $\Omega = [-1,1]\times[-1,1]$ (see Figure 1). It is divided into four sub-domains according to the permeability K (see the Figures 1 and 2). The permeability K is a positive constant in each of the sub-domains and is discontinuous across the surfaces of sub-domains. Let the permeability in the sub-domain $\Omega_i$ be $\mathrm{K}_i$. Assuming that

V.N. Alexandrov et al. (Eds.): ICCS 2006, Part II, LNCS 3992, pp. 102–105, 2006.

$K_1 = K_3 = R$ and $K_2 = K_4 = 1.0$. $K_1$, $K_2$, $K_3$ and $K_4$ refers to the permeabilities in the subdomains $\Omega_1$, $\Omega_2$, $\Omega_3$ and $\Omega_4$, respectively. The parameter $R$ is given below. Let the exact solution in the polar form be [5]

$$p(r,\theta) = r^{\gamma}\,\eta(\theta)\ , \tag{2}$$

where the parameter $\gamma$ denotes the singularity in the solution [5] and it depends on the permeability distribution in the domain (see Figure 2 for the permeability for the singularity $\gamma = 0.1$). $\eta(\theta)$ is given as

$$\eta(\theta) = \begin{cases} \cos[(\pi/2-\sigma)\gamma]\ \cos[(\theta-\pi/2+\rho)\gamma]\ , & \theta \in [0,\pi/2]\ , \\ \cos(\rho\gamma)\ \cos[(\theta-\pi+\sigma)\gamma]\ , & \theta \in [\pi/2,\pi]\ , \\ \cos(\sigma\gamma)\ \cos[(\theta-\pi-\rho)\gamma]\ , & \theta \in [\pi,3\pi/2]\ , \\ \cos[(\pi/2-\rho)\gamma]\ \cos[(\theta-3\pi/2-\sigma)\gamma]\ , & \theta \in [3\pi/2,2\pi]\ . \end{cases} \tag{3}$$

It can be shown that solution $p$ (given by equation (2)) barely belongs in the fractional Sobolev space $\mathbf{H}^{1+\kappa}(\Omega)$ with $\kappa < \gamma$ (cf. [6]).

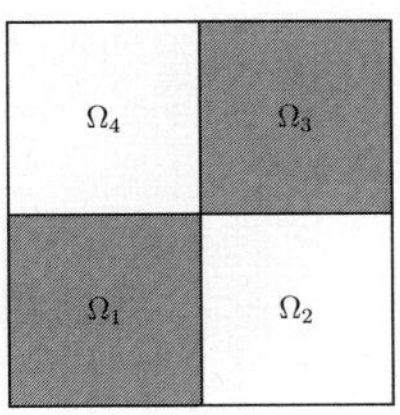


**Fig. 1.** Domain

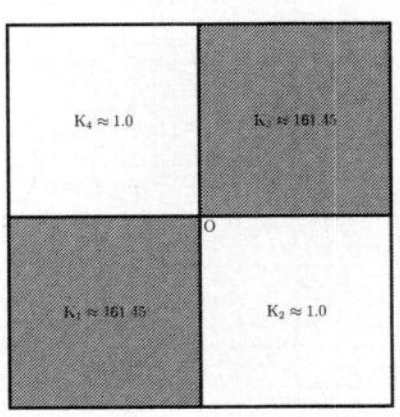


**Fig. 2.** Permeability distribution

For the singularity $\gamma = 0.1$, the parameters are

$$R \approx 161.4476\ , \quad \rho \approx 0.7854 \quad \text{and} \quad \sigma \approx -14.9225\ .$$

We solve the problem (1) on the four meshes. The exact solution is given by the equation (2). We enforce the solution inside the domain by the Dirichlet boundary condition and the source term. For solving discrete system of equations formed on the meshes, we use the Conjugate Gradient (CG) solver (see [4]). Table 1 presents eigenvalues and condition numbers of the matrix systems associated with the different meshes. Note that in this table, the largest eigenvalue on all four meshes is approximately same. However, the smallest eigenvalue associated with the adaptive mesh is greater than the smallest eigenvalues associated with other three meshes. When solving the Symmetric Positive Definite (SPD) linear system $\boldsymbol{A}\mathbf{p}_h = \mathbf{b}$ with the CG, the smallest eigenvalues of the matrix slowes down the convergence (cf. [4]). Several techniques have been proposed in the literature to remove bad effect of the smallest eigenvalue (see [2, 4, and references therein]). Convergence of the CG solver for these the four systems are shown in the Figure 7. It is clear from the Table 1 and the Figure 7 that it is easier to solve a matrix system associated with an adaptive mesh than to solve systems associated with uniform, localised and locally refined meshes.

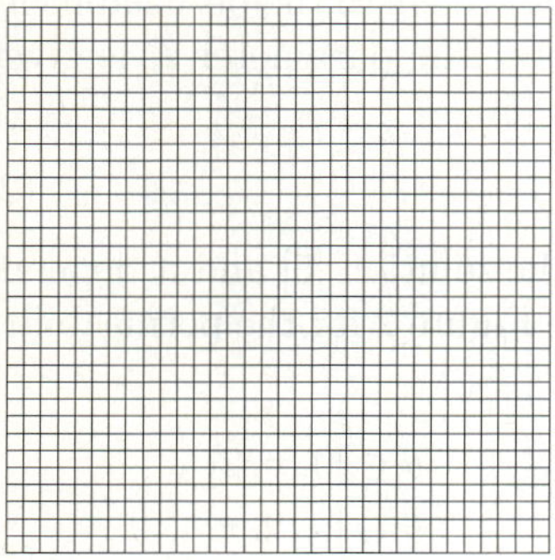

Fig. 3. Uniform mesh

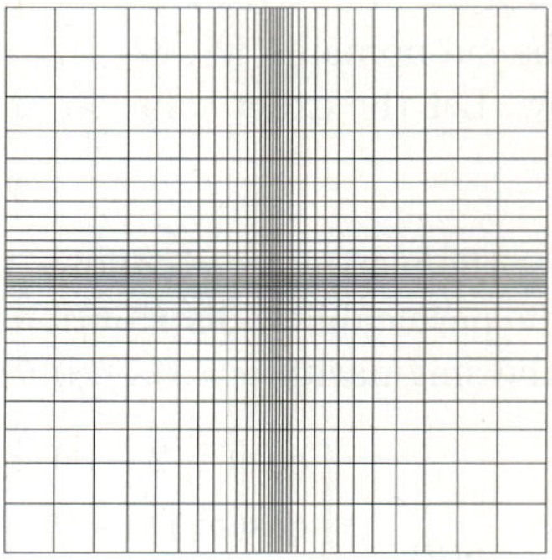

Fig. 4. Localised mesh

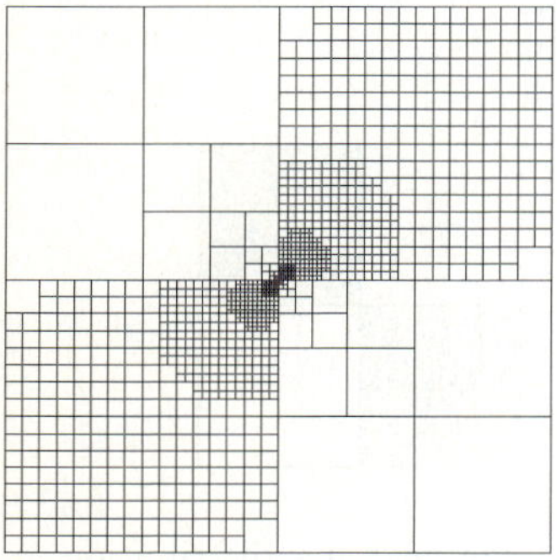

Fig. 5. Adaptive mesh

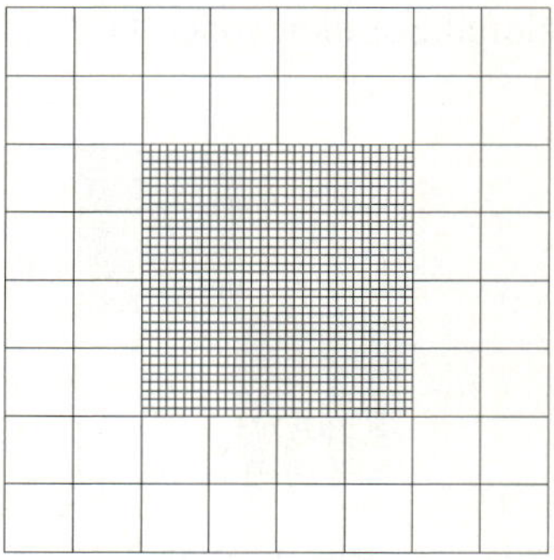

Fig. 6. Locally refined mesh

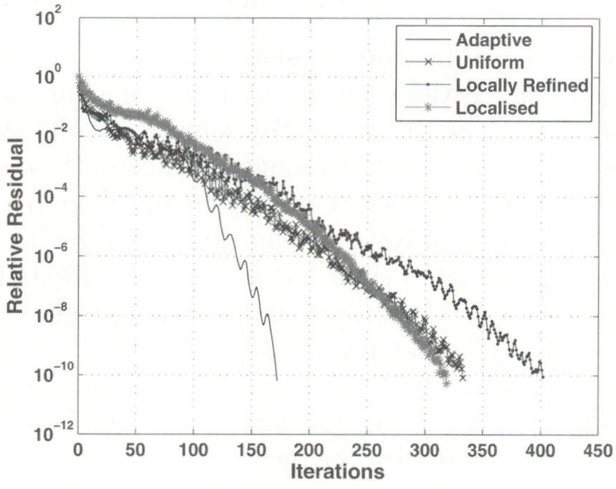


Fig. 7. Convergence curves for the matrix system formed on the different meshes

**Table 1.** Eigenvalues and condition numbers of different matrix systems

| Mesh | Smallest eig. | Largest eig. | Cond. Num. |
|---|---|---|---|
| Adaptive | $\mathbf{4.15 \times 10^{-1}}$ | $1.28 \times 10^{3}$ | $\mathbf{3.10 \times 10^{3}}$ |
| Localised | $5.50 \times 10^{-2}$ | $0.78 \times 10^{3}$ | $1.42 \times 10^{4}$ |
| Uniform | $7.62 \times 10^{-2}$ | $1.28 \times 10^{3}$ | $1.69 \times 10^{4}$ |
| Locally Refined | $3.94 \times 10^{-2}$ | $1.28 \times 10^{3}$ | $3.25 \times 10^{4}$ |

## 2 Conclusions

We have shown that it is easier to solve a matrix system associated with an adaptive mesh than solving systems associated with uniform, localised and locally refined meshes. The adaptive mesh is generated by equal distribution of the fluxes over all the cells in the mesh. Why do equal distribution of fluxes is create meshes which are better conditioned ? Or, why do equal distribution of fluxes remove bad effect of small eigenvalue ? Answers to these questions can help in designing new preconditioners or improving existing preconditioners.

## References

1. Aavatsmark, I.: An introduction to multipoint flux approximations for quadrilateral grids. Comput. Geosci., 6(3-4), (2002) 405–432.
2. Carpentieri, B., Duff, I.S. and Giraud, L.: A Class of Spectral Two-Level Preconditioners. SIAM Journal on Scientific Computing, 25(2), (2003) 749–765.
3. Ewing, R.E., Lazarov, R.D. and Vassilevski, P.S.: Local refinement techniques for elliptic problems on cell-centered grids. I. Error analysis. Math. Comp., 56(194), (1991) 437–461.
4. Giraud, L., Ruiz, D. and Touhami, A.: A comparative study of iterative solvers exploiting spectral information for SPD systems. Technical Report TR/PA/04/40, CERFACS, Toulouse, France, (2004).
5. Khattri, S.K.: Numerical Tools for Multiphase, Multicomponent, Reactive Transport: Flow of $CO_2$ in Porous Medium. PhD Thesis, The University of Bergen, Norway, (2006).
6. Strang, G. and Fix, G.J.: An analysis of the finite element method. Wiley New York, 1, (1973).
7. Süli, E.: Convergence of finite volume schemes for Poisson's equation on nonuniform meshes. SIAM J. Numer. Anal., 28(5), (1991) 1419–1430.
8. Weiser, A. and Wheeler, M.F.: On convergence of block-centered finite differences for elliptic-problems. SIAM J. Numer. Anal., 25(2), (1998) 351–375.

# Parallel Simulation of Three–Dimensional Bursting with MPI and OpenMP*

S. Tabik[1], L.F. Romero[2], E.M. Garzón[1] and J.I. Ramos[2]

[1] Depto de Arquitectura de Computadores y Electrónica, Universidad de Almería, 04120 Spain
[2] Room I-320, E.T.S. Ingenieros Industriales, Universidad de Málaga, 29080 Spain

**Abstract.** This work presents a mathematical model and its parallel implementation via two parallel paradigms for the simulation of three–dimensional bursting phenomena. The mathematical model consists of four nonlinearly coupled partial differential equations and includes fast and slow subsystems. The differential equations have been discretized by means of a linearly–implicit finite difference method in equally–spaced grids. The resulting system of equations at each time level has been solved by means of an optimized Preconditioned Conjugate Gradient (PCG) method. The proposed mathematical model has been implemented via: (1) a message passing paradigm based on the standard MPI and (2) a shared address space paradigm based on SPMD OpenMP. The two implementations have been evaluated on two current parallel architectures, i.e., a cluster of biprocessors Xeon and an SGI Altix 3700 Bx2 based on Itanium. It is shown that better performance and scalability are obtained on the second platform.

## 1 Introduction

Bursting phenomena occur in physiology and biology. For example, neurons communicate by firing and transmitting action potentials. Usually, action potentials occur in a periodic fashion, as in response to a constant applied current of sufficiently large magnitude. In addition, many cell types, e.g., pancreatic $\beta$-cells, exhibit more complex behavior characterized by brief bursts of oscillatory activity interspersed with quiescent periods during which the cell membrane potential changes slowly. Models of bursting electrical activity can be classified into two main groups. The first and earliest one was based on the assumption that bursting was caused by an underlying slow oscillation in the intracellular $Ca^{2+}$ concentration [1, 2]; however, recent experiments indicate that this assumption is not entirely correct and, as a consequence, models relying on alternative mechanisms have been developed [3]. The different known bursting mechanisms can be classified into three main groups. In type I, bursts arise from hysteresis and bistability as in the pancreatic $\beta$–cell model. In type II, bursts arise from an underlying slow oscillation, while, in type III, bursting arises from a subcritical

* This work was supported by the Ministerio de Educación y Ciencia of Spain under Projects TIN2005-00447, FIS2005–03191 and Fondos FEDER.

V.N. Alexandrov et al. (Eds.): ICCS 2006, Part II, LNCS 3992, pp. 106–113, 2006.

Hopf bifurcation [1, 2, 4, 5]. This classification is by no means complete, for it is based on knowledge acquired with nonlinear ordinary differential equations. In extended systems, i.e., spatio–temporal systems, one can easily show that bursting occurs through one of the three types mentioned above provided that the system is homogeneous. However, when the system is not initially homogeneous and the diffusion coefficients are not nil, the homogeneous bursting solution may bifurcate and result in spatio–temporal nonhomogeneities and/or quenching.

The objective of this paper is several–fold. First, a three–dimensional model of bursting consisting of four, nonlinearly coupled partial differential equations is proposed. This model has been selected so that it exhibits bursting under homogeneous conditions, and both bursting and quenching for non–homogeneous initial conditions. The source or reaction terms of these partial differential equations can be classified into two subsystems exhibiting fast and slow behavior. The accurate simulation of the fast processes demands the use of small time steps, while the extended system considered here requires the use of sufficiently small spatial step sizes in order to accurately resolve the steep gradients of the dependent variables.

Second, a parallel implementation of the finite difference equations resulting from the discretization of the three–dimensional bursting model equations is developed and analyzed using two parallel paradigms: (1) a message passing paradigm where all communications between processors are established via MPI and (2) a shared address space paradigm based on SPMD OpenMP. An efficient parallel conjugate gradient (CG) solver has been used to solve the system of linear equations. Both parallel implementations are then evaluated on two parallel platforms: (1) a cluster of biprocessors Intel(R) Xeon(TM) and (2) a SGI Altix 3700 Bx2 system based on Itanium.

## 2 Formulation and Numerical Method

Bursting in extended systems has been simulated by means of the following nonlinearly coupled system of partial differential equations

$$\frac{\partial \mathbf{U}}{\partial t} = \mathbf{D}\left(\frac{\partial^2 \mathbf{U}}{\partial x^2} + \frac{\partial^2 \mathbf{U}}{\partial y^2} + \frac{\partial^2 \mathbf{U}}{\partial z^2}\right) + \mathbf{S}(\mathbf{U}) \tag{1}$$

where $\mathbf{U} = (u, v, w, p)^T$, $\mathbf{D}$ is a diagonal matrix with components equal to $D_u$, $D_v$, $D_w$ and $D_p$, $t$ is time, $x$, $y$ and $z$ are Cartesian coordinates, $\mathbf{S} = (S_u, S_v, S_w, S_p)^T$ is the vector of the source/reaction terms given by

$$\begin{cases} S_u = f(u) - v - gw(u - u_0) \\ S_v = \frac{1}{5}(v_\infty(u) - v) \\ S_w = f_w(w) + \alpha_w(p - 0.3) \\ S_p = \beta_p((1 - p)H(u) - p) \end{cases} \tag{2}$$

and

$$\begin{cases} f(u) = 1.35u(1 - \frac{u^2}{3}) \\ f_w(w) = -0.2(w - 0.2)(w - 0.135)(w - 0.21) \\ v_\infty(u) = tanh(5u) \\ H(u) = \frac{3}{2}(1 + tanh(5u - 2.5)) \end{cases} \tag{3}$$

with $u_0 = 2$, $\alpha_\omega = 0.002$, $\beta_z = 0.00025$ and $g = 0.73$. The functions $f(u)$ and $f_\omega(\omega)$ have three zeros each, whereas $v_\infty(u)$ and $H(u)$ are monotonically increasing functions of their arguments.

Eq. 1 is a simplified model of bursting electrical activity in cells, where $u$ and $w$ are the fast system and represent the currents of activated and voltage–dependent channels, respectively, whereas $w$ and $p$ constitute the slow subsystem and represent a current and its activation, respectively.

Eq. 1 have been solved in a parallelepiped $\Omega \equiv [-L_x, L_x] \times [-L_y, L_y] \times [-L_z, L_z]$ and homogeneous Neumann boundary conditions have been used on all the boundaries. The initial conditions used in the computations are

$$u = -2.5, v = -0.2, \omega = -0.5, z = 0.5 \text{ in } \omega$$
$$u = 2.5, v = -0.2, \omega = 0.5, z = -0.5 \text{ in } \Omega - \omega$$

where $\omega = [-\frac{L_x}{2}, \frac{L_x}{2}] \times [-\frac{L_y}{2}, \frac{L_y}{2}] \times [-\frac{L_z}{2}, \frac{L_z}{2}]$. These initial conditions were selected so that, under homogeneous conditions, Eq. 1 exhibits bursting oscillations. By discretizing the time variable in Eq. 1 by means of the (second–order accurate) Crank-Nicolson technique, one can obtain a nonlinear system of elliptic equations at each time step. These elliptic equations can be linearized with respect to time in order to obtain a system of linear elliptic equations at each time step, i.e., the nonlinear terms $\mathbf{S}^{n+1}$ are approximated by means of the second-order accurate terms $\mathbf{S^n} + \mathbf{J}^n \Delta\mathbf{U}$ where $\mathbf{J}$ denotes the Jacobian of the source terms, the superscript $n$ denotes the $n$–th time level, i.e., $t^n = n\Delta t$, $n = 0, 1, 2, 3, \ldots$, $\Delta t$ is the time step, and $\Delta\mathbf{U} \equiv \mathbf{U}^{n+1} - \mathbf{U}^n$. The resulting system of linear elliptic equations was discretized in space by means of second–order accurate central finite difference formulae in a regular grid of $N_x \times N_y \times N_z$ points, and the resulting system of linear algebraic equations can be written as

$$\begin{array}{l} (\mathbf{I} - \frac{k}{2}\mathbf{J}^n_{i,j,k})\Delta\mathbf{U}_{i,j,k} \\ -\alpha_x(\Delta\mathbf{U}_{i+1,j,k} - 2\Delta\mathbf{U}_{i,j,k} + \Delta\mathbf{U}_{i-1,j,k}) \\ -\alpha_y(\Delta\mathbf{U}_{i,j+1,k} - 2\Delta\mathbf{U}_{i,j,k} + \Delta\mathbf{U}_{i,j-1,k}) \\ -\alpha_z(\Delta\mathbf{U}_{i,j,k+1} - 2\Delta\mathbf{U}_{i,j,k} + \Delta\mathbf{U}_{i,j,k-1}) = \mathbf{T}^n_{i,j,k} \end{array} \tag{4}$$

where $\mathbf{I}$ is the identity or unit matrix, $\alpha_x = \frac{k}{2\Delta x^2}\mathbf{D}$, $\alpha_y = \frac{k}{2\Delta y^2}\mathbf{D}$, $\alpha_z = \frac{k}{2\Delta z^2}\mathbf{D}$, $\Delta x$, $\Delta y$ and $\Delta z$ are the grid spacings in the $x$, $y$ and $z$ directions, respectively, and

$$\begin{array}{l} \mathbf{T}^n_{i,j,k} = 2\alpha_x(\mathbf{U}_{i+1,j,k} - 2\mathbf{U}_{i,j,k} + \mathbf{U}_{i-1,j,k}) \\ +2\alpha_y(\mathbf{U}_{i,j+1,k} - 2\mathbf{U}_{i,j,k} + \mathbf{U}_{i,j-1,k}) \\ +2\alpha_z(\mathbf{U}_{i,j,k+1} - 2\mathbf{U}_{i,j,k} + \mathbf{U}_{i,j,k-1}) + \mathbf{S}^n_{i,j,k} \end{array} \tag{5}$$

Since the finite difference discretization of a system of $N_e (= 4)$ partial differential equations results in $N_e \times N_x \times N_y \times N_z$ nodal variables and algebraic equations, a main issue when solving systems of linear algebraic equations such as Eqs. 4 is the ordering of the equations and variables. A natural ordering of the grid points and blocking of nodal variants has been chosen here to obtain a well structured matrix, and because of its better cache behavior. For this ordering, Eq. 1 can be expressed in matrix form as $Ax = b$, where $A$ is an heptadiagonal block matrix, with $4 \times 4$ blocks, because $N_e = 4$ in Eq 1, and $x = \Delta \mathbf{U}$. We have used the conjugate gradient method with Jacobi preconditioner to solve the linear system of equations $Ax = b$. The sequential algorithm can be described as follows:

Initialize the source/reaction terms according to Eqs. 2 and 3.
Initialize the solution $U^0$ according to the initial conditions considered.
do n=1,2,3,...
- Update the matrix $A$ and right–hand–side $b$ according to Eqs. 4 and 5, respectively.
- Solve the system of equations $A \cdot \Delta U = b$.
- Update the solution $U^n = U^{n-1} + \Delta U$.

end do

Preliminary profiling of the serial simulation has shown that solving the system of equations is the most computationally consuming step. Therefore, the efficiency of the whole code depends mainly on the efficiency of the solution of the system of linear equations. In addition, the storage format of the matrix has a large influence on the performance of the solver used [12]. For this reason, an optimized preconditioned Conjugate Gradient (PCG) method for banded matrices, designed by Ortigosa et al. [7], has been used. The optimized PCG is based on a maximum exploitation of the locality of data and message-computation overlap. Moreover, The matrix has been stored using a compressed diagonal format which takes into account the sparsity pattern of the Jacobian term, and the implementation of the sparse matrix-vector product includes a consequent hand-unrolled product which exploits the pipelined floating point units.

In this paper, all computations have been performed on (coarse) $51 \times 51 \times 51$ and (fine) $101 \times 101 \times 101$–point equally-spaced meshes. The coarse mesh represents the largest step size for which accurate results can be obtained, and the time step is equal to 0.0001 $s$ .

Some sample results corresponding to $L_x = L_y = L_z = 15$ are presented in Fig. 1 which shows the time history of $u$ at two monitor points: one (identified with asterisks) in $\omega$ and the other (identified with squares) in $\Omega - \omega$. This figure shows that, for the initial conditions considered in this paper, the time history of $u$ in $\Omega - \omega$ exhibits an oscillatory behavior characterized by periods of intensive variations in $u$ starting from $t = 0$, whereas the time history of $u$ in $\omega$ takes a while before it starts oscillating. The oscillations observed in Fig. 1 are typical of relaxation oscillations.

Fig. 1 also shows that the oscillations of $u$ in $\omega$ and $\Omega - \omega$ are not in phase, but nearly coincide at about $t = 1370$ a.u.; thereafter, the one that was lagging overcomes the other, and, after some time, both time histories reach the same slowly varying value. Similar trends to those shown in Fig. 1 have also been observed at other six monitoring locations.

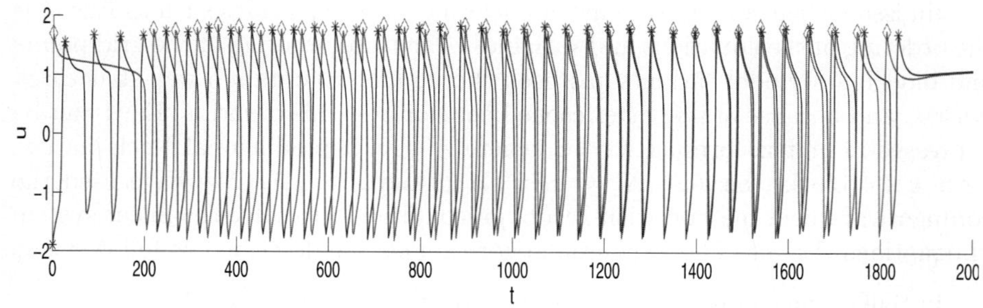


**Fig. 1.** $u$ versus time $t$ (in a.u.), in $\omega$ $(*)$ and $\Omega - \omega$ $(\diamond)$

Although not shown here, three–dimensional, time–dependent visualizations of $u$ indicate that there are inward and outward propagating waves in $\omega$ before the outward ones propagate in $\Omega - \omega$, i.e., there is an interplay between the values of $u$ in $\omega$ and $\Omega - \omega$ caused by the diffusion of the dependent variables. This back–and–forth play between the inner and outer domains is a consequence of the initial conditions, the source/reaction terms, and the diffusion coefficients of Eqs. 1.

## 3 Parallel Implementation

CLUsters of MultiProcessors (CLUMPs) have become today's dominant platforms. Such architectures support three parallel programming paradigms, i.e., message passing paradigm, shared address space paradigm and hybrid paradigm. Here, we propose two parallel implementations of the three–dimensional bursting model described above; the first is based on the message passing paradigm using MPI, while the second is based on a shared address space using OpenMP. Both implementations are based on SPMD style, whereby each processor works on its own sub-domain.

### 3.1 MPI Code

At each time-step, each processor updates its sub-matrix and local right–hand–side. Then, the parallel solution of the system $A \cdot \Delta U = b$ is performed by the PCG solver as in [7]. Using a Jacobi preconditioner, each iteration of the PCG method includes two inner products and one sparse matrix vector product; therefore, three communications and their corresponding synchronizations are needed. In this implementation, computations and communications have been overlapped using asynchronous messages in order to minimize the communications overhead, specially in the matrix vector product. In this way, after the iterations of the PCG are completed, a message with the boundary values of $U$ is sent to the neighboring processors and this is overlapped with the last update of $\Delta U$.

### 3.2 OpenMP Code

We have chosen the SPMD style instead of the loop level one because it reduces overhead and results in better scalability [10]. In this implementation, only one parallel section covers the whole dynamic extent of the code and, therefore, only OpenMP synchronization directives have been used. The ordering of the computations employed in the MPI implementation to minimize the communication overhead is also employed here to reduce the waits in the synchronizations points. For the inner product, the synchronization events have been implemented by using counters protected by lock variables. Additional flags have been included in order to grant permission for accessing shared data. As soon as a processor has computed the data on its borders, it enables the flag. If other processor requires these data, it waits for this flag to be enabled. Reset of counters and flags has been carefully implemented by means of odd and even sense-reversing flags to enhance performance and avoid data race conditions. In both massage-passing and shared space address paradigms, global barriers have been avoided to minimize waiting times.

## 4 Evaluation of the Parallel Implementation

We have assessed the performance of both parallel codes for the coarse and fine grids on the architectures described below:

- A cluster of Intel(R) Xeon(TM) biprocessors at 3.06 GHz with 2 GB RAM and 512 KB cache. The nodes of the cluster are interconnected via two gigabit Ethernet networks: one for data (NFS) and the other for computation.
- An SGI Altix 3700 Bx2 of 64 processors at 1600 MHz Intel Itanium 2 Rev with 128 GB RAM and 6 MB L3. The Altix 3700 computer system is based on a Distributed Shared Memory architecture [8] and uses a cache–coherent Non-Uniform Memory Access (NUMA) where the latency of processors to access the local memory is lower than the latency to access the global (or remote) memory [9].

Preliminary evaluation has shown that for both codes, the computational cost is similar for each time step; therefore, in this section, we show and analyze the average execution time for only one time step.

On the cluster of biprocessors, the parallel performance and the scalability of the code based on MPI paradigm are better when the dimension of the matrix is large as shown in Table 1. One of the underlying reasons is the increase in the computation–communication ratio, i.e., a finer grid or a large matrix involves more computation, but also requires more communications. The evaluation of the code based on OpenMP cannot be completed in more than two processors; therefore, no significative conclusions can be made for this case.

On a SGI Altix 3700 Bx2 system, both paradigms yield very good performance for both coarse and fine grids as shown in Table 2. This is due to the high speed interconnection technology used in this architecture [9] and to its good shared memory features. Besides, both codes includes pipelining techniques, which allow to reach very good performance on VLIW architectures such those based on Itanium [12]. Moreover, the scalability for both codes is slightly better for the coarse grid; this can be explained by two reasons, first, the good behavior of the

**Table 1.** Performance and speedup of the parallel simulation per time step on a cluster of biprocessors, for the coarse and fine grids

| | coarse grid | | | | fine grid | | | |
|---|---|---|---|---|---|---|---|---|
| | openMP | | MPI | | openMP | | MPI | |
| proc | runtime (s) | speedup | runtime (s) | speedup | runtime (s) | speedup | runtime (s) | speedup |
| 1 | 0.61 | 1.00 | 0.61 | 1.00 | 6.20 | 1.00 | 6.20 | 1.00 |
| 2 | 0.50 | 1.20 | 0.58 | 1.05 | 5.05 | 1.23 | 3.02 | 1.99 |
| 4 | | | 0.32 | 1.88 | | | 1.64 | 3.66 |
| 6 | | | 0.24 | 2.52 | | | 1.26 | 4.76 |
| 8 | | | 0.20 | 3.03 | | | 1.03 | 5.83 |

**Table 2.** Performance and speedup of the parallel simulation per time step on a SGI Altix 3700 Bx2, for the coarse and fine grids

| | coarse grid | | | | fine grid | | | |
|---|---|---|---|---|---|---|---|---|
| | openMP | | MPI | | openMP | | MPI | |
| proc | runtime (s) | speedup | runtime (s) | speedup | runtime (s) | speedup | runtime (s) | speedup |
| 1 | 0.258 | 1.00 | 0.258 | 1.00 | 2.001 | 1.00 | 2.001 | 1.00 |
| 2 | 0.135 | 1.91 | 0.135 | 1.91 | 1.010 | 1.98 | 1.025 | 1.97 |
| 4 | 0.070 | 3.68 | 0.072 | 3.59 | 0.591 | 3.39 | 0.855 | 2.36 |
| 6 | 0.047 | 5.50 | 0.044 | 5.91 | 0.407 | 4.91 | 0.459 | 4.43 |
| 8 | 0.035 | 7.44 | 0.035 | 7.42 | 0.2789 | 6.19 | 0.336 | 6.01 |

memory management, since the percentage of data that fits in the cache memory in this case is higher than that for the fine grid; second, both implementations use a reorganization of the computations in order to minimize the communications overhead (in MPI-code) and the synchronizations overhead in (OpenMp-code), the penalty of this reordering on the locality accesses is higher for the fine grid.

Finally, it should be mentioned that the performance of both sequential codes on the platform based on Itanium is 2 to 3 times better than that on the platform based on Xeon, and this is in accord with the results of the Standard Performance Evaluation Corporation (SPEC) [11].

## 5 Conclusions

A mathematical model for three–dimensional bursting phenomena and two parallel implementations of it have been presented. The model is described by four, nonlinearly coupled partial differential equations which have been discretized by means of a second–order accurate, linearly–implicit finite difference method in equally–spaced grids. The resulting system of linear algebraic equations at each time level has been solved by means of the PCG solver optimized for banded matrices and implemented using two parallel paradigms: (1) a message passing paradigm with a message-computation overlap and (2) a SPMD OpenMP style. Both implementations have been evaluated on two parallel platforms: a DSM platform based on Itanium and a cluster of biprocessors Xeon. The performance of both implementations depends on the mesh size, the parallel paradigm used and the architecture of the platform. It has been shown that both parallel paradigms are suitable for the platform based on Itanium, especially for coarse grids, whereas for clusters based on Xeon processors, only with fine grids and using the message passing paradigm one can achieve the expected performance.

## References


1. Rinzel, J. and Lee, Y. S., Dissection of a model for neuronal parabolic bursting, *J. Math. Biol.* **25** (1987) 653–675.
2. Rinzel, J., Electrical excitability of cells, theory and experiment: review of the Hodgkin–Huxley foundation and an update, *Bull. Math. Biol.* **52** (1990) 5–23.
3. Smolen, P. and Keizer, J., Slow voltage inactivation of $Ca^{+}2$ currents and bursting mechanisms for the mouse pancreatic $\beta$–cell, *J. Membr. Biol.* **127** (1992) 9–19.
4. Bertram, R., Buttle, M. J., Kiemel, T. and Sherman, A., Topological and phenomenological classification of bursting oscillations, *Bull. Math. Biol.*, **57** (1995) 413–439.
5. Keener, J. and Sneyd, J., *Mathematical Physiology*, Springer, New York, 1998.
6. Ramos, J. I., Linearization methods for reaction–diffusion equations: Multidimensional problems. *Appl. Math. Comput.* **88** (1997) 225-54.
7. Ortigosa, E. M., Romero, L. F. and Ramos, J. I., Parallel scheduling of the PCG method for banded matrices rising from FDM/FEM, *J. Parall. Distr. Comput.* **63** (2003) 1243 - 1256.
8. Protic, J., Tomasevi, M. and Milutinovic, V., *Distributed Shared Memory: Concepts and Systems*, Wiley, New York, 1997.
9. Dunigan, T., Vetter, J. and Worley, P., Performance evaluation of the SGI Altix 3700, In *Proc. of the IEEE Int. Conf. Parallel Proc., ICPP*, 231–240, 2005.
10. Krawezik, G. and Cappello, F., Performance comparison of MPI and OpenMP on shared memory multiprocessors, *Concurr. Comput.: Practice and Experience*, **18** (2006) 29–61.
11. http://www.spec.org/
12. Mellor–Crummey, J. and Garvin, J., Optimizing sparse matrix–vector product computations using unroll and jam. Int. J. High Perfor. Comput. Applic., 18 (2004) 225–236.

# Numerical Simulation of Phase Transformations in Shape Memory Alloy Thin Films

Debiprosad Roy Mahapatra and Roderick V.N. Melnik

Mathematical Modelling and Computational Sciences,
Wilfrid Laurier University, Waterloo, ON, N2L3C5, Canada

**Abstract.** A unified variational framework and finite element simulations of phase transformation dynamics in shape memory alloy thin films are reported in this paper. The computational model is based on an approach which combines the lattice based kinetics involving the order variables and non-equilibrium thermodynamics. Algorithmic and computational issues are discussed. Numerical results on phase nucleation under mechanical loading are reported.

## 1 Introduction

Phase transforming solids, in particular metallic alloys with large differences in lattice constants of their crystollographic variants have interesting properties. The variants (phases) which exist at low temperature are called martensites and the parent state which exists at high temperature is called austenite. When temperature is increased, the martensites are transformed to austenite at certain critical temperature. Further, when mechanical force is applied at constant temperature, the structure undergoes a different transformation path from austenite to martensites and recovers the original shape. This is known as shape memory effect. The associated strain can be quite large ($6\% - 50\%$) depending upon the size of the sample. Shape memory effect has wide range of applications in mechanical, bio-medical and micro-device engineering. Due to thermomechanically coupled phase transformation which is not diffusive but of first-order type, and which produces microstructures with sharp interfaces with large rotations of lattice vectors, the experimental, analytical and computational characterization of the material properties and overall structural responses become highly challenging. There are three important mathematical and computational issues for reliable numerical simulation of shape-memory alloys, which are (1) an accurate description of the free energy density, its frame-invariance and material symmetry properties [1, 2], (2) prediction of the microstructures [2, 3] and their evolutions [4, 5, 6] and (3) prediction of the thermo-mechanical hysteresis at the macroscopic scale, which is important in the numerical simulation based design of shape-memory alloy devices [7]. Therefore, a challenging task is to address all of the above objectives systematically within a *unified* modeling, analysis and computational framework. In this paper we report a finite element based numerical simulation of phase transformations in shape memory alloy thin films. The computational model is based on a Ginzburg-Landau free energy description,

V.N. Alexandrov et al. (Eds.): ICCS 2006, Part II, LNCS 3992, pp. 114–121, 2006.

detailed analytical studies of which can be found in [8, 9, 10, 11]. Applications of the Landau theory to the description of first-order martensitic phase transitions can be found in [12]. Also, numerical simulations of microstructure due to cubic to tetragonal transformation in thin films based on continuum theory of lattice have been reported [3], where quadratic polynomial of the strain invariants have been employed to construct the free energy densities separately for the austenite and the martensites. In the above work, a free-boundary type variational formulation was employed for quasi-static analysis of microstructure.

In order to perform numerical simulations, we have developed a variational framework and a finite element code, wherein a general thermo-mechanical loading can be handled while studying the lattice-based kinetics of the microstructures. The computational model developed here employs a variational framework, where the thermodynamic conservation law couples the mechanical deformation, temperature and the order variables. The order variables describe the type of phase (parent austenite phase or a martensitic variant) at a material point. The paper is structured as follows. In Sec. 2, we summarize the Ginzburg-Landau free energy coupled model, which has been implemented in the variational framework. The variational framework and the finite element formulation are discussed in Sec. 3. Computational issues are discussed in Secs. 4 and 5. Numerical results for cubic to tetragonal phase transformations in Ni-Al thin films are reported in Sec 6.

## 2 Ginzburg-Landau Free Energy Model

We denote the order variables $\eta_k \in [0,1]$, where $k = 1, \cdots, N$ indicates the number of martensitic variants, $\eta_k = 0 \, \forall k$ defines the austenite and $\eta_k = 1\,, \eta_j = 0\,, k \neq j$ defines the $k$th martensitic variant at a material point. According to the point group of crystallographic symmetry, only one type of martensite is allowed to exist at a material point. Denoting the vector of the order variables as $\boldsymbol{\eta} = \{\eta_1, \cdots, \eta_N\}^T$, the Gibbs free energy density is defined as $G(\boldsymbol{\sigma}, \theta, \boldsymbol{\eta})$. Here $\boldsymbol{\sigma}$ is the stress tensor, $\theta$ is the temperature. The finite strain tensor $\varepsilon$ is decomposed into the elastic part and the transformation-induced part as

$$\varepsilon = \varepsilon_{\mathrm{el}} + \sum_{k=1}^{N} \varepsilon_k^t \varphi(\eta_k) \,, \tag{1}$$

where $\varphi(\eta_k)$ is a polynomial in $\eta_k$ and $\varepsilon_k^t\,, k = 1, \cdots, N$ are the transformation strain tensors [2] obtained using experiments. The structure of the initially unknown polynomial $\varphi(\eta_k)$ is such that it satisfies the following two conditions:

$$\varphi(0) = 0\,, \quad \varphi(1) = 1\,. \tag{2}$$

The frame-invariance property of the free energy density is imposed by the polynomial structure in $\eta_k$ such that interchanges between two indices produce identical structure of $G$. Material symmetry under proper rotation of the lattice vector is preserved due to the decomposition in Eq. (1). The Gibbs free energy density then takes the following form.

$$G(\eta) = -\frac{1}{2}\boldsymbol{\sigma} : \left[\boldsymbol{\lambda}_0 + \sum_{k=1}^{N}(\boldsymbol{\lambda}_k - \boldsymbol{\lambda}_0)\varphi(\eta_k)\right] : \boldsymbol{\sigma} - \boldsymbol{\sigma} : \sum_{k=1}^{N} \boldsymbol{\varepsilon}_k^t \varphi(\eta_k)$$

$$- \boldsymbol{\sigma} : \left[\boldsymbol{\varepsilon}_{\theta 0} + \sum_{k=1}^{N}(\boldsymbol{\varepsilon}_k^\theta - \boldsymbol{\varepsilon}_0^\theta)\varphi(\eta_k)\right] + \sum_{k=1}^{N} f(\theta, \eta_k) + \sum_{i=1}^{N-1}\sum_{j=i+1}^{N} F_{ij}(\eta_i, \eta_j) \, , \quad (3)$$

where $\boldsymbol{\lambda}_k$ is the second-order forth-rank compliance tensor for the $k$th martensitic variant ($M_k$ phase), $\boldsymbol{\lambda}_0$ is for austenite phase ($A$ phase), $\boldsymbol{\varepsilon}_0^\theta = \boldsymbol{\alpha}_0(\theta - \theta_e)$, $\boldsymbol{\varepsilon}_k^\theta = \boldsymbol{\alpha}_k(\theta - \theta_e)$. $\theta_e$ is the temperature at which the stress-free martensite losses stability. $\boldsymbol{\alpha}_0$ and $\boldsymbol{\alpha}_k$ are the thermal expansion tensors for $A$ and $M_k$ phases, respectively. $f(\theta, \eta_k)$ is the chemical part of the free energy of the $M_k$ phases and assumed in the form of a polynomial which is to be determined. $F_{ij}$ is an interaction potential required to preserve the frame-invariance of $G$ with respect to the point group of symmetry and uniqueness of the multivariant phase transformation at a given material point. The description of the order variables can now be generalized with three sets of order parameters: $\bar{0} = \{0, \eta_k = 0, 0\}$ for $A$ phase, $\bar{1} = \{0, \eta_k = 1, 0\}$ for $M_k$ phase and $\bar{\eta}_k = \{0, \eta_k, 0\}$, $\eta_k \in (0, 1)$ for diffused $A - M_k$ interface. The role of the first-order kinetics in the order variables is to assist in reaching the bottom of the energy well, i.e.,

$$\frac{\partial G}{\partial \eta_k} = 0 \, , \quad \boldsymbol{\eta} = \bar{0}, \bar{1} \, , \quad (4)$$

$$\frac{\partial^2 G}{\partial \eta_k^2} \leq 0 \, , \quad \boldsymbol{\eta} = \bar{0} \quad (A \rightarrow M_k), \quad (5)$$

$$\frac{\partial^2 G}{\partial \eta_k^2} \leq 0 \, , \quad \boldsymbol{\eta} = \bar{1} \quad (M_k \rightarrow A). \quad (6)$$

The transformation energy associated with $A \leftrightarrow M_k$ transformation is

$$G(\boldsymbol{\sigma}, \theta, \bar{0}) - G(\boldsymbol{\sigma}, \theta, \bar{1}) = \boldsymbol{\sigma} : \boldsymbol{\varepsilon}_k^t - \Delta G^\theta \, , \quad (7)$$

where $\Delta G^\theta$ is the jump in the free energy due to phase transformation. With the help of Eqs. (2)-(7), we determine $\varphi(\eta_k)$ and $f(\theta, \eta_k)$ (see [10] for the details). According to Landau theory, a quadratic polynomial in strain components can be adequate to describe the free energy. Therefore, following Eq. (1), one finds that for cubic to tetragonal transformation, the interaction potential has the following form (see [9] for the details)

$$F_{ij} = \eta_i \eta_j (1 - \eta_i - \eta_j) \left[B \left\{(\eta_i - \eta_j)^2 - \eta_i - \eta_j\right\} + D \eta_i \eta_j\right] + \eta_i^2 \eta_j^2 (\eta_i Z_{ij} + \eta_j Z_{ji}) \quad (8)$$

where the material constants $B$ and $D$ are obtained from experiments or numerical estimation. The matrix elements $Z_{ij}$ are obtained as functions of the constants $B$ and $D$ and an energy scale factor in $f(\theta, \eta_k)$.

## 2.1 Thermodynamic Conservation

Note that a jump in the free energy $\Delta G^{\theta}$ has been introduced in Eq. (7). The consequence of this jump, as well as the jump in the total strain across the $A-M_j$ interface, is the thermodynamic forcing as a source of dissipation. The forcing term would eventually be balanced by the kinetic force. Therefore, one has to establish a link between the evolution of the phases and the non-negativity of the thermodynamic potential (Helmholtz free energy). This is unlike the notion in plasticity-based framework (see e.g.[13]), where the non-negativity of the rate of phase fraction is directly enforced.

For the present problem, the kinetic equation is derived by balancing the thermodynamic force with the kinetic force as

$$C\frac{\partial \eta_k}{\partial t} + \frac{\partial G'}{\partial \eta_k} = 0 \,, \tag{9}$$

where $C$ is a constant and $G' = G + \tilde{G}(\nabla \boldsymbol{\eta})$ describes the modified Gibbs free energy including the gradient terms to account for the non-local nature of the interface energy. By rearranging Eq. (9) and expanding the forcing terms, we get the Ginzburg-Landau equation for phase kinetics, which is given by

$$\frac{\partial \eta_k}{\partial t} = -\sum_{p=1}^{N} L_{kp}\left[\frac{\partial G}{\partial \eta_p} + \boldsymbol{\beta}_p : \boldsymbol{\nabla\nabla}\eta_p\right] + \theta_k \,, \tag{10}$$

where $L_{kp}$ are positive definite kinetic coefficients, $\boldsymbol{\beta}_p$ are positive definite second rank tensor. $\theta_k$ is the thermal fluctuation satisfying the dissipation-fluctuation theorem. Eq. (10) is complemented by the macroscopic energy conservation law

$$\frac{\partial}{\partial t}\left[\mathcal{W} - \theta\frac{\partial \mathcal{W}}{\partial \theta}\right] - \boldsymbol{\nabla}\cdot(\boldsymbol{\sigma}\cdot\dot{\boldsymbol{u}} - \boldsymbol{q}) = h_\theta \,, \tag{11}$$

and the momentum balance equation

$$\rho\frac{\partial^2 \boldsymbol{u}}{\partial t^2} = \boldsymbol{\nabla}\cdot\boldsymbol{\sigma} + \boldsymbol{p} \,, \tag{12}$$

where $\mathcal{W}$ is the Helmholtz free energy given by

$$\mathcal{W} = G + c_v\theta + \frac{1}{2}\boldsymbol{\sigma} : \left[\boldsymbol{\lambda}_0 + \sum_{k=1}^{N}(\boldsymbol{\lambda}_k - \boldsymbol{\lambda}_0)\varphi(\eta_k)\right] : \boldsymbol{\sigma}$$
$$+\, \boldsymbol{\sigma} : \left[\boldsymbol{\varepsilon}_{\theta 0} + \sum_{k=1}^{N}(\boldsymbol{\varepsilon}_{\theta k} - \boldsymbol{\varepsilon}_{\theta 0})\varphi(\eta_k)\right] \,, \tag{13}$$

$\boldsymbol{q}$ is the heat flux, $h_\theta$ is the heat source and $\boldsymbol{p}$ is body force.

## 3 Variational Framework and Finite Element Discretization

We relate the elastic part of displacements to the elastic strain $\boldsymbol{\varepsilon}_{\text{el}}$ via the linear strain-displacement relation, i.e. $\boldsymbol{\varepsilon}_{\text{el}} = ((\nabla \boldsymbol{u}) + (\nabla \boldsymbol{u})^T)/2$. With this assumption of strain and Eq. (1), it is now obvious that the order variables $\eta_k, k = 1, \cdots, N$, are to be treated as internal variables in the variational formulation. We want to interpolate the fields $\boldsymbol{u}(x, y, z, t)$, $\theta(x, y, z, t)$ and $\eta_k(x, y, z, t)$ over the domain $\Omega(x, y, z) \subset R^2$ with Lipschitz continuous boundary $\partial\Omega$, using fixed-order finite elements with $h$-refinement. We consider the Lagrangian isoparametric interpolation function $\boldsymbol{N}$,

$$\{u_1,\ u_2,\ u_3\}^T = \boldsymbol{N}_u \boldsymbol{v}^e\ , \quad \theta = \boldsymbol{N}_\theta \boldsymbol{v}^e\ , \quad \eta = \boldsymbol{N}_\eta \boldsymbol{v}^e\ , \tag{14}$$

$$\boldsymbol{v} = \{u_1,\ u_2,\ u_3,\ \theta,\ \eta_1\ , \cdots\ , \eta_n\}^T\ . \tag{15}$$

Here, the superscript $e$ indicates element nodal quantities. Introducing admissible weights $\{\bar{u}_i, \bar{\theta}, \bar{\eta}_k\}$ chosen from the linear span of $\boldsymbol{v}^e$, the variational formulation of the problem can be stated as follows

$$\delta\Pi = \delta\Pi_{\text{PT}} + \delta\Pi_\theta + \delta\Pi_u + \delta W = 0\ , \quad t \in [0, +\infty] \tag{16}$$

where

$$\delta\Pi_{\text{PT}} = \int_\Omega \sum_{k=1}^{N} \sum_{p=1}^{N} \delta\bar{\eta}_k \left[ L_{kp} \left( \frac{\partial G}{\partial \eta_p} + \boldsymbol{\beta}_p : \boldsymbol{\nabla\nabla}\eta_p \right) \right] \mathrm{d}\mathbf{x}$$

$$\int_\Omega \sum_{k=1}^{N} \delta\bar{\eta}_k \left[ \frac{\partial \eta_k}{\partial t} - \theta_k \right] \mathrm{d}\mathbf{x} - \int_{\partial\Omega} \sum_{k=1}^{N} \sum_{p=1}^{N} \delta\bar{\eta}_k L_{kp} \frac{\partial G}{\partial \eta_k} \mathrm{d}s(\mathbf{x})\ , \tag{17}$$

$$\begin{aligned} \delta\Pi_\theta = & \int_\Omega \delta\bar{\theta} \left[ \frac{\partial}{\partial t} \left( \mathcal{W} - \theta \frac{\partial \mathcal{W}}{\partial \theta} \right) - \boldsymbol{\nabla} \cdot (\boldsymbol{\sigma} \cdot \frac{\partial \boldsymbol{u}}{\partial t}) \right] \mathrm{d}\mathbf{x} \\ & + \int_\Omega \delta\bar{\theta} \left[ \boldsymbol{\nabla} \cdot \left( -\kappa \boldsymbol{\nabla}\theta - \alpha' \kappa \boldsymbol{\nabla} \frac{\partial \theta}{\partial t} \right) \right] \mathrm{d}\mathbf{x} - \int_{\partial\Omega} \delta\bar{\theta} \boldsymbol{q}_\perp \mathrm{d}s(\mathbf{x})\ , \end{aligned} \tag{18}$$

$$\delta\Pi_u = \int_\Omega \delta\bar{\boldsymbol{u}}^T \left[ \rho \frac{\partial^2 \boldsymbol{u}}{\partial t^2} - \boldsymbol{\nabla} \cdot \boldsymbol{\sigma} \right] \mathrm{d}\mathbf{x} - \int_{\partial\Omega} \delta\bar{\boldsymbol{u}}^T \boldsymbol{\sigma}_\perp \mathrm{d}s(\mathbf{x})\ , \tag{19}$$

and $W$ is the external work done over the sample. Integrating Eq. (16) by parts, we obtain the finite element approximation

$$\boldsymbol{M} \frac{\partial^2 \boldsymbol{v}}{\partial t^2} + \boldsymbol{D} \frac{\partial \boldsymbol{v}}{\partial t} + \boldsymbol{K}\boldsymbol{v} = \boldsymbol{f}\ , \tag{20}$$

with initial state and microstructure

$$\boldsymbol{v}(\boldsymbol{t}) = \boldsymbol{v}(0)\ , \quad \frac{\partial}{\partial t}\boldsymbol{v}(t) = \mathbf{0}\ . \tag{21}$$

## 4 Deformation, Phase Kinetics and Multiple Scales

The problem of simulating the microstructure in shape memory alloys has been extensively discussed in the literature [1], primarily under static loadings. Dynamic loadings present a major remaining challenge in the field. While simulating the macroscopic deformation under dynamic loadings, resolving the sharp interfaces and the dendritic microstructure with multiscale features are some of the key computational difficulties that need to be addressed. A smoothed version of the transformation conditions in Eqs. (4)-(6) leads to the time-dependent Ginzburg-Landau phase kinetics, where the fast time scale has to be controlled depending on the global time stepping (in slow time scale) in the finite element time integration. Note that $\eta_k$ represents the reordering of the atomic ensemble within a box (or in other words within a finite element). Choice of the length-scale for defining this box can be arrived at by applying a finite-difference scheme to the phase kinetic equation. The chosen time scale for the kinetics dictates the limit to the coarse graining of $\eta_k$. This, in turn, restricts the order of interpolation for a given size of the finite element. Since the elastic part of the strain is defined in the lattice coordinate for the current phase, special care is necessary to ensure continuity of elastic strain and hence displacement within that phase. For example, if linear interpolation is used for displacement and there exists a number of elements having same phase connected to a finite element node, then we average out the displacement for that particular node. For higher-order interpolation, this is not necessary, except when additional intermediate nodes are used in an element for interpolation of $\eta_k$.

## 5 Computational Scheme

We have implemented the finite element model discussed in Sec. 3 in a general three-dimensional finite element code. For numerical simulation of cubic to tetragonal transformation, we deal with three different phases, that is $N = 3$. We employ an 8-node, 7 d.o.f/node hexahedral element with tri-linear isoparametric interpolation and reduced Gauss-quadrature integration for in-plane shear terms. Since the energy minimization process can take a different and unphysical path, a special care should be taken in organizing nonlinear iterations. The implemented algorithm is given below.

(1) Form the matrix $\boldsymbol{M}$ and the linear parts of the matrices $\boldsymbol{D}$ and $\boldsymbol{K}$.
(2) Time step: $t_i = t_0$; Form the nonlinear system. Solve: $\boldsymbol{K}(\boldsymbol{v}_i)\boldsymbol{v}_{i+1} = \bar{\boldsymbol{f}}(\boldsymbol{v}_i)$.
(3) Iteration step : $j = 1$; From (2), $\boldsymbol{v}^j_{i+1} = \boldsymbol{v}_{i+1}$. Form $\Delta\bar{\boldsymbol{f}}^j_{i+1} = \bar{\boldsymbol{f}}(\boldsymbol{v}^j_{i+1})_{\text{int}} - \bar{\boldsymbol{f}}(\boldsymbol{v}_i)$. Solve [14]: $\Delta\boldsymbol{v}^{j+1}_{i+1} = -\boldsymbol{K}(\boldsymbol{v}^j_{i+1})^{-1}\Delta\bar{f}^j_{i+1}$.
(4) Update: $\boldsymbol{v}^{j+1}_{i+1} = \boldsymbol{v}^j_{i+1} + \Delta\boldsymbol{v}^{j+1}_{i+1}$. Check stability conditions Eqs. (5)-(6). If not consistent, update $\eta_k$ only and repeat (3).
(5) Update: $\boldsymbol{\varepsilon}^{j+1}_{i+1}$, $\boldsymbol{\sigma}^{j+1}_{i+1}$. Check convergence. If not converged $j \leftarrow j+1$, repeat (3) else go to (6).
(6) Update: $\dot{\boldsymbol{v}}_{i+1} = \dot{\boldsymbol{v}}^{j+1}_{i+1}$, $\ddot{\boldsymbol{v}}_{i+1} = \ddot{\boldsymbol{v}}^{j+1}_{i+1}$. $i \leftarrow i+1$, repeat (2).

## 6 Numerical Simulations

A Ni-Al thin film is considered for numerical analysis in this section as an example. The material properties are taken from [9]. $5Hz$ sinusoidal stress is applied in the longitudinal direction throughout the left and right edges of the rectangular film. The stress distribution over the edge has equilateral triangular shape. The other two parallel edges of the film are kept free. Constant temperature of $300K$ with cubic phase (austenite) is assumed to be the initial state. Figures 1 and 2, respectively, show the displacements and the nucleated phases near the stressed left and right edges of the film.

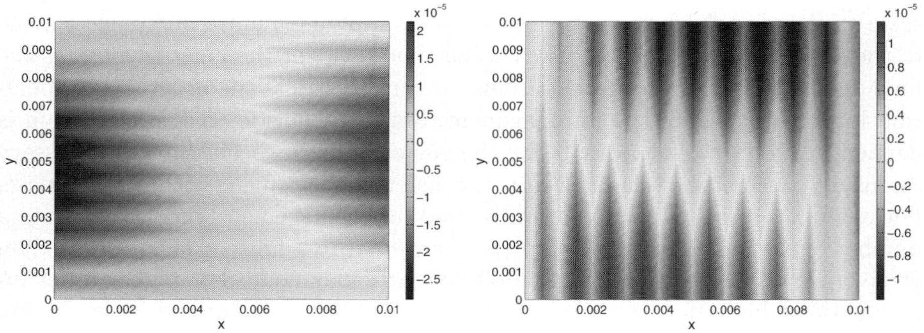


(a) (b)

**Fig. 1.** (a) Longitudinal displacement $u_1$ and (b) lateral displacement $u_2$ during nucleation of the martensites from the right edge at the first one-fourth of the loading cycle ($t = 25ms$)

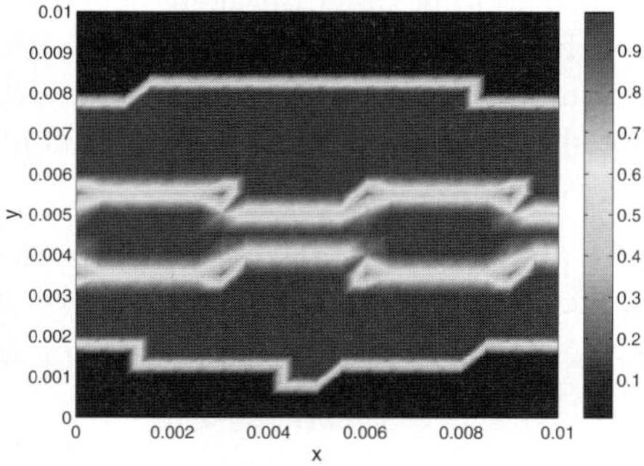


**Fig. 2.** Martensitic variant $\eta_1 \rightarrow 1$ (seen as the horizontal strips in the middle) during nucleation from the left and right edges at the first one-fourth of the tensile loading cycle ($t = 25ms$)

## 7 Conclusions

A computational model has been developed and the simulations based on this model capture a qualitative behaviour of the Ni-Al microstructure observed experimentally (see [15]).

## References

1. J.M. Ball and C. Carstensen, Compatibility conditions for microstructures and the austenite-martensite transition, *Mater. Sci. and Eng. A*, 273, 231-236 (1999).
2. K. Bhattacharya, *Microstructure of Martensite*, Oxford University Press (2003).
3. P. Belik and M. Luskin, Computational modeling of softening in a structural phase transformation, *Multiscale Model. Simul.*, 3(4), 764-781, (2005).
4. R. Abeyaratne, C. Chu and R.D. James, Kinetics of materials with wiggly energies: The evolution of twinning microstructure in a Cu-Al-Ni shape memory alloys, *Phil. Mag.*, 73A, 457-496 (1996).
5. A. Artemev, Y. Wang, A.G. Khachaturyan, Three-dimensional phase field model and simulation of martensitic transformation in multilayer systems under applied stresses, *Acta Mater.*, 48, 2503-2518 (2000).
6. T. Ichitsubo, K. Tanaka, M. Koiwa and Y. Yamazaki, Kinetics of cubic to tetragonal transformation under external field by the time-dependent Ginzburg-Landau approach, *Phy. Rev. B*, 62, 5435 (2000).
7. F. Auricchio and L. Petrini, A three-dimensional model describing stress-temperature induced solid phase transformations: solution algorithm and boundary value problems, *Int. J. Numer. Meth. Engng.*, 61, 807-836, (2004).
8. V.I. Levitas and D.L. Preston, Three-dimensional Landau theory for multivariant stress-induced martensitic phase transformations. I. Austenite $\leftrightarrow$ martensite, *Phys. Rev. B*, 66, 134206 (2002).
9. V.I. Levitas, D.L. Preston and D.W. Lee, Three-dimensional Landau theory for multivariant stress-induced martensitic phase transformations. III. Alternative potentials, critical nuclei, kink solutions, and dislocation theory, *Phys. Rev. B*, 68, 134201 (2003).
10. D.R. Mahapatra and R.V.N. Melnik, A dynamic model for phase transformations in 3D samples of shape memory alloys, *LNCS* Springer-Verlag, 3516, 25-32 (2005).
11. D. Roy Mahapatra and R.V.N. Melnik, Finite element approach to modelling evolution of 3D shape memory materials, *Math. Computers Simul.* (submitted) Sep 2005.
12. F. Falk and P. Kanopka, Three-dimensional Landau theory describing the martensitic phase transformation of shape-memory alloys, *J. Phys.: Condens. Matter*, 2, 61-77 (1990).
13. J.G. Boyd and D.C. Lagoudas, A thermodynamical constitutive constitutive model for shape memory materials. Part I. the monolithic shape memory alloy, *Int. J. Plasticity*, 12(6), 805-842, 1996.
14. J.C. Simo and T.J.R. Hughes, Computational Inelasticity, Springer-Verlag, 1997.
15. P. Boullay, D. Schryvers and J.M. Ball, Nanostructures at martensite macrotwin interfaces in $Ni_{65}Al_{35}$, *Acta Mater.*, 51, 1421-1436, 2003.

# A Virtual Test Facility for Simulating Detonation-Induced Fracture of Thin Flexible Shells

Ralf Deiterding, Fehmi Cirak, Sean P. Mauch, and Daniel I. Meiron

California Institute of Technology, MC 158-79, Pasadena CA 91125, USA
{ralf, cirak, sean, dim}@cacr.caltech.edu
http://www.cacr.caltech.edu/asc

**Abstract.** The fluid-structure interaction simulation of detonation- and shock-wave-loaded fracturing thin-walled structures requires numerical methods that can cope with large deformations as well as topology changes. We present a robust level-set-based approach that integrates a Lagrangian thin shell finite element solver with fracture and fragmentation capabilities with an Eulerian Cartesian detonation solver with optional dynamic mesh adaptation. As an application example, the rupture of a thin aluminum tube due to the passage of an ethylene-oxygen detonation wave is presented.

## 1 Introduction

The Center for Simulation of Dynamic Response of Materials at the California Institute of Technology has developed a virtual test facility (VTF) for studying the three-dimensional dynamic response of solid materials subjected to strong shock and detonation waves propagating in fluids [7]. The fluid flow is simulated with a Cartesian finite volume upwind method in Eulerian coordinates that considers the solid as a moving embedded body by utilizing a ghost-fluid approach (see Sec. 2). In the present paper, the fracturing thin-walled solid is modeled as a Kirchhoff-Love type thin-shell in Lagrangian coordinates. The finite element discretization of the underlying energy functional is accomplished with subdivision finite elements (see Sec. 3).

The evolving shell geometry is implicitly represented on the Cartesian fluid mesh by a scalar level set function, which stores the unsigned distance to the solid surface. As the solid deforms, the level set function is updated on-the-fly by a highly efficient auxiliary algorithm based on geometric characteristic reconstruction and scan conversion (see Sec. 4). In the VTF, we apply a loosely coupled temporal splitting method for updating the boundary's position and velocity between time steps. The fluid-structure coupling approach and its implementation on distributed memory computing platforms are briefly described in Sec. 5. As a large-scale three-dimensional computational example, we present in Sec. 6 the simulation of the rupture of a thin aluminum tube due to an internal detonation wave arising from the combustion of an ethylene-oxygen mixture. Simulations of

V.N. Alexandrov et al. (Eds.): ICCS 2006, Part II, LNCS 3992, pp. 122–130, 2006.

this type will be compared directly to available experimental results [1] and will serve as a validation case for the VTF.

## 2 Eulerian Detonation Solver

The governing equations of detonation wave propagation in gases are the inviscid Euler equations [9]. Throughout this paper, we consider only the simplified case of a single exothermic chemical reaction $A \longrightarrow B$ with a progress variable $Y$ corresponding to the mass fraction ratio between the partial density of the reactant $A$ and the total density $\rho$, i.e. $Y = \rho_A/\rho$. The governing equations of the hydrodynamic model are

$$\begin{aligned} \partial_t \rho + \nabla \cdot (\rho \boldsymbol{u}) &= 0 \,, & \partial_t(\rho \boldsymbol{u}) + \nabla \cdot (\rho \boldsymbol{u} \otimes \boldsymbol{u}) + \nabla p &= 0 \,, \\ \partial_t(\rho E) + \nabla \cdot ((\rho E + p)\boldsymbol{u}) &= 0 \,, & \partial_t(Y\rho) + \nabla \cdot (Y \rho \boldsymbol{u}) &= \psi \,. \end{aligned} \tag{1}$$

Herein, $\boldsymbol{u}$ is the velocity vector and $E$ the specific total energy. The hydrostatic pressure $p$ is given by $p = (\gamma-1)(\rho E - \frac{1}{2}\rho \boldsymbol{u}^T\boldsymbol{u} - \rho Y q)$ with $\gamma$ denoting the ratio of specific heats and $q$ the heat release due to the chemical reaction per unit mass. A one-step reaction would typically be modeled with an Arrhenius law such as $\psi = -kY\rho \exp(-E_A\rho/p)$ [9], but in the specific case considered here, we utilize the constant volume burn model suggested by Mader [10]. This model neglects the detailed chemical depletion, and therefore the internal detonation structure, but ensures the right propagation speed and the correct state in chemical equilibrium at all grid resolutions. The model is intended to be applied together with the fractional step method that numerically decouples chemical reaction and hydrodynamic transport. First, the *homogeneous* system (1) is advanced at a full time step, then the reactant density $\rho_A$, pressure $p$, and total energy $E$ are modified locally in each cell; the total density $\rho$ and the velocity vector $\boldsymbol{u}$ remain unaltered. The algorithm for the detonation model reads:

$V := \rho^{-1}$, $V_0 := \rho_0^{-1}$, $V_{CJ} := \rho_{CJ}$, $Y' := 1 - (V - V_0)/(V_{CJ} - V_0)$
if $0 \le Y' \le 1$ and $Y > 10^{-8}$
  if $Y < Y'$ and $Y' < 0.9$ then $Y' := 0$
  if $Y' < 0.99$ then $p' := (1 - Y')p_{CJ}$ else $p' := p$
  $\rho_A := Y'\rho$, $E := p'/(\rho(\gamma - 1)) + Y'q_0 + \frac{1}{2}\boldsymbol{u}^T\boldsymbol{u}$

In the latter, the index 0 indicates the unreacted state (assumed to be constant), while CJ refers to the equilibrium values that can be calculated in advance following Chapman-Jouguet theory [9, 6] for a given detonation velocity.

For the shock-capturing finite volume upwind scheme, we utilize a straightforward extension of the flux-vector splitting method by Van Leer (cf. [6]). Second-order accuracy in smooth solution regions is achieved with the MUSCL-Hancock variable extrapolation technique [12]. Geometrically complex moving boundaries are incorporated in the upwind scheme by using some of the finite volume cells as ghost cells for enforcing immersed moving wall boundary conditions [8]. The boundary geometry is mapped onto the Cartesian mesh by employing a scalar level set function $\phi$ that stores the unsigned distance to the boundary surface and allows the efficient evaluation of the boundary outer normal in every mesh point as $\boldsymbol{n} = -\nabla\phi/|\nabla\phi|$. Note that for topologically closed boundary surfaces the signed distance may be used instead of the unsigned distance [5]. A cell is considered to be a valid fluid cell, if the distance at the cell *midpoint* satisfies the condition $\phi > h/2$ and as an exterior ghost cell otherwise. The mesh received from the shell solver corresponds to a two-dimensional manifold surface mesh (cf. Sec. 3) and the utilization of condition $\phi > h/2$ is a straightforward, unambiguous solution to achieve the mandatory thickening of this mesh by the shell thickness $h$. The contour line $\phi = h/2$ effectively represents the embedded boundary for the fluid solver (depicted as dotted line around shell elements in Fig. 1). The hydrodynamic load on each shell element is evaluated as the difference between the approximated pressure values at $\phi = h/2$ in the positive and negative direction of each element's normal, i.e. $p^F := p^+ - p^-$.

For the governing equations (1), the boundary condition at a rigid wall moving with velocity $\boldsymbol{w}$ is $\boldsymbol{u} \cdot \boldsymbol{n} = \boldsymbol{w} \cdot \boldsymbol{n}$. Enforcing the latter with ghost cells, in which the discrete values are located at the cell centers, requires the mirroring of the primitive values $\rho$, $\boldsymbol{u}$, $p$, $\rho_{\mathrm{A}}$ across the embedded boundary. The normal velocity in the ghost cells is set to $(2\boldsymbol{w} \cdot \boldsymbol{n} - \boldsymbol{u} \cdot \boldsymbol{n})\boldsymbol{n}$, while the mirrored tangential velocity remains unmodified. Mirrored values are constructed by calculating spatially interpolated values in the point $\tilde{\boldsymbol{x}} = \boldsymbol{x} + 2\phi\boldsymbol{n}$ from neighboring interior cells. We employ a dimension-wise linear interpolation for this operation, but it has to be emphasized that directly near the boundary the number of interpolants needs to be decreased to ensure the monotonicity of the numerical solution. This property is essential in simulating hyperbolic problems with discontinuities, like detonation waves. Figure 1 also highlights the necessary reduction of the interpolation stencil for some exemplary cases. The interpolation locations are indicated by the origins of the arrows normal to the contour line that defines the embedded boundary. After the application of the numerical scheme, cells that have been used to impose internal boundary conditions are set to the entire state vector of the nearest cell in the fluid interior. This operation ensures proper values in case such a cell becomes a regular interior cell in the next step due to boundary movement. The consideration of $\boldsymbol{w}$ in the ghost cells guarantees that the embedded boundary propagates at most one cell in every time step.

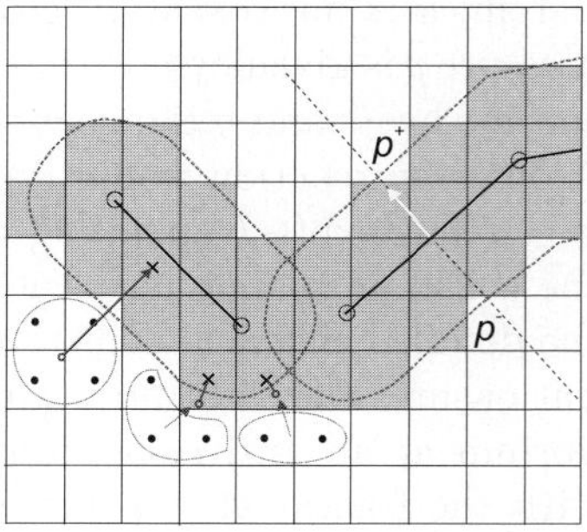


**Fig. 1.** Ghost cells (shaded gray) around shell elements and construction of mirrored values

In order to supply a fine local temporal and spatial resolution efficiently, the finite volume scheme described above has been incorporated into a block-oriented hierarchical mesh adaptation method. The numerical scheme is then applied as a single-grid routine in a loop over all subgrids (see [6] or [7] for details).

# 3 Lagrangian Thin-Shell Solver

The Kirchhoff-Love type thin-shell model applied in this study takes the membrane as well as bending response of the surface into account, and has been discretized with subdivision finite elements [2, 3]. The underlying kinematic assumptions allow for arbitrarily large displacements as well as rotations of the shell. Fracture initiation and propagation is considered as a progressive failure phenomenon in which the separation of the crack flanks is resisted by cohesive tractions. The relationship between the crack-opening displacements and the tractions is given by a cohesive model. Cohesive interface elements are inserted at inter-element edges and constrain the opening of the crack flanks to the deformation of the shell middle surface and its normal. This approach allows for fracture in an in-plane or tearing mode, a shearing mode, or a "bending of hinge" mode.

To kinematically describe a fractured thin-shell as sketched in Fig. 2, we consider a shell of uniform thickness $\overline{h}$ occupying an undeformed configuration $\overline{V}$. The position vector $\overline{\boldsymbol{\varphi}}$ of a material point on the undeformed shell body is assumed to be

$$\overline{\boldsymbol{\varphi}} = \overline{\boldsymbol{x}} + \theta^3 \overline{\boldsymbol{n}} \tag{2}$$

with $-h/2 \leq \theta^3 \leq h/2$. The position vector of the shell middle surface is denoted by $\overline{\boldsymbol{x}}$ and its out-of-surface unit normal by $\overline{\boldsymbol{n}}$. In other words, the shell middle surface represents a two-dimensional manifold in $\mathbb{R}^3$. The deformation mapping $\boldsymbol{\varphi}$ maps the shell body into the deformed configuration $V$ and is discontinuous across the crack

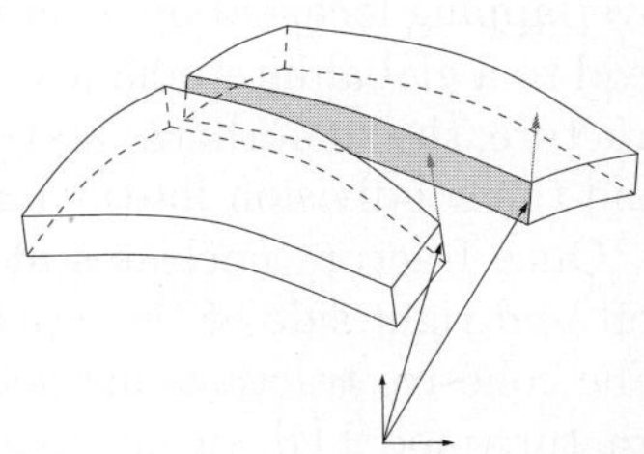

**Fig. 2.** Fractured shell body: Opposite crack flanks and corresponding normals.

$$[\![\boldsymbol{\varphi}]\!] = \boldsymbol{\varphi}^+ - \boldsymbol{\varphi}^- = [\![\boldsymbol{x}]\!] + \theta^3 [\![\boldsymbol{n}]\!] \tag{3}$$

where the superscripts + and − refer to the opposing crack flanks. Further, the first term describes the discontinuity of the deformation of the middle shell surface, and the second term the discontinuity in the shell out-of-surface normal. The discontinuities in the deformations can also be interpreted as the *opening displacement* of the crack.

A standard semi-inverse approach is followed for obtaining the shell equilibrium equations in weak from. To this end, the assumed reduced kinematic equations for the shell body (Equations (2) and (3)) are introduced into the conventional virtual work expression for the three-dimensional body. As previously mentioned, we consider fracture as a gradual separation phenomenon, resisted by cohesive tractions. Consequently, the internal virtual work expression contains the virtual work of the cohesive interface ($\delta\Pi_{C,\text{int}}$) in addition to the virtual work of the bulk material ($\delta\Pi_{S,\text{int}}$)

$$\delta\Pi_{S,\text{int}} + \delta\Pi_{C,\text{int}} - \delta\Pi_{\text{ext}} = 0 \tag{4}$$

with

$$\delta \Pi_{S,\text{int}} = \int_{\overline{\Omega}} \int_{-\overline{h}/2}^{\overline{h}/2} \boldsymbol{P} : \delta \boldsymbol{F} \, \mu \, d\theta^3 d\overline{\Omega} \; , \quad \delta \Pi_{C,\text{int}} = \int_{\overline{\Gamma}_C} \int_{-\overline{h}/2}^{\overline{h}/2} \boldsymbol{T} \cdot [\![\boldsymbol{\varphi}]\!] \, \mu \, d\theta^3 d\overline{\Gamma}_C$$

where $\boldsymbol{P}$ is the first Piola-Kirchhoff stress tensor, $\boldsymbol{T}$ the related traction vector at the cohesive surface, and $\boldsymbol{F}$ the deformation gradient. The virtual work expression for the bulk material is integrated over the undeformed shell middle surface $\overline{\Omega}$ and for the cohesive interface over the crack path $\overline{\Gamma}_C$. The scalar factor $\mu$ accounts for the curvature of the shell in the volume computation [3].

Next, we briefly outline the discretization of the governing equation (4). A detailed presentation of the used subdivision finite element discretization technique can be found in [2] and [3]. In this approach, the reference ($\overline{\boldsymbol{x}}$) and deformed ($\boldsymbol{x}$) shell surfaces are approximated using smooth subdivision surfaces belonging to the Sobolev space $H_2$ with square-integrable curvatures. The subdivision interpolation within one element is accomplished with shape functions, which have support on the element as well as on the one-ring of neighboring elements. The overlapping local subdivision interpolants, each defined over one patch, together lead to a global interpolant with square-integrable curvatures. In the presence of fracture, the smoothness and/or continuity of the interpolation has to be relaxed and the subdivision interpolant needs to be modified (see [4] for details).

Once fracture nucleates along an element edge, the element patches on the left and right side of the cracked edge interact only through cohesive tractions. The cohesive tractions are self-balanced internal forces derived from a cohesive fracture model [4]. In this model, the opening displacement $[\![\boldsymbol{\varphi}]\!]$ plays the role of a deformation measure while the traction $\boldsymbol{T}$ is the conjugate stress measure.

Finally, the inelastic behavior of the bulk material, i.e. the relation between $\boldsymbol{P}$ and $\boldsymbol{F}$, is described with a conventional $J_2$ plasticity model with isotropic power-law hardening. The rate-dependent behavior is described with a power viscosity law and constant rate sensitivity.

## 4 Efficient Level Set Evaluation

In Sec. 2, we have sketched the concept of employing a distance function to represent a complex embedded boundary on a Cartesian mesh. While distance functions are easily prescribed for single elementary geometric objects, their evaluation can be cumbersome for complex shapes. In coupled Eulerian-Lagrangian simulations, this complex shape is defined by the deforming shell surface mesh.

One can efficiently compute the distance on a grid by solving the eikonal equation with the method of characteristics and utilizing polyhedron scan conversion [11]. For a given grid point, the relevant closest point on the triangular mesh lies on one of the primitives (faces, edges and vertices) that comprise the surface. The characteristics emanating from each of these primitives form polyhedral shapes. Such a *characteristic polyhedron* contains all of the points which are possibly closest to its corresponding face, edge or vertex. The closest points to a triangle face must lie within a triangular prism defined by the face and

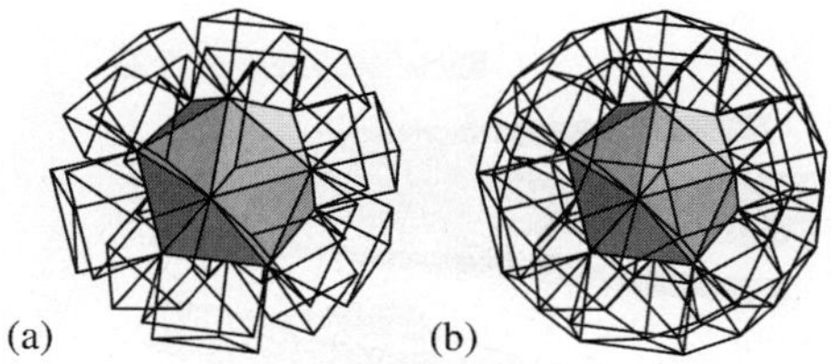


**Fig. 3.** The characteristic polyhedra for faces and edges of an icosahedron

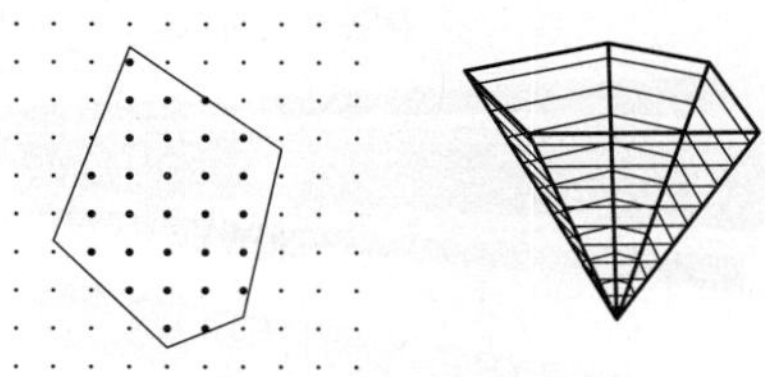

**Fig. 4.** Scan conversion of a polygon in 2-D and slicing of a polyhedron to form polygons

its normal; the closest points to an edge lie in a cylindrical wedge defined by the line segment and the normals to the two incident faces (see Fig. 3 for face (a) and edge (b) polyhedra as an example). Analogously, polygonal pyramids emanating from the vertices are also possible (not shown). We then determine the grid points that lie inside a characteristic polyhedron with polyhedron scan conversion. The polyhedron is first sliced along each sheet of the grid lattice to produce polygons, cf. Fig. 4. Simple geometric formulas are finally used to calculate the distance.

By utilizing the outlined techniques, and evaluating the distance exactly only within a small distance around the surface, a highly efficient algorithm can be formulated that has linear computational complexity both in the number of Cartesian mesh points and the surface triangles [11, 7].

## 5 Fluid-Structure Coupling

The explicit fluid and solid solvers are weakly coupled by successively applying appropriate boundary conditions in a time-operator splitting technique. In the case of inviscid flows, the compatibility conditions are simply the continuity of the velocity component normal to the embedded boundary $u_n$ in solid (S) and fluid (F), i.e. $u_n^S = u_n^F$, and the continuity of the component normal of the solid's Cauchy traction vector, $p^S = (\boldsymbol{\sigma n})\boldsymbol{n}$ with $\boldsymbol{\sigma} = 1/det(\boldsymbol{F})\boldsymbol{FP}$ , and the hydrodynamic pressure $p^F$, i.e $p^S = p^F$. We use the update algorithm of Fig. 5 to implement these conditions numerically. After evaluating the distance function $\phi$ for the currently available shell surface mesh, the embedded wall boundary velocities for the fluid solver are set to the solid velocities in the nearest shell element midplane. The same velocity $\boldsymbol{w}$ is enforced in the fluid on upper ($^+$) and lower ($^-$) side of each element. After setting embedded rigid wall boundary conditions as sketched in Sec. 2 and the fluid update, a new hydrodynamic pressure load $p^F := p^+ - p^-$ on each shell element is derived (compare Fig. 1). With these new boundary conditions, the cycle is completed by advancing the solid by $\Delta t$,

update $\phi(t)$
$\boldsymbol{w}_F^{+/-} := \boldsymbol{u}^S(t)$
update_fluid( $\Delta t$ )
$p^S := p^F(t + \Delta t)$
update_solid( $\Delta t$ )
$t := t + \Delta t$

**Fig. 5.** Basic coupling algorithm

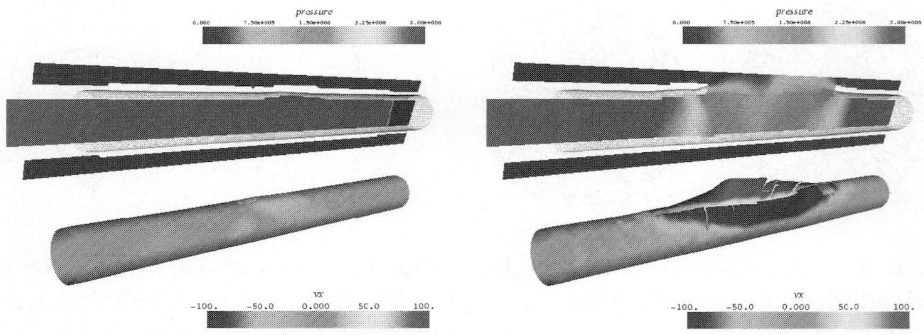


**Fig. 6.** Coupled simulation of detonation-driven rupture of a thin aluminum tube. Two snapshots of the fracturing tube with velocity iso-contours (lower row); cuts through the fluid domain show the resulting hydrodynamic venting (upper row).

which in practice is typically done by taking multiple, smaller time steps in the solid solver to effectively accommodate the more restrictive stability condition in the solid. An extended version of above coupling scheme compatible with the optional fluid mesh adaptation method with recursive time step refinement has also been developed [7].

In our current implementation, fluid and shell solver are parallelized separately for distributed memory machines using rigorous domain decomposition methods. The two independent solvers run on two disjoint set of processors. To facilitate the efficient exchange of the distributed fluid-shell boundary information (surface mesh and velocities to the fluid; pressure loadings to the solid) we have implemented a non-blocking high-level communication library that determines the necessary point-to-point communication patterns by intersecting Cartesian bounding boxes enclosing the local domains.

## 6 Detonation-Driven Fracture Example

We present one representative simulation that corresponds to an experimental configuration studied by Chao [1]. The setup consists of a detonation tube of 1.52 m, filled with $C_2H_4 + 3\,O_2$ at $p_0 = 180\,\mathrm{kPa}$ and $\rho_0 = 2.33\,\mathrm{kg/m^3}$, to which a thin-walled aluminum (Al6061-T6) test tube is attached. The test specimen has a length of 45.7 cm, an inner radius of 1.975 cm, and a wall thickness of 0.89 mm. While the lower end of the entire device is closed, a thin diaphragm seals the upper end. To ensure a reproducible fracture pattern Chao's specimen has a central longitudinal notch of 6.32 cm at the middle, which is modeled as an initial crack in the computations.

The mixture is thermally ignited at the closed end. A combustion wave arises that has the characteristics of a freely propagating self-sustained detonation when it enters the test specimen. Utilizing the full GRI 3.0 reaction mechanism Chapman-Jouguet theory predicts its velocity to be $D_{\mathrm{CJ}} = 2404\,\mathrm{m/s}$ and an adiabatic mixture coefficient for the fully reacted state of 1.24 [1]. For the simplified detonation model of Sec. 2, we choose $\gamma = 1.24$ and evaluate $q_0 \approx 5.20\,\mathrm{MJ/kg}$, $p_{\mathrm{CJ}} \approx 6.10\,\mathrm{MPa}$, and $\rho_{\mathrm{CJ}} \approx 4.16\,\mathrm{kg/m^3}$. A one-dimensional simulation is used to calculate the flow field at the moment the detonation enters the specimen. The one-dimensional data is then taken as initial conditions for the three-dimensional reactive Euler equations. Fig. 7 shows the detonation propagation in the one-dimensional case and the development of a Taylor rarefaction wave due to the closed end is clearly visible. To model this wave correctly, the Cartesian fluid domain also encompasses 0.92 m of the downstream end of the detonation tube as a static embedded boundary. Fig. 6 shows computational results of a typical coupled fluid-structure interaction simulation with a shell mesh of 8665 elements and a uniform Cartesian fluid mesh of $40 \times 40 \times 725$ cells that required $\approx 900$ h CPU on 27 nodes of a Pentium-4-2.4 GHz dual processor system (21 fluid and 33 solid processes). 1300 coupled time steps with fixed step size to a final time of 0.26 ms have been calculated (20 solid solver sub-steps in each fluid time step). The left graphic of Fig. 6 shows the beginning of the crack opening at $t \approx 0.15$ ms (detonation propagating from left to right). The snapshot on the right shows the rupture at the final time $t = 0.26$ ms. The venting of high pressure reacted gas out of the tube is clearly visible. During this simulation, level set evaluation and finite volume scheme (excluding parallel synchronization operations) have about the same computational costs on each fluid processor.

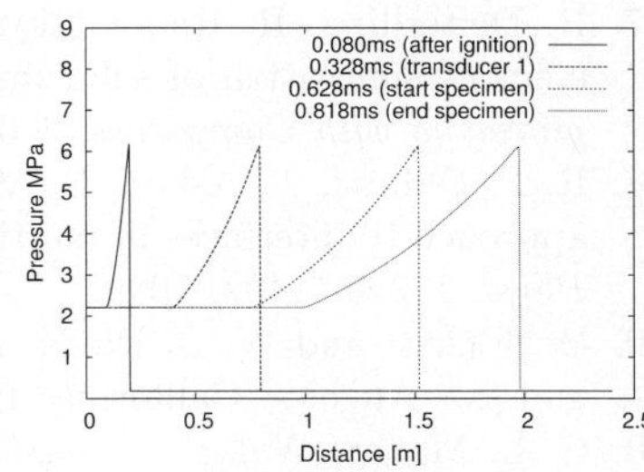


**Fig. 7.** 1d detonation propagation through tube and specimen

## 7 Conclusions

A parallel level-set-based fluid-structure coupling method for the time-accurate simulation of thin flexible shells dynamically responding to gaseous detonation waves has been described. The approach has been demonstrated to handle arbitrary topology changes and large deformations and is computationally very efficient. Its implementation in the software framework "Virtual Test Facility" is freely available for research purposes (cf. http://www.cacr.caltech.edu/asc).

## References

1. T.-W. Chao. *Gaseous detonation-driven fracture of tubes.* PhD thesis, California Institute of Technology, 2004.
2. F. Cirak, M. Ortiz, P. Schröder. Subdivision surfaces: a new paradigm for thin-shell finite-element analysis. *Int. J. Numer. Meth. Engineering*, 47:2039–2072, 2000.
3. F. Cirak and M. Ortiz. Fully $C^1$-conforming subdivision elements for finite deformation thin-shell analysis *Int. J. Numer. Meth. Engineering*, 51:813–833, 2001.

4. F. Cirak, M. Ortiz, and A. Pandolfi. A Cohesive Approach to Thin-Shell Fracture and Fragmentation. *Computer Methods in Appl. Mechanics and Engineering*, 194:2604–2618, 2005.
5. F. Cirak and R. Radovitzky. A Lagrangian-Eulerian Shell-Fluid Coupling Algorithm Based on Level Sets. *Computers & Structures*, 83:491–498, 2005.
6. R. Deiterding. *Parallel adaptive simulation of multi-dimensional detonation structures.* PhD thesis, Brandenburgische Technische Universität Cottbus, Sep 2003. Available at http://www.cacr.caltech.edu/~ralf/html/pub.htm.
7. R. Deiterding, R. Radovitzky, S. P. Mauch et al. A virtual test facility for the efficient simulation of solid materials under high energy shock-wave loading. *Engineering with Computers*, 2005. Invited submission.
8. R. P. Fedkiw, T. Aslam, B. Merriman, and S. Osher. A non-oscillatory Eulerian approach to interfaces in multimaterial flows (the ghost fluid method). *J. Comput. Phys.*, 152:457–492, 1999.
9. W. Fickett and W. C. Davis. *Detonation.* University of California Press, Berkeley and Los Angeles, California, 1979.
10. C. L. Mader. *Numerical modeling of detonations.* University of California Press, Berkeley and Los Angeles, California, 1979.
11. S. P. Mauch. *Efficient Algorithms for Solving Static Hamilton-Jacobi Equations.* PhD thesis, California Institute of Technology, 2003.
12. E. F. Toro. *Riemann solvers and numerical methods for fluid dynamics.* Springer-Verlag, Berlin, Heidelberg, 2nd edition, 1999.

# Data-Driven Inverse Modelling of Ionic Polymer Conductive Composite Plates

John G. Michopoulos[1] and Moshen Shahinpoor[2]

[1] Special Projects Group, Code 6390.2, Center for Computational Material Science, Naval Research Laboratory, USA
john.michopoulos@nrl.navy.mil

[2] Artificial Muscle Research Institute School of Engineering and School of Medicine University of New Mexico Albuquerque, NM 87131, USA
shah@unm.edu

**Abstract.** Analytical solutions of the partial differential equations (PDEs) governing the behavior of ionic polymer plates have not been yet obtained and therefore only time consuming discrete numerical methods can be used instead. To avoid the computational cost of numerical solutions this paper introduces a solution construction method that exploits analytical approximation basis functions borrowed from solutions of single physics formulations associated with rectangular ionic polymer plates for artificial muscle applications. This is achieved by utilizing an inverse approach that exploits global optimization. An objective function is constructed to express the error between the experimental and analytical values of the selected state variables. Minimization of this objective function yields an efficient determination of the unknown free coefficients. Comparisons between the determined approximations and the experimental data along with computational efficiency improvements conclude this paper.

## 1 Introduction

Recent progress on processing and development of various multi-field activated materials such as electro-active polymers (EAP) and ionic polymer conductive composites (IPCC) for artificial muscle applications [1, 2, 3] has underlined two main needs. First, the general need for a rigorous, complete and preferably automated modelling methodology that will eventually accelerate the design, certification and prototyping of these materials and the devices or systems built around them. Second, the need for efficient exercise and application of the models derived from the previous methodology. We have already addressed the first need elsewhere [4, 5, 6] by following standard continuum multiphysics modeling approaches [7, 8, 9, 10]. The present paper focuses on the second need. Thus, here an attempt to increase the computational efficiency of the process required to exercise models of rectangular plates constructed from various multi-field activated materials is presented.

Previous modeling efforts of EAP or IPCC plates [4, 5] have resulted to a system of partial differential equations governing their multi-field behavior in space

V.N. Alexandrov et al. (Eds.): ICCS 2006, Part II, LNCS 3992, pp. 131–138, 2006.

and time. This non-linear system is an extended Von-Karman set of equations. We have demonstrated [4, 5, 6] that this system is solvable via the the use of the Finite Element Method (FEM). However, the time required to obtain FEM solutions has been hindered by two main difficulties. First, a general couple PDE system solver code is required by the investigator at hand. There are commercial codes (i.e. "flexPDE" [11], "COMSOL Multiphysics" [12]) and open source free codes (i.e. "freeFEM++" and "freeFEM3D" [13]) that may exceed the user's monetary and/or time (for learning how to use it) investment capability. Second, -assuming that a user owns a code of this type-, the computational cost can me prohibiting for specific applications especially because the non-linear and dynamic character of the associated PDEs.

The goal of this paper is to develop a method that alleviates the second of these problems by introducing a methodology that avoids the discretization and the numerical integration of the associated PDEs. The proposed method has been inspired by the overall "data-driven" framework associated with the development of a general computational environment for multiphysics applications (DDEMA) that has been preliminarily described elsewhere [14, 15].

The novelty of this paper is three fold. First, it avoids the need to integrate the multiphysics PDEs describing the dynamics of the system all together. Second, it proposes a technique based on approximating the solution fields by sums of basis functions borrowed from well known single-physics approximations. Third, it casts the problem as an optimization problem where an objective function is constructed that expresses the norm of the error vector between the experimental values of the fields involved in the problem and their respective analytical approximations constructed as previously described.

The paper presents the coupled and the partially uncoupled multiphysics formulations for a rectangular IPCC plate in Section 2. It subsequently describes the experimental procedure for acquiring the data to be utilized in Section 3. The corresponding global optimization problem is being setup and applied and validated in section 4. Conclusions provide the closure of the paper.

## 2 Multi-field Coupled and Uncoupled System of PDEs

We have demonstrated in the past [4, 5, 6] that introduction of Lagrange strains (to account for large deformations) into the general theory of multi-component species diffusion of electro-thermo-elastic continua, with insignificant ionic currents and chemical reactivity, excited by generalized loading conditions of multi-component mass concentrations in the boundaries, electric potential, temperature and mechanical deformation yields the following set of modified Von-Karman nonlinear PDEs :

$$\nabla^2\nabla^2 w + (1+\nu)\sum_k \nabla^2 X^k = \frac{h}{N}\left(\frac{q}{h} + F_{,22}w_{,11} - 2F_{,12}w_{,12} + F_{,11}w_{,22}\right), \quad (1a)$$

$$\nabla^2\nabla^2 F + E\sum_k \nabla^2 X^k = E[(w_{,12})^2 - w_{,11}w_{,22}], \quad (1b)$$

$$\frac{1-\nu}{2E}(\nabla^2\delta_{ij} - \partial_i\partial_j)F\delta_{ij} + \nabla^2 X^k = 0. \tag{1c}$$

Here $w, F, X^k$ are the deflection, the Airy stress function and the additional generalized field state variables and $\delta_{ij}, \nu, E, h, N, q$ are Kronecker's delta, Poisson's ratio, the modulus of elasticity, the plate thickness, the flexural rigidity and the distributed load respectively.

When all non-mechanical fields $X^k$ are not present, this system reduces to the traditional system of the large deflection Von-Karman equations. Analytical approximation solutions of this system are to be used for solving Eqs. (1).

## 3 Experimental Procedure for Data Collection

To acquire preliminary experimental data reflecting the electrically activated bending of IPCC plates, an IPCC square specimen was cut with dimensions $50 \times 50 \times 0.3$ mm, where the small number corresponds to the the thickness of the plate while the other two numbers correspond to the length and width of the plate. Figure 1(a) presents a view of the IPCC plate mounted on conductive frame that was used to apply voltage boundary conditions along the entire boundary. In Fig. 1(b) the experimental setup shown including a load cell used to determine the deflection at the mid-point of the plate surface. The load cell is mounted on a precision screw type platform capable of applying measurable displacement towards and away from the center of the plate. Consecutive turns of the screw were applied all corresponding to a displacement of a about a micron. For each one of these turns, when zero contact force was first observed (the load cell was not touching the plate) the distance traveled by the load cell becomes equal to the deflection of the plate.

Figure 2(a) shows the experimental results of applying a sinusoidally varying voltage and the corresponding deflection at the middle of the plate as a function of time. Figure 2(b) shows the voltage vs. the mid-point deflection for the first two seconds. These are the data to be used in the process that follows.

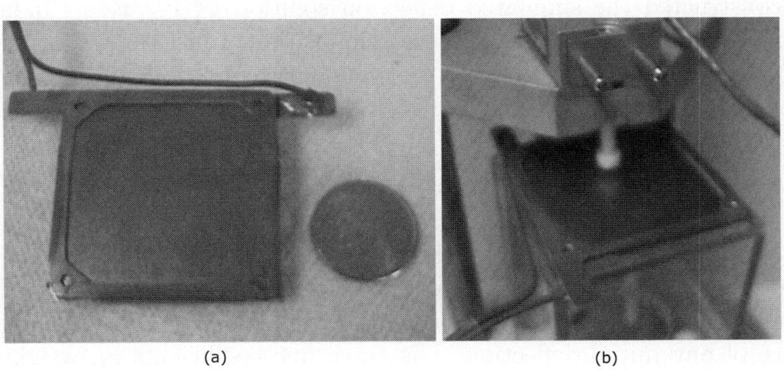

**Fig. 1.** Closeup view of Square IPCC plate mounted on conductive frame (a), and experimental setup of loading frame with loading cell in place (b)

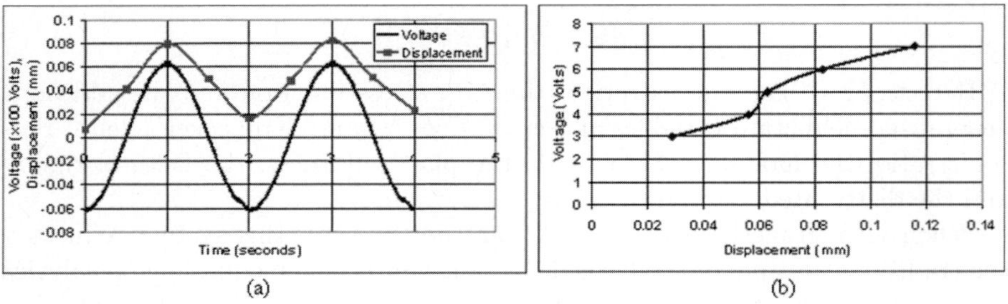


**Fig. 2.** Voltage and mid-point deflection vs. time (a), and voltage vs. mid-point deflection (b)

## 4 Global Optimization for Analytical Approximation of Simulated Behavior

The inverse approach character of the proposed methodology is based on minimizing in the least-squares sense the objective function

$$f^o(c_{ij}) = \{\sum_{i=1}^{n}[w_i^s(c_{1j}) - w_i^e]^2 + \sum_{i=1}^{n}[F_i^s(c_{2j}) - F_i^e]^2 + \sum_{i=1}^{n}[V_i^s(c_{3j}) - V_i^e]^2\}, \quad (2)$$

where $w_i^s(c_{1j}), F_i^s(c_{2j}), V_i^s(c_{3j})$ are the unknown variables of the simulated fields corresponding to the deflection, Airy stress function and voltage distributions; $w_i^e, F_i^e, V_i^e$ are the corresponding experimental values of the same variables respectively. Here, the unknowns to be determined by the minimization of Eq. 2 are the free coefficients $c_{ij}$.

In our preliminary experimental implementation the Airy stress function and the voltage distribution over the domain of the space variables $(x, y)$ were not measured. Therefore, only the first term of Eq. 2 (corresponding to the deflections) was used in the actual numerical analysis. This implies that $c_{ij} = c_{1j} = c_j$.

We constructed the simulated deflection solution $w_i^s(c_{1j}; x_i, y_i)$ indexed for specific locations $(x_i, y_i)$ as an additive composition of just three basis functions $\tau_j$ with, $(j = 1, 2, 3)$, that satisfy the boundary conditions of zero deflection along the edges according to

$$w_i^s(x_i, y_i) = \frac{4q^{equiv}a^4}{\pi^5 D}[c_1\tau_1(x_i, y_i) + c_2\tau_2(x_i, y_i) + c_3\tau_3(x_i, y_i)] + w_i^{offset}, \quad (3)$$

where $D$ is the flexural rigility of the plate, $q^{equiv}$ is an equivalent lateral load distribution per unit of area of a small deflections plate that generates deflections identical to our multi-filed plate, and $w_i^{offset}$ was added to capture the possible existence of any initial deflection. The three free coefficients $c_j$, weighting the basis functions are the unknowns to be determined.

Based on the infinite series analytical solutions constructed for approximating a solution satisfying the biharmonic equation governing the single physics small deflection bending of rectangular plates given in the past, we chose our basis functions to consist of the first three terms from one of them [16] according to

$$\tau_m(x_i,y_i)=\cos\frac{m\pi x_i}{a}(1-\frac{\alpha_m\tanh\alpha_m+2}{2\cosh\alpha_m}\cosh\frac{m\pi y_i}{a}+\frac{1}{2\cosh\alpha_m}\frac{m\pi y_i}{a}\sinh\frac{m\pi y_i}{a}) \tag{4}$$

with $\alpha_m = m\pi b/2a$ and where $a, b$ are the dimensions of the plate along the $x, y$ axes respectively.

Since the experimental data for the deflection are collected only at the mid-point of the plate we used another solution of an equivalent single physics approximating solution of the biharmonic equation to generate pseudo experimental data for the rest of the points on the plate according to,

$$w_i^e(x_i,y_i) = \frac{16q^{equiv}}{\pi^6 D}\sum_{m=1}^{100}\sum_{n=1}^{100}\frac{a^4b^4}{mn(b^2m^2+a^2n^2)^2}\sin\frac{m\pi(2x_i+a)}{2a}\sin\frac{m\pi(2y_i+b)}{2b}+ \\ + w_i^{offset} + w_i^{gaussian\_noise}, \tag{5}$$

where $w_i^{gaussian\_noise}$ represents a term injecting noise for the purpose of emulating the error due to the experimental data-acquisition methodology. It is worthwhile mentioning here that the fact that we use only the first 100 terms of the double series in the relation above is due to the fact that we are not concerned to capture accurately the deflection of the plate but rather in gross manner. The injected noise is expected and has been chosen to be of equal or higher magnitude than the approximation error introduced from not considering more terms.

To eliminate the presence of the equivalent mechanical load distribution from Eqs. 3 and 5 the Voltage vs. mid-point (maximum) deflection observed in Fig. 2(b) has been approximated by the following second order polynomial

$$w^{\max} = 0.00000137903V_o^2 + 0.00000635696V_o + 0.000000234504. \tag{6}$$

Equating this deflection with that provided by Eq. 3 at the midpoint, and solving the resulting equations with respect of $q^{equiv}$ we obtain the following voltage depended solution for

$$q^{equiv} = \frac{\pi^5 D(0.00000137903V_o^2+0.00000635696V_o+0.000000234504 - w_i^{offset})}{4a^4\,[c_1\tau_1(0,0)+c_2\tau_2(0,0)+c_3\tau_3(0,0)]}. \tag{7}$$

Equations 3-7 fully define the quantities participating in Eq.2 and therefore determination of the unknown parameters $c_j$ can be achieved by using any quadratic programming methodology for minimizing Eq. 2. In the context of this work the implementation of the optimization procedure was formulated in Mathematica [17] via the algorithms that were available in the package Global Optimization "GO-5.0" [18]. The resulting solution is presented in Table 1 and the three coefficients are compared with the known values of the first 3 terms of the infinite series that approximates the biharmonic solution. The computational time for determining these coefficients such that the objective function

**Table 1.** Computed solution coefficients $c_i$ and its comparison with know terms of the single physics solution

| | $c_1$ | $c_2$ | $c_3$ |
|---|---|---|---|
| Solution | 1.00000 | 0.00432335 | 0.00000310941 |
| Known Terms | 1.00000 | 0.00411523 | 0.00032 |

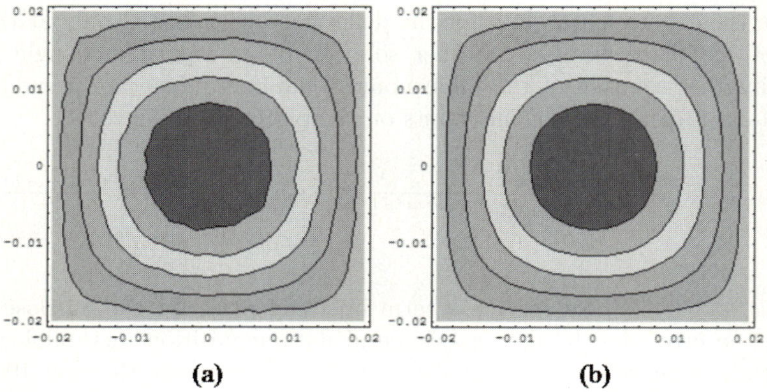


**Fig. 3.** Deflection field distributions for the pseudo-experimental case (a), and the analytically computed one (b)

given by Eq. 2 is minimized, for a problem that evaluates the simulated and experimental values of deflection at 25 points of the plate, on a Pentium IV 3.2 GHz machine was just under 53 seconds. Now that $c_j$s are known the deflection given by Eq. 3 is fully determined. For comparison purposes Fig. 3 shows the experimental (left) and simulated (right) distributions of the displacement field over the square plate at the same arbitrary time moment.

Clearly, the computationally determined simulated solution tracks the pseudo-experimental one very well. The jaggedness of the contours in the pseudo-experimental deflection distributions shown in Fig. 3(a) is due to the noise introduced in Eq. 5.

The computational cost of such a distribution via Eq. 3 is about 0.005 seconds while the respective solution due to the appropriate finite element analysis takes an average of 195 seconds. The speedup factor is therefore $195/0.005 = 39,000$. Even if one includes the time for applying the optimization process to determine the coefficients $c_j$ the speedup factor is $195/(53 + 0.005) = 3.67$. The advantage of using a data-driven approximation of the unknown deflection field in place of a numerical discretization solution of the associated PDEs via FEM is obvious.

## 5 Conclusions

In this effort, we have described a methodology for determining partial analytical representations approximating the solution of the multi-field PDEs governing the behavior of the large deflection generalized Von-Karman plates constructed from IPCC. This is achieved by exploiting single physics (just mechanical) solutions of the biharmonic PDE governing small deflection plates to construct analytical approximations for the simulated deflection field. An optimization scheme for minimizing an objective function constructed in the least square sense as a measure of the square difference between the simulated and experimentally acquired (and pseudo-experimentally constructed) deflection field. The resulting solution tracks very well the experimental fields. The realized speedup factors are significant and justify the usage of this approach.

An essential limitation of this approach is that due to lack of experimental measurements for the spatial distribution of the Airy stress function and the voltage fields, no approximations for these fields has been derived. Future efforts will address this limitation. Once this is done then the PDEs themselves can further restrict the coefficients of the approximating solutions and thus further simplify their form. In addition, instead of only considering approximation solution of the small deflection plate theory this method can be extended to include approximation solutions of the large deflection theory of Von-Karman equations [19, 20]. Another, needed activity is a sensitivity analysis to classify how the increase of the number of the basis functions participating in the solution effects both the accuracy and the computational time of the entire process.

**Acknowledgement.** The authors acknowledge the support by the National Science Foundation under grants EIA-0205663 and CNS-0540419. Partial support from NRL's 6.1 core-funding is also greatly acknowledged.

## References

1. Kim K. J. and Shahinpoor M., Ionic Polymer-Metal Composites II. Manufacturing Techniques, Smart Materials and Structures (SMS), Institute of Physics Publication, **12**(1) (2003) 65-79.
2. Kim K. J. and Shahinpoor M., Ionic Polymer-Metal Composites III. Modeling and Simulation As Biomimetic Sensors, Actuators, Transducers and Artificial Muscles, Smart Materials and Structures Int. J., **13**(4), (2004) 1362-1388.
3. Shahinpoor M. and Kim K.J., Ionic Polymer-Metal Composites IV. Industrial and Medical Applications, Smart Materials and Structures Int. J., Smart Mater. Struct. **14**(1), (2005)197-214.
4. Michopoulos, J.G., Shahinpoor, M., Towards a Multiphysics Formulation of Electroactive Large Deflection Plates Made from Ionic Polymeric Artificial Muscles, Proceedings of First World Congress on Biomimetics and Artificial Muscles, 9-11 December 2002, Albuquerque NM.
5. John G. Michopoulos, Generalized multifield Von-Karman equations for large deflection of artificial muscle plates, Proc. SPIE Vol. 5387, Smart Structures and Materials 2004: Active Materials: Behavior and Mechanics; Dimitris C. Lagoudas; Ed., (2004), 12-23.

6. John Michopoulos, "Computational Modeling of Multi-Field Continuum Systems," Proc., International Conference Computational Science - ICCS 2004, Krakw, Poland, June 6-9, 2004, Part IV Series: Lecture Notes in Computer Science, Vol. 3039 - ICCS 2004 Bubak, M.; Albada, G.D.v.; Sloot, P.M.A.; Dongarra, J. (Eds.), **LNCS 3039**, (2004), 621-628.
7. Green, A. E., Naghdi, P. M., A unified procedure for construction of theories of deformable media. I. Classical continuum physics, Proc. Roy. Soc. London Ser. **A 448 /1934**, (1995), 335-356.
8. Eringen, A.C., Maugin., G.A., Electrodynamics of Continua I. Foundations and Solid Media, Springer-Verlag, New York, 1990.
9. Sih, G.C., Michopoulos, J.G., Chou S.C., Hygrothermoelasticity, Martinus Nijhoff Publishers (now Kluwer Academic), 1986.
10. Michopoulos, J.G., Charbel Farhat, Jacob Fish, Modeling and Simulation of Multiphysics Systems, Journal of Computing and Information Science in Engineering – September , **5**(3), (2005), 198-213.
11. PDE Solutions Inc., FlexPDE 5 Reference manual, PDE Solutions Inc., URL: http://www.pdesolutions.com/, (2005).
12. COMSOL Inc., COMSOL Multiphysics 3.2 Reference manual, COMSOL Inc. , URL: http://www.comsol.com/products/multiphysics/, (2006).
13. Ionut Danaila, Frdric Hecht, Olivier Pironneau, freeFEM++ and freeFEM3D distributions at URL: http://www.freefem.org/, (2005).
14. Michopoulos, J., Tsompanopoulou, P., Houstis, E., Rice, J., Farhat, C., Lesoinne, M., Lechenault, F., DDEMA: A Data Driven Environment for Multiphysics Applications,in: Proceedings of International Conference of Computational Science - ICCS'03, Sloot, P.M.A., et al. (Eds.) Melbourne Australia, June 2-4, **LNCS 2660**, Part IV, Springer-Verlag, Haidelberg, (2003) 309-318.
15. Michopoulos, J., Tsompanopoulou, P., Houstis, E., Farhat, C., Lesoinne, M., Rice, J., Joshi, A., On a Data Driven Environment for Multiphysics Applications, Future Generation Computer Systems, **21**(6), (2005), 953-968.
16. Timoshenko, Stephen P.; Woinowsky-Krieger, S., "Theory Of Plates And Shells," Engineering Societies Monographs, New York: McGraw-Hill, (1959).
17. Wolfram, S.,1999, The Mathematica Book, 4th Edition, Wolfram Media, Cahmpaign and Cambridge University Press, Cambridge.
18. Loehle Enterprises, Global Optimization 5.0 for Mathematica, Loehle Enterprises, (2004).
19. vonKarman, T., 1910, Festifkeitsprobleme im Maschinebau, Encyk. der Math. Wiss., pp. 349.
20. Levy, S., 1942, Bending of Rectangular Plates with Large Deflections, NACA Technical Report No. 737.

# Exploiting Real-Time 3d Visualisation to Enthuse Students: A Case Study of Using Visual Python in Engineering

Hans Fangohr

University of Southampton, Southampton SO17 1BJ, United Kingdom
hans.fangohr@physics.org

**Abstract.** We describe our experience teaching programming and numerical methods to engineering students using Visual Python to exploit three dimensional real time visualisation. We describe the structure and content of this teaching module and evaluate the module after its delivery. We find that the students enjoy being able to visualise physical processes (even if these have effectively only 1 or 2 spatial degrees of freedom) and this improves the learning experience.

## 1 Introduction

Computer controlled devices are used in virtually every area of science and technology, and often carry business and safety critical [1, 2] roles. A substantial part of the work of engineers and scientist – both in academia and industry – is to use and to develop such devices and their controlling software. This requires designing and maintaining software in subject areas including control, data analysis, simulations and design optimisation.

While students are generally given a broad education on mathematics which includes basics as well as advanced material, it is often the case that the subject of software engineering is not taught. Instead, it is assumed that engineers and scientists will be able to pick up software engineering skills by learning a programming language (for example by reading a book) at the point of their career where software development is required. While this may be justified for very small programs it is inappropriate for larger projects. It is unrealistic to assume someone would be able to master solving differential equations if they have only learned about polynomials (but not differential operators).

There are two cost factors attached to not having appropriate software engineering skills: (i) often not the most efficient approach to solve a given problem is chosen, and (ii) the written code is unlikely to be re-usable.

However, there are practical constraints on the amount of time that can be dedicated to learning software engineering in non-computer science degree programmes such as mathematics, physics, chemistry, biology and engineering. Thus it is important to provide a thorough introduction to the fundamental ideas in very short time. As always in education, this will be most efficient if students enjoy the material and the way in which it is learned. That summarises the motivation for the work presented in this paper.

V.N. Alexandrov et al. (Eds.): ICCS 2006, Part II, LNCS 3992, pp. 139–146, 2006.

In section 2 we explain what motivates us to use Visual Python. Section 3 reports from teaching a module using Visual Python, and evaluates feedback from students and teachers. We suggest that virtual reality tools should be used more broadly in section 4 and close with a summary in section 5.

## 2 Background

### 2.1 What is Python?

Python [3] is an interpreted, interactive, object-oriented programming language. It is often compared to Tcl, Perl, Scheme or Java and combines remarkable power with very clear syntax. It has been argued that Python is an excellent choice of programming language for beginners both in computer science [4] and in other disciplines [5, 6, 7].

### 2.2 What is Visual Python?

Visual Python (VPython) [8] is a 3d graphics system that is an extension of the Python language. Its main usage has been in the area of demonstration of physical systems in physics, chemistry, and engineering. VPython was initially written by David Scherer under the supervision of Bruce Sherwood and Ruth Chabay and is is released under the GNU Public License.

### 2.3 Motivation for Using Visual Python

Scientists in all stages of their career like solving puzzles: curiosity and the strong desire to understand the world drive them. We argue that this is not too distant from the keen young student who might like to play computer games: here, too, the player needs to solve a puzzle (to win the game). Some of the attractiveness of computer games stems from (i) their interactivity and (ii) the real time virtual reality graphics.

It is well known that the learning process is more successful if learners enjoy the learning activity and even more so if they start to explore the subject following their own ideas. We therefore strive to inspire the learners' imagination and provide them tools that encourage experimentation in computational science. We pick up the two points made above and make computational science (i) interactive and (ii) employ virtual reality graphics.

The interactivity can be achieved in programming by using an interpreted language (such as Python): a student's command is dealt with immediately, and the computer "responds" with an answer or action (or an error message). It can, of course, be argued that the edit-compile-run cycle is interactive, too (although slightly less direct than using an interpreted language). The use of 3d-graphical programming (often referred to as virtual reality) can be used to make exercises far more attractive (if the 3d programming is easy to use). Visual Python provides such an environment.

In a historical context, 2d visualisation has been introduced into teaching programming (and mathematics) via the "turtle" within the LOGO programming language more than 25 years ago [9]. More recently, "Alice" [10] provides an environment which allows the student to be the director of a movie or the creator of a video game and to create highly sophisticated 3d scenes with very little effort. Again, the aim is to allow traditional programming concepts to be more easily taught and more readily understood [11].

We have chosen Visual Python because it offers true 3d graphics (with simple building blocks such as spheres, boxes, cones etc), it is extremely easy to learn, and it is sufficiently "serious" to be used in real-life tasks in engineering and science.

## 3 Case Study

### 3.1 Module Layout and Structure

In 2004/2005 the 85 second year Aerospace students at the University of Southampton were taught a computing module [12] as outlined and evaluated in this section.

The course consists of 12 lectures (one a week) and 6 associated practical sessions in a computer laboratory equipped with standard PCs running MS Windows. The practical laboratories take place every two weeks, last 3 hours, and every student is doing one self-paced assignments using their computer. Demonstrators are available in the laboratories (approximately one demonstrator per 10 students) to provide help if necessary. In each of the first four laboratory sessions students have to complete one self-paced assignment. When completed, they are asked to explain their work (both computer programmes and any other notes) to a demonstrator in a mini-viva between one demonstrator and one student lasting between 10 and 15 minutes (in front of the student's computer). The fifth assignment is slightly larger and students have to submit a written report to be marked off-line by the lecturer (without the student being present). The fifth assignment is also used as a report writing exercise.

### 3.2 Content

**Students' background in numerical methods:** This module is preceded by 12 lectures (each lasting 45 minutes) providing a theoretical introduction to numerical methods (including elementary linear algebra, root finding, numerical integration of ordinary differential equations) without practical exercises. Due to time-tabling constraints it has not been possible to combine the more problem-solving based module described here with these theory lessons (although this is desirable).

**Students' background in programming:** The students have had some exposure to the Matlab [13] environment and programming language in their first year. This included four computer based exercises introducing fundamental concepts such as if-then statements, for-loops and functions on a very basic level.

*A sphere at position* $\mathbf{r} = (r_x, r_y, r_z)$ *of mass* $m = 1$kg *is subject to a horizontal force* $\mathbf{F}_{\text{spring}} = (-kr_x, 0, 0)$ *and to a vertical force due to gravity* $\mathbf{F}_{\text{grav}} = (0, -mg, 0)$. *The initial position is* $\mathbf{r}(t_0) = (3, 5, 0)$m, *initial velocity* $\mathbf{v}(t_0) = (0, 0, 0)$m/s *and* $k = 5$N/m. *Compute the time development of the system, assuming that the sphere will bounce elastically when it touches the ground at* $r_y = 0$.

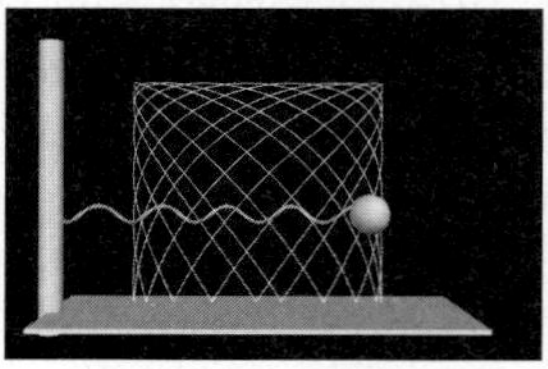

**Fig. 1.** (Left) An example problem and (right) a snapshot of an animation of the solution. The faint line which starts filling a rectangular is the trajectory of the sphere and updated as the sphere moves.

**The content of this module** is given in table 1 and ordered by lectures (numbered from 1 to 12) and laboratory sessions (numbered from 1 to 6).

As can be seen, we have combined a repetition of numerical methods with the introduction of the Python programming language (laboratories 1 and 2) and Visual Python (laboratories 3, 4 and 5).

**Required Software:** Apart from `Python` [3], we require `Numeric` [14], `Scientific Python` [15], `Pylab` (formerly `Matplotlib`) [16] and `Visual Python` [8]. All of these can be installed on the major three platforms MS Windows, Linux and Mac OS X.

For MS Windows, we have found the "Enthought Python" edition [17] of great value which bundles Python with all the extra packages we need (apart from Visual Python which can be installed afterwards) and simplifies the installation of the software both for IT personnel at the university as well as for students at home.

### 3.3 Results and Discussion

**A visualisation example** is shown in figure 1 and has been created by the program shown in figure 2. It is outside the scope of this paper to explain the workings of the program detail. However, hopefully this demonstrates that it encourages experimentation to support understanding the system. (This problem could be tackled by the students in laboratory session 4).

**Feedback from students** was obtained using an anonymous questionnaire to evaluate the student experience at the end of this module. One of the questions posed was "Did Visual Python improve the learning process?" and possible ratings range from 1 ("Not at all") to 5 ("Very much"). The average student rating was 4.2 with a standard deviation of 0.68. This is clear evidence that the students considered the use of Visual Python useful or very useful to improve the learning process.

**Feedback from the teacher** includes the following observations:

▷ The overall reception of the module was very good, in particular taking into account that within the student body the subject of "computing" is often regarded as difficult and boring.

**Table 1.** Overview of material taught in lectures and rehearsed in laboratory sessions

| Lecture | Lab. | Content |
|---|---|---|
| 1 & 2 | | Introduction & formalities, Using IDLE, basic data types: strings, floats, ints, boolean, lists, type conversion, `range`, for-loop, if-then, importing modules, the `math` module, the `pylab` module, plotting simple functions $y = f(x)$, defining python functions, basic printing, importing python files as modules. |
| | 1 | Programs to write:<br>1. computer chooses random integer, user has to guess<br>2. finding the plural of (regular) English nouns automatically<br>3. plotting mathematical functions $y = f(x)$<br>4. retrieve current weather conditions in Southampton from Internet (i.e. processing of text file) |
| 3 & 4 | | Ordinary Differential Equations (ODEs), Euler's method in Python, Use of `Numeric` and `scipy`, use of `scipy.integrate.odeint` to solve ODEs |
| | 2 | Programs to write:<br>1. proving that $\sum_{i=1}^{n} i = \frac{1}{2}n(n+1)$ for $n = 1000$<br>2. currency conversion (exercise functions)<br>3. implement composite trapezoidal rule for integration of $f(x)$ and evaluate convergence properties empirically<br>4. use of `scipy`'s `quad` for integration<br>5. automatic integration of function and plotting of integrand |
| 5 & 6 | | Introduction to Visual Python, finite differences for differentiation, Newton method for root finding. Calling Python functions with keyword arguments, name spaces, exceptions. Example code for dealing with 3d vectors and scalars. |
| | 3 | Programs to write:<br>1. implement a 2nd order Runge Kutta integrator for ODEs<br>2. solve given 1d ODE using `scipy.integrate.odeint`<br>3. visualise $\mathbf{r}(t) \in \mathbb{R}^3$ in real-time using Visual Python<br>4. compute and visualise solution to 2nd order ODE with two degrees of freedom using Visual Python |
| 7 & 8 | | Finding ODEs to describe a given system. Example code dealing with time dependent 3d problems and visualisation. |
| | 4 | Programs to write:<br>1. Use `scipy`'s root finding tools (`bisect`) to find root of $f(x)$<br>2. Use root finding and integration of ODE to solve boundary value problem ("shooting method") visualised with Visual Python<br>3. (Exercise on LaTeX– therefore only 2 other tasks.) |
| 9 & 10 | | Explanation of laboratory assignment 5 |
| | 5 | Larger assignment requiring written report. Tasks include implementing root finding using Newton's method, making Newton's method safe, integrating ODEs, visualising 3d time-dependent data. All examples from space exploration (mainly trajectories). |
| 11 & 12 | | Introduction to Object Orientation |
| | 6 | Time available to complete assignment 5 |

```
from Numeric import array, concatenate
import scipy, visual

def rhs(y,t):
    """function that returns dy/dt(y,t) for system of ODE"""
    vx, vy, vz, rx, ry, rz = y
    mass = 1.0              #mass of object in kg
    g = 9.81                #acceleration from Earth in N/kg
    F_grav = array([0,-g, 0])*mass
    spring_x = 0            #spring equilibrium at x=0 (in m)
    k = 5                   #spring stiffness in N/m
    F_spring = k*array([spring_x-rx,0,0])
    dvdt = (F_spring + F_grav)/mass
    drdt = (vx,vy,vz)
    return concatenate([dvdt,drdt])

#main program starts here
r = array([3,5,0])          #initial position of object in m
v = array([0,0,0])          #initial velocity of object in m/s
t = 0                       #current time in s
dt = 1/30.0                 #time step in seconds, to match framerate

#visualisation
visual.scene.autoscale = False #don't zoom in and out
visual.scene.center = (0,3,0)  #focus camera it this point
base     = visual.box(pos=(0,-0.5,0), length=10, height=0.1, width=4)
pole_x   = -5                  #position arbitrarily chosen
cylinder= visual.cylinder(pos=(pole_x,-1,0),axis=(0,7,0),radius=0.3)
ball     = visual.sphere(pos=r,radius=0.5,color=visual.color.white)
spring   = visual.helix(pos=r,axis=(pole_x-r[0],-1,0),thickness=0.1)
path     = visual.curve()   #initiate drawing path of trajectory

while True: #infinite time loop starts
    y = concatenate([v,r])
    #integrate the system from time t to t+dt
    y = scipy.integrate.odeint(rhs,y,array([t,t+dt]))
    t = t + dt              #advance time
    v = y[-1,0:3]           #extract last row from output from odeint
    r = y[-1,3:6]           #for visualisation

    if r[1] < 0:            #if below base plate
        v[1] = -v[1]           #then reverse velocity (elastic bounce)

    visual.rate(30)         #visualisation: keep frame rate constant
    ball.pos = r            #               update position of ball
    path.append(r)          #               update trace
    spring.pos=r            #               update spring head
    spring.axis = (pole_x-r[0],0,0)        #update spring tail
```

**Fig. 2.** Source code to generate animated visualisation shown in figure 1

- ▷ A number of students wrote programs unrelated to the module and in their spare time because they enjoyed the process. These included an analog clock (using Visual Python for the central knob, hands and hour-ticks) and a visualisation of the orbits of planets within the solar system.
- ▷ Visual Python can render its 3d graphics for coloured anaglyph glasses (here red-cyan) with the following command `visual.scene.stereo='redcyan'`. This allows seeing the scene with spacial depth and simple glasses can be bought for about one US dollar or Euro. This proved popular with students and has clearly boosted the motivation.
- ▷ This module was previously taught in Matlab [13] using a similar content and structure. However, it turned out that it was not possible to simple 're-write' the learning materials and lectures to be delivered in Python. For example, previously a complete laboratory session was used to introduce the correct syntax (in Matlab) to pass a function to a function. This is required if, say, general purpose integrators are written. With Python, a function is an object as is any other object, and can be passed to a function and then used as if it was defined elsewhere in the code (see also [6]). This change immediately freed up several hours of time in the laboratory sessions and allowed us to proceed further in terms of numerical methods.
- ▷ The real-time integration of ordinary differential equations (ODEs) and the real-time display of the simulated process allows students to gain a deeper understanding of the physical process in comparison to looking at 2d graphs showing the displacement of an object against time. We are not able to quantify this but have got the impression that this was supporting the learning process.

## 4 Outlook

There are a number of topics in Physics [18], Mathematics [19] and elsewhere which could benefit from being taught together with virtual reality tools such as Visual Python. For example, when the topic of Vibrations (or "normal modes") is taught, the lecturer often demonstrates with an experimental set up how energy transfers from one pendulum to another when these are coupled by a spring. The students could be given a Visual Python program which allows to repeat this experiment and to vary the initial conditions and system parameters as much as they like outside the lecture.

While often there is good reason for fundamental mathematics being taught by mathematicians, programming being taught by computer scientists and design being taught by engineers, in many real-life situations engineers and scientists need to combine knowledge from all areas to solve problems. Programming, computational methods and visualisation provide opportunities to bring the disciplines together already in the teaching curriculum.

## 5 Summary

In summary we describe our experiences of using Visual Python to enthuse engineering students when learning the fundamental concepts of software engineering and programming. Feedback from students and staff shows that the real-time virtual reality environment has improved the learning experience. We link exercises to other parts of their curriculum (such as ordinary differential equations) to contextualise the new skills. We argue that there are a number of areas which naturally allow (and would benefit from) the integration of computation and visualisation into the existing curriculum.

## References

1. Je'ze'quel, J.M., Meyer, B.: Design by contract: The lessons of Ariane. IEEE Computer **30**(2) (1997) 129–130
2. Leveson, N.G., Turner, C.S.: An investigation of the Therac-25 accidents. IEEE Computer **26**(7) (1993) 18–41
3. van Rossum, G.: Python tutorial. Centrum voor Wiskunde en Informatica (CWI), Amsterdam. (1995) `http://www.python.org`.
4. Downey, A., Elkner, J., Meyers, C.: How to Think Like a Computer Scientist: Learning with Python. Green Tea Press (2002) `http://www.greenteapress.com/thinkpython/html`.
5. Donaldson, T.: Python as a first programming language for everyone. In: Western Canadian Conference on Computing Education. (2003) `http://www.cs.ubc.ca/wccce/Program03/papers/Toby.html`.
6. Fangohr, H.: A comparison of C, Matlab and Python as teaching languages in engineering. Lecture Notes on Computational Science **3039** (2004) 1210–1217
7. Roberts, S., Gardner, H., Press, S., Stals, L.: Teaching computational science using vpython and virtual reality. Lecture Notes on Computational Science **3039** (2004) 1218–1225
8. Scherer, D., Sherwood, B., Chabay, R.: (2005) `http://www.vpython.org`.
9. Papert, S.: Mindstorms: Children, Computers and Powerful Ideas. Prentice Hall Europe (1980)
10. Carnegie Mellon University: (2005) `http://www.alice.org`.
11. Dann, W., Cooper, S., Pausch, R.: Learning to Program with Alice. Prentice Hall (2005)
12. Fangohr, H.: Computing module SESA2006, Aerospace Engineering, University of Southampton (2004) The complete learning materials are available from the author on request.
13. The Mathworks: Matlab (2005) `www.mathworks.com`.
14. `http://numeric.scipy.org`.
15. `http://scipy.org`.
16. `http://matplotlib.sourceforge.net`.
17. `http://www.enthought.com/python`.
18. Chabay, R.W., Sherwood, B.A.: Matter and Interactions: Modern Mechanics and Electric and Magnetic Interactions. John Wiley and Sons (2003)
19. Urner, K.: Python in the mathematics curriculum. (2004) `http://www.python.org/pycon/dc2004/papers/15`.

# Involving Undergraduates in Computational Science and Engineering Research: Successes and Challenges

R.M. Kirby, C.R. Johnson, and M. Berzins

School of Computing and Scientific Computing and Imaging Institute,
University of Utah, Salt Lake City, UT, USA

**Abstract.** The undergraduate years of one's educational career are among the most formative in terms of education and research outlook. The university education should both broaden the student's perspective, while simultaneously sharpening critical thinking skills. Though classroom experience is beneficial, it is incomplete. The purpose of this paper is to address some of the issues associated with providing computational science and engineering education at the undergraduate level by surveying efforts made at the University of Utah. Specifically, we will describe a program that allows students to become involved in research during their undergraduate years, and will provide two success stories demonstrating its efficacy. We also discuss efforts being made at associated undergraduate curriculum reform.

## 1 Introduction

The mission of the university at the undergraduate level is to educate and train. The first effort attempts to develop critical thinking skills that transcend any specific discipline; the second effort attempts to instill discipline-specific practical skills that help students to be productive members of the work-force and of society. It is well understood in many fields that although the classroom experience is necessarily, is it incomplete. It is our contention that education in the sciences and engineering can be made complete by introducing research opportunities at the undergraduate level. Research experience for undergraduates provides a means of honing critical thinking and research skills under the tutelage of faculty interested in both the student's research and academic development. Traditionally project-work towards the end of an undergraduate degree provides some experience of research for students. However there are a multitude of benefits at the student, faculty and university level for undergraduate students to be immersed in a research environment for a lengthy period of time. For instance, at the student level, the benefits are:

- Many students find being part of a research cohort to be intellectually stimulating. As part of a mentor's research group, they are able to participate with others in the process of discovery; in group meetings to update members involved on the research project; and to obtain access to faculty who often treat the student more like a graduate student.

V.N. Alexandrov et al. (Eds.): ICCS 2006, Part II, LNCS 3992, pp. 147–154, 2006.

- They are able to advance their personal/professional goals (i.e. enhancement of cognitive/communication skills, letters of recommendation, graduate school training, etc.)
- Integration within a research laboratory often helps motivate students to study harder in their classes by giving them context for what they are learning. Many of us have had the experience of being presented with a seemingly abstract topic with little motivation, e.g. infinite series expansions or functional analysis, and thus did not expend much time on the subject, only to find out later that this material is now key to understanding many important methods. Students not stimulated by the university curricula often find their research experience to be a far more invigorating blend of application and theory.

The student is not the only one to benefit. Benefits of this experience for the faculty and university are:

- It helps to expand the university's commitment to scholarly activity and/or scientific inquiry, and as such provides a good source of "advertisement" to attract new students.
- The undergraduate helps the advisor in advancing his/her career and/or research environment, and many undergraduates are capable of functioning at a level similar to that of many graduate students for lower funding needs.
- In the end, we believe that the students that participate in a research experience receive a superior education, and that is good for all concerned.

In this paper we describe two efforts at expanding computational science and engineering at the University of Utah. The first effort we will discuss is the Engineering Scholars Program, a College of Engineering program geared to get undergraduate students into the research environment as early as possible. We will present some details of the program and will discuss two student success stories. The second effort is our attempts to revamp the view of scientific computing undergraduate education through new undergraduate courses that present the computational science pipeline.

## 2 Engineering Scholars Program

The focus of the Undergraduate Engineering Scholars Program (ESP) is the engagement of between ten and twenty *first year* undergraduate engineering students in research opportunities. This early exposure to the research environment enhances the engineering education experience and better trains these future engineers for positions in industry, government, and advanced graduate training.

The quest of the ESP is to leverage the exciting engineering-based research at the University by exposing first year students to actual research activities. The addition of a research component to the first year student's experience allows the student to see over the horizon offered by prerequisite courses. By giving the first year student a more in-depth look into engineering, it is hoped that we will increase the enthusiasm level of the students with the results being significantly improved retention rates and better trained students. Higher retention rates will translate into more, better-trained and much-needed engineers and computer scientists.

The program is designated as a one-year program during the freshmen year where the students are exposed to engineering research through a set of tours and demonstrations in the Fall semester. During Spring semester, students are placed in Engineering and Computer Science labs where they work between 10-20 hours per week participating in ongoing research projects. Although they may not be able to do the most complicated parts of the research, the opportunity to participate in associated tasks allow them to gain an understanding of the research process.

A unique aspect to the ESP is that students are engaged in research their freshman year. Several universities have programs that engage undergraduates during their senior year (in the form of a senior project or senior thesis). While we think that any engagement in research during undergraduate training is positive, we have noted that ESP students, as is demonstrated below by the student examples, often continue to work within research laboratories throughout their entire undergraduate training. There have been multiple instances where ESP students have co-author journal and conference papers, as well as given presentations at research conferences. It is almost always the case that people are astonished that undergraduate students are doing such a level of research. We too have been impressed at the level of research that some of the ESP students can tackle when properly challenged and motivated.

An additional benefit of the Engineering Scholars Program is that the students participating in the program are introduced to fellow peers in the program, and indeed some are even housed in the same building. Events such as social gatherings and field trips are regularly scheduled for both current members of the program and the program alumni. A highlight is the annual inaugural luncheon where students and parents are brought together with participating professors, top-level College of Engineering administrators and program sponsors.

### 2.1 Engineering Scholar – Lindsey Healy

Lindsey started her research experience working in a Bioengineering lab, but then migrated to working with Radiology Professor Sam Browd, on a project involving finding a way to visualize the cervical vertebral body in children with Down Syndrome.

Children with Down Syndrome are more prone to congenital abnormalities of the cervical spine than their normal age-matched counterparts. These abnormalities of the cervical spine leave the Down Syndrome population with an increased susceptibility to spinal injury from normal activity. In order to analyze the origin of spinal instability of the atlanto-occipital joint, Lindsey performed segmentation on CT images taken prior to surgical fixation of the unstable joint complex. After segmenting and creating a reconstruction of the first cervical vertebral body for the patient and control population, the superior particular facet was

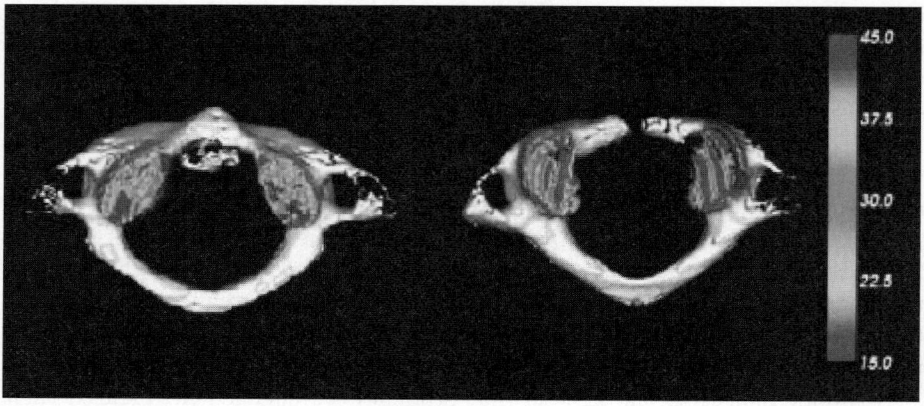


**Fig. 1.** Representative data comparing the normal age-matched control (left), to the Down syndrome patient (right). Numbers on color scale indicate degrees between the perpendicular at any particular point and the vertical. Blue therefore indicates areas closer to parallel (more sloped) with the vertical while red indicates areas closer to perpendicular (less sloped) with the vertical.

color mapped according to the slope to visualize differences between subjects and age-matched controls as shown in Figure 1. In addition, quantitative measurements were taken and length over depth ratios were calculated. The results showed a significant difference (p = 0.01) in the morphometric measurements between the two groups as shown in Table 1. As part of this research, Lindsey worked closely with members of the Pediatric Neurosurgery Department at Primary Children's Medical Center in Salt Lake. This research was accepted for an oral presentation at the AANS Pediatric Section1 annual meeting in December 2004 and has been submitted for publication. It is the hope that this research will provide an avenue for further research and software development for improved and preventative screening within the Down Syndrome population.

Because of Lindsey's positive experience, she has gone on to work on multiple research projects during her undergraduate education, including a segmentation and visualization project with geneticists Mario Capecchi and Charles Keller that resulted in a paper that was recently submitted to Nature Genetics, and then worked with Radiology Professor Norman Foster, who directs the Center for Alzheimer's Care, Imaging, and Research at the University of Utah.

Lindsey graduated in 2005 with a B.S. in Bioengineering. As an undergraduate, she was awarded nine scholarships. At the time of this paper, she is applying for graduate programs in Bioengineering.

### 2.2 Engineering Scholar – Curtis Hamman

Curtis Hamman began his Mechanical Engineering studies at the University of Utah in the Fall of 2003. By the end of that first Fall semester, Curtis had obtained a research position directed by Professor Mike Kirby at the Scientific

Computing and Imaging Institute. Curtis has been active in two interrelated projects presented below.

**Project 1 – Visualization and Analysis of Turbulent Flows:** The study of fluid flows is typically based on the three vector velocity components $\boldsymbol{u} = (u, v, w)^T$ and the scalar pressure field $p$. These quantities can adequately describe the state of any incompressible fluid flow, nevertheless, the computation and visualization of derived quantities such as vorticity $\boldsymbol{\omega}$ can improve our understanding about the flow dynamics.

In the same spirit, Curtis has been involved in the study of a newly appreciated derived quantity, the divergence of the lamb vector. At a time in their student careers when most are attempting to grasp velocity and vorticity as concepts, Curtis was using research experience to solidify his understanding of fluid mechanics concepts and their applications within engineering.

As an example – Curtis was challenged with the task of understanding both mathematically and through numerical evidence the differences between isosurfaces of vorticity and isosurfaces of the divergence of the lamb vector. His working hypothesis was that the divergence of the lamb vector is a far more localized quantity than vorticity, and that as such is far more amenable to modeling. To support his hypothesis, he post-processed and visualized data from a fully-developed turbulent channel flow with two homogeneous directions with $Re_\tau \approx 180$ and $Re_\delta \approx 3300$ as shown in Figures 2 and 3. These figures demonstrate that divergence of the lamb vector is a far more localized quantity (as shown in Figure 3) than vorticity (as shown in Figure 2). Presentation of these results to faculty in the Mechanical Engineering department here at the University stimulated an ongoing inter-disciplinary research effort to understand the differences between vorticity and the lamb vector.

**Project 2 – High-Performance Computing Applied to Channel Flow Simulations:** This project investigated the mathematical and computational

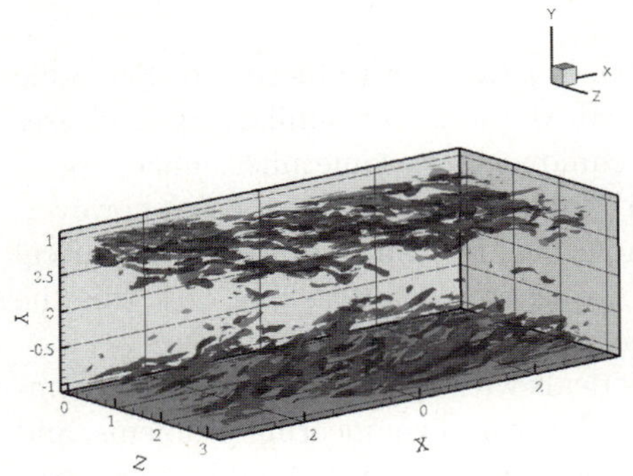


**Fig. 2.** Streamwise vorticity isocontours at $|w_x| = 2.5$. Red is positive, and blue is negative

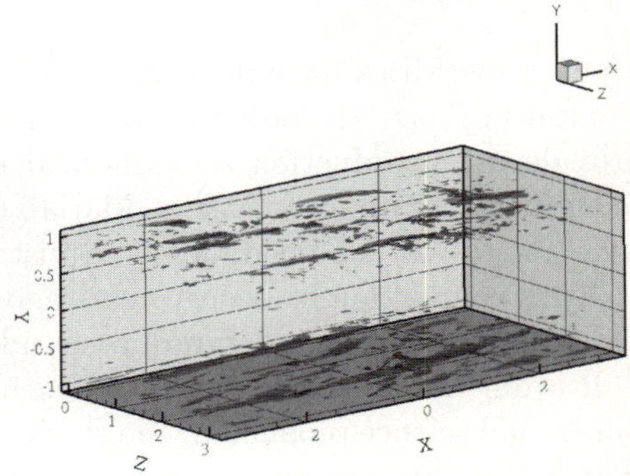


**Fig. 3.** Lamb vector divergence isocontours at $|n| = 250$. Red is positive, and blue is negative

**Table 1.** Wall clock time, speedup, and efficiency for varying numbers of processors for a simulation with $N_e = 32$, $P = 8$, $N_x = 64$, and $N_z = 64$. The predicted efficiency is given by an equation derived as part of the study.

| # Processors | Wall clock time (s) | Speedup | Efficiency | Predicted Efficiency |
|---|---|---|---|---|
| 1 | 20.47 | 1 | 1 | 1 |
| 2 | 10.72 | 1.91 | 0.96 | 0.98 |
| 4 | 5.74 | 3.57 | 0.90 | 0.92 |
| 8 | 3.08 | 6.65 | 0.83 | 0.83 |
| 16 | 1.90 | 10.78 | 0.67 | 0.67 |

properties of a one-dimensional Helmholtz solver with high-order spectral element discretizations. The principal benefit of this study is to elucidate the salient aspects of this particular parallel implementation in order to deduce its effectiveness in the Direct Numerical Simulation (DNS) of turbulent channel flow where wall-normal Helmholtz solves are conjugate to the streamwise-spanwise Fourier transforms. As part of the project, a parallel channel code was developed and computational models were formulated to predict efficiency. The results of a study of one of the stages of the flow solver are shown in Table 1.

Although these results are for relatively small numbers of processors they are significant in that they also show a close relationship between the predicted scalability and the actual scalability. More significantly the same code has been extended and shown to scale nicely on as many as 1024 processors [2]. Although the underlying parallel modeling methodology was that discussed in [1], the study was independent of a taught course, moved more quickly, and used many more processors by being linked to the existing DOE CSAFE research program at Utah.

## 3 Computational Science and Engineering Course Development

Many universities have undergraduate classes titled "Introduction to Scientific Computing" or "Introduction to Numerical Methods" or similar. Such classes provide an introduction to numerical techniques and sometimes make use of scientific software tools such as Matlab or Mathematica. In some cases however, such classes would be difficult to distinguished from a traditional numerical analysis course taught twenty years ago. A question we are asking is: is this the right way to expose undergraduate students to scientific computing?

It is our thesis that scientific computing deals with the application of the computational science pipeline to the study of science and engineering problems, and that a scientific computing class at the undergraduate level should expose students to this process. This idea has precedents in which problems and solutions were both considered [3], but has been extended to the complete computational science pipeline.

What distinguishes the computational science pipeline? Historically, the scientific method was formulated around the idea of postulating a model of natural phenomenon, making observations to validate one's model, and correcting the model based upon discrepancies between the phenomenon and nature. Later, the scientific process was extended to include the idea of the controlled experiment. No longer was the scientist limited to passively observing the world around him to deduce the correctness of the model. This gave rise to the idea of devising controlled experiments designed to evaluate the correctness of the hypothesis in a systematic manner. This systematic process allowed the model to be updated based upon the lessons learned through the experiment. With the advent of modern computing, a new paradigm called computational science has emerged, in which *experiment* now employed within the scientific method consists of the computational solution of the model. The computational science pipeline consists of the following stages:

- Scientific Problem of Interest (Problem Identification): Statement of the scientific or engineering problem of interest. Questions should be developed in such a way that quantifiable metrics for determining the level of success of the computational science endeavor can be evaluated.
- Modeling: The development of a model which abstracts the salient features of the problem of interest in such a way that exploration and evaluation of the model allows an answer to the questions specified concerning the problem of interest. Modeling techniques include, but are not limited to, deterministic or probabilistic, discrete or continuous mathematical models. Means of validating the model (determining the error introduced due to the model abstraction of the real phenomenon) should be established.
- Computation: The generation of algorithms and implementations which accurately and efficiently evaluate the model over the range of data needed to answer the questions of interest. This simulation of the physical phenomenon by computational expression of the model provides the experiment upon which the computational science pipeline hinges.
- Evaluation: The distillation and evaluation of the data produced through computational simulation to answer the questions of interest and to provide quantifiable determination of the success of the experiment. Methods such as, but not limited to, scientific visualization provide a means of tying the simulation results back to the problem of interest. Typically this stages leads to a revised model or computation and the process being repeated.

We are undertaking the creation of a new undergraduate version of semester long Introduction to Science Computing course that will expose students to the computational science method and the associated tools and techniques. As such, we will reduce the amount of numerical methods and add sections on mathematical and geometric modeling, visualization, and statistical evaluation. In addition we will stress the entire process by a few well chosen driving computational science problems.

An example of such a problem is the classic chemical morphogenesis work of Turing [4]. Following Turing, a basic idea of the chemical processes can be presented to the students, followed by the mathematical modeling tools which allow one to simplify the problem of studying morphogenesis down to understanding a system of ordinary differential equations. Once can then introduce basic time-stepping methods and concepts such as stability, accuracy and error control. Visualization is then used as a means of understanding the numerical results of the system, with the goal of connecting it back to the original problem of interest – chemical morphogenesis. This is just one of a multitude of such examples which help students to see the entire computational science pipeline in action.

## 4 Summary and Discussion

Multidisciplinary research has become an integral part of the research landscape, and its importance will continue to grow in the future. How discipline-centered university programs adapt to the changing nature of research will directly impact scientific and engineering progress in this next century. More tightly coupled integration of research and teaching is mandatory – especially at the undergraduate level. The University of Utah's Engineering Scholars Program and undergraduate computational science course as described in this paper demonstrate how undergraduates can be exposed to research ideas through both the classroom and laboratory environment.

## References

[1] M. Berzins, R.M. Kirby, and C.R. Johnson, "Integrating Teaching and Research in HPC: Experiences and Opportunities", Proceedings of the International Conference on Computational Science, 2005.

[2] C.W. Hamman, R.M. Kirby and M.Berzins, "Parallelization and scalability of spectral element solver", SCI Institute Report UUSCI-2005-011.

[3] T.A. Grandine, "The numerical methods programming projects book", *Oxford University Press*, 1990.

[4] A.M. Turing, "The chemical basis of morphogensis", *Phil. Trans. Roy. Soc. Lond.*, B237:37-72, 1952.

# A Project Based Approach to Teaching Parallel Systems

Alistair P. Rendell

Department of Computer Science, Australian National University
Canberra ACT0200, Australia
alistair.rendell@anu.edu.au

**Abstract.** For several years we have delivered advanced undergraduate courses related to computational science using a traditional approach of lectures, laboratory exercises and assignments. In recent years, however, we have moved away from this towards project based approaches. In this paper we discuss our attempts to structure a course in parallel systems around a group project that required the students design, build and evaluate their own message passing environment.

## 1 Introduction

In 2001 under a joint initiative between the Departments of Mathematics and Computer Science, and with funding from the Australian Partnership in Advanced Computing (APAC) [1], an undergraduate degree in computational science was established at the Australian National University. This degree involves students taking roughly a quarter of their courses in computer science, another quarter in mathematics and the remaining half in their chosen field of specialization [2]. Two courses that are offered by the Dept. of Computer Science as part of this degree are a third year course in "High Performance Scientific Computing" [3] and a fourth year course in "Parallel Systems" [4]. In the last two years, and for both these courses, we have adopted a project centered approach to course delivery. The aim is to give the students a better understanding of the hardware and software that they are using, a sense of achieving something significant, and to provide a natural springboard for in depth discussions and subsequent research projects. In the high performance scientific computing course this has involved getting the students to develop a simple parallel molecular dynamics application code that they then run on a cluster that they have built. Some details of the molecular dynamics case study, and how it relates to modern concepts in software design, are given in reference 5. In this paper we focus instead on the parallel systems course, and our efforts to build this course around a group project in which the students write their own basic message passing environment.

## 2 COMP4300: A Course in Parallel Systems

Within the Australian system a science, IT or computational science degree can be completed in three years full-time study. Students who achieve a mid-credit or above

V.N. Alexandrov et al. (Eds.): ICCS 2006, Part II, LNCS 3992, pp. 155 – 160, 2006.

average (≥65%) are then permitted to study for one extra year and gain an honours degree. In addition to this path some other degrees, such as software engineering, are four year honours programs from the outset. The parallel systems course targets undergraduate students who are in their fourth year, so by in large these are science, IT or computational science honours students or software engineers in their final year of study. Technically it is possible for a standard 3 year degree student to enroll in the course, but this requires them to have undertaken a non-standard enrolment pattern, or be a part-time student. Suffice it to say that the general standard of the students embarking on this course is quite high, so the course is designed to be challenging.

The course is offered in alternate years and aims to cover several aspects of parallel systems, including hardware architecture, programming paradigms, parallel algorithms and sample applications of parallel computers. It is based loosely on the text book *"Parallel Programming: techniques and applications using networked workstations and parallel computers"*, by Barry Wilkinson and Michael Allen [6], augmented with additional material covering, for example, one-sided communications in MPI-2 [7] and Global Arrays [8]. It also usually includes one or two guest lectures and a tour of the APAC National Facility [1]. The course has roughly 30 hours of lectures distributed over a 13 week semester and includes 12 hours of supervised laboratory work. Course prerequisites include a second year course in "Concurrent and Distributed Systems" and a third year course in either "High Performance Scientific Computing" (the course mentioned above) or "Algorithms".

In 2004 after having delivered the parallel systems course 2-3 times in a fairly standard lecture format, the opportunity arose to trial a radical new format. Part of the motivation for this was the fact that in this year the course would only be open to a small group of 5 honours students; making it much easier to map out a new course structure on the fly. And so it was decided to drive much of the course around a group project where the students developed their own message passing environment based solely on the use of simple UNIX utilities; essentially the students were required to write their own limited version of the Message Passing Interface (MPI) [9].

The project, codename `mympi`, began after 2 weeks of introductory lectures and one laboratory session. It was broken down into five 2-week segments that were each assigned to one of the students. Logistically the main lecture time was a block of 2 hours each Friday afternoon. In the first week of each 2-week segment the relevant student was required to come and discuss their part of the project in private a few days before the Friday lecture. At this meeting they were asked to outline what they thought was required, and how they proposed to tackle it. After clarifying any misconceptions, and ensuring that the student was on the right track, they made a formal 10 minute presentation of their proposed work to the rest of the class during the Friday lecture. This would invariably evolve into a class discussion and further refinement of their ideas. In the second week the student would discuss their progress in private before giving a formal presentation and demonstration of their code to the class during the Friday lecture. The student was required to handover their work to the next student before the following Monday. Both presentations were peer marked, with more detailed marking and feedback given after the student had submitted a formal write-up.

The five project stages, their requirements and some comments on the objectives and outcomes are given below:

## 2.1 Stage 1 – Basic Process Creation and Communication

***Requirements:*** Develop an elementary message passing capability using UNIX processes created using the `fork` and `exec` system calls with inter-process communication performed using TCP/IP sockets. Demonstrate operation on a multiprocessor shared memory system with simple byte stream data transfers. The environment developed should be similar to other MPI implementation with program initiation taking place via the following command:

```
mympirun -n n_proc a.out
```

where `n_proc` is the number of copies of the executable (`a.out`) that will be created. The program that gave rise to executable `a.out` should include calls to functions `mympi_init` and `mympi_finalize` that are part of the `mympi` library and are responsible for initializing and terminating the message passing environment.

***Comments:*** As mentioned above a second year course in concurrent and distributed systems is a prerequisite for enrolment. This provided the students with a basic understanding of process creation using `fork` and `exec`, and some exposure to buffered asynchronous message passing using pipes (not sockets). Fundamental design decisions relating to performance, extensibility and understandability were discussed, with particular attention given to the topology of the connections between the processes. In lectures the latter was related back to the more general issue of network topology on parallel computers. For simplicity a completely connected network was chosen, although this was highlighted as a potential bottleneck for very large process count. How to communicate the various socket port numbers between the different processes was discussed, as was ensuring that the network of processes was established in a deadlock free manner. These issues were solved by augmenting the command line arguments that were passed to the user program in the `exec` call and by imposing a specific ordering when the connections were established. From this work the roles of `MPI_Initialize` and `MPI_Finalize` in a real MPI implementation were immediately apparent to the students. The final demonstration involved sending a simple byte stream around a ring of processes – a so called "communication snake" or "com-snake" program.

## 2.2 Stage 2 – Rank, Size, Typed and Tagged Communication with Multihosts

***Requirements:*** Write the equivalent of `MPI_Comm_rank`, `MPI_Comm_size`, `MPI_Send`, and `MPI_Recv`, but without the use of communicators and requiring the send and receive calls only to support `int`, `double`, and `byte` data types. Specifically, the send and receive calls should be tagged, and there should be wild cards that permit receive calls to match messages from any sending process or any incoming tag. Extend the original implementation to run on multiple platforms of the same type.

***Comments:*** Inclusion of message types and tags requires some additional information beyond the message content to be transferred between processes. The concept of a message header naturally follows from this. The ability to receive a message from any process prompts discussion of non-determinism, busy waiting, and use of the select system call. How to match a message with a specific tag requires the receiver to interrogate an incoming message, read its header, and then potentially receive the message. This clearly shows the need for buffers to store header information (and maybe more) for messages that have been interrogated but found not to match the required message tag. The transition from asynchronous to synchronous message passing that occurs in most MPI implementations as the message size increases (and can no longer be held in the intermediate buffer) is now very obvious. Expanding mympi to involve multiple hosts requires thought as to how the hosts will be specified (e.g. via command line list, environment variable, or other means) and how to create processes on remote machines. In particular the option of having daemons on each host responsible for creation of application (a.out) processes, versus direct remote initiation was discussed. Different policies for mapping a.out processes to available hosts were considered, introducing concepts like round-robin and blocked allocation. Some security issues associated with the creation of processes on remote machines was also discussed. The final demonstration was a modification of the com-snake demo, but with typed data and across multiple hosts.

### 2.3 Stage 3 – Heterogeneous Hosts and Global Operations

***Requirements:*** Use XDR (external data representation) to extend the above code to run between heterogeneous UNIX computers (specifically between UltraSPARC/Solaris and x86/Linux systems). Using the mympi rank, size, send and recv routines developed above construct higher level functions for performing collective operations equivalent of MPI_Barrier, MPI_Bcast, MPI_Reduce, MPI_Allreduce, MPI_Gather and MPI_Scatter (again without communicators and only for reduction calls involving summation). Provide theoretical and observed performance characteristics for all of these functions.

***Comments:*** Moving to a heterogeneous environment requires different binaries to be run on each system type, and considering how these locations should be specified. The difference between big and little endian, learnt during earlier courses, was now very obvious. While use of XDR was mandated, its impact on performance was discussed, as was the use of lighter weight alternatives. Various alternative approaches to constructing barriers and other collective operations were discussed and the cost analyzed as a function of number of processes and length of message. (For the student the primary objective was to implement correct collective operations and understand their performance characteristics. This invariably resulted in binary tree type implementations, although more elaborate schemes such as those outlined by Rabenseifner [10] were discussed in class.) The demonstration involved running the com-snake program over a heterogeneous environment, and then using other programs to perform a variety of collective operations.

### 2.4 Stage 4 – Shared Memory

***Requirements:*** Modify the code to use UNIX shared-memory segments and semaphores for message transfers that occur between processes on the same shared-memory node. Compare performance using shared-memory transfers with socket-based transfers.

***Comments:*** The students had at this stage completed a laboratory class that covered basic use of shared-memory segments. They were also familiar with the concept of a semaphore through the second year course in concurrent and distributed systems, although this had not specifically covered UNIX semaphore arrays. To enable easy switching between use of shared-memory and socket based intra-node communications, a command line option was added to `mympirun`. The concept of a group of processes, defined as all processes running on the same host, comes naturally when using clusters of shared-memory processors. The number of shared-memory segments and semaphore arrays to be used was discussed in the context of contention (for shared resources) and possible system wide limits. In the end a model that used one shared-memory segment divided up to provide unique read and write buffers for each pair of processes on the same host was used. How to handle wild card options that may involve data being received either in the shared memory segment or on a socket was solved, rather inefficiently, using a busy wait loop that monitored all possible sockets and semaphores in a round robin fashion. Superior performance was demonstrated by running a simple ping-pong benchmark with and without shared-memory transfers.

### 2.4 Stage 5 – Performance Evaluation and General Critique

***Requirements:*** Perform extensive performance evaluation for all functions in `mympi`. Consider possible TCP/IP tuning options [11]. Download and install at least one other version of MPI (eg LAM-MPI or MPI-CH [12]) and compare its performance with that of `mympi`. Give consideration to issues raised by your fellow developers during the earlier stages of this project and comment on where `mympi` might go from here.

***Comments:*** Since the performance evaluation was undertaken on a rather noisy student computing environment no performance data will be given here, just details of what was evaluated. The Nagle algorithm, used to concatenate small messages on a TCP/IP network, was identified as a possible performance bottleneck raising latencies for small message transfers. Some tests were run to compare transfers with and without the Nagle algorithm invoked. The effect of changing the size of the shared-memory buffers, identified in stage 4 as a performance issue, was investigated. MPI-CH was installed and its performance compared to `mympi` for simple ping-pong transfers and for collective operations. Results both within a shared-memory node and between nodes were obtained. While MPI-CH was found to be slightly faster, the students were pleased to find that it was not hugely faster (noting that this conclusion was drawn from experiments run on a non-dedicated environment). Error handling and error detection (e.g. when a remote process dies) were identified as the two areas that most urgently required further work.

## 3 Discussion

The group project as outlined above constituted 25% of the students final course mark. Another 25% was associated with a more conventional assignment requiring parallelization of a 1-D fast Fourier transform using MPI on a cluster and pthreads or OpenMP (the student could chose) on a shared memory system. The final exam was worth 50%. With just 5 students in the course feedback was informal, and along the lines of the course being hard and requiring considerable work, but that they all greatly enjoyed the challenge and the effort made to run a very different style of course. Of the five students who completed the course, 3 went on to obtain first class honours, while 2 ended up leaving university to take up full-time employment before they had completed their honours year. Of the students who obtained first class honours one is now pursuing a PhD in work related to cluster computing.

As with any group project that comprises an incremental set of steps the danger is that someone in the chain fails to deliver. Fortunately this did not occur, and what the students actually achieved was quite impressive. Likewise for any software development project it is also important that the first student makes wise design decisions, as this will intricately affect all later students.

Having trialed this project once with a small group of able students it would be relatively easy to adapt it to a much larger diverse class by, e.g. dividing the students into groups of mixed ability and having each group work on their own version of `mympi`. The exercise could then be run as a challenge between groups to produce the best performing message passing software.

Perhaps the biggest limitation in the project as carried out was the lack of a dedicated teaching cluster so that the students could obtain reliable performance data. In future, however, this will not be a problem, since due to the success of this course and related project work in the high performance scientific computing course we were awarded in mid 2005 a teaching grant that has enabled us to purchase an 8 node dual core Athlon 64 cluster.

**Acknowledgements.** The author gratefully acknowledge support from the Computational Science Education Program of the Australian Partnership in Advanced Computing.

## References

1. The Australian Partnership in Advanced Computing, see http://www.apac.edu.au
2. ANU Bachelor of Computational Science degree, see http://bcomptlsci.anu.edu.au/
3. COMP3320: *High Performance Scientific Computing*, see http://cs.anu.edu.au/student/comp3320
4. COMP4300: *Parallel Systems*, see http://cs.anu.edu.au/student/comp4300
5. J. Roper and A.P. Rendell, *Introducing Design Patterns, Graphical User Interfaces and Threads within the Context of a High Performance Computing Application*, LNCS **3515**, 18 (2005).
6. *Parallel Programming: techniques and applications using networked workstations and parallel computers*, Barry Wilkinson and Michael Allen. Prentice Hall 1999. ISBN 0-13-671710-1.
7. *Using MPI-2: Advanced Features of the Message-Passing Interface*, William Gropp, Ewing Lusk and Anthony Skjellum, MIT Press, 1999; ISBN 0-262-57132-3.
8. Global Arrays, www.emsl.pnl.gov/docs/global
9. Message Passing Forum, see http://www.mpi-forum.org
10. R. Rabenseifner, *Optimization of Collective Reduction Operations*, LNCS **3036**, 1 (2004)
11. See for example: *Enabling High Performance Data Transfers* http://www.psc.edu/networking/projects/tcptune/
12. LAM-MPI, see http://www.lam-mpi.org, MPI-CH, see http://www-unix.mcs.anl.gov/mpi/mpich

# Learning by Doing: Software Projects in CSE Education

Martin Bernreuther and Hans-Joachim Bungartz

[1] IPVS, Universität Stuttgart, Universitätsstr. 38, D-70569 Stuttgart, Germany
[2] Institut für Informatik, TU München, Boltzmannstr. 3, D-85748 Garching, Germany

Martin.Bernreuther@ipvs.uni-stuttgart.de

**Abstract.** Software development is one of the main routine activities in Computational Science and Engineering (CSE). Nevertheless, there is a huge gap between software engineering techniques available and established today in most fields where mainstream software is developed on the one hand and the typical extent of their application in a CSE context on the other hand. CSE curricula often reflect this tendency by not including software engineering topics adequately. This contribution reports experiences with a new course format called "student project" in the CSE master's program at TU München. There, for about half a year, a group of 4-8 students cooperate on a software development project – this time dealing with molecular dynamics. Although it is one objective to get a well performing code, the project's focus is on the consequent application of software engineering and project management practices.

**Keywords:** computational science and engineering, software engineering, education, molecular dynamics.

## 1 Introduction

If we look at both CSE curricula and CSE practice, there is, frequently, an obvious lack of software-related issues. This seems to be somewhat strange, esp. against the background that a significant part of a CSE researcher's or practitioner's everyday work is about making software (or programs, at least). The reasons of this gap are diverse: On the one hand, today, the focus of software engineering is a completely non-numerical one, while on the other hand most CSE people are strongly convinced that software engineering methods might be helpful elsewhere, but are definitely harmful to code performance. Recently, finally, the opinion that this continuing alienation might lead to a real simulation software crisis has got more and more support from the CSE side, too. First, experience and statistical investigations say that, typically, more than two thirds of software overall lifecycle cost is due to maintenance, only. Second, in the memorandum [1], classical software issues such as valdiation, verification, or quality management are identified as key issues for large-scale software in a CSE context, too. And third, the 2005 PITAC report on Computational Science [2]

V.N. Alexandrov et al. (Eds.): ICCS 2006, Part II, LNCS 3992, pp. 161–168, 2006.

even speaks of a "crisis in computational science software" we are going to face or already facing, resp.

As a consequence of this development, CSE curricula must react and invest more in teaching selected aspects of software engineering, at least. Hence, at Technische Universität München (TUM), we decided to do a first step and to integrate a so-called student project into TUM's international CSE master's program [3]. The student project is a format adopted from Universität Stuttgart, where such projects have been a very successful part of the Software Engineering curriculum for several years [4] and have, recently, also been offered by the authors with a CSE-related topic (*Computational Steering - The Virtual Wind Tunnel* [5]). Of course, normally, planning to add something new to an existing study program is a quite complicated endeavour, since hardly anyone will be eager to give up some course as a countermove. However, we were lucky to have the necessary freedom, since our new honours program *Bavarian Graduate School of Computational Engineering (BGCE)* [6], a special offer to the best students of the CSE program and of two other master's programs, has a ten credit slot for some project work that had to be filled with life anyway, and we used exactly that slot for the new student project.

What are the essentials of this format? First of all, such a student project shall reflect the real-life process of making software as much as possible in an academic environment. For that, a team of, here, 4-8 students of, typically, different background gets the task of designing and developing a precisely specified program system in a limited amount of time (actually 6-8 months). Furthermore, not just the implementation, but several steps of the software development cycle have to be covered – from the very first concepts and design considerations up to the final handing-over of the well-documented code to the customer. This implies that there are different parts to be filled: the advisor, the examiner, and a customer on the instructors' side, as well as a project manager, a configuration manager, and several specialists such as a parallel programmer etc. on the students' side. Here, the explicit existence of a customer as a person not that much involved in the project, but nevertheless deciding about the amount of success at the end, has turned out to be a crucial feature. The students have to present their design and concept to the customer at the beginning and to make a formal contract concerning functionalities, milestones, and deadlines. They have to organize both their work and themselves in a way that ensures the final success of the project. Of course, there are accompanying courses in software engineering or project management – but nevertheless, there is an intended high degree of "learning by doing" in this concept.

As a topic for the first run of this format at TUM, we chose molecular dynamics for nanofluidic simulations. There were several reasons for this decision. First, nanofluidics is a very interesting and active field of research; second, particle methods such as molecular dynamics are not in the primary focus of the CSE program and are, hence, a nice supplement to the program; third, the underlying mathematical models as well as the numerical algorithms needed are not too complicated, compared with an implementation of a Navier Stokes solver, for example; finally, our group is involved in a research project dealing with nanofluidic simulations of condensation processes in real fluids on the molecular level, which provides a nice environment for this student project.

The remainder of this paper is organized as follows: In Sect. 2, we briefly present the necessary fundamentals of molecular dynamics simulations for nanofluidics. Then, in Sect. 3, the issue of parallel computing – as far as relevant for this student project – is addressed. Afterwards, Sect. 4 is dedicated to the outline of the student project *MolDyn* itself. Finally, in Sect. 5, the current state of the still running project as well as outcomes and future perspectives are discussed.

## 2 Molecular Dynamics Basics

The classical Molecular Dynamics (MD) simulation is well-established for the simulation of fluids on an atomic scale [7, 8, 9]. For real simple fluids, a molecule is modelled as a set of atoms or sites, resp. Molecules are rigid, which means that there are no internal degrees of freedom and only non-bonded interactions have to be considered. The student project has to implement a single site type only: the well-known Lennard-Jones (LJ) sphere. Sites interact with each other and there's a potential associated to this interaction. Only binary interactions and hence pair potentials are considered.

The LJ 12-6 potential $u_{ij}(r) = 4\varepsilon\left[(\sigma/r_{ij})^{12} - (\sigma/r_{ij})^{6}\right]$ for two LJ centers $i$ and $j$ with positions $\boldsymbol{r}_i$, $\boldsymbol{r}_j$ and distance $r_{ij} := \left|\boldsymbol{r}_j - \boldsymbol{r}_i\right|$ is a semi-empiric function. It covers repulsion through the empiric $r_{ij}^{-12}$ term and dispersive or van der Waals attraction, resp., through the physically based $r_{ij}^{-6}$ term. Therefore, it can be used to model the intermolecular interactions of non-polar or weakly polar fluids. The reference length parameter $\sigma$ corresponding to an atom diameter and the energy parameter $\varepsilon$ are adjustable parameters. For mixtures, the modified Lorentz-Berthelot rules provide these for the interaction between molecules of different components $A$ and $B$: $\sigma_{AB} := (\sigma_A + \sigma_B)/2$ and $\varepsilon_{AB} := \xi\sqrt{\varepsilon_a \varepsilon_B}$. The parameter $\xi$ is heuristic and adjustable with a value close to 1 [10]. The force between two sites is given by the gradient $\boldsymbol{F}_{ij} = -\nabla u_{ij}$ and fulfils Newton's third law (*actio* = *reactio*). Typically, Hamiltonian mechanics is used, and the phasespace consists of all molecule positions $r_i$ and velocities $v_i$. Newton's equations of motion for the atoms set up a system of ordinary differential equations: $\dot{\boldsymbol{v}}_i = \sum_{j \neq i} \boldsymbol{F}_{ij}/m_i$. The resulting initial value problem is solved numerically with some suitable time integration scheme. Regarding a canonical ensemble, the number of molecules $N$, the volume $V$, and the temperature $T$ are kept constant. The cuboid domain with periodic boundary conditions is used and during the simulation run, the temperature is controlled with a thermostat. This is achieved through a simple isokinetic scaling of the velocities here. Regarding multicentered molecules, the phasespace also has to contain the orientations and angular velocities. The intermolecular force between two molecules is the sum of all site-site forces, which also generate torque. An enhanced time integration scheme takes care of that [11].

## 3 Parallel Algorithms

Due to the enormous computational requirements resulting from the large number of molecules to simulate and the small time steps, MD still abuts against the limits of today's possibilities even on supercomputers.

First, the sequential algorithm has to be optimized. Due to the property of the used potential and force to decay very fast with increasing distance, there are typically many very small values to be found in the force matrix, which corresponds to the interaction graph. Considering only interactions of molecules below a certain distance, the cut-off radius $r_c$, these values will be neglected and set to zero, and the matrix will get sparse. To avoid unneccessary distance calculations, the linked cell method is applied. It, first, bins all molecules using regular cells and then only examines molecules within a cell and neighbouring cells within the cut-off radius. The cells approximate the sphere-shaped interaction volume, and Newton's third law can be incorporated, including only half of the neighbours and adding/subtracting a calculated force to both molecules at once. The algorithm achieves linear complexity for short-range MD fluid simulations and provides a basis for parallel versions.

The target platform are clusters of workstations, which belong to the class of distributed memory (NORMA) machines. There are three common parallelisation strategies known for parallel MD simulations [12]: Replicated Data (RD), Force Decomposition (FD), and Spatial Decomposition (SD).

Replicated Data, which is also known as Atom Decomposition, is the easiest, but least efficient strategy: each Processing Element (PE) will store all data, but calculates only one part of the molecules without taking into account their positions. In contrast to a shared memory parallelisation where all the data can be accessed from every PE by definition and are stored only once, the redundant replicated data has to be synchronised. The communication-computation ratio is usually quite large for massively parallel systems, since each PE has to communicate with all other PEs.

Force Decomposition not only distributes the molecules to update among the PEs but also assigns parts of the force matrix to each PE (cf. Fig. 1), where each PE calculates only partial forces of its molecules. Rearranging the columns reduces the communication partners for each PE. Regarding e.g. PE 7 in Fig. 1, molecule positions are needed from PEs 3,11,15,5,6,8 and partial sums have to be exchanged with 5,6,8. Each PE communicates with $O(\sqrt{p})$ PEs, where $p$ denotes the number of PEs. But this is depending on the number itself. A square number is favourable whereas a prime number leads to the RD approach.

Spatial Decomposition is used for the parallelisation of various applications. The whole domain is divided into subdomains and distributed among the PEs. Regarding MD, each PE calculates the associated molecules. Special care has to be taken regarding interactions between two molecules belonging to different PEs. These occur only at the boundaries, and the molecules concerned have to be replicated and synchronized. This is equivalent to enlarging each subdomain with an overlapping halo region. A heterogeneous distribution of molecules entails unequal molecule quantities for the PEs and requires load balancing.

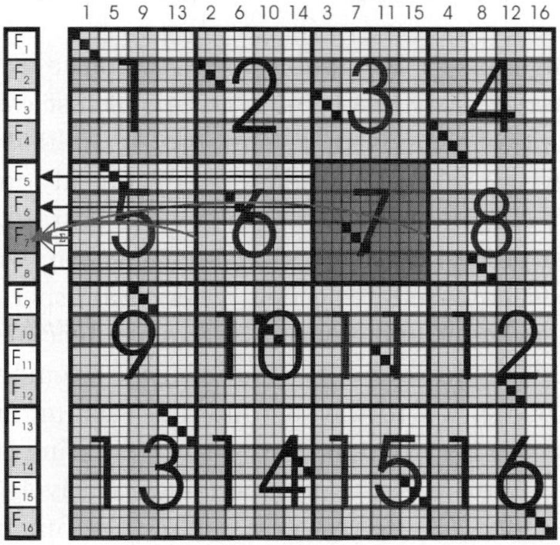


**Fig. 1.** Force Decomposition for 16 PEs (The force matrix calculation is distributed among the PEs, which have to communicate molecule positions and partial sums.)

# 4 The *MolDyn* Project

## 4.1 Task

The student project is about a software system to perform molecular dynamics simulations and visualise the results. The simulations have to deal with mixtures of rigid multiple sites molecules. To organise the molecule types used, the component definitions are stored in an extensible library. Tools shall help the user with the set up and administration. There's also a need for utilities to generate the necessary input data for multicomponent simulations from macroscopic parameters, as mentioned in the first example of Sect. 4.2. Since a simulation run typically takes a long time, an interruption has to be possible, whereas the program will save all necessary data to a file needed for a restart. Platform independence is required to some extent to run the simulation on various supercomputing environments offering MPI communication and also on forthcoming systems, since the rapid product cycle for computer hardware is expected to be substantially shorter than for the simulation software. The targeted user is an experienced researcher. It has to be considered that the program structure must be flexible and should allow to include innovations such as new site types or evaluation functions in the future. Due to the large computational requirements, the modularity has to be combined with efficiency, which refers to the chosen algorithms as well as to implementation issues. Documentation is always important, but particulary if the code shall serve as a base for further development by the customer.

It's not only the result that counts, but also the fulfilment of conditions defined within the contract. Besides the detailed system requirements, which have to be negotiated with the customer, mainly time restrictions have to be obeyed. A time plan including milestones is mandatory and part of the contract.

### 4.2 Test Cases

The simulation package has to be capable to simulate the dynamics of multicentered molecules. It particularly has to pass the given test cases, presented next. Fig. 2 shows an example of a canonical ensemble of $N = 40000$ atoms in a domain of volume $V^* = 97^3$ with periodic boundary conditions at temperature $T^* = 0.7$. A face-centered cubic lattice serves as the initial configuration (cf. Fig. 2a). The observable nucleation process initiates a phase transition, which also has consequences for the computational effort needed. There is another $NVT$-ensemble test case, where spherical methane and ethane clusters of molecules in a domain of volume $V^* = 120^3$ with periodic boundary conditions at a temperature $T^* = 1.0739$ are colliding. The vaporisation to the surrounding vacuum and the collision will spread the molecules over the domain, but at the beginning the clusters cause an inhomogeneous distribution over the domain and a relatively large amount of interactions to be calculated. These test cases define the functionality of the code and constitute a requirement for the acceptance of the final product.

**(a)** $t^* = 0$ **(b)** $t^* = 100$ **(c)** $t^* = 500$ **(d)** $t^* = 3000$

**Fig. 2.** Nucleation process

### 4.3 Related Courses

Of course, although there is a strong component of "learning by doing" in this format, the participating students are not left alone – concerning neither MD nor the software and project issues. For example, the customer provided a detailed description of the product expected before the project started, and there was a block course on MD tailored to the project's needs as well as to the students' knowledge and held esp. for the participants of this project. Furthermore, as a part of the regular CSE course program, there is an accompanying lecture on software engineering, which provides the basic background for this part of the project. Hence, the student project is well embedded into the CSE and BGCE programs at TUM.

## 5 Current State and Outcomes

The first action of the new team was to self-organize. Each of the five members got his/her own role with corresponding responsibilities, such as project manager, configuration manager, quality assurance, development, or documentation.

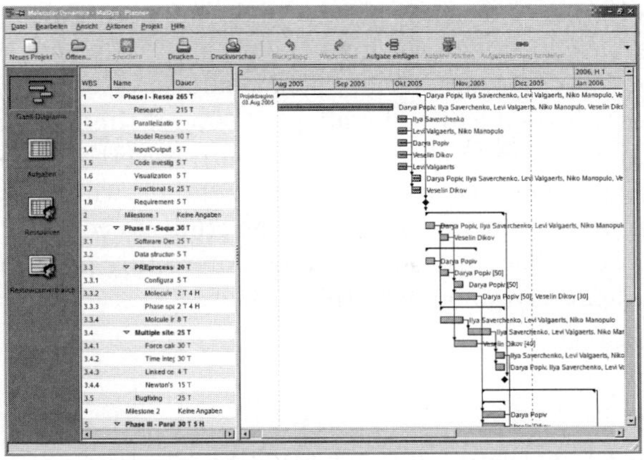

**Fig. 3.** Time planning with Gantt-diagram

The team also had to agree upon a common environment in terms of tools to be used. Besides a document processor and a project management system also tools such as an IDE and a CVS had to be chosen. The students also had to get acquainted with MD and MD implementation details, which was further supported by the advisor and additional literature. Before the design and implementation phase, the requirements and functional specification was written. Customer meetings to clarify the requirements were supplemented by electronical communication via email for renegotiation.

An important issue for the accomplishment of a project is time management. The students had to set up a work plan. Working packages to carry out the different tasks had to be identified and associated to a time schedule with related milestones. The first of these mandatory milestones included the work plan itself. Fig. 3 shows a Gantt-diagram created by the students with some work packages assigned to the team members.

The simulation was divided into the classical parts: preprocessing, computation, and postprocessing. The preprocessing programs dealing with the XML and binary input files are written in Java and responsible to set up an initial configuration for the phasespace to start a simulation. The team was free to use adequate programming languages for different tasks and decided to use C++ for the parallel simulation code. The simulation results have to be evaluated, and a visualisation of the molecules is required. The open-source visualisation framework OpenDX was integrated here to save work. Recently, the second milestone was obtained and a sequential version of the program, including the interface to and customization of the visualisation, was delivered to the customer. Right now, the students struggle with parallelisation issues and work on the implementation of the FD method combined with the linked cells algorithm.

From the very beginning, all five participants have been enthusiastic and highly motivated. Although they all have a bachelor's degree in engineering or science and, hence, more a CSE than a computer science background, they are esp. interested in the software engineering aspects. They organized the project and themselves in a very disciplined way, and they now really experience that applying software engineering principles is not an overhead not worthwhile, but is inevitable for larger projects and, then, really pays off. At the beginning, the original time schedule had to be adapted several times (mainly due to the fact that their MD background was close to zero). Nevertheless, after the initial phase, all deadlines were met, and the team certainly will be able to deliver a quite impressive product at the end. Since the students consider this project to be one of the most challenging, but also most interesting parts of their master's studies, we will definitely go on with this format.

## References


1. D. E. Post and L. G. Votta. *Computational Science Demands a New Paradigm.* Physics Today, 2005 (no. 1), pp. 35-41.
2. M. R. Benioff and E. D. Lazowska. *Computational Science: Ensuring America's Competitiveness.* PITAC report, 2005.
3. Technische Universität München. *International Master's Program CSE – Computational Science and Engineering.* `http://www.cse.tum.de/`.
4. J. Ludewig and R. Reißing: *Teaching what they need instead of teaching what we like – the new Software Engineering curriculum at the Universität Stuttgart.* Information and Software Technology 40 (4), 1998, pp. 239-244.
5. M. Bernreuther and H.-J. Bungartz: *Experiences with Software Projects for CSE Education.* Submitted to Computing in Science & Engineering.
6. Technische Universität München, Friedrich-Alexander-Universität Erlangen-Nürnberg. *BGCE – The Bavarian Graduate School of Computational Engineering.* `http://www.bgce.de`.
7. M.P. Allen and D.J. Tildesley: *Computer Simulation of Liquids.* Oxford University Press, 2003 (reprint).
8. R.J. Sadus: *Molecular Simulation of Fluids - Theory, Algorithms and Object-Orientation.* Elsevier, 1999.
9. D. Frenkel and B. Smit: *Understanding Molecular Simulation - From Algorithms to Applications.* Academic Press, 2002 (2nd ed.).
10. J. Stoll, J. Vrabec and H. Hasse: *Vapor-Liquid Equilibria of Mixtures Containing Nitrogen, Oxygen, Carbon Dioxide, and Ethane.* AIChE J. 49: 2003, 2187–2198.
11. D. Fincham: *Leapfrog rotational algorithms.* Molec. Sim. 8 1992, 165–178.
12. S. Plimpton: *Fast parallel algorithms for short-range molecular dynamics.* J. Comp. Phys. 117 1995, 1–19.

# Computational Math, Science, and Technology (CMST): A Strategy to Improve STEM Workforce and Pedagogy to Improve Math and Science Education*

O. Yaşar, L. Little, R. Tuzun, K. Rajasethupathy, J. Maliekal, and M. Tahar

CMST Institute, State University of New York, College at Brockport, NY 14420
{oyasar, llittle, rtuzun, kraja, jmalieka, mtahar}@brockport.edu

**Abstract.** Integrated approach to education in science, technology, engineering, and mathematics (STEM) has recently become a common practice. Technologies such as wireless graphing calculators, electronic whiteboards, and computers help engage students in classroom activities. Modeling and simulation software allow experimentation without usual prerequisites while enabling a pedagogical approach to display interconnectedness of math and science. Controlled simulations and layered approach to learning seem to offer long-sought support for inquiry-based curricula in public schools. Today's jobs require multiple skills, necessitating a broader college education. While students with multiple majors can be seen on college campuses, the overall number of degree holders in technical fields has gone down steadily. A handful of institutions have established multidisciplinary programs to adjust to these concerns and to better serve diverse interests of students.

## 1 Introduction

Estimates of the size of the STEM workforce in the USA varies between 3-10 million individuals, depending on the definition and perspective used. For the period from 2000 to 2010, Bureau of Labor Statistics predict that employment in science and engineering occupations will increase about 3 times faster than the growth rate for all occupations [1]. The economy as a whole is expected to provide approximately 15 % more jobs over this decade, with employment opportunities for STEM jobs expected to increase by 2.2 million jobs, or about 47 percent. Approximately, 86 percent of the projected increase in STEM jobs is in computer-related occupations.

This projected trend is supported by the current and past data released by the American Institute of Physics [2] and the U.S. National Science Foundation. According to AIP survey, physics graduates indicated that the three most important skills in their jobs were scientific problem solving, synthesizing information, and mathematical skills. The importance of math and computer skills

* Authors acknowledge support by National Science Foundation's Math and Science Partnership grant (EHR-0226962) and Cyberinfrastructure grant (SCI-0520036).

V.N. Alexandrov et al. (Eds.): ICCS 2006, Part II, LNCS 3992, pp. 169–176, 2006.

is also evidenced in National Science Board's 1996 report [3]: 74 % of math and computer science doctorates worked in the same field as their degree versus 57 % in physical sciences. A similar trend is seen at the bachelor's level (35 % versus 22 %). While there is a great demand for math and computer skills, this data also shows the importance of a science education that may open doors to a variety of jobs. An education combining math and computer skills with a science or an engineering field would be a valuable investment. This is particularly important for undergraduate students because the likelihood of being employed outside of one's field is high for most disciplines. Over the past several years, some science and engineering disciplines have added more math and computing courses in their curricula while math and computer science disciplines increased student exposure to areas of applications (see the ABET accreditation requirements for Computing programs). A more dramatic approach, yet still in infancy, has been creation of a new interdisciplinary field, namely computational science and engineering, which offers an integrated approach to STEM education.

While there seems to be a great demand for STEM workers, there is a fundamental problem with STEM education within the USA. The percentage of bachelor's degrees in STEM has steadily gone down; from 35% in 1966 to 31% in 2001 [4]. The same trend is seen with master's (29% to 21% ) and doctoral (64% to 62% ) degrees. When one looks at the pre-college student achievement in math and science, the lack of interest in STEM becomes obvious. The United States ranks at the bottom of the 12th grade TIMSS (Trends in International Mathematics and Science Study) exam. Math and science are perceived to be difficult topics and only 3% of boys and 1% of girls in grade schools go on to pursue a STEM-related education in college. To make matters worse, there is a shortage of high school students nationwide [5]. Several reports have been issued to draw attention to this crisis. The Business-Higher Education Forum [6] declared that competence in mathematics and science are essential to all individuals. Realizing the seriousness of the crisis, NSF created a new program [7] to promote a strong partnership of stakeholders (college faculty and school teachers) and subject areas. A recent report [8] by President Bush's Science Advisor introduced Computational Science as one of the most important disciplines of the 21st Century. It further stated that progress in Computational Science is seen as a way to ensure America's Competitiveness in a new world of increasing global competition.

## 2 CMST Pedagogy: Secondary Schools

The CMST Institute promotes a strong partnership of higher education and school districts as well as of subject areas (math, science, and technology). While colleges demand better students from school districts (SD), school districts demand better teachers. Their success is interdependent. The partnership of subject areas comes from their interconnectedness as intended by nature. Through computation, one can see the interplay of math, science, and technology to run controlled simulations and scientific inquiries.

At secondary school level, students cannot be expected to construct mathematical and computer models to run scientific simulations, unless there are tools to make transparent the underlying mathematics, programming, and the detailed scientific content of the experiment. Latest technologies have brought powerful tools to the aid of researchers and educators. It is now possible, indeed, to construct simulations without usual tedious task of programming (FORTRAN, C++, or JAVA) and advanced mathematics. We have used tools such as Interactive Physics [9], AgentSheets [10], Stella [11], Geometer's SketchPad [12], Excel, Project Interactivate (www.shodor.org) and graphing calculator software along with laptops, TI-84 graphing calculators [13] and electronic smartboards [14].

Since 2003, we offered CMST training to more than 265 teachers, with majority being from Rochester City and Brighton Central School Districts. 148 teachers received summer training once (60 hours), 42 teachers twice (108 hours), and 15 teachers three times (156 hours). We also offered, through collaboration with Texas Instruments, TI-certification training (60 hours) to 38 MST teachers. Following the summer workshop, teachers are assigned coaches among college faculty and experienced teachers. They implement CMST-based lessons during the academic year and also participate with their students in the CMST Challenge program. We hold bi-monthly meetings for all faculty, teachers and coaches. CMST Teachers receive laptops, graphing calculators, viewscreens and share electronic smartboards and LCD projectors. Testimonies from teachers show that electronic smart-boards are very effective, as are LCD projectors. According to teacher testimonies, while it usually takes 6 months for a student to come to a blackboard for an in-class demonstration, it only takes 6 hours with the smart board.

Although we rely on teacher and faculty training to impact classroom learning, we reach out directly to middle and high school students as well. In 2005, we reached more than 347 students through Interactive Physics Week and the Challenge Competition. A ceremony was held to display project posters and demonstrations. Outside evaluators judged the projects and selected projects were presented and displayed during the 2005 summer institute. These projects set good examples for CMST integration into classroom teaching as well as how modeling can be applied to solution of real world problems. Evidence thus far has shown that students even at middle school can indeed learn modeling tools and use them to further their knowledge of science and appreciation of math.

Our project evaluators have found a correlation between the number of CMST Teachers in a school and the improvements in student achievement. At the same time, there seems to be a strong correlation between the amount of teacher training and student achievement. As part of institutionalizing our project activities, these trainings are now formalized as credit-bearing college courses and offered outside the summer institute. The CMST training has also been integrated into the professional development days in school districts. Hundreds of CMST-based lesson plans posted on project web site [15] are enhancing our outreach efforts, as they enable CMST teachers to demonstrate to others the effectiveness of the new pedagogy. Instructors from our suburban partner [16] helped us train our urban school [17] teachers. Integration of CMST approach and tools has been promoted within the context of NSF-supported Connected Math [18] and Core Plus [19] curricula.

Student achievement improved considerably at Rochester City School District (RCSD) since the beginning of our project in 2003. The district average passing rate reached 90% in Math-A Regents exam, a 39-point jump from June 2003. Similarly, the district passing rate in state 8th grade math exam improved by 8 percentage points, from 11% (2003) to 19% (2004). In 2005, the district maintained the gains made since 2003.

Findings from RCSD show a consistent multi-year trend, including: a) students receiving instruction from multiple CMST teachers outperformed students receiving instruction from only one CMST teacher, b) data suggest a correlation between the number of CMST teachers in a given school and the performance of participating students, c) teachers indicate with greater frequency that student performance is improving throughout the school year in their classrooms, d) student performance in technology classrooms exceeded performance in science and math, which probably indicates that students in technology classes have higher access and exposure to CMST tools than others in theoretical math and science classes, e) students taught by beginner-level teachers in 2004 did better than those in 2003, which supports other data that the quality of summer institute got better as years passed as a result of several factors including faculty experience as well as inclusion of peer-teaching, f) academic performance of students participated in the Challenge activity exceeded those who did not, g) attendance of students increased with exposure to CMST pedagogy, and h) dropout rates were lower for students exposed to CMST pedagogy.

Technology access is not a problem at Brighton, contrary to RCSD. A stark contrast between these two districts is that the number of courses and the curriculum seem to be fixed at BCSD while there are many variations and courses within RCSD schools. A similarity between BCSD and RCSD is that Brighton technology treatment classrooms had the highest acceptable grade rate, followed by science and math. This again points out the fact that treatment is maximized in classrooms with adequate technology infrastructure. Another similarity is that although the acceptable grades (A/B/C) rate among the suburban-based staff far exceeded that for the urban staff, the trend of the more comprehensively trained teachers from the cohort having more successful students continued. The achievement by students in the Challenge Program in BCSD also exceeded other peer groups within the same school. Three Brighton students [20] in the 2005 Challenge program wrote the following in their report:

> *Our motivation to prove Kepler's laws has historical significance. As 9th graders, we have not taken any physics course and we are not fully knowledgeable about laws of universe that govern planetary motion. That is not different from the situation of Kepler; as no one quite knew how gravitational forces worked until Newton came. Kepler had access to data compiled by Tycho Brahe and he looked for patterns. We had access to CMST tools and experienced teachers and we looked for miracles! We learned how to transfer visuals images from Interactive Physics*

*into Geometer's Sketchpad, which was a tough endeavor all throughout. We also understood better, through Excel, the physical laws and the computational steps involved in Interactive Physics simulations. While we did not make a discovery in physics, we certainly discovered that physics is not a threatening or boring subject after all. The foreboding nature of complicated physics was abolished as we worked on this project and now we look forward to taking AP physics classes offered in our school.*

## 3 CMST Education: Workforce Preparation

SUNY Brockport launched the first Computational Science (CPS) undergraduate degree program in the USA in 1998. The department has developed BS, MS, and combined BS/MS degrees in this area, the curricula of which integrate math, computer science, and application sciences. The department has also played an active role in the establishment of other similar programs in the United States and elsewhere. The CPS department has generated workers with practical knowledge and problem solving skills in 3 technical fields (math, science, and technology) to raise quality, adaptability, and versatility of our regional and national workforce. Recruiting and sustaining a sizable student body in a multidisciplinary STEM program is a tough challenge. Thus far, however, the department graduated more than 35 students with both BS and MS degrees; all gainfully employed.

The CPS program is built on seven (7) student learning outcomes, including computational tools, high performance computing, applied and computational math, simulation and modeling, visualization, application sciences, and technical communication [21-31]. It seeks high parameter students who are interested in modeling, programming, and applied mathematical skills in industrial, governmental, and academic settings. Through the minor, the department also seeks students who are interested in learning these skills but not necessarily making them the major focus of their careers. Through general education courses the department seeks to educate the more general college population about the applications and relevance of computational modeling. Through the CMST Institute the department seeks to teach high school and middle school teachers the use of computational modeling software in the classroom, in order to help bring up a more science- and math-literate population.

Preparing a student to credibly learn 3 subject areas (mathematics, computing, and sciences) in 4 years has been a tough challenge. Advising a student body with a broad education base is a challenge for faculty members. The department has gone through 3 major revisions to balance its diversified course work. Even though there is no accreditation in this new field, the department sought peer review from local, regional, and national communities via presentations in conferences and articles in journals [21-31]. Within just 7 years, the department developed and taught 27 new courses, including Introduction to Computational Science, Computational Science Tools I-II, Internet and Technology Ethics, Forensic Computing, History of Science and Technology, High Performance Computing, Simulation and Modeling, Computational Mathematics, Embedded Computing, Computational Chemistry, Computational Physics, Computational Finance, Scientific Visualization, Computational Fluid Dynamics, Computational Biology, Computational Methods for Teachers I-II-III, Advanced Software Tools, Computational Methods in the Physical Sciences, Deterministic and Dynamical Systems, Supercomputing Applications, and Society, Technology, and Science.

Recognizing the difficulties student might face in a multidisciplinary program, the department sought to offer scholarships and internships for students, which created a pipeline of high quality students with internship experiences, employment opportunities, and doctoral assistantships in places such as Harvard, Carnegie-Mellon, Lockheed Martin, Ford, Renselear Polytechnique Institute, George Mason, Sandia National Labs, Eastman Kodak, to name a few. Department faculty members have been very active in recruitment, both domestically and internationally. They developed brochures, posters, and other promotional materials using personal time and took a program from 3 undergraduate students in its first semester to a steady enrollment of 50 undergraduate and graduate students over several years.

As described earlier, the department formed a multi-disciplinary (CMST) institute in 2002, with members from eight departments and two school districts. The institute sponsored funds and team work, and provided a platform for close collaboration between departments and between the college faculty and school teachers. CMST faculty developed several in-service courses for teachers and teacher candidates. These include four natural science courses, two liberal studies courses, and three General Education courses (Internet Ethics, Forensic Computing, and History of Science and Technology). Courses such as Computational Chemistry and Computational Biology (Bioinformatics) are now being taught by their respective departments.

Besides 27 new courses in the department, CPS faculty also taught 15 courses outside the department, particularly in math, computer science, education, and liberal studies programs. The department has continuously and consistently strived to serve the entire college community, and beyond, through curriculum development in which computational modeling is used as a vehicle for understanding. Institute activities led to modifications in more than 20 existing courses in the college, including Forensic Computing, History of Science and Technology, College Geometry, College Algebra, Pre-calculus, Calculus I-II-III, Business Calculus, Finite Math, Math for Elementary Teaching I-II, History of Math, College Physics I-II, Field Methods in Earth Sciences, and Anatomy and Physiology.

Through the grants and equipment donations, the CPS department provided access to supercomputers, high-performance servers, laptops, LCD projectors, graphing calculators, and electronic smart boards for more than 15 faculty members in support of their professional development, scholarly activities, best-practices, and teaching methodologies. The multidisciplinary aspect of the computational science program attracted large scale funding from the government agencies. The CPS faculty brought the largest single grant ($ 3.8 million) in the history of our college. Although the CPS department is rather small (3 faculty members), it has provided sponsored funding for more than 12 faculty members in math, computer science, physics, chemistry, earth sciences, biology, and education departments.

CMST, or briefly computational science, has gone through several stages of development as witnessed from the history of our program at Brockport. Leading the nation in CMST education, Brockport program was first to introduce computational science as a discipline of its own, raising it from a mere overlap of mathematics, computer science, and sciences as depicted in [25]. Availability of high performance computer power and wide use of computational techniques in many fields in later stages led to creation of subfields within the computational science community such as computational physics, finance, and others. Today, as witnessed from our own experience, computational science has become a core for many new scientific fields; a role similar to the one played by mathematics. From a new pedagogy to teach math and science to tools for industrial design and to a methodology to perform scientific research, computational science has proven to be a high-return investment area for academia, industry, and public education.

## 4 Conclusion

CMST education is an innovative approach to prepare new generations for the challenges of the information age. The amount of new knowledge is doubling every year and it is a challenge to manage exabytes of digital information created every year. Today's jobs require multiple skills and multiple domains of knowledge. Professionals no longer are specialized in one domain, but in multiple domains. How do we filter this tremendous amount of information and present it in a hierarchical way? The growth of Google (a search engine) is a good indication that filtering and presentation of volumes of data to those who need it is very important. How do we educate new generations?

Technology has changed the way we learn and teach. Campuses nowadays discuss about 'active learning' and seek ways to deliver more content with less cost. We need to better prepare students with general mathematical and computational skills as well as disciplinary knowledge. We need to constantly assess whether we are preparing tomorrow's workforce and leaders. College faculty needs to constantly renew itself in the sea of information. The best way to deal with this is to have a mechanism that will deal with newly created multi-faceted knowledge in a collaborative way. In our CMST experience, we have seen concrete results, ranging from faculty/student quality and quantity in STEM to teacher/student quality in public schools. However, we also recognize a major difficulty to be the lack of necessary human capacity on a single campus. Courses that integrate CMST tools into application fields (science, engineering, business, and humanities) need to be fully developed. Development of student learning outcomes and well-designed assessment strategies, a well-defined curriculum, and a collection of tools and application courses is essential to successfully implement a CMST program. Computational science programs need to engage more disciplinary faculty in their campuses or collaborate with each other via distance learning and other digital repositories to enrich resources offered locally.

## References


1. Bureau of Labor Statistics, 2001, *http://www.bls.gov/emp/home.htm*
2. Courtesy The American Institute of Physics, *http://www.aip.org*
3. National Science Board, Science and Engineering Indicators, National 1996 Report
4. National Science Foundation, *http://www.nsf.gov/statistics*
5. New York Times, Feb 5 (2005)
6. Responding to the Crisis in Mathematics & Science Education, The Business-Higher Education Forum, *http://www.bhef.com*
7. NSF Math and Science Partnership Program, *http://www.ehr.nsf.gov/msp*
8. PITAC Report, Computational Science: Ensuring America's Competitiveness, *http://www.nitrd.gov/pitac.*
9. Interactive PHysics Website, *http://www.interactivephysics.com*
10. AgentSheets Website, *http://www.agentsheets.com*
11. Stella Website, *http://www.iseesystems.com*
12. Geometer's Sketchpad Website, *http://www.keypress.com/sketchpad*
13. Texas Instruments Website, *http://www.ti.com*
14. Blue Ribbon Learning Sites on the Web, *http://www.kn.sbc.com/wired/bluewebn*
15. CMST Website, *http://www.brockport.edu/cmst*
16. Brighton School District, *http://www.bcsd.org*
17. Rochester City School District, *http://www.rcsdk12.org*
18. NSF-supported Connected Math Project, *http://www.math.msu.edu/cmp*
19. NSF-supported Core Plus Project, *http://www.wmich.edu/cpmp*
20. S. Kashyap, J.P. Yaşar, R. Ravago, *http://www.brockport.edu/cmst*
21. O. Yaşar, J. Maliekal, L. J. Little, D. Jones, Computational Technology Approach to Math and Science Education, IEEE CiSE (to appear)
22. H. Dag, G. Soykan, S. Piskin, and O. Yaşar, CSE Education at Istanbul Technical University. IEEE Comp. in Sci. & Eng (2005), Vol. 7 (1), 72-77
23. O. Yaşar, CMST Pedagogical Approach to Math and Science Education. LNCS, Vol. 3045 (2004), 807-816
24. H. Dag, G. Soykan, S. Piskin, O. Yaşar, CSE Education, LNCS, Vol. 3045 (2004), 795-806
25. O. Yaşar and R. Landau, Elements of CSE Education. SIAM Review (2003), Vol. 45 (4), 787-805
26. O. Yaşar and R. Landau, Elements of Computational Science Education. Proceedings of SIAM CSE Conference, Feb 2003, San Diego, CA.
27. O. Yaşar, The Bachelor of Science Degree in Computational Science at SUNY Brockport. Proceedings of SIAM CSE Conference. Feb 2003, San Diego, CA.
28. L. Little, The Computational Science Major at SUNY Brockport. FGCS (2003), Vol. 19 (8), 1285-1292
29. O. Yaşar, Computational Science Education: Standards, Learning Outcomes and Assessment. LNCS, Vol. 2073 (2001), 1159-1169
30. O. Yaşar, K. Rajasethupathy, R. Tuzun, A. McCoy, and J. Harkin, A New Perspective on Computational Science Education. IEEE CiSE (2000), Vol. 5 (2), 74-79
31. O. Yaşar, Computational Science Program at SUNY Brockport. Proceedings of the First SIAM CSE Conference, September 21-24 (2000), Wash, D.C., *http://www.siam.org/meetings/cse00*

# The School of Computational Science at Florida State University

Gordon Erlebacher and Janet Peterson

Florida State University, Tallahassee, FL 32312
erlebach@csit.fsu.edu
http://www.scs.fsu.edu/~erlebach

**Abstract.** Florida State University has recently launched a new program in Computational Science. In this paper, we describe the need for computational science, how we defined computational science, and explain the rational behind our computational science curriculum.

## 1 Introduction

Over the last few decades, computations have joined theory and experimentation to form the three pillars of scientific discovery and technological design. Moreover, in many cases scientific computations have superseded both theory and experimentation. Thus, whether one is studying subatomic particles or galaxies, whether one is designing minute nano-composites or huge skyscrapers, whether one is sequencing the human genome or protecting fragile ecosystems, whether one is studying the flow of blood in capillaries or predicting the winds in a hurricane, computations play a central role. The computations that enable these and a myriad of other studies depend on the invention, implementation, and testing of algorithms and software that computers use to solve scientific and engineering problems. This is the work of computational scientists.

The public has little awareness of the important and ubiquitous role of computational science in their everyday lives. Especially here in Florida, people hear on the Weather Channel that computer models predict that hurricane X will make landfall at city Y at time Z as a category W storm but they are completely unaware that CAT (computer aided tomography) and MRI (magnetic resonance imaging) scans are completely reliant on the contributions of computational scientists that allow for the transformation of measurements taken by the scanning device into images that a radiologist can read and interpret. Nor do they realize as they look out the window of a Boeing 777 that the wing they see was designed using algorithms developed by computational scientists that predict both its aerodynamical and structural behavior, with experiments used just to ensure the computations were right. Nor do they know that large retail companies such as Walmart and Target use sophisticated computational algorithms to manage the distribution and delivery of goods to their many stores in an efficient manner so that costs are kept down. In these and countless other settings, computational scientists are indispensable contributors.

V.N. Alexandrov et al. (Eds.): ICCS 2006, Part II, LNCS 3992, pp. 177–184, 2006.

The high national priority of computational science is amply illustrated by the fact that the President's Information Technology Advisory Committee (PITAC) in 2003 chose it, along with health care information technology and cyber security, as the three areas of greatest national importance related to information technology. The Committee's June 2005 report on Computational Science [2] states that the most scientifically important and economically promising research frontiers in the 21st century will be conquered by those most skilled with advanced computing technologies and computational science applications.

Based on this and earlier trends, Florida State University has established the School of Computational Science with the following missions:

1. Develop innovative interdisciplinary graduate training programs in scientific computing and its applications;
2. Foster research in scientific computing and its applications in a variety of disciplines;
3. Provide a supportive environment for high performance computing on the providing the best possible education and training in computational science should be an item of high priority at any research university.

In this paper, we concentrate on the first mission.

## 2 What Is Computational Science

There is no accepted definition of computational science. Many institutions equate computational science with computational frameworks, problem solving environments, etc. While that is certainly viable, we define it somewhat differently. At Florida State University (FSU), we state that Computational Science lies at the intersection of Mathematics, Computer Science, Applied Science and Engineering. Thus, the common themes that thread all these areas constitutes computational science (see Figure 1). Examples of such commonalities include algorithm development, multiscale techniques, transform methods, an appreciation for frameworks and problem solving environments, scientific visualization, stochastic methods, optimization, and others. Each of these subject areas encompass techniques that are used in almost all applied areas of science. Advances in multiscale algorithms and theory will further the development of the physics of turbulence, improve understanding of geological and climate processes, lead to better modeling of biological processes in the human body, etc. This definition is consistent with the following quote from the PITAC report[2]: "While it is itself a discipline, computational science serves to advance all of science."

Many of the important problems facing society today can only be solved by teams of individuals from several disciplines. Among these teams, one or more computational scientists, according to our definition, will serve as the glue that will help members from these different teams communicate with each other, translate each others languages, and facilitate the transfer of technology, whether algorithmic or deliverables.

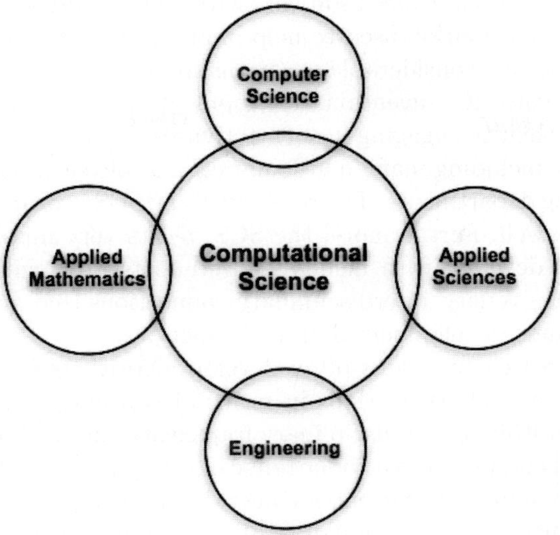


**Fig. 1.** The interdisciplinary nature of computational science

## 3 Multi- and Inter-disciplinary Research

In the past, the inter- and multi- disciplinary nature of computational science has been largely ignored in the training of computational scientists. That training has been mono-disciplinary in nature, with students largely confined to courses and research within just one of the many disciplines that intersect with the world of computational science. Thus, despite the fact that the need to improve the training of computational scientists is universally recognized and despite the fact that numerous reports (see references within [2]) have stated that the United States is facing a national crisis due to the its diminishing leadership role in computational science, not much has been done to improve the situation. A list of existing graduate programs in computational science is available at [1].

The Computational Science program at Florida State University is committed to improving the training of future computational scientists at the Ph.D. level through an innovative curriculum with courses designed to function across disciplines rather than *within* a single discipline.

Computational methods for solving problems in science and engineering are ubiquitous within the University. The vast majority of the natural and social science and engineering departments, centers, schools and institutes in the University include faculty, postdoctoral associates, and students who use computational methods in their research. Moreover, several non-scientific departments such as the film, dance and business departments have a need for generic tools, e.g., for the creation of animated films, for choreographing dances, for marketing and finance decision making, etc. Even the athletic department makes use of (although does not develop) advanced software with feature detection algorithms to extract important information used to help improve their athletes.

The University also has considerable strengths in those aspects of computational science that involve the invention, development, and testing of new computational methods. Faculty engaging in such activities are scattered throughout the campus, certainly including many mathematical and physical science departments and engineering departments. The research of the SCS is primarily devoted to these activities as well. Furthermore, the SCS, by its very mission, seeks to exploit, explore, and develop, with faculty and students both within the SCS and throughout the University, interdisciplinary connections that synergistically can enhance algorithmic development and deployment.

These different research programs often develop algorithms in isolation, although they might have strong overlaps in terms of concepts and even implementation. Working within a interdisciplinary framework can lead to algorithms that are more general and useful to a wider range of disciplines. The training provided to our future computational scientists must therefore impart the knowledge in those areas of science that are common most of the ongoing research activies in Applied Sciences, Engineering, and even the social sciences.

Programs at other universities also try to emphasize the interdisciplinary nature of computational science. For example, the Institute for Computational Engineering and Sciences (ICES) at the University of Texas at Austin is a consortium of research centers and groups that includes nine university based centers and programs ranging from Computational Materials to Computational Finance. George Mason University has a School of Computational Sciences organized into several semi-autonomous units with their own faculty and chairs. In both cases, the idea is to foster an environment that invites students to develop graduate students to do develop skills across multiple disciplines.

## 4 SCS Program

The stated goal of the Ph.D. program in Computational Science is to train of graduate students in the art of computational science and provide them with the opportunity to acquire expertise in a particular area of applied science or engineering. Thus the degree provides the student with breadth as well as depth. Graduates should be able to successfully collaborate with scientists in other disciplines. Ideally, students should learn to develop and analyze new computational procedures for use in a variety of fields.

The identifying characteristic of a Ph.D. degree is the ability of its graduates to perform independent research in a selected field rather than just accumulating credit hours. Although, there is much to learn, and some feel that more coursework is indicated, we have chosen to impose a number of credit hours consistent with the University's residence requirements. As a result, our primary emphasis is breadth across topics rather than depth in a given scientific discipline.

Students entering our degree program have a strong desire to do computational science. We broadly classify the students into two groups. The first group is more interested in developing and analyzing scientific algorithms for problems arising in a wide range of application areas, developing techniques for managing large volumes of digital information regardless of the application area, or perhaps in visualizing their data from areas as vastly different as medical imaging or weather prediction. The second group is interested in similar goals except that the students have already identified a science or engineering area of interest, in which they want to concentrate their research.

We offer two computational tracks within the SCS Ph.D. program: 1) a major (computational) track for students of the first group, and 2) computational tracks with a specialization within one of the disciplines from either the applied sciences or engineering for students in the second group. All tracks have the same number of course hours and the required (core) courses are identical. A specialization degree requires that the student register for a minimum of 9 credit hours in the specific area of specialization. Students in the first group are free to choose credits in more than one applied field. Based on the current expertise at the SCS, we currently offer major tracks in Computational Science, and Computational Science tracks with specialties in the areas of Atmospheric Science, Biochemistry, Biological Science, Geological Science, Materials Science, and Physics.

### 4.1 Curriculum

Since computational science is an interdisciplinary program, programs of study can be quite varied. Without a flexible Ph.D. program, it will be difficult to adapt to the fast-paced change in the scientific research community. For example, opportunities will surely exist in the future to combine nano-technology or material science with biological sciences, or imaging with structural biology. With the flexibility built into our program, it becomes possible to respond to these trends in a rapid and flexible manner. This translates directly into a robust set of course requirements, to be determined on an individual basis by the curriculum committee and the advisory committee of the student. The former has knowledge of the large course base available and can offer advise to the student and his advisory committee, while the later has detailed knowledge of the student goals and can offer advice and how to achieve them.

Every student must take three core courses, for a total of 11 credit hours (see next section). These courses are designed to cut across disciplines. A second set of courses, called core electives, provide training in computational science, in more detail. Students choose three courses among the electives. Three examples are scientific visualization, numerical solution of partial differential equations, numerical methods for earth and environmental sciences. Finally, 9 additional credit hours are selected from existing departmental courses in computer science, engineering, mathematics or an applied science. The department is chosen based on the stated specialization of the student (second group). There are a total of 29 credit hours of coursework in addition to a minimum of 9 credits of dissertation hours. Additional credit hours may be obtained through dissertation hours, or coursework, under the guidance of the student's supervisory committee.

### 4.2 The backbone of SCS: The Core Courses

The required core elements in our program are three new courses that provide the student with the programming skills, tools and algorithms necessary to tackle a wide range of real world problems. After completion of these courses, students are expected to have mastered the course content, and more importantly, understand the inter-relationships between the various concepts and feel comfortable utilizing the tools discussed. The goal we seek to achieve is to have students develop the instinct to choose the appropriate computational methods when faced with a new task.

The primary course on which the others depend is *Scientific Programming*, offered for the first time in Fall 2005. Programming languages form the backbone of virtually all areas of scientific research involving computers. Although many students have an elementary knowledge of a single compiled language such as C/C++ or Fortran, or an interpreted language such as Matlab, this knowledge is not sufficient for the development of large-scale scientific codes. Indeed, these programs are usually strewn over multiple files, multiple languages, and multiple operating systems. Students are confronted with these realities from the onset of the course. In addition, they are exposed to benchmarking, profiling and documentation. These concepts are integrated into the language concepts from day one and must be used in all their projects. The hope is that by the time the students are involved in their own research projects, these skills will be integrated into their work habits during program development, rather than as an afterthought. This course covers the common elements of Fortran 90, C++ and Java, three languages of widespread use in scientific computing. The students are taught to solve problems at the conceptual level using mainstream object-oriented programming ideas. The course has been well received with fourteen students evenly distributed across four departments (Mechanical Engineering, Physics, Mathematics, and Meteorology). SCS is planning to offer the course on an ongoing basis, both as a service course to science students in general and as a required course to the students enrolled in our program.

The remaining two required courses, Applied Computational Science I & II, are innovative new courses that combine classwork (4 hours) with a weekly computational lab (4 hours). We have already established that researchers in an interdisciplinary environment must possess a working knowledge of a wide variety of tools and algorithms. Even more important is the researcher's to combine these tools and understand their inter-relationships. These tools span the domains of computer science, mathematics and applied science. A major goal behind these courses is to foster a frame of mind where the student's first instinct is to use tools in combination rather than in isolation. We hope to achieve this through a combination of more conventional lectures, in which algorithms, software, tools are described and demonstrated, and laboratory work where the students get the chance to try the tools out and apply them to practical problems. They will also be required to write the software for some simple tools.

Achieving this objective necessitates a delicate balance between theory and practice. In the theory section of the course, the students are exposed to a wide range of algorithms in fields the span differential equations, visualization, stochastic methods, imaging, optimization, eigenvalues, clustering, etc. While not describing each of these fields in every detail detail (after all, each of the aforementionned fields is easily a course onto itself, which in many cases can be taken as part of course electives), representative algorithms are described, pseudo-code is given, and the students are expected to generate working code that executes the algorithm. The novel part of the course is that whatever the project the student is working on, the results generated are analyzed/visualized, or further processed with other tools (visualization, clustering, etc.). These "other" tools are provided to the student (either obtained from the web or developed by students in the courses given in previous years). Naturally, code benchmarking and profiling are mandatory, regardless of the project underway. Some of the helper tools might themselves be developed from first principles during later sections of the course. Thus, to run a PDE solver, the student might be required to first generate a grid with some existing grid generation package. The lab work provides an environment to put theory into practice under the guidance of the instructors. This is where coding, debugging, and tool exploration happens. The lab exercises are designed to expand knowledge of many existing tools. In a fast changing world, it is unrealistic to teach students to use individual tools. More importantly, students must learn to adapt to new tools, learn how to use tools in isolation, and most importantly with each other. The course will teach students to understand the tools, construct the tools, and use the tools. In addition, they will learn to explore the web to find new tools knowing only some of the required specifications.

### 4.3 Support Courses: Electives

Elective courses are existing departmental courses and courses under development that relate to computational science, and serve to complement the core courses and provide more in depth of computational science concepts. Currently these courses are Computational Biophysics I&II, Computational Evolutionary Biology, Computational Finite Element Methods, Introduction to Bioinformatics, Molecular Dynamics, Monte Carlo/Markov Chain Simulations, Numerical Linear Algebra I&II, Numerical Methods for Earth and Environmental Sciences, Parallel Programming, Algorithms and Architectures, and Survey of Methods for Numerical PDEs. Additional courses, on topics that cut across disciplines will be added as needed. Examples might include Grid Generation, Multi-Scale Analysis, Transform Methods, Wavelet Algorithms, Applied Database Techniques, and more.

## 5 What Is Different?

Within any new degree program (and ours is no exception), the question arises: why could a student enrolled in our degree program not obtain essentially the same degree from an existing department. The reason is simple: computational science cuts across departments, concentrating on the sub-disciplines that are common to all: programming, algorithm development, visualization, statistics, just to give three examples. Providing students with knowledge in these various disciplines generates breadth rather than breadth. Both types of scientists are in demand in today's world. The computational scientist serves as a buffer between domain experts who have most of their knowledge in a single field. he also serves as a conduit to transfer technology from one discipline to another. Without our program, a student in a science or engineering degree program is only trained in computational methods appropriate to their chosen discipline. In the SCS, students are more involved with algorithm development across disciplines, rather than with the use of an existing numerical method to solve an outstanding problem in some applied area. While computations do play a large role in the training of a least some students in the Mathematics or Computer Science departments, algorithm development usually plays a minor role. It is the application itself that drives the research. In our school however, it is the need for algorithms that are applicable *across* disciplines that is the driver. Nonetheless, we recognize the need to expose students to one or more applied disciplines to gain an appreciation for the algorithms being developed: to keep a foot on the ground rather than get lost in abstractions. The SCS is not in competition with existing programs. Rather, the objective is to train students in the emerging discipline of computational science, which combines aspects of computational mathematics, computer science, computational statistics as well as application areas in science and engineering.

## 6 Conclusions

We aim to train students for careers that focus on interdisciplinary research on the invention, development, and testing of new computational methods. Thus, it builds on the existing strengths, found within narrower disciplinary avenues, in the use and development of such methods. The program also recognizes the truly interdisciplinary nature of computational science and the need to train future generations of computational scientists in ways that take advantage of that nature. Our program does not intend to replace the training of computational scientists in a specialized discipline within a department. Rather, it aims to offer students a novel and needed type of training with a broadness that they cannot provide. Instead the departments inculcate any additinal specialization not provided by the SCS.

## References

1. Computational Science Graduate Programs. http://www.cs.mun.ca/~wlodek/comp-sci-links.html .
2. PITAC Report to the President on Computational Science: Ensuring America's Competitiveness (June 2005). http://www.nitrd.gov/pitac/reports/20050609_ computational/computational.pdf .

# Teaching the Foundations of Computational Science on the Undergraduate Level

C. Freundl, H. Köstler, and U. Rüde

Chair for System Simulation, University of Erlangen-Nürnberg, Germany
Christoph.Freundl@informatik.uni-erlangen.de
Harald.Koestler@informatik.uni-erlangen.de
Ulrich.Ruede@informatik.uni-erlangen.de

**Abstract.** This paper describes a new course introduced at University of Erlangen-Nürnberg to teach the foundations of computational science for undergraduates at the end of their second year. The course is a mandatory part of the curriculum in three different computer science-related degree programs. The paper discusses the motivation for introducing the course, the topics chosen for it, and the teaching scheme developed for this unique course.

## 1 Motivation

The University of Erlangen-Nürnberg offers programs in Computational Engineering (Bachelor, Master, PhD), Computer Science (Diplom Informatik), and Information and Communication Technology (Diplom). All three programs rely on a joint set of core course modules for the first two years of study. Beginning this year, a new required course was introduced for students in the fourth semester of these programs, entitled *Data Structures and Algorithms for Continuous Systems.*

The birth of this new course originates form a discussion within the department that many current computer applications deal with objects that are fundamentally of continuous nature. Example applications include the analysis and synthesis of images, processing of audio signals and movie sequences, and simulation. Arguably, these topics are gaining in importance for computer scientists and computational engineers. However, these topics were not covered in the curriculum of the first two years of any of these programs.

Continuous models have also become important to model extremely large data sets and are being used for data mining or web search engines. Many modern algorithms in this area depend on intricate techniques from numerical linear algebra and numerical analysis.

This, however, stands in contrast to the state to which our computer science curriculum had evolved over the past two decades. Topics, such as numerical linear algebra had essentially disappeared from the undergraduate curriculum and would only appear within optional courses for higher semesters, e.g. as part of taking a minor in mathematics.

V.N. Alexandrov et al. (Eds.): ICCS 2006, Part II, LNCS 3992, pp. 185–192, 2006.

Of course, advanced classes on computer graphics, visualization, image processing, pattern recognition, signal processing, and simulation require these topics as a prerequisite, and therefore many of these courses included introductory material to cover their mathematical prerequisites. Clearly, this setup did not exploit that there is a shared theoretical basis. As a consequence, this had lead to a significant overlap in advanced level courses.

The new course *Data Structures and Algorithms for Continuous Systems* has been designed to address this problem and to make the overall curriculum more efficient by teaching the foundations in a unified form, with the goal of saving time and enable the higher level courses to progress more quickly to advanced material.

An early analysis of the desired topics for the course showed that a simple revitalization of a conventional numerical analysis course would not satisfy our requirements (see [1, 2, 3]). We have therefore decided to develop a completely new course based on the material extracted from existing advanced courses in pattern recognition, computer graphics, and simulation.

This approach was also chosen in the hope that it would be easier in this way to motivate students to study topics that many of them perceive as difficult and highly abstract. On the other hand, the designer of the course and its instructors face a challenge to balance the presentation of fundamental theory with the motivation through practical examples.

In summer 2005, the new course was given for the first time. Since no text books for such a unique course were available, the course has been taught based on a set of self-developed power point presentations which were made available to the students as a substitute for lecture notes and as a guideline to find the relevant material from the various textbook sources.

The remainder of this paper reports on our first experience with teaching this course. It is organized as follows. In section 2 we will outline the material taught in the lecture itself. Section 3 describes the design of the exercise classes, selected assignments are presented in section 4. A concept for instructing unexperienced tutors is shown in section 5. Finally, section 6 discusses evaluation results for this course.

## 2 Contents of the Lecture

The course *Data Structures and Algorithms for Continuous Systems*[1] consisted of a set of 26 two-hour lectures corresponding to the typical length of the summer semester in Germany.

The course began with an introduction to continuous data sets with examples such as audio or video signals, covering topics such as quantization, discretization, the curse of dimensionality, rounding, floating point arithmetic, stability, and the condition of a problem.

[1] Website of the course: `http://www10.informatik.uni-erlangen.de/Teaching/Courses/SS2005/AlgoIII/`

As a second major topic area, the course covered basic algorithms from numerical linear algebra, like the LU- and Cholesky factorization, QR decompositions, and also data structures for matrices, including sparse matrix formats. For iterative methods, it was decided to present only elementary relaxation methods, and neither Krylov space methods nor multigrid methods. This part of the course also included a review of norms for vectors and matrices.

Wherever possible, this material was motivated by applications as mentioned in Section 1, e.g. by emphasizing the geometric interpretation of rotations, reflections, and orthogonalization. When possible, the use of these methods in applications was explicitly worked out, e.g. the use of linear algebra for the coordinate transformations in computer graphics. Similarly, the lecture on least squares methods was given in the context of examples from pattern recognition and a more involved example was taken from the algebraic reconstruction technique within computed tomography, including a brief excursion to Tikhonov regularization. An introduction to vector space techniques for text retrieval and the page rank model of Google was also presented.

Generally speaking, the linear algebra part of the lecture was driven less by the mathematical analysis of the algorithms, but rather by their application. Nevertheless, the goal of the lecture was to present these techniques as a general toolbox for a wide variety of problems and to make the students aware of issues like the algorithmic complexity or numerical stability. Additionally, the course attempted to give guidelines for the practical implementation of the algorithms by discussing data structures and tools such as Matlab.

The next topic in the class revolved about data structures and algorithms for geometric objects, such as curves, surfaces, and volumes. Consequently, this part of the course started from polynomial approximation and interpolation, to introduce spline functions and Bézier curves. From there the lecture progressed to (tensor-product) surfaces, transfinite interpolation, and Bézier surfaces, and finally volume models, such as Octrees or constructive solid geometry, as used in computer aided geometric design. This part also included basic methods for numerical quadrature, motivated by computing volumes and weights of geometric objects.

A further topic covered in the course was optimization, and nonlinear equations. This included bisection, Newton's method, the Nelder and Mead simplex method, and steepest descent. Because of their practical importance, the course also covered some methods for discrete optimization, such as dynamic programming, backtracking (with heuristics), simulated annealing, and genetic algorithms.

Finite differences and finite elements were covered briefly in a two hour lecture each, and together with the discussion of the FFT algorithm, they concluded the course.

## 3 Design of the Exercise Classes

Typical for the German system, the course had the format of a lecture (given to the whole class of about 150 students) and was accompanied by tutorial sessions (*exercises*) which were broken up to limit the number of students to at most twenty in each group. These exercises consisted of two parts:

- class room sessions where the mathematically oriented assignments were presented and discussed, and
- programming assignments.

Organizing the exercise sessions required a major effort. As another peculiarity of the German university system, the active participation in the exercises cannot be required of the students, since they only need to pass the final exam successfully. The success of the students is solely determined by their grade in the final exam which is formally independent of the lecture, and which — as another anachronism of the system — is given at earliest two months after the end of the semester. As an option the students can even choose to take this exam even at a later time, e.g. eight months after the end of the course. This ridiculous situation is one reason for the poor motivation of many students in such courses and the ineffective learning of students lacking a sufficient degree of self-discipline and self-motivation. As an example, more than a third of the participants of our course have chosen to defer the exam by more than half a year.

These constraints led us to develop a special strategy to motivate students: the exercises had to be sufficiently interesting and self-motivating as possible. On the other hand, with only limited teaching capacity and the high student numbers, the exercises had to be given in an efficient way.

### 3.1 Classroom Exercise Sessions

The exercise classes were given weekly as two-hours sessions and in groups of at most twenty students. The assignments were handed out and posted on the web in the week prior to the classes. Students were expected to prepare the assignments, but — as described above — were not formally required to do the assignments. Therefore, in the class itself the solution of the tasks was presented, but some time in each session was reserved for the students to work out the solutions on their own with the additional possibility of asking the exercise tutors for help. Depending on the difficulty of a task, the solution was then presented by either a student or the tutor on the whiteboard.

### 3.2 Programming Assignments

To additionally motivate students to do the work intensive programming assignments, they were cast in the form of a contest. To reduce the teaching load, students were grouped in teams of three. There were only four, but then somewhat larger programming assignments. Furthermore we set up a system based on the versioning software *Subversion* [4] which allowed the students to hand in their solutions electronically. Using Python [5] scripts, we provided mechanisms to evaluate the correctness of a solution automatically such that the students got immediate feedback on the quality of their solution. All the tasks required solving a numerical problem so the correctness of a solution could be determined by measuring norms of residuals or errors. Also, we rewarded an efficient programming style by incorporating the runtime of a program into the score. The system provided a high score of the best solutions on the course's web site, thus promoting a competition among the strongest student teams. This did help to keep the contest interesting and has proven quite effective in giving especially the good students an additional motivation.

### 3.3 Results and Insights

The classroom exercise sessions were frequently and steadily attended by many students. This may also have been caused by the fact that we did not hand out or post master solutions. In all classes, at least some of the students showed high effort and enthusiasm in solving the given assignments. Nevertheless, as any of the more mathematically demanding classes, the course was found quite difficult by a majority of the students and many of them gave up early on finding their own solution.

The programming assignments showed different results. Although the level of participation was quite high in the beginning (34 groups with a total of about 75 students), only four groups (with nine students altogether) handed in solutions for all of the given tasks. Given the effort to create and present these assignments, this is of course quite disappointing. We believe that this unsatisfactory participation was caused by:

1. Since each assignment was quite work intensive, many students were discouraged from working on them.
2. The evaluation software initially had some technical problems, and was not always quick enough to provide feed back to the students, especially when the deadline was approaching.

Summarizing, the system used for the course was favoring the strong students who put into the course a large effort and who then commented favorably on the learning experience. The disadvantage was that many of the average and below average students were quickly getting frustrated and — not quite unexpectedly — essentially gave up on many of the assignments.

This could of course be corrected most easily by making participation in the exercises a requirement for passing the course, but the mentioned peculiarities of the German university system currently prohibit a more effective teaching setup. This anachronism is partly inherited from the traditional German university structure, and is additionally aggravated by the yet unfinished Bologna reform[2].

## 4 Selected Assignments

Over the whole semester the students had to do four programming assignments. For a somewhat more detailed exposition, we pick here the third one that was

[2] The Bologna Process aims at introducing a Europe-wide unification of the various national University systems. Besides introducing a Bachelor/Master degree structure it also requires the introduction of a standardized credit point system by year 2010.

motivated from medical image processing. The goal was to register two non-aligned images from the same patient e.g. in order to locate the position of a tumor in each image. This led us to use a variational approach for non-rigid image registration and thus mathematically to a system of nonlinear partial differential equation representing the deformation between the images [6]. The students were given parts of a Matlab program that performs the image registration for a simple test case. Their task was to implement the registration algorithm in C or C++ and to present some results for arbitrarily chosen images. The assessment in this case was on the one hand done by measuring the error of the registration algorithm and on the other hand every group had to hand in a video sequence showing the transformation from one image into the other. These sequences were then rated by the other students. The best ten groups got additional bonus points. Figure 1 shows the image sequence of a deformation process to register the first with the last image in the sequence.

**Fig. 1.** Reference images (top) and template images (bottom) after the deformation

## 5 Supervised Teaching

Usually several student tutors from higher semesters are employed to teach some of the exercise classes. These students are supervised by regular university staff — e.g. an advanced graduate student or post doc. Within the context of the Elite Network of Bavaria [7] we have recently been awarded special funds to do this in a more systematic way and to introduce a scheme to *teach the teachers*.

The student tutors were participating in the elite program and thus they could earn credit for participating in our setup of *supervised teaching*. For this, they were required to prepare each assignment set and the corresponding exercise class fro presentation and discuss this before they gave the class with their supervisor. The supervisor additionally sat in the class itself and provided feedback to the instructor after each class. Typical discussions included e.g. how the instructor managed to explain the material to the audience or to what extent he or she could motivate the younger students to participate actively in the class.

*Supervised teaching* additionally included a one-day seminar given by the head of the Didaktikzentrum Nürnberg [8]. Here the focus lay again on general teaching skills, e.g. a discussion of several presentation techniques and types of leadership. Many practical tips and tricks for teaching were given, e.g. how to deal with students who try to disturb a lesson or how to motivate them.

In addition to to the individual supervison and the seminar, we had arranged a professional coaching of one of the exercise classes by a professional trainer. Here, one class session by the student tutor was videotaped and discussed in detail after class. The goal of this was to help young instructors with technical teaching skills like the right use of the blackboard, but also to systematically develop their teacher personality and to improve their interaction with the students.

The feedback from the first student participating as an instructor in "supervised teaching" was very positive. He felt that he himself had learned intensively since teaching required a high level of abstraction and a deep knowledge of the underlying material in order to be able to teach it well for the younger students. He felt that the responsibility to express material technically correctly and precisely presented a special challenge in itself, and he has found that he could already profit from these skills on several other occasions.

It remains to note that also the participants in the classes taught by student tutors were very positive, since this may have lead to better prepared exercise sessions overall.

## 6 Evaluation

The Faculty of Engineering Sciences of the University Erlangen-Nürnberg conducts a systematic evaluation of all courses through a web-based questionnaire of all enrolled students.

We cite a few student comments.

- "Mr Rüde tries hard to arouse interest in the fairly complex subject, and this with success. Particularly, I enjoyed the references to practical applications in image processing and the shown videos."
- "Absolutely interesting, I will probably choose this as a major subject. The lecturer is very capable, friendly, and stands relatively close to us students. Points out, what his subject is, what you can do with it, and which perspectives it offers."
- "Subject is way too difficult, very many topics. Better leave out a few issues and go more into details, give more examples or else comprehension will be a torture."

Altogether, the evaluation showed average ratings compared to similar classes. Since this class was given for the first time and was still in a prototype stage, we believe that this is a sign that we are on the right track to develop a good course. More in detail, the students have strongly acknowledged and commented positively on our effort as instructors both for the central lecture and for the exercise sessions. Some of their constructive suggestions, such as an encouragement to work more practical examples on the blackboard during in the central lecture (as compared to using only power point slides) could even be realized in the ongoing semester and this was generally well received. However, the students commented in their majority quite critically on the topical composition of the course as too theoretical and too mathematical, and generally as too difficult. The disappointing turnout in the programming exercises together with this feedback will require a re-evaluation of the course goals and possibly an adaption of its contents.

## Acknowledgments

The authors wish to thank Dr. Ch. Alberternst for the professional coaching of the supervised teaching students, Dipl.-Päd. J. A. Wendorff for giving the seminar of didactics and Q. Meyer for giving feedback on the supervised teaching.

## References

1. Rüde, U.: Computational Engineering Programs at University Erlangen-Nuremberg. In: Proc. of the 2002 International Conference on Computational Science (ICCS2002), Part III. Volume 2331 of Lecture Notes in Computer Science., Amsterdam, The Netherlands, Springer (2002) 852–857
2. Bungartz, H.J.: Some remarks on CSE education in Germany. In: Proceedings of the 2004 International Conference on Computational Science: ICCS 2004. Volume 3039 of Lecture Notes in Computer Science., Heidelberg, Springer (2004) 1180–1187
3. Fabricius, U., Freundl, C., Köstler, H., Rüde, U.: High performance computing education for students in Computational Engineering. In Sunderam, V., Albada, G., Sloot, P., Dongarra, J., eds.: Computational Science - ICCS 2005. Volume 3515 of LNCS., Springer (2005) 27–35 ISBN-10 3-540-26043-9, ISBN-13 978-3-540-26043-1, ISSN 03-2-9743.
4. CollabNet: Subversion version control system (2000–2005) `http://subversion.tigris.org`.
5. Python Software Foundation: Python (2001–2005) `http://www.python.org`.
6. Modersitzki, J.: Numerical methods for image registration. Oxford University Press (2004)
7. Bavarian State Ministry of Sciences, Research and Arts: Elite Network of Bavaria web page (2005) `http://www.elitenetzwerk-bayern.de`.
8. DiZ: Zentrum für Hochschuldidaktik der bayerischen Fachhochschulen (2005) `http://www.diz-bayern.de`.

# Computational Science: An Intermingling of Science, Mathematics, and Computer Science

Frederick R.W. McCourt

Departments of Chemistry and Physics, University of Waterloo, 200 University Avenue West, Waterloo, Ontario, Canada N2L 3G1

**Abstract.** The development of a full undergraduate program in computational science, with both a broad-swath academic plan and four specialization plans, one each in biochemistry, chemistry, earth sciences, and physics, is described. The reasons behind the establishment of the program are presented, and the evolution of the program over the past five years is described. Some of the initial difficulties encountered in setting up such a cross-disciplinary and relatively flexible program of this type are also pointed out.

## 1 Background Information

Some background information regarding the University of Waterloo (UW) will be useful prior to discussing our program in Computational Science. UW is what can be referred to as a medium–sized comprehensive university, with six Faculties: Applied Health Sciences, Arts, Engineering, Environmental Studies, Mathematics, and Science. It was founded in 1957 as an engineering school with auxiliary faculty members for the teaching of those arts and science courses required for graduates of the academic programs to meet the specifications required for accreditation as professional engineers. All academic programs were offered only under a concept referred to as 'cooperative education', in which students alternate academic terms of four months duration with 'work terms' of four months duration, during which they are employed by a company or agency participating with the University of Waterloo in one of its 'work–study' programs.

Independent Arts and Science Faculties had evolved by 1960, each offering academic programs, leading to BA and BSc degrees, respectively. They both began by offering programs only in the usual mode of sequential four–month Fall/Winter terms per year; not much later, however, the Science Faculty saw the advantages of offering some of its programs in the cooperative education mode developed by the Faculty of Engineering. It was not for another fifteen or so years that the Faculty of Arts also began offering some of its academic programs in the cooperative mode. Today, the University of Waterloo has over 10,000 coop students, spread over all six Faculties. Graduate programs in most areas of endeavour had also emerged by the mid-1960s.

This new university recognized quite early on that digital computers were going to revolutionize our society and, consequently, in the mid–1960s a decision

V.N. Alexandrov et al. (Eds.): ICCS 2006, Part II, LNCS 3992, pp. 193–198, 2006.

was taken to create a Faculty of Mathematics centered on the discipline of Computer Science, but offering both graduate and undergraduate degrees in all areas of mathematics as well as in the area of computer science. This initiative was a bold one forty years ago: its success can be seen in the approximately 3500 undergraduate students who register annually in the Faculty of Mathematics, with about two–thirds of them working towards a baccalaureate in computer science.

The University of Waterloo is probably best known for its Faculties of Engineering and Mathematics, specifically for their electrical and computer engineering and computer science academic programs, respectively. The Faculty of Science is the fourth largest (or third smallest) Faculty at the University of Waterloo and is made up of four Departments and two Schools: Departments of Biology, Chemistry, Earth Sciences, and Physics, and Schools of Optometry and, most recently, of Pharmacy. Each academic unit has its own degree program(s); in addition, some units partner with others to offer joint programs, such as Biochemistry, Bioinformatics, Chemical Physics, and Geochemistry, or participate in the broader Faculty of Science programs, such as Honours Science, Honours Science and Business, and Honours Computational Science, via specializations in their core disciplines.

The Computational Science program is the youngest full program in the Faculty of Science and is also one of the youngest programs at the Univeristy of Waterloo. A set of program plans was borne out of the perception that there was a need in modern society for graduates of an academic program in which the combination of a thorough grounding in a conventional area of science, coupled with a significant degree of mathematical sophistication and a strong computer science background, would create graduates who would be prepared to tackle a wide variety of problems arising in today's world.

## 2 The UW Computational Science Program

The Computational Science program at UW has been designed to provide a series of academic plans for students who either wish to attain fairly intensive coverage of one of the traditional areas of science while simultaneously acquiring a broad background in mathematics together with an in–depth knowledge of computer science or who wish a broader but less intensive coverage of all of the traditional areas of science but with the same coverage of mathematics and computer science. The program has four academic plans that focus upon specific traditional areas of science, namely biochemistry, chemistry, earth sciences, and physics, as areas of specialization, plus one academic plan for those who do not wish to specialize in one specific area of science. The Biology Department did not participate in this program as it had already joined forces with the Computer Science Department in mounting a four–year degree program in Bioinformatics.

A typical full program plan at UW consists of forty term lecture courses (each comprised of 33 lectures) plus any other specified laboratory and/or field courses. All Computational Science program plans have a common set of seventeen core

**Table 1.** Core courses for all Computational Science program plans

| Faculty | Course Number | Basic Course Content |
|---|---|---|
| Science | CHEM 120/120L | Properties of matter |
| | CHEM 123/123L | Chemical reaction rates, equilibrium |
| | EARTH 121/121L | Natural geological processes |
| | EARTH 122/122L | The Earth's history and resources |
| | PHYS 121/121L | Particle dynamics, conservation laws |
| | PHYS 122/122L | Wave motion, gravitation, and heat |
| Engineering | ENVE 224 | Probablilty and statistics |
| Mathematics | MATH 114 | Linear algebra |
| | MATH 127 | Differential calculus |
| | MATH 128 | Integral calculus |
| | MATH 227 | Multivariate and vector calculus |
| | MATH 228 | Ordinary differential equations |
| | CS 123 | Programming principles using JAVA |
| | CS 126 | Principles of software development |
| | CS 230 | Computers and computer systems |
| | CS 234 | Data types and structures |
| | CS 370 | Numerical computation |

term courses, consisting of six freshman science courses with accompanying laboratories, six mathematics courses, and five computer science courses. This set of courses can be found in Table I. Each student in the program must also select two additional elective junior or senior level computer science courses to complete the basic core component of the program. The remaining twenty–one term courses depend upon the specific plan chosen by a student: however, because of the need to acquire a reasonable depth of knowledge in mathematics and computer science in addition to the basic areas of science, plans in this program contain at most six free electives.

Each specialization plan contains a set of courses that are considered core for the particular specialization, a number of courses from which to select specialization electives, plus a few free electives, while the non-specialized plan has an additional core set of seven chemistry, four earth science, and four physics courses, for a total of fifteen additional core science courses, plus six free electives. The structures of the various plans are shown in Table II.

We believe that the current structures of the Computational Science program plans allow students a reasonable level of flexibility for choosing courses that are of interest to them and to tailor an academic plan that suits their perceived needs. The program is challenging, and has so far attracted a number of very capable, even gifted, students, judging from the number from this program who make the Dean's Honours list each term.

The seven specialization electives associated with the Biochemistry Specialization have been structured to allow students to obtain three–course sequences at the sophomore level and above in at least two of the traditional analytical,

**Table 2.** Course structures for Computational Science plans at UW

| Plan | Number of term courses | | | |
|---|---|---|---|---|
| | Basic Core | Other Core | Electives | |
| | | | Spec$^n$ | Free |
| No Specialization | 19 | 15 | – | 6 |
| Specialization: | | | | |
| Biochemistry | 19 | 10 | 7 | 4 |
| Chemistry | 19 | 8 | 7 | 5 |
| Earth Sciences | 19 | 11 | 5 | 5 |
| Physics | 19 | 12 | 4 | 5 |

inorganic, and organic chemistry subdisciplines. Similarly, the seven specialization electives associated with the Chemistry Specialization have been structured so as to ensure that students will complete three–course sequences in physical chemistry plus two of the other three traditional subdisciplines of chemistry.

## 3 Growing Pains with the Initial Version of the Program

Two problems were encountered when this program was first offered in 2001. First, students were being asked to choose a particular plan of study upon entry into their freshman year: this had the consequence that changes in plan were difficult to handle after the first term had been completed, as each specialization plan already contained at least one required specialization core course by the end of the Winter term. Second, the initial version of the program had calculus, chemistry, computer science, and physics/earth science courses in each of the two freshman terms, plus a linear algebra course in the first term and either a free elective or a specialization core course in the second term. This turned out to be the more serious problem for the initial group of 35 students: not only did they find all these courses rather demanding, but also there seemed to them to be almost no overlap of content. Moreover, as they were also adapting to the significant differences between life in high school and university environments, many of the students found the program overwhelming, with the end result being that a number of them fell by the wayside.

To make our Computational Science program more flexible, and at the same time to help students undergo a less demanding adaptation into the university environment, all five plans were rearranged to allow the creation of a common first year. This not only alleviated much of the pressure on the freshman group, but also allowed the decision of whether to specialize or not to be made at the beginning of the sophomore year rather than prior to starting university. At present, all students entering this program share a common first year of study with two term courses each of freshman calculus, chemistry, earth sciences, and physics, and one term course in each of linear algebra and introductory computer science programming principles. Upon completion of this set of common introductory courses, students are asked either to choose one of the four areas of specialization or to indicate that they wish to remain in the broader non-specialized plan.

An ongoing concern for students in this program is the number of timetable conflicts that can arise due both to the cross–disciplinary nature of the program itself and to its crossing of Departmental and, especially, Faculty lines. Because the number of students in the program is still relatively small (roughly fifteen to twenty students in each year of the program) and a smaller still number of students take any particular course following the freshman year, there is not much opportunity to negotiate time changes for lecture courses that are in conflict. This requires the program advisor to be thoroughly conversant with all available antirequisite courses and to be able to suggest and/or approve changes to individual plans to allow similar, though not necessarily equivalent, courses to be substituted as needed in order to facilitate successful completion of a particular student's academic plan.

## 4 Some Comments and Observations

After five years of operation, despite having had to go through a number of growing pains, we have concluded that our Computational Science program can be considered a success. While it does not yet draw as many students as we had originally envisaged, this may be in part due to some of the specific growing pains mentioned above and in part to the fact that we have not been overly aggressive in promoting our program, especially in Ontario high schools. Support for this latter observation comes from anecdotal evidence based upon discussions with students who have transferred into our program from other Faculty programs on the UW campus, during which comments have been made to the effect that had they been aware of the UW Computational Science program when they graduated from high school, they would have entered it directly at that time. This may in large part be because while the Computational Science program has equal billing with Departmental programs (such as those in Chemistry and Physics) in the Faculty of Science brochure sent out to all Science applicants, students who have applied to other Faculties would not normally receive information about this particular program.

We haven't been able to obtain any significant statistics on post–graduate employment or studies as we have yet to graduate a full class from this relatively new program. However, one potentially relevant measure of success of the Computational Science program lies in the success rate for placement of the roughly two–thirds of our students who follow the cooperative mode of study: there have been almost no work placement problems, even though students in our program compete directly for these work–term positions against students from both the Computer Science and Computer Engineering programs at UW. This observation is further corroborated by the level of satisfaction with the performance of our students expressed by the coop work–term employers. Further, we are aware of two of our forthcoming graduates who have applied to carry out graduate work in chemistry and physics departments in the USA and in Canada: one has already informed us that he has been accepted by a major Canadian graduate school in chemistry, conditional upon successful completion of his Computational Science with Chemistry Specialization degree at UW. Of course, Computational Science students and graduates have different overall skill sets than do CS students and graduates: on the one hand, they take roughly half the number of CS courses taken by CS students, and on the other hand, they have both a broader and deeper coverage of traditional Science subjects.

It is worth pointing out that a number of computationally intensive courses have been created in the Faculty of Science at the University of Waterloo over the past four years, mainly arising because of the existence of a core of computationally literate students, some of whom are associated with the Computational Science program itself. Courses that fall into this category are: BIOL 366 (Introduction to Bioinformatics), BIOL 382 (Computational Modelling of Cellular Systems), CHEM 355 (Computational Physical Chemistry), CHEM 452 (Computational Quantum Chemistry), CHEM 465 (Molecular Modelling and Protein Folding), PHYS 239 (Scientific Computation 1; numerical analysis applied to physics problems), PHYS 339 (Scientific Computation 2; Monte Carlo and molecular dynamics methods). There are also a number of other courses that have become available through a new complementary Computational Mathematics program introduced very recently by the Faculty of Mathematics. While this newer computationally oriented program has a number of aims in common with our Computational Science program, it has a stronger mathematics focus and does not have the strong science background of our our program. Finally, it should be said that additional computationally intensive courses will be created in the next few years, all of which can enrich the experience of students in our Computational Science program.

## 5 Conclusion

Academic plans of the type offered under the aegis of a Computational Science program should appeal to a select group of modern students who have a desire to couple their interests in science with their fascination for digital computers and the power of numerical procedures and simulation methods. Computational Science provides both a sufficient challenge to good students and a recognition of what can be accomplished when the power of digital computation is combined with in–depth scientific training.

# A Framework for Conceptually Modelling the Domain Knowledge of an Instructional System

Emilia Pecheanu, Luminita Dumitriu, Diana Stefanescu, and Cristina Segal

University 'Dunarea de Jos' of Galati, Department of Computer Science and Engineering, 800008 Galati, Romania
{Pecheanu.Emilia, Luminita.Dumitriu, Diana.Stefanescu, Cristina.Segal}@ugal.ro

**Abstract.** This paper presents a solution for conceptually modelling the teaching domain knowledge for computer-assisted instructional systems. The model consists of a theoretical framework and a knowledge representation approach. This model might be adopted in order to build a well-defined conceptual structure for the domain knowledge of a CAI system. The paper also presents an authoring system which implements the features of the modelling methods. This approach can offer a better solution to the problem of knowledge structuring and organising for computer-assisted instructional systems.

## 1 Knowledge Modelling in Computer Assisted Instruction

Increasing the assistance capability of instructional environments has lately become an important research subject in the area of computer–assisted instruction. A direction pursued by many systems designers is to build flexible, well defined models of the domain knowledge (the knowledge belonging to the training area), which is presented to users as instructional content.

In many instructional theories [3] [4] [6] it is considered that learning occurs as a result of accumulating and reorganising the knowledge within the human cognitive–mental structures. This vision of human knowledge acquisition, corroborated with representational properties of network-like architectural structures, leads to the idea that the domain knowledge might be organised in the form of "conceptual networks".

A conceptual network's node has to contain a description of some of the domain's concepts (main ideas), and the links between the network's nodes should be kept for the multiple relationships between these concepts.

The organisation model of the instructional content, which is tightly connected to the structure of the domain knowledge, has to provide the user with multiple views or presentation patterns of the concepts in the training domain.

Studies and assessments of computer-assisted instructional systems (CAI systems) have shown that if they contain a structural model, they facilitate learning. A well-structured architecture of the instructional content can also improve the efficiency of any guidance method the instructional system might use.

V.N. Alexandrov et al. (Eds.): ICCS 2006, Part II, LNCS 3992, pp. 199 – 206, 2006.

## 2 A Theoretical Framework for Knowledge Modelling in Computer Assisted Instruction

### 2.1 Main Design Principles

The modelling methods described herein might be adopted in order to build and to represent the knowledge space of an instructional system. The resulting models integrate the domain knowledge pertaining to a CAI system in a single structure. The modelling methods have been developed and fully described in [8]. The modelling methods have been mainly derived from Formal Concepts Analysis, [2] and Logical Concept Analysis, [1].

The theoretical framework and the knowledge modelling methods presented in this paper comply with the following principles, taken from [5]:

1. Domain knowledge modelling is based on the didactic experience of educators, who know how a discipline should be taught within institutional frameworks
2. Knowledge modelling is realised in preparation for the act of instruction
3. Knowledge modelling is a way of organising the resources that are necessary in the instructional process, in this approach.

Another design principle should be highlighted. Most CAI systems belong to one of the following categories, depending on the main features they possess [5]:

1. Pedagogy-oriented systems, focused on implementing well-defined (explicit) teaching and organising/sequencing methods for the instructional content.
2. Performance-oriented instructional systems, focused on assessing/diagnosing the knowledge acquired by the learners.

The modelling framework described herein is aimed at developing conceptual structures for pedagogy-oriented instructional systems.

### 2.2 Key Elements in Modelling the Domain Knowledge

The key elements in modelling the domain knowledge of a CAI system are the following: the Conceptual Unit, the Conceptual Structure, and the Conceptual Transition Path. These three elements are defined below.

**Definition 1:** A **Conceptual Unit $C_i$** is a group of related notions (concepts, basic ideas) belonging to the domain knowledge of an instructional system. Conceptual Units are obtained by applying a mathematical relation (or a set of relations) over a set of notions belonging to that domain knowledge.

**Definition 2:** A **Conceptual Structure** $\mathcal{S}$ is a set ($N$, $R_N$, $C_S$, $L_S$), where N is a set of notions belonging to the domain knowledge of a course or teaching discipline, $R_N$ is an order relation over N, $C_S$ is the set of Conceptual Units obtained by applying the relation $R_N$ over the set N, and $L_S$ is the set of traversing paths of the structure $\mathcal{S}$.

**Definition 3:** A **Conceptual Transition Path** $\mathcal{T}_s \in L_S$ is a set of conceptual units ($C_{1s}$, $C_{2s}$, ..., $C_{Ns}$) connected one after the other within a Conceptual Structure $\mathcal{S}$. A conceptual transition path $\mathcal{T}_s$ has as origin a the Conceptual Unit $C_{1s}$, considered initial, and as destination a Conceptual Unit $C_{Ns}$, which encapsulates a set of notions comprising a study goal within a teaching course.

A Conceptual Structure is a model meant to represent the domain knowledge of a teaching course. A Conceptual Structure should map the cognitive structure of the domain knowledge, and should also reflect the pedagogical vision of the teacher-author of that course. The model has to allow for flexibility, i.e. to provide as many transition paths as possible in order to learn the domain's main concepts.

A Conceptual Transition Structure should offer various solutions to traverse a group of notions in the interactive course. This requirement comes from the fact that the users of an instructional system have various learning styles, as well as various behavioural attitudes in learning. As such, an educational software-system should offer its users a flexible support for the learning act.

In computer-assisted instruction, a teaching-domain's notion or group of notions is transmitted to users through a presentational software-object, which can be: an explanatory text, an image, an animated image or a combination thereof. The presentational object can be as well a full-fledged application or software tool, intended to mediate learning through more complex methods of instructing: tutoring, exploring, practising, and so on. One can observe that a presentational object belonging to an instructional system could constitute the medium to learn/assimilate several related domains' notions.

The present approach assumes that for each main notion or group of notions belonging to the teaching-domain, there is a corresponding presentational object within the instructional software system. Presentational objects are supposed to possess a consistent internal description, permitting storage and retrieval.

The previous introductory ideas have led to the following developments, setting up several significant elements in computer-assisted instructional activity.

### 2.3 Notions and Relationships Within the Instructional Content

The notions (concepts, basic ideas) existing in the knowledge space of a domain are interconnected through multiple relationships. Within the frame of an instructional activity, a sense of knowledge in a domain might be acquired only if one considers the domain's notions and relationships as a whole. The most common relationships between notions within a teaching content can be considered as following:

1. **the relationship of precedence between notions**, i.e. the order or the sequence of these notions in the teaching process;
2. **the relationship of contribution of a notion in teaching another notion,** this relationship, which can also be expressed using phrases such as "significant contribution" or "reduced contribution", can then be quantified within a well stated range of values.

Applying any of the above relationships over the notions within an instructional content is always generating a specific structure. This structure might be a simple sequence of notions, a hierarchy or - most commonly - a network of interconnected notions. In this approach, the process of learning/assimilating a course's notions can be considered as a way of traversing the network of notions. This way of traversing should be carried out by respecting a specific order: it should begin with the "basic" notions and end up with the notion (or group of notions) considered final, or the instructional target.

## 2.4 The Modelling Approaches

A lattice-like model has been developed in [8] in order to be the Conceptual Transition Structure previously defined. This model is named **COUL-M** (COnceptual Units' Lattice Model). It might be adopted in order to represent the relationships between notions within the space of knowledge of a teaching domain in a comprehendsible way.

The COUL-M model has been built upon the mathematical formalisation of the precedence and contribution relationships. The mathematical equivalents of precedence and contribution relationships are the Precedence Relation and the Contribution Relation. These relations have several properties permitting to transform them by applying sequences of aggregation and decomposition operations into one-to-one incidence relations.

The **formal contexts** and **the formal concepts' complete lattice** can further be built for these relations [2]. Then, the pairs of sets (intent, extent) which compose **formal concepts** can be extracted from these formal contexts and **the formal concepts' complete lattice** can further be built [2]. The formal concepts are mathematical, abstract representations for sets of related notions within the teaching material of the course. The formal concepts' intent part (or the intent sets of the formal concepts) will stand as the "Conceptual Units" related to the teaching material. Several lattices of Conceptual Units, standing as Conceptual Structures for the domain knowledge can eventually be derived.

Thus, the COUL-M model for the domain's knowledge of a CAI system is defined as a complete lattice $C_S$ . The latter is generated by rewriting the relations between the course notions as relationships of incidence, where the formal concepts are an abstract representation for sets of related notions within the teaching content.

The subposition operation [2] has been applied in order to compose the formal contexts of the Precedence and Contribution Relations. The results of this operation were integrated into the model named **COUL-RM**, defined as a set (N, $R_N$ , $C_S$, $L_S$), where:

1. N is a set of notions from teaching-domain of a course,
2. $C_S$ is the complete lattice,
3. $R_N$ is the relation of order within lattice $C_S$, established by set inclusion,
4. $L_S$ is the set of elementary chains to traverse the lattice $C_S$ , by respecting the order subconcept –concept.

The COUL-M model has the following properties:

1. it is a complete lattice - a well defined mathematical structure - able to represent the interactive course's notions, as specified by the teacher-author of the course
2. it is a set of Conceptual Units
3. each Conceptual Unit in the COUL-M model constitutes a formally expressed description of several course's notions
4. within the COUL-M model the "new" notions to assimilate within each instructional sequence depend on what has been learned in the previous sequence (they depend on the selected instructional path) and cannot be listed beforehand.

The COUL-RM model attempts to capture the semantic links between the notions of an instruction subject. It also attempts to corroborate them with the way in which the instruction can be carried out, taking into account the resources that are available at a given time within the interactive course.

## 2.5 The Knowledge Compiler

Let's consider the domain knowledge of a system meant for assisted-instruction in an IT course. This course might be "Operating Systems" for example. Furthermore, let's consider a set of basic notions belonging to a specific chapter in this course, i.e. "Mutual Exclusion". Starting from the pedagogical relationships, which are connecting this chapter's notions, the COUL-M model can be built.

In the teacher's pedagogical approach, the most important concepts of the "Mutual Exclusion" chapter are considered to be the following ones: mex, cr, cs, sem, sap, wap, pca, lra, where:

A = mex = mutual exclusion
C = cs = critical section
E = sap = strict-alternation of processes
F = pca = producer-consumer algorithms
B = cr = critical resource
D = sem = semaphore
F = wap = wait-activation of processes
G = lra = lecturer-redactor algorithms

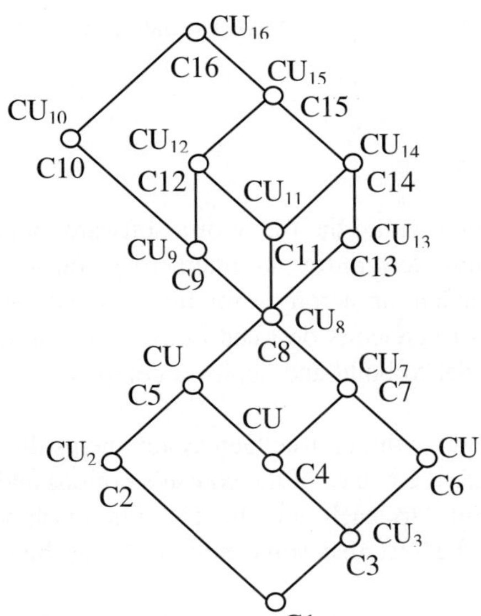


Conceptual Units

$CU_{16}$ = ABCDEFGH
$CU_{15}$ = ABCDEFG
$CU_{14}$ = ABCDEF
$CU_{13}$ = ABCDE
$CU_{12}$ = ABCDFG
$CU_{11}$ = ABCDF
$CU_{10}$ = ABCDGH
$CU_9$ = ABCDG
$CU_8$ = ABCD
$CU_7$ = ABC
$CU_6$ = AB
$CU_5$ = BCD
$CU_4$ = BC
$CU_3$ = B
$CU_2$ = D
$CU_1$ = $\Phi$

**Fig. 1.** The COUL-M model for the teaching-knowledge within "Mutual Exclusion" chapter

It is assumed that the teacher who has authored the interactive course has specified the pedagogical precedence relationships between these notions (depending on their meaning and on this teacher's pedagogical approach) as an ordered list:

cr, mex, cs, sem, sap, wap, pca, lra

The contribution relationship between notions has been specified by the following expressions:

cr → mex and wap → (pca, lra) and sap → wap
(cr, mex, cs, sem) → (sap, wap, pca, lra) and (cr, sem) → cs

The resulting COUL-M lattice-like structure for these notions and relationships - as described by the teacher - is depicted in Figure 1.

The COUL-M model has been implemented by means of a software tool - a knowledge compiler - named COUL-COMP (COnceptual Units' Lattice – knowledge COMPiler). The compiler has been developed so as to test the validity of the theoretical model and of the modelling methods carried out by the authors.

Based-upon the relationships specifications the compiler will produce the Conceptual Transition Path in the following form:

1. a list of the various learning paths the user can adopt in order to assimilate the notions of the interactive course
2. the list of the corresponding learning stages for each learning path.

In this approach, the learning paths calculated by the compiler are ordered sets of learning stages. The learning stages are standing for collections of interconnected notions that succeed in a well-defined order.

Various tests have been deployed and thus the authors have validated the modeling methods.

## 3 The Authoring System

The COUL-M model has been implemented within the frame of a software tool able to realise the teaching-content's structure design for an interactive course. The software-system has been named InStructGen, an acronym for Instructional-content Structure Generation and Deployment. InstructGen is destined to play the role of an authoring system and might be used in order to build and deploy a computer-assisted instructional environment.

Through a user-friendly interface (Figure 2) the InStructGen system helps the user (the teacher-author of the course) to specify the names of the course's notions and the location of software-objects corresponding to each notion. The relationships of contribution between these notions, i.e. the fact that some notions "contribute" to teach other notions must also be specified.

After the course notions and relationships specifications the conceptual structure can be generated. This structure is built according to the design principles previously described in this paper and can be presented to the teacher author in the shape of a lattice-like diagram (Figure 3).

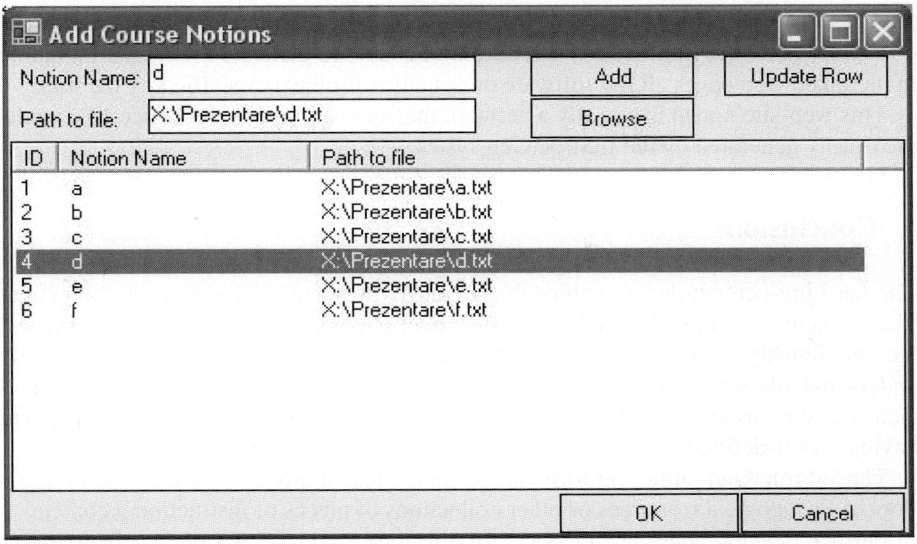


**Fig. 2.** An InsStructGen user interface screen-shot: specifications of software objects corresponding to the course's notions

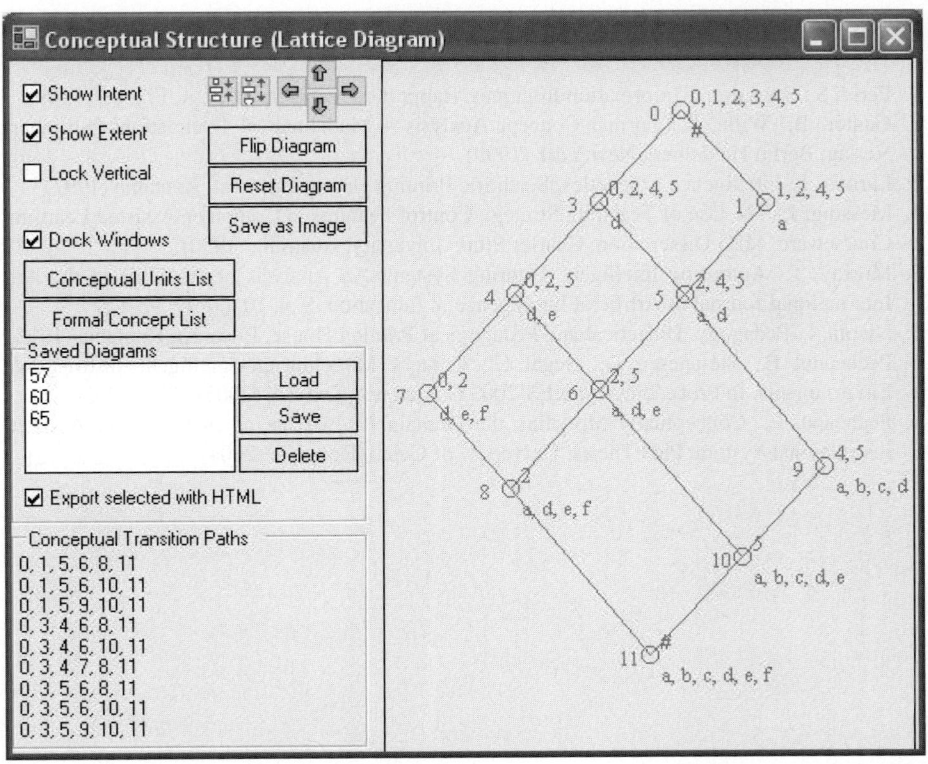


**Fig. 3.** A lattice-like model for the teaching content as generated by the InsStructGen system

Furthermore, the InStructGen software tool is able to generate a web-site including in its linked web pages all the software presentational-objects specified by the user.

This web-site's map is actually a network that strictly follows the lattice-like model previously generated by the InStructGen system.

## 4 Conclusions

The teaching-content design approach previously presented is based upon the lattice-like mathematical model COUL-M. This model is considered to be able to represent the relationships between notions within a training domain in a coherent, easily understandable way. The model has been implemented in a software tool – InStruct-Gen- standing as an authoring system that can generate an instructional environment having a well-defined conceptual structure.

The InStructGen authoring tool can be integrated in any environment including a Pool of Pedagogical resources or other collections of pieces of instructional content.

Further developments are planned for this project in the future: a client server architecture for the instructional environment generated by InStructGen and more extensive live-testing sessions carried out by users enrolled in instructional activities.

## References

1. Ferré, S.: Systèmes d'Information Logiques. Rapport de recherche. IRISA, France, (1999)
2. Ganter, B., Wille, R.: Formal Concept Analysis – Mathematical Foundations. Springer-Verlag, Berlin Heidelberg New York (1999)
3. Larmat, J.: Intelligence's Genetics, Scientific Printing House, Bucarest, Romania (1997)
4. Messing, J.: The Use of Teaching Strategy Control Features in Computer Assisted Learning Courseware. MEd Dissertation. Charles Sturt University, Australia (1990)
5. Murray, T.: Authoring Intelligent Tutoring Systems-An Analysis of the State of the Art. International Journal of Artificial Intelligence in Education, Vol. 10, (1999) 98-129
6. Nicola, I.: Pedagogy. Didactical and Pedagogical Printing House, Bucarest, Romania (1992)
7. Pecheanu, E., Stefanescu D., Segal C.: Content Modeling in Intelligent Instructional Environments. In Proceedings of KES'2003 Conference, Oxford, (2003) 1229–12351
8. Pecheanu, E.: Conceptually Modeling the Domain Knowledge of an Computer Assisted Instructional System. PhD Thesis. University of Galati, Romania (2004)

# Platyhelminthes Are [Re]constructed Recursively*

Alberto de la Encina[1], Mercedes Hidalgo-Herrero[2], and Olga Marroquín-Alonso[1]

[1] Dpto. Sistemas Informáticos y Programación, Facultad de Informática
[2] Dpto. Didáctica de las Matemáticas, Facultad de Educación
Universidad Complutense de Madrid, Spain
albertoe@sip.ucm.es, mhidalgo@edu.ucm.es, alonso@sip.ucm.es

**Abstract.** In this paper we propose a progressive technique for teaching recursion with the aim of making easier the understanding of this concept by students of other areas than computer science. Since knowledge is intended to be actively constructed by the students and not passively absorbed from textbooks and lecturers, the adopted teaching technique is derived from constructivism. This paper presents a study of the results obtained in 2005 by two groups of students from Mathematics and Statistics degrees.

## 1 Introduction

During the last years, computational science has been proved to be a powerful tool in many areas such as economy, mathematics, statistics, biology, and so on. Therefore, the corresponding curricula contain basic courses on computer science, since the students involved in these areas should learn its basic concepts. This applied aspect of computer science has led us to study its methodology in Mathematics and Statistics degrees, in order to check the understanding level of its foundations. In our opinion, recursion should be included in such degrees, owing to its suitability in program design [10]. In fact, recursive-functional languages have proven their usefulness in simplifying programming efforts when dealing with complex problems in computational science (see e.g. [6]). In this respect, there exist two generalized tendencies in teaching recursion: In a basic course or as an advanced tool [1]. We consider that it should be taught as soon as possible, so that students can both take profit from it during their studies, and by doing so, strengthen their knowledge about this useful technique. In this paper, we have taken both approaches: Our first choice was considering recursion as a basic tool, but its treatment as an advanced one is also present in our experiment, because the individuals in the group of Mathematics hardly received training in recursion during their first years.

In addition to this, we pretended to introduce recursion as a natural tool for solving problems whose iterative solution is highly complex. Students were previously taught *if* statements, loops, and subprograms. The new approach was conceived by considering not only our teaching experience, but also the unfruitful traditional methodology [11] and the initial iterative tendency of students.

* Work partially supported by the Spanish MCYT project TIC2003–07848–C02–01.

V.N. Alexandrov et al. (Eds.): ICCS 2006, Part II, LNCS 3992, pp. 207–214, 2006.

### 1.1 Didactical Framework: Constructivism

The traditional methodology for teaching consists in transmitting knowledge by means of teacher explanations. By contrast, we will consider *constructivism* [2], where the teacher task is to pose problems for being solved by the students. In such a case, the teacher just observes the strategies employed by the scholars, but without revealing the solutions of the proposed questions [5].

Besides, thought, as an essential mechanism for solving logical/mathematical questions, comes from acting over an environment [9]. Moreover, the experimentation must be both constant and organized [4], and as a consequence, the teacher will guide the learning of the students by organizing the posed problems. The first one must be easily solvable by modifying the previous knowledge, and the remaining sequence will be proposed by introducing *didactic variables*, that is, planned items which can be modified by the teacher in order to sway the students hierarchy of strategies for solving problems [3, 7].

Following this constructivist methodology, we have used questions dealing with platyhelminthes, that is, flatworms with two kinds of reproduction which allow us to recursively study their family tree. Our proposal comprises four sections: "General definition", "base case/recursive case", "final recursion/non-final recursion", and "inefficiency of several recursive calls".

## 2 Our Proposal

This section comprises the problems posed to the students by paying special attention to the used didactic variables and their purposes. The solving of such problems was performed in the following way:

1. Students were given the formulations.
2. Problems were solved by the individuals by following the given instructions. The teacher behaved as constructivism states.
3. Students explained their solutions. Afterwards, the most suitable ones were selected. In this framework, the discussion among students is extremely important so that the validation of the results is carried out by themselves.

The use of a programming language was not compulsory, so specifications in natural language were also admitted. The problems were the following: Let us consider a colony of platyhelminthes which is characterized by both its way of reproduction (every worm has at least one children) and the immortality of its individuals. Answer concisely the proposed questions. The following functions[1] can be used:

1. `funhermaphrodite(w:worm)returns(b:bool)` where `hermaphrodite(w)` $\leftrightarrow$ `w` is an hermaphrodite worm (sexual reproduction)[2].
2. `fun parent(w:worm) returns (`$\mathtt{w_r}$`:worm)` where $\mathtt{w_r}$ either generated `w` by fragmentation (asexual reproduction) or is the male worm which generated `w` by internal fecundation (sexual reproduction).

[1] They are written by using the notation defined in [8].

[2] The way of reproduction of a worm is constant in its lifetime.

3. fun children(w:worm) returns ($ch_s$:v) where type v=array [1..n] of worm and $ch_s$ contains the children of worm w.
4. fun firstborn_child(w:worm) returns ($w_r$:worm) where $w_r$ is the firstborn child of w (generated from either w's head or its first fertilized egg).
5. fun name(w:worm) returns (s:string) where s is the first name of w.
6. action write(s:string) writes the string s.

## 2.1 General Definition

The first problem deals with the concept of *recursion* by means of two questions which ask for the ancestors and the descendants of a given worm, respectively.

In order to avoid several ascending branches in the family tree, the didactic variable *way of reproduction* takes the value *by fragmentation*. In the second question, fragmentation also sets the number of descendants at two.

1. Let us consider that there did not exist an original worm in the colony such that the remaining worms come from it. Define the set of ancestors of a given flatworm w if all the reproductions leading to it have been by fragmentation. Similarly, define the set of descendants of platyhelminth w if its reproduction and those of its descendants have been by fragmentation.

Because of the way of reproduction of the ancestors, the solution for the first question is a *linear* or *simple recursive function*, whereas the second one is solved by using a *non-linear* or *multiple recursive function.*

Notice that the problem asks for neither the cardinal of ancestors nor the one of descendants of w. Defining a function to calculate the number of ancestors is trivial, whereas the number of descendants can be expressed by the following geometric progression: $\sum_{n=1}^{\infty} 2^n$.

Anchor cases are not possible in solving these questions because the colony features make all the branches in the family tree infinite.

## 2.2 Base Case/Recursive Case

In order to introduce the *base* or *anchor case* concept, we modify the previous formulation by combining both values for the variable *way of reproduction.*

2. Let us consider a non-hermaphrodite flatworm w. Define a function to work out the number of ancestors of w whose reproduction is by fragmentation. The calculation must finish when getting the youngest ancestor whose reproduction has been sexual. Similarly, define a function to calculate the number of descendants of w which have bred asexually and whose parent and sibling are asexual as well.

The first function base case is satisfied by any ascending branch where a sexual worm exists. Similarly, the second function anchor case is met by a descending branch as long as there exists an asexual worm with an hermaphrodite child. The calculation of descendants requires two recursive calls and a base case, whose definition is simplified by the imposed condition on the descendants of w.

## 2.3 Final Recursion/Non-final Recursion

In order to solve the following problem a *non-final recursive function* is needed. Even though the previous questions required non-final recursion, now our purpose goes further: The order between the result processing and recursive calls becomes crucial when defining the required sets. This significance is due to the introduction of the didactic variable *age order* with two possible values: *From the eldest to the youngest* and *from the youngest to the eldest*. Besides, the variable *way of reproduction* allows to obtain both a simple sequence of platyhelminthes (ancestors of `w`) or a graph/tree of flatworms (descendants of `w`).

3a. Let us assume that all the flatworms in the colony have reproduced by fragmentation. In these conditions, the firstborn child of a platyhelminth is the one generated from its head. Besides, let us consider that all the flatworms have a first name which identifies them univocally. Given a platyhelminth `w`, enumerate its firstborn descendants from the eldest one to the youngest one.

3b. Let us suppose a combination of reproduction types where every flatworm has at least one hermaphrodite descendant. In these conditions, the firstborn child of a platyhelminth is the one either generated from its head (asexual reproduction) or born from its first fertilized egg (sexual reproduction). Besides, let us suppose that all the flatworms have a first name which identifies them univocally. Given a non-hermaphrodite platyhelminth `w`, enumerate its firstborn non-hermaphrodite descendants. The calculation must finish when getting the first sexual one and the enumeration must be from the eldest one to the youngest one.

3c. Let us assume the conditions in 3b. Given a non-hermaphrodite platyhelminth `w`, enumerate the first names of its non-hermaphrodite firstborn descendants. The calculation must finish when a sexual worm is reached and the enumeration must be from the youngest one to the eldest one.

It is easy to see that questions 3a and 3b[3] are *final recursion* cases, whereas the solution of question 3c is a non-final recursive algorithm.

### 2.4 Inefficiency of Several Recursive Calls

Recursion drawbacks encompass inefficiency because the usual overload produced by algorithm invocations is increased by using this mechanism. The reason is that an external call may generate a lot of internal invocations.

The following problem will make students face this drawback as they are asked to design a method for improving recursion efficiency.

4a. Let us consider that in the colony there existed an original flatworm which is the eldest ancestor of all the platyhelminthes there. Besides, each worm is bigger than its ancestors as shown by the following expression: `rings(w)=rings((parent(w))+(rings(parent(parent(w)))div2)`. Calculate the number of times each platyhelminth number of rings is computed when the final task is calculating the number of rings of a great-great-grandchild of the original flatworm[4].

By analyzing the simplest solution of this question it is clear that some recursive algorithms are inefficient owing to the used design method. In such a case, the same values are calculated many times, though these repeated computations can be avoided by using auxiliary parameters.

---

[3] In question 3b, the writing action cannot be included in the base case.

[4] The students were given the following data: `rings(firstborn_child(w`$_0$`))=4` and `rings(w`$_0$`)=2` where `w`$_0$ is the original flatworm.

4b. Let us assume the conditions in question 4a. Calculate the number of rings of a flatworm in such a way that each ancestor is processed only once.

## 3 Obtained Results: Mathematics

The detailed analysis of the solutions proposed by the students of the subject *Java Programming* allowed us to check the fact that they hardly know the concept of recursion except from the idea that a recursive function or procedure calls itself repeatedly. As a consequence, many students answer with iterative algorithms to inherently recursive questions, such as the enumeration of both the ancestors and descendants of a given flatworm. This iterative tendency yields never-ending loops whenever the recursion we are dealing with lacks a base case. Next, we discuss such design issues by analyzing some of the obtained results.

### 3.1 General Definition

More than half of the students (seven of 13) defined iterative algorithms based on loops instead of recursive functions. A representative fragment of (Java) code extracted from these algorithms is the following, where the boolean condition is trivially satisfied because there did not exist an original worm in the colony.

```
worm wr=w; while(parent(wr)!=null){wr=parent(wr);}
```

In addition to this, just one of the six remaining students deemed the question too difficult, but the rest of solutions were correctly formulated in a recursive way. Next, we show the definition of the set of ancestors of `w` provided by the only student who learnt Haskell in his first year of Computer Science. The analysis of his answers confirms the idea that teaching functional programming languages before imperative ones makes easy the understanding of the concept of recursion:

$$\mathtt{y} \in \mathtt{A(w)} \leftrightarrow \mathtt{y} = \mathtt{parent(w)} \vee \mathtt{y} \in \mathtt{A(parent(w))}$$

### 3.2 Base Case/Recursive Case

In this case, nine students defined a function to work out the number of ancestors of `w` and seven provided a function to calculate the number of descendants of the same worm. In spite of this low level of response, the proposed algorithms such as the following show a clear iterative tendency in defining recursive concepts.

```
worm wr=w; boolean asexual=true;
while (asexual){
 if (hermaphrodite(parent(wr))) asexual=false else wr=parent(wr);}
```

Seven (of nine) students provided similar algorithms to the one shown above, whereas there was just one suitable solution recursively defined. Concerning the calculation of the descendants of `w`, many functions were not properly defined owing to the fact that they were based on the expression

$\mathtt{hermaphrodite(parent(w))} \vee \mathtt{hermaphrodite(sibling(w))}$, where

$\mathtt{sibling(w)} = \mathtt{w_r} \leftrightarrow \mathtt{w_r} \in \mathtt{children(parent(w))} \wedge \mathtt{name(w_r)} \neq \mathtt{name(w)}$,

while a closer analysis of the question yields the boolean condition

$\mathtt{hermaphrodite(children(g)[1])} \vee \mathtt{hermaphrodite(children(g)[2])}$.

Therefore, two (of seven) students provided an iterative algorithm by using the first expression, and five of them defined a recursive implementation, two of which were correctly designed. Finally, we note that students tend either to solve both questions (seven of 13) or to leave blank both of them (four of 13). The two remaining students provided only an algorithm to work out the number of ancestors of `w`.

### 3.3 Final Recursion/Non-final Recursion

As expected, the number of students who tried to solve a problem is inversely proportional to the level of difficulty of such problem. In this way, seven students provided solutions corresponding to questions 3a and 3b, whereas question 3c was solved by four students. In any case, loops were less used than before, though it was easy to find in them many mistakes quite similar to those previously described. The proposed algorithms dealing with final recursion were correct in most cases, as shown by the following functional definition:

```
names x = name x : names firstborn_child x
```

Nevertheless, non-final recursion proved to be hard to understand, since in order to solve the last question, students defined a double path through the family tree: First the youngest sexual descendant of `w` is found, and then its ancestors are properly displayed. In short, four students solved properly questions 3a and 3b, one of whom provided a correct algorithm in Haskell for question 3c.

### 3.4 Inefficiency of Several Recursive Calls

The question was solved by only three students, who also provided algorithms for the previous problems. They correctly calculated the number of rings of a great-great-grandchild of the original worm in the colony, but the provided solutions for computing such number in the general case were not accurate. To be exact, just one student proposed a correct solution in Haskell for the whole problem, whereas his schoolmates, who learnt Pascal in their first year of Computer Science, did not handle accumulators properly.

## 4 Obtained Results: Statistics

The results in this section show the difficulties in teaching recursion owing to the iterative tendency of human mind. As a consequence, our main purpose is to make students understand recursion as a natural mechanism for problem solving.

Besides, these individuals, who study statistics, are not familiarized with programming techniques, so they are not supposed to give formal solutions for the posed questions but informal specifications of the corresponding algorithms.

### 4.1 General Definition

The first approach for the first question (ancestors of `w`) given by the 14 students was iterative, so that the following hint was given: "Let us assume that one of your ancestors knows the set of his/hers. Could you answer this question by using such set?". This clue led the students to a correct recursive algorithm, though it was fairly suggested by the teacher. Concerning the question about the descendants of `w`, one student provided an iterative algorithm, whereas the remaining 13 proposed a solution defined in a recursive way, one of which was correct. The rest of recursive algorithms called themselves just once, maybe because in those days statistics students had not learnt arrays in depth.

### 4.2 Base Case/Recursive Case

The number of clues given by the teacher diminished considerably for this problem, owing to the fact that recursion was introduced in the previous one. In spite of this, not only three students provided iterative algorithms, but also another two defined solutions by combining loops with recursive calls in the non-trivial case. The following fragment of code shows such combination.

```
repeat anc:=parent(w)+ances(parent(w)) (*recursive call*)
until (hermaphrodite(parent(w))=true)
```

Nine students proposed recursive solutions, two of which lacked a base case.

### 4.3 Final Recursion/Non-final Recursion

According to the constructivist method, the teacher did not gave any hint in this case, since it was assumed that recursion was introduced enough. In some cases, the traverse of the family tree was wrongly accomplished, due to the fact that just two generations were reached, as shown by the following implementation.

```
nameW:=name(w)+name(firstborn_child(w))
```

However, other recursive solutions were quite clear. We show the most representative one corresponding to question 3b.

```
fun list(w:worm) returns (n:string)
 n=name(w)+if (hermaphrodite(firstborn_child(w)))
             list(firstborn_child(w))
```

Finally, the analysis of the obtained results confirms that students tend to consider recursion when solving those problems whose iterative solutions have an increasing complexity. The ratio between the number of iterative solutions and the total number of answers were: Ten to 14 in question 3a, five to ten in question 3b, and one to eight in question 3c. Besides, we noticed that half of the recursive algorithms provided for questions 3b and 3c did not consider a base case, though these solutions were more correct than the iterative ones.

### 4.4 Inefficiency of Several Recursive Calls

As expected, only few students tried to solve this problem, as it shows the fact that only six individuals provided a solution for the first question. Besides, every student except one mistaken the calculations in the following way: Given the expressions

$\texttt{rings}(\texttt{w}_4)=\texttt{rings}(\texttt{w}_3)+(\texttt{rings}(\texttt{w}_2)\texttt{div2})$,
$\texttt{rings}(\texttt{w}_3)=\texttt{rings}(\texttt{w}_2)+(\texttt{rings}(\texttt{w}_1)\texttt{div2})$, and
$\texttt{rings}(\texttt{w}_2)=\texttt{rings}(\texttt{w}_1)+(\texttt{rings}(\texttt{w}_o)\texttt{div2})$

they did not consider that platyhelminth $w_2$ is consulted three times, owing to the number of rings of $w_2$ is used for processing $w_4$ and $w_3$. This mistake is due to the fact that students did not develop the whole invocations tree, that is, they just worked out for each flatworm its appearance in the right hand-side of the equations. Finally, the last question was answered by four students, who failed to solve the involved recurrence and did not propose any recursive solution.

## 5 Conclusions and Current Work

The main purpose of this test was the study of teaching recursion in other areas than computer science. It can be observed that recursion can be successfully taught in a basic course in order to provide a useful technique for developing computational systems by students of either Mathematics or Statistics degrees. Concerning our platyhelminth problem, it is easy to see that students tend to provide iterative algorithms to solve the simplest questions, whereas recursion is used in the complex ones, because recursive programming needs fewer low level details than loop programming. In this way, the teacher's role as a simple guide has proved to be very useful, since students have evolved by themselves from iterative to recursive solutions. Besides, since the posed problems deal with family trees (a familiar concept) instead of the typical mathematical items (e.g. the Fibonacci series), its use in areas such as biology or economy is more plausible.

As current work, we pretend to pose again these questions by simplifying their formulations with the purpose of making them suitable for other programming subjects and areas.

## References

1. O. Astrachan. Self-reference is an illustrative essential. In *Proceedings of SIGCSE 1994, Technical Symposium on Computer Science Education*, pages 238–242. ACM Press, 1994.
2. M. Ben-Ari. Constructivism in computer science education. *Journal of Computers in Mathematics and Science Teaching*, 20(1):45–73, 2001.
3. J. Briand and M. C. Chevalier. *Les Enjeux Didactiques dans l'Enseignement des Mathématiques.* Hatier, 1995.
4. J.S. Bruner. *On Knowing:Essays fortheLeftHand.* MA:HarvardUniversityPress,1962.
5. M. Chamorro, J. M. Belmonte Gómez, S. Llinares, M. Ruiz Higueras, and F. Vecino-Rubio. *Didáctica de las Matemáticas para Primaria.* Pearson, 2003.
6. A. Encina, I. Rodríguez, and F. Rubio. Testing spculative work in a lazy/eager parallel functional language. In *Languages and Compilers for Parallel Computing (LCPC'05)*, LNCS. Springer-Verlag, 2006. In press.
7. M. Hidalgo-Herrero, I. Rodríguez, and F. Rubio. Testing learning strategies. In *Int. Conference on Cognitive Informatics (ICCI'05)*, pages 212–221. IEEE, 2005.
8. R. Peña. *Diseño de Programas. Formalismo y Abstracción.* Prentice Hall, 1998.
9. J. Piaget. *Introduction à l'Épistemologie Genetique.* PUF, 1973.
10. A. B. Tucker. Computing curricula 1991. *Communication of the ACM*, 34(6):68–84, 1991.
11. J. Velázquez-Iturbide. Recursion in gradual steps (is recursion really that difficult?). In *Proceedings of SIGCSE 2000, Technical Symposium on Computer Science Education*, pages 310–314. ACM Press, 2000.

# Extensions for 3D Graphics Rendering Engine Used for Direct Tessellation of Spline Surfaces

Dr. Adrian Sfarti, Prof. Brian A. Barsky, Todd J. Kosloff, Egon Pasztor, Alex Kozlowski, Eric Roman, and Alex Perelman

University of California, Berkeley

**Abstract.** In current 3D graphics architectures, the bus between the triangle server and the rendering engine GPU is clogged with triangle vertices and their many attributes (normal vectors, colors, texture coordinates).

We have developed a new 3D graphics architecture using data compression to unclog the bus between the triangle server and the rendering engine. This new architecture has been described in [1]. In the present paper we describe further developments of the newly proposed architecture.

The current paper shows several interesting extensions of our architecture such as backsurface rejection, NURBS real time tesselation and a description of a surface based API. We also show how the implementation of our architecture operates on top of the pixel shaders.

## 1 Introduction

In [1] we described a new graphics architecture that exploits faster arithmetic such that the CPU will serve parametric patches and the rendering engine (GPU) will triangulate those patches in real time. Thus, the bus sends control points of the surface patches, instead of the many triangle vertices forming the surface, to the rendering engine. The tesselation of the surface into triangles is distance-dependent, it is done in real time inside the rendering engine. The implementation of the architecture is through reprogramming one or more vertex processors inside the GPU.

## 2 Previous Work

There have been very few implementations of real time tesselation in hardware. In the mid-1980's, Sun developed an architecture for this that was described in [2] and in a series of associated patents. The implementation was not a significant technical or commercial success because it did not exploit triangle based rendering; instead it attempted to render the surfaces in a pixel-by-pixel manner [3]. The idea was to use adaptive forward differencing to interpolate infinitesimally close parallel cubic curves imbedded into the bicubic surface.

[4] describe hardware support for adaptive subdivision surface rendering.

Recently, NVIDIA has resurrected the real time tesselation unit [5]. The NVIDIA tesselation unit is located in front of the transformation unit and outputs triangle databases to be rendered by the existent components of the 3D graphics hardware.

More recently Bolz and Schroder have attempted to evaluate subdivision surfaces via programmable GPUs [6].

V.N. Alexandrov et al. (Eds.): ICCS 2006, Part II, LNCS 3992, pp. 215–222, 2006.

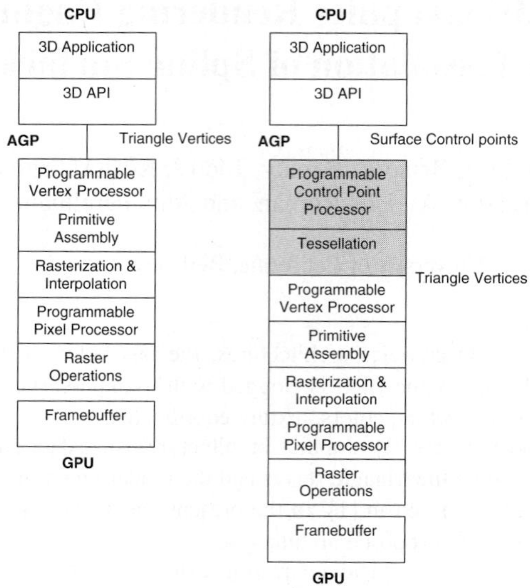


**Fig. 1.** Conventional programmable architecture (left), new architecture (right). The new architecture adds two stages to the GPU pipeline, which are shown in grey.

## 3 Tesselator Unit Back-Facing Surface Removal and Trivial Clipping

Since surfaces that are entirely back-facing should be discarded before attempting tesselation, we first test to determine if any portion of the patch is facing towards the viewer. Although back-facing triangles are discarded in the triangle rendering portion of the pipeline, the discarding of entire surfaces as early as possible in the pipeline reduces time spent on tesselation. To determine whether or not a patch is back-facing, we use the convex hull of the patch. Specifically, we test the polyhedral faces of the convex hull to determine if they are forward-facing. If any face is forward-facing, then the patch is not discarded. Note that not all faces need to be tested; as soon as a forward-facing polyhedral face is found, then we know that the patch cannot be discarded.

If the 16 control points of a surface all lie outside the same clipping plane, the surface can be trivially clipped out of the database.

We produced five different animations in order to prove our backface removal and surface trivial clipping algorithm.

## 4 Tesselator Unit Extension to NURBS

The algorithm described above has a straightforward extension for rational surfaces such as non-uniform rational B-spline (called "NURBS") surfaces.

A nonuniform rational B-spline surface of degree $(p,q)$ is defined by

$$S(s,t) = \frac{\sum_{i=1}^{m}\sum_{j=1}^{n} N_{i,p}(s)N_{j,q}(t)w_{i,j}P_{i,j}}{\sum_{i=1}^{m}\sum_{j=1}^{n} N_{i,p}(s)N_{j,q}(t)w_{i,j}}$$

where $N_{i,p}()$ denotes the B-spline basis of degree $p$, $w$ denotes the matrix of weights, and $P$ denotes the matrix of control points.

Each patch of such a surface lies within the convex hull formed by its control points. To illustrate the concept, consider $p = q = 4$ and $m = n = 4$. There are 16 control points, $P_{11}$ through $P_{44}$ (similar to the surfaces described in the previous paragraph). The surface lies within the convex hull formed by $P_{11}$ thru $P_{44}$.

Now consider any one of the curves:

$$C(s) = \frac{\sum_{i=1}^{m} N_{i,p}(s) w_i P_i}{\sum_{i=1}^{m} N_{i,p}(s) w_i}$$

where the parameter $p$ denotes the order, $N_{i,p}(s)$ the B-spline basis functions, $P_i$ dentoes the control points, and $w_i$ is the weight of control point $P_i$ and is represented as the last coordinate of the homogeneous point. The curve lies within the hull formed by the control points. Such a curve can be obtained by fixing one of the two parameters $s$ or $t$ in the surface description. For example holding $t = 0$ while letting $s$ vary produces such a curve. As in the case of Bézier surfaces, there are eight such curves; that is, four boundary curves and four internal curves. The tesselation of the surface reduces to the subdivision of the hull of the boundary curves or of the internal curves as described in the case of the Bézier surfaces. By subdividing the B-spline surface we reduced the problem to a problem that we have already solved. Trimmed NURBS are treated as the intersection of the triangle meshes resulting from the tesselation of the untrimmed NURBS used in the intersection. This simple method outputs the trimming loops rather than requiring them as an input in a drastic departure from the current approaches.

## 5 Tesselator Unit and Programmable Shaders

While we can readily build GPUs according to the proposed architecture from existing designs, there are some issues surrounding the transition that we will identify and resolve. One concern for vertex shader programming under this new architecture, is how to pass custom per-vertex information to the vertex shader. Since the 3D

**Fig. 2.** Per pixel lighting using programmable shaders

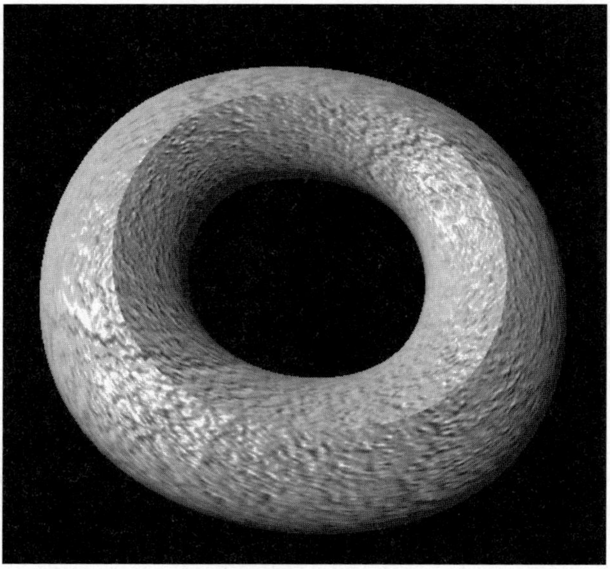

**Fig. 3.** Bump mapped ridged torus using programmable shaders

application is sending a stream of control points to the GPU, all of the vertices generated by the tesselator and subsequently fed to the programmable vertex processor, have properties calculated by the Tesselation Processor as described in [1], not by the application. This complication can be resolved quite naturally using a technique already familiar to shader programmers. Property maps for custom attributes can be encoded as 2D-textures, and vertex shaders could sample the texture to extract interpolated values. This allows each vertex generated within the surface to assume a meaningful custom attribute value (could use s/t values, position, texture coordinates etc. for sampling location). Many existing vertex shaders rely on heavily (and uniformly) tessellated models. The greater number of vertices can increase the quality of per-vertex lighting calculations, and facilitate displacement mapping of vertices. We support this class of shader despite our distance-dependent subdivision, to the degree required by the vertex shader program.

- Firstly, we might increase the overall fineness of the resulting tessellation by controlling the termination criterion.
- Secondly, we could invoke an adaptive mesh refinement as part of our tesselation algorithm This could be sufficient for displacement mapping, where the vertex shader merely needs more points to play with. More sophisticated mesh refinements would benefit lighting and texturing without having to evaluate more points using the surface description.

Given the introduction of the control point shader, for maximum effectiveness the vertex shader should only permute geometry in subtle ways. Following this convention will ensure that vertex shaders do not invalidate the perceptual accuracy of our distance-dependent tessellation. Lastly, one should remember that the Tesselation Processor already incorporates a minimum level of tessellation . The use of vertex/pixel shaders was a natural extension in our software demonstration, and required no special enhancements. Figures 2 and 3 show the combination of real-time tessellation and programmable shaders to bump map bicubic surfaces.

## 6 A Prototype for a Graphics Library Utility

To facilitate the design of drivers for the proposed architecture, we must develop a Graphics Library Utility (GLU). The primitives of the GLU are strips, fans, meshes and indexed meshes. Current rendering methods are based on triangle databases (strips, fans, meshes) resulting from offline tesselation via specialized tools. These tools tessellate the patch databases and ensure that there are no cracks between the resulting triangle databases. The tools use some form of zippering. The triangle databases are then streamed over the AGP bus into the GPU. There is no need for any coherency between the strips, fans, etc., since they are, by definition, coherent (there are no T-joints between them). The net result is that the GPU does not need any information about the entire database of triangles, which can be quite huge. Thus, the GPUs can process virtually infinite triangle databases. Referring to Figure 4., in a strip, the first patch will contribute sixteen vertices, and each successive patch will contribute only twelve vertices because four vertices are shared with the previous patch. Of the sixteen vertices of

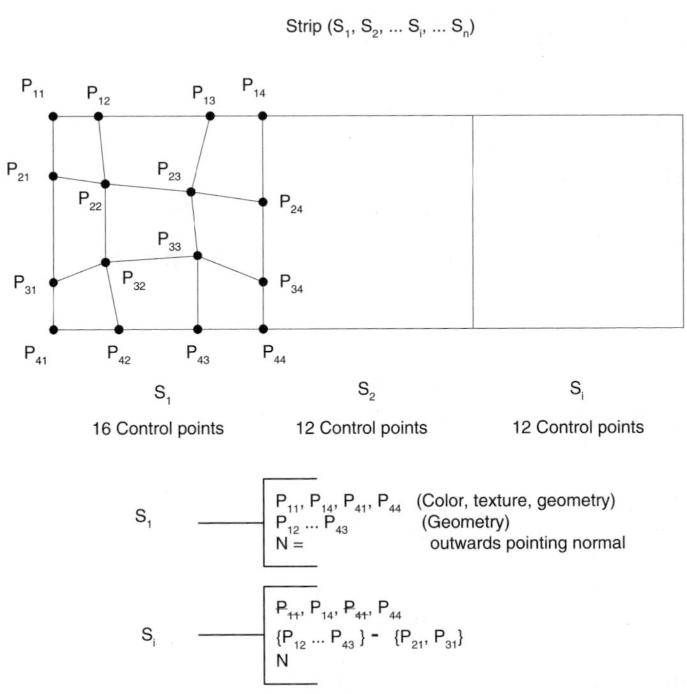


**Fig. 4.** Strip

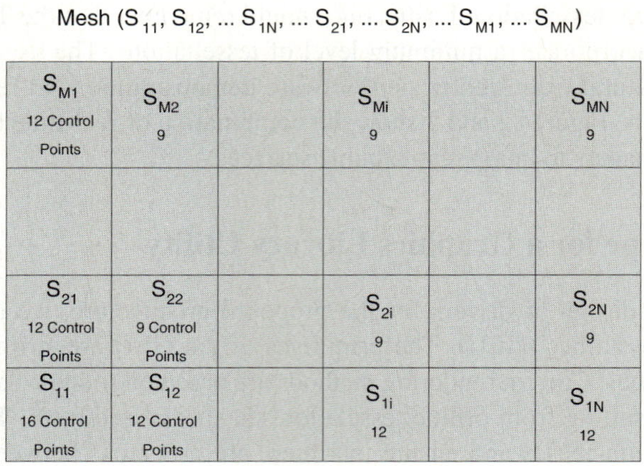


**Fig. 5.** Mesh

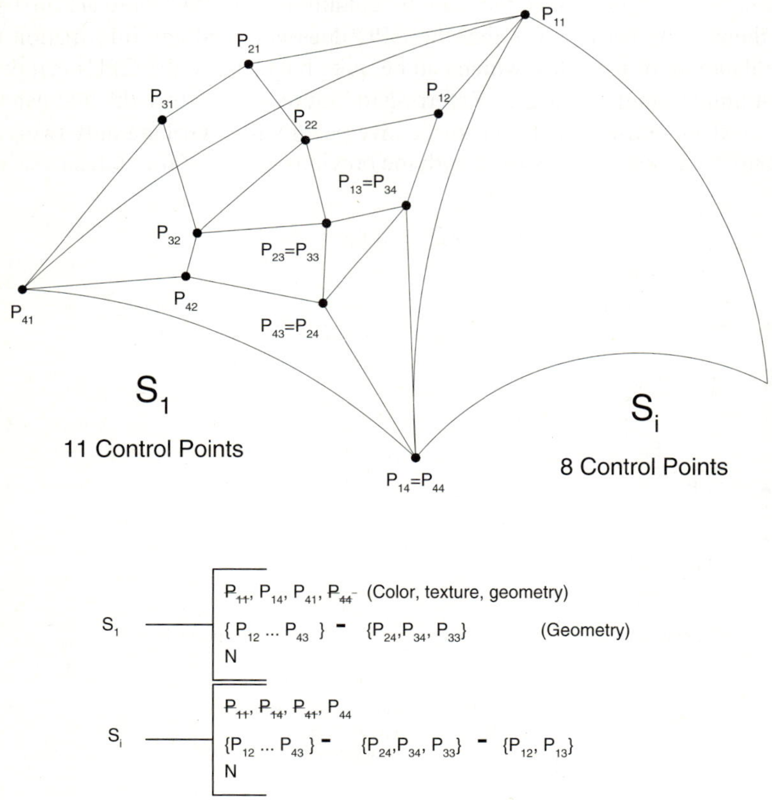


**Fig. 6.** Fan

the first patch, $S_1$, there will only be four vertices (namely, the corners $P_{11}$, $P_{14}$, $P_{41}$, $P_{44}$) that will have color and texture attributes; the remaining twelve vertices will have only geometry attributes. Of the twelve vertices of each successive patch, $S_i$, in the strip, there will only be two vertices (namely, $P_{14}$ and $P_{44}$) that will have color and texture attributes. It is this reduction in the number of vertices that will have color and texture attributes that accounts for the reduction of the memory footprint and for the reduction of the bus bandwidth necessary for transmitting the primitive from the CPU to the rendering engine (GPU) over the AGP bus. Further compression is achieved because a patch is expanded into potentially many triangles by the Tessellator Unit inside the GPU. Referring to Figure 5, in a mesh, the anchor patch, $S_{11}$ has sixteen vertices, all the patches in the horizontal and vertical strips attached to $S_{11}$ have twelve vertices and all the other patches have nine vertices. Each patch has an outward pointing normal. Referring to Figure 6, each patch has only three boundary curves, the fourth boundary having collapsed to the center of the fan. The first patch in the fan enumeration has eleven vertices; each successive patch has eight vertices. The vertex $P_{11}$, which is listed first in the fan definition, is the center of the fan and has color and texture attributes in addition to geometric attributes. The first patch, $S_1$, has two vertices with color and texture attributes, namely $P_{41}$ and $P_{14}$; the remaining nine vertices have only geometric attributes. Each successive patch, Si, has only one vertex with all the attributes.

The meshed curved patch data structures introduced above are designed to replace the triangle data structures used in the conventional architectures. Within one patch strip, the edge database must be retained for zippering reasons but no information needs to be stored between two abutting patches. If two patches bounding two separate surfaces share an edge curve, they share the same control points and they will share the same tesselation. By doing so we ensure the absence of cracks between patches that belong to data structures that have been dispatched independently and thus our method scales the exactly the same way the traditional triangle based method does.

## 7 Conclusion

We developed a new 3D graphics architecture that replaces the conventional idea of a 3D engine (GPU) that renders triangles with a GPU that will tesselate surface patches into triangles. Thus, the bus sends control points of the surface patches, instead of the many triangle vertices forming the surface, to the rendering engine. The tesselation of the surface into triangles is distance-dependent, it is done in real time inside the rendering engine.

## Acknowledgements

The authors would like to thank Tim Wong, Grace Chen, Clarence Tam, and Chris Lai for their collaboration in the programming of the demos.

## References

1. Sfarti, A., Barsky, B., Kosloff, T., Pasztor, E., Kozlowski, A., Roman, E., Perelman, A.: New 3d graphics rendering engine architecture for direct tessellation of spline surfaces. In: International Conference on Computational Science (2). (2005) 224–231

2. Lien, S.L., Shantz, M., Pratt, V.R.: Adaptive forward differencing for rendering curves and surfaces. In: SIGGRAPH '87 Proceedings, ACM (1987) 111–118
3. Lien, S.L., Shantz, M.: Shading bicubic patches. In: SIGGRAPH '87 Proceedings, ACM (1987) 189–196
4. Boo, M., Amor, M., Dogget, M., Hirche, J., Stasser, W.: Hardware support for adaptive subdivision surface rendering. In: ACM SIGGRAPH/Eurographics workshop on Graphics Hardware. (2001) 30–40
5. Moreton, H.P.: Integrated tesselator in a graphics processing unit. U.S. patent (2003) #6,597,356.
6. Bolz, J., Schroder, P.: Evaluation of subdivision surfaces on programmable graphics hardware. http://www.multires.caltech.edu/pubs/GPUSubD.pdf (2003)
7. Clark, J.H.: A fast algorithm for rendering parametric surfaces. In: Computer Graphics (SIGGRAPH '79 Proceedings). Volume 13(2) Special Issue., ACM (1979) 7–12
8. Moreton, H.P.: Watertight tesellation using forward differencing. In: Proceedings of the ACM SIGGRAPH/Eurographcs workshop on graphics hardware. (2001)
9. Chung, A.J., Field, A.: A simple recursive tesselator for adaptive surface triangulation. JGT **5(3)** (2000)
10. Moule, K., McCool, M.: Efficient bounded adaptive tesselation of displacement maps. In: Graphics Interface 2002. (2002)
11. Boo, M., Amor, M., Dogget, M., Hirche, J., Strasser, W.: Hardware support for adaptive subdivision surface rendering. In: Proceedings of the ACM SIGGRAPH/Eurographics workshop on Graphics Hardware. (2001) 33–40
12. Hoppe, H.: View-dependent refinement of progressive meshes. In: Proceedings of the 24th annual conference on computer graphics and interactive techniques. (1997)
13. Sfarti, A.: Bicubic surface rendering. U.S. patent (2003) #6,563,501.
14. Sfarti, A.: System and method for adjusting pixel parameters by subpixel positioning. U.S. patent (2001) #6,219,070.
15. Barsky, B.A., DeRose, T.D., Dippe, M.D.: An adaptive subdivision method with crack prevention for rendering beta-spline objects. Technical Report, UCB/CSD 87/384, Computer Science Division, Electrical Engineering and Computer Sciences Department, University of California, Berkeley, California, USA (1987)
16. Lane, J.F., Carpenter, L.C., Whitted, J.T., Blinn, J.F.: Scan line methods for displaying parametrically defined surfaces. In: Communications of the ACM. Volume 23(1)., ACM (1980) 23–24
17. Forsey, D.R., Klassen, R.V.: An adaptive subdivision algorithm for crack prevention in the display of parametric surfaces. In: Proceedings of Graphics Interface. (1990) 1–8
18. Velho, L., de Figueiredo, L.H., Gomes, J.: A unified approach for hierarchical adaptive tesselation of surfaces. In: ACM Transactions on Graphics. Volume 18(4)., ACM (1999) 329–360
19. Kahlesz, F., Balazs, A., Klein, R.: Nurbs rendering in opensg plus. In: OpenSG 2002 Papers. (2002)

# An Evolution Computation Based Approach to Synthesize Video Texture

Yu Meng, Wen-hui Li, Yan Wang, Wu Guo, and Wei Pang

College of Computer Science and Technology, Key Laboratory of Symbol Computation and Knowledge Engineering of the Ministry of Education, Jilin University, Changchun, 130012, P.R.China
myu_jlu@126.com

**Abstract.** Texture synthesis is one of the hottest areas in computer graphics, computer vision and image processing fields, and video texture synthesis is one subset of it. We bring forward a new method on video texture synthesis, in which evolution computing technique is introduced into the processes of synthesizing videos. In the method, by analyzing and processing a finite source video clip, Infinite video sequences obtained can be played smoothly in vision. Comparing with many existing video texture synthesis algorithms, this method can not only get high-quality video results without complicated pre-processing of source video, but also improve the efficiency of synthesis.

## 1 Introduction

Texture synthesis is of great importance in computer vision and graphics. Many methods on texture synthesis were introduced in the last decade. Heeger & Bergen [1] use color histograms across frequency bands as a texture description. Bonet[2], who first brought non-parameter texture synthesis method, uses a multiresolution filter-based approach to generate textures. Efors & Leung[3] are first to copy pixels directly from input textures to generate new textures. Portilla & Simoncelli[4] use a synthesis procedure by decomposing texture images to complex wavelets and synthesize new images by matching the joint statistics of these wavelets. Wei & Levoy[5] have improved Efros & Leung's method by using a multiresolution image pyramid based on a hierarchical statistical method. Liang [6] and Efors & Freeman[7] copy whole patches from input textures to synthesize new textures.

The work described above mainly deals with static scenes. Recently has been appeared many video texture synthesis methods. Arno Schödl creates long videos by rearranging original frames from a short source video[8], and extends video textures to allow for control moving objects in video[9]. Wang&Zhu[10] models the motion of texture particles in source video to synthesize new videos. Doretto[11] uses Auto-Regressive filters to model and edit the complex motion of fluids in video. Kwatra[12] combines volumes of pixels along minimum error seams to create new sequences that are longer than the original video. Bhat[13] analyzes the motion of textured particles in the input video along user-specified flow lines, and synthesizes seamless video of arbitrary length by enforcing temporal continuity long a second set of user-specified flow lines.

V.N. Alexandrov et al. (Eds.): ICCS 2006, Part II, LNCS 3992, pp. 223 – 230, 2006.

## 2 Evolution Computation-based Video Texture Synthesis

We suggest a video texture synthesis method using Genetic Algorithm[14], which is inspired by Schödl's algorithm[8]. Compared with other algorithm, this method uses an appropriate fitness function in the processes of creating video sequences instead of much complex pre-processing of source video clip. Additionally, constraints of selection operator, crossover operator and mutation operator ensure that this algorithm can get high-quality video results quickly.

### 2.1 Video Textures Algorithm

The original idea of Video Texture Algorithm supposes that there are two similar frames in a given video clip: frame *i* and frame *j*. Naturally, the switch from frame *j* to frame *j*+1 is very smooth, we can gain the unaffected video from switching frame *i* to frame *j*+1. Here *i* is called "transition" point. When the video is played frame *i*, the next frame can be frame *i*+1 or frame *j*+1. If there are several such transition points in the source clip, videos with arbitrary length can be synthesized from it by switching at these points continuously.

The original procedure of Video Textures Algorithm is described as follows:

1. Calculate the similarities between all pairs of frames in the input sequence *S*. Calculate frame-to-frame distances and stored them in the similarity matrix of *D* with the following formula:

$$D_{ij}=\sum_{k=1}^{m}\sum_{l=1}^{n} sqrt\left\{\left[R\left(P_{k,l}^{i}\right)-R\left(P_{k,l}^{j}\right)\right]^2+\left[G\left(P_{k,l}^{i}\right)-G\left(P_{k,l}^{j}\right)\right]^2+\left[B\left(P_{k,l}^{i}\right)-B\left(P_{k,l}^{j}\right)\right]^2\right\} \quad (1)$$
$$0<i,j\le len;\quad 0<k\le m;\quad 0<l\le n;$$

   Where $P_{k,l}^{i}$ denotes the pixel of coordinates (*k*,*l*) in the *i*-th frame with size m×n in video sequence *S*. *R()*, *G()*, *B()* represent the RGB values of a pixel. And *len* is the frame number of *S*;
2. Calculate the transition cost matrix between two frames. Transition cost is a sum of image differences around a transition. Transition cost from frame $F_i$ to frame $F_j$ can be calculated as below and stored in a transition cost matrix of $D_{ij}^{'}$ :

$$D_{ij}^{'}=\sum_{k=-m}^{m-1} w_k D_{i+k+1,j+k};\quad 0<i,j\le len; \quad (2)$$

   Where m=1 or 2, corresponding to a sum over 2 or 4 image differences with weights $w_k$ (1 1) or (1 2 2 1), and *len* is the frame number of *S*;
3. Avoid dead-ends and prune part of transitions. The former operation aim at preventing the video from being lead to regions without transitions, and the latter operation can save on storage space and improve the quality of result videos.

4. Calculate future cost for each transition using the following formula, that reflects the expected average cost of future transitions and store it in future cost matrix of $D''_{ij}$.

$$D''_{ij} = (D'_{ij})^p + \alpha \min_k D''_{jk}; \qquad 0 < i, j, k \le len; \tag{3}$$

   Here, $p$ is a constant controlling the tradeoff between multiple good (low-cost) transitions and a single poorer one. The constant $\alpha$ is used to control the relative weight of future transitions in the metric. In order to reach convergence, it should be limited to (0,1). *len* is the frame number of *S*;
5. Create video loops, called compound loops, which can be combined to create additional cyclic sequences. The final video sequences are composed of several compound loops.

Applying Video Textures Algorithm to process not only random textures but also structured textures can obtain high-quality result video. However, the pre-processing to source clip before creating video sequences is too much and complex. Furthermore, the Dynamic Programming Algorithm(DPA) used to create video sequences only allows backward transition so that its synthesizing results have less diversity. We find that creating of video sequences can be considered as a combinatorial optimum problem and solved by Genetic Algorithm. Using Genetic Algorithm to synthesize videos only need an appropriate fitness function to preserve dynamics and continuity instead of much complex pre-processing. In addition, the GA-based algorithm allows not only backward transition but also forward transition so that its synthesizing results have better diversity.

## 2.2 Genetic Algorithm (GA)

GA is based on the biological phenomenon of genetic evolution. The GA maintains a set of solutions which is known as a population or a generation. GA operates in an iterative manner and evolves a new generation from the current generation by application of genetic operators. The process of basic Genetic Algorithm is as below:

1. Generate random population of n chromosomes (suitable solutions for the problem)
2. Evaluate the fitness f(x) of each chromosome x in the population
3. Create a new population by repeating following steps until the new population is complete
   3.1 Select two parent chromosomes from a population according to their fitness. For each chromosome, the better fitness it has, the bigger chance it would be selected;
   3.2 Cross over the parents to form new offspring with a crossover probability. If no crossover was performed, offspring is the exact copy of parents.
   3.3 Mutate new offspring at each position in chromosome with a mutation probability;
   3.4 Place new offspring in the new population

If the end condition is satisfied, stop, and return the best solution in current population; otherwise, go to step 2.

### 2.3 Evolution Computation-Based Video Texture Synthesis

Now we apply GA to Synthesize videos:

**Encoding.** Given a video sequence $S$ with $l$ frames, S={$F_1$, $F_2$...$F_l$}, a solution to video texture synthesis will be represented as an ordered list of numbers. Define each decimal number $i(i\in[1,l])$as a gene and let it denote a frame $F_i$. For instance, (3 4 5 1 2 3 6 7) is the chromosome representation of video sequence {$F_3$, $F_4$, $F_5$, $F_1$, $F_2$, $F_3$, $F_6$, $F_7$}. In the following algorithm description, we assume that there are $m$ chromosomes in the population and each chromosome $C_k$ ($k\in[1,m]$) denotes a solve to video texture synthesis. The simple permutation representation is a common and popular representation for solving the order-based problems.

**Fitness.** In order to make new video sequence continuous in vision, we should compute similarities among frames and cost of transitions. First we should calculate similarity matrix of D and transition cost matrix of $D'_{ij}$. Then we use formula(4) to calculate mathematical expectation $E_k$ and standard error $\sigma_k$ of transition cost for each chromosome $C_k$.

$$E_k=\left(\sum_{j=1}^{n-1}D'_{c_j,c_{j+1}}+D'_{c_n,c_1}\right)\times n^{-1}\,,$$
$$\sigma_k=\sqrt{\left(\sum_{j=1}^{n-1}(D'_{c_j,c_{j+1}}-E)^2+(D'_{c_n,c_1}-E)^2\right)\times n^{-1}} \tag{4}$$

Here, $n$ is the number of frames in $C_k$, $c_j$ represents the $j$-th($j\in[1,n]$) gene of $C_k$ For each chromosome $C_k$ if its $E_k$ is small and its $\sigma_k$ is large, there are less transitions with high transition costs in the video sequence corresponding to $C_k$. And thus there are some obvious abruptions in the video sequence; if its $E_k$ is large and its $\sigma_k$ is small, there are more transitions with low transition costs in the video sequence corresponding to $C_k$. The video sequence is regarded as repeatedly playing a small clip. It seems that when its $E_k$ and $\sigma_k$ are both relative small, chromosomes have better continuity and diversity. Therefore, we can define fitness function as a weighted sum of $E_k$ and $\sigma_k$. Furthermore, considering that $E_k$ and $\sigma_k$ may be not at the same quantity-level, we should normalize them at first. Formula of fitness function is as below:

$$Fit(C_k)=\left(\alpha\cdot E_k+\beta\cdot\frac{\sigma_k\cdot\overline{E}}{\overline{\sigma}}\right)^{-1}, \tag{5}$$

Where $\overline{E}=\left(\sum_{t=1}^{m-1}E_t\right)\times m^{-1}$ , $\overline{\sigma}=\left(\sum_{t=1}^{m-1}\sigma_t\right)\times m^{-1}$ ,and $\alpha+\beta=1$

$m$ is the number of chromosomes in the population. $\alpha$ and $\beta$ are weight coefficients. In our experiments, we obtain the best result videos in vision when $\alpha$ =0.7 and $\beta$ =0.3.

**Selection Operator.** Selection Operator gives preference to better individuals, allowing them to pass on their genes to the next generation. In our experiments, we use roulette wheel method for selection operation. Survivable probability $Ps$ of each individual $C$ is directly determined by its fitness.

$$Ps=\frac{Fit(C)}{\sum_{k=1}^{m}Fit(C_k)} \tag{6}$$

Here, $m$ is the amount of individuals in population. $C_k$ means the $k$-th individual in population.

**Crossover Operator.** This process recombines portions of good existing individuals to create even better individuals, it leads population to move towards the best solution. Single-point crossover is adopted and each time two genes from different individuals are exchanged. Crossover probability is set as a random float in [0.5,1].

**Mutation Operator.** The purpose of Mutation Operation is to introduce new gene and maintain diversity within the population and inhibit premature convergence. Selection of Mutation probability is much important for GA. If its value is too great, the population will not be converged, if its value is too small, the population may be led to premature convergence. In this paper, we set the value to 0.05.

During each iteration of GA, the individual having the largest fitness within all individuals in every generation is selected and checked whether its fitness is larger than a given threshold value of convergence $V_t$. If the fitness is larger or the number of generations is beyond a specified value $G_{max}$, the algorithm stops iterations and reports the individual having the largest fitness as the best solution.

## 3 Results and Analysis

### 3.1 Results

The current video synthesizing system is implemented in Visual C++6.0. Some video clips designed using the system are shown in the following figures. The system maintains reasonably smooth frames on a PC (Pentium4 2.0GHz). These video clips demonstrate several different video textures produced by the above-mentioned methods.

**Fig. 1.** a 63-frame source video of a man speaking, result video synthesized by the algorithm in this paper maintains continuity of his head and mouse moving.

**Fig. 2.** an 89-frame source video of flame, result video synthesized by the algorithm in this paper maintains continuity of flame jumping

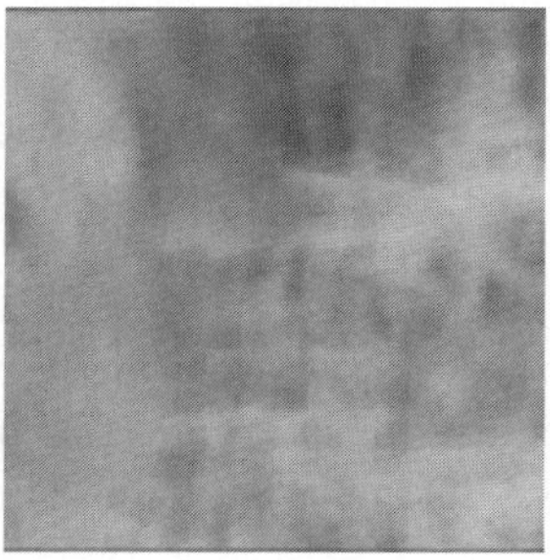

**Fig. 3.** a 61-frame source video of waterfall, result video synthesized by the algorithm in this paper maintains continuity of water flowing.

**Fig. 4.** a 63-frame cartoon source video, result video synthesized by the algori-thm in this paper maintains flying regularity of hair and scarf in the source video.

### 3.2 Analysis of Population Size

Population size is a main key to influence the efficiency and quality of our results. If its value is too small, the algorithm may fail to give us a satisfied solution of video texture synthesis; On the other hand, if its value is too large, the algorithm may waste much of calculating time. Therefore, it is more important to set a proper population size for the efficiency and quality of our algorithm.

In our experiments for each source clip, we set different population sizes to test. When processing the example in Fig.1, we set the number of chromosomes to 30, 50, 100, 150, 300, 500, 800, 1000 and 1500 respectively. Results of our experiments show that the algorithm quickly converges to an unsatisfied solution when the number of chromosomes is tiny; while the number of chromosomes is 500, the algorithm is able to do a fairly work; as the population size is larger than 500, fitness of the best solution has unobvious change. For the rest figures, satisfactory solutions can be obtained when the population sizes are 800, 500 and 500 respectively.

### 3.3 Analysis of Max Generation Number

Number of Max generation is another main key to influence the efficiency of our results. When processing the example in Fig.1, we specify that the population size is 500, and set the number of Max generations to 150, 300, 500, 800, 1000, 1500, 2000 and 3000 respectively. Results of our experiments show that when the number of Max generations is 1000, satisfied solutions can be obtained. For the rest Figures, we specify that the three population sizes are 800, 500 and 500 respectively, satisfied solutions can be obtained when the number of Max generations are 1500, 1000, and 1000.

### 3.4 Analysis of Computation Complexity

In Schödl's algorithm[8], the computation complexity of pre-processing is much more than that of creating video sequence. Without computing the part time of pre-processing, the total computation complexity of Video Textures Algorithm is $O(L^2N^2)$, where $L$ is the length of result video sequence and $N$ is the number of transitions. However the total computation complexity of our GA-based algorithm is $O(CG)$ ,where $C$ is the population size and $G$ is the number of Max generation. Only when the computation complexity of pre-processing is ignored, the two algorithms have comparable efficiency. Obviously, GA-based method is more efficient.

## 4 Conclusions

In this paper, evolution computation technique is introduced into the processes of synthesizing videos, and a new video texture synthesis method is suggested. This method uses an appropriate fitness function instead of much complex pre-processing of source video clip. Additionally, constraints of selection operator, crossover operator and mutation operator can ensure this algorithm to get high-quality video results quickly. Compared with other existing representative video texture synthesis algorithms, this algorithm has less computational complexity and improves the efficiency of synthesis.

Although we put more emphasis on the continuities and regularities of video sequences in our algorithm, we need to care for the structures of them. A more appropriate fitness function is anticipated to be found in future research work.

## References

1. Heeger D.J. and Bergen, J.R.: Pyramid-based texture analysis/synthesis. Computer Graphics, ACM SIGGRAPH. (1995), 229-238
2. Bonet J.S.D.: Multiresolution sampling procedure for analysis and synthesis of texture images. Computer Graphics, ACM SIGGRAPH. (1997) 361-368
3. Efros A. and Leung T.: Texture synthesis by non-parametric sampling. International Conference on Computer Vision, Vol 2. (1999) 1033-1038

4. Portilla J. and Simoncelli E.P.: A parametric texture model based on joint statistics of complex wavelet coefficients. International Journal of Computer Vision. vol.40, No.1. (2000) 49-70
5. Wei L.Y. and Levoy M.: Fast texture synthesis using tree-structured vector quantization. Computer Graphics, ACM SIGGRAPH. (2000) 479-488
6. Liang L., Liu C., Xu Y.Q., Guo B. and Shum H.Y.: Real-time texture synthesis by patch-based sampling. ACM Transactions on Graphics, Vol. 20, No. 3. (2001) 127-150
7. Efros A.A. and Freeman W.T.: Image quilting for texture synthesis and transfer. Proceedings of SIGGRAPH 2001. (2001) 341-346.
8. Schödl A., Szeliski R., Salesin D.H. and Essa I.A.: Video textures. Proceedings of SIGGRAPH 2000. (2000) 489-498.
9. Schödl A., ESSA I. A. Controlled animation of video sprites. Proceedings of ACM Symposium on Computer Animation 2002, (2002) 121~127
10. Wang Y., Zhu S. C. A generative model for textured motion: Analysis and synthesis. In: Proceedings of European Conf. on Computer Vision (ECCV)2002, Copenhagen, (2002) 583~598
11. Doretto G., Chiuso A., Soatto S., And Wu Y. Dynamic textures. International Journal of Computer Vision2003, (2003) 51(2): 91~109.
12. Kwatra V., Schödl A., Essa I. A., Turk G., and Bobick A. Graphcut Textures: Image and Video Synthesis Using Graph Cuts. In Proceedings of ACM SIGGRAPH 2003. (2003) 277~286
13. Bhat K.S., Seitz S.M., Hodgins J.K., and Khosala P.K., Flow-based video synthesis and editing. In Proc. of ACM SIGGRAPH 2004. (2004), 360~363
14. Holland, J.: Adaptation in Natural and Artificial Systems. University of Michigan Press. (1975)

# Deformation of Dynamic Surfaces

L.H. You and Jian J. Zhang

National Centre for Computer Animation
Bournemouth University
BH12 5BB
United Kingdom
{lyou, jzhang}@bournemouth.ac.uk

**Abstract.** Based on dynamic fourth order partial differential equations, we present an iterative finite difference algorithm. With C++ language and OpenGL graphics library, we implement the finite difference algorithm into a user interface and develop shape control parameters, density, damping coefficient, boundary tangents and external forces into user handles for dynamic manipulation of surface deformations. Using the developed user interface, we investigate how these user handles influence the deformation of dynamic surfaces.

**Keywords:** Surface deformation, dynamic fourth order partial differential equations, iterative finite difference algorithm, influences of user handles.

## 1 Introduction

Deformation of surfaces can be purely geometric or based on some physical laws. Purely geometric deformation tools change surface shapes mainly by tweaking control points of surfaces represented by Bézier, B-spline or NURBS [1].

In order to follow underlying laws of surface deformations, physics based surface deformation methods were introduced. Depending on whether dynamic effects and time variable are considered, these methods can be classified into static and dynamic.

Static deformation methods of surfaces only consider static material properties. Based on the concepts of elasticity, Kang and Kak presented a finite element analysis method at two resolutions: a coarse resolution for calculations of gross deformations, and a finer resolution for a better and smoother delineation of surface layers [2]. Léon and his co-workers presented an approach for parametric deformations of free-form surfaces which uses a bar network to represent the control polyhedron of a surface and changes the surface shape by deforming the bar network [3, 4].

Dynamic deformation approaches consider the effects of density and damping on deformations and motion and are time dependent. In general, these approaches are based on the equation of motion and solved by some numerical methods such as the finite element method and finite difference method. Terzopoulos and his co-workers proposed a deformable model derived from the theory of elasticity [5, 6] and further incorporated viscoelasticity, plasticity and fracture [7]. Later on, Metaxas and Terzopoulos gave a technique developed from parametrically defined solid primitives, global geometric deformations and local physics-based deformations and used it to build and animate constrained, nonrigid, unibody or multibody objects [8]. Celniker

V.N. Alexandrov et al. (Eds.): ICCS 2006, Part II, LNCS 3992, pp. 231 – 238, 2006.

and Gossard solved the deformable model based on the minimization of an energy functional subjected to user controlled geometric constraints and loads [9]. Güdükbay and Özgüç described a system for the animation of deformable models [10]. Deriving the control equations of motion from Lagrangian mechanics, Guan et al. introduced a deformable curve and surface modeling module and applied it in construction of N-sided patches, surface smooth joining, curve and surface fairing and so on [11]. Terzopoulos and Qin described an interactive modeling system called dynamic NURBS which incorporates mass distributions, internal deformation energies and other physical quantities into the popular NURBS geometric substrate and demonstrated the flexibility of their models in a variety of applications [12]. Qin and Terzopoulos also presented a dynamic NURBS swung surface model which can be formulated constructively from two NURBS profile curves or by applying a nonlinear constraint to a dynamic NURBS surface [13].

Partial differential equation (PDE) based modeling was firstly proposed by Bloor and Wilson [14] and applied in surface generation[15], surface blending[16], shape parameterization of a two-stroke engine [17] and fluid membranes and vesicles[18], and aerodynamic design of a flying wing [19] etc. In recent years, this modeling method has become more and more active. For example, Ugail and Wilson discussed shape parameterization for automatic design optimization based a PDE formulation [20]. Ugail also used the spine of a PDE surface to parameterize the shape of a complicated object [21]. You and Zhang proposed some new resolution methods and applied them in surface blending [22, 23]. Du and Qin combined PDE with the equation of motion and developed a novel modeling approach using popular numerical methods [24, 25].

In this paper, we will present an iterative finite difference algorithm to dynamic fourth order partial differential equations, implement it into a user interface, and investigate how different user handles affect dynamic deformations of surfaces.

## 2 Dynamic PDEs

Considering the influences of density and damping on deformations and motion, dynamic surfaces can be described with the solution to the following dynamic fourth order partial differential equations together with the corresponding boundary conditions

$$\left\{b_i \frac{\partial^4}{\partial u^4} + c_i \frac{\partial^4}{\partial u^2 \partial v^2} + d_i \frac{\partial^4}{\partial v^4} + \rho \frac{\partial^2}{\partial t^2} + \eta \frac{\partial}{\partial t}\right\} x_i(u,v,t) = f_i(u,v,t) \quad (i=1,2,3) \tag{1}$$

where $u$ and $v$ are parametric variables, $t$ is a time variable, $x_1$, $x_2$ and $x_3$ are position functions, $b_i$, $c_i$ and $d_i$ are shape control parameters, $\rho$ is the density, $\eta$ is the damping coefficient, $f_1(u,v,t)$, $f_2(u,v,t)$ and $f_3(u,v,t)$ are force components acting in the directions $x_1$, $x_2$ and $x_3$, respectively.

Surface patches can be defined by four boundary curves and the boundary tangents of the surfaces at these boundary curves. Therefore, the boundary conditions used for solving the above dynamic partial difference equations are

$$
\begin{array}{lll}
u=0 & x_i = b_{i0}(v,t) & \dfrac{\partial x_i}{\partial u} = b_{i1}(v,t) \\
u=1 & x_i = b_{i2}(v,t) & \dfrac{\partial x_i}{\partial u} = b_{i3}(v,t) \\
v=0 & x_i = b_{i4}(u,t) & \dfrac{\partial x_i}{\partial v} = b_{i5}(u,t) \\
v=1 & x_i = b_{i6}(u,t) & \dfrac{\partial x_i}{\partial v} = b_{i7}(u,t) \\
(i=1,2,3) & &
\end{array}
\tag{2}
$$

where $b_{i0}(v,t)$, $b_{i2}(v,t)$, $b_{i4}(u,t)$ and $b_{i6}(u,t)$ are boundary curves, and $b_{i1}(v,t)$, $b_{i3}(v,t)$, $b_{i5}(u,t)$ and $b_{i7}(u,t)$ are boundary tangents.

## 3 Iterative Finite Difference Algorithm

Partial differential equations (1) subjected to boundary conditions (2) are difficult to solve. For most cases, closed form analytical solutions do not exist. In order to address this problem, in this section, we introduce an iterative finite difference algorithm. Using the mesh indicated in Fig. 1 and the central difference approximation, the difference equations of different orders of partial derivatives of the position functions $x_i$ with respect to the parametric variables $u$ and $v$ in Eqs. (1) and (2) can be written as

$$
\begin{aligned}
&\left(\frac{\partial x_i}{\partial u}\right)_0^{t+\Delta t} = \frac{x_{i1}^{t+\Delta t} - x_{i3}^{t+\Delta t}}{2h} \\
&\left(\frac{\partial x_i}{\partial v}\right)_0^{t+\Delta t} = \frac{x_{i2}^{t+\Delta t} - x_{i4}^{t+\Delta t}}{2h} \\
&\left(\frac{\partial^4 x_i}{\partial u^4}\right)_0^{t+\Delta t} = \frac{1}{h^4}\left[6x_{i0}^{t+\Delta t} - 4(x_{i1}^{t+\Delta t} + x_{i3}^{t+\Delta t}) + x_{i9}^{t+\Delta t} + x_{i11}^{t+\Delta t}\right] \\
&\left(\frac{\partial^4 x_i}{\partial u^2 \partial v^2}\right)_0^{t+\Delta t} = \frac{1}{h^4}\left[4x_{i0}^{t+\Delta t} - 2(x_{i1}^{t+\Delta t} + x_{i2}^{t+\Delta t} + x_{i3}^{t+\Delta t} + x_{i4}^{t+\Delta t})\right. \\
&\qquad\qquad \left. + x_{i5}^{t+\Delta t} + x_{i6}^{t+\Delta t} + x_{i7}^{t+\Delta t} + x_{i8}^{t+\Delta t}\right] \\
&\left(\frac{\partial^4 x_i}{\partial v^4}\right)_0^{t+\Delta t} = \frac{1}{h^4}\left[6x_{i0}^{t+\Delta t} - 4(x_{i2}^{t+\Delta t} + x_{i4}^{t+\Delta t}) + x_{i10}^{t+\Delta t} + x_{i12}^{t+\Delta t}\right] \\
&(i=1,2,3)
\end{aligned}
\tag{3}
$$

where $\Delta t$ is a time increment, $h$ is the interval between two adjacent nodes, the subscript 0 on the left-hand side of the equations represents the typical node 0, the first subscript $i$ on the right-hand side of the equations indicates the ith component of the position functions $x_i$, and the second subscript stands for the index of nodes.

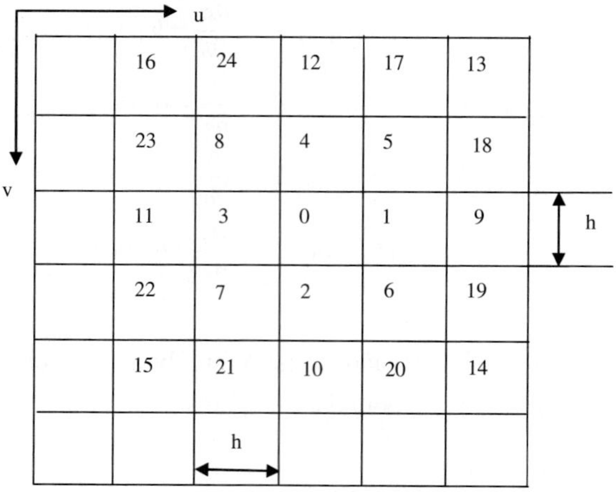


**Fig. 1.** Typical node of finite difference approximation

Similarly, the first and second partial derivatives of the position functions $x_i$ with respect to the time variable t are transformed into

$$\left(\frac{\partial x_i}{\partial t}\right)_0^t = \frac{x_{i0}^{t+\Delta t} - x_{i0}^{t-\Delta t}}{2\Delta t}$$

$$\left(\frac{\partial^2 x_i}{\partial t^2}\right)_0^t = \frac{x_{i0}^{t+\Delta t} - 2x_{i0}^t + x_{i0}^{t-\Delta t}}{\Delta t^2} \tag{4}$$

$$(i = 1, 2, 3)$$

Substituting Eqs. (3) and (4) into (1), formulating the finite difference equations for all nodes excluding those on the boundaries, and removing $x_{i0}^{t-\Delta t}$ by means of the relation between the velocity and the displacement

$$V_{i0}^t = \frac{x_{i0}^t - x_{i0}^{t-\Delta t}}{\Delta t} \qquad (i = 1, 2, 3) \tag{5}$$

we obtain the following linear algebra equations

$$\begin{aligned} &\left[4(3b_i + 2c_i + 3d_i)\Delta t^2 + (2\rho + \eta\Delta t)h^4\right] x_{i0}^{t+\Delta t} - 4(2b_i + c_i)\Delta t^2 (x_{i1}^{t+\Delta t} + x_{i3}^{t+\Delta t}) \\ &- 4(c_i + 2d_i)\Delta t^2 (x_{i2}^{t+\Delta t} + x_{i4}^{t+\Delta t}) + 2c_i\Delta t^2 (x_{i5}^{t+\Delta t} + x_{i6}^{t+\Delta t} + x_{i7}^{t+\Delta t} + x_{i8}^{t+\Delta t}) + 2b_i\Delta t^2 \\ &(x_{i9}^{t+\Delta t} + x_{i11}^{t+\Delta t}) + 2d_i\Delta t^2 (x_{i10}^{t+\Delta t} + x_{i12}^{t+\Delta t}) = 2\Delta t^2 h^4 f_{i0}^t + (2\rho + \eta\Delta t)h^4 x_{i0}^t + (2\rho \\ &- \eta\Delta t)\Delta t h^4 V_{i0}^t \end{aligned} \tag{6}$$

$$(i = 1, 2, 3)$$

Substituting Eq. (3) into (2), the boundary conditions are changed into the finite difference equations below

$$
\begin{array}{lll}
u=0 & x_{i0}^{t+\Delta t}=b_{i0}(v_0,t+\Delta t) & \dfrac{x_{i1}^{t+\Delta t}-x_{i3}^{t+\Delta t}}{2h}=b_{i1}(v_0,t+\Delta t) \\
u=1 & x_{i0}^{t+\Delta t}=b_{i2}(v_0,t+\Delta t) & \dfrac{x_{i1}^{t+\Delta t}-x_{i3}^{t+\Delta t}}{2h}=b_{i3}(v_0,t+\Delta t) \\
v=0 & x_{i0}^{t+\Delta t}=b_{i4}(u_0,t+\Delta t) & \dfrac{x_{i2}^{t+\Delta t}-x_{i4}^{t+\Delta t}}{2h}=b_{i5}(u_0,t+\Delta t) \\
v=1 & x_{i0}^{t+\Delta t}=b_{i6}(u_0,t+\Delta t) & \dfrac{x_{i2}^{t+\Delta t}-x_{i4}^{t+\Delta t}}{2h}=b_{i7}(u_0,t+\Delta t) \\
(i=1,2,3) & &
\end{array}
\tag{7}
$$

where the subscript 0 stands for the typical node on the boundaries.

Solving Eqs. (6) and (7) for each time instant, the values of the position functions $x_i$ at all nodes are determined and the deformed surface is defined by the new positions of these nodes. This process is repeated until the whole time period is completed.

## 4 Influences of User Handles

The above finite difference algorithm has been implemented into a user interface, and shape control parameters, density, damping coefficient, force functions and boundary

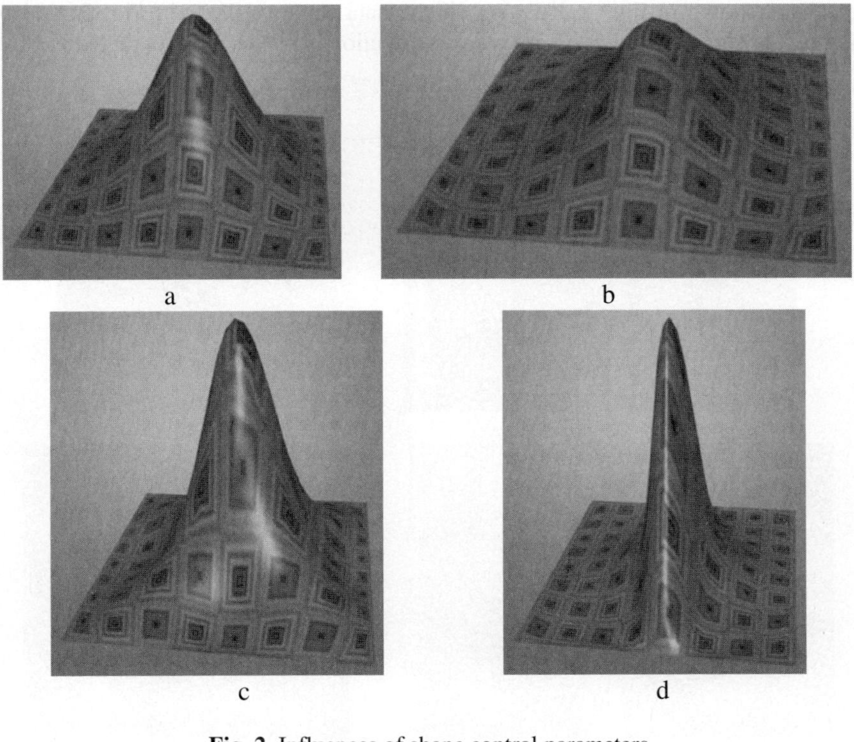

**Fig. 2.** Influences of shape control parameters

tangents are developed into user handles. In the following, we investigate how these user handles influence shapes of dynamic surfaces.

The basic data of user handles are taken to be: shape control parameters $b_i = d_i = 1$ and $c_i = 2$ $(i = 1, 2, 3)$, density $\rho = 0.1$, damping coefficient $\eta = 1$, force functions $f_1 = f_2 = 0$ and $f_3 = 800$, and all the boundary tangents are set to zero. The obtained surface was given in Fig. 2a.

Firstly, we use different shape control parameters and see how the surface shape varies. Setting $b_i = 10$ only leads to the shape in Fig. 2b, changing $c_i$ to -1.5 causes that in Fig. 2c, and taking $d_i = -0.001$ generates the surface in Fig. 2d. It can be concluded that a big value of shape control parameters $b_i$ decreases the deformation, minus values of $c_i$ and $d_i$ raise the size of the deformation, but reduce the deformation region. Between them, the impact of $d_i$ is more obvious.

Next, the effects of boundary tangents on deformations are discussed. Taking different values of boundary tangents, the images in Fig. 3 were generated where Fig. 3a is from $b_{i1}(v,t) = 3$, Fig. 3b is from $b_{i3}(v,t) = 10$, Fig. 3c is from $b_{i1}(v,t) = b_{i3}(v,t) = b_{i5}(u,t) = 3$, and Fig. 3d is from $b_{i1}(v,t) = b_{i5}(u,t) = -3$ and $b_{i3}(v,t) = b_{i7}(u,t) = 3$. Clearly, these different combinations of boundary tangents create quite different appearances of the surface.

Finally, how force functions, density and damping coefficient affect the surface shape were shown in Fig. 4. Keeping all the data same as those in Fig. 2a and only changing the sign of the force, a reverse deformation is produced as depicted in Fig. 4i.

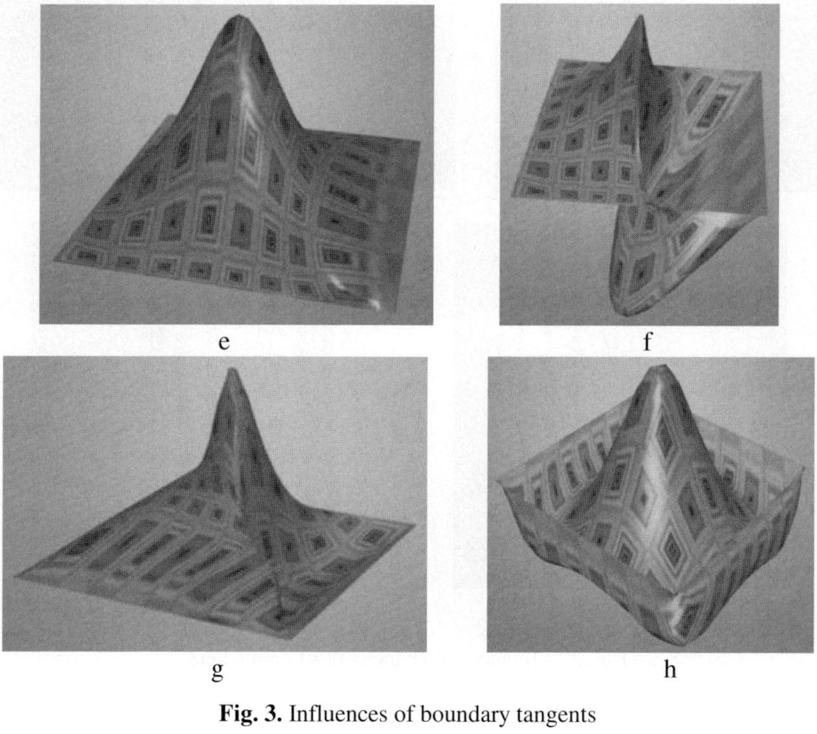


**Fig. 3.** Influences of boundary tangents

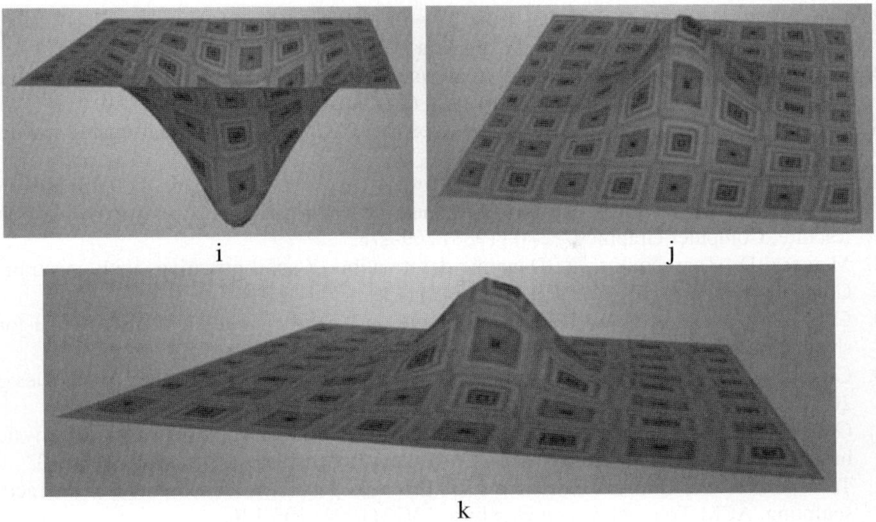


**Fig. 4.** Influences of force functions, density and damping

Increasing the damping coefficient from 1 to 20, the deformation of the surface becomes more difficult resulting in a much smaller deformation in Fig. 4j. Raising the density of the surface can also decrease the deformation of the surface. Compared the image in Fig. 2a where the density is 0.1, the deformation in Fig. 4k is noticeably reduced due to a rising of the density from 0.1 to 1.

## 5 Conclusions

In this paper, we have presented an iterative finite difference approximation for dynamic deformations of surfaces which is based on dynamic fourth order partial differential equations. We also implemented the finite difference algorithm into a user interface using C++ and OpenGL graphics library.

With the developed user interface, we examined the influences of user handles on surface deformations. These user handles include shape control parameters, density, damping coefficient, force functions and boundary tangents. They are found to be very effective in shape manipulation of dynamic surface deformations.

## Acknowledgements

The authors are very grateful for the donation of the Maya licenses from Alias.

## References

1. Farin, G.: Curves and Surfaces for CAGD: A Practical Guide, 5th Edition, Morgan-Kaufmann (2001).
2. Kang, H., Kak, A.: Deforming virtual objects interactively in accordance with an elastic model, Computer-Aided Design, 28(4) (1996) 251-262.

3. Léon, J.C., Veron, P.: Semiglobal deformation and correction of free-form surfaces using a mechanical alternative, The Visual Computer, 13 (1997) 109-126.
4. Guillet, S., Léon, J.C.: Parametrically deformed free-form surfaces as part of a variational model, Computer-Aided Design, 30(8) 1998 621-630.
5. Terzopoulos, D., Platt, J., Barr, A., Fleischer, K.: Elastically deformable models, Computer Graphics, 21(4) (1987) 205-214.
6. Terzopoulos,D.,Fleischer,K.:Deformable models,The Visual Computer, 4 (1988) 306-331.
7. Terzopoulos, D., Fleischer, K.: Modeling inelastic deformation: viscoelasticity, plasticity, fracture, Computer Graphics, 22(4) (1988) 269-278.
8. Metaxas, D., Terzopoulos, D.: Dynamic deformation of solid primitives with constraints, Computer Graphics (SIGGRAPH '92), 26(2) (1992) 309-312.
9. Celniker, G., Gossard, D.: Deformable curve and surface finite-elements for free-form shape design, Computer Graphics (SIGGRAPH '91), 25(4) (1991) 257-266.
10. Güdükbay, U., Özgüç, B.: Animation of deformable models, Computer-Aided Design, 26(12) (1994) 868-875.
11. Guan, Z.D., Ling, J., Tao, N., Ping, X., Tang, R.X.: Study and application of physics-based deformable curves and surfaces, Computers & Graphics, 21(3) (1997) 305-313.
12. Terzopoulos, D., Qin, H.: Dynamic NURBS with geometric constraints for interactive sculpting, ACM Transactions on Graphics, 13(2) (1994) 103-136.
13. Qin, H., Terzopoulos, D.: Dynamic NURBS swung surfaces for physics-based shape design, Computer-Aided Design, 27(2) (1995) 111-127.
14. Bloor, M.I.G., Wilson, M.J.: Generating blend surfaces using partial differential equations, Computer-Aided Design, 21(3) (1989) 165-171.
15. Bloor, M.I.G., Wilson, M.J.: Complex PDE surface generation for analysis and manufacture, Computing Suppl., (10) (1995) 61-77.
16. Bloor, M.I.G., Wilson, M.J., Mulligan, S.J.: Generating blend surfaces using a perturbation method, Mathematical and Computer Modelling, 31(1) (2000) 1-13.
17. Mimis, A.P., Bloor, M.I.G., Wilson, M.J.: shape parameterization and optimization of a two-stroke engine, Journal of Propulsion and Power, 17(3) (2001) 492-498.
18. Bloor, M.I.G., Wilson, M.J.: Method for efficient shape parametrisation of fluid membranes and vesicles, Physical Review E (Statistical Physics, Plasmas, Fluids, and Related Interdisciplinary Topics), 61(4) (2000) 4218-429.
19. Sevant, N.E., Bloor, M.I.G., Wilson, M.J.: Aerodynamic design of a flying wing using response surface methodology, Journal of Aircraft, 37(4) (2000) 562-569.
20. Ugail, H., Wilson, M. J.: Efficient shape parametrisation for automatic design optimisation using a partial differential equation formulation, Computers and Structures, 81 (2003) 2601-2609.
21. Ugail, H.: Spine based shape parameterisation for PDE surfaces, Computing, 72 (2004) 195-206.
22. You, L.H., Zhang, J.J., Comninos, P.: Blending surface generation using a fast and accurate analytical solution of a fourth order PDE with three shape control parameters, The Visual Computer, 20 (2004) 199-214.
23. You, L.H., Comninos, P., Zhang, J.J.: PDE blending surfaces with $C^2$ continuity, Computers & Graphics, 28 (2004) 895-906.
24. Du, H., Qin, H.: A shape design system using volumetric implicit PDEs, Computer-Aided Design, 36 (2004) 1101-1116.
25. Du, H., Qin, H., Dynamic PDE-based surface design using geometric and physical constraints, Graphical Models, 67 (2005) 43-71.

# A New Smoothing Algorithm for Quadrilateral and Hexahedral Meshes

Sanjay Kumar Khattri

Department of Mathematics, University of Bergen, Norway
sanjay@mi.uib.no
http://www.mi.uib.no/~sanjay

**Abstract.** Mesh smoothing (or r-refinement) are used in computer aided design, interpolation, numerical solution of partial differential equations, etc. We derive a new smoothing called parallelogram smoothing. The new smoothing tries to fit a given domain by the parallelograms. We present several numerical examples and compare our results against the traditional Laplacian smoothing. Presented numerical work shows that the new approach is superior to the Laplacian smoothing.

## 1 Introduction

Mesh smoothing such as the Laplacian smoothing [13] and the Winslow smoothing [14] has many applications. For example, computer graphics [7, 8, 9, and references therein], interpolation [10, and references therein], numerical solution of partial differential equations [11, 12, 13, 14, 17, and references therein], Arbitrary Lagrangian Eulerian Methods [1] and mesh generation [3, 4, 5]. It is known (see [2, 13, 17]) that mesh smoothing improves accuracy of the solution and also reduces the overall computational effort. Smoothing is effective in improving distorted and inverted elements in automatically generated meshes [13]. Mesh smoothing adjusts the positions of the interior vertices in the mesh while preserving its topology [13, 17]. Due to simplicity and computationally efficiency, the Laplace smoothing is the most popular method for improving mesh quality. It repositions the vertices of the mesh by moving each interior node to the geometric center of its neighbours [13]. However, the Laplacian smoothing can produce inverted elements (See the Section 2) and such meshes cannot be used in applications. We present a new mesh smoothing algorithm called the parallelogram smoothing. Numerical work shows that the new smoothing is performing better than the Laplacian smoothing. Parallelogram smoothing tries to fit a given domain with parallelograms in 2D and parallelopipeds in 3D.

Let us mention advantages of the parallelogram smoothing. The error of the Streamline Methods on parallelogram and parallelopiped mesh is minimum (see the Figures 7, 11 and 15 in [15]). Many discretization methods such as the Multi Point Flux Approximation (MPFA) results in a symmetric discrete system on a parallelogram or parallelopiped mesh [16]. So, efficient solvers such as the Conjugate Gradient can be used. Quadrilateral and hexahedral elements are expressed

V.N. Alexandrov et al. (Eds.): ICCS 2006, Part II, LNCS 3992, pp. 239–246, 2006.

by bi-linear and tri-linear mappings from a reference square and cube respectively. These mappings are widely used with discretization techniques such as Finite Element Methods, Finite Volume Methods [16] and Streamline Methods [15]. One of the major problem in numerical computations is ensuring the invertibility of the bi-linear and tri-linear mappings. Since these mappings are not linear, the positivity of the Jacobian at the vertices does not ensure the global invertibility of the mappings (or ensuring the invertibility of the mapping is computationally intensive). One encouraging fact is that for parallelograms and parallelepipeds elements these mappings become linear.

An outline of the article is as follows. In the Section 2, a measure of parallelogram property of quadrilateral and hexahedral elements, and a quadratic parallelogram functional for quadrilateral and hexahedral meshes are presented. Optimization of the functional leads to a new smoothing algorithm. Section 2 presents numerical work and comparison with the Laplacian smoothing. Finally Section 2 concludes the article.

## 2 Measure of Parallelogramness and Parallelogram Functional

Figure 1(a) shows a quadrilateral element with vertices $\mathbf{v}_1$, $\mathbf{v}_2$, $\mathbf{v}_3$ and $\mathbf{v}_4$. The midpoints of the diagonals $\mathbf{v}_1\mathbf{v}_3$ and $\mathbf{v}_2\mathbf{v}_4$ are $\mathbf{d}_1$ and $\mathbf{d}_2$ respectively. Lets call this quadrilateral element k. Euclidean distance between the midpoints $\mathbf{d}_1$ and $\mathbf{d}_2$ is

$$Q(k) = \left\| \frac{\mathbf{v}_1 + \mathbf{v}_3}{2} - \frac{\mathbf{v}_2 + \mathbf{v}_4}{2} \right\|_{L_2} . \qquad (1)$$

Here k means that we are referring to the quadrilateral element named k. Since the mid points of two diagonals of a parallelogram co-incide; i.e., $\mathbf{v}_1 + \mathbf{v}_3 = \mathbf{v}_2 + \mathbf{v}_4$. If Q(k) approaches zero than the element k approaches a parallelogram

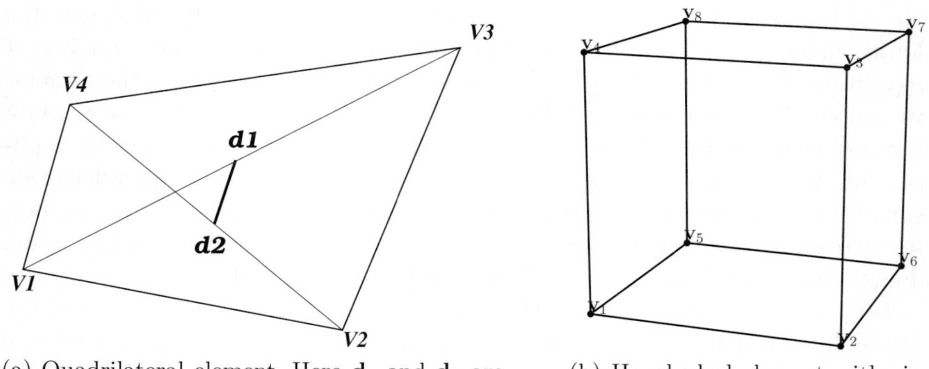


(a) Quadrilateral element. Here $\mathbf{d}_1$ and $\mathbf{d}_2$ are the midpoints of the diagonals.

(b) Hexahedral element with six quadrilateral faces

**Fig. 1.** Quadrilateral and Hexahedral Elements

and for Q(k) equal to zero the element k will be a parallelogram. Q(k) will be called a measure of the parallelogramness of the quadrilateral element k.

Figure 1(b) shows a planar hexahedral element (a hexahedra with planar surfaces) with vertices $\mathbf{v}_1$, $\mathbf{v}_2$, $\mathbf{v}_3$, $\mathbf{v}_4$, $\mathbf{v}_5$, $\mathbf{v}_6$, $\mathbf{v}_7$ and $\mathbf{v}_8$. Lets call this hexahedra k. Let us further denote quadrilateral surfaces of the hexahedra k by $s_i$ where $i = 1 \ldots 6$. The surface $s_i$ will be a parallelogram if $Q(s_i)$ is equal to zero. Parallelogramness of the hexahedral element k can be expressed as a sum of parallelogramness of the six quadrilateral faces

$$Q(k) = \sum_{i=1}^{6} \|Q(s_i)\|_{L_2} \ . \qquad (2)$$

A hexahedral element is a parallelogram if the six surfaces are parallelograms. Thus the hexahedral element k will be a parallelogram if Q(k) is equal to zero. Figure 2 shows a structured 2 × 2 quadrilateral mesh. The mesh shown in the Figure 2 consists of four quadrilateral elements lets call them $k_i$, $i = 1 \ldots 4$. For the 2 × 2 mesh shown in the Figure, 2 $\mathbf{r}_{14}$, $\mathbf{r}_1$, $\mathbf{r}_{12}$, $\mathbf{r}_2$, $\mathbf{r}_{23}$, $\mathbf{r}_3$, $\mathbf{r}_{34}$, $\mathbf{r}_4$ are the boundary nodes and these nodes are fixed during smoothing process. Let the coordinates of the node $\mathbf{r}_i$ be $(x_i, y_i)$. The Laplacian smoothing moves grid points to the geometric center of incident vertices. The Laplacian smoothed position of the node $\mathbf{r}_0 = (x_0, y_0)$ is

$$\mathbf{r}_0 = \frac{\mathbf{r}_1 + \mathbf{r}_2 + \mathbf{r}_3 + \mathbf{r}_4}{4.0} \ . \qquad (3)$$

For deriving an alternative smoothing, let us first define a functional $f$ for a mesh shown in the Figure 2 as

$$f(x_0, y_0) = \sum_{i=1}^{4} Q(k_i)^2 \ . \qquad (4)$$

Here, index i is for the four surrounding quadrilaterals around the node 0 and $Q(k_i)$ is the measurement of the parallelogramness of the quadrilateral element named $k_i$ in the mesh. Similar functional can also be defined for unstructured quadrilateral meshes. It should be noted that $f(x_0, y_0)$ will be zero if $(x_0, y_0)$

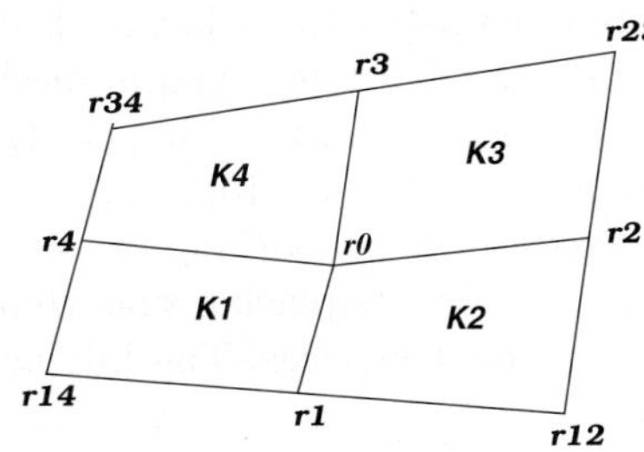


**Fig. 2.** A 2 × 2 structured quadrilateral mesh

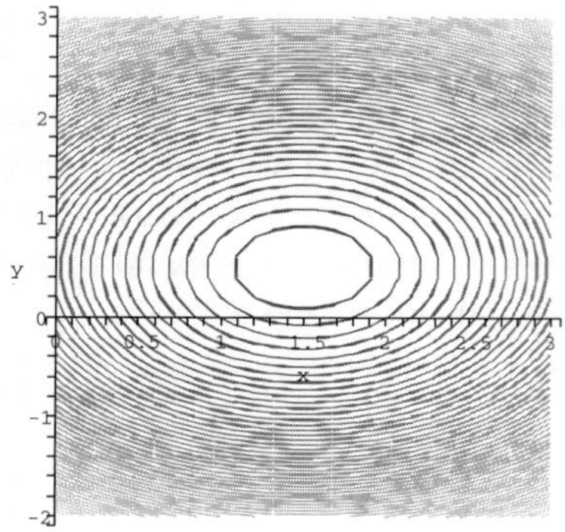


(a) Contours of the functional f

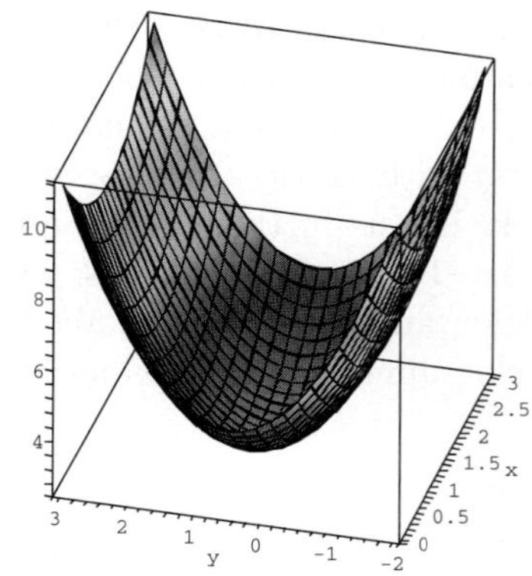


(b) Surface plot the functional f

**Fig. 3.** Convex nature of the functional f

is situated such that the four surrounding quadrilaterals are parallelograms. Functional f is quadratic in nature and it can be easily shown that it is convex; i.e., the Hessian of the functional f is positive definite. Figure 3(a) shows the elliptic contours of this functional for a quadrilateral mesh and Figure 3(b) shows the surface plot of the functional f. Since functional f is convex so it has a unique minimum. The minimization of this functional f will provide the proper nodal position for the node $\mathbf{r}_0$ and the newly formed surrounding cells around this node will be the best parallelogram fit (if squares or rectangles can be fit, the minimization will provide the nodal position for squares or rectangles fit since square and rectangle are a special kind of parallelograms). It can be shown that the independent variables $x_0$ and $y_0$ ($\mathbf{r}_0$) that minimizes the functional f given by the equation (4) are

$$\mathbf{r}_0 = \frac{\mathbf{r}_{14} + \mathbf{r}_{12} + \mathbf{r}_{23} + \mathbf{r}_{34}}{4.0} - \frac{\mathbf{r}_1 + \mathbf{r}_2 + \mathbf{r}_3 + \mathbf{r}_4}{2.0} \ . \tag{5}$$

Our new smoothing approach for a structured quadrilateral mesh is given by the equation (5). Thus, our new approach will move internal nodes according to the equation (5). In the next Section several interesting numerical examples are explored for verifying the robustness of the new smoothing given by equation (5) compared to the Laplacian smoothing given by equations (3). Similar smoothing can be derived for structured or unstructured hexahedral meshes.

Figure 4 shows a structured $2 \times 2 \times 2$ hexahedral mesh. The mesh 4 consists of eight hexahedral elements lets call them $k_i$ with $i = 1, \ldots, 8$. For the mesh 4, 14 is the internal node while $1, \ldots, 13$ and $15, \ldots, 27$ are the boundary nodes. Boundary nodes are kept fixed during smoothing process. Let the coordinate of a node i is given as $\mathbf{r}_i = (x_i \ , y_i)$. The Laplacian smoothing moves grid points to the geometric center of the incident vertices. The Laplacian smoothed position of the node 14 is given as

$$\mathbf{r}_{14} = \frac{\mathbf{r}_{11} + \mathbf{r}_{15} + \mathbf{r}_{17} + \mathbf{r}_{13} + \mathbf{r}_5 + \mathbf{r}_{23}}{6.0} \ , \tag{6}$$

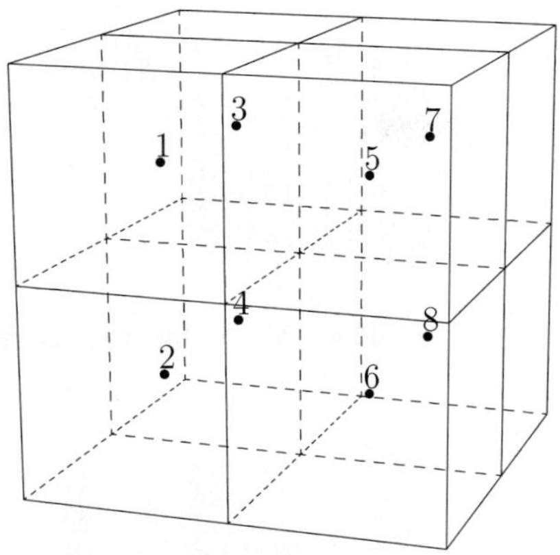


(a) Central node is surrounded by 8 hexahedrals

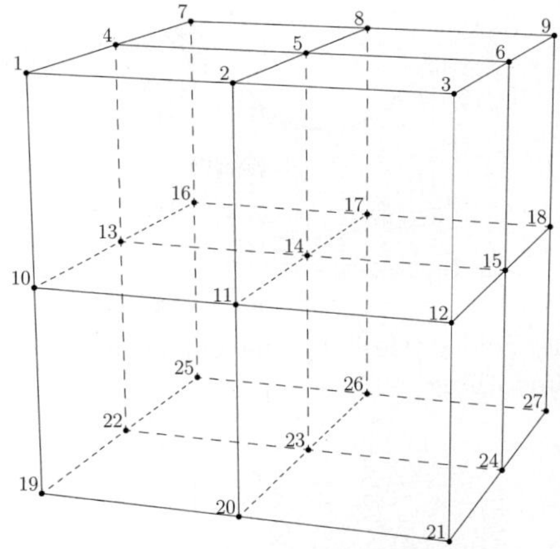


(b) Mesh consists of eight hexahedrals

**Fig. 4.** A $2 \times 2 \times 2$ structured hexahedral mesh

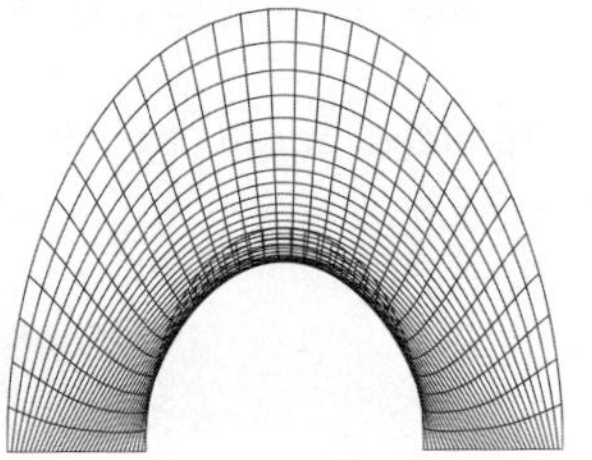

(a) Folded Mesh by the Laplacian Smoothing

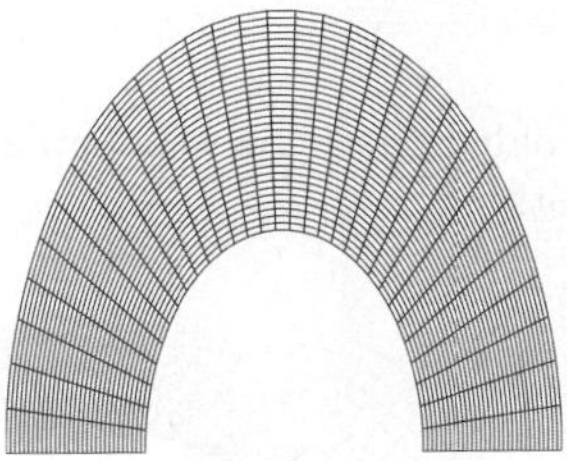

(b) Unfolded Mesh by the New Method

**Fig. 5.** Comparison of Laplacian and New Method

For deriving an alternative smoothed position for the node 14 shown in Figure 4, we define a functional f as follows

$$f(x_{14}, y_{14}, z_{14}) = \sum_{i=1}^{8} Q(k_i) \ . \qquad (7)$$

Here index i runs over eight surrounding hexahedral elements around the node 14 and $Q(k_i)$ is measurement of the parallelogramness of the $i^{th}$ hexahedra defined by equation (2). Similar functional can also be defined for unstructured hexahedral meshes. It can be seen that $f(x_{14},y_{14},z_{14})$ will be zero if the coordinates $(x_{14},y_{14},z_{14})$ of the node 14 are situated such that the surrounding hexahedrals are parallelepiped. Again the functional f is quadratic and convex with a positive definite Hessian. It can be shown that independent variables $(x_{14},y_{14},z_{14})$ that minimizes (7) are given as (solution of the equations $\partial f/\partial x_{14} = 0$, $\partial f/\partial y_{14} = 0$ and $\partial f/\partial z_{14} = 0$).

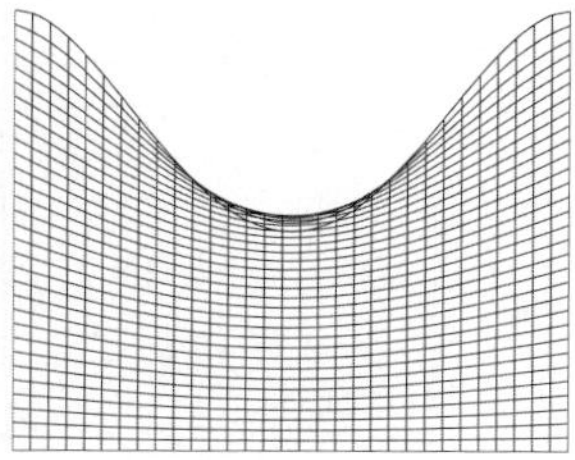

(a) Folded Mesh by the Laplacian Smoothing

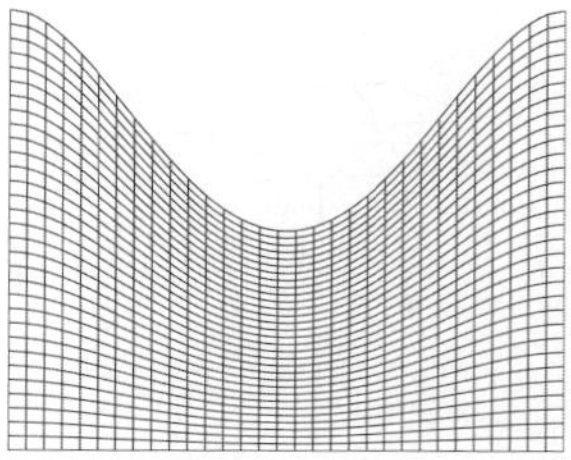

(b) Unfolded Mesh by the New Method

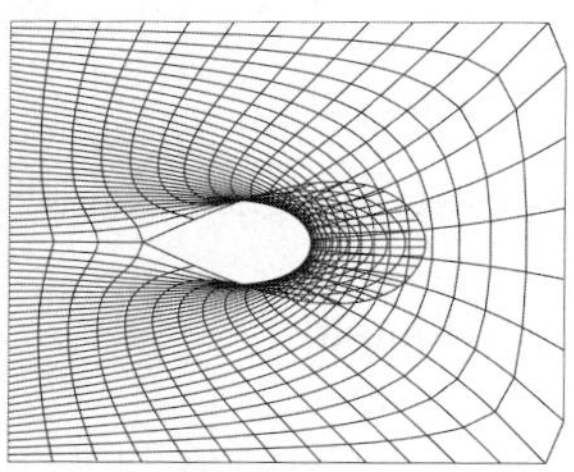

(c) Folded Mesh by the Laplacian Smoothing

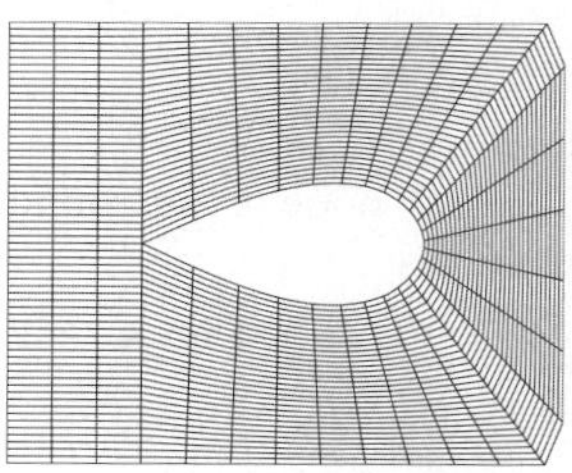

(d) Unfolded Mesh by the New Method

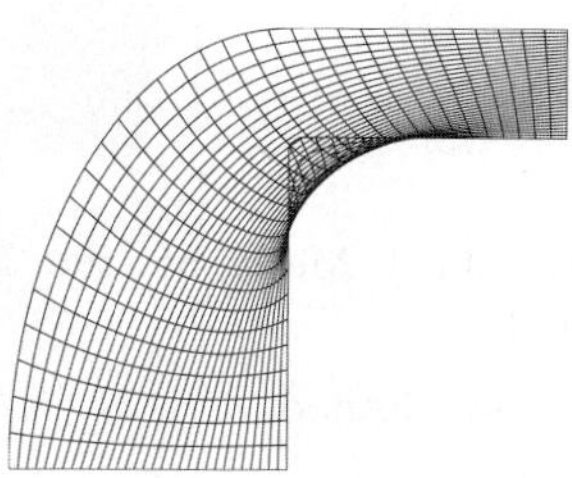

(e) Folded Mesh by the Laplacian Smoothing

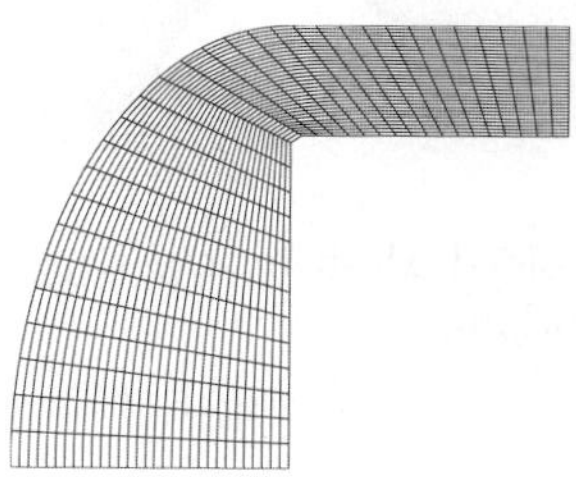

(f) Unfolded Mesh by the New Method

**Fig. 6.** Comparison of Laplacian and New Method

$$\mathbf{r}_{14} = \frac{\mathbf{r}_{11} + \mathbf{r}_{15} + \mathbf{r}_{17} + \mathbf{r}_{13} + \mathbf{r}_{5} + \mathbf{r}_{23}}{3.0} - \frac{\mathbf{r}_{10} + \mathbf{r}_{12} + \mathbf{r}_{18} + \mathbf{r}_{16}}{24.0} - \frac{\mathbf{r}_{24} + \mathbf{r}_{26} + \mathbf{r}_{2} + \mathbf{r}_{6} + \mathbf{r}_{8} + \mathbf{r}_{4}}{48.0} \,. \quad (8)$$

Equation (8) forms a new smoothing algorithm for structured hexahedral mesh. Similar formulas can be derived for unstructured hexahedral meshes.

## 3 Numerical Experiments

We are using both the new approach and the Laplacian method for quadrilateral mesh smoothing. Figures 6 and 5 report outcome of our numerical experiments. Initial grids are generated by Transfinite Interpolation. It is clear from the Figures that for 2-D geometries the Laplacian smoothing can generate folded grids and the resulting mesh can not be used in numerical simulations. It can be seen in the Figures 6 and 5 that the Laplacian smoothing move nodes outside the boundary and it results in inverted elements. Numerical experiments demonstrate robust nature of the new approach.

## 4 Conclusions

A simple alternative to the traditional Laplacian smoothing is presented. Reported numerical work shows that the new approach is superior to the Laplacian smoothing. This method is simple to understand and can be implemented in an existing mesh generation package. Smoothing is derived for the structured quadrilaterals and hexahedral meshes. The real use of this approach can be in the area of hexahedral mesh generation where many meshing algorithms are capable of generating inverted elements.

## References


1. Hermansson, J. and Hansbo, P.: A variable diffusion method for mesh smoothing, Comm. Numer. Methods Engrg., **19**, (2003), 897-908.
2. Daniel, W. and Garry, R.: Improved vector FEM solutions of Maxwell's equations using grid pre-conditioning, Internat. J. Numer. Methods Engrg., **40**, (1997), 3815–3837.
3. Ding, Y., Xia, J., Wang, Y. and Xiao, J.: Fully automatic generation of finite element mesh, Progr. Natur. Sci. (English Ed.), **5**, (1995), 93–98.
4. O'Sullivan, F.: Discretized Laplacian smoothing by Fourier methods, J. Amer. Statist. Assoc., **86**, (1991), 634–642.
5. Hansbo, P.: Generalized Laplacian smoothing of unstructured grids. Communications in Numerical Methods in Engineering, **11**, (1995), 455-464.
6. Field, D.A.: Laplacian smoothing and Delaunay triangulations. Communications in Applied Numerical Methods. **4**, (1998), 709–712.
7. Floater, M.S.: Mean value coordinates. Computer Aided Geometric Design. **20**, (2003), 19–27.
8. Khodakovsky, A., Litke, N., and Schröder, P.: Globally smooth parameterizations with low distortion. ACM Transactions on Graphics (TOG). **22**, (2003).
9. Sheffer, A. and Sturler, E.D.: Smoothing an overlay grid to minimize linear distortion in texture mapping. ACM Transactions on Graphics (TOG). **21**, (2002), 874–890.
10. Mallet, J.L.: Discrete smooth interpolation. ACM Transactions on Graphics (TOG). **8**, (1998), 121–144.
11. Bank, R.E. and Smith, R.K.: Mesh smoothing using a posteriori error estimates. SIAM J. Numer. Anal. **34**, (1997), 979–997.

12. Frey, W.H. and Field, D.A.: Mesh relaxation: a new technique for improving triangulations. Internat. J. Numer. Methods Engrg. **31**, (1991), 1121–1133.
13. Shontz, S.M. and Vavasis, S.A.: A linear weighted laplacian smoothing framework for warping tetrahedral meshes. Submitted to SIAM Journal on Scientific Computing on July 17, 2004 for publication. Available on line at http://arxiv.org/abs/cs.NA/0410045.
14. Knupp, P.M.: Winslow smoothing on two-dimensional unstructured meshes. Proceedings, 7th International Meshing Roundtable, Sandia National Lab. (1998), 449-457.
15. Hægland, H., Dahle, H.K., Eigestad, G.T., Lie, K.A. and Aavatsmark, I.: Improved Streamlines and Time of Flight for Streamline Simulation on Irregular Grids. Submitted in Journal. Available on line at http://heim.ifi.uio.no/~kalie/papers/cvi-sl.pdf. October (2005).
16. Khattri, S.K.: Analyzing Finite Volume for Single Phase Flow in Porous Media. Journal of Porous Media. Accepted for Publication, (2006).
17. Persson, P.-O. and Strang, G.: A Simple Mesh Generator in MATLAB. SIAM Review, 46 (2), (2004), 329-345.

# The Calculation of Parametric NURBS Surface Interval Values Using Neural Networks

Erkan Ülker and Ahmet Arslan

Selçuk University, Computer Engineering Department,
42075 Konya, Turkey
{Eulker, Ahmetarslan}@selcuk.edu.tr
http://www.mmf.selcuk.edu.tr/bilgisayar/akademik/akademik1.html

**Abstract.** Three dimensional coordinate values of parametric NURBS (Non-Uniform Rational B-Splines) surfaces are obtained from two dimensional parameters *u* and *v*. An approach for generating surfaces produces a model by giving a fixed increase to *u* and *v* values. However, the ratio of three dimensional parameters increases and fixed increase of *u* and *v* values is not always the same. This difference of ratio costs unrequired sized breaks. In this study an artificial neural network method for simulation of a NURBS surface is proposed. Free shaped NURBS surfaces and various three dimensional object simulations with different patches can be produced using a method projected as network training with respect to coordinates which are found from interval scaled parameters. Experimental results show that this method in imaging modeled surface can be used as a simulator.

## 1 Introduction

Customer's preferences for aesthetic appearance may vary and this causes increase of using free surfaces in product design. Therefore using Computer Aided Design and Computer Aided Manufacturing (CAD/CAM) became inevitable in modern industry. NURBS (Non-Uniform Rational B-Spline) surfaces are known as common parametric surface definition method. Moreover, coordinate calculating about NURBS surface is rather complex because of conversion to a basic type of surface. Since each surface has a different mathematical expression, complete surface conversion is not easy and conversation procedure always require casualty of surface data [1, 2].

Non-parametric methods require big amount of surface data for storage surfaces [3]. Three dimensional coordinate values of parametric NURBS surfaces are produced from *u* and *v* two dimensional parameters. The main advantage of the NURBS surfaces is their ability to reproduce second-degree basic surfaces, such as cylindrical, spherical, parabolic and hyperbolic surfaces. Furthermore, usage of NURBS in computer graphic and CAD increase rapidly [4-11]. Another type of parametric surface type is double cubic polynomial surface, which is out of the scope of this study.

A common approach to produce the surface is forming a model by fixed increments of *u* and *v* values. Despite the fixed increments in parameters, 3-D coordinates do not change in the same amount, and this causes unwanted fractures in the surface. As a

V.N. Alexandrov et al. (Eds.): ICCS 2006, Part II, LNCS 3992, pp. 247 – 254, 2006.

result the surface will not be aesthetic enough. Another approach is to make calculations with the derivations of the parametric variables to determine the interval values of the surface. However, this approach increases cost and complexity of surface calculation [12].

Artificial Neural Networks (ANN) is a machine learning algorithm which simulates human brain cells. ANN is efficient method to solve complex and non-linear problems. They are successfully used in many areas especially in image processing and pattern recognition. Once the Neural network is trained, it can solve the problem efficiently [13-18]. ANN is formed by a number of connected neurons in layers. A neuron in ANN puts a linear combination of its input(s), either from the previous layer of neurons or from user input and its weight into an activation function. This function returns the value that the neuron will pass on to the next layer. The ANN learns by adjusting these weights by some pre-chosen algorithm. To solve a non-linear problem, a multi-layer ANN should be used which contains one or more hidden layer [19-23].

In this study we present an alternative method based on an ANN model, to produce three dimensional surfaces as an image. The ANN model is developed to calculate intermediate values of a NURBS surface which is produced by giving a fixed increase to *u* and *v* parameters.

## 2 Parametric NURBS Surfaces

NURBS is an industry standard for geometric design and visualization. Some advantages of using NURBS surfaces are listed below [24-28]: (i) It is a common form that is suggested for both Standard Analytical Form shapes and free form shapes; (ii) It provides flexibility for designing shapes in many different types; (iii) They can be processed efficiently by digital and stable algorithms; (iv) They have an unchangeable character under affine like perspective transformations. NURBS are generalization of rational and non-rational Bezier curves and surfaces. One of the disadvantages of the NURBS is, even defining a basic shape like a circle, requirement of additional memory. In addition to control points, some other parameters appear in the definition of NURBS. But after the operation it supplies wanted. NURBS shapes can not be defined only with control points; weight of each control point is also required. A NURBS curve, *C(u)*, is a vector valued and piecewise rational function which can be expressed as;

$$C(u) = \frac{\sum_{i=0}^{n} W_i * P_i * N_{i,k}(u)}{\sum_{i=0}^{n} W_i * N_{i,k}(u)} \tag{1}$$

where $W_i$ represents *weights*, $P_i$ represents *control points* (vector) and $N_{i,k}$ is the normal *k degree B-spline basis functions* and $N_{i,k}$ can be defined as a recursive function as follows:

$$N_{i,k}(u) = \frac{u - t_i}{t(i+k) - t_i} * N_{(i,k-1)}(u) + \frac{t_{(i+k+1)} - u}{t_{(i+k+1)} - t_{(i+1)}} * N_{(i+1,k-1)}(u) \tag{2}$$

$$N_{i,0} = \begin{Bmatrix} 1, & if \quad t_i \le u \le t_{i+1} \\ 0, & t_i \le u \le t_{i+1}, \quad otherwise \end{Bmatrix}$$

where $t_i$ represents knots that are shaped by a knot vector and $U= \{t_0,t_1,\dots,t_m\}$.

A definition of a NURBS curve in equation (1) can be organized using rational basic functions as follows:

$$C(u) = \sum\nolimits_{i=0}^{n} P_i * R_{i,k}(u), \quad and\ R_{i,k}(u) = \frac{W_i * N_{i,k}(u)}{\sum\nolimits_{j=0}^{n} W_j * N_{j,k}(u)} \tag{3}$$

Any NURBS surface is defined similarly by:

$$S(u,v) = \sum\sum P_{i,j} * R_{i,k,j,l}(u,v), \quad and\ R_{i,k,j,l}(u,v) = \frac{W_{i,j} * N_{i,k}(u) * N_{j,l}(v)}{\sum\nolimits_{r=0}^{n} \sum\nolimits_{s=0}^{m} W_{r,s} * N_{r,k}(u) * N_{s,l}(v)} \tag{4}$$

## 3 Artificial Neural Network Model

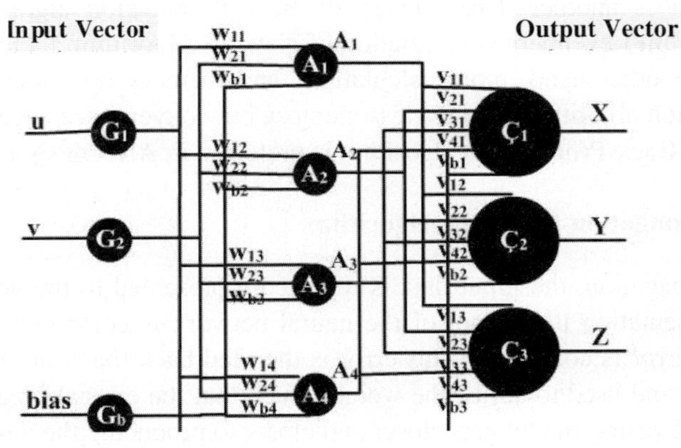


**Fig. 1.** Desired model of Artificial Neural Network

### 3.1 Major Construction of Network

ANN structure is shown in *Fig 1*. Network is constructed in two layers. First layer has two neurons *($G_1$, $G_2$)* and second layer has three ($C_1$, $C_2$, $C_3$). Second layer works as a shape producer and function of the first layer is key point distributor. Network's input is a 3 dimensional $X=$ *(u, v, bias)* vector. Bias input is 1, and network's output is 3 dimensional $Y=(x, y, z)$ vector. Each neuron of first layer is one-to-one related to each input and first two layers are completely related to each other. The power of relation between $m^{th}$ input and $n^{th}$ neuron of first layer is represented by $W_{mn}$. $W$ power matrix is formed from $m$ X $n$ power matrix and the power of relation between first layer's $n^{th}$ output and $p^{th}$ nerve under second layer is represented by $v_{np}$. $V$ power matrix has $nXp$ size. In the network design, $m=3$, $n=4$, and $p=3$.

A learning period is needed by artificial neural network similar to the human brain. During the calculation process in designed ANN, $x$, $y$ and $z$ coordinate values are calculated and saved into a file according to the addition of $u$ and $v$, then generalization is done by using weights which obtains relation between neurons according to the these data (NURBS surface coordinate $x$, $y$ and $z$ that match $u$ and $v$ values). In the learning process back-propagation algorithm is used to obtain desired output.

### 3.2 Training Algorithms

In general, training algorithms for neural networks can be classified into three basic groups: controlled learning, non-controlled learning, and reinforcement learning. Learning in the neural network determines the weight set that gives the desired output of the network which is tested for inputs and outputs. This weight vector is formed by increasing or decreasing the values of the weights in a loop until the differences between desired output and network output is minimal. Back-Propagation learning algorithm is one of the most common of them. It has been used in several areas. However its learning time is long because its approach ratio is very poor. Catching local minimum is another disadvantage of Back-Propagation algorithm. Another learning algorithm Levenberg is a variation of Newton Algorithm. Its learning time is shorter. On the other hand, more calculations and memory resources are required. Back-Propagation algorithm requires less memory but convergence to result is slower. For this reason Back-Propagation algorithm is preferred in ANN design.

### 3.3 Back-Propagation Learning Algorithm

With back-propagation, the input data is repeatedly presented to the neural network. With each presentation the output of the neural network is compared to the desired output and an error is computed. This error is then fed back (back propagated) to the neural network and used to adjust the weights such that the error decreases with each iteration and the neural model gets closer and closer to producing the desired output.

In this study, particularly, delta rule is used which is commonly used in training of multi-layered networks. Modification of weight between $j^{th}$ input and $i^{th}$ neuron for $k$ sample pair of input-output is expressed as follows:

$$\Delta w_{i,j}^{k} = \alpha (T_i^k - O_i^k) x_j^k \tag{5}$$

where $T_i$ is desired target output, $O_i$ is real output, $\alpha$ is ratio of learning, and $x_i$ is $j^{th}$ input of network. For a sample pair on $k^{th}$ layer, total error and average error function is defined as:

$$E = \frac{1}{2}\sum_k (T_k - O_k)^2 = \frac{1}{2P}\sum_p \sum_j (T_j - O_j)^2 \tag{6}$$

where $T_k$ is desired target output, $O_j$ is calculated output, and $p$ is the total number of input-output pair in the training set. In the case of not converging, the formula can be generalized as in (7) for setting weights as it includes $\beta$ momentum term:

$$\Delta w_{i,j}(t+1) = \alpha \, \delta_i^k \, O_j^k + \beta \Delta w_{ij} \tag{7}$$

where $\beta$ is momentum term, $\delta_i^k$ is $i^{th}$ neuron's error value for $k$ sample pair and contains derivative of used threshold function. The threshold function determines the output of process element. Usually, a function which can derivative is preferred. Although there are many types of threshold functions, sigmoid function is used in this study because it is used commonly in back-propagation.

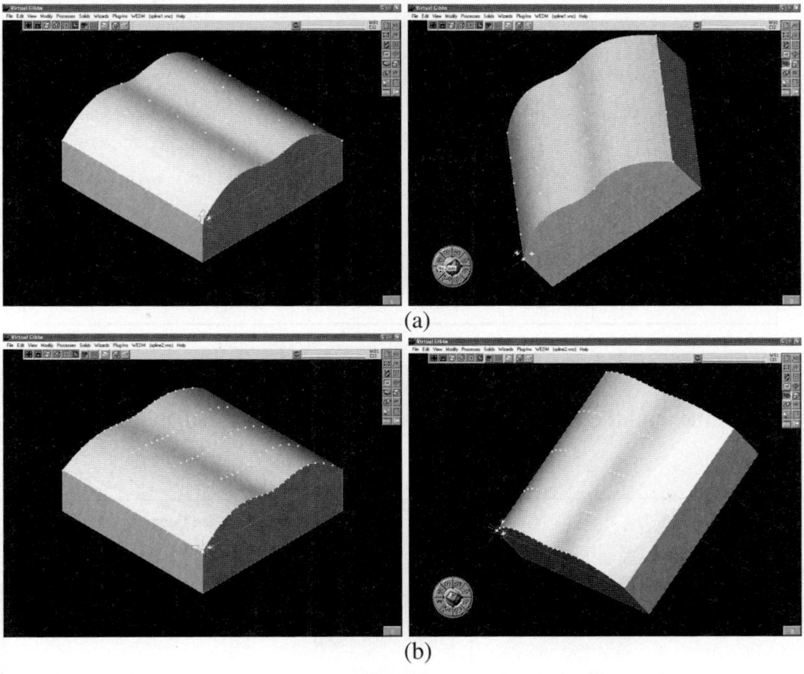

**Fig. 2.** Two different sample surface; (a) First two Surfaces (above) (b) Second Surfaces (below) (are produced by Solidworks and Neural Network)

## 4 Experimental Results

Two different surfaces produced by commercial software *Solid Works 2004* and displayed in *Fig 2* are considered in this study for experiments. Coordinates on the layer ($S(u,v)$) are generated by fixed increments of *0.1* on $u$ and $v$ values as the surface is created. After network is trained according to these values, weight values are calculated by *Back Propagation Learning Algorithm*. Then surface coordinates is produced by giving $u$ and $v$ the values that are multiples of *0.1*. Parametric equation of surface, real coordinates which is produced according to this addition, and coordinate values which are produced by suggested method are compared and a part of comparison results for one of these surfaces is given in *Table 1*. Experiments show that our model is flexible enough and able to model any desired shape. A comparison of the computation time shows that our approach yields faster results than the other algorithms that have fixed *u* and *v* increment after the first training of network.

**Table 1.** Surface Coordinates Values that are Produced by Solid Works 2004 and Back Propagation Algorithm

| u | v | Solid Works Outputs | | | ANN Outputs | | |
|---|---|---|---|---|---|---|---|
| | | X | Y | Z | X | Y | Z |
| 0 | 0.2 | 0 | 40 | 0 | 2 | 42 | 11 |
| 0 | 0.8 | 0 | 160 | 0 | 10 | 167 | 9 |
| 0.2 | 0.8 | 49 | 160 | 30 | 34 | 125 | 24 |
| 0.4 | 0.2 | 84 | 40 | 27 | 77 | 30 | 20 |
| 0.6 | 0 | 115 | 0 | 27 | 110 | 9 | 27 |
| 0.6 | 0.2 | 115 | 40 | 27 | 111 | 20 | 27 |
| 0.8 | 0.8 | 163 | 141 | 30 | 150 | 160 | 30 |
| 1 | 0.6 | 200 | 120 | 0 | 199 | 170 | 0 |
| 1 | 0.9 | 200 | 190 | 0 | 200 | 197 | 0 |

The graphical relationship between the number of iterations in training the ANN, the computation time, and the $r^2$ values for the surface of *Fig 3.a* is displayed in *Fig 3.b*.

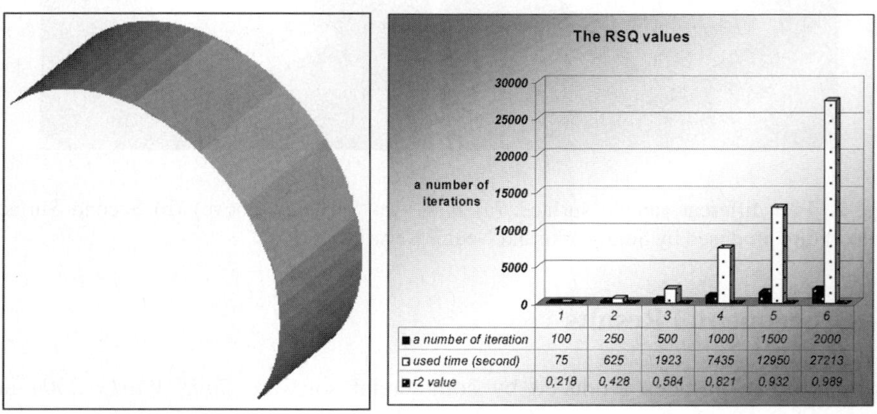


(a) (b)

**Fig. 3.** (a) The test surface saved as NURBS for simplicity. (b) The relation between a number of iterations and used time for training of ANN in terms of surface demonstrated in Fig. 3.a.

## 5 Conclusion and Future Work

An approach is defined by using ANN's for getting intermediate coordinates of three dimensional NURBS surfaces. Free form shaped NURBS surfaces which do not contain pockets and various three dimensional NURBS models which are formed from several patches can be produced by using this method. Experimental results show that this method can be used as an alternative simulation tool for visualization of the modeled surface.

As a future work, other Artificial Intelligence techniques can be employed to calculate intermediate values of these surfaces and results can be compared.

## Acknowledgements

This study has been supported by the Scientific Research Projects of Selcuk University (in Turkey).

## References

1. Zeid, İbrahim , CAD/CAM Theory and Practice, McGraw-Hill, (1991).
2. Hyunbo Shim and Euikwon Suh, Contact treatment algorithm for the trimmed NURBS surface Journal of Materials Processing Technology, Vol. 104, p. 200-206, (2000).
3. Hearn, Donald, Computer graphics, Prentice Hall, New Jersey, (1994).
4. Aziguli W., Goetting M. And Zeckzer D., Approximation of NURBS curves and surfaces using adaptive equidistant parameterizations, Tsinghua Science&Technology, Vol. 10, pp. 316-322, (2005).
5. Che X., Liang X., and Li Q., G1 continuity conditions of adjacent NURBS surfaces, Computer Aided Geometric Design, Vol. 22, pp. 285-298, (2005).
6. Yang J., and Abdel-Malek K., Approximate swept volumes of NURBS surfaces or solids, Computer Aided Geometric Design, Vol. 22, pp. 1-26, (2004).
7. Ma W., But W.-C., and He P., NURBS based adaptive slicing for efficient rapid prototyping, Computer Aided Geometric Design, Vol. 36, pp. 1309-1325, (2004).
8. Yau H.-T., Kuo M.-J., NURBS machining and feed rate adjustment for high-speed cutting of complex sculptured surfaces, International Journal of Production Research, Taylor & Francis, Vol. 39, (2001).
9. Kumar S., Manocha D., Lastra A., Interactive display of large NURBS models, Visualization and Computer Graphics, IEEE Transactions on Vol. 2, pp. 323-336, (1996).
10. Wang Q, Hua W., Li G., Bao H., Generalized NURBS curves and surfaces, Geometric Modeling and Processing, Proceedings p 365-368, (2004).
11. Hatna A., Grieve B., Cartesian machining versus parametric machining: a comparative study, International Journal of Production Research, Taylor & Francis, Vol 38, (2000).
12. L. Piegl, W. Tiller: Geometry-based triangulation of trimmed NURBS Surfaces. Computer Aided Design, Vol. 30(1), pp. 11-18, (1998).
13. E. Castillo, A. Iglesias, J.M. Gutiérrez, E. Alvarez, J.I., Functional Networks. An application to fitting surfaces, World Multiconference on Systemics, Cybernetics and Informatics, Proceedings of the ISAS-98 Fourth International Conference on Information Systems, Analysis and Synthesis Vol. 2, pp. 579-586, (1998).

14. A. Iglesias, A. Galvez, Applying Functional Networks to CAGD: the Tensor Product Surface Problem, Fourth International Conference on Computer Graphics and Artificial Intelligence, pp. 105-115, (2000).
15. A. Iglesias, A. Galvez, Fitting 3D data points by extending the neural networks paradigm, Computational Methods and Experimental Measurements, WIT Press /Computational Mechanics Publications, Southampton-Boston (Series: Computational Engineering) Vol. 3, 809-818., (2001).
16. A. Iglesias, A. Gálvez, Applying Functional Networks to Fit Data Points From B-spline Surfaces, Fitting 3D data points by extending the neural networks paradigm, Proceedings of the Computer Graphics International, CGI'2001, Hong-Kong (China), IEEE Computer Society Press, Los Alamitos, California 329-332, (2001).
17. G. Echevarría, A. Iglesias, A. Gálvez, Extending Neural Networks for B-spline Surface Reconstruction, Computational Science-ICCS'2002, Springer-Verlag (Series: Lectures Notes in Computer Science), Berlin Heidelberg, Vol. 2330, pp. 305-314 (2002).
18. A. Iglesias, G. Echevarria, A. Galvez, Functional Networks for B-spline Surface Reconstruction, Future Generation Computer Systems, Special Issue on "Computer Graphics and Geometric Modeling", Vol. 20, Issue 8, pp. 1337-1353, (2004).
19. Barhak, J.,Fischer, A.: Parameterization and reconstruction from 3D scattered points based on neural network and PDE techniques. IEEE Trans. on Visualization and Computer Graphics, 7(1), pp 1-16, (2001).
20. Mishkoff H. C. Understanding Artificial Intelligence, Radio Shack, (1986).
21. Gevarter W. B. Intelligence Machines: An Introductory, Prentice-Hall, (1985).
22. Hoffman, M., Varady, L.: Free-form surfaces for scattered data by neural networks. Journal for Geometry and Graphics, 2, pp. 1-6, (1998).
23. Iglesias, A., Galvez, A.: A new Artificial Intelligence paradigm for Computer-Aided Geometric Design. Lecture Notes in Artificial Intelligence, Vol. 1930, pp. 200-213, (2001).
24. Piegl Les, On NURBS: A Survey, IEEE Computer Graphics and Applications, Vol. 11, No. 1, pp. 55 – 71, (1991).
25. Rogers David F., Rae A. Earnshaw (editors), State of the Art in Computer Graphics - Visualization and Modeling, New York, Springer-Verlag, pp. 225 – 269, (1991).
26. C. deBoor, A Practical Guide to Splines, New York, Springer-Verlag, (1978).
27. Watt Alan, Watt Mark, Advanced Animation and Rendering Techniques, New York, AMC press Addision-Wesley, (1992).
28. Foley James D. et al, Introduction to Computer Graphics, Addision-Wesley, (1994).

# Characterizing and Covering Some Subclasses of Orthogonal Polygons*

Ana Mafalda Martins[1] and António Leslie Bajuelos[2]

[1] Escola Superior de Ciências e Tecnologia, Universidade Católica Portuguesa, Campus Viseu, Portugal
ammartins@crb.ucp.pt

[2] Dept. of Mathematics & CEOC - Center for Research in Optimization and Control, University of Aveiro, Portugal
leslie@mat.ua.pt

**Abstract.** A *grid n-ogon* is a $n$-vertex orthogonal polygon that may be placed in a $\frac{n}{2} \times \frac{n}{2}$ unit square grid and that does not have collinear edges. Given a grid $n$-ogon $P$, let $|\Pi(P)|$ be the number of rectangles that results when we partition $P$ by extending the edges incident to reflex vertices towards its interior. $P$ is called FAT if $|\Pi(P)|$ is maximal for all grid $n$-ogons; $P$ is called THIN if $|\Pi(P)|$ is minimal for all grid $n$-ogons. THINs with area $2r+1$ are called MIN-AREA. We will show that $\lceil \frac{n}{6} \rceil$ vertex guards are necessary to guard a MIN-AREA grid $n$-ogon and present some problems related to THINs.

## 1 Introduction

Art Gallery problems represent a classic and very interesting field of Computational Geometry. The original art gallery problem was introduced by V. Klee in 1973 in a conference of Mathematics. He posed the following problem to V. Chvtal: How many stationary guards are needed to cover an art gallery room with $n$ walls? Informally the floor plan of the art gallery room is modeled by a *simple polygon* (simple closed polygon with its interior) $P$ and a guard is considered a fixed point in $P$ with $2\pi$ range visibility. We say that a point $x$ *sees* point $y$ (or $y$ is visible to $x$) if the line segment connecting them lies entirely in $P$. A set of guards covers $P$, if each point of $P$ is visible by at least one guard. Thus, the Art Galery Problem deals with setting a minimal number of guards in a gallery room whose floor plan has polygonal shape, so that they could see every point in the room. Two years later Chvtal established the well known *Chvátal Art Gallery Theorem*: $\lfloor \frac{n}{3} \rfloor$ *guards are occasionally necessary and always sufficient to cover a simple polygon of n vertices.*

Many variants of the original art gallery problem have been considered and studied over the years, see [4, 5, 9] for comprehensive surveys. An interesting variant is the *Orthogonal Art Gallery Theorem.* This theorem was first formulated

* Partially funded by *Fundação para Ciência e Tecnologia (FCT)* and by CEOC (Univ. of Aveiro) through *Programa POCTI, FCT*, co-financed by *EC* fund *FEDER.*

V.N. Alexandrov et al. (Eds.): ICCS 2006, Part II, LNCS 3992, pp. 255–262, 2006.

and proved by Kahn et al, in 1983. It states that $\lfloor \frac{n}{4} \rfloor$ *guards are occasionally necessary and always sufficient to cover an orthogonal simple polygon of n vertices. Orthogonal simple polygons* (simple polygons whose edges meet at right angles) are an important subclass of polygons. Indeed, they are useful as approximations to polygons; and they arise naturally in domains dominated by Cartesian coordinates, such as raster graphics, VLSI design, or architecture. Efficient algorithms, based on the proofs of the above theorems, were developed to cover both arbitrary and orthogonal simple polygons with $\lfloor \frac{n}{3} \rfloor$ and $\lfloor \frac{n}{4} \rfloor$ guards, respectively. While this number of guards is necessary in some cases, often it is far more than it is needed to cover a particular simple polygon. For instance, it is known that convex polygons only require one guard. Similarly, depending on the structure of a simple polygon the minimum number of guards may be smaller than the estimated. A variant of this problem is the MINIMUM VERTEX GUARD (MVG) problem, that is the problem of finding the minimum number of guards placed on vertices (*vertex guards*) needed to cover a given simple polygon. This is a NP-hard problem both for arbitrary and orthogonal simple polygons [2, 6].

*Our Contribution.* This paper has as intention to introduce a subclass of orthogonal polygons that presents sufficiently interesting characteristics that we are studying and formalizing, in particular the way they can be guarded. Of these polygons two classes stand out: the FATs and THINs. We think FATs and THINs are representative of extremal behavior and they are used experimentally to evaluate some approximated methods of resolution of the MVG problem [8]. The paper is structured as follows: in the next section we will present some introductory definitions and useful results. In section 3, we will study the MVG problem for a subclass of THIN grid $n$-ogons (the MIN-AREA) and in section 4 we will refer to some problems related to THINs.

## 2 Conventions, Definitions and Results

For every $n$-vertex orthogonal simple polygon ($n$-*ogon* for short), $n = 2r + 4$, where $r$ denotes the number of reflex vertices, e.g. [4]. A *rectilinear cut* ($r$-cut) of an $n$-ogon $P$ is obtained by extending each edge incident to a reflex vertex of $P$ towards the interior of $P$ until it hits $P$'s boundary. We denote this partition by $\Pi(P)$ and the number of its pieces by $|\Pi(P)|$. Each piece is a rectangle and so we call it a $r$-*piece*. A $n$-ogon that may be placed in a $\frac{n}{2} \times \frac{n}{2}$ square grid and that does not have collinear edges is called *grid* $n$-*ogon*. We assume that the grid is defined by the horizontal lines $y = 1, \ldots, y = \frac{n}{2}$ and the vertical lines $x = 1, \ldots, x = \frac{n}{2}$ and that its northwest corner is $(1, 1)$. Grid $n$-ogons that are symmetrically equivalent are grouped in the same class [1]. A grid $n$-ogon $Q$ is called FAT iff $|\Pi(Q)| \geq |\Pi(P)|$, for all grid $n$-ogons $P$. Similarly, a grid $n$-ogon $Q$ is called THIN iff $|\Pi(Q)| \leq |\Pi(P)|$, for all grid $n$-ogons $P$. Let $P$ be a grid $n$-ogon with $r$ reflex vertices, in [1] is proved that, if $P$ is FAT then $|\Pi(P)| = \frac{3r^2+6r+4}{4}$, for $r$ even and $|\Pi(P)| = \frac{3(r+1)^2}{4}$, for $r$ odd; if $P$ is THIN then $|\Pi(P)| = 2r + 1$. There is a single FAT $n$-ogon (except for symmetries of the grid) and its form is illustrated in fig. 1(a). However, the THIN $n$-ogons are not unique (see fig. 1(b)).

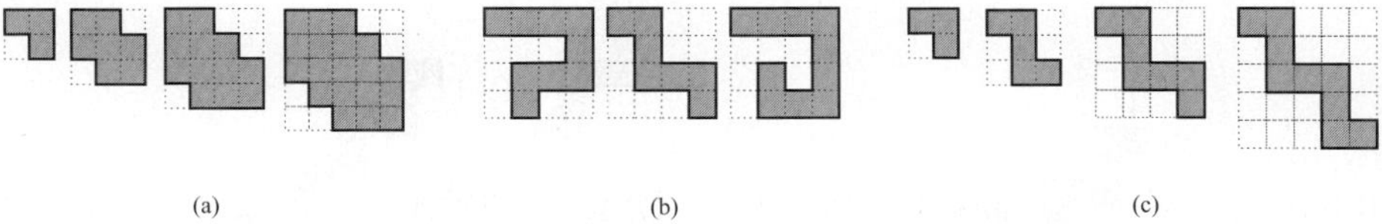

**Fig. 1.** (a) The unique FAT $n$-ogon, for $n = 6, 8, 10, 12$; (b) Three THIN 10-ogons; (c) The unique MIN-AREA grid $n$-ogons, for $n = 6, 8, 10, 12$

The area of a grid $n$-ogon is the number of grid cells in its interior. In [1] it is proved that for all grid $n$-ogon $P$, with $n \geq 8$, $2r + 1 \leq A(P) \leq r^2 + 3$. A grid $n$-ogon $P$ is a MAX-AREA grid $n$-ogon iff $A(P) = r^2 + 3$ and it is a MIN-AREA grid $n$-ogon iff $A(P) = 2r + 1$. There exist MAX-AREA grid $n$-ogons for all $n$, but they are not unique. However, there is a single MIN-AREA grid $n$-ogon (except for symmetries of the grid) and it has the form illustrated in fig. 1(c). Regarding MIN-AREA grid $n$-ogons, it is obvious that they are THIN $n$-ogons, because $|\Pi(P)| = 2r + 1$ holds only for THIN $n$-ogons. However, this condition is not sufficient for a grid $n$-ogon to be a MIN-AREA grid $n$-ogon.

Our aim is to study the MVG problem for grid $n$-ogons. Since THIN and FAT $n$-ogons are the classes for which the number of $r$-pieces is minimum and maximum, we think that they can be representative of extremal behavior, so we started with them. We already proved that to cover any FAT grid $n$-ogon it is always sufficient two $\frac{\pi}{2}$ vertex guards (*vertex guards with $\frac{\pi}{2}$ range visibility*) and established where they must be placed [3]. THINs are much more difficult to cover, on the contrary of what we might think once they have much fewer pieces than FATs. Since THIN grid $n$-ogons are not unique, we intend to characterize structural properties of classes of THINs that allow to simplify the problem study. Up to now the only quite characterized subclass is the MIN-AREA grid $n$-ogons: the subclass for which the number of grid cells is minimum.

## 3 Guarding Min-Area Grid $n$-ogons

Given $P$, a MIN-AREA, we will denote by $g(P)$ the minimum number of vertex guards that is needed to cover $P$. We will show not only that $g(P) = \lceil \frac{r+2}{3} \rceil$ but also in which vertices these guards must be placed.

**Lemma 1.** *Two vertex guards are necessary to cover the Min-Area* grid 12-ogon. *Moreover, the only way to do so is with the vertex guards $v_{2,2}$ and $v_{5,5}$.*

*Proof (Sketch).* This demonstration is based on the fact that the unit squares $Q_0$ and $Q_1$ will have to be guarded and that the only vertex guards that can do it and simultaneously guard all the polygon are $v_{2,2}$ and $v_{5,5}$ (see fig. 2(a)). □

**Proposition 1.** *If we "merge" $k \geq 2$* MIN-AREA *grid* 12*-ogons, we will obtain the* MIN-AREA *grid $n$-ogon with $r = 3k + 1$. More, $k + 1$ vertex guards are necessary to cover it, and the only way to do so is with the vertex guards: $v_{2+3i,2+3i}, i = 0, 1, \ldots, k$.*

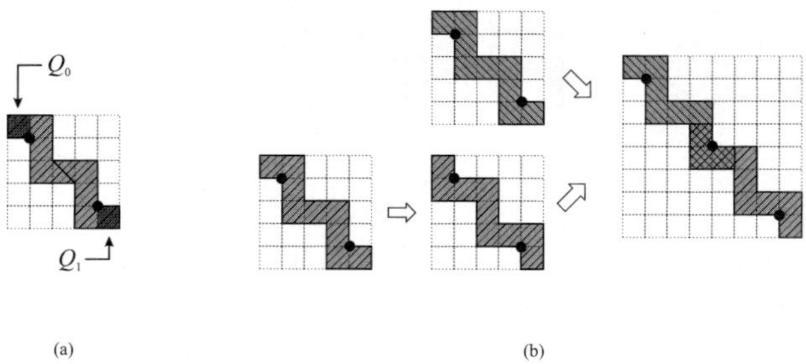


**Fig. 2.** (a) MIN-AREA grid 12-ogon; (b) Construction of the Min-Area grid 18-ogon from two Min-Area grid 12-ogons

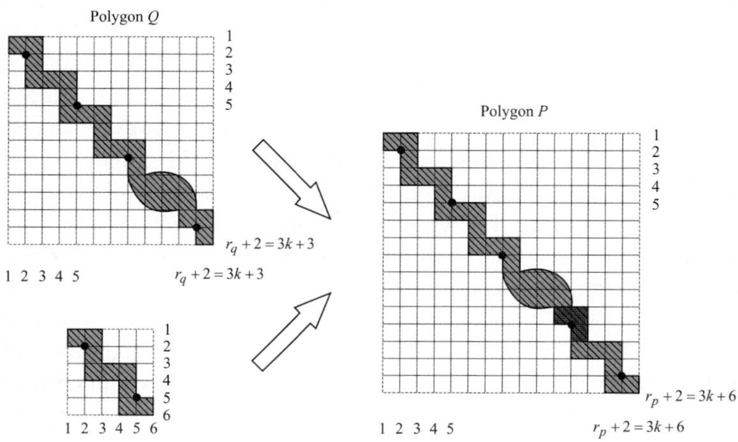


**Fig. 3.** Polygon $P$ ("merge" of $Q$ with the Min-Area grid 12-ogon)

*Proof.* Let $P$ be the Min-Area *grid n-ogon* with $r = 7$ reflex vertices. $P$ can be obtained "merging" two Min-Area *grid* 12-*ogons* (see fig. 2(b)).

By lemma 2 and as we can see, 3 vertex guards are necessary to cover $P$, and the only way to do that is with $v_{2,2}$, $v_{5,5}$ and $v_{8,8}$. Thus, for $k = 2$, the proposition is true. Let $k \geq 2$, we will show that the proposition is true for $k+1$ (*induction thesis*), assuming that it is true for $k$ (*induction hypotheses*).

By induction hypothesis, "merging" $k$ MIN-AREA grid 12-ogons we obtain $Q$, the MIN-AREA grid $n$-ogon with $r_q = 3k + 1$ reflex vertices. If we "merge" $Q$ with the MIN-AREA grid 12-ogon, we will obtain a polygon $P$ (see fig. 3). $P$ has $r_p = 3k+4$ reflex vertices and $A(P) = 2r_p + 1$. Therefore, "merging" $k+1$ MIN-AREA grid 12-ogons we obtain $P$, the MIN-AREA grid $n$-ogon with $r = 3k + 4$. Furthermore, by induction hypotheses and from what we can observe in fig. 3, we can conclude that $k + 2$ vertex guards are necessary to cover $P$. Moreover, the only way to do so is with the vertex guards: $v_{2+3i,2+3i}, i = 0, 1, \ldots, k+1$. □

**Proposition 2.** $\lceil \frac{r+2}{3} \rceil$ *vertex guards are always necessary to guard a* MIN-AREA *grid n-ogon with r reflex vertices.*

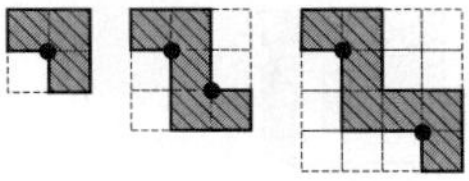

**Fig. 4.** Min-Area grid $n$-ogons with $r = 1, 2, 3$

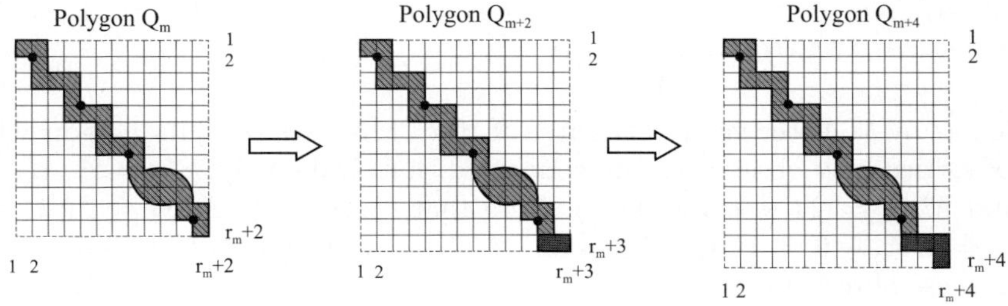


**Fig. 5.** Min-Area grid $n$-ogons $Q_m$, $Q_{m+2}$ and $Q_{m+4}$

*Proof.* Let $P_n$ be a MIN-AREA grid $n$-ogon with $r_n = \frac{n-4}{2}$ reflex vertices. We may easily check that 1, 2 and 2 vertex guards are necessary to guard MIN-AREA grid $n$-ogons with $r_n = 1, 2, 3$, respectively (see fig.4).

Let $r_n \geq 4$. If $r_n \equiv 1\ (mod\ 3)$ then, by proposition 1, the $\lceil \frac{r_n+2}{3} \rceil$ vertex guards $v_{2+3i,2+3i}, i = 0, 1, \ldots, \frac{r_n-1}{3}$, are necessary to cover $P_n$. Thus, we just need to prove the following cases: $r_n \equiv 2\ (mod\ 3)$ and $r_n \equiv 0\ (mod\ 3)$.

In any case, $P_n$ can be obtained, by INFLATE-PASTE (a complete process to generate grid $n$-ogons, well described in [7]), from a Min-Area $Q_m$ with $r_m = \frac{m-4}{2}$ and such that $r_m = 3k_m + 1$ (see fig. 5). The first case corresponds to polygon $Q_{m+2}$, in fig. 5, and $r_n = r_m + 1$. The second case corresponds to polygon $Q_{m+4}$, in fig. 5, and $r_n = r_m + 2$.

As we can see, in any case, is always necessary one more vertex guard, which can be $v_{r_n+1,r_n+1}$. Thus, $\lceil \frac{r_m+2}{3} \rceil + 1 = \lceil \frac{r_n+2}{3} \rceil$ vertex guards are necessary to guard $P_n$. □

Proposition 2 not only gives the guarantee of that $\lceil \frac{r+2}{3} \rceil$ vertex guards are required to guard the MIN-AREA grid $n$-ogon with $r$ reflex vertices, but also establishes a possible positioning.

## 4 Some Problems Related to Thin $n$-Ogons

As we saw in section 1, on the contrary of the FAT the THIN grid $n$-ogons are not unique. In fact, 1 THIN 6-ogon exists, 2 THIN 8-ogons exist, 30 THIN 10-ogons exist, 149 THIN 12-ogons exist, etc. Thus, it is interesting to evidence that the number of THIN grid $n$-ogons ($|\text{THIN}(n)|$) grows exponentially. Will it exist some expression that relates $n$ to $|\text{THIN}(n)|$? Also, we can question on the value of the area of the THIN grid $n$-ogon with maximum area (MAX-AREA-THIN $n$-ogon) and if the MAX-AREA-THIN $n$-ogon is unique.

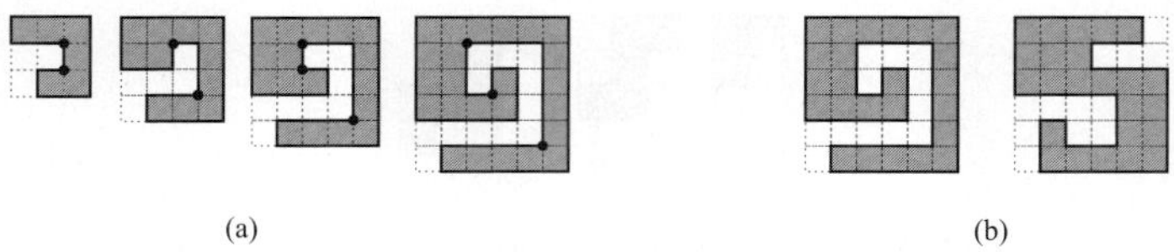

Fig. 6. (a) From left to right $MA_2 = 6, MA_3 = 11, MA_4 = 17, MA_5 = 24$; (b) Two THIN 14-ogons with area 24, $MA_5 = 24$

Denote by $MA_r$ the value of the area of "the" MAX-AREA-THIN $n$-ogon with $r$ reflex vertices. By observation we concluded that $MA_2 = 6, MA_3 = 11, MA_4 = 17$ and $MA_5 = 24$ (see Fig. 6(a)). Note that, $MA_2 = 6$, $MA_3 = MA_2 + 5$, $MA_4 = MA_3 + 6 = MA_2 + 5 + 6$ and $MA_5 = MA_4 + 7 = MA_2 + 5 + 6 + 7$. From these observations it follows:

*Conjecture 1.* $MA_r = MA_2 + 5 + 6 + 7 + \ldots + (r + 2) = \frac{r^2+5r-2}{2}$.

If conjecture 1 is true we can say the THIN grid $n$-ogon with maximum area is not unique (see Fig. 6(b)).

**Definition 1.** *A* THIN *$n$-ogon is called* SPIRAL-THIN *if its boundary consists of two polygonal chains: a chain of reflex vertices and a chain of convex vertices.*

From left to right in fig. 6(a), the second SPIRAL-THIN can be obtained from the first by INFLATE-PASTE, the third SPIRAL-THIN can be obtained from the second... So we believe that a MAX-AREA-THIN grid $(n + 2)$-ogon can always be obtained from a MAX-AREA-THIN grid $n$-ogon. We intend to use the following results and the SPIRAL-THIN grid $n$-ogons illustrated in fig. 6(a) to prove conjecture 1.

The dual graph of $\Pi(P)$ captures the adjacency relation between pieces of $\Pi(P)$. Its nodes are $r$-pieces and its non-oriented edges connect adjacent $r$-pieces, i.e., $r$-pieces with a common edge. We will denote the dual graph of $\Pi(P)$ by $DG(\Pi(P))$.

**Lemma 2.** *Let $P$ be a* THIN *$(n + 2)$-ogon. Then every grid $n$-ogon that yields $P$ by* INFLATE-PASTE *is also* THIN.

The proof of this lemma is strongly based on the reverse process of INFLATE-PASTE.

**Proposition 3.** *Let $P$ be a* THIN *grid $n$-ogon with $r = \frac{n-4}{2} \geq 1$ reflex vertices, then $DG(\Pi(P))$ is a path graph (i.e., a tree with two nodes of vertex degree 1, called leaves, and the other nodes of vertex degree 2) (see examples in fig. 7(a)).*

The proof of this proposition is done by induction on $r$ and uses lemma 2.

**Proposition 4.** *Let $P$ be a grid $n$-ogon. If $P$ is not* THIN *then $DG(\Pi(P))$ is not a tree (see example in fig 7(b)).*

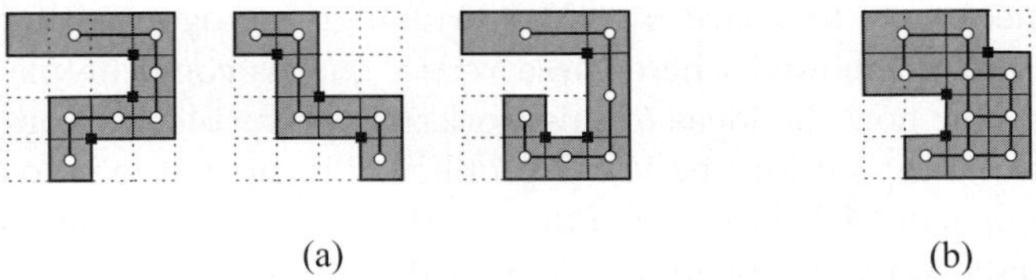


**Fig. 7.** (a) Three THIN grid 10-ogon and respective dual graphs; (b) A grid 10-ogon and respective dual graph

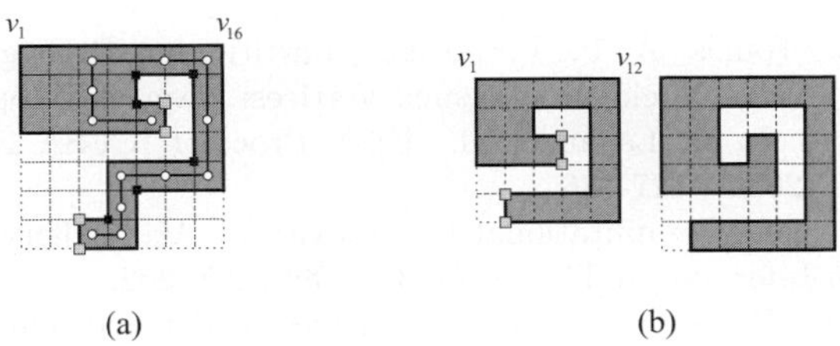


**Fig. 8.** (a) The only convex vertices that could yield, by INFLATE-PASTE, THIN grid 14-ogons are $v_3$, $v_4$, $v_{11}$ and $v_{12}$ (in CCW order); (b) The only convex vertices, from the first SPIRAL-THIN, that could yield the second are $v_3$, $v_4$, $v_9$ and $v_{10}$

**Proposition 5.** *The unique convex vertices of a* THIN *grid $n$-ogon that could be used to yield a* THIN *grid $(n+2)$-ogon, by Inflate-Paste, are those which belong to the $r$-pieces associated to the leaves of $DG(\Pi(P))$.*

Lemma 2 and proposition 5 can be very useful in the generation, by INFLATE-PASTE, of THIN grid $n$-ogons ($n \geq 8$). Lemma 2 says that we must take a THIN grid $(n-2)$-ogon, and proposition 5 establishes that the only convex vertices that can "work" are those which belong to the $r$-pieces associated to the leaves of $DG(\Pi(P))$ (which are in number of 4). In this way we do not need to apply INFLATE-PASTE to all the convex vertices of a THIN and then to check which of the produced polygons are THINs. We just need to apply INFLATE-PASTE to 4 convex vertices and then check which of the produced polygons are THINs. So the number of case analysis is significantly reduced (see fig 8).

*Conjecture 2.* There exists at least a THIN grid $n$-ogon for which $\lfloor \frac{r}{2} \rfloor + 1$ vertex guards are necessary to cover it.

It seems to us, with some certainty, that the SPIRAL-THIN grid $n$-ogons, illustrated in fig. 6(a), require $\lfloor \frac{r}{2} \rfloor + 1$ vertex guards.

## 5 Conclusions and Further Work

We defined a particular type of polygons - grid $n$-ogons - and presented some results and problems related to them. Of these problems, the guarding problems are the ones that motivate us more. We proved that $\lceil \frac{r+2}{3} \rceil$, i.e., $\lceil \frac{n}{6} \rceil$ vertex guards are necessary to guard any MIN-AREA grid $n$-ogon with $r$ reflex vertices. Moreover, we showed where these vertex guards could be placed. We are investigating now how the ideas of this work may be further exploited to obtain better approximate solutions to MVG problem. The next step is to characterize structural properties of classes of THINs with the aim of simplifying our next objective: study MVG problem for THIN grid $n$-ogons.

## References


1. Bajuelos, A. L., Tomás, A. P., F.Marques: Partitioning Orthogonal Polygons by Extension of All Edges Incident to Reflex Vertices: lower and upper bounds on the number of pieces. In A. Lagan et al. (Eds): Proc. of ICCSA 2004, LNCS 3045, Springer-Verlag (2004), 127-136.
2. Lee, D. and A. Lin: Computational Complexity of Art Gallery Problems, IEEE Transactions on Information Theory IT-32 (1996) 276-282.
3. Martins A.M and Bajuelos A.: Some Properties of Fat and Thin grid n-ogons, in Proc. of International Conference of numerical Analisys and Applied Mathematics (ICNAAM 2005), Wiley-VCH Verlag, (2005). 361-365.
4. O'Rourke, J.: Art Gallery Theorems and Algorithms, Oxford U. Press (1987).
5. Shermer T.: Recent results in art galleries. Proc. IEEE, 80 (1992) 1384-1399.
6. Schuchardt D., Hecker H.: Two NP-Hard Art-Gallery Problems for Ortho-Polygons. Math. Logiv Quart 41 (1995) 261-267
7. Tomás, A.P., Bajuelos, A.L.: Quadratic-Time Linear-Space Algorithms for Generating Orthogonal Polygons with a Given Number of Vertices.In A. Lagan et al. (Eds): Proc. of ICCSA 2004, LNCS 3045, Springer-Verlag (2004), 117-126.
8. Tomás, A.P., Bajuelos, A.L., Marques F.: On Visibility Problems in the Plane-Solving Minimum Vertex Guard Problems by Successive, in online Proceedings of Artificial Intelligence and Mathematics, Florida, USA, http://anytime.cs.umass.edu/aimath06/, (2006).
9. Urrutia, J.: Art Gallery and Illumination Problems. In J.-R. Sack and J. Urrutia, editores, Handbook of Computational Geometry, Elsevier (2000).

# Techniques for Computing Viewpoint Entropy of a 3D Scene

Pascual Castelló[1], Mateu Sbert[2], Miguel Chover[1], and Miquel Feixas[2]

[1] Departamento de Lenguajes y sistemas Informáticos, Universitat Jaime I, Campus Riu Sec, 12080 Castellón de la Plana, Spain
{castellp, chover}@lsi.uji.es
http://graficos.uji.es
[2] Institut d'Informàtica i Aplicacions, Universitat de Girona, Campus Montilivi, 17071 Girona, Spain
{Mateu.Sbert, Miquel.Feixas}@ima.udg.es

**Abstract.** Viewpoint entropy is a metric that allows measuring the visibility goodness of a scene from a camera position. In this work, we analyze different software and hardware assisted techniques to compute the viewpoint entropy. The main objective of this study is to identify which of these techniques can be used in real time for 3D scenes of high complexity. Our results show that interactivity can be obtained with occlusion query technique and that for real time we need a hybrid software and hardware technique.

## 1 Introduction

Recently, several methods have been developed to compute the goodness of a viewpoint. These methods have in common the use of the viewpoint complexity concept [2, 3, 15, 7, 8, 9, 13]. The notion of viewpoint complexity is used in several areas of Computer Graphics such as scene understanding and virtual world exploration, radiosity and global illumination, image-based modelling and rendering, etc.

In scene understanding and virtual world exploration, viewpoint entropy is used to automatically calculate suitable positions and trajectories for a camera exploring a virtual world [2, 3, 6, 1, 12].

In Monte Carlo radiosity and global illumination, viewpoint complexity is used to improve the scene subdivision in polygons and the adaptive ray casting [4, 5, 14].

In image-based modelling, viewpoint entropy is used to compute a minimum optimized set of camera positions.

Among the metrics that have been introduced for the complexity calculation, the viewpoint entropy has been the most fruitful metric up to date [10]. Recently, it has been embedded to the field of volume visualization to compute the best n-views of a volumetric object. However, the viewpoint entropy computation can be very expensive, especially when a very complex scene and multiple viewpoints have to be evaluated.

In this paper, we will assess different alternatives for its calculation on several geometric models with increasing complexity. For our calculation, we will make use of the facilities of modern hardware cards such as OpenGL histogram and Occlusion query as well as the new symmetric bus PCI Express [16, 17].

V.N. Alexandrov et al. (Eds.): ICCS 2006, Part II, LNCS 3992, pp. 263 – 270, 2006.

## 2 Viewpoint Entropy

The viewpoint entropy, based on the Shannon entropy definition, was introduced in [7, 9] and is a measure of the information provided by a point of view.

The Shannon entropy of discrete random variable $X$, with values in the set $\{a_1, a_2, ..., a_n\}$, is defined as

$$H(X) = -\sum_{i=1}^{n} p_i \log p_i, \tag{1}$$

where $p_i = Pr[X = a_i]$, the logarithms are taken in base 2 and $0 \log 0 = 0$ for continuity. As $-\log p_i$ represents the information associated with the result $a_i$, the entropy gives the average information (or the uncertainty) of a random variable. The unit of information is called a bit.

To define viewpoint entropy we use as probability distribution the relative area of the projected faces (polygons) over a sphere of directions centered in the viewpoint. Thus, given a scene $S$ and a viewpoint $p$, the entropy of $p$ is defined as

$$I(S, p) = -\sum_{i=0}^{N_f} \frac{A_i}{A_t} \log \frac{A_i}{A_t}, \tag{2}$$

where $N_f$ is the number of polygons of the scene, $A_i$ is an approximation on a plane of the projected area of polygon $i$ over the sphere, $A_0$ represents the projected area of background in open scenes, and $A_t$ is the total area of the sphere. In a closed scene, the whole sphere is covered by the projected polygons, and thus $A_0=0$.

The maximum entropy is obtained when a certain point can see all the polygons with the same relative area. So, in an open scene, the maximum entropy is $\log (N_f + 1)$ and, in a closed scene it is equal to $\log N_f$. Among all the possible viewpoints, the best is the one that has maximum entropy, i.e. maximum information captured.

## 3 Techniques for Computing the Entropy

In order to compute the viewpoint entropy, we need the number of pixels covered for each visible triangle from a particular camera position. This number will give us the projected area. Next we analyze several techniques that allow us to compute those areas.

### 3.1 OpenGL Histogram

The OpenGL histogram was first used to compute the entropy in [11]. The OpenGL histogram let us analyze the colour information of an image. Basically, it counts the appearances of a colour value of a particular component. However, we can also use it to calculate the area of triangles that are visible from a viewpoint, without reading the buffer. Since version 1.2, OpenGL includes an extension called *glHistogram*. This extension is part of the image processing utilities. The OpenGL histogram is hardware-accelerated, although there are just a few graphics cards that actually support it (for instance, 3DLabs WildCat) and often is implemented in software.

In order to obtain the area of each visible triangle, we need to assign a different colour to each triangle. An important limitation is that histograms have a fixed size, normally of 256 different values. This is the most common value in many graphics cards. The *glGetHistogram* command returns a table that counts each colour value separated into channels. If we use the 4 RGBA colour channels, a 256 item table of 4 integer values will be returned, where each integer is the number of pixels this component has. Thus, if we want to detect a triangle, this should be codified using one single channel. This gives us a total of 1020 different values. That is to say, for channel R (1,0,0,0) up to (255,0,0,0), for channel G (0,1,0,0) up to (0,255,0,0), for channel B((0,0,1,0) up to (0,0,255,0) and finally for channel A (0,0,0,1) up to (0,0,0,255). The value (0,0,0,0) is reserved for the background.

Obviously the main drawback of this technique is that we need several rendering passes for objects with more than 1020 triangles. In each pass, we will obtain the area of 1020 different triangles. Using histograms with a higher number of items and making a rendering off-screen, will increase the number of colours and therefore making necessary less rendering passes. However, this possibility is outside the OpenGL specification and is hardware dependent. It was not possible for us to use a larger size histogram in the several graphics cards tested.

### 3.2 Hybrid Software and Hardware Histogram

The OpenGL histogram allows us to obtain the area of each visible triangle. However, as we said in the previous section, several rendering passes are needed for objects with more than 1020 triangles. Currently, new symmetric buses have appeared such as the PCI Express. In this new bus the buffer read operation is not as expensive as before. Therefore, it is possible to obtain a histogram avoiding making several rendering passes. The way to get it is very simple. A different colour is assigned to each triangle and the whole object is sent for rendering. Next, a buffer read operation is done, and we analyze this buffer pixel by pixel retrieving data about its colour. Using a RGBA colour codification with a byte value for each channel, up to 256*256*256*256 triangles can be calculated with only one single rendering pass. In Figure 1 we show an example of the entropy calculation using this method.

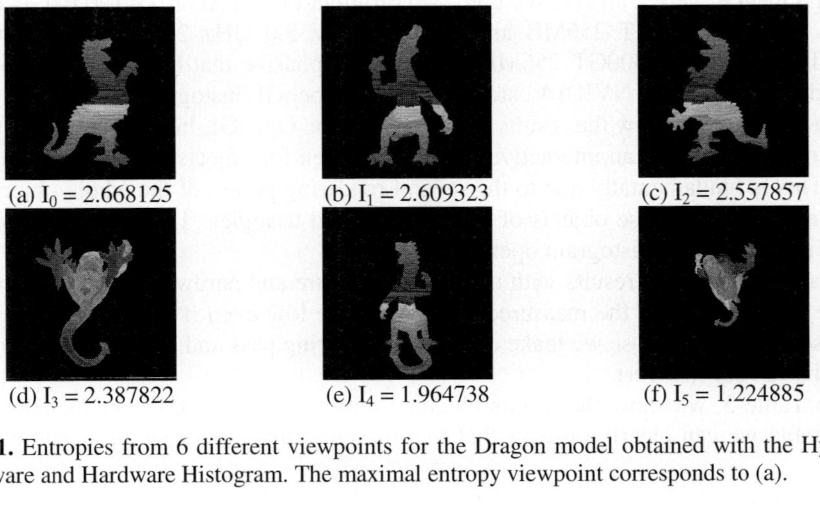

(a) $I_0 = 2.668125$ (b) $I_1 = 2.609323$ (c) $I_2 = 2.557857$

(d) $I_3 = 2.387822$ (e) $I_4 = 1.964738$ (f) $I_5 = 1.224885$

**Fig. 1.** Entropies from 6 different viewpoints for the Dragon model obtained with the Hybrid Software and Hardware Histogram. The maximal entropy viewpoint corresponds to (a).

### 3.3 Occlusion Query

This OpenGL extension is normally used to identify which scene objects are hidden by others, and therefore we shouldn't send them to render. In fact, what we do is just to render the bounding box of an object and, if it is not visible, the object is not sent for rendering. However, it can also be used to compute the area of the triangles that are visible from a particular camera position.

The OpenGL ARB_occlusion_query extension returns the number of visible pixels. In order to compute the area of each visible triangle from an object with this technique we will proceed as follows. First, the whole object is sent for rendering and the depth buffer is initialized. Second, we independently send each triangle for rendering. With this procedure it is necessary to make n + 1 rendering passes, n being the number of triangles in an object. We must mention that only in the first pass the whole geometry is rendered. In the following passes, one single triangle is rendered. However, a high number of renderings can significantly penalize this technique. In order to improve the results, this extension can be used asynchronously in contrast to its predecessor HP_occlusion_query. That is to say, it does not use a "stop-and-wait" execution model for multiple queries. This allows us to issue many occlusion queries before asking for the result of any one. But we must be careful with this feature because, as we mentioned above, this extension was not designed to deal with thousands of multiple queries. Thus, we can have some limitations depending on the graphics card.

## 4 Comparison

We calculated the viewpoint entropy from 6 camera positions, regularly distributed over a sphere that covers the object, using the different techniques we described above. In order to compare them, we measured the time needed to compute the entropy from those cameras. As test models, we used several models of different complexities (see Figure 2). All models were rendered in a 256x256 pixels resolution using OpenGL vertex arrays. We used two different PCs: a Xeon 2.4 GHz 1GB RAM with an ATI X800XT 256MB and a Pentium IV 3.0 GHz 2 1GB RAM with an NVIDIA GeForce 6800GT 256MB. We must emphasize that between the two analyzed GPUs, only the NVIDIA card supports the OpenGL histogram.

In Table 1, we show the results obtained with the OpenGL histogram. These times are too high to allow an interactive calculation, even for objects with a low complexity. This is fundamentally due to the several rendering passes of the whole object that we make when we use objects of several thousand triangles. The main cost component is the OpenGL histogram operation.

Table 2 shows the results with the hybrid software and hardware histogram. In this table we can see that the measured times are quite low even if the complexity is increased, mainly because we make one single rendering pass and the buffer read operation has a very low cost.

In Table 3, we show the results obtained with the Occlusion Query technique. In this table we can clearly observe that the measured times increase proportionally in

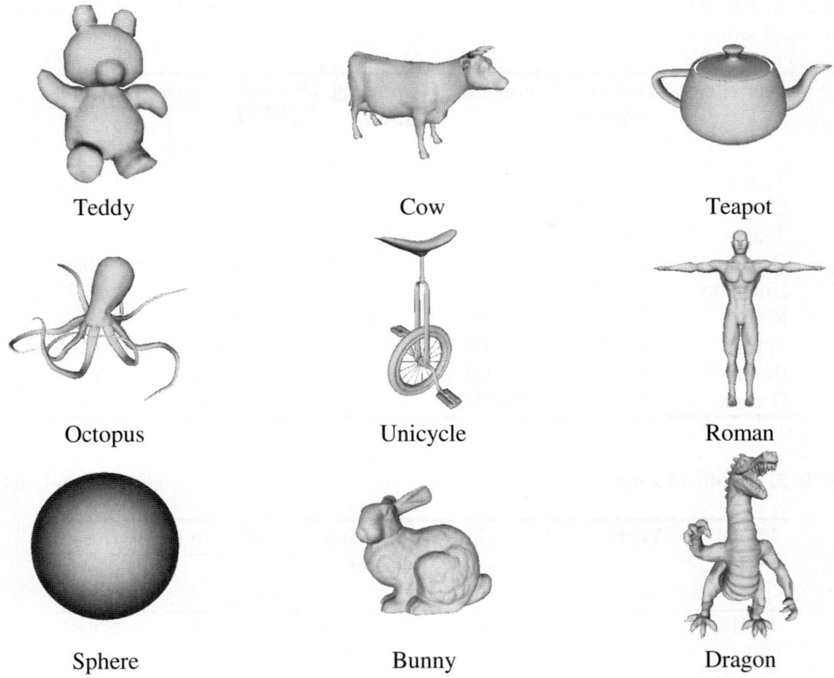


**Fig. 2.** Models used in our experiments

**Table 1.** Results obtained for the viewpoint entropy calculation with the OpenGL Histogram

| **Model** | **Vertices** | **Triangles** | **Rendering passes** | **OpenGL Histogram(ms)** | |
|---|---|---|---|---|---|
| | | | | GeForce 6800 GT | Radeon X800 XT |
| Teddy | 1,598 | 3,192 | 4 | 2,811.45 | - |
| Cow | 2,904 | 5,804 | 6 | 4,227.28 | - |
| Teapot | 3,644 | 6,320 | 7 | 4,927.65 | - |
| Octopus | 4,242 | 8,468 | 9 | 6,339.67 | - |
| Unicycle | 6,973 | 13,810 | 14 | 9,886.66 | - |
| Roman | 10,473 | 20,904 | 21 | 14,888.38 | - |
| Sphere | 15,314 | 30,624 | 31 | 22,136.10 | - |
| Bunny | 34,834 | 69,451 | 69 | 50,445.86 | - |
| Dragon | 54,296 | 108,588 | 107 | 80,029.94 | - |

relation to the complexity of the analyzed model. In the same way as the previous technique, the ratio remains unchanged here because the number of rendering passes is proportional to the number of triangles. A complete rendering of the object is only done at the first pass.

Finally, in Figure 3 we show as a summary a performance comparison among the different techniques. These results were obtained with the previously described NVIDIA card. We used an NVIDIA card because it fully supports all the techniques.

**Table 2.** Results obtained for the viewpoint entropy calculation with the Hybrid Software and Hardware Histogram

| **Model** | **Vertices** | **Triangles** | **Rendering passes** | **SW+HW Histogram(ms)** GeForce 6800 GT | Radeon X800 XT |
|---|---|---|---|---|---|
| Teddy | 1,598 | 3,192 | 1 | 11.66 | 16.62 |
| Cow | 2,904 | 5,804 | 1 | 13.36 | 19.10 |
| Teapot | 3,644 | 6,320 | 1 | 14.84 | 19.37 |
| Octopus | 4,242 | 8,468 | 1 | 17.28 | 20.69 |
| Unicycle | 6,973 | 13,810 | 1 | 18.53 | 23.24 |
| Roman | 10,473 | 20,904 | 1 | 24.12 | 29.85 |
| Sphere | 15,314 | 30,624 | 1 | 36.65 | 38.09 |
| Bunny | 34,834 | 69,451 | 1 | 57.91 | 67.04 |
| Dragon | 54,296 | 108,588 | 1 | 79.35 | 88.75 |

**Table 3.** Results obtained for the viewpoint entropy calculation with the Occlusion Query

| **Model** | **Vertices** | **Triangles** | **Rendering passes** | **Occlusion Query(ms)** GeForce 6800 GT | Radeon X800 XT |
|---|---|---|---|---|---|
| Teddy | 1,598 | 3,192 | 3,193 | 26.88 | 25.19 |
| Cow | 2,904 | 5,804 | 5,805 | 47.49 | 44.41 |
| Teapot | 3,644 | 6,320 | 6,321 | 50.88 | 48.31 |
| Octopus | 4,242 | 8,468 | 8,469 | 67.17 | 64.48 |
| Unicycle | 6,973 | 13,810 | 13,811 | 109.88 | 100.78 |
| Roman | 10,473 | 20,904 | 20,905 | 162.75 | 151.31 |
| Sphere | 15,314 | 30,624 | 30,625 | 238.09 | 221.42 |
| Bunny | 34,834 | 69,451 | 69,452 | 553.36 | 460.74 |
| Dragon | 54,296 | 108,588 | 108,589 | 829.75 | 665.33 |

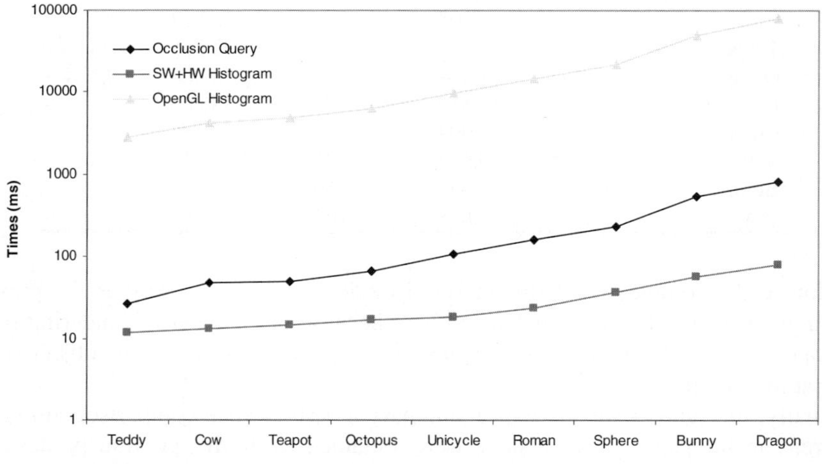


**Fig. 3.** Comparison of results obtained with the different analyzed techniques

Anyway, if we examine the proportions among the techniques with the ATI card, we can see that they are practically the same as with the NVIDIA card.

These results show clearly that by using the hybrid software and hardware histogram we can calculate the entropy in real time and even for complex objects (100,000 triangles), because times increase very slowly as complexity goes up. The next best technique is the Occlusion Query. Note that its cost grows as the object complexity increases, being unapproachable for complex objects for real time. Lastly, the OpenGL histogram technique is worst than the two others. This technique is useless for real time, unless we use objects of low complexity (1,000 triangles).

## 5 Conclusions

The viewpoint entropy is a metric that has been mainly used to determine the best viewpoint of a 3D object. In this paper we studied several hardware assisted techniques to allow computing the viewpoint entropy in an efficient way. Among the different analyzed techniques, the viewpoint entropy calculation with the hybrid software and hardware histogram has the best performance, followed by the occlusion query based technique. By using the hybrid software and hardware histogram technique we can practically achieve the entropy calculation in real time even for complex objects, while occlusion query technique allows us to obtain only interactivity.

We must take into account that the performance of the hybrid software and hardware histogram technique depends on the analysis of pixels done by the CPU and the read operation of the PCI Express bus. We also did some tests using higher resolutions, for example: 960x960, and we observed that the times for the occlusion query are constant, but even in higher resolutions the hybrid software and hardware histogram technique gets better results than occlusion queries. The proportion is not as higher as before but still is significantly better. For our goal, we think that the resolution used in our experiments (256x256) is enough to obtain accurate results in the viewpoint entropy calculation.

## References


1. C. Andújar, P. P. Vázquez, M. Fairén, Way-Finder: guided tours through complex walkthrough models, Computer Graphics Forum (Eurographics 2004), 2004
2. P. Barral, G. Dorme, D. Plemenos. Scene understanding techniques using a virtual camera. Eurographics 2000, Interlagen (Switzerland), August 20-25, 2000, Short papers proceedings
3. P. Barral, G. Dorme, D. Plemenos. Visual understanding of a scene by automatic movement of a camera. International Conference GraphiCon'99, Moscow (Russia), August 26 – September 3, 1999
4. V. Jolivet, D. Plemenos, M. Sbert, A pyramidal hemisphere subdivision method for Monte Carlo radiosity, Eurographics 99 Short Papers proceedings
5. M. Feixas, An Information Theory Framework for the Study of the Complexity of Visibility and Radiosity in a Scene. PhD thesis, Technical University of Catalonia, 2002
6. D. Plemenos. Exploring Virtual Worlds: Current Techniques and Future Issues. International Conference GraphiCon'2003, Moscow (Russia), September 5-10, 2003

7. P. P. Vázquez, M. Feixas, M. Sbert, and W. Heidrich. Viewpoint Selection Using Viewpoint Entropy. Vision, Modeling, and Visualization 2001 (Stuttgart, Germany), pp. 273-280, 2001
8. P. P. Vázquez, M. Feixas, M. Sbert, and A. Llobet. Viewpoint Entropy: A New Tool for Obtaining Good Views for Molecules. VisSym '02 (Eurographics - IEEE TCVG Symposium on Visualization) (Barcelona, Spain), 2002
9. P. P. Vázquez, PhD thesis, On the Selection of Good Views and its Application to Computer Graphics. Technical University of Catalonia, 2003
10. P. P. Vázquez, M. Feixas, M. Sbert, and W. Heidrich. Automatic View Selection Using Viewpoint Entropy and its Application to Image-Based Modeling. Computer Graphics Forum, December 2003
11. P. P. Vázquez, M. Sbert. On the fly best view detection using graphics hardware 4th IASTED International Conference on Visualization, Imaging, and Image Processing (VIIP 2004)
12. P. P. Vázquez and M. Sbert. Automatic indoor scene exploration. In International Conference on Artificial Intelligence and Computer Graphics, 3IA, Limoges, May 2003
13. J. Rigau, M. Feixas, and M. Sbert. Information Theory Point Measures in a Scene. IIiA-00-08-RR, Institut d'Informàtica i Aplicacions, Universitat de Girona (Girona, Spain), 2000
14. J. Rigau, M. Feixas, and M. Sbert. Entropy-Based Adaptive Sampling. Graphics Interface 2003 (Halifax, Canada), June 2003
15. M. Sbert, M. Feixas, J. Rigau, F. Castro, and P. P. Vázquez. Applications of Information Theory to Computer Graphics. Proceedings of 5th International Conference on Computer Graphics and Artificial Intelligence, 3IA'02 (Limoges, France), pp. 21-36, 2002
16. M. Segal, Kurt Akeley. The OpenGL Graphics System: A Specification (Version 2.0 – October 22, 2004). Silicon Graphics, Inc., 2004
17. A.Wilen, J.Schade, R.Thornburg. Introduction to PCI Express. A Hardware and Software Developer's Guide. Intel Press, 2003

# 3D Object Repair Using 2D Algorithms

Pavlos Stavrou[1,*], Pavlos Mavridis[1], Georgios Papaioannou[2],
Georgios Passalis[1], and Theoharis Theoharis[1]

[1] National Kapodistrian University of Athens , Department of Informatics
{p.stavrou, grad0751, passalis, theotheo}@di.uoa.gr
http://graphics.di.uoa.gr

[2] Athens University of Business and Economics, Department of Informatics
gepap@aueb.gr
http://www.aueb.gr/users/gepap/

**Abstract.** A number of three-dimensional algorithms have been proposed to solve the problem of patching surfaces to rectify and extrapolate missing information due to model problems or bad geometry visibility during data capture. On the other hand, a number of similar yet more simple and robust techniques apply to 2D image data and are used for texture restoration. In this paper we make an attempt to bring these two-dimensional techniques to the 3D domain due to their obvious advantage of simplicity and controllability. Creating a depth image with the help of a voxelisation algorithm will allow us to apply a variety of image repair algorithms in order to mend a 3D object. The use of three variations of the texture synthesis algorithm is investigated. Constrained texture synthesis and its variations using the Haar wavelet and image decomposition methods are also proposed in order to preserve patterns appearing on the object while trying to maintain its geometry intact.

## 1 Introduction

The repair of incomplete three-dimensional objects is a common problem in computer graphics. We subjectively regard a 3D model as incomplete when it exhibits missing surfaces and patterns. Incomplete three-dimensional objects may result from a 3D scanning procedure or any other three-dimensional data acquisition method or may be models of actually damaged real objects, in which case we seek to rectify their original shape. Data corruption or bad modelling are two other problems where shape repair may be applied to fix the model geometry.

In cases, where detail preservation is important, a more general and efficient method is required which simultaneously preserves the geometry of the object while detecting and reproducing patterns appearing on its surface. We, therefore, propose the use of 2D algorithms on 2D mapping of 3D objects to enhance performance and retain patterns and geometry.

Given an incomplete 3D object, a method for its repair based on two-dimensional algorithms is introduced in this paper. In order to apply 2D algorithms for the repair of a 3D object we need to represent the object in a two-dimensional form, which is

* The Authors would like to acknowledge financial support from the Hellenic General Secretariat for Research and Technology (PENED).

V.N. Alexandrov et al. (Eds.): ICCS 2006, Part II, LNCS 3992, pp. 271 – 278, 2006.

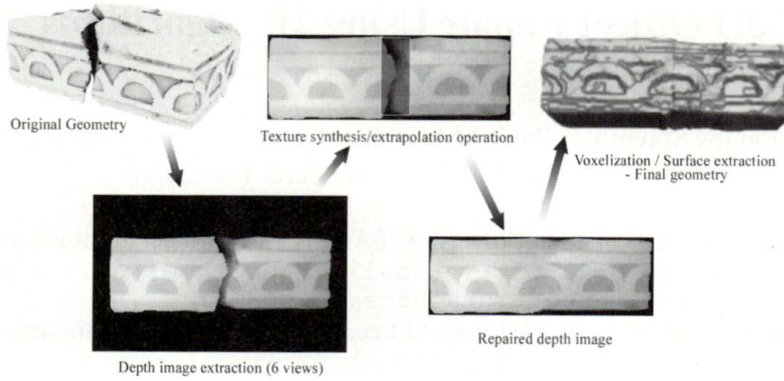


**Fig. 1.1 .** Procedure of a 3D object repair using a 2D algorithm

capable of preserving gradient information for both high and low spatial frequency bands. We extract depth (Z-range) images for this task and then apply algorithms for image repair to them. Subsequently the depth information is transformed back into three-dimensional data via voxelisation [10,12]. The final volumetric model represents the repaired 3D object (Fig 1.1).

### 1.1 Related Work in Two Dimensions

A variety of algorithms have been proposed for image repair. The *in-painting* algorithm, presented by Bertalmio et al. [1], is probably the best known. Given an initial image, which contains areas where information is corrupt, the goal of this algorithm is to produce an image that contains a fair approach of the missing information. Bertalmio et al. [2] suggests an improved method that simultaneously deals with both structure and texture. The initial image is decomposed into two coefficients for each pixel, the first representing the structure of the image and the second storing high detail texture information. The in-painting algorithm is then applied to the structure coefficient and texture synthesis is applied to the texture coefficient before the image is composed back together. Texture synthesis [3] is a common method for image reconstruction and is based on filling the missing pixels of the corrupt image by searching for similar patterns on the image.

### 1.2 Related Work in Three Dimensions

In the three-dimensional case, the proposed solutions mainly rely on two different approaches, one being hole filling and the other being surface blending.

The most popular hole-filling techniques use the implicit surface approach [4] to represent objects that are to be repaired. A local approximation of the surface is performed by fitting a function that best describes the geometry, including the missing areas. One way to achieve this is through a sum of *Radial Basis Functions* (RBF). Once a RBF is found, the surface is automatically completed. Davis at al. [5] introduced volumetric diffusion for hole filling. Their algorithm first constructs a volumetric signed distance function around the surface samples. Then an iterative Gaussian convolution propagates adjacent distance values to fill the holes. Verdera et al. [6] propose the use of the in-painting method mentioned above to fill holes in a mesh by first representing the surface of interest in implicit form.

Surface blending techniques aim to finding a blend surface that smoothly connects two or more surfaces. One of the most interesting approaches for surface blending is the combined subdivision schemes method by Levin [7]. This particular approach allows boundary constraints to be taken into account during the generation of the blend surface points. High quality blend surfaces can be constructed using variational subdivision schemes [8]. Such surfaces minimize an energy metric of the surface relying on its curvature. Application of this method to triangular surfaces is called Discrete Fairing [9].

## 2 Method Overview

As mentioned in the introduction, we have opted to work with two-dimensional algorithms to repair three-dimensional objects. In order to work with such algorithms a two-dimensional representation of an object is required. The use of depth images of three-dimensional objects is adopted in this paper. The depth image creation method we selected to use is based on the setup of the voxelisation method described by Karabassi et al. [10], acquiring depth images from the 6 $z$-buffers, a pair for each of the three orthocanonical directions. Depth images encode geometric detail as intensity values, allowing a fast implementation of the selected two-dimensional algorithms. In order for the repair method to work correctly, we must acquire at least six depth images, which correspond to the faces of the bounding cube, and the object must be aligned so as to maximally expose the missing part.

After the acquisition of the depth image, the constrained texture synthesis algorithm is used for the repair of the object. We have experimented with three different variations of the method:

- Constrained texture synthesis
- Constrained texture synthesis using Haar wavelets
- Constrained texture synthesis using image decomposition

In the first and simplest variant, the algorithm is directly applied to the raw depth-image values without any pre-processing. In the second variant, the Haar wavelet decomposition [11] was applied to the depth image prior to using the texture synthesis algorithm. Texture synthesis is then performed separately on selected bands of the resulting image. The reconstructed depth image is produced by the inverse Haar wavelet transform of the synthesised sub-images. Finally, in the third variant, image decomposition transform is applied to the depth image in advance, thus acquiring the $U$ and $V$ coefficients of the image, where the $U$ coefficient corresponds to the structure of the image and the $V$ coefficient corresponds to the texture. Texture synthesis is applied to the $V$ coefficient and an interpolation method to the $U$ coefficient before the image is reconstructed using the inverse image decomposition transform. Detailed descriptions of the object are presented in section 3 of the paper.

Having obtained the restored depth-images, the repaired three-dimensional object is calculated by voxelisation according to the range information stored in them, using the hardware-based voxelisation of Passalis et al. [12].

# 3 Texture Synthesis for Image Repair

The first algorithm proposed for texture synthesis is *texture synthesis by non-parametric sampling* by Efros and Leung [3]. This algorithm composes one pixel at a time. For each pixel *p*, it extracts its neighbourhood and searches in the texture sample for a similar pattern. The texture sample is defined as a part of the initial image or the whole image itself and the neighbourhood of a pixel is defined as a square window, which contains all the known neighbouring pixels of *p*. The search is exhaustive as all neighbourhoods in the image are compared and a distance factor is calculated. Next, the algorithm randomly selects one of the minimum-distance neighbourhoods and copies the corresponding pixel into pixel *p*. The distance factor uses a number of metrics. One of the most popular is the *mean square distance* metric:

$$d = \frac{1}{N}\sum \left\| x_{ij} - y_{ij} \right\|^2 \tag{1}$$

where $x_{ij}$ ($y_{ij}$) is the value of the pixel at position $(i,j)$, in neighbourhood $x(y)$ and $N$ is the number of $(i,j)$ positions in which both pixels $x_{ij}$ and $y_{ij}$ are not missing from the image.

A further improvement of the mean square distance metric is the weighted mean square distance metric, which weights the differences in Eq. 1 according to the distance of the elements from the central pixel. Therefore, pixels that are closest to the centre have greater contribution than the ones further away. As a weighting function we use the Gaussian distribution function. This is the distance metric adopted in this paper.

## 3.1 Constrained Texture Synthesis

Constrained texture synthesis is an improvement to the texture synthesis method by non-parametric sampling. Given the sample of a texture, which contains gaps, we attempt to fill these gaps in such a manner that the produced texture and resulting image do not exhibit any discontinuities, especially when dealing with well-defined shapes as patterns. The main idea is to apply texture synthesis as above, introducing a limit to the values that certain pixels may acquire. The algorithm processes the whole texture sample and composes only the missing pixels. For better results, the pixel with the largest number of neighbours is composed first. In this paper, a new constraint is introduced, which permits only tileable new textures to be produced. This means that a uniform texture without discontinuities can be formed by repetition of the reconstructed image. Instead of repeating a square-window pattern search at the current composed pixel, a symmetric search around the current location is performed up to a pre-defined distance in order to detect and bridge textural details across the gap. Results of using this method are presented in figure 3.1.

## 3.2 Constrained Texture Synthesis Using Haar Wavelets

Let $H$ be the Harr wavelet transform [11]. For four adjacent pixels in the image aligned as $[\mathbf{L},\mathbf{R}]:\mathbf{L}=[a,c]^T,\mathbf{R}=[b,d]^T$, a vector $[a,b,c,d]^T$ is constructed and is multiplied with matrix $H$ to produce a vector $[ll,lh,hl,hh]^T$ . Tilling and applying this square window across the image produces four bands (images) LL, LH, HL and HH each of which has the resolution of one quarter of the original image. This is one step of the Haar wavelet transform. We can choose to continue applying the transform recursively either only to the LL band or to all resulting bands (Fig 3.2a).

The HH, HL and LH bands of the image now only contain repeated patterns of the image and are free of smooth (slow) intensity transitions where the texture synthesis algorithm would fail. The LL band works as a box filter and contains only the low frequencies of the image. Increasing the number of steps of the transform will result in the LL band containing very few details of the original image and only the color gradient will appear, allowing us to use a simple interpolation method for the missing pixels. We apply the constrained texture synthesis algorithm to the other three bands to fill the missing textures.

Finally we apply the inverse Haar wavelet transform to the synthesized image to acquire the reconstructed depth image (Fig 3.2b).

## 3.3 Constrained Texture Synthesis Using Image Decomposition

The image decomposition technique allows us to decompose the original image ($f$) into two coefficients. The two coefficients represent different attributes of the original image. The $v$ coefficient holds the noise and object texture of the image, while the $u$ coefficient represents the shape and large intensity variations, which correspond to the main geometric shape in our case (Fig 3.3). Combined, the two coefficients give the original image:

$$f = u + v \tag{2}$$

Rubin, Osher and Fatemi proposed [13] the minimization of the following equation for the construction of the $u$ and $v$ coefficients, where $\lambda$ is a free parameter which corresponds to the desired coherency between the input image and the $u$ image.

$$\inf_u F(u) = \iint \|\nabla u\| + \lambda \iint \|f - u\|^2 \, dxdy \tag{3}$$

Vese et al. [14] proposed an improvement of the above model, which achieves better separation of texture and noise:

$$\inf_{u,g_2,g_2} F(u,g_1,g_2) = \iint \|\nabla u\| + \lambda \iint \|f - u - \partial_x g_1 - \partial_y g_2\|^2 dxdy + \mu \iint \|\nabla u\| \sqrt{g_1^{\ 2} + g_2^{\ 2}} \, dxdy \tag{4}$$

where $g_1 = \frac{1}{2\lambda} \frac{u_x}{\|\nabla u\|}$ and $g_2 = \frac{1}{2\lambda} \frac{u_y}{\|\nabla u\|}$.

The minimization of equation (4) with respect to $u$ , $g_1$ and $g_2$ results in an Euler-Lagrange differential equation system.

Having now decomposed the depth image into the structure ($u$) and texture ($v$) coefficients, we apply the constrained texture synthesis algorithm to the texture coefficient in order to fill out the missing detail information. As far as the structure coefficient is concerned, since it contains only slow variations, a simple interpolation

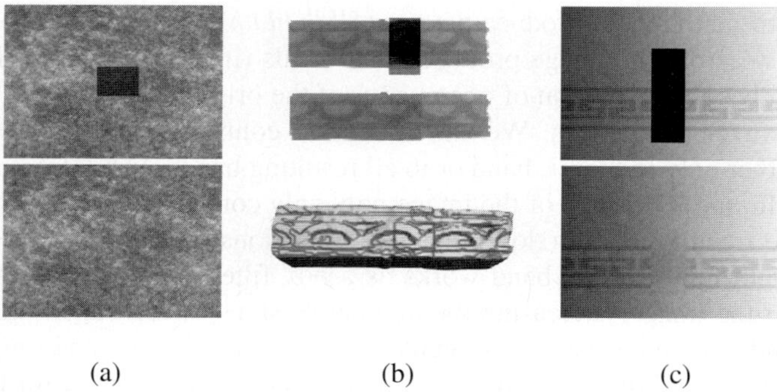

Fig. 3.1. Constrained texture synthesis. (a) Input and repaired image. Black area represents the missing pixels. (b) Repair of an ancient artifact. One set of input and output depth images is shown. The object is voxelised from the rectified depth images. (c) Repair of a curved wall depth image. The gradient prevents the algorithm from detecting similar pixel regions.

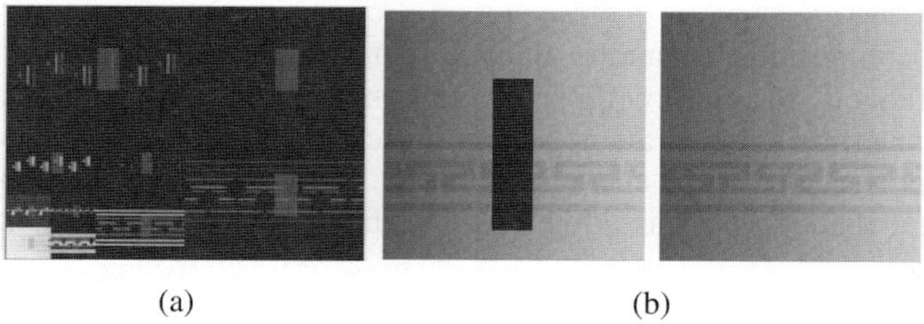

Fig. 3.2. Constrained texture synthesis using Haar wavelets. (a) Three-step Haar wavelet transform of an image. (b) Image reconstruction using Haar wavelet method.

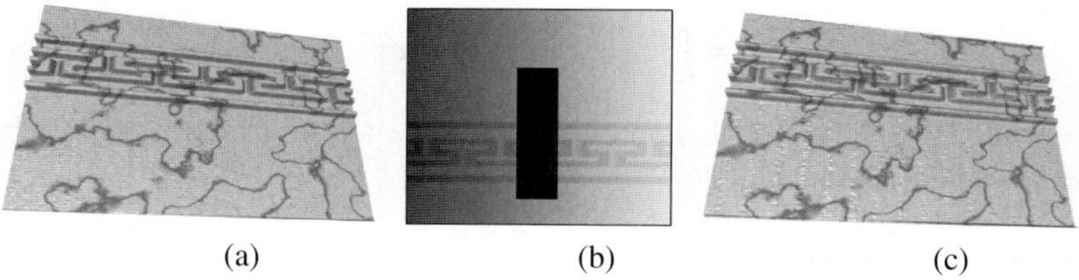

Fig. 3.3. Constrained depth image synthesis using image decomposition. (a) The original 3D surface. (b) A damaged depth image of (a). (c) The final reconstructed surface after the voxelisation of the repaired and re-composed depth images.

method is adequate for the restoration of the missing information, provided the gaps are not disproportionately large compared to the object. For this task, we can use either the RBFs or the image in-painting method [4][2]. After the two coefficients have been repaired, the image is re-composed (Eq. 2) to form the restored depth information (Fig 3.3).

## 4 Results

The constrained texture synthesis algorithm produces perfect results when applied to images with repeated structural patterns, which conform to the Markov random fields model. However when applied to depth images (Fig 3.1b), the resulting image exhibits discontinuities at the depth values. These are of course translated as slope discontinuities on the reconstructed three-dimensional surface, after voxelisation. This is an inherent problem of the texture synthesis algorithm, which blindly copies pixel values to complete the missing information and disregards changes in the overall structure. In Fig 3.1c, where the depth image of a curved wall is shown, we observe another failure of the algorithm. Following a vertical path to complete missing pixels and following a horizontal path (both follow the rule of the missing pixel with the most neighbours to be completed first) can produce different results due to the gradient direction of the image, which is a result of depth value variance. This issue was addressed by simply using both approaches and each time selecting the most suitable.

When applied to the same test case, the constrained texture synthesis using wavelets achieves far better results due to the isolation of the patterns in the high-frequency Haar coefficients. However, we can still observe a small amount of noise in the reconstructed image (Fig 3.3). Texture synthesis tends to perform worse in high levels of wavelet decomposition due to the decrease of the spatial resolution of the wavelet bands. At the same time, for the proper application of the interpolation algorithm to repair the LL, many iterations of the wavelet analysis are required. The problem of automatically selecting the level of the wavelet decomposition, so that both texture synthesis and interpolation methods work, needs further investigation.

Finally, the constrained texture synthesis method using image decomposition presents the best results so far, as the texture of the $v$ coefficient is properly synthesized and the patterns are correctly pieced together. Slight discontinuities appear when repairing the $u$ coefficient using our interpolation, indicating that a more accurate interpolation method is needed to produce more consistent results.

It must be noted that in all cases the produced results were symmetrically correct to the original object. Patterns exhibited on the input object were coherently reproduced, during the repair procedure, to form an object whose texture was smoothly reconstructed (Figures 3.1b, 3.2b, 3.3).

## 5 Conclusion and Future Work

We presented a method that uses texture synthesis on depth images to repair a 3D object. We utilized three methods of texture synthesis which worked well on specific cases. None of the three produced very high quality results for all cases of 3D objects. However, texture synthesis results were impressive, considering their simplicity and speed of execution. The combined use of these methods, based on the characteristics. of each object, will greatly increase the efficiency of 3D object repair using 2D algorithms. In addition, one should consider extensions of the presented algorithms, such as more accurate interpolation techniques, to increase the accuracy of the results.

## References


1. Marcelo Bertalmio, Guillermo Sapiro, Vicent Caselles, and Coloma Ballester. Image inpainting. In Kurt Akeley, editor, *SIGGRAPH 2000, Computer Graphics Proceedings*, pages 417.424. ACM Press / ACM SIGGRAPH / Addison Wesley Longman, 2000.
2. M. Bertalmio, L. Vese, G. Sapiro, and S. Osher. Simultaneous structure and texture image inpainting. UCLA CAM Report, 02(47), 2002.
3. Alexei A. Efros and Thomas K. Leung. Texture synthesis by non-parametric sampling. In *ICCV (2)*, pages 1033.1038, 1999.
4. Jonathan C. Carr, Richard K. Beatson, Jon B. Cherrie, Tim J. Mitchell, W. Richard Fright, Bruce C. McCallum, and Tim R. Evans. Reconstruction and representation of 3D objects with radial basis functions. *SIGGRAPH 2001, Computer Graphics Proceedings*, pages 67.76. ACM Press / ACM SIGGRAPH, 2001.
5. Davis, J., Marschner, S. R., Garr, M., and LevoyY, M. 2002. Filling holes in complex surfaces using volumetric diffusion. In Proceedings of the 1st International Symposium on 3D Data Processing Visualization and Transmission (3DPVT-02), IEEE Computer Society, Los Alamitos, CA, G. M. Cortelazzo and C. Guerra, Eds., 428.438.
6. Verdera, J., Caselles, V., Bertalmio, M., and Sapiro, G. 2003. Inpainting surface holes. In 2003 International Conference on Image Processing 2003 ICIP.
7. Adi Levin. Combined subdivision schemes - an introduction. In *SIGGRAPH course notes*, Subdivision for Modeling and Animation, pages 129.132, 2000.
8. Leif Kobbelt. Variational subdivision schemes. In *SIGGRAPH course notes*, Subdivision for Modeling and Animation, pages 159.164, 2000.
9. Leif Kobbelt. Discrete fairing and variational subdivision for freeform surface design. The Visual Computer, 16(3-4):142.158, 2000.
10. Karabassi E.A., Papaioannou G., and T. Theoharis. A fast depth-buffer-based voxelisation algorithm. ACM Journal of Graphics Tools, pages 5.10, 1999.
11. E. Stollnitz, T. DeRose and D. Salesin. Wavelets for Computer Graphics : Theory and Applications. Morgan Kaufmann Publishers Inc., 1996
12. G. Passalis, I. A. Kakadiaris, and T. Theoharis. Efficient Hardware Voxelisation. In *Proceedings of Computer Graphics International*, pp. 374-377, Crete, Greece, June 2004.
13. L. Rudin, S. Osher, and E. Fatemi. Nonlinear total variation based noise removal algorithms. Phys.~D, 60 (1992), pp. 259--268.
14. L. Vese and S. Osher. Modeling textures with total variation minimization and oscillating patterns in image processing, 2002. UCLA CAM Report, 02(20), 2002.

# Extraction of Ridges-Valleys for Feature-Preserving Simplification of Polygonal Models

Soo-Kyun Kim[1], Sun-Jeong Kim[2], and Chang-Hun Kim[3,*]

[1]Samsung Electronics
nicesk@gmail.com
[2] Div. of Info. Eng. and Telecom, Hallym University
sunkim@hallym.ac.kr
[3] Dept. of Computer Science and Engineering, Korea University
chkim@korea.ac.kr
Tel.: +82-2-3290-3199; Fax: +82-2-953-0771

**Abstract.** We propose a new method for simplifying polygonal models while retaining salient features. We extract ridge and valley features defined by means of curvature derivatives on the model surface. We combine this extraction with a simplification technique so as to construct coarse models with features preserved. Experimental results have better quality and smaller geometric error than these of previous methods.

## 1 Introduction

Over the last decade, many simplification algorithms have been developed that transform a 3D polygonal model into a simpler version, while preserving its original shape and appearance. Although existing techniques produce plausible results in many cases, the semantic or high-level meaning of the detail of models are often disregarded, because perception of these features is subjective. For example, in the dragon model of Figure 3, regions such as the sharp backbone can be semantically crucial, although they have which will affect small geometric errors any attempt to simplify them.

Some approaches [10, 13] allow users to guide the simplification of semantic features; but such methods require users to select a threshold, which cannot be found intuitively, and users have to discover the threshold value which allow a reasonable set of semantic features to be found. This tedious process requires a lot of skill and time.

We believe that it is possible to get a better result by using a new feature-extraction technique to find semantic features in dense polygonal models. This technique identifies well-connected ridges and valleys that represent salient features. Then we compute a simplifying criterion based on quadric and feature quadric error. Finally, a new quadric error metric is generated by combining these two metrics.

* The Corresponding author.

V.N. Alexandrov et al. (Eds.): ICCS 2006, Part II, LNCS 3992, pp. 279–286, 2006.

## 2 Previous Work

Many successful mesh-simplification algorithms [14, 16] are based on iterative edge contraction [4, 5]. They allow vertices to be moved so as to be controlled to preserve the shape or appearance of the original model. Garland [5] and Hoppe [6] use an extended error metric that combines geometric and surface attributes such as normals, colors and texture coordinates. This extended method produces a more accurate and a higher-quality mesh than geometric information alone. Olano et al.[3] use a texture error measurement algorithm to control simplification.

Kho and Garland [10] introduced a user-guided simplification technique and Pojar et al. [13] have presented a method for the user-controlled creation of multiresolution meshes. But these methods still require the user to select regions of geometry to be preserved.

Avioding the need for user-guided features, Watanabe and Belyaev [17] introduced a method for the stable detection of salient curvature extrema on surfaces approximated by dense meshes, which preserves triangles of significant curvature using the Garland-Heckbert method [4]. Hubeli et al. [7] proposed a method to find feature edges in 3D meshes that gives weights to edges based on the difference between two normals, and then fits polynomials to the lines of intersection between a parameter plane and a mesh. A drawback of this method is that users must select a threshold value that cannot be found intuitively. Besides, neither method eliminates the necessity of extracting feature points from the 3D mesh. Yoshizawa et al. [19] have extracted feature lines by using local polynomial fitting of triangulated meshes, which they combined with a quadric-based simplification technique. Kim et al. [8] showed that discrete curvature can be a good criterion for a simplification that preserves the shape of an original model. Walter and Healey [18] have proposed applying feature-preserving mesh simplification to a multidimensional dataset.

## 3 Feature Extraction

Our approach to finding line-type features is based on the method of extracting ridges and valleys due to Kim et al. [9]. As a foundation for feature extraction, we will briefly discuss the basics of finding ridge and valley lines, which are curves on a surface formed by the set of loci of points at which the principal curvature has an extremum along its curvature line.

Consider a dense smooth oriented surface. Let $\kappa_{max}$ and $\kappa_{min}$ be the maximal and minimal curvatures respectively ($\kappa_{max} > \kappa_{min}$), with $\mathbf{t}_{max}$ and $\mathbf{t}_{min}$ as the corresponding principal directions. The derivatives of $\kappa_{max}$ and $\kappa_{min}$ in directions $\mathbf{t}_{max}$ and $\mathbf{t}_{min}$ are given by the zero-crossing of $e_{max} = \nabla\kappa_{max} \cdot \mathbf{t}_{max}$ and $e_{min} = \nabla\kappa_{min} \cdot \mathbf{t}_{min}$. Note that we exclude umbilical points, at which $\kappa_{max} = \kappa_{min}$. The extrema of principal curvatures along their curvature directions can be used to define ridges and valleys. The ridges are given by

$$\nabla e_{max} \cdot \mathbf{t}_{max} < 0, \quad \kappa_{max} > |\kappa_{min}|$$

and valleys by

$$\nabla e_{min} \cdot \mathbf{t}_{min} > 0, \quad \kappa_{min} < -|\kappa_{max}|.$$

We aim to detect ridges and valleys from an input polygonal mesh, not a smooth surface. First we estimate the curvatures and their derivatives directly from the mesh by employing a local MLS (moving-least-squares) approximation technique to fit the discrete surface. MLS approximations to surfaces usually involve two phases: determining the local tangent plane and then computing a local polynomial over that plane. But our modified technique uses a substitute for the tangent plane, with a corresponding reduction in time-complexity. We quickly estimate curvatures and their derivatives at each vertex using a local polynomial that moves over the plane. Then, we detect ridge and valley vertices by checking for zero-crossings of the extrema of principal curvatures at mesh edges. Finally, the ridge and valley vertices that we have found are connected in the direction of principal curvature.

We shall omit specific references to valleys in this paper, since the methods for detecting ridges and valleys are essentially duals of one another.

## 4 New Quadric Error Metric

It is hard to measure an accurate feature edge on a highly curved surface using only a quadric error metric. We therefore introduce a new quadric error metric (NQEM), which is a combination of a basic error metric with a feature quadric error metric. It preserves features which are removed by a basic error metric.

### 4.1 Basic Quadric Error Metric

The quadric error metric uses a quadric form [4] as its distance error metric which is computed as the weighted sum of squared distances. On each face $f$ of the original mesh, a quadric $Q^f(\mathbf{p})$ is defined, which equal to the squared distance from a point $\mathbf{p} \in \mathbf{R}^3$ to the plane containing the face $f$. Each vertex $v$ of the original mesh is assigned the sum of quadrics on its adjacent faces weighted by face area:

$$Q^v(\mathbf{p}) = \sum_{v \in f} area(f) \cdot Q^f(\mathbf{p}).$$

Let us now derive $Q^f(\mathbf{p})$ for a given face $f = (v_1, v_2, v_3)$. The signed distance of $\mathbf{p}$ to the plane containing $f$ is $\boldsymbol{n}^T\mathbf{p}+d$, where the face normal $\boldsymbol{n} = (\mathbf{p}_2 - \mathbf{p}_1) \times (\mathbf{p}_3 - \mathbf{p}_1)/\|(\mathbf{p}_2 - \mathbf{p}_1) \times (\mathbf{p}_3 - \mathbf{p}_1)\|$ and the scalar $d = -\boldsymbol{n}^T\mathbf{p}_1$. These parameters can be obtained by solving the linear system,

$$\begin{pmatrix} \mathbf{p}_1^T & 1 \\ \mathbf{p}_2^T & 1 \\ \mathbf{p}_3^T & 1 \end{pmatrix} \begin{pmatrix} \boldsymbol{n} \\ d \end{pmatrix} = \begin{pmatrix} 0 \\ 0 \\ 0 \end{pmatrix}$$

with the additional constraint that $\|\boldsymbol{n}\| = 1$.

Therefore, the squared distance between point $\mathbf{p}$ and the plane containing $f$ is

$$Q^f(\mathbf{p}) = (\boldsymbol{n}^T\mathbf{p} + d)^2 = \mathbf{p}^T(\boldsymbol{n}\boldsymbol{n}^T)\mathbf{p} + 2d\boldsymbol{n}^T\mathbf{p} + d^2,$$

which can be represented as a quadric functional $\mathbf{p}^T\mathbf{A}\mathbf{p} + 2\mathbf{b}^T\mathbf{p} + c$, where $\mathbf{A}$ is a symmetric $3 \times 3$ matrix, $\mathbf{b}$ is a column vector of size 3, and $c$ is a scalar. Thus,

$$Q^f = (\mathbf{A}, \mathbf{b}, c) = (\boldsymbol{n}\boldsymbol{n}^T, d\boldsymbol{n}, d^2)$$

which can be stored using 10 coefficients (6+3+1). The advantage of this representation is that the quadric is obtained as a simple linear combination of these coefficient vectors.

Finally, a quadric for the distance error metric of a new vertex $v$ after an edge collapse $(v_1, v_2) \rightarrow v$ is formulated as the sum of the quadrics on its adjacent faces, weighted by face area for vertex $v_1$ and $v_2$:

$$Q^v = \sum_{v_1 \in f} area(f) \cdot Q^f + \sum_{v_2 \in f} area(f) \cdot Q^f. \tag{1}$$

### 4.2 Feature Edge Error Metric

We extend Lindstrom's approach [11] to reconstructing the feature lines of an original model by preserving surface boundaries. A feature error metric is defined to be the distance between the plane of the associated feature edge $e_{fe}$ and a point. We define the normal to the feature edge $e_{fe}$ by averaging the normals of two neighboring triangles that share that edge. We then construct a plane $m_{e_1}$ which is close to orthogonal to the edge. This plane is spanned by the edge $e_1^n = (n_1^f + n_2^f)/2$, and $m_{e_1} = e_1 \times e_1^n$, where $e_1^n$ is the computed normal at the edge $e_1$ (see Fig. 1).

Using these estimates, we define the feature edge error metric for feature edge $e_{fe}$ of a vertex as the sum of the squared distances from a point $v$ to the planes containing the neighboring feature edges:

$$Q^{fe} = \sum_{e \in fe(v)} (m_e)(m_e)^T. \tag{2}$$

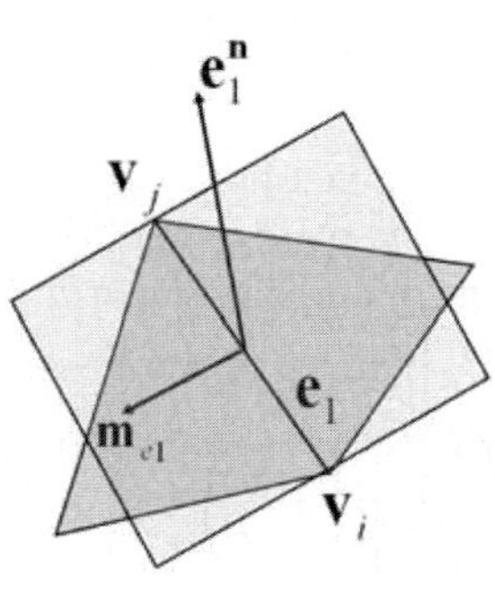


**Fig. 1.** Computation of feature edge error metric

## 5 Simplification Algorithm

Our simplification algorithm is based on iterative edge collapses using an error metric which is a combination of the surface error and the feature edge error, based on geometric correspondence in 3D. The error metric is defined as the sum:

$$Q^v = \sum_{v_n \in ngbh(v)} Q^{v_n} + \sum_{fe_n \in fe(v)} Q^{fe_n}. \tag{3}$$

where the geometric error $Q^{v_n}$ is the squared distance from a vertex on the plane $\mathbf{v} \subset \mathbf{R}^3$ containing the incident triangle, and the feature edge error $Q_{fe_n}$ is the distance between a point and orthogonal vector calculated from edge $e_{fe}$. We start with a polygonal model and perform a simplification algorithm. For each edge collapse $(v_1, v_2) \rightarrow \bar{v}$, the simplification algorithm determines the error of $v$ and then moves $v$ to a new position that minimizes the error. The edge collapse with the lowest error is chosen and carried out. The algorithm can be summarized as follows:

**Algorithm:** The simplification algorithm.

Step 1 Extract ridge and valley lines.
Step 2 Compute the new quadrics for the initial polygonal model using Eq. (3).
Step 3 Compute the new vertex position for each edge collapse $(v_1, v_2) \rightarrow \bar{v}$. The error metric in the portion of the vertex $v$ represents the cost of that collapse.
Step 4 Place all the edge collapses in a heap with the minimum-cost edge at the top.
Step 5 Iteratively remove the edge collapse of least cost from the heap, collapse this edge, and update the costs of all edge collapses involving $v_1$ and $v_2$.

## 6 Implementation and Results

All models were simplified on a PC with a Pentium IV 3.0Ghz processor and 1GB of main memory. We simplified a number of models, and used MeshDev [12, 15] to measure the surface deviation after simplification. The MeshDev software [1] compares two triangular meshes according to geometrical data or appearance attributes. It returns visual and numerical results. The method is based on the geometric error used by Metro tool [2].

Figs. 2 show the mean geometric deviations between the original and the simplified models. There is the comparison with four graphs representing geometric errors of simplified models using with feature edge and without it. It can be seen from these graphs that the simplified meshes generated using our algorithm have smaller mean geometric errors than meshes produced using without feature error metrics. In Figs. 3 and 4, our method preserves the silhouettes of the back and teeth more accurately.

Fig. 5 shows various degrees of simplification. When simplified to 700 faces without feature error quadrics, the model lost one propeller. But our simplified result did not lose a propeller until the model was simplified to 536 faces.

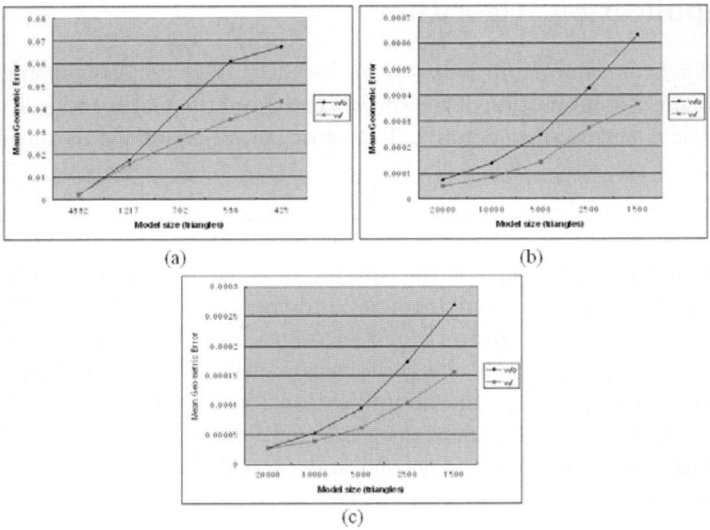


**Fig. 2.** Mean geometric error measured by MeshDev [1]: (a) Cessna, (b) Dragon and (c) Dinosaur model. $w/$:with feature edge error metric, $w/o$: without feature edge error metric.

**Fig. 3.** Simplified results for the Dragon model. (a) original model (320,000 faces), (b) red ridges and blue valleys, c) with feature edge error metric, (d) without feature edge error metric.

**Fig. 4.** Simplified results for the Dinosaur model. (a) original model (112,623 faces), (b) red ridges and blue valleys, c) with feature edge error metric, (d) without feature edge error metric.

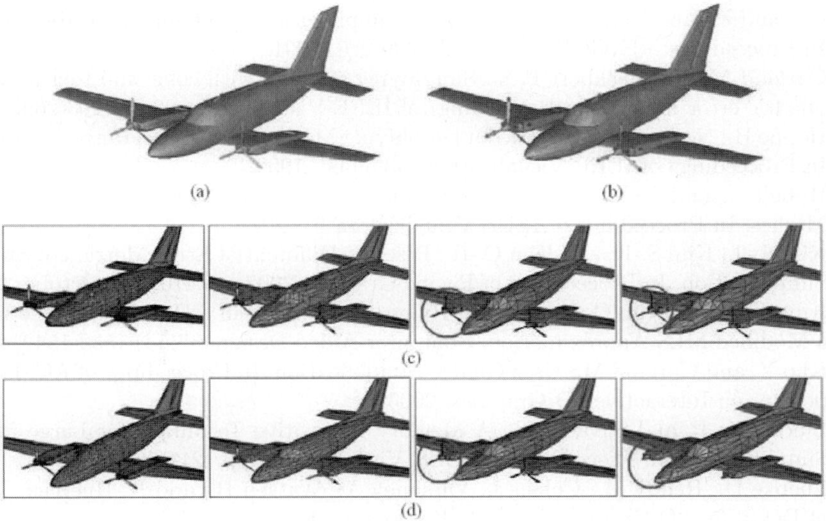


**Fig. 5.** Simplified results of the Cessna model. (a) original model (13,546 faces), (b) red ridges and blue valleys, (c) and (d) are each simplified to 4332, 1217, 702, and 536 faces.

## 7 Conclusion and Future Work

We have proposed a method of simplifying polyhedral models which extracts and preserves salient features. Simplification is controlled by a feature error metric combined with a basic error metric. The new error metric overcomes the drawback of a basic error metric which is inaccurate on highly curved regions, and preserves sharp or prominent regions despite a drastic simplification.

In future work, it would be desirable to measure the error metrics of attributes for surface simplification and provide a more theoretical analysis.

**Acknowledgments.** This work was supported by grant No.(R01-2005-000-10120-0) from Korea Science and Engineering Foundation in Ministry of Science & Technology, and Three-Dimensional Digital Real Object Duplication System (RODS) Development Project from the ministry of commerce, industry and energy. This research was also supported by University IT Research Center Project.

## References

1. http://meshdev.sourceforge.net/
2. Cignoni P., Rocchini C., and Scopigno R.: Metro: measuring error on simplification surfaces. Computer Graphics Forum, 17(2):167-174, 1998.
3. Cohen J., Olano M., and Manocha D.: Appearance-Preserving Simplification. In Proceedings of SIGGRAPH 98, pp.115-122, 1998.

4. Garland M. and Heckbert P. S.: Surface simplification using quadric error metrics. In Proceedings of SIGGRAPH 97, pp.209-216, 1991.
5. Garland M. and Heckbert P. S.: Simplifying surfaces with color and texture using quadric error metrics. In Proceedings of IEEE Visualization 98, pp.263-269, 1998.
6. Hoppe H.: New Quadric Metric for simplifying Meshes with Appearance Attributes. In Proceedings of IEEE Visualization, pp.59-66, 1999.
7. Hubeli A. and Gross M.: Multiresolution Feature Extraction from Unstructured Meshes. In Proceedings of IEEE Visualization 01, 2001.
8. Kim S.-J., Kim S.-K. and Kim C.-H.: Discrete Differential Error Metric For Surface Simplification. In Proceedings of Pacific Graphics 2002, pp.276-283, October 2002.
9. Kim S.-K. and Kim C.-H.: Finding Ridges and Valleys in a Discrete Surface Using a Modified MLS Approximation. Computer-Aided Design, 37(14):1533-1542, 2005.
10. Kho Y. and Garland M.: User-Guided Simplification. In Proceedings of ACM Symposium on Interactive 3D Graphics. 2003.
11. Lindstrom P. and Silva C. T.: A Memory Insensitive Technique for Large Model Simplification. In Proceedings of IEEE Visualization, pp.121-126, 2001.
12. Luebke D., Reddy M., Cohen J., Varshney A., Watson B., and R. Huebner: Level of Detail for 3D Graphics. Morgan Kaufmann, 2003.
13. Pojar E. and Schmalstieg D.: User-controlled creation of multiresolution meshes. In Proceedings of ACM Symposium on Interactive 3D Graphics, pp.127-130, 2003.
14. Rossignac J. and Borrel P.: Multi-resolution 3D approximations for rendering complex scenes. Modeling in Computer Graphics: Methods and Application, pp.455-465, 1993.
15. Roy, M., Foufou, S., and Truchetet, F. Mesh Comparison using Attribute Deviation Metric. International Journal of Image and Graphics (IJIG), 4(1):127-140, 2004.
16. Schroeder, W. J., Zarge, J. A, and Lorensen, W. E. Decimation of Triangle Meshes. Computer Graphics (In Proceedings of SIGGRAPH 92), 26(2):65-70, 1992.
17. Watanabe, K., and Belyaev, A. G. Detection of Salient Curvature Features on Polygonal Surfaces. Computer Graphics Forum (In Proceedings of Eurographics 01), 20(3):385-392, 2001.
18. Walter, J. D., and Healey, C. G. Attribute Preserving Dataset Simplification. In Proceedings of IEEE Visualization 01, pp.113-120, 2001.
19. Yoshizawa, S., Belyaev, A. G., and Seidel, H.-P. Fast and robust detection of crest lines on meshes. In Proceedings of ACM Symposium on Solid and Physical Modeling, pp.227-232, 2005.

# Multiresolution 3D Rendering on Mobile Devices

Javier Lluch, Rafa Gaitán, Miguel Escrivá, and Emilio Camahort

Computer Graphics Section
Departament of Computer Science
Polytechnic University of Valencia
Camino de Vera s/n, 46022. Valencia, Spain

**Abstract.** We present a client/server system that is able to display 3D scenes on handheld devices. At the server we extract the geometry that is visible for each client and send it. We also extract texture and material information. The clients, running on mobile devices, use that information to render realistic images. Our geometry extraction algorithm employs multiresolution and view-dependent simplification. We present results of our system running on PocketPC 2003 PDAs.

## 1 Introduction

Over the last few years mobile computing platforms have made important advances. We can see handheld devices with increasing processor speed and integrated wireless technologies. Computer graphics are also rapidly advancing in mobile devices. Still, these devices have limitations of memory and processing power. Also, most of them do not offer hardware accelerated graphics. This makes it difficult to render and interact with large 3D scenes on these devices.

A different issue is the size of the geometric data sets. Recent advances in 3D design, acquisition and simulation have led to larger and larger geometric data sets that exceed the size of the main memory and the rendering capabilities of current graphics hardware. Therefore, it is necessary to apply techniques like multiresolution and simplification in order to reduce the amount of geometry. Still, multiresolution models can only be applied to data sets that do not exceed the size of the main memory.

To solve this problem new models have been developed that store large scenes in secondary memory (*out-of-core*). We can still render these models at interactive speeds using view-dependent simplification. One of the immediate applications of this type of models is rendering of three-dimensional scenes on mobile devices.

We have developed a client-server system for remote rendering on mobile devices. Our system uses multiresolution, view-frustum culling and out-of-core techniques to deliver geometry from a PC server to one or more mobile devices. Our system has the advantage that it only sends the changes in geometry from the server to the client. Additionally, our system can handle and send textures to the clients.

In this paper, we present the system and the algorithm used to extract and send geometry from the client to the server. The paper is structured as follows. First, we introduce OpenSceneGraph(OSG) and multiresolution modeling, two components used in our system. Then, we present our client-server system. In the following Section we introduce an algorithm to efficiently extract geometry differences at the server. We also

V.N. Alexandrov et al. (Eds.): ICCS 2006, Part II, LNCS 3992, pp. 287–294, 2006.

explain how textures are sent to the client. Finally, we present some results, conclusions and directions for future work.

## 2 Background

In this section we present the background of our system and our implementation. First we summarize OSG and the benfits of its use. Next we explain how we have merged OSG and the Multitesselation(MT) library.

### 2.1 OpenSceneGraph

OSG [8] is an open source, cross platform graphics toolkit for the development of high peformance graphics applications such as flight simulators, games, virtual reality and scientific visualization. Based around the concept of a scene graph, it provides an object oriented framework on top of OpenGL, freeing the developer from implementing and optimizing low level graphics calls. OSG provides many additional utilities for rapid development of graphics applications.The key strengths of OSG are its performance, scalability, portability and the productivity gains associated with using a fully featured scene graph.

### 2.2 Multiresolution

Multiresolution meshes are a common basis for building representations of geometric shapes at different levels of detail. The use of the term multiresolution means that the accuracy (or level of detail) of a mesh in approximating a shape is related to the mesh resolution, i.e., to the density (size and number) of its cells. A multiresolution mesh provides several alternative mesh-based approximations of a spatial object (e.g., a surface describing the boundary of a solid object, or the graph of a scalar field).

A multiresolution mesh is a collection of mesh fragments, describing usually small portions of a spatial object with different accuracies. It also includes suitable relations that allow selecting a subset of fragments (according to user-defined accuracy criteria), and combining them into a mesh covering part or the whole object. Existing multiresolution models differ in the type of mesh fragments they consider and in the way they define relations among such fragments. The reader is referred to the surveys [4, 7, 2] for a detailed description of multiresolution mesh representations.

### 2.3 Rendering on Mobile Devices

In [5] we address the problem of rendering complex three-dimensional scenes on mobile computer devices. The issue is that the scene does not fit in the devices's main memory. So we use out-of-core storage and view-dependent simplification to keep the smallest possible number of polygons stored in the mobile device for rendering. To achieve this goal we employ multiresolution meshes.

We propose applying current multiresolution, viewing frustum culling and out-of-core [3] techniques to 3D graphics rendering on mobile devices [1, 6]. We present a client-server system that delivers simplified geometry to a PDA over a wireless connection. The server stores the scene graph and extracts levels of detail to be rendered at the client. Careful selection of appropriate levels of detail allows rendering at interactive rates on a handheld device.

Our system uses a geometry cache that manages two copies of the set of visible objects. One copy is located at the server and the other at the client. The server updates the cache with the visible geometry when the scene or the viewing parameters change. A synchronization process mantains the coherence between the server and the client copies of the cache. With this approach we only send geometry updates from the server to the client, thus reducing the latency and bandwidth requirements of our system. It also has the added advantage that we can use advanced geometry culling methods at the server side.

Our server system also supports multiresolution and view-dependent simplification, thus allowing the selection of a suitable level of detail for rendering each desired view. The system can render scenes with any number of polygons. It improves on previous systems because we do not store the entire geometry at the mobile device. Neither do we render at the server and send the rendered frames to the client. Instead we make sure that the client only receives those parts of the scene graph that are visible. These parts are currently rendered using a software library. Once hardware accelerated rendering is available we will be able to render larger amounts of geometry on the mobile device.

## 3 Geometry and Texture Extraction

The system just described caches geometry at the mobile client so that only new geometries need to be sent from the server. Still, the architecture needs to send all the multiresolution geometry every time there is a change in the resolution level. This behavior causes the latency to increase. To solve this problem we propose an algorithm that efficiently computes differences between two extracted levels of detail. This reduces the latency between client and server with multiresolution geometries. We also show how to extract textures at the server in order to send them to the client for realistic rendering.

### 3.1 Geometry Extraction

We have a library that binds OSG and the MT library. We have improved the library by adding new methods that extract only the differences between two different geometry resolutions.

MT works by extracting two arrays of triangle identifiers. One for the geometry inside the interest volume and the other for the geometry outside the interest volume. Until now our server was extracting all the geometry for every change in the resolution and sending all the new geometry information to the client. Between two extractions of geometry the information differs only in a few vertices and of course in their triangle configuration. We propose to reuse this information to avoid sending all the geometry information to the client.

We use an algorithm that is made of three steps:

1. First we extract the geometry and save the triangle identifiers and the index pointer (of each vertex) to the vertex array inside the triangle. The data structure that we use

to save the information is a hash map(*index hash map*). This step is necessary because the next step uses the saved information to calculate the differences between geometries.
2. The second step obtains the new arrays of triangle identifiers from the MT geometry, and compares each triangle with the triangles that are already stored at the server. For those triangles we generate the index array with the index pointer saved in the *index hash map* during the previous step. For those triangles not saved in the hash map we use a different hash map (*difference hash map*) to save the triangle identifier, and the vertex and normal information. We must have the difference information available because the next step will use it to generate the new geometry.
3. The third step takes the *difference hash map* with the triangles not yet cached and appends the triangles to the vertex and normal arrays. The index hash map is also updated with the new added triangle identifier and the indices of its vertices in the vertex and normal arrays. It is also necessary to update the geometry index array with the index pointers added to the *index hash map*. Finally we clean the difference hash map and go to step 2 to calculate differences again.

Once steps 2 and 3 have been run, we can draw the complete geometry with a different level of detail. Algorithm 1 shows the methods used for computing the geometry differences (left) and reconstructing the geometry from those differences (right).

The server system has been modified to extract differences between two continuous levels of detail. When the cull process traverses the scene, if the geometry is a multiresolution geometry then we run (i) step one if is the first time we find the geometry or (ii) step two if the geometry had already been created. Next we add the geometry to the local geometry cache to make the syncronization between client and server.

The cache has been modified in order to support this kind of new geometries. We have extended our cache to save the differences for multiresolution geometries and to calculate the geometry using step three of the algorithm.

**Algorithm 1.** Calculate Differences and Geometry Extraction by Differences

```
procedure CALCULATEDIFFERENCES
    act_tri ⇐ MT_ExtractActiveTiles()
    for all t in act_tri do
        if t is in indexHashMap then
            indices.add(indexHashMap[t].i0)
            indices.add(indexHashMap[t].i1)
            indices.add(indexHashMap[t].i2)
        else
            diffHashMap[t].v0 ← t.v0
            diffHashMap[t].v1 ← t.v1
            diffHashMap[t].v2 ← t.v2
        end if
    end for
end procedure
```

```
procedure CALCGEOMBYDIFFERENCES
    act_tri ⇐ MT_ExtractActiveTiles()
    for all t in diffHasMap do
        vertices.add(t.v0,t.v1,t.v2)
        lenv ← vertices.size()
        indices.add(lenv − 3)
        indices.add(lenv − 2)
        indices.add(lenv − 1)
        leni ← indices.size()
        indexHashMap[t].i0 ← (leni − 3)
        indexHashMap[t].i1 ← (leni − 2)
        indexHashMap[t].i2 ← (leni − 1)
    end for
    diffHashMap.clean()
end procedure
```

### 3.2 Texture Extraction

In order to improve the realism of our system we implement functionality to extract textures at the server and send them to the client for rendering. To accomplish this task, we had to modify the **geometry extractor**, the **client** and the **scene cache**. Figure 2 contains some screenshots of our system.

When the server traverses the scene to collect all the geometries to send to the client, we need to accumulate the state of OpenGL (*textures, lights, ...*). So, when a geometry is found, the current accumulated state is added to the server cache with its associate geometry. Then the cache is synchronized with the client cache, and the new geometry and the associated OpenGL state are sent to the client.

The client synchronizes the scene cache with the server's cache. When this happens, the new geometry and the material state are received by the client. Then, the client adds the received geometry to its cache.

Both parts, client and server, have a scene cache that must be synchronized. The scene cache has been modified to support the material state and textures. We need to send those materials through the network, but only new materals are sent in order to improve cache synchronization speed. The new geometries and textured materials are received at the *back buffer cache*, so when the synchronization process finishes, the system swaps the cache and the new geometries, so that they can be used for rendering.

## 4 Results

To test the implementation of our system we have run the server on a desktop PC and the client on different devices. We run the client on a laptop with wireless access and on an HP ipaq 4150 running PocketPC 2003.

We used different scenes to test our system. Figure 2(b) shows a screenshot of the system running our win32 client. Figure 2(c) shows a scene with several multiresolution objects rendered using OSG. Figures 2(e) and 2(f) show the client running on a PDA.

To analyze our system we compare the evolution of the time of extraction between the difference algorithm and the complete extraction. We also show the latency of the system with multiresolution geometries and textured materials. The comparisons have been made moving the camera along a couple of pre-generated paths.

Table 1(a) shows a comparison between the average, maximum and minimum of extraction time of one multiresolution geometry. We observe that the average extraction time is substantially better using the difference algorithm. This allows us to greatly reduce the time needed to prepare the geometry to send it to the client. Table 1(b) shows a comparison of the transmission times required by the difference algorithm and the algorithm that sends all the geometry. We observe that the average transmission time of the difference algorithm is 2.5 times faster than sending all the geometry. Table 1(c) shows a comparison of sending static geometries with textured materials through two different types of network.

Plot 1(b) shows latency time with multiresolution geometries. Note that the difference algorithm runs faster. Reducing latency time allows more extractions thus reducing popping artifacts. Plot 1(c) shows the evolution of latency time when rendering static geometry and textured materials. The textured scene has three geometries with the same

| | Recreate | Differences |
|---|---|---|
| Average | $0.0057sec$ | $0.0026sec$ |
| Max | $0.0187sec$ | $0.0078sec$ |
| Min | $0.0038sec$ | $0.0021sec$ |

(a) Comparison of extraction time of a single multiresolution geometry when using the difference algorithm and the complete reconstruction algorithm. The test scene has roughly 69000 triangles, viewing as many as 4000 triangles in the PDA.

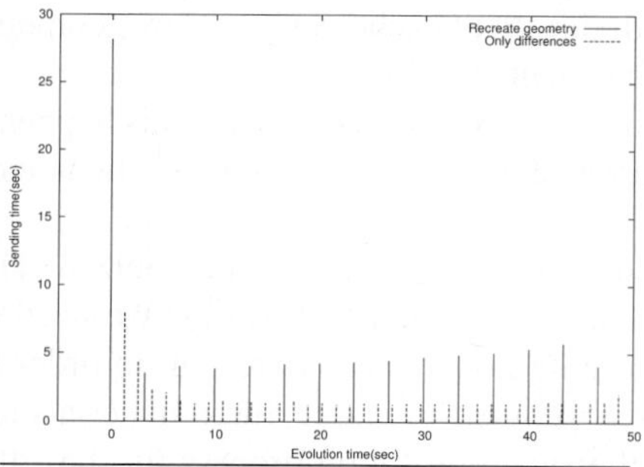


| | Complete | Differences |
|---|---|---|
| Average | $6.053sec$ | $2.412sec$ |
| Max | $28.173sec$ | $27.791sec$ |
| Min | $3.525sec$ | $1.236sec$ |

(b) Latency times for multiresolution geometries. The test scene has more than 960000 triangles, viewing as many as 10000 in the PDA. Maximum values are obtained at the initial time.

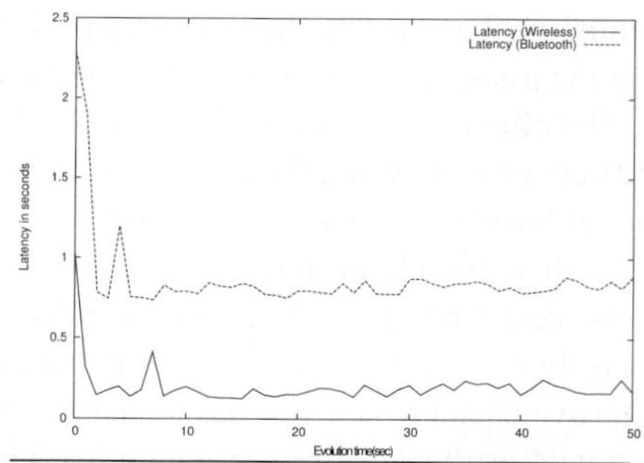


| | Bluetooth | Wireless |
|---|---|---|
| Average | $0.870sec$ | $0.211sec$ |
| Max | $2.292sec$ | $1.035sec$ |
| Min | $0.736sec$ | $0.124sec$ |

(c) Evolution of latency time when rendering static geometry and textured materials across two different types of network. The test scene has 1500 triangles.

**Fig. 1.** Results obtained with our system running on a PocketPC 2003 PDA

material. Note that the high latency at the beginning occurs because we send two geometries and the associated material. The following high latency in the plot appears at roughly second 8 is the sending of the third geometry.

## 5 Conclusions and Future Work

In this paper we present a client-server system for remote rendering on mobile devices. Our system uses multiresolution, view-frustum culling and out-of-core techniques to deliver geometry from a PC server to one or more mobile devices. Our system has the advantage that it only sends the changes in geometry from the server to the client. Additionally, our system can handle and send textures to the clients.

The system's server uses *OSG* and multiresolution meshes for scene management. Given the camera parameters delivered by the client, the server culls the scene graph and sends to the client the geometry to be rendered. We propose a new algorithm that ex-

(a) A container terminal (b) Detailed view of the same terminal

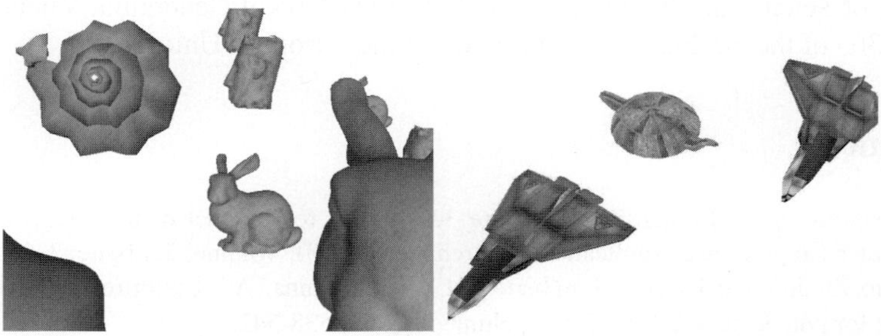

(c) A large scene with several multiresolution models. The test scene used in 1(b).

(d) A scene with several texture-mapped objects.

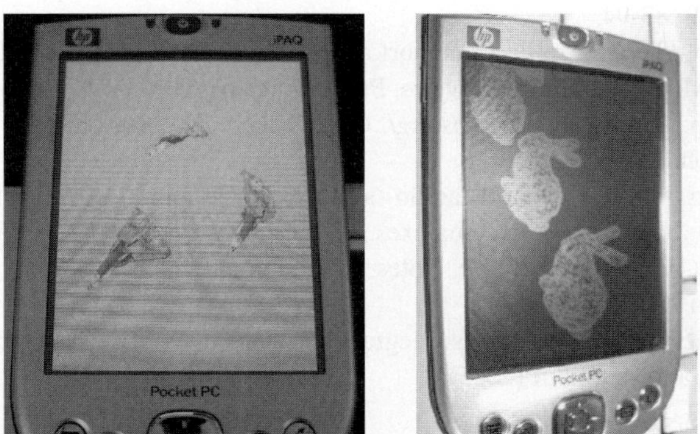


(e) Texture-mapped objects. (f) A scene with multiresolution objects.

**Fig. 2.** Example renderings of our client system running on *top* and *middle* a PC, and *bottom* a PDA

tracts the differences between two geometry levels of detail. With this algorithm only the difference in geometry needs to be sent to the client, thus substantially reducing latency.

Our algorithm also allows extracting textures in addition to geometry. Texture and material data can be extracted, cached and delivered at the client exactly like geometry. This produces an increased realism in the renderings generated at the mobile client.

We need to improve the cache in order to avoid overloading if we continuously append differences.

We want to implement an out-of-core multiresolution model for PDA systems and we want to compare it with the system just described. We are also preparing some benchmarks to measure the transmisssion efficiency of different network connections.

## Acknowledgement

This work was partially supported by grants: TIN2005-08863-C03-01 of the Spanish Ministry of Science and Technology, GV04B-497 04-05 of Generalitat Valenciana and IST-004363 of the 6th Framework Program of the European Union.

## References

1. B. D'amora and F. Bernardini. Pervasive 3d viewing for product data management: IEEE Computer Graphics and Applications (March/April 2003), Volume: 23, Issue: 2, 14-19.
2. Claudio Zunino and Fabrizio Lamberti and Andrea Sanna: A 3d multiresolution rendering engine for pda devices. In SCI 2003, volume 5 (2003) 538-542.
3. J. El–Sana and Y.-J. Chiang: External memory view-dependent simplification. In EuroGraphics volume 19 (2000), 139-150.
4. J. El–Sana and A. Varshney: Generalized view-dependent simplification. In Proceedings EuroGraphics (1999), 83-94.
5. J. Lluch, Rafael Gaitán, Emilio Camahort and Roberto Vivó: Interactive Three-Dimensional Rendering on Mobile Computer Devices. Proceedings of ACM ACE (2005) 254–256.
6. Khronos Group: http://www.khronos.org/. OpenGLES - The Standard for Embedded Accelerated 3D Graphics (2004).
7. Leila De Floriani and Paola Magillo and Enrico Puppo: Multiresolution representation of shapes based on cell complexes. In Discrete Geometry for Computer Imagery, number 1568 (1999) 3-18.Lecture Notes in Computer Science. http:// www.disi.unige.it/person/MagilloP/MT.
8. OSG Community: http://www.openscenegraph.org. OpenSceneGraph - Open Source high performance 3D graphics toolkit (2005).

# Multiresolution Remeshing Using Weighted Centroidal Voronoi Diagram

Chao-Hung Lin[1], Chung-Ren Yan[2], Ji-Hsen Hsu[2], and Tong-Yee Lee[2]

[1] Dept. of Geomatics, National Cheng Kung University, Taiwan
[2] Dept. of Computer Science and Information Engineering,
National Cheng Kung University, Taiwan
http://couger.csie.ncku.edu.tw/~edirvr/

**Abstract.** We present a novel method for multiresolution remeshing of irregular mesh. First, the original mesh (two-manifold any genus) is decomposed into several patches, each patch is homeomorphic to a 2D triangle. The goal of this decomposition process is that the decomposed patches are size-equally. First, a mesh is manually cut into a few disk-like patches. With the help of weighted centroidal Voronoi diagram (WCVD), each patch is then automatically partitioned into more triangular patches with nearly equal size. Recursively subdividing these triangular patches, we finally get a semi-regular mesh.

**Keywords:** Parameterization, Remeshing, Voronoi Diagram, Resampling.

## 1 Introduction

In computer graphics and geometric modeling, 3D models are often represented by triangular meshes. The connectivity of triangular mesh produced by modeling software is usually irregular. Irregular meshes are not appropriate to handle in such tasks as model texturing and computation of finite element. Therefore, the remeshing process, i.e., modifying the sampling and connectivity of a geometry to generate a new mesh, is a fundamental step for efficient geometric processing.

The majority of previous works propose semi-regular remeshing techniques [1, 2, 3] are based on an initial step of constructing a parameterization of the original model over a base domain, i.e., so-called base mesh, consisting of a small number of triangles. A triangle in the base mesh maps to a triangular patch in the original mesh. This base mesh is then recursively subdivided, i.e., vertex resampling, to accurately fit the original mesh over the parametric domain.

Mesh re-tiling technique is proposed by [4, 5, 6] to resample the original mesh based on a particle system approach. A new set of vertices are first distributed over the surface of a given mesh and are then triangulated these vertices to create a surface that accurately fits the original mesh. A global relaxation procedure balances the distribution of the vertices by shifting them on the surface. In [6], to exactly find the sharp features, the authors extend the relaxation procedure by attracting the vertices to feature edges.

V.N. Alexandrov et al. (Eds.): ICCS 2006, Part II, LNCS 3992, pp. 295–301, 2006.

The concept of interactive user control in geometry remeshing has been recently proposed in [7]. In [7], the original mesh is substituted by a series of 2D maps in parameter space. The user can combine these maps to create a control map which is used to control the vertex sampling over the surface.

The weighted centroidal Voronoi diagram (WCVD) has been applied to many applications such as non-photorealistic rendering (stippling) [13], and surface remeshing [14]. In [14], an adaptive surface remeshing technique is presented. The vertex resampling is achieved by a given density function. Utilizing the WCVD, the mesh can be adaptive tessellated. This method focuses on generating a precise isotropic surface remeshing, but it is difficult to generate a multiresolution remeshing using their remeshing scheme. Multi-resolution remeshing technique is important and can be applied to many applications such as morphing [8, 9] and LOD. In this paper, a multiresolution remeshing technique for arbitrary geometry is proposed. Unlike the previous works that create a base mesh using simplification algorithm [1, 2, 3], we decompose the original mesh into several triangular patches utilizing WCVD in the parametric domain. These decomposed triangular patches are nearly equal-size and similar to equilateral triangle. Once the decomposition is accomplished, these triangular patches are re-embedded onto the parametric domain to obtain a low distortion parameterization. This decomposition process can also be treated as a vertex resampling step. The next step is to recursively subdivide the triangular patches to produce a new semi-regular mesh as the result.

## 2 Methodology

The basic idea of the proposed method is to decompose the original mesh $M$ into several triangular patches using the WCVD. These triangular patches are then subdivided to produce a new mesh with new connectivity that accurately fit the mesh $M$. The proposed remeshing algorithm consists of the following steps.

1. First pass decomposition: Manually partition the mesh $M$ into a few disk-like patches $\{P^1, P^2, ..., P^m\}$.
2. Parameterization: Each individual surface patch is then embedded into an isomorphic planar quadrilateral. The boundary of the parameterization is fixed to be a square so that a geometric map can be efficiently stored and processed as a regular floating point image.
3. Construct WCVD: A geometric map is computed over the parameterization and stored in 2D image as the weight of WCVD. For each quadrilateral, given $n$ sites, we construct a WCVD. The construction of WCVD will be described in more detail in Section 2.1.
4. Second pass decomposition: After triangulating the WCVD, the disk-like patch $P^i$is further decomposed into several nearly equal-size triangular patches $\{P^i_1, P^i_2, ..., P^i_m\}$. This step will be also described in more detail in Section 2.1.
5. Reparameterization: For each individual triangular patch, we re-embed it into an isomorphic planar triangle for a lower distortion parameterization.

6. Resampling: Perform recursive 4-to-1 subdivisions on all triangular patches $P_j^i$ over the parametric domain. Finally, we can obtain a new semi-regular mesh.

### 2.1 Mesh Decomposition Using Weighted Centroidal Voronoi Diagram

Once the mesh $M$ has been decomposed into a few disk-like patches, i.e., the first pass decomposition, and embedded into planar quadrilateral, we utilize the WCVD to further decompose each disk-like patch into several triangle-like patches.

**Voronoi Diagram.** Given a set of points, called Voronoi sites, a Voronoi diagram partition space into several regions. For each region, any position in this region is closer to the Voronoi site of this region than to another one. In this paper, we unitize the space partitioning ability of Voronoi diagram to decompose surface patch. We adopt a fast 3D graphics hardware-based algorithm [10] to compute a Voronoi diagram. This algorithm draws a set of cones with their apexes at each site. Each cone is given a unique color as the site's identity. For each pixel, the z-buffer determines which cone is closer to the viewer and assigns the color of this cone, i.e., the site's identity, to it. Finally, we can obtain each region by the colors. This technique allows us to compute a discrete Voronoi diagrams extremely fast.

**Centroidal Voronoi Diagram.** To uniformly decompose the surface patch, a centroidal Voronoi diagram (CVD) is more suitable. The CVD has the property that each site lies exactly on the centroid of its Voronoi region. The centroid of a region is defined as equation 1.

$$C = \frac{\sum_{x_i \in R} x_i}{n} \tag{1}$$

Where $R$ is the Voronoi region, is the pixel in the region $R$ and $n$ is the number of pixels in the region . Specifically, a centroidal distridution of the sites is useful because the sites are evenly distributed on the defined plane and the Voronoi region are nearly size-equally (the CVD is an optimization in some sense). These properties are appropriate to handle uniform resampling or uniform decomposition. In this paper, we adopt a relaxation algorithm [11] to produce a CVD. This algorithm can simply be stated as follows: compute Voronoi Diagram, find centroid of each region and move site to the centroid. This process is terminated when the positions of these sites are convergent. Figure 1 shows an example of CVD with 100 sites randomly placing in an equilateral quadrilateral. Figure 1(a) is the Voronoi diagram of the given sites. After 23 iterations, we obtain the CVD as shown in Figure 1(e).

**Weighted Centroidal Voronoi.** Diagram The CVD is created in the parametric domain which causes the regions in the planar quadrilateral to be distorted.

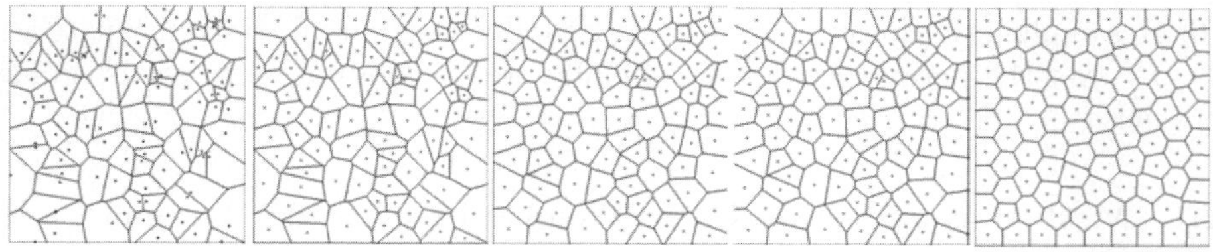

(a) (b) (c) (d) (e)

**Fig. 1.** Centroidal Voronoi Diagram. (a) voronoi diagram ;(b) 5 iterations ; (c) 10 iterations ; (d) 15 iterations ; (e) 23 iterations.

In this paper, we apply a weight function to the centroidal Voronoi diagram to solve this problem. The Eq. 1, i.e., calculate the Voronoi centroid, is rewritten as follows:

$$C = \frac{\sum_{x_i \in R} x_i w_i}{\sum w_i} \tag{2}$$

Where $w_i$ is the weight of $x_i$. We identify a relevant geometrical value: area distortion map $M_A$ to serve as the weights of CVD. The map $M_A$ is easily computed using the ratio $A_{3D}/A_{2D}$ of each triangle in the parameterization. Where $A_{2D}$ and $A_{3D}$ are the triangle area in the parameterization and its corresponding area on the 3D surface, respectively. After normalizing the scale $A_{3D}/A_{2D}$, the geometric map $M_A$ can be treated as a grayscale image range from 0 to 255. For a better visual effect, we set the pixel with largest area distortion to be 0 and the pixel with the lowest area distortion to be 255. Following the relaxation algorithm by S. Lloyd [11], the process of constructing WCVD can be described as the pseudo-code below. In Figure 2, we show an example of the surface patch decomposition using WCVD. During the relaxation process, the darker regions of the geometry map have lager attraction for sites. With the help of WCVD, the surface patch is uniformly decomposed (as shown in Figure 2(e)). One disadvantage of the proposed method is the use of discrete calculation of the Voronoi diagram that is affected by the resolution of the diagram and the distortion of the parameterization. A related problem is that if the resolution is too low, the sites will not distribute over the areas with an extremely high distortion. A solution described in [10] is to split the diagram into tiles and compute each tile at

```
Procedure WCVD ()
{
  Give n sites
  Repeat
     Compute Voronoi diagram
     For each region R_i of site s_i
        Find the weighted centroid s'_i of the region R_i using Eq. (2)
        Move the site s_i to the weighted centroid s'_i
  Until convergence
}
```

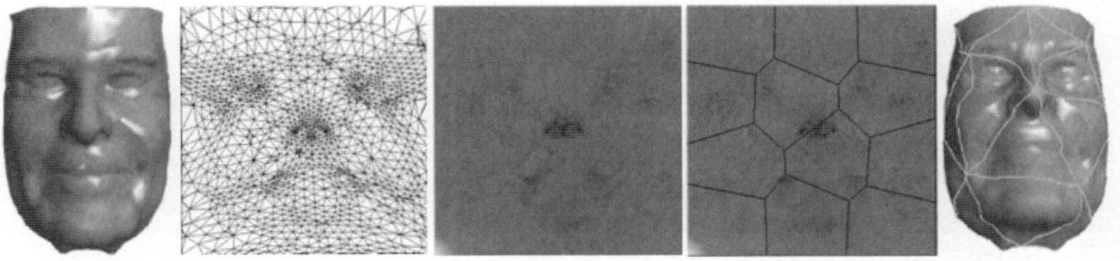

(a) (b) (c) (d) (e)

**Fig. 2.** Surface patch decomposition using WCVD. (a) Surface patch; (b) The parameterization of (a); (c) The area distortion map of (b); (d) The WCVD with 12 sites; (e) The surface decomposition result.

the higher resolution (increase arbitrary) and then stitch the full diagram back together. Another solution is to repartition the surface patch and then re-embed these sub-patches to parametric domain until the distortion is less than a defined threshold. In this paper, we compute a WCVD in a high resolution plane. If the problem described above still occurs, we will repartition the surface patch.

### 2.2 Mesh Remeshing

Once the decomposition process is completed, we re-embed each triangle patch to parametric domain to obtain a lower distortion parameterization. Then, we produce a new mesh with subdivision connectivity from the parameterization. The simple strategy we adopted is to perform a 4-to-1 split that the position of the split point is at the middle of an edge.

## 3 Experimental Results

In Figure 3 and 4, we show the preliminary remeshing results using the proposed method. Follow the evaluate function in [12]. Accuracy is measured as Peak Signal to Noise Ratio $PSNR = 20 * log(peak/d)$, where $peak$ is the diagonal length of the bounding box and $d$ is the symmetric RMS Hausdorff error (geometric distance) between the original mesh and the remesh. The Table 1 shows the Statistics of PSNR for the remeshing in the Figure 3 and 4. Roughly speaking, the PSNR about 70dB is considered to be a nice approximation.

## 4 Conclusion and Future Work

We have introduced a novel approach for multiresolution uniform remeshing. Utilize the weighted centroidal Voronoi diagram to uniformly distribute the vertices in the surface and then decompose the surface to several triangular patches with nearly equal size. In the further work, we will address the processing of consistent remeshing of several different poses models. This would be very useful in the morphing and animation issues.

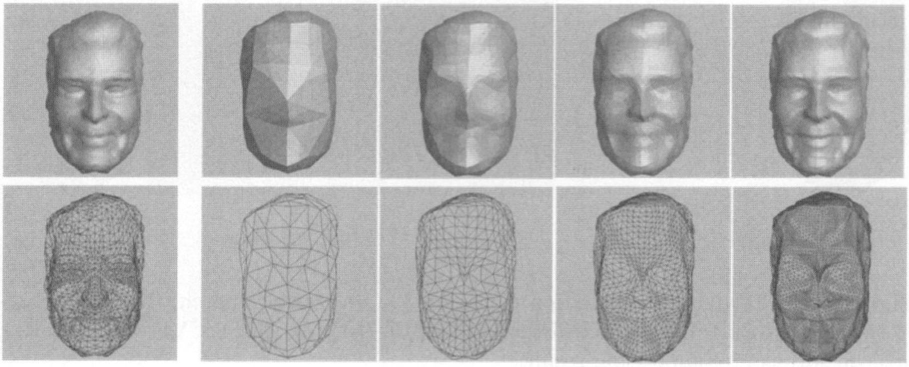

(a) (b) (c) (d) (e)

**Fig. 3.** Mesh remeshing of a man's head model. (a) The original mesh; (b) The decomposition result (the coarsest level of subdivision: 353 triangles); (c),(d),(e) Remeshing with 1408 triangles (level 1), 5632 triangles (level 2) and 22528 triangles (level3), respectively.

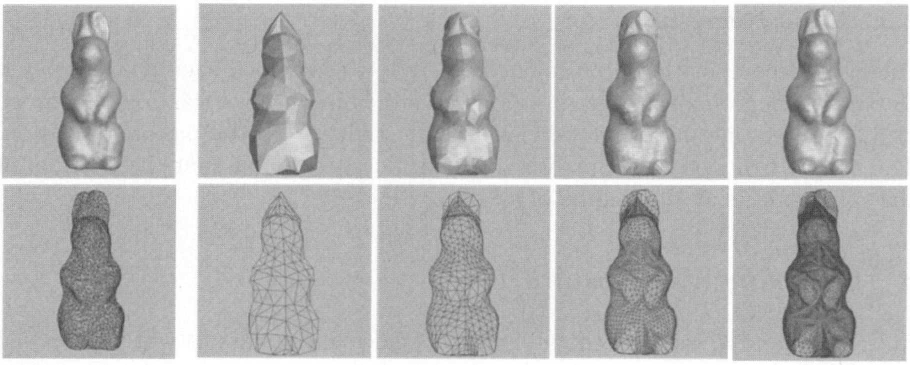

(a) (b) (c) (d) (e)

**Fig. 4.** Mesh remeshing of a rabbit model. (a) The original mesh; (b) The decomposition result (the coarsest level of subdivision: 472 triangles); (c),(d),(e) Remeshing with 1888 triangles (level 1), 7552 triangles (level 2) and 30208 triangles (level3), respectively.

**Table 1.** Statistics for PSNR

| Model | Level | Lv 0 | Lv 1 | Lv 2 | Lv 3 | Lv 4 |
|---|---|---|---|---|---|---|
| Man's Head | PSNR(dB) | 44.42 | 52.81 | 60.44 | 68.40 | 85.71 |
| | no. of face | 352 | 1,408 | 5,632 | 22,528 | 90,112 |
| Rabbit | PSNR(dB) | 44.40 | 50.5 | 58.71 | 61.27 | 73.96 |
| | no. of face | 472 | 1,888 | 7,552 | 30,208 | 120,832 |

**Acknowledgements.** This project is supported by National Science Council, Taiwan, ROC, under contract No. 94-2213-E-006-063 and 94-2213-E-006-005.

## References

1. A. W.F. Lee, D. Dobkin, W. Sweldens and P. Schroder, "MAPS: Multiresolution Adaptive Parameterization of Surfaces." ACM SIGGRAPH '98, pp. 95-104, 1998.
2. I. Guskov, W. Sweldens and P. Schroder, "Multiresolution Signal Processing for Meshes", ACM SIGGRAPH '99, pp. 325-334, 1999.
3. A. Khodakovsky, N. Litke and P. Schroder, "Globally Smooth Parameterizations with Low Distortion", ACM SIGGRAPH'03, pp. 350-357, 2003.
4. L. Kobbelt, T. Bareuther, and H.-P. Seidel "Multiresolution shape deformations for meshes with dynamic vertex connectivity", Computer Graphics Forum volume 19(3), pp. 249-260, 2000.
5. G. Turk, "Re-tiling polygonal surfaces", ACM SIGGRAPH'92, vol. 26, pp. 55-64, 1992.
6. J. Vorsatz, C. Rossl, Leif P. Kobbelt and Hans-Peter Seidel. "Feature Sensitive Remeshing", Proceedings of EUROGRAPHICS'01, pp. 393-401, 2001.
7. P. Alliez, M. Meyer and M. Desbrun, "Interactive Geometry Remeshing", ACM SIGGRAPH '02, pp. 347-354, 2002.
8. C.-H. Lin and T.-Y. Lee, "Metamorphosis of 3D Polyhedral Models Using Progressive Connectivity Transformations", IEEE Transactions on Visualization and Computer Graphics, Vol. 11, No.1, pp. 2-12, Jan./Feb., 2005.
9. T.-Y. Lee and P.-H. Hung, "Fast and Intuitive Metamorphosis of 3D Polyhedral Models Using SMCC Mesh Merging Scheme", IEEE Transactions of Visualization and Computer Graphics, Vol. 9, No.1,pp. 85-98, January-March 2003.
10. K. E. Hoff, T. Culver, J. Keyser, M. Lin and D. Manocha, "Fast Computation of Generalized Voronoi Diagrams Using Graphics Hadware", ACM SIGGRAPH '99, pp. 277-286, 1999.
11. S. Lloyd, "Least Square Quantization in PCM", IEEE Transactions on Information Theory, volume 28, pp.129-137, 1982.
12. P. Sender, J. Snyder, S. Gortler and H. Hoppe, "Texture Mapping Progressive Meshes", ACM SIGGRAPH '01, pp. 409-416, 2001.
13. A. Secord, "Weighted Voronoi Stippling", In Proceedings of NPAR, 2002.
14. V. Surazhsky, P. Alliez and C. Gotsman, "Isotropic Remeshing of Surfaces: A Local Parameterization Approach", Proceedings, 12th International Meshing Roundtable, Sandia National Laboratories, pp.215-224, Sept, 2003.

# Metric 3D Surface Mesh Generation Using Delaunay Criteria

Tomasz Jurczyk and Barbara Głut

AGH University of Science and Technology, Kraków, Poland
{jurczyk, glut}@agh.edu.pl

**Abstract.** This paper presents a technique of incorporating anisotropic metric into the Delaunay triangulation algorithm for unstructured mesh generation on 3D parametric surfaces. Both *empty circumcircle* and *inner angles* criteria of Delaunay retriangulation can be successfully used with the developed method of coordinate transformation with little adjustments. We investigate the efficiency of mesh generation process for different criteria and the quality of obtained meshes.

## 1 Introduction

The concept of a non-Euclidean metric is commonly used for generation of unstructured anisotropic meshes on 3D surfaces [1, 2, 3, 4]. In typical approach the *metric tensor* is used for evaluation of edge length, which is the main geometrical calculation governing the meshing process.

We consider parametric 3D surfaces which can be treated as two-dimensional domains with additional distortion introduced by the parameterization. The triangulation process is performed entirely in this parametric space basing on the Delaunay incremental insertion algorithm[1] with elements evaluation and nodes placement governed by metric. The metric description is stored in a *control space* (covering the whole domain), which can have various structure (e.g. background mesh or quadtree grid) depending on the source of the sizing data[6].

The method of mesh generation is based on Delaunay triangulation which can be obtained using criteria of *empty circumcircle* or *inner angles* [1]. In this paper we present a technique of adaptation of these criteria for generation of meshes with anisotropic metric using coordinate transformation. Utilization of both criteria is investigated with respect to the influence on triangulation efficiency and quality of obtained meshes.

## 2 Metric Definition

The metric is usually defined as a *metric tensor* $\mathcal{M} = \mathcal{R}\Lambda\mathcal{R}^{-1}$, where $\mathcal{R}$ is the eigenvector matrix (responsible for defining main directions) and $\Lambda = \text{diag}(\lambda_i)$ is the eigenvalue matrix, defining the required length of element edges along the main directions, and $\mathcal{M}$ is a symmetric positive-definite matrix.

V.N. Alexandrov et al. (Eds.): ICCS 2006, Part II, LNCS 3992, pp. 302–309, 2006.

Another method which was used in this work, is to introduce the non-Euclidean metric through an appropriate transformation of the coordinates of points within the domain, using the *transformation matrix* $\mathbf{M}_t$:

$$P_i' = \mathbf{M}_t(P_i)P_i \ . \tag{1}$$

Similar approach was presented in [5] where authors use transformation of coordinates in a different algorithm of constrained Delaunay triangulation.

The relation between the *metric tensor* and the *transformation matrix* is given by formula

$$\mathcal{M}(P) = \mathbf{M}_t(P)\mathbf{M}_t^T(P) \ . \tag{2}$$

Equation (2) defines a family of matrices, but the symmetric one is the most desirable, being much better fitted for metric-related operations[7]. For two-dimensional parametric space such symmetric matrix $\mathbf{M}_t^{(\mathrm{sym})}$ can be calculated as:

$$\mathbf{M}_t^{(\mathrm{sym})} = \begin{bmatrix} m_{11} & m_{12} \\ m_{12} & m_{22} \end{bmatrix} = \begin{bmatrix} \frac{1}{h_1}\cos^2\alpha + \frac{1}{h_2}\sin^2\alpha & (\frac{1}{h_1} - \frac{1}{h_2})\sin\alpha\cos\alpha \\ (\frac{1}{h_1} - \frac{1}{h_2})\sin\alpha\cos\alpha & \frac{1}{h_1}\sin^2\alpha + \frac{1}{h_2}\cos^2\alpha \end{bmatrix} , \tag{3}$$

where $\alpha$ denotes the direction of element stretching, $h_1$ and $h_2$ are the required lengths of element along the base directions.

During meshing process all geometrical formulas are calculated in metric space (i.e. basing on points coordinates transformed using the transformation matrix). The optimal final mesh should have all edges of unit length in local metric space (so called *unit mesh*[3]).

In order to account for distortion introduced by an arbitrary parameterization of the given surface patch $\mathbf{p} : (u, v) \rightarrow (x, y, z)$, we use the matrix $\mathbf{G}$ of the first fundamental form:

$$\mathbf{G} = \begin{bmatrix} g_{11} & g_{12} \\ g_{12} & g_{22} \end{bmatrix} = \begin{bmatrix} \langle \mathbf{p}_u, \mathbf{p}_u \rangle & \langle \mathbf{p}_u, \mathbf{p}_v \rangle \\ \langle \mathbf{p}_u, \mathbf{p}_v \rangle & \langle \mathbf{p}_v, \mathbf{p}_v \rangle \end{bmatrix} . \tag{4}$$

However, instead of using it directly, the symmetric *parameterization matrix* $\mathbf{M}_P$, where $\mathbf{M}_P\mathbf{M}_P^T = \mathbf{G}$, has to be calculated. Then, the final matrix used for transformation of coordinates is calculated as a product of sizing and parameterization matrices:

$$\mathbf{M}_t = \mathbf{M}_t^{(\mathrm{sym})}\mathbf{M}_P \ . \tag{5}$$

## 3 Metric in Meshing Algorithms

Introducing the metric into the meshing process requires small changes in all procedures which use geometrical properties. In most cases the only requirement is to calculate the local *transformation matrix* and to translate the coordinates of vertices used in the geometrical equations. The algorithms using the mesh topology information only may remain unchanged.

The coordinate transformation technique depends on the metric to be locally constant (or close to) in the neighborhood from which the points are being taken for calculation in any given meshing step. Usually this condition of small local variation is fulfilled. However, there are situations which can pose problems, mostly at the beginning of the meshing process (when the single mesh transformations cover large areas) or when the control space has (incorrectly) large metric variations. In order to avoid errors, an additional check for metric invariance has to be performed in some algorithms.

The workflow of used procedure of 3D surface mesh generation:

1. For discretization of contours of surface patches, an iterative placement of nodes for creating edges of unitary length in local metric is used. Procedure runs in parametric space of selected patch, and obtained discretization may be further refined according to control space defined in other patches incident to the given contour.
2. Triangulation of boundary vertices is created separately for each surface patch. Boundary nodes are being inserted one by one with local retriangulation of the mesh in order to maintain the Delaunay property. The retriangulation procedure adjusts the connectivity in the neighborhood of the node inserted into the mesh in order to reinstate the property of Delaunay triangulation. Both *empty circumcircle* and *inner angles* criteria[1] can be successfully adapted to produce proper results with metric introduced via the technique of coordinate transformation. Finally, the missing edges are recovered and obsolete triangles are removed.
3. The mesh is refined according to the prescribed sizing map by introducing new nodes. The selection of triangles to improve is performed basing on theirs metric quality coefficient. The new node is being inserted in the middle of the circumcircle and local retriangulation is performed.
4. Several iterations of mesh smoothing procedures are used (e.g. Laplacian smoothing).
5. If required, the triangular mesh can be further converted to quadrilateral, using one of the implemented algorithms of advancing front triangle merging[8], adapted for the developed technique of coordinate transformation.

### 3.1 Empty Circumcircle Criterion

The introduction of new node into the existing mesh can be accomplished using the empty circumcircle criterion, which requires locating and removing of all triangles having this node in theirs circumcircle and adding new triangles incident to the new node. Procedure starts with locating the triangle containing the new node. Adjacent triangles are iteratively checked for containing the new node in theirs circumcircles (Fig. 1). Selected elements are removed, and the vertices of the created empty cavity are being connected with the new node.

For each inserted node local metric is retrieved from control space once. The coordinates of new node and vertices of triangles being checked for empty circumcircle test have to be transformed into metric space (on average 7 triangles are checked and 16 vertices are transformed). If this procedure is used during the phase of insertion of inner nodes, the quality of all new triangles (on average

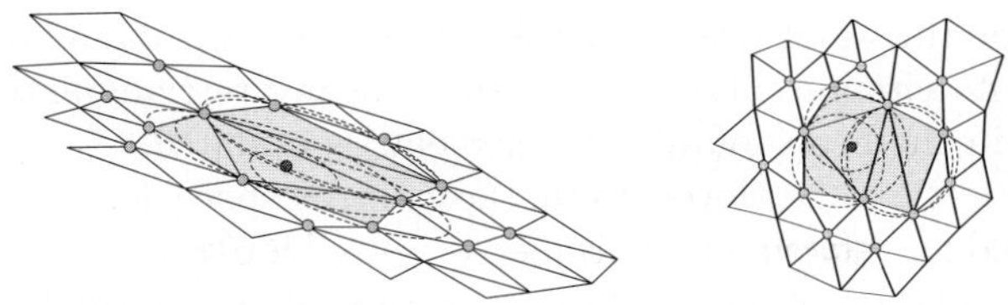

(a) insertion of new node (b) metric space

**Fig. 1.** Insertion of new node using empty circumcircle criterion

6) has to be recalculated which requires additional calls to control space and transformation of vertices coordinates (in total about 24 calls and 18 transformations).

### 3.2 Inner Angles Criterion

Procedure starts with locating the triangle containing the new node. This triangle is then substituted by three new triangles (Fig. 2). If the node is located at some edge, two triangles have to be removed and four new elements are inserted instead. Then the edges in the adjacent elements are swapped according to inner angles criterion. The swapping procedure starts with checking the new triangles and theirs neighbors for fulfillment of the criterion, then in case of swap further pairs of triangles have to be checked iteratively.

For each inserted node local metric is retrieved from control space once. The coordinates of new node and vertices of each pair of adjacent triangles being checked for inner angles test have to be transformed into metric space (on average 13–16 pairs of triangles are checked and 2–4 edges are swapped which requires about 9–12 transformation of coordinates). If this procedure is used during the phase of insertion of inner nodes, the quality of all modified triangles (on average 6) has to be recalculated which requires additional calls to control space and transformation of vertices coordinates (in total about 24 calls and 18 transformations).

### 3.3 Insertion of Inner Nodes

This phase starts with reorganizing the triangles into heap-list which requires calculation of quality (proper size and shape) of all elements. Then, in each step

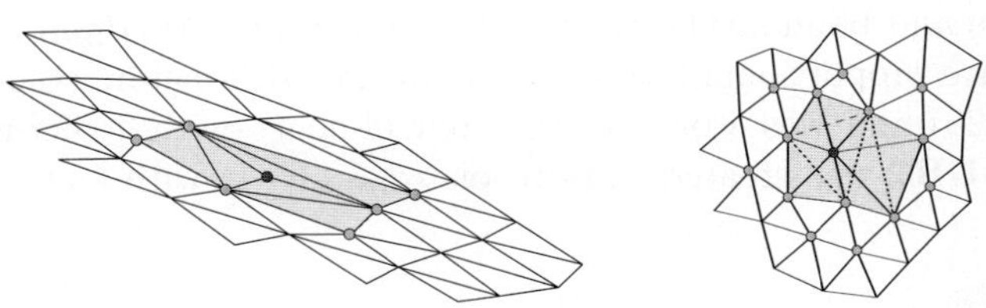

(a) insertion of new node (b) metric space

**Fig. 2.** Insertion of new node using inner angles criterion of the main loop the worst triangle is taken from the heap and an additional node is inserted in its circumcenter. The mesh is locally retriangulated and quality (and position in the heap) of affected elements is updated.

The local metric is retrieved from the control space for barycenter of the triangle selected for refinement. This metric is used for transformation of triangle vertices for calculation of new node and for reverse-transformation of this node before insertion into mesh.

Introduction of new nodes in circumcenter of triangle allows to obtain good quality of elements, but may be unacceptable in some cases (e.g. if the new node lies too close to boundary or outside of it). Using metric approach, additional conditions have to be checked. First, the new node can't be too close to one of vertices of the containing triangle. Then it must be checked whether the insertion of the node and retriangulation according to the metric defined in this node would at all affect the improved triangle. In order to reduce number of canceled insertions, at the beginning phase of this procedure inner nodes are inserted in the middle of the longest edge of triangle.

The quality of triangle is calculated basing on the the area (in metric space) of the circle circumscribed on the given triangle, which controls both size and shape of created triangles. All vertices of the evaluated triangle have to be transformed to the metric space. The transformation matrix is calculated as intersection[7] of metric matrices retrieved from the control space for each vertex of the triangle and for its barycenter.

## 4 Examples

Several meshes with different size and structure are presented. Meshes **AN0** (Fig. 3(a)) and **AN1** were generated on a plane with rectangular boundary $u \in [0,2], v \in [0,1]$ and metric defined by functions $f_m^i : (u,v) \rightarrow (\alpha, h_1, h_2)$:

$$\begin{aligned} f_m^0 &= (\pi/6, 0.01, 0.11 - 0.05u) \\ f_m^1 &= (\pi/6, 0.01, 0.0101 - 0.005u) \end{aligned} \tag{6}$$

Meshes **SUR0** (Fig. 3(b)), **SUR1** and **SUR2** were created on surface patch

$$\mathbf{p} : (u,v) \rightarrow (u, v, 2.5e^{-0.1(u^2+v^2))\sin(2u)\cos(2v)}) \tag{7}$$

where $u, v \in [-6, 6]$. The metric was automatically recognized from the curvature of the surface and boundary[6] (maximum anisotropy ratio $d_{\max} = 10$, lengths of elements are proportional to radii of curvature with factors respectively $c_0 = 0.1, c_1 = 0.025$) and was stored in the control space with structure of regular grid. Mesh **SUR2** was created as isotropic ($d_{\max} = 1$) with factor $c_2 = 0.1$.

### 4.1 Results

Table 1 presents metric utilization statistics for all tested meshes. As can be seen in columns $\overline{N}_c$, $\overline{N}_{pm}$ and $\overline{N}_{mp}$ the average number of control space calls

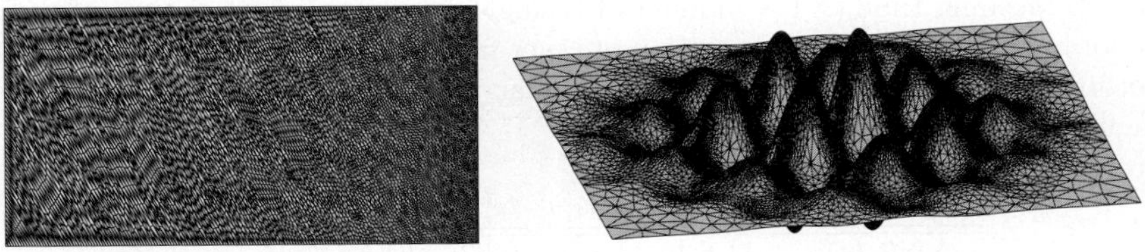

(a) planar mesh AN0 (b) surface mesh SUR0

**Fig. 3.** Anisotropic meshes

**Table 1.** Utilization of control space and metric during mesh generation using (c) – *empty circumcircle* and (a) – *inner angles* criteria. NP is number of points in the final mesh, $N_c$ – number of calls to control space, $N_{pm}$ – number of transformation of coordinates to metric space, $N_{mp}$ – reverse transformations. $\overline{N}_c$, $\overline{N}_{pm}$ and $\overline{N}_{mp}$ are additionally divided by number of points.

| | NP | $N_c$ | $N_{pm}$ | $N_{mp}$ | $\overline{N}_c$ | $\overline{N}_{pm}$ | $\overline{N}_{mp}$ |
|---|---|---|---|---|---|---|---|
| **AN0** (c) | 5069 | 115321 | 217728 | 4099 | 22.8 | 42.95 | 0.8 |
| (a) | 5139 | 117899 | 183426 | 4174 | 22.9 | 35.69 | 0.8 |
| **AN1** (c) | 104447 | 2475058 | 4661639 | 81994 | 23.7 | 44.63 | 0.8 |
| (a) | 105013 | 2543917 | 3934650 | 82994 | 24.2 | 37.47 | 0.8 |
| **SUR0** (c) | 23854 | 580469 | 1136499 | 20674 | 24.3 | 47.64 | 0.9 |
| (a) | 23707 | 599628 | 974884 | 21039 | 25.3 | 41.12 | 0.9 |
| **SUR1** (c) | 250031 | 5773210 | 11052111 | 214788 | 23.1 | 44.20 | 0.9 |
| (a) | 249628 | 5874957 | 9324457 | 215831 | 23.5 | 37.35 | 0.9 |
| **SUR2** (c) | 53404 | 1225174 | 2350930 | 45785 | 22.9 | 44.02 | 0.9 |
| (a) | 53265 | 1244957 | 1982100 | 45580 | 23.4 | 37.21 | 0.9 |

and metric transformations per point of the final mesh is fairly constant for all meshes.

Using *empty circumcenter* criterion tends to require lower number of control space calls, while for *inner angles* there are noticeably less transformation of coordinates. However, it must be noted that the operation of control space call for local metric retrieving is much more costly, especially for structures like background mesh.

Meshing times (2.6GHz Intel P4) for created meshes are presented in Table 2. The triangulation speed is close to linear for larger meshes and the triangulation process with *empty circumcircle* criterion is consistently faster, although the difference is not that large.

Table 3 presents quality evaluation of generated meshes (with two iterations of Laplacian smoothing in metric space). All presented coefficients ($\mu_{EL}$, $\mu_T$ and $\mu_\alpha$) should be equal to 1 for ideal mesh, $\mu_T$ evaluates only size and $\mu_\alpha$ only shape of created triangles.

Both criteria of retriangulation allow to produce meshes with comparably good quality. In case of example **SUR0** the metric variation supposedly is too large in comparison with the prescribed sizes of element, which results in creation smaller elements than expected.

**Table 2.** Meshing time (NT – number of triangles in final mesh, $t_b$ – time of triangulation of boundary nodes, $t_i$ – insertion of inner nodes, $t_s$ – two iterations of Laplacian smoothing, $\tau_t$ – speed of triangulation (boundary and inner nodes) and $\tau_{ts}$ – speed of triangulation with smoothing)

| | | NT $[10^3]$ | $t_b[s]$ | $t_i[s]$ | $t_s[s]$ | $\tau_t[10^3/s]$ | $\tau_{ts}[10^3/s]$ |
|---|---|---|---|---|---|---|---|
| **AN0** | (c) | 11.6 | 0.02 | 0.17 | 0.03 | 61.5 | 52.8 |
| | (a) | 11.7 | 0.02 | 0.19 | 0.05 | 57.7 | 46.9 |
| **AN1** | (c) | 237.6 | 5.64 | 4.81 | 0.89 | 22.7 | 20.9 |
| | (a) | 238.9 | 6.09 | 4.91 | 0.89 | 21.7 | 20.1 |
| **SUR0** | (c) | 53.5 | 1.49 | 1.81 | 0.39 | 16.2 | 14.5 |
| | (a) | 53.6 | 1.49 | 1.91 | 0.39 | 15.8 | 14.2 |
| **SUR1** | (c) | 589.6 | 0.95 | 20.27 | 4.45 | 27.8 | 23.0 |
| | (a) | 589.8 | 0.97 | 20.52 | 4.42 | 27.5 | 22.8 |
| **SUR2** | (c) | 126.6 | 1.49 | 3.47 | 0.91 | 25.6 | 21.6 |
| | (a) | 126.9 | 1.49 | 3.56 | 0.89 | 25.1 | 21.4 |

**Table 3.** Mesh quality ($\mu_{EL}$, $\sigma_{EL}$ – average length and standard deviation of edges length in metric space, $N_{EL}[\%]$ – number of edges shorter than 0.5 or longer than 2, $\mu_T$ and $\sigma_T$ – triangle size quality (ratio of optimal area to current one, in metric space), $\mu_\alpha$ and $\sigma_\alpha$ – alpha criterion[9] of triangle shape quality in metric space)

| | | $\mu_{EL}$ | $\sigma_{EL}$ | $N_{EL}[\%]$ | $\mu_T$ | $\sigma_T$ | $\mu_\alpha$ | $\sigma_\alpha$ |
|---|---|---|---|---|---|---|---|---|
| **AN0** | (c) | 1.079 | 0.165 | 0.03 | 0.912 | 0.244 | 0.961 | 0.042 |
| | (a) | 1.072 | 0.167 | 0.03 | 0.927 | 0.252 | 0.959 | 0.042 |
| **AN1** | (c) | 1.042 | 0.175 | 0.14 | 0.969 | 0.325 | 0.952 | 0.054 |
| | (a) | 1.039 | 0.175 | 0.14 | 0.974 | 0.326 | 0.952 | 0.054 |
| **SUR0** | (c) | 0.802 | 0.255 | 16.66 | 1.577 | 1.275 | 0.913 | 0.098 |
| | (a) | 0.806 | 0.255 | 16.39 | 1.568 | 2.151 | 0.913 | 0.101 |
| **SUR1** | (c) | 0.989 | 0.195 | 1.69 | 1.045 | 0.503 | 0.953 | 0.054 |
| | (a) | 0.990 | 0.195 | 1.68 | 1.046 | 0.659 | 0.952 | 0.055 |
| **SUR2** | (c) | 0.999 | 0.173 | 0.22 | 0.982 | 0.292 | 0.957 | 0.042 |
| | (a) | 1.000 | 0.173 | 0.22 | 0.979 | 0.293 | 0.958 | 0.042 |

## 5 Conclusion

In this paper there were described details of the developed technique of metric introduction into meshing process via coordinate transformation. The presented examples show potential of this method to generate unstructured, possibly anisotropic meshes on 3D parametric surfaces. The running time is close to linear which allows to create meshes with very large number of elements. The quality of meshes in most cases is good with respect to both size and shape of elements, although results for the mesh **SUR0** suggest that additional processing of control space would be advisable to adjust its variation to prescribed size of elements.

**Acknowledgments.** The partial support of the Polish Ministry of Scientific Research and Information Technology (MNII) Grant No. 3T11F00828 is gratefully acknowledged.

## References


1. George, P.L., Borouchaki, H.: Delaunay Triangulation and Meshing. Applications to Finite Elements. Hermès, Paris (1998)
2. Dolejsi, V.: Anisotropic mesh adaptation for finite volume and element methods on triangular meshes. Comput. Visual Sci. **1** (1998) 165–178
3. Frey, P.: Anisotropic metrics for mesh adaptation. In: Proc. ECCOMAS 2004, Jyvaskyla, Finland (24–28 July 2004)
4. Bottasso, C.: Anisotropic mesh adaptation by metric-driven optimization. Int. J. Numer. Meth. Engng. **60** (2004) 597–639
5. Vigo, M., Pla, N.: Computing directional constrained Delaunay triangulations. Computers & Graphics **24** (2000) 181-190
6. Głut, B., Jurczyk, T.: Definition and interpolation of discrete metric for mesh generation on 3d surfaces. Computer Science, Annual of University of Mining and Metallurgy (2005) in press.
7. Jurczyk, T., Głut, B.: Metric 3D surface mesh generation using coordinate transformation method. In: Proc. of CMS'05 Conference on Computer Methods and Systems. Volume 1., Kraków, Poland (2005) 395–405
8. Jurczyk, T., Głut, B.: Generation of good quality quadrilateral elements for anisotropic surface meshes. In: Proc. of ADMOS2003 Conference on Adaptive Modeling and Simulation, Goeteborg, Sweden (2003)
9. Lo, S., Lee, C.: Generation of gradation meshes by the background grid technique. Computers & Structures **50**(1) (1994) 21–32

# A Multiresolution Model for Non-photorealistic Rendering of Trees

Celso Campos[1], Ricardo Quirós[2], Joaquin Huerta[2], Emilio Camahort[3], Roberto Vivó[3], and Javier Lluch[3]

[1] Departamento de Lenguajes y Sistemas Informáticos, Universidad de Vigo, Spain
ccampos@ei.uvigo.es
[2] Departamento de Lenguajes y Sistemas Informáticos, Universitat Jaume I, Spain
{quiros, huerta}@lsi.uji.es
[3] Departamento de Sistemas Informáticos y Computación, Universidad Politécnica de Valencia, Spain
{camahort, rvivo, jlluch}@dsic.upv.es

**Abstract.** Modeling and rendering of trees has recently received a lot of attention. Models have been developed that allow photorealistic rendering of trees at interactive frame rates. However, little attention has been devoted to expressive rendering of these models. In this work we present a multiresolution model designed specifically to speed up painterly rendering of trees. Our method proposes a novel clustering technique based on the computation of nested convex hulls. We employ variable multiresolution to obtain efficient models that contain higher resolution representations for the outside of the tree and lower resolution representations for the inner parts. This variable multiresolution method mimics the techniques used by traditional artists to paint trees.

## 1 Introduction

Due to the intrinsic complexity of trees, the models that represent them are made of large amounts of polygons. Therefore, we can not achieve interactive rates when rendering these models. Current research in the literature used acceleration techniques based on replacing geometry by images. Only a few methods work with geometry.

Geometry, however, is necessary in some cases like in painterly rendering of trees. In this case, the most popular method of representation is polygonal models. The mathematical simplicity of this type of representation makes it possible to render a great number of polygons with the current graphics hardware. However, due to the vast amount of polygons that compose the tree models, it is necessary to use some method that reduces the number of polygons without loss of visual appearance. The most common techniques are multiresolution models or discrete level of detail (LOD) models.

Most of the existing generic multiresolution models can not be applied to our problem, since these models work with polygonal surfaces, whereas the leaves of a tree must be handled as independent polygons. Other specific models for vegetation also present important limitations. The model proposed by Lluch et al. [3] uses precomputed images that replace the leaves, which eliminates the geometry necessary to generate the brush strokes. The work presented by Remolar et al. [11] proposes a

V.N. Alexandrov et al. (Eds.): ICCS 2006, Part II, LNCS 3992, pp. 310 – 317, 2006.

simplification algorithm (LSA: Leaf Simplification Algorithm) that collapses pairs of leaves preventing more efficient approaches.

In this work we present a geometry-based multiresolution model that allows the expressive rendering of trees using brush strokes. We introduce a clustering algorithm to organize the leaves of a tree. Then we propose a simplification algorithm suitable for rendering trees expressively. Our proposal extends the LSA algorithm and adapts it to the requirements of this type of rendering. Our models render at interactive rates on modern graphics hardware.

## 2 Previous Work

Our work is primarily related to two different areas of computer graphics: non-photorealistic rendering of plants and trees and multiresolution modeling.

The first methods for automatic illustration of vegetable species were introduced by Yessios [5] and Sasada [6]. They both produce tree sketches for architectural applications. Kowalski et al. [7] and Markosian et al. [8] also create abstract tree sketches using geometric primitives that approximate the tree's foliage. Deussen [9] presents a method that creates pen-and-ink illustrations of plants and trees. His method supports the representation of specific plants and trees and not just generic trees like those in [7] and [8]. More recently Di Fiore [10] proposed a method that renders cartoon shaded trees from models generated from L-systems.

In previous work [4] we presented a stroke-based method for non-photorealistic rendering of plants and trees. This method improves on Meier's algorithm [1] by supporting interactive frame rates. The method can be applied to videogames, virtual walkthroughs and other real-time computer graphics applications. It models trees using random parametric L-Systems (RL-Systems) [2]. This approach has several advantages over the surface patch algorithm of Meier. For example, it supports the simultaneous generation of both the tree's geometry and its stroke particles for future rendering. The stroke particles are distributed using the same RL-system that is used for modeling the tree's geometry. To achieve this goal we use a *shape instantiation* process. This process represents every instantiable object - a branch or leaf - using both 3D geometry and a cloud of strokes.

Multiresolution is a modeling technique that was first introduced to accelerate rendering of complex geometries [14][16]. Additionally, multiresolution techniques have been proposed for image representations [17], curve and surface modeling [12], and volumetric data sets [13].

Geometry-based simplification methods have been successfully applied to many areas of computer graphics. However, they fail to maintain the general structure of a tree. During the process, the tree's volume diminishes and loses its visual appearance. Some techniques have been proposed for multiresolution models for plants and trees, like degradation at range, space partitioning, layered depth images, volumetric textures, bidirectional textures, leaf impostors and leaf collapsing [15].

However, the proposed models only work for realistic purposes and we need specific techniques to render them expressively.

# 3 Multiresolution Model

RL-Systems allow modeling trees with realistic appearance preserving their structural complexity. This amount of detail can be suitable for photorealistic rendering, but it may be excessive for stroke-based rendering. We propose solving this problem using a multiresolution representation for the trees.

One of the main components of a multiresolution model is its data structure, because it affects the quality of the final rendering of the model. We start with an initial data structure that maintains the structural complexity of the tree. Then a procedure builds the tree's multiresolution model. The model is made of two different data structures: one for the branch brush strokes and the other for the leaf brush strokes.

The extended data structure stores the data implicitly, both for the branches and for the leaves. To limit the size of the data structure we do not store all possible LODs of the tree. Implicit storage also avoids sudden changes (jumps) between contiguous LODs. Although using implicit storage generally increases the cost of LOD extraction, in our case the increase is minimum.

## 3.1 Branch Structure

A tree modeled using an RL System undergoes an instantiation process that generates both a geometric and a brush-stoke representation for branches, leaves, flowers and fruit [4]. The brush-stroke representation is stored in a linked list to optimize memory management. The process generates a branch representation made of multiple LODs. Each LODs has a certain number of strokes starting at three strokes per branch. The minimum LOD provides a low quality result for close views of the tree. The quality can be improved by rendering a larger number of brush strokes for each branch.

To add new brush strokes, we use an adaptive scheme [4] based on the branching level. For each branch we may have several LODs. The LOD finally selected for each branch depends on its branching level. The amount of detail (number of strokes) decreases from the trunk to the outer branches. Strokes are stored in a list containing all LODs organized sequentially. At rendering time the list is traversed until the desired LOD is reached.

## 3.2 Leaf Structure

Each leaf is represented using a brush stroke. For large trees this may require so many brush strokes that rendering may not be interactive. We propose using a model that stores a list of nodes. Each node contains the brush strokes associated to a cluster of leaves. The model stores an implicit representation of the different LODs. This implies that we do not store all computed LODs.

To classify the original leaves in the different nodes we define criteria for clustering and simplification. These criteria are described with detail in the following section. The width of the leaf stroke data structure (see Figure 1) depends on the number of clusters, whereas the depth depends on the number of LODs selected by the user. Width and depth are computed in the clustering process (Section 4). For our test tree we use a structure of four LODs and thirty five clusters.

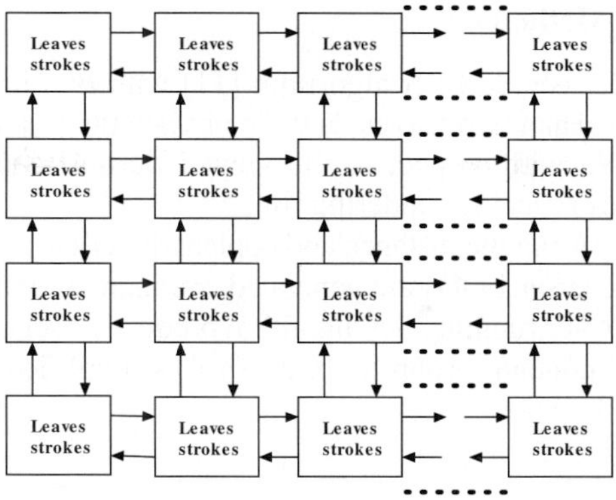


**Fig. 1.** Data structure of our foliage representation: the top row contains the clusters of leaves; lower rows contain lower LOD representations of the leaf clusters

## 4 Clustering and Simplification

The generation of the different LODs of the tree leaves requires a simplification method to decide which information is less relevant and remove it from the model. To implement a simplification method we first cluster the brush strokes of the leaves.

There are two possible approaches. We can cluster the leaves in pairs and then collapse each pair into a single object [11]. Or we can use information about the topology of the tree to cluster the leaves into groups that, for example, share the same branch [10]. This second approach preserves the natural layout of the leaves but it produces a suboptimal simplification.

Our simplification method could benefit from this clustering method, since it gives better visual consistency to certain trees like coniferous. However, we propose a new clustering method based on nested convex hulls.

### 4.1 Clustering Method

To cluster the brush strokes of the leaves we compute a set of nested convex hulls. Then we fill the clusters with the strokes between pairs of neighboring convex hulls. This method is based on techniques used by traditional artists to paint trees. The painter uses initially large rough strokes to paint the tree, and then adds detail using fine strokes over the original ones.

To compute the set of convex hulls, we take the cloud of points formed by the centers of the polygons of all leaves. We obtain their convex hull and remove the points belonging to the convex hull from the initial cloud. We repeat this process until there are three or less points left. This produces a large set of convex hulls. So we only keep a user-selected number of them. They are chosen equally spaced between the largest convex hull and the smallest convex hull that contains at least 10% of the total number of leaf strokes. Given this set of leaf clusters we apply our simplification algorithm to each of them. This approach supports simplifying inner clusters more than outer clusters.

### 4.2 Simplification Method

For simplification we use the LSA algorithm [11] with two improvements: we allow leaf collapses of more than two leaves, and we obviate the requirement to compute the orientation of the tree with respect to the viewer during rendering. This speeds up stroke extraction, thus reducing rendering time.

We collapse leaves using the distance and coplanarity criteria of the LSA. As simplification proceeds and coarser LODs are generated, we increase the distance and coplanarity angle thresholds. After running, our algorithm produces a set of LODs for each cluster of strokes. Each LOD contains groups of brush strokes simplified to single strokes.

## 5 Rendering

Multiresolution modeling requires a data representation, a simplification algorithm and an LOD extraction algorithm for rendering. In this Section we describe how to extract LODs for both branch and leaf brush strokes.

### 5.1 Extraction of Brush Strokes Associated to Branches

The brush strokes associated to the branches are stored in a linked list (see Section 3.1). To extract an LOD we traverse the list rendering all brush strokes we find until a maximum number is reached. This number depends on the branching level and the distance from the viewer to the tree. Our method includes a pruning step that removes branches whose projected length is so short that they can not be perceived.

### 5.2 Extraction of Brush Strokes Associated to Leaves

After applying the simplification algorithm, we know the number of strokes stored at each node of the resulting data structure. For each row of the structure we store the total number of strokes. We use these numbers to decide how to traverse the structure to extract the desired LOD.

The process begins computing the distance from the observer to the model. Based on this distance we determine the number of brush strokes to display. With that number we select the two rows that contain the closest numbers of strokes. We traverse the two rows and select the combination of nodes that better fits the desired number of strokes. Figure 2 shows two example rows with all possible traversals in different

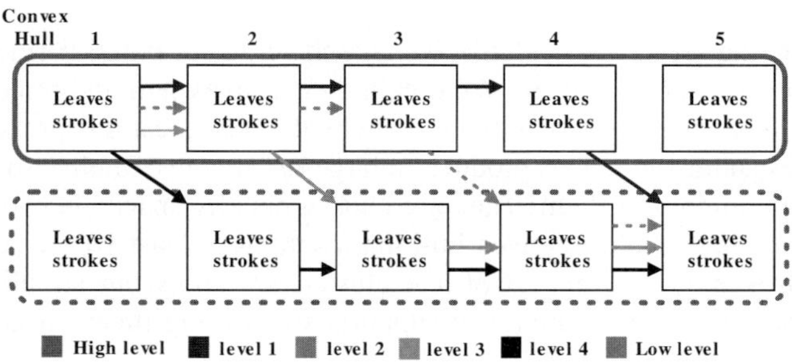


**Fig. 2.** LOD extraction from two rows of the leaf stroke data structure

colors. Note that the first nodes in the selection always belong to the top row and the last nodes to the bottom row.

## 6 Results

We have applied our algorithms to a generic ternary tree with 9841 branches and 8748 leaves. We have rendered the tree using OpenGL in a PC with Windows 2000, a 2.8 GHz AMD Athlon processor and an nVidia GeForce FX 5200 with 128 Mb. Figure 3 shows four renderings of the branches of the tree at different LODs.

We have clustered the leaves into 35 convex hulls each of them simplified four imes. Figure 4 shows four LODs of the leaves of our test tree. Simplified leaf trokes have been drawn in red to show how simplification proceeds from the nside to the outside of the tree. Figure 5 shows a front view of the tree endered at four different

**Fig. 3.** The branches of the test tree and three finer LODs generated with our method. The trees contain 33663, 3754, 1159 and 363 branch strokes with pruning.

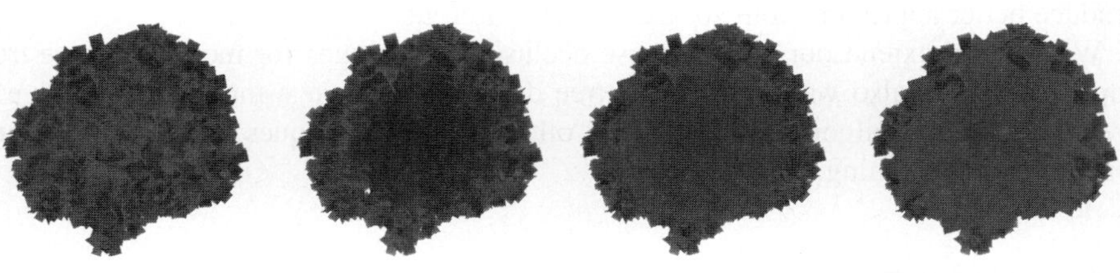

**Fig. 4.** The test tree and three finer LODs. They have 8748, 4156, 2463 and 761 leaf strokes.

**Fig. 5.** The test tree and three finer LODs. They have 42411, 7910, 3622 and 1124 strokes.

**Table 1.** Rendering times of the tree models shown in Figures 3 and 5 (in msec)

| **Only branches (Fig. 3)** | Test Tree | LOD 1 | LOD 2 | LOD 3 |
|---|---|---|---|---|
| Number of strokes | 33663 | 3754 | 1159 | 363 |
| Rendering time | 66.7 | 27.8 | 13.5 | 13.5 |
| **Full tree (Fig. 5)** | Test Tree | LOD 1 | LOD 2 | LOD 3 |
| Number of strokes | 42411 | 7910 | 3622 | 1124 |
| Rendering time | 250 | 125 | 91 | 66.7 |

LODs. The top row shows the LODs at the same size, he bottom row shows real rendering sizes. This view shows better the results f our simplification algorithm.

Table 1 shows the rendering times of the models depicted in Figures 3 and 5.

## 7 Conclusions and Future Work

We have presented a multiresolution model specially designed for painterly rendering of trees. For each tree we generate and render brush strokes. Our solution separately handles the branches and the leaves of a tree. The storage of the LODs is implicit, avoiding an excessive increase in the spatial cost of the data structure. Our method allows efficient extraction and rendering of LODs. Rendering runs at interactive rates and sudden changes between contiguous LODs are not noticeable.

We can extend our method to include viewer-distance dependent adaptive simplification. We are also considering applying our model to photorealistic rendering and including the structural information of the trees in the clustering process. This would produce better approximations for trees like coniferous.

We want to extend our model to use occlusion techniques for modeling large tree populations. We also want to create a tree database, and we want to extend our expressive rendering algorithms to support other artistic techniques like sketching, engraving and half-toning.

## Acknowledgements

This work was partially supported by grant 05VI-1C02 of the University of Vigo, rant TIN2005-08863-C03 of the Spanish Ministry of Education and Science and by STREP project IST-004363 of the 6th Framework Program of the European Union.

## References

1. B. J. Meier, "Painterly rendering for animation", Proceedings of SIGGRAPH 96, pp 477-484, Agosto 1996. New Orleans, Louisiana.
2. J. Lluch, M. J. Vicent, R. Vivó, and R. Quirós, "GREEN: A new tool for modelling natural elements", Proceedings of WSCG'2000 International Conference on Computer Graphics and Rendering, Plzen, 2000.
3. J. Lluch, E. Camahort, R. Vivó, "An Image-Based Multiresolution Model for Interactive Foliage Rendering", The 12-th International Conference in Central Europe on Computer Graphics, Rendering and Computer Vision 2004, Czech Republic.

4. C. Campos, R. Quirós, J. Huerta, M. Chover, J. Lluch, and R. Vivó, "Non Photorealistic Rendering of Plants and Trees", International Conference on Augmented, Virtual Environments and Three-Dimensional Imaging, ICAV3D, Greece, 2001.
5. C. I. Yessios, "Computer drafting of stones, wood, plant and ground materials." Proceedings of SIGGRAPH'79 , pp 190-198, 1979.
6. T. T. Sasada, "Drawing Natural Scenery by Computer Graphics", Computer-Aided Design, vol. 19, pp 212-218, 1987.
7. M. A. Kowalski, L. Markosian, J. D. Northrup, L. D. Bourdev, R. Barzel, L. S. Holden, and J. F. Hughes, "Art-Based Rendering of Fur, Grass and Trees", Proceedings of SIGGRAPH 99, pp 433-438, Agosto 1999. Los Angeles, California.
8. L. Markosian, B. J. Meier, M. A. Kowalski, L. S. Holden, J. D. Northrup, and J. F. Hughes, "Art-based Rendering with Continous Levels of Detail", Proceedings of NPAR 2000, Annecy, France, 2000.
9. O. Deussen and T. Strothotte, "Computer-Generated Pen-and-Ink Illustration of Trees", Proceedings of SIGGRAPH 2000, pp 13-18, Julio 2000.
10. F. Di Fiore, W. Van Haevre, and F. Van Reeth, "Rendering Artistic and Believable Trees for Cartoon Animation", Proceedings of CGI2003, 2003.
11. I. Remolar, C. Rebollo, M. Chover, J. Ribelles, Real-Time Tree Rendering, Lecture Notes in Computer Science, Proc. of Computational Science ICCS 2004, Springer, ISBN/ISSN 3-540-22129-8, krakow (Poland), vol. 3039, pp. 173-180, June, 2004.
12. Matthias Eck, Tony DeRose, Tom Duchamp, Hugues Hoppe, Michael Lounsbery, and Werner Stuetzle. Multiresolution analysis of arbitrary meshes. Proceedings of SIGGRAPH 95, Computer Graphics Proceedings, Annual Conference Series, August 1995, pages 173-182.
13. T. He, L. Hong, A. Kaufman, A. Varshney and S. Wang. Voxel-based object simplification. Proceedings of Rendering 95, IEEE Computer Society Press, 1995.
14. Paul Heckbert and Michael Garland. Multiresolution Modeling For Fast Rendering. In Proceedings of Graphics Interface '94, pages 43-50.
15. O.Deussen, P.Hanrahan, B.Lintermann, R.Měch, M.Pharr, P.Prusinkiewicz. "Realistic modeling and rendering of plant ecosystems." In Proceedings of SIGGRAPH 98, 275-286.
16. E. Puppo and R. Scopigno. Simplification, LOD and Multiresolution Principles and Applications. In Proceedings of EUROGRAPHICS 97. CG Forum, vol. 16, no. 3, 1997.
17. Azriel Rosenfeld. Multiresolution Image Processing and Analysis. Springer-Verlag, Berlin, 1984.

# Model Creation by Velocity Controlled Surface Deformation

Risto Rangel-Kuoppa[1,2] and David Mould[1,⋆]

[1] University of Saskatchewan, Saskatoon SK S7N 5C9, Canada
rir785@mail.usask.ca, mould@cs.usask.ca
[2] Universidad Autónoma Metropolitana - Azc., Av. San Pablo No. 180,
C.P. 02200, México, D.F., México
rrk@correo.azc.uam.mx

**Abstract.** We present a scheme for the semiautomatic creation of 3D models through velocity-controlled surface deformations. Our surface representation consists of oriented points with volatile inter-neighboring point linkage. The surface is resampled in order to maintain an even distribution of points. Points created during resampling inherit their characteristics from their progenitors. Our surface representation and resampling behavior support detailed irregular surfaces with smooth transitions such as those of organic entities. Surface velocities are set by the combination of two types of operators, selection and velocity assignment, with operator application managed by a finite state machine. We demonstrate our scheme with the creation of some branched, fruit-like and mushroom-like models.

## 1 Introduction

We present a velocity-based surface deformation control scheme for model creation. Modeling is achieved by beginning with a simple surface such as a sphere and progressively deforming it to add shape and detail. Our surfaces are represented by a collection of oriented points; each point has a link to every other nearby point. The surface deformation is specified by a set of velocity operators applied to its points, and these operators are in turn managed by a velocity control scheme. Our surface deformation system includes automatic spatial density resampling; surface resampling creates new surface elements that inherit their attributes from the surface elements which spawned them. Our scheme is suited to creating models consisting of detailed surfaces with smoothly varying irregularities, especially organic models such as plants. Figure 1 depicts some of the above concepts and shows a model created by our method.

⋆ We thank and acknowledge the *Universidad Autónoma Metropolitana U. Azc.-D. CBI* and *CONACyT (Mexican Council for Science and Technology)* for sponsoring Risto Rangel-Kuoppa PhD studies. This work was supported in part by NSERC RGPIN299070-04.

V.N. Alexandrov et al. (Eds.): ICCS 2006, Part II, LNCS 3992, pp. 318–325, 2006.

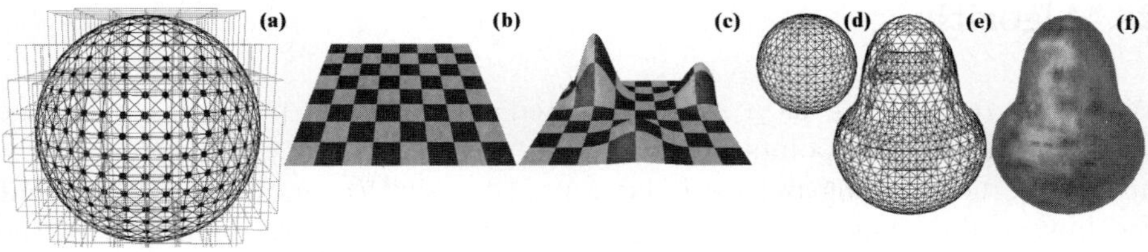


**Fig. 1.** (a) A spherical point cloud rendering with its inter-point neighboring linkage and space partitioning. (b,c) Flat surface deformation to demonstrate surface elements' feature inheritance and normal computation during resampling. (d) An initial sphere. (e) Sphere deformed to a pear shape. (f) Resulting model with pear texture.

### 1.1 Previous Work

Surface deformation has been previously investigated for both polygonal mesh and point-based surface representations. One approach for modifying polygonal meshes is to deform the mesh components: edges and vertices. For example, Combaz and Neyret [1] "paint" stressing forces onto the surface, leading to the elongation or contraction of the edges, and generating folds. Similarly, Lawrence and Funkhouser [2] "paint" velocities on the surface elements and simulate the resulting movement to obtain various models. Both their approaches and ours use the concept of applying velocities to surface elements. We avoid the self-intersection problem, present in both Combaz and Neyret's work and Lawrence and Funkhouser's, by using spatial subdivision to detect self-intersections and correct them by fusing samples.

Techniques for rendering point-based surfaces were reported by Pfister et al. [6] and Rusinkiewicz and Levoy [7] who used oriented points and a space partitioning structure to control the level of detail. This type of surface has also been used as a platform for model creation. Pauly et al. [5], Zwicker et al. [10], and Szeliski and Tonnesen [9] used oriented particles to define their surfaces and provided deformation mechanisms where local or inter-point information was used for resampling purposes. Our approach builds on some of the concepts provided by the previous authors. We determine our surface deformation based on a collection of velocity operators, rather than requiring a user to paint velocity or (in the case of PointShop3D [10]) to manipulate the points interactively. Unlike Zwicker et al., but like Lawrence and Funkhouser, we allow points to have velocities in arbitrary directions. Unlike pure point-based methods, we include in our surface model neighboring inter-point connectivity information that is used to generate the final surface model and to inherit point characteristics during resampling. The surface model of Szeliski and Tonnesen [9] considers the resampling of the surface when its components are too far away from each other, but their surface elements suffer from additional displacements because of the inter-point attracting and repelling forces they used. This situation does not occur in our framework since the inter-point connectivity information is used only to resample the surface and not to alter the positions of surface elements.

## 2 Algorithm

We define our surface as a set of oriented points. Each point has a linkage to its neighbors; two points $p_i$ and $p_j$ are neighbors if they are closer than $\rho$ units. The neighboring linkage is later used for surface resampling. Note that the linkage does not define a polygonal mesh because a triangulation of the points and their linkage would include intersecting and overlapping triangles; nevertheless, a manifold surface would be a subset of this triangulation. Our surface definition lies between the pure point surface definitions [5, 6, 7, 9, 10] and the more traditional polygonal mesh representations.

We accomplish surface deformation by associating velocities with individual points of the surface. These velocities are used to update the points' positions at each simulation step (see Fig. 2) and since the velocities may produce stretching and compressing displacements, resampling is required. Dynamic resampling occurs by reacting to the distance between registered neighbors: if the neighbors are farther than $R_{\text{Max}}$ or closer than $R_{\text{Min}}$ units then a *splitting* or *fusion* operation is respectively triggered.

The resampling operations generate new intermediate points between the pair of points whose distance value triggered the operation. The position of the new point is obtained by linearly interpolating between the old points. Note that our resampling scheme is different from a polygonal mesh resampling [4] since the relations between points may not correspond to a triangulated mesh. Our resampling is also different from those of point-only surface representations [5, 8, 10] because we do not approximate the new points from a subset of points; rather, we only use the spawning points' information.

The points created due to a resampling operation inherit all other characteristics from their parents (including position and velocity), but the normal (orientation) requires a different computation, described as follows and depicted in Fig. 3. First, we create a plane $P_l$ whose normal is the difference between the spawning points positions, $p_{\text{diff}} = \text{norm}(p_j - p_i)$. Next, we take an approximate normal $N_{\text{le}}$ by linearly interpolating the spawning points' normals. Finally, the

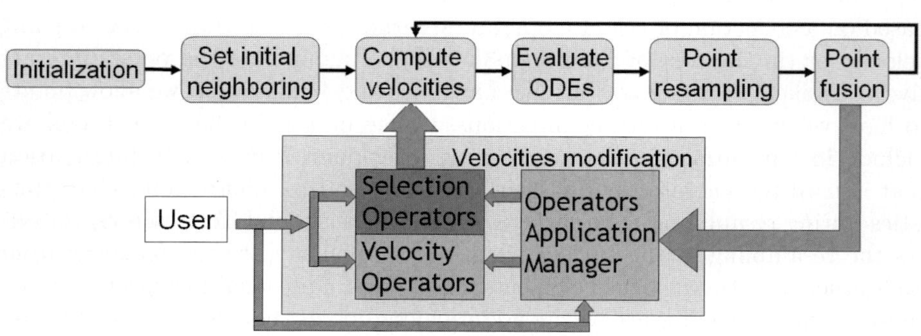


**Fig. 2.** Surface deformation simulation cycle; note that the individual point velocity evaluations and surface resampling are executed in parallel

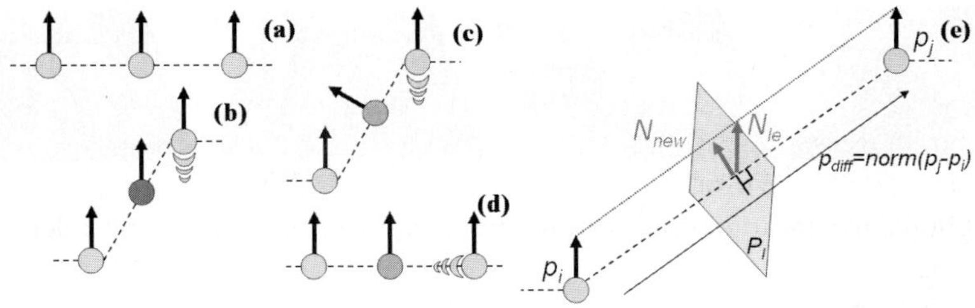


**Fig. 3.** (a) Initial surface, (b) Wrong normal Computation, (c) and (d) proper normal, (e) How normal is computed

normal of the new point is the normalized projection of $N_{\mathrm{le}}$ on the plane $P_{\mathrm{l}}$, i.e., $N_{new} = \mathrm{norm}(N_{\mathrm{le}} - c \cdot p_{\mathrm{diff}})$ where $c = N_{\mathrm{le}} \cdot p_{\mathrm{diff}}$, as shown in Fig. 3.

We will next discuss assigning velocities to the points. We define two types of operators: Selection Operators (SO) and Velocity Operators (VO); the SO selects a group of points and the VO changes the velocities of points to which it is applied. Both types of operators are specified in terms of any measurable geometric property of the surface or its points, as well as accepting numerical parameters; for example, "All points reachable by traversing $n$ links from point $p_{\mathrm{k}}$" would be an SO, and "A point's velocity is twice its distance to the surface centroid" a VO. Typically we apply a VO to the points selected by an SO.

We have used two operators in the examples in this paper. The first one is the "Scaled Unitary Gaussian" operator represented by $SUG(p_{\mathrm{k}}, \sigma, d)$ where $p_{\mathrm{k}}$ is a surface point, $\sigma$ a standard deviation value, and $d$ a distance. The SUG is a two stage operator. The first stage is an SO that chooses all points whose Euclidean distance is less than $d$ units from $p_{\mathrm{k}}$. The second stage modifies the chosen points' velocities by adding a velocity in the normal direction, scaled by a normalized Gaussian distribution centered in $p_{\mathrm{k}}$, i.e., $v_{p_i} = v_{p_i} + n_{p_i} \cdot e^{-\frac{\mathrm{dist}(p_k, p_i)^2}{2 \cdot \sigma^2}}$. The overall effect of the SUG operator is to increase the velocity of points surrounding $p_k$ along their normal direction by a magnitude that falls off as $e^{-x^2}$.

The second operator is the Voronoi Regionalization $VR(m, t)$ operator, which is an SO only. Using a variant of Lloyd's algorithm [3], this operator creates $m$ subsets of points making $t$ passes of the algorithm; the subsets correspond to centroidal Voronoi regions. Applying this operator to a model results in a partition of the model, which can later be used by another operator. For example, in Fig. 4.a we can see the results of applying $VR(13, 1)$ and in Fig. 4.b we see $VR(13, 10)$. In Fig. 4.c we have used the centers of the regions created with $VR(13, 10)$, say $p_1, \ldots, p_{13}$, and applied $SUG(p_i, 0.9, 8.0)$ $i = 1, \ldots, 13$. In Fig. 4.c the points color intensities are proportional to their velocities. Finally, Fig. 4.d shows the surface after it has evolved into a surface with Voronoi-distributed bumps; each bump has a smooth transition to the flat surface.

Even though just having a library of operators would be very useful for the creation of different models, a user would still need to apply them individually. As part of our scheme we provide an Operator Application Manager (OAM) to control which operators are applied and when and where they are applied.

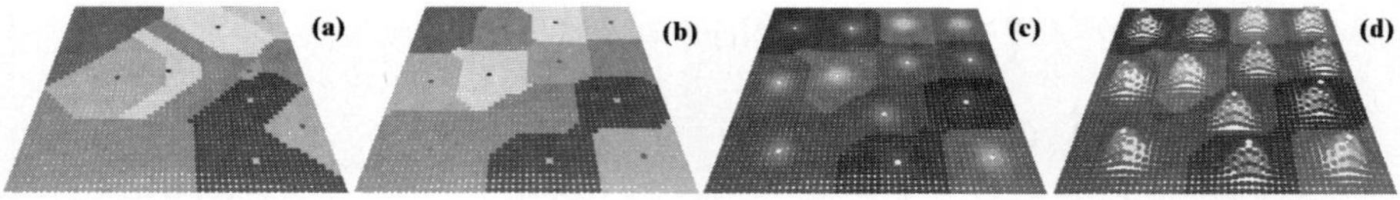


**Fig. 4.** Simple deformation operators used in series to achieve a complex deformation

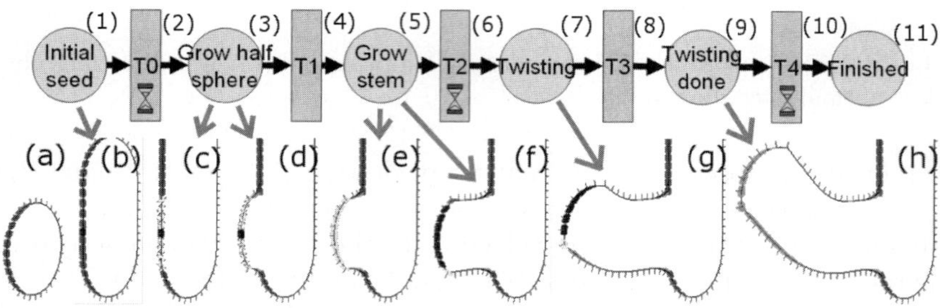


**Fig. 5.** Above, an OAM net controlling the operators to be applied. Below, application of the OAM net

The OAM controls the operators through a finite state machine. An OAM is similar to a Petri Net, the difference being that the transition elements in the OAM can be fired when only one of the input states contains a token. The OAM net transitions are constantly monitoring a specified point cloud metric; possible metrics include maximum height, number of subregions, or time elapsed. The OAM net states can be associated with a VO. Figure 5 shows an OAM net that is used to grow a branch that later twists to a given angle. With the proper application of the operators and the OAM nets, our system can simulate a variety of effects. Global effects like gravity, local effects such as tropism, or local modifications to the mesh such as branches can all be described within our system.

We next describe the behavior of the OAM net in Fig. 5. This net contains only one token, initially placed in the state *Initial seed.*

1. State *Initial seed* – Applies the SO CreateRadialSubset$(P_i, r)$ which selects the reachable points in a radius $r$ from $P_i$ and records the current position of $P_i$, $\overrightarrow{p_0} = \overrightarrow{p_i}$.
2. Transition *T0* – Timed transition firing after 0 seconds.
3. State *Grow half sphere* – Applies a VO that assigns to each of the points selected in the previous state a normalized velocity proportional to the difference of their distance to $P_i$ and $r$, i.e., $\overrightarrow{V_{\text{new}}} = \overrightarrow{N} \cdot (r - \|\overrightarrow{p_x} - \overrightarrow{p_i}\|)$
4. Transition *T1* – Fires when the distances from all selected points to the original position of $P_i$, $\overrightarrow{p_0}$, are greater than $r$.
5. State *Grow stem* – Applies a VO setting the selected points' velocities to have the same direction as their normals.

6. Transition *T2* – Use an VO to make the selected points' velocities equal to zero and compute a pivoting place in space $\overrightarrow{T} = \overrightarrow{p_x} - (Y_{\text{offset}} \cdot \widehat{j})$ where $Y_{\text{offset}} \geq r$.
7. State *Twisting* – Rotates the selected points upward using pivot $\overrightarrow{T}$.
8. Transition *T3* – Fires when all the selected points have rotated $\gamma$ degrees.
9. State *Twisting done* – Applies the same operator used in state *Grow stem.*
10. Transition *T4* – Fires after *t4* seconds
11. State *Finished* – Marks the end.

All of the concepts described in this section have been implemented in a software tool. This tool has an interface similar to that of Lawrence and Funkhouser [2] with which a user can specify the parameters of an initial surface. The surface can be deformed by either applying velocities to individual points selected by the user or applying the deformation operators. The deformation is controlled by the user and can be paused at any time; also the user can apply new velocities or operators at any moment. This software tool was used to create the models shown in the following section.

## 3 Results

Using the surface specification, operators, and OAM defined previously, we have created a set of models. Figure 6.a-f depicts the steps involved in the creation of an apple model. First, following a ring path, an SUG operator is applied to the upper half of the initial sphere. Next, four instances of the same operator are applied on the lower half plus a negative weighted SUG is applied to the lowermost point, creating a crown-like section. Figure 6.c shows the surface evolved in time. Two additional Gaussian-like velocity distributions are applied to the upper half to mimic the heart-shaped contour that some apples present (Fig. 6.d). Figure 6.e-f shows the final point cloud and its connectivity mesh. More complex models can be obtained if the operators are applied to subsets of the point cloud. Figure 6.g-l shows the steps to create the basic structure of mushroom-like models.

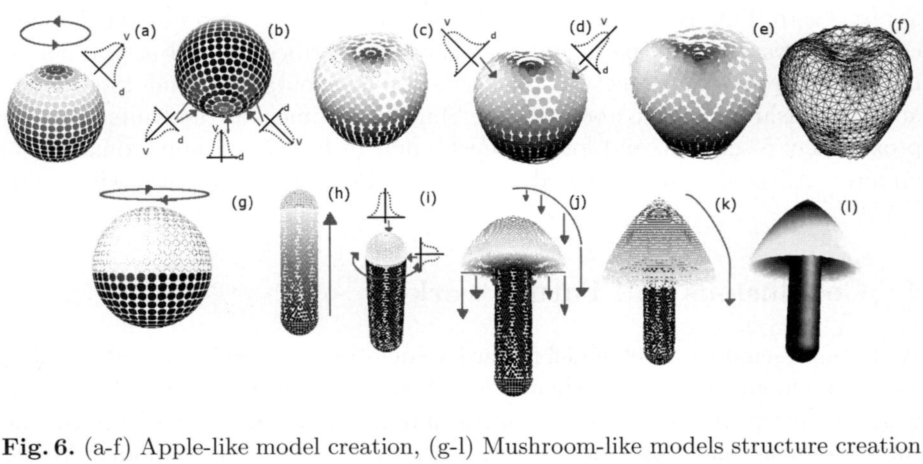


**Fig. 6.** (a-f) Apple-like model creation, (g-l) Mushroom-like models structure creation

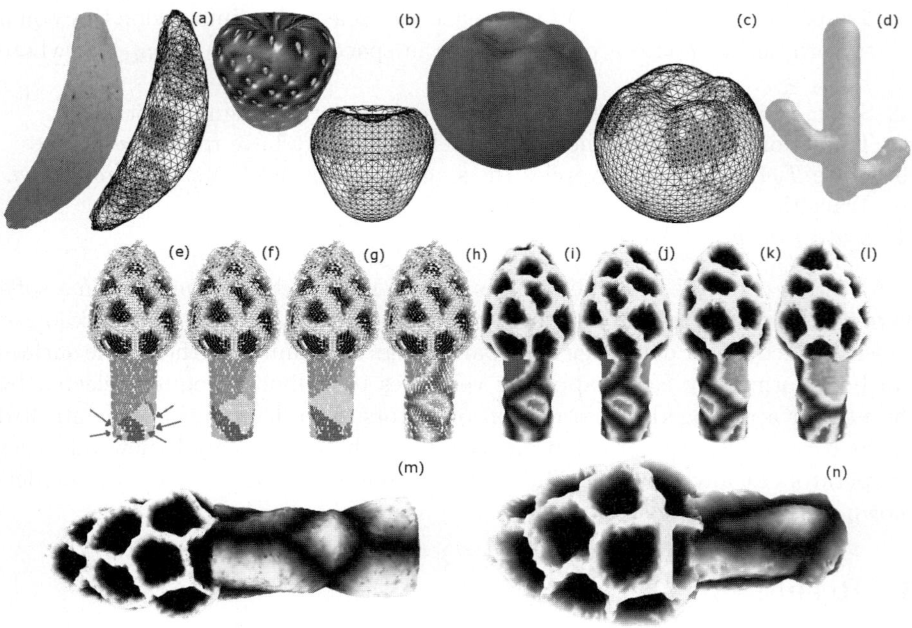


**Fig. 7.** (a-d) Fruit-like models, (e-h) Mushroom model detail creation, (i-l) Rotated views of mushroom model, (m-o) Close ups of mushroom model

Additional examples created with similar techniques are shown in Fig. 7.a-c. Figure 7.d was generated with two of the OAM nets shown Fig. 5. One of our favorite models is shown in Fig. 7.e-n. Initially we have a mushroom model with a stem and an egg-shaped cap to which a Voronoi regionalization operator is applied. Figure 7.e-g shows the result of the Voronoi regionalization operator with initial seeds randomly placed on the stem's base. In Fig. 7.h, the points' velocity is set as a function of their distance to the center of their respective Voronoi region; if the velocity is greater than a certain threshold, then the velocity to apply is zero. Finally, in Fig. 7.m and Fig. 7.n we can see two different models created by varying the parameters previously described as well as a different initial mushroom-like structure. The fruit models contain around 1500 points and the mushrooms have about 27000. Simulation times for the fruits were approximately 5 seconds, and for the mushrooms of Fig. 7.e-n, approximately 30 minutes. All models were created on a 3 GHz Pentium 4 computer with 1 GB RAM.

## 4 Conclusions and Future work

We have presented a novel model for surface deformation and its use with a semi-automatic model generation scheme. The surface specification can be considered a hybrid between point based surfaces and triangulated meshes because the surface is represented by a point cloud with connections between points. We do not have to maintain consistency as with a mesh, but we do have neighborhood information for performing local operations on the surface, unlike a pure point cloud. We have defined some basic operators and provided specific detail on two specialized operators: application of a Gaussian velocity distribution and Voronoi regionalization of point subsets. We showed some models whose surfaces have irregularities with smooth transitions, the type of surfaces commonly present in natural models. As part of our continuing research we are looking to implement different operators that reflect other natural phenomena of growing entities, such as bark and lichen growth, and we are also aiming to simulate aging effects on surfaces like cracking and erosion.

## References

1. Combaz J. and Neyret F., *Painting Folds using Expansion Textures*, Pacific Graphics, 2002, vol. 1, no. 1, pp. 176-183, October
2. Lawrence J. and Funkhouser T., *A Painting Interface for Interactive Surface Deformations*, 11th Pacific Conference on Computer Graphics and Applications (PG'03), 2003, vol. 1, no. 1
3. Lloyd S., *Least squares quantization in PCM*, IEEE Transactions on Information Theory, 1982, vol. 28, no. 2, pp. 129-137
4. Mandal Chandomay, Qin H. and Vemuri B., *Dynamic modeling of butterfly subdivision surfaces*, IEEE Transactions on Visualization and Computer Graphics, 2000, vol. 6, no. 3, pp. 265-287
5. Pauly M., Keiser R., Kobbelt L.P. and Gross M., *Shape modeling with point-samples geometry*, Proceedings of the ACM SIGGRAPH 2003, 2003, vol. 22, no. 3, pp. 641-650, July
6. Pfister H., Zwicker M., van Baar J. and Gross M., *Surfels: Surface elements as rendering primitives*, Proceedings of the 27th Annual Conference on Computer Graphics and Interactive Techniques SIGGRAPH'00, 2000, vol. 1, no. 1, pp. 335-342
7. Rusinkiewicz S. and Levoy M., *QSplat: A multiresolution point rendering system for large meshes*, Proceedings of the 27th Annual Conference on Computer Graphics and Interactive Techniques, SIGGRAPH'00, 2000, vol. 1, no. 1, pp. 343-352
8. Sethian J., *Level Set Methods*, United States of America: Cambridge University Press
9. Szeliski R. and Tonnesen D., *Surface Modeling with Oriented Particle Systems*, Computer Graphics, 1992, vol. 2, no. 26, pp. 185-194, July
10. Zwicker M., Pauly M., Knoll O. and Gross M., *Pointshop3D: An interactive system for point-based surface editing*, Proceedings of the 29th Annual Conference on Computer Graphics and INteractive Techniques (SIGGRAPH'02), 2002, vol. 1, no.1, pp. 322-329

# Rendering of Unorganized Points with Octagonal Splats

Sun-Jeong Kim and Chang-Geun Song

Hallym University,
Division of Information Engineering and Telecommunications,
Chuncheon-si, Gangwon-do 200-702 Korea

**Abstract.** Point sets from range scanners are noisy and do not contain additional data such as normals, which makes it difficult to create a mesh for rendering. We use the moving least-squares (MLS) approximation to get normals and differential information, and approximate the surface near each point by an MLS surface. From this we obtain a splat of a new octagonal shape which approximates the local curvature at a point and allows fast and high-quality rendering. We show several results which demonstrate the practicality and benefits of our algorithm.

## 1 Introduction

Recent advances in scanning technology have led to a rapid increase in the availability and size of geometric data sets. Some consist of hundreds of millions of points. Point data are usually considered by generating and simplifying a triangular mesh. However some of the meshes obtained from scan data are too large to be displayed in real time, and the running-time and space requirements of mesh reduction and progressive display algorithms are impractical at this scale.

Point-based rendering schemes have evolved as an alternative to the use of intermediate triangle-based representations. Their major advantage is that point sets do not require any connectivity information, and no topological consistency conditions have to be satisfied. Moreover these schemes involve a hierarchical data structure which allows the efficient control of frame rates and visual quality.

Most point rendering algorithms handle points sampled from a large mesh, which means that the normals of points can be easily obtained from the input mesh. Grossman and Dally [4] converted geometric models into point-sampled data sets and then rendered them. They addressed the issues of sampling rate and gaps in the rendered images. The *Surfel* concept [8] follows from this work. This technique uses an octree-based structure and forward warping. A significant challenge for point rendering techniques is to reconstruct continuous surfaces from irregularly spaced point samples. In order to eliminate any gaps in the displayed objects, splat-based schemes are used. *QSplat* [9] replaces points by ellipses or rectangles in image space and uses a hierarchy of bounding spheres. Sphere radii and normal cones are saved to perform time-critical rendering. *Surface splatting* [11] is an object-space technique and focuses on texture filtering. Point rendering with Gaussian resampling kernels, also called *EWA splatting*, has provided

V.N. Alexandrov et al. (Eds.): ICCS 2006, Part II, LNCS 3992, pp. 326–333, 2006.

the highest image quality so far, with antialiasing capabilities. This technique has been applied to interactive shape design [12]. To reduce the computational complexity of EWA splatting, Botsch et al. [3] proposed an approximation using look-up tables. The above point rendering schemes store a limited amount of information derived from the point primitives, such as normals, bounding spheres, and tangent plane disks. But Kalaiah and Varshney [5] store local differential geometric information with every point. Their approach renders a surface as a collection of local neighborhoods, consisting of a rectangle for each point, with per-pixel shading. Although the rendering quality is very high but their scheme cannot be applied to an irregular point set because they sample points uniformly in the parametric domain of a NURBS surface. Wu and Kobbelt [10] presented sub-sampling technique for dense point sets. They compute a minimal set of splats that covers a point set within a given error tolerance. To reduce the splat count, point simplification [7] is applied.

Levin [6] has recently introduced a new point-based surface representation called moving least-squares (MLS) surfaces. Based on this representation, Alexa et al. [1] have proposed new modeling and rendering techniques, and Amenta and Kil [2] have developed a new explicit definition of point-set surfaces and variants of the MLS surface.

In this paper, we propose a new algorithm to render completely unorganized points which have been obtained from a range scanning system. For each point, the normal vector is computed by the eigen-analysis of the covariance matrix and then with this normal an MLS surface is generated to approximate the points in a small neighborhood around it. After approximation, we construct a splat of an octagonal shape by extracting the positions and normals of vertices of octagon from an MLS surface. At last the points are rendered using these octagonal splats. Our contributions are the follows:

– We handle point sets straight from a range scanning system. Previous rendering algorithms make use of sampling order or other pre-existing information such as normals and differential information. But our technique allows the display of point sets that are genuinely unorganized.
– We render points with octagonal splats which allow for the curvatures at a point. The two orthogonal axes and normals of each octagon correspond to the estimated principal curvature directions and normals on an approximated smooth surface. And an octagon is faster to draw than a circle or ellipse.

## 2 Normal Estimation

Normal vectors are estimated by analyzing the local neighborhood of each point. Our computation is similar to [1]. Since there is no connectivity information in a point set, the local neighborhood of each point is constructed by $k$-nearest neighbors. Let $r$ be a point and $\{r_0, \ldots, r_{k-1}\}$ its $k$-nearest neighbors. The matrix of weighted covariances be defined by

$$C := \sum_{i=0}^{k-1} \theta_i (r_i - r)(r_i - r)^T \in \mathcal{R}^{3\times3}, \tag{1}$$

with the weights $\theta_i = \theta(\|r_i - r\|)$. $\theta$ is a non-negative weight function, usually a Gaussian of the form:

$$\theta(d) = e^{-\frac{d^2}{h^2}},$$

where $h$ is a fixed parameter reflected the anticipated spacing between neighboring points. Finding a global value of $h$ is difficult for non-uniformly distributed point clouds. Similar to [7], for each point we choose different $h$ value as the distance to the seventh closest neighbor point from it.

The eigenvector of $C$ corresponding to the smallest eigenvalue gives an estimate for the normal direction. Since there is no consistent orientation over all points, we correct the normal directions by propagation along a minimal spanning tree. We also use the other two eigenvectors for computing a splat shape because they are approximately equal to the principal curvature directions $\boldsymbol{\kappa}_1$ and $\boldsymbol{\kappa}_2$.

## 3 MLS Approximation

We begin with the brief review of the definition of MLS surfaces [6]. An MLS surface is implicitly defined as points that project on to themselves under the MLS projection. The procedure to approximate this surface involves two steps: the approximation of a local reference plane and fitting a local bivariate polynomial to points projected on to the reference plane.

Let $\{r_i\}_{i\in I}$ be points on a surface $S$ in $\mathcal{R}^3$ (possibly subject to noise). Given a point $r$, we first find a local reference plane $H = \{x|\langle \boldsymbol{n}, x\rangle - d = 0, x \in \mathcal{R}^3\}, \boldsymbol{n} \in \mathcal{R}^3, \|\boldsymbol{n}\| = 1$ in $\mathcal{R}^3$, and a point $q$ on $H$ (i.e. $\langle \boldsymbol{n}, q\rangle = d$ – see Figure 1) such that the following quantity is minimized:

$$\sum_{i\in I} (\langle \boldsymbol{n}, r_i\rangle - d)^2 \theta(\|r_i - q\|), \tag{2}$$

where $\langle \cdot, \cdot \rangle$ is the standard inner product in $\mathcal{R}^3$, $\|\cdot\|$ is the Euclidean distance in $\mathcal{R}^3$.

Now we compute a local bivariate polynomial approximation of $S$ over the reference plane. Let $\{x_i\}_{i\in I}$ be the orthogonal projections of the points $\{r_i\}_{i\in I}$ on to $H$, represented in a specific orthonormal coordinate system on $H$, and let $f_i = \langle n, r_i\rangle - d, i \in I$ be the heights of the points $\{r_i\}_{i\in I}$ over $H$. We will choose the origin of the orthonormal coordinate system on $H$ to be at $q$. We define a local approximation to $S$ of degree $m$ as a bivariate polynomial $p$ that minimizes the weighted least-squares error:

$$\sum_{i\in I} (p(x_i) - f_i)^2 \theta(\|r_i - q\|). \tag{3}$$

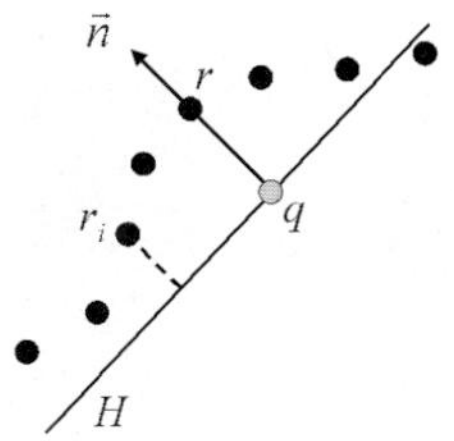


**Fig. 1.** The MLS projection. Given a point $r$, a local reference plane $H$ and a point $q$ on $H$ are computed. An MLS surface is defined implicitly by the MLS projection as the set of points that project on to themselves.

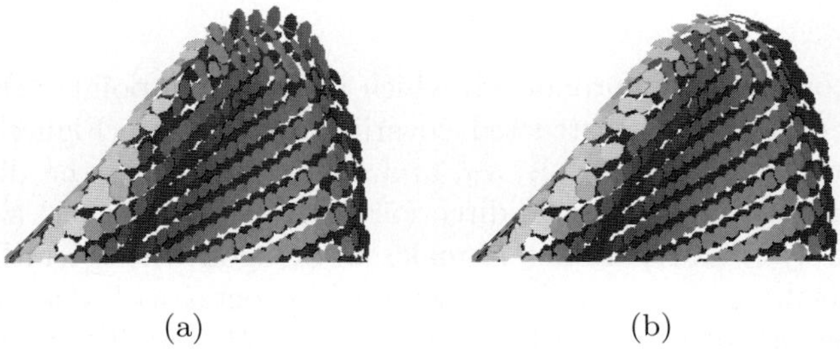


**Fig. 2.** Splat shapes with (a) the wrong-estimated normals on the top of a point-set by eigen-anlaysis and (b) the corrected normals by minimizing the MLS error

In this paper, we use a third-degree polynomial, because it produces a good fit, does not oscillate, and is quick to compute.

To obtain an MLS approximation surface, we should solve non-linear minimization problem. However, for simplifying this computation, we use an estimated normal as $\boldsymbol{n}$ and set $t = 0$ (i.e. $q = r$), since in our experiments an estimated normal is reliable and the position of a reference plane is not important after the normal direction is fixed. Even if an estimated normal is wrong, it can be corrected by choosing the one to minimize the MLS error among three eigenvectors (see Figure 2).

## 4 Octagonal Splat

After the MLS approximation, we construct an octagonal splat shape and its normal vectors for each point. The two orthogonal axes and normals of each octagonal splat correspond to the estimated principal curvature directions and normals on an approximated MLS surface. The octagonal shape is faster to render than the elliptical shape which many previous schemes constructed. For hole-free rendering, the Voronoi diagram is used. Because an octagonal splat covers the Voronoi cell which includes the point, there is no gap between splats.

On the reference plane $H$ of a point $r$ (Figure 3(a)), the Voronoi diagram of orthogonal projected points is computed (Figure 3(b)). A bounding square is

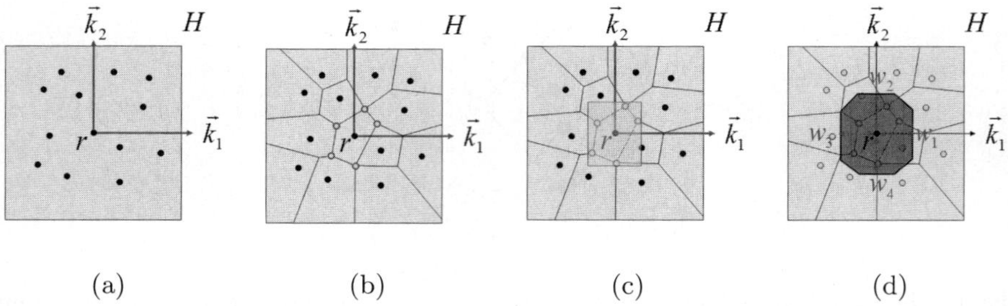

Fig. 3. Construction of an octagonal splat shape: (a) the orthogonal projections of neighbor vertices for each point $r$ on to $H$, (b) the Voronoi diagram, (c) a bounding square which is constructed covering the Voronoi cell, (d) four parameters $\omega_1$, $\omega_2$, $\omega_3$, and $\omega_4$

constructed covering the Voronoi cell which includes the point $r$ (Figure 3(c)), and then an octagon is constructed covering this square (Figure 3(d)). Now we decide the parameters $\omega_1$, $\omega_2$, $\omega_3$, and $\omega_4$ by measuring the distance from the point $r$ to the octagon in the direction of principal axes $\boldsymbol{\kappa}_1$, $\boldsymbol{\kappa}_2$, $-\boldsymbol{\kappa}_1$, and $-\boldsymbol{\kappa}_2$ respectively. These parameters make an octagon whose width is $\omega_1 + \omega_3$ and whose height is $\omega_2 + \omega_4$. The center of an octagonal splat is the point $O(0,0)$ in the orthogonal coordinate system on $H$ and its eight end-points are $x_i = \{(\omega_1, -s\omega_4), (\omega_1, s\omega_2), (s\omega_1, \omega_2), (-s\omega_3, \omega_2), (-\omega_3, s\omega_2), (-\omega_3, -s\omega_4), (-s\omega_3, -\omega_4), (s\omega_1, -\omega_4)\}$, where $s = \sqrt{2} - 1$. Therefore the vertices $v_i$ of the octagon for each point $r$ in $\mathcal{R}^3$ are computed as follows:

$$v_i = r + \alpha_i \boldsymbol{\kappa}_1 + \beta_i \boldsymbol{\kappa}_2 + \gamma_i \boldsymbol{n}, \tag{4}$$

where $x_i = (\alpha_i, \beta_i)$ and $\gamma_i = p(x_i) - p(O)$.

The octagon is not planar, but distorted to correspond to the estimated curvature at the point. In the last step, we compute the normals of vertices of the octagon. Because we have already an MLS polynomial function, in order to obtain the normals we just differentiate the polynomial partially and then transform them into the global coordinate system. The normals $\boldsymbol{n}_i$ of the vertices $v_i$ in the local coordinate system is

$$\boldsymbol{n}_i = \Big( -\frac{p(x_i)}{\partial \boldsymbol{\kappa}_1}, \ -\frac{p(x_i)}{\partial \boldsymbol{\kappa}_2}, \ 1 \Big). \tag{5}$$

## 5 Implementation and Results

All our tests were run on a PC with a Pentium IV 2.4 GHz processor and 2GB of main memory. Figure 4 shows the pictures produced by our point rendering algorithm. These figures show how our octagonal splat shape avoids gaps in the rendered image.

We have compared our rendering results with those of QSplat [9] and Pointshop 3D [12] (Figure 5 and 6). QSplat uses flat splats such as circles, ellipses, or

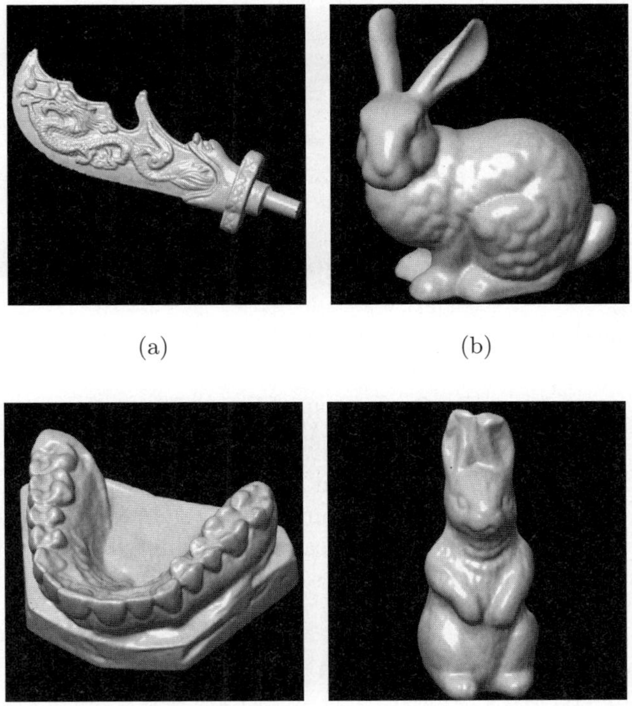

(a) (b)

(c) (d)

**Fig. 4.** Differential rendering of various point-sets: (a) the "Chinese Scimitar" (441,113 points) (b) the "Stanford Bunny" (34,834 points), (c) the "Teeth" (116,604 points), (d) the "Rabbit" (67,039 points)

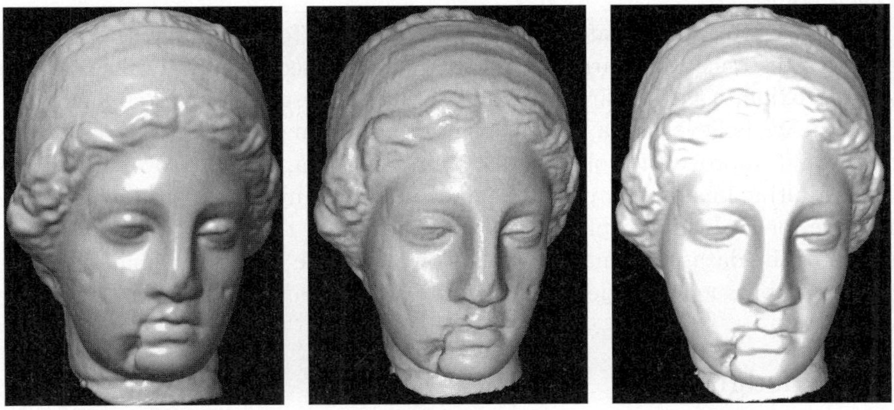

(a) Our method (b) Qsplat (c) Pointshop 3D

**Fig. 5.** Comparison of our technique with QSplat [9], and Pointshop 3D [12]: the "Igea" model (134,345 points)

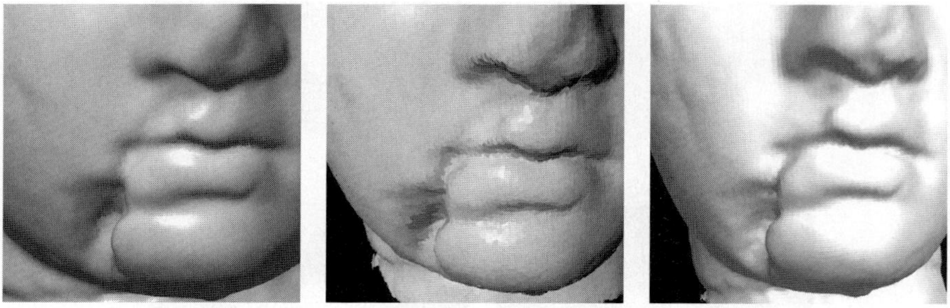

(a) Our method (b) Qsplat (c) Pointshop 3D

**Fig. 6.** Close-up of Figure 5

rectangles, while ours is a curved octagon. Consequently, the visual quality of our results is higher than these of QSplat. Pointshop 3D samples the image on a regular grid based on Surfels. Therefore, the sampling technique may lead to aliasing. The close-ups in Figure 6 show that our rendering quality is higher than others.

## 6 Conclusion and Discussion

We have proposed a new scheme to render unorganized points. While previous schemes render sampled-point data, ours can handle scanned data which have no additional pre-existing information such as normals. To construct the splat shape and obtain the differential information, we made use of the MLS approximation. Also we built up octagonal splats for each point which reflected the curvature at each point.

During MLS approximation, it is very difficult to determine the nearest-neighbors distance in an irregular point-set. We used adaptive MLS surfaces that approximate the $k$ nearest neighbors, but the best value of $k$ remains undetermined. We believe that a preprocessing step to thin out over-dense data or fill holes would improve the robustness of our algorithm, and improve the quality of the resulting image. Many point rendering algorithms construct a hierarchical structure for real-time display. But the efficiency of this structure depends on the size of the dataset: it may actually slow down the display of small datasets. We need to pursue an efficient compromise.

## Acknowledgments

This research was supported by Hallym University Research Fund, 2005 (HRF-2005 - 31) and the project for developing innovative local industries which was conducted by Regional Innovation Division in the Ministry of Commerce Industry and Energy of the Korean Government.

## References


1. Alexa, M., Behr, J., Fleishman, S., Cohen-Or, D., Levin, D., and Silva, C. T. Computing and rendering point set surfaces. *IEEE Transactions on Visualization and Computer Graphics 9*, 1 (2003), pp. 3–15.
2. Amenta, N. and Kil, Y. J. Defining point-set surfaces. *ACM Transactions on Graphics 23*, 3 (SIGGRAPH 2004), pp. 264–270.
3. Botsch, M., Wiratanaya, A., and Kobbelt, L. Efficient high quality rendering of point sampled geometry. *Proceedings of 13th Eurographics Workshop on Rendering* (2002), pp. 53–64.
4. Grossman, J. P. and Dally, W. J. Point sample rendering. In *Proceedings of the 9th Eurographics Workshop on Rendering* (1998), pp. 181–192.
5. Kalaiah, A. and Varshney, A. Differential point rendering. In *Proceedings of the 12th Eurographics Workshop on Rendering* (2001), pp. 139–150.
6. Levin, D. Mesh-independent surface interpolation. *Geometric Modeling for Scientific Visualization* (2004), Brunnett, G., Hamann, B., Muller, H., and Linsen, L. (eds.), Springer-Verlag, pp. 37–50.
7. Pauly, M., Gross, M., Kobbelt, L. Efficient simplification of point-sampled surfaces. In *Proceedings of IEEE Visualization 2002* (2002), pp. 163–170.
8. Pfister, H., Zwicker, M., Baar, J. V., and Gross, M. Surfels: surface elements as rendering primitives. In *Proceedings of SIGGRAPH 2000* (2000), pp. 335–342.
9. Rusinkiewicz, S. and Levoy, M. QSplat: a multiresolution point rendering system for large meshes. In *Proceedings of SIGGRAPH 2000* (2000), pp. 343-352.
10. Wu, J. and Kobbelt, L. Optimized sub-sampling of point sets for surface splatting. *Computer Graphics Forum 23*, 3 (Eurographics 2004), to appear.
11. Zwicker, M., Pfister, H., Baar, J. V., and Gross M. Surface splatting. In *Proceedings of SIGGRAPH 2001* (2001), pp. 371–378.
12. Zwicker, M., Pauly, M., Knoll, O., and Gross M. Pointshop 3D: An interactive system for point-based surface editing. *ACM Transactions on Graphics 21*, 3 (SIGGRAPH 2002), pp. 322–329.

# Statistical Based Vectorization for Standard Vector Graphics

Sebastiano Battiato, Giovanni Maria Farinella, and Giovanni Puglisi

Dipartimento di Matematica e Informatica, University of Catania, Italy
Image Processing Laboratory
{battiato, gfarinella, puglisi}@dmi.unict.it
http://www.dmi.unict.it/~iplab

**Abstract.** In this paper a novel algorithm for raster to vector conversion is presented. The technique is mainly devoted to vectorize digital picture maintaining an high degree of photorealistic appearance specifically addressed to the human visual system. The algorithm makes use of an advanced segmentation strategy based on statistical region analysis together with a few ad-hoc heuristics devoted to track boundaries of segmented regions. The final output is rendered by Standard Vector Graphics. Experimental results confirm the effectiveness of the proposed approach both in terms of perceived and measured quality. Moreover, the final overall size of the vectorized images outperforms existing methods.

## 1 Introduction

The vectorial format is useful for describing complex graphical scenes using a collection of graphic vectorial primitives, offering the typical advantages of vectorial world: scalability, resolution independence, etc. In this paper we propose a novel technique to cover the gap between the graphical vectorial world and the raster real world typical of digital photography. Standard Vector Graphics (SVG) is a standard language for describing two-dimensional graphics in XML ([1]). This format could be very useful in the world of mobile imaging device where the typical capability of a camera needs to match with limited colour/dimension resolutions display. Another potential application is presented in [2] where an ad-hoc method for icon vectorization is applied emphasizing the good performances in terms of resolution scalability.

Recently, some commercial and free software have been developed using some solution to the "raster to SVG" problem (Vector Eye [3], Autotrace [4], Kvec [5]). Almost all of them are devoted to SVG rendering of graphic images (e.g. clip art, etc.) showing in such case good performances but also several perceptual drawbacks when applied to digital pictures.

Considering photorealistic vectorization, a few advanced approaches are described in SWaterG [6], SVGenie and SVGWave [7], where ad-hoc segmentation strategies, making use respectively of watershed decomposition and adaptive triangulation, allow to obtain good performances both in terms of perceptual accuracy and overall compression size. An useful review is presented in [8]. We

V.N. Alexandrov et al. (Eds.): ICCS 2006, Part II, LNCS 3992, pp. 334–341, 2006.

also mention the work presented in [9] where a region-contour segmentation scheme that takes into account contours resulting from colour gradients in an image, as well as their interrelationships in terms of bounding a common region, is described. One of the main drawback of almost all cited approaches is the parameters tuning with respect to the final application. The correct trade-off between final perceived quality, scalability and the corresponding file size is a very challenging task, mainly for user that doesn't know the algorithmic details of each solution. Our proposed technique is based on a single input parameter, that fixes the degree of "coarseness" to be used in the segmentation step. An efficient algorithm has been used for the preliminary analysis of input raster data: Statistical Region Merging (SRM) [10]. Such segmentation algorithm is able to capture the main structural components of a digital image using a simple but effective statistical analysis.

Our technique consists of two main steps: the image is partitioned in polygonal regions using SRM; the borders of segmented regions are properly tracked and described by SVG primitives making also use of some smart tips.

The final results show how the proposed strategy outperforms state-of-art techniques in terms of visual perception, measured image quality and compression size. The overall performances are described in terms of Rate-Distortion plots, by varying the unique involved quality parameter.

The rest of the paper is organized as follows. In Section 2 we briefly review the main details of SRM. The successive Section describes how boundary representation is computed while Section 4 reports the generation of SVG representation by using suitable mapping strategies. In Section 5 several experimental results together with some comparisons with other techniques are reported. A final Section closes the paper tracking also direction for future works.

## 2 Statistical Region Merging

Segmentation is the process of partitioning an image into disjoint and homogeneous regions. A more formal definition can be given in the following way [11]: let $I$ denote an image and let $H$ define a certain homogeneity predicate; the segmentation of $I$ is a partition $P$ of $I$ into a set of $N$ regions $R_1$,$R_2$,...,$R_N$, such that: $\bigcup_{n=1}^{N} R_n = I$ with $R_n \cap R_m = \emptyset$, $n \neq m$; $H(R_n) = true \quad \forall n$; $H(R_n \cup R_m) = false \quad \forall R_n$ and $R_m$ adjacent.

Recently, thanks to the increasing speed and decreasing cost of computation, many advanced techniques have been developed for segmentation of colour images. In particular we used the Statistical Region Merging [10] algorithm that belongs to the family of region growing techniques with statistical test for region fusion. SRM is based on the follow model of image: $I$ is an image with $|I|$ pixels each containing three values (R, G, B) belonging to the set {1, 2, ..., g}. The model considers image $I$ as an observation of perfect unknown scene $I^*$ in which pixels are represented by a family of distributions from which each colour level is sampled. In particular, every colour level of each pixel of $I^*$ is described by a set of Q independent random variables with values in [0, g/Q]. In $I^*$ the optimal regions satisfy the following homogeneity properties: inside any statistical region and for any colour channel, statistical pixels have the same expectation value for this colour channel; the expectation value of adjacent regions is different for at least one colour channel. From this model a suitable merging predicate is applied as described in [10].

The order in which the tests of merging were done follows a simple invariant $A$: when any test between two true regions occurs, that means that all tests inside each region have previously occurred.

In the experiments, $A$ is approximated by a simple algorithm based on gradient of nearby pixels. In particular Nielsen and Nock consider a function $f$ defined as follow:

$$f(p,p') = \max_{a \in R,G,B} f_a(p,p') \ . \tag{1}$$

A simple choice for $f_a$ is:

$$f_a(p,p') = |p_a - p'_a| \ . \tag{2}$$

More complex function that extends classical edge detection convolution kernels could be used to define $f_a$. In our case we used the Sobel gradient mask.

The pseudo-code of algorithm is the following:

```
INPUT: an image I
Let S_I the set of the 4-connexity couples of adjacent pixels in image I
S'_I=orderIncreasing(S_I, f);
for i = 1 to |S'_I| do
 if((R(p_i) ≠ R(p'_i)) and P(R(p_i), R(p'_i)) == true) then
  merge(R(p_i), R(p'_i));
```

The set of the pairs of adjacent pixel ($S_I$) is sorted according to the value of (1). Afterwards the algorithm takes every couple of pixels $(p,p')$ of $S_I$ and if the regions to which they belong ($R(p)$ *and* $R(p')$) were not the same and satisfactory the merging predicate, it merges the two regions. SRM algorithm gives, for each segmented region, the list of pixel belonging to it and the related mean colour. We use this output as starting point to create a vectorial representation of image.

## 3 Contouring

To obtain a vectorial representation we have to find the border of segmented regions. This could be done more easily if we consider the pixels belonging to several groups (Fig. 1(a)). First of all pixels are divided in:

- *internal pixels*: pixels with all neighbours (in a 4-connexity scheme) belonging to the same region;
- *border pixels*: remaining pixels.

Due to the overall complexity of border regions a further setting into two groups is required:

- *close pixels*: pixels with at least an internal pixel as neighbour (in a 8-connexity scheme);
- *open pixels*: remaining pixels.

After having assigned each pixel to the corresponding category we describe regions in vectorial form. In particular there are two types of curves: *close curves* and *open curves*. In both cases we could approximate their boundaries through segments with eight possible directions (Fig. 1(d)).

A *close curve* is a curve made up of only *close pixels*. Initially we consider a simple configuration (Fig. 1(b)) to explain how segments could be found from *close pixels* list. The pseudo-code that describes the algorithm is the following:

```
initialPixel = findFirstPixel();
currentPixel = findNextPixel(initialPixel);
direction = findDirection(initialPixel, currentPixel);
segment = createNewSegment(initialPixel, direction);
while (currentPixel != initialPixel){
 oldCurrentPixel = currentPixel;
 oldDirection = Direction;
 currentPixel = findNextPixel(oldCurrentPixel);
 direction = findDirection(oldCurrentPixel, currentPixel);
 if (direction != oldDirection){
  setFinalCoordinate(segment, oldCurrentPixel);
  insertInSegmentsList(segment);
  segment = createNewSegment(oldCurrentPixel, direction);}}
setFinalCoordinate(segment, currentPixel);
insertInSegmentsList(segment);
```

The functions are:

- `findFirstPixel()`: it chooses the top-left pixel as initial pixel.
- `findNextPixel(currentPixel)`: it looks for the next pixel in the neighbourhood following a counter clockwise direction.
- `createNewSegment(oldCurrentPixel, direction)`: it creates a new segment with first coordinate `oldCurrentPixel` and direction `direction`.
- `setFinalCoordinate(segment, oldCurrentPixel)`: it sets the final coordinate of segment `segment` at `oldCurrentPixel`.
- `insertInSegmentList(segment)`: it adds `segment` in the list of segments that describes the *close curve*.

Our algorithm chooses the top-left pixel of curve as initial pixel and following the boundary in counter clockwise direction creates the segments necessary for a vectorial description. There are a series of complex configuration: regions with internal curve; ambiguity in the choice of next `currentPixel`; several *close curve* belonging to the same region.

To properly consider these cases it is needed to slightly modify the algorithm described above. The full description and details are reported at `http://www.-dmi.unict.it/~iplab`.

### 3.1 Open Curve

Even if a good segmentation algorithm should create regions with simple boundaries and not ragged this is not always the case. For this reason we have divided border pixels into two groups: *close pixels* and *open pixels*. The last are the pixels devoted to describe the ragged above mentioned. For simple configurations (Fig. 1(c)) we could use the following algorithm:

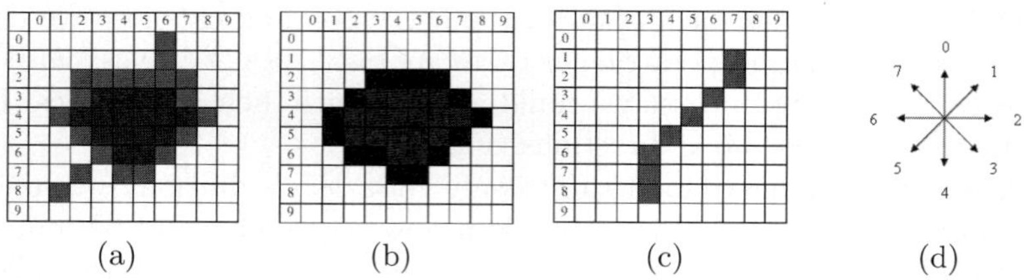


**Fig. 1.** An example of different kind of pixels: internal (blue), close (red), and open (green) (a). Example of simple close (b) and open curve (c). Possible directions of segments that approximate the boundaries of regions (d).

```
initialPixel = findFirstPixel();
currentPixel = findNextPixel(initialPixel);
direction = findDirection(initialPixel, currentPixel);
segment = createNewSegment(initialPixel, direction);
while (it is possible to find new nearby pixels){
 oldCurrentPixel = currentPixel;
 oldDirection = direction;
 currentPixel = findNextPixel(oldCurrentPixel);
 direction = findDirection(oldCurrentPixel, currentPixel);
 if (direction != oldDirection){
  setFinalCoordinate(segment, oldCurrentPixel);
  insertInSegmentsList(segment);
  segment = createNewSegment(oldCurrentPixel, direction);}}
setFinalCoordinate(segment, currentPixel);
insertInSegmentsList(segment);
```

It is very similar to the one described for the *close curve*, but the following function has a different behaviour:

- `findFirstPixel()`: it choices a pixel with only a neighbour as initial pixel. Moreover when a pixel is chosen it is deleted from the list of *open pixels*.

For complex configuration we used a more sophisticated algorithm. Full description with more details are available at the following web address `http://www.dmi.unict.it/~iplab`.

# 4 SVG Generation Code

After tracking the curve boundaries is necessary to map the data by SVG primitives. In particular a generic *close curve* could be represented in the following way:

```
<path d="M x1,x1 L x2,y2 Lx3,y3 Z" stroke ="#RRGGBB" fill"#RRGGBB" />
```

where `x1,y1,x2,y2,x3,y3` are the vertexes coordinates and `RR, GG, BB` are respectively the hex representation of red, green, blue mean value of the region that *close curve* belong to. An *open curve* could be represented in the following way:

```
<path d="M x1,x1 L x2,y2 Lx3,y3" stroke ="#RRGGBB" fill="none" />
```

*Open curves* are no filled (`fill = "none"`) and start point is not equal to final point (`Z` parameter is absent). In order to obtain small file size some optimization could be done [12]:

- `path` element permits to eliminate some separator characters, to use relative coordinate (`m, l` command) and `h, v` command for horizontal and vertical lines respectively;
- `<g>` elements is used to properly ensemble common graphic properties of various primitives.

## 5 Experimental Results

We have done several experiments in order to study the behaviour of our solution as the Q parameter increases. In particular we used two indexes of performance: PSNR and bit per pixel. The first index is tied with the quality of the image, the second with the file size.

We use "Lena" image to show the relation between Q parameter and some indexes of performance (PSNR, bpp). Analogous results have been found for other images. Figures (2(a)) and (2(b)) show that both quality and image size grow as the Q parameter increases. In fact increasing Q, SRM produces more regions and more details need to be represented. Moreover it is not useful increasing the Q parameter over a certain threshold (see Fig. 2(c)).

For sake of comparison SVG images obtained with our algorithm have been compared with other techniques such as Vector Eye [3], SWaterG [6], SVGenie and SVGWave [7]. The set of images used for comparison are of different kinds: graphics, textual, colours and grey levels. The parameters of different techniques have been set in order to obtain images visually agreeable. Figure (3(a)) shows

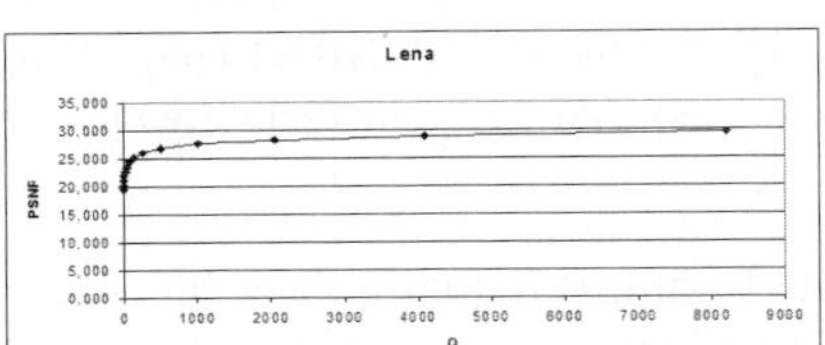


(a) Relation between PSNR and Q.

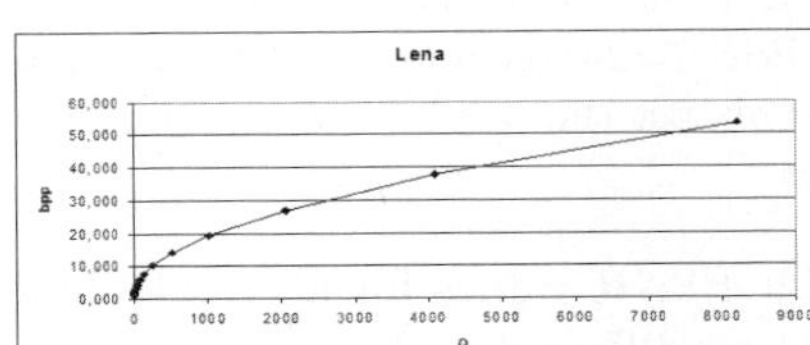


(b) Relation between bpp and Q.

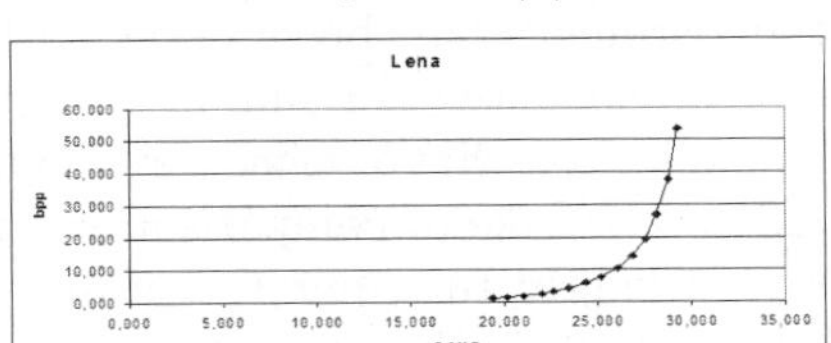


(c) Relation between PSNR and bpp.

**Fig. 2.** Relation between Q parameter and some indexes of performance (PSNR, bpp) for Lena image

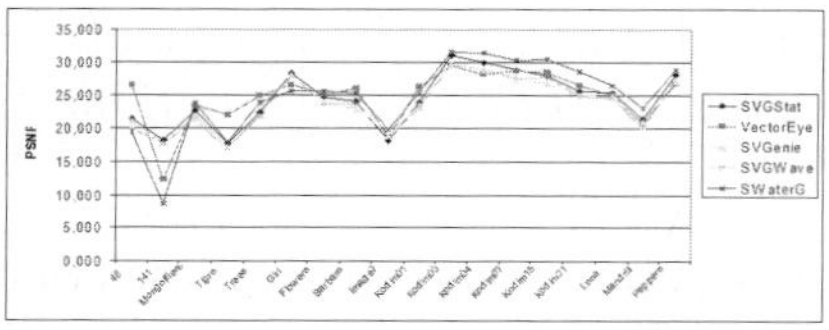


(a) PSNR comparison of dataset images outputs.

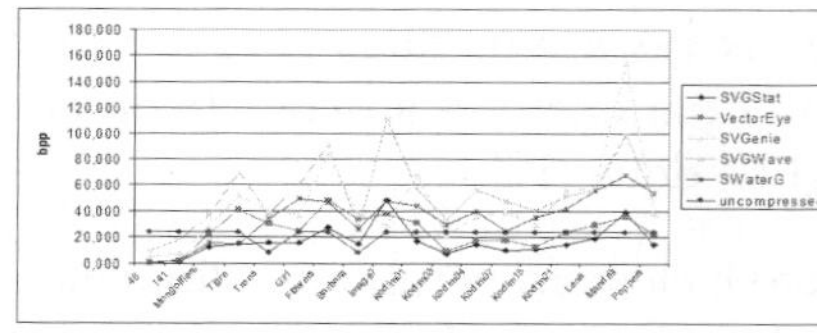


(b) Bit per pixels comparison of dataset images outputs.

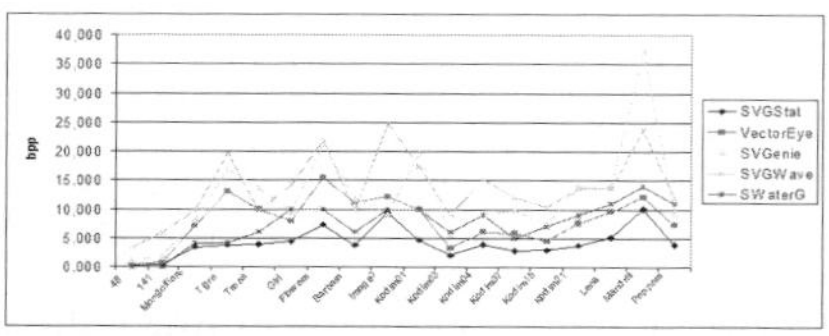


(c) Bit per pixels comparison of dataset images compressed outputs (SVGZ embedded coding).

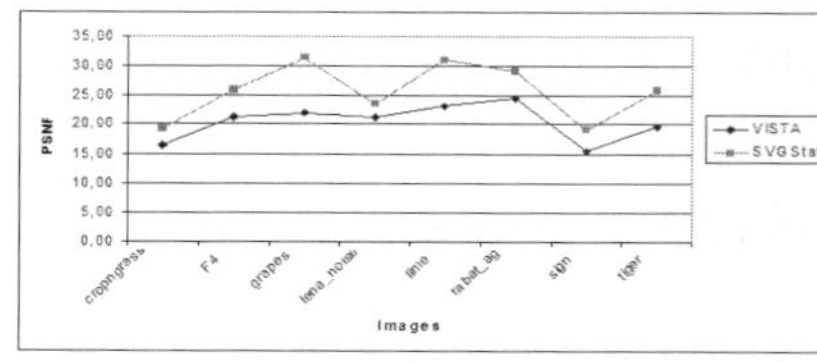


(d) PSNR comparison of our solution with respect to VISTA dataset images outputs.

**Fig. 3.** Comparison between different techniques

**Fig. 4.** Visual comparison between tiger image vectorized with our solution (left) (99 KB, PSNR=26,51) and VISTA SVG image (right) (94 KB, PSNR=19,69). The file size is almost the same but the quality of image produced with our technique is clearly better.

similar PSNR values for all the techniques. Fixing the quality, the file size produced by different techniques changes. In particular our algorithm outperforms the others (see Fig. 3(b) and 3(c)) producing files of lower size.

A different kind of experiment has been done with the recent VISTA technique [9]. In particular we have used for suitable comparison the data (original, SVG code) downloadable from `http://www.svgopen.org/2005/papers-/Prasad_Abstract_R2V4SVG`. For sake of comparison we fix the overall bit size of each image in order to properly tune our Q parameter. The final quality, measured (PSNR) and perceptual, is clearly better then corresponding VISTA images as showed in Fig. (3(d)) and (4). It is useful to consider that our algorithm has only a parameter (Q) that is easily tuneable to obtain the desired quality-size trade-off. Further experiments can be found at the following web address: `http://www.dmi.unict.it/~iplab`.

## 6 Conclusion and Future Works

In this paper we have proposed a novel technique able to convert raster images in a SVG format. Moreover, we have carried out several experiments showing that our algorithm outperforms other similar techniques. Future researches will be devoted to studying advanced region merging heuristics, to use of Bezier curves and filter enhancement.

## References

1. Duce, D., Herman, I., Hopgood, B.: Web 2D Graphics File Format. Computer Graphics forum **21**(1) (2002) 43–64
2. Rabaud, V., Belongie, S.: Big Little Icons. In: CVAVI, San Diego CA (2005)
3. Vantighem, C., Laurent, N., Deckeur, D., Plantinet, V.: Vector eye 1.0.7.6 (2003) Copyright SIAME e Celinea, `http://www.siame.com`, `http://www.celinea.com`.
4. Weber, M.: Autotrace 0.31 (2002) GNU General Public License, http://www.autotrace.sourceforge.net.
5. Kuhl, K.: Kvec 2.99 (2003) Copyright KK-Software, `http://www.kvec.de`.
6. Battiato, S., Costanzo, A., Di Blasi, G., Nicotra, S.: SVG Rendering by Watershed Decomposition. In: Proceeding of SPIE Electronic Imaging-Internet Imaging VI-. Volume 5670.3. (2005)
7. Battiato, S., Barbera, G., Di Blasi, G., Gallo, G., Messina, G.: Advanced SVG Triangulation Polygonalization of Digital Images. In: Proceeding of SPIE Electronic Imaging-Internet Imaging VI-. Volume 5670.1. (2005)
8. Battiato, S., Di Blasi, G., Gallo, G., Messina, G., Nicotra, S.: SVG Rendering for Internet Imaging. In: Proceeding of IEEE CAMP'05, International Workshop on Computer Architecture for Machine Perception, Palermo(Italy) (2005) 333–338
9. Prasad, L., Skourikhine, A.: Raster to Vector Conversion of Images for Efficient SVG Representation. In: Proceedings of SVGOpen'05, NL (2005)
10. Nock, R., Nielsen, F.: Statistical Region Merging. IEEE Transaction on Pattern Analysis and Machine Intelligence **26**(11) (2004) 1452–1458
11. Lucchese, L., Mitra, S.: Color Image Segmentation: A State-of-the-Art Survey. In: Proc. of the Indian National Science Academy(INSA-A). Volume 67 A. (2001) 207–221
12. World Wide Web Consortium: Scalable Vector Graphics (SVG) 1.1 Specification. (2003) `http://www.w3.org/TR/2003/REC-SVG11-20030114/`.

# Robustly Computing Intersection Curves of Two Canal Surfaces with Quadric Decomposition

Jinyuan Jia[1], Ajay Joneja[2], and Kai Tang[2]

[1] Zhuhai College of Jilin University,
Jinwan District, Zhuhai, P. R. China, 519041
csjyjia@yahoo.com.cn
[2] The Hong Kong University of Science and Technology,
Clear Water Bay, Kowloon, Hong Kong, P. R. China
Joneja@ust.hk, mektang@ust.hk

**Abstract.** This paper revisits the intersection problems of two canal surfaces with a new quadric decomposition we proposed for canal surfaces. It reduces computing intersection curves of two canal surfaces to computing intersection curves of two revolute quadrics. Furthermore, Bounding Cylinder Clipping is proposed for efficient intersection determination. Compared to the existing method, our method can (*i*) run more robustly and efficiently; (*ii*) represent the final intersection curves as a piecewise closed-form RQIC; and (*iii*) give a simple shape analysis.

## 1 Introduction

Surface intersection is a fundamental issue in CAGD and geometric modeling. Robustness, accuracy and efficiency are used to evaluate surface intersection algorithms. Several important algorithms that were developed over few decades have been summarized in [16]. To design accurate, robust and efficient algorithms of computing intersection curves of two surfaces, even two special surfaces, e.g. quadrics, cyclides and canal surfaces still remains an open challenge. Although methods for general surface intersections can be applied to special surface intersection problems, they are inefficient. Considering that special surfaces usually have good geometric properties, it desirable to develop more efficient intersection algorithms for them. Therefore many papers have addressed specific intersection problems for CSG primitives, e.g. plane, sphere, cylinder, cone, quadric and tori [1, 2, 3, 5, 13, 14, 17, 18, 19] and some potential geometric primitives e.g. cyclides, surfaces of revolution, ruled surfaces and ringed surfaces [4, 6, 7, 8, 9, 10, 11, 12, 17].

Canal surfaces are one of important geometric primitives in solid modeling, VR, CG, CAD and CAM. Examples of canal surfaces include natural quadrics, revolute quadrics, tori, Dupin cyclides, surfaces of revolution and pipe surfaces. Canal surfaces are very useful in representing long thin objects, for instance, pipes, poles, 3D fonts, brass instruments, internal body organs, and a variety of filleted surfaces. Therefore, it is essential to devise robust and efficient intersection algorithms for canal surfaces.

V.N. Alexandrov et al. (Eds.): ICCS 2006, Part II, LNCS 3992, pp. 342–349, 2006.

## 2 Related Works

Subdivision is an important approach to solve surface intersection problems. Heo *et al* use *circle decomposition* to solve intersection problem of two canal surfaces in [6], which subdivides a canal surface into a dense set of characteristic circles, and reduces intersection problem of two canal surfaces to a zero-set searching problem of a bi-variant function $f(u,v) = 0$, that is much simpler than the original intersection problem. However, (*i*) the numerical behavior of zero-set searching of $f(u,v) = 0$ is both time and memory consuming at high precision; (*ii*) it outputs the intersection curves with a set of discrete sampling points, that is not easy to concatenate; and (*iii*) no shape analysis is performed on the intersection curves, e.g. loops and singularity.

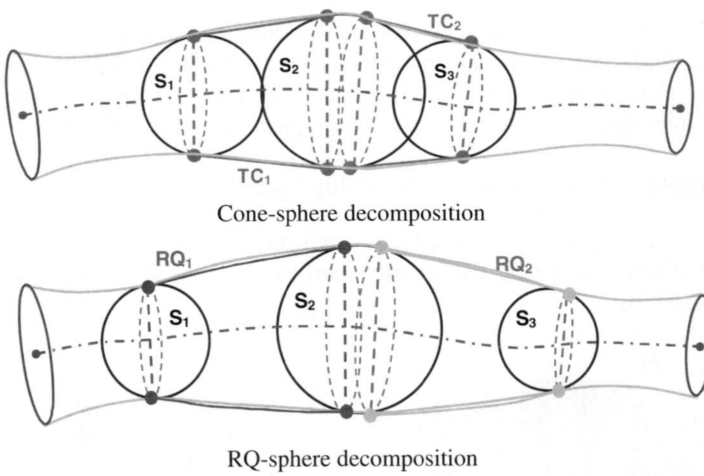


**Fig. 1.** Two quadric decomposition schemes of same canal surface: (a) Cone-sphere decomposition; (b) RQ-sphere decomposition

We revisit this problem by proposing more suitable subdivision scheme for canal surfaces. Except for circle decomposition, there are three other subdivision schemes for canal surfaces, *cyclide* [20], *cone-sphere* [15] and *RQ-sphere* [11]. The first one approximates a canal surface with a set of $G^1$ truncated cyclides [20]. The second one approximates a canal surface with a series of sampling spheres and associated tangential truncated cones (see Fig. 1(a)). However, the density of cone-spheres increases for good approximation quality at high precision and easily causes self-intersection of two neighboring truncated cones when the spine curve has high curvature. The last one approximates canal surfaces with a series of sampling spheres and associated tangential revolute quadrics (see Fig. 1(b)), instead of truncated cones. Apparently, not only less RQ-sphere pairs are required than cone-sphere scheme for same approximation quality, but the self-intersection problem can be avoided as well. In fact, cone-sphere is a special case of RQ-sphere, *cyclide* decomposition reduces canal/canal intersection computing to cyclide/cyclide intersection, that has to solve order 8 polynomial equation numerically. Both *cone-sphere* and *RQ-sphere* can reduce computing canal/canal intersection curves to computing RQ/RQ intersection curves, that has closed-form solutions [1, 2, 5, 16, 17, 19, 20]. However, cone-sphere may yield incorrect intersection curve if it has self-intersection. So we decided to employ RQ-sphere decomposition to solve canal/canal intersection problem. In this paper, we propose a new conception, *canal valid intersection intervals* (CVII), a good hierarchical data structure, *cylindrical bounding volume* (BCT), and a new method, *bounding cylinder clipping* for efficient intersection determination in Section 3. The rough idea of canal/canal intersection algorithm is described in Section 4. Experimental examples given in Section 5 show the robustness and efficiency of our method. Conclusions and future work are presented in Section 6.

## 3 Bounding Cylinder for Canal Surfaces

Compared to traditional bounding volumes, bounding box and bounding sphere, *bounding cylinder* BC is chosen for canal/canal intersections because (*i*) it can enclose canal surfaces more closely; (*ii*) it can be constructed rather easily; (*iii*) more important, intersections of two BCs can be computed geometrically [4].

### 3.1 Hierarchical Construction of Bounding Cylinder Tree

It is uneasy to construct a smallest BC for canal surfaces theoretically. In practice, it is acceptable to construct a near BC for canal surfaces, as long as it can reasonably close to the smallest BC.

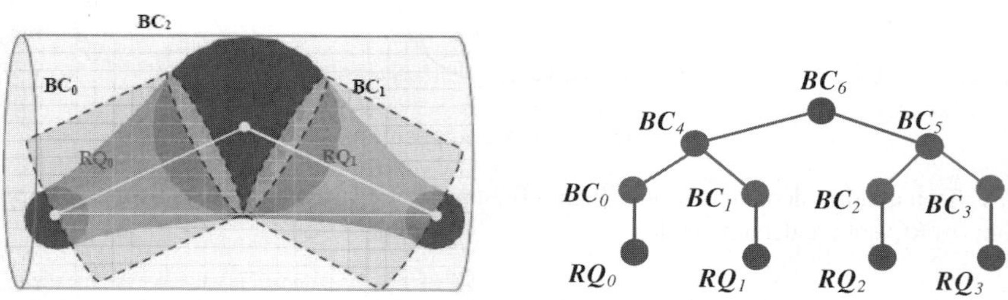


**Fig. 2.** A binary bounding cylinder tree BCT for a canal surface

Considering that canal surfaces have good geometric properties, they should have simpler but more efficient BC construction methods. We use a hierarchical BC construction method for canal surfaces. In our *RQ-sphere* decomposition, a canal surface is subdivided as a set of $n$ $G^1$ *RQ-spheres*. It is easy to construct a bounding cylinder BC for each *RQ*. For two neighboring $RQ_0$ and $RQ_1$, we can have two bounding cylinders $BC_0$ and $BC_1$ in a straightforward way, and as shown in Fig. 2, a bigger bounding cylinder $BC_2$ can be constructed geometrically to enclose $BC_1$ and $BC_2$. In the same manner, a binary tree of bounding cylinders HBCT can be organized hierarchically.

### 3.2 Canal Valid Intersection Intervals (CVII)

Suppose that two canal surfaces are subdivided into $n$ and $m$ *RQ-sphere* pairs respectively. A brute force method for canal/canal intersection would require invoking $n*m$

*RQ/RQ* intersection computations. However, it becomes quite inefficient when two canal surfaces only intersect each other within a small overlapping region (see Fig. 3(1)), in this case, only few *RQ/RQ* pairs of the two canal surfaces within their overlapping regions need to be examined. Therefore, by filtering out those *RQ/RQ* pairs outside the overlapping regions, the expected performance of the algorithms should improve. To describe it conceptually, *canal valid intersection intervals* CVII is defined as the interval pairs (*u-interval, v-interval*) corresponding to the overlapping regions of the two canal surfaces, hereafter denoted by ($CVII_u$, $CVII_v$). The idea, *bounding cylinders clipping*, hereafter denoted by *BC clipping*, is also proposed to detect ($CVII_u$, $CVII_v$), or to find all those potential intersecting *RQ/RQ* candidates for more efficient intersection determination.

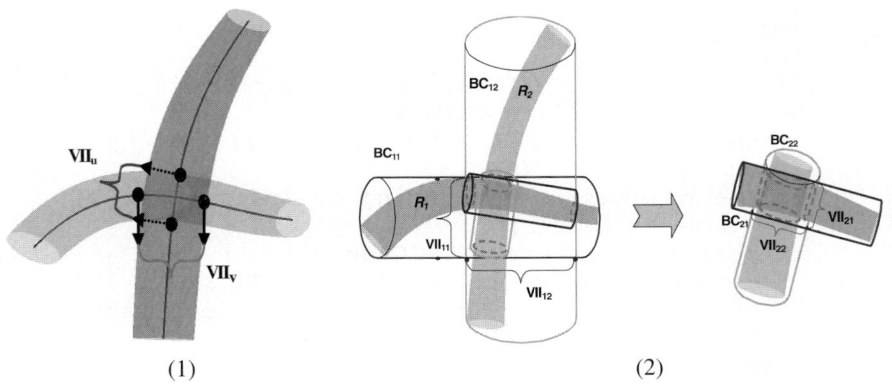


**Fig. 3.** The valid intersection intervals of two canal surfaces: (1) Definition of VII of two overlapping canal surfaces; (2) Two consecutive rounds of BC clipping

### 3.3 Clipping of Two Bounding Cylinders

It is difficult to determine the exact interval ($CVII_u$, $CVII_v$) of two canal surfaces $R_1$ and $R_2$ directly by solving complicated equations numerically. Therefore, it is acceptable to estimate ($CVII_u$, $CVII_v$) of $R_1$ and $R_2$ only approximately but more efficiently.

Similar to computing RVII of two surfaces of revolution [9, 10], we propose *BC clipping* to estimate ($CVII_u$, $CVII_v$) of two canal surfaces approximately by computing the overlapping regions of their respective bounding cylinders $\mathbf{BC_{11}}$ and $\mathbf{BC_{12}}$, then refining the regions recursively as shown in the right part of Fig. 3(b). The first round of *BC clipping* for $\mathbf{BC_{11}}$ and $\mathbf{BC_{12}}$ yields the initial intersection interval ($CVII_{11}$, $CVII_{12}$). It is a very rough approximation to the real CVII, within which there are still some *RQ-sphere* pairs of $R_1$ and $R_2$ that have no intersection, since $\mathbf{BC_{11}}$ and $\mathbf{BC_{12}}$ enclose $R_1$ and $R_2$ rather loosely. Furthermore, two smaller bounding cylinders $\mathbf{BC_{21}}$ and $\mathbf{BC_{22}}$ are constructed respectively for those *RQ-sphere* pairs within $CVII_{11}$, and $CVII_{12}$, then, the second round *BC clipping* is taken on $\mathbf{BC_{21}}$ and $\mathbf{BC_{22}}$ as shown in the left part of Fig. 3 (b) (amplified version of second round of $2^{nd}$ *BC clipping*), giving a smaller intersection interval pair ($CVII_{21}$, $CVII_{22}$). Usually, very few rounds of such *BC clipping* output very close to the real ($CVII_u$, $CVII_v$) of two canal surfaces.

# 4 Computing Intersections of Two Canal Surfaces

The intersection curve of two canal surfaces, hereafter denoted *CSIC*, is computed in three steps: (*i*) computing CVII ($CVII_u$, $CVII_v$) of the two canal surfaces; (*ii*) computing all the intersection curve segments of all potential *RQ*/*RQ* pairs, RQIC, within ($CVII_u$, $CVII_v$) by using Goldman's method [5]; (*iii*) concatenating all the individual *RQIC* into *CSIC* as a set of independent components (open branches or closed loops).

## 4.1 Computing Intersection Curves of Two Canal Surfaces

The procedure, Find_CVII ($BCT_1$, $BCT_2$), is to compute ($CVII_{11}$, $CVII_{12}$) of two canal surfaces. There are three possible cases which should be treated differently:

```
procedure FindCVII (BCT1, BCT2)
  begin
    If both BCT1 and BCT2 are only two bounding cylin-
    ders (0-level BCT) [Case 1]
      compute their CVII (CVII11, CVII12);
      return (CVII11, CVII12);
    If one of them is a bounding cylinder [Case 2]
      assume it be BC1;
    If both of them are BCT [Case 3]
      assume the lower one of them be BC1;
    if BC1 and the root BC of BCT2 do not overlap
      return an empty CVII;
    else
      [check if BC1 and two children BCT21 and BCT22
      overlap recursively]
      Find_CVII(BC1, BCT21);
      Find_CVII(BC1, BCT21);
      add all the individual sub-CVIIs to CVII;
    end if;
    return (CVIIu, CVIIv);
  end.
```

The main idea of our computing intersection curves of two canal surfaces *CSIC* based on *RQ-sphere* decomposition can be sketched as follow:

```
program ComputeCSIC (BCT1, BCT2)
  begin
    (CVIIu CVIIv) = FindCVII (BCT1, BCT2);
    assume CVIIu be shorter one;
    for each RQui (BCui) within CVIIv
      construct BCTvj for all the RQs within CVIIv;
      CVIIij = Find_CVII (BCui, BCTvj);
      for each RQvj within CVIIij
        if RQui and RQvi overlap
           RQICij = ComputeRQIC (RQui, RQvi);
           if RQICij is a closed loop itself
              output it as a new loop;
           else
```

```
                Check coincidence of RQIC_ij and prior
                RQIC_i-1,j along their common bounding
                circle C_i-1;
                Check coincidence of RQIC_ij and prior
                RQIC_i,j-1 along their common bounding
                circle C_j-1;
                if no coincidence between them
                  output it as a new branch;
                else
                   concatenate RQIC_i,j with RQIC_i-1,j and
                   RQIC_i,j-1 respectively;
                   if a closed loop is formed,
                     output it as a new closed loop;
                   end if;
                end if;
             end if;
end.
```

### 4.2 Computing Self-intersections of a Single Canal Surface

Similarly, self-intersection of a single canal surface *R* can be solved. Assume that the bounding cylinder tree of *R* is $BCT_0$, *R* is decomposed *n RQ-sphere* pairs $\mathrm{RQS}_i$ ($i = 1, 2, \ldots, n$), $BCT_0$ has two sub-trees $BCT_{\mathrm{Left}}$ (enclosing the first half $\mathrm{RQS}_i$, $i = 1, 2, \ldots, [n/2]$) and $BCT_{\mathrm{Right}}$ (enclosing the other half $\mathrm{RQS}_i$, $i = [n/2], [n/2]+1, \ldots, n$). If $BCT_{\mathrm{Left}}$ and $BCT_{\mathrm{Right}}$ overlap, then, *R* may intersect itself and its self-intersection curves can be computed by calling $CSIC(BCT_{\mathrm{Left}}, BCT_{\mathrm{Right}})$. Otherwise, conquer each half (sub-tree) recursively. This idea is sketched roughly as follows.

```
procedure Canal_Self_Intersection (BCT_0)
  begin
    if BCT_left and BCT_right, two children of BCT_0 overlap
        ComputingCSIC(BCT_Left, BCT_Right);
    else
        Canal_Self_Intersection(BCT_Left);
        Canal_Self_Intersection(BCT_Right);
    Output all the open branches or loops;
  end.
```

## 5 Illustrative Examples

The proposed algorithms have been implemented with *C++* and *OpenGL* under Windows *XP* and PC (Pentium III, 512M RAM, 512M HZ). Two examples are given in Fig. 4, one is for computing general *CSIC*, the other is for self-intersection. Their spine curves and radii are represented with cubic Beizer form. Both of them are computed within one second.

Comparing determination method of global self-intersection for a single pipe surface in [16], it is simpler algorithmically, easier for implementation and also more robust, since both fundamental intersection computations of BC/BC and *RQ*/*RQ* have closed form solutions.

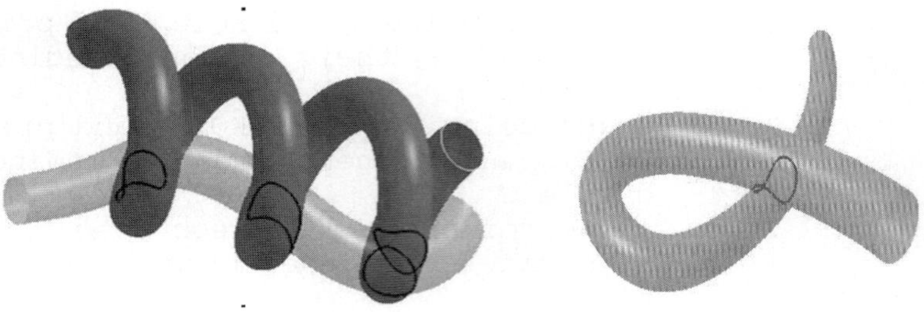

(1) (2)

**Fig. 4.** Intersection Curves on canal surfaces: (1) The intersection curves of two canal surfaces; (2) Self-intersection curves of a single canal surface

# 6 Conclusion

Robustness is one of most important factors for surface intersection algorithms. We have shown and analyzed the instability of the method [6] in our previous work [9]. Our *RQ-sphere* decomposition based method reduces computing intersections of two canals to computing intersections of two *RQ*s, which can be solved by Goldman's method [5] robustly and efficiently. Further, *BC clipping* makes computing the *CSIC* relatively efficient. In fact, the efficiency, accuracy and robustness of two revolute quadrics *RQ*/*RQ* intersection can be further enhanced by more recent algorithms [1, 2, 18, 19]. Also, *RQ-sphere* decomposition facilitates tracing the intersection curves because it is easy to recognize the closed loops and singular points on the intersection curves. Therefore, both theoretical analysis and practical implementation show the robustness and efficiency of our proposed method.

The *RQ-sphere* decomposition of canal surfaces also can be extended to solving other geometric problems of canal surfaces, e.g. collision detection, isophotes, silhouette, bisector, distance computing and so on.

# References

1. Dupont L., Lazard D., Lazard S., Petitjean S.: Towards The Robust Intersection of Implicit Quadrics. In Proc. of Workshop on Uncertainty in Geometric Computations, Sheffield, UK, (2001).
2. Dupont L., Lazard D., Lazard S., Petitjean S.: Near-Optimal Parameterization of the Intersection of Quadrics. In *ACM Symposium on Computational Geometry*, San Diego, USA, (2003).
3. Elber G., IRIT Software, http://www.cs.technion.ac.il/~irit/, The Technion-IIT, Haifa, Israel, (1993).
4. Eberly David: Intersection of two cylinders. Magic Software Company. http://www.magic -software. com/ Intersection.html (2000).
5. Goldman R. N.: Quadrics of Revolution. IEEE Computer Graphics and Applications. 3(2), (1983) 68-76.
6. Heo H. S., Hong S. J., Seong J. K., Kim M. S.: The Intersection of Two Ringed Surfaces and Some Related Problems. Graphical Model. 63(4), (2001) 228-244.

7. Johnstone J.: A New Intersection Algorithm for Cyclides and Swept Surfaces Using Circle Decomposition. CAGD, 10(1). (1993) 1-24.
8. Jia J., Tang K., Joneja A, Kwok K. W.: New Algorithm of Computing Planar Sections of Surfaces of Revolution Based on Revolute Quadric Subdivision". Proc. of Shape Modeling and Applications. Genova. Italy. (2004).
9. Jia J., Baciu G., Kwok K. W.: Revolute Quadric Decomposition for Computing Intersection Curves of Surfaces of Revolution. Graphical Model. 55(5). (2004) 363-383.
10. Jia J., Tang K, Joneja A.: Bi-Conic Subdivision for Surfaces of Revolution and Its Applications to Intersection Problems. The Visual Computer. 20(7), (2004) 457-478.
11. Jia J., Joneja A. and Tang K., "Revolute Quadric Decomposition of Canal Surfaces and Its Application". LNCS. Vol. 3514. Springer-Verlag. (2005) 213-222.
12. Kim M. S.: Intersecting Surfaces of Special Types. Proc. of Shape Modeling and Processing. University of Aizu, Japan. (1999) 122-128.
13. Levin J. Z.: A Parametric Algorithm for Drawing Pictures of Solid Objects Composed of Quadrics. Communications of the ACM. 19(10). (1976) 555-563.
14. Levin J. Z..: Mathematical Models for Determining The Intersections of Quadric Surfaces. Computer Graphics and Image processing. Vol. 1, (1979) 73-87.
15. Nelson Max: Cone-Spheres. ACM SIGGRAPH Computer Graphics, Vol. 24, (1990) 59-62.
16. Patrikalakis N. M., Maekawa T.: Shape Interrogation in Computer Aided Design and Manufacturing. Springer Verlag. (2002).
17. Seong J. K., Kim K. J., Kim M. S., Elber G.: Intersecting a Freeform Surface with a Ruled or a Ringed Surface. Proceedings of Geometric Modeling and Processing. Beijing, (2004) 38-45.
18. Wang W., Goldman R., Tu C.: Enhancing Levin's Method for Computing Quadric-Surface Intersections. CAGD, 20(7). (2003) 401-422.
19. Wang W., Joe B., Goldman R.: Computing Quadric Surface Intersections Based on An Analysis on Plane Cubic Curves. Graphical Model. 64(6), (2003) 335-367.
20. Paluszny, M., Bühler, K.: Canal Surfaces and Inversive Geometry. Mathematical Methods for Curves and Surfaces II. In Daehlen M., Lyche T., Schumaker L. (eds.), (1998) 367-375.

# Triangle Strip Multiresolution Modelling Using Sorted Edges*

Ó. Belmonte Fernández, S. Aguado González, and S. Sancho Chust

Department of Computer Languages and Systems
Universitat Jaume I
12071 Castellon, Spain
oscar.belmonte@uji.es

**Abstract.** This paper presents a new multiresolution model based exclusively on the triangle strip primitive. The model is independent of the initial set of triangle strips and the simplification method used to obtain the lower resolution versions of the original model. The information stored in the data structures is sorted to minimise the time needed to recover a Level of Detail (LoD). The orientation of triangles in each strip is maintained as the model is simplified, so back-face culling can be used. The main result is a dramatic improvement in the rendering time.

## 1 Introduction

Polygonal meshes are widely used to describe surfaces in Computer Graphics, especially in Real Time Computer Graphics, and triangle meshes are the most popular representation of such models. Triangle meshes currently used in Computer Graphics are composed of thousands of triangles. A good example of this are the polygonal models used in cultural heritage [7]. Rendering these highly detailed polygonal models is a challenging problem in interactive Computer Graphics applications, where a minimum rate of 15 frames per second is needed to mimic continuous movement.

To address this problem, multiresolution models try to adapt the number of triangles in a scene according to some criterion [8, 5, 11]. The basic primitive in these models is the triangle. Only a few models use the triangle strip primitive in the rendering stage [6, 2], and only recent works have used this primitive both in the data structures and in the rendering stage [1, 9].

This work presents a continuous multiresolution model based exclusively on the triangle strip primitive. The model is independent of the initial collection of triangle strips. Unlike [9], the model is also independent of the simplification method used to obtain the lowest resolution of the original model. The inner edges of each triangle strip are sorted to achieve quick recovery of a demanded

* This work was partly supported by grant P1 1B2005-17 of Fundació Caixa Castelló-Bancaixa, and grant IST-2-004363 of the European Community.

V.N. Alexandrov et al. (Eds.): ICCS 2006, Part II, LNCS 3992, pp. 350–357, 2006.

LoD. Another characteristic that distinguishes this model from [1, 9] is that triangle orientation is maintained at any level of detail, so back-face culling can be used. The result is a dramatic improvement over models based on triangles and better performance than models based exclusively on the triangle strip primitive.

The rest of the paper is organised as follows, Section 2 includes a review of previous work Section 3 presents the new multiresolution model. Section 4 shows the experiments carried out and their results. Finally, section 5 summarises the main conclusions.

## 2 Previous Work

In [6] H. Hoppe presented a view-dependent multiresolution model in which data is stored in a hierarchy similar to that presented in [12]. After the recovery algorithm has recovered a set of triangles for a given LoD, triangle strips are searched over them. This model used triangle strips only in the rendering stage.

In [2] El-Sana et al. presented a view-dependent multiresolution model for polygonal models. The basic data structure of the model is the hierarchical tree presented in [12]. Triangle strips are searched over the model using the STRIPE algorithm [3], with the constraint of not generating strips with swap operations. These triangle strips are coded as skip-list data structures. In this case triangle strips are more closely related to the basic data structures of the model, but they are not an essential compound of it.

F. Ramos et al. [9] presented a view-dependent multiresolution model whose only data structure is the triangle strip. The building of the model used an ad hoc sequence of vertex pair contractions, so the model losses quality at lower resolutions. This model does not maintain the orientation of the triangles in the strip for every LoD, and hence back-face culling can not be applied.

In [1] Ó. Belmonte et al. presented a multiresolution model based exclusively on the triangle strip primitive. Although this model provides a frame rate that doubles those provided by multiresolution models based on triangles, its recovery time is in general higher than that offered by this latter type of model. Moreover, this model does not maintain the orientation of the triangles in the strip for every LoD, and so back-face culling can not be applied.

## 3 The Sorted Edge Triangle Strip Model

The information required to begin the building of a SETS multiresolution model is a set of triangle strips over the polygonal model. Any triangle strip search algorithm can be used to do this, for example an algorithm that takes advantage of the vertex cache size to minimise vertex cache misses or a general triangle strip search algorithm. We have chosen the STRIPE algorithm [3] because it minimises the number of swap operations in the strips.

The triangle strip search algorithm does not impose any restriction upon the simplification method that can be used to obtain the simplified version of the original model. Furthermore, no restrictions are imposed upon the triangle strip search algorithm by the simplification method.

Any simplification method can be used on the initial set of triangle strips to obtain the simplified versions of the original model. We have opted for the QSlim algorithm [4]. This algorithm works by collapsing pairs of vertices and, given a target number of triangles, it is able to simplify the original polygonal model to reach a number of triangles near the target number. The output of this algorithm is an ordered sequence of the pairs of collapsed vertices.

If a vertex in a triangle strip collapses, the sequence that describes the strip must then be updated. A special case occurs when the vertex at the beginning of the strips changes and the previous beginning is no longer valid. Figure 1 shows a series of pairs of vertex collapses and the updated sequences of vertices.

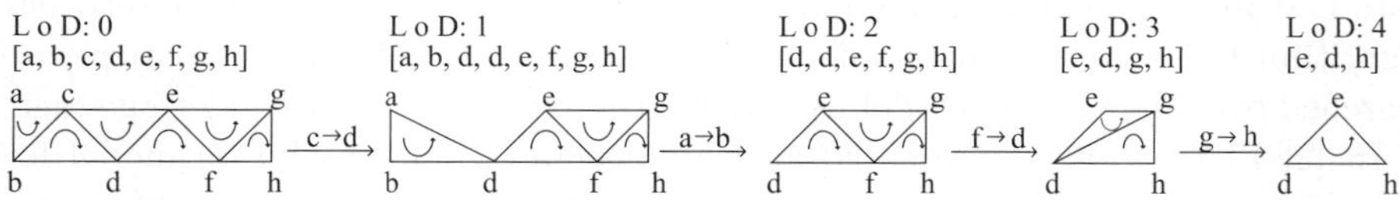


**Fig. 1.** After each pair of vertices collapses the sequence that describes the triangle strip must be updated to properly describe the new strip

For instance, let us first take vertex $b$. At LoD 0, vertex $b$ is followed by vertex $c$, which we denote by $b \to c(0)$. At LoD 1, the sequence of vertices that describes the strip has changed, and vertex $b$ is followed by vertex $d$, which we denote by $b \to d(1)$. Vertex $b$ does not appear at any other LoD.

Let us now take vertex $d$ as an example. At LoD 0, vertex $d$ is followed by vertex $e$ ($d \to e(0)$). At LoD 1, the first occurrence of vertex $d$ is followed by vertex $d$ and the second occurrence is followed by vertex $e$, which we denote as $d \to d, e(1)$. At LoD 2, vertex $d$ is followed by vertices $d$, $e$ and $g$, which is denoted as $d \to d, e, g(2)$. Finally, at LoD 3, vertex $d$ is followed by vertices $d$ and $g$ ($d \to d, g(3)$.

So, the information that needs to be store, for each level of detail, in the multiresolution model is: a) The vertex at the beginning of the strip and the maximum LoD until it is valid; b) The sorted sequence of vertices after a given one and the maximum LoD until it is valid.

### 3.1 Data Structures

A labelled graph is used to efficiently store the information needed to recover the sequence of the strip for each LoD. In this graph, each node (`ColElm`) represents a vertex in the strip, each arc (`RawElm`) is the series of vertices following on from that vertex, and the label in the arc (`RawElm.resolution`) means the maximum LoD at which the series is valid.

The vertices at the beginning of the strip (`StripBeginning`) and the maximum LoD until these become valid are stored in an array. Table 1 shows the data structure used in the model.

**Table 1.** Data structures of the model

| StripBegenning | RawElm | ColElm | MultiresolutionStrip |
|---|---|---|---|
| Integer resolution | Integer resolution | Integer idVertex | Array aStripBeginning |
| Integer idColElm | Array aIdColElm | Array aRawElm | Array aColElm |
| | | Integer shift | Array aResolution |

## 3.2 Building Process

The building process is performed just once for each polygonal model, and the resulting information is stored as the multiresolution representation of the model.

If vertex $a$ collapses over vertex $b$, all occurrences of vertex $a$ must be replaced with vertex $b$ in the sequence of vertices that describes the strip. In some cases, the sequence can be optimised by removing subsequences of the main sequence while maintaining the orientation of the triangles in the strip. The two conditions that a sub-sequence must comply with in order to be removed are: a) The number of vertices that are removed must be even, b) The remaining vertices must not add new information, that is, they must not describe triangles that do not exist before removing.

The initial sequence of vertices in Figure 1 is $[a, b, c, d, e, f, g, h]$, and the first triangle in the sequence is oriented counter-clockwise, so the triangle at the beginning of the sequence must be counter-clockwise oriented for every LoD. The first contraction moves vertex $c$ over vertex $d$ ($c \rightarrow d$) and the updated sequence is $[a, b, d, d, e, f, g, h]$; thus, no optimisation can be performed over this sequence. The next contraction is ($a \rightarrow b$), the updated sequence is $[b, b, d, d, e, f, g, h]$, and in this case it can be optimised to $[d, d, e, f, g, h]$ which is the correct counter-clockwise sequence. Finally, the third contraction ($f \rightarrow d$) yields the sequence $[d, d, e, d, g, h]$, and this can be optimised to $[e, d, g, h]$, the first triangles of which have the correct orientation.

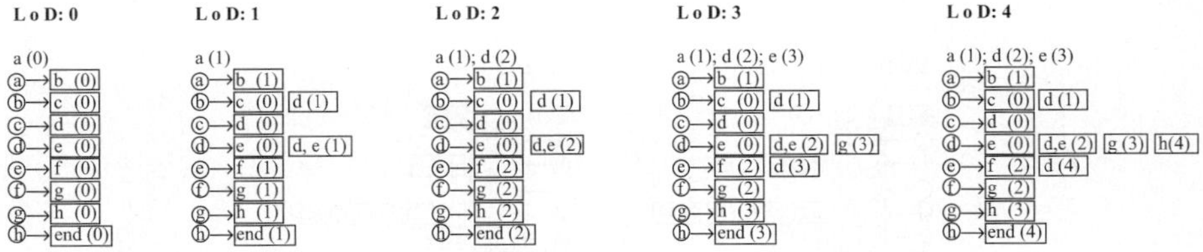


**Fig. 2.** Building process of a multiresolution triangle strip

Following with the example in Figure 1, the first LoD is 0, and for this vertex $a$ is the vertex at the beginning of the strip. This information is stored in the array of vertices at the beginning of the strip, as shown in Figure 2. Vertex $a$ has just one successor in the sequence, vertex $b$, so this is added to vertex $a$ as an arc with the current resolution. This was noted this by $a \rightarrow b(0)$. In the same way, vertex $b$ has only one successor in the sequence of vertices, vertex $c$, so this is added to $b$ as an arc with the current resolution $b \rightarrow c(0)$. The building process continues until the last vertex in the strip, which is vertex $h$, is reached. In this case, the special arc $END$ is added to $h$, and noted by $h \rightarrow END(0)$.

At the next LoD 1, vertex $a$ is still the vertex at the beginning of the strip, so its maximum LoD must be updated. In the same way, if the successors of a vertex are the same as in the previous LoD, only the information relative to the resolution must be updated in the corresponding arc; otherwise, a new arc with the new vertices and the current resolution must be added.

### 3.3 LoD Recovery Algorithm

The recovery algorithm traverses the data structures to recover the model at the demanded LoD. The elements in the array of nodes at the beginning of the strip and the arcs for each node in the graph are sorted according to the LoD, and then a dichotomic search can be used over them. In addition, once an arc is found, this is valid for all successors of the node at the same resolution.

```
algorithm LoDrecover(res)
begin
id=aStripBeginning.dichotomicSearch(res)
while(id != END)
  storeCoords(aColElm[id].shift)
  shift = vECol[id].shift
  if shift == 0
    aColElm[id].dichotomicSearch(res)
  aColElm[id].shift++
  id = aColElm[id].aRawElm.
    current(shift)
endWhile
end
```

**Fig. 3.** Level of detail recovery algorithm

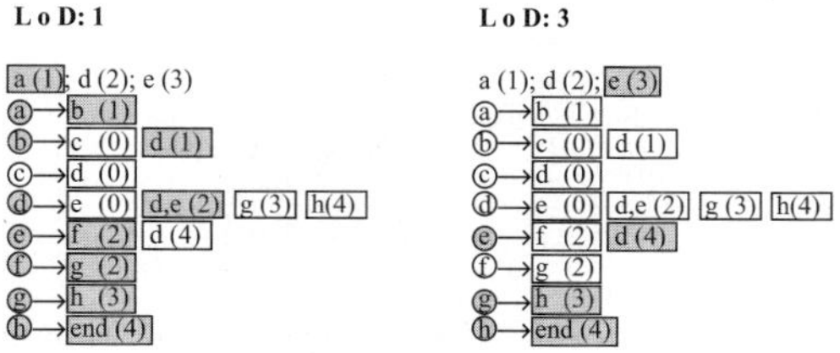


**Fig. 4.** LoD 1 is demanded on the left, LoD 3 on the right

Figure 3 summarises the recovery algorithm. The recovery algorithm starts by performing a binary search over the array of vertices at the beginning of the strip `MultiresolutionStrip.aStripBeginning` at a given resolution. If a vertex is found, its `StripBeginning.idColElm` is the identifier of the node from which the traversal of the graph must be begun. The traversal is done through arcs whose valid resolution is strictly lower than the given resolution. If the node has not yet been visited (`ColumnElm.shift == 0`), a binary search is carried out over the arcs; otherwise, the binary search is not needed because the field `ColElm.shift` points to the next vertex in the sequence. The algorithm stops when an arc with the label $END$ is reached. Figure 4 shows two examples of the recovery process. The grayed nodes and arcs are traversed during the recovery process.

### 3.4 Coherence

Coherence means recovering only data that has changed from one LoD to another, while maintaining data that remains in both LoDs. In the SETS model this implies traversing just those graphs whose vertex sequence has changed.

To achieve this, a validity interval is defined for each strip, the bounds of which represent the minimum and maximum LoD until the current sequence is valid. This is encoded as an array containing LoDs at which a change in the sequence happens. This array is filled up when the model is loaded, so that two consecutive LoDs in the array are the bounds of the validity interval. At the beginning, the bounds in the validity interval are LoDs at position 0 and 1 of the array. When a new LoD is demanded, if it is within the validity interval the sequence that represents the strip is valid, otherwise the graph must be traversed. Finally, the bounds of the interval are updated, the maximum bound being the first element in the array greater than the current LoD, and the element in the previous position is the minimum bound.

## 4 Experimental Results

This section begins by showing the spatial cost of the multiresolution model. The, visualisation time of the SETS model is then compared with a modification of the MTS model [1] in which the orientation of the triangles in the strip is preserved as the model is simplified.

The tests were conducted using the models whose characteristics are shown in Table 2. The platform used was a Pentium IV processor at 1.8 GHz. with 256 Kb cache, 1 Gb RAM and an nVidia GeForce3 Ti 200 graphics card. The compiler used was gcc 3.3.3 version in a Linux 2.6.5 kernel.

Table 2 also shows the memory size of the SETS multiresolution models compared to those for the MTS multiresolution models. In all cases the size of the SETS models are bigger than the MTS models, but we think that the better performance of the recovery algorithm offsets this disadvantage.

**Table 2.** Polygonal data and spatial cost of the multiresolution models in Mb

| Model | #Vertices | #Triangles | #Strips | SETS | MTS |
|---|---|---|---|---|---|
| Cow | 2904 | 5804 | 136 | 0.338 | 0.253 |
| Bunny | 34834 | 69451 | 1229 | 4.150 | 2.964 |
| Phone | 83044 | 165963 | 1747 | 9.832 | 6.766 |

The tests conducted to measure the performance of the recovery algorithm were those defined in [10]. Each test recovers a percentage of the total number of LoDs, and the total time is averaged over the number of LoDs recovered. The difference between the tests is the distribution taken to recover the LoDs: linear or exponential. The linear distribution takes LoDs with the same distance in between; the exponential distribution takes close LoDs when they are next to the maximum resolution, and distant LoDs when they are next to the minimum resolution. These two tests can be interpreted as the model moves towards or away from the observer.

In both tests 1, 3, 6, 10, 15 and 20% of the total number of LoDs present in the multiresolution model were recovered. To avoid possible influences of the direction of the tests, these start at the maximum LoD, descend towards the minimum LoD and then return to maximum LoD.

Table 3 shows that the more LoDs are recovered the lower the average time is, regardless of the model and the test used. This behaviour is due to the use of coherence, the lower the distance between LoDs is, the more LoDs are recovered.

**Table 3.** Visualisation time results for the tests. Time in ms

| | **Cow** | | | | **Bunny** | | | | **Phone** | | | |
|---|---|---|---|---|---|---|---|---|---|---|---|---|
| | *Linear* | | *Exponential* | | *Linear* | | *Exponential* | | *Linear* | | *Exponential* | |
| *LoD* | MTS | SETS | MTS | SETS | MTS | SETS | MTS | SETS | MTS | SETS | MTS | SETS |
| 1% | 1.186 | 1.694 | 1.186 | 1.525 | 11.951 | 11.922 | 12.253 | 11.520 | 26.683 | 25.189 | 28.194 | 24.346 |
| 3% | 1.200 | 1.371 | 1.086 | 1.428 | 10.010 | 8.544 | 10.761 | 8.645 | 22.914 | 18.089 | 25.156 | 18.107 |
| 6% | 0.974 | 1.146 | 0.974 | 1.117 | 9.143 | 7.071 | 10.012 | 7.401 | 21.427 | 15.107 | 23.918 | 15.645 |
| 10% | 0.929 | 0.963 | 0.947 | 0.998 | 8.715 | 6.342 | 9.700 | 6.741 | 20.625 | 13.604 | 23.224 | 14.380 |
| 15% | 0.861 | 0.861 | 0.873 | 0.872 | 8.468 | 5.908 | 9.497 | 6.352 | 20.206 | 12.757 | 22.881 | 13.691 |
| 20% | 0.818 | 0.775 | 0.861 | 0.792 | 8.334 | 5.659 | 9.383 | 6.182 | 19.940 | 12.359 | 22.664 | 13.301 |

The visualisation time is higher for the exponential than for the linear test. This is due to the fact that the number of vertices recovered is higher in the exponential than in the linear test, and so more time is used in rendering.

The SETS model yields better results than the MTS model, as it has more triangles. This is due to the fact that data is stored in the data structures of the SETS model in a sorted way, and in this way the sequence that represents a triangle strip can be recovered faster than in the MTS model. The SETS model also yields better results as more LoD are recovered in all cases.

## 5 Conclusions

A new multiresolution model based only on the triangle strip primitive has been presented. The main characteristic of the model is that data is stored in a sorted fashion, taking into account the maximum LoD at which any inner edge exists. The building of the models is independent both of the algorithm used to find a collection of triangle strips and the simplification method use to obtain simplified versions of the original model. In addition, the orientation of the triangles in a strip is preserved as it is simplified so that back-face culling can be used to speed up the rendering stage.

The main result is a speeding up of the visualisation time, 70% faster in the best case, as compared to that offered by models that do not use sorted data.

The main drawback is an increase in the spatial cost of the models in main memory, but we believe that this is well justified taking into account the better performance offered by the model.

## References

1. Ó. Belmonte, I. Remolar, J. Ribelles, M. Chover, M. Fernández. Efficiently Using Connectivity Information between Triangles in a Mesh for Real-Time Rendering. *Future Generation Computer Systems*, 20(8), pp: 1263-1273, 2004.
2. J. El-Sana, E. Azanli, A. Varshney, Skip Strips: Maintaining Triangle Strips for View-dependent Rendering, *IEEE Visualisation '99*, pp:131-138, 1999.
3. F. Evans, S. Skiena, A. Varshney, Optimising triangle strips for fast rendering, *IEEE Visualisation '96*, pp: 319-326, 1996.
4. M. Garland, P. Heckbert, Surface Simplification Using Quadric Error Metrics, *Proceedings of SIGGRAPH '97*, pp: 209-216, 1997.
5. M. Garland. Multiresolution modelling: survey & future opportunities, *State of the Art Reports of EUROGRAPHICS '99*, pp: 111-131, 1999.
6. H. Hoppe, View-dependent refinement of progressive meshes, *Proceedings of SIGGRAPH '97*, pp: 189-197, 1997.
7. M. Levoy et al, The digital Michelangelo project: 3D scanning of large statues. *Proceedings of SIGGRAPH 2000*, 2000.
8. D. Luebke, M. Reddy, J. D. Cohen, A. Varshney, B. Watson, R. Huebner. Level of detail for 3D graphics. *Morgan-Kaufmann*, 2003.
9. F. Ramos, M. Chover, LodStrips, *Proceedings of Computational Science ICCS 2004*, pp: 107-114, 2004.
10. J. Ribelles, M. Chover, A. López, J. Huerta, A First Step to Evaluate and Compare Multiresolution Models, *Short Papers and Demos of EUROGRAPHICS'99*, pp: 230-232, 1999.
11. J. Ribelles, A. López, O. Belmonte, I. Remolar, M. Chover. Multiresolution modeling of arbitrary polygonal surfaces: a characterization. *Computers & Graphics*, 26(3), pp: 449-462. 2002.
12. J. Xia, A. Varshney. Dynamic View-Dependent Simplification for Polygonal Models, *Visualization '96 Proceedings*, pp: 327-334, 1996.

# Improvement Construction for Planar G2 Transition Curve Between Two Separated Circles

Zhong Li[1,2], Lizhuang Ma[2], Mingxi Zhao[2], and Zhihong Mao[2]

[1] Department of Mathematics and Science, Zhejiang Sci-Tech University, Hangzhou, 310018, China
lizhongzju@hotmail.com
[2] Department of Computer Science and Engineering, Shanghai Jiao Tong University, Shanghai, 200030, China
ma-lz@cs.sjtu.edu.cn, {Zhaomx, mzh_yu}@sjtu.edu.cn

**Abstract.** In this paper, we use the undetermined coefficient method to find a desirable pair of cubic Bezier spirals and a desirable pair of quintic PH spirals to generate planar G2 transition curve between two separated circles. The G2 transition curve can be gotten by the rooting formula, which simplifies the computation, and the ratio of two radii has no restriction, which extends the application area.

## 1 Introduction

Transition curves are useful for several Computer Graphics and CAD applications. They may be used for blending in the plane, i.e. to round corners, or for smooth transition between two curves, usually two circular arcs. For some applications, such as the design of highways, railways or robot trajectories, it is desirable that the curvature varies monotonically with arc-length. So the research on G2 transition curve is important [1-3].

There have been some studies for the G2 transition curve when two circles have different positions. The Clothoid or Cornu spiral has traditionally been used in highway design for many years [2,3], but it can not be represented exactly as a NURBS curve, so it is not used easily in standard graphics packages. For two tangent circles, Li and Meek [9] gave the smooth method of arc spline by a pair of rational quadratic Bezier curves or one cubic Bezier curve. For two separated circles, Walton and Meek [4,6] found a family of cubic Bezier spirals and used one of them to construct the transition curve, but it requires the numerical method to solve, and in order to make sure there is one solution, the ratio of two radii sometimes has some restriction. Walton and Meek [5,7] also found a family of quintic PH spirals and used one of them to construct the transition curve, but it needs the numerical computation, and the ratio of two radii has some limitation too.

In this paper, we give the improvement construction for G2 transition curve between two separated circles. Based on the undetermined coefficient method, we find a pair of cubic Bezier spirals and a pair of quintic PH curve spirals to generate G2 transition curve. The advantage of improvement method is that

V.N. Alexandrov et al. (Eds.): ICCS 2006, Part II, LNCS 3992, pp. 358–361, 2006.

the transition curve can be gotten by the rooting formula, which simplifies the computation. And the ratio of two radii has no limitation, which extends the application area. The improvement construction of G2 transition curve is more convenient for applications in the fields of CAD and Computer Graphics.

## 2 G2 Transition Curve by a Pair of Cubic Bezier Spirals

The standard cubic Bezier curve can be written as

$$B(t) = (1-t)^3 B_0 + 3t(1-t)^2 B_1 + 3t^2(1-t)B_2 + t^3 B_3, t \in [0,1],$$

where $B_0, B_1, B_2, B_3$ are control points of the cubic Bezier curve.

Denote $h_0 = ||B_1 - B_0||, h_1 = ||B_2 - B_1||, h_2 = ||B_3 - B_2||$, $\theta$ is the angle from $B_0B_1$ to $B_1B_2$, $\phi$ is the angle form $B_1B_2$ to $B_2B_3$. From R[8], we know when $\theta = 0, h_1 = h_0$, and $h_2 \leq \frac{6}{5}h_1 \cos\phi$, the cubic Bezier curve is a spiral, the curvature at one endpoint is zero, the curvature at another endpoint is a constant, the curvature is monotone for $t \in [0,1]$. Here, we denote $h_2 = \frac{6}{5}xh_1 \cos\phi$, where $x$ is a real number and $0 < x \leq 1$. Helped by the undetermined coefficient method to find a desirable $x$, we can use the quadratic equation to get a pair of Bezier spirals for planar G2 transition curve in S shape between two separated circles.

In this case, assume that two circles $\Omega_0, \Omega_1$ centred at $C_0, C_1$ with radius $r_0, r_1, r_0 \geq r_1$, see Fig 1. We use a pair of Bezier spirals $B_0(t), B_1(t)$ to construct the G2 transition curve. Suppose $B_0(0), B_1(0)$ are the joint of two spirals at $B(0)$, points $B_0(1), B_1(1)$ connect two circles, $T, N$ are the unit tangent vector and the unit normal vector at the joint of two spirals, $T_a, T_b$ are the unit tangent vector at another endpoints of two curves, the angle from $T$ to $T_b$ and the angle from $T_a$ to $T$ are both $\phi$.

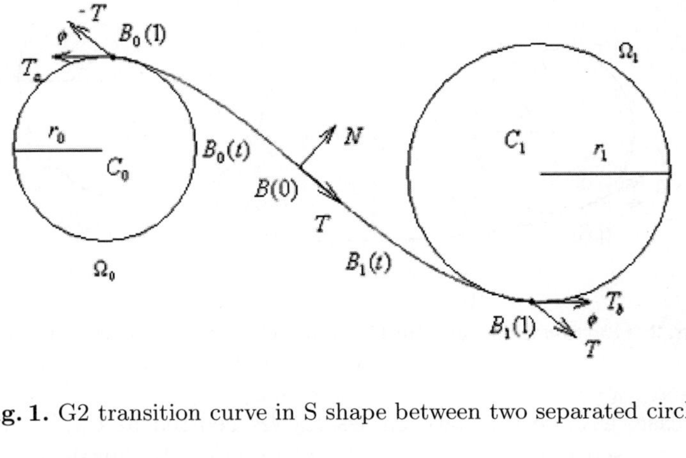


**Fig. 1.** G2 transition curve in S shape between two separated circles

Let $x = \frac{5}{9}$ and use the geometry relation, we can obtain a quadratic equation

$$q(\phi) = -||C_1 - C_0||2\cos\phi^2 - 8(r_1 + r_0)^2 \cos\phi + 9(r_1 + r_0)^2.$$

We can guarantee there is the unique root of $q(\phi) = 0$ when $\phi \in (0, \frac{\pi}{2})$. This root can be obtained by the rooting formula of quadratic equation. At the same time, since $x = \frac{5}{9} < 1$, the cubic Bezier curve is a spiral. And in the computation process, we find the ratio of two radii has no restriction.

Similarly, we can use a pair of cubic Bezier spirals to construct the G2 transition curve in C shape between two separated circles. The solution can also be gotten by the rooting formula of quadratic equation and the ratio of two radii has no restriction.

## 3 G2 Transition Curve by a Pair of Quintic PH Spirals

Consider the curve $Q(t) = x(t)T_0 + y(t)N_0$ and its derivative with respect to $t$, where $x(t)$ and $y(t)$ are polynomials in $t$, $T_0$ and $N_0$ are two orthonormal vectors. The curve $Q(t)$ is said to be a PH curve if $\{x'(t)\}^2 + \{y'(t)\}^2$ can be expressed as the square of a polynomial in $t$. To ensure $Q(t)$ is indeed a PH curve, define $x'(t)$ and $y'(t)$ as $x'(t) = U^2(t) - V^2(t)$ and $y'(t) = 2U(t)V(t)$, where $U(t)$ $V(t)$ are polynomials. A PH quintic Bezier form of $Q(t)$ is obtained by defining $U(t)$ and $V(t)$ as $U(t) = i(1-t)^2 + 2jt(1-t) + kt^2, V(t) = u(1-t)^2 + 2vt(1-t) + wt^2$.

From R[8], we know when $u = v = 0, i = j$, and $i \geq \frac{7k}{4(1+\cos\theta)}$, where $\theta$ is the angle between tangents ate two endpoints, the quintic PH curve is a spiral. Here, we denote $i = j = \frac{7xk}{4(1+\cos\theta)}, k = \frac{y}{2}\sqrt{7r\sin\theta}, u = v = 0, w = k\sqrt{\frac{1-\cos\theta}{1+\cos\theta}}$, where $x, y$ are real numbers, $x \geq 1, y > 0$. Based on the undetermined coefficient method to find the desirable $x, y$, we can use the rooting formula to construct the G2 transition curve in C shape between two separated circles.

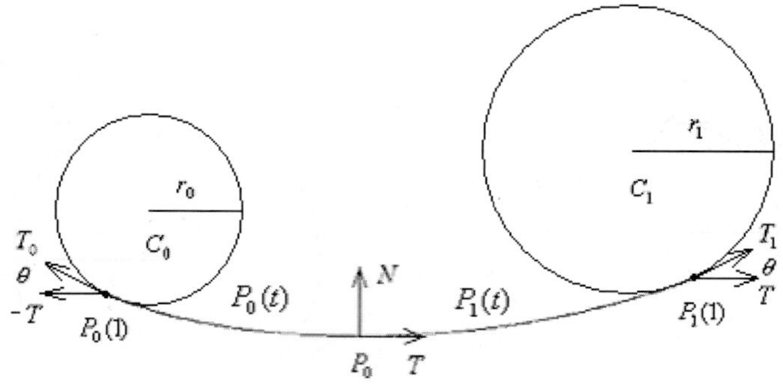


**Fig. 2.** G2 transition curve in C shape between two separated circles

In this case, assume that two circles $\Omega_0, \Omega_1$ centred at $C_0, C_1$ with radius $r_0, r_1, r_0 \geq r_1$, see Fig 2. We use a pair of quintic PH Spirals $P_0(t), P_1(t)$ to construct the transition curve. Suppose $P_0(0), P_1(0)$ are the joint of two spirals at $P_0$. Points $P_0(1), P_1(1)$ connect two circles, $T, N$ are the unit tangent vector and the unit normal vector at $P_0$, points $T_0, T_1$ are the unit tangent vector at another endpoints of two curves, the angle from $T$ to $T_1$ and the angle from $T_0$ to $T$ are both $\theta$.

Let $\varepsilon = \frac{6}{7}(\frac{||C_1-C_0||}{||r_1-r_0||} - 1)$, $x = \begin{cases} 1 \text{ when } x \leq 1 \\ \varepsilon \text{ when } x > 1 \end{cases}$, $y = \sqrt{\frac{10}{7}}$, by the geometry relation, we can obtain the following cubic equation

$$h(\theta)=||C_1-C_0||^2(1+\cos\theta)^3-\frac{(r_1+r_0)^2}{120^2}(1-\cos\theta)g_1^2(\theta)-\frac{(r_1-r_0)^2}{60^2}(1+\cos\theta)g_2^2(\theta),$$

where $g_1(\theta) = (140x - 120)\cos\theta + 490x^2 + 140x - 120$, $g_2(\theta) = (60 - 70x)\cos\theta + 70x + 60$.

we can guarantee that $h(0) > 0, h(\frac{\pi}{2}) < 0$ and $h'(\theta) < 0$ for $\theta \in (0, \frac{\pi}{2})$. So there is the unique real root of $h(\theta) = 0$ when $\theta \in (0, \frac{\pi}{2})$. This root can be gotten by the rooting formula of cubic equation. In the computation process, we find the ratio of two radii has no restriction.

For the G2 transition curve in S shape between two separated circles, we can also use a pair of quintic PH spirals to construct the transition curve. It can be gotten by the rooting formula of cubic equation and the ratio of two radii has no restriction.

## Acknowledgements

This paper was supported by National Natural Science Foundation of China (No.60373070 and No.60573147), Research Foundation of Microsoft China (Project-2004-Image-01), Postdoctor Foundation of Shanghai (05R214129), Education Foundation of Zhejiang (20050786).

## References

1. Guggenheimer H.W.: Differential geometry. New York: McGraw-Hill (1963)
2. Baass K.G.: The use of clothoid templates in highway design. Transportation Forum (1984) 1, 47-52
3. Meek D.S., Walton D.J.: The use of Cornu spirals in drawing planar curves of controlled curvature. Journal of Computational and Applied Mathematics (1989) 25, 69-78
4. Walton D.J., Meek D.S.: A planar cubic Bezier spiral. Journal of Computational and Applied Mathematics (1996) 72, 85-100
5. Walton D.J., Meek D.S.: A Pythagorean hodograph quintic spiral. Computer Aided Design (1996) 28, 943-950
6. Walton D.J., Meek D.S.: Planar G2 transition curves composed of cubic Bezier spiral segments. Journal of Computational and Applied Mathematics, (2003) 157, 453-476
7. Walton D.J., Meek D.S.: Planar G2 transition with a fair Pythagorean hodograph quintic curve. Journal of Computational and Applied Mathematics (2002) 138, 109-126
8. Walton D.J., Meek D.S.: G2 curve composed of cubic and Pythagorean hodograph quintic spirals. Computer Aided Geometry Design (1998) 15, 547-566
9. Li Z., Meek D.S.: Smoothing an arc spline. Computers & Graphics (2005) 29, 576-587

# B-Spline Curve Fitting Using Dominant Points*

Hyungjun Park[1] and Joo-Haeng Lee[2]

[1] Chosun University, Gwangju 501-759, Korea
hzpark@chosun.ac.kr
[2] Electronics and Telecommunications Research Institute, Daejeon 305-350, Korea
joohaeng@etri.re.kr

**Abstract.** Motivated by an insight that properly selected points, called *dominant* points, can play an important role in producing better curve approximation, we propose a new approach for B-spline curve fitting to a sequence of points. The approach is substantially different from the conventional ones in knot placement and dominant point selection. It can generate a B-spline curve in good quality with less deviation. Adopted in the error-bounded curve approximation, it can play an important role in generating B-spline curves with much less control points.

## 1 Previous Work

Least-squares B-spline curve fitting has been used as a basic tool to construct a B-spline curve $\mathbf{c}(t)$ from points $\mathbf{p}_i$ ($i$=0,..,$m$), which takes three steps: parameterization, knot placement, and the least-squares minimization [1]. In parameterization, we select the parameters $\bar{t}_i$ of the points $\mathbf{p}_i$. The chord length or centripetal methods have been widely adopted for this task. In knot placement, we determine a knot vector $\mathbf{T}=\{t_0,t_1,...,t_{n+p-1},t_{n+p}\}$ after specifying the order (degree+1) $p$ and the index $n$. In the least-squares minimization, we determine control points $\mathbf{b}_j$ ($j$=0,..,$n$) of a B-spline curve $\mathbf{c}(t)$ by minimizing the least-squares error $E(\mathbf{b}_0,..,\mathbf{b}_n)=\sum_{i=0}^{m}\|\mathbf{c}(\bar{t}_i)-\mathbf{p}_i\|^2$. This minimization problem is transformed into the problem of solving a linear system. The knots are determined to reflect the distribution of the parameters. The interior knots $t_i$ of the knot vector $\mathbf{T}$ can be spaced as follows [1]:

$$t_{p+i-1}=\begin{cases}\dfrac{1}{p-1}\sum_{j=i}^{i+p-2}\bar{t}_j & for\ m=n\\ (1-u)\bar{t}_{k-1}+u\bar{t}_k & for\ m>n\end{cases}\qquad (i=1,..,n-p+1) \tag{1}$$

where $k=\text{int}(i\times d)$, $d=(m+1)/(n-p+2)$, and $u=i\times d-k$. This is called the averaging technique (AVG) for $m=n$, and the knot placement technique (KTP) for $m>n$. These techniques give a stable system of equations [1]. Nonetheless, not all stable equations yield acceptable curves. When $m$ is nearly greater than $n$ ($|m-n|$ is small),

* This work was supported in part by research funds from Chosun University, 2004.

V.N. Alexandrov et al. (Eds.): ICCS 2006, Part II, LNCS 3992, pp. 362–366, 2006.

it is liable to generate undesirable results. To avoid this, Piegl and Tiller [2] suggested another knot placement technique (NKTP).

## 2 Proposed Approach

The proposed approach takes four main steps: parameterization, dominant point selection, knot placement, and least-squares minimization. The approach is substantially different from the conventional ones in knot placement and dominant point selection.

### 2.1 Determination of Knots

Consider that dominant points $\mathbf{d}_j$ $(j=0,..,n)$ have been selected among the given points $\mathbf{p}_k$ ($k$=0,..,$m$). Generally, $n$ is not greater than $m$ ($n \le m$). With the points $\mathbf{d}_j$, interior knots $t_i$ can be computed as follows:

$$t_{p+i-1} = \frac{1}{p-1}\sum_{j=i}^{i+p-2} \bar{t}_{f(j)} \qquad (i=1,..,n-p+1) \tag{2}$$

where $\bar{t}_k$ are the parameters of the points $\mathbf{p}_k$ and $f(j)$ is a simple function that returns the index of the point $\mathbf{p}_k$ corresponding to a dominant point $\mathbf{d}_j$. The knots are computed by averaging the parameters of the dominant points, so the knot vector coincides with the one for B-spline curve interpolation to the dominant points [1].

### 2.2 Selection of Dominant Points

Eventually, the quality of the resulting curve depends on how to select dominant points $\mathbf{d}_j$ ($j$=0,..,$n$) among the points $\mathbf{p}_i$ ($i$=0,..,$m$). We can devise an algorithm realizing that fewer dominant points are chosen at flat regions but more at complex regions. Its main steps are given as follows:

**(1)** Set $n_d = -1$ where $n_d$ denotes the highest index of current dominant points.

**(2)** Select *seed* points from the point set, and store them in a list **S** in decreasing order of *significance*. While $\mathbf{S} \ne \varnothing$ AND $n_d < n$, pop a point from the list **S**, make it a *new* dominant point, and $n_d \leftarrow n_d + 1$.

**(3)** While $n_d < n$, select dominant points $\mathbf{d}_j$ adaptively as follows:

(3.1) Find two successive dominant points $\mathbf{d}_k = \mathbf{p}_s$ and $\mathbf{d}_{k+1} = \mathbf{p}_e$ ($|e-s|>1$) whose corresponding segment incurs the largest deviation.

(3.2) Choose a point $\mathbf{p}_w$ $(s<w<e)$ as a *new* dominant point, and $n_d \leftarrow n_d + 1$.

We herein consider two kinds of seed points: two end points and local curvature maximum points (LCM). We detect LCM points from the curvatures estimated at the given points. Two end points are the most significant. For LCM points, the greater their curvatures are, the more significant they are.

With the current dominant points, we can generate a current curve by knot placement and least-squares minimization. For two neighboring dominant points $\mathbf{d}_j = \mathbf{p}_s$ and $\mathbf{d}_{j+1} = \mathbf{p}_e$, we can build a segment $\mathbf{S}_{s,e} = \{\mathbf{p}_k \mid k = s,..,e\}$. Among the segments consisting of at least three points, we find the segment with the largest deviation to the current curve. We then choose a *new* dominant point from the segment. A simple approach for the choice is to select the median of the points of the segment, but a more promising approach is to select a point $\mathbf{p}_w$ that minimizes $|\lambda_{s,w} - \lambda_{w,e}|$ for $s < w < e$ where $\lambda_{s,e}$ denotes the shape complexity of a segment $\mathbf{S}_{s,e}$ that can be estimated by combining the total curvature and the arc length of the segment.

## 3 Experimental Results

Fig. 1 shows cubic B-spline curve fitting to a point set that consists of 251 points. The points are enclosed in an 84×126 rectangle. In Fig. 1(d), KTP selects the knots such that each knot span contains almost the same number of parameters. In Fig. 1(e), NKTP shows a similar trend except at the first and the last knot spans. In Figs. 1(c) and 1(f), the proposed approach creates fewer dominant points (curve segments) at flat regions but more at complex regions.

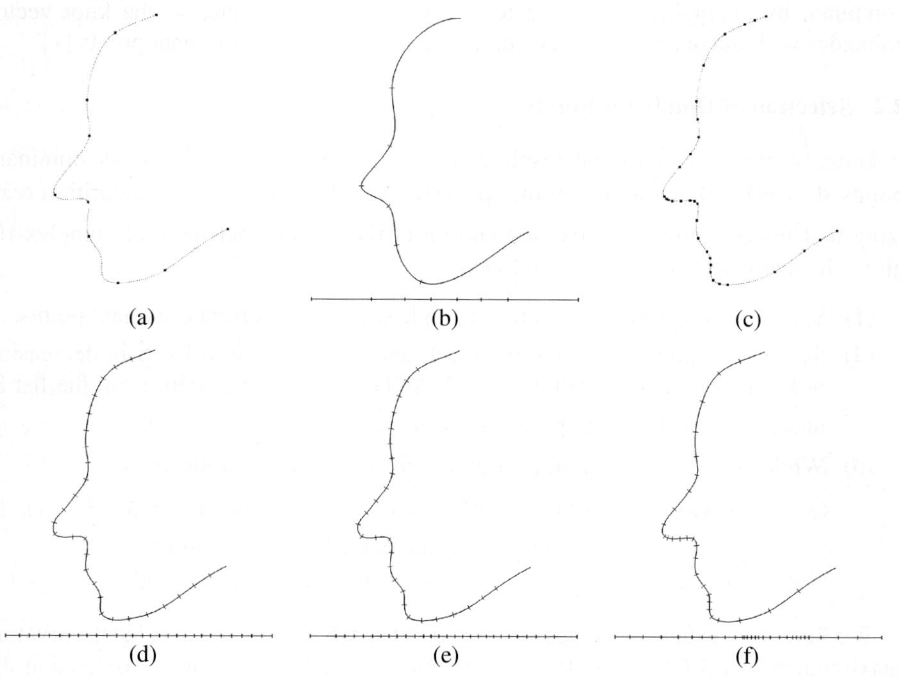


**Fig. 1.** B-spline curve fitting: (a) point set with 10 seed points; (b) curve obtained using 10 seed points as dominant points; (c) point set with 31 dominant points; (d) curve obtained by KTP ($n$=31); (e) curve obtained by NKTP ($n$=31); (f) curve obtained with 31 dominant points

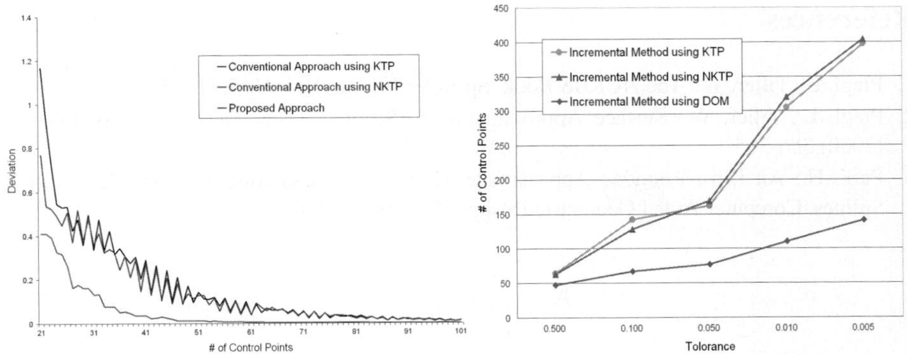


(a) (b)

**Fig. 2.** Plots for comparison: (a) Deviation plot; (b) Plot of the number of control points

Fig. 2(a) shows the deviation caused by B-spline curve fitting to the point set in Fig. 1(a). The deviation is computed as we increase the number of control points from 16 to 251. Deviations for more than 101 control points were omitted from the plot since they are not differentiable visually. The deviation caused by the proposed approach is 69% less than the one caused by the approach using KTP, and the deviation caused by the approach using NKTP is nearly the same as (actually 3% greater than) the one caused by KTP. Additionally, we adopted three approaches KTP, NKTP, and DOM in the incremental method for the error-bounded curve approximation [3]. DOM denotes the proposed approach. The incremental method starts with few control points and adds control points repeatedly to maintain the error bound [1,3]. We compared the three approaches by the number of control points required to approximate the point set within a tolerance. Fig. 2(b) shows the comparison result of the number of required control points. Note that DOM outperforms the others in data reduction.

## 4 Concluding Remarks

Because of the page limit of the paper, we had to shorten the description of the proposed approach to the minimum. The approach can realize the concept of adaptive refinement that fewer curve segments are generated at flat regions but more at complex regions. It can generate a B-spline curve in good quality with less deviation. Adopted in the error-bounded curve approximation, the approach can play an important role in generating B-spline curves with much less control points.

There still remain important issues not covered in this paper: properties of the knot placement using dominant points; non-deterministic algorithms such as genetic algorithms for the optimal selection of dominant points; expansion of the proposed approach to B-spline surface fitting. We expect that all the issues will be handled by the future research.

## References

1. Piegl, L., Tiller, W.: The NURBS Book. Springer-Verlag, New York (1995)
2. Piegl, L., Tiller, W.: Surface Approximation to Scanned Data. The Visual Computer 16 (2000) 386-395
3. Park, H.: An Error-Bounded Approximate Method for Representing Planar Curves in B-Splines. Computer Aided Geometric Design 21 (2004) 479-497

# Quality and Performance Evaluation of Ray-Space Interpolation for Free Viewpoint Video Systems

Fan Liangzhong[1,2], Yu Mei[1,3], Yu Zhou[1], and Jiang Gangyi[1,2]

[1] Faculty of Information Science and Engineering, Ningbo University, 315211, Ningbo, China
[2] Institute of Computer Technology, Graduate School of Chinese Academy of Sciences, 100871, Beijing, China
[3] National Key Lab of Software New Technology, Nanjing University, Nanjing 210093, China

**Abstract.** Ray-space is the main technology to realize Free Viewpoint Video (FVV) systems with complicated scenes. One of the key technologies to make the ray-space feasible is interpolation. In this paper, two fast interpolation methods based on feature points and directionality detection are compared with block based matching method, and the experimental results show that the two methods improve visual quality as well as *PSNR*s of rendered virtual viewpoint image greatly while computational burden is also reduced significantly.

## 1 Ray-Space Interpolation for Free Viewpoint Systems

In the application scenario of free viewpoint video (FVV) systems, the viewers experience the free viewpoint navigation within the range covered by the shooting cameras. In recent years, image based rendering techniques have been developed to generate novel views of an environment from a set of pre-acquired images. Ray-space representation, as a new technique of image based rendering, renders arbitrary viewpoint images without complicated object segmentation and 3D modeling. However, in real world situation, it is difficult to set many cameras very closely, and the ray data obtained by the camera setup is too sparse in viewpoint axis to apply the ray-space method [1]. Therefore, it is necessary to generate dense ray data by interpolation techniques. It should be noted that ray-space slice has strong directionalities, so that conventional interpolation methods developed for natural image are not suitable anymore. Directionality detection of each pixel to be interpolated becomes very important in ray-space interpolation. Because ray-space interpolation is in fact associated with arbitrary intermediate view rendering, it is one of the key technologies in ray-space based FVV systems.

The block-based matching method (BMI) consists of searching for corresponding blocks in the two EPI lines (epipolar lines) to find the best-corresponding pixel pair for interpolation in the assigned maximum disparity [2]. Our directionality detection based method (DDI) [3] and feature point growth based method (FGI) [4] are two feature based interpolation methods. In DDI and FGI, feature pixels are first extracted from sparse ray-space slice, and their directionalities are determined by block matching technique, then directionality of each pixel to be interpolated in dense ray-space

V.N. Alexandrov et al. (Eds.): ICCS 2006, Part II, LNCS 3992, pp. 367 – 370, 2006.

slice is linear interpolated with the directionalities of these feature pixels, and then linear interpolation is done according to directionality of the pixel to be interpolated. In fact, the so-called feature pixel is the one near edge or within area with fine texture. Directionality searching is only done for these feature pixels, because these searching usually can obtain more reliable results compared with processing of non-feature pixels around which there is no enough texture information for marching. Moreover, the computational burden is also reduced because the matching is done only for feature pixels but not all pixels to be interpolated. Different from DDI method, FGI method is proposed to solve the occlusion problem further. In FGI, each of the feature points grows in the nearby epipolar lines until the direction of the grown feature point departures from the direction of the initial point, which means that occlusion occurs or the feature point is in fact a noise.

## 2 Experimental Results and Analysis

To compare the above three methods, we perform experiments on two test sequences of real data called "Xmas" and "Cup" as given in Fig.1 on a PC with CPU of 1.8G Hz and RAM of 256M. "Xmas" sequence is provided by Tanimoto Laboratory, Nagoya University, 101 viewpoint images are captured synchronously with camera interval of 3mm, the image resolution is 640×480, and the distance between cameras and object is about 30cm. The "Cup" sequence is captured by a shifted camera with interval of 1mm, the number of viewpoint images and the resolution are the same as "Xmas" sequence, distance between camera and object is about 20cm, so the maximum disparity in "Cup" sequence is much larger than "Xmas" sequence.

(a) "Xmas" (b) "Cup"

**Fig. 1.** The original left-most and right-most viewpoint images of the two test sequences

The EPI obtained from the 101 images is adopted as the dense EPI, while the EPI transformed from some images with same camera interval is used to simulate the sparsely sampling. To evaluate the performance of ray-space interpolation method, the interpolated EPI is compared with the corresponding dense one, and the rendered intermediate viewpoint images are also compared with the real captured images.

Table 1 gives the average *PSNR*s of the 480 interpolated EPIs as well as the corresponding average interpolating times. It is clear that the DDI and FGI methods achieve much higher *PSNR*s than BMI method. The interpolating time of the BMI method increases as camera interval increases, but the proposed methods consume approximately constant time under different camera intervals, and the interpolating time is much saved compared with BMI method especially under large camera interval.

**Table 1.** Average *PSNR* / Interpolating time of 480 EPIs unit: dB / ms

| Test sets | Camera interval | BMI | DDI | FGI |
|---|---|---|---|---|
| Xmas | 15mm | 39.31 / 97 | 40.69 / 28 | 41.03 / 34 |
| | 30mm | 38.81 / 185 | 39.59 / 28 | 40.27 / 36 |
| | 60mm | 37.72 / 354 | 38.60 / 29 | 39.72 / 37 |
| | 75mm | 36.60 / 438 | 38.07 / 29 | 39.50 / 38 |
| Cup | 5mm | 33.84 / 165 | 34.27 / 26 | 34.89 / 36 |
| | 10mm | 33.11 / 335 | 33.60 / 25 | 34.23 / 39 |
| | 20mm | 31.58 / 672 | 32.68 / 26 | 33.50 / 41 |
| | 25mm | 30.35 / 839 | 31.36 / 25 | 31.99 / 41 |

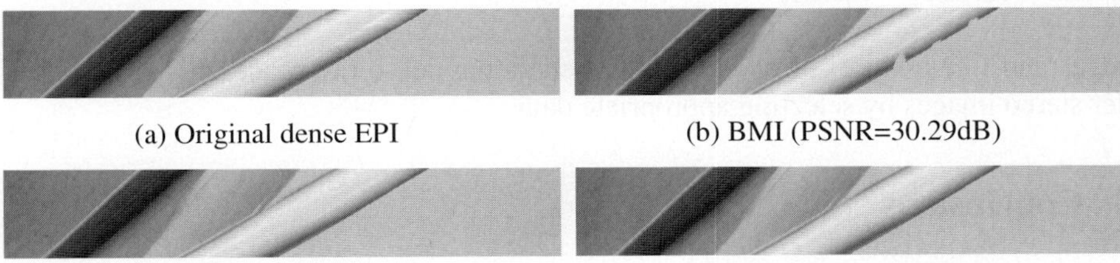

(a) Original dense EPI (b) BMI (PSNR=30.29dB)

(c) DDI (PSNR=32.03dB) (d) FGI (PSNR=32.29dB)

**Fig. 2.** Results of the interpolated 155th EPI with camera interval 25mm (from "Cup" sequence)

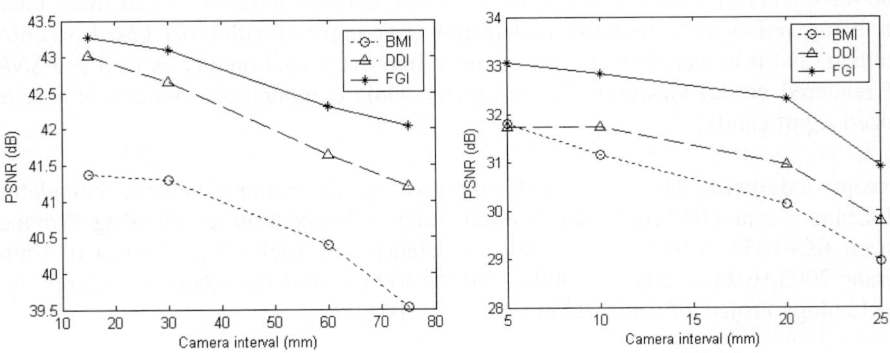


(a) "Xmas" (b) "Cup"

**Fig. 3.** Average *PSNR*s of rendered intermediate viewpoint images

Fig.2 gives an example of ray-space interpolation with the three different methods. Fig.3 shows average *PSNR*s of intermediate viewpoint images rendered from the interpolated ray-space. In the figures, the proposed interpolation methods outperform BMI method, because it can obtain more accurate directions. Moreover, the proposed methods run much faster than BMI method because it only searches some feature points' directions but not all of the pixels to be interpolated. Fig.4 shows parts of the rendered viewpoint images with respect to the above three interpolation methods. The text and the door's edges on the cup are well kept by the proposed methods, it is obvious that the proposed methods are much better than BMI method in keeping

(a) Original image (b) BMI (29.35dB) (c) DDI (30.99 dB) (d) FGI (32.74dB)

**Fig. 4.** Part of the rendered 40th virtual view from interpolated ray-space with camera interval 25 mm (from "Cup" sequence)

edges and fine textures of objects. After generating dense ray-space, it is easy to render stereo images by selecting appropriate data.

## 3 Conclusions

Ray-space representation has superiority in rendering arbitrary viewpoint images of complicated scene in real-time. Interpolation is one of the key techniques to make the ray-space based FVV system feasible, and it also determines the cost of application and the quality of rendered image. In this paper, two fast interpolation methods based on feature points and directionality detection are compared with block based matching method, and it is seen that the two methods improve visual quality as well as *PSNR*s of rendered virtual viewpoint image greatly while computational burden is also reduced significantly.

**Acknowledgment.** This work was supported by the Natural Science Foundation of China (grant 60472100), the Natural Science Foundation of Zhejiang Province (grant RC01057, Y105577), the Ningbo Science and Technology Project of China (grant 2003A61001, 2004A610001, 2004A630002), and the Zhejiang Science and Technology Project of China (Grant 2004C31105).

## References

1. Fujii T., Tanimoto M.: Free-viewpoint TV system based on Ray-Space representation. In: Proceedings of SPIE, vol. 4864. Boston (2002) 175-189
2. Tehrani, M. P., Fujii, T., Tanimoto, M.: Offset Block Matching of Multi-View Images for Ray-Space Interpolation. The journal of the Institute of Image Information and Television Engineers. Vol. 58. (2004) 540-548
3. Jiang, G., Yu, M., Ye, Xien.: New method of Ray-Space interpolation for free viewpoint video. In: Proceedings of International Conference on Image Processing. (2005) 1138-1141
4. Jiang, G., Fan, L.,Yu, M.: Fast Ray-Space Interpolation based on Occlusion Analysis and FeaturePoints Detection. In: Proceedings of International Conference on Computational Intelligence and Security. Vol. 3802. Xi an, China (2005) 935-940

# Framework for Adaptive Sampling of Point-Based Surfaces Using Geometry and Color Attributes


Duck Bong Kim[1], Eui Chul Kang[1], Kwan H. Lee[1], and Renato B. Pajarola[2]

[1] Gwangju Institute of Science and Technology (GIST), Intelligent Design and Graphics laboratory, Department of Mechatronics, 1 Oryong-dong, Buk-gu, Gwangju, 500-712, Korea
{eorka333, eckang, lee}@kyebek.gist.ac.kr

[2] Univ. of Zurich, Department of Informatics, Winterthurerstr. 190 8057, Zurich, Switzerland
pajarola@acm.org



**Abstract.** Point-based rendering has offered a powerful alternative to triangle meshes when it comes to the rendering of highly complex objects consisting of densely sampled point clouds due to its flexibility and simplicity. The technological advance of 3D scanners has made it possible to acquire color as well as geometry data of highly complex objects. However, scanning and acquisition systems often produce surfaces that are much more dense than actually required for the intended application. Mobile devices, computer games and distributed virtual environments must often operate on systems where rendering and transmission capacity is highly constrained and therefore require strict control over the level of detail used in models. In this research, we present a framework for adaptive sampling of point-based surfaces using both geometry and color information.


## 1 Introduction

The recent advancement of 3D scanners has made it possible to acquire color as well as geometry information of highly complex objects with very high speed and with good accuracy. However, the acquired data poses great challenges in storage, editing, transmission, and rendering due to the heavy data set. Therefore, simplification of highly detailed objects is necessary for the real time implementation and has become an important issue in many application fields such as entertainment, industrial design, virtual reality and other related fields.

Polygonal-based simplification has been studied by many researchers. In spite of the simplicity and flexibility of polygonal-based simplification, it has some limitations and drawbacks in some applications due to the need of establishing connectivity. However, a point-based method does not need the connectivity information during the simplification process. Efficient simplification of point-sampled surfaces has been researched [1,2]. However, many existing point-based simplification methods focus only on decimating an original model into simplified models considering only the geometry information.


V.N. Alexandrov et al. (Eds.): ICCS 2006, Part II, LNCS 3992, pp. 371–374, 2006.

(a) Real object (b) Scanned data (c) Adapt. sampling (d) Display

**Fig. 1.** 3D discrete LOD models of point-based surface creation pipeline

In our research, there are mainly three steps in adaptive sampling of the generation of point-based surfaces. Fig. 1 illustrates the whole process of generating 3D discrete LOD (level of detail) models of point-based surfaces acquired from a 3D scanning system. First, a real object is digitized using a 3D scanning device. The raw data from the 3D scanner is to be sampled adaptively according to the required level or desired application. Finally, the model is displayed using a point based rendering system based on blending and visibility splatting [3].

The main goal of this work is to present a framework for an adaptive sampling of point-based surfaces based on 3D grids using both color and geometry information to create the same visual quality and to speed up the rendering with lower sampling density.

## 2 Adaptive Sampling of Point-Based Surface

In this stage, the raw data from the 3D scanner is to be sampled adaptively according to the required level. The proposed method uses a 3D grid algorithm [4]. It subdivides the point cloud into a number of sub grids according to normal and color standard deviation of points, each of which is replaced by one representative point sample. Using the 3D grid algorithm, we can sample more points in the regions of high curvature and big color difference, as shown in Fig. 2(c) and 2(d). Before using 3D grid algorithm, normal estimation is performed locally using an algorithm [3] based on covariance matrix for each point and its neighbors. Fig. 2 illustrates the whole procedure of the adaptive point sampling method based on 3D grids from the unorganized point data set. Fig. 2(a) shows the model Nefertiti. Initial grids are generated by a user defined input, as shown in Fig. 2(b). Each initial grid is subdivided according to the standard deviation of normal values, as shown in Fig. 2(c). The subdivided grids are repeatedly subdivided according to the standard deviation of color values, as shown in Fig. 2(d). A Refinement step is performed in order to avoid holes and intervention between discrete point samples, as shown in Fig. 2(e). Upon completion of the 3D grid subdivision, the decimated discrete point samples are converted into surface splats in order to fill the gap between the neighboring samples. In order to estimate the extent of each splat, another calculation is required. However, we estimate the extent of each splat according to the depth of each grid.

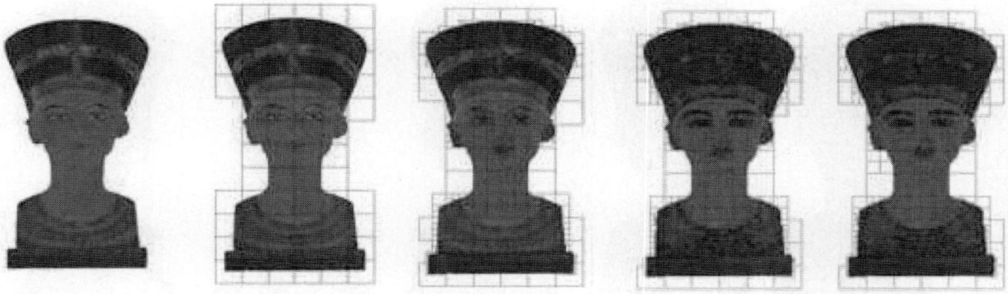

**Fig. 2.** Adaptive sampling method procedure: (a) Nefertiti (b) Initial grids (c) Subdivided by normal variation (d) Subdivided by color variation (e) Refinement

## 3 Rendering Result

Fig. 3 shows the visual quality comparison between the original model (left), the model simplified by uniform sampling method (left middle), the model simplified by only using geometry variation (right middle), and the model simplified by geometry and color variation (right). The model, Nefertiti, is simplified from the original point set of 448k to 14k by the uniform sampling method and the proposed algorithm and rendered by the proposed point-based rendering technique. This rendering result demonstrates that the eyes of Nefertiti simplified by geometry and color variation look sharp, while the eyes of Nefertiti, simplified by uniform sampling method and only geometry variation, are blurred.

Fig. 4 illustrates the visual quality according to 3D discrete LOD (level of detail) point-based surfaces. The visual quality of each point-based surface is poorer in accordance with the smaller 3D discrete LOD (level of detail) point-based surfaces. However, the rendering speed time becomes faster as the level of detail decreases.

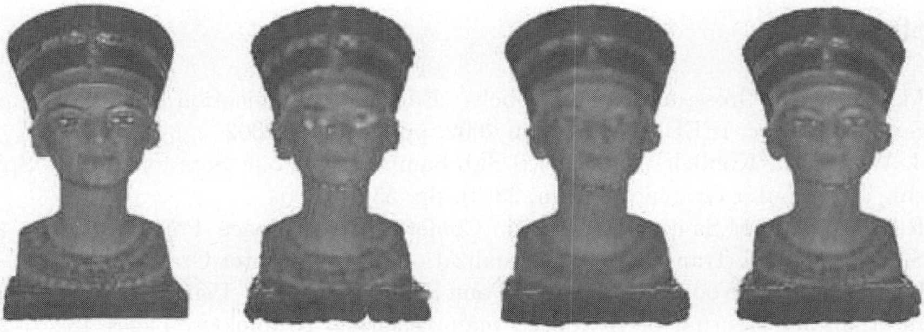

**Fig. 3.** Visual comparison of Model Nefertiti: (a) Original (b) sampled by uniform (c) using geometry variation (d) using geometry and color variation

**Fig. 4.** Nefertiti at different levels of detail. From left to right, 450k, 110k, 45k, 14k and 5k points of the original model, rendered with a point splatting renderer.

## 4 Conclusion

In this paper, we proposed a framework for adaptive sampling of point-based surfaces using both geometry and color attributes. Experimental rendering results demonstrate that the visual quality of the model simplified by using geometry and color variation has the best quality compared with other approaches and a trade off exists between the visual quality and the speed of rendering time. In future we will define new mathematical criteria for determining the extent of each splat to generate point-based surfaces without holes and intervention.

## Acknowledgements

This work was supported by the Immersive Contents Research Center and by the Realistic Broadcasting Research Center at GIST, Gwang-ju, South Korea and by the Korea Science and Engineering Foundation (KOSEF). The authors would like to thank Inus technology, Inc. for providing the geometric model Nefertiti.

## References

1. M. Pauly, M. Gross, and L.P. Kobbelt, "Efficient Simplification of Point-Sampled Surfaces," Proc. IEEE Visualization 2002, pp. 163-170, 2002
2. J. Wu and L. Kobbelt, "Optimized Sub-Sampling of Point Sets for Surface Splatting", Computer Graphics Forum, 23(3), pp. 53-64. 2004
3. R.B.Pajarola, M.Sainz, P.Guidotiti. Confetti Object-Space Point Blending and Splatting. IEEE Transaction on Visualization and Computer Graphics, 2004
4. K. H. Lee, H. Woo and T. Suk, "Point Data Reduction Using 3D Grids", The International Journal of Advanced manufacturing Technology, pages 18:201-210, 2001

# Normalizing Relational Database Schemas Using Mathematica

Ali Yazici[1] and Ziya Karakaya[2]

[1] Computer Engineering Department, TOBB University of Economics & Technology, Ankara - Turkey
aliyazici@etu.edu.tr
[2] Computer Engineering Department, Atilim University, Ankara - Turkey
ziya@atilim.edu.tr

**Abstract.** In this paper, basic relational database (DB) normalization algorithms are implemented efficiently as Mathematica modules. It was observed that, Mathematica provided a straightforward platform as opposed to previous ones, mainly Prolog based tools which required complex data structures such as linked list representations with pointers. A Java user interface called JMath-Norm was designed to execute the Mathematica modules in a systematic way. For this purpose, Mathematica's Java link facility (JLink) is utilized to drive the Mathematica kernel. JMath-Norm provides an effective interactive tool in an educational setting for teaching DB normalization theory.

## 1 Introduction

Conceptual modeling is one of the most important steps in the design of DB systems. It is in a way the transformation of the real life problem or business rules into a set of entities with certain attributes and relationships among them. The ER, EER or UML data modeling tools are utilized for the purpose of representing this transformation graphically. The entity set in the conceptual model is eventually mapped into a set of relation schemas $\{R_1, R_2, R_3, ..., R_n\}$ where each $R_i$ represents one of the relations (tables) in the DB schema. This mapping is accomplished by using a set of informal rules [1].

The next step in the design process is normalization in which each $R_i$ along with its attribute set and a set of functional dependencies defined are considered and the relation schema is decomposed into sub relations to satisfy the so called normalization criteria. In the literature 1NF, 2NF, 3NF, BCNF, 4NF and 5NF are commonly cited even though some other special normal forms do exist.

Although automatic normalization has been a point of attraction for DB researchers from the very beginning of the relational DB theory, commercial DB tools do not provide a complete solution for automatic normalization.

Computational problems and some algorithmic problems in the design of normalized relation schemas has been considered in [2]. A Prolog implementation of the normalization process can be found in [3]. Another Prolog based normalization tool is described in [4]. Some new and fast algorithms for DB normalization is

V.N. Alexandrov et al. (Eds.): ICCS 2006, Part II, LNCS 3992, pp. 375–382, 2006.

provided in [5, 6] where a new method for computing minimal covers is presented using a modified definition of closure for normalizing relations. Normalization and some associated problems have been investigated from a semantical point of view in [7]. A polynomial time algorithm for finding all possible candidate keys in a relation schema is given in [8].

Decomposing relations into sub relations during the normalization process normally introduces some overhead in the DB querying. Cost effectiveness of normalization process has been undertaken by some researches [9, 10].

A practical solution to 1NF problem was given in [11] where the repeating groups are handled by creating an extra superfluous repeater table. A discussion on relations which are in 3NF but not in BCNF is provided in [12].

In this paper, the symbolic computational power of Mathematica and the Java are combined to design a normalization tool for DB relation schemas. Basic normalization algorithms are implemented efficiently in Mathematica and Java is utilized to design a windows based application. This tool aims to provide an educational setting for the undergraduate students studying DB systems. All standard normalization algorithms are implemented and can be tested individually or in an integrated fashion to provide 3NF relations. The design approach in this study is similar to the one given in [13] although the underlying environment and the purpose of the study are totally different.

The basic data structure used in all implementations given here are just lists together with the standard Mathematica set operations as opposed to the linked lists and pointers and complicated programming structures used in the previous implementations (see for example [3, 4, 13]).

The tool designed in this work utilizes JLink [14], Java link mechanism of the Mathematica.

Section 2 briefly discusses the normalization theory. In Section 3 Mathematica implementation of basic normalization algorithms is given. Java interface via JLink is given in Section 4. Final remarks and conclusions are provided in the final section.

## 2 Theoretical Background

As mentioned in the preceding section, the relational DB schema as an outcome of the conceptual modeling should be normalized to sustain the integrity of the DB. The normalization process relies on the concept of functional dependency (FD) between set of attributes of the DB schema. Formally, a FD between two sets of attributes X and Y, denoted by, $X \rightarrow Y$ specifies that for any two tuples $t_1$ and $t_2$ in a relation state r of $R = \{A_1, A_2, \cdots, A_n\}$ with n attributes that have $t_1[X] = t_2[X]$, they must also have $t_1[Y] = t_2[Y]$. One says that Y is functionally dependent on X.

For example, for a DB schema named *STUDENT*= {***studentNo***, *studentName, instructorId, officeNo*}, with **studentNo** as the primary key, one can write the following FD's:

$FD1$ : *studentNo* $\rightarrow$ {*studentName, instructorId, officeNo*}
$FD2$ : *instructorId* $\rightarrow$ *officeNo*

Other set of FD's can be obtained using the concept of attribute set closure. For each set of attributes X, we determine $X^+$, the set of attributes that can be determined by X based on the set of FD's F of the schema. An algorithm to do so is given below [1]:

**Algorithm ClosureX:** Determining $X^+$ under FD
Given a DB schema R, a set of attributes X and FD set F calculate $X^+$

1. $X^+ := X$;
2. repeat
   $tempX^+ := X^+$
   for each FD $Y \rightarrow Z$ in $F$ do
     if $X^+ \supseteq Y$ then $X^+ := X^+ \cup Z$
   until $(X^+ = tempX^+)$

Given a set of FD's F, an attribute B is said to be extraneous [5] in $X \rightarrow A$ with respect to F if $X = ZB$, $X \neq Z$, and $A \in Z^+$. A set of FD's H is called a minimal cover[1, 5] for a set F if each dependency in H as exactly one attribute on the right-hand side, if no attribute on the left-hand side is extraneous, and if no dependency in H can be derived from the other dependencies in H. All normalization algorithms uses the minimal cover of a given set of FD's.

In this study, the closure and minimal cover algorithms are implemented in Mathematica. An algorithm to calculate the minimal cover is given in [1]. Actually, the calculation of a minimal cover consists of "Elimination of Extraneous Attributes" followed by the "Elimination of Redundant Dependencies". Normalization is a step by step procedure in which the original DB schema is decomposed into a set of relations to satisfy the so called normalization conditions. 1NF relations are those without any multi-valued attributes. In a 2NF relation there are no partial dependencies. 3NF relations do not contain any transitive dependencies among its attributes. This paper utilizes Bernstein's algorithm below [1, 15] to provide 3NF schemas for a given set of attributes and a set of FD's F. 2NF condition is automatically satisfied and dependencies are preserved. However, lossless join property [1] is not guaranteed.

**Algorithm ThirdNormalForm:** Bernstein's Synthesis Algorithm [15]
Given a DB schema R and FD set F, find 3NF decompositions D

1. Find a Minimal Cover G for F
2. For each a left-hand side X of a FD that appears in G, create a relation schema in D with attributes $\{X \cup \{A_1\} \cup \{A_2\} \cup \ldots \cup \{A_k\}\}$, where $X \rightarrow A_1$, $X \rightarrow A_2, \ldots, X \rightarrow A_k$ are the only dependencies in G with X as the left-hand side (X is the key)
3. Place any remaining attributes (that have not been placed in any relation) in a single relation schema to ensure the attribute preservation

# 3 Mathematica Implementation and Results

## 3.1 3NF Normalization with Mathematica

In this section implementation of the Bernstein's 3NF algorithm in Mathematica is discussed. The use case diagram in Fig.1 shows the overall structure of the subtasks of the Third Normal Form algorithm. As mentioned earlier, normalization codes encountered in the literature mainly make use of linked list structures with pointers and requires complex coding. In this work, the same algorithms are effectively implemented using only the Mathematica's basic list structure and the well-known set operations. These operations are *Union*[], *Complement*[], *Intersection*[], *MemberQ*[], *Extract*[], *Append*[], and *Length*[].

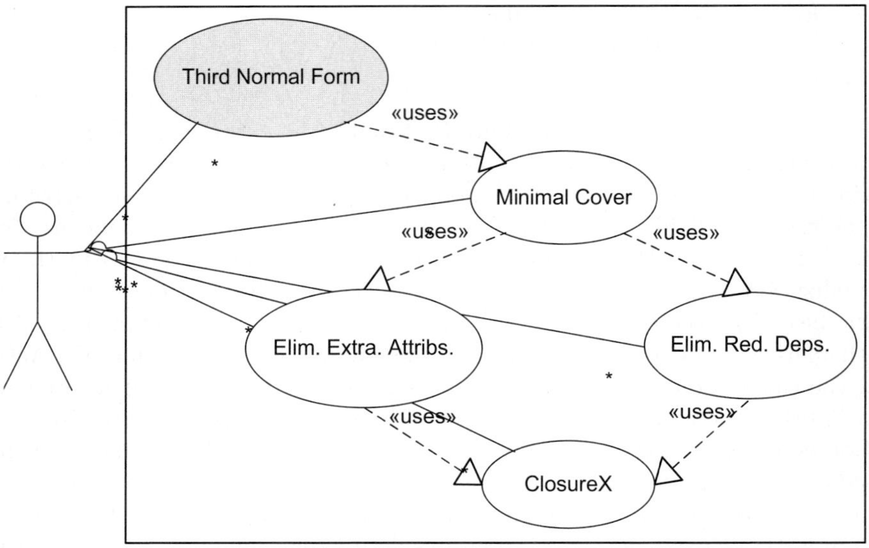


**Fig. 1.** Use Case Diagram for 3NF Modules

The tasks shown in Fig.1 are implemented effectively as Mathemamatica modules. As an illustration, the Mathematica code for the ThirdNormalForm only is given below. This program takes as input the minimal cover $[AGL, AGR]$ of the original set of FD's and generates a set of decompositions DL's each in 3NF.

```
ThirdNormalForm[AGL_,AGR_]:=
Module[{i,k,m,NewGL,NewGR,DR,DL,Index,TempGL,TempDL,TempGR,count},
   GL=AGL; GR=AGR; i=1; count=1;
   While[count<=Length[AGL] && Not[GL=={}],
     Index={i};
     If[Length[GL[[i]]]<=1, DL={GL[[i]]}, DL=GL[[i]]];
     TempDL=DL; DR={GR[[i]]};
```

```
      TempGR={GR[[i]]}; DL=Union[DL,TempGR];
      For[k=1, k<=Length[GL], k++,
         If[Length[GL[[k]]]<=1, TempGL={GL[[k]]},TempGL=GL[[k]]];
         DR={GR[[k]]};
         If[i!=k && TempGL==TempDL, Index = Union[Index,{k}];
           DL=Union[DL,DR]];
        ];
      NewGL={}; NewGR={};
      For[m = 1, m <= Length[GL], m++,
         TempGL=GL[[m]]; TempGR=GR[[m]];
         If[Not[MemberQ[Index, m]], NewGL=Append[NewGL, TempGL];
           NewGR=Append[NewGR, TempGR]];
        ];
      GL=NewGL; GR=NewGR;
      i=1; count=count+1; Print["DL=", DL];
      ];
   ]
```

### 3.2 Sample Runs

A set of experiments are performed to check the correctness of the programs using benchmark tests found in the literature. In this section, the results of running the programs with two different benchmark schemas only are given.

**Test 1.** Consider the STUDENT DB schema and the FD's given by FD1, and FD2 above. Here, 3NF is violated for the *STUDENT* schema above, because of the transitive dependency between *studentNo* and *officeNo*. First of all, a minimal cover is to be obtained which requires the elimination of extraneous attributes followed by the elimination of redundant dependencies. In this case, there are no extraneous attributes. However, the redundant dependency $studentNo \rightarrow officeNo$ is automatically eliminated from FD2. The new set of FD's which is the minimal cover is obtained as follows:

$FD1'$**:** *studentNo* $\rightarrow$ {*studentName, instructorId*}
$FD2'$**:** *instructorId* $\rightarrow$ *officeNo*

By calling the *ThirdNormalForm* program with the minimal cover set above as input, the following 3NF decompositions are produced:

$STUDENT_1$ ={*studentNo, studentName, instructorId*},
$STUDENT_2$ ={*instructorId, officeNo*}

**Test 2.** This benchmark test is artificially produced in order to check the correctness of all the subtasks involved. The DB schema is $BM = \{a, b, c, d, e, g\}$ and a set of eleven FD's are given as follows:

$\{a,b\}\rightarrow c,\ c\rightarrow a,\ \{b,c\}\rightarrow d,\ \{a,c,d\}\rightarrow b,\ d\rightarrow e,\ d\rightarrow g,\ \{b,e\}\rightarrow c,\ \{c,g\}\rightarrow b,$
$\{c,g\}\rightarrow d,$ *and* $\{c,e\}\rightarrow g$

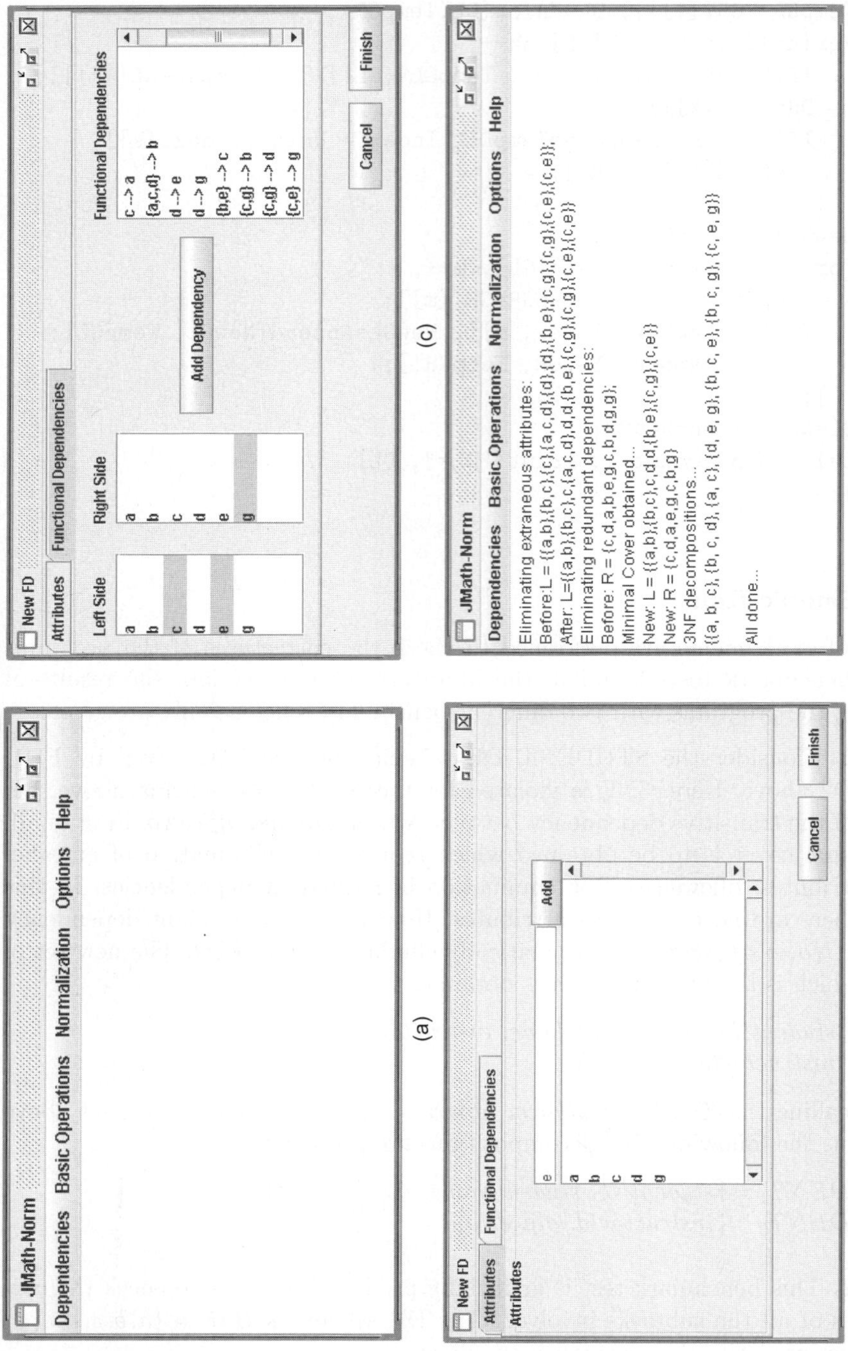


**Fig. 2.** JMath-Norm in Action (a) Main menu (b) Defining attributes (c) Specification of FD's (d) Normalization into 3NF

The minimal cover module eliminates the extraneous attribute $a$ from $\{a,c,d\}\rightarrow b$, replacing it by $\{c,d\}\rightarrow b$. This FD set has 3 redundant dependencies, namely, $c\rightarrow a$, $\{c,d\}\rightarrow b$, and $\{c,g\}\rightarrow d$. Hence, all three is eliminated giving a minimal cover set with eight dependencies. 3NF decomposition sets as an output from $ThirdNormalForm$ are obtained as follows:

$\{a,b,c\}$, $\{b,c,d\}$, $\{d,e,g\}$, $\{b,c,e\}$, $\{b,c,g\}$, $\{a,c\}$, and $\{c,e,g\}$

## 4 JMath-Norm: A Java Tool with Mathematica and JLink

### 4.1 System Architecture

In this section, a prototype Java tool called JMath-Norm, developed by the authors is going to be illustrated. This tool is written in Java language with Java Link facility (JLink) of Mathematica which allows linking the Java code with the Mathematica kernel and execute the modules written in Mathematica. The main user interface is shown in Fig.2(a) which consists of submenus to manage FD's, basic normalization tasks, normalization to ThirdNormalForm, and others. Using JMath-Norm's *Dependencies* pull-down menu, the users may define a new set of FD's, open an existing one, save or edit FD's using a data entry dialog box. Basic normalization tasks such as *Set Closure*, and *Minimal Cover* can be carried out individually by utilizing the operations menu. 3NF form of a DB schema can be obtained from *Normalization* submenu.

### 4.2 Using JMath-Norm

The example given in Test.2 above will be used to illustrate some of the features of JMath-Norm. The user starts by selecting New FD from the *Dependencies* pull-down menu to enter the attribute names one by one. This process is shown in Fig.2(b) below. The user then selects *Functional Dependencies* tab from the dialogue box, to efficiently define FD set by selecting attributes from the left and right pane respectively to enter the lhs and rhs of the FD set using the *Add Dependency* button(see Fig.2(c)).

After completing the specification of FD's, the user may either choose to select from a set of aforementioned basic normalization processes or directly obtain the 3NF decompositions by selecting 3NF from the *Normalization* tab of the main menu. The output for the later is shown in Fig.2(d).

## 5 Conclusions

In this paper, an efficient Mathematica implementation of basic DB normalization algorithms is presented. It was observed that, the implementation uses only Mathematica's collection operations. Previous implementations, however, are heavily based on linked lists and pointer structures which produces computationally inefficient systems.

The Mathematica modules developed for the purpose are put together into a window based system JMath-Norm which provides interaction between the user and Mathematica. JMath-Norm is written in Java and utilizes JLink to derive the Mathematica kernel. By using the system user can arrange I/O into files which is displayed on user-friendly screens for which no Mathematica knowledge is actually required.

This prototype provides an interactive tool that is useful for teaching the normalization theory in a DB management course. The system, when completed, will provide modules for the other normalization forms and enable users to create DB tables automatically for a specific DBMS chosen.

## References

1. Elmasri, R. and Navathe, S.B.: Fundamentals of Database Systems, 4th Ed., Addison Wesley (2003).
2. Beeri, C. and Bernstein, P.A.: Computational Problems Related to the Design of Normal Form Relational Schemas, ACM Transactions on Database Systems, Vol.4, No.1 (1979) 30-45.
3. Ceri, S. and Gottlob, G.: Normalization of Relations and Prolog, Communications of the ACM, Vol.29, No.6 (1986)
4. Welzer, W., Rozman, I. and Gyrks, J.G.: Automated Normalization Tool, Microprocessing and Microprogramming, Vol.25, (1989) 375-380.
5. Diederich, J. and Milton, J.: New Methods and Fast Algorithms for Database Normalization, ACM Transactions on Database Systems, Vol.13, No.3(1988) 339-365.
6. Diederich, J.: Minimal Covers Revisited: Correct and Efficient Algorithms, SIGMOD Record, Vol.20, No.1 (1991)
7. Hallasy, B.: Normal Forms and Normalization: Practical Designer's View, Information and Software Technology, Vol.25, No.6 (1991) 451-461.
8. Lucchesi, C.L. and Osborn, S.L.: Candidate Keys for Relations, Journal of Computer and System Sciences, Vol.17, No.2 (1978) 270-279.
9. Westlend, J.C.: Economic Incentives for Database Normalization, Information Processing and Management, Vol.28, No.5 (1992)647-662.
10. Lee, H.: Justifying Database Normalization: A Cost/Benefit Model, Information Processing and Management, Vol.31, No.1 (1995) 56-67.
11. Millet, I.: Normalization on the Fly, Journal of Systems Software, Vol.40 (1998) 85-88.
12. Vincent, M.W. and Srivinasan, B.: A Note on Relation Schemas Which Are in 3NF But Not in BCNF, Information Processing Letters, Vol.48 (1993) 281-283.
13. Du, H. and Wery, L.: Micro: A Normalization Tool for Relational Database Designers, Journal of Network and Computer Applications, Vol.22, (1999) 215-232.
14. Wolfram, S.: The Mathematica Book, 4th Ed., Wolfram Media/Cambridge University Press (1999).
15. Bernstein, P.A.: Synthesizing Third Norm Relations from Functional Dependencies, ACM Trans. Database Systems, Vol.1, No.4 (1976) 277-298.

# Extending Maple Capabilities for Solving and Displaying Inequalities

A. Iglesias[1,*] and R. Ipanaqué[2]

[1] Department of Applied Mathematics and Computational Sciences,
University of Cantabria, Avda. de los Castros,
s/n, E-39005, Santander, Spain
iglesias@unican.es
http://personales.unican.es/iglesias

[2] Department of Mathematics, National University of Piura,
Urb. Miraflores s/n, Castilla, Piura, Perú
robertipanaquechero@yahoo.es
http://www.unp.edu.pe/pers/ripanaque

**Abstract.** Solving inequalities is a very important topic in computational algebra. This paper presents a new *Maple* package, `IneqGraphics`, for displaying the two-dimensional solution sets of several inequalities of real variables. The package also deals with inequalities involving complex variables by displaying the corresponding solutions on the complex plane. The package provides graphical solutions to many inequalities (such as those involving polynomial, rational, logarithmic, exponential and trigonometric functions) that cannot be solved by using the standard *Maple* commands, which are basically reduced to linear inequalities. In addition, our outputs are consistent with *Maple*'s notation and results in the sense that the package provides a similar output for those cases that can also be solved with *Maple*. To show the performance of the package, several illustrative and interesting examples are described.

## 1 Introduction

Solving inequalities is a very important topic in Mathematics, with outstanding applications in many problems of theoretical and applied science [1, 2, 4, 6, 7]. However, since there is not a general methodolody for solving inequalities, their symbolic computation is still a challenging problem in computational algebra [3]. For example, *Maple* includes some commands for dealing with inequalities, but they are drastically limited to the linear case [5]. This drawback motivated us to create a *Maple* package, `IneqGraphics`, that is introduced in this paper. The package displays the two-dimensional solution sets of several inequalities. Typically, in our package the input might be comprised of any sequence of inequalities (connected by either the `and` or `or` operators) involving nonlinear expressions - such as polynomial, rational, trigonometric, exponential and logarithmic functions - that cannot be solved by using the standard *Maple* commands. The

* Corresponding author.

V.N. Alexandrov et al. (Eds.): ICCS 2006, Part II, LNCS 3992, pp. 383–390, 2006.

package also deals with inequalities involving complex variables by displaying the corresponding solutions on the complex plane.

The structure of this paper is as follows: Section 2 describes the main standard *Maple* tools for solving inequalities. Then, Section 3 introduces the new *Maple* package, `IneqGraphics`, and describes the commands implemented within. The performance of the package is also discussed in this section by using some illustrative examples. Finally, Section 4 closes with the main conclusions of this paper and some further remarks.

## 2 Standard *Maple* Tools for Solving Inequalities

*Maple* incorporates only a few commands for solving inequalities. The command `solve(ineq,var)`, solves one inequality in one variable, `var`, for all the cases where the equality points of `ineq` can be properly ordered. While it is not required for the inequality to be linear (for instance, you can solve $e^x > x+1$) it only accepts one inequality, thus strongly limiting its applicability.

In addition, the *Maple* 10 package `plots` provides the command `inequal` that plots regions defined by logical combinations of linear inequalities in the $\mathbb{R}^2$ space. Although its input allows us to enter more than one equality or inequality, they must necessarily be linear. For example, the package fails to display the solution sets of each of the inequalities $x^2+y<1$ and $x^2+y^2 \leq 16$ on the set $[-5,5]\times[-5,5]$:

```
>with(plots):
```

*Warning, the name changecoords has been redefined*

```
> inequal(x^2+y<1,x=-5..5,y=-5..5);
```

*Error, (in inequal) the inequalities should be linear in 'x' and 'y'*

```
> inequal(x^2+y^2<16,x=-5..5,y=-5..5);
```

*Error, (in inequal) the inequalities should be linear in 'x' and 'y'*

This problem has been partially overcome in [8] where the author presents an interesting extension of the command `inequal` to the case of nonlinear expressions. In particular, his new command `inequalities` admits systems of nonlinear inequalities, including the case of expressions given in polar coordinates. However, the package does not solve the problem in all its generality as the inequalities can only be connected by the `and` operator, not all nonlinear expressions can be considered and the polar form fails in presence of loops, multivalued curves and self-intersection scenarios, to mention some of the limitations we noticed.

The package `IneqGraphics`, described in the next section, overcomes many of those limitations and allows the user to solve a large family of real and complex inequality systems and equations and display their two-dimensional solution sets accordingly.

## 3 The Package IneqGraphics: Some Illustrative Examples

As shown in the previous section, inequalities involving nonlinear functions cannot be visualized by means of the `inequal` command. They can be solved, however, by loading the package developed by the authors:

```
> libname:="C:\\Program files\\Maple 10/mypackages", libname:
> with(IneqGraphics);
```

*[complexineqplot, ineqplot]*

which includes the command

```
ineqplot(ineqs, x=xmin..xmax, y=ymin..ymax, ops)
```

for displaying the two-dimensional region of the set of points satisfying the inequalities `ineqs` of real numbers inside the square `[xmin, xmax]`×`[ymin, ymax]`. For example, the inequalities indicated in the previous section can be solved as follows:

```
> ineqplot(x^2+y<1,x=-5..5,y=-5..5);
```

*See Figure 1(left)*

```
> ineqplot(x^2+y^2<=16,x=-5..5,y=-5..5);
```

*See Figure 1(right)*

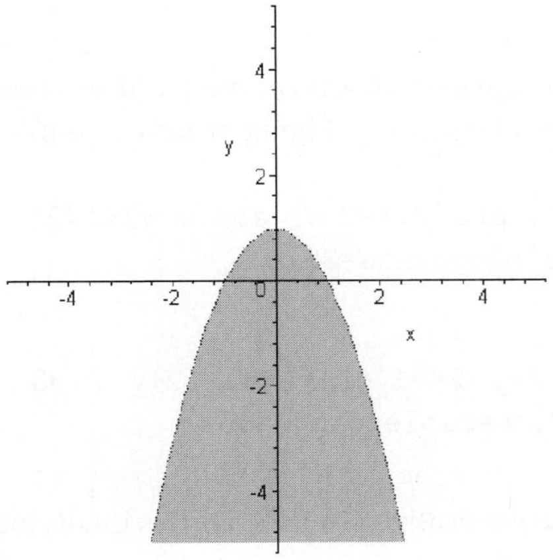


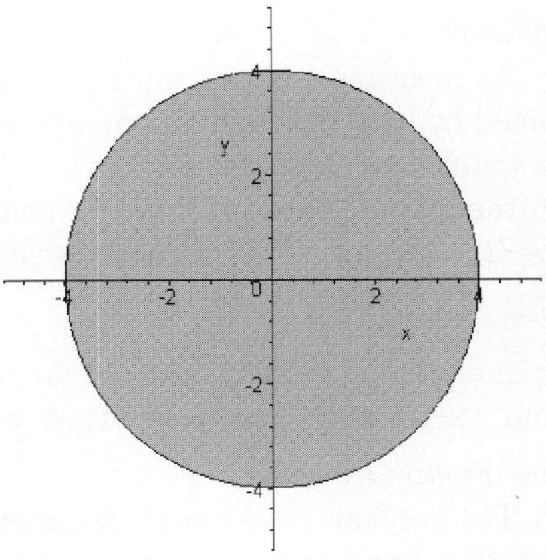


**Fig. 1.** Some examples of inequality solutions on the square $[-5, 5] \times [-5, 5]$: (left) $x^2 + y < 1$; (right) $x^2 + y^2 \leq 16$

Similarly, Fig. 2 displays the solution sets of the inequalities $1 < x^2 + y^2 < 4$ and $sin(x^2 + y^2) \geq \frac{1}{2}$ on the squares $[-2.5, 2.5] \times [-2.5, 2.5]$ and $[-4, 4] \times [-4, 4]$, respectively. Note that the *Maple* command `inequal` does not provide any solution for these inequalities either.

```
> ineqplot({1<x^2+y^2,x^2+y^2<4},x=-2.5..2.5,y=-2.5..2.5);
```

*See Figure 2(left)*

```
>ineqplot(sin(x^2+y^2)>=1/2,x=-4..4,y=-4..4,
feasiblepoints=50,linespoints=2500);
```

*See Figure 2(right)*

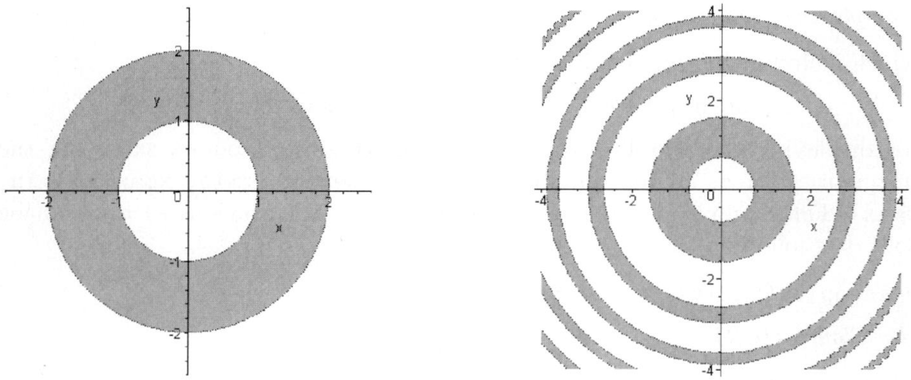

**Fig. 2.** Some examples of inequality solutions: (left) $1 < x^2 + y^2 < 4$ on the square $[-2.5, 2.5] \times [-2.5, 2.5]$; (right) $sin(x^2 + y^2) \geq \frac{1}{2}$ on the square $[-4, 4] \times [-4, 4]$

As mentioned above, our input can be comprised of several inequalities combined by using the `and` and `or` operators simultaneously. The next two examples are aimed to show this fact:

```
> ineqplot(1<abs(x)+abs(y) and abs(x)+abs(y)<=2 or abs(x+y)<1/2,
x=-3..3,y=-3..3,linespoints=400,feasiblepoints=35);
```

*See Figure 3(left)*

```
> ineqplot((x+2)^2+y^2<=1 or (x-2)^2+y^2<=1 or ((x+2)^2+y^2>=3
and (x-2)^2+y^2>=3),x=-4..4,y=-2..2,feasiblepoints=30);
```

*See Figure 3(right)*

The previous command, `ineqplot`, can be generalized to inequalities involving complex numbers. The new command:

```
complexineqplot(ineqs,z=(Rezmin..Rezmax,Imzmin..Imzmax),opts)
```

displays the solution sets of the inequalities `ineqs` of complex numbers inside the square in the complex plane given by `[Rezmin, Rezmax]` × `[Imzmin,Imzmax]`. In this case, the functions appearing within the inequalities need to be real-valued functions of a complex argument, e.g. *Abs*, *Re* and *Im*. For example:

```
> complexineqplot({1<abs(z^2-z+1),abs(z^2-z+1)<4},
z=(-2...3.,-3...3.),feasiblepoints=25);
```

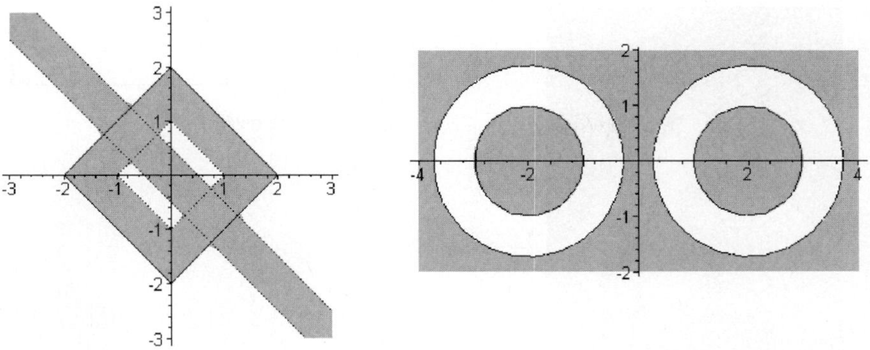

**Fig. 3.** Some examples of inequality solutions connected with the `and` and `or` operators: (left) $1 < |x| + |y| \leq 2$ `or` $|x+y| < \frac{1}{2}$ on the square $[-3,3]\times[-3,3]$; (right) $(x+2)^2+y^2 \leq 1$ `or` $(x-2)^2+y^2 \leq 1$ `or` $(x+2)^2+y^2 \geq 3$ `or` $(x-2)^2+y^2 \geq 3$ on the square $[-4,4]\times[-2,2]$

*See Figure 4(left)*

```
> complexineqplot({1<abs(sin(z-2)+z),abs(sin(z-2)+z)<4},
z=(-5...8.,-4...4.),feasiblepoints=35,linespoints=550);
```

*See Figure 4(right)*

We would like to point out that the output obtained from the package is consistent with *Maple*'s notation and results in the sense that the package provides a similar output for those cases already solved by *Maple*. Figure 5 illustrates this fact: on the left, the solutions provided by the standard *Maple* command

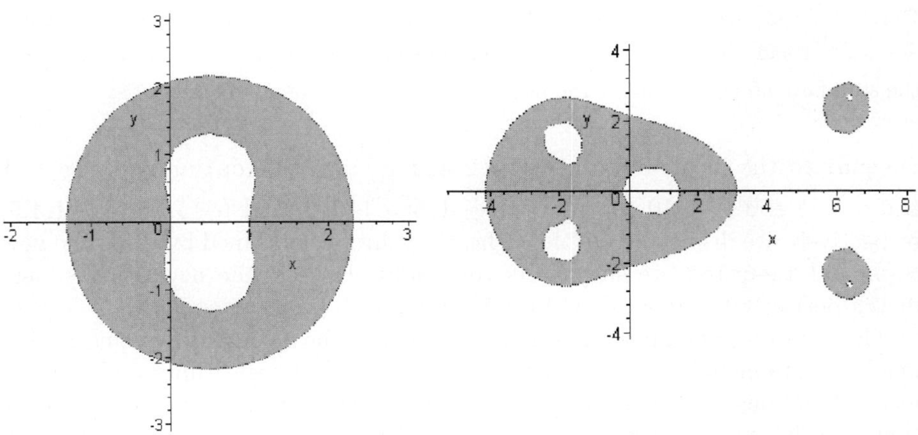


**Fig. 4.** Some examples of inequality solutions for $z \in \mathbb{C}$: (left) $1 < ||z^2 - z + 1|| < 4$, $Re(z) \in [-2,3]$, $Im(z) \in [-3,3]$; (right) $1 < ||sin(z-2) + z|| < 4$, $Re(z) \in [-5,8]$, $Im(z) \in [-4,4]$

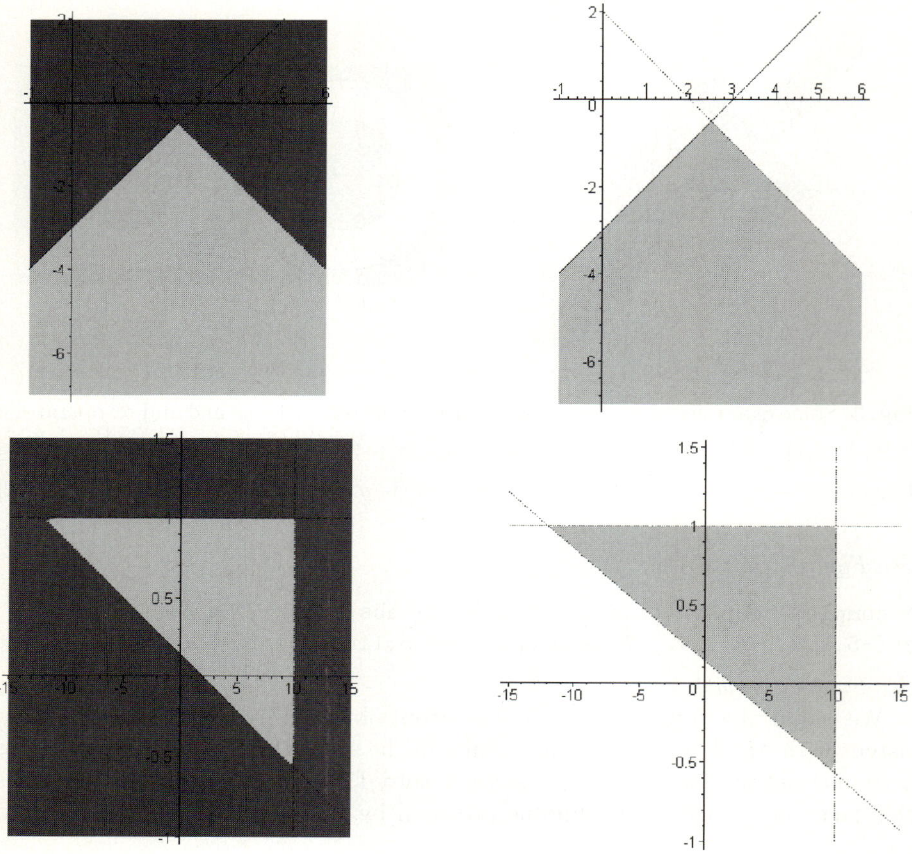

**Fig. 5.** Solution sets for the inequality systems: (top) $x+y \leq 2$ and $x-y \geq 3$; (bottom) $\frac{x}{2}+7y>1$ and $y<1$ and $x<10$. The solutions have been obtained by applying: (left) the standard *Maple* command `inequal`; (right) the new command `ineqplot`.

`inequal` to the inequalities $x+y \leq 2$ and $x-y \geq 3$ (top) and $\frac{x}{2}+7y>1$ and $y<1$ and $x<10$ (bottom) on $[-1,6]\times[-7,2]$ and $[-15,15]\times[-1,1.5]$ respectively are displayed. On the right, the solutions obtained by using the new command `ineqplot` are shown. As the reader can see, the new solution sets match perfectly those obtained from the standard *Maple* commands.

The last examples aim to show how complicated the inequality systems can be: in addition to include polynomial, rational, exponential, logarithmic and trigonometric functions, combinations and even compositions of these (and other) functions can also be considered. In Figures 6 and 7 the solutions sets of the inequality systems:

$$x^2+y^2<5 \text{ and } |log(x^2+1)+y|>1 \quad (1)$$

$$x^2-y<2 \text{ and } e^x+e^y>1 \text{ and } e^x+e^y<4 \quad (2)$$

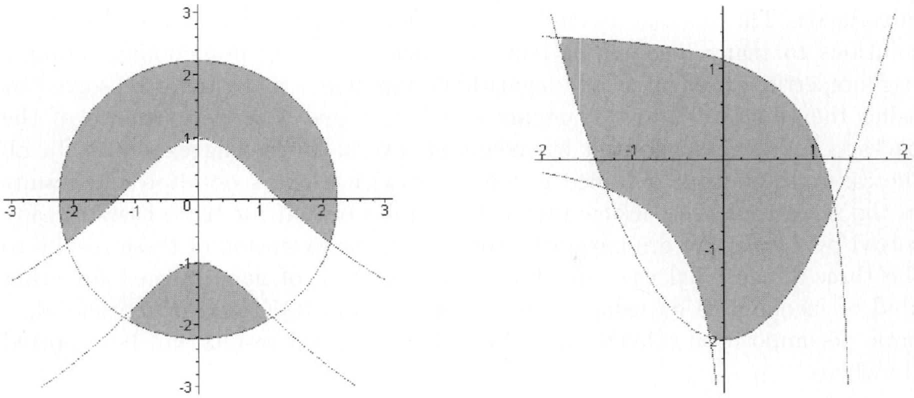


**Fig. 6.** (left, right) Solution sets for the inequality systems given by (1) and (2) respectively

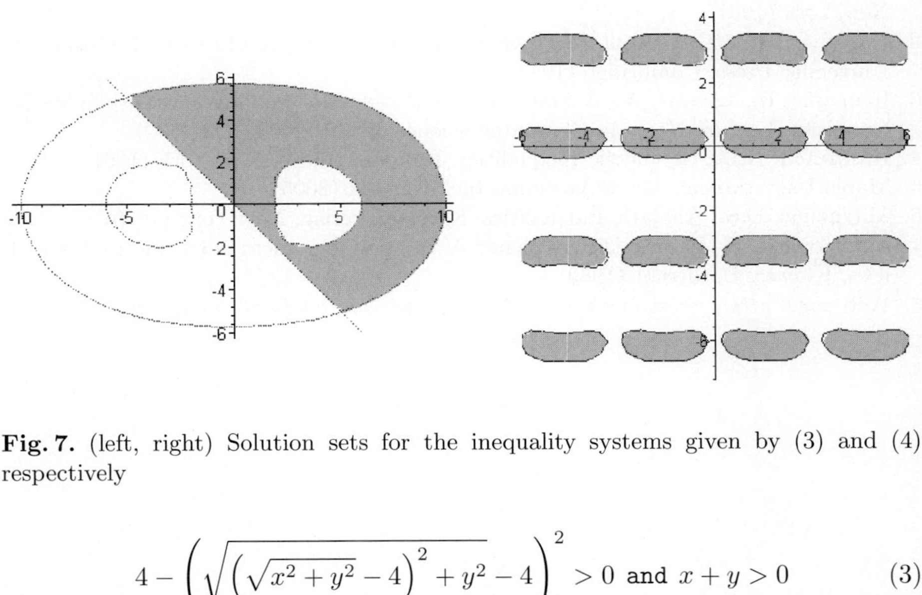


**Fig. 7.** (left, right) Solution sets for the inequality systems given by (3) and (4) respectively

$$4-\left(\sqrt{\left(\sqrt{x^2+y^2}-4\right)^2+y^2}-4\right)^2>0 \text{ and } x+y>0 \tag{3}$$

$$\sqrt{sin^2(x)+(sin(y)-2)^2}\sqrt{(sin(x)+1)^2+sin^2(y)}\sqrt{(sin(x)-1)^2+sin^2(y)}\leq\frac{1}{8} \tag{4}$$

on $[-3,3]\times[-3,3]$, $[-2,2]\times[-2.5,1.6]$, $[-10,10]\times[-6,6]$ and $[-6,6]\times[-7,4]$ respectively, are displayed.

## 4 Conclusions and Further Remarks

In this paper a new *Maple* package, `IneqGraphics`, to solve real and complex inequalities and display their associated two-dimensional solution sets is introduced. The package extends *Maple*'s capabilities by providing graphical solutions to many inequalities (such as those involving polynomial, rational, trigonometric, exponential and logarithmic functions) that cannot be solved by using the standard *Maple* commands and packages. The performance of the package has been illustrated by means of several interesting examples. In all the cases, the output obtained is consistent with *Maple*'s notation and results in the sense that the package provides a similar output for those cases already solved by *Maple*. We are currently working on the extension of these results to the three-dimensional case and the implementation of new routines for other kind of inequalities by using quantifier elimination (QE) and cylindrical algebraic decomposition (CAD) algorithms. The obtained results will be reported elsewhere.

## References

1. Beckenbach, E.F., Bellman, R.E.: An Introduction to Inequalities. Random House, New York (1961)
2. Hardy, G.H., Littlewood, J.E., Pólya, G.: Inequalities (Second Edition). Cambridge University Press, Cambridge (1952)
3. Ipanaque, R., Iglesias, A.: A Mathematica Package for Solving and Displaying Inequalities. Lecture Notes in Computer Science, **3039** (2003) 303-310
4. Kazarinoff, N.D.: Geometric Inequalities. Random House, New York (1961)
5. Maple User Manual. Maple Waterloo Inc., Canada (2005)
6. Mitrinovic, D.S.: Analytic Inequalities. Springer-Verlag, New York (1970)
7. Mitrinovic, D.S., Pecaric, J.E., Volenec, V.: Recent Advances in Geometric Inequalities. Kluwer, Dordrecht (1989)
8. Web page: *http://www.math.ubc.ca/~israel/advisor/advisor6/h9r1.htm*

# Phase Response Curves, Delays and Synchronization in MATLAB

W. Govaerts and B. Sautois

Department of Applied Mathematics and Computer Science, Ghent University,
Krijgslaan 281-S9, B-9000 Ghent, Belgium
{willy.govaerts, bart.sautois}@ugent.be

**Abstract.** MATCONT is a Matlab software package for the study of dynamical systems. We extended this package with the functionality of computing the phase response curve or PRC of a neural model. An important application lies in the study of synchronization of neurons. In synchronization studies, often the delays in a network are discarded or set to zero. We show in this paper that the delay can be crucial for the ability of neurons in an excitatory network to synchronize. We also show that, by studying the PRC of a neural model, one can compute the necessary delay to allow synchronization or phase-locking in a network of such neurons.

## 1 Introduction

CL_MATCONT [1] and its GUI version MATCONT [2] are MATLAB packages for the study of dynamical systems and their bifurcations. Among other things, they support the numerical continuation of equilibria, limit cycles, limit points, Hopf points, fold, flip, and torus bifurcations of cycles. Both packages are freely available at `http://www.matcont.ugent.be/`. The MATLAB platform, for which the packages are written, is attractive for both developer and user. The problem of speed, which arises since MATLAB is not compiled, is solved by incorporating C-code, while preserving the program's portability and user-friendly features.

In neural modeling, a phase response curve is a function that indicates the response of a spiking neuron to an input pulse. A positive value means that the next spike will be fired sooner, a negative value means that the next spike is delayed. In CL_MATCONT and MATCONT we have incorporated the functionality of calculating the phase response curve (PRC) and its derivative (dPRC) of limit cycles during continuation.

In the present paper, we show the importance of the PRC and dPRC for synchronization and phase-locking properties of neurons in a network with excitatory connections. In studies of such networks, very often delays are regarded as being of less importance, or simply set to zero ([3],[4]). We show that in fact, the delays of connections play a crucial role in the ability of neurons to phase-lock or synchronize. This necessary delay can be computed using the PRC.

We restrict ourselves to networks of identical neurons, with uniform excitatory connections. More chaotic synchronization studies have been made, using other software, e.g. in Mathematica [5], [6].

V.N. Alexandrov et al. (Eds.): ICCS 2006, Part II, LNCS 3992, pp. 391–398, 2006.

## 2 The MATLAB Software

CL_MATCONT and MATCONT are packages for the study of dynamical systems and their bifurcations. They are successor packages to packages written in compiled languages (Fortran, C, C++). The full Matlab platform of CL_MATCONT and MATCONT is attractive because it makes them more user-friendly than other packages, and portable to all operating systems, as MATLAB is platform-independent. Also, MATLAB has GUI-development tools built-in and allows a standard handling of data files, graphical output, etcetera. A typical screenshot, showing several of the windows that can be opened in MATCONT, is shown in Figure 1.

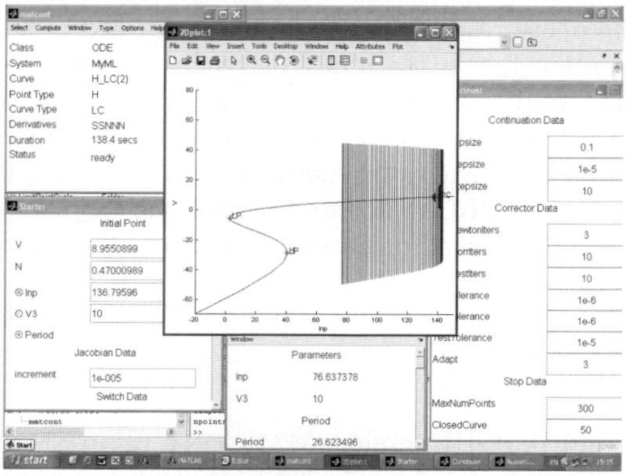


**Fig. 1.** Typical screenshot from MATCONT

MATCONT uses the MATLAB standard ODE-solvers for time-integration of dynamical systems. Also, it makes intensive use of the linear algebra routines that form the core of MATLAB. Further, it uses the sparse solver routines that are provided in MATLAB.

On the other hand, the code is inevitably slower than a compiled programming language. It is possible however, through some intermediate C-code, to have the C-code and MATLAB-code communicate with each other. Because there is a C-compiler present in every version of MATLAB, we can supply the C-source-code and let the program compile the C-code at runtime. This causes a slight delay at startup of the program, negligible when compared to the time needed to do further computations. The advantage of this approach is that it keeps the program completely platform-independent. The platform-dependent compiled C-code is only generated at runtime, and is thus adapted to the machine.

## 3 Phase Response Curve of a Model

The phase response curve or PRC of a periodically spiking neuron expresses its reaction to an input pulse given at a certain time during its cycle period.

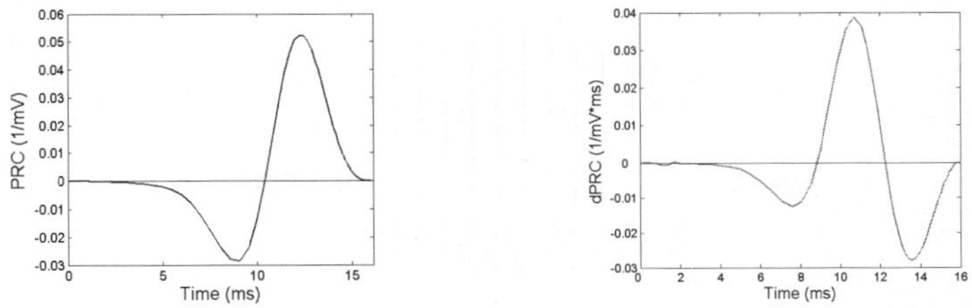


**Fig. 2.** Left: PRC ($\phi$) for the Hodgkin-Huxley system. Right: dPRC ($\phi'$) for the same system. The equations and parameter values for this model are given in [7].

More formally, the phase response curve is a function, which is defined over the timespan of one cycle in the spike train of a firing neuron with no extra input. As starting point of the period we take the peak of a spike. At each point of the cycle, the phase response curve (PRC) indicates the effect on the oncoming spikes of an input pulse to the neuron at that particular time in its cycle.

If we denote by $T_{old}$ the period of the unperturbed cycle, and by $T_{new}$ the period when the input pulse $\alpha$ is given at time $t$, the linearized phase response curve is defined by

$$\mathtt{PRC}(t) = \phi(t) = \lim_{\alpha \to 0} \frac{G(t,\alpha)}{\alpha}\,,$$
$$G(t,\alpha) = \frac{T_{old} - T_{new}(\alpha)}{T_{old}}. \tag{1}$$

Note that time $t$ is expressed in ms, and the input pulse $\alpha$ in mV. Thus, $G(t,\alpha)$ is dimensionless, and the actual PRC, $\phi$, has units 1/mV.

It is also possible to compute the derivative of the PRC (dPRC or $\phi'$), which can be useful, as we show in this paper. This function has units 1/mV*ms. An example of both curves is given in Figure 2. For a more detailed discussion of linearized PRCs and how we compute them using our software package MATCONT, we refer to [7].

## 4 1:1 Phase-Locking to Repeated Inputs

Consider an oscillator with period $T$, which gets repeated external inputs with a fixed period which is close to $T$, say $T_\Delta$. The ability of the oscillator to phase-lock with this input is an important concept, as discussed in [8] and [9].

Say the oscillator or neural model fires a spike at time $t = 0$, and gets an input with strength $\alpha$ at time $t = t_1$. Then the next spike of the neuron will occur at time $t = T - \alpha T\ \phi(t_1)$. If we call $t_2$ the distance between the following spike and input, then

$$t_2 = (t_1 + T_\Delta) - T(1 - \alpha\phi(t_1)) \tag{2}$$

So

$$t_2 = (t_1 + T_\Delta) + \alpha T\ \phi(t_1)\ \mathtt{mod}\, T \tag{3}$$

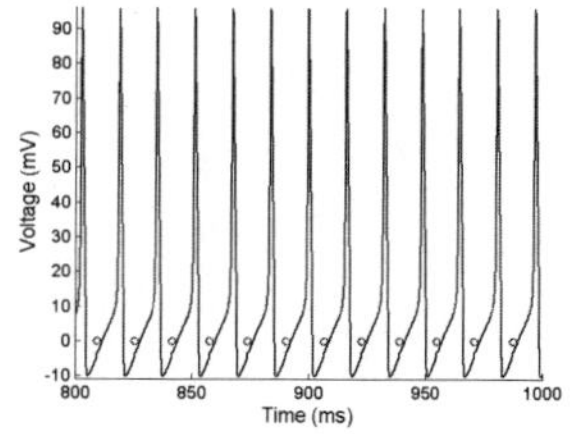


**Fig. 3.** The spikes show the spiking regime of the Hodgkin-Huxley neuron. The circles indicate when the repetitive input comes in. It is clear that this is a phase-locked state, for $t_1 \approx 7$ ms, with period 16.28 ms.

Any fixed point of this map has the property that

$$T_\Delta + \alpha T\ \phi(t_1) = 0 \ \mathtt{mod}\, T \tag{4}$$

Thus, in such points, 1:1 phase locking is possible. Since the linearized PRC of a neuron can only be used reliably for $\alpha$ small enough, the period of the inputs is not too different from the oscillator's original period.

The formula can be understood as follows: when the PRC for a certain delay is positive (negative), then the next spikes will be sped up (delayed), therefore there can only be synchronization if $T_\Delta < T$ ($T_\Delta > T$).

The so obtained fixed point of the map $t_1 \to t_1 + T_\Delta + \alpha T \phi(t_1) - T$ is stable if and only if its unique multiplier $1 + \alpha T \phi'(t_1)$ has modulus less than 1, i.e.

$$\frac{-2}{\alpha T} < \phi'(t_1) < 0 \tag{5}$$

So, if an oscillator gets a series of inputs of which the period is close to its own original period ($T_\Delta$ satisfies (4)), and the delay $t_1$ meets constraint (5), then one can get stable synchronization of the oscillator to the inputs.

These results were tested on the Hodgkin-Huxley equations, with inputs of strength 1, for which the PRC and dPRC are shown in Figure 2. For an input with a longer period, both $\phi$ and $\phi'$ will have to be negative at the input time to allow phase-locking, so the inputs will come earlier than about 8.5 ms after the last spike. In all cases with negative $\phi$ and negative $\phi'$, $\phi' > -2/T = -0.125$, so constraint (5) is always satisfied. And finally, the exact timing of input and spike at the phase-locked state will depend on the period of the spike. Since $T_\Delta > T$, we know that $T_\Delta - T = T\ \phi(t_1)$. E.g. when we give inputs with period $T_\Delta = T(1 + 0.01) \approx 16.28$, the neuron will always phase-lock with the inputs in such a timing that the inputs will reach the neuron at time $t_1$, such that $\phi(t_1) = -0.01$, and in the negative regime of $\phi'$, this is the case for $t_1 = 6.747$ ms. The experimental results are shown in Figure 3, with circles indicating the times of arriving input. An analogous analysis can be done for inputs arriving with a shorter period than that of the uncoupled model.

# 5 Two Single-Spiking Neurons

## 5.1 Synchronization

Consider two identical neurons with excitatory connections to each other, and suppose that the connection has a delay of duration $t_1$ and a connection strength of size $\alpha$. Suppose the neurons, without input, have period $T$, and the linearized PRC of the neurons is given by the function $\phi$. Finally assume that neuron 1 fires at time $t = 0$, and neuron 2 at time $t = \Delta$, $\Delta < t_1$.

Then neuron 1 gets an input signal at time $\Delta + t_1$ and neuron 2 gets a signal at absolute time $t_1$, which is in relative time, relative to its spike, $t_1 - \Delta$ (due to the phase difference). For the next spikes we conclude:

- neuron 1 will fire its next spike at $t = T - \alpha T\ \phi(t_1 + \Delta)$.
- neuron 2 will fire its next spike at $t = (\Delta + T) - \alpha T\ \phi(t_1 - \Delta)$.

The time difference map (for the time difference between spikes of the two neurons)

$$\Delta \to \Delta + \alpha T \phi(t_1 + \Delta) - \alpha T \phi(t_1 - \Delta) \tag{6}$$

has the trivial fixed point $\Delta = 0$ which is stable if and only if its unique multiplier $1 + 2\alpha\phi'(t_1)$ has modulus less than 1, i.e.

$$\frac{-1}{\alpha T} < \phi'(t_1) < 0 \tag{7}$$

Once the neurons are synchronized, the new period is $T - \alpha T\ \phi(t_1)$.

This shows that for two identical neurons with excitatory connections to synchronize, the connection delay plays a crucial role. Moreover, the delays that allow synchronization, can be determined from the period of the uncoupled neuron, its PRC and its dPRC.

Consider two identical Hodgkin-Huxley neurons, connected through excitatory connections with strength 1. From the plot of $\phi'$ in Figure 2, we notice that $\phi'(t_1)$ is always greater than $-1/T \approx -0.0624$. So constraint (7) is always met when $\phi'(t_1) < 0$. So whenever we initiate the neurons with a connection delay $t_1$ such that $\alpha \leq 1$, $\phi'(t_1) < 0$, the neurons will move to synchronization with new period $T - \alpha T\ \phi(t_1)$.

An example is shown in Figure 4. The first plot shows the start, and the second one the result, when starting with 2 HH-neurons, firing 1 ms apart, and with a connection delay of 6 ms. They clearly converge, with a new period of 16.1 ms, which matches our formula.

Analogously, experiments support the theory that there is synchronization of the two neurons with a shorter emerging period when $\phi'(t_1) < 0$ and $\phi(t_1) > 0$, and there is no synchronization if $\phi'(t_1) > 0$.

## 5.2 1:1 Phase-Locking

If a neuron has a phase response curve $\phi$ such that $\phi(t_1 + \Delta) = \phi(t_1 - \Delta)$, for a certain connection delay $t_1$ and a phase difference $\Delta$, then this point is a fixed point of the map (6).

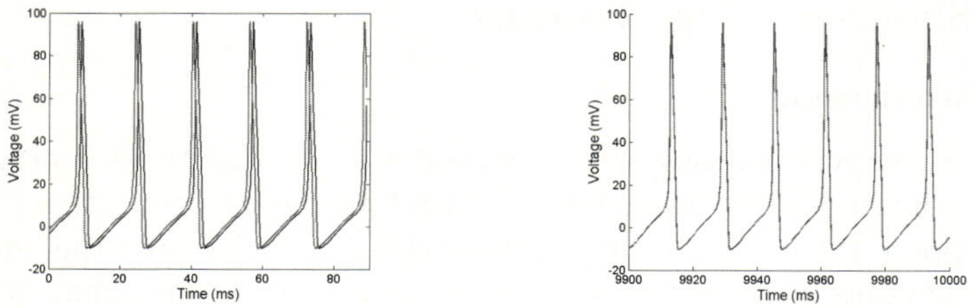


**Fig. 4.** Left: 2 HH-neurons spiking with 1 ms difference in timing. Right: The 2 neurons have converged.

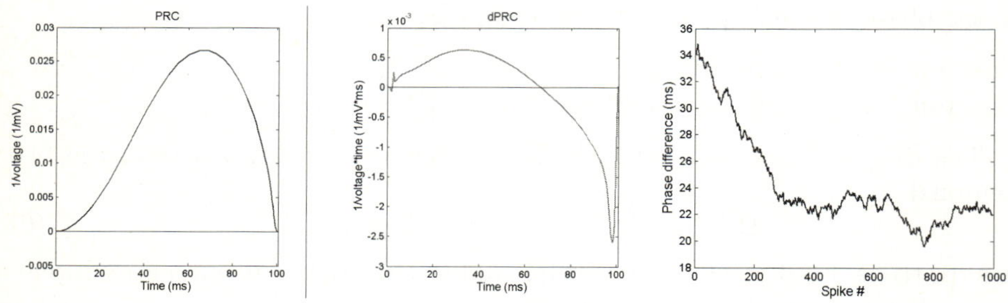


**Fig. 5.** Left: PRC for the Connor system. Middle: dPRC for the same system. Right: phase difference between two spiking neurons with excitatory connection, evolving to a stable state.

This fixed point is stable when

$$
\begin{aligned}
|1 + \alpha T \phi'(t_1 + \Delta) + \alpha T \phi'(t_1 - \Delta)| &< 1 \\
\frac{-2}{\alpha T} < \phi'(t_1 + \Delta) + \phi'(t_1 - \Delta) &< 0
\end{aligned}
\tag{8}
$$

So the fall of $\phi$ must be steeper than its rise.

As an example, this is the case in the Connor model. The equations for this model are given in [10]. Figure 5 shows the PRC (left) and dPRC (middle) of the Connor model at parametervalue $I = 8.49$. When the connection delay $t_1 = 64$, then $\Delta \approx 22$ is a fixed point, such that $\phi(t_1 + \Delta) = \phi(t_1 - \Delta) = 0.019$. Also $\phi'(t_1 + \Delta) = -8e - 4$ and $\phi'(t_1 - \Delta) = 6e - 4$, so constraint (8) is met. A result is shown in the rightmost picture in Figure 5; it shows how the phase difference between two spiking neurons moves from its initial value, chosen at 35 ms, to 22 ms, the stable fixed point.

## 6 Synchronization of an N-Neuron Single-Spiking Network

It is possible to generalize the theory for a 2-neuron network to an N-neuron network. Consider a network of N identical oscillators with period $T$ and PRC $\phi$. Suppose this is an all-to-all connected excitatory network, with all connections having a delay $t_1$ and strength $\alpha$.

Let $X = (x_1\ x_2\ ...\ x_N)^T$ be a vector of time variables close to each other such that oscillator 1 spikes at time $x_1$, oscillator 2 at time $x_2$, etc. Then the next spikes will be fired at times

$$A(X) = \begin{pmatrix} x_1 + T - \alpha T\phi(t_1 + x_2 - x_1) - ... - \alpha T\phi(t_1 + x_N - x_1) \\ x_2 + T - \alpha T\phi(t_1 + x_1 - x_2) - ... - \alpha T\phi(t_1 + x_N - x_2) \\ ... \\ x_N + T - \alpha T\phi(t_1 + x_1 - x_N) - ... - \alpha T\phi(t_1 + x_{N-1} - x_N) \end{pmatrix} \quad (9)$$

If there is synchronization, i.e. $x_1 = x_2 = ... = x_N$, then all components of $A(X)$ are equal to $x_1 + T - \alpha(N-1)T\phi(t_1)$. So the period of the synchronized state must be $T - \alpha(N-1)T\phi(t_1)$.

We want to know whether this synchronized state is stable and attracting. Consider the map

$$F : X \mapsto A(X) - \begin{pmatrix} T(1-\alpha(N-1)\phi(t_1)) \\ T(1-\alpha(N-1)\phi(t_1)) \\ ... \\ T(1-\alpha(N-1)\phi(t_1)) \end{pmatrix} \quad (10)$$

$F$ has a fixed point $(0\ ...\ 0)^T$ and its Jacobian $F_X$ evaluated there is

$$\begin{pmatrix} 1 + \alpha T\phi'(t_1)(N-1) & -\alpha T\phi'(t_1) & ... & -\alpha T\phi'(t_1) \\ -\alpha T\phi'(t_1) & 1 + \alpha T\phi'(t_1)(N-1) & ... & -\alpha T\phi'(t_1) \\ ... & ... & ... & ... \\ -\alpha T\phi'(t_1) & ... & ... & 1 + \alpha T\phi'(t_1)(N-1) \end{pmatrix} \quad (11)$$

To determine the asymptotic stability of the synchronized state, we consider the difference vector $Y = (y_2\ ...\ y_N)^T = (x_2 - x_1\ ...\ x_N - x_1)^T$.

The difference vector of $F(X)$ is $G(Y) = (F_2(X) - F_1(X)\ \ ...\ \ F_N(X) - F_1(X))^T$ where $F_i$ is the $i$ - th component of $F$.

So the stability of the synchronized state is determined by the map $G$. Its Jacobian $G_Y$ can easily be obtained from $F_X$ and is given by

$$\begin{pmatrix} 1 + \alpha T\phi'(t_1)(N-1) & ... & -\alpha T\phi'(t_1) \\ ... & ... & ... \\ -\alpha T\phi'(t_1) & ... & 1 + \alpha T\phi'(t_1)(N-1) \end{pmatrix} + \begin{pmatrix} \alpha T\phi'(t_1) & ... & \alpha T\phi'(t_1) \\ ... & ... & ... \\ \alpha T\phi'(t_1) & ... & \alpha T\phi'(t_1) \end{pmatrix} \quad (12)$$

This is a diagonal matrix with $(N-1)$ equal diagonal elements and thus all eigenvalues equal to $1 + \alpha N\ T\phi'(t_1)$. So the synchronization is asymptotically stable if

$$\frac{-2}{\alpha N\ T} < \phi'(t_1) < 0 \quad (13)$$

This is the generalization of constraint (7) to the synchronization of an N-neuron network with all-to-all excitatory connections.

This is again confirmed by tests on a network of Hodgkin-Huxley neurons. Figure 6 shows on the left the initial state: 8 HH neurons, firing spikes 1 ms

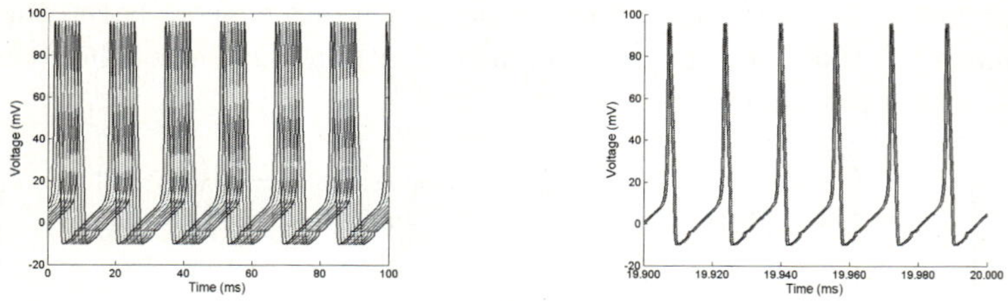


**Fig. 6.** Left: 8 HH-neurons spiking with 1 ms difference in timing. Right: The 8 neurons have converged.

apart. They are connected through excitatory connections, giving each other pulses of size 1 (mV). The connections have a delay of 6 ms. After 20000 ms, the result is shown on the right of the figure: all 8 neurons have synchronized strongly.

**Acknowledgement.** Bart Sautois is a research assistant of the Research Foundation - Flanders (FWO - Vlaanderen).

## References

1. Dhooge, A., Govaerts, W., Kuznetsov, Yu.A., Mestrom, W., Riet, A.M.: A continuation toolbox in MATLAB. Manual (2003), `http://www.matcont.ugent.be/doc_cl_matcont.zip`
2. Dhooge, A., Govaerts, W., and Kuznetsov, Yu.A.: MATCONT: A MATLAB package for numerical bifurcation analysis of ODEs. ACM TOMS 29,2 (2003) 141-164
3. Hansel, D., Mato, G., Meunier, C.: Synchrony in excitatory neural networks. Neural Comput. 7 (1995) 307-337
4. van Vreeswijk, C., Hansel, D.: Patterns of synchrony in neural networks with spike adaptation. Neural Comput. 13 (2001) 959-992
5. Gutiérrez, J.M., Iglesias, A.: A Mathematica package for the analysis and control of chaos in nonlinear systems. Comp. Phys. 12,6 (1998) 608-619
6. Iglesias, A., Galvez, A.: Revisiting some control schemes for chaotic synchronization with Mathematica. LNCS 3516 (2005) 651-658
7. Govaerts, W., Sautois, B.: Computation of the phase response curve: a direct numerical approach. To appear in Neural Comput. 18,4 (2006)
8. Brown, E., Moehlis, J., Holmes, P.: On the phase reduction and response dynamics of neural oscillator populations. Neural Comput. 16 (2004) 673-715
9. Izhikevich, E.M.: Dynamical systems in neuroscience: the geometry of excitability and bursting. To be published by the MIT press (2006)
10. Connor, J.A., Walter, D., McKown, R.: Modifications of the Hodgkin-Huxley axon suggested by experimental results from crustacean axons. Biophys. J. 18 (1977) 81-102

# A Hybrid Approach for Normal Factorization of Polynomials

Nicos Karcanias[1], Marilena Mitrouli[2,⋆], and Dimitrios Triantafyllou[2]

[1] Control Engineering Centre, School of Engineering and Mathematical Science, City University Northampton Square, London EC1V 0HV, UK
N.Karcanias@city.ac.uk

[2] Department of Mathematics, University of Athens, Panepistemiopolis 15784, Athens, Greece
{mmitroul, dtriant}@math.uoa.gr

**Abstract.** The problem considered here is an integral part of computations for algebraic control problems. The paper introduces the notion of normal factorization of polynomials and then presents a new hybrid algorithm for the computation of this factorization. The advantage of such a factorization is that it handles the determination of multiplicities and produces factors of lower degree and with distinct roots. The presented algorithm has the ability to specify the roots of the polynomials without computing them explicitly. Also it may be used for investigating the clustering of the roots of the polynomials. The developed procedure is based on the use of algorithms determining the greatest common divisor of polynomials. The algorithm can be implemented symbolically for the specification of well separated roots and numerically for the specification of roots belonging in approximate clusters.

**Keywords:** Polynomials, factorization, approximate computations, greatest common divisor.

## 1 Statement of the Problem

Some of the key problems of algebraic computations are the computation of the greatest common divisor (GCD), the computation of the least common multiple (LCM) of a set of polynomials and the computation of the factors of a polynomial. The problem of finding the GCD of a set $\mathcal{P}_{m,d}$, of $m$ polynomials of $\Re[s]$ of maximal degree $d$, is a classical problem that has been considered before, see [2, 6, 8]. The numerical computation of GCD has been considered so far by transforming it to an equivalent problem of real matrix computations (see methods such as Extended Row Equivalence and Shifting (ERES) [6], Matrix Pencil see [2] and [8] for other methods). The advantage of real matrix computations is that we can discuss the problem of approximate solutions and thus introduce the notion of "approximate GCD". In several engineering computations it is useful to define an approximate GCD of the set within a specified accuracy. This

⋆ This research was financially supported by PENED 03ED740 (Greek Secretary of Research and Technology).

V.N. Alexandrov et al. (Eds.): ICCS 2006, Part II, LNCS 3992, pp. 399–406, 2006.

approximate GCD can be used in several applications such as the definition of the almost zeros of a given polynomial system see [1]. The ERES method carries out succesfully the computation of approximate GCD. In [7] other methods for computing approximate GCD are also proposed.

The problem of factorizing polynomials is within the general framework of algebraic computations and it is crucial for determining Smith forms, solving Diofantine equations and general polynomials, or rational matrix equations. Of special interest is the problem of factorizing polynomials without resorting to root finding, as well as handling issues of approximate factorizations, when there is uncertainty on the exact values of the coefficients. This paper deals with the definition, the symbolic and numerical aspects of computation of a special factorization of polynomials, which is within the general factorization of polynomial, and can be performed without resorting to procedures based on finding roots.

For every polynomial $f(s) \in \Re[s]$ there exist positive integers $d_1, ..., d_\sigma$ where $d_1 > d_2 > ... > d_\sigma \geq 1$, such that f(s) may be expressed as

$$f(s) = e_1(s)^{d_1} e_2(s)^{d_2} ... e_\sigma(s)^{d_\sigma} \quad (1)$$

where the polynomials $\{e_1(s), e_2(s), \ldots, e_\sigma(s)\}$ are pairwise coprime and the polynomial $\hat{f}(s) = e_1(s)e_2(s)...e_\sigma(s)$ has distinct roots. □

The above factorization will be called a *normal factorization* of f(s) [3], $(d_i,\ i \in \widetilde{\sigma})$ will be called the *power set* and the set of polynomials $\{e_i(s),\ i \in \widetilde{\sigma}\}$ will be referred to as the *base* of the factorization. Such factorizations indicate the clustering of sets of roots, as this is indicated by the power set. Computing such factorizations is the problem considered in this paper. More precisely, we develop an algorithm specifying the set $\{e_i(s), i = 1, \ldots, k\}$ of polynomials and the corresponding degrees $\{d_i(s), i = 1, \ldots, k\}$. The advantage of such a factorization is that it handles the determination of multiplicities and produces factors of lower degree and with distinct roots. The use of algorithms for computing GCD is central to the current method and methods such as ERES [6], Matrix Pencil [2] etc may be used for this computation, and thus frequently lead to "approximate" rather than exact evaluations of the GCD. It is clear that with the notion of "approximate GCD" we also have the problem of defining "almost factorizations" of a polynomial. For approximate values of the GCD we define the order of approximation as a factor of the polynomial. The new computational procedure is demonstrated finally by a number of examples.

Throughout the paper $\Re[s]$ denotes the ring of real polynomials. The symbol $\partial\{f(s)\}$ denotes the degree of a polynomial. If a property is said to be true for $i \in \widetilde{n},\ \ n \in \mathcal{Z}^+$, this means it is true for all $1 \leq i \leq n$. If $g_1(s), g_2(s) \in \Re[s]$ and $g_1(s)$ divides $g_2(s)$ then we denote it by $g_1(s)/g_2(s)$.

## 2 Background of Approximate Algebraic Computations

In engineering computations, the numerical procedures must have the ability to work on data with numerical inaccuracies and thus lead to "approximate algebraic computations". In the sequel we study the notions of "approximate GCD" and "approximate normal factorization".

Let $\mathcal{P}_{m,d} = \{p_i(s) : p_i(s) \in \mathcal{R}[s],\ i = 1, 2, \ldots, m,\ d_i = deg\{p_i(s)\},\ d = max\{d_i,\ i = 1, 2, \ldots, m\}\}$ be the set of $m$ polynomials of $\mathcal{R}[s]$ of maximal degree $d$. Given a set of polynomials $t_i(s) \in \mathcal{R}[s]$, $i = 1, 2, \ldots, k$, define a numerical procedure for the computation of their "approximate" GCD, and associated factorization avoiding root finding.

**Approximate GCD**

The notion of the GCD of many polynomials is characterized by the property that its computation is nongeneric; in fact, the set of polynomials for which a nontrivial GCD ($\neq 1$) may be defined is a set of measure zero. However, the need for defining notions such as "almost zeros" and "approximate GCD" has been recognized as important in many applications. Methods computing the GCD of the set $\mathcal{P}$, which deploy relaxation of the exact conditions for GCD evaluation, such as the ERES method [6] lead to expressions for the "approximate GCD". In [7] the "approximate GCD" problem has been considered in the context of Euclidean division and for the case of two polynomials. Recently Karcanias etc [4], introduced formally the notion of the "approximate GCD" and then developed a computational procedure that allows the evaluation of how good is the given "approximate GCD" by estimating its strength of approximation.

## 2.1 Algorithm Approximate GCD

The Resultant Matrix Pencil Algorithm [5] appropriately modified allows the capturing of "approximate GCDs" and this is summarized by the following algorithm applied on the Modified Sylvester Resultant [9] of the given polynomial set and which exploits the notion of "near nullspace" of a matrix. The basic steps are:

**Step1:** Define a threshold $t > 0$. Apply the SVD algorithm to $S^*$ to define a basis $\widetilde{M}$for the right 'near nullspace' of the Sylvester matrix $S^*$ constructed from the given polynomial set.
**Step2:** Define the GCD Matrix Pencil $\widetilde{Z}(s) = s\widetilde{M}_1 - \widetilde{M}_2$ ,where $\widetilde{M}_1$ and $\widetilde{M}_2$ are the matrices obtained from $\widetilde{M}$ by deleting the last and the first row of $\widetilde{M}$ respectively.
**Step 3:** Construct a matrix with elements all nonzero minor determinants $d(s)$ of $\widetilde{Z}(s)$ and compute its basis $\widetilde{B}$.
**Step 4:** Apply the SVD algorithm to matrix $\widetilde{B} : \widetilde{B} = \widetilde{U}^T \widetilde{\Sigma} \widetilde{V}$. The corresponding to the largest singular value column of $\widetilde{V}$ defines the approximate GCD.
**Step 5:** Compute the angle of the spaces $\widetilde{M}_1$ and $\widetilde{M}_2$ as an indicator for the strength of approximation.

The next example demonstrates the implementation of the above algorithm to a computer.

*Example 1.* Consider the following set of polynomials :
$\{p_1(s) = s^3 - 6\,s^2 + 11\,s - 6, p_2(s) = s^2 - 3\,s + 2, p_3(s) = s^2 - 2\,s + 0.9999,$
$p_4(s) = s^2 - 1\}$

The above set of polynomials does not have an exact GCD. We have the following modified Sylvester matrix :

$$S^* = \begin{bmatrix} 1 & -6 & 11 & -6 & 0 \\ 1 & -3 & 2 & 0 & 0 \\ 1 & -2 & 0.9999 & 0 & 0 \\ 1 & 0 & -1 & 0 & 0 \\ 0 & 1 & -6 & 11 & -6 \\ 0 & 1 & -3 & 2 & 0 \\ 0 & 1 & -2 & 0.9999 & 0 \\ 0 & 1 & 0 & -1 & 0 \\ 0 & 0 & 1 & -3 & 2 \\ 0 & 0 & 1 & -2 & 0.9999 \\ 0 & 0 & 1 & 0 & -1 \end{bmatrix} \in \Re^{11\times 5}$$

Applying the SVD algorithm to $S^*$ we obtain as numerical results the following singular values and the matrix V are :

$$\Sigma = \begin{bmatrix} 19.2719 & 0 & 0 & 0 & 0 \\ 0 & 8.6015 & 0 & 0 & 0 \\ 0 & 0 & 2.6935 & 0 & 0 \\ 0 & 0 & 0 & 1.1627 & 0 \\ 0 & 0 & 0 & 0 & 0.0001 \end{bmatrix} and V = \begin{bmatrix} -0.0426 & -0.1206 & -0.4488 & 0.7630 & 0.4472 \\ 0.2834 & 0.5225 & 0.6347 & 0.2096 & 0.4472 \\ -0.6642 & -0.4355 & 0.3864 & -0.1408 & 0.4472 \\ 0.6516 & -0.4942 & -0.0828 & -0.3526 & 0.4472 \\ -0.2282 & 0.5279 & -0.4895 & -0.4793 & 0.4472 \end{bmatrix}$$

The tolerance we use is $10^{-15}$. The last singular value is $10^{-4}$. If the threshold t=0.0002 then the last singular value is smaller than t. The corresponding column of V is the last one and so $\widetilde{M}$ is :

$$\widetilde{M} = \begin{bmatrix} 0.4472 \\ 0.4472 \\ 0.4472 \\ 0.4472 \\ 0.4472 \end{bmatrix}, \widetilde{M}_1 = \begin{bmatrix} 0.4472 \\ 0.4472 \\ 0.4472 \\ 0.4472 \end{bmatrix}, \widetilde{M}_2 = \begin{bmatrix} 0.4472 \\ 0.4472 \\ 0.4472 \\ 0.4472 \end{bmatrix}$$

The angle of $\widetilde{M_1}$ and $\widetilde{M_2}$ acts as an indicator for the approximation. Their angle is: 0 (very good approximation).

The GCD Pencil and its basis matrix are correspondingly :

$$\widetilde{Z}(s) = s\widetilde{M}_1 - \widetilde{M}_2 = \begin{bmatrix} 0.4472s - 0.4472 \\ 0.4472s - 0.4472 \\ 0.4472s - 0.4472 \\ 0.4472s - 0.4472 \end{bmatrix} and \widetilde{B} = \begin{bmatrix} 0.4472 & -0.4472 \\ 0.4472 & -0.4472 \\ 0.4472 & -0.4472 \\ 0.4472 & -0.4472 \end{bmatrix}$$

Applying the SVD algorithm to $\widetilde{B}$ we have :

$$\widetilde{\Sigma} = \begin{bmatrix} 1.2649 & 0 \\ 0 & 0 \end{bmatrix} \; and \; \widetilde{V} = \begin{bmatrix} 0.7071 & -0.7071 \\ -0.7071 & -0.7071 \end{bmatrix}$$

The maximal singular value is the first one and the corresponding column of V is also the first. This columns gives the coefficients of the GCD : 0.7071 and -0.7071, which divided by 0.7071 give 1 and -1. So the GCD is s-1.

**$\epsilon$-Normal Factorization**
Next we introduce the definition of the approximate normal factorization with accuracy $\epsilon$, or the $\epsilon$-normal factorization.

For every polynomial $f(s) \in \Re[s]$ there exist positive integers $d_1, ..., d_\sigma$ where $d_1 > d_2 > ... > d_\sigma \geq 1$, such that f(s) may be expressed as

$$f(s) = \prod_{1}^{\sigma} e_i(s)^{d_i} + \Delta\ P(x) \tag{2}$$

where the polynomials $\{e_1(s), e_2(s), \ldots, e_\sigma(s)\}$ are pairwise coprime , the polynomial $\hat{f}(s) = e_1(s)e_2(s)...e_\sigma(s)$ has distinct roots and $||\Delta\ P|| \leq \epsilon$, where $\epsilon$ a selected accuracy for the computation. □

## 3 The Algorithm

Let $f(s) \in \Re[s]$ be a given polynomial. The following algorithm computes the normal factorization of $f(s)$. More specifically $f(s) = e_1(s)^{d_1} e_2(s)^{d_2} ... e_\sigma(s)^{d_\sigma}$.

### 3.1 Algorithm NF

STEP 0: Compute the GCD of $\{f^{(0)}(s), f^{(1)}(s)\}$.
If the polynomials are coprime
$f^{(0)}(s)$ has well separated roots. Perform numerically or symbolically STEPS 1 - 4
Else
compute numerically the approximate GCD.

STEP 1: Construct the associated set of $f(s)$
$G_f^{d_1} = \{g_0(s), g_1(s), ..., g_{d_1-1}(s), g_{d_1}(s)\}$, :
$g_k(s) = \text{GCD}\{f^{(0)}(s), f^{(1)}(s), ..., f^{(k)}(s)\}$
where the index $d_1$ is defined from
$d_1 = min_k g_k(s) = 1$

STEP 2: Define the prime set of $f(s)$
$\mathcal{T}_f = \{t_1(s), t_2(s), ..., t_{d_1}(s)\}$, where
$t_i(s) = \frac{g_{i-1}(s)}{g_i(s)}, \quad i = 1, \ldots, d_1$

STEP 3: Define the factorization set of $f(s)$
$P_f = \{p_i(s), i = 1, 2, \ldots, d_1\}$, where
$p_j(s) = \frac{t_j(s)}{t_{j+1}(s)}, \quad j = d_1, d_{1-1}, \ldots, 1, \quad t_{d_1+1} = 1$

STEP 4: From $\mathcal{T}_f$ construct the Ferrer diagram
Specify the index set
$\mathcal{I}_f = \{(d_i, v_i),\ i \in \widetilde{\sigma},\ d_1 > d_2 > ... > d_\sigma\}$
Form the normal factorization $f(s) = e_1(s)^{d_1} ... e_\sigma(s)^{d_\sigma}$ as follows:

Construct the essential factorization $\mathcal{P}_f^*$ of $f(s)$
$\mathcal{P}_f^* = \{p_{d_i}(s), \ \ i = 1, 2, ..., \sigma, \ d_1 > d_2 > ... > d_\sigma \geq 1\}$
base=$\{e_i(s) = p_{d_i}(s) : p_{d_i}(s) \in \mathcal{P}_f^*, \ \ i = 1, 2, \ldots, \sigma\}$
power set= $\{d_i, \ i = 1, 2, \ldots, \sigma\}$
$v_i$=number of all e.d. of $f(s)$ over $\mathcal{C}$ with degree $d_i$.

### 3.2 Numerical-Symbolical Results

The above method for $NF$ evaluation has a hybrid nature. Its numerical part was programmed in Matlab. The complexity is of order concerning mostly the execution of the Matrix Pencil algorithm [2]. The algorithm was tested on a Pentium machine over several polynomials characterized by various properties. The symbolical implementation of the method was performed using Maple. According to the nature of the examples the numerical or the symbolical implementation will be chosen.

*Example 2.*
Compute the normal factorization of the following polynomial with real coefficients:
$f(s) = s^7 - 24.35s^6 + 225.2s^5 - 1021.225s^4 + 2526.7875s^3 - 3494.896875s^2 + 2544.75s - 761.4$

Implementing the algorithm numerically with internal required tolerance for the GCD computation of order $O(10^{-8})$ we obtain the following factors: $(s-1.5)^4(s-2.35)(s-8)^2$ (real roots well separated). The same result gives also the symbolical implementation of the algorithm with increase of the required time.

*Example 3.*
This example demonstrates the ability of the method to distinguish between "almost equal factors". Let us suppose that we are given the following polynomial:

$$f(s) = (s+1)^2(s+5)^2(s+3)(s+\epsilon)^k$$

(i) For $\epsilon = 1.001$ and $k = 1$ implementing the algorithm numerically with internal required accuracy for the GCD computation of order $O(10^{-12})$ we obtain the following factors : $(s^2+6s+5)^2(s^2+4.001s+3.003)$ We see that the first factor contains the product of $(s+1)(s+5)$ whereas the second factor contains the product of $(s+1.001)(s+3)$. If we require to know the exact values of the roots of $f(s)$ we can now proceed computing the roots of the appearing factors which as we can see are of rather lower degree i.e. 2nd degree each instead of 6th degree that is the original polynomial.
(ii) For $\epsilon = 1.001$ and $k = 2$ implementing the algorithm numerically with internal required accuracy for the GCD computation of order $O(10^{-12})$ we obtain the following factors : $(s^3 + 7.001s^2 + 11.006s + 5.005)^2(s+3)$ We see that the first factor contains the product of $(s+1)(s+5)(s+1.001)$.
(iii) For $\epsilon = 1.000001$ and $k = 1$ implementing the algorithm numerically with internal required accuracy for the GCD computation of order $O(10^{-12})$ using

he modified LU or QR factorization we obtain the following factors : $(s+1.00000029520885)^3(s+5)^2(s+3)$ We see that the product $(s+1)^2(s+1.000001)$ is the six digit approximation of the first appearing factor. Thus 1.00000029520885 is the approximate value of the three roots of $f(s)$ that they belong to the same cluster.

If we have polynomials with roots belonging in same clusters we can apply the following Criterion:

The Clustering Criterion [3]: A polynomial $f(s) \in \mathcal{R}(s)$ has distinct roots if and only if $f(s)$ and $f^{(1)}(s)$ are coprime.

When we want to apply the numerical method for polynomials of rather high degree problems might appear due to rounding errors since it will be required the computation of several derivatives and their GCDs. Thus for such cases might be better the symbolical computation as illustrated by the following example:

*Example 4.*
Compute the normal factorization of the following polynomial:

$$f(s) = (s-2)^{15}(s-4)^5$$

Applying algorithm NF symbolically in Maple we get exactly as basic factors of the above expanded polynomial the terms: $(s-2)$ and $(s-4)$ with degrees 15 and 5 respectively. From this example it is evident that implementing the method symbolically we can determine the normal factorization of any polynomial of any degree with well separated roots.

*Example 5.* Specify the root clustering of the following polynomial:

$$f(s) = s^3 - 5.000001s^2 + 7.000004s - 3.000003$$

Applying the Clustering Criterion we get that $\{f(s), f^{(1)}(s)\}$ have "almost common factor" the polynomial $s - 1.0000005$ whereas $\{f(s), f^{(1)}(s), f^{(2)}(s)\}$ are coprime. Thus $f(s)$ has an "almost common factor" of degree 2 and therefore its 2 roots belong in the same cluster. A representative of this cluster is the value 1.0000005.Thus we attain the following almost factorization: $f(s) = (s-1.0000005)^2(s+\lambda)$ where $\lambda$ is defined by division or almost factorization. Note that theoretically the above polynomial is the product of the following factors: $f(s) = (s-1)(s-1.000001)(s-3)$. Thus $f(s)$ has "almost common factors" (or "almost zeros"). Using the Clustering Criterion we determined them. Applying the approximate resultant matrix pencil we get the the factor $s - 0.99999949999911$ of degree 2.

Root clustering can also be applied to polynomials with complex conjugate roots.

## 4 Conclusions

An algorithm achieving the normal factorization of polynomials has been introduced. The algorithm may be worked out using only algebraic tools, such as determination of GCD and symbolic computation of derivatives. The normal factorization can be considered as a first step into the factorization of polynomials, where multiplicities are computed first and then root finding is reduced to smaller degree and distinct root polynomials. The nature of the presented algorithm is hybrid. It can be applied symbolically or numerically according to the given data and the required results. The comparison of the current approach with other procedures that use different approximate GCD computation techniques is under study. The current approach also allows the study of root clustering, since the approximate GCD allows the grouping of roots which are close to each other.

## References

1. Karcanias, N., Giannakopoulos, C., Hubbard, M.: Almost zeros of a set of polynomials of $\mathcal{R}[s]$. Int. J. Control, **38**, (1983) 1213–1238.
2. Karcanias, N., Mitrouli M.: A Matrix Pencil Based Numerical Method for the Computation of the GCD of Polynomials. IEEE Trans. Autom. Cont., **39**, (1994) 977–981.
3. Karcanias, N., Mitrouli, M.: Normal Factorization of Polynomials and Computational Issues. Computers and Mathematics with Applications, **45**, (2003) 229–245.
4. Karcanias, N., Mitrouli, M., Fatouros, S., Halikias, G.: Approximate Greatest Common Divisor of many polynomials and generalised resultants. Computers and Mathematics with Applications, to appear (2006).
5. Karcanias, N., Mitrouli, M., Fatouros, S.: A resultant based computation of the Greatest Common Divisor of two polynomials. Proc. of 11th IEEE Med. Conf on Control and Automation, Rodos Palace Hotel, Rhodes, Greece, (2003b).
6. Mitrouli, Mm., Karcanias, N.: Computation of the GCD of polynomials using Gaussian transformation and shifting. Int. Journ. Control, **58**, (1993) 211–228.
7. Noda, M., Sasaki, T.: Approximate GCD and its applications to ill-conditioned algebraic equations. Jour. of Comp. and Appl. Math., **38**, (1991) 335–351.
8. Pace, I., S., Barnett, S.: Comparison of algorithms for calculation of GCD of polynomials. Int. Journ. System Scien., **4**, (1973) 211–226.
9. Triantafyllou, D., Mitrouli, M.: Two Resultant Based Methods Computing the Greatest Common Divisor of Two Polynomials. In: Li, Z., Vulkov, L., Wasniewski, J. (eds.): Numerical Analysis and Its Applications. Lecture Notes in Computer Science. Vol. 3401, (2005) 519–526.
10. Turnbull, H., W., Aitken, A., C.: An Introduction to the Theory of Canonical Matrices. Dover Publ., New York (1961).

# Computer Algebra for the Formation of Structural Matrices of Piezoceramic Finite Elements

Algimantas Čepulkauskas, Regina Kulvietienė, and Genadijus Kulvietis

Vilnius Gediminas Technical University
Saulėtekio 11, Vilnius 2040, Lithuania
{algimantas_cepulkauskas, regina_kulvietiene, genadijus_kulvietis}@gama.vtu.lt

**Abstract.** This paper deals with the description of a theoretical background of systematic computer algebra methods for the formation of structural matrices of piezoceramic finite elements. The efficiency of computer algebra application was compared here with the numerical integration methods of forming the structural matrices of the finite elements. To this end, the computer algebra system VIBRAN was used. Two popular finite elements for modelling piezoceramic actuators of the sector-type and the triangular one are discussed. All structural matrices of the elements were derived, using the computer algebra technique with the following automatic program code generation.

## 1 Introduction

Recent advances in the development, theory and applications of new smart materials, structures and devices, including the materials with extremely high piezoelectric or magnetostrictive properties (e.g., Terfenol-D or PMN – lead magnesium niobate) have extended the area of mechatronics, providing systems with very high levels of integration and multifunctionality [3]. Some concepts, e.g., actuators with an infinite number of degrees of freedom, transmission of energy to actuators through some distance, active bearings, etc., could be treated as solutions that define a problem. In some cases, the introduction of piezomechanical systems creates a new synergistic effect and, as in all integrated systems, the problem of maximum interaction between subsystems is the key to optimal design. Although the introduction of new piezoactive materials, that have found application in the areas of actuators, transducer technology, energy transformers, control devices, etc., has been very intensive in recent years [1, 7, 12, 16], the main concepts, ideas and effects are relatively unknown to design engineers. The holistic complex interaction of dynamic effects, energy transformation and devices based on them are described, while the physics involved in the piezoelectric theory may be regarded as the coupling between Maxwell's equations of electromagnetism and the elastic stress equations of motion [9, 11, 13, 14].

The finite element method (FEM) is widely used for modelling complicated structures [17]. In the FEM the continuum is digitized and thus the numerical

V.N. Alexandrov et al. (Eds.): ICCS 2006, Part II, LNCS 3992, pp. 407–413, 2006.

approximation is chosen. By means of numerical integration the result of sufficient accuracy could often be obtained, and using more effective computer facilities, this technique is mostly chosen. However, the order of numerical integration implies thorough adjustment to the order of polynomial expressions to be integrated, and this fact is not generally known. The use of a more precise formula for numerical integration will not produce more accurate results, but will somewhat 'stiffen' the examined structure [5, 6]. On the other hand, even the use of the exact numerical integration scheme always yields some numerical error that can become important in particular finite element applications, requiring a high accuracy of analysis [10, 15].

Moreover, it is known that the formation of the structural matrices is one of the most computationally expensive procedures in FEM [1], and the use of numerical integration for these purposes is not the optimal solution. This fact becomes particularly keen in case of finite element applications in the structures of piezoactive materials. Besides the conventional stiffness matrix there appear three additional ones: matrices of mass, electro-elasticity, and capacity [1, 4]. The numerical integration procedure is especially expensive for the formation of the latter matrices, because the power of integral expressions is higher. The computer algebra offers wide opportunities in the formation of finite element structural matrices. Regretfully we see that much routine algebraic work in the FEM, is usually done by hand or by means of numerical approximations, while it could be done by the computer algebra technique much faster, reliably and better. Another most important aspect is that the results, obtained in this way, often make considerable economies of computer resources and assure a proper accuracy of evaluations [5]. Despite that, even the classical books on FEM [17] do not mention any possibility of symbolic manipulations in the formation of element structural matrices.

There is one advantage of the numerical integration mentioned: for a particular type of finite elements, the structural matrices are always to be expressed in the same way through interpolating functions and their derivatives, and therefore the universal computational subroutines can be applied to various finite elements [8, 17]. However, the same extent of generality is a characteristic feature of computer algebra methods, too. Moreover, the user of programs, generated by computer algebra, does not have to consider the powers of polynomial expressions to be integrated, which is necessary in order to select the proper numerical integration scheme [2]. Even the users of modern commercial codes (such as ALGOR, ANSYS, etc.) have determined the necessary number of integration points for some types of finite elements at the stage of initial data reading [5].

## 2 Application of the Finite Element Method to Piezoceramic Structures

The mechanical and electrical phenomena as well as their coupling complicate the application of FEM to piezoceramic structures, such as actuators or other devices mentioned above. Thus the state of each finite element nodal point can be represented by the values of the nodal displacement $\{\delta\}$ and potential $\{\varphi\}$ [1]. When describing a finite element, the displacements and potentials at any point are expressed via nodal values as follows:

$$\begin{cases} \{U\} = [N]\{\delta\}, \\ \{\Phi\} = [L]\{\varphi\}, \end{cases} \tag{1}$$

where $[N]$ and $[L]$ are, in the general case, interpolating functions, and$\{\delta\}, \{\varphi\}$ are nodal displacements and potentials, respectively.

The strain vector could be expressed in the form [17]

$$\{\varepsilon\} = [B]\{\delta\}. \tag{2}$$

Similarly, the electric field could be expressed in the form [1]

$$\{E\} = [B_E]\{\varphi\}. \tag{3}$$

Thus, the equations of piezoelectric effect on the elementary volume are expressed as [16]

$$\begin{cases} \{\sigma\} = \left[c^E\right]\{\varepsilon\} - [e]^T\{E\}, \\ \{D\} = [e]\{\varepsilon\} + \left[\ni^S\right]\{E\}, \end{cases} \tag{4}$$

where $[c^E]$, $[e]$, $[\ni^s]$ are the matrix of stiffness for a constant electric field, the matrix of the piezoelectric constant, and the matrix of the dielectric constant evaluated under constant strain, respectively; $\{\sigma\}, \{\varepsilon\}$, {D}, {E} are the vectors of stress, strain, electric induction, and the electric field, respectively.

The analysis of piezoelectric actuators must be made taking into consideration the electric occurrence in the system. Based on the FEM, each node of the element has an additional degree of freedom used for electric potentials in FEM modelling. The solution applied to the equations of motion, fit for the actuator, can be derived from the principle of the minimal potential energy by means of the variation functional [4]. The basic dynamic FEM equation of motion for piezoelectric transducers that are entirely covered with electrodes can be expressed as follows [1]:

$$\begin{cases} [M]\left\{\ddot{\delta}\right\} + [C]\left\{\dot{\delta}\right\} + [K]\{\delta\} - [T]\{\varphi\} = \{R(t)\}, \\ [T]^T\{\delta\} + [S]\{\varphi\} = \{Q\}, \end{cases} \tag{5}$$

where [M], [K], [T], [S], [C] are the matrices of mass, stiffness, electro elasticity, capacity, and damping, respectively;$\{\delta\}, \{\varphi\}, \{R\}, \{Q\}$ are the vectors of node displacements, potentials, external mechanical forces, and charges coupled on the electrodes, respectively.

Here the structural matrices of the piezoceramic finite element of volume $V$ are:

$$[K] = \int_V [B]^T \left[c^E\right][B]\,\mathrm{d}V, \tag{6}$$

$$[T] = \int_V [B]^T [e] [B_E] \, \mathrm{d}V, \tag{7}$$

$$[S] = \int_V [B_E]^T [^s] [B_E] \, \mathrm{d}V, \tag{8}$$

$$[M] = \int_V [N]^T \gamma [N] \, \mathrm{d}V, \tag{9}$$

$$[C] = \alpha [M] + \beta [K] . \tag{10}$$

The damping matrix [C] is derived using the mass and stiffness matrices by denoting the constants $\alpha$ and$\beta$. $\gamma$ is the mass density.

The procedure to obtain the structural matrices of finite elements consists of the following steps:

1. Definition of interpolating functions, which are inverse to the configuration matrix obtained from interpolating polynomials.
2. Obtaining the derivatives of interpolating functions.
3. Multiplication of all the necessary submatrices.
4. Integration of these products over the volume of a finite element.

## 3 Implementation of Computer Algebra

Usually piezoceramic actuators are of a simple form, e.g., plates, rings, cylinders, etc. [3, 7]. Therefore the simple form finite elements are used. Two popular finite elements will be presented for modelling piezoceramic actuators of a sector-type and the triangular one. The interpolating functions for the sector-type finite element with four nodal points in polar coordinates $(\rho, \theta)$ could be expressed as follows [1]:

$$\{N\} = \frac{1}{(R_2 - R_1)(\theta_2 - \theta_1)} \left\{ \begin{array}{c} (R_2 - \rho)(\theta_2 - \theta) \\ (\rho - R_1)(\theta_2 - \theta) \\ (\rho - R_1)(\theta - \theta_1) \\ (R_2 - \rho)(\theta - \theta_1) \end{array} \right\}, \tag{11}$$

where $R_1, R_2, \theta_1, \theta_2$ are the values of sector radius and the angle boundary.

And the matrix $[B_i]$ for the $i$-th node of the element is [1]

$$[B_i] = \begin{bmatrix} \frac{\partial N_i}{\partial \rho} & 0 \\ \frac{N_i}{\rho} & \frac{1}{\rho}\frac{\partial N_i}{\partial \theta} \\ \frac{1}{\rho}\frac{\partial N_i}{\partial \theta} & \frac{\partial N_i}{\partial \rho} - \frac{N_i}{\rho} \end{bmatrix}, \quad i = 1, 4. \tag{12}$$

Analogously the matrix $[B_{Ei}]$ for the $i$-th node is of the same form [16]

$$[B_{Ei}] = \begin{bmatrix} \frac{\partial N_i}{\partial \rho} \\ \frac{1}{\rho}\frac{\partial N_i}{\partial \theta} \end{bmatrix}, \quad i = 1, 4. \tag{13}$$

Thus, the expression for the stiffness matrix of the sector-type finite element becomes

$$[K] = h \int_{\theta_1}^{\theta_2} \int_{R_1}^{R_2} [B]^T \left[c^E\right] [B] \, \rho \, \mathrm{d}\rho \, \mathrm{d}\theta, \tag{14}$$

where $h$ is thickness of the element.

Analogously, formulas (7)–(10) of the other matrices have the same structure.

The structural matrices of the sector-type finite element were derived using the computer algebra system VIBRAN, that is a FORTRAN preprocessor for analytical perturbation with polynomials, rational functions, and trigonometric series [3]. A special VIBRAN procedure can generate an optimized FORTRAN code from the obtained analytical expressions [2], which can be directly used in the programs for numerical analysis. The computer algebra technique is more convenient than numerical for this purpose, because the polynomials under integrals are of the negative power of order $-1$ for the sector-type finite element in the polar coordinates. Consequently, numerical integration is not very useful. Table 1 illustrates the number of products necessary to form the structural matrices of the piezoceramic sector finite element. For the sake of clarity, we consider only three matrices of the mass, stiffness, and electro-elasticity, because their formation is more expensive.

**Table 1.** The number of products to form the sector-type element matrices

| Matrix | Numerical integration | Computer Algebra |
|---|---|---|
| $K$ | 1296 | 820 |
| $T$ | 972 | 634 |
| $M$ | 2792 | 926 |
| Total: | 5060 | 2380 |

Another geometrically suitable finite element is triangle, which allows modelling various surfaces [17]. The first member of a triangular membrane or a bending element family is the element with three nodes and the interpolating polynomial of second order. The triangular finite element is also suitable to determine the configuration of optimal electrodes of the piezoceramic actuators [4]. The interpolating functions for the triangular finite element are expressed in the area of $L$-coordinates [5, 17]

$$\{N_i\} = \left\{ \begin{array}{c} L_1 + L_1^2 L_2 + L_1^2 L_3 - L_1 L_2^2 - L_1 L_3^2 \\ b_2 \left(L_3 L_1^2 + 0.5 L_1 L_2 L_3\right) - b_3 \left(L_1^2 L_2 + 0.5 L_1 L_2 L_3\right) \\ c_2 \left(L_3 L_1^2 + 0.5 L_1 L_2 L_3\right) - c_3 \left(L_1^2 L_2 + 0.5 L_1 L_2 L_3\right) \end{array} \right\}, \tag{15}$$

$i = 1, 2, 3,$

where $b_i, c_i$ are nodal coordinate differences with cyclic permutation of indices $1 \rightarrow 2 \rightarrow 3 \rightarrow 1$ for the remaining coefficients [5]. The matrix $[B_i]$ of the triangular element for the $i$-th node is [17]

$$[B_i] = \begin{bmatrix} -\frac{\partial^2}{\partial x^2} N_i \\ -\frac{\partial^2}{\partial^2 y} N_i \\ \frac{\partial^2}{\partial x \partial y} N_i \end{bmatrix}, \quad i = 1, 2, 3. \tag{16}$$

Thus, the expression of the stiffness matrix for the triangular finite element becomes like this

$$[K] = h \int_0^1 \int_0^{1-L_1} [B]^T \left[c^E\right] [B] \, \mathrm{d}L_2 \, \mathrm{d}L_1, \tag{17}$$

where $h$ is thickness of the element.

Analogously, the formulas (7)–(10) of the other matrices are of the same structure. Integration of the products (17) in algebraic form over the area of a finite element in $L$-coordinates according to [17] could be expressed as follows:

$$\int_{\Delta} L_1^n L_2^m L_3^k \, \mathrm{d}\Delta = \frac{n!m!k!}{(n+m+k+2)!} 2\Delta, \tag{18}$$

where $2\Delta$ is the triangle area.

Table 2 illustrates the number of products necessary to form the structural matrices of the piezoceramic triangular finite element.

**Table 2.** The number of products used to form triangular element matrices

| Matrix | Numerical integration | Computer Algebra |
|---|---|---|
| $K$ | 2430 | 1204 |
| $T$ | 1660 | 946 |
| $M$ | 3240 | 794 |
| Total: | 7330 | 2944 |

## 4 Conclusions

The proposed analytical formation of the structural matrices of piezoceramic finite elements reduces the number of floating point operations more than twice, particularly for the mass matrix of the elements. The highest economy is gained during the integration procedure, because, after the integration, the expressions of matrix elements become shorter due to disappearance of variables. The distributed mass, electro-elasticity, and damping matrices always yield relatively simple final analytical expressions even for complex finite elements. Due to the fact that the integral expressions to be integrated, in these cases, have the power at least twice higher than that of the ordinary stiffness matrix, the use of computer algebra remarkably saves the required computer time as compared to the numerical integration technique. It should be also emphasized that the computer time economy is followed by an increased accuracy of computations, which is, in many cases, the most important gain achieved.

## References


1. Bansevicius, R., Barauskas, R., Kulvietis, G., Ragulskis, K.: Vibromotors for Precision Microrobots. Hemisphere Publishing Corp., USA (1988)
2. Bansevicius, R., Cepulkauskas, A., Kulvietiene, R., Kulvietis, G.: Computer Algebra for Real-Time Dynamics of Robots with Large Number of Joints, Lecture Notes in Computer Science, Vol. 3039. Springer-Verlag, Berlin Heidelberg New York (2004) 278–285
3. Bansevicius, R., Parkin, R., Jebb, A., Knight, J.: Piezomechanics as a Sub-System of Mechatronics: Present State of the Art, Problems, Future Developments. IEEE Transactions on Industrial Electronics, vol. 43 (1) (1996) 23–30.
4. Bansevicius, R., Kulvietis, G., Mazeika, D.: Optimal Electrodes Configuration of Piezodrives. Solid Mechanics and Its Applications, N 73, Kluwer Academic Publ. (2000) 77–83
5. Belevicius, R., Pedersen, P.: Analysis and Sensitivity Analysis by Computer Algebra for a Third Order Plate Finite Element. Comp. & Struct. 49 (1993) 243–252
6. Besseling, J.F.: Finite Element Properties, Based upon Elastic Potential Interpolation. Hybrid and Mixed Finite Element Meth. Int. Symp. Atlanta, 1981. John Wiley, Chichester (1983) 253–266
7. Han, J.M., Adriaens, T.A., Willem, L., de Koning, Reinder Banning: Modelling Piezoelectric Actuators. IEEE/ASME Transactions on Mechatronics, Vol. 5 (4) (2000) 331–337
8. Laouafa, F., Royis, P.: An Iterative Algorithm for Finite Element Analysis. Journal of Computational and Applied Mathematics, Vol. 164–165 (2004) 469–491
9. Liew, KM, He, XQ, Kitipornchai, S.: Finite Element Method for the Feedback Control of FGM Shells in the Frequency Domain via Piezoelectric Sensors and Actuators. Computer Methods in Applied Mechanics and Engineering, Vol. 193 Issue 3 (2004) 257–273
10. Marczak, R.J.: Object-Oriented Numerical Integration – a Template Scheme for FEM and BEM Applications. Advances in Engineering Software, *In Press, Corrected Proof*, Available online 11 July 2005
11. Oliveira, A., Sousa P., Costa Branco, P.J.: Surface Deformation by Piezoelectric Actuator: from Park and Agrawal Models to a Simplified Model Formulation Sensors and Actuators. A: Physical, Vol. 115 (2–3) (2004) 235–244
12. Parashar, S., Das Gupta, A., von Wagner, U. & Hagedorn, P.: Non-linear Shear Vibrations of Piezoceramic Actuators International Journal of Non-Linear Mechanics, 40 (2005) 429–443
13. Samal, M.K. and oths.: A Finite Element Model for Nonlinear Behaviour of Piezoceramics under Weak Electric Fields Finite Elements in Analysis and Design, Vol. 41 (15) (2005) 1464–1480
14. Storck, Heiner, Wallaschek, Jörg: The Effect of Tangential Elasticity of the Contact Layer between Stator and Rotor in Travelling Wave Ultrasonic Motors. International Journal of Non-Linear Mechanics 38 (2003) 143–159
15. Tenchev, R.T.: A Study of the Accuracy of some FEM Stress Recovery Schemes for 2D Stress Concentration Problems. Finite Elements in Analysis and Design, Vol. 29 (2) (1998) 105–119
16. Tzou, H.S.: Piezoelectric Shells (Distributed Sensing and Control of Continua), Kluwer Academic Publishers (1993)
17. Zienkiewicz, O.C., Taylor, R.L.: The Finite Element Method. McGraw-Hill (1991)

# Symbolic Analysis of Economical Models with Mathematica

A. Gálvez and A. Iglesias

Department of Applied Mathematics and Computational Sciences, University of Cantabria, Avda. de los Castros, s/n, E-39005, Santander, Spain

**Abstract.** Functional equations is a very powerful technique to obtain consistent models in Economics. Unfortunately, there is only a few computer tools for solving functional equations. In this paper a *Mathematica* package, `FSolve`, is applied to the symbolic analysis of some economical models for price and advertising policies. The procedure is as follows: firstly, we make some assumptions about the functional structure of the functions describing the models. Such assumptions are given in terms of functional equations that account for some economical properties. Then, the package is applied to compute the solutions of these equations and check for inconsistencies. The interesting cases of the monopoly and duopoly models are discussed.

## 1 Introduction

One of the most appealing applications of the computer algebra systems (CAS) is the symbolic analysis of economical models [8, 9]. By this we mean to determine the symbolic solutions of the equations governing those models, which must be based on reasonable properties. The starting point of this process is to establish the conditions - which must be meaningful from the viewpoint of economical sciences - that those functions must satisfy necessarily. Then, the derived equations must be solved. However, it is quite hard in general to determine which are the conditions characterizing the problem and thus, some assumptions are usually taken. For instance, the functional structure of the functions describing the models is sometimes arbitrarily chosen, based either on convenience or on empirical considerations. Very often, this fact leads to absurd solutions and/or inconsistencies.

Our experience is that the functional equations is an optimal technique to overcome these undesirable situations. They provide powerful and consistent methods to describe the common sense properties of the economical functions and, simultaneously, the mathematical tools for solving the resulting equations [1, 2, 3, 4, 5]. The drawback of this picture is that most of this work must be performed by hand, as there is is only a few computer tools for solving functional equations. One remarkable exception is a *Mathematica* package, `FSolve`, developed by one of the authors and described in [6]. In this paper, the package will be successfully applied to the analysis of some monopoly and duopoly models for price and advertising policies.

V.N. Alexandrov et al. (Eds.): ICCS 2006, Part II, LNCS 3992, pp. 414–421, 2006.

The structure of this paper is as follows: in Section 2 we describe briefly the package FSolve. Then, in Section 3 we analyze some economical models for price and advertising policies. The procedure is as follows: firstly, we make some assumptions about the functional structure of the functions describing the models. Such assumptions are given in terms of functional equations that account for some economical properties. Then, the package is applied to compute the solutions of these equations and check for inconsistencies. The interesting cases of the monopoly and duopoly models are discussed.

## 2 The Package FSolve

In this section the *Mathematica* package FSolve is briefly described. The package has been implemented a few years ago in *Mathematica 3.0* [6] but updated versions for *Mathematica 4.0, 4.1, 4.2* and *5.0* are also available. We start our discussion by loading the package:

```
In[1]:=<<FunctionalEquations`FSolve`
```

which includes the command

FSolve[*eqn*, {*functions*}, {*variables*}, *options*]

where *eqn* denotes the functional equation to be solved, {*functions*} is the list of unknown functions, {*variables*} is the list of variables and *options* allows the users to consider different domains for the variables (see Table 1) and classes of feasible functions (see Table 2).

For instance, we can calculate the solution of the functional equation $f(x+y) = g(x) + h(y)$ where $x, y \in \mathbb{R}$ and $f, g, h$ are continuous functions as:

```
In[2]:= FSolve[f[x+y]==g[x]+h[y],{f,g,h},{x,y},Domain->Real,
                 Class->Continuous]
```

$$Out[2] := \begin{array}{l}\{f(x) \to C(1)\,x + C(2) + C(3), g(x) \to C(1)\,x + C(2),\\ h(x) \to C(1)\,x + C(3)\}\end{array}$$

where $C(1)$, $C(2)$ and $C(3)$ are arbitrary constants. Note that the general solution can depend on one or more arbitrary constants and even on arbitrary functions (see $Out(3)$ and $Out(4)$ for two examples). Note also that a single equation can determine several unknown functions (such as $f, g$ and $h$ in this example). The reader is referred to [7] for a general introduction to the theory of functional equations and their applications.

## 3 Price and Advertising Policies

In this section the package FSolve is applied to analyze some economical models for price and advertising policies. In particular, we focus on the problem of modeling the sales $S(p, v)$ of a single-product firm such that they depend on the price $p$ of its product and on the advertising expenditure $v$. Because of limitations of space, we will restrict our discussion to the cases of monopoly and duopoly models.

**Table 1.** List of all valid domains used in the `FSolve` package

| *Domains* | *Package names* | *Definition* |
|---|---|---|
| $\mathbb{R}$ | `Arbitrary` | $\mathbb{R}$ |
| $\mathbb{R}$ | `Real` | $\mathbb{R}$ |
| $\mathbb{I}$ | `Irrational` | $\mathbb{R} - \mathbb{Q}$ |
| $\mathbb{Q}$ | `Rational` | $\mathbb{Q}$ |
| $\mathbb{Z}$ | `Integer` | $\mathbb{Z}$ |
| $\mathbb{R} - \{0\}$ | `NonZeroReal` | $\mathbb{R} - \{0\}$ |
| $\mathbb{Q} - \{0\}$ | `NonZeroRational` | $\mathbb{Q} - \{0\}$ |
| $\mathbb{Z} - \{0\}$ | `NonZeroInteger` | $\mathbb{Z} - \{0\}$ |
| $\mathbb{R}_+$ | `RealPositiveZero` | $\{x \in \mathbb{R}/x \geq 0\}$ |
| $\mathbb{Q}_+$ | `RationalPositiveZero` | $\{x \in \mathbb{Q}/x \geq 0\}$ |
| $\mathbb{Z}_+$ | `IntegerPositiveZero` | $\{x \in \mathbb{Z}/x \geq 0\}$ |
| $\mathbb{R}_{++}$ | `RealPositive` | $\{x \in \mathbb{R}/x > 0\}$ |
| $\mathbb{I}_{++}$ | `IrrationalPositive` | $\{x \in \mathbb{R} - \mathbb{Q}/x > 0\}$ |
| $\mathbb{Q}_{++}$ | `RationalPositive` | $\{x \in \mathbb{Q}/x > 0\}$ |
| $\mathbb{Z}_{++}$ | `IntegerPositive` | $\{x \in \mathbb{Z}/x > 0\}$ |
| $\mathbb{R}_-$ | `RealNegativeZero` | $\{x \in \mathbb{R}/x \leq 0\}$ |
| $\mathbb{Q}_-$ | `RationalNegativeZero` | $\{x \in \mathbb{Q}/x \leq 0\}$ |
| $\mathbb{Z}_-$ | `IntegerNegativeZero` | $\{x \in \mathbb{Z}/x \leq 0\}$ |
| $\mathbb{R}_{--}$ | `RealNegative` | $\{x \in \mathbb{R}/x < 0\}$ |
| $\mathbb{I}_{--}$ | `IrrationalNegative` | $\{x \in \mathbb{R} - \mathbb{Q}/x < 0\}$ |
| $\mathbb{Q}_{--}$ | `RationalNegative` | $\{x \in \mathbb{Q}/x < 0\}$ |
| $\mathbb{Z}_{--}$ | `IntegerNegative` | $\{x \in \mathbb{Z}/x < 0\}$ |

**Table 2.** Classes of valid functions used in the `FSolve` package

| *Name* | *Classes of functions* |
|---|---|
| `Differentiable` | Differentiable |
| `Invertible` | Invertible |
| `Increasing` | Increasing |
| `Decreasing` | Decreasing |
| `Continuous` | Continuous |
| `ContinuousPoint` | Continuous at a point |
| `Monotonous` | Monotonous |
| `Arbitrary` | Arbitrary |

### 3.1 The Monopoly Model

Let us assume a firm such that the sales $S$ of a single product depend on the unitary price $p$ and on the advertising expenditure $v$, that is, $S = S(p, v)$. The function $S$ cannot be arbitrary, but it must satisfy the following properties:

(M1) The $S(p, v)$ function is continuous in both arguments.
(M2) $\forall\, v$, the $S(p, v)$ function, considered as a function of $p$ only, must be convex from below and decreasing. This implies that, for the same advertising

expenses, any increment in the unit price of the product leads to a reduction in sales, and that its derivative decreases with $p$.

(M3) $\forall\, p$, the $S(p,v)$ function, considered as a function of $v$ only, must be concave from below and increasing. This implies that, for the same unit price, an increment in the advertising expenses leads to an increment in sales.

(M4) A multiplicative change in the advertising expenditure leads to an additive change in sales, that is,

$$S(p,v\,w) = S(p,v) + T(p,w), \tag{1}$$

where $p \geq 0$, $v \geq 0$, $w \geq 0$, $T(p,1) = 0$ and $T(p,w)$ is increasing with $w$.

(M5) The sales due to an increment $q$ in price are equal to the previous sales times a real number, which depends on $q$ and $v$, that is,

$$S(p+q,v) = S(p,v)\,R(q,v) \tag{2}$$

where $p \geq 0$, $p+q \geq 0$, $v \geq 0$, $R(0,v) = 1$ and $R(q,v)$ is decreasing in $q$.

Eq. (1) can be solved by using the package `FSolve` as follows:

```
In[3]:= em1=FSolve[S[p,v*w]==S[p,v]+T[p,w],{S,T},{p,v,w},
        Domain->RealPositiveZero,Class->Continuous]
```

$Out[3] := \{S(p,v) \to Arb1(p)\,Log(v) + Arb2(p), T(p,w) \to Arb1(p)\,Log(w)\}$

where $Arb1(p)$ and $Arb2(p)$ denote two arbitrary functions depending on the variable $p$. Similarly, we can solve eq. (2) as:

```
In[4]:= em2=FSolve[S[p+q,v]==S[p,v]*R[q,v],{S,R},{p,q,v},
        Domain->RealPositiveZero,Class->Continuous]
```

$Out[4] := \left\{S(p,v) \to Arb3(v)\,e^{p\,Arb4(v)}, R(q,v) \to e^{q\,Arb4(v)}\right\}$

where $Arb3(v)$ and $Arb4(v)$ denote two arbitrary functions depending on the variable $v$. Once we have solved functional equations (1) and (2) separately, the general solution of the system (1)-(2) is given by:

```
In[5]:= FSolve[Equal @@ (S[p,v] /. First[#]& /@ {em1,em2}),
                    {Arb1,Arb2,Arb3,Arb4},{p,v}];
In[6]:= S[p,v] /. %
```

$Out[6] := (C(1) + C(2)\,Log[v])\,Exp[-C(3)\,p]$

where $C(1)$, $C(2)$ and $C(3)$ are arbitrary constants. Note that in *Out[6]* we have no longer arbitrary functions, but arbitrary constants. This means that the parametric model is completely specified and that we can estimate its parameters $C(1)$, $C(2)$ and $C(3)$ using empirical data.

The obtained solution shows a logarithmic increment of sales with advertising expenditures and an exponential decrease with price, in agreement with assumptions (M4) and (M5). One justification of this model of sales is the so-called Weber-Fechner law, that states that the stimuli of the intensity of perception is a linear function of the logarithm of the intensity of the stimulus.

The previous model was proposed by Eichhorn in 1978 [5]. It can be argued, however, that the function $R$ should depend on the price $p$, instead of $v$. Thus, we can replace (M5) by:

(M6) The sales due to an increment $q$ in price are equal to the previous sales times a real number, which depends on $q$ and $p$, that is,

$$S(p+q,v) = S(p,v)\,R(q,p) \tag{3}$$

where $p \geq 0$, $p+q \geq 0$, $v \geq 0$, $R(0,p) = 1$ and $R(q,p)$ is decreasing in $q$.

The general solution of (3) can be obtained by using the package `FSolve` as:

```
In[7]:= em3=FSolve[S[p+q,v]==S[p,v]*R[q,p],{S,R},{p,q,v},
        Domain->RealPositiveZero,Class->Continuous]
```

$$Out[7] := \left\{ S(p,v) \to Arb5(p)\,Arb6(v),\; R(q,p) \to \frac{Arb5(p+q)}{Arb5(p)} \right\}$$

Alternatively, we can assume a multiplicative, instead of an additive, change in the price $p$ and we can question whether or not choosing between one of these assumptions influences the resulting model. In other words, we can assume:

(M7) The sales due to a multiplicative change ($w$ times) in the price are equal to the previous sales times a real number, which depends on $w$ and $p$, that is,

$$S(p\,w,v) = S(p,v)\,R(w,p) \tag{4}$$

where $p \geq 0$, $w \geq 0$, $v \geq 0$, $R(1,p) = 1$ and $R(w,p)$ is decreasing in $w$.

```
In[8]:= FSolve[S[p*w,v]==S[p,v]*R[w,p],{S,R},{p,v,w},
        Domain->RealPositiveZero,Class->Continuous]
```

$$Out[8] := \left\{ S(p,v) \to Arb5(p)\,Arb6(v),\; R(w,p) \to \frac{Arb5(p\,w)}{Arb5(p)} \right\}$$

Note that the $S$ functions in *Out[7]* and *Out[8]* are identical. Thus, equations (3) and (4) are equivalent. Consequently, the above mentioned two assumptions (M6) and (M7) lead to the same model.

Now, the solution of the system (1)-(3) can be obtained as:

```
In[9]:= FSolve[Equal @@ (S[p,v] /. First[#]& /@ {em1,em3}),
                {Arb1,Arb2,Arb5,Arb6},{p,v}]
```

$$Out[9] := \begin{array}{l} \{Arb2(p) \to \dfrac{C(4)}{C(3)}\,Arb1(p), Arb5(p) \to \dfrac{-Arb1(p)}{C(3)}, \\ Arb6(p) \to -C(3)\,Log(p) - C(4)\} \end{array}$$

which leads to the model:

```
In[10]:= S[p,v] /. %
```

$$Out[10] := Arb1(p) \left[ Log(v) + \frac{C(4)}{C(3)} \right]$$

where the function $Arb1(p)$ and the constants $C(3)$ and $C(4)$ are arbitrary. For this solution to satisfy assumptions (M2) and (M3) above, $Arb1(p)$ must be convex from below and decreasing. Note that $Log(v) + \frac{C(4)}{C(3)}$ is increasing. We also remark that model in $Out[10]$ is more general than model in $Out[6]$. In fact, the resulting model is not completely specified because it depends on arbitrary functions. This means that new requirements might be established.

### 3.2 The Duopoly Model

Assume now that we have two different firms that compete in the market. Assume also that the sales $S$ of the product by firm 1 depend on the unit prices $p$ and $q$ and on the advertising expenditures $u$ and $v$, of the two firms, that is, $S = S(p, q, u, v)$. The function $S(p, q, u, v)$ must satisfy the following properties:

(D1) The $S(p, q, u, v)$ function is continuous in all arguments.
(D2) $S(p, q, u, v)$ is increasing in $q$ and $u$.
(D3) $S(p, q, u, v)$ is decreasing in $p$ and $v$.
(D4) A multiplicative change in the advertising expenditure of firm 1 leads to an additive change in sales, that is,

$$S(p, q, u\,w, v) = S(p, q, u, v) + T(p, q, w, v) \tag{5}$$

(D5) The sales due to an increment $r$ in price of firm 1 are equal to the previous sales times a real number, which depends on $r$ and $p$, that is,

$$S(p + r, q, u, v) = S(p, q, u, v)\, R(r, p, q, v) \tag{6}$$

where $p \geq 0$, $p + r \geq 0$, $v \geq 0$ and $R(0, p, q, v) = 1$.

The general solution of the system (5)-(6) is given by the following sequence of calculations: firstly, we compute the functions $S$, $T$ and $R$ of the previous equations, and then we apply the outputs to calculate the functional structure of function $S$.

```
In[11]:= ed1=FSolve[S[p,q,u*w,v]==S[p,q,u,v]+T[p,q,w,v],{S,T},
                    {p,q,u,v,w},Domain->RealPositiveZero,
                    Class->Continuous];

In[12]:= ed2=FSolve[S[p+r,q,u,v]==S[p,q,u,v]+R[r,p,q,v],{S,R},
                    {p,q,r,u,v},Domain->RealPositiveZero,
                    Class->Continuous];

In[13]:= FSolve[Equal @@ (S[p,q,u,v] /. First[#]& /@ {ed1,ed2}),
                   {Arb1,Arb2,Arb3,Arb4},{p,q,u,v}];
In[14]:= S[p,q,u,v] /. %
```

$Out[14] := Arb1(p, q, v)\ [Log(u) + Arb2(q, v)]$

where $Arb1(p, q, v)$ and $Arb2(q, v)$ are arbitrary functions. In addition we can consider the following assumption:

(D6) The total sales of both firms is a constant $K$, that is,

$$S(p, q, u, v) + S(q, p, v, u) = K \tag{7}$$

which, using the previous output, leads to

```
In[15]:= FSolve[((S[p,q,u,v]+S[q,p,v,u]) /. %)==K,{Arb1,Arb2},
         {p,q,u,v},Domain->RealPositiveZero,Class->Continuous]
         // FSimplify;
```

```
In[16]:= S[p,q,u,v] /. %
```

$$Out[16] := \frac{1}{Arb7(p) + Arb7(q)} \left[ Log\left(\frac{u}{v}\right) + K\, Arb7(q) \right]$$

where $Arb7(p)$ is an arbitrary but increasing function of $p$. The physical interpretation of this model is as follows: if the advertisement expenditures of both firms are coincident, the sales are proportional to the ratios $\frac{Arb7(q)}{Arb7(p) + Arb7(q)}$ and $\frac{Arb7(p)}{Arb7(p) + Arb7(q)}$ for firms 1 and 2, respectively. On the other hand, the advertisement expenditures influence sales directly proportional to the logarithm of the ratio $\frac{u}{v}$ and inversely proportional to $Arb7(p) + Arb7(q)$.

We can now consider two additional assumptions:

(D7) The sales $S(p + r, q + s, u, v)$ of firm 1 due to increments $r$ and $s$ in the prices of firms 1 and 2, respectively, are the initial sales $S(p, q, u, v)$ of firm 1 times two factors which consider the associated reduction and increments due to these two changes, that is,

$$S(p + r, q + s, u, v) = S(p, q, u, v)\, U(r, p, q)\, V(s, p, q) \qquad (8)$$

(D8) The sales $S(p, q, u + r, v + s)$ of firm 1 due to increments $r$ and $s$ in the advertisement expenditures of firms 1 and 2, respectively, are the initial sales $S(p, q, u, v)$ of firm 1 times two factors which consider the associated increments and decrements due to these two changes, that is,

$$S(p, q, u + r, v + s) = S(p, q, u, v)\, U(r, u, v)\, V(s, u, v) \qquad (9)$$

Combining now (D7) and (D8) and solving the system of equations (8)-(9), we get:

```
In[17]:= FSolve[S[p+r,q+s,u,v]==S[p,q,u,v]*U[r,p,q]*V[s,p,q],
                {S,U,V},{p,q,r,s,u,v},Domain->RealPositiveZero,
                Class->Continuous];
```

```
In[18]:= FSolve[S[p,q,u+r,v+s]==S[p,q,u,v]*U[r,u,v]*V[s,u,v],
                {S,U,V},{p,q,r,s,u,v},Domain->RealPositiveZero,
                Class->Continuous];
```

```
In[19]:= FSolve[Equal @@ (S[p,q,u,v] /. First[#]& /@ {%%,%}),
                 {Arb1,Arb2,Arb3,Arb6,Arb7,Arb8},{p,q,u,v}];
```

```
In[20]:= S[p,q,u,v] /. %
```

$$Out[20] := Arb1(p)\, Arb2(q)\, Arb6(u)\, Arb7(v)$$

where the functions $Arb1(p)$ and $Arb7(v)$ are decreasing and the functions $Arb2(q)$ and $Arb6(u)$ are increasing, but otherwise arbitrary. The physical interpretation of this model is that all the factors (prices and advertisement expenditures) act independently and contribute to the total sales of firm 1 as a factor which is less than 1 and decreasing for $p$ and $v$ and greater than 1 and increasing for $q$ and $u$.

## References

1. Aczél, J.: On a system of functional equations determining price and productivity indices. Utilitas Math. **7** (1975) 345-362
2. Aczél, J.: Measurement in economics: theory and applications of economic indices. Physica, Heidelberg (1988) 3-17
3. Aczél, J., Eichhorn, W.: Systems of functional equations determining price and productivity indices. Utilitas Math. **5** (1974) 213-226
4. Boulding, W., Lee, E., Staelin, R.: Mastering the mix: do advertising, promotion and sales force activities lead to differentiation?. Journal of Marketing Research **31** (1994) 159-172
5. Eichhorn, W.: Functional Equations in Economics. Addison Wesley, Redwood City, CA (1978)
6. Castillo, E., Iglesias, A.: A package for symbolic solution of real functional equations of real variables. Aequationes Mathematicae, **54** (1997) 181-198
7. Castillo, E., Iglesias, A., Ruiz-Cobo, R.: Functional Equations in Applied Sciences. Elsevier Pub., Amsterdam (2005)
8. Montoro, J.D., Paz, J.V.: Detecting macroeconomic chaos. In: Keranen, V., Mitic, P., Hietamaki, A. (eds): Innovations in Mathematics. Proceedings of the Second International Mathematica Symposium, IMS'97. Computational Mechanics Publications, Southampton (1997) 353-360
9. Paz, J.V., Montoro, J.D.: Classic economic dynamics: a neo-ricardian approach to economic growth and income distribution by means of Mathematica. In: Keranen, V., Mitic, P. (eds): Mathematics with Vision. Proceedings of the First International Mathematica Symposium, IMS'95. Computational Mechanics Publications, Southampton (1995) 299-308

# Polarizable Theta-Stable Parabolic Subalgebras and $K_{\mathbb{C}}$-Saturation in the Non-compact Real Forms of $G_2$ and $F_4$

Steven Glenn Jackson and Alfred G. Noël

Department of Mathematics
University of Massachusetts
Boston, MA 02125-3393, USA
jackson@math.umb.edu
anoel@math.umb.edu

**Abstract.** A general method for finding theta-stable parabolic subalgebras was given in 2004 . In this paper, we develop *LiE* subroutines to find representatives of conjugacy classes of polarizable theta-stable parabolic subalgebras. Using a theorem of Tauvel, we implement algorithms for computing the $K_{\mathbb{C}}$-saturation of the nilradical of such parabolic subalgebras. Furthermore, we provide a tool for testing a long standing conjecture that relates the wave front set of a representation of a real or p-adic group to special nilpotent orbits. Incidently, the conjecture is true for classical groups.

## 1 Introduction

Let $\mathfrak{g}$ be a real semisimple Lie algebra with adjoint group $G$ and Cartan decomposition $\mathfrak{g} = \mathfrak{k} \oplus \mathfrak{p}$ relative to a Cartan involution $\theta$. We will denote by $\mathfrak{g}_{\mathbb{C}}$ the complexification of $\mathfrak{g}$. Let $\sigma$ be the conjugation of $\mathfrak{g}_{\mathbb{C}}$ with respect to $\mathfrak{g}$. Then $\mathfrak{g}_{\mathbb{C}} = \mathfrak{k}_{\mathbb{C}} \oplus \mathfrak{p}_{\mathbb{C}}$ where $\mathfrak{k}_{\mathbb{C}}$ and $\mathfrak{p}_{\mathbb{C}}$ are obtained by complexifying $\mathfrak{k}$ and $\mathfrak{p}$ respectively. $K$ will be a maximal compact Lie subgroup of $G$ with Lie algebra $\mathfrak{k}$ and $K_{\mathbb{C}}$ will be the connected subgroup of the adjoint group $G_{\mathbb{C}}$ of $\mathfrak{g}_{\mathbb{C}}$, with Lie algebra $\mathfrak{k}_{\mathbb{C}}$. It is well known that $K_{\mathbb{C}}$ acts on $\mathfrak{p}_{\mathbb{C}}$ and the number of nilpotent orbits of $K_{\mathbb{C}}$ in $\mathfrak{p}_{\mathbb{C}}$ is finite. Furthermore, for a nilpotent $e \in \mathfrak{p}_{\mathbb{C}}$, $K_{\mathbb{C}}{\cdot}e$ is a connected component of $G_{\mathbb{C}}{\cdot}e \cap \mathfrak{p}_{\mathbb{C}}$ and $\dim K_{\mathbb{C}}{\cdot}e = 1/2 \dim G_{\mathbb{C}}{\cdot}e$.

Let $\mathfrak{q}$ be a parabolic subalgebra of $\mathfrak{g}_{\mathbb{C}}$ with Levi decomposition $\mathfrak{q} = \mathfrak{l} \oplus \mathfrak{u}$. Denote by $Q$ the connected Lie subgroup of $G_{\mathbb{C}}$ with Lie algebra $\mathfrak{q}$. Then there is a unique nilpotent orbit $\mathcal{O}_{\mathfrak{g}_{\mathbb{C}}}$ of $G_{\mathbb{C}}$ on $\mathfrak{g}_{\mathbb{C}}$ meeting $\mathfrak{u}$ in an open dense set. The intersection $\mathcal{O}_{\mathfrak{g}_{\mathbb{C}}} \cap \mathfrak{u}$ consists of a single $Q$-orbit under the adjoint action of $Q$ on $\mathfrak{u}$. These facts were first proved by Richardson [9].

The Kostant-Sekiguchi correspondence is a bijection between nilpotent orbits of $G$ in $\mathfrak{g}$ and nilpotent orbits of $K_{\mathbb{C}}$ on $\mathfrak{p}_{\mathbb{C}}$. Thus, the correspondence allows us to study certain questions about real nilpotent orbits by looking at nilpotent orbits of $K_{\mathbb{C}}$ on the symmetric space $\mathfrak{p}_{\mathbb{C}}$. Therefore the following is a natural

V.N. Alexandrov et al. (Eds.): ICCS 2006, Part II, LNCS 3992, pp. 422–429, 2006.

question: Maintaining the above notations and assuming that $\mathfrak{q}$ is $\theta$-stable, *when does $Q \cap K_{\mathbb{C}}$ admit an open dense orbit on $\mathfrak{u} \cap \mathfrak{p}_{\mathbb{C}}$?* This would be the equivalent of the Richardson theorem for the real case. It turns out that this statement is not true in general. Patrice Tauvel gave a counter example in [12] on page 652 for $\mathfrak{g}_{\mathbb{C}} = D_4$. However, he was able to prove a version of Richardson's theorem for polarizable parabolic subalgebras.

**Definition 1.** *Let $\mathfrak{q} = \mathfrak{q} \cap \mathfrak{k}_{\mathbb{C}} \oplus \mathfrak{q} \cap \mathfrak{p}_{\mathbb{C}}$ be a $\theta$ stable parabolic subalgebra of $\mathfrak{g}_{\mathbb{C}}$ with Levi decomposition $\mathfrak{q} = \mathfrak{l} \oplus \mathfrak{u}$. Let $e$ be a nilpotent element of $u \cap \mathfrak{p}_{\mathbb{C}}$ . We say that $\mathfrak{q}$ is a polarization of $\mathfrak{g}_{\mathbb{C}}$ at $e$ if*

$$2\dim\mathfrak{q} = \dim\mathfrak{g}_{\mathbb{C}}^{e} + \dim\mathfrak{g}_{\mathbb{C}}$$

*where $\mathfrak{g}_{\mathbb{C}}^{e}$ is the centralizer of $e$ in $\mathfrak{g}_{\mathbb{C}}$.*

*A theta-stable parabolic $\mathfrak{q} = \mathfrak{l} \oplus \mathfrak{u}$ is said to be polarizable if there exists $e \in \mathfrak{u} \cap \mathfrak{p}_{\mathbb{C}}$ such that $\mathfrak{q}$ is a polarization of $\mathfrak{g}_{\mathbb{C}}$ at $e$.*

Observe that since $\dim\mathfrak{g}_{\mathbb{C}}^{e} = \dim\mathfrak{g}_{\mathbb{C}} - \dim G_{\mathbb{C}}{\cdot}e$ and $\dim K_{\mathbb{C}}{\cdot}e = 1/2\dim G_{\mathbb{C}}{\cdot}e$ the above formula becomes:

$$\dim\mathfrak{q} = \dim\mathfrak{g}_{\mathbb{C}} - \dim K_{\mathbb{C}}{\cdot}e.$$

This is the criterion that we will use to test for polarization.

The next proposition could be seen as a version of Richardson's theorem for complex symmetric spaces.

**Proposition 1.** *(P. Tauvel). Maintaining the above notations, suppose that there exists $z$ in $\mathfrak{p}_{\mathbb{C}}$ such that $\mathfrak{q}$ is a polarization of of $\mathfrak{g}_{\mathbb{C}}$ at $z$. Then*

*i. There exists a unique $K_{\mathbb{C}}$-nilpotent orbit, $\mathcal{O}_k$, in $\mathfrak{p}_{\mathbb{C}}$ such that $\mathcal{S} = \mathfrak{u} \cap \mathfrak{p}_{\mathbb{C}} \cap \mathcal{O}_k$ is an open and dense in $\mathfrak{u} \cap \mathfrak{p}_{\mathbb{C}}$.*

*ii. $\mathcal{S}$ is a $Q \cap K_{\mathbb{C}}$-orbit.*

*iii. if $x \in \mathcal{S}$ then $[x, \mathfrak{k}_{\mathbb{C}} \cap \mathfrak{q}] = \mathfrak{u} \cap \mathfrak{p}_{\mathbb{C}}$, $[x, \mathfrak{p}_{\mathbb{C}} \cap \mathfrak{q}] = \mathfrak{u} \cap \mathfrak{k}_{\mathbb{C}}$ and $\mathfrak{q}$ is a polarization of $\mathfrak{g}_{\mathbb{C}}$ at $x$.*

*Proof.* See [12] proposition 4.6. □

The orbit $\mathcal{O}_k$ is said to be the $K_{\mathbb{C}}$-saturation of $\mathfrak{u} \cap \mathfrak{p}_{\mathbb{C}}$. We use the notation $\mathcal{O}_k = K_{\mathbb{C}}(\mathfrak{u} \cap \mathfrak{p}_{\mathbb{C}})$. A study of the proof of the above proposition reveals that if $e$ is in $\mathfrak{u} \cap \mathfrak{p}_{\mathbb{C}}$ and $\mathfrak{q}$ is a polarization of $\mathfrak{g}_{\mathbb{C}}$ at $e$ then $\mathcal{O}_k = K_{\mathbb{C}}{\cdot}e$. This observation allows us to design an algorithm for computing the $K_{\mathbb{C}}$-saturation of $\mathfrak{u} \cap \mathfrak{p}_{\mathbb{C}}$ when $\mathfrak{q}$ is polarizable.

## 2 Algorithms

### 2.1 Root Decomposition

Let $\mathfrak{h}$ be a fundamental Cartan subalgebra of $\mathfrak{g}$. Then $\mathfrak{h} = \mathfrak{t} \oplus \mathfrak{v}$, where $\mathfrak{t}$ is a Cartan subalgebra of $\mathfrak{k}$ and $\mathfrak{v} \subseteq \mathfrak{p}$. We have the following decompositions in the root spaces of $\mathfrak{g}_{\mathbb{C}}$ generated by the roots of $\mathfrak{h}_{\mathbb{C}}$:

$$\mathfrak{k}_{\mathbb{C}} = \mathfrak{t}_{\mathbb{C}} \oplus \sum_{\alpha \ compact\, imaginary} \mathbb{C}X_\alpha \oplus \sum_{(\alpha,\theta\alpha) \ complex\ pairs} \mathbb{C}(X_\alpha + \theta(X_\alpha)),$$

$$\mathfrak{p}_{\mathbb{C}} = \mathfrak{v}_{\mathbb{C}} \oplus \sum_{\alpha \ non\, compact\ imaginary} \mathbb{C}X_\alpha \oplus \sum_{(\alpha,\theta\alpha) \ complex\ pairs} \mathbb{C}(X_\alpha - \theta(X_\alpha)).$$

Here $X_\alpha$ is a nonzero vector of the root space $\mathfrak{g}_{\mathbb{C}}^\alpha$. An imaginary root $\alpha$ is compact (noncompact) if its root space $\mathfrak{g}_{\mathbb{C}}^\alpha$ lies in $\mathfrak{k}_{\mathbb{C}}$ ($\mathfrak{p}_{\mathbb{C}}$). See ([7]) for more details.

## 2.2 Algorithms

**Algorithm I.** The following algorithm generates all the $\theta$-stable parabolic subalgebras associated with a system of simple roots $\Delta$.

**Input:**

$\Delta$: set of simple roots
$\Upsilon$: set of compact imaginary roots
$\Gamma$: set of non-compact imaginary roots
$\Lambda$: set of complex roots
$\Phi^+$ : set of positive roots

**Computation:**

1. $\mathfrak{h}_{\mathbb{C}} = \bigoplus_{\alpha \in \Delta} \mathbb{C}H_\alpha$
2. $\mathfrak{u} \leftarrow \sum_{\alpha \in \Phi^+} \mathbb{C}X_\alpha$
3. $\mathfrak{q} \leftarrow \mathfrak{h}_{\mathbb{C}} \oplus \mathfrak{u}$ [ This is the Borel subalgebra defined by $\Phi^+$ ]
4. **For** each subset $S$ of $\Delta$ **Do**
5. **For** each $\alpha \in S$ of **Do**
 **If** $\alpha \in \Lambda$ **Then** $S \leftarrow S \cup \{\theta(\alpha)\}$ **EndIf-EndFor**
6. $\Phi \leftarrow < S >$ [ The root subsystem generated by the positive roots in $S$ ]
7. $\mathfrak{l} \leftarrow \mathfrak{h}_{\mathbb{C}} \oplus \sum_{\alpha \in \Phi \cap \Upsilon} \mathbb{C}X_\alpha \oplus \sum_{\alpha\ \Phi \cap \Gamma} \mathbb{C}X_\alpha \oplus \sum_{\alpha \in \Phi \cap \Lambda} \mathbb{C}(X_\alpha + \theta(X_\alpha))$
 $\oplus \sum_{\alpha \in \Phi \cap \Lambda} \mathbb{C}(X_\alpha - \theta(X_\alpha))$
8. $\hat{\Phi} = \Phi^+ \setminus (\Phi \cap \Phi^+)$
9. $\mathfrak{u}' \leftarrow \bigoplus_{\mu \in \hat{\Phi}} \mathbb{C}X_\mu$
10. $Next(\mathfrak{q}) \leftarrow \mathfrak{l} \oplus \mathfrak{u}'$ [ build the next parabolic subalgebra in the list ]

**EndFor**

**End.**

**Algorithm II.** The next algorithm determines whether or not a given parabolic is polarizable. If it is polarizable then the $K_{\mathbb{C}}$-saturation of its niradical in $\mathfrak{p}_{\mathbb{C}}$ is returned. Although we use the algorithm on exceptional Lie groups it does handle the classical cases also. The dimensions of the nilpotent orbits of the exceptional complex Lie groups are given in [1] and such data are used in order to check the polarization condition. In [5] we gave algorithms for computing representatives of $K_{\mathbb{C}}$-nilpotent orbits for exceptional Lie groups. Djoković [2] has computed such representatives also. Each representative of a nilpotent orbit is given as a linear combination of non zero vectors of root spaces.

We write $e = c_{\mu_1}X_{\mu_1} + \ldots c_{\mu_m}X_{\mu_m}$ where the $\mu_i$'s are roots to denote a representative. Furthermore, $\mu_1, \ldots, \mu_m$ are simple roots of a reductive Lie subalgebra of $\mathfrak{g}_{\mathbb{C}}$. For example in $F_{4(4)}$, $e = \sqrt{2}(X_{\alpha_1+2\alpha_2+3\alpha_3+\alpha_4} + X_{-\alpha_1-\alpha_2-\alpha_3})$ is a representative of orbit #9 in Djoković's tables and these two short roots generate a subalgebra of type $A_2$. We write $< e >$ to denote the subalgebra of rank $m$ generated by $\mathcal{R}_e = \{\mu_1, \ldots, \mu_m\}$.

**Input:**
$L$ : List of of lists of representatives of each $K_{\mathbb{C}}$ orbits.
$q$: A theta-stable parabolic subalgebra $\mathfrak{q} = \mathfrak{l} \oplus \mathfrak{u}$ of $\mathfrak{g}_{\mathbb{C}}$.
$\mathcal{R}_e$: defined above

**Computation:**
0. $\mathcal{R} \leftarrow \{\alpha | X_\alpha \in \mathfrak{u} \cap \mathfrak{p}_{\mathbb{C}}\}$
1. **For** each $K_{\mathbb{C}}$-orbit $\mathcal{O}_k$ **Do**
2. **If** $\dim \mathfrak{q} = \dim \mathfrak{g}_{\mathbb{C}} - \dim \mathcal{O}_k$ **Then** [ q may be polarizable ]
  3. **For** each representative $e$ of $\mathcal{O}_k$ **Do**
  4. $r \leftarrow$ rank $< e >$ [$< e >$ is the subalgebra generated by $e$]
  5. **For** each subset $S \subseteq \mathcal{R}$
    such that $|S| = r$ and $Type(< S >) = Type(< e >)$ **Do**
    6. **If** $S$ is $K_{\mathbb{C}}$- conjugate to $\mathcal{R}_e$ **Then**
      $q$ is polarizable and $\mathcal{O}_k = K_{\mathbb{C}}(\mathfrak{u} \cap \mathfrak{p}_{\mathbb{C}})$
    7. **Else** go back to step 3.
  **EndIf-EndFor-EndFor**
**EndIf-EndFor**
**End.**

**Comments.** We implement both algorithms in *LiE*. The first algorithm is an extension of the procedure discussed in [6]. A proof of correctness was given also in that paper. A careful examination of the second algorithm should reveal that step 6 is indeed the most costly condition on average because it may require traversing the Weyl group of $K_{\mathbb{C}}$.

The number of $K_{\mathbb{C}}$-conjugacy classes of systems of simple roots is given by $\frac{|W(G_{\mathbb{C}})|}{|W(K_{\mathbb{C}})|}$ where $W(G_{\mathbb{C}})$ and $W(K_{\mathbb{C}})$ are the Weyl groups of $G_{\mathbb{C}}$ and $K_{\mathbb{C}}$ respectively. One can enumerate all such classes by successive reflections through non compact imaginary roots. Thus for each class the algorithms will give a list of theta-stable polarizable parabolic subalgebras. There could be redundancies as some parabolic subalgebras could lie in different classes of simple roots. However, this problem can be easily resolve by keeping a list of distinct parabolic subalgebras. A more subtle problem is to decide when two theta parabolic subalgebras are conjugate under $K_{\mathbb{C}}$ which may require traversing $W(K_{\mathbb{C}})$. Keep in mind that in general $W(K_{\mathbb{C}})$, although finite, could be large. This situation is under investigation.

Finally, we shall discuss an important conjecture in Representation theory. Certain complex nilpotent orbits are said to be *special*. For a definition of the term special in this context see [1], [10]. Special nilpotent orbits are the building blocks of the celebrated Springer's correspondence which relate nilpotent orbits to Representation theory. The Wave Front theory (See [4]) suggests that the $K_{\mathbb{C}}$-saturation is always a connected component of a special nilpotent complex orbit. Moëglin [4] proved that the conjecture is true for classical p-adic groups. Trapa [13] seems to have a proof for classical real Lie groups. The results below support the conjecture for polarizable parabolic subalgebras; all the $K_{\mathbb{C}}$-saturations obtained come from special complex nilpotent orbits of $G_{\mathbb{C}}$ on $\mathfrak{g}_{\mathbb{C}}$.

# 3 Results

Because of space constraints we will only give a small subset of our results.

## 3.1 $G_2$

$\mathfrak{g} = G_{2(2)}$ and dim $G_2 = 14$.

$\Delta = \{\alpha_1, \alpha_2\}$, $\mathfrak{h}_{\mathbb{C}} = \mathbb{C}H_{\alpha_1} \oplus \mathbb{C}H_{\alpha_2}$

**Table I**

1. $\mathfrak{q} = \mathfrak{h}_{\mathbb{C}} \oplus \mathbb{C}X_{\alpha_1+\alpha_2} \oplus \mathbb{C}X_{-\alpha_2} \oplus \mathbb{C}X_{\alpha_1} \oplus \mathbb{C}X_{2\alpha_1+\alpha_2} \oplus \mathbb{C}X_{3\alpha_1+\alpha_2} \oplus \mathbb{C}X_{3\alpha_1+2\alpha_2}$
dim $\mathfrak{q} = 8$ $\mathcal{O}_k = K_{\mathbb{C}}(\sqrt{6}X_{\alpha_1+\alpha_2} + \sqrt{10}X_{-\alpha_2})$ dim $\mathcal{O}_k = 6$

2. $\mathfrak{q} = (\mathfrak{h}_{\mathbb{C}} \oplus \mathbb{C}X_{\alpha_1} \oplus \mathbb{C}X_{-\alpha_1}) \oplus \mathbb{C}X_{\alpha_1+\alpha_2} \oplus \mathbb{C}X_{\alpha_2} \oplus \mathbb{C}X_{2\alpha_1+\alpha_2} \oplus \mathbb{C}X_{3\alpha_1+\alpha_2} \oplus \mathbb{C}X_{3\alpha_1+2\alpha_2}$ dim $\mathfrak{q} = 9$ $\mathcal{O}_k = K_{\mathbb{C}}(X_{\alpha_1+\alpha_2} + X_{3\alpha_1+\alpha_2})$ dim $\mathcal{O}_k = 5$

3. $\mathfrak{q} = (\mathfrak{h}_{\mathbb{C}} \oplus \mathbb{C}X_{\alpha_1+\alpha_2} \oplus \mathbb{C}X_{-\alpha_1-\alpha_2}) \oplus \mathbb{C}X_{\alpha_1} \oplus \mathbb{C}X_{-\alpha_2} \oplus \mathbb{C}X_{2\alpha_1+\alpha_2} \oplus \mathbb{C}X_{3\alpha_1+\alpha_2} \oplus \mathbb{C}X_{3\alpha_1+2\alpha_2}$ dim $\mathfrak{q} = 9$ $\mathcal{O}_k = K_{\mathbb{C}}(X_{2\alpha_1+\alpha_2} + X_{-\alpha_2})$ dim $\mathcal{O}_k = 5$

## 3.2 $\mathfrak{g}_{\mathbb{C}} = F_4$

We order the roots of $F_4$ as in the table below and we use the Bourbaki system of simple roots $\Delta = \{\alpha_1, \alpha_2, \alpha_3, \alpha_4\}$. The Cartan involution $\theta$ with $+1$-eigenspace $\mathfrak{k}$ and $-1$-eigenspace $\mathfrak{p}$ depends on the real forms. If $\mathfrak{g} = F_{4(4)}$ then $\mathfrak{k}_{\mathbb{C}} = sp_3(\mathbb{C}) \oplus sl_2(\mathbb{C})$ and the vector space $\mathfrak{p}_{\mathbb{C}}$ is the complex span of non-zero root vectors $X_\beta$ where $\beta = c_1\alpha_1 + c_2\alpha_2 + c_3\alpha_3 + c_4\alpha_4$ with $c_1 = \pm 1$. If $\mathfrak{g} = F_{4(-20)}$ then

$\mathfrak{k}_{\mathbb{C}} = so_9(\mathbb{C})$ and the vector space $\mathfrak{p}_{\mathbb{C}}$ is the complex span of non-zero root vectors $X_\beta$ where $\beta = c_1\alpha_1 + c_2\alpha_2 + c_3\alpha_3 + c_4\alpha_4$ with $c_4 = \pm 1$.

**Positive roots of $F_4$**

| | | |
|---|---|---|
| 1. [1,0,0,0] | 9. [0,1,2,0] | 17. [1,2,2,1] |
| 2. [0,1,0,0] | 10. [0,1,1,1] | 18. [1,1,2,2] |
| 3. [0,0,1,0] | 11. [1,1,2,0] | 19. [1,2,3,1] |
| 4. [0,0,0,1] | 12. [1,1,1,1] | 20. [1,2,2,2] |
| 5. [1,1,0,0] | 13. [0,1,2,1] | 21. [1,2,3,2] |
| 6. [0,1,1,0] | 14. [1,2,2,0] | 22. [1,2,4,2] |
| 7. [0,0,1,1] | 15. [1,1,2,1] | 23. [1,3,4,2] |
| 8. [1,1,1,0] | 16. [0,1,2,2] | 24. [2,3,4,2] |

In the preceding table each vector indexed by $i$ represents the coefficients of the $i^{th}$ postive root in the Bourbaki base $\Delta$. We use $X_i$ or $X_{-i}$ to denote a non zero root vetor in the root spaces of $i$ and $-i$ respectively. The maximum Cartan subalgebra is $\mathfrak{h}_{\mathbb{C}} = \mathbb{C}H_{\alpha_1} \oplus \mathbb{C}H_{\alpha_2} \oplus \mathbb{C}H_{\alpha_3} \oplus \mathbb{C}H_{\alpha_4}$.

$\mathfrak{g} = F_{4(4)}$ and dim $F_4 = 52$.

**Table II**

1. $\mathfrak{q} = \mathfrak{h}_{\mathbb{C}} \oplus (2,3,4,6,7,9,10,13,16,24,14,15,17,18,$
$19,20,21,22,23,-1,-5,-8,-11,-12)$
dim $\mathfrak{q} = 28$ $\quad \mathcal{O}_k = K_{\mathbb{C}}(\sqrt{22}X_{16}+\sqrt{30}X_{15}+\sqrt{42}X_{-11}+4X_{-12})$ dim $\mathcal{O}_k = 24$

2. $\mathfrak{q} = (\mathfrak{h}_{\mathbb{C}} \oplus \pm 3) \oplus (2,4,6,7,9,10,13,16,24,12,14,15,17,18$
$,19,20,21,22,23,-1,-5,-8,-11)$
dim $\mathfrak{q} = 29$ $\quad \mathcal{O}_k = K_{\mathbb{C}}(\sqrt{12}X_{12}+\sqrt{14}X_{14}+\sqrt{10}X_{-11}+\sqrt{20}X_{-5})$ dim $\mathcal{O}_k = 23$

3. $\mathfrak{q} = (\mathfrak{h}_{\mathbb{C}} \oplus \pm 14) \oplus (2,3,4,6,7,9,10,13,16,24,15,17,18,19,20$
$,21,22,23,-1,-5,-8,-11,-12)$
dim $\mathfrak{q} = 29$ $\quad \mathcal{O}_k = K_{\mathbb{C}}(\sqrt{14}X_{14}+\sqrt{18}X_{18}+\sqrt{8}X_{-11}+\sqrt{10}X_{-12})$ dim $\mathcal{O}_k = 23$

4. $\mathfrak{q} = (\mathfrak{h}_{\mathbb{C}} \oplus \pm(14,15)) \oplus (2,3,4,6,7,9,10,13,16,24,17,18,19,20$
$,21,22,23,-1,-5,-8,-11,-12)$
dim $\mathfrak{q} = 30$ $\quad \mathcal{O}_k = K_{\mathbb{C}}(\sqrt{5}X_{17}+3X_{18}+X_{-11}+\sqrt{8}X_{-12})$ dim $\mathcal{O}_k = 22$

5. $\mathfrak{q} = (\mathfrak{h}_{\mathbb{C}} \oplus \pm(2,4)) \oplus (3,6,7,9,10,13,16,24,11,14,15,17$
$,18,19,20,21,22,23,-1,-5,-8,-11)$
dim $\mathfrak{q} = 30$ $\quad \mathcal{O}_k = K_{\mathbb{C}}(\sqrt{5}X_{17}+X_{18}+3X_{11}+\sqrt{8}X_{-8})$ dim $\mathcal{O}_k = 22$

6. $\mathfrak{q} = (\mathfrak{h}_{\mathbb{C}} \oplus \pm(2,11,14)) \oplus (4,-12,15,-8,17,18,3,20,6,7,24,10,-5,19$
$,-1,21,9,13,16,22,23)$
$\dim \mathfrak{q} = 31 \quad \mathcal{O}_k = K_{\mathbb{C}}(\sqrt{5}X_{17} + 3X_{18} + \sqrt{8}X_{-12}) \ \dim \mathcal{O}_k = 21$

7. $\mathfrak{q} = (\mathfrak{h}_{\mathbb{C}} \oplus \pm(3,12,15)) \oplus (2,4,6,7,9,10,13,16,24,-3,14,17,18$
$,19,20,21,22,23,-1,-5,-8,-11)$
$\dim \mathfrak{q} = 31 \quad \mathcal{O}_k = K_{\mathbb{C}}(\sqrt{10}X_{14} + \sqrt{6}(X_{18} + X_{-8})) \ \dim \mathcal{O}_k = 21$

8. $\mathfrak{q} = (\mathfrak{h}_{\mathbb{C}} \oplus \pm(3,4,7)) \oplus (2,6,9,10,13,16,24,5,8,11,$
$12,14,15,17,18,19,20,21,22,23,-1)$
$\dim \mathfrak{q} = 31 \quad \mathcal{O}_k = K_{\mathbb{C}}(\sqrt{10}X_{-1} + \sqrt{6}(X_{18} + X_8)) \ \dim \mathcal{O}_k = 21$

9. $\mathfrak{q} = (\mathfrak{h}_{\mathbb{C}} \oplus \pm(3,12,14,15)) \oplus (2,4,6,7,9,10,13,16,24,17,18$
$,19,20,21,22,23,-1,-5,-8,-11)$
$\dim \mathfrak{q} = 32 \quad \mathcal{O}_k = K_{\mathbb{C}}(\sqrt{3}X_{17} + X_{18} + 2X_{-5} + X_{-11}) \ \dim \mathcal{O}_k = 20$

10. $\mathfrak{q} = (\mathfrak{h}_{\mathbb{C}} \oplus \pm(2,4,8,12)) \oplus (3,6,7,9,10,13,16,24,11,14$
$15,17,18,19,20,21,22,23,-1,-5)$
$\dim \mathfrak{q} = 32 \quad \mathcal{O}_k = K_{\mathbb{C}}(\sqrt{3}X_{17} + X_{18} + 2X_{-5} + X_{11}) \ \dim \mathcal{O}_k = 20$

11. $\mathfrak{q} = (\mathfrak{h}_{\mathbb{C}} \oplus \pm(3,4,7,24)) \oplus (2,6,9,10,13,16,14,17,$
$19,20,21,22,23,-1,-5,-8,-11,-12,-15,-18)$
$\dim \mathfrak{q} = 32 \quad \mathcal{O}_k = K_{\mathbb{C}}(\sqrt{2}X_{14} + X_{20} + \sqrt{3}X_{22} + X_{-8}) \ \dim \mathcal{O}_k = 20$

12. $\mathfrak{q} = (\mathfrak{h}_{\mathbb{C}} \oplus \pm(2,4,24,11,14,15,17,18,20)) \oplus (3,6,7,10,19,21,$
$-8,-12,9,13,16,22,23,-1,-5)$
$\dim \mathfrak{q} = 37 \quad \mathcal{O}_k = K_{\mathbb{C}}(\sqrt{2}(X_{19} + X_{-8})) \ \dim \mathcal{O}_k = 15$

13. $\mathfrak{q} = (\mathfrak{h}_{\mathbb{C}} \oplus \pm(3,4,7,5,8,11,12,15,18)) \oplus (2,6,9,10,13,$
$16,24,14,17,19,20,21,22,-1,23)$
$\dim \mathfrak{q} = 37 \quad \mathcal{O}_k = K_{\mathbb{C}}(X_{14} + X_{21} + X_{-1}) \ \dim \mathcal{O}_k = 15$

14. $\mathfrak{q} = (\mathfrak{h}_{\mathbb{C}} \oplus \pm(2,3,6,9,24,12,15,17,19)) \oplus (4,7,10,13,$
$18,20,21,22,23,-1,-5,-8,-11,-14,16)$
$\dim \mathfrak{q} = 37 \quad \mathcal{O}_k = K_{\mathbb{C}}(\sqrt{2}(X_{18} + X_{-1})) \ \dim \mathcal{O}_k = 15$

15. $\mathfrak{q} = (\mathfrak{h}_{\mathbb{C}} \oplus \pm(2,3,4,6,7,9,10,13,16)) \oplus (1,5,8,11,12,14,$
$15,17,18,19,20,21,22,23,24)$
$\dim \mathfrak{q} = 37 \quad \mathcal{O}_k = K_{\mathbb{C}}(\sqrt{2}(X_{23} + X_1)) \ \dim \mathcal{O}_k = 15$

$\mathfrak{g} = F_{4(-20)}$.

**Table III**

1. $\mathfrak{q} = (\mathfrak{h}_{\mathbb{C}} \oplus \pm(1,2,3,5,6,8,9,11,14)) \oplus (16,18,20,21,22,,$
$23,24,-4,-7,-10,-12,-13,-15,-17,-19)$
$\dim \mathfrak{q} = 37 \quad \mathcal{O}_k = K_{\mathbb{C}}(\sqrt{2}(X_{19} + X_4)) \ \dim \mathcal{O}_k = 15$

## 4 Conclusion

The results presented in this paper are the output of an implementation of the above algorithms in the language *LiE* [14]. We choose this language because it contains many basic functions that can be used in such investigations. We are also quite familiar with the accuracy and the speed of its underlying algorithms. We have not implemented our algorithms in other languages. It would be an interesting exercise to produce implementations in GAP, Mathematica or Mapple. One of the reviewers wanted to know whether or not our program can be used to produce counter-examples similar to the one given by Tauvel. Although this is a clever suggestion, we did not pursue it. However, we think that it is worth pointing out that Richardson's theorem holds for $sl_n(\mathbb{R})$.

## References


1. Collingwood, D. H. and McGovern, W. M. *Nilpotent orbits in semisimple Lie algebras.* Van Nostrand Reinhold Mathematics Series, New York (1992).
2. Djoković D. *The closure diagrams for nilpotent orbits of real forms of $F_4$ and $G_2$*, J. Lie Theory **10** (200), 491-510.
3. Djoković D. *Classification of nilpotent elements in the simple exceptional real Lie algebras of inner type and description of their centralizers* J. Alg. **112 (2)** (1988) 503-524.
4. Moëglin, C *Front d'onde des représentations des groupes classiques p-adiques* American Journal of Mathematics **118** (1996) 1313-1346.
5. Noël, A. G. *Computing maximal tori using LiE and Mathematica* Lectures Notes in Computer Science, Springer-Verlag **2657** (2003) 728-736.
6. Noël, A. G. *Computing theta-stable parabolic subalgebras using LiE.* Lectures Notes in Computer Science, Springer-Verlag **3039** (2004) 335-342.
7. Knapp, A. W. *Lie groups beyond an introduction.* Second edition, Birkhaüser Progress in Mathematics **140** (2002)
8. Kostant, B. *The principal three dimensional subgroup and the Betti numbers of a complex Lie group.* Amer. J. Math., **81**, (1959) 973-1032 .
9. Richardson R. *Conjugacy classes in parabolic subgroups of semisimple algebraic groups*, Bull. London Math. Soc. **6** (1974), 21–24.
10. Spaltenstein N. *Classes unipotentes et sous-groupes de Borel* Lecture Notes in Math. **946**, Springer-Verlag, New York, 1982.
11. Sekiguchi J. *Remarks on real nilpotent orbits of a symmetric pair* J. Math. Soc. Japan **39**, No. 1 (1987), 127-138.
12. Tauvel P. *Quelques résultats sur les algèbres de Lie symétriques*, Bull. Sci. math. **125** 8 (2001), 641-665.
13. Trapa, P. *Richardson orbits for real classical groups* Journal of Algebra, **286** (2005), 386-404.
14. Van Leeuwen, M. A. A., Cohen, A. M., and Lisser, B. *LiE : A package for Lie group computations*, Computer Algebra Nederland, Amsterdam, Netherlands (1992)

# Dynamic Load Balancing with MatlabMPI*

Ricolindo L. Cariño[1], Ioana Banicescu[1,2], and Wenzhong Gao[3]

[1] Center for Computational Sciences ERC, PO Box 9627
[2] Department of Computer Science and Engineering, PO Box 9637
[3] Department of Electrical and Computer Engineering, PO Box 9571
Mississippi State University, Mississippi State MS 39762, USA
{rlc@erc, ioana@cse, wgao@ece}.msstate.edu

**Abstract.** A number of packages have recently been developed to enable the execution of MATLAB programs on parallel processors. For many applications, an integrated load balancing functionality is also necessary to realize the full benefits of parallelization. This paper describes a toolkit based on MatlabMPI to ease the parallelization and integration of load balancing into MATLAB applications that contain computationally-intensive loops with independent iterations. Modifications to existing code to incorporate the toolkit are minimal. Performance tests of two nontrivial MATLAB programs with the toolkit on a general-purpose Linux cluster indicate that significant speedups are achievable.

## 1 Introduction

MatlabMPI [1] is a set of MATLAB [2] scripts that implements a subset of the Message-Passing Interface (MPI) standard [3]. MatlabMPI allows multiple copies of a MATLAB program executing on parallel processors to communicate. MatlabMPI utilizes file I/O for communications and exploits the built-in I/O capabilities of MATLAB to read/write (receive/send) any MATLAB variable. While MatlabMPI enables parallelization of a MATLAB program, the program developer is still responsible for ensuring that the parallel version achieves high performance. Thus, additional MatlabMPI functionality is necessary to reduce the burden of developers in writing efficient parallel MATLAB code.

The MATLAB programs that benefit most from MatlabMPI are those that are embarrassingly parallel. These programs typically contain computationally-intensive parallel loops. The iterations of these loops can be executed independently and in any order on several processors without affecting the correctness of the computations. However, the parallel execution of iterations gives rise to a load balancing problem, especially if the iterations have nonuniform execution times which may not be known *a priori*, or if the processors are heterogeneous or are unpredictably loaded. To our knowledge, as of the time of writing this paper, there are no visible research efforts addressing the dynamic load balancing problem in MatlabMPI.

* This work was partially supported by the NSF grants #9984465, #0313274 and #0132618.

V.N. Alexandrov et al. (Eds.): ICCS 2006, Part II, LNCS 3992, pp. 430–437, 2006.

The contribution of this paper is a toolkit to address the potential load imbalance that arises while using MatlabMPI. The toolkit simplifies the parallelization *and* load balancing of MATLAB programs that contain loops with independent iterations, requiring only very minor revisions to the programs. Owing to the overhead of using file I/O for communications, the toolkit achieves significant speedups only when the loop iterations are more costly than file I/O. The toolkit implements a scheduler–worker load balancing strategy, where the scheduler decides the worker loads and also participates in executing iterations. The worker loads are determined by a user-selected dynamic loop scheduling technique. The toolkit has been successfully integrated into a number of nontrivial MATLAB applications that exhibit significant performance improvement when executed on a general-purpose Linux cluster. Thus, we believe the toolkit is useful for load balancing many other MATLAB applications, thereby increasing the productivity of users and improving the utilization of parallel computing resources.

The rest of this paper first reviews related work on parallel MATLAB in Section 2, and then describes our implementation of the load balancing toolkit based on MatlabMPI in Section 3. Section 4 describes the integration of the toolkit into actual MATLAB programs, and their resulting parallel performance on a general-purpose Linux cluster. Section 5 presents our conclusions and describes future work.

## 2 Parallel MATLAB

MATLAB is a high-level language and interactive environment for algorithm development, data visualization, data analysis, and numeric computation [2]. It is very popular and is widely used in scientific and technical computing. Arguments for not developing parallel MATLAB were offered in 1995 [4]; however, a recent survey [5] found 27 parallel MATLAB projects, some defunct and some implemented in commercial packages. The survey summarizes the approaches to making MATLAB parallel, including: compiling MATLAB programs into native parallel code; providing a parallel back end to a graphical MATLAB front end; and providing communication and coordination facilities for MATLAB processes. MatlabMPI [1] uses the last approach. It is a set of MATLAB scripts implementing a subset of MPI. With MatlabMPI, a MATLAB program may use the six basic functions (MPI_Init, MPI_Comm_size, MPI_Comm_rank, MPI_Send, MPI_Recv and MPI_Finalize) to carry out parallel processing. Additionally, MPI_Abort, MPI_Bcast and MPI_Probe are implemented for convenience.

In order to realize the full benefits of parallelization in MATLAB programs, load balancing is usually required especially if the parallel tasks have nonuniform computational requirements or if the processors are heterogeneous. *Dynamic* load balancing becomes a necessity when the nonuniformity of tasks or heterogeneity of processors are not known *a priori*. Two of the authors have previously designed a dynamic load balancing routine based on MPI for C or Fortran applications containing parallel loops [6]. The toolkit described in this paper is based on similar design principles.

## 3 A Dynamic Load Balancing Toolkit

We developed a toolkit based on MatlabMPI to reduce the burden of developing parallel MATLAB programs that require dynamic load balancing. In addition to reducing programming efforts, the toolkit improves the productivity of users through shorter program running times, achieved through parallelization and load balancing. And as a consequence of decreased running times, parallel computing resources are utilized more efficiently. This section briefly describes the usage of the toolkit, the MATLAB scripts, and the shell script to submit the parallel MATLAB job to a Linux cluster managed by the Portable Batch System (PBS) [7].

### 3.1 Toolkit Usage

The toolkit inherits the target programs of MatlabMPI. These targets are MATLAB programs that contain embarrassingly parallel loops. These loops must be of the form `for i=1:N` {`i-iteration`} `end` or the equivalent `while` form, where the iterations are independent and computationally-intensive. The integration of the toolkit into a target program is illustrated by Figure 1. Essentially, the original `for` loop is converted to a `while` loop where chunks of iterations can be executed concurrently on different processors. Since the `i-iteration` invokes CPU-intensive computations which may require tens or hundreds of MATLAB statements, the additional code to utilize the toolkit constitutes a small percentage of the total number of lines of code for the application.

```
% serial version    |  % parallel version
                    |  MPI_Init;
...                 |  ...
for i=1:N           |  scheduler = 0; method = 4; dispLevel = 2;
  {i-iteration}     |  DLBinfo = DLB_StartLoop (MPI_COMM_WORLD, ...
end                 |             scheduler, 1, N, method, dispLevel);
...                 |  while ( ~DLB_Terminated(DLBinfo) )
                    |    [iStart iSize DLBinfo] = DLB_StartChunk(DLBinfo);
                    |    for i=iStart:(iStart+iSize-1)        % was i=1:N
                    |      {i-iteration}
                    |    end
                    |    DLBinfo = DLB_EndChunk (DLBinfo);
                    |  end
                    |  [iIters iTime] = DLB_EndLoop (DLBinfo); % (OPTIONAL)
                    |  ...
                    |  MPI_Finalize;
```

**Fig. 1.** Parallelization and dynamic load balancing of a MATLAB for-loop

Results of the `i-iterations`, if any, will be scattered among the processors in MPI_COMM_WORLD. The `scheduler` maintains information about all the chunks of iterations in the `chunkMap` field of `DLBinfo`. The other processors keep the information about chunks they execute in their own `DLBinfo.chunkMap`. After loop execution, the workers can send results to the scheduler. An example of how this can be accomplished is provided in Section 4.

### 3.2 Descriptions of Functions

The toolkit consists of the following functions:

`function DLBinfo = DLB_StartLoop ( iComm, master, firstIter, lastIter, iMeth, dispLevel)` is the synchronization point at the start of loop execution. A dynamic load balancing environment is created on `iComm`. The information about this environment is maintained in `DLBinfo`. The scheduler will have the rank specified by `master`. The loop range is [`firstIter, lastIter`]. The loop scheduling method that will be used is indicated by `iMeth` (see `DLB_StartChunk`). The amount of debugging information to be displayed is `dispLevel` (0 = no tracing; 1 = trace the sends and receives; and 2 = trace the sends, receives and probes).

`function done = DLB_Terminated(DLBinfo)` returns true if all loop iterations have been executed.

`function [iStart iSize DLBinfo] = DLB_StartChunk (DLBinfo)` returns a range for a chunk of iterations. This range starts with iteration `iStart` and contains `iSize` iterations. Let `N` and `P` denote the number of iterations and the number of processors, respectively, and let `remaining` denote the number of unscheduled iterations at any time during loop execution. The following loop scheduling methods are used by the scheduler for determining the chunk sizes:

- `iMeth=0: iSize=N/P` (static scheduling);
- `iMeth=1: iSize=1` (self scheduling);
- `iMeth=2: iSize=(N/P)/log2(N/P)` (fixed size chunking, with the same number of chunks as in factoring; see `iMeth=4`);
- `iMeth=3: iSize=(remaining/P)` (guided self scheduling [8]); and
- `iMeth=4: P` chunks, each of `iSize=remaining/(2*P)` (factoring [9]).

Other loop scheduling methods such as variants of adaptive weighted factoring [10, 11, 12, 13] and adaptive factoring [14, 15] will be added in the future. Internally, the scheduler probes for requests for chunk data from workers and responds by sending `[iStart iSize]` if `remaining`>0; otherwise, the scheduler sends `iSize`=0, which will be interpreted by workers as the termination signal. If there are no requests, the scheduler sets `iSize`=1 and proceeds to execute one iteration.

`function DLBinfo = DLB_EndChunk(DLBinfo)` signals the end of execution of a chunk of iterations. Internally, a worker processor requests from the scheduler the start and size of the next chunk.

`function [iIters iTime] = DLB_EndLoop(DLBinfo)` (optional) marks the end of loop execution. `iIters` is the number of iterations done by the calling processor, and `iTime` is the cost measured using the MATLAB function `etime`. `iIters` and `iTime` are useful for assessing the performance gains achieved by dynamic load balancing. For example, the sum of the `iTimes` from all participating processors gives an estimate of the cost of executing the loop on a single processor.

### 3.3 Submission to a Linux Cluster

MatlabMPI provides a MPI_Run.m script to initiate the parallel execution of a MATLAB program on a collection of machines. On a Linux cluster managed by PBS, the PBS script for the parallel MATLAB job must first create the list of machines to be supplied as an argument to MPI_Run, and then start a MATLAB session to execute MPI_Run. In order to avoid initiating this extra MATLAB session on a Linux cluster, we moved the essential functionality of MPI_Run to the PBS script. Thus, the PBS scripts for our MatlabMPI jobs first create the parallel MATLAB scripts, and then start the remote MATLAB sessions to run these scripts. In contrast to those generated by MPI_Run, our parallel MATLAB scripts include the statements for the remote MATLAB sessions to create MPI_COMM_WORLD, instead of loading it from disk.

## 4 Evaluation

We have integrated the toolkit into nontrivial MATLAB programs. This section presents results of testing the toolkit on two such programs. The tests were conducted on the heterogeneous general-purpose EMPIRE cluster of the Mississippi State University [16]. The cluster has a total of 1038 processors organized into 16 racks. A rack contains 32 nodes of dual 1.0GHz or 1.266GHz Pentium III processors and 1.25GB RAM. Each node is connected to a 100Mb/s Ethernet rack switch. The rack switches are connected by a gigabit Ethernet cluster switch. Installed software includes RedHat Linux and PBS. The general submission queue allows 64-processor, 48-hour jobs; a special queue allows 128-processor, 96-hour jobs from a restricted set of users. According to the Top 500 Supercomputer Sites list published in June 2002, EMPIRE then was the 126th fastest computer in the world.

The first example is a MATLAB program used to verify the correctness of a filtered back projection algorithm for tomographic imaging [17]. An outline of the computationally-intensive section of the program and its parallel version with the toolkit is illustrated by Figure 2. The other sections of the program are executed redundantly by all participating processors.

Figure 3 compares the performance of the imaging application on 1, 4, 8 and 12 processors. In addition to the parallelization, the application variable `rays` was increased in value from 16 to 64 to generate a parallel loop with 256 iterations. For this problem with a short runtime, the speedups on the loop execution (loop time plus gather time) are 1.83, 3.43 and 4.94 for 4, 8 and 12 processors, respectively.

The second example is a program to benchmark several methods for solving nonnegative quadratic programming problems (http://www.cis.upenn.edu/feisha/codes/nqpBench.m). In particular, the program compares the convergence

```
%serial version           | % parallel version
for x=x_min:x_max2,       | comm = MPI_COMM_WORLD;
   for y=y_min:y_max2,| info = DLB_StartLoop(comm,0,x_min,x_max2,4,2);
     sum=0.0;             | while ( ~DLB_Terminated(info) )
     for i=1:angles,      |   [iStart iSize info] = DLB_StartChunk(info);
       ...                |   for x=iStart:(iStart+iSize-1)
       sum = sum + ...|     (The "for y=y_min:y_max2..." is unchanged)
     end;                 |   end;
     f(x,y)=sum;          |   info = DLB_EndChunk(info);
   end;                   | end;
 end;                     |         % gather results
                          | if (comm.rank==0) % 0 receives remote chunks
                          |    for i=1:info.numChunks
                          |       iStart = info.chunkMap{1,i}(1,1);
                          |       iSize = info.chunkMap{1,i}(1,2);
                          |       src = info.chunkMap{1,i}(1,3);
                          |       f(iStart:iStart+iSize-1,:) = ...
                          |          MPI_Recv(src,iStart,comm);
                          |    end
                          | else % workers send executed chunks to 0
                          |    for i=1:info.numChunks
                          |       iStart = info.chunkMap{1,i}(1,1);
                          |       iSize = info.chunkMap{1,i}(1,2);
                          |       MPI_Send(0,iStart,comm,...
                          |          f(iStart:iStart+iSize-1,:));
                          |    end
                          | end;
```

**Fig. 2.** Outline of the computionally-intensive section in a MATLAB program for tomographic imaging (left) and its parallel version (right) with code to gather results into one of the processors

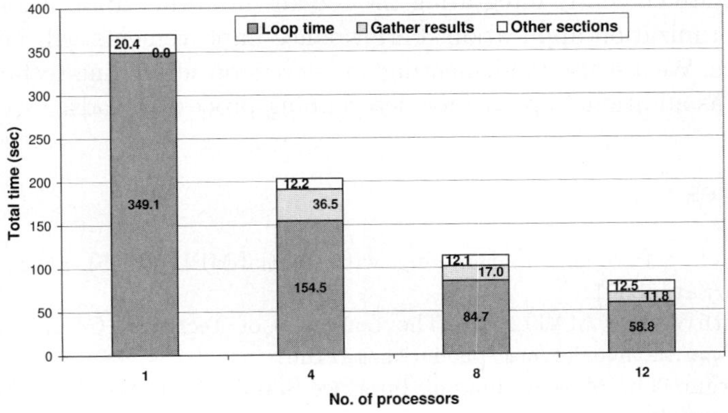


**Fig. 3.** Total time of tomographic imaging application

**Table 1.** Performance of benchmark program

| Processors | 1 | 3 | 4 | 5 |
|---|---|---|---|---|
| Wall time (sec) | 59096.5 | 23495.1 | 17655.5 | 11854.5 |
| Speedup | - | 2.52 | 3.35 | 4.99 |

rate of a recently developed multiplicative updates technique [18] to the rates of MATLAB's quadprog() with preconditioned conjugate gradients and quadprog() with direct solver. The program generates `numTrial=1000` random problems and computes the average solution time for each technique. Table 1 summarizes the performance of the program on 1, 3, 4 and 5 processors when `numTrial=10000`. The 5-processor limit is due to the number of concurrent users allowed for a MATLAB toolbox required by the benchmark program.

## 5 Conclusions and Future Work

MatlabMPI and similar packages enable the execution of MATLAB programs on parallel processors. However, load balancing is necessary to fully exploit the advantages of parallelization and to make more productive use of expensive parallel computing resources.

We developed a toolkit based on MatlabMPI to ease the parallelization and integration of load balancing into MATLAB programs that contain computionally-intensive parallel loops. The toolkit implements a dynamic scheduler-worker load balancing strategy where loads are determined through loop scheduling techniques. The toolkit drastically reduces the burden of program developers in writing efficient parallel MATLAB code.

We successfully tested the toolkit by integrating it into a number of nontrivial MATLAB programs and executing these on a general-purpose Linux cluster. Results of sample tests indicate significant speedups are achievable with the toolkit. We are currently integrating the toolkit into other applications such as a design optimization application. Performance improvements will be reported in the future. We are also implementing adaptive loop scheduling techniques into the toolkit as additional options for determining processor loads.

## References


1. Kepner, J.: Parallel Programming with MatlabMPI. (2005) `http://www.ll.mit.edu/MatlabMPI`.
2. The MathWorks: MATLAB - The Language of Technical Computing. (2005) `http://www.mathworks.com/products/matlab`.
3. MPI Forum: The Message Passing Interface Standard. (1995) `http://www-unix.mcs.anl.gov/mpi`.
4. Moler, C.: Why ther isn't a parallel MATLAB (1995) `http://www.mathworks.com/company/newsletters/news_notes/pdf/spr95cleve.pdf` .

5. Choy, R., Edelman, A.: Parallel MATLAB: Doing it right. Proceedings of the IEEE **93** (2005) 331–341
6. Cariño, R.L., Banicescu, I.: A load balancing tool for distributed parallel loops. Cluster Computing: The Journal of Networks, Software Tools and Applications **8**(4) (2005) To appear
7. Altair Grid Technologies: Portable Batch Systems. (2003) `http://www.openpbs.org`.
8. Polychronopoulos, C., Kuck, D.: Guided self-scheduling: A practical scheduling scheme for parallel supercomputers. IEEE Transactions on Computers **C-36**(12) (1987) 1425–1439
9. Hummel, S.F., Schonberg, E., Flynn, L.E.: Factoring: A method for scheduling parallel loops. Communications of the ACM **35**(8) (1992) 90–101
10. Banicescu, I., Velusamy, V.: Performance of scheduling scientific applications with adaptive weighted factoring. In: Proceedings of the 15th IEEE International Parallel and Distributed Processing Symposium - 10th Heterogenous Computing Workshop (IPDPS-HCW 2001) CDROM, IEEE Computer Society (2001)
11. Banicescu, I., Velusamy, V., Devaprasad, J.: On the scalability of dynamic scheduling scientific applications with adaptive weighted factoring. Cluster Computing: The Journal of Networks, Software Tools and Applications **6** (2003) 215–226
12. Cariño, R.L., Banicescu, I.: Dynamic scheduling parallel loops with variable iterate execution times. In: Proceedings of the 16th IEEE International Parallel and Distributed Processing Symposium - 3rd Workshop on Parallel and Distributed Scientific and Engineering Computing With Applications (IPDPS-PDSECA 2002) CDROM, IEEE Computer Society (2002)
13. Cariño, R.L., Banicescu, I.: Load balancing parallel loops on message-passing systems. In Akl, S., Gonzales, T., eds.: Proceedings of the 14th IASTED International Conference on Parallel and Distributed Computing and Systems (PDCS 2004), ACTA Press (2002) 362–367
14. Banicescu, I., Liu, Z.: Adaptive factoring: A dynamic scheduling method tuned to the rate of weight changes. In: Proceedings of the High Performance Computing Symposium (HPC) 2000. (2000) 122–129
15. Banicescu, I., Velusamy, V.: Load balancing highly irregular computations with the adaptive factoring. In: Proceedings of the 16th IEEE International Parallel and Distributed Processing Symposium - 11th Heterogeneous Computing Workshop (IPDPS-HCW 2002) CDROM, IEEE Computer Society (2002)
16. Mississippi State University ERC: EMPIRE: The ERC's Massively Parallel Initiative for Research and Engineering. `http://www.erc.msstate.edu/computing/empire/`(2001)
17. Rao, R.P., Kriz, R.D., Abbott, A.L., Ribbens, C.J.: Parallel implementation of the filtered back projection algorithm for tomographic imaging. `http://www.sv.vt.edu/xray_ct/parallel/Parallel_CT.html` (1995)
18. Sha, F., Saul, L.K., Lee, D.D.: Multiplicative updates for nonnegative quadratic programming in support vector machines. In S. Becker, S.T., Obermayer, K., eds.: Advances in Neural Information Processing Systems 15. MIT Press, Cambridge, MA (2003) 1041–1048

# Maple Implementation of the Chor-Rivest Cryptosystem

L. Hernández Encinas, J. Muñoz Masqué, and A. Queiruga Dios

Dpt. Information Processing and Coding, Applied Physics Institute, CSIC
C/ Serrano 144, E28006-Madrid, Spain
{luis, jaime, araceli}@iec.csic.es

**Abstract.** A Maple implementation of the Chor-Rivest cryptosystem is presented. The main problems in this implementation are: to determine discrete logarithms over $GF(q^h)$, $q \approx 200$ and $h \approx 25$, and to use the arithmetic of the finite fields in order to encrypt and decrypt messages.

## 1 Introduction

As it is known, the goal of Cryptography is to guarantee the secrecy and confidentiality of communications between two or more users, who use an insecure channel ([6], [7]). On the other hand, the goal of cryptanalysis is to break the security and privacy of communications. Public Key Cryptography (PKC) was born in 1976 when Diffie and Hellman proposed a protocol for key agreement using an insecure channel ([4]). In PKC two keys are used: The encryption (or public) key is used by the sender to encrypt a message; and the decryption (or private) key is kept in secret by the receiver and it allows him to decrypt the message. The security of these schemes is based on the computational intractability of some Number Theory problems. In 1985, a cryptosystem based on the knapsack problem was proposed (see [1], [2]). This system has been recently broken by Vaudenay ([12]), but only for the parameters originally proposed. The Chor-Rivest cryptosystem is based on the arithmetic of finite fields and it needs to compute discrete logarithms. The discrete logarithm problem (DLP) can be posed as follows: Given a prime integer $p$, a generator $\alpha$ of the cyclic group $\mathbb{Z}_p^*$, and an element $\beta \in \mathbb{Z}_p^*$, the DLP consists in finding an integer $x$, $0 < x \leq p-1$, such that $\beta = \alpha^x$. This problem is considered to be difficult because the best algorithm known for solving it is the number field sieve ([11]) which has a subexponential expected running time: $O\left(e^{\left((64/9)^{1/3}+o(1)\right)(\ln p)^{1/3}(\ln \ln p)^{2/3}}\right)$.

The cryptosystem has the interesting property that its security depends on the knapsack problem but not on the DLP. In fact, the more tractable the DLP, the easier to implement Chor-Rivest cryptosystem. So, results on how to compute discrete logarithms more efficiently help to improve the implementations of Chor-Rivest cryptosystem.

In this paper, we present a Maple implementation of the Chor-Rivest cryptosystem. This implementation includes several procedures to compute discrete logarithms in the finite field $GF(q^h)$, $q \approx 200$ and $h \approx 25$. The rest of the paper

V.N. Alexandrov et al. (Eds.): ICCS 2006, Part II, LNCS 3992, pp. 438–445, 2006.

is organized as follows: In section 2, we introduce the Chor-Rivest cryptosystem to encrypt and decrypt messages. In section 3, the Maple procedures to compute discrete logarithms in finite fields and the implementation of the cryptosystem are presented. Finally, in section 4, the conclusions are stated.

# 2 The Chor-Rivest Cryptosystem

## 2.1 System Generation

The parameters of the cryptosystem are chosen as follows (for more details, see [1] and [2]).

1. Let $q$ be a prime and let $h \leq q$ be an integer so that the DLP can be efficiently solved in the finite field $GF(q^h)$. This property is crucial because the user has to compute discrete logarithms in $GF(q^h)$ to determine his keys. It is known that the DLP can be efficiently solved if the order of the multiplicative group considered, $\mathrm{GF}(q^h)^*$, factorizes as a product of small prime factors ([8]). The running time of Pohlig-Hellman algorithm is proportional to the largest prime factor of the order of the group.
2. A random element $T \in GF(q^h)$ is chosen such that $T$ is algebraic of degree $h$ over $GF(q)$, by considering $GF(q^h) \approx GF(q)[T]/(f(T))$, where $f(T)$ is a random irreducible monic polynomial of degree $h$ in $GF(q)[T]$. Note that the elements in $GF(q^h)$ are polynomials of degree $\leq h-1$ with coefficients in $GF(q)$, and the operations are done modulo $q$ and $f(T)$.
3. Pick a generator $g$ of the group $GF(q^h)^*$. To determine such generator, one can choose a random element $g \in GF(q^h)^*$ until it verifies $g^{(q^h-1)/s} \neq 1$ for all prime divisors, $s$, of $q^h - 1$. Note that $q$ an $h$ are chosen so that $q^h - 1$ has small prime factors, hence the above property is easy to verify.
4. The discrete logarithms, $a_i = \log_g(T + \alpha_i)$, $\forall \alpha_i \in GF(q)$, are computed.
5. The values $b_i = a_{\pi(i)}$ are obtained by means of a random permutation, $\pi : \{0, 1, \ldots, q-1\} \to \{0, 1, \ldots, q-1\}$.
6. A random noise, $0 \leq r \leq q^h - 2$, is added to obtain the public knapsack: $c_i \equiv (b_i + r)(\mathrm{mod}(q^h - 1))$, $i = 0, 1, \ldots, q-1$.
7. The public key of the user is the set $(c_0, c_1, \ldots, c_{q-1})$.
8. The private key is the set $(T, g, \pi, r)$.

## 2.2 Encryption

The sender encrypts the binary message $M = (m_0, m_1, \ldots, m_{q-1})$, $m_i \in \{0, 1\}$, $i = 0, \ldots, q-1$ as follows:

$$E = \sum_{i=0}^{q-1} m_i \cdot c_i (\mathrm{mod}(q^h - 1)).$$

In section 3 we will show how to transform a binary text without constrains into a binary vector of length $q$ and weight $h$ ([3]) and conversely.

### 2.3 Decryption

To decrypt an encrypted message, $E$, the receiver executes the following steps:

1. Compute $s' \equiv E - h \cdot r(\text{mod}(q^h - 1))$.
2. Determine the polynomial of degree $h-1$, $Q(T) = g^{s'}(\text{mod}(f(T))$.
3. Compute the $h$ roots $\alpha_{\pi(i)}$ of $f(T) + Q(T)$ over the $GF(q)$:

$$f(T) + Q(T) = \prod_{i \in I}(T + \alpha_{\pi(i)}).$$

4. Apply $\pi^{-1}$ in order to recover the coordinates of the original message $M$ having the bit 1.

## 3 Implementation and Procedures in Maple

In this section we present the Maple procedures and commands needed to encrypt and decrypt messages by using the Chor-Rivest system, with real parameters.

### 3.1 Procedures to Compute Discrete Logarithms in $GF(q^h)$

To generate the system parameters, two procedures to compute discrete logarithms over $GF(q^h)$ are needed: The Pohlig-Hellman algorithm ([6], [8]) which calls to the Baby-step Giant-step algorithm ([6]). The input of the `PohligHellman` procedure is a generator ($\alpha$) of $GF(q^h)^*$ and an element ($\beta$) of the group. The output is $\log_\alpha \beta$. Our procedure is specifically designed to compute discrete logarithms in finite fields and it is suppose that $q$, $h$ and $f = f(T)$ are known.

```
PohligHellman := proc(beta,alpha)  local NN, FF, RR, DD, ii, PP, EE, XX,
       QQ, EX, GG, LL, AA, jj, GG1, BB, OUT:
  NN:=(q^h)-1: FF:=convert(factorset(NN),list):
  RR:=nops(FF): DD:=ifactors(NN)[2]:
  for ii from 1 to RR do PP[ii]:=DD[ii,1]: EE[ii]:=DD[ii,2]: end do:
  for ii from 1 to RR do
    XX[ii]:=0: QQ:=PP[ii]: EX:=EE[ii]: GG:=1: LL[-1]:=0:
    AA:=Powmod(alpha,NN/QQ,f,T) mod q:
    for jj from 0 to EX-1 do
      GG:=Powmod(GG*Powmod(alpha,LL[jj-1]*QQ^(jj-1),f,T) mod q,1,f,T)
        mod q: GG1:=Powmod(GG,-1,f,T) mod q:
      BB:=Powmod(Powmod(beta*GG1,1,f,T) mod q,NN/(QQ^(jj+1)),f,T) mod q:
        LL[jj]:=BabyGiant(BB,AA,QQ^EX): XX[ii]:=XX[ii]+LL[jj]*QQ^jj:
    end do:
  end do:
  OUT:=chrem([seq(XX[ii],ii=1..RR)],[seq(PP[ii]^EE[ii],ii=1..RR)]):
  return(OUT):
end:
```

The above procedure calls to BabyGiant procedure:

```
BabyGiant := proc(pB, pA, pI) local MM, TT, TT1, GG, ii, jj, kk, AA, OUT:
  MM:=ceil(evalf(sqrt(pI))):
  TT:=[seq(Powmod(pA,jj,f,T) mod q,jj=0..MM-1)]:
  TT1:=[seq([jj-1,TT[jj]],jj=1..nops(TT))]: AA:=Powmod(pA,-MM,f,T) mod q:
  for ii from 0 to MM-1 do
    GG:=Powmod(pB*Powmod(AA,ii,f,T) mod q,1,f,T) mod q:
    if member(GG,TT,'kk') then OUT:=ii*MM+TT1[kk][1]: end if:
  end do:
  return(OUT):
end:
```

Both procedures (preceded by the command `with(numtheory)`) can be saved in the file `ProcsDLP.mpl` to be loaded when it be necessary.

### 3.2 Message Transformation

The procedures to transform a binary message without constraints, $T$, into a binary vector of length $q$ and weight $h$, and its inverse, are the following. $T$ is divided into blocks of length $\lfloor \log_2 \binom{q}{h} \rfloor$ bits. Each block can be considered as the binary representation of an integer $n$, where $0 \leq n < \binom{q}{h}$. Then, if $n \geq \binom{q-1}{h}$, the first bit of the output, $v$, is 1; otherwise, the bit is 0. Then $q$ and $h$ are updated and the procedure is iterated $q$ times.

```
Transformation := proc(int::integer,qq::integer,hh::integer)
  local NN, QQ, HH, LL, ii:
  NN:=int: QQ:=qq: HH:=hh: LL:=array(1..QQ):
  for ii from 1 to QQ do
    if NN >= binomial(QQ-ii,HH) then
      LL[ii]:=1: NN:=NN-binomial(QQ-ii,HH): HH:=HH-1:
    else LL[ii]:=0:
    end if:
  end do:
  return([seq(LL[ii],ii=1..QQ)]):
end:
ITransformation:=proc(pL::list,pQ::integer,pH::integer)
  local iN, ii, iQ, iH, lL:
  iN:=0: iQ:=pQ: iH:=pH: lL:=pL:
  if nops(lL) <> iQ or numboccur(lL,1) <> iH or
    apply(max,seq(lL[i],i=1..iQ)) > 1 or
    apply(min,seq(lL[i],i=1..iQ)) < 0 then
    return('Error'):
  end if:
  for ii from 1 to iQ do
    if lL[ii] = 1 then iN:=iN+binomial(iQ-ii,iH): iH:=iH-1: end if:
  end do:
  return(iN):
end:
```

Both procedures (preceded by the command `with(ListTools))`) can be saved in the file `ProcsMessT.mpl` to be loaded when it be necessary.

### 3.3 The Cryptosystem

In this subsection we present the Maple commands to generate the keys, encrypt and decrypt messages using the Chor-Rivest cryptosystem.

First, the receiver, $R$, restarts Maple to clear the memory and variables, loads the procedures to compute discrete logarithms, and determine the parameters of the system, which are saved in a file.

```
>restart: read "C:/CASA2006/ProcsDLP.mpl":
>q:=197; h:=24; G:=GF(q,h):
> f:=128+18*T+121*T^2+107*T^3+132*T^4+110*T^5+88*T^6+107*T^7+97*T^8+
159*T^9+131*T^10+95*T^11+73*T^12+168*T^13+130*T^14+171*T^15+40*T^16+
70*T^17+28*T^18+69*T^19+19*T^20+ 172*T^21+139*T^22+88*T^23+T^24:
>save q, h, "C:/CASA2006/Parameters.crc";
```

$R$ computes and saves his private key (noise, permutation and generator).

```
>noise:=rand(1..(q^h)-2): r:=noise();
with(combinat,randperm): pii:=map(x->x-1,randperm(q));
factrs:=convert(factorset((q^h)-1),list); nfactrs:=nops(factrs);
>rg:=rand(0..1): rg1:=rand(1..q-1):
expg:=[seq(rg(),i=1..3)]: coefg:=[seq(rg1(),i=1..3)]:
g:=coefg[1]*expg[1]*T^2+coefg[2]*expg[2]*T+coefg[3]*expg[3]:
for i from 1 to nfactrs do
 if Powmod(g,((q^h)-1)/factrs[i],f,T) mod q = 1 then
   print ("Please, choose a different value for g"):break:
 end if:
end do:
>save f, r, pii, g, "C:/CASA2006/PrivateKey.prk":
```

Finally, $R$ computes the DL $a_i$ and determines the elements of the public key (the knapsack), $c_i$, $i = 0, \ldots, 196$, which are saved and made public.

```
>for i from 0 to q-1 do
 PH:=PohligHellman(T+i,g): a[i]:=PH: b[i]:=a[pii[i+1]]:
 c[i]:=(b[i]+r) mod ((q^h)-1):
end do:
PublicKey:=[seq(c[i],i=0..q-1)]:
save PublicKey, "C:/CASA2006/PublicKey.puk":
```

Some partial results obtained in this example are:

$$
\begin{aligned}
r &= 59417724624775040217124916789513066102507311261970443 92, \\
\pi &= [81, 146, 160, \ldots 64, 111, 169], \\
g &= T + 8, \\
PublicKey &= \{76253232057300822171301002758956182898356846840230205 28, \\
&\quad \ldots, \\
&\quad 10145577571184720190572346163115496979342053810009806448\}.
\end{aligned}
$$

In the encryption process, the sender, $S$, restarts Maple and loads some procedures and the public key of the receiver.

```
>restart: with(StringTools): read "C:/CASA2006/ProcsMessT.mpl":
read "C:/CASA2006/Parameters.crc": read "C:/CASA2006/PublicKey.puk":
```

$S$ writes down the text of his message, computes the length and the number of blocks of the message, transforms and divides the message into blocks of the same length, and transforms it in blocks of length $q$ and weight $h$.

```
>Text:="Computer Algebra Systems and Their Applications, CASA'2006";
>ltext:=nops(Explode(Text)); lengt:=floor(log[2](binomial(q,h)));
lblock := floor(lengt/8); lmess:=lblock*ceil(ltext/lblock);
nblocks:=lmess/lblock; mascii:=map(Ord,Explode(Text)):
for i from 1 to lmess do
 if (i<=ltext) then masciic[i]:=mascii[i]: else masciic[i]:=0: end if:
end do:
>for i from 1 to nblocks do
 mess256[i]:=(convert([seq(masciic[j],j=(i-1)*lblock+1..i*lblock)],
   base,256,10)):
 messblock[i]:=sum(mess256[i]['j']*10^('j'-1),'j'=1..nops(mess256[i])):
 M[i]:=Transformation(messblock[i],q,h):
end do:
```

Finally, $S$ encrypts the message and saves it in a file to be sent to $R$.

```
>for i from 1 to nblocks do
 E[i] := sum(M[i]['j']*PublicKey['j'],'j'=1..q) mod ((q^h)-1):
end do:
EncryptedMessage:=[seq(E[j],j=1..nblocks)]:
save EncryptedMessage, "C:/CASA2006/EncryptedMessage.mes":
```

The encrypted message corresponding to the original one is:

$$\{10855811452733482391430481217627249373290172810833537037,\\ 3202585566164216102458406919307132366444257340079736316,\\ 11323380025066712989876028216475460861617165984770898397,\\ 211917943989265378754649813365968072528466655688931543,\\ 5620567254380493400571633868218080569011665365895027971\}.$$

To recover the original message, the receiver restarts Maple and loads some Maple procedures, the parameters of the system, his private key, and the encrypted message.

```
>restart: with(StringTools): with(numtheory):
read "C:/CASA2006/Parameters.crc": read "C:/CASA2006/PrivateKey.prk":
read "C:/CASA2006/EncryptedMessage.mes":
read "C:/CASA2006/ProcsMessT.mpl":
```

Then, $R$ computes the polynomials of degree $h$ over $GF(q^h)$, $Q_i(T)$, and determines the roots of $f(T) + Q_i(T)$, by factoring those polynomials. In the example, some partial results are:

$$Q_1(T) = 119T^{23} + T^{22} + 53T^{21} + \ldots + 84T^2 + 17T + 80,$$
$$f(T) + Q_1(T) = (T + 69)(T + 111)(T + 1) \cdots (T + 50)(T + 96)(T + 159),$$
$$\{\alpha_{\pi(i)}\} = \{128, 15, 16, 147, 36, 158, \ldots, 76, 79, 82, 86, 91, 101\}.$$

Later, $R$ applies the inverse permutation to recover the coordinates of the original message, obtains the partial messages and computes the integer numbers corresponding to each partial message of length $q$ and weight $h$.

```
>nblocks:=nops(EncryptedMessage); alias(alpha = RootOf(f)):
for j from 1 to nblocks do
 sprime[j]:=EncryptedMessage[j]-h*r mod ((q^h)-1):
 Q[j]:=Powmod(g,sprime[j],f,T) mod q:
end do:
>parc:=[]: R:=[]:
for j from 1 to nblocks do
 pol[j]:=Factor(f+Q[j],alpha) mod q: sol[j]:=[msolve(pol[j]=0,q)]:
 for i from 1 to h do
  parc:=[op(parc),-rhs(op(sol[j][i])) mod q]:
 end do:
 if member(0,parc) then parc:=Rotate(parc,1): end if:
 R:=[op(R),parc]: parc:=[]:
end do:
>Pos:=[]: psc:=[]:
for j from 1 to nblocks do
 for i from 1 to h do
  member(R[j][i],pii,'pos'): psc:=sort([op(psc),pos]):
 end do:
 Pos:=[op(Pos),psc]: psc:=[]:
end do:
>for j from 1 to nblocks do
 m[j]:=array(1..q):
 for i from 1 to h do m[j][Pos[j,i]]:=1: end do:
 for i from 1 to q do
  if m[j][i]<>1 then m[j][i]:=0: end if:
 end do:
end do:
>for j from 1 to nblocks do
 Me[j]:=ITransformation(convert(m[j],list), q, h):
end do:
Dm:=[seq(Me[j],j=1..nblocks)]:
```

Finally, the receiver decrypts the message and obtains the original message.

```
>for i from 1 to nblocks do
 Mes:=(convert(Dm[i],base,256)): Message[i]:=map(Char,Mes):
end do:
RecoveredText:=Implode(Flatten([seq(Message[i],i=1..nblocks)],1));
```

"*Computer Algebra Systems and Their Applications*, $CASA'2006$"

Most of the time of CPU is spent in the calculation of discrete logarithms. In this example, the process of generation of the parameters and keys has required 19.37 minutes on a Pentium 4 CPU, 3.00Ghz. Note that this work is executed only when the keys have to be stated. The CPU time used for encrypting and decrypting the message have been of 0.124 and 0.108 seconds, respectively.

## 4 Conclusions

We have presented a Maple implementation of the Chor-Rivest cryptosystem over finite fields by using commands and procedures. The main problems in this implementation are to determine discrete logarithms over a finite field $GF(q^h)$, where $q \approx 200$ and $h \approx 25$, and to use the arithmetic in these finite fields.

**Acknowledgements.** This work has been supported by Ministerio de Educación y Ciencia (Spain), under grants SEG2004-02418 and MTM2005-00173. Authors thank anonymous referees for their valuable suggestions.

## References

1. B. Chor, *Two issues in public key cryptography. RSA bit security and a new knapsack type system*, The MIT Press, Cambridge, MS, 1985.
2. B. Chor and R.L. Rivest, A knapsack-type public key cryptosystem based on arithmethic in finite fields, *IEEE Trans. Inform. Theory* **34**, 5 (1988), 901–909.
3. T.M. Cover, Enumerative source encoding, *IEEE Trans. Inform. Theory* **19** (1973), 73–77.
4. W. Diffie and M. E. Hellman, New directions in Cryptography, *IEEE Trans. Inform. Theory* **22** (1976), 644–654.
5. T. ElGamal, A public-key cryptosystem and a signature scheme based on discrete logarithm, *IEEE Trans. Inform. Theory* **31** (1985), 469–472.
6. A. Menezes, P. van Oorschot, and S. Vanstone, *Handbook of applied cryptography*, CRC Press, Boca Raton, FL, 1997.
7. R.A. Mollin, *An introduction to cryptography*, Chapman & Hall/CRC, Boca Raton, FL, 2001.
8. R.C. Pohlig and M.E. Hellman, An improved algorithm for computing logarithms over $GF(p)$ and its cryptographic significance, *IEEE Trans. Inform. Theory* **24** (1978), 106–110.
9. R.L. Rivest, A. Shamir, and L. Adleman, A method for obtaining digital signatures and public-key cryptosystems, *Commun. ACM* **21** (1978), 120–126.
10. A. Shamir, A polynomial-time algorithm for breaking the basic Merkle-Hellman cryptosystem, *IEEE Trans. Inform. Theory* **30**, 5 (1984), 699–704.
11. O. Schirokauer, D. Weber and T. Denny, Discrete logarithms: the effectiveness of the index calculus method, *Algorithmic Number Theory*, LNCS **1122** (1996), 337–361, Springer-Verlag, Berlin.
12. S. Vaudenay, Cryptanalysis of the Chor-Rivest cryptosystem, *J. Cryptology* **14** (2001), 87-100.

# Development of TRIP: Fast Sparse Multivariate Polynomial Multiplication Using Burst Tries

Mickaël Gastineau and Jacques Laskar

Paris Observatory - IMCCE - CNRS UMR8028
77, avenue Denfert Rochereau
75014 Paris
gastineau@imcce.fr, laskar@imcce.fr

**Abstract.** Flat vector representation of sparse multivariate polynomials is introduced in the computer algebra system TRIP with specific care to the cache memory. Burst tries are considered as an intermediate storage during the sparse multivariate polynomial multiplication by paying attention to the memory allocations. Timing and memory consumption are examined and compared with other recursive representations and other computer algebra systems.

## 1 Introduction

A large number of celestial mechanics applications, such as the classical expansions of the Kepler problem and several expansions of disturbing functions, requires to handle multivariate generalized power series. We develop a general symbolic and numeric computer algebra system TRIP [1] dedicated to celestial mechanics. It handles generalized power series (1) : Coefficients $C_{j,k}$ can be numerical coefficients (fixed and multi-precision rational number, double and quadruple precision floating-point) or rational functions. Complex expressions of the form $\exp^{\imath\lambda_m}$ are encoded as variables, so negative exponents are permitted.

$$S(X_1, ...X_n, \lambda_1, ..., \lambda_m) = \sum_{j_1 ... j_n, k_1 ... k_m \in \mathbb{Z}} C_{j,k} X_1^{j_1} .... X_n^{j_n} \exp^{\imath k_1 \lambda_1 + .... + k_m \lambda_m} \quad (1)$$

TRIP is tuned to compute large series with millions of terms depending on a large number of variables. Computation of large series needs fast in-memory data storage and fast algorithms. Computer algebra systems have different internal representations of polynomials and series. Internal representations should depend on the sparsity of polynomials in the studied problem. Some symmetries are present in celestial mechanics problems, such as d'Alembert relations in the planetary motion (see Laskar, [2]). This implies that sparse series appear during computations. Specialized computer algebra systems, called Poisson Series Processors, handle Poisson series as list of objects (e.g., see [3], [4], or [5]). General

V.N. Alexandrov et al. (Eds.): ICCS 2006, Part II, LNCS 3992, pp. 446–453, 2006.

computer algebra systems represent sparse polynomial as a recursive list or a recursive vector but they always have only one of these types to handle sparse polynomials. TRIP have three internal representations for sparse polynomials, discussed in Section 1.

An efficient multiplication algorithm is critical for most computations on series. Hida [6] demonstrates the effects of polynomials representation and blocking loops on the execution time to perform a sparse polynomial multiplication. We investigate here the usage of burst tries [7] as an intermediate representation during the sparse polynomial multiplication. Memory management for this intermediate representation is investigated in order to reduce memory consumption and improve scalability on SMP hardware. In section 6, we compare performance of polynomials multiplication in TRIP with other computer algebra systems.

## 2 Series Representations

TRIP supports multiple memory representations of a polynomial in order to handle it efficiently. Polynomials are always stored in an expanded form. In the two first representations described in (Laskar, [8]), multivariate polynomials are stored as a recursive list or a recursive vector in memory. The recursive list is a recursive single linked-list containing only the non-zero coefficients with the associated exponents. The recursive vector stores the minimal and maximal exponents and all coefficients between the minimal and maximal exponents are stored in this array. Depending on the polynomial, the recursive list representation consumes more or less memory than the recursive vector.

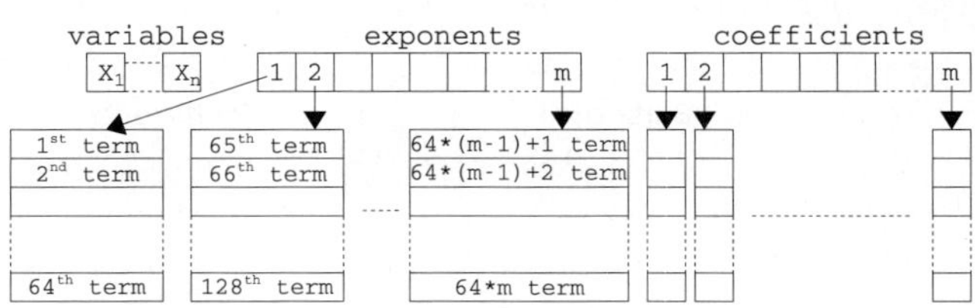


**Fig. 1.** Flat vector representation of a polynomial with variables $X_1, ...X_n$ containing $64m$ terms. Blocks of exponents are viewed in uncompressed form.

These two previous representations contain many indirections (pointers) to handle large sparse polynomials. Modern computers are less efficient with indirection because load instructions from memory or cache could take long time to complete. In these cases, a processor could stall until the data are available. Hida's experiments [6] to improve cache locality, such as blocking, show a speedup up to 47% on the sparse polynomial multiplication. It shows that flat structure are aware of cache locality.

We introduce a flat vector representation (Fig. 1) in TRIP : Exponents and coefficients of each term are stored in two arrays subdivided in short blocks. Terms are always sorted on exponents. A short block contains at most 64 terms. Operations on blocks of exponents and on coefficients, when fixed-size coefficients are used, take advantage of cache and SIMD pipelines available on some processors, such as Altivec or SSE extensions. Storing all exponents could consume much more memory than recursive representation. To reduce memory footprint of this representation, we compress blocks of exponents using shift and mask operations when exponents become unused. Compression factor is good because exponents have many times the same values for the most factorized variables in a block. Compression and decompression operations, when polynomial multiplications or additions are performed, show a negligible runtime overhead because it requires only integer arithmetics.

## 3 Multiplication Using Burst Tries

Multiplication of two sparse polynomials requires to use the naive algorithm of the multiplication because Karatsuba algorithm and FFT methods are only well adapted for dense polynomials. When polynomials are stored in the recursive list or vector, a simple naive recursive algorithm is performed. If the recursive vector or list is viewed as a tree, multiplication consists only as insertion of nodes in a tree. Using the flat vector representation, multiplication requires to sort terms on exponents. Hida [6] uses an hash table to sort terms during multiplication of its flat vectors in order to reduce execution time. To have good performance with hash table, it requires to find an efficient hashing function and the size of the hash table must be sufficiently large in memory. The massively parallel Deprit's algorithm [9] requires much more memory : Their first step needs memory for $2nm$ coefficients to perform the product of two series with $n$ and $m$ terms.

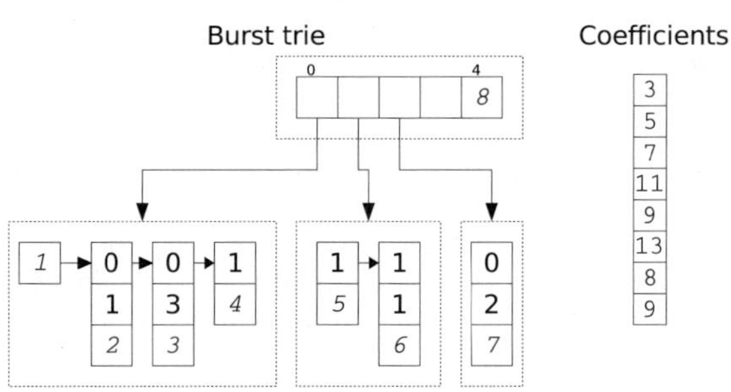


**Fig. 2.** Burst trie holding $3 + 5z + 7z^3 + 11y + 9zy + 13zyx + 8z^2x^2 + 9x^4$. Here, the burst threshold is set to the value 5 instead of 10. Italic numbers in the burst trie are indices of coefficients in the array of coefficients.

We prefer to find another method to sort terms during multiplication. Exponents in a term could be viewed as a string not of characters but of integers. Efficient data structure to sort strings of characters are binary search trees (BST), splay trees, judy array and tries. Heinz has developed a new data structure, burst trie [7], to sort large strings in memory. A burst trie consists in an access trie whose leaves are containers. A container, which could be a sorted list or BST, has few data. The access trie can be viewed as a recursive vector whose leaves are list or BST. The trie node contains only the minimal and maximal exponents. The container has all trailing exponents. The trailing zero exponents are not stored in the burst trie. Figure 2 shows an example of burst trie which contains the polynomial $3 + 5z + 7z^3 + 11y + 9zy + 13zyx + 8z^2x^2 + 9x^4$. A burst trie is built during the multiplication and a final step is performed to copy exponents into the vector. When terms are inserted into the burst trie and the number of data in a container reaches a threshold, the container requires to be converted in a trie node with new containers as leaves. This process is called *bursting*. In our implementation, containers are a sorted list and experiments show that bursting must be performed when containers have 10 elements. Coefficients associated with the terms in the burst trie are stored in large blocks and their indexes are stored with the $k^{th}$ exponent if we multiply 2 polynomials of $k$ variables. Figure 3 shows a stable speedup about to 60% over recursive vector when a flat vector representation and multiplication using burst trie as intermediate data structure are used.

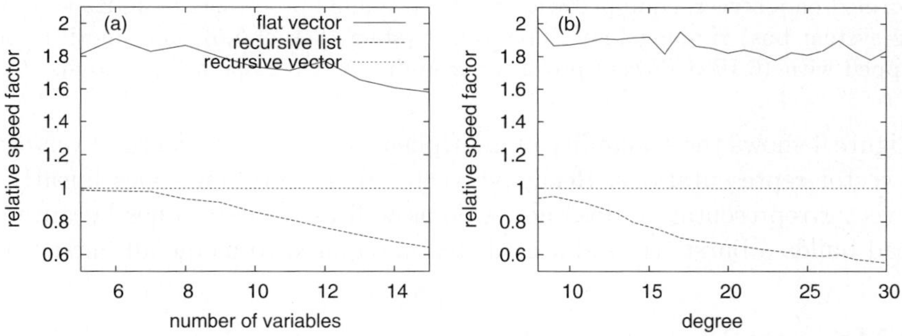


**Fig. 3.** (a). Relative computation time of $s_1 * (s_1 + 1)$ with $s1 = (1 + x_1 + x_2 + ... + x_p)^{15}$ using different representations versus computation time for the recursive vector. (b). Relative computation time of $s_1 * (s_1 + 1)$ with $s1 = (1 + x + y + z)^p$ using different representations versus computation time for the recursive vector.

## 4 Parallel Multiplication on SMP Hardware

We decide to parallelize the multiplication operation because it takes most of the time in our computations. Currently, Symmetric Multi-Processing or Multi-Threading Systems are becoming more and more popular and cheaper. TRIP uses threads to implement parallelism in shared memory multiprocessor architectures. All threads within a process share the same address space. Communication between threads is more efficient and easier to use than communication between processes. Threads must take care when they write to shared data : it requires synchronization mechanisms. Thread mechanisms could be implemented using OpenMP API or POSIX Threads API. For POSIX Threads, we adopt a manager-worker queue model in TRIP : a single thread, the manager assigns work to other threads, the workers. The manager thread uses a dynamic scheduling to split the multiplication operation. All workers are created at the beginning and accept work from a common queue. The number of workers is dynamic : the user can fix the number of workers by setting a global variable in its interactive TRIP session.

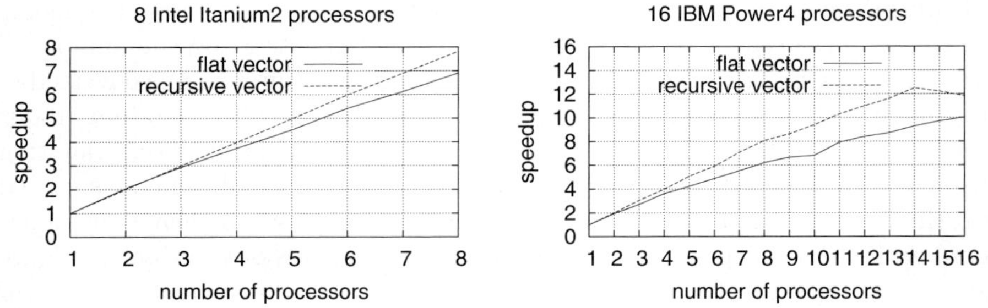


**Fig. 4.** Speedup of computation $(1+x+3y+5z+7t)^{23}*(1+11x+13y+17z+23t)^{23}$ using different representations with a manager-worker queue model. Computations are performed on a server, equipped with 8 Intel Itanium2 processors (1.5GHz with a 400 MHz system bus) running Linux operating system (kernel 2.6), and another server, equipped with 16 IBM Power4 processors running AIX 5.2 operating system.

Figure 4 shows the scalability of multiplication using the recursive vector and flat vector representations. Recursive vector representation scales linearly but flat vector representation does not scale as well as recursive ones because each thread builds a burst trie and a final step is required to merge all burst tries.

## 5 Memory Management

Most computer algebra systems use a garbage collector to perform memory operations instead of explicit memory management. TRIP have two objectives for memory management : reduce memory consumption and have efficient performance. Garbage collection generally results in poorer performance or higher space consumption. Garbage collectors require much more memory to be faster than explicit memory management (Zorn, [10] and Hertz, [11]). We decide to use explicit memory management (allocation with system call `malloc` and deallocation with `free`).

The frequency of allocation and the size of allocated objects depend on the representations of series used in TRIP. Flat vectors allocate large blocks of memory but it doesn't happen very often. This representation uses operating system system calls (`malloc` and `free`). A recursive list representation requires many allocations of very small blocks of memory for the elements in the list. These blocks always have the same size. A recursive vector representation often allocates blocks of different sizes but these sizes are smaller than flat vectors ones.

### 5.1 Memory Management for a Recursive List Representation

Most explicit memory managers add an overhead to each allocated block of memory. For example, Doug Lea's Malloc [12], used in the GNU C library, adds size and status information before the available user space data returned by the system call `malloc`. In addition, it aligns data on 8 bytes even on 32-bit operating systems. So each element in the list of the recursive list representation will have an unnecessary overhead. We designed a custom memory manager for the single threaded version of TRIP, called **rlalloc** in this paper, in order to reduce the overhead in the allocation of elements of lists. The design of an efficient custom memory manager is an hard task as shown in the Berger's review [13]. Our memory manager splits a large memory chunk requested from the operating system in individual blocks of the size of the element of the recursive list representation. The only overhead is a single linked-list which contains the address of the large memory chunk. Free blocks are stored in a single linked-list: The pointer of this list is stored in the user data space. So it doesn't consume extra memory. The free operation sums up to add the freed block in this single linked-list.

**Table 1.** Computation time and memory consumption of $s_1 * (s_1 + 1)$ with $s1 = (1+x+y+z+t)^{25}$ using different memory managers. Polynomial is stored as a recursive list. Measured time (sum of user CPU time and system CPU time) is expressed in seconds and space used (size of the trip process in memory) in kilobytes.

| Machine | time (s) | | memory (Kb) | |
|---|---|---|---|---|
| | system | rlalloc | system | rlalloc |
| Intel Xeon 3.06Ghz - Linux 32-bits - glibc 2.3.4 | 108 | 83 | 22360 | 19688 |
| Intel Xeon 3.6Ghz - Linux 64-bits - glibc 2.3.4 | 87 | 63 | 44656 | 33920 |
| Apple G5 2.0Ghz - Mac os X 10.4 64-bits | 225 | 186 | 28748 | 22504 |

Table 1 presents timing and memory consumption with our custom memory manager and the system memory manager (memory manager located in the C library of the system). Computations are performed on several operating system and hardware with a single cpu to see the effect of the system memory manager on time and memory used. The numerical coefficients in the polynomial $s1$ are double-precision floats in order to reduce computation time in coefficient arithmetic. Our memory manager reduces memory footprint about 25%. We observe increased speed up to 27% with our allocator. Most system memory managers, like "Doug Lea's Malloc" on Linux, use a best-fit algorithm even for small allocated blocks which produces many split and merge operations on their internal chunks. Furthermore, this memory manager is used for fixed-size numerical coefficient, like quadruple-precision float or arbitrary size rational numbers.

### 5.2 Memory Management for the Burst Tries

When multiplication of two sparse polynomials using burst tries is performed, the containers in burst tries have many small lists. The memory manager described in the previous paragraph is used to allocate memory for this list in a multi-thread context with no lock mechanism because burst tries are always allocated and freed by the same thread. This provides better performance on SMP hardware.

## 6 Benchmarks with Other Computer Algebra Systems

Previous benchmarks on sparse polynomial multiplication, such as Fateman's review [14], show that PARI/GP and Singular have very good performance. So we compare the speed and memory consumption of TRIP, PARI/GP [15] and Singular [16] for sparse polynomial multiplication. This benchmark is performed on a single-core processor server to disable parallelization multiplication in TRIP. As the benchmark depends on coefficient arithmetic, the same version (4.1.4) of GNU Multiple Precision Arithmetic Library [17] is used for all computations. The current stable version 2.1.7 of PARI/GP supports only a native kernel for multi-precision kernel, a comparison with the current development version of this library with a GMP kernel is performed in order to have the same multi-precision kernel. Table 2 presents timing results and memory used for computing $s_1 * (s_1 + 1)$ with $s1 = (1 + x + y + z + t)^{20}$ and $s_2 * (s_2 + 1)$ with $s2 = (1/3 * x + 3/5 * y + 5/7 * z + 7/11 * t)^{20}$. Our recursive and flat vector have better execution time than PARI/GP and Singular but the recursive representation consume more memory than Singular in some cases.

**Table 2.** Comparison of time and memory consumption with other computer algebra systems. Measured time (sum of user and system CPU time) is expressed in seconds and space used in kilobytes. Computations are performed on an Intel Xeon processor with Linux operating system running at 3.06Ghz with a 533 MHz system bus.

| Computer algebra systems | $s_1 * (s_1 + 1)$ | | $s_2 * (s_2 + 1)$ | |
|---|---|---|---|---|
| | time (s) | memory (Kb) | time (s) | memory (Kb) |
| PARI/GP 2.1.7 | 82.1 | N/A | 225.6 | N/A |
| PARI/GP 2.2.11 (GMP kernel) | 67.8 | N/A | 177.6 | N/A |
| Singular 3.0.1 | 101.5 | 8383.5 | 42.8 | 982.9 |
| TRIP 0.98 (recursive list) | 54.6 | 7138.5 | 15.5 | 1370.0 |
| TRIP 0.98 (recursive vector) | 42.9 | 5953.7 | 16.5 | 1831.4 |
| TRIP 0.98 (flat vector) | 24.1 | 3609.1 | 13.3 | 721.2 |

## 7 Conclusion

Flat vector representation of sparse multivariate polynomials improves the execution time and memory usage when multiplications using burst tries are performed. The multiplication using burst tries of sparse polynomials is efficient for a large number of variables and also for a large degree and it scales almost linearly on SMP architectures. Internal representations of sparse polynomials in the computer algebra system TRIP are sufficiently tuned to compute power series and polynomials up to a high degree with a large number of variables.

**Acknowledgement.** Part of the computations were made at IDRIS-CNRS.

## References


1. Gastineau, M., Laskar, J.: TRIP 0.98. Manuel de référence TRIP, IMCCE, Paris Observatory (2005) http://www.imcce.fr/Equipes/ASD/trip/trip.html.
2. Laskar, J.: Accurate methods in general planetary theory. Astronomy Astrophysics **144** (1985) 133–146
3. San-Juan, F., Abad, A.: Algebraic and symbolic manipulation of poisson series. Journal of Symbolic Computation **32** (2001) 565–572
4. Ivanova, T.: A New Echeloned Poisson Series Processor (EPSP). Celestial Mechanics and Dynamical Astronomy **80** (2001) 167–176
5. Henrard, J.: A Survey of Poisson Series Processors. Celestial Mechanics and Dynamical Astronomy **45** (1988) 245–253
6. Hida, Y.: Data structures and cache behavior of sparse polynomial multiplication (2002) Class project CS282.
7. Heinz, S., Zobel, J., Williams, H.E.: Burst tries: a fast, efficient data structure for string keys. ACM Trans. Inf. Syst. **20** (2002) 192–223
8. Laskar, J.: Manipulation des séries. In: Modern Methods in Celestial Mechanics, Comptes Rendus de la 13ieme Ecole Printemps d'Astrophysique de Goutelas (France), 24-29 Avril, 1989, Gif-sur-Yvette, Editions Frontieres (1990) 63–88
9. Deprit, T., Deprit, E.: Processing Poisson series in parallel. Journal of Symbolic Computation **10** (1990) 179–201
10. Zorn, B.G.: The measured cost of conservative garbage collection. Software - Practice and Experience **23** (1993) 733–756
11. Hertz, M., Berger, E.D.: Quantifying the performance of garbage collection vs. explicit memory management. In: OOPSLA '05: Proceedings of the 20th annual ACM SIGPLAN conference on Object oriented programming systems languages and applications, New York, NY, USA, ACM Press (2005) 313–326
12. Lea, D.: A memory allocator (1998) http://g.oswego.edu/dl/html/malloc.html.
13. Berger, E.D., Zorn, B.G., McKinley, K.S.: Reconsidering custom memory allocation. SIGPLAN Not. **37** (2002) 1–12
14. Fateman, R.: Comparing the speed of programs for sparse polynomial multiplication. SIGSAM Bull. **37** (2003) 4–15
15. The PARI Group Bordeaux: PARI/GP, version 2.1.5. (2004) available from http://pari.math.u-bordeaux.fr/.
16. Greuel, G.M., Pfister, G., Schönemann, H.: SINGULAR 3.0. A Computer Algebra System for Polynomial Computations, Centre for Computer Algebra, University of Kaiserslautern (2005) http://www.singular.uni-kl.de.
17. Granlund, T.: GNU multiple precision arithmetic library 4.1.4 (2004) http://swox.com/gmp/.

# A Symbolic Approach to Quantum Computation Simulation*

António Pereira and Rosália Rodrigues

Department of Mathematics,
University of Aveiro -Portugal,
{antoniop, rosalia}@mat.ua.pt

**Abstract.** This paper presents SQCS: a *Mathematica* package for the symbolic simulation of quantum computation. It can be a useful tool in the development and testing of quantum algorithms as well as in the teaching and learning of quantum computation.

The strength of this symbolic approach is here illustrated with Grover's quantum algorithm for database search.

## 1 Introduction

Simulating quantum computations on a classical computer is, in general, a hard task. Every classical simulator of the dynamics of a quantum computer faces two problems: the representation of quantum states and its unitary evolution. (See, for example, Nielsen and Chuang [1] or Kitaev [2] for introductory readings on the basic concepts of quantum computation.)

In fact, a pure state of a composite closed quantum system with $n$ $d$-level qudits can be seen as a unit vector on a Hilbert space of dimension $d^n$, i.e., a normalized vector with $d^n$ complex components. Also every unitary operator defined in a Hilbert space of dimension $d^n$ can be described by a $d^n \times d^n$ complex unitary matrix. Working directly with these vectors and matrices becomes unfeasible even when the number of qudits is small.

Most of the existing quantum computation simulators follow the linear algebra matrix approach, ranging from simple applets that handle computations with a maximum of 4 qudits to heavy clusters of computers capable of solving problems with a maximum of 30 qubits.

The complexity of simulating the state evolution of a quantum computer, for a given quantum algorithm, is strongly related with the number of entangled qudits in the system. We expect a quantum algorithm to admit an efficient simulation when, along the simulation process, only small subsystems of the whole system become entangled.

* This work was partially supported by the R&D unit Center for Research in Optimization and Control (CEOC) of the University of Aveiro – Portugal.

V.N. Alexandrov et al. (Eds.): ICCS 2006, Part II, LNCS 3992, pp. 454–461, 2006.

## 2 The Symbolic Quantum Computation Simulator

On trying to build a simulator over an environment like *Mathematica*[3], the first observation to take into account is that not the content of a quantum object need to be stored, but only its identifier. To our knowledge, even the existing quantum computation applications developed with *Mathematica* fail to take advantage of this fact.

For instance, any basis state of a qudit on a $d$-dimensional Hilbert space need not be represented as a full length $d$ list, not even as a packed array, but simply as a symbolic labelled object with head `ket`. Subsequent operations on such objects should work only symbolically. Also every unitary operator can be represented by a list of rules that define its symbolic action on quantum states.

The basic objects and operators defined in SQCS are introduced next. Most have external notations for the *Mahematica* frontend associated with the internal representations.

### 2.1 Kets

For $d \geq 2$ let $\mathbb{Z}_d$ denote the set $\{0, \ldots, d-1\}$. A *qudit* is a quantum system whose state is a unit vector in a $d$-dimensional Hilbert space. We assume that the underlying Hilbert space is always $\mathcal{H}_d = \mathbb{C}^d$. It has become common practice to identify the canonical computational basis of $\mathcal{H}_d$ with the set $\{|k\rangle, k \in \mathbb{Z}_d\}$, in Dirac notation.

The general state of a qudit is $|\psi\rangle = \sum_{k \in \mathbb{Z}_d} a_k |k\rangle$, where $a_k \in \mathbb{C}$, for $k \in \mathbb{Z}_d$ and $\sum_{k \in \mathbb{Z}_d} |a_k|^2 = 1$. The case $d = 2$ gives the usual definition of a *qubit* with general state, $|\psi\rangle = a|0\rangle + b|1\rangle$, $|a|^2 + |b|^2 = 1$, $a, b \in \mathbb{C}$.

In SQCS each basis state of a Hilbert space is represented by a symbolic object with head `ket`. Figure 1 shows some examples and basic properties of `ket` objects.

```
<< SQCS`

{ket[0], ket[1], ket[0, 3], ket[1, 3], ket[2, 3]}

{|0⟩, |1⟩, |0₃⟩, |1₃⟩, |2₃⟩}

α |ψ⟩ + α |ϕ⟩ // Simplify

α (|ϕ⟩ + |ψ⟩)

α |ϕ⟩ + β |ϕ⟩ // Simplify

(α + β) |ϕ⟩

|ϕ⟩ - |ϕ⟩

0
```

```
<< SQCS`

|ψ₀⟩ = 1 / Sqrt[2] |0⟩ + I / Sqrt[2] |1⟩

|0⟩/√2 + i |1⟩/√2

|ψ₁⟩ = I |ψ₀⟩ + 1 / Sqrt[2] |1⟩

|1⟩/√2 + i (|0⟩/√2 + i |1⟩/√2)

|ψ₁⟩ // Simplify

i |0⟩/√2
```

**Fig. 1.** Some algebraic properties of ket objects

### 2.2 Bras

The set of linear functionals on a Hilbert space $\mathcal{H}$ is also an Hilbert space, called the dual space of $\mathcal{H}$ and denoted by $\mathcal{H}^\dagger$. According to Riesz theorem, the inner product in $\mathcal{H}$ induces a one to one correspondence between the elements of $\mathcal{H}$ and $\mathcal{H}^\dagger$. In Dirac notation, the dual of the ket state $|\psi\rangle \in \mathcal{H}$ is the *bra* state $\langle\psi| = |\psi\rangle^\dagger \in \mathcal{H}^\dagger$.

In SQCS the dual of a `ket` object is another object with head `bra`. Figure 2 displays some basic properties of `bra` objects and of the `dagger` function.

### 2.3 BraKets and the Inner Product

Considerer a `bra` object $\langle\psi| \in \mathcal{H}^\dagger$. Since $\langle\psi|$ is a linear functional in $\mathcal{H}$, the action of $\langle\psi|$ on a `ket` object $|\phi\rangle \in \mathcal{H}$, $\langle\psi|\,(|\phi\rangle)$ gives a complex number, denoted by the `braKet` $\langle\psi|\phi\rangle$.

Objects of type `braKet` can also have a different meaning. Recall that the Hilbert space $\mathcal{H}$ is equipped with a inner product, assumed to be conjugate-linear in the first argument and linear in the second. If we denote the inner product of $|\psi\rangle$ and $|\phi\rangle$ by $\langle\psi|\cdot|\phi\rangle$, then $\langle\psi|\cdot|\phi\rangle = \langle\psi|\phi\rangle = \langle\psi|\,(|\phi\rangle)$.

Some properties of the inner product of kets are shown in fig. 3.

### 2.4 The Kronecker Product

The underlying Hilbert space of a composite system, say a system composed of two qubits, is the tensor product space of the underlying Hilbert spaces of the

```
<< SQCS`

{bra[0], bra[1], bra[0, 3], bra[1, 3], bra[2, 3]}
```

$\{\langle 0|, \langle 1|, \langle 0_3|, \langle 1_3|, \langle 2_3|\}$

```
(α|ψ⟩ + β|ϕ⟩)†
```

$\langle\psi|\,\alpha^* + \langle\phi|\,\beta^*$

```
<< SQCS`

|ψ0⟩ = 1/Sqrt[2] |0⟩ + I/Sqrt[2] |1⟩
```

$\frac{|0\rangle}{\sqrt{2}} + \frac{i\,|1\rangle}{\sqrt{2}}$

```
|ψ0⟩†
```

$\frac{\langle 0|}{\sqrt{2}} - \frac{i\,\langle 1|}{\sqrt{2}}$

**Fig. 2.** Some algebraic properties of bra objects

```
<< SQCS`

{⟨0|1⟩, ⟨1|1⟩, ⟨1₃|2₃⟩}
```

{0, 1, 0}

```
{⟨0|·|1⟩, ⟨1|·|1⟩, ⟨1₃|·|2₃⟩}
```

{0, 1, 0}

```
<< SQCS`

(α⟨ψ| + β⟨ϕ|)·|η⟩ //. expandInnerProduct
```

$\langle\psi|\eta\rangle\,\alpha^* + \langle\phi|\eta\rangle\,\beta^*$

```
⟨η|·(α|ϕ⟩ + β|ψ⟩) //. expandInnerProduct
```

$\alpha\,\langle\eta|\phi\rangle + \beta\,\langle\eta|\psi\rangle$

**Fig. 3.** Some algebraic properties of the inner product of kets

```
<< SQCS`

| ϕ⟩ ⊗ (| ψ⟩ ⊗ | η⟩) === (| ϕ⟩ ⊗ | ψ⟩) ⊗ | η⟩

True

(α | ϕ⟩) ⊗ | ψ⟩

α | ϕ⟩ ⊗ | ψ⟩

| ϕ⟩ ⊗ (α | ψ⟩)

α | ϕ⟩ ⊗ | ψ⟩
```

```
<< SQCS`

(α | ϕ⟩ + β | ψ⟩) ⊗ | η⟩ //. expandKron

α | ϕ⟩ ⊗ | η⟩ + β | ψ⟩ ⊗ | η⟩

| η⟩ ⊗ (α | ϕ⟩ + β | ψ⟩) //. expandKron

α | η⟩ ⊗ | ϕ⟩ + β | η⟩ ⊗ | ψ⟩

(| 0⟩ + | 1⟩)^⊗2 /. expandPow //. expandKron

| 0⟩ ⊗ | 0⟩ + | 0⟩ ⊗ | 1⟩ + | 1⟩ ⊗ | 0⟩ + | 1⟩ ⊗ | 1⟩
```

**Fig. 4.** Some algebraic properties of the tensor product

subsystems, $\mathcal{H}_2^{\otimes 2} = \mathcal{H}_2 \otimes \mathcal{H}_2$. Thus the state of a quantum system with $n$ qudits is a unit vector of ${\mathcal{H}_d}^{\otimes n}$. This space has a computational basis with $d^n$ states denoted by $|x_{n-1}\rangle \otimes \cdots \otimes |x_1\rangle \otimes |x_0\rangle$, where $x_j \in \mathbb{Z}_d$, $j = 0, 1, \ldots, n-1$.

Each basis state in ${\mathcal{H}_d}^{\otimes n}$ naturally associates with the radix $d$ representation of the integer $\sum_{j=0}^{n-1} x_j d^j \in \mathbb{Z}_{d^n}$, hence the following notations are equivalent: $|x_{n-1}\rangle \otimes \cdots \otimes |x_1\rangle \otimes |x_0\rangle \equiv |x_{n-1} \cdots x_1 x_0\rangle \equiv \left|\sum_{j=0}^{n-1} x_j d^j\right\rangle$.

A state $|\psi\rangle \in {\mathcal{H}_d}^{\otimes n}$ is called a *product state* if it can be decomposed as $|\psi\rangle = |\psi_{n-1}\rangle \otimes \cdots \otimes |\psi_1\rangle \otimes |\psi_0\rangle$, for some sequence of states $|\psi_i\rangle \in \mathcal{H}_d$, $i = 0, \ldots, n-1$. If no such sequence exists, $|\psi\rangle$ is said to be an *entangled state*.

The tensor product, also known as the Kronecker product, is associative, noncommutative and distributive with respect to linear combinations. Figure 4 displays some examples.

### 2.5 Operators

In SQCS every linear operator is represented by an object `op[name_,n_,f_]`, where `name` is a label for the operator, `n` is the number of qudits on which the operator acts, and `f` is the function that defines the action of the operator on the basis qudits.

The efficient simulation of the action of a unitary operator on a given quantum state is directly related to the possibility of decomposing it as a tensor product of operators acting on a small number of qudits. Thus a major concern when defining a new operator in SQCS is to represent it as factored as possible.

Several operators are already available in SQCS. In particular, the identity operator $\mathrm{I}_n$, where $n$ is the number of qudits on which the operator acts (thus diverging from the usual notation where subscripts represent the dimension of the underlying Hilbert space of the operator). Other examples of provided operators are:

**The Hadamard Operator.** One of the fundamental operators in quantum computation for creating superpositions of states is the Hadamard operator, H.

It acts on one qubit by $\mathrm{H}\,|i\rangle = \frac{1}{\sqrt{2}}\left(|0\rangle + (-1)^i\,|1\rangle\right)$, for $i = 0, 1$. Figure 5 illustrates the fact that H is its own inverse.

```
<< SQCS`
numericOff
H·|0⟩
(|0⟩ + |1⟩)/√2
H·|1⟩
(|0⟩ - |1⟩)/√2
```

```
<< SQCS`
numericOff
H·(|0⟩ + |1⟩)/√2 //. expandLinearInside // Simplify
|0⟩
H·H
I₁
```

**Fig. 5.** The Hadamard operator in action

```
<< SQCS`
numericOff;
H^⊗4 · |0⟩^⊗4
1/4 (|0⟩ + |1⟩)^⊗4
H^⊗3 · |0⟩⊗|1⟩⊗|0⟩
(|0⟩ + |1⟩)⊗(|0⟩ - |1⟩)⊗(|0⟩ + |1⟩) / (2 √2)
```

```
<< SQCS`
numericOff;
H^⊗2 · |0⟩⊗|1⟩ /. expandKron
1/2 (|0⟩⊗|0⟩ - |0⟩⊗|1⟩ + |1⟩⊗|0⟩ - |1⟩⊗|1⟩)
H^⊗5 · H^⊗5
I₅
```

**Fig. 6.** The Walsh-Hadamard operator in action

**The Walsh-Hadamard Operator.** Let $\mathrm{H}^{\otimes n}$ denote the Walsh-Hadamard operator on $n$ qubits. It is defined by

$$\mathrm{H}^{\otimes n} \left|i\right\rangle = \frac{1}{2^{n/2}} \sum_{j=0}^{2^n-1} (-1)^{i \cdot j} \left|j\right\rangle , \tag{1}$$

for each basis state $|i\rangle$ of the Hilbert space $\mathcal{H} = \mathcal{H}_2{}^{\otimes n}$ and where $i \cdot j$ denotes the bitwise inner product of $i$ and $j$ modulo 2.

Using (1) directly in *Mathematica* would require exponentially increasing memory resources. This can be avoided by observing that, given $|\psi\rangle = \otimes_{i=0}^{n-1} |\psi_i\rangle$, where each $|\psi_i\rangle \in \mathcal{H}_2$, the state $\mathrm{H}^{\otimes n} |\psi\rangle$ can be computed in linear time by $\mathrm{H}^{\otimes n} |\psi\rangle = \bigotimes_{i=0}^{n-1} \mathrm{H} |\psi_i\rangle$.

Some properties of the Walsh-Hadamard operator are displayed in fig. 6. Its implementation is based on the general concept of tensor powers of quantum objects, which is fully supported in SQCS. Note that the internal evaluation is a purely symbolic process, hence taking no significant memory resources.

**The Outer Product.** Consider the Hilbert space $\mathcal{H} = \mathcal{H}_d{}^{\otimes n}$ and two basis states $|i\rangle, |j\rangle \in \mathcal{H}$. The outer product operator $|i\rangle \langle j|$ is defined by the action on each basis state $|k\rangle$, $k = 0, \ldots, d^n - 1$, by $(|i\rangle \langle j|) |k\rangle = \langle j|k\rangle |i\rangle$.

```
<< SQCS`

(|ψ⟩ · ⟨ϕ |) · |η⟩
⟨ϕ | η⟩ | ψ⟩

(| 0⟩ · ⟨1 |)^⊗4 · (| 0⟩⊗| 0⟩⊗| 1⟩⊗| 1⟩)
0

(| 0⟩ · ⟨1 |)^⊗4 · (| 1⟩⊗| 1⟩⊗| 1⟩⊗| 1⟩)
| 0⟩⊗| 0⟩⊗| 0⟩⊗| 0⟩
```

```
<< SQCS`

(Sum[| i_4⟩ · ⟨i_4 |, {i, 0, 3}]) · | 0_4⟩ //. expandLinearOutside
| 0_4⟩
```

**Fig. 7.** The outer product operator in action

Given a state $|\psi\rangle = \otimes_{s=0}^{n-1}|\psi_s\rangle$, where $|\psi_s\rangle \in \mathcal{H}_2$, the outer-product action $(|i\rangle\langle j|)|\psi\rangle$ can be computed in time $\mathcal{O}(n)$ by first computing radix-2 decompositions of the integers $i$ and $j$, i.e., $|i\rangle = \otimes_{s=0}^{n-1}|i_s\rangle$, $\langle j| = \otimes_{s=0}^{n-1}\langle j_s|$, and then use the identity $(|i\rangle\langle j|)|\psi\rangle = \bigotimes_{s=0}^{n-1}(|i_s\rangle\langle j_s|)|\psi_s\rangle$.

An important property involving the outer product operator is the completeness relation $\sum_{i=0}^{d^n-1}|i\rangle\langle i| = \mathrm{I}_n$. Figure 7 exhibits several properties of this operator.

## 3 Simulation of Grover's Algorithm in SQCS

Consider the problem of searching through an unstructured database. Let $N = 2^n$ be the number of records in a database labelled by the integers $0, 1, \ldots, N-1$. Let $x_0$ be the label of the only record in the database to be searched for.

Any classical algorithm solving this problem must take, on average, $\mathcal{O}(N)$ steps to find $x_0$. However, Lov Grover [4] devised a quantum algorithm that finds the unknown label, with probability not less than $1 - \frac{1}{N}$, in just $\mathcal{O}\left(\sqrt{N}\right)$ steps.

Grover's algorithm was simulated for databases sizes varying from $2^2$ to $2^{32}$. All the simulations were performed on a Pentium IV, 3.0 GHz processor, 1 GB of RAM, personal computer.

In all the simulations the final measurement step was ignored, in order to overcome the problem of generating a probability distribution over an exponential number of values and then sampling from it (although an efficient implementation of the measurement operator is possible due to the simple structure of the representation of the final state in Grover's algorithm).

Next we briefly review Grover's algorithm. For a more detailed analysis and generalizations see for example [5].

### 3.1 A Glimpse of Grover's Algorithm

We are given a blackbox function, which we can query as many times as we like, defined by

$$f(x) = \begin{cases} 1 \text{ if } x = x_0 \\ 0 \text{ otherwise} \end{cases} . \tag{2}$$

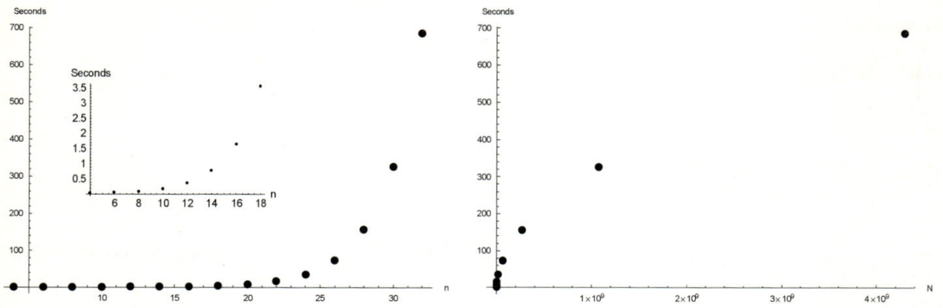


**Fig. 8.** Simulation times for the quantum database case

This function, for which no knowledge of the internal computational work is assumed, will be called the oracle.

In the quantum perspective, the oracle is given as a blackbox unitary operator $U_f$ defined by $|x\rangle \otimes |y\rangle \mapsto |x\rangle \otimes |f(x) \oplus y\rangle$, where $\oplus$ denotes the addition modulo 2. $U_f$ is equivalent to the unitary operator $I_{|x_0\rangle} = I - 2\,|x_0\rangle\,\langle x_0|$.

Grover's algorithm is based on the operator $Q = -\,H^{\otimes n}\, I_{|0\rangle}\, H^{\otimes n}\, I_{|x_0\rangle}$ which performs a rotation of the state of the system towards the desired state $|x_0\rangle$.

**Algorithm[Grover].**

1. Initialize the state of the system to $|\psi_0\rangle = |0\rangle^{\otimes n}$.
2. Apply the Walsh-Hadamard operator, $|\psi_1\rangle = H^{\otimes n}\,|\psi_0\rangle$.
3. For $i = 1$ to $k \approx \lfloor \frac{\pi}{4}\sqrt{N} \rfloor$ do $|\psi_{i+1}\rangle = Q\,|\psi_i\rangle$.
4. Measure $|\psi_{k+1}\rangle$ with respect to the basis $\{|0\rangle, |1\rangle, \ldots, |N-1\rangle\}$.

### 3.2 Simulations for a Quantum Database

In this case it is assumed that the oracle is given by the operator $I_{|x_0\rangle}$. The index $x_0$ is randomly picked up from $\{0, \ldots, 2^n - 1\}$ and no database is in fact generated. When the goal is to test the behaviour and correctness of Grover's algorithm this layout seems adequate.

**Simulation Results.** Grover's algorithm was simulated over instances of size $n \in \{2, \ldots, 32\}$ qubits (searching on a database with $k = 2^n$ elements). We were able to simulate Grover's algorithm for $n = 30$, using a common PC, in about 5 minutes. As expected, the running times display an exponential growth when taken as a function of the number of qubits, whereas when taken as a function of the database size they follow the square-root curve, as depicted in fig. 8.

### 3.3 Simulations for a Classical Database

In this case a classical database is first generated and the oracle operator defined by $\text{Oracle} \cdot |x\rangle = (-1)^{f(x)}\,|x\rangle$ where $f(x)$ is the classical oracle call function (2). When applying Grover's operator, we had to fully expand some intermediary states since no useful tensor product decomposition of the oracle operator was found.

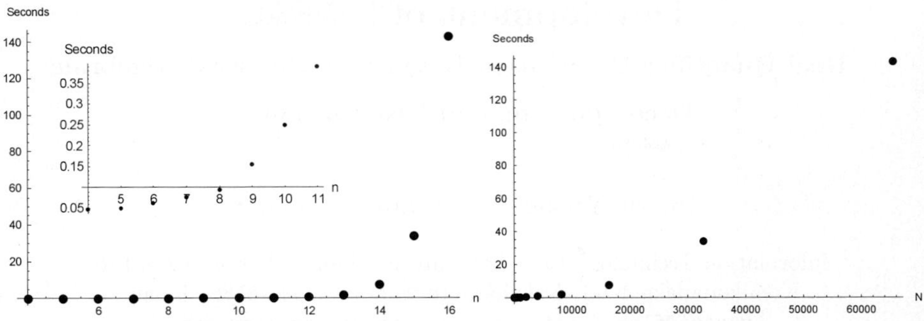


**Fig. 9.** Simulation times for the classical database case

**Simulation Results.** Grover's algorithm was simulated over instances of size $n \in \{4, 5, \ldots, 16\}$ qubits. Figure 9 displays the simulation running times as a function of the number of qubits, as well as a function of the database size. Both graphics follow an exponential curve, explained by the inherent inefficiency of the oracle calls in the algorithm.

## 4 Conclusions

*Mathematica* can provide a powerful environment for the implementation of a quantum computer simulator, as long as we are able to keep, as much as possible, the implementation at the symbolic level.

A new symbolic approach for the simulation of quantum computations has been proposed. SQCS is an ongoing project still lacking many features. Measuring operators, an improved quantum register address manager and simulations of more quantum algorithms are currently being developed.

## References

1. Nielsen, M.A., Chuang, I.L.: Quantum Computation and Quantum Information. Cambrige University Press (2000)
2. Kitaev, A.Y., Shen, A., Vyalyi, M.: Classical and quantum computation. Volume 47 of Graduate Studies in Mathematics. American Mathematical Society (2002)
3. Wolfram, S.: TheMathematica Book, Fifth Edition. Wolfram Media, Inc. (2003)
4. Grover, L.K.: A fast quantum mechanical algorithm for database search. In: Proc. 28th Annual ACM Symposium on the Theory of Computing. (1996) 212–219
5. Biham, E., Biham, O., Biron, D., Grassl, M., Lidar, D.A.: Grover's quantum search algorithm for an arbitrary initial amplitude distribution. Physical Review A **60** (1999) 27–42

# Development of SyNRAC

## —Real Quantifier Elimination Based on Cylindrical Algebraic Decomposition and Visialization—

Hitoshi Yanami[1,2] and Hirokazu Anai[1,2]

[1] Information Technology Core Laboratories, Fujitsu Laboratories Ltd., Kamikodanaka 4-1-1, Nakahara-ku, Kawasaki 211-8588, Japan
yanami@flab.fujitsu.co.jp, anai@jp.fujitsu.com
[2] CREST, Japan Science and Technology Agency, Kawaguchi Center Building, 4-1-8, Honcho, Kawaguchi 332-0012, Japan

**Abstract.** We present newly implemented functions in SyNRAC, which is a Maple package for solving real algebraic constraints derived from various engineering problems. The current version of SyNRAC has added quantifier elimination (QE) by cylindrical algebraic decomposition (CAD), a general QE procedure. We also show a visualization tool for representing the possble region of an output quantifier-free formula for the two-dimensional case.

## 1 Introduction

Nowadays symbolic computation methods have been widely applied to solving scientific and engineering problems, which has been caused by the efficient symbolic algorithms introduced or improved for these few decades and by the advancement of computer technology that has hugely increased the CPU power and memory capacity.

We have been developing a Maple toolbox, called SyNRAC, for solving real algebraic constraints, or first-order formulas over the reals. SyNRAC stands for a Symbolic-Numeric toolbox for Real Algebraic Constraints and is aimed at being a comprehensive toolbox including a collection of symbolic, numerical, and symbolic-numeric solvers for real algebraic constraints derived from engineering problems and the biological sciences. Our main method is quantifier elimination (QE), which is a procedure removing the quantified variables from a given formula to return a quantifier-free equivalent.

In this paper we present newly implemented procedures in SyNRAC. In [1] two types of special QE methods as well as some simplification procedures of quantifier-free formulas had been implemented in SyNRAC. In [2] QE by virtual substitution for weakly parametric linear formulas and a preliminary version of QE by CAD were implemented. The newly implemented functions in the current version of SyNRAC are following:

- general QE by cylindrical algebraic decomposition
- visualization of a resulting quantifier-free formula (plotting of 2-D CAD)

V.N. Alexandrov et al. (Eds.): ICCS 2006, Part II, LNCS 3992, pp. 462–469, 2006.

Against the inherent computational complexity of QE based on a CAD algorithm, several researchers have focused on QE algorithms specialized to particular types of input formulas; see [3, 4, 5, 6, 7]. This direction is quite promising in practice since a number of important problems in engineering have been successfully reduced to a certain type of input formulas and resolved by using specialized QE algorithms. For the concrete applications of these schemes, see [8, 9, 10, 11, 12, 13]. However, there still remain many significant problems in engineering that cannot be recast as such particular formulas. Therefore, it is strongly desired to develop an efficient algorithm to realize CAD. These new features extend the applicability and tractability of SyNRAC for solving real algebraic constraints in science and engineering.

This paper is organized as follows. We explain some changes in expressing formulas in SyNRAC in Section 2. We briefly describe CAD in Section 3 and in Section 4 we show some exapmle commands of QE by CAD and our plan for combining QE by CAD with numerical methods. In Section 5, a tool for visualizing two-dimensional possible region are shown. We state a conclusion and our future work in Section 6.

## 2 Formula Description on Maple 9.5

SyNRAC has been developed on the Maple software. When the SyNRAC project started, we were doing our implementation on Maple 8. We were using relational and logical operators bundled in Maple in the very first stage, but it turned out that some of those operators were unsuitable for our purpose. Here is a simple example. Let $x$ be just an indeterminate (called a name in Maple). The `evalb` command, which evaluates a Maple expression in Boolean context, returns `false` if $x = 0$ is input. This behavior does not meet our expectation, because we want to remain $x = 0$ unchanged and undecided until some real value is assigned to $x$. We introduced a user-defined operator `&=` to avoid such a reaction and replaced it for the Maple's equal sign '='. We have also defined `&and` and `&or`, as taking a list of atomic formulas as an argument because binary logical operators `and` and `or` in Maple had a disadvantage when dealing with long formulas.

Since then Maple 9, Maple 9.5 and Maple 10 have been released. We are now developing SyNRAC on Maple 9.5. The most influential change is that the Logic Package has been introduced in Maple 9.5. (We have not used Maple 10 yet, but we hear that the Logic Package in Maple 10 is the same as in Maple 9.5.)

The package has its own set of logical operators such as `&and`, `&or`, and `&not`, as well as several commands for manipulating expressions. For example, `Normalize` converts the input into a disjunctive or conjunctive normal form and `BooleanSimplify` changes a given Boolean expression into a minimal sum of prime implicants. `Import` and `Export` transform a certain type of Maple expression into a form available in the Logic Package, which enables users to convert a Boolean expression represented by built-in logical operators into one suitable for the Logic Package, and vice versa.

The names `&and` and `&or` in the Logic Package are coincidentally the same as those defined in SyNRAC on Maple 8. One of the differences was that `&and` (or `&or`) in the Logic Package takes a sequence as input, while our counterpart had been defined as taking a list of atomic formulas as input, like `&and([a>0, b>0, c>0])`. We have modified our `&and` and `&or` to receive a sequence of atomic formulas as input so that we can apply commands in the Logic Package such as `Logic[Normalize]` to quantifier-free formulas in SyNRAC.

## 3 Cylindrical Algebraic Decomposition

Cylindrical algebraic decomposition (CAD) was discovered by Collins in 1973; see [14] for his monumental work. Collins also proposed a general QE algorithm based on CAD, which has provided a powerful method for solving real algebraic constraints.

Let $A$ be a finite subset of $\mathbb{Z}[x_1, \ldots, x_n]$. An *algebraic decomposition* for $A$ is a collection of mutually disjoint, semi-algebraic, $A$-invariant sets that partitions the Euclidean $n$-space $E^n$. To define the term *cylindrical,* we explain three parts of a CAD procedure—the projection phase, the base phase, and the lifting phase.

In the projection phase of a CAD, the PROJ function plays a central role. Let $r$ be an integer greater than 1. PROJ maps a finite set of integral polynomials in $r$ variables to a finite set of integral polynomials in $r-1$ variables: for $A_r \subset \mathbb{Z}[x_1, \ldots, x_r]$, $\mathrm{PROJ}(A_r) \subset \mathbb{Z}[x_1, \ldots, x_{r-1}]$. For a given $A \subset \mathbb{Z}[x_1, \ldots, x_n]$, we obtain a list

$$A = A_0 \overset{\mathrm{PROJ}}{\mapsto} A_1 \overset{\mathrm{PROJ}}{\mapsto} A_2 \overset{\mathrm{PROJ}}{\mapsto} \cdots \overset{\mathrm{PROJ}}{\mapsto} A_{n-1},$$

where $A_i \subset \mathbb{Z}[x_1, \ldots, x_{n-i}]$.

In the base phase we partition $E^1$ by using a set of univariate polynomials $A_{n-1} \subset \mathbb{Z}[x_1]$; we find all the real zeros of $A_{n-1}$ and partition $E^1$ into $A_{n-1}$-invariant regions that consist of the zeros of $A_{n-1}$ and the remaining open intervals. These points and intervals are called sections and sectors, respectively.

The lifting phase inductively constructs a decomposition of $E^{i+1}$ from the decomposition of $E^i$, $i = 1, 2, \ldots, n-1$. Suppose $D$ is a decomposition of $E^i$. A lifting of $D$ is a decomposition $\bar{D}$ of $E^{i+1}$ obtained by decomposing the space $R \times E^1$ by using $A_{n-i-1}$ for each region $R \in D$ and putting all of them together. Let $R$ be a region of a decomposition $D$ of $E^i$. $R \times E^1$ is decomposed by the following; Take a point $(p_1, \ldots, p_i)$ in $R$ and substitute it for $(x_1, \ldots, x_i)$ in each polynomial in $A_{n-i-1}$ to obtain a set of univariate polynomials in $x_{i+1}$; Partition $E^1$ into, say, $L_0$, $L_1$, …, $L_{2k+1}$ by using the roots of the polynomials in $x_{i+1}$; Regard $R \times L_0$, $R \times L_1$, …, $R \times L_{2k+1}$ as the resulting decomposition. The condition for this process to work is that $A_{n-i-1}$ is *delineable* on $R$, in other words, every pair of polynomials in $A_{n-i-1}$ has no intersections on $R$. In such a case the decomposition is independent of the choice of a sample point. A decomposition $D$ of $E^r$ is *cylindrical* if it is constructed by iterating the above lifting method, i.e., $r = 1$ and $E^1$ is decomposed as in the base phase, or $r > 1$ and $D$ is a lifting of some cylindrical decomposition $D'$ of $E^{r-1}$.

Given a formula $\varphi$ one can construct a CAD for the polynomials of the atomic formulas in $\varphi$. The point for CAD-based QE is that the truth value of $\varphi$ is determined regionwise because each region in the CAD is $A$-invariant. See [14] for details.

It is the PROJ function that is crucial in a CAD procedure. The fewer polynomials PROJ produces, the more efficient the CAD program becomes. But PROJ must be constructed to maintain the delineability and make the lifting phase possible. Some improvements in the projection phase of CAD are found in [15, 16, 17]. In the next section, we show some QE commands based on CAD. Our present implementation of QE by CAD is based on Collins's original method; there is still a long way to catch up with latest tools such as QEPCAD.

# 4 Quantifier Elimination Based on CAD

## 4.1 Commands for QE by CAD

Here we show an example of the QE by CAD command in SyNRAC. Consider $\psi = \exists y\ (x^2 + y^2 \le 1 \ \wedge\ x^3 - y^2 < 0)$. To solve this formula, we first construct a CAD for $A = \{x^2+y^2-1, x^3-y^2\} \subset \mathbb{Z}[x, y]$. The graph of these two polynomials is shown in Fig. 1. The `Projection` command repeats PROJ and returns `P[1]`= $\mathrm{PROJ}^0(A) = A$ and `P[2]`= $\mathrm{PROJ}(A) = \{x^2 - 1, x^3 + x^2 - 1, x\}$.

The `Base` command partitions $E^1$ by using `P[2]` and returns a list of points that represent respective sections or sectors. A rational point is taken as a sample point for a sector, and a unique vanishing polynomial and an isolated interval are taken for a section. There are four real roots (sections) in `P[2]` and they make five open intervals (sectors).

```
> Base(P[2], x);

  [-2, &algn(x+1,-1,-1), -1/2, &algn(x,0,0), 3/8,
   &algn(x^3+x^2-1,3/4,7/8), 15/16, &algn(x-1,1,1), 2]
```

where `&algn(`$f(x), l, r$`)` represents the algebraic number that is a unique root of $f(x)$ lying between the two rational numbers $l$ and $r$ ($l \le r$).

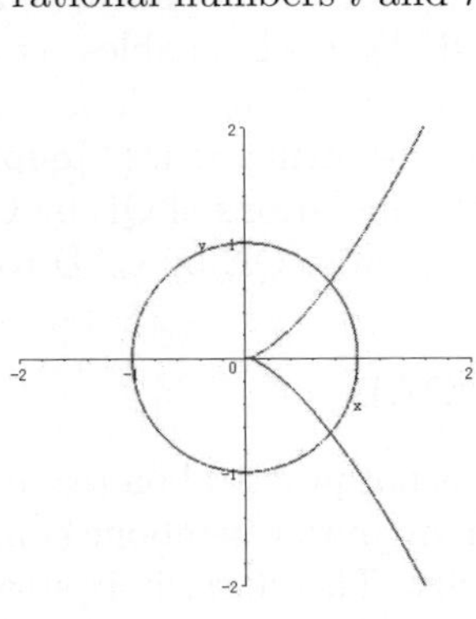


**Fig. 1.** The graph of A

The `Lifting` command makes a stack for each section or sector. Out of the nine regions, we have the fifth one displayed. The fifth region is a sector with a rational sample point `[3/8]` and the stack on it is represented in a list of nine sample points of sections/sectors.

```
> L:=Lifting(P, [x,y]):
> op(L[5]);

   [3/8, -2], [3/8, &algn(64*y^2-55,-1,-1/2)], [3/8, -1/2],
   [3/8, &algn(512*y^2-27,-1/2,0)],[3/8, 0], [3/8, &algn(512*y^2-27,0,1/2)],
   [3/8, 1/2], [3/8, &algn(64*y^2-55,1/2,1)], [3/8, 2]
```

And the `QEbyCAD` command returns a quantifier-free equivalent to the input formula $\psi$ by using the sample points:

```
> QEbyCAD(&Ex([y],&and(x^2 + y^2 <= 1, x^3 - y^2 < 0)));

  &or(&and(0 < -x^2+1, 0 < -x, 0 <= x+1, x <= 0),
      &and(x^2-1 = 0, 0 <= x+1, x <= -1), &and(x = 0, 0 <= x, x <= 0),
      &and(0 < x, 0 <= x, 0 < -x^3-x^2+1, 8*x <= 7))
```

### 4.2 Advantage of QE Based on CAD

Though QE by CAD has a bad computational complexity (doubly exponential in terms of the number of variables), it has some advantages against specialized QE algorithms that have already been implemented in SyNRAC, such as QE by the Sturm-Habicht sequence for positive definiteness of an univariate polynomial with parametric coefficients and QE by virtual substitution for linear and quadratic formulas with respect to quantified variables:

- QE by CAD is a general-purpose QE algorithm and can handle any first-order formulas.
- Moreover, one of the big advantages of QE by CAD is its ability to produce simple solution formula.

Large quantifier-free formulas can be simplified well by using QE by CAD due to the second advantage. QE by CAD enables us to simplify the output formulas generated by a specialized QE algorithm, which are mostly too large to understand (see `http://www.cs.usna.edu/{~}qepcad/SLFQ/Home.html`). For a very large formula, repeated operations of QE by CAD to its subformulas may succeed when a direct application of QE by CAD to the entire formula fails.

### 4.3 Symbolic-Numeric CAD

Unfortunately, QE by CAD is not practical on real computers, since a CAD procedure usually consists of many purely symbolic computations and has inherent bad computational complexity. Therefore, it is strongly desired to develop an efficient algorithm to realize CAD. One way of tackling this is, as was stated in the previous subsection, to consider a specialized algorithm for restricted input formulas. Another effective way for more efficient CAD is achieved by utilizing numerical computation and by making use of the derived numerical information on algebraic numbers to the full extent—until it violates the correctness of the results—instead of by handling symbolic computation. So far there are some related works to introduce numerical computation into CAD construction [18].

This strategy greatly improves the efficiency, while this causes uncertainty of the computed results depending on accuracy of numerical computation. Hence we need an effective validation scheme of the results. Now we are developing a symbolic-numeric procedure using as small a number of symbolic computations as possible—they are used only when validating unreliable numerical results. Symbolic reconstruction procedure in the lifting phase is improved if we utilize a dynamic evaluation (DE) technique [19, 20, 21] combined with successive representations of algebraic extensions.

## 5 Visualization

Here we show a newly introduced function for visualizing output of QE computation, in particular, in the two-dimensional case. By QE computaiton we obtain a quantifier-free formula in the unquantified variables. We can use the command `DrawDnf2d(dnf,xmin,xmax,ymin,ymax,grid)` to visualize the CAD of the two-dimensional space with colored feasible regions if the output formula is given by a disjunctive normal form (DNF). The range where the result is shown is given by the rectangle $[\texttt{xmin}, \texttt{xmax}] \times [\texttt{ymin}, \texttt{ymax}]$. This process is done by analyzing the topology of the CAD, sweeping through from left to right with repeated increments of a stride, and marking the appropriate points to the model of the CAD. This does not change the topology of the model; it makes the picture look prettier. In the plotting where the view is $[-2, 2]\times[-2, 2]$, its stride is given by $1/grid$. The DNF given as $(x^2+y^2-1 > 0\wedge -x+y > 0)\vee(x^2-y > 0)$ is drawn as follows;

```
> DrawDnf2d := define_external('DrawDnf2d','MAPLE','LIB'="DrawDnf2d.dll");
  DrawDnf2d := proc () option call_external;
  call_external(0, 22024192, true, args) end proc

> dnf := "[[x^2+y^2-1>0,-x+y>0]][[x^2-y>0]]";

> DrawDnf2d(dnf, -2, 2, -2, 2, 50);
  "ok..."
```

Our visualizer, as you see, is an external library written in C, hence we call it from maple via "DrawDnf2d.dll." Though QE by CAD can generate "adjacency information" for two-dimensional CADs which allows it to generate topologically correct plots, we do not care for the adjacency in this implementation.

## 6 Conclusion and Future Work

We have presented newly developed functions in Maple-package SyNRAC. The current version of SyNRAC provides QE by CAD and visualization of 2-D resulting

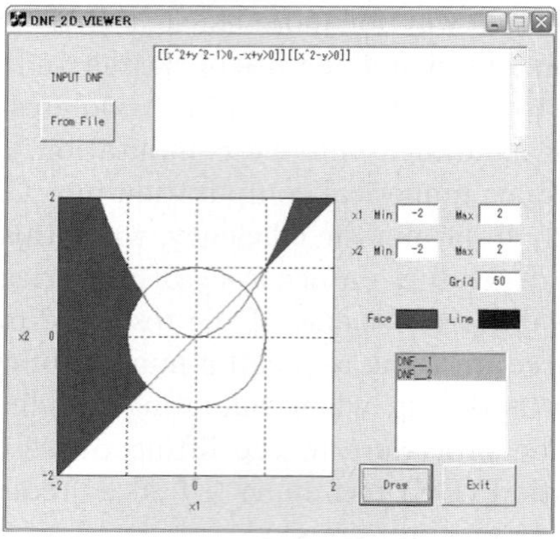


**Fig. 2.** 2-D visualization

quantifier-free formulas. The new features greatly extend the applicability and tractability of SyNRAC for solving real algebraic constraints in engineering.

We are planning to combine QE by CAD with some numerical methods to improve efficiency. We proceed to implement other known QE algorithms and improve them, and are setting about developing symbolic-numeric algorithms. We also plan to advance our toolbox for parametric robust control design on MATLAB using SyNRAC as a core engine [13], which we keep upgrading.

**Acknowledgements.** The authors thank Hong Myunghoon for the C code of the 2-D visualization tool in the present paper.

## References

1. Anai, H., Yanami, H.: SyNRAC: A maple-package for solving real algebraic constraints. In: Proceedings of International Workshop on Computer Algebra Systems and their Applications (CASA) 2003 (Saint Petersburg, Russian Federation), P.M.A. Sloot et al. (Eds.): ICCS 2003, LNCS 2657, Springer (2003) 828–837
2. Yanami, H., Anai, H.: Development of synrac – formula description and new functions. In: Proceedings of International Workshop on Computer Algebra Systems and their Applications (CASA) 2004 : ICCS 2004, LNCS 3039, Springer (2004) 286–294
3. Weispfenning, V.: The complexity of linear problems in fields. Journal of Symbolic Computation **5** (1988) 3–27
4. Loos, R., Weispfenning, V.: Applying linear quantifier elimination. The Computer Journal **36** (1993) 450–462 Special issue on computational quantifier elimination.
5. Weispfenning, V.: Quantifier elimination for real algebra—the quadratic case and beyond. Applicable Algebra in Engineering Communication and Computing **8** (1997) 85–101
6. Hong, H.: Quantifier elimination for formulas constrained by quadratic equations. In Bronstein, M., ed.: Proceedings of the International Symposium on Symbolic and Algebraic Computation (ISSAC 93), Kiev, Ukraine, ACM, ACM Press (1993) 264–274

7. González-Vega, L.: A combinatorial algorithm solving some quantifier elimination problems. In Caviness, B., Johnson, J., eds.: Quantifier Elimination and Cylindrical Algebraic Decomposition. Texts and Monographs in Symbolic Computation. Springer, Wien, New York (1998) 365–375
8. Weispfenning, V.: Applying quantifier elimination to problems in simulation and optimization. Technical Report MIP-9607, FMI, Universität Passau, D-94030 Passau, Germany (1996) To appear in the Journal of Symbolic Computation.
9. Sturm, T., Weispfenning, V.: Rounding and blending of solids by a real elimination method. In Sydow, A., ed.: Proceedings of the 15th IMACS World Congress on Scientific Computation, Modelling, and Applied Mathematics (IMACS 97). Volume 2., Berlin, IMACS, Wissenschaft & Technik Verlag (1997) 727–732
10. Dolzmann, A., Sturm, T., Weispfenning, V.: Real quantifier elimination in practice. In Matzat, B.H., Greuel, G.M., Hiss, G., eds.: Algorithmic Algebra and Number Theory. Springer, Berlin (1998) 221–247
11. González-Vega, L.: Applying quantifier elimination to the Birkhoff interpolation problem. Journal of Symbolic Computation (1996) To appear.
12. Anai, H., Hara, S.: Fixed-structure robust controller synthesis based on sign definite condition by a special quantifier elimination. In: Proceedings of American Control Conference 2000. (2000) 1312–1316
13. Anai, H., Yanami, H., Sakabe, K., Hara, S.: Fixed-structure robust controller synthesis based on symbolic-numeric computation: design algorithms with a cacsd toolbox (invited paper). In: Proceedings of CCA/ISIC/CACSD 2004 (Taipei, Taiwan). (2004) 1540–1545
14. Collins, G.E.: Quantifier elimination for real closed fields by cylindrical algebraic decomposition. In Caviness, B., Johnson, J., eds.: Quantifier Elimination and Cylindrical Algebraic Decomposition. Texts and Monographs in Symbolic Computation. Springer, Wien, New York (1998) 85–121
15. Hong, H.: An improvement of the projection operator in cylindrical algebraic decomposition. In Caviness, B., Johnson, J., eds.: Quantifier Elimination and Cylindrical Algebraic Decomposition. Texts and Monographs in Symbolic Computation. Springer, Wien, New York (1998) 166–173
16. McCallum, S.: An improved projection operation for cylindrical algebraic decomposition. In Caviness, B., Johnson, J., eds.: Quantifier Elimination and Cylindrical Algebraic Decomposition. Texts and Monographs in Symbolic Computation. Springer, Wien, New York (1998) 242–268
17. Brown, C.W.: Improved projection for cylindrical algebraic decomposition. Journal of Symbolic Computation **32** (2001) 447–465
18. Anai, H., Yokoyama, K.: Numerical cylindrical algebraic decomposition with certificated reconstruction. In: Proceedings of 11th GAMM - IMACS International Symposium on Scientific Computing, Computer Arithmetic, and Validated Numerics (Fukuoka, Japan). (2004) 31
19. Duval, D.: Algebraic numbers : an example of dynamic evaluation. Journal of Symbolic Computation **18** (1994) 429–445
20. Duval, D., González-Vega, L.: Dynamic evaluation and real closure. Math. Comput. Simulation **42** (1996) no. 4-6, 551–560
21. Hong, H.: An efficient method for analyzing the topology of plane real algebraic curves. Selected papers presented at the international IMACS symposium on Symbolic computation, new trends and developments. Elsevier Science Publishers B. V., Springer, Amsterdam, The Netherlands (1996) 571-582

# Automated Discovery in Elementary Extrema Problems

Francisco Botana* and José L. Valcarce

Departamento de Matemática Aplicada I, Universidad de Vigo, Campus A Xunqueira, 36005 Pontevedra, Spain
fbotana@uvigo.es
Departamento de Matemáticas, IES Pontepedriña, 15702 Santiago, Spain
jvalcarce@edu.xunta.es

**Abstract.** This paper describes an enhancement of GDI, a dynamic geometry environment that uses symbolic techniques for performing automatic proof and discovery in elementary geometry. We report a successful add–on that automatically discovers critical points in elementary extrema problems specified via diagrams. Based in classical one variable calculus, the new module, Optimus, is a step ahead in the cooperation between dynamic geometry and computer algebra systems. It can be used in technology rich environments for mathematics education, adding new calculus abilities to dynamic geometry software.

## 1 Introduction

Dynamic geometry software refers to computer programs where accurate construction of geometric configurations can be done. The key characteristic of this software is that unconstrained parts of the construction can be moved and, as they do, all other elements automatically self–adjust, preserving all dependent relationships and constraints. Commercial versions of dynamic geometry environments, namely Cabri Geometry [9] and The Geometer's Sketchpad [8], appeared ending the 80's, and, since they experienced a fast and wide dissemination in computer aided mathematical education, remain today as market dominant tools. Nevertheless, their approach heavily relies on numerical strategies that have originated some problems which are creating a real bottleneck for future developments: the so called continuity problem, the loci generation and the proof capabilities [5].

The continuity problems refers to dynamic diagrams where a small move of an element causes a sudden jump of a dependent element. It appears in constructions where a dependent point is located on a intersection comprising at least a quadratic object.

The loci generation problem deals with the way geometric loci are computed in standard dynamic geometry. The sampling approach taken in most systems

* Partly supported by grant MTM2004-03175 from the Spanish MEC and by the Technical University of Eindhoven, as a visiting scholar during March 2004.

V.N. Alexandrov et al. (Eds.): ICCS 2006, Part II, LNCS 3992, pp. 470–477, 2006.

only allows the generation of parametrically defined loci, and their equations are also unknown [2].

Regarding proof abilities, it must be noted that the accuracy of the constructions and the possibility to experiment with different instances of them have led to the extensive use of an activity that has been termed as 'visual proving' (of properties and theorems). Furthermore, Cabri incorporates an (unexplained) 'property checker' that can give an answer to a reduced list of geometric statements about a diagram.

Partly reacting to this numerical approach last generation dynamic geometry programs such as Cinderella [11], GDI [1, 3, 4], GEOTHER [13] and GEX [7] start including formal tools. Based on projective geometry and Cayley-Klein geometries (Cinderella), on the Groebner bases method (GDI), or on Ritt–Wu characteristic sets (GEX and GEOTHER), these systems connect dynamic geometry with solid mathematical theories. Besides that, GDI and GEOTHER show up two successful examples of cooperation between an interactive environment and a computer algebra system (Mathematica and CoCoA [6] in GDI, Maple in GEOTHER).

This simple note describes a new way for computing, in an automatic manner, critical points in elementary extrema points described through dynamic diagrams. Section 2 states the problem and illustrates the Optimus module by means of a single exercise. The limitations and some extensions of the approach are discussed in Section 3. Section 4 deals with a possible connection of Optimus with other dynamic geometry environments via OpenMath.

## 2 User Interface and Architecture

We will restrict the study to optimization problems in one variable. Furthermore, we will only consider problems that admit a plane geometric modelization in a dynamic geometry setting, where the magnitude to optimize is defined by means of a dependent point, or can be reduced to it.

Let us illustrate the use of Optimus with a simple case: *Two vertical poles OA and PB are secured by a rope AXB going from the top A of the first pole to a point X on the ground between the poles and then to the top B of the second pole. If the length of poles are 1 and 2, find the position of X that requires the least rope.*

The first task consists of building a dynamic model of the problem (Fig. 1). Note that, besides the coordinate axes based on points $O(0,0)$ and $U(1,0)$ (being $U$ hidden in the diagram), there are only three free points $A(0,1)$, $P(2,0)$, $B(2,2)$ and a dependent point $X(x_1, x_2)$, which lies on the line passing through $O$ and $U$. Since GDI uses a coordinate–based approach, the whole construction can be algebraically described by the coordinates of points and the condition $x_2 = 0$. Once the construction is done, the user must specify the magnitude to optimize and the point on which it depends through an ad–hoc designed window (Fig. 2). The magnitude *distance(A,X)+distance(X,B)* is internally translated as the sum of a pair of polynomial constraints $x_3 + x_4$, whose definitions are

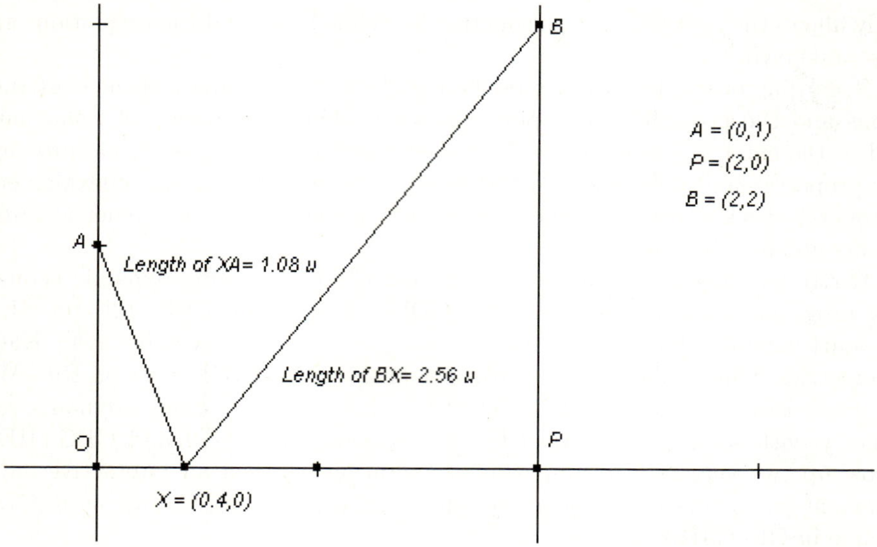


**Fig. 1.** The construction for the two poles problem

$x_3^2 = (x_1^2 + (1 - x_2)^2)$ and $x_4^2 = ((2 - x_1)^2 + (2 - x_2)^2)$. Although taking the square distances will be a source of imprecision, the need of coherence with other GDI agents (prover, discoverer, ...) compelled us to maintain the computations in a strict algebraic environment.

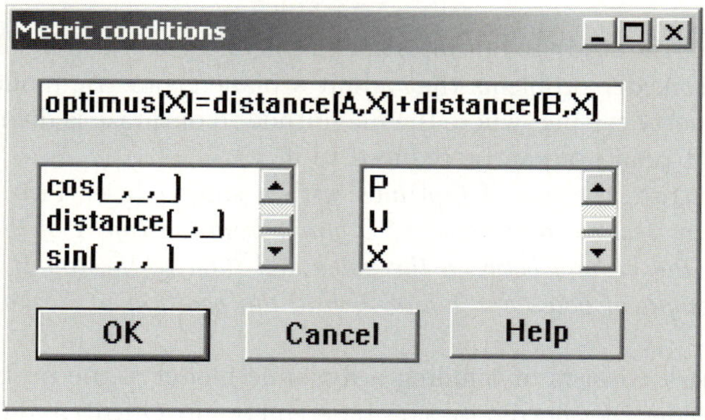


**Fig. 2.** Specifying the point and the magnitude to optimize

Once given the preceding information, GDI launches a local copy of Mathematica, if any, or calls a remote kernel via webMathematica. A small Mathematica package then performs the following operations:

- read the construction points and constraints and the optimization items,
- solve the polynomial constraints ($x_2 \to 0$),

– solve the optimization constraints
  $(x_3 \to \pm\sqrt{x_1^2 + (1 - x_2)^2}, x_4 \to \pm\sqrt{(2 - x_1)^2 + (2 - x_2)^2})$,
– for each combination of the preceding solutions
  - solve the derivative of the optimization condition with respect to the unbound variable of the optimization point $(x_1)$,
  - compute the coordinates of the construction dependent points,
  - add this set of points to a list of possible solutions,
– return this list.

The list of critical and other dependent points is shown as a window in the main screen of GDI or, if a remote Mathematica kernel is being used, by means of a browser (Fig. 3). A numerical approximation of the coordinates is also given in order to help the student placing the points in the dynamic construction. Note also that, of course, the first point $X(-2, 0)$ does not optimize the length of the rope, and it has been found due to the polynomial restriction above mentioned.

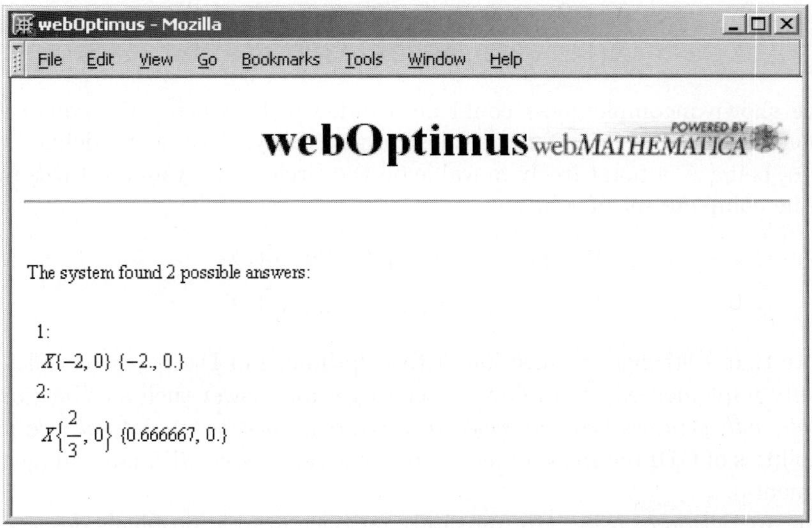


**Fig. 3.** The critical points found by the system

## 3 Limitations and Extensions of the Approach

Apart from the introduction of spurious solutions due to the polynomial restriction, it must be noted that the approach is not complete. Furthermore, it is recommended that the system will be used under teacher guidance in elementary levels, since it sometimes poses subtle questions requiring mature mathematical skills or a deep knowledge of the system design.

Consider, for instance, finding which isosceles triangle inscribed in a given circle has the greatest area. A common behavior in a dynamic geometry setting would be drawing a circle and placing two points $A(x_1, x_2)$ and $X(x_3, x_4)$ on the circle. The third triangle vertex $B(x_7, x_8)$ can be easily constructed through the perpendicular bisector of segment $AX$ and its intersection with the circle. With this construction Mathematica finds 32 solutions to the set of polynomial constraints when solving respect to an unbound coordinate of $X$, say $x_3$. Eight of them are degenerate cases ($A = X$), since they include the rules $x_1 \to x_3$ and (taken with equal sign) $x_2 \to \pm\sqrt{1-x_3^2}$, $x_4 \to \pm\sqrt{1-x_3^2}$. So, point $B$ is ill–defined and no derivative is computed because the optimization condition is null. Other sixteen solutions have two degrees of freedom: no rules for $x_1, x_3$ are found, according to the fact that $A$ and $X$ have just been defined as arbitrary points on the circle. The triangle area is then a two–argument function, and the algorithm aborts the a posteriori computations. The remaining eight solutions include the rules $x_1 \to x_3$, $x_2 \to \pm\sqrt{1-x_3^2}$, $x_4 \to \mp\sqrt{1-x_3^2}$, generating the four solutions returned by the system

$$X(-1/2,-\sqrt{3}/2), A(-1/2,\sqrt{3}/2), B(1,0)$$
$$X(-1/2,\sqrt{3}/2), A(-1/2,-\sqrt{3}/2), B(1,0)$$
$$X(1/2,-\sqrt{3}/2), A(1/2,\sqrt{3}/2), B(-1,0)$$
$$X(1/2,\sqrt{3}/2), A(1/2,-\sqrt{3}/2), B(-1,0)$$

The shown incompleteness could be avoided just reducing the construction freedom. Paying the price for some lost of generality, $A$ could be defined to be $U(1,0)$, being $X$ a point freely movable on the circle. The system returns in this case the complete set of solutions

$$X(-1/2,-\sqrt{3}/2), B(-1/2,\sqrt{3}/2)$$
$$X(-1/2,\sqrt{3}/2), B(-1/2,-\sqrt{3}/2)$$

Note that a stronger connection of the Optimus and Discovery modules (not yet fully implemented) will allow a user to get an answer such as *The isosceles triangle with greatest area inscribed in the circle must be equilateral* (see [5] for the abilities of GDI finding geometric properties expressed with natural language statements).

The restriction of using Optimus only for one variable problems deserves additional explanation. GDI offers an option for specifying extra constraints to a construction. This option is the key for finding geometric loci with a posteriori conditions, or for discovering facts about essentially non constructive situations (see [10] for a paradigmatic example, the MacLane $8_3$ theorem). Through the introduction of extra constraints the system can be extended to deal with optimization problems in several variables. Consider finding the right triangle of given perimeter and maximum area: defining a vertex to be the origin $O$ and placing the remaining vertices $A, X$ as dependent points on both axes, the user can declare the perimeter to be a constant, say 14,

$$distance(O,X) + distance(O,A) + distance(A,X) = 14$$

and ask about the maximum area of the triangle, taking $X$ as the dependent point. Two of the eight found solutions are shown in Fig. 4, being the remaining ones symmetrical. Again, the consideration of signed distances leads up to the second group of solutions.

Another possible extension of Optimus has been studied. A construction involving an optimization condition can be done without any reference to numerical coordinates, that is, the coordinates of the free points or the values of the constants can be symbolic (as they are when using the proving abilities of GDI). In this case, the obtained solutions will exhibit more generality, but the price to pay is, in general, an unmanageable increment of the expressions of dependent points. So, since these values are not easily readable, we have restricted the domain of Optimus to constructions referred to a coordinate system.

4:

$$X\{0,\ 7(2-\sqrt{2})\}\ \{0.,\ 4.10051\}\quad A\{\frac{14(-7+7(2-\sqrt{2}))}{-14+7(2-\sqrt{2})},\ 0\}\ \{4.10051,\ 0.\}$$

8:

$$X\{0,\ 7(2+\sqrt{2})\}\ \{0.,\ 23.8995\}\quad A\{\frac{14(-7+7(2+\sqrt{2}))}{-14+7(2+\sqrt{2})},\ 0\}\ \{23.8995,\ 0.\}$$

**Fig. 4.** A pair of solutions of the two variable problem

## 4 Using Optimus with Other Dynamic Geometry Systems

A goal of this short note consists of illustrating the connection of a dynamic geometry environment with a computer algebra system [12] for performing exact computations about optimization problems. Since the inputs to the algorithm described in Section 2 are the list of construction points, the constraints, a dependent point and the magnitude to optimize, any other dynamic geometry system could be easily modified to deal with such problems. Nevertheless, the character of proprietary software of most programs would convert this task in a time consuming work, unless developed by their own designers. The Geometer's Sketchpad, for instance, can export its constructions to html code, for being visualized using the JavaSketchpad applet [15]. In this code, the value of an applet parameter contains a readable description of the construction. Similarly, Cabri has an option, *Show Figure Description*, that roughly describes a construction. Cinderella and GEX have also a mechanism to see the textual description of constructions. These tools can be used to analyse the construction and to get the list of points and constraints. A little piece of code could be then developed in order to select the point and optimization conditions and to launch Optimus. Nevertheless, in our opinion, a more promising future involves the use of a common protocol for intercommunication such as OpenMath [16] or the newest proposal GOOL [14]. Using OpenMath and the experimental `plangeo` Content Dictionaries [17], a construction such as the one used in Section 2 can be described by

$$point(O(0,0)), point(U(1,0)), point(A(0,1)), point(P(2,0)), point(B(2,2)),$$

$$line(l, incident(O,l), incident(U,l)), point(X, incident(l,X))$$

and the optimization condition, defining a new OpenMath symbol *optimus*, by

$$optimus(X, distance(A,X) + distance(X,B)$$

The first part of this OpenMath code can be easily translated into a set of points and a list of polynomials, as shown in the preceding Section. The optimization condition carries up the rest of the required information for Optimus. So, any OpenMath–capable system could connect to a remote Mathematica kernel for performing this kind of computations.

## 5 Conclusion

We extend a dynamic geometry environment with the ability of computing critical points in elementary one variable optimization problems diagrammatically described. To this end, we connect our dynamic tool with a computer algebra system (Mathematica), where standard calculus techniques are employed. Furthermore, this connection can be either local, either via a remote Mathematica kernel.

While optimization problems are usually managed by means of approximation techniques and empirical testing in educational uses of common dynamic geometry systems, our proposal gives exact solutions to the exercises and, being fully integrated with the geometry reasoning system, can be also exploited for proving or discovering tasks.

## References

1. Botana, F., Valcarce, J.L.: A dynamic-symbolic interface for geometric theorem discovery. *Computers and Education*, 38(1-3), 21–35 (2002)
2. Botana, F.: Interactive versus symbolic approaches to plane loci generation in dynamic geometry environments. *Proc. I Int. Workshop on Computer Graphics and Geometric Modeling* (CGGM 2002), LNCS, 2330, 211-218 (2002)
3. Botana, F., Valcarce, J.L.: A software tool for the investigation of plane loci. *Mathematics and Computers in Simulation*, 61(2), 141–154 (2003)
4. Botana, F., Valcarce, J.L.: Automatic determination of envelopes and other derived curves within a graphic environment. *Mathematics and Computers in Simulation*, 67(1–2), 3–13 (2004)
5. Botana, F., Recio, T.: Towards solving the dynamic geometry bottleneck via a symbolic approach. *Proc. 5th Int. Workshop on Automated Deduction in Geometry* (ADG 2004), LNAI, 3763, 92–110 (2006)

6. Capani, A., Niesi, G., Robbiano, L.: CoCoA, a system for doing Computations in Commutative Algebra. Available via anonymous ftp from: `cocoa.dima.unige.it`
7. Gao, X.S., Zhang, J.Z., Chou, S.C.: *Geometry Expert.* Nine Chapters, Taiwan (1998)
8. Jackiw, N.: *The Geometer's Sketchpad v 4.0.* Key Curriculum Press, Berkeley (2002)
9. Laborde, J. M., Bellemain, F.: *Cabri Geometry II.* Texas Instruments, Dallas (1998)
10. Recio, T., Botana, F.: Where the truth lies (in automatic geometry theorem proving). *Proc. II Technical Session on Computer Graphics and Geometric Modeling* (TSCG 2004), LNCS, 3044, 761–770 (2004)
11. Richter–Gebert, J., Kortenkamp, U.: *The Interactive Geometry Software Cinderella.* Springer, Berlin (1999)
12. Roanes–Lozano, E., Roanes–Macías, E., Villar, M.: A bridge between dynamic geometry and computer algebra. *Mathematical and Computer Modelling*, 37(9–10), 1005-1028 (2003).
13. Wang, D.: GEOTHER: A geometry theorem prover. *Proc. 13th International Conference on Automated Deduction* (CADE 1996), LNCS, 1104, 166–170 (1996)
14. Wang, D., Liang, T.: Towards a Geometric–Object–Oriented Language. *Proc. 5th Int. Workshop on Automated Deduction in Geometry* (ADG 2004), LNAI, 3763, 130–155 (2006)
15. `http://www.keypress.com/sketchpad/javasketchpad/`
16. `http://www.openmath.org`
17. `http://www.win.tue.nl/~amc/oz/om/cds/geometry.html`

# Stabilizing Second-Order Linear Dynamic Systems Via Hybrid Output Feedback Controls

Liguo Zhang, Yangzhou Chen, and Pingyuan Cui

School of Electronic and Control Engineering,
Beijing University of Technology,Beijing, 100022, China
zhangliguo@emails.bjut.edu.cn

**Abstract.** This paper considers the open problem whether there exists a finite-state hybrid output feedback control to asymptotically stabilize a second-order linear dynamic system. More precisely, for second-order linear time-invariant systems which are not stabilizable via a single static output feedback, we find two different output feedback gains and a switching law orchestrating the feedback gains such that the closed-loop system is asymptotically stable.

## 1 Introduction

Given a linear time-invariant dynamic system

$$\begin{aligned} \dot{x}(t) &= Ax(t) + Bx(t) \\ y(t) &= Cx(t) \end{aligned} \tag{1}$$

where $x \in R^n$, $u \in R^m$, $y \in R^p$, and $A$, $B$, $C$ are matrices of suitable dimensions, if system (1) is stabilizable and detectable, then there exists a linear dynamic output feedback law that asymptotically stabilizes the system [1]. In practice, however, such continuous dynamic feedback might not be implemental, and a hybrid version of the controller is often desired. We present a motivation example, which was also discussed in [2], [3] and [4].

*Example 1.* Consider the harmonic oscillator model, which can be written in the form of (1) with

$$A = \begin{pmatrix} 0 & 1 \\ -1 & 0 \end{pmatrix}, \quad B = \begin{pmatrix} 0 \\ 1 \end{pmatrix}, \quad C = \begin{pmatrix} 1 & 0 \end{pmatrix}. \tag{2}$$

Although this system is both controllable and observable, it cannot be stabilized by a single static output feedback [2]; however, it is stabilizable by a hybrid output feedback [3], [4]. Letting $u = -y$ and $u = \frac{y}{2}$ we obtain the following systems, respectively,

$$\dot{x}(t) = \begin{pmatrix} 0 & 1 \\ -2 & 0 \end{pmatrix} x(t), \tag{3}$$

and

$$\dot{x}(t) = \begin{pmatrix} 0 & 1 \\ -\frac{1}{2} & 0 \end{pmatrix} x(t). \tag{4}$$

V.N. Alexandrov et al. (Eds.): ICCS 2006, Part II, LNCS 3992, pp. 478–485, 2006.

Define $V(x) = x_1^2 + x_2^2$. If system (3) is active in the first and third quadrants, while system (4) is active in the second and fourth quadrants, we will have $\dot{V} < 0$ whenever $x_1 x_2 \neq 0$, which implies that the entire switched system is asymptotically stable by LaSalle's Principle (e.g., [15]).

Therefore, the hybrid output feedback stabilization problem can be formulated as follows [5]: find a finite collection of $m \times p$ gain matrices $\{K_1, \cdots, K_l\}$, and a switching law $\sigma : [0, \infty) \to \{1, \cdots l\}$, such that the switched linear system $\dot{x}(t) = A_{\sigma(t)} x(t)$ is asymptotically stable, where $A_{\sigma(t)} = A + BK_{\sigma(t)}C$, and the output feedback control is given by $u(t) = K_{\sigma(t)} y(t)$. To rule out trivial cases, we assume the system (1) cannot be static output feedback stabilized.

Stability analysis for switched systems has achieved many results, which mostly based on the common Lyapunov function or the multiple Lyapunov function approach [6], [7], [8], [9], [10], [11]. Stabilization problem is the control design for stability, while available techniques that can be used in the hybrid output feedback control have been very few. The difficulty led researchers to study it first in the low-order systems. In [12], it has been shown that if system (1) is controllable and observable, then it admits a stabilizing hybrid output feedback that uses a countable number of discrete states. More recently, [5] constructs a 2-state output feedback for a class of second-order linear systems incorporated with a conic switching law [13], [14], which provide a description of the switching law based on the angles of subsystem vector fields and the geometric properties of the phase plane. However, their approach may be complicated to work with or determine the exact parametric value for which the switching law yields instability of the feedback system.

In this paper, we propose an approach for constructing a 2-state hybrid output feedback control to stabilize a second-order linear dynamic system. This approach is composed of two important parts. First, we consider the asymptotic stabilization problem of second-order autonomous switched linear systems consisting of two subsystems with unstable focus equilibrium. Using the "most stabilizing" switching law, we translate the switched system into a piecewise linear system that yields an easily verifiable sufficient condition for stabilization of this system. Secondly, 2-state output feedback gains are constructed in terms of control system matrices $(A, B, C)$, which extends the results in [5].

The rest of the paper is organized as follows. Section 2 analyzes the asymptotic stabilization for an autonomous switched linear system. Section 3 designs the 2-state hybrid output feedback control. Numerical examples are given in section 4. Section 5 summarizes.

## 2 Stabilization for Switched Linear Systems

Consider an autonomous switched linear system given by

$$\begin{aligned} &\dot{x}(t) = A_{\sigma(t)} x(t) \\ &\sigma : [0, \infty) \to \{1, 2\} \end{aligned} \tag{5}$$

where $x \in R^2$ and $\sigma(t)$ is the switching law indicating the active subsystem at each instant. Let $\lambda_{i,\pm} = \alpha_i \pm \beta_i$ be the eigenvalues of $A_i$, respectively.

**Lemma 1.** *[16]. For the two subsystems of the switched linear system (5), there exists a nonsingular matrix Q, such that*

$$QA_1Q^{-1} = \begin{pmatrix} \alpha_1 & \beta_1/E \\ -E\beta_1 & \alpha_1 \end{pmatrix}, \quad QA_2Q^{-1} = \begin{pmatrix} \alpha_2 & \beta_2 \\ -\beta_2 & \alpha_2 \end{pmatrix}. \tag{6}$$

Without loss of generally, we only discuss the results for $|E| < 1$. In the case of $0 < E < 1$, the system trajectory diverges to $\infty$ clockwisely, while in the case $-1 < E < 0$, then counter-clockwisely. Therefore, in the following discussion, we assume the subsystems' state matrices have the form (6).

We first observe that the locus of both subsystems circle around the origin and diverge to clockwisely or counter-clockwisely. We will show that the "most stabilizing" switching law corresponds to switching between the two subsystem's vector fields are parallel. Here, by "most stabilizing" case, we mean if the switched system is stabilizable, then the system's trajectory converges to the origin under this switching law, that is, select an active subsystem which would drive the trajectory closer to the origin.

Define $l_-$ and $l_+$ as the lines on which the subsystem's vector field $A_1x$and $A_2x$ are parallel. Then on $l_\pm$ we have

$$\frac{-\beta_2 x_1 + \alpha_2 x_2}{\alpha_2 x_1 + \beta_2 x_2} = \frac{-E\beta_1 x_1 + \alpha_1 x_2}{\alpha_1 x_1 + (1/E)\beta_1 x_2}. \tag{7}$$

Solving this equation we obtain that if

$$D \geq 0,$$

then

$$x_2 = \nu_\pm x_1, \tag{8}$$

where, when

$$\rho_1 - \frac{\rho_2}{E} \neq 0,$$

we have,

$$\nu_\pm = \frac{(E - (1/E)) \mp 2\sqrt{D}}{2(\rho_1 - (\rho_2/E))}; \tag{9}$$

or, when

$$\rho_1 - \frac{\rho_2}{E} = 0,$$

we have,

$$\nu_- = \infty$$

$$\nu_+ = \frac{\rho_1 - E\rho_2}{E - (1/E)}. \tag{10}$$

We denote

$$D = R^2 + 2\rho_1\rho_2 R - (1 + \rho_1^2 + \rho_2^2)$$

and

$$R = \frac{1}{2}(E + \frac{1}{E}), \quad \rho_1 = \frac{\alpha_1}{\beta_1}, \quad \rho_2 = \frac{\alpha_2}{\beta_2}.$$

Define the regions

$$R_1 = \{(x_1, x_2)|\nu_- x_1 > x_2 > \nu_+ x_1\},$$

$$R_2 = \{(x_1, x_2)|x_2 \geq \nu_- x_1 \ \ and \ \ x_2 \geq \nu_+ x_1\},$$

where $-R_i$ is the opposite region of $R_i$ against the origin on the phase plane, $i = 1, 2$, see Fig.1.

In the case $0 < E < 1$, the trajectories of subsystems 1, 2 are all clockwise logarithmic spirals. Then the "most stabilizing" switching law would be:

Switch to subsystem 1 whenever the system trajectory enters the regions $R_1$ and $-R_1$, and switch to subsystem 2 whenever the system trajectory enters the regions $R_2$ and $-R_2$.

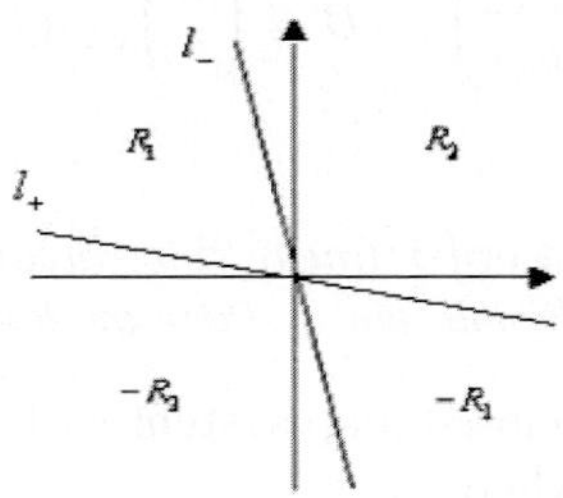


**Fig. 1.** Region partition when $D > 0$

Under the "most stabilizing" switching law, we have translated system (5) into a piecewise linear system

$$\dot{x} = \begin{cases} A_1 x & x \in R_1 \bigcup -R_1 \\ A_2 x & x \in R_2 \bigcup -R_2 \end{cases}. \tag{11}$$

Now, if the piecewise linear system (11) is asymptotically stable, then the switched linear system (5) asymptotically stabilizable.

**Theorem 1.** *[16]. The autonomous switched linear system (5) consisting of two subsystems with unstable focus equilibrium is asymptotically stabilizable if $D > 0$, $0 < E < 1$ and $\rho < 1$, where*

$$\rho = exp[\rho_1 arctan(\frac{\rho_1 R - \rho_2}{\sqrt{D}}) + \rho_2 arctan(\frac{\rho_2 R - \rho_1}{\sqrt{D}}) - \frac{\pi}{2}(\rho_1 + \rho_2)]$$

$$\times \sqrt{\frac{(\rho_1\rho_2 + R) + \sqrt{D}}{(\rho_1\rho_2 + R) - \sqrt{D}}}. \tag{12}$$

*Remark 1.* For $D > 0, 0 < E < 1$, the trajectories of subsystems 1, 2 are counter-clockwise, clockwise logarithmic spirals, respectively. In the case we could also define the "most stabilizing" switching law and get the similar conclusion as Theorem 1. When $D > 0$, then at least on the half phase plane, one subsystem's vector field always points on the same side of other. Form the proof of Theorem 1 we know in this case system (5) cannot be asymptotically stabilizable.

## 3 Hybrid Output Feedback Control

In this section, we construct 2-state hybrid output feedback gains in terms of control system matrices $(A, B, C)$, which deduced closed-loop system matrices satisfy the conditions of Theorem 1. For simply, we only focus discussion on second-order single-input-single-output (SISO) systems, i.e., $m = 2$, $n = p = 1$ in system (1). Since a nonsingular linear transformation does not change the stabilization property, we consider system (1) in the following form:

$$A = \begin{pmatrix} a_{11} & a_{12} \\ a_{21} & a_{22} \end{pmatrix}, \quad B = \begin{pmatrix} 0 \\ 1 \end{pmatrix}, \quad C = \begin{pmatrix} 1 & c \end{pmatrix}, \tag{13}$$

where $a_{ij}, c \in R$, $i, j = 1, 2$.

**Theorem 2.** *For the second-order linear time-invariant system (13), if $c = 0$, and $a_{12} \neq 0$, then it is stabilizable via a static or 2-state hybrid output feedback.*

*Proof.* For system (13), the closed-loop system with output feedback $u = k_i y$ is given by $\dot{x} = A_i x$, $i = 1, 2$, where

$$A_i = A + k_i BC = \begin{pmatrix} a_{11} & a_{12} \\ a_{21} + k_i & a_{22} \end{pmatrix}. \tag{14}$$

Since $a_{12} \neq 0$ , if $a_{11} + a_{22} < 0$, obviously there exists output feedback gain $k_i$ such that $A_i$ is Hurwitz stable, that is, system (13) could be static output feedback stabilizable. As for the case $a_{11} + a_{22} \geq 0$ , let $\Delta_i$ is the discriminant of the characteristic equation for the coefficient matrices $A_i, i = 1, 2$, If we choose$k_i$ such that

$$\Delta_i = 4(\alpha^2 - |A| + a_{12}k_i) < 0 \tag{15}$$

then $A_i$'s eigenvalues are express as $\lambda_{i\pm} = \alpha + j\beta_i$ , where

$$\alpha = \frac{a_{11} + a_{22}}{2},$$

$$\beta_i = \sqrt{|A| - \alpha^2 - a_{12}k_i}.$$

From Lemma 1, we know there exists a coordinate transformation $y = Qx$, $Q = (q_{ij})_{2\times2}$, changes the closed-loop system into $\dot{y} = A_i' y$, where $A_i' = QA_iQ^{-1}$, $i = 1, 2$, and

$$A_1' = \begin{pmatrix} \alpha & \beta_1/E \\ -E\beta_1 & \alpha \end{pmatrix}, \quad A_2' = \begin{pmatrix} \alpha & \beta_2 \\ -\beta_2 & \alpha \end{pmatrix}. \tag{16}$$

Solving the equation $QA_i = A_i^{'}Q$, $i = 1, 2$, we obtain that $E = \frac{\beta_1}{\beta_2}$, $q_{12} = 0$, and

$$(a_{11} - a_{22})q_{22} = 2a_{12}q_{21},$$

$$\beta_2 q_{22} = a_{12}Q_{11}.$$

If we further assume that

$$\beta_1 < \beta_2, \tag{17}$$

then according the discussion in Section 2, we have

$$0 < E < 1, \quad D = \frac{1}{4}(E - \frac{1}{E})^2 > 0,$$

and

$$\rho = \prod_{i=1}^{2} \exp \rho_i (\arctan \rho_i - \frac{\pi}{2}) \times \sqrt{\frac{\alpha^2 + \beta_2^2}{\alpha^2 + \beta_1^2}}. \tag{18}$$

Under the coordinate transformation $y = Qx$, we obtain that at the parallel lines

$$\nu_{\pm}^{'} = \frac{y_2}{y_1} = \frac{q_{21}x_1 + q_{22}x_2}{q_{11}x_1 + q_{12}x_2},$$

that is,

$$\nu_{\pm} = \frac{q_{21} - \nu_{\pm}^{'}}{q_{12}\nu_{\pm}^{'} - q_{12}}.$$

Therefore

$$\rho_1 - \frac{\rho_2}{E} = 0, \quad \nu_{-}^{'} = \infty, \quad \nu_{+}^{'} = -\frac{\alpha}{\beta_2}.$$

In view of the discussion thus far, we should choose the constants $k_i$ satisfies (15), (18), and such that $\beta_i$ satisfies $\rho < 1$. It is not difficult to show that such $k_i$ always exist. Therefore, we conclude that in this case system (13) always stabilizable via the 2-state hybrid output feedback control

$$u = \begin{cases} k_1 y & x_2 \leq -\frac{a_{22}}{a_{12}}x_1, and \;\; x_1 \geq 0, or \\ & x_2 \geq -\frac{a_{22}}{a_{12}}x_1, and \;\; x_1 \leq 0 \\ k_2 y & otherwise \end{cases}. \tag{19}$$

The proof is complete.

*Remark 2.* For $c \neq 0$, and $a_{12} \neq 0$, we could also design the output feedback gains such that closed-loop switched system satisfies the condition of Theorem 1. Of course they might be nonexistence. When $a_{12} = 0$, system (13) is uncontrollable, then unstabilizable.

## 4 A Numerical Example

*Example 2.* Consider the linear system of type (13) with

$$A = \begin{pmatrix} -1 & 2 \\ -2 & 3 \end{pmatrix}, \quad B = \begin{pmatrix} 0 \\ 1 \end{pmatrix}, \quad C = \begin{pmatrix} 1 & 0 \end{pmatrix}. \tag{20}$$

We can easily verify that the pair $(k_1, k_2) = (-1, -38)$ satisfies the required conditions (15), (18), and $\rho = 0.8697 < 1$. Then the 2-state hybrid output feedback stabilizable control is

$$u = \begin{cases} -y & 0 \leq x_1 \leq -\frac{2}{3}x_2, or - \frac{2}{3}x_2 \leq x_1 \leq 0 \\ -38y & otherwise \end{cases}. \tag{21}$$

Fig. 2 shows the trajectory of the system with the initial condition $x(0) = (2, 2)$.

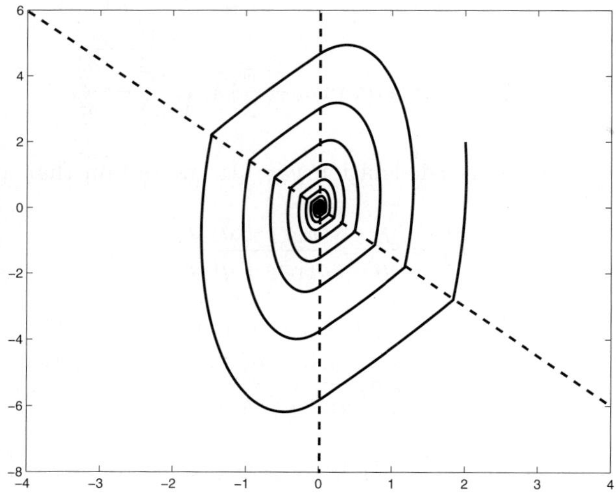


**Fig. 2.** The trajectory for Example 2

## 5 Conclusion

This paper studies the hybrid output feedback stabilization of second-order linear dynamic systems. For a class of such systems, we showed that the system is stabilizable via a 2-state static output feedback incorporated with an appropriate switching law. Output feedback gains and switching law are constructed in terms of control system matrices. Extension of the proposed results to general high-dimensional linear systems is considered to be an interesting future work.

## Acknowledgements

This paper was supported in part by the National Natural Science Foundation of China (No.60374007), and in part by the Beijing Natural Science Foundation of China (No.4042006).

## References


1. Syrmos, V.L., Abdallah, C.T., Dorato, P., and Grigoriadis, K.: Static output feedback: a survey. Automatica. **33** (1997) 125-137
2. Artstein, Z.: Examples of stabilization with hybrid feedback, in: R. Alur et al. (Eds.), Hybrid Systems III: Verification and Control, (1996) 173-185
3. Liberzon, D.: Stabilizing a linear system with finite-state hybrid output feedback, in: Proceedings of the 7th IEEE Mediterranean Conference on Control and Automation, (1999) 176-183
4. Liberzon, D., Morse, A.S.: Basic problems in stability and design of switched systems, IEEE Control Systems Magazine **19** (1999) 59-70
5. Hu, B., Zhai, G., Michel, A.N.: Hybrid output feedback stabilization of two-dimensional linear control systems, in: Proceedings of the 2000 American Control Conference, (2000) 2184-2188
6. Branicky, M.S.: Multiple Lyapunov functions and other analysis tools for switched and hybrid systems, IEEE Trans. Automat. Control **43** (1998) 475-482
7. DeCarlo, R., Branicky, M., Pettersson, S., Lennartson,B.: Perspectives and results on the stability and stabilizability of hybrid systems, Proc. IEEE **88** (2000) 1069-1082
8. Johansson, M., Rantzer, A.: Computation of piecewise quadratic Lyapunov functions for hybrid systems, IEEE Trans. Automat. Control **43** (1998) 555-559
9. Michel, A.N.: Recent trends in the stability analysis of hybrid dynamical systems, IEEE Trans. Circuit and Systems-I: Fundamental Theory and Applications **45** (1999) 120-134
10. Morse, A.S.: Control using logic-based switching, in: A. Isidori (Ed.), Trends in Control: a European Perspective, Springer, Berlin, (1995) 69-113
11. Peleties, P., DeCarlo, R.: Asymptotic stability of m-switched systems using Lyapunov-like functions, in: Proceedings of the 1991 American Control Conference, (1991) 1679-1684
12. Litsyn, E., Nepomnyashchikh, Y.V., and Ponosov, A.: Stabilization of linear differential systems via hybrid feedback controls, SIAM J. Control Optim. **38** (2000) 1468-1480
13. Xu, X., Antsaklis, P.J.: Design of stabilizing control laws for second-order switched systems, in: Proceedings of the 14th IFAC World Congress, vol. C, (1999) 181-186
14. Xu, X., Antsaklis, P.J.: Stabilization of second-order LTI switched systems, Int. J. Control **73** (2000) 1261-1279
15. Sastry, S.: Nonlinear Systems, Springer, New York, (1999)
16. Zhang, L., Chen, Y., and Cui, P.: Stabilization of a Class of Switched Linear Systems, Nonlinear Analysis: Theory, Methods and Applications, Special Issue on Hybrid Systems and Applications, **62** (2005) 1527-1535

# Computation of the Adjoint Matrix

Alkiviadis Akritas[1] and Gennadi Malaschonok[2]

[1] University of Thessaly, Department of Computer and Communication Engineering, GR-38221 Volos, Greece
akritas@uth.gr
[2] Tambov State University, Laboratory for Algebraic Computations, Internatsionalnaya 33, 392622 Tambov, Russia
malaschonok@math-iu.tstu.ru

**Abstract.** The best method for computing the adjoint matrix of an order $n$ matrix in an arbitrary commutative ring requires $O(n^{\beta+1/3} \log n \log\log n)$ operations, provided that the complexity of the algorithm for multiplying two matrices is $\gamma n^\beta + o(n^\beta)$. For a commutative domain – and under the same assumptions – the complexity of the best method is $6\gamma n^\beta/(2^\beta - 2) + o(n^\beta)$. In the present work a new method is presented for the computation of the adjoint matrix in a commutative domain. Despite the fact that the number of operations required is now 1.5 times more, than that of the best method, this new method permits a better parallelization of the computational process and may be successfully employed for computations in parallel computational systems.

## 1 Statement of the Problem

The adjoint matrix is a transposed matrix of algebraic complements. If the determinant of the matrix is nonzero, then the inverse matrix may be computed as the adjoint matrix divided by the determinant. The adjoint matrix of a given matrix $A$ will be denoted by $A^*$: $A^* = \det(A)A^{-1}$.

The best method for computing the adjoint matrix of an order $n$ matrix in an arbitrary commutative ring requires $O(n^{\beta+1/3} \log n \log\log n)$ operations (see [1] and [2]). For a commutative domain the complexity of the best method is $6\gamma n^\beta/(2^\beta - 2) + o(n^\beta)$ (see [3]). It is asssumed that the complexity of the algorithm for multiplying two matrices is $\gamma n^\beta + o(n^\beta)$.

In a commutative domain the algorithm is based on applications of determinant identities [3]. It generalizes in a commutative domain the following formula for the inverse matrix $\mathcal{A}^{-1}$:

$$\mathcal{A}^{-1} = \begin{pmatrix} I & -A^{-1}C \\ 0 & I \end{pmatrix} \begin{pmatrix} I & 0 \\ 0 & (D - BA^{-1}C)^{-1} \end{pmatrix} \begin{pmatrix} I & 0 \\ -B & I \end{pmatrix} \begin{pmatrix} A^{-1} & 0 \\ 0 & I \end{pmatrix},$$

where $\mathcal{A} = \begin{pmatrix} A & C \\ B & D \end{pmatrix}$ – is an invertible matrix with invertible block $A$.

In the present work a new method is proposed for the computation of the adjoint matrix in a commutative domain. Despite the fact that the number of

V.N. Alexandrov et al. (Eds.): ICCS 2006, Part II, LNCS 3992, pp. 486–489, 2006.

operations required is now 1.5 times more, than that of the algorithm described in [1], this new method permits a better parallelization of the computational process.

This new method generalizes in a commutative domain the following factorization of the inverse matrix $\mathcal{A}^{-1}$:

$$\mathcal{A}^{-1} = \begin{pmatrix} I & -A^{-1}C \\ 0 & I \end{pmatrix} \begin{pmatrix} I & 0 \\ 0 & (B^{-1}D - A^{-1}C)^{-1} \end{pmatrix} \begin{pmatrix} I & 0 \\ -I & I \end{pmatrix} \begin{pmatrix} A^{-1} & 0 \\ 0 & B^{-1} \end{pmatrix}.$$

In the next section we present without proofs the determinant identity of column replacement, which is used as the basis of the proposed method for computing the adjoint matrix, along with several additional theorems, which are fundamental for the new method. In the final section the algorithm and a small example are presented for the computation of the adjoint matrix.

## 2 Identity of Column Replacement

Let $B$ be a matrix of order $n$ and assume two different columns are fixed. We denote by $B_{\{x,y\}}$ the matrix which is obtained from $B$ after replacing the two fixed columns by the columns $x$ and $y$, respectively.

**Theorem 1.** *(Identity of column replacement.)*
*For every matrix $B \in R^{n\times n}$ and columns $a, b, c, d \in R^n$ the following identity holds*

$$\det B_{\{ab\}} \det B_{\{cd\}} = \begin{vmatrix} \det B_{\{ad\}} & \det B_{\{db\}} \\ \det B_{\{ac\}} & \det B_{\{cb\}} \end{vmatrix} \tag{1}$$

For example, for the matrix of order 2 the identity of column replacement is as follows

$$\begin{vmatrix} a & c \\ b & d \end{vmatrix} \begin{vmatrix} x & u \\ y & v \end{vmatrix} = \begin{vmatrix} a & u \\ b & v \end{vmatrix} \begin{vmatrix} x & c \\ y & d \end{vmatrix} - \begin{vmatrix} a & x \\ b & y \end{vmatrix} \begin{vmatrix} u & c \\ v & d \end{vmatrix}.$$

**Theorem 2.** *Let $R$ be a commutative domain, $\mathcal{A} = \begin{pmatrix} A & C \\ B & D \end{pmatrix}$ a matrix of order $2n$ over $R$, $A, B, C, D$ square blocks, $\alpha = \det A \neq 0$, $\beta = \det B \neq 0$, $F = \alpha B^* D - \beta A^* C$.*

*Then every minor of order $k$ of the matrix $F$ $(k \leq n)$ is divisible by $(\alpha\beta)^{k-1}$ and the following identity holds*

$$\det F = (\alpha\beta)^{n-1} \det \mathcal{A}. \tag{3}$$

**Theorem 3.** *Let $R$ be a commutative domain, $\mathcal{A} = \begin{pmatrix} A & C \\ B & D \end{pmatrix}$ a matrix of order $2n$ over $R$, $A, B, C, D$ square blocks, $\alpha = \det A \neq 0$, $\beta = \det B \neq 0$ and $F = \alpha B^* D - \beta A^* C$.*

*Then $(\alpha\beta)^{-n+2} F^* \in R^n$ and*

$$\mathcal{A}^* = \begin{pmatrix} \alpha^{-1}|\mathcal{A}|I & -\alpha^{-2}\beta^{-1}A^*C \\ 0 & \alpha^{-1}\beta^{-1}I \end{pmatrix} \begin{pmatrix} I & 0 \\ 0 & (\alpha\beta)^{-n+2}F^* \end{pmatrix} \begin{pmatrix} I & 0 \\ -\beta I & \alpha I \end{pmatrix} \begin{pmatrix} A^* & 0 \\ 0 & B^* \end{pmatrix}. \tag{4}$$

**Theorem 4.** *Let $R$ be a commutative domain, $0 \neq \gamma \in R$, $\mathcal{A} = \begin{pmatrix} A & C \\ B & D \end{pmatrix}$ a matrix of order $2n$ ($n \geq 2$) over $R$, such that every minor of order $k$ is divisible by $\gamma^{k-1}$, $A, B, C, D$ square blocks, $\alpha = \gamma^{1-n} \det A \neq 0$, $\beta = \gamma^{1-n} \det B \neq 0$, $\mathbf{A}^* = \gamma^{2-n} A^*$, $\mathbf{B}^* = \gamma^{2-n} B^*$ and $F = (\alpha\gamma^{-1}\mathbf{B}^* D - \beta\gamma^{-1}\mathbf{A}^* C)$.*

*Then,*

$$\gamma^{2-2n}\mathcal{A}^* = \begin{pmatrix} \alpha^{-1}\gamma^{1-2n}|\mathcal{A}|I & -(\alpha^2\beta\gamma)^{-1}\mathbf{A}^*C \\ 0 & (\alpha\beta\gamma)^{-1}I \end{pmatrix} \times$$

$$\begin{pmatrix} I & 0 \\ 0 & (\alpha\beta)^{-n+2}F^* \end{pmatrix} \begin{pmatrix} I & 0 \\ -\beta I & \alpha I \end{pmatrix} \begin{pmatrix} \mathbf{A}^* & 0 \\ 0 & \mathbf{B}^* \end{pmatrix}. \qquad (5)$$

*Here $\gamma^{2-2n}\mathcal{A}^*$ and the last three factors on the right-hand side of (5) are matrices over $R$.*

## 3 The Algorithm

Using the theorems we proved about the factorization of the adjoint matrix we now introduce the algorithm for computing it along with the determinant of a given matrix.

Let $R$ be a commutative domain, $0 \neq \gamma \in R$, $\mathcal{A} = \begin{pmatrix} A & C \\ B & D \end{pmatrix}$ a matrix of order $2n = 2^N$ over $R$, such that every minor of order $k$ is divisible by $\gamma^{k-1}$. Moreover, we assume that all minors, on which a division is performed during the computation of the adjoint matrix, are non-zero.

The inputs to the algorithm are the matrix $\mathcal{A}$ and the number $\gamma = 1$.

The outputs from the algorithm are $\gamma^{1-2n}|\mathcal{A}|$ and $\gamma^{2-2n}\mathcal{A}^*$. Note here that the determinant of the matrix has been divided by $\gamma^{2n-1}$, and that the adjoint matrix has been divided by $\gamma^{2n-2}$.

**Algorithm ParAdjD**

$$\{\ \gamma^{1-2n}|\mathcal{A}|, \gamma^{2-2n}\mathcal{A}^*\ \}\textbf{=ParAdjD}(\mathcal{A}, \gamma)$$

*Input:* $\mathcal{A} = \begin{pmatrix} A & C \\ B & D \end{pmatrix}$, and $\gamma$. $A, B, C, D \in R^{n \times n}$, $\gamma \in R$.
*Output:* $\{\gamma^{1-2n}|\mathcal{A}|,\ \gamma^{2-2n}\mathcal{A}^*\ \}$.

1. If the matrix $\mathcal{A}$ is of order two, then
*output:*

$$\left\{\gamma^{-1}(AD - BC), \begin{pmatrix} D & -C \\ -B & A \end{pmatrix}\right\}.$$

otherwise, proceed to the next point.
2. Concurrently compute
$\{\ \alpha, \mathbf{A}^*\}$**=ParAdjD**$(A, \gamma)$ and $\{\ \beta, \mathbf{B}^*\}$**=ParAdjD**$(B, \gamma)$.

3. Concurrently compute
   $N = \gamma^{-1}\mathbf{B}^* D$ and $M = \gamma^{-1}\mathbf{A}^* C$, and then
   $F = \alpha N - \beta M$.
4. Compute
   $\{\varphi, \mathbf{F}^*\}=$**ParAdjD**$(F, \alpha\beta)$.
5. Concurrently compute
   $\varphi' = \gamma^{-1}\varphi$, $H = \alpha^{-1}\gamma^{-1}\mathbf{F}^*\mathbf{A}^*$ and $L = \beta^{-1}\gamma^{-1}\mathbf{F}^*\mathbf{B}^*$.
6. Concurrently compute
   $H' = \alpha^{-1}(\varphi'\mathbf{A}^* + MH)$ and $L' = -\alpha^{-1}ML$.

*Output:*

$$\left\{\varphi', \begin{pmatrix} H' & L' \\ -H & L \end{pmatrix}\right\}.$$

### 3.1 Example

*Input:*

$$\mathcal{A} = \begin{pmatrix} A & C \\ B & D \end{pmatrix} = \begin{pmatrix} 0 & 2 & -2 & 2 \\ 1 & -3 & 1 & -2 \\ 3 & 0 & -3 & 0 \\ -1 & 3 & -1 & 1 \end{pmatrix}, \ \gamma = 1.$$

*Output:*

$$\left\{\varphi', \begin{pmatrix} H' & L' \\ -H & L \end{pmatrix}\right\} = \left\{6, \begin{pmatrix} -9 & -12 & 4 & -6 \\ -6 & -6 & 2 & 0 \\ -9 & -12 & 2 & -6 \\ 0 & -6 & 0 & -6 \end{pmatrix}\right\}.$$

## References

1. Kaltofen E.: On Computing Determinants of Matrices Without Divisions. Proc. Internat. Symp. Symbolic Algebraic Comput. ISSAC'92. ACM Press, (1992) 342–349.
2. Kaltofen E., Villard G.: On the complexity of computing determinants. Proc. Fifth Asian Symposium on Computer Mathematics, ASCM 2001. Extended abstract. (2001) 13–27.
3. Malaschonok G.I.: Effective Matrix Methods in Commutative Domains. Formal Power Series and Algebraic Combinatorics. Springer (2000) 506–517.

# MathBlackBoard as Effective Tool in Classroom

Deguchi Hiroaki[1] and Hashiba Hirokazu[2]

[1] Kobe University, 2-1 Rokkodai-cho, Nada-ku, Kobe 657-8501, Japan
[2] Akashi Junior High School attached to Kobe University,
3-4 Yamashita-cho, Akashi 673-0878, Japan

**Abstract.** MathBlackBoard is a Java program that has unique GUI (Graphical User Interface). By using GUI, MathBlackBoard users can input, edit and manipulate mathematical expressions easily. And MathBlackBoard has its own kernel. However, it has functions to use other computer algebra system as computational engine.

In this paper, we describe an example of using MathBlackBoard. We have added a new function which is designed to be used in junior high school classroom. And we have used it in classrooms.

## 1 Introduction

MathBlackBoard is based on a Java applet named "blackboard applet" which was written in 1997 by Matsushima J. ([1]). The blackboard applet has a GUI where users can input, edit and manipulate mathematical expressions easily. And GUI of the blackboard applet is inherited to MathBlackBoard. Moreover, a kernel of MathBlackBoard is developed from some objects of the blackboard applet.

The purpose of the blackboard applet was to create an "ease to use" computer algebra system. Matsushima argued that difficult matters about the educational use of computer algebra systems are "difficulties of input by using CUI (Character-based User Interface or Command-line User Interface)" and "dependence to OS." The blackboard applet was produced to try to resolve these issues. "GUI buttons for mathematical operation" and "drag & drop with the meaning of mathematical operation" were mentioned as resolutions to the first issue. And second issue has been resolved by "use of Java applet on web browser."

## 2 GUI of MathBlackBoard

We have improved the blackboard applet after the master thesis([1]), and decided to rename it MathBlackBoard ([2][3]). Now MathBlackBoard is able to run not only as Java applet, but also as Java application. And it is reconstructed as a set of objects. Then, MathBlackBoard has been enriched with new functions which can connect to some computer algebra systems and use them as computational engine ([4]). In addition, its GUI operations have been improved.

V.N. Alexandrov et al. (Eds.): ICCS 2006, Part II, LNCS 3992, pp. 490–493, 2006.

### 2.1 GUI Windows

MathBlackBoard has three GUI windows: Blackboard window (Fig. 1), Keyboard window and GraphScreen window. Blackboard window is main window that has editing space. Mathematical expressions are shown in this window. And users can input, edit and manipulate expressions on Blackboard window.

Keyboard window is used to input and edit expressions. And MathBlackBoard provides GraphScreen window where graphs are drawn.

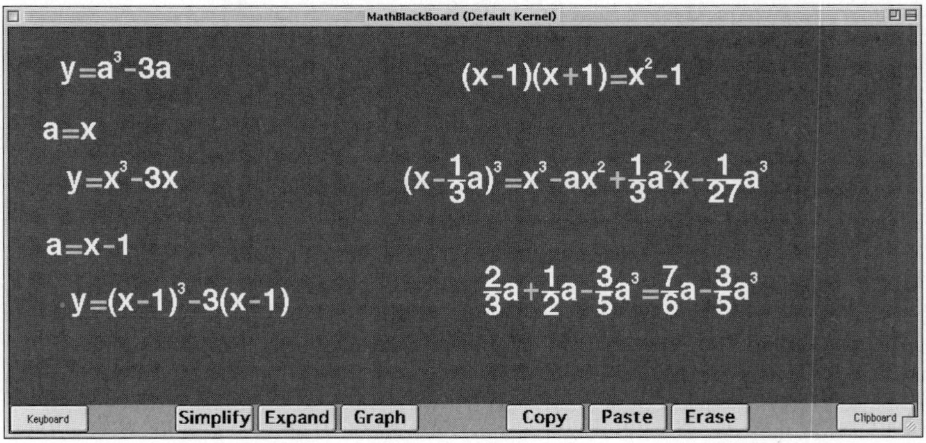


**Fig. 1.** Blackboard Window

The function buttons are placed at the bottom of Blackboard window (Fig. 1). These GUI buttons are provided to execute functions instead of typing commands.

### 2.2 Drag & Drop

One of characteristics of MathBlackBoard is flexibility of position of mathematical expressions displayed. Expressions can be moved to anywhere on the editing area of Blackboard window. And MathBlackBoard provides "drag & drop" to manipulate mathematical expressions. Editorial operations and mathematical operations are assigned to "drag & drop."

Editorial operations are as follows: **Move** expression to another point, **Connect** grabbed expression to the other expression, **Copy** selected expression to clipboard, and **Paste** an expression from clipboard.

Mathematical operations are as follows: **Substitute**, **Draw** graph, and **Perform** functions assigned to GUI buttons.

## 3 Using MathBlackBoard in Classroom

We have had an opportunity of testing MathBlackBoard in classrooms, at Akashi Junior High School attached to Kobe University. The classes were in the unit

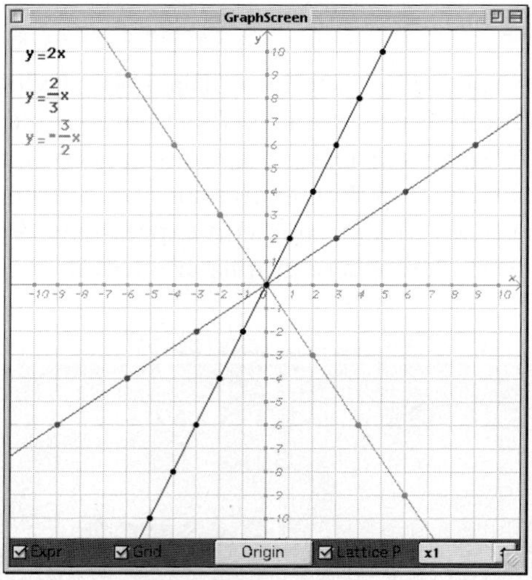


**Fig. 2.** New GraphScreen Window for the Class

"direct & inverse proportion." MathBlackBoard has been used in the lessons to make graphs of direct proportion. The purposes of the lesson were as follows:

1. To understand the difference of graphs between plus-constant and minus-constant of direct proportion
2. To make graphs of direct proportion of fraction-constant, by using pen and paper

### 3.1 Preparations for the Class

The teacher of the classes requested to improve MathBlackBoard as follows:

1. The range of GraphScreen window must be "-10 to 10" which is about the same of student's graph paper
2. Mathematical expressions of graphs are shown by default
3. Grid lines are drawn by default
4. Lattice points of graphs are described, if it could be

It is the reason for the request "4" that students of the classes had learned to make graphs by connecting dots. The teacher expected that students read out graphs on MathBlackBoard strictly, and compare them with graphs on students' paper.

### 3.2 Scenes on the Class

MathBlackBoard was used as follows:

1. The teacher present how to make a graph of direct proportion by using MathBlackBoard
2. By using pen and paper, students make graphs of some expressions presented

3. Students use MathBlackBoard, to make graphs of presented expressions
4. Students compare graphs of their paper with graphs of MathBlackBoard

## 4 Conclusion

With those implementation above, MathBlackBoard has been students' easily accessible tool. Students were able to acquire how to use MathBlackBoard after teacher's presentation. The teacher presented how to use MathBlackBoard just once per a class. It was enough time for students to acquire how to use MathBlackBoard. They were able to grade their graphs on paper easily.

After these classes, we discussed about improvement of MathBlackBoard. The teacher remarked that he required new functions. And he showed a plan to use these new functions in other scene.

We consider "substitution by using drag & drop" can help student's understanding of concept of substitution. But it is necessary to bring MathBlackBoard close to the "pen and paper" style. We have to implement new style similar to student's work in their notebook, instead of mechanical replacement.

The goal of MathBlackBoard development is to provide new environment. It is not an environment that solve problems, but an environment where users can solve problems.

## References

1. Matsushima J.: *Useful Computer Algebra System by Using Java.* Master Thesis, Kobe University. (1998) [in Japanese]
2. Deguchi H.: Blackboard Applet, *Journal of Japan Society for Symbolic and Algebraic Computation* **9**(1) (2002) 32–37 [in Japanese]
3. Deguchi H.: MathBlackBoard, *Journal of Japan Society for Symbolic and Algebraic Computation* **11**(3,4) (2005) 77–88 [in Japanese]
4. Deguchi H.: MathBlackBoard as User Interface of Computer Algebra Systems, *Proceedings of the 10th Asian Technology Conference in Mathematics* (2005) 246–252

# Finding Inefficiencies in OpenMP Applications Automatically with Periscope*

Karl Fürlinger and Michael Gerndt

Institut für Informatik,
Lehrstuhl für Rechnertechnik und Rechnerorganisation
Technische Universität München
{Karl.Fuerlinger, Michael.Gerndt}@in.tum.de

**Abstract.** Performance optimization of parallel programs can be a time-consuming and difficult task. Therefore, tools are desirable that help application developers by automatically locating inefficiencies. We present Periscope, a system for automated performance analysis based on the notion of performance *properties*.

We present the overall architecture of Periscope, which consists of a set of analysis agents and show how properties of OpenMP applications are detected. We describe the set of OpenMP properties we have defined so far and the data model used in the specification of these properties. Practical tests on the feasibility of our approach are performed with a number of OpenMP applications.

## 1 Introduction

With creating scientific parallel programs comes the question of efficiency. Does the program make optimal use of the available resources? Is the effort of parallelization paying off or is performance lost somewhere? Assisting programmers in answering such questions by automated performance analysis is important because performance analysis can be a difficult and time-consuming task.

In this paper we present Periscope, a tool for automated performance analysis of OpenMP and MPI codes and evaluate it on finding inefficiencies in OpenMP applications. Periscope detects inefficiencies by automatically searching for performance problems that are specified in terms of *performance properties*. We present the overall design and implementation of Periscope and test it on a number of OpenMP applications.

The rest of this paper is organized as follows: In Sect. 2 we describe the overall structure of Periscope and our monitoring approach for OpenMP applications. We describe how properties are specified with respect to a data model and specify the data model for OpenMP applications. Sect. 3 then describes a number of properties we have defined for OpenMP. In Sect. 4 we test our approach on a several applications from the NAS benchmark suite, in Sect. 6 we summarize and discuss ideas for future work.

* This work was partially funded by the Deutsche Forschungsgemeinschaft (DFG) under contract GE1635/1-1.

V.N. Alexandrov et al. (Eds.): ICCS 2006, Part II, LNCS 3992, pp. 494–501, 2006.

## 2 Periscope

Periscope is a system for automated performance analysis on large-scale cluster-like systems. The analysis is performed by a set of *agents* distributed over the machine. The agents are arranged in a hierarchy, node-level agents form the lowest level of the hierarchy, a master-agent integrates the performance data of the whole system and connects to the front-end of the tool with is the interface to the user. Intermediate agents form a tree with the master-agent as root and the node-level agents as leaves. All components use a registry service to register themselves and to discover higher or lower level agents, see Fig. 1.

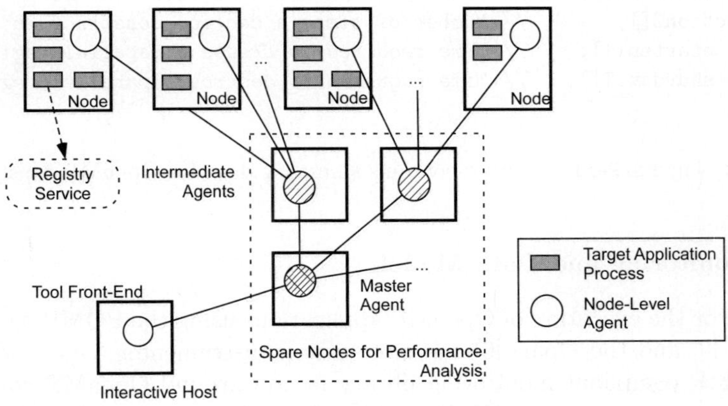


**Fig. 1.** The arrangement of the Periscope agents forms a hierarchy. At the lowest level, node-agents detect performance properties. Intermediate agents integrate the results of the node-level agents. A single master agent forms the connection to the tool's front-end. All components register themselves with a registry service.

While parts of the Periscope system are still under development (e.g., the higher-level agents) the functionality for performance analysis of OpenMP applications is already available in the node-level agents and this is what we evaluate in this paper. Periscope's agent hierarchy mainly serves to enable scalable analysis on large machines (hundreds of nodes), where the central collection of performance data can lead to problems related to the scalability of data management, analysis and visualization.

Periscope's node-level agents perform an automated search for performance properties of the target application. The properties are formally specified in a language called ASL (APART specification language) [1] with respect to a certain data-model that depends on the particular programming model used (message passing or shared memory programming). The data model for OpenMP is described in the next section, while the properties we have defined for OpenMP are described in Sect. 3

```
ParPerf {
  Region *reg,              // The region for which the summary is collected
  Experiment *exp,          // The experiment where this data belongs to
  int threadC,              // Number of threads that executed the region
  double execT[],           // Total execution time per thread
  double execC[],           // Total execution count per thread
  double exitBarT[],        // Time spent in the implicit exit barrier
  double singleBodyT[],     // Time spent inside a single construct
  int singleBodyC[],        // Execution count in a single construct
  double enterT[],          // Time spent waiting to enter a construct
  int enterC[],             // Number of times a threads enters a construct
  double exitT[],           // Time spent to exit a construct
  int exitC[],              // Number of times a thread exits a construct
  double sectionT[],        // Time spent inside a section construct
  int sectionC[],           // Number of times a section construct is entered
  double startupT[],        // Time required to create a parallel region
  double shutdownT[],       // Time required to destroy a parallel region
}
```

**Fig. 2.** The `ParPerf` structure contains summary data for OpenMP constructs

### 2.1 Monitoring and Data Model

We monitor the execution of OpenMP applications using the POMP monitoring interface [5] and the Opari [6] source-to-source instrumenter. Opari adds calls to a POMP compliant monitoring library in and around OpenMP constructs. Periscope implements the POMP interface in a library that is linked with the target application and is thus able to observe the execution of OpenMP applications.

Our monitoring library observes the events the application generates and writes event records to a buffer that is located in a shared memory segment. The event records are processed and analyzed by a node-level agent executing on the same node as the target application. This way, we try to minimize the perturbation of the target application, as any more time-consuming analysis is performed by the node-level agent executing on a processor set-aside for performance analysis. More details about our monitoring system as outlined above can be found in [2].

The node-level analyzes the monitoring events and generates performance data corresponding to the ASL data model. For OpenMP applications, the data model is represented by the `ParPerf` data structure shown in Fig. 2.

The `ParPerf` structure holds summary (profiling) data for individual OpenMP regions (i.e., the extent of OpenMP regions and user-defined regions). The `reg` member of the `ParPerf` structure points to the region for which the data is collected. The meaning of the other members is briefly explained in the form of "comments" in Fig. 2. In [3] we describe this data model in detail. Note that not all data members are meaningful for all OpenMP regions. For example, `sectionT` and `sectionC` are only defined for a `section` construct.

## 3 Performance Properties

Performance properties are formally specified in ASL (APART specification language). The property specification has three major components: condition, confidence and severity. Severity describes the impact on the performance of the application that a property represents. Confidence is a measure of the certainty that a property holds and condition specifies how the property can be checked. An example ASL specification for the `ImbalanceAtBarrier` property is shown in Fig. 3.

In this example, and in general, the severity depends on the computed imbalance which is divided by a value returned by a 'ranking basis' function (`RB`). The ranking basis allows a scaling of the severity with respect to the experiment conducted. In all results presented in this study, the ranking basis corresponds to the overall time used by all threads (wall-clock time $\times$ number of threads). Hence the severity of a property corresponds to improvement in runtime a program would experience, if the reason for the inefficiency could be eliminated entirely.

```
property ImbalanceAtBarrier(ParPerf pd) {
  let
    imbal = pd.execT[0]+...+pd.execT[pd.threadC-1];

  condition  : (pd->reg.type==BARRIER) && (imbal > 0);
  confidence : 1.0;
  severity   : imbal / RB(pd.exp);
}
```

**Fig. 3.** The ASL specification of the `ImbalanceAtBarrier` property

Among others we have defined the following properties for OpenMP applications: `ImbalanceAtBarrier`, `ImbalanceInParallelSections`, `Imbalance-InParallelLoop`, `ImbalanceInParallelRegion`, `UnparallelizedInSingleRe-gion`, `UnparallelizedInMasterRegion`, `ImbalanceDueToNotEnoughSections`, `ImbalanceDueToUnevenSectionDistribution`, `LockContention`, and `Criti-calSectionContention`.

The `ImbalanceAtBarrier` and `ImbalanceIn*` properties refer to time spent waiting at explicit or implicit barriers, respectively. For explicit (programmer-added) barriers the barrier wait time of thread $n$ is available in the `execT[`$n$`]` member of the `ParPerf` structure, see Fig. 3. Implicit barriers are added to work-sharing constructs and parallel regions by Opari in order to measure the load imbalance. The waiting time is available in the `exitBarT` member in this case (the barrier is added at the end of the respective construct, hence the name).

The `ImbalanceDueTo*` properties give the reason for the discovered imbalance in a more detailed fashion. `NotEnoughSections` refers to the fact that there where too few sections for the available number of threads, whereas `Uneven-SectionDistribution` indicates that some threads executed more sections than others. As an example consider a `section` construct that contains six individual `sections`. When the work per section is approximately equal and four threads execute the construct, two threads will execute 2 sections and two threads will only execute one. If, on the other hand, eight threads are used, six threads will each execute one section and the remaining two will be idle. The `sectionC` member of `ParPerf` can be used to find out the number of section constructs executed by each thread.

The `UnparallelizedIn*` properties capture the situation that time was lost due to a single thread executing a construct. For `single` constructs the severity is measured by the summed time in `exitBarT`, for `master` region, the severity is approximated by the master's time divided by the number of threads.

`CriticalSection` and `LockContention` sum up the time lost due to threads waiting to acquire a lock or to enter a critical section, respectively. The properties are based on the `enterT` and `exitT` times.

### 3.1 Implementation

In Periscope properties are implemented as C++ classes compiled to dynamically loadable objects (.so files). To simplify development, only the `condition` code has to be written for each property and a script generates a complete C++ class. A severity value is computed in the condition code and is returned by the `severity` method of the C++ class, the confidence is fixed to 1.0 for our prototype implementation.

Having the properties available as dynamically loadable modules allows the development and deployment of the tool separately from the "knowledge base". Without changing the main tool, new performance properties can be implemented and tested or existing ones can be modified.

## 4 Test Setup and Results

We tested Periscope on the OpenMP version of the NAS parallel benchmarks version 3.2. The programs in the NAS benchmark suite are derived from CFD applications, consists of five kernels (EP, MG, CG, FT, IS) and three simulated CFD applications (LU, BT, SP).

We executed the applications on a 32-CPU SGI Altix system based on Itanium-2 processors with 1.6 GHz and 6MB L3 Cache using a batch system. The number of threads was set to eight and the Periscope node-level agent was executed on a separate CPU (i.e., nine CPUs were requested for the batch runs).

The Periscope node-level agents have the ability to conduct the search for performance properties at any time during the execution of the target application (on-line performance analysis). In this study, however, we use the node-level agents in a post-mortem mode, i.e., the search for performance properties is triggered when the application finishes.

The table in Figure 4 shows all properties identified by Periscope for the NAS benchmarks. Note that for completeness this table shows all properties without applying a severity cutoff threshold. Some properties in Figure 4 have very

| Property | BT | CG | EP | FT | IS | LU | MG | SP |
|---|---|---|---|---|---|---|---|---|
| ImbalanceAtBarrier | | | | | 1 | 3 | | |
| ImbalanceInParallelSections | | | | | | | | |
| ImbalanceInParallelLoop | 12 | 13 | 1 | 8 | 2 | 9 | 12 | 16 |
| ImbalanceInParallelRegion | 6 | 9 | 1 | | 2 | 8 | 2 | 5 |
| UnparallelizedInSingleRegion | | | | | | 3 | | |
| UnparallelizedInMasterRegion | 4 | | | | | 13 | 2 | 5 |
| ImbalanceDueToNotEnoughSections | | | | | | | | |
| ImbalanceDueToUnevenSectionDistribution | | | | | | | | |
| CriticalSectionContention | | | | | | | | 1 |
| LockContention | | | | | | | | |

**Fig. 4.** Performance Properties identified by Periscope. This table lists all discovered performance properties, even such with very low severity values.

| Benchmark | Property | Region | Severity |
|---|---|---|---|
| BT | ImbalanceInParallelLoop | rhs.f 177--290 | 0.0446 |
| BT | ImbalanceInParallelLoop | y_solve.f 40--394 | 0.0353 |
| BT | ImbalanceInParallelLoop | rhs.f 299--351 | 0.0347 |
| CG | ImbalanceInParallelLoop | cg.f 556--564 | 0.0345 |
| CG | ImbalanceInParallelRegion | cg.f 772--795 | 0.0052 |
| CG | ImbalanceInParallelRegion | cg.f 883--957 | 0.0038 |
| EP | ImbalanceInParallelRegion | ep.f 170--230 | 0.0078 |
| EP | ImbalanceInParallelLoop | ep.f 129--133 | 0.0001 |
| FT | ImbalanceInParallelLoop | ft.f 606--625 | 0.0676 |
| FT | ImbalanceInParallelLoop | ft.f 653--672 | 0.0304 |
| FT | ImbalanceInParallelLoop | ft.f 227--235 | 0.0269 |
| IS | ImbalanceAtBarrier | is.c 526 | 0.0272 |
| IS | ImbalanceInParallelRegion | is.c 761--785 | 0.0087 |
| IS | ImbalanceInParallelLoop | is.c 397--403 | 0.0020 |
| LU | ImbalanceAtBarrier | ssor.f 211 | 0.0040 |
| LU | ImbalanceAtBarrier | ssor.f 182 | 0.0032 |
| LU | ImbalanceInParallelLoop | rhs.f 189--309 | 0.0011 |
| MG | ImbalanceInParallelLoop | mg.f 608--631 | 0.0831 |
| MG | ImbalanceInParallelLoop | mg.f 779--815 | 0.0291 |
| MG | ImbalanceInParallelLoop | mg.f 536--559 | 0.0248 |
| SP | ImbalanceInParallelLoop | x_solve.f 27--296 | 0.0285 |
| SP | ImbalanceInParallelLoop | y_solve.f 27--292 | 0.0265 |
| SP | ImbalanceInParallelLoop | z_solve.f 31--326 | 0.0239 |

**Fig. 5.** The three most severe performance properties with source-code location and severity value, identified by Periscope (only two properties were found for EP)

low severity values and do not actually represent severe performance problems. Figure 5 shows the three most severe properties identified by Periscope. All properties have severity values below nine percent. Most properties are in the range of three or four percent.

## 5 Related Work

A number of tools try to automate the process of performance analysis. Expert [9] performs automated search for inefficiencies in trace files. The execution of instrumented programs generates traces in the Epilog format that are analyzed by Expert. Expert presents the results in a viewer with three hierarchies, one hierarchy contains the type of inefficiency, one hierarchy shows the machine and one hierarchy shows the program's resources (files, functions,...). Recent improvements of Expert include an algebra for performing cross-experiment analysis and support for virtual topologies.

Paradyn [4] is a tool for automated on-line performance analysis based on dynamic instrumentation. In the running application Instrumentation is added and removed as required by the currently tested *hypothesis* (which performance problem exist) and *focus* (where the problem exists). An infrastructure for the efficient collection of performance data called MRNet [7] has been developed and integrated in Paradyn, while the search for performance problems is still performed centrally by Paradyn's Performance Consultant. Recently, a distributed search methodology supporting a partially distributed approach and a fully distributed approach has been added to Paradyn, called the Distributed Performance Consultant [8].

## 6 Summary and Future Work

Periscope is a tool for performance analysis based on the automated search for performance properties. Properties are formally specified with respect to a data-model that depends on the programming model employed. We described the data model used for OpenMP applications and a set of properties based on that data model.

We have tested Periscope and the performance properties on the OpenMP NAS benchmark suite. The study shows that Periscope is an efficient tool for the automated detection of inefficiencies. Inefficiencies were detected in each of the applications. In some applications the detected inefficiencies were not signification (e.g., EP); for MG an inefficiency of about 8 percent was discovered.

An attractive feature of Periscope lies in the fact that the specification of the tool's knowledge base is separate from the implementation of the tool itself. Performance properties are specified as C++ objects that are compiled into shared objects that are dynamically loaded by Periscope at tool startup. This allows for an easy extension of Periscope without the need to re-compile the entire tool and allows a performance expert to easily experiment with new property specifications.

Future work is planned along several directions: the data model described in this paper and the properties specification based on it only use timing data. For the future we plan to integrate hardware-counter data as well. This will enable the detection of important inefficiencies, related to cache usage for example.

## References


1. Thomas Fahringer, Michael Gerndt, Bernd Mohr, Felix Wolf, Graham Riley, and Jesper Larsson Träff. Knowledge specification for automatic performance analysis. APART technical report, revised edition. Technical Report FZJ-ZAM-IB-2001-08, Forschungszentrum Jülich, 2001.
2. Karl Fürlinger and Michael Gerndt. Distributed application monitoring for clustered SMP architectures. In Harald Kosch, László Böszörményi, and Hermann Hellwagner, editors, *Proceedings of the 9th International Euro-Par Conference on Parallel Processing*, pages 127–134. Springer, August 2003.
3. Karl Fürlinger and Michael Gerndt. Performance analysis of shared-memory parallel applications using performance properties. In *Proceedings of the 2005 International Conference on High Performance Computing and Communications (HPCC-05)*, pages 595–604, September 2005. Accepted for publication.
4. Barton P. Miller, Mark D. Callaghan, Jonathan M. Cargille, Jeffrey K. Hollingsworth, R. Bruce Irvin, Karen L. Karavanic, Krishna Kunchithapadam, and Tia Newhall. The Paradyn parallel performance measurement tool. *IEEE Computer*, 28(11):37–46, 1995.
5. Bernd Mohr, Allen D. Malony, Hans-Christian Hoppe, Frank Schlimbach, Grant Haab, Jay Hoeflinger, and Sanjiv Shah. A performance monitoring interface for OpenMP. In *Proceedings of the Fourth Workshop on OpenMP (EWOMP 2002)*, September 2002.
6. Bernd Mohr, Allen D. Malony, Sameer S. Shende, and Felix Wolf. Towards a performance tool interface for OpenMP: An approach based on directive rewriting. In *Proceedings of the Third Workshop on OpenMP (EWOMP'01)*, September 2001.
7. Philip C. Roth, Dorian C. Arnold, and Barton P. Miller. MRNet: A software-based multicast/reduction network for scalable tools. In *Proceedings of the 2003 Conference on Supercomputing (SC 2003)*, November 2003.
8. Philip C. Roth and Barton P. Miller. The distributed performance consultant and the sub-graph folding algorithm: On-line automated performance diagnosis on thousands of processes. In *Proceedings of the ACM SIGPLAN Symposium on Principles and Practice of Parallel Programming (PPoPP'06)*, March 2005. Accepted for Publication.
9. Felix Wolf and Bernd Mohr. Automatic performance analysis of hybrid MPI/OpenMP applications. In *Proceedings of the 11th Euromicro Conference on Parallel, Distributed and Network-Based Processing (PDP 2003)*, pages 13–22. IEEE Computer Society, February 2003.

# Analysis of the Spatial and Temporal Locality in Data Accesses

Jie Tao, Siegfried Schloissnig, and Wolfgang Karl

Institut für Technische Informatik
Universität Karlsruhe (TH), 76128 Karlsruhe, Germany
{tao, schloissnig, karl}@ira.uka.de

**Abstract.** Cache optimization becomes increasingly important for achieving high computing performance, especially on current and future chip-multiprocessor (CMP) systems, which usually show a rather higher cache miss ratio than uni-processors. For such optimization, information about the access locality is needed in order to help the user in the tasks of data allocation, data transformation, and code transformation which are often used to enhance the utilization of cached data towards a better cache hit rate.

In this paper we demonstrate an analysis tool capable of detecting the spatial and temporal relationship between memory accesses and providing information, such as access pattern and access stride, which is required for applying some optimization techniques like address grouping, software prefetching, and code transformation. Based on the memory access trace generated by a code instrumentor, the analysis tool uses appropriate algorithms to detect repeated address sequences and the constant distance between accesses to the different elements of a data structure. This allows the users to pack data with spatial locality in the same cache block so that needed data can be loaded into the cache at the same time. In addition, the analysis tool computes the push back distance which shows how a cache miss can be avoided by reusing the data before replacement. This helps to reduce cache misses increasing therefore the temporal reusability of the working set.

## 1 Introduction

Due to the widen gap between memory and CPU speed caches were introduced into the computer systems for buffering reused data and for providing a low access latency that matches the processor speed. However, many applications still suffer from excessive data accesses in the main memory. This problem is more challenged on current and future chip-multiprocessor systems, since these machines show a higher cache miss ratio, which can be up to four folds of that on uni-processor systems, according to some research reports [7]. As a consequence, optimization with respect to cache locality has been regarded as a critical issue in raising the overall performance of modern processors.

Currently, this kind of optimization is primarily based on two approaches: compiler-level automatic optimization and user-level manual optimization. For the former, the optimization is performed during the compiling time through code transformation

V.N. Alexandrov et al. (Eds.): ICCS 2006, Part II, LNCS 3992, pp. 502–509, 2006.

[10, 8, 9], array padding [1, 14], or both of them [4], while for the later optimization is done directly within the source code via manually rewriting the program code using the same techniques [5, 12]. In contrast to the compiler approach, the user-level optimization is more common due to its straight-forward manner. However, for both of them detailed information about the access pattern of applications is needed.

This information can be achieved using profiling tools, performance counters, simulation systems, or heuristic analysis models; however, we proposed and developed an analysis tool which performs the data analysis directly on memory accesses performed at the runtime during the execution of an application. Hence, this kind of analysis is more accurate and allows the acquisition of some information that can not be obtained using other approaches.

The memory accesses for this tool are provided by a code instrumentor, which inserts additional instructions in the assembly codes. This results in the generation of a memory trace containing all memory references, each of them shows the access address, access type, location in the source file, and the ID of the process issuing this access. Based on this trace, the analysis tool applies appropriate algorithms to detect the regularity among the references and give hits about optimization possibilities.

Overall, our tool makes the following contributions:

- Detecting access group, a group of neighboring accesses with different target address and occurring repeatedly. This helps to allocate these accesses in the way that they are held in the same cache block with a runtime effect of the access to the first data causing the load of other data into cache.
- Detecting access stride, a constant distance between accesses to elements of a data array. This gives the knowledge about which data will be required for further computing and directs the user or the system to prefetch data which is really needed. This also helps programmers to allocate data structures in the manner of presenting the spatial locality.
- Computing cache parameters, including reuse distance, set reuse distance, cache hit/miss, and push back distance. This shows whether an access is a cache hit or a cache miss and in case of misses a push back distance indicates how many steps an access must be pushed back in order to change the miss to a cache hit. This directs the users to use code transformation to enable an earlier issuing of an access before the data is evicted from the cache due to conflicts.
- Mapping between virtual address and data structures in source codes. The analysis tool is capable of providing all information described above in both virtual addresses and data structures. This enables the users to directly locate the optimization object in the source code.

The rest of the paper is organized as follows. Section 2 introduces the algorithms for detecting access patterns. This is followed by a detailed description of the functionality of this analysis tool, especially the implementation details, in Section 3. Section 4 shows some verification results based on applications from standard benchmark suites. The paper concludes in Section 5 with a short summary and some future directions.

## 2 Algorithms for Pattern Analysis

The main goal of our analysis tool is to provide information needed for applying cache optimization techniques like grouping and software prefetching. For this, we provide both address group and access stride, which is defined as:

**Address Group.** A repeated sequence of memory accesses with different target addresses and/or of position holders, where a position holder represents any address. The following is a simple example:

Assuming a memory access trace containing references with target address
100 200 300 400 ... 100 200 301 400 ... 100 200 302 400 ... 100 200 303 400 ...
Then two Address Groups exist ("-" is the position holder)
$< 100, 200 >$
$< 100, 200, -, 400 >$

**Stride.** A constant distance among neighboring accesses to different elements of a data array. A Stride is presented using the form $< start, distance, length >$, where start is the address of the first access, distance is the access stride, and length is the number of accesses with this stride. An example:

Assuming a memory access trace containing the following references
402 102 200 300 900 400 104 899 106 108 898 200 500 897 110
Then three Strides exist
$< 102, 2, 5 >$ $(102, , , , 104, , 106, 108, , , , , 110)$
$< 200, 100, 4 >$ $(200, 300, , 400, , , , , , , 500)$
$< 900, -1, 4 >$ $(900, , , 899, , , 898, , , 897)$

### 2.1 Detecting Address Groups

The work of detecting Address Groups is similar to the work of searching for repeated pattern from a DNA, RNA, or protein sequence in Bioinformatics. Because this is a hot research area, several algorithms have been proposed for efficient pattern discovery. Well-known examples are Sequence Alignment [3], Suffix-Tree [2], and Teiresias [13]. The Sequence Alignment algorithm is usually used to order two or more sequences to detect the maximal identity and similarity, while Suffix-Tree detects patterns and their positions from a sequence through building a tree with a single branch for each suffix of the sequence. However, due to the high requirement on memory space and overhead, neither of them is appropriate for detecting Address Groups from a large access trace.

Teiresias is another algorithm for pattern recognition. It first finds small patterns and then reconstructs them into larger ones. For small patterns two parameters are needed to specify both the maximal length of the pattern (L) and the number of letters (N), written in form $< L, N >$. The difference between L and N specifies the number of position holders. For example, the following are patterns with L=5 and N=3:

ABC
A-B-C
A–BC

For pattern discovery, the algorithm first performs a *scan/extension* phase for generating all small patterns of $< L, N >$. This is achieved by first building patterns of length 1 (one letter with a minimal number of occurrence) and then extending them to patterns of length L with maximal N position holders starting and ending with a letter, i.e. no position holder at the first and the last position of a pattern.

For each detected pattern the Prefixes and Suffixes are then computed. A Prefix is a prefix of a pattern having to be ended with a letter. For instance, the possible Prefixes of the three patterns above are: AB, A-B, and A–B, where each Prefix must contain at least two letters and the last position must be a letter. Similarly, a Suffix is defined as a suffix of a pattern with a letter at the starting position. For example, the Suffixes of the three patterns are BC, B-C, and BC.

In the following, the algorithm performs a *convolution* phase to those pairs of detected small patterns with the feature of the Prefix of one pattern and the Suffix of the other being the same. Such pairs are combined to larger patterns. For example, from pattern DF-A-T and A-TSE pattern DF-A-TSE is generated. In addition, the *convolution* phase is also responsible for finding other patterns, like those overlapped. The following is an example:

pattern 1: abcdf
pattern 2: abcef
overlapped pattern: abc-f

In contrast with Sequence Alignment and Suffix-Tree, Teiresias has lower memory requirement. According to some researchers, Teiresia has also a good runtime behavior. Hence, we deploy this algorithm to detect Address Groups.

## 2.2 Detecting Strides

For detecting Strides we deploy an algorithm similar to that described in [6]. As shown in Figure 1, the algorithm uses a search window to record the difference between two references. The head line demonstrates the access addresses in a sample access trace, while the differences are presented in the columns corresponding to each address, with the first difference (to the previous access) in line 1, the second difference (to the second previous) in line 2, and so on. The number -297 in line 4 of the last column, for example, shows the difference between address 103 and 400.

| Access trace: | 200 | 300 | 501 | 923 | 400 | 102 | 881 | 500 | 103 | ...... |
|---|---|---|---|---|---|---|---|---|---|---|
| 1 | | **100** | 201 | 422 | −523 | −298 | 779 | −381 | −397 | |
| 2 | | | 301 | 623 | −101 | −821 | 481 | 398 | −778 | |
| 3 | | | | 723 | **100** | −399 | −42 | **100** | 1 | |
| 4 | | | | | 200 | −198 | 380 | −423 | −297 | |
| 5 | | | | | | −98 | 581 | −1 | −820 | |
| 6 | | | | | | | 681 | 200 | −398 | |
| 7 | | | | | | | | 300 | −197 | |
| 8 | | | | | | | | | −97 | |

**Fig. 1.** Algorithm for detecting Strides

To discover a Stride, pairs with the same difference value are searched within the window. An example is difference 100 which is highlighted in the figure. The first position lies in line 1 of address 300 (difference between 300 and its direct left neighbor 200) and the second position is line 3 of address 400 (difference between 400 and its

3rd left neighbor 300). Then a stride $< 200, 100, 3 >$ can be concluded that describes an access stride beginning with address 200, of length 100, and repeating for three accesses. In the following a further difference 100 in line 3 of address 500 is observed. While this is the difference to address 400, the last address in the detected Stride, the Stride can be extended to $< 200, 100, 4 >$.

## 3 Implementation

For developing this analysis tool, we first need a memory trace that records all references performed at the runtime. We rely on a code instrumentor to generate the trace. This is followed by the implementation of the chosen algorithms for both Address Groups and Strides. In the next step we implement algorithms to determine whether an access is a hit or a miss and to give hints about how to transform a cache miss to a cache hit.

**Memory Trace.** The memory trace is generated by a code instrumentor, called Doctor. Doctor is originally developed as a part of Augmint [11], a multiprocessor simulation toolkit for Intel x86 architectures, and is used to augment assembly codes with instrumentation instructions that generate memory access events. For every memory reference, Doctor inserts code to pass its address, pid, and type to the simulation subsystem of Augmint. For this work, we slightly modified Doctor allowing to generate a trace file which stores all memory accesses performed by the application at runtime.

**Teiresias and the Algorithm for Strides.** The implementation of Teiresias follows actually the description of this algorithm in Section 2. First in the *scan/extension* phase small Address Groups $< L, N >$ are detected. The number of addresses and the length of an Address Group can be specified through command line parameters. Due to the large size of the memory trace, we slightly modified the functionality of this phase in order to reduce the execution time. In addition, we use a parameter K to specify the minimal occurrence of an address sequence and this reduces the number of small groups and thereby the time for generating larger groups in the *convolution* phase. As mentioned in Section 2, small patterns are combined in this phase and additionally overlapped patterns are detected and generated.

The implementation of the algorithm for Strides also follows the description in Section 2. First the search window is initialized and then for each reference the differences to all former accesses are calculated and stored in the window. It is clear that the size of the window is restricted and in case that the window is full the differences related to further accesses cover the columns in the window from left to right. Strides are generated and extended during the building and update of the search window.

**Cache Parameters.** In order to know if an access is a cache hit or a cache miss, our analysis tool models the architecture of the specified cache. The configuration information, such as cache size, block size, and the associativity, is delivered to the tool via command line parameters. According to the cache organization, reuse distance and set reuse distance are computed, where the former is the number of different addresses between two accesses to the observed address and the latter has the same meaning but considers only addresses lying in the same set. Using both cache parameters it is possible to determine whether an access lands in cache or not. It is also possible to compute the Push Back Distance, which shows how many steps a miss access must be shifted in order to achieve a hit. Due to the high overhead, we do not compute the Push Back Distance for all variables in the program. Rather only user-specified variables are handled. For hit/miss estimation, however, all variables are covered in order to allow a combined analysis of the access behavior and the detection of optimization strategies.

**Output.** At the end of the analysis, our tool provides for each process three XML files, one for Address Groups, one for Strides, and the other for Cache Parameters. Each record in the Address Group file shows the addresses in the group and number of occurrence of this group. The Stride file gives all detected regular access stride together with the start address and the number of elements that hold this stride. In the Parameter file, the hit/miss and Push Back Distance, together with access address, variable, and source information, are demonstrated.

## 4 Verification with Standard Applications

We use several shared memory applications from the SPLASH-II benchmark suite [15] and a few small easy-to-understand codes to test this analysis tool for examining its feasibility in detecting access patterns and optimization possibilities.

**Address Group.** We first use the SPLASH-II applications to examine the ability of the tool for detecting frequently occurring Address Groups. Table 1 depicts the results with FFT (Fast Fourier Transformation), Barnes-Hut (solution of n-body problem), and Radix (integer radix sort).

**Table 1.** Results of Address Group detection with SPLASH-II applications

| | Parameters | | | Number of patterns | |
|---|---|---|---|---|---|
| | L | W | K | process 1 | process 2 |
| FFT | 3 | 5 | 500 | 46 | 31 |
| | 5 | 8 | 500 | 30 | 18 |
| | 5 | 15 | 5000 | 4 | 4 |
| Barnes-Hut | 3 | 5 | 500 | 111 | 587 |
| | 5 | 8 | 500 | 60 | 366 |
| | 5 | 15 | 5000 | 3 | 3 |
| Radix | 3 | 5 | 500 | 1492 | 1484 |
| | 5 | 15 | 5000 | 1255 | 48 |
| | 5 | 15 | 10000 | 21 | 22 |

It can be observed, and it is also clear, that for each application less Address Groups are detected with the increase of the number of addresses in the group (L), the group length (W), and the number of minimal occurrence (K). However, some applications still show a good behavior with large groups. For example, Radix holds more than 20 Address Groups of length 15 that repeat more than 10000 times. This indicates how well a program can be optimized with the grouping strategy. Applications, which hold many smaller Address Groups, would benefit from caches with small cache blocks, while applications with many longer Address Groups benefit more from caches of large blocks.

It is also interested to see that some applications show an identical behavior among processes. For Barnes-Hut, for example, we have examined the corresponding Address Group of the same ID in the detected Groups of all processes and found that they map to each other and have the same structure: position holders at the same position and the same frequency of occurrence. We also detected that these mapping Groups target the same code region. This means that for these applications address grouping can be generally done without special handling for each process.

**Access Stride.** For verifying the profit capable of being achieved by the information about access strides we use a small code containing mainly the following loop:

$for(n1 = 0; n1 < DIM; n1++)$
$\quad for(n2 = 0; n2 < DIM; n2++)$
$\quad\quad arr[n2 * DIM + n1] = 1;$

The result of analysis shows that all accesses to the array introduce a cache miss. In the XML file for detected Strides we found a set of records like:

............
$< stride - id = 18\ start = 536883260\ step = 128\ length = 32 >$
$< stride - id = 18\ start = 536883188\ step = 128\ length = 32 >$
............

We conclude that the misses are caused by the large stride between accesses to different elements of the array. With this stride the elements, which are loaded into cache together with the required element, can not be used before the cache block is evicted from cache due to conflict. With this observation we exchange the loops of n1 and n2 and this time only one Stride is reported:

$< stride - id = 18\ start = 536883140\ step = 4\ length = 1024 >$

The result shows significantly less cache misses with one miss every eight accesses. This is caused by the improvement in spatial locality, where data close to the accessed one is required in the next step of computing.

In summary, the experimental results demonstrate that our analysis tool can provide information which is needed and even necessary for conducting optimizations with respect to cache locality.

## 5 Conclusions

In this paper we introduce an analysis tool capable of providing information, like sequence of repeated access addresses and stride between accesses to elements of a data structure. It also exhibits the hit/miss behavior of each access and gives hints about how to transform a cache miss to a cache hit.

Currently we are developing a programming environment that both visualizes the information provided by the analysis tool and establishes a platform for analyzing, optimizing, compiling, and executing the applications. This environment also shows the influence of any optimization with the program codes. After this work we will starts with the optimization process first using benchmark applications and then realistic codes.

## References

1. David F. Bacon, Jyh-Herng Chow, Dz ching R. Ju, Kalyan Muthukumar, and Vivek Sarkar. A Compiler Framework for Restructuring Data Declarations to Enhance Cache and TLB Effectiveness. In *Proceedings of CASCON'94 – Integrated Solutions*, pages 270–282, October 1994.
2. A. Chattaraj and L. Parida. An Inexact-suffix-tree-based Algorithm for Detecting Extensible Patterns. *Theoretical Computer Science*, 335(1):3–14, 2005.
3. A. Delcher, S. Kasil, R. Fleischmann, O White J. Peterson, and S. Salzberg. Alignment of Whole Genomes. *Nucelic Acids Research*, 27(11):2369–2376, 1999.
4. C. Ding and K. Kennedy. Improving Cache Performance in Dynamic Applications through Data and Computation Reorganization at Run Time. *ACM SIGPLAN Notices*, 34(5):229–241, May 1999.
5. C. C. Douglas, J. Hu, M. Kowarschik, U. Rüde, and C. Weiss. Cache Optimization for Structured and Unstructured Grid Multigrid. *Electronic Transaction on Numerical Analysis*, 10:21–40, 2000.
6. T. Mohan et. al. Identifying and Exploiting Spatial Regularity in Data Memory References. In *Supercomputing 2003*, Nov. 2003.
7. S. Fung. Improving Cache Locality for Thread-Level Speculation. Master's thesis, University of Toronto, 2005.
8. Somnath Ghosh, Margaret Martonosi, and Sharad Malik. Precise Miss Analysis for Program Transformations with Caches of Arbitrary Associativity. *ACM SIGPLAN Notices*, 33(11):228–239, November 1998.
9. Somnath Ghosh, Margaret Martonosi, and Sharad Malik. Automated Cache Optimizations using CME Driven Diagnosis. In *Proceedings of the 2000 International Conference on Supercomputing*, pages 316–326, May 2000.
10. N. Megiddo and V. Sarkar. Optimal Weighted Loop Fusion for Parallel Programs. In *Proceedings of the 9th Annual ACM Symposium on Parallel Algorithms and Architectures*, pages 282–291, New York, June 1997.
11. A-T. Nguyen, M. Michael, A. Sharma, and J. Torrellas. The augmint multiprocessor simulation toolkit for intel x86 architectures. In *Proceedings of 1996 International Conference on Computer Design*, October 1996.
12. J. Park, M. Penner, and V. Prasanna. Optimizing Graph Algorithms for Improved Cache Performance. In *Proceedings of the 16th International Parallel and Distributed Processing Symposium*, pages 32–33, April 2002.
13. I. Rigoutsos and A. Floratos. Combinatorial Pattern Discovery in Biological Sequences: the TEIRESIAS Algorithm. *Bioinformatics*, 14(1):55–67, January 1998.
14. G. Rivera and C. W. Tseng. Data Transformations for Eliminating Conflict Misses. In *Proceedings of the ACM SIGPLAN Conference on Programming Language Design and Implementation*, pages 38–49, Montreal, Canada, June 1998.
15. S. C. Woo, M. Ohara, E. Torrie, J. P. Singh, and A. Gupta. The SPLASH-2 Programs: Characterization and Methodological Considerations. In *Proceedings of the 22nd Annual International Symposium on Computer Architecture*, pages 24–36, June 1995.

# A Performance Profile and Test Tool for Development of Embedded Software Using Various Report Views*

Yongyun Cho and Chae-Woo Yoo

School of Computing, Soongsil University,
1-1 Sangdo-dong, Dongjak-gu, Seoul 156-743, Korea
yycho@ss.ssu.ac.kr, cwyoo@comp.ssu.ac.kr

**Abstract.** In this paper, we suggest a performance profiling and testing tool that offers developers convenient environments to profile or test an embedded software's performance and to analyze the results through various graphic report views. Because the suggested tool does not need any additional hardware, it is efficient in aspects of costs and management for profiling or testing an embedded software's performance. The tool consists of a code analyzer, a test suite generator, and a report generator. The code analyzer expresses a software's structure as a parse tree and decides a position that additional codes for profiling must be inserted into. The test suite generator offers a test script wizard for users to easily make a test driver. The report generator converts a string-type result to an XML-based class instance in order to raise reusability for the result. To offer various report views we divide the instance into two sections, which are for result data and for visual views. Therefore, users can get various report views by associating the two sections according to their intention.

## 1 Introduction

Because an embedded system generally offers less computing resources than a general-purpose computer system does, developers make every effort to improve the quality of their embedded software and make it to always have a good performance in resource usage[3, 4]. To do this, developers occasionally use embedded software evaluation tools to increase development efficiency for embedded softwares. With a software evaluation tool, developers know whether the developed software is efficiently optimized for embedded system's restricted resources. Because embedded software is commonly developed in the cross-platform, in which the test execution occurs on a target-side but the result analysis in a host-side existing embedded software evaluation tools are based in the environments. Some of them need an additional hardware to profile and test embedded software. An additional hardware is more or less profitable in a view of times, but may be financial burden on middle-sized embedded software developers. Many embedded

* This work was supported by the Soongsil University Research Fund.

V.N. Alexandrov et al. (Eds.): ICCS 2006, Part II, LNCS 3992, pp. 510–517, 2006.

software vendors include profile and test tools in their products. However, because many of them produce the testing results that occasionally are text-based strings, to analyze the string data to find where to be revised becomes often very tiresome and time-consuming work. Of course, many embedded software vendors support profiling or testing through graphical result views. However, they do not offer various report views enough to increase the analysis efficiency.

In this paper, we suggest a graphic tool based on pure software without any additional hardware for profiling and testing embedded software's performance. The tool includes a code analyzer, a test suite generator, and a report generator. The code analyzer inserts profile codes into a target source through parsing and generates an execution file including profiling codes. The test suite generator makes test scripts and generates test drivers after parsing the scripts. In this paper, we design an XML-based test script DTD to easily make a test script. To generate various report views, the report generator uses a result converter to represent string-typed profile or test results to object instances through XML-based classes. The XML-based class consists of two parts. One is to represent graphical views, and the other is to describe result data. Because the two parts are separated, users can get several graphical views by associating one result data with various graphical values according to the users' preference.

## 2 Related Work

### 2.1 Existing Profile and Test Tools for Embedded Softwares

Telelogic's Tau TTCN Suite is a system to test telecom and datacom equipment ranging from built-in communication chips to huge switches and intelligent network services. It includes various tools such as script editors, compilers and simulators, but it is not suitable for testing embedded software because it is a test tool for telecommunication vendors. It also is very expensive because it is mostly additional hardware equipment to test telecom tools. AstonLinux's CodeMaker is an IDE (Integrated Development Environment) to develop embedded software based on Linux in Windows. It supports remote debugging and source-level debugging, but it doesn't offer any function to test and analyze embedded software's performance because it is only an IDE for specific RTOS/chip vendor. Rational's TestRealTime is a target-based performance evaluation system for real-time software. It offers various result views so that users can easily analyze real-time software's performance. It can also execute various performance tests ranging from memory usage, memory leak, cpu usage to code coverage. However, it is somewhat difficult for developers to understand the meaning of the result view at a glance because it is not a performance system for embedded software based in the cross-platform environment. Additionally, it offers a somewhat difficult script language for users to make a test script. Because the script language is a perfectly new script language, users must spend a lot of time to study how to use for making a test script. To solve the problems, in this paper we design an XML-based script language that common users can intuitively understand and easily use.

## 3 The Proposed Performance Evaluation Tool

### 3.1 A System Architecture

In this paper, we suggest a tool for profiling and testing embedded software's performance that consists of pure software without additional hardware equipment and offers such various performance tests as memory, code coverage, code trace and function performance [3, 5, 6]. The evaluation tool offers users graphical report views that they can easily and intuitively analyze the test result. Figure 1 is the proposed architecture for a performance evaluation tool.

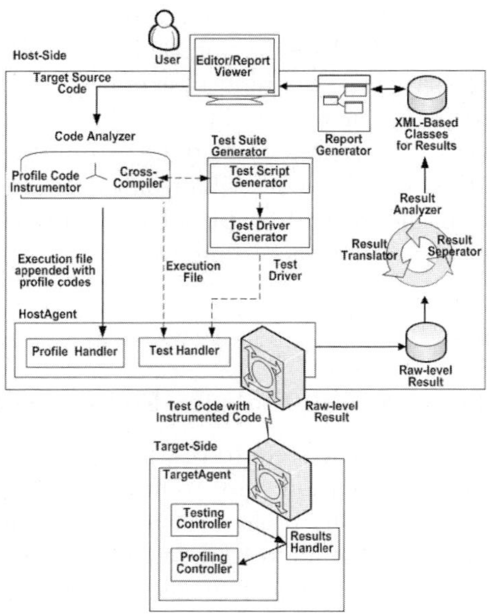


**Fig. 1.** The Architecture of The Proposed Performance Evaluation Tool

In Fig. 1, the proposed evaluation tool is composed of a GUI, host/target-side agents, a code analyzer, a result analyzer, and a report viewer. The code analyzer consists of the code corrector to insert additional code into source code and a cross-compiler to create target-executable file for the source code. The evaluation tool is a client/server model based in host-target architecture. Because an embedded system offers insufficient memory and an inconvenient user interface, the suggested tool places an agent not only on the host-side to offer users convenient GUI but also on the target-side to execute performance testing of the software in the target board. The agents keep a communication connection to deliver source files and test result to each other. First, the host-side agent transfers an inputted source to the target-side agent through a LAN cable or wireless network. Then, the target-side agent executes a testing process, gains results from the test events, and sends the results to its host-side counterpart.

Consequently, the host-side agent stores the string-typed result received from target-side one into the result DB.

### 3.2 Profiling for Embedded Softwares

Generally, embedded software must use minimum process and memory resources. To make embedded software to meet those requirements, the suggested tool tests software's performance for 4 items, which are trace, memory, performance, and code coverage profile [4, 5]. Through trace profiling, users can trace what functions are executed according to the software's execution process and find what functions are unnecessarily called. Through memory profiling, users can know about memory allocation/de-allocation, memory leaks, and code sections frequently to use memory. Users can use performance profiling to estimate how much time it takes to execute the whole or part of embedded software and it confirms whether it becomes optimized in embedded system. Code coverage profiling offers users information about used or unused code sections, and frequently or infrequently used code sections. Users can make embedded software more efficient by using information profiled according to the 4 items. In this paper, we classify string-typed profiling results according to the items and converts them into instances objectified by the classes. The report generator uses the instances to make various report views according to the user's requirements. The instances consist of a section to represent result data and a section to describe graphical elements that construct a report view. For example, when a user wants to know how many memory are used in a test target embedded software through a pie graph or a bar graph, the user can get the information by combining the software's memory profile result with class pre-defined for the graphical elements, pie or bar.

### 3.3 Testing for Embedded Softwares

To enhance the performance of embedded softwares, testing is very important. The suggested tool supports testing for embedded softwares. Commonly, to test embedded softwares users need a test source code and a test driver [7]. A test driver calls code units that must be tested and collects results of the execution. A user can make a test driver in a program language that a test target program is written in. However, because it needs a lot of times, a user commonly uses a tool that automatically translates a test script that a user makes with a script language to a test driver. With existing tools, a user must use a specific script language which is never easy for common users to understand and use. In this paper, we suggest an XML-based test script language. Because the script language is based on XML, users can easily understand its grammar and learn how to use it. Figure 2 shows the schema of the suggested script language.

In Figure 2, <testDriver> is a test script's root. <test> can occur repeatedly and describes test cases. <element> is to set initial values and expected values which are <var> element's attributes. <run> describes the test execution. It executes <runTest> or a branch routine by using <if>, <then>, and <else> according to the test results. Figure 3 shows an example test script for add() function to add two integer values using the suggested script schema.

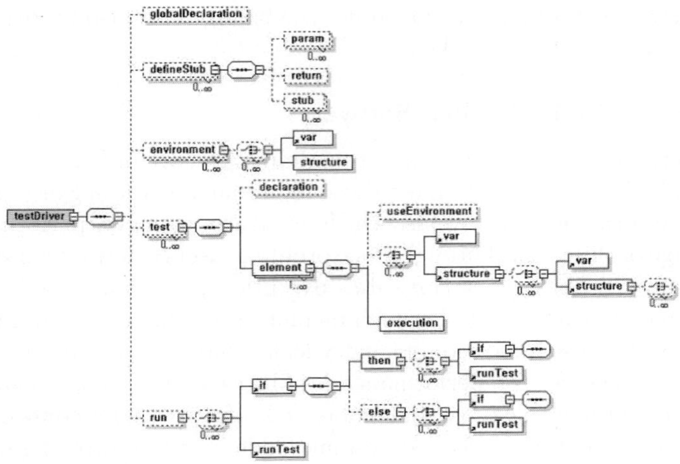


**Fig. 2.** The Suggested XML-Based Script Language's Schema

```
<testDriver>
        <test id="add">
                <declaration>
                        <![CDATA[
                        int x, y, rval_add;
                        ]]>
                </declaration>
                <element>
                        <var name="x" init="3">
                        <var name="y" init="4">
                        <var name="rval_add" ev="7">
                        <execution>
                                <![CDATA[
                                rval_add = add(x, y);
                                ]]>
                        </execution>
                </element>
        </test>
        <run>
                <runTest id="add"/>
        </run>
</testDriver>
```

Test Definition Section (<test id="add"> … </test>)

Test Execution Section (<run> … </run>)

**Fig. 3.** A Sample Test Script for add(x, y) function

## 4 Experiments and Results

The suggested evaluation tool is implemented in the Java 2 platform and we use an HRP-SC2410 (Ami) launched strong ARM chip and embedded Linux. We use a calculator program in C language as input source code. Its code size is about 520 lines and consists of three modules. We will profile them depending on the 4 items, which are trace, memory, performance, and code coverage profile, and execute unit testing for them. As results for the experiments, we will show various graphical report views for the profiling and testing process.

Figure 4 shows various report views generated by the suggested tool after profiling the sample C program. Each report view's visual element is not static, because the suggested tool divides profile results with visual elements for construction of report views. Therefore, users can make various report views according to their intention. Figure 4(a) shows a trace view that is a style of UML

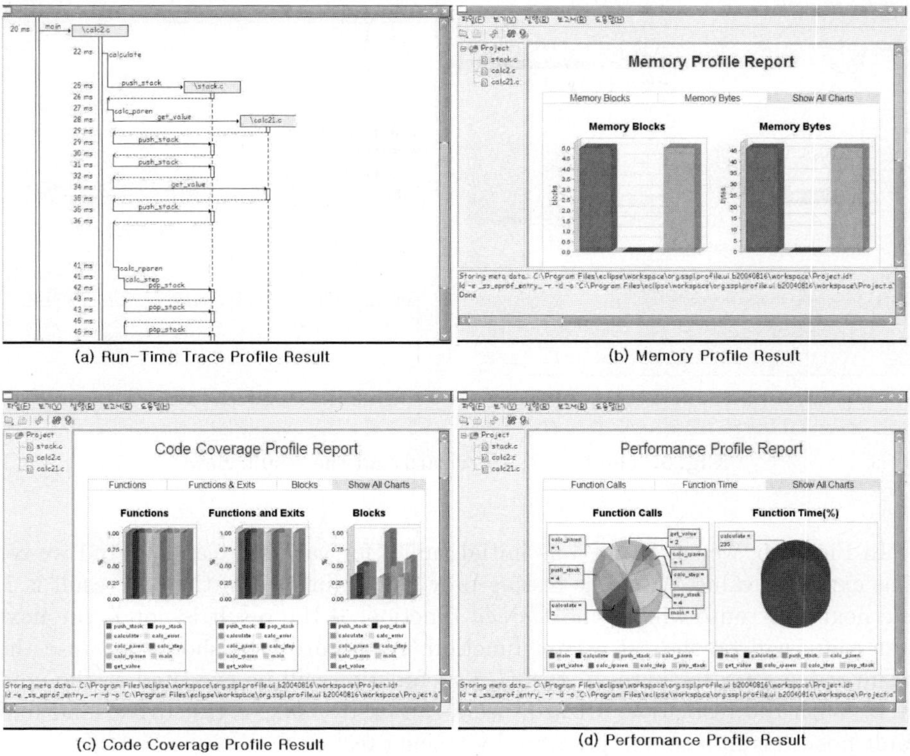


**Fig. 4.** Various Report Views for Profile Results

sequence diagram [9, 10]. With the view, developers can analyze a program's execution by tracing functions' call orders. Figure 4(b), (c), and (d) show respectively memory, code coverage, and a performance report view. With Figure 4(b), we can know how much memory software uses and can find whether any freed memory section is called for freeing, or any unallocated memory section is called for freeing. With that, developers can find where memory leaks or illegal memory usages happen. With Figure 4(c), we can know whether functions in source code were executed and how many blocks those functions were executed. Through the information, developers can decide which function must be revised to enhance the entire efficiency of the software. With Figure 4(d), we can know the call times for each function and the function's execution time with its lower functions or without them. We can also find the average execution time of any function against total execution time. Through the result, developers can know which function is most busy, and they can divide the burden of the function to other function in order to raise the execution efficiency of the software.

Figure 5 shows a test driver wizard for testing and a result view after testing.

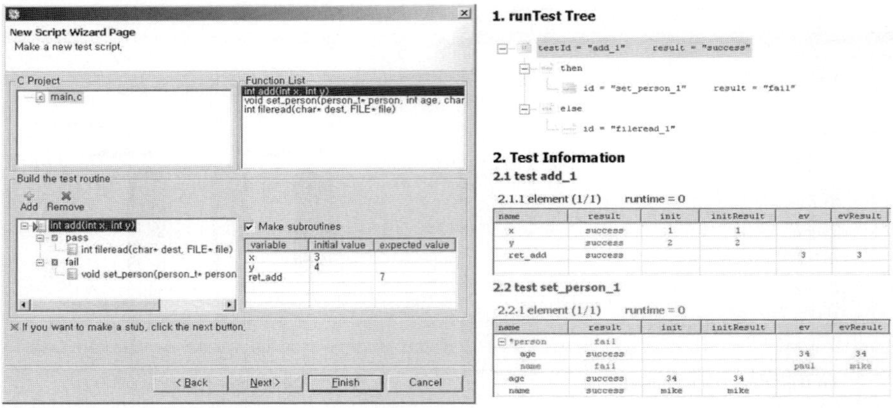

(a) Test Driver Wizard for Making Test Driver (b) Test Result View after Testing with Test Driver

**Fig. 5.** The test driver wizard and the result view

In Figure 5, we set 3 and 4 as initial values for variable x and y. And we set 7 as expected value. After the testing process has finished, if the test result is 7, the next test route will be the *fileread* function. If the result is not 7, the next test route will be the *set_person* function. As Figure 5(b) shows, because the result value after testing is 7, the *fileread* function will be executed as the next step. With the test driver wizard, users can easily make a test driver and find fault positions in a source program by tracing test path.

## 5 Conclusion

In the development of embedded software, developers have always to consider good efficiency not only for resource usage but also for the development time. In this paper, we suggest a GUI-based tool which easily profiles and tests embedded software's performance and intuitively analyzes the results. For this, we suggested a code analyzer that produces a parse tree as a result of parsing and uses a parse tree handler to insert profile codes into aimed positions of the source code. With the parse tree handler, the code analyzer can decide exact instrumentation points not only in compilation time but also in execution time. Additionally, we also suggested the test suite generator that makes a test script and a test driver. For that, we have designed an XML-based test script DTD for easy understanding and using. By using the DTDs, the report generator converts string-typed results to XML class instances and generates various report views through which developers can easily understand the meaning of the results and revise the inefficient portions of the source codes. In the experiments with the suggested tool, we profiled the performance of some C source codes for the 4 items mentioned in Section 3.2, and showed the results graphically through the report viewer. We also showed that developers can easily make a test driver by using the test driver wizard. Therefore, through the suggested tool, developers can clearly know what must be fixed in software's source code and can improve development efficiency of embedded software.

## References

1. Roper, Marc, Software Testing, London, McGraw-Hill Book Company, 1994.
2. Boris Beizer, Software Testing Techniques 2nd edition, New York: Van Nostrand Reinhold, 1990.
3. Bart Broekman and Edwin Notenboom, Testing Embedded Software, Addisson-wesley, Dec. 2002
4. Dr. Neal Stollon, Rick Leatherman and Bruce Ableidinger,Multi-Core Embedded Debug for Structured ASIC Systems, proceedings of DesignCon 2004, Feb, 2004.
5. David B. Stewart, Gaurav Arora, A Tool for Analyzing and Fine Tuning the Real-Time Properties of an Embedded System. IEEE Trans. Software Eng., Vol.TSE-29, No.4, April 2003, pp.311-326.
6. Ichiro Satoh, A Testing Framework for Mobile Computing Software. IEEE Trans. Software Eng., Vol.TSE-29, No.12, December 2003, pp.1112-1121.
7. Paul Anderson, Thomas W. Reps, Tim Teitelbaum, Design and Implementation of a Fine-Grained Software Inspection Tool. IEEE Trans. Software Eng., Vol.TSE-29, No.8, August 2003, pp.721-733.
8. John Joseph Chilenski and Steven P. Miller, Applicability of Modified Condition/Decision Coverage to Software Testing, Software Engineering Journal, September 1994, Vol. 9, No. 5, pp. 193-200.
9. Robert B. France, Dae-Kyoo Kim, Sudipto Ghosh, Eunjee Song, A UML-Based Pattern Specification Technique, IEEE Trans. Software Eng., Vol.TSE-30, No.4, April 2004, pp. 193-206.
10. Ludovic Apvrille, Jean-Pierre Courtiat, Christophe Lohr, Pierre de Saqui-Sannes, TURTLE: A Real-Time UML Profile Supported by a Formal Validation Toolkit. IEEE Trans. Software Eng., Vol.TSE-30, No.7, July 2004, pp. 473-487.

# SCE Toolboxes for the Development of High-Level Parallel Applications*

J. Fernández, M. Anguita, E. Ros, and J.L. Bernier

Departamento de Arquitectura y Tecnología de Computadores,
Universidad de Granada, 18071-Granada, Spain
{jfernand, manguita, eros, jlbernier}@atc.ugr.es
http://atc.ugr.es

**Abstract.** Users of Scientific Computing Environments (SCE) benefit from faster high-level software development at the cost of larger run time due to the interpreted environment. For time-consuming SCE applications, dividing the workload among several computers can be a cost-effective acceleration technique. Using our PVM and MPI toolboxes, MATLAB® and Octave users in a computer cluster can parallelize their interpreted applications using the native cluster programming paradigm — message-passing. Our toolboxes are complete interfaces to the corresponding libraries, support all the compatible datatypes in the base SCE and have been designed with performance and maintainability in mind. Although in this paper we focus on our new toolbox, MPITB for Octave, we describe the general design of these toolboxes and of the development aids offered to end users, mention some related work, mention speedup results obtained by some of our users and introduce speedup results for the NPB-EP benchmark for MPITB in both SCE's.

## 1 Introduction

GNU Octave [1] is a high-level language, primarily intended for numerical computations. It provides a convenient command line interface for solving linear and nonlinear problems numerically, and for performing other numerical experiments using a language that is mostly compatible with MATLAB.

MATLAB [2] is a high-level technical computing language and interactive environment for algorithm development, data visualization, data analysis, and numeric computation. Decades of experience, a careful selection of algorithms and a number of improvements to accelerate the interpreted environment (JIT compiler, etc.), can make MATLAB faster for computationally intensive tasks than traditional programming languages such as C, C++, and FORTRAN, if the programmer is not so careful in the algorithm selection or implementation in terms of native machine instructions.

Both environments can be extended, either with scripts written in the interpreted language — the so-called "M-files", or external (FORTRAN, C, C++) programs using

* This work was supported by the Spanish CICYT project SINTA-CC (grant TIN2004-01419) and the European FP-6 integrated project SENSOPAC (grant IST-028056).

V.N. Alexandrov et al. (Eds.): ICCS 2006, Part II, LNCS 3992, pp. 518 – 525, 2006.

the MATLAB API or Octave library — called "MEX-files" or "DLD functions", respectively. This latter choice lets us build an interface to a message-passing library.

Fig. 1 is an overview of how the end-user application would sit on top of the SCE (Octave, in the example shown) and our toolbox (MPITB in this case), being thus able to communicate with other SCE instances running in other cluster nodes.

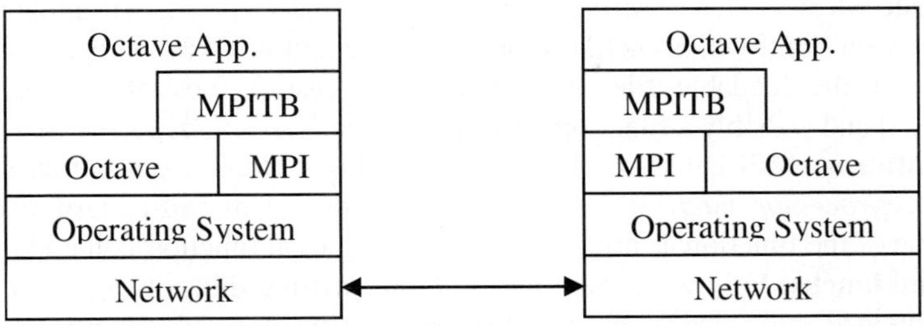


**Fig. 1.** High-level overview diagram of Toolbox role. The Toolbox invokes message-passing calls and SCE API calls to intercommunicate SCE processes running on different cluster nodes.

PVM (Parallel Virtual Machine) [3] was built to enable users to exploit their existing computer hardware to solve much larger problems at minimal additional cost. PVM can of course be used as message-passing mechanism in a dedicated cluster.

MPI (Message Passing Interface) [4,5] is a standard, and as such, it only defines the routines that should be implemented in a compliant implementation.

LAM/MPI (Local Area Multicomputer) [6,7] implements MPI-1.2 and large portions of the MPI-2.0 standard. In addition to high performance, LAM provides a number of usability features key to developing large scale MPI applications.

Our PVMTB and MPITB toolboxes interface the PVM and LAM/MPI libraries from within the MATLAB or Octave environments, thus letting SCE users parallelize their own applications by means of explicit message-passing. Since users are expected to modify their sequential code to adapt it to parallel execution, a number of utility M-files have also been developed on top of this direct, "low-level" library interface, to facilitate the parallelization process and reduce those modifications to a minimum.

To further clarify the idea, MEX or DLD interface is required, in order to be able to call PVM or LAM from C code. That cannot be accomplished from the interpreted language. Octave's *interpreted language* is compatible with MATLAB, but the MEX version of MPITB will simply *not load* as a valid DLD under Octave, let alone work correctly. Each version uses a different API to access SCE data and pass it from/to the message-passing library. For instance, to return without errors we would either code `*mxGetPr(plhs[0])=0` for MATLAB or `return octave_value(0)` for Octave.

Section 2 summarizes the design principles for these toolboxes, compares them to other related work, describes the utilities built on top of them and shows how to use these to ease the parallelization of sequential SCE code. Section 3 shows performance measurements for the NPB-EP benchmark. Section 4 describes speedup results from some of our users. Section 5 summarizes some conclusions from this work.

## 2 Design Principles

Although the explanation and examples below are focused on our most recent MPITB for Octave, the same guidelines hold for the other toolboxes. The key implementation details are:

- **Loadable files:** For each MPI call interfaced to the Octave environment, an **.oct** file has been built, which includes the doc-string and invokes the call-pattern. The source for the loadable file includes at most just 2 files: the global include (`mpitb.h`) and possibly a function-specific include (`Send.h`, `Topo.h`, `Info.h`, etc).
- **Call patterns:** MPI calls are classified according to their input-output signature, and a *preprocessor macro* is designed for each class. This call-pattern macro takes the name of the function as argument, substituting it on function-call code.
- **Building blocks:** Following the criterion of not writing things twice, each piece of self-contained code that is used at least twice (by two call patterns or building blocks) is itself promoted to building block and assigned a *macro name* to be later substituted where appropriate. This feature makes MPITB maintenance tractable.
- **Include files:** build-blocks and call patterns are grouped for affinity, so that only the functions that use them have to preprocess them. A building block or call pattern shared by two or more such affine groups is moved to the global include file.
- **Pointers to data:** MPI requires pointers to buffers from/to where the transmitted info is read/written. For arrays, pointer and size are obtained using methods **`data()`** and **`capacity()`** common to all Octave array types. For scalars, C++ data encapsulation is circumvented, taking the object address as starting point from which to obtain the scalar's address. Shallow-copy mechanisms are also taken into account by invoking method **`make_unique()`** if the Octave variable is shared.
- **RHS value return:** Octave variables used as MPI buffers must appear in the right-hand-side (RHS) of the equal sign in MPITB function definitions; that is, we rather define **`[info,stat]=MPI_Recv(buf,src,tag,comm)`** than return **`buf`** as Octave left-hand-side (LHS) such as in **`[buf,info,stat]=MPI_Recv(src,tag,comm)`**. That would be a low-performance design decision, since the returned **`buf`** object should be constructed. Depending on **`buf`** type and size, that could be a very expensive operation, particularly in the context of a ping-pong test or an application with a heavy communication pattern. We instead force buffers to be Octave symbols (named variables), so the output value can be returned in the input variable itself.

According to our experience [8], these last two principles guarantee that performance will be superior to any toolbox lacking of them. This is particularly true for the mentioned object construction in LHS return, which is not required for MPI operation.

### 2.1 Related Work

Perhaps due to the different design principles used, other parallel Octave toolboxes no longer work, proving that the task is not trivial. The interface is elaborate except for a few of the simplest calls, and the design must keep an eye on performance and maintainability. We have surveyed the Octave mailing list in search for related work.

**Table 1.** Survey on parallel Octave prototypes

| Package | based on | #cmds | date | status |
|---|---|---|---|---|
| PVMOCT | PVM | 46 | Feb. 1999 | developed for octave-2.0.13 would require major edition |
| Octave-MPI | LAM/MPI | 7 | Nov.2000 | prepatched for octave-2.1.31 would require major edition |
| P-Octave | LAM/MPI | 18 | Jun. 2001 | works with octave-2.1.44 after minor source code edit |
| D-Octave | LAM/MPI MPICH | 2+6 | Mar. 2003 | prepatched for octave-2.1.45 proof-of-concept only |
| MPITB | LAM/MPI | 160 | Apr. 2004 | recompiled for 2.1.50-57-69-72 supported |

Our MPITB interfaces all MPI-1.2 and some MPI-2.0 calls — 160 commands. It does not rely so heavily on the Octave version — no Octave patching is required, just minor code changes to adapt to changes in Octave API and internal representation of data, or to new Octave datatypes. The observed design principles convert MPITB maintenance in a feasible task. Our approach of pointers to data and RHS return is unique, so all other toolboxes would suffer from data copy and object creation costs.

## 2.2 Utilities

End-users are the ones knowing their own applications, and hence the most appropriate to design the message-passing protocol among the parallel SCE instances. However, many users benefit from some simple examples, either to ease the initial learning curve or to save development time by customizing very frequently used code.

This has been the origin for the startup files and protocol and instrumentation utilities. Again, although we concentrate here on our most recent MPITB for Octave, the same discussion is applicable to our other parallel toolboxes.

**Startup Files**

The default Octave run-command (`.octaverc`) can be replaced or modified to add the toolbox path, and to discriminate spawned SCE instances by checking for existence of the LAMPARENT environment variable. This use has evolved to a collection of startup files, each customized to a particular situation.

- **Path:** When starting multiple SCE instances via `mpirun`, the instances are expected to `MPI_Init` and `_Finalize` by themselves, without further support. This is most similar to a normal C-MPI application. The NPB-EP demo works this way.
- **Communicator Merge:** When starting slave instances from a master via the MPITB command `MPI_Comm_spawn`, the slaves' and master's communicators could be merged to simplify rank addressing. Slave `MPI_Init` and `_Intercomm_merge` are included in this startup. The master Octave is expected to `_Intercomm_merge` some time after `MPI_Comm_spawn`. The MPITB tutorial examples work this way.
- **Broadcast Protocol:** In many simple applications, slaves are sent just one and the same Octave command (an MPI string), which must be evaluated on all slaves. The

master Octave is expected to `MPI_Bcast` the required string. Slave `_Bcast` is included in this startup. MPITB demos *Pi*, *PingPong* and *Spawn* use this protocol.

- **NumCmds Protocol:** For more complex applications, users might want to send a fixed or undetermined number of commands, and possibly related variables, to slave SCE's; or perhaps the same command is repeated over and over on different data. A more complex startup is provided to cope with all these situations, as well as supporting scripts for the master. The MPITB demo *Wavelets* uses this protocol.

**Protocol Utilities**

The NumCmds protocol works as depicted in Fig. 2. The above mentioned supporting scripts are shown in the left SCE instance. The user would log in the headed cluster node, start an SCE (master) session, and define a cell-array of cluster hostnames.

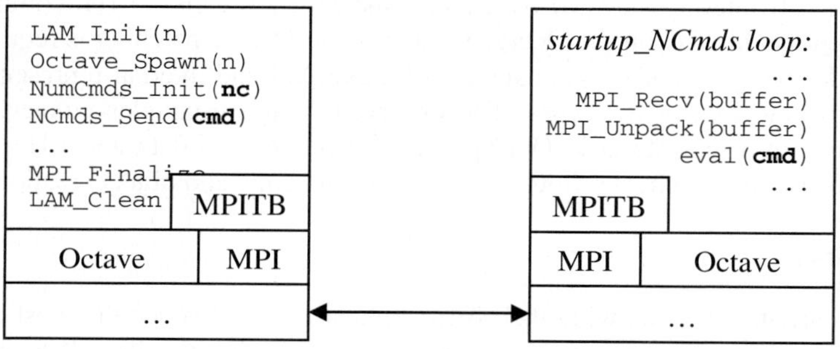


**Fig. 2.** NumCmds protocol overview. *LAM_Init* boots *n* LAM daemons. *Octave_Spawn* starts *n* Octave instances on those cluster nodes. These instances are engaged inside *startup_NumCmds* in a receive-unpack-eval loop. The loop is here fixed at *nc* epochs due to *NumCmds_Init* on the master side. On each epoch, *NumCmds_Send* is used in the master Octave to send a different command string and optional data to each slave (null string "" to repeat previous command).

**`n`** of these hosts are used by the `LAM_Init` and `Octave_Spawn` utilities to build the parallel Octave session. The slave instances are expected to source the `NumCmds` startup file, which expects an initial number **`nc`** from the master, with 0 meaning an undetermined number of commands —slaves will proceed until a `quit` command is received. Slaves then enter a receive-unpack-eval loop, executing on each iteration the command sent by the master instance.

**Instrumentation Utilities**

Since each user using our toolboxes will need to know how time is spent on the sequential SCE application, a number of simple to use routines are provided to generate, manipulate and graphically display the instrumentation data:

- init_stamp: Allocates (n annotations) and initializes the timestamp data structure.
- time_stamp: Annotates the current time together with a descriptive label.
- save/load_instr: Saves/loads the collected instrumentation data to/from a file.
- look_stamp: Lets the user graphically browse through the instrumentation data.

Showing the data in an intuitive way is vital to correctly parallelize applications.

# 3 NPB-EP Benchmark

The NAS (Numerical Aerodynamic Simulation) Program has been developing for two decades or so [9,10] a series of test codes aimed to benchmark highly parallel computers. They have tried to avoid both restricted kernels and full scale applications, arriving to a set of 5 kernels (EP, MG, CG, FT, IS) and 3 application benchmarks (LU, SP, BT), collectively known as NPB (NAS Parallel Benchmarks). Together, they mimic the computation and data movement characteristics of large scale CFD applications. These codes are well-known. Particularly the kernels have been ported to many programming languages, including Java, SAC, UPC and ZPL — see [11] for a review.

The EP (embarrassingly parallel) kernel measures floating point performance by tabulating statistics on pseudorandom data. It exhibits the simplest possible communication pattern among processes, hence the name. The MPITB version of this code uses the *path* startup file and is run using `mpirun`.

The cost of porting EP from the original FORTRAN version to MATLAB/Octave is mainly in the random number generator, adapted from the NAS-provided `randdp` double precision version. Particularly, the few MPI calls can be almost literally ported from the original FORTRAN bindings to the MPITB bindings. The `common` FORTRAN storage for the pseudo-randoms is ported to a `global` MATLAB variable. A number of performance-related details have been considered, including the use of real math in MATLAB (`reallog` & `realsqrt` instead of the normal complex versions `log` & `sqrt`) and the use of vectorization in Octave —not in MATLAB, where the JIT compiler on the iterative version is faster than the vectorized version.

Fig. 3 shows runtimes and relative speedups when EP is run in up to 8 computers, for all three versions, FORTRAN, MATLAB and Octave.

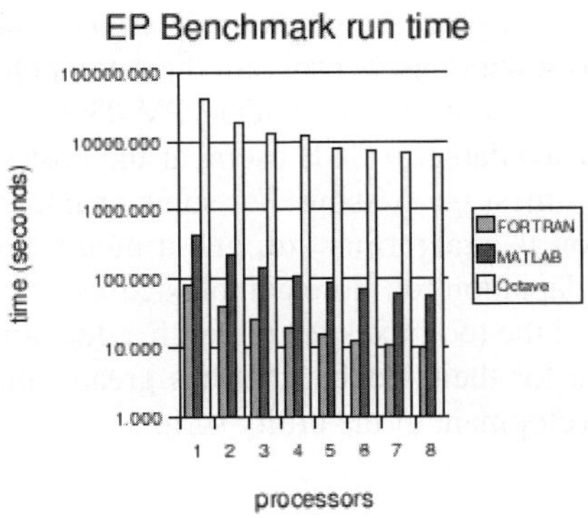


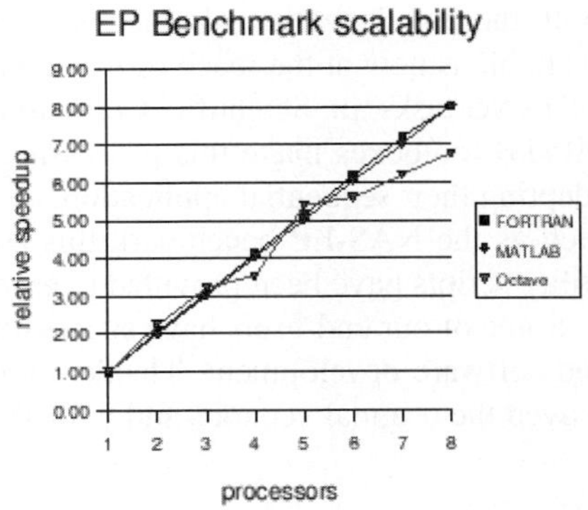


**Fig. 3.** Scalability test for the NPB-EP benchmark on an 8-node AMD Athlon 2400+ cluster. MATLAB run times in the left are one order of magnitude greater than FORTRAN ones, and two smaller than Octave. For all three versions, scalability is practically linear. The large computation time for Octave-2.1.69 made it difficult to measure it in a clean, zero-user situation.

## 4 End-Users Examples

Sebastián Dormido (UNED, Spain) has used PVMTB for MATLAB in order to implement DP (Dynamic Programming) in clusters of computers [12]. His group already had a large background on DP and MATLAB programming, but no previous knowledge on message-passing. They have developed several *parallel serial monadic DP* algorithms, some of them with very good scalability and speedup results, near to 12 using 15 slaves (80% efficiency). The referenced paper includes plots of these results.

Sébastien Goasguen (then at Purdue, IN, USA) used both PVMTB and MPITB for MATLAB in order to model quantum transport in nanoscale transistors [13,14]. He reduced simulation time from 3 days to 45 minutes (speedup 101x with 120 CPUs, 84% efficiency) after parallelizing the NanoMOS simulator. Incidentally, NanoMOS has been the subject for experiments on dynamic Grid filesystem sessions [15] where the whole MATLAB or just MPITB were updated.

Michael Creel (now at UAB, Spain) is an active developer of Octave, and has used MPITB for Octave in order to parallelize Monte Carlo simulations, Maximum Likelihood estimation and Kernel Regression [16]. He has included MPITB in his Parallel-Knoppix Linux distribution [17] with which he obtained speedups from 9 to 11 when using 12 computers (efficiency 75-92%). The code is available from the OctaveForge site (econometrics package [18]), so other users of this package can immediately benefit from cluster computing without previous knowledge of message-passing.

Morris Law (HKBU, Hong Kong) is the IT coordinator for the Science Faculty in HKBU. One of their clusters (TDG) ranked #300 in Top500 on 6/2003. Morris prepared some coursework related to MPITB for MATLAB [19]. MPITB has also been involved in research in HKBU [20].

## 5 Conclusions

With the popularization of clusters of computers, a processing power previously unavailable is now at the reach of normal users. But many users program their compute-intensive tasks in Scientific Computing Environments (SCE's). Our PVMTB and MPITB toolboxes make this processing power available to SCE users, at the cost of adapting their sequential applications to explicit message-passing. For some examples such as the NAS-EP benchmark this adaptation is straightforward, and a number of utility scripts have been provided to ease this adaptation for more complex cases.

Some of our end-users have successfully used the toolboxes for research, education and software development. Thanks to all them for their feedback; it has greatly improved the original versions and guided the development of the utility scripts.

## References

1. Eaton, J. W.: GNU Octave Manual. Network Theory Ltd. (2002) ISBN: 0-9541617-2-6.
2. Moler, C. B.: Numerical Computing with MATLAB. SIAM (2004) ISBN: 0-89871-560-1.
3. Geist, A., Beguelin, A., Dongarra, J., Jiang, W., Manchek, R., Sunderam, V.: PVM: Parallel Virtual Machine. A Users' Guide and Tutorial for Networked Parallel Computing. The MIT Press (1994). ISBN: 0-262-57108-0.

4. MPI Forum: MPI: A Message-Passing Interface Standard. Int. J. Supercomput. Appl. High Perform. Comput. Vol.8, no.3/4 (1994) 159–416. See also the MPI Forum Documents: MPI 2.0 standard (2003) University of Tennessee, Knoxville. Web http://www.mpi-forum.org/.
5. Gropp, W., Lusk, E., Skjellum, A.: Using MPI: Portable Parallel Programming with the Message Passing Interface. 2nd Edition. The MIT Press (1999) ISBN: 0262571323.
6. Burns, G., Daoud, R., Vaigl, J.: LAM: an Open Cluster Environment for MPI. Proceedings of Supercomputing symposium (1994), 379–386
7. Squyres, J., Lumsdaine, A.: A Component Architecture for LAM/MPI. Proceedings of the 10th European PVM/MPI Users' Group Meeting, Lecture Notes in Computer Science, Vol.2840 (2003) 379–387.
8. Fernández, J., Cañas, A., Díaz, A.F., González, J., Ortega, J., Prieto, A.: Performance of Message-Passing MATLAB Toolboxes, Proceedings of the VECPAR 2002, Lecture Notes on Computer Science, Vol.2565 (2003), 228–241. Toolboxes URL http://atc.ugr.es/ ~ javier.
9. Bailey, D. et al: The NAS Parallel Benchmarks. RNR Technical Report RNR-94-007 (1994)
10. Bailey, D. et al: The NAS Parallel Benchmarks 2.0. Report NAS-95-020 (1995). Reports and software available from http://www.nas.nasa.gov/Software/NPB/.
11. Buss, B.J.: Comparison of serial and parallel implementations of Benchmark codes in MATLAB, Octave and FORTRAN. M.Sc. Thesis, Ohio State University (2005). Thesis and software available from http://www.ece.osu.edu/~bussb/research/.
12. Dormido C., S., de Madrid, A.P., Dormido B., S.: Parallel Dynamic Programming on Clusters of Workstations, IEEE Transactions on Parallel and Distributed Systems, vol.16, no.9 (2005), 785–798
13. Goasguen, S., Venugopal, R., Lundstrom, M.S.: Modeling Transport in Nanoscale Silicon and Molecular Devices on Parallel Machines, Proceedings of the 3rd IEEE Conference on Nanotechnology (2003), Vol.1, 398–401. DOI 10.1109/NANO.2003.1231802.
14. Goasguen, S.; Butt, A.R.; Colby, K.D.; Lundstrorn, M.S.: Parallelization of the nano-scale device simulator nanoMOS-2.0 using a 100 nodes linux cluster, Proceedings of the 2nd IEEE Conference on Nanotechnology (2002) 409–412. DOI 10.1109/ NANO. 2002. 1032277
15. Zhao, M., Chadha, V., Figueiredo, R.J.: Supporting Application-Tailored Grid File System Sessions with WSRF-Based Services, Procs of the 14th IEEE Int. Symp. on High Perf. Distributed Computing HPDC-14 (2005), 24–33. DOI 10.1109/HPDC.2005.1520930.
16. Creel, M.: User-Friendly Parallel Computations with Econometric Examples, Computational Economics, Springer (2005) 26 (2): 107–128. DOI 10.1007/s10614-005-6868-2
17. Creel, M.: Parallel-Knoppix Linux, http://pareto.uab.es/mcreel/ParallelKnoppix/.
18. Creel, M.: Econometrics Octave package at OctaveForge, http://octave.sf.net/. See package index at http://octave.sourceforge.net/index/extra.html#Econometrics.
19. Law, M.: MATLAB Laboratory for MPITB (2003), Coursework resource for MATH-2160 HKBU, available from http://www.math.hkbu.edu.hk/math2160/materials/MPITB.pdf. See also Guest Lecture on Cluster Computing (2005), Coursework resource for COMP-3320, available http://www.comp.hkbu.edu.hk/~jng/comp3320/ClusterLectures-2005.ppt.
20. Wang, C.L.: Grid Computing research in Hong Kong, 1st Workshop on Grid Tech. & Apps. (2004), http://www.cs.hku.hk/~clwang/talk/WoGTA04-Taiwan-CLWang-FNL.pdf.

# Introducing the Open Trace Format (OTF)

Andreas Knüpfer, Ronny Brendel, Holger Brunst,
Hartmut Mix, and Wolfgang E. Nagel

ZIH, TU Dresden, Germany
{andreas.knuepfer, ronny.brendel, holger.brunst
hartmut.mix, wolfgang.nagel}@tu-dresden.de

**Abstract.** This paper introduces the new Open Trace Format. The first part provides a small overview about Trace Format Libraries in general and existing Formats/Libraries and their features. After that the important requirements are discussed. In particular it concerns efficient parallel and selective access to trace data. The following part presents design decisions and features of OTF comprehensively. Finally, there is some early evaluation of OTF. It features comparison of storage size for several examples as well as sequential and parallel I/O benchmarks. At the end, a conclusion will summarize the results and give some outlook.

## 1 Introduction

Today, there are many established trace formats already. Let's name for example Epilog [11], Pajé's trace format [3], Pablo Self-Defining Data Format (SDDF) [1], Structured Trace Format (STF) [4], TAU trace format [2, 10] and Vampir Trace Format 3 (VTF3) [8]. All of them share the basic principle storing event information in so called *event records* which are sorted by time. Special definition records deliver auxiliary information and allow more efficient storage. Event types usually supported by all formats are: function call events, point to point message events, collective communication events, performance counter samples and few more. Usually, there are only minor differences in representation. The general functionality is the same.

To all *trace formats* there is a *trace format library* as well[1]. It serves as read/write layer and passes trace records from/to files. Although all formats could be written or read directly with more or less effort it is not recommended. Instead, one should always use the supplied library. It will perform parameter checking, parsing, data filtering etc. Most of the time this will be easier, safer, faster and more complete.

In High Performance Computing (HPC) there are some special requirements for trace formats and their support libraries. Access speed and storage efficiency are important. Furthermore, both properties must scale well with very huge amounts of data. Parallel I/O is one necessary concept to accomplish this on today's HPC platforms. To incorporate parallel I/O efficiently the trace format library must provide special support.

It turns out, that even the trace formats are affected. Only with data management and storage scheme adapted towards parallel I/O maximum performance becomes achievable. This feature is missing from all but one of the trace file formats mentioned above.

[1] The terms "trace format" and "trace format library" are often used synonymously.

V.N. Alexandrov et al. (Eds.): ICCS 2006, Part II, LNCS 3992, pp. 526–533, 2006.

STF [4] implements parallel I/O and advanced selective access. Unfortunately, that format is not available on all platforms and for all purposes (anymore).

This paper presents the OTF format itself and especially the trace format library. It is not intended as comprehensive programming documentation but an overview of functionality. In the first part of the paper there will be a definition of requirements. It is followed by a presentation of the design considerations and main features. Finally, there are some (early) performace data as well as comparisons with the state of the art.

## 2 Requirements to a Modern Trace Format

First of all, a complete set of record types for the usual purposes has to be provided. The interface should use the same basic form as all existing trace format libraries: It provides record type specific write functions and a call-back mechanism for reading.

The format must be platform independent, especially with respect to endianness and word size (32bit/64bit). Of course, economical storage size is very important. Last but not least, read/write access needs to be fast and efficient. This involves both, selective access and sequential linear read/write operations. The former is the key to parallel I/O which relies on selecting disjoint parts of a trace to be read independently. The latter is relevant for sequential and parallel performance as it determines sustained I/O speed.

Selective access is to be provided with respect to three criteria: First, selection per record type as supported by all existing trace formats. This simply means to ignore unwanted record types. Second, selection per trace process. This must not read everything dropping unwanted parts because it would overstress I/O. Loading only worthwhile data saves a lot of effort. In case a single process contains still too much data the third selection facility can reduce this further. It allows to select a certain time interval (per process). Again, it is important that only the requested data is loaded rather than the whole process trace.

The last part of selective access requires another new facility. In order to start reading at an arbitrary time stamp additional helper information is necessary. Otherwise, the initial state at that time stamp would be unknown, rendering all following events useless. Therefore, a special class of records has to provide resumption points, i.e. full state information at selectable time stamps.

In order to provide a clue what processes resp. time intervals to select *before* reading the whole trace there should be another kind of auxiliary information. This is to provide statistical overview information about what happens on certain processes and in certain time intervals. Those information should be small and concise and yet allow arbitrary granularity.

## 3 Design Decisions and Features

Based on above requirements the following basic design was chosen: OTF uses *multiple streams per trace*. It uses an *ASCII encoding* which allows *binary search on files* for time intervals. Furthermore, it supports *optional auxiliary information*. Please, see below for a comprehensive description.

### 3.1 Multiple Streams

The most important point regarding parallel I/O is the distribution of a single trace to multiple *streams*. Those streams are separate files that can be accessed independently. A stream contains events of one or more processes in temporal order. But every trace process/thread is mapped to one stream exclusively, i.e. a process cannot span over multiple streams. This is necessary to maintain the original order of events with identical time stamp in one process. Thus, a trace of $n$ processes can be stored with $1 \dots n$ streams.

In addition, there is an index file with suffix .otf that stores the mapping of processes to streams. It can be set or queried via the trace format library, too. During tracing it might be very convenient to use $n$ independent streams. Simply adding the index file as well as a global definitions file will produce a perfectly valid trace without explicit merging[2]. Later, it might be convenient to reduce the number of streams. Perhaps several thousands of files are undesirable for massive parallel traces. The **otfmerge** tool is able to translate this to any number $1, \dots, n$ of streams or to a fixed number of processes per stream.

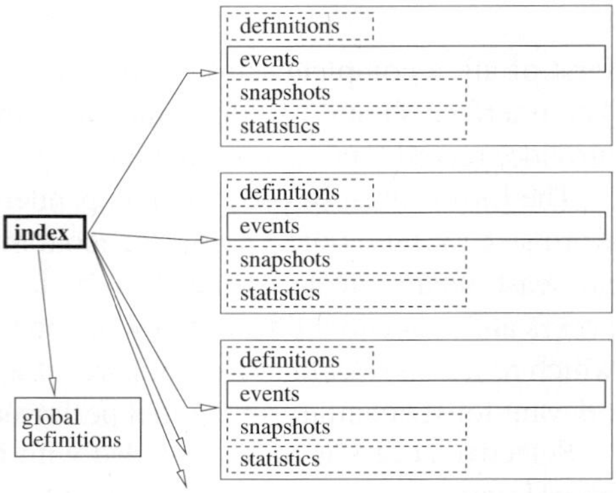


**Fig. 1.** OTF storage scheme. Index file, global definitions file as well as event files are mandatory. All local definitions, snapshots and statistics are optional.

When writing a trace, the process to stream mapping can either be specified explicitly or created automatically. With respect to reading event records the process-to-stream mapping is completely transparent. The user specifies which processes are to be read. Then, the OTF library accesses only streams containing those processes. If multiple streams are read they are merged on-the-fly to provide a single sorted input stream to the caller. Furthermore, an arbitrary number of streams can be handled with limited numbers of file handles. Figure 1 gives an impression of OTF's storage scheme.

### 3.2 ASCII Encoding

The internal representation of records in OTF uses an ASCII encoding. This makes OTF platform independent in the purest way. Of course, this is a slight disadvantage in terms of storage size which is compensated later. This encoding makes OTF human readable and accessible via tools like grep, awk, perl, etc.

With this ASCII representation every record is stored as a single line of text. Information about time stamp and process of event records is separated from the records

[2] Merging and post-processing of trace data are frequently used for several purposes. It might still be necessary for timer-synchronization, token renaming, etc.

and managed by a *state machine* where every stream keeps track of current time and process. Special records update those values which then remain until the next update. Ordinary event records contain no time stamp or process but simply refer to the current state of the stream. By this means redundant time stamp and process specifications are eliminated within streams.

Another advantage of the ASCII representation is the ability to recover reading at arbitrary file positions. After a **`fseek()`** operation to any file position a simple forward search for the next newline character will produce a valid resumption point. Further search will deliver an update of the stream's state machine such that reading can be started at that point in a reliable way. This is most important for selective access as introduced below. Unlike for pure binary formats there is no need for anchor information, escape sequences, padding, etc.

### 3.3 Efficient Selective Access

OTF provides selective access to traces according to three aspects: by record types, by processes and by time intervals.

Selection by record types is possible by disabling a certain record type's call-back handler. The OTF library will then skip all records of that type. It will even avoid fully parsing such records.

For selective reading by processes the user requests a set of processes to be read. This may range from a single process to all. Based on that, OTF accesses only affected streams. If a stream contains enabled as well as disabled processes, it is read and filtered. Thus, only records of the enabled process(es) are passed to the consumer. Again, records of disabled processes are skipped without parsing.

The OTF interface allows only to select by processes but not by streams. However, the user can query the process-to-stream mapping and derive an optimal selection. This is highly recommended for parallel reading.

By this means, multiple processes or threads of a parallel application can read/write disjoint parts of a trace concurrently. OTF does not do any parallelism internally but adapts to the parallel paradigm of the user application. The OTF library is thread-safe but doesn't contain any explicit multi-threading or multi-process parallelism internally. User applications need to take care of race-conditions during parallel OTF output!

Finally, OTF also supports selective access with respect to arbitrary time intervals. From the user interface point of view this works by simply restricting the time interval to be read. Internally, OTF performs a binary search for the start time stamp on all according files. With $O(\log n)$ effort for total record count $n$ this scales very well for large files. Once the start position is found the OTF library performs standard reading until the end of the time interval is reached or all streams are exceeded.

### 3.4 ZLib Support

For the sake of efficient storage size OTF supports transparent ZLib [6] data compression as an experimental feature. This will compensate for the ASCII encoding which is not most economical by itself.

Every file that is part of an OTF trace can be compressed by itself, it is automatically detected when reading. OTF uses a rather small compression level which allows fast compression and yet good data reduction to $20\%$ - $35\%$ of the original size on average.

ZLib compression is achievable directly during trace file creation or later via the **`otfcompress`** tool, which also supports de-compression. In order to support random access even for compressed files OTF uses a blockwise compression scheme as directly supported by ZLib. Default block size is $10\ mathrmKB$ which allows good compression and fine-grained access at the same time.

So far, this paper mentioned event records only. Of course, there are definition records, too. They specify certain global properties, like timer resolution and token definitions. Such global definitions are never performance critical and are stored in a single global file (see Figure 1).

OTF supports local definitions for convenience. They are attached to streams but not to individual processes inside a stream. This is provided for trace libraries which might want to store local definitions at trace time. However, local definitions should be translated into preferred uniform global definitions.

Besides definition records there are two more classes of auxiliary records. First, there are so called *snapshot records* which assist user applications when reading arbitrary time intervals. They provide information about the complete state of a trace process at a certain time. This includes the function call stack, pending messages, etc.

Second, so called *summary records* allow a reasonable guess about which processes resp. time intervals to chose for selective access. They provide a variety of properties for certain event record types. For example number of calls and exclusive/inclusive runtime per function. Alternatively, it can be summarized for groups of functions instead of single functions.

Summary information at a time stamp $t$ covers the time interval $[0, t)$, i.e. from the beginning of the trace to the current time. All properties $S$ can be derived for the interval $[a, b)$ by the simple transformation

$$S\big([a,b)\big) = S\big([0,b)\big) - S\big([0,a)\big). \tag{1}$$

By this scheme of representation only $n$ summary points are sufficient to cover $n \cdot (n-1)$ intervals of time.

Both, snapshot and summary records are optional and reside in separate files per stream (see Figure 1). This allows to generate, re-create or remove them without changing the event information. Snapshots and summaries can be supplied by a trace library, by any user application or by the **`otfaux`** tool which adds both to an existing trace with selectable granularity.

## 4 Early Evaluation

Below early performance evaluation is shown. It was performed on a commodity workstation (AMD Athlon 64 3200+, 2GB main memory) and on a brand-new super computer (SGI Altix 3700, 128x Intel Itanium2 1500MHz, 490GB shared memory, very fast I/O).

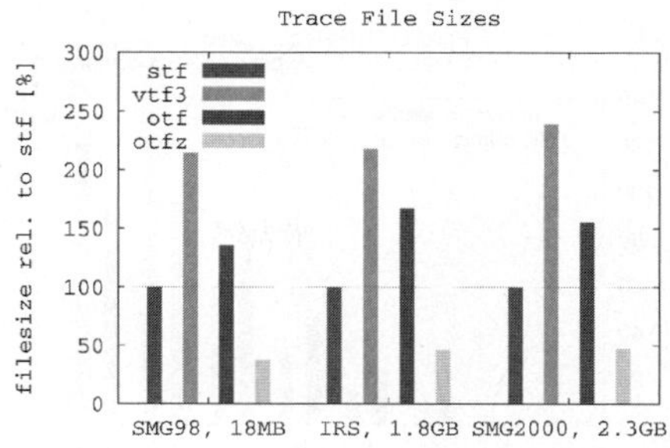


**Fig. 2.** Storage sizes for STF and VTF3 vs. plain and compressed OTF

**Storage Size.** The first interesting issue concerns storage size which is a platform independent issue. Fig. 2 compares storage size for identical traces in STF [4] and VTF3 [8] formats to plain and compressed OTF. The examples are original STF traces with 18 MB, with 1.8 GB and with 2.3 GB size from well known ASCI benchmarks [5, 9].

Both, compressed and even plain OTF traces are much smaller than binary VTF3, i.e. even a pure ASCII encoding results in a quite efficient storage scheme. Yet, plain OTF is 30% to 60% larger than STF. However, ZLib compressed OTF format turns out to use $< 50\%$ of STF's storage volume. Thus, in terms of storage size compressed OTF provides a notable advantage over two well established binary trace formats.

**Read Performance.** The second issue of interest is raw read performance of the trace format libraries. Traces are read, parsed and send to dummy handlers measuring the formats' read performance including internal overhead.

Figure 3 shows reading speed in million records per second for the 3 example traces in the 4 traces formats. On the AMD 64 platform (Figure 3 left) there is an obvious difference between small traces (18 MB), where I/O is noncritical, and huge traces ($>$ 1 GB), where I/O becomes the limiting factor.

For the small example uncompressed OTF speed is almost double of STF despite the larger storage size. Reading compressed OTF is no advantage here. Instead, the additional computation decreases performance while reduced input amount ($\approx 1/3$) brings only minor benefit.

For large traces the situation is different. Now, the run-time behavior looks perfectly reciprocal to the storage size diagrams. The I/O effort determines the read performance and thus, compressed OTF is fastest by far.

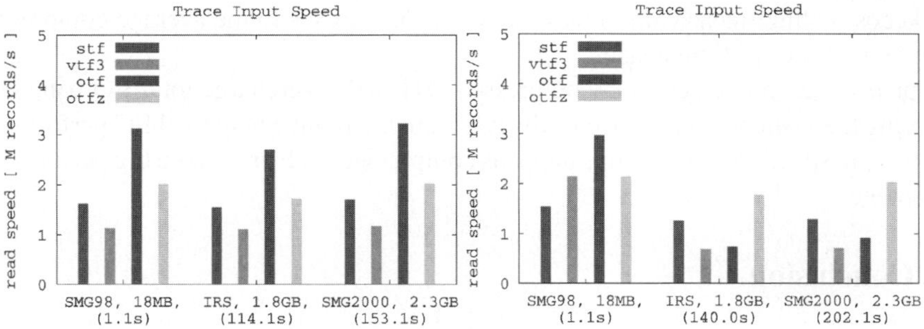


**Fig. 3.** Read speed for three real-life traces on AMD Athlon 64 3200+ (left) and on SGI Altix 3700 (right). Figures are normalized relative to STF.

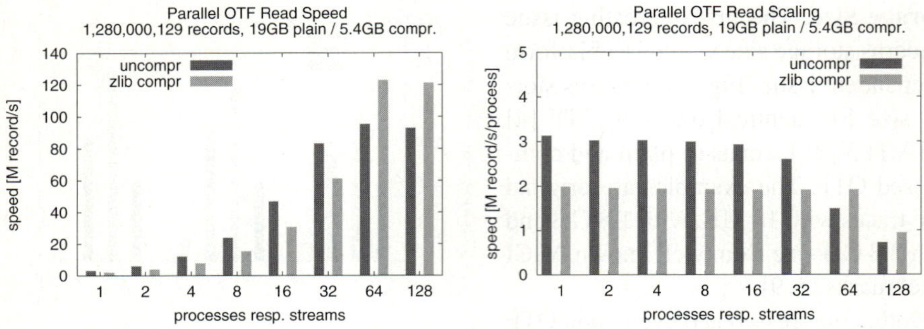


**Fig. 4.** Parallel input speed on SGI Altix 3700: total (left) and normalized (right)

On the Altix platform (Figure 3 right) all three examples behave similar to the small case on AMD 64 platform. Because of this machine's excellent I/O capabilities input speed is just not critical, not even for traces of 2.3 GB size. Because of that, uncompressed OTF is fastest for all three examples. It is followed by compressed OTF which is still faster than STF and VTF3. Nevertheless, OTF compression should be used by default because of the reduced storage footprint.

**Parallel Read Performance.** Finally, Figure 4 shows parallel performance for plain and compressed OTF on the Altix platform. It was evaluated with a very huge trace containing 128 processes and 1 280 000 000 records. This results in $\approx$ 19 GB uncompressed or $\approx$ 5.4 GB compressed total storage size. This same trace was stored with $n = 1 \dots 128$ streams and read with $n$ processes.

The left hand side of Figure 4 shows the aggregated total read performance (in million records per second), the right hand side the normalized counterpart (in million records per second per process).

The diagrams reveal a very nice scaling parallel input speed when reading uncompressed OTF. The sustained input speed of 3 000 000 records/s/process scales almost linearly up to 16 processes. At 32 processes the total input speed is saturated and the uncompressed read performance per process drops notably.

The normalized speed when reading compressed OTF reaches only $\approx 2/3$ of the uncompressed speed for small processor counts. When scaling to larger processor counts uncompressed OTF reading speed declines at $> 16$ processes. Compressed OTF reading keeps scaling linearly up to 64 processes! This relates to the average compression ratio $\approx 4:1$, i.e. $1/4$ input size.

For $n = 128$ processes even compressed OTF will overcharge total I/O bandwidth reducing the sustained input speed. Always, there is a point where total I/O performance limits input speed. As soon as this happens compressed OTF provides a big performance advantage.

## 5 Conclusion

OTF is a fast and efficient trace format library with special support for parallel I/O. It provides a convenient interface similar to existing trace format's ones. In addition it offers extended selective access along with some support features.

It achieves very good performance on single processor workstations as well as on massive parallel super computers. It delivers fast sequential and parallel access which scales very well depending on raw I/O capacity.

Experimental transparent blockwise ZLib compression allows to reduce storage size notably and improves parallel scalability on the same time.

OTF is available at [7] under BSD open source license that allows free usage for academic and commercial applications. Future work will include porting and tuning for additional platforms. Furthermore, a carefully extention of record types supported by OTF will be done according to user feedback. This might involve new MPI-2 related record types, for example.

**Acknowledgment.** Design and development of OTF has been done in cooperation with University of Oregon and Lawrence Livermore National Laboratory (LLNL) of University of California. Especially we would like to thank Prof. Allen D. Malony and Dr. Sameer S. Shende.

Furthermore, we are very grateful to the High Performance Computer group of ZIH, TU Dresden for their support with early I/O performance measurements on SGI Altix 3700. Special thanks to Michael Kluge and Thomas Blümel.

## References

1. Ruth A. Aydt. The Pablo Self-Defining Data Format. Technical report, Department of Computer Science, University of Illinois, Urbana/Illinois, USA, 1994. `http://wotug.kent.ac.uk/parallel/performance/tools/pablo/`.
2. Holger Brunst, Allen D. Malony, Sameer S. Shende, and Robert Bell. Online Remote Trace Analysis of Parallel Applications on High-Performance Clusters. In *Proceedings of ISHPC'03 Conference*, Springer LNCS 2858, pages 440–449, 2003.
3. J. Chassin de Kergommeaux, B. de Oliveira Stein, and G. Mouni. Paje Input Data Format. Technical report, 2003.
4. Intel GmbH, Brühl, Germany. *Intel Trace Collector User's Guide*.
5. Guohua Jin and John Mellor-Crummey. Experiences Tuning SMG98 - a Semicoarsening Multigrid Benchmark based on the Hypre Library. In *Proceedings of the International Conference on Supercomputing*, New York, June 2002.
6. Jean loup Gailly and Mark Adler. *zlib 1.1.4 Manual*, March 2002. `http://www.zlib.net/manual.html`.
7. Hartmut Mix. Open Trace Format. `http://tu-dresden.de/zih_otf/`, 2006.
8. Stephan Seidl. VTF3 - A Fast Vampir Trace File Low-Level Library. personal communications, May 2002.
9. The ASCI Project. The IRS Benchmark Code: Implicit Radiation Solver. `http://www.llnl.gov/asci/purple/benchmarks/limited/irs/`, 2003.
10. University of Oregon. *TAU User's Guide*. `http://www.cs.uoregon.edu/research/tau/docs/newguide/`.
11. Felix Wolf and Bernd Mohr. EPILOG Binary Trace-Data Format. Technical report, Forschungszentrum Jülich, University of Tennessee, 2004. FZJ-ZAM-IB-2004-06.

# Dynamic Instrumentation of Distributed Java Applications Using Bytecode Modifications

Włodzimierz Funika* and Pawel Swierszcz

Inst. Comp. Science, AGH-UST, al. Mickiewicza 30, 30-059 Krakow, Poland
funika@uci.agh.edu.pl
Tel.: (+48 12) 617 44 66; Fax: (+48 12) 633 80 54

**Abstract.** Java's features such as system platform independence, dynamic and network oriented architecture, robustness as well as growing number of common standards make it a language of choice for many projects. However an increasing complexity of created software and requirement for high stability and high quality of applications make it desirable for a developer to inspect, monitor, debug or in any way alter Java programs behaviour on-the-fly. The main goal of this paper is to present the design of a system for instrumenting Java classes at runtime. This system is to aid developer in modifying program by adding fragments of code at specific locations that implement some new functionality. This allows programmer to enhance classes with logging, monitoring, caching or any other capabilities that are required at run-time.

**Keywords:** Java, instrumentation, bytecode, J-OMIS, J-OCM.

## 1 Introduction

With the fast increasing power of modern computers equipped with more and more advanced processing units, memory and disk resources more complex and larger applications are created. Rapidly developing networking made distributed systems common as network-enabled computers are standard. Additionally, modern systems are often assembled from components coming from different vendors, sometimes running on quite different platforms and not fully compliant to implemented standards. This makes applications prone to errors, unpredicted behaviour, runtime flaws, deadlocks and all kinds of programmer mistakes.

Java is a leading solution for developing modern, object-oriented software, being a platform for a great number of advanced systems in various use cases both scientific projects and business oriented applications. Together with this comes greater effectiveness of software production in many domains from mobile devices to large scale systems and middle-tier integration of heterogeneous legacy solutions. Java is also successfully applied to massive computational tasks executed on a farm of high performance servers or quite contrary in distributed environment or for example grid network.

* Corresponding author.

V.N. Alexandrov et al. (Eds.): ICCS 2006, Part II, LNCS 3992, pp. 534–541, 2006.

The increasing complexity of developed software makes the whole process of testing, bugs detecting and fixing difficult and time-consuming. Large hardware potential made solving much complicated problems possible nowadays but this has a side-effect of making optimizing and debugging systems even more difficult. Although Java platform provides with many options of debugging, testing or in other ways producing stable applications it does not include ready-to-use kits that can be helpful when dealing with advanced, parallel and distributed programming. This leads to a demand for more sophisticated tools intended to aid the developer in creating stable end efficient software, e.g. monitoring systems, debuggers, logging systems, profilers and other. For realizing their tasks in the distributed environment, such tools might exploit various techniques and interfaces, in this paper we address one of these techniques - instrumentation.

*Instrumentation* does not modify original application's behaviour in terms of which instrumented program should work in the exact same way is was working. Obviously, in order to perform code instrumentation, some knowledge about how application works is indispensable. This could include some code structure details like locations of interesting fragments of code or runtime behaviour information (for example how is application using it's external resources). However these application meta-data could be collected in automatic or semi-automatic way through code (source or binary) analysis.

In this paper we address the issue of dynamic instrumentation of distributed Java applications. In order to achieve this goal, we combine techniques, which enable the analysis of application code with run-time bytecode modification and communication with distributed JVMs which are running the target application.

## 2 Related Work

There are other systems that exploit various instrumentation techniques in application monitoring or manipulation. One example might be the `JSpy` system for runtime analysis of Java applications [10]. It reads instrumentation specification (a set of rules in a predicate/action form) and generates logging statements in the target code. The general concept of the system is quite similar to ours, binary Java classes are also extended with additional bytecode instructions (low-level library used for that is in this case `JTrek`) and there is interoperability with external monitoring system (`JPax`). Our system is more generic while `JSpy` is entirely analysis/monitoring oriented.

Another example of instrumentation tool capable of instrumenting and modifying programs at the time of execution is the `DynInst` library [14]. It provides functions for building instrumentation tools by creating *mutator* programs that connect to the target application at runtime and change the program's behaviour on-the-fly. Although DynInstAPI is machine independent it is also highly C++ oriented and thus practically unusable for instrumenting Java applications. On the other hand, DynInst allows for fine-grained instrumentation while our system produces new versions of whole classes.

A slightly different approach is used in `Java Instrumentation Suite` (`JIS`), where instead of modifying class methods prior to executing them in JVM the actual runtime environment is instrumented [11]. Calls to dynamic native methods are wrapped that allows to react on runtime events like starting thread or entering monitor. This makes bytecode modifications needless which is good as they are quite error prone. On the other hand, it limits the set of events the instrument can react on to those which include calling native methods from within the JVM runtime libraries. The whole system has functionality strictly limited to monitoring multithreaded applications, debugging deadlocks, etc.

An example of a tool that gives a developer a greater extent of freedom could be the `BIT` tool [12], being a framework for creating custom instrumentation applications, which provides a set of interfaces for enhancing applications with arbitrary code. BIT operates somewhere between an instrumentation system like ours and a low-level bytecode altering library. It provides no direct support for distributed applications nor monitoring - at the cost of making instrumentation process more complicated and time consuming it gives almost unlimited power to the programmer when it comes to inserting custom code at a fine-grained level.

## 3 Motivation and Goals

The main goal of our research is to design and implement a system that will support the developer in instrumenting Java applications. This support consists of automating some of the tasks that are the required steps of instrumentation process. In this way, the developer can concentrate on a higher abstraction level - the design of instrumentation elements, choice of instrumentation spots, combining additional functionality with the existing application classes to create functional execution units.

The designed system has to preserve maximum flexibility that means it must not limit instrument's functionality so that it can perform any task that the original program could do. This is to be achieved by using bytecode manipulation that allows for arbitrary code to be executed as an instrument is invoked. An alternative approach could be taken - that is to change the JVM behaviour rather than an application's code. Although this frees us from error-prone (and sometimes costly in execution time) bytecode modifications it requires a custom virtual machine. This is a major downside as we don't want to rely on a single JVM implementation [4]. Operating on bytecode has an advantage that it puts minimal constraints on the instrumentation process - no source code is needed and rewritten classes are undistinguishable from those produced by the original Java compiler. Therefore this is another goal of the work - to create methods for automatic bytecode modification from an instrumentation design, possibly with some help from third party libraries.

As mentioned above, distributed applications are increasingly widely used and thus some kind of tool support for them is highly required. An instrumentation system must be able to act in a distributed environment and communicate with remote virtual machines and instruments running on them. To achieve this, another tool is used - the `J-OCM` distributed monitoring system [1, 2]. It allows to discover execution nodes, collect various data on running machines and what is more important - to interact with JVMs through low-level debugging interfaces. Therefore the main project goals include inspection of the possibilities to combine the generic instrumentation engine with the monitoring system.

The secondary goals include creating structures and algorithms for representing the instrumented application's code in an object-oriented fashion, examining libraries for the modification of binary classes in the context of instrumentation and creating a universal, generic model of instrumentation design - specifying common instrument features and ways of binding instruments to applications.

## 4 System Concept

The system being described comprises a few components. The main element of the architecture is an instrumentation engine with graphical user interface allowing for choosing classes to be instrumented, designing instruments to be applied and controlling the actual process of dynamic instrumentation. Other elements include low-level libraries for class manipulations (e.g. Bytecode Engineering Library), instruments definitions (class files realizing a proper interface and holding data about a specific instrumentation pattern) and last but not least the monitoring system for communicating with distributed virtual machines.

The process of instrumentation performed by the described system consists of several stages. The first one is choosing which application classes are to be modified and analysing their contents. A class model is created in the form of `SIR` tree which is a format for representing an applications' code structure [7]. Then it comes to the design of additional functional units called *instruments*. Each instrument implements some feature (e.g. writes out the current class field value) and is connected to program hooks which are places in the code where the instrument code can be injected. Having prepared the design of instrumentation, the developer can advance to the third stage and perform actual class files modifications. New fields are added to the classes, their internal structures are modified and new bytecode instruction sections are introduced to classes' methods - accordingly to previously crafted instruments and their binding to SIR elements.

After all classes that need to be changed are changed, they are distributed to (possibly remote) Java Virtual Machines running an application that is to be operated on. The original program classes are replaced in JVMs with their instrumented versions. Every object of instrumented class instantiated from that point on as well as static calls to class methods will possess and perform extended functionality. For adding network-awareness to the design, we used the J-OCM distributed Java applications monitoring system. It is used as a communication and transportation framework, provides information about hosts (nodes) and JVMs running the application as well as implements low-level class hot-swapping service using Java's new (`Java Virtual Machine Tool Interface`) (`JVMTI`) [5] . This is where the described instrumentation system differs mostly from traditional solutions. Typically, instrumentation tools focus on either enhancing the application's code or interoperating with a target virtual machine, to gather required data or perform some operations. Using J-OCM in the described system gives the developer a better view on the environment where the instrumented application is going to be executed as well as allows to interact with it. Since one can write custom services in the J-OCM monitoring system, this functionality of interacting with the distributed environment can be further extended as new features are required.

The usability of the described system is not limited to monitoring or profiling only. Instruments can implement almost any kind of service including object state persistence (for example to relational database), caching with an external caching system or even workarounds or quick bugfixes of encountered application's problems. Practically, the developer has only to be careful about not damaging original program's functionality and not using excessive amounts of system resources which could affect application's performance or stability.

The architecture of the system is shown in Fig. 1.

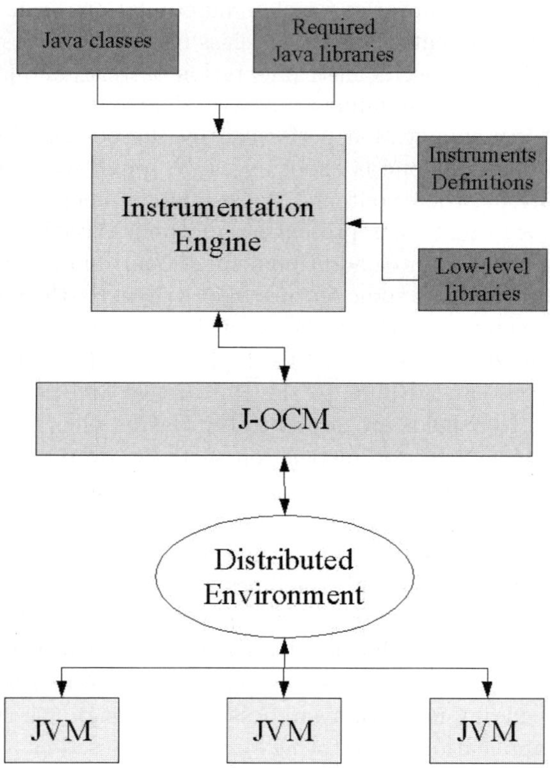


**Fig. 1.** Instrumentation system architecture

## 5 Implementation Issues

The system was implemented in Java using `Java Developers Kit` version 5.0. It was created with a free and open developer's environment Eclipse [13]. Graphic user interface (`GUI`) uses `SWT` (`Standard Widget Toolkit`) library which is an alternative to standard libraries included in Java distribution such as `AWT` (`Abstract Windowing Toolkit`) or `Swing`. Alternatively, a scripting approach would be possible, but we believe that it's easier for the user to interact with graphical interface. For example it is more convenient to examine the application's structure in a point-and-click widget than to read large XML file [6], also joining instruments with application elements is less error-prone and more user-friendly when done using a GUI. Class introspection and manipulation was performed mostly with `Java Object Instrumentation Engine` [9] v. 0.9 beta.

Like in many other applications, XML format was used for storing instrument definitions, SIR structures, and application configuration between sessions. In our project, however, instead of using the standard `SAX` or `JDOM` parser, an alternative solution is chosen. `Castor` tool was used to transform plain Java objects into XML structures based on the mapping files created by the developer [8]. Castor was chosen because it frees from writing custom XML processing code. Keeping the conversion policy in an XML file makes it also more maintainable and easier to apply any potential changes.

The most challenging task was to implement bytecode modifications to be performed based on instrument definitions and the original class structure. Not only bytecode sections have to be changed but also other static structures of the class like the constant pool containing all constant values the class uses or the stack size field that could result in a runtime error if set to an insufficient value. The actual bytecode instructions inserted have also to be valid and correct - not only from the syntax point of view but also should not collide with original instructions, change a field or variable values or in any other way alter the original runtime behaviour. A small mistake in an instruction parameter, index argument or similar details will end up in crashing VM or will cause the class verifier to refuse loading the instrumented class. Unfortunately, there is no easy way to debug instrumented classes at run-time. There is no source code for the instrumented version of the class and instrumented classes contain no valid line numbering information. What is even worse a debugger cannot be used in case of a low level class validation error or runtime VM crashes, at least no debugger capable of warning in such conditions is known to the authors. The most effective yet inconvenient way of debugging is bytecode instruction analysis. By comparing the original bytecode with the bytecode of instrumented classes, it is possible to reveal what instructions were added by the instrumentation system - it is up to the developer to find what's wrong with it.

## 6 Case Study

For testing purposes, a few instruments were created (designed as Java classes implementing proper interface) and later instantiated in the instrumentation system and applied to simple test classes and methods. Later, bytecode implementations were compared between original and modified versions. The instruments used in these tests were:

– `HeapMemoryInstrument` - measures the amount of heap memory available at the time it is invoked. It can be configured to react on any runtime event but its most natural use is to inject such a measurement into those places in application code where new objects are created.
– `MethodTimeInstrument` - is intended to react on the `method execution` and `return from method` events. The first one causes instrument to get a current system time and store it as a field value. The second event makes the instrument check the system time again and compare it to a saved value. In this way, one can calculate the time spent in a method.
– `VariableWatchingInstrument` - enables to monitor changes to a variable value or class/object field value.
– `ExceptionLoggingInstrument` - reacts on the `throw exception` runtime event, helpful in debugging applications; with a few enhancements it can be used to preserve important data when a system error occurs.
– `GarbageCollectingInstrument` - is more of a utility than a monitoring probe, forcing garbage collection in VM.

The system proved to be producing a code that is working stable and correct. Instrument implementations were injected into the original code and actually executed by VM just as they were written in the original source code and compiled with the rest of the class. The user is enabled to define instrumentation with GUI and observe the whole instrumentation process with it.

## 7 Conclusions

While most of the above research goals have been achieved, some work still remains. The dynamic class reloading of classes after bytecode instrumentation via monitoring system is left for future development.

The instrumentation system discussed in this paper is found to be capable of performing all the tasks mentioned above with the exception of distributed class redefinition. First of all, it allows for the in-depth introspection of binary classes of the instrumented application and the creation of a model representing its structure. It provides an easy and extendable way of defining instruments as Java classes which implement a special interface. The instrument can perform arbitrary actions and what actually is done in the instrument depends only on the instrument creator. Binary modifications are performed fully automatically and the user performing instrumentation does not need to know any bytecode-level details.

As for examining the possibilities of combining the generic instrumentation system with J-OCM - using techniques defined in the J-OMIS specification [3], monitoring system services can be defined and called to perform on-line dynamic instrumentation through Java agent plugged into the target VM. This requires, however, knowledge about how J-OCM works as well as some native coding for implementing services.

Instrumentation is often used as a means of monitoring. In this context, instead of actually modifying classes low-level interfaces provided by the `JVMTI` can be used to obtain information about the running application and even perform a sort of instrumentation for the purposes of gathering data to be utilized by tools. The general concept stays the same with the difference that the J-OCM system requests are not used for class redefinition but rather for calling proper tool interface methods via the agent.

**Acknowledgements.** This research was partially supported by the EU IST K-Wf Grid project and the AGH grant.

## References

1. M. Bubak, W. Funika, M.Smętek, Z. Kilianski, and R. Wismüller: Architecture of Monitoring System for Distributed Java Applications. In: Proc. EuroPVMMPI'2003, LNCS 2840, pp. 447-454, Springer Verlag, 2003
2. M. Smetek: OMIS-based Monitoring System for Distributed Java Applications, M.Sc. Thesis, AGH, Krakow, 2003
3. Bubak, M., Funika, W., Wismüller, R., Mętel, P., Orłowski. Monitoring of Distributed Java Applications. In: Future Generation Computer Systems, 2003, no. 19, pp. 651-663. Elsevier Publishers, 2003
4. Tim Lindholm, Frank Yellin: The Java Virtual Machine Specification, 1999 `http://java.sun.com/docs/books/vmspec/`
5. JVM Tool Interface: `http://java.sun.com/j2se/1.5.0/docs/guide/jvmti/`
6. W3Org: Extensible Markup Language (XML) 1.0 (Second Edition) `http://www.w3.org/TR/REC-xml`
7. C. Seragiotto, Jr., Hong-Linh Truong, B. Mohr, T. Fahringer, M. Gerndt: Standarized Intermediate Representation for Fortran, Java, C and C++ (to be published)
8. S. Gignoux, K. Visco: Castor XML Mapping `http://www.castor.org/xml-mapping.html`
9. G. A. Cohen, J. S. Chase, D. L. Kaminsky: Automatic Program Transformation with JOIE, 2002
10. A. Goldberg, K. Havelund: Instrumentation of Java Bytecode for Runtime Analysis, 2003 `http://ase.arc.nasa.gov/havelund/Publications/jspy-final.pdf`
11. J. Guitart, J. Torres, E. Ayguade, J. Oliver, J. Labarta: Java Instrumentation Suite: Accurate Analysis of Java Threaded Applications, 2000 `http://citeseer.ist.psu.edu/guitart00last.html`
12. H.B. Lee, B. G. Zorn: BIT: A Tool for Instrumenting Java Bytecodes, 1997 `http://citeseer.ist.psu.edu/lee97bit.html`
13. Object Technology International, Inc.: Eclipse Platform Technical Overview, 2003 `http://www.eclipse.org/whitepapers/eclipse-overview.pdf`
14. B. Buck,J.K. Hollingsworth: An API for Runtime Code Patching, 2000 `http://www.dyninst.org/`

# Fine-Grained Instrumentation and Monitoring of Legacy Applications in a Service-Oriented Environment*

Bartosz Baliś[1], Marian Bubak[1,2], and Krzysztof Guzy[1]

[1] Institute of Computer Science, AGH, al. Mickiewicza 30, 30-059 Kraków, Poland
[2] Academic Computer Centre – CYFRONET, Nawojki 11, 30-950 Kraków, Poland
{bubak, balis}@uci.agh.edu.pl
Tel.: (+48 12) 617 39 64; Fax: (+48 12) 633 80 54

**Abstract.** Legacy applications are still heavily used in Grid systems. In modern service-oriented Grid environments such applications are adapted to work as services. Monitoring of applications is often necessary for various purposes, such as debugging, performance analysis, fault-tolerance, etc. In this paper we present a framework for monitoring of legacy applications wrapped as services. We adapt the OCM-G system for monitoring of distributed legacy applications (such as MPI) to work as part of a broader service-oriented monitoring infrastructure GEMINI. We address monitoring and instrumentation at both service-level and legacy-level. Our instrumentation is fine-grained, i.e. we support instrumentation and monitoring at the level of individual code regions.

**Keywords:** grid, monitoring, instrumentation, legacy applications, service-oriented architecture.

## 1 Introduction

Grid systems based on modern service-oriented architecture (SOA) still heavily use legacy code. For needs of debugging or performance measurement there is necessity to monitor such legacy applications both at the level of service invocations and legacy code. The basis of monitoring of applications is instrumentation: we insert additional instructions into the application's code to generate events and pass monitoring information to the monitoring system. Instrumentation in a service-oriented environment cannot be statically inserted whenever we need to monitor an application, since we usually do not have the opportunity to change the source code, compile and deploy the application on demand.

We propose a solution for instrumentation which is: (1) dynamically enabled and disabled, (2) fine-grained to enable monitoring at the level of code regions, (3) accessible through a standardized instrumentation service to expose instrumentation functionality to arbitrary tools and services. This feature involves

* This work is supported by EU-IST Project K-WfGrid Knowledge Based Workflow System for Grid Applications, IST-2002-511385, http://www.kwfgrid.net

V.N. Alexandrov et al. (Eds.): ICCS 2006, Part II, LNCS 3992, pp. 542–548, 2006.

a standardized high-level representation of the application to let the user easily pick code regions to be instrumented, and a standardized language for expressing instrumentation requests.

For legacy applications wrapped as services, it is important to address monitoring at both service-level and legacy-code level in a coordinated way, e.g. to relate delays at the service-level to respective code regions at the legacy-level. However, those two aspects involve completely different execution environments. If different monitoring and instrumentation tools are used for those two levels, we are likely to fail in providing a unified view of the monitoring data. Therefore, a generic monitoring system is desirable to collect data from various sources.

To our knowledge, though some grid application monitoring approaches do exist, e.g. Mercury [8], none of the current efforts addresses the described problems in a comprehensive way.

We present a framework to instrument and monitor legacy applications wrapped as services. In this effort, we employ several existing systems and specifications. The existing system OCM-G [1] [2] is used for monitoring of MPI applications. We extend the OCM-G to support the concept of Standard Intermediate Representation (SIR) [3] to provide an abstract view of the application as a convenient way for the user to pick individual code regions to be instrumented. We have designed an instrumentation service compliant with a standardized language WIRL (Workflow Instrumentation Request Language) for specifying instrumentation requests. The mentioned functionality is integrated with the GEMINI monitoring infrastructure [4] [5], which provides us with opportunity to build custom sensors in order to collect information about any entity we want to monitor (legacy application in our case), and ensures the transport of the monitoring data over the Grid.

## 2 Legacy Monitoring Framework

Our framework used to monitor legacy applications is depicted in Fig. 1. The main monitoring system is GEMINI – Generic Monitoring Infrastructure for Grid resources and applications. GEMINI accepts requests for monitoring data and instrumentation, and also transports those requests and monitoring data over the network. GEMINI also deals with instrumentation at the service-level of the application. For this we used application sensors developed with GEMINI. However, in this paper we will focus on the legacy-level aspect. More information about GEMINI can be found in section 4.

The actual instrumentation and monitoring of the legacy application is done by the OCM-G monitoring system [1] [2]. We adapt the OCM-G to work as a GEMINI sensor. The OCM-G which supports multi-site parallel applications is based on the OMIS specification. The OCM-G custom sensor is an adaptation layer between GEMINI and the OCM-G. From the viewpoint of the OCM-G, the sensor is a tool connected to the OCM-G which sends monitoring requests expressed in OMIS. For GEMINI, however, this component works as a sensor compliant to the Generic Sensor Infrastructure. In OCM-G, the requests for

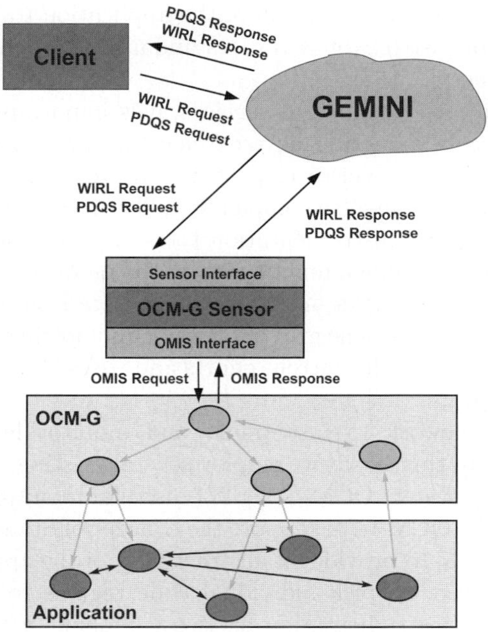


**Fig. 1.** Legacy Monitoring Framework

data can be expressed in a powerful imperative manner using OMIS[7]. GEMINI, on the other hand, employs an less complex, declarative monitoring request language PDQS combined with the instrumentation request language WIRL. GEMINI also uses a different data representation than the OCM-G. Thus, the main task of the high-level OCM-G sensor is to process queries from monitoring service of GEMINI (expressed in WIRL and PDQS) and convert them to appropriate OMIS monitoring requests supported by OCM-G. And also to convert OMIS responses containing monitoring data to the data representation compliant with GEMINI.

Additionally, we have also adopted the OCM-G to fully support fine-grained instrumentation of legacy applications (see section 3).

Clients of GEMINI are provided with service interface which enable them to control the instrumentation and monitoring and to obtain the monitoring data. Client thus can request SIR to have an abstract view of application and then select individual code regions to be instrumented.

## 3 Fine-Grained Instrumentation

Our approach to instrumentation is to combine source code instrumentation and binary wrapping with the dynamic control of the measurement process at runtime. The instrumentation is inserted statically via patching of source code or binary libraries, while activation and deactivation of the instrumentation is done at runtime. In our opinion, this approach could work for services: while developers of applications might be required to instrument their code, the cost of inactive instrumentation will be insignificant, and it will be possible to activate or deactivate required parts of instrumentation at runtime. Future implementations of instrumentation may be based on a fully dynamic approach. We provide a tool to automatically insert probe functions at defined places into source files and to generate SIR descriptions of the code. We include SIR into the application executable.

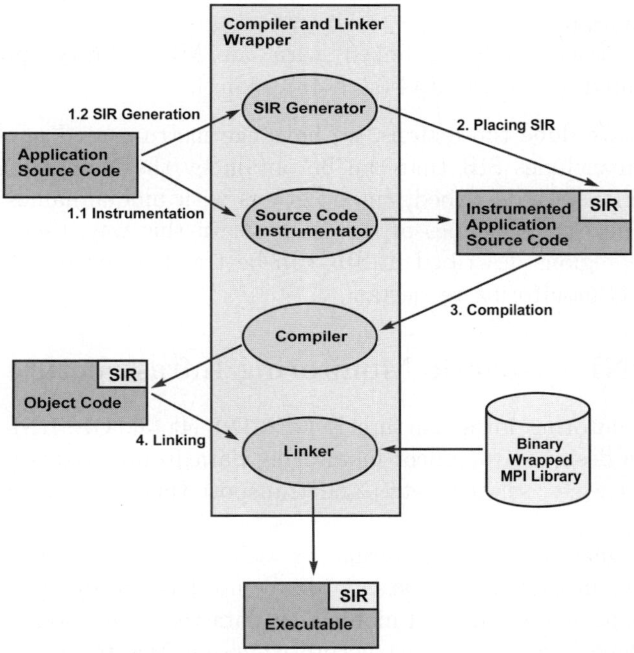


**Fig. 2.** Instrumentation process

To specify instrumentation requests, GEMINI exposes a service interface compliant with WIRL language (Workflow Instrumentation Request Language). WIRL is a XML-based language in which the user specifies which code regions are to be instrumented and what metrics to compute for those code regions (e.g. *wall-clock time for code region A*). The WIRL requests are translated into OCM-G's OMIS requests.

Our goal is to support monitoring at the level of individual code regions. Currently, the OCM-G does not allows this, we can only monitor *all* calls to a specified function. However, the OCM-G offers the mechanism of probes which can be inserted at arbitrary places into the source code of the application to generate custom events. We have extened the OCM-G with the capability to extract the SIR description from the application. With this and the usage of probes, the OCM-G can fully support instrumentation at the level of individual code regions. The detailed process of instrumentation and SIR generation is presented in Fig. 2.

1. First we take the application's source file and run a tool to instrument all code regions in it (currently function invocations) – step 1.1 Instrumentation. The tool, based on the parsing of the code, also generates the SIR description of the code – (step 1.2 SIR Generation).
2. Then SIR is inserted (as a static string variable) into the instrumented source file (step 2. Placing SIR).
3. Instrumented code is next compiled by the OCM-G compiler wrapper (step 3. Compilation).
4. Received object code is linked with an MPI library, probably pre-instrumented by the OCM-G (step 4. Linking).

After we have done those steps, we have the instrumented application executable which includes SIR that can be obtain by the monitoring system. In SIR, code regions are described, among others with unique names that can be transformed into probe names of the OCM-G. In this way, user "clicks" that indicate code regions described in SIR can be transformed into GEMINI and finally OCM-G monitoring requests.

## 4 GEMINI – Generic Monitoring Infrastructure

Our main monitoring infrastructure is GEMINI [4] [5]. GEMINI is a generic infrastructure designed to collect monitoring data from arbitrary sources (applications, resources, services, etc.) and transport this data over a distributed network.

GEMINI is generic not only in terms of a variety of supported monitoring data types, but also in that if offers standardized interfaces to query and subscribe for the data and a standardized monitoring data representation. GEMINI also provides a Generic Sensor Infrastructure which enables to easily develop and deploy sensors which produce monitoring data to GEMINI.

Fig. 3 presents the peer-to-peer architecture of the GEMINI Framework. This is a planned architecture; the current prototype provides a full monitoring functionality though only multiple separate Monitors can be deployed. GOM is an external registry that is used to publish GEMINI services and monitoring data provided. D-Monitors and Monitors form a super-peer architecture to transport monitoring data and data requests. D-Monitors expose interfaces to query and subscribe for monitoring data in PDQS and also to issue instrumentation requests in WIRL. Sensors are data producers which are connected to Monitors and D-Monitors.

To specify instrumentation requests in GEMINI, WIRL (Workflow Instrumentation Request Language) language is used. In WIRL one can request to attach application to be ready for instrumentation, enable/disable or finalize instrumentation and get the SIR of an application. The second interface exposed by GEMINI is PDQS (Performance Data Query and Subscription). It is employed to specify query and subscription requests for monitoring data.

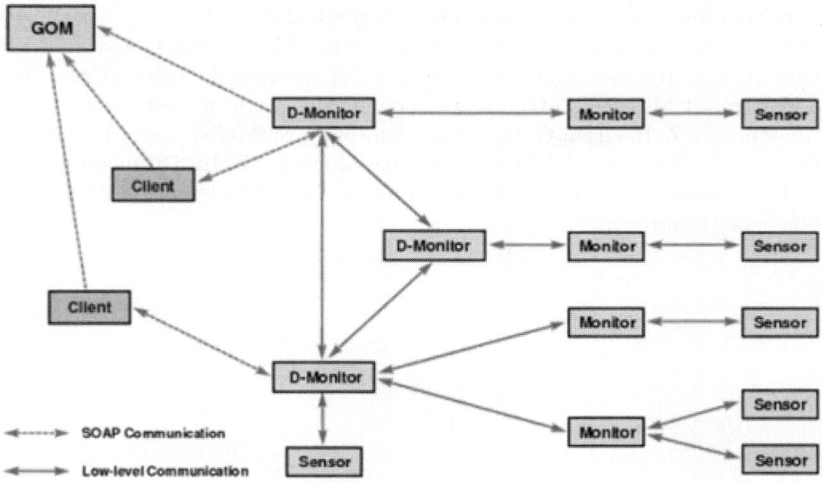


**Fig. 3.** Architecture of the GEMINI Monitoring Infrastructure

## 5 Summary

We have presented a framework for monitoring legacy applications wrapped as services and deployed in a grid. The framework is composed of a generic monitoring infrastructure GEMINI to transport the monitoring data and also data and instrumentation requests and a legacy system for monitoring distributed applications the OCM-G. The OCM-G was adapted to be compliant with GEMINI. A completely novel aspect of instrumentation was the support to instrument the application at the level of code regions. Thanks to GEMINI, unified view on monitoring data from the service-level and from the legacy-level of application is possible. This paper was focusing on the legacy level. In the future we would like to evaluate the presented solution on some real-world application scenarios.

## References

1. B. Baliś, M. Bubak, M. Radecki, T. Szepieniec, and R. Wismüller. Application Monitoring in CrossGrid and Other Grid Projects. In Dikaiakos M., editor, Grid Computing. Proc. Second European Across Grids Conference, pages 212-219, Nicosia, Cyprus, January 2004. Springer.
2. The OCM-G homepage: http://www.icsr.agh.edu.pl/ocmg
3. C. Seragiotto Jr., H.-L. Truong, B. Mohr, T. Fahringer, M. Gerndt, T. Li. Standardized Intermediate Representation for Fortran, Java, C and C++ Programs. APART Technical Report Workpackage 1, http://www.fz-juelich.de/apart
4. B. Balis, M. Bubak, J. Dziwisz, H.-L. Truong, and T. Fahringer. Integrated Monitoring Framework for Grid Infrastructure and Applications. In P. Cunningham and M. Cunningham, editors, Innovation and the Knowledge Economy. Issues, Applications, Case Studies, pages 269-276, Ljubljana, Slovenia, October 2005. IOS Press.
5. The GEMINI homepage: http://gemini.icsr.agh.edu.pl

6. The K-Wf Grid Project homepage, http://www.kwfgrid.net
7. OMIS – On-line Monitoring Interface Specification. Version 2.0. Lehrstuhl fur Rechnertechnik und Rechnerorganisation Institut fur Informatik (LRR-TUM), Technische Universitat Munchen. http://wwwbode.informatik.tu-muenchen.de/ omis
8. N. Podhorszki, Z. Balaton, G. Gombas. Monitoring Message-Passing Parallel Applications in the Grid with GRM and Mercury Monitor. In: In: Dikaiakos M., editor, Grid Computing. Proc. Second European Across Grids Conference, Nicosia, Cyprus, January 2004, Springer.

# Monitoring of WS-Based Applications

Lechoslaw Trebacz[1], Piotr Handzlik[2], Wlodzimierz Funika[3,⋆], and Marcin Smetek[3]

[1] Department of Computer Methods in Metallurgy, AGH, Krakow, Poland
[2] Department of Physical Chemistry and Electrochemistry, AGH, Krakow, Poland
[3] Inst. Comp. Science, AGH, Krakow, Poland
ltrebacz@metal.agh.edu.pl, piohandz@interia.pl
{funika, smetek}@uci.agh.edu.pl
Tel.: (+48 12) 617 44 66; Fax: (+48 12) 633 80 54

**Abstract.** The paper presents a Java-related monitoring platform for distributed applications which are oriented towards the use of Web Services (WS). We focus on the building a monitoring platform based on a well-defined interface (OMIS) between a monitoring system organized as middleware and tools that use the facilities provided by the monitoring middleware for WS-based applications. We aim at the observation and analysis of SOAP messages (data) exchanged with a Web Service (requests and responses). Our system is intended to monitor the characteristics of the remote call, especially: request time, transport time, cache access time, response time.

**Keywords:** monitoring tools, Web Services, OMIS, J-OCM, Java.

## 1 Introduction

With the increased possibilities of distributed Java programming, e.g. Web Services (WS), the demand for tool support (performance analyzers, debuggers etc.) for efficient programming increases as well. Nowadays, there are few environments which provide efficient monitoring support for distributed Java programs. One of the opportunities to solve this problem is an approach exploited in the On-line Monitoring Interface Specification [1]. A universal, open interface between tools and a monitoring system and an OMIS compliant monitoring system (OCM) allowed to specify such a Java-oriented monitoring interface (J-OMIS) [2] and monitoring infrastructure, which enables an extensible range of functionality intended for various kinds of tools and programming paradigms.

In this paper we consider extending the functionality of the J-OCM system [3] by the monitoring of WS-based applications. This extension is intended to:

– provide information about running Web Services and their number,
– provide information about the main stages of a WS life cycle, e.g.: receiving a request from the client side, starting particular services, time used to parse an SOAP message, etc.

⋆ Corresponding author.

V.N. Alexandrov et al. (Eds.): ICCS 2006, Part II, LNCS 3992, pp. 549–556, 2006.

- make possible to manipulate particular services, for example: stopping any operation at any moment, calling a method of a Web Service from within the monitoring system etc.
- enable access to available information about a Web Service,
- make possible to check, which stages of a WS life cycle take most of time, to analyze the performance of this application,
- provide information about errors when running a Web Service.

The paper is organized as follows: In Section 2 we focus on the principal features of WS, interesting from the performance point of view. In Section 2 (Related work) we show what issues connected withe the operation of WS are not covered by the existing tools. In Section 3 we briefly describe the J-OCM monitoring system with showing its some features aimed at the monitoring of distributed applications. In Section 4 we focus on the concept of support for WS monitoring in the J-OCM monitoring system and further on implementation details of the WS-related extension to it. We give also some results of overhead measurements when monitoring sample WS-based applications.

## 2 Related Work

There are a number of monitoring tools aimed at providing performance information and testing of the operation of WS. Such tools as `internetvista` [1] or `Parasoft SOAtest` [2] provide for the user a lot of useful features. They allow the user to verify all aspects of WS, from WSDL validation, to unit and functional testing of the client and server, to performance testing. SOAtest addresses key Web service issues such as interoperability, security, change management, and scalability. On the other side, each of these tools runs as a client application. They test WS by sending requests and waiting for a response. However, an issue is that they do not allow the user to get insight into what really happens *inside* the WS. There is another group of monitoring tools like TAU [15] which allow for advanced performance visualization of distributed applications, but do not provide monitoring information on WS.

Our goal was to overcome this constraint and to provide that our approach can enable to locate which part of WS (initialization, request processing, operation invocation or response) is responsible for performance problems. The system under discussion should be easily extended by new functionality to meet the emerging needs of the WS user. Moreover, we aim at adapting the existing performance analysis mechanisms like this exploited in TAU for the goals of WS monitoring.

## 3 Web Service

Web Services are a programming paradigm of distributed systems [4, 5, 6], being a programmable application logic accessible using standard Internet protocols.

[1] [`http://www.internetvista.com`]
[2] [`http://www.parasoft.com/`]

Web Services combine the best aspects of component-based development and the Web. As components, Web Services represent the functionality that can be easily reused without knowing how the service is implemented. Web Service features the following:

- WS is accessible over the Web. Web Services communicate using platform-independent and language-neutral Web protocols,
- WS supports loosely coupled connections between systems, by passing messages to each other.
- WS provides an interface that can be called from within another program. The WS interface acts as a liaison between the Web and the actual application logic that implements the Service.

WS can communicate by the Extensible Markup Language (XML) [7]. Web Services use XML to describe their interfaces and to encode their messages. XML underlies three standards used by WS: SOAP which defines a standard invocation protocol for WS [8], WSDL which defines a standard mechanism to describe a WS [9], and UDDI which provides a standard mechanism to register and discover WS [10].

The needs in the monitoring of Web Services are mainly motivated by the performance problems when using this programming paradigm. So the main goal of the monitoring of WS is to obtain as many information as possible to improve WS performance, to discover places where a WS-based application has bottlenecks, memory leaks, and errors.

## 4 J-OCM Monitoring System vs. Accessing JVM

J-OCM is a monitoring system for Java applications[3], compliant with the OMIS specification [1] extended by support for distributed Java programs. The idea of OMIS (On-line Monitoring Interface Specification) is to separate the functionality of a monitoring system from monitoring tools. J-OMIS is a monitor extension to OMIS for Java applications intended to support the development of Java distributed applications with on-line tools. The work dates back to 1995 when OMIS, a monitor/tool interface specification was released. J-OMIS specifies three types of services: *information services* (providing information about an object), *manipulation services* (allowing to change the state of an object), *event services* (trigger arbitrary actions whenever a matching event is raised).

J-OCM comprises the following components: (1) *Node Distribution Unit* is part, which is responsible for distributing requests and assembling replies; (2)*Local Monitor* is a monitor process, which resides on a node. LM's extensions provide new services defined by J-OMIS, which control Java Virtual Machine via agent. LM stores information about the target Java application's object, such as JVMs, threads, classes, interfaces, objects, methods, etc., referred to by tokens; (3) *Agent* uses JVM native interfaces to access a low-level mechanism for interactive monitoring of JVM.

In Java 1.5 Sun Microsystems has incorporated a new native programming interface for use by tools, JVMTI (JVM Tool Interface) [11], to replace less efficient interfaces JVMPI and JVMDI, which were used formerly in J-OCM.

JVMTI is intended to provide a VM interface for the full range of tools that need access to VM state, including but not limited to: profiling, debugging, monitoring, thread analysis, and coverage analysis tools. It is a two-way interface. A client of JVMTI, hereafter called an *agent*, can be notified of interesting events. JVMTI can query and control the application through many functions, either in response to events or independent of them. Agents run in the same process with and communicate directly with the virtual machine executing the application being examined. A native in-process interface allows maximal control over the application with minimal intrusion on the part of a tool. Agents can be controlled by a separate process which implements the bulk of a tool's function without interfering with the target application's normal execution. JVMTI provides support for bytecode instrumentation which can be applied in three ways:

- Static Instrumentation: The class file is instrumented before it is loaded into the VM - e.g., by creating a duplicate directory of `*.class` files which have been modified to add the instrumentation.
- Load-Time Instrumentation: When a class file is loaded by the VM, the raw bytes of the class file are sent for instrumentation to the agent. The `ClassFileLoadHook` event provides this functionality.
- Dynamic Instrumentation: A class which is already loaded (and possibly even running) is modified. This feature is provided by the `RedefineClasses` function. Classes can be modified multiple times and can be returned to their original state.

## 5 Concept and Implementation of WS Monitoring in J-OCM

Prior to coming to the proper discussion of performance issues of WS, we will mention the software platform used to build WS. Next, we present our approach to building a WS-oriented monitoring mechanism, followed by some details on the implementation aspects, usage, and the overhead induced by the system.

**Software Platform for Building WS.** For building WS to be monitored, we use Jakarta Tomcat [14] and AXIS[12, 13]. AXIS (the Apache eXtensible Interaction System) is an Open-Source product from the Apache Software Foundation and its tools for writing client Java programs that use WS (including Microsoft `.NET` WS) and tools for deploying Java programs as WS. AXIS provides transparent access to WS for Java programmers, which allows to focus on the business logic of their applications rather than to worry about low-level network protocols (like SOAP), to use WS. AXIS also allows for automated deployment of Java programs as WS by generating a WSDL description from Java code directly.

To use a WS, a client program (typically, another WS-based application in Java, C++, C#, etc.) sends a request to the WS and receives a reply much like a web browser requests a web page and receives HTML in reply. However, the requests and replies are encoded by SOAP. With AXIS, the developer can build client programs in Java that use WS as if they used methods in any Java class. AXIS automatically generates an additional "glue" code that hides the details of SOAP from the client program. In AXIS, existing Java programs can be deployed quickly as WS by installing AXIS e.g. on a J2EE application server engine.

Tomcat, an official reference implementation for the Java Servlet 2.2 and JavaServer Pages 1.1 technologies , is a servlet container with in a JSP environment, which is a runtime shell that manages and invokes servlets on behalf of users. Tomcat runs with any web server that supports servlets and JSPs. The Tomcat servlet engine often appears in combination with an Apache webserver. Tomcat can also function as an independent web server: it operates in development environments with no requirements for speed and transaction handling.

**WS Performance Issues.** In WS-based applications, many issues need monitoring, especially, those directly connected with the SOAP protocol and with the life-cycle of WS activities. Usually, WS performs the following basic operations:

1. receives a request via SOAP,
2. parses XML contained within the SOAP request,
3. executes the functionality specified by the XML,
4. formats the results in XML,
5. transmits the reply via SOAP.

The above operations which may cause performance problems are intended to be monitored by a WS extension to J-OCM.

As a basis of our system, we use J-OCM , but prior to using this system we needed to reimplement it from Java 1.4 to Java 1.5, where the issue of monitoring of Java applications is realized via the JVMTI interface. This enables to capture much more information from VM compared to the earlier interfaces of JVM (JVMPI, JVMDI).

**Architecture of the J-OCM/WS System.** The monitoring system (Fig. 1) comprises three layers of components (darked elements in Fig. 2 are parts of the system): (1)*application monitor* (AM) - agent like in J-OCM, (2)node's *local monitor* (LM), and (3)*service monitor* (SM). AM is embedded into the AXIS, which manages SOAP messages. It is used to perform monitoring activities in the context of the application. A node's LM is created on each node, where WS to be monitored resides. LM receives requests from SM and distributes them to AMs. LM assembles replies from AMs to send an integrated reply to SM. SM is a permanent component and exposes monitoring services to tools.

**Metrics.** The J-OCM/WS system is intended to provide metrics on particular operations performed within a WS life-cycle, e.g., the time used to parse a SOAP message, time of computation, the whole time of the activity of a particular WS, etc. An important element to be monitored is the SOAP message and its content. It is done in an event-driven fashion. At the start and the end of each

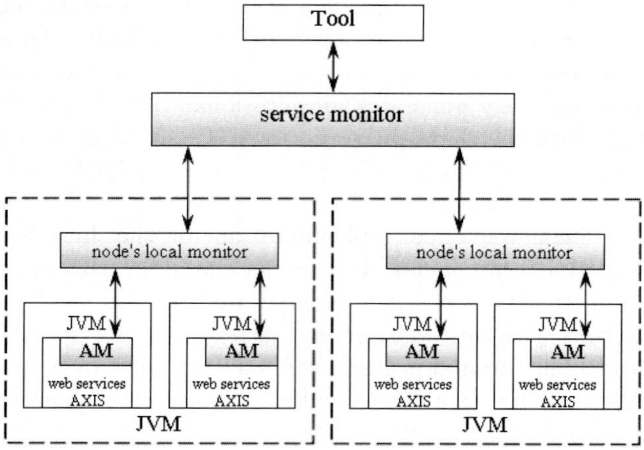


**Fig. 1.** Architecture of J-OCM/WS

above mentioned stage of WS operation, there are placed sensors which generate events and convey them to the AM. The sensors are dynamically placed into the class images of AXIS classes. The time stamps obtained from these events are used to produce *dynamic metrics*, by which we understand metrics related to the life-cycle of the WS-based application. The second group of metrics comprise *static metrics*, i.e. metrics related to the WS provider. There will be monitored how many WS are used at any moment, how many calls to the same WS are performed. The system provides information about the usage of WS (e.g. name, end-point URL) and about the used method (operation) of WS, e.g. the name of this operation.

J-OCM/WS provides information about:

- start/end of the whole WS execution or of its operation,
- start/end of a coming in/going out request/response message,
- start/end of parsing an SOAP message,
- calls to non-existing operation by the user or application,
- errors during an operation execution,
- number and names of WS,
- number, names, and signatures of WS operations,
- number of currently running WS,
- number of calls to each WS.

**Use of the Monitoring System.** In order to start the monitoring of a WS-based application, one must perform the following commands:

- to start AXIS on Tomcat with the agent in JVM:

```
java -jar agentlib:jvmlm bootstrap.jar
```

where:

- `jvmlm` - application monitor
- `bootstrap.jar` - `jar` with AXIS classes

A query from a tool to J-OCM/WS can look like the following (in compliance to the OMIS interface specification):

– a request for getting all WS tokens[3] on the JVM that runs the WS-container (AXIS):

```
:jvm_ws_get_tokens([jvm_j_1_n_1])
```

– a request for information on events (with the time, name, and token of WS and the name and token of an operation of WS), when any operation of any WS on JVM 3 on node 2 begins:

```
jvm_ws_event_operationStart([jvm_j_3_n_2]):print(["Operation
Starts at",$time, $wsStr, $ws, $opStr, $op])
```

**Monitoring Overhead.** The monitoring overhead measured in our tests on Intel 2.4 GHz platform can be broken down into a start-up overhead and that on a per WS basis. The start-up overhead occurs only when the software platform Tomcat and Axis), resulting from the dynamic instrumentation of all classes in JVM, which is performed by JOCM (Tomcat includes ca. 1400 classes). The start-up time increases from 6.343 s to 12. 17 s (increase by ca. 90%). The influence of monitoring on the duration of WS operations depends on the duration of the operations. In case of a 'Hello World' WS which lasted 0.142 s the overhead was 30%, while in case of a WS which lasted 1.741 s the overhead was more negligible, only 5%. Thus, the longer is WS, the less is the relative overhead.

## 6 Concluding Remarks

The monitoring system J-OCM/WS we presented in this paper is an implementation of our concept of the monitoring of Web Services [16], based on the inherent extendibility of the OMIS specification and its Java-bound implementation, the J-OCM monitoring system, towards new paradigms and new tools.

The WS-related monitoring system under discussion is open source and offers a functionality required in case of the monitoring of distributed Java applications and also Web Services implemented in Java.

For monitoring Web Services we used the dynamic instrumentation of AXIS classes. At the start and the end of each stage of processing SOAP message there are placed sensors (in AXIS) which generate events and convey them to the agent. We used the JVM Tool Interface and JNI to obtain information about the state of Java Virtual Machine (e.g. information about a class of WS). As a result we keep to a minimum a overhead to the monitored application performance.

[3] *Token* is a string reference to a monitored object, i.e. in this case it is a reference to JVM 1 on node 1.

Due to the fact that our system extends J-OCM, commands of this extension comply to the same syntax as the commands of J-OCM, our system should be easy to use for the users of tools which are compliant with OMIS. Within our recent research we have adapted the J-OCM/WS monitoring system to cooperate with the SCIRUN/TAU framework [17, 18]

**Acknowledgements.** This research was partially supported by the EU Core-GRID IST-2002-004265 project and the corresponding SPUB-M grant.

## References

1. Ludwig, T., Wismüller, R., Sunderam, V., and Bode, A.: OMIS – On-line Monitoring Interface Specification (Version 2.0). Shaker Verlag, Aachen, vol. 9, LRR-TUM Research Report Series. 1997.
`http://wwwbode.in.tum.de/~omis/OMIS/Version-2.0/version-2.0.ps.gz`
2. Bubak, M., Funika, W., Wismüller, R., Mętel, P., Orłowski. Monitoring of Distributed Java Applications. In: Future Generation Computer Systems, 2003, no. 19, pp. 651-663. Elsevier Publishers, 2003.
3. W. Funika, M. Bubak, M.Smętek, and R. Wismüller. An OMIS-based Approach to Monitoring Distributed Java Applications. In: Yuen Chung Kwong (ed.) Annual Review of Scalable Computing, volume 6, chapter 1. pp. 1-29, World Scientific Publishing Co. and Singapore University Press, 2004.
4. http://www.webservices.org/index.php/ws/content/view/full/1390/
5. http://www.xml.com/pub/a/ws/2001/04/04/webservices/index.html
6. http://www.w3.org/TR/2002/WD-ws-arch-20021114/
7. http://www.w3.org/XML/
8. $http : //www.w3schools.com/soap/soap_intro.asp$
9. http://www.w3.org/TR/wsdl
10. http://www.uddi.org/
11. http://java.sun.com/j2se/1.5.0/docs/guide/jvmti/jvmti.html
12. http://ws.apache.org/axis/java/user-guide.html
13. http://ws.apache.org/axis/java/architecture-guide.html
14. http://jakarta.apache.org/tomcat/
15. A. D. Malony, S. Shende, and R. Bell, "Online Performance Observation of Large-Scale Parallel Applications", Proc. Parco 2003 Symposium, Elsevier B.V., Sept. 2003.
16. P. Handzlik, L. Trebacz, W. Funika, M.Smętek, Performance Monitoring of Java Web Service-based Applications. Proc. Cracow Grid Workshop'2004, December 2004, Cracow, CYFRONET, Krakow, 2005
17. Funika, W., Koch, M., Dziok, D., Malony, A. D., Shende, S., Smetek, M., Wismuller, R. An Approach to the Performance Visualization of Distributed Applications, in: Bubak, M., Turaa, M., Wiatr, K. (Eds.), Proceedings of Cracow Grid Workshop - CGW'04, December 13-15 2004, ACC-Cyfronet UST, 2005, Krakw, pp. 193-199.
18. Funika, W., Koch, M., Dziok, D., Smetek, M., Wismuller, R. Performance Visualization of Web Services Using J-OCM and SCIRun/TAU, in: Laurence Tianruo Yang, Omer F. Rana, Beniamino Di Martino, Jack Dongarra (Eds.), Proc. HPCC 2005, pp. 666-671, Lecture Notes in Computer Science, no. 3726, Springer, 2005.

# Using Sequential Debugging Techniques with Massively Parallel Programs

Christian Schaubschläger, Dieter Kranzlmüller, and Jens Volkert

GUP, Joh. Kepler University Linz
Altenbergerstr. 69
A-4040 Linz, Austria, Europe
schaubschlaeger@gup.uni-linz.ac.at

**Abstract.** Debugging is a crucial part of the software development process. Especially massively-parallel programs impose huge difficulties to program analyis and debugging due to their higher complexity compared to sequential programs. For debugging and analysing parallel programs there are several tools available, but many of these fail in case of massively-parallel programs with potentially thousands of processes.

In this work we introduce the *single process debugging strategy*, a scalable debugging strategy for massively-parallel programs. The goal of this strategy is to make debugging large scale programs as simple and straight-forward as debugging sequential programs. This is achieved by adapting and combining several techniques which are well known from sequential debugging. In combination, these techniques give the user the possibility to execute and investigate small fractions of a possibly huge parallel program, without having to (re-)execute the entire program.

## 1 Introduction

In the past years we can see a drastic increase of hardware and software complexity, especially in the field of high performance computing. In order to satisfy the steadily increasing needs of the supercomputing community, computer manufacturers try to increase the degree of parallelism of the hardware. This can be observed on several system layers, starting at processor level. With the integration of more than one processor core on one CPU very compact SMP systems with a relatively high number of processor cores can be built.

Since scalability of SMP systems is limited to several dozens of processor cores, NUMA systems or clustering techniques are used to increase the number of processor cores further. NUMA systems scale well up to a few thousand of processor cores. Using cluster computing techniques, SMPs or NUMAs are connected to larger systems with possibly more than a hundredthousand processor cores.

In recent years we see also software-sided approaches to build systems which exceed the borders of large clusters. Such ideas evolved originally from distributed computing, where distributed resources were used to solve a given problem, like in the SETI@home Project or in the GIMPS Project, just to name two popular. Researchers work on advanced middleware to build so called GRID Systems [1]. Such systems improve and

V.N. Alexandrov et al. (Eds.): ICCS 2006, Part II, LNCS 3992, pp. 557–564, 2006.

extend the idea of distributed computing in terms of resource allocation, scheduling, security, and many others. One goal is to build a possibly global metacomputer with hundreds of thousands of processors.

With the increase of the available computing power, one can also observe an increase of software complexity. To utilise the hardware resources on such large scale computers as good as possible, usually massively-parallel programs are used. Due to their size (in terms of number of processes, memory consumption, disk space, etc.) such programs impose big problems especially when it comes to program analysis and debugging. Most of the available tools in that area cannot handle such huge programs well enough. A common approach is to debug not the full scale program, but try to work on a smaller problem size, use less processors, etc. However, this approach does not work always, since some errors occur only in the full scale case. Therfore there is a strong need for new strategies in program analysis and debugging of massively-parallel programs. In this paper we propose a novel strategy, which tries to hide the high complexity of the program under analysis from the user, thus making it easier to handle.

The rest of the paper is organised as follows. In section 2 we describe important aspects of program analysis in general and problems of analysing massively-parallel program in particular. In section 2.3 we give a brief overview about related work in this area. In section 3 the *single process debugging strategy* is introduced and described. Section 4 concludes the paper.

## 2 Program Analysis

### 2.1 Debugging Techniques

Program analysis is an important part of the software life-cycle, because an effective analysis and proper use of analysis results can reduce the costs of software development and maintainance enormously. Three main aspects of program analysis are performance analysis, testing, and error debugging. In our work the focus lies on debugging, and debugging strategies respectively.

Many well known techniques and methods for program debugging have been developed originally for debugging sequential programs. Amongst others we mention the following, which are important for our debugging strategy: *Single-step debugging*, *Cyclic debugging*, *Checkpointing*, *Breakpointing*, and *Control flow analysis and Program slicing*. The reasons why these techniques cannot be applied to parallel or massively parallel programs, respectively, ad-hoc, and why such programs impose enormous difficulties on the debugging process, is described in the following sections.

### 2.2 Massively Parallel Programs

The difficulties in debugging massively parallel programs stem from their sheer size, both in execution time, as well as in space (number of processes, memory consumption, etc). A common practice to get information about a program's execution is to instrument the program, i.e. to insert additional monitoring code, which collects information during execution. This data is subject to analysis, either post-mortem or online.

This approach works well for small and medium-size programs, but it becomes impractical for large programs, since the amount of collected data is too big. There are hardly any analysis techniques to extract useful information, and even more important, there are almost no proper techniques to visualize such big programs in a meaningful way.

Another obstacle in debugging parallel programs is introduced by nondeterminism, which plays a bigger role in the parallel case than in sequential programs. This is due to the fact that in addition to the sources of nondeterminism that we know from sequential programs, there are several nondeterministic program constructs that are specific to parallel programs. Examples are wild-card receive function calls in message passing programs or unordered access to shared data in shared memory programs. The problems that arise from nondeterministic behaviour during the debugging process are severe, and are usually subsummed as *probe effect*, *completeness problem*, and *irreproducibility effect* [2]. For debugging the irreproducibility effect is the more problematic, since it inhibits cyclic debugging. For this reason several methods have been developed to enable deterministic re-executions, one of the first being the instant replay mechanism proposed by LeBlanc and Mellor-Crummey [3]. In principle all these replay approaches work in two phases. In the record phase, information about what happened during the program's execution is recorded and saved to stable memory. This information, the so-called trace of the program, is used in the second phase, the replay phase, where the program is re-executed under the control of a replay system. The recorded information is used to enforce the same decisions at nondeterministic events as in the original program run, hence permitting any number of equivalent re-executions of the program.

### 2.3 Related Work

Many existing debugging tools, commercial and non-commercail, for sequential and parallel programs, support some or all of the above described techniques. Well known examples for sequential debuggers are gdb and dbx, all with similar and mature functionality. In the parallel case the situation is more wide-spread. Some approaches extend the sequential case and use multiple instances of a sequential debugger like gdb simultaneously [4, 5]. This class of parallel debugger does not scale very well, and they do not address the pitfalls of nondeterminism. Therfore they are useful only for small deterministic programs.

Another class of parallel debuggers uses the record&replay approach. Such debuggers can handle nondeterminism and hence can provide cyclic debugging. Some provide distributed checkpoints and breakpoints, which is useful for long running programs. Examples are [6, 7, 8]. Unfortunately they don't scale well on the process axes, i.e. they are difficult to use for massively parallel programs.

Our approach is in some sense similar to program slicing in general [9] and the work of Dusterwald et al. on slicing distributed programs in particular [10]. However, their work is based on control flow analysis and the distributed dependence graph, while we use the event graph (as described below) as fundament. Also, they don't support breakpoints and checkpoints, which is a big drawback for long running programs.

## 3 Single Process Debugging

In this section we describe our debugging strategy, which we call *single-process debugging*. The target applications of our work are massively-parallel message-passing programs. Without loss of generality we confine ourselves to SPMD message-passing programs written in the C programming language. For inter-process communication we use only two functions defined in the MPI-Standard [11], namely MPI_Send for sending, and MPI_Recv for receiving data.

There are several prerequisites which must be met for our approach to work. Firstly, we need to be able to replay parallel programs deterministically. We can do this in a very efficient manner using our Record&Replay approach called RON. Furthermore we describe how we can isolate a process (or a group of processes) from a parallel program, in a sense, that we can (re-)execute this process stand-alone, without having to execute the entire parallel program. We use checkpoints as well as breakpoints; both techniques have to be adapted to work in the case of parallel programs. Finally we describe how we combine these techniques to the single process debugging strategy.

### 3.1 Modelling Program Executions

For analysis of the recorded trace data we need a formalism in order to be able to apply mathematical methods like graph operations on the data. As a formal representation of a program's execution we use the event graph model:

**Definition Event Graph [12].** An event graph is a directed graph $G = (E, \rightarrow)$ , where $E$ is the non-empty set of events $e \in E$, while $\rightarrow$ is a relation connecting events, such that $x \rightarrow y$ means that there is an edge from event $x$ to event $y$ in $G$ with the "tail" at event $x$ and the "head" at event $y$.

The events $e \in E$ are the recorded events observed during a program's execution, like for example send or receive events. The relation connecting the events of an event graph is the so called happened-before relation [13].

### 3.2 Deterministic Replay

A crucial aspect of any monitoring or Record&Replay approach, respectively, is to keep the influence of the monitoring system on the program as low as possibe. Given the fact, that in MPI non-overtaking communication is guaranteed, one can show that it is sufficcient to record only the process-id of the corresponding sending process for every nondeterministic call to MPI_Recv. This can be done with very little effort during program execution and results in very small tracefiles. In subsequent replay iterations equivalent program runs are guaranteed by replacing every occurance of a wild-card receive with a deterministic receive, where the process-id of the message source is read from the previously recorded tracefiles. Once the execution path is fixed one can extend the amount of recorded data without running the risk to perturb the order of message reception. This Record&Replay mechanism has been implemented in the monitoring tool NOPE [14], which is part of the DeWiz debugging environment [15].

With deterministic replay we are now able to do cyclic debugging. However, as stated before, due to the size of massively-parallel programs, cyclic debugging is by far not sufficient as debugging strategy. The next step is therefore to reduce the amount of analysis data. We can do this by narrowing the investigated parts of the program in two dimensions, namely along the process axes using process isolation, as well as along the time axes using breakpoints and checkpoints.

### 3.3 Process Isolation

During debugging of a massively-parallel program the user might get the suspicion that there is a bug on a certain process. At this point, one would start the program again, possibly several times, paying special attention on that particular process. During all these debugging iterations the whole program, which consists of many processes, must be executed all the times, although the user focuses only on one or a few processes. This has many disadvantages, it wastes CPU time, which is a cost issue, unneccessary trace data is generated (and possibly analysed in vain), and so on. Therefore we use the idea of process isolation, where we "isolate" one process (or a group of processes) from the rest of the parallel program, making them executable stand-alone.

Isolating process $P_i$ from its context requires three steps:

1. Initial execution of the entire program: generation of trace data.
2. Build the event graph.
3. Isolation step: re-execute process $P_i$ and all (directly or indirectly) dependent processes. Save the contents of all incoming messages on process $P_i$.

During subsequent isolated executions of $P_i$ incoming messages are not received by MPI_Recv, but the contents of the message is read from the previously generated tracefiles. The number of processes that have to be re-executed in step 3 is not fixed, but depends on the program being analysed. Based on the event graph we use a recursive algorithm to determine which processes have to be re-executed.

### 3.4 Breakpoints in Parallel Programs

Distributed breakpoints (and checkpoints) are used to reduce the program's complexity along the time axis. A distributed breakpoint is a set of sequential breakpoints, one on each process of the parallel program: $B_i = (S_{i,0}, ..., S_{i,(n-1)})$, with $n$ being the number of processes in the parallel program. In our approach we give the user two possibilities to define a distributed breakpoint: manually, by modifying the source code, or semi-automatically, based on the event graph.

Using the manual approach the user can instert special breakpoint code at any position in the source code. However, in this case it lies in the responsibility of the user that this code is placed properly so that the sequential breakpoints are reached by every process during program execution (which of course is crucial for a distributed breakpoint to be useful).

The second possibility to specify a distributed breakpoint is situated on a higher level of abstraction (or program representation) compared to the source code level. A recorded program run, or rather the event graph, can be visualized as space-time diagram, where - amongst others - there are send and receive events as nodes in the graph, which are pairwise connected by arcs. The user can select any event $e_{i,p}$ (the $i^{th}$ event on process $p$) in the graph by simply clicking on it; this event will be the origin of the distributed breakpoint. After selection of such an event, we can automatically determine which events on all other processes have to be part of the distributed breakpoint, by using the happened-before relation.

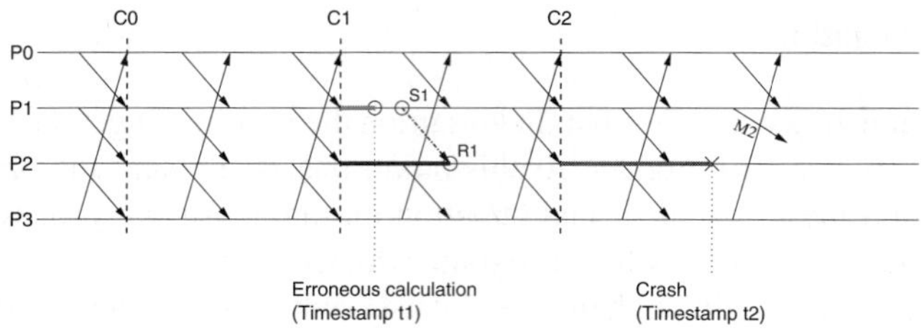


**Fig. 1.** Example debugging scenario

### 3.5 Checkpoints in Parallel Programs

Another approach to reduce the cycle time of the debugging iterations is to use checkpointing. Similar to a distributed breakpoint a distributed checkpoint consists of one sequential checkpoint for each process of the parallel program, which form a so called checkpoint line. An important aspect is the placement of checkpoints, which can be done either semi-automatically or automatically. The semi-automatic approach is done via the visual representation of the event graph of the parallel program. To place a checkpoint the user can select any arbitrary point on one process, the corresponding checkpoints on the remaining processes are calculated automatically, based on the event graph. During the next replay of the program at all user specified checkpoint locations the current state of the process is saved to stable storage.

The semi-automatic approach has several disadvantages, namely placing checkpoints on a regular base would become a tedious task. Therefore we have the possibility to place checkpoints automatically. In principle each process saves it's state periodically to stable storage. However, several requirements must be met to ensure so-called consistent checkpoints, e.g. one must take care of messages that would cross a checkpoint line, and others. Details on that can be found in [16].

### 3.6 Putting It All Together

All these techniques have been applied in debugging programs, sequential and parallel as well. Due to the difficulties introduced by massively-parallel programs as described in section 2.2 these techniques are limited to relatively small parallel programs. Therefore we have tried to combine them in a proper way, giving the user the possibility to handle even the biggest parallel programs.

The two main ideas are to keep the user's focus only on relevant parts of the program on the one hand, i.e. on parts where errors are suspected, and on the other hand to save computing resources during debugging, which is often an important cost factor. We can achieve this by re-executing only small parts of the massively parallel program during the debugging cycles. The following simple scenario in Figure 1 illustrates the idea.

1. A small parallel program consisting of four processes $P_0$ to $P_3$ was executed and crashed after a while, the user starts a debugging session. Unless the initial execution was not monitored, the user would initiate a monitored re-execution in order to permit deterministic re-executions and to generate trace data. Also checkpoints are generated periodically.

2. The event graph of the previous execution is generated and visualized as space-time diagram, as displayed in Figure 1. We see message exchanges indicated by arrows, and three distributed checkpoints $C_0$ to $C_2$.
3. The user suspects a crash on $P_2$ at timestamp $t_2$ (based on various indications, eg. error messages from the MPI environment, or - in our example - by the fact that message $M_2$ is pending, etc.), therefore debugging starts on $P_2$. The cause of the crash is an erroneous calculation on process $P_1$ at timestamp $t_1$, which of course is unknown to the user at that time.
4. In a first step only a small part of $P_2$ is selected and debugged: $P_2$ is isolated starting at the closest checkpoint to $t_2$, in our case $C_2$. The small section $[C_2 : t_2]$ can now be re-executed and inspected any arbitrary number of times stand alone.
5. Since the cause of the crash could not be found in section $[C_2 : t_2]$, the investigated interval is extended to $[C_1 : t_2]$, again using process isolation, etc. Here the user finds that an erroneous value is received at receive event $R_1$. Therefore the debugging session is continued on process $P_1$ in the interval $[C_1 : C_2]$, since this is the location of the corresponding send event $S_1$. For that purpose process $P_1$ must be isolated and re-executed in the mentioned interval.
6. Finally the user finds the location of the erroneous calculation on process $P_1$ at timestamp $t_1$, which caused the crash later on $P_2$.

It is important to note that all steps necessary for process isolation, all transitions from one process to another, etc. can be performed automatically and transparently in the background. The user simply selects a (small) region on a process for debugging, and this section is isolated automatically. In case of the reception of an erroneous value as in the example above, the user just initiates a switch to the sending process; which section (or checkpointing interval, respectively) on the sending process has to be isolated can be calculated automatically using the relations in the event graph.

## 4 Conclusion and Future Work

We have presented a scalable debugging strategy for massively parallel programs. The main idea is to narrow the range of the investigated parts of the program during debugging in an intelligent way, both in time and in space. This keeps the focus of the user on the relevant parts of the program, and, since re-execution during debugging is much less ressource intensive, may actually save money.

We are working on several improvements of this approach. For example, the integration of control flow analysis could improve the understanding of error propagation in the program, hence making the tracking of errors more efficient. Several work has been done in this field, for example the work on slicing parallel programs described in [10]. Also an extension to shared memory programs and distributed programs in general seems useful and will be investigated further.

## References

1. Foster, I., Kesselman, C., Tuecke, S.: The anatomy of the grid: Enabling scalable virtual organizations. The International Journal of High Performance Computing Applications **15** (2001) 200–222

2. Schaubschläger, C.: Automatic testing of nondeterministic programs in message passing systems. Masters Thesis, GUP, Johannes Kepler University, Linz, Austria. http://www.gup.uni-linz.ac.at/~cs/thesis (2000)
3. LeBlanc, T.J., Mellor-Crummey, J.M.: Debugging parallel programs with instant replay. IEEE Trans. Comput. **36** (1987) 471–482
4. Balle, S.M., Brett, B.R., Chen, C.P., LaFrance-Linden, D.: Extending a traditional debugger to debug massively parallel programs. Journal of Parallel and Distributed Computing **64** (2004) 617–628
5. Cunha, J., Lourenco, J., Antao, T.: A debugging engine for parallel and distributed environment (1996)
6. Kacsuk, P.: Systematic macrostep debugging of message passing parallel programs. Future Gener. Comput. Syst. **16** (2000) 609–624
7. Etnus: Totalview debugger. http://www.etnus.com/ (2005)
8. Absoft, Corp.: DDT - Distributed Debugging Tool (2005)
9. Weiser, M.: Program slicing. In: ICSE '81: Proceedings of the 5th international conference on Software engineering, Piscataway, NJ, USA, IEEE Press (1981) 439–449
10. Duesterwald, E., Gupta, R., Soffa, M.L.: Distributed Slicing and Partial Re-execution for Distributed Programs. In: Languages and Compilers for Parallel Computing. (1992) 497–511
11. Message Passing Interface Forum: MPI: A Message-Passing Interface Standard - Verion 1.1. http://www.mcs.anl.gov/mpi/ (1995)
12. Kranzlmüller, D.: Event graph analysis for debugging massively parallel programs. PhD thesis, GUP, Joh. Kepler Univ. Linz. http://www.gup.uni-linz.ac.at/~dk/thesis (2000)
13. Lamport, L.: Time, clocks, and the ordering of events in a distributed system. In: Communications of the ACM, Vol. 21, No. 7. (1978) 558–565
14. Kranzlmüller, D., Volkert, J.: Nope: A nondeterministic program evaluator. In: Proc. of ACPC99, 4th Intl. ACPC Conference, LNCS, Vol. 1557. (1999) 490–499
15. Kobler, R., Schaubschläger, C., Aichinger, B., Kranzlmller, D., Volkert, J.: Examples of monitoring and program analysis activities with dewiz. In: Proc. DAPSYS 2004 (5th Austrian-Hungarian Workshop On Distributed And Parallel Systems). (2004)
16. Thoai, N.: Checkpointing techniques for minimizing the waiting time during debugging long-running parallel programs. PhD. Thesis, GUP, Johannes Kepler University, Linz, Austria (2003)

# Workflow for Integrated Object Detection in Collaborative Video Annotation Environments

Lars Grunewaldt, Kim Möller, and Karsten Morisse

University of Applied Sciences Osnabrück, D-49076 Osnabrück, Germany
http://www.diva.fh-osnabrueck.de

**Abstract.** Annotation tools for media are getting more and more important. The application for these kind of applications are very manifold. This paper describes some ongoing work on a video annotation environment supporting the media production and post-production process. Based on a former approach, some new development like integrated audio conferencing with recording facilities for audio based work instructions and an automatized video segmentation module are presented. Beside these technical improvements an object based approach for the video annotation process is presented.

## 1 Introduction

Annotation tools for media, e.g. audio, video or animation are getting more and more attractive and important. Existing approaches and tools are very manifold. There are tools for a collaborative learning approach using hyper-video functionality [3]. Other tools are extending the video information with metadata to make it searchable in a media database, e.g. IBM MPEG-7 Annotation Tool [12]. Moreover, annotation tools can be a useful support in the production process of media. An approach for this kind of annotation has been shown with DiVA [9]. In [7] some important requirements and an approach to use annotations in the production process has been presented. However, during the usage of DiVA some things attracted negative attention. This paper describes some technical improvements in the existing software and shows a way to the next step in video annotation that is an approach to object-based video annotation. All of the annotation tools known so far are using an manual-geometric approach, i.e. annotation in the video plane are defined manually with elementary geometric objects, e.g. circles, rectangles or polygons. It would be very desirable to have an object-based approach, where one can activate a single object in the video layer, e.g. a person or a table. Figure 1 shows the difference between these two annotation approaches.

The organization of this article is as follows. In section 2 the annotation approach of DiVA in video production is briefly described and an overview about some technical improvements is given. In sections 3 and 4 known technology for object-detection from 2D-picture is considered and an approach for object based annotations is presented. Finally in section 5 some concluding remarks and some prospectives for future work are given.

V.N. Alexandrov et al. (Eds.): ICCS 2006, Part II, LNCS 3992, pp. 565–572, 2006.

(a) Rectangle as a manual-geometric annotation

(b) Object contour highlighting for object-based annotation

**Fig. 1.** Geometric vs. object-based annotation

## 2 DiVA – Video Annotation in Post-production

DiVA[1] [9] is a collaboration tool which supports some basic features for synchronized video annotation. Some general requirements for such a tool (see [7]) are: synchronized navigation for several users, definition of textual and graphical annotations which can be archived for work in the subsequent production phases and usage of high-quality video material.

Annotations in DiVA can be defined manually as a text or a graphic object directly on the video layer (manual-geometric approach from figure 1). By a client-server approach several users can navigate in a synchronized manner through the video sequence to define categorized annotations, which are in fact graphical objects combined with textual metadata (see figure 2).

### 2.1 Technical Improvements for Video Annotation

During the usage of DiVA some things attracted negative attention. These were: manual definition of shots within the video sequence, proprietary format for annotation storage and the lack of an integrated audio conferencing component.

**Integrated Audio-Conferencing.** As a collaboration tool DiVA has to provide ways of communication to the users while annotating the video content. In a first prototype of the system, such an audio-visual communication was only possible using third party products like external VoIP software or a standard phone line. As this creates a cumbersome overhead of providing a seperate tool next to DiVA itself, integrating a VoIP component is an important feature for synchronized collaboration tools as well for asynchronous tools, if audio recording is considered as well.

[1] DiVA - **Di**stributed **V**ideo **A**nnotation.

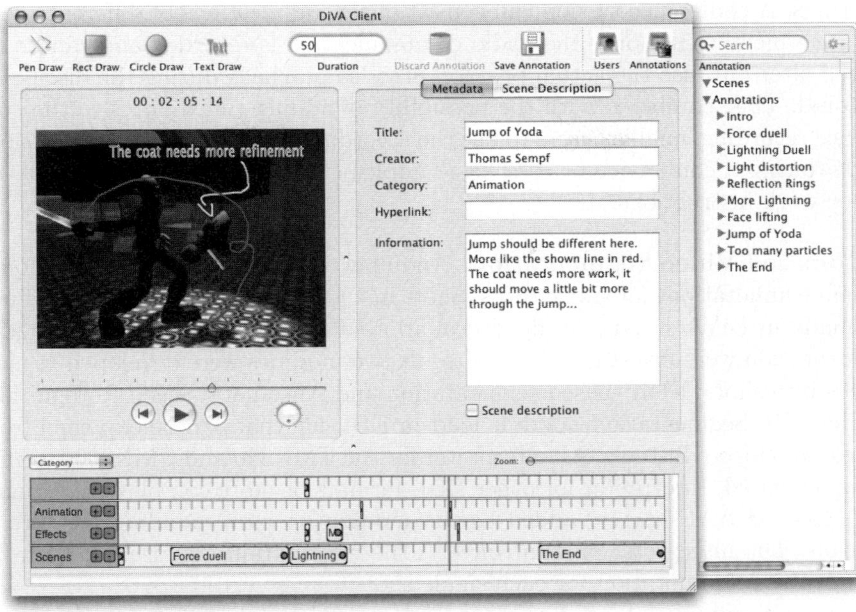


**Fig. 2.** DiVA - Scene annotation with graphic and text objects, synchronized navigation and annotation archiving

Therefore in [4] a Voice-over-IP component has been integrated into DiVA. Right now, several types of VoIP communication systems and protocols are in wide-spread use; mainly the H.323 protocol suite and the SIP protocol can be found in common VoIP software applications such as Netmeeting or hardware-implemented VoIP components like the Sony PCS series. Some applications are based either on H.323 or SIP, some support both protocols. Additionally, there are several products that are based on propriatary protocols, e.g. Skype. When implementing the audio conference plug-in for DiVA, the ability to easily replace the audio conference unit with another one was very important. Development of VoIP technology is still swift, and it seemed useful to be able to switch between different audio conference implementations to rapidly adopt current communication methods. The current conference plug-in supports the H.323 protocol using a H.323 conference server to manage conferences on the DiVA server side. To match bandwidth requirements different audio codecs are available, e.g. GSM-codecs for lower bandwiths, and G.711 for higher bandwidths. The DiVA client automatically connects to the audio conference system when joining a DiVA annotation session. Once set up, no additional user interaction is needed to activate audio conferencing when connecting to the DiVA system.

For a tool supporting the production workflow management, it is important to track the discussion process about the video content. Therefore it is desirable to provide a functionality to record audio comments of the users. This has been realized by extending the conceptual video annotation process to sessions and meetings. A conference session can consist of several meetings of different users. For each meeting not only the audio communication is recorded and can be replayed later but also the actual position of video playback during the discussion is stored. This, combined with the possibility of adding timed notes to the conference recording, enables users to use the recordings for later checks of decisions made during a conference or as a basic implementation of a audio-based to-do list or work instructions.

**Automated Video Segmentation.** An important issue in video annotation is to define annotations for single shots. Shots in a video sequence had to be defined manually in DiVA so far.. In [6] several approaches for an automation in video segmentation were investigated. As a result two modules were developed to split videos into shots: EDL[2]-based segmentation and Automatic Scene Recognition.

The EDL-Segmentation- module loads a EDL-file that can be exported from almost all video-edit tools. An interpreter for the CMX340 and CMX3600 format is implemented. Interpreter for other EDL-formats or more complex file formats like MXF or AAF can be added to the module by corresponding interpreter plug-ins. The interpreter plug-in converts the information found in an EDL-file and creates an annotation for each single shot.

A second and much more flexible module provides a shot segmentation without additional information like an EDL-file. It analyzes the video content for cuts and can be used on all video formats which are compatible to the Apple Quicktime Framework. The price for this very flexible approach in content segmentation is the obligation to analyze each single frame of the video sequence[3]. Off course, for special video codecs dedicated algorithms can be used[4]. They can be added to the segmentation module by a flexible plug-in approach. In [6] several known and some new algorithms for shot segmentation had been investigated. Some of them were integrated in the DiVA-system. With the Contrast Compare algorithm (see [5]) it is even possible to find fades and dissolves. In general, the histogram systems returned equal or even better results than more complex methods like color coherence vectors [11] or principal coordinate systems [10]. A newly developed algorithm is the 2 x 2 histogram, which creates four independent histograms of the four quarters of a frame and compares them with the next frame. The high retrieval rate of shots is a major advantage of this algorithm. However, like in other systems too, rapid light changes can still lead to over-found cuts. But in an application like DiVA this is not a real shortage. The collaborative video annotation is a highly interactive process where the

[2] EDL - Edit Decision List.

[3] This is done by rendering each single frame in a bitmap image and to analyze all pixel information. These pixel information, e.g. histogram, is then compared to the previous frame. In case of a large difference a new shot or segment has been found. The ratio of deviation can be used as a parameter of the segmentation algorithm.

[4] For example codecs based on interframe coding like the codecs of the MPEG-family with techniques like motion-compensation and predictive coding. Here one can also consider the GOP-structure of the video sequence to design much more powerful algorithms for this special case.

users define new annotations or delete existing annotations and shots found by the segmentation module are just an indicator for the collaboration process and can be changed easily.

**Standardized Annotation.** In the prototype of the software, the annotations were saved in a proprietary XML-format. This circumstance causes a lack of integration and exchange with other software tools. Hence, the MPEG-7 standard had been evaluated for the description of video annotations. This part is still under development.

## 3 Related Work for Annotation and Object Detection and Recognition

### 3.1 Annotation Tools

Most of the available systems for video annotation are systems for single user usage and not intended for collaboration with several persons involved concurrently. So far video annotation systems are not known to assist the video production process. The known systems are tools to describe the actual plot of a video. Most of these systems offer a bundle of tools for navigating through the video sequence and adding metadata to it. This metadata is then stored on a server for later retrieval. In some systems MPEG-7 [16] is used for the metadata, e.g. in the IBM MPEG-7 Annotation Tool [12]. This metadata is then saved and could be viewed with a MPEG-7 conformable player.

[3] describes a tool for hyper-video functionality. This approach as well as the Hyperfilm project [13] are extending the hypertext concept to continuous media by defining mouse sensitive regions for one or several video frames which are linked to other information (e.g. pictures, text or other videos). Some other systems for hyper-video are described in [2]. Although hyper-text functionality is an interesting feature, it is not the main focus of the annotation approach considered here.

An interesting project is the filmEd [8] research project of the University of Queensland, Australia. Within filmEd tools for annotating, discussing, and resuming videos are developed. Moreover, software tools to discuss and annotate video sequences in real-time over the Internet in a distributed manner are considered. A prototype called Vannotea has been developed, which enables the collaborative indexing, annotation and discussion of audiovisual content over high bandwidth networks. However, the navigation system is master/client based, that means that only one person can navigate through the video sequence and all clients will be automatically synchronized.

### 3.2 Object Recognition and Interaction in 2D Pictures

In [1] an overview is given about object detection and recognition in 2D pictures. The presented systems work with a sequence of model points of an object, a degree of invariance (e.g. through rotation or scale) and a description of the image data, that is e.g. independent to photometrically invariant. Some standard objects like characters or faces can be build in an object database, more complex objects have to be determined by the user in an initial sequence, showing what and where an object is. By-and-by the size of the object database increases and the software, often used in neural networks, can work independent and detect or recognize objects.

But, the enlarging database leads to a slower detection. In movies it is of course possible to use a faster working object tracker to find the object in the next frame again. Nevertheless, to work with objects, the user has to catalog them in a time consuming process. Also it takes time to get a good working system. But even then there will be still errors caused by e.g. multiple objects or partial hidden objects in a frame.

A very interesting approach for object based video annotation will be made possible by a full implementation of the MPEG-4 standard. It offers a lot of additional features for interactive video applications. For instance the user even has the option to move or turn objects directly in a movie. But at the moment these features are usually not implemented. Only a few implementations of MPEG-4 beyond simple audio and video playback are known. Most of them are not available to the public. One interesting approach is the GPAC project [15], that supports at least some of these features in a MPEG-4 production tool and a player. Certainly these features could be used effectively in a collaborative video discussion tool like DiVA.

## 4 An Approach for Object Based Video Annotation

Recognizing objects in 2D pictures is time consuming and, without any additional information also inaccurate and erroneous. Therefore a new approach is considered in DiVA, because it is a tool supporting the production process. The basic idea is thus not to find objects in existing movies or TV contents, but to get as much as possible information about objects while creating a video sequence or animation with 3D-, videoedit- and/or posteffect-tools.

With a 3D modeling tool like Cinema 4D or a procedural approach (e.g. scenegraph-based systems like OpenSG or Java), one creates a virtual 3D scene but also makes shots of this scene from different camera angles and camera flights in between. Of course, when rendering these different shots, one has all information available for object detection later on. There are two possibilities to store the needed information: coordinates and contours of objects in a rendered frame can be stored as metadata in a MPEG-7 file or objects can be stored directly, when using the MPEG-4 BIFS[5] format. Using the stored data in a video annotation tool, an object contour highlighting, as shown in figure 1(b), can be realized.

In real video sequences with natural actors and objects, only the case when working with more than one video-layer and perhaps the use of chroma keys in editing tools is considered. The contents of the different layers can also be stored

[5] **BI**nary **F**ormat for **S**cenes.

as objects to MPEG-7 or MPEG-4. Certainly, when one video-layer contains more than one object, this method does not guarantee the identification of every single object in that layer.

When rendering frames firstly, then arrange and mix them afterwards with natural video in an editing tool and finally adding some posteffects, the object information, collected during the rendering process, can become invalid if stored to a MPEG-7 file. Thus naturally, the editing- and posteffect-tool also need extension, ensuring the object positions and contours stay valid during the workflow.

## 5 Summary and Further Work

DiVA is a collaboration platform for video production that supports different steps for a distributed video production process. The distributed approach for navigation, annotation and collaboration on video content makes DiVA unique among other systems.

This paper presents some important improvements like integrated audio conferencing and automation in video segmentation. Especially the later one is a must have feature for these kind of tools. By implementation of a new shot detection algorithm based on partial RGB-histogram values, good automatized recognition results for shots could be reached. The audio-based work instructions are an innovative approach in the workflow of media production. This has been realized by the integration of a audio conferencing and storage component.

Moreover a new approach for object based video annotation has been presented. Work on this feature is still under development and its effectiveness in practice must be shown. But it opens the door for a wide range of new applications. An interesting application of annotated video sequences by an open standard like MPEG-7 is in broadcasting. Regarding DiVA in the production workflow of videos, the created MPEG-7 file can be used for example in interactive MHP applications, when the video is broadcasted. A similar approach is also considered in the GMF4ITV-project [14].

## References

1. Amit, Y.: 2D Object Detection and Recognition MIT Press, Cambridge, USA (2002)
2. Dommasch, C.: Entwurf und Realisierung einer komponentenbasierten Anwendung für den kooperativen Wissenserwerb auf Basis von interaktiven Videoinhalten. Diplomarbeit, Fachhochschule Osnabrück, 2005.
3. Finke, M., Balfanz, D.: A reference architecture supporting hypervideo content for ITV and the internet domain. Computers & Graphics 28 (2004).
4. Grunewaldt, L.: Konzeption und Realisierung einer Audio-Konferenz- und Aufzeichnungskomponente im Rahmen eines Video-Annotationswerkzeugs Diplomarbeit, Fachhochschule Osnabrück (2005)
5. Lienhart, R.: Verfahren zur Inhaltsanalyse, zur Indizierung und zum Vergleich von digitalen Videosequenzen, Shaker Verlag, Aachen (1998)

6. Möller, K.: Softwareentwicklung zur automatischen Segmentierung von digitalem Videomaterial, Diplomarbeit, Fachhochschule Osnabrück (2005)
7. Morisse, K., Sempf, T.: An Approach for Video Annotation in Post-Production. In Proceedings of Int. Conference on Computational Science ICCS 2005, Atlanta, GA, USA, May 2005, LNCS 3516, Springer Verlag, 2005
8. Schroeter, R., Hunter, J., Kosovic, D.: FilmEd - Collaborative Video Indexing, Annotation and Discussion Tools Over Broadband Networks. International Conference on Multi-Media Modeling, Brisbane, Australia, January 2004
9. Sempf, T., Morisse, K.: Video Annotation in der Postproduktion. Digital Production. **1** (2005) 103-105
10. Yilmaz, A., Ali Shah, M.: Shot Detection using Principal Coordinate System. IASTED Intl Conference, Internet and Multimedia Systems and Applications Las Vegas, USA (2000)
11. Zabih, R., Pass, G., Miller, J.: Comparing Images Using Color Coherence Vectors Proceedings of the Forth ACM International Conference on Multimedia 96, ACM Press, p. 65-73 Boston, USA (Nov 1996)
12. IBM MPEG-7 Annotation Tool: http://www.alphaworks.ibm.com/tech/videoannex
13. Hyperfilm - Extending hypertext in the video realm: http://www.hyperfilm.it/eng/index.html
14. Generic Media Framework for Interactive Television: http://www.gmf4itv.org
15. GPAC Project: http://gpac.sourceforge.net
16. MPEG (Moving Picture Experts Group): http://www.chiariglione.org/mpeg/

# RMIX: A Dynamic, Heterogeneous, Reconfigurable Communication Framework*

Christian Engelmann and Al Geist

Computer Science and Mathematics Division,
Oak Ridge National Laboratory,
Oak Ridge, TN 37831-6164, USA
{engelmannc, gst}@ornl.gov
http://www.csm.ornl.gov

**Abstract.** RMIX is a dynamic, heterogeneous, reconfigurable communication framework that allows software components to communicate using various RMI/RPC protocols, such as ONC RPC, Java RMI and SOAP, by facilitating dynamically loadable provider plug-ins to supply different protocol stacks. With this paper, we present a native (C-based), flexible, adaptable, multi-protocol RMI/RPC communication framework that complements the Java-based RMIX variant previously developed by our partner team at Emory University. Our approach offers the same multi-protocol RMI/RPC services and advanced invocation semantics via a C-based interface that does not require an object-oriented programming language. This paper provides a detailed description of our RMIX framework architecture and some of its features. It describes the general use case of the RMIX framework and its integration into the Harness metacomputing environment in form of a plug-in.

## 1 Introduction

Collaborative environments for heterogeneous distributed computing strive to enable research institutions and universities world-wide to pool their computing and storage resources in order to enable joint research in computational sciences, such as nanoengineering and quantum chemistry. Frameworks for metacomputing, data and computational grids, and peer-to-peer environments help to facilitate resource and data sharing among collaborating sites using standardized interfaces and interoperable software components.

Remote Method Invocation (RMI) is the most important communication paradigm for heterogeneous distributed collaborative environments as it extends the semantics of local method calls to networked systems. RMI is an object oriented analogy to the Remote Procedure Call (RPC) concept. It enables client components to invoke methods of objects that have been previously exported on

* This research is sponsored by the Mathematical, Information, and Computational Sciences Division; Office of Advanced Scientific Computing Research; U.S. Department of Energy. The work was performed at the Oak Ridge National Laboratory, which is managed by UT-Battelle, LLC under Contract No. De-AC05-00OR22725.

V.N. Alexandrov et al. (Eds.): ICCS 2006, Part II, LNCS 3992, pp. 573–580, 2006.

local or remote server components using a protocol stack that defines connection management, message formats and data encoding.

Traditional RMI communication frameworks typically implement one specific protocol stack, where connection management and message formats are defined by client and server libraries, and data encoding is defined using client-side and server-side stubs for each class of objects.

With this paper, we present a native (C-based), flexible, adaptable, multi-protocol RMI/RPC communication framework that complements the Java-based RMIX variant [1, 2, 3, 4] previously developed by our partner team at Emory University as part of the Harness research effort [5, 6, 7, 8, 9].

RMIX is a dynamic, heterogeneous, reconfigurable communication framework that allows software components to communicate using various RMI/RPC protocols, such as ONC RPC, Java RMI and SOAP, by facilitating dynamically loadable provider plug-ins to supply different protocol stacks. While the RMIX base library contains functions that are common to all protocol stacks, like networking and thread management, RMIX provider plug-ins contain protocol stack specific functions for connection management, message formats and data encoding. Since it is up to the provider plug-ins to reuse base library functions, implementations may range from lightweight to heavyweight. Moreover, client- and server-side object stubs are very lightweight and protocol independent as they only perform an adaptation to the RMIX system.

Furthermore, RMI semantics have been expanded within RMIX to support the RPC paradigm and its protocols. In addition to standard synchronous RMI/RPC mechanisms, RMIX also offers advanced RMI/RPC invocation semantics, such as asynchronous and one-way.

This paper is structured as follows. First, we briefly discuss related past and ongoing research. We continue with a detailed description of the RMIX framework architecture and some of its features. We describe the general use case of the RMIX framework and its integration into the Harness lightweight metacomputing environment in form of a plug-in. This paper concludes with a short summary of the presented research and its current status.

## 2 Related Work

The work in heterogeneous, adaptable, reconfigurable, networked systems (Harness) is a collaborative research effort among Oak Ridge National Laboratory (ORNL), University of Tennessee, Knoxville, and Emory University focusing on the design and development of technologies for flexible, adaptable, reconfigurable, lightweight environments for heterogeneous distributed computing.

The C-based RMIX framework developed at ORNL and described in this paper follows a similar architectural design approach to the Java-based variant by offering the same interface and similar functionality. However, the C-based implementation is not a one-to-one translation of the Java-based solution as this would require reimplementing many of the Java language features in a non-Java language, i.e. reinventing the Java Virtual Machine (JVM) and parts of the Java Standard Development Kit (SDK).

Our approach is to offer the same multi-protocol RMI/RPC services and advanced invocation semantics via a C-based interface that does not require an object-oriented programming language. Instead of mapping RPC to RMI calls, like in the Java-based variant, we actually map RMI to RPC calls by emulating an object-oriented interface. Any object-oriented application just needs to implement the C-based stubs as they were method calls, with the exception that the first call argument is always a reference to the object itself. Polymorphism is guaranteed as the server-side stub calls the real object method.

The following related work has already been previously discussed in the context of the Java-based RMIX research [1, 2, 3, 4].

XSOAP [10] (a.k.a. SoapRMI) is a RMI system based on the SOAP protocol and offers Java and C++ implementations to create and access Web Services. While it supports multiple transports and custom data encoding, it is not a universal multiprotocol RMI system.

JavaParty [11] is a drop-in replacement for standard Java RMI, written in pure Java and exploiting optimized serialization. While JavaParty supports non-TCP/IP communication networks, e.g. Myrinet, it is not interoperable with ordinary RMI applications and services.

The Manta [12] approach sacrifices Java portability and uses native code to achieve the best possible performance. Manta is a native Java compiler that compiles Java source code to Intel x86 executables. It does not offer any multi-protocol support nor does it provide a native equivalent.

Web Services [13, 14] have become a de facto standard for simplifying integration and access of heterogeneous networked services by presenting the user with a much lower level of abstraction for distributed computing. Web Service protocols are XML-based and are not designed for efficient data transfer. However, a RMI layer on top of Web Services, as exemplified by XSOAP and the Java-based RMIX variant, can offer simple and elegant access to heterogeneous resources over the Web.

## 3 RMIX Framework Architecture

The main goal of the RMIX communication framework is to provide efficient and interoperable RMI/RPC capabilities to system software, middleware and applications in a heterogeneous, distributed, collaborative computing environment. Conceptually, the RMI (and RPC) paradigm is based on a client-server architecture, where a client invokes a method (or function) at a server-side object. The RMIX approach allows each client-server pair to choose at compile time or even to negotiate at runtime the most efficient RMI protocol stack that is supported by both sides, while client-side and server-side object stubs remain the same as they only perform an adaptation to the RMIX framework and are not involved in the protocol stack.

The RMIX framework architecture (Figure 1) consists of two parts: a base library and a set of provider plug-in software modules. While the base library contains functions that are common to all protocol stacks, like advanced RMI

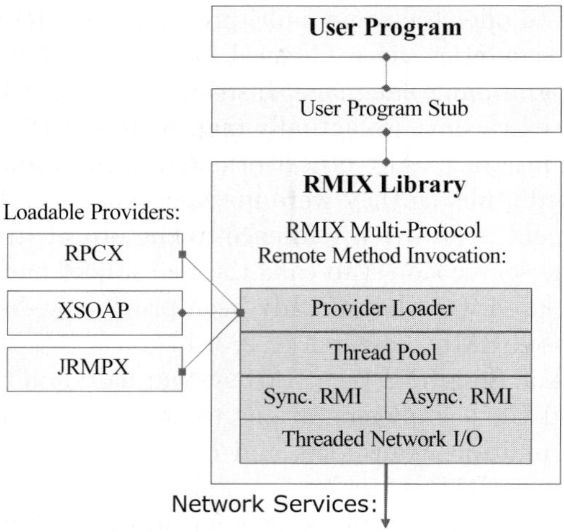


**Fig. 1.** RMIX Framework Architecture

semantics, networking and thread management, provider plug-ins contain protocol stack specific functions for connection management, message formats and data encoding. Since it is up to the provider to reuse base library functions, implementations may range from lightweight to heavyweight.

The base library reuses technology developed earlier for the lightweight Harness run time environment [15], such as a thread pool to simplify thread management and a provider plug-in loader to allow dynamic adaptation. It also provides TCP/IP based networking support and advanced invocation semantics, such as asynchronous and one-way. Due to the pluggable nature of RMIX, provider plug-ins are able to extend these basic functions by supplying their own implementation or by loading other plug-ins.

RMI calls are mapped to RPC calls in part by the base library using an object registry that stores necessary object interface information at the server side. When exporting an object, the user has to specify the object interface including method names, signatures and respective stub function pointers. Server-side object stubs adapt incoming RPC calls to method invocations and client-side stubs adapt outgoing RMI calls to RPC calls. The adaptation performed by both stubs also includes transposing the RMI/RPC call arguments between C-function and argc/argv style in order to be able to pass them through the RMIX base library for encoding and decoding in provider plug-ins. The mapping of RMI to RPC calls is independent from the protocol stack.

### 3.1 General Use Case

In order to enable an application to perform remote method invocations using RMIX, it just needs to link the RMIX base library and supply the necessary lightweight client- and server-side object stubs.

First, an object needs to be exported at the server-side base library with protocol parameters, object interface and object pointer in order to accept incoming RMI calls. The base library registers the object, loads the appropriate provider using the specified protocol, and calls the provider via a standardized interface to evaluate the protocol parameters and to export the object. The provider returns a local reference for the exported object back to the user. Part of this local object reference is the remote object reference.

Remote object references may be dynamically stored at name servers, such as the RMIX registry, or statically assigned at compile time using advanced protocol parameters to force a specific export behavior.

When calling a remote method, the user invokes the client-side method function using the remote object reference as first argument followed by the original arguments. The client-side stub transposes the arguments to an argc/argv style and forwards them and additional object interface information to the base library, which loads the appropriate provider using the protocol specified in the remote object reference and calls it via a standardized interface.

Providers are free to use the base library components for encoding and networking, but are not forced to do so. On the server side, the provider receives and decodes the RMI call and looks up the server-side object interface and object pointer at the object registry in the base library. The provider calls the server-side method function with object pointer and argc/argv style arguments either directly or using a separate thread. The server-side stub transposes the arguments back to C-function style and calls the appropriate function or method. Any return values are passed from the server-side stub and provider back to the client-side provider and stub in a similar fashion.

Provider plug-ins may be preloaded before exporting an object or calling a remote method to improve performance. Furthermore, a provider registry file is used to store information about providers and supported protocols. Protocol parameters specified for exporting an object may be used to configure protocol stack parameters, such as type mapping. They are also part of the remote object reference to ensure symmetric protocol stack configuration.

### 3.2 Advanced RMI Semantics

The RMI paradigm is based on a request/response model, where the request message contains the call input and the response message holds the call output. However, applications are not interested in response messages if they communicate using the message passing paradigm. Furthermore, each RMI call takes a certain amount of time for remote processing causing the caller to wait idle for the response message.

RMIX supports one-way invocations via a separate invocation interface that allows the caller to continue after the request message has been sent and accepted. Any response is eliminated at the client-side provider plug-in in a separate thread to maintain RMI/RPC protocol compliance.

Asynchronous invocations are offered by RMIX via another separate interface that also allows the caller to continue after the request has been sent and accepted. The caller obtains an invocation reference in order to retrieve the response later. The client-side provider uses a separate thread to wait for the response and to store it locally. Multiple method invocations may be interleaved, i.e. called in succession without retrieving the response in between. The server-side protocol plug-in guarantees the invocation order.

Asynchronous and one-way invocations have to be explicitly supported by a provider plug-in using separate implementations for each invocation style.

### 3.3 Remote Object Registry

Our C-based RMIX variant provides a name server style registry to dynamically associate remote object references with names similar to the Java RMI registry. However, in contrast to the Java RMI registry, the RMIX registry has multi-protocol support to improve interoperability. In fact, the RMIX registry is itself just a name server object that is exported using RMIX.

## 4 Harness Integration

Harness is a pluggable heterogeneous Distributed Virtual Machine (DVM) environment for parallel and distributed scientific computing. Conceptually, the Harness software architecture consists of two major parts: a runtime environment (RTE) and a set of plug-in software modules. The multi-threaded userspace RTE manages the set of dynamically loadable plug-ins. While the RTE provides only basic functions, plug-ins may provide a wide variety of services needed in fault-tolerant parallel and distributed scientific computing, such as messaging, scientific algorithms and resource management. Multiple RTE instances can be aggregated into a DVM.

The C-based RMIX variant has been integrated into the C-based lightweight Harness RTE [15] in form of a plug-in (Figure 2) to provide multiprotocol RMI/RPC capabilities to the RTE and to plug-ins. This effort also complements earlier work in integrating the Java-based RMIX solution into the Java-based Harness run time environment H2O [16].

The use case scenario of the C-based RMIX variant within the Harness context has already been discussed in [15]. While the RMIX base library and Harness RTE stubs are wrapped into a Harness-RMIX plug-in, stubs for other plug-ins are also implemented as plug-ins. Since the Harness RTE supports plug-in dependencies, a plug-in requiring RMIX automatically loads its stub plug-in(s), which subsequently load the RMIX plug-in.

Ongoing work in this area focuses on using both lightweight Harness RTEs to provide adaptability and interoperability in a heterogeneous collaborative environment for distributed scientific computing. Furthermore, we are also currently investigating parallel plug-in programming paradigms using the Harness/RMIX combination as a backbone for adaptive, fault tolerant, distributed, component-based scientific applications.

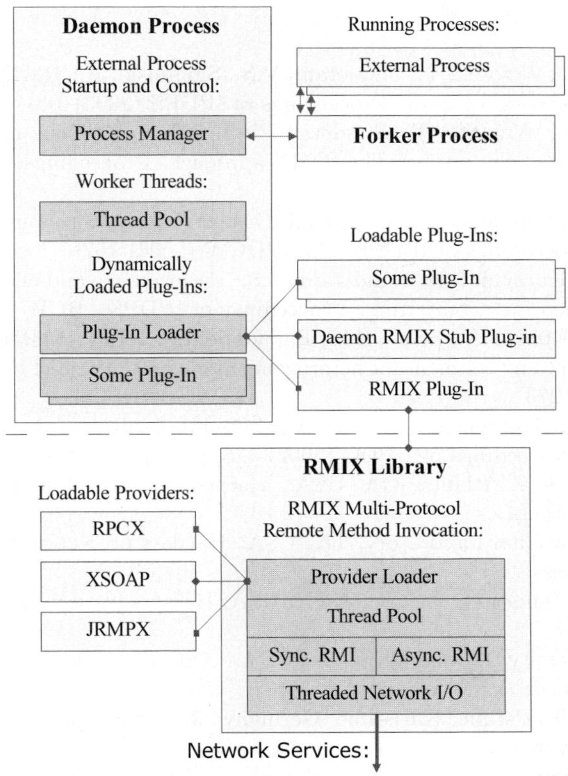


**Fig. 2.** RMIX Plug-in for the Harness Metacomputing System

## 5 Conclusions

With this paper, we presented a native (C-based), flexible, adaptable, multi-protocol RMI/RPC communication framework that complements the Java-based RMIX solution. We described the RMIX framework architecture, its general use case and some of its features, such as advanced invocation semantics, in more detail. We also explained recent integration efforts with the Harness lightweight metacomputing environment.

RMIX is part of the Harness software distribution package from Oak Ridge National Laboratory. Currently, we supply the Harness runtime environment together with the RMIX base library and the fully functional RPCX provider plug-in, which offers a ONC RPC compliant protocol stack using XDR encoding. The RPCX provider also supports one-way and asynchronous invocations. Ongoing work focuses on SOAP, IIOP and Java RMI (JRMP) provider plug-ins to further improve heterogeneity. Future work will target security related issues, such as authentication, authorization and encryption.

## References

1. Kurzyniec, D., Wrzosek, T., Sunderam, V.S., Slominski, A.: RMIX: A multiprotocol RMI framework for Java. Proceedings of IPDPS (2003) 140
2. Kurzyniec, D., Wrzosek, T., Sunderam, V.S.: Heterogeneous access to service-based distributed computing: The RMIX approach. Proceedings of IPDPS - HCW (2003) 100
3. Kurzyniec, D., Sunderam, V.S.: Semantic aspects of asynchronous rmi: The RMIX approach. Proceedings of IPDPS - JavaPDCW (2004) 157
4. Wrzosek, T., Kurzyniec, D., Sunderam, V.S.: Performance and client heterogeneity in service-based metacomputing. Proceedings of IPDPS - HCW (2004) 113
5. Geist, G.A., Kohl, J.A., Scott, S.L., Papadopoulos, P.M.: HARNESS: Adaptable virtual machine environment for heterogeneous clusters. Parallel Processing Letters **9** (1999) 253–273
6. Sunderam, V., Kurzyniec, D.: Lightweight self-organizing frameworks for metacomputing. Proceedings of HPDC (2002) 113–124
7. Emory University, Atlanta, GA, USA: Harness project at http://www.mathcs.emory.edu/harness
8. Oak Ridge National Laboratory, TN, USA: Harness project at http://www.csm.ornl.gov/harness
9. University of Tennessee, Knoxville, TN, USA: Harness project at http://icl.cs.utk.edu/harness
10. Indiana University, Bloomington, IN, USA: XSOAP project at http://www.extreme.indiana.edu/xgws/xsoap
11. University of Karlsruhe, Karlsruhe, Germany: JavaParty project at http://www.ipd.uka.de/javaparty
12. Maassen, J., van Nieuwpoort, R., Veldema, R., Bal, H.E., Plaat, A.: An efficient implementation of Java's remote method invocation. Proceedings of PPoPP (1999) 173–182
13. Vasudevan, V.: A Web Services primer. http://www.xml.com/pub/a/2001/04/04/webservices (2001)
14. Christensen, E., Curbera, F., Meredith, G., Weerawarana, S.: Web Services Description Language (WSDL). http://www.w3.org/TR/wsdl (2001)
15. Engelmann, C., Geist, G.A.: A lightweight kernel for the Harness metacomputing framework. Proceedings of IPDPS - HCW (2005) 120
16. Kurzyniec, D., Drzewiecki, D., Sunderam, V.S.: Towards self-organizing distributed computing frameworks: The H2O approach. Parallel Processing Letters **13** (2003) 273–290

# Developing Collaborative Social Software

Ismail Bhana and David Johnson

Advanced Computing and Emerging Technologies Centre,
School of Systems Engineering,
The University of Reading,
Reading, RG6 6AY, United Kingdom
{i.m.bhana, d.johnson}@reading.ac.uk

**Abstract.** The Internet has an increasing role in facilitating communication between people and groups of people. As access to the Internet and World Wide Web is widely available, collaborative services enabled over the Internet are also burgeoning. In this paper, we present the current issues and our techniques for developing collaborative social software. We discuss online communities in the context of social collaborative systems. We then describe our approach to the development of supporting software for online communities and collaboration.

## 1 Introduction

The Internet has an increasing role in facilitating communication between people and groups of people. Traditional forms, such as telephones, mail and even face-to-face meetings, are gradually being assumed by computer based forms where socialisation is enhanced through rich media sharing, reflection on past experiences and greater availability of services. Communication software allows for audio conferencing, video conferencing, as well as real-time application sharing. Reflection is supported by allowing the extensive logging of communications and the ability to annotate these logs. Content sharing can provide instant publishing of and access to information. As access to the Internet and World Wide Web (WWW) is widely available, collaborative services enabled over the Internet are also burgeoning.

In this paper, we present the issues and techniques for developing collaborative social software. In section 3 we discuss online communities and our approaches to supporting such communities. We then go on to describe our approach to the development of supporting software for online interaction and collaboration.

## 2 Related Work

ABC [1] is an exploration into how to conceptualise and model human activities in the context multi-user interactions. ABC consists of a Java-based cross-platform development framework and supporting runtime infrastructure on which applications supporting both synchronous (i.e. real-time) and asynchronous cooperation or

V.N. Alexandrov et al. (Eds.): ICCS 2006, Part II, LNCS 3992, pp. 581 – 586, 2006.

collaboration can be built. ABC focuses on activity sharing rather than the application sharing that is more commonly found in CSCW.

Friendster [2] is a Web-based Internet social networking service that supports online communities with shared content (pictures, videos, and Blogs) and online profiles that are accessible to other community members. The objective is to enable individuals to connect, form social networks and interact in virtual communities that are linked by shared acquaintances. Friendster is based on the circle of friends technique for networking individuals in virtual communities and has its roots in Ryze [3], a business networking site founded in 2001.

The DreamTeam framework [4] consists of an environment that is designed to facilitate the rapid development of collaborative distance learning applications. The motivation behind DreamTeam is to hide the complexity of the underlying network and to enable developers to build collaborative applications transparently, using the facilities and services of the DreamTeam runtime infrastructure.

## 3 Online Community

Several common features characterise an online community or 'networked communities' [6]:

- Technological mediation of communications.
- Persistence of the overall context or activity of the community.
- Multiple interaction styles allowing entities to communicate in a range of ways.
- Real-time interaction capabilities facilitating spontaneous collaborations.
- Supporting multiple users as opposed to one-to-one contacts.

We identify three interaction styles for communication that encompass all of the aforementioned factors:

- One-to-one - Direct communication from one participant to another. Examples of this include email, audio/video chat and instant messaging (IM).
- One-to-many - A single participant communicating with multiple participants. Most commonly seen examples of this are publishing information on the WWW in the form of personal web-pages or Blogs. Other examples include audio/video conferencing.
- Many-to-many - Multiple entities communicating with each other in a larger context, for example collaborative authoring in Wikis [7] and knowledge sharing in discussion forums.

In building online communities, both synchronous and asynchronous communication can play a role. Ad hoc collaborations enable entities to self organise and communicate in real-time, as is the norm in audio/video/IM chat and conferencing. The ability to edit or annotate content allows entities to communicate over greater lengths of time where organisation of real-time meetings cannot be realised, or the nature of the task requires drawn-out periods to complete. Examples of persistence include publishing information on the WWW using Wikis, discussion forums and Blogs.

Issues of identity also come into play as community building, whether online or not, require contributions from and collaboration between identifiable individuals. Establishing an identity is pivotal in building relationships with others within an online community, especially if there is no pre-existing relationship outside the community context. For example, an online community may bring together entities with no experience of each other but on the other extreme, a community might be formed that mirrors existing relationships in the 'real' world. In the latter case, establishing identity is not always necessary.

We identify three enablers for supporting identity in an online community:

- Profiles – A profile represents a persona that a participant wishes to project to the community.
- Portfolios – A record of participant's experiences. A portfolio represents achievements, experience, and evidence, including reflections on each of these. Dependant on the context, parts of portfolios are publishable to the community for others to form opinions.
- Participations – The process of interacting with other community entities. By communicating and collaborating with others, one builds an identity purely through interaction, rather than publishing some record of identity. As community participants form opinions of each other, social norms, reputations and individual's roles can be defined. Norms and roles are not static once identified, and can change as collective experiences go on.

As an individual builds up a persona defined by their experiences through contributing, communicating and collaborating with others in a community, issues regarding intellectual property arise. Typically information is centralised. Users' profiles, portfolios and content tend to reside in one place so that access may be tightly controlled; this is normally the case with Web-based community services. However, there is a growing need to be able to share and exchange information to support community interactions, such as personal data and content, between multiple different community systems (Websites or applications) transparently and provide services that unify various user profiles, portfolios and participations across multiple systems.

In most cases, access to profile, portfolio and participation data is controlled by the service provider for a particular community system, forum or community website (with limited user control). Most service providers require users to register and agree to certain terms and conditions of service that specify how user information is handled by that provider alone. This has the affect of hindering the process of interoperation of community service where there is no dialog between providers (or where services are hosted by competing organisations). Options here do exist, such as Microsoft Passport [8], but there is a need for open specifications and technology to facilitate the transference of user data (as we have defined above) transparently between services and community systems – ideally where it is managed and controlled by the participants themselves.

Depending on the needs of a service provider, we submit that a balance between the needs of individual community systems to enable meaningful contextual interactions and freedom of the participants to control access to profile, portfolio and participation data transparently across multiple community systems is needed.

## 4 Collaborative Communities with Coco

Coco [9] consists of a set of peer-to-peer (P2P) Internet web services that support the creation of spontaneous multi-user collaborative sessions, allowing users to self-organise and communicate, share tasks, workloads, and content, and interact transparently using multiple different computing devices, networks, and operating systems. Coco is designed to handle both heterogeneity (interoperability between systems) and to scale from very small to very large communities using P2P technologies.

Coco is built on a services-based architecture with collaborative community services hosted by JXTA peers and peer groups. Community data (such as profiles, portfolios, and participations) is represented using RDF data structures and exchanged by peers using the Coco metadata content service [10]. Synchronous interactions are supported by the Coco Messaging Service which allows groups of participants to engage in real-time video, audio, and IM interactions and the Interaction Service, which supports real-time presence notification within group interactions. The objective in providing distinct services is to allow collaborative community applications to access either all or a subset of Coco services hosted by participating peers.

To fulfill the need for enabling collaboration across a range of devices, we are developing a version of Coco aimed at mobile devices, such as PDAs and mobile phones. MicroCoco [11] is being developed using Java 2 Micro Edition (J2ME) [12] and JXTA for J2ME (JXME) [13] technologies. JXME interoperates with JXTA protocols allowing for mobile devices to participate in JXTA networks. The services provided by MicroCoco are a subset of the full platform. This is because the hardware constraints allow a limited amount of bandwidth for communication with other full Coco peers. For example, a group of Coco peers may be able to participate in video conferencing; however a MicroCoco peer running on a mobile phone may not necessarily have the bandwidth to deal with an incoming video stream, or the hardware capabilities to display that stream. A MicroCoco peer could however participate in sessions that require more asynchronous communications such as instant messaging, content sharing and co-authoring.

We assume that the mobile device will not be a user's primary computing device and have designed MicroCoco accordingly. To overcome some of the problems with mobile devices, we pair mobile peers with desktop peers. The mobile peer can then be used for particular tasks whilst away from the desktop, such as searching for content, whilst actually redirecting downloads to the paired desktop peer. This content redirection allows the user to still consume large amounts of content whilst using a constrained mobile device.

To support seamless collaboration across devices, we are implementing a session handoff mechanism that allows a user collaborating on one device to transfer session information to another device. The concept of Application Session Handoff (ASH) has been identified and practically demonstrated by UCLA's Pervasive Computing Laboratory project, iMASH [14]. In [14], the authors discuss the need for continuous computing – the ability to migrate application sessions between heterogeneous platforms. We plan on implementing this concept for collaborative sessions in Coco where some of the issues of dealing with platform heterogeneity are inherently solved through the use of Java and JXTA technologies.

## 5 Conclusions

This paper has presented a number of issues that reflect our experiences in the development of collaborative community systems. We submit that the goal of future community systems should be to support transparency via interoperability in the underlying architecture. Such transparency will enable a diverse range of community systems (including websites and applications) to access profile data, portfolios, and historic participations of participants and built community facilities by consuming collaborative services from a range of sources.
There are a number of potential areas of future work, including:

- Managing context switching – the implementation of a facility (possibly by via the incorporation of a new service into the design) that allows users to dynamically migrate from one device to another without affecting the continuity of collaboration.
- Support for user roles – Coco does not currently provide mechanisms for defining user roles within a collaborative context. Roles are, however, fundamental to effective collaborative and in future the platform needs to provide extended support for complex roles.
- Support for user activities – research into user activities, via case studies, will enable us to look more closely into the social aspects of online communities, particularly where mobile technologies are involved.
- Technological performances. Evaluating consistency of centralised vs. decentralised communities in terms of user-experience (speed of forming communities, development of roles, and socialisation).

## References

1. Bardram, J. E.: Supporting Mobility and Collaboration in Ubiquitous Computing. Technical Report CfPC 2003–PB–38, Center for Pervasive Computing, Aarhus, Denmark, (2003).
2. Friendster, online community of friends: http://www.friendster.com, (2005)
3. Ryze, business networking: http://www.ryze.com/, (2005)
4. Marsic, I.: An Architecture for Heterogeneous Groupware Applications, Piscataway, NJ, USA: Rutgers University, (2000).
5. Mynatt, E. D., Adler, A., Ito, M., and O'Day, V. L.: Design for network communities. In Proceedings of the SIGCHI Conference on Human Factors in Computing Systems (Atlanta, Georgia, United States, March 22 - 27, 1997). S. Pemberton, Ed. CHI '97. ACM Press, New York, NY, 210-217.
6. Wiki, What Is Wiki: http://wiki.org/wiki.cgi?WhatIsWiki, (2005)
7. Microsoft Passport: http://www.microsoft.com/net/default.mspx , (2005)

8. Bhana, I. Johnson, D. Alexandrov, N.: Supporting ad hoc Collaboration in Peer-to-Peer Networks", Proc. of 17 International Conference on Parallel and Distributed Computing, San Francisco (2004)
9. Bhana, I. Johnson, D.: A Peer-to-Peer Approach to Content Dissemination and Search in Collaborative Networks", Proc. of International Conference on Computational Science 2005, Atlanta (2005)
10. Bhana, I. Johnson, D.: Knowledge Management for Collaborative Learning, Proc. of the Interactive Computer-aided Learning 05 conference, Villach, Austria (2005)
11. Java 2 Platform Micro Edition (J2ME), http://java.sun.com/j2me/, (2005)
12. JXME: JXTA Platform Project, http://jxme.jxta.org/proxied.html, (2005)
13. Bagrodia, R., Bhattacharyya, S., Cheng, F., Gerding, S., Glazer, G., Guy, R., Ji, Z., Lin, J., Phan, T., Skow, E., Varshney, M., and Zorpas, G. 2003. iMASH: interactive mobile application session handoff. In Proceedings of the 1st international Conference on Mobile Systems, Applications and Services (San Francisco, California, May 05 - 08, 2003). MobiSys '03. ACM Press, New York, NY, 259-272.

# An Efficient and Reflective Event Filtering for Context-Awareness in Ubiquitous Computing

Kyu Bong Cho[1], Sung Keun Song[1], Hee Yong Youn[1], and Gyung Leen Park[2]

[1] School of Information and Communications Engineering
Sungkyunkwan University, Suwon, Korea
mongbong@skku.edu, skk103@devg.org, youn@ece.skku.ac.kr
[2] Department of Computer Science and Statistics
Cheju National University, Cheju, Korea
gpark@cheju.ac.kr

**Abstract.** In the ubiquitous computing system fast and reliable event delivery is important to properly adapt to the dynamically changing environment. Such requirement could be satisfied by message-based middleware based on the publish/subscribe model. In this paper we propose a scheme that allows reflective filtering through the reasoning process. The objective of the reflective filtering is to prevent waste in the space of the event queue of the event channel due to useless events and reduce the network load. Performance evaluation reveals that the proposed scheme displays a significant improvement over the existing event service, Java messaging service (JMS) and omnievent service, in terms of event delivery time.

**Keywords:** Context-oriented filtering, event service, message-oriented middleware, publish/subscribe, reflective filtering.

## 1 Introduction

The ubiquitous computing for any-time, any-where, any-network, any-device, and any-service is becoming a core paradigm of the modern information-oriented society. In such environment the post PC-type computers provide various services useful to human by using real-time context information of the objects [1]. Here reliable and fast communication must be guaranteed to adapt to the dynamically changing environment and receive the context data of the environment including the event data [2,3].

In the ubiquitous environment the basic mechanisms employed for event transmission are remote procedure call (RPC), remote method invocation (RMI), and message- oriented middleware (MOM), etc. The RPC offers an abstraction of high level, but it does not guarantee reliability in addition to static binding and limited scalability. To solve the problem while dealing with the heterogeneous environment of various programming languages, hardware platforms, and operating systems, the MOM based on message queuing was developed. Later, the distributed object computing paradigm providing more abstraction and transparency of users appeared. There also exist peer-to-peer, publish/subscribe, and hierarchical channel model as the messaging scheme for event transmission. The common characteristics of various MOM

V.N. Alexandrov et al. (Eds.): ICCS 2006, Part II, LNCS 3992, pp. 587 – 594, 2006.

products and the publish/subscribe model based on message queuing and filtering are presented in [4,5].

The publish/subscribe model among various messaging models is a representative event transmission model used widely. The current popular event-based transmission services are JMS (Java Message Service) [7], CORBA event service [9], and CORBA Notification service [6]. They employ filtering to filter unwanted events out. However, the overhead of filtering in the Notification service has been recognized to be very high because filtering occurs in several proxies. Note that the JMS needs JVM, which is an additional overhead of Java-based system with respect to the speed of event delivery. The JMS handles message filtering in the server like CORBA Notification service, and thus the overhead in the processing and managing the messages in the queue is high [8].

In this paper we propose a reflective filtering approach where the filtering criteria of the client are sent to the supplier through a reasoning process. The approach can enhance the capability of event service using the algorithm of highly efficient and scalable reflective message filtering [9]. It implements context-awareness for handling the requests generated from the environment the users belong to. The objective of the proposed reflective filtering is to prevent the waste in the space of the event queue of the event channel due to useless events and reduce the network load. Extensive experimentation reveals that the proposed approach allows consistently higher performance than the JMS and CORBA event service in terms of event delivery time for various numbers of servers and data sizes. Also, it reveals that the performance significantly varies according to where the filtering is applied.

The rest of the paper is organized as follows. Section 2 presents the related work and Section 3 introduces the proposed event filtering approach. The proposed approach is evaluated in Section 4, and conclusion is made in Section 5.

## 2 The Related Work

### 2.1 Classification of Filtering Algorithms

Several algorithms for distributed filtering have been proposed. Evaluation and comparison of the algorithms are difficult due to the diversity of the approaches. What is needed is a concise classification of the approaches employed for distributed filtering [10].

**Location of Filtering:** Filtering can be performed at the subscribers (flooding of events) or providers (flooding of profiles), or at certain broker nodes. Flooding of events results in high network traffic, but requires small memory space. Flooding of profiles, however, results in the opposite: less network traffic but large memory space. Filtering at fixed brokers allows the advantage of filtering control according to available resources, but has the disadvantage of high load at the filtering brokers in terms of network traffic and memory usage.

**Spreading of Filtering Complexity:** The filtering complexity can be spread over several brokers by exclusive filtering at certain brokers or by distributed filtering. Exclusive filtering can be implemented with some control overhead. A disadvantage of exclusive filtering is the danger of multiple notifications for a single event because the event notification may be forwarded to several neighbor brokers.

**Communication with the Subscribers:** In direct communication only the filtering broker and the subscriber are involved in the communication. A disadvantage here is that either a connectionless protocol has to be used (resulting in unreliable communications) or new connections have to be established over the time. When forwarding messages via the network of brokers, only the neighbor brokers and local clients communicate directly. Local clients are the publishers and subscribers that are directly connected to a broker. A drawback here is high memory usage. When using brokers as the proxies, the brokers act as the subscribers to their neighbor nodes and thus the number of subscribers each broker node has to deal with in limited. Exploiting the coverage between the profiles of several subscribers is possible and beneficial in several aspects. A disadvantage, though, is the necessity of post-filtering in the notification to the client subscribers.

## 2.2 Java Message Service (JMS)

The Java Message Service is a Java API that allows the applications to create, send, receive, and read the messages. A JMS provider is a messaging system that implements the JMS interfaces and provides administrative and controlling features. The two kinds of administered objects are destinations and connection factories. The technology underlying these objects is likely to be very different from one implementation of the JMS API to another. Therefore, management of the objects belonging to different administrative tasks varies from provider to provider. The JMS clients access the objects through portable interfaces, and therefore a client application can be run with little or no change on more than one implementation of the JMS API. Figure 1 illustrates the way the components in the JMS interact with each other.

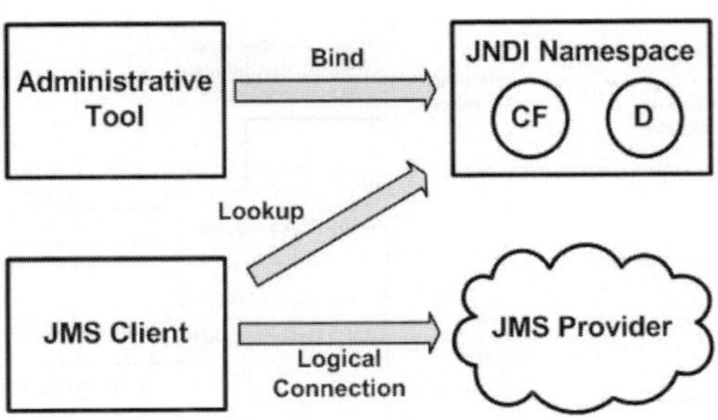


**Fig. 1.** The JMS API architecture

The JMS message is composed by a header, properties, and body, and it is typed based on the data contained in the message body. The class hierarchy of the JMS message is organized such that the Message interface, which is the root interface for all the JMS message types, defines the header fields, property facility, and associated accessor methods. A JMS message selector is used by a JMS client to specify which messages it is interested in. It is based on the values contained in the message header, which includes standard header fields as well as optional fields that are added via application, standard, or provider-specific properties. As noted earlier, a message selector cannot refer to the value of the message body. A message selector is a Boolean expression that, when it evaluates to be true, results in the matched message to be passed to the client. It is defined as a string, and its syntax is based on a subset of the SQL92 conditional syntax. Creating a selector involves the generation of a string that conforms to the defined syntax [7].

## 3 The Proposed Event Filtering

The proposed event filtering is a part of a new reflective context-oriented event service embedded in the CALM (Component-based Autonomic Layered Middleware) research project [11]. We first briefly introduce the CALM.

### 3.1 CALM

The CALM adopts the reflective paradigm and agent system to construct a flexible and efficient middleware platform for ubiquitous computing. It also provides useful development tools. It consists of two internal layers, one external layer, and various tools forming an efficient agent-based platform. The internal layer consists of the communication platform layer and agent platform layer. The communication platform layer is constructed such that it can provide various distributed services based on the context and situation using appropriate communication protocols and devices in wired and wireless environment. The agent platform layer is composed using the components so that it can maximize the efficiency of the service, adapt itself to the environment, and accommodate the advantage of diverse agent systems. The external layer is composed by self-growing engine and ontology-based situation-awareness engine required for providing intelligent services.

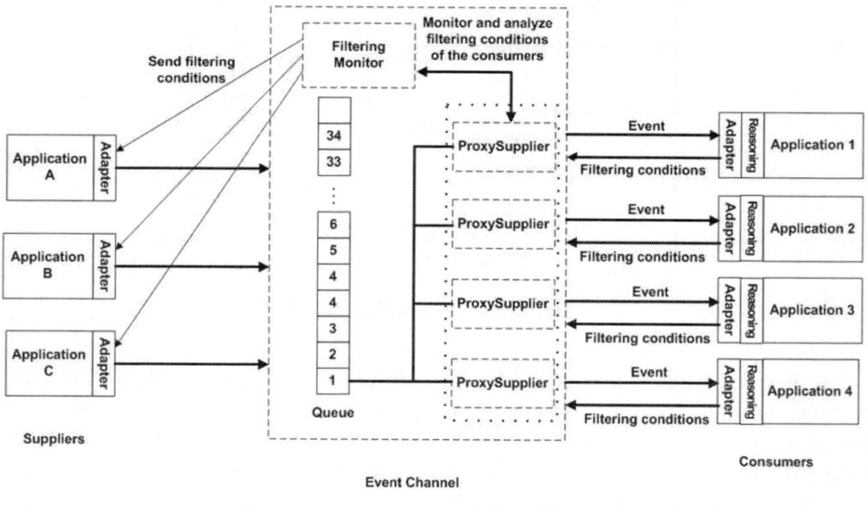


**Fig. 2.** The structure of the proposed reflective filtering

### 3.2 Reflective Filtering

The objective of reflective filtering is to prevent waste in the space of the event queue of the event channel due to useless events and reduce the network load. The concept of reflective filtering process is shown in Figure 2. The consumers decide the filtering criteria with which each consumer declares what kinds of context information are desired to be received through reasoning. The adapter of the consumers connects it to the event channel and sends its own filtering criteria to relevant proxysuppliers. The filtering monitor of the event channel monitors and analyzes the filtering criteria of the consumers and sends the result of analysis to the adapters of the suppliers. The filtering monitor is normally in sleeping mode and activates itself upon receiving a new filtering criterion from a consumer or a new consumer connects to the event channel. The adapter of the suppliers receives the filtering criteria from the filtering monitor of the event channel. The adapter of the suppliers filters out the events that the application has not indicated to receive from the event channel.

### 3.3 Reliable Communication

The proposed event service employs event numbering and event queues to support reliable communication. The format of an event of the proposed event service is as follows;

```
struct EventHeader{
        bool                ACK;
        unsigned short      EventID;
        unsigned short      EventSize;
        long                Timestamp;
        long                Expiration;
        bool                DeliveryMode;
        bool                Priority;
        string              EventType;
        struct              MessageProperty;
};
```

The event channel and adapter have an event queue, respectively. Reliable communi-cation is provided as follows; the adapter of the supplier stores the events at its own event queue before sending them to the event channel. Thereafter, it sends the events to the event channel. It waits for an acknowledgment from the event channel after sending the events. If the adapter receives an ACK from the event channel and the event queue is full, it overwrites the Acked events with new events until the events are Acked. In case that the adapter receives a retransmission request from the event channel, it resends the events stored in the event queue.

Efficient message filtering needs an effective design of header and payload of the message. Message is the most important part in message filtering and the message format is shown in Figure 3. The events communicate with each other using the messages carrying the data of applications, and different components of the adapter are used in message transmission. The components compose an alive system.

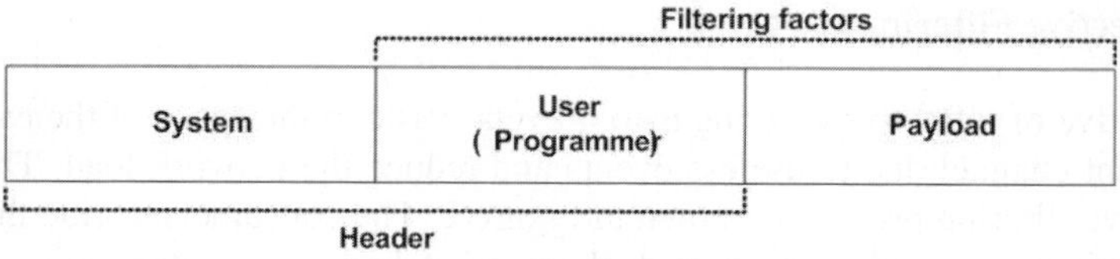


**Fig. 3.** The message format

A message object is divided into two parts, message header and message data itself (payload). The messages have several different types decided based on the payload of the message. The payload is StreamMessage, MapMessage object, or takes a standardized TextMessage, ObjectMessage and BytesMessage type. A message carries data, and it is also used as a simple means of notification of the events occur-ing in the system.

The message header supplies the MetaData describing who created the message, when it was created, and how long it is valid. Also, a message can include the attribute that can be defined and set by the client. The subscribers can filter the message through this attribute and select the messages according to the value of the specification header and payload. The message attribute is put in the additional header that can be attached to the message, and it allows the application developer to include more information in a message.

## 4 Performance Evaluation

The performance of the proposed approach is carried out concerning message delivery time, and it is compared with the omnievent service and JMS. The test platform includes twelve identical Windows XP-based PCs, each with one 2.1-Ghz CPU and

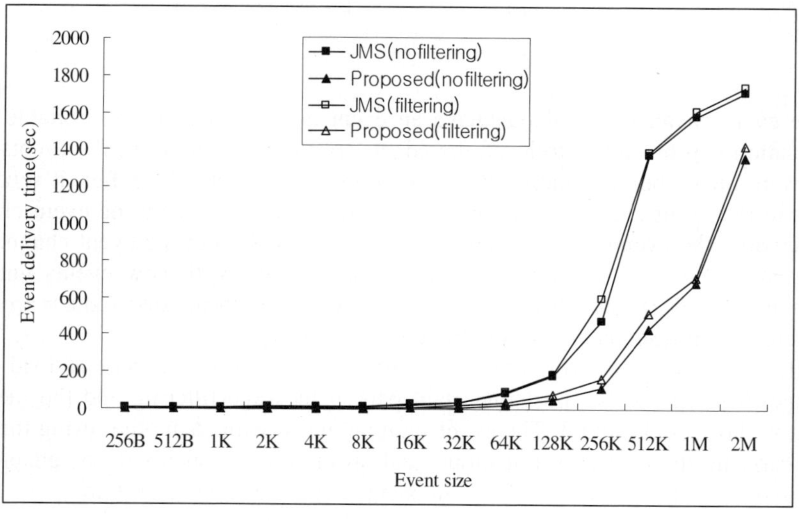


**Fig. 4.** The comparisons of event delivery time for one-to-one connection

512-Mbyte RAM. In case of the proposed reflective filtering and omnievent service ten PCs host suppliers, one PC hosts the naming server, and one PC hosts the consumer. In case of the JMS ten PCs are senders (suppliers), which connect to their local server and a PC is their subscriber (consumer). Figure 5 and 6 show the average event delivery time of 1000 test runs in case of one-to-one and many-to-one connection, respectively.

In Figure 4 each supplier generates 2000 context information while the event size is increased up to 2Mb. Notice from the figure that filtering allows faster delivery than the case of nonfiltering for both the systems. Also, notice that the proposed event service displays always smaller delivery time while the improvement gets more significant as the event size grows. Figure 5 shows the filtering efficiency when the number of suppliers is increased. Again, the proposed scheme shows a significant improvement because filtering occurs in the JMS server while the suppliers transmit data without considering the consumer's preference. Note that the proposed event service applies filtering in the supplier. We obtain the same result with the omnievent service, which is not included here due to space limit.

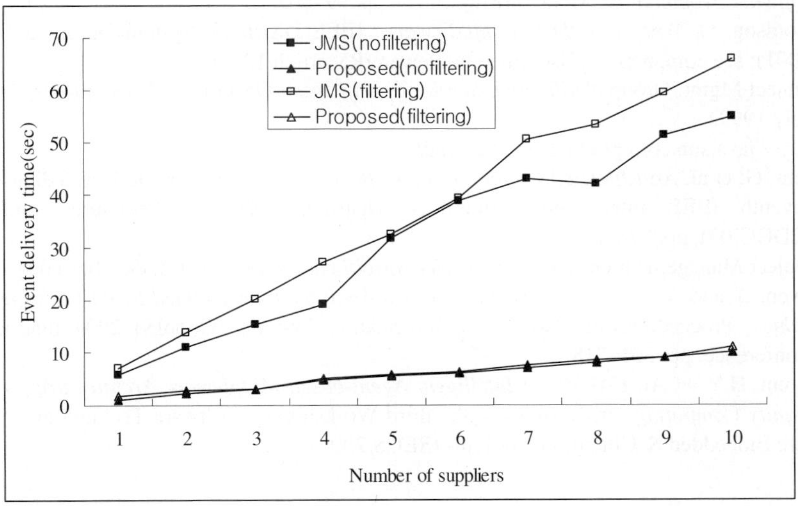


**Fig. 5.** The comparisons of event delivery time for many-to-one connection

## 5 Conclusion and Future Work

In the pervasive computing the importance of event filtering is very high since a large number of events are generated in the environment. Various distributed filtering algorithms have been proposed for the publish/subscribe based systems, but systematic analysis and comparison of them is difficult. In this paper we have proposed a context-oriented event filtering approach supporting reliable communica-tion and reflective filtering through the adapters. The proposed event filtering is implemented in the CALM developed by the authors. It also reduces the network load of CORBA-based event servers which is an important requirement for large scale ubiquitous environment. Extensive computer simulation reveals that the proposed approach allows consistently higher performance than other approaches in terms of event delivery time for various number of servers and data sizes. We will further investigate the filtering approaches dynamically adaptable for various operational conditions and requirements.

## References


1. Ha, W. *Ubiquitous IT revolution and the third space* Korean Elictrinic, 2002.
2. Karen, H. and Jadwiga, I. *A Software Engineering Framework for Context-Aware Pervasive computing*. Proceedings of the Second IEEE Annual Conference on Pervasive Computing and Communications (PERCOM'04), pp. 77.
3. Schantz, R. E. and Schmidt, D. C. (2001) *Middle-ware for Distributed Systems: Evolving the Common Structure for Network-centric Application* Encyclopedia of Software Engineering, New York, Wiley & Sons pp. 801-813.
4. Astley, M., Sturman, D. and Agha, G. *Customizable middleware for modular distributed software. Commun* ACM 44, 5 (May 2001), pp. 99–107.
5. Coulson, G. *What is reflective middleware*? IEEE Distribute System Online 2, 8 (Dec. 2001); see computer.org/dsonline/middleware/RMarticle1.htm.
6. Object Mgmt. Group *Notification Service Specification* OMG Doc. Telecom/99-07-01 ed., July 1999.
7. http://java.sun.com/products/jms/tutorial/
8. Lan, G. et al. *An Efficient, Scalable Content-Based Messaging System.* Proceedings of the seventh IEEE International Enterprise Distributed Object Computing conference (EDOC'03), pp. 278–285.
9. Object Management Group *Event Service Specifi cation* Version 1.2, October 2004.
10. Sven, B. and Annnika, H. *Classification and Analysis of Distributed Event Filtering Algorithms.* Proceedings of Cooperative Information Systems (CoopIS) 2004 International Conference, pp. 301–318.
11. Youn, H.Y. et Al. *CALM: An Intelligent Agent-based Middleware Architecture for Community Computing,* Proceedings of the third Workshop on Software Technologies for Future Embedded & Ubiquitous Systems (SEUS 2006).

# Creation and Control of Interactive Virtual Environments

Adrian Haffegee, Priscilla Ramsamy, Ronan Jamieson, and Vassil Alexandrov

Advanced Computing and Emerging Technologies Centre,
The School of Systems Engineering, University of Reading,
Reading, RG6 6AY, United Kingdom
sir04amh@reading.ac.uk

**Abstract.** Within the confines of a Virtual Environment (VE) almost anything is possible. It is easy to establish the benefits such an application could provide throughout the many walks of life, and yet current VE development remains within the domain of Virtual Reality application programmers. We describe methods that enhance VE development, first by providing scene creation for non-programmers, and second through a scene management entity that controls interaction within the environment. We explore methods for interacting through the scene to enable multiuser collaboration, and detail sample applications making use of this approach.

## 1 Introduction

The field of Virtual Reality (VR) concerns itself with the study of how the senses and perception can by fooled into believing that virtual objects exist, or that a non existing experience has occurred. This ability to alter what each of the senses is telling the brain, combined with the way the brain is translating them, provides vast scope in exploring the different aspects of reality. The uses of a perfect VR system are only limited by the imagination provided in development of its applications. Those depicting objects could have commercial uses in previewing or analyzing products. Alternatively medical based objects could provide surgeons a preview of what they will be facing without the prior need of investigative surgery. Indeed a virtual representation of any object could be created and then limitlessly studied and modified, thereby negating a large dependence on real world models or prototypes.

Rather than just depicting individual objects, an entire Virtual Environment (VE) could also be constructed. Representations of real environments could be used to experience products or locations, such as a new house or a holiday destination, while viewing from a different location.

Adding user interaction creates a more dynamic environment that could react to certain user provided stimulii. Examples here would include training exercises where users could be repeatedly placed inside simulations of otherwise difficult to reproduce situations, thereby learning as though they were in the real environment. However VEs need not just be constrained to environments that

V.N. Alexandrov et al. (Eds.): ICCS 2006, Part II, LNCS 3992, pp. 595–602, 2006.

humans would naturally find themselves in. For instance animal behavior biologists may benefit from experiencing an environment as though they were seeing through the eyes of a different creature, or for entertainment purposes a user could become a hero in some fictional fantasy world.

The natural extension to having one person using a VE is to allow multiple users to share the environment. Ideally this should be possible regardless of the users' physical location, allowing geographically dispersed users to meet together in a Networked Virtual Environment (NVE)[1].

Once multiple users are connected together within a VE it is possible to use such an environment to replace current forms of remote user to user communication and interaction. These current methods make use of limited resources such as telephones, videoconferencing systems through to simple pens and paper. All of this functionality can brought into the almost limitless domain of the VE, making it a perfect medium supporting inter-person communications. However such technology extends beyond the boundaries imposed through the conventional resources; rather than just conveying their views to distant participants, users can interact with each other within their shared virtual world, dynamically manipulating it to convey their thoughts.

One issue effecting the usage of VEs is the effort taken in their development. While tools for their creation do exist [2, 3] they require programming experience along with a detailed knowledge of the area. This coupled with limited accessibility to VR hardware has resulted in restricted VE development. Current environments have little interaction between their users, with them often only sharing head and hand positioning and simple action events. Although some work has been undertaken in bringing realistically mixed audio into the environment [4, 5], generally external audio applications [6, 7] are used by VE participants.

Recently, the AGJuggler project [8] has attempted to bring the AccessGrid [9] conferencing technology into the virtual world. This is still a work in progress, and beyond the inclusion of remote video streams into the VE, currently has limited user interaction. However it does demonstrate areas where real world communications technology is beginning to merge into the VE.

Assuming the complexities of VE implementation can be ignored there also exists the problem of designing the environment itself. Whilst almost anyone could populate a blank environment with walls, floors, windows and doors, it would still take someone with architectural knowledge to be able to construct a practical and realistic virtual building. Similarly a psychologist would be needed to develop environments for psychological testing, or a doctor for simulating medical procedures. It is clear that the development of these advanced environments require skills that are not usually available for VR programmers.

This paper describes a method for creating and controlling interactive virtual environments that can be used by most people regardless of their computer programming background. Section 2 details the tools that achieve this functionality before section 3 describes how they can be used for different types of interaction. Section 4 describes sample applications built around this methodology and then the paper concludes with section 5.

## 2 Controlling the Virtual World

This work builds on top of a lower level VR application development toolset [10], that described a library of infrastructure tools providing functionality such as display management and user representation/control. A key development from this work was the creation of a common architectural platform for writing VR applications. Since they share the same source code, these applications could be used for cross system development from desktop PCs through to CAVEs [11].

The new additions presented here consist of a configuration tool that provides non-programmers with a method of constructing a VE from various virtual objects, and the scene controller which is a management entity that administers dynamic aspects of a scene.

### 2.1 Scene Components

Generally speaking immersive VR can be used to represent 2 different types of data; visualization and realistic, lifelike scenes. Visualization concerns itself with the visual representation and manipulation of data sets. It does not try to persuade the user that the data is anything more than what it is, and immersion is only used to provide improved representation. Lifelike scenes however are used to persuade the user that they really are in an alternate place. The greater the user feels attached to the scene and has a presence within it, the more successful the VE and the greater its likelihood of achieving its desired results. This split does not mean visualization cannot be used in lifelike scenes; on the contrary data sets are often used in such scenes, however then the visualization object is just a part of the full scene.

Similarly to the number of objects that surround us in real life, a virtual scene consists of a heterogeneous array of different virtual objects. Creating a believable lifelike scene depends on incorporating the right objects for the VE being built.

Objects created from primitive geometries have basic shapes such as spheres, boxes and cylinders. They are easy to define but lack the complexity required for many scenes. However they often have uses representing basic man-made objects such as balls, walls, tubes and so on.

More complex complex objects can be created through CAD or modeling packages. Assuming they are available in, or can be converted to an acceptable format, they can be brought into the VE as realistic representations. For example, an architect could create a CAD model of a house he is designing and add it life size to an environment. He could then enter the environment and examine or demonstrate his design before any materials have even been ordered, allowing potential modifications if required.

Textures can be added to objects to improve realism, and range from simple patterns to high quality images. Such textures can greatly enhance the virtual objects, for instance a basic box can be covered in a texture of a previously photographed item to create a realistic virtual representation (see figure 1). Being image based, textures can be enhanced for providing additional effects

**Fig. 1.** CAVE applications demonstrating virtual shopping (left), multimedia conferencing (middle) and a virtual control room (right)

such as areas of transparency, lighting artifacts (reflections/shadows), aging or fogging. Textures do not need to remain static, and useful effects such as the illusion of passing time can be created from a dynamic sequences of different images.

While static objects are sufficient in a number of cases, it is often desirable for some degree of animation. In its simplest form objects can be made to move, either independently or resulting from user interaction. More advanced transformations could include scaling changes or morphing between objects.

## 2.2 Scene Configuration

The scope for potential VE development areas is vast. Its successful implementation would open huge avenues in many different fields. For instance it would become the perfect blank canvas for psychologists, artists and architects to name but a few. Using it they could create their environments tailored to their own specifications. However, these people generally do not possess the necessary VR programming skills to directly access the scene creation libraries. Simplified VR scripting languages (e.g VRML [12]) do exist, although being originally developed for creating/browsing 3D content on the Internet, have limited functionality in an immersive environment and lack widespread support for the various VR platforms. In an attempt to address the ease of creation issue, the Scene Configurator tool has been developed.

The Scene Configurator is a utility that creates a VE following some simple commands from a human readable configuration file. These commands define the objects that exist within the scene (geometries, models and any textural information), along with their location, animation and interaction characteristics. Using this approach non-programmers can create and edit the configuration files that manage their scenes. Currently proprietary plain text is used for the simplification of the commands, however it is anticipated that an additional text formating such as XML may be considered.

Having a common format for scene configuration enables further development of scene generation tools, enabling applications to store the state of their VE for backup or scene editing purposes. It also allows for automation in generating the scenes from from external data sources such as architectural design plans, shop-floor layouts and so on.

### 2.3 Scene control

An optional scene controller module has been added to the library of tools to free the application developer from some of the maintenance tasks within an environment. When used it creates and manages a registered hierarchy of virtual objects within the scene and can be used to provide animation and interaction. In addition to their geometries the objects have specific personalities effecting their behavior and enabling them to interact. Predefined personalities exist for these objects such as those that are manipulable (can attach to and be moved by a user or another object), mobile (they have their own rules determining how they move locally or around the scene) or are container objects (which can be used to hold/carry other objects). This base level of scene control can be provided to objects defined through the scene configurator. Additionally, further objects with extended characteristics and behavior can be derived from these base personalities.

Due to the underlying nature of the OpenSG scenegraph [13] that this work is built on, the object hierarchy has good support over distributed or multi-processor architectures. This coupled with state storage in persistent or shared memory allows for more complex objects, or those that have additional processing requirements to exist without hampering rendering performance in the scene. An simple example of this could involve the dynamic motion and interaction of objects exposed to multiple forces. The scene controller may only need positional information from the objects, and this could be retrieved with minimal overhead. However additional threads (and processors) could be constantly calculating the physics behind the objects' motion to determine the resultant positions.

## 3 User Interactions Within the Environment

An extension of a single user interactive environment would involve multiple users, using aspects of the environment as a communication medium. Current applications sometimes use avatars [14, 15] as virtual world representations of remote users as they share a VE. Here tracking devices follow user actions and map these onto their avatar counterpart. Using this method remote users can be seen to gesture or otherwise interact with the virtual world. However the expense of current tracking technology tends to result in a maximum of 2-3 tracked points for each person, thereby limiting its effectiveness.

A more acceptable method for interaction, and one that would help drive the uptake of VEs would be one that uses readily available technologies alongside the virtual world. Currently, most media formats revert to a 2D image such as those seen in books, televisions or displayed on PCs. While in the future it would be desirable for all users to be able to share and interact with 3D objects and models, right now these existing 2D formats can be incorporated into the 3D world. An example of this could be a virtual television or screen upon which could be displayed a traditional 2D image.

To facilitate this a particular type of virtual object has been developed with changeable textures each of which can represent a single frame/page of 2D data. This texture is read as needed from a shared memory segment attached to the object. The shared memory is written to by an external process based around the Multicast Application Sharing Tool (MAST) [16], which receives and decodes streams of application display data generated during group collaborations. More recently this tool has also been used for streaming face video. The resulting virtual object is therefore able to display a flat plane within the environment upon which can displayed live streams of video or application data. Since MAST supports multiple channels such an object could easily be used to represent a virtual monitor that could switch its display from one of many sources. Several of these could then be combined in the VE to form the basis of a multiple user multimedia conferencing system.

## 4 Applications

Several applications have already been developed from the described technologies, and demonstrate just some of the possible uses of these utilities. Figure 1 shows three of these applications running in a CAVE like environment.

The first of these is a virtual supermarket which allow users to pick up and examine different products. The whole scene has been developed with the scene configurator, so could easily have been created by a non-programmer. The store, the shelves and all the items on them as well as how they can be manipulated have been defined within the configuration file. The products and signs are simple geometries upon which textures created from photographs of the real objects have been added. Similar environments could just as easily be created to demonstrate alternative scenes, maybe museums containing artifacts that users could examine, or architectural designs to prototype new houses.

The second diagram shows a potential use of multimedia conferencing, in this case a virtual art class. MAST has been used to distribute application and face video from number of users. The teacher within the VE can view the students progress as they develop their drawing regardless of the application that each of them is using. The video and audio streams are provided to enhance communication between participants. This type of interaction could be used in many types of virtual meetings where audio, video and application data may wish to be shared amongst distributed users.

The final application shows a virtual environment being used as a security centred virtual control room. Multiple video streams from security cameras would be fed into the VE, and remote sensors could also be used to trigger alarm notification. The user from within the VE could be made aware of any intrusion, and enhance the respective video stream for further examination. Should it be necessary the user could notify the authorities while also having access to virtual models of the real environment to assist with any remedial action. Since it is unlikely that a user will be permanently required to monitor the same location, a virtual control centre could be established where a user could monitor many sites, each with their own virtual representation. If any of these sites require servicing then that VE could be brought up for the user's attention.

## 5 Conclusion

This work describes extensions to the VE development process that first open up the field to non-VR experts, and second provide a control entity that help in the management of dynamic virtual worlds. It discusses how virtual objects can be derived alongside these extensions to populate the environment with interactive items, and how these items can be used for multiuser collaboration and interaction. Finally it demonstrates applications that make use of this technology and suggests how they could be further enhanced.

Future work should extend the range of behaviors of the virtual objects and provide support for these within the scene configurator. Additional scene generation tools could provide automated or assisted development, potentially allowing extensive scene editing from within the VE. Improved collaborative interaction could also be developed through closer integration with MAST, potentially allowing bidirectional messaging for remote application communication and control.

## References


1. Macedonia, M.R., Zyda, M.J.: A taxonomy for networked virtual environments. IEEE MultiMedia **4** (1997) 48–56
2. Rohlf, J., Helman, J.: IRIS performer: A high performance multiprocessing toolkit for real-time 3d graphics. In: SIGGRAPH, ACM Press (1994) 381–394
3. Bierbaum, A.D.: VR Juggler: A virtual platform for virtual reality application development. Master's thesis, Iowa State University, Ames, Iowa (2000)
4. Radenkovic, M., Greenhalgh, C., Benford, S.: A scaleable audio service for CVEs. In: Proc of the sixth conference of the UK VRSIG. (1999)
5. Neumann, T., Fünfzig, C., Fellner, D.W.: TRIPS - a scalable spatial sound library for OpenSG. Technical Report TUBSCG-2003-02, Institute of ComputerGraphics (2003)
6. Robust Audio Tool (RAT) website. Available on: http://www-mice.cs.ucl.ac.uk/multimedia/software/rat
7. Teamspeak website. Available on: http://www.goteamspeak.com
8. Gonzalez, D.: AGJuggler: An architecture for virtual reality within a collaboration environment. Master's thesis, Purdue University (2005)
9. Childers, L., Disz, T., Olson, R., Papka, M., Stevens, R., Udeshi, T.: Access grid: Immersive group-to-group collaborative visualization. In: 4th International Immersive Projection Technology Workshop. (2000)
10. Haffegee, A., Jamieson, R., Anthes, C., Alexandrov, V.: Tools for collaborative VR application development. In: International Conference on Computational Science, Springer Verlag (2005) 350–358
11. Cruz-Neira, C., Sandin, D.J., Defanti, T.A., Kenyon, R.V., Hart, J.C.: The CAVE: Audio visual experience automatic virtual environment. Communications of the ACM **35** (1992) 64–72

12. Carey, R., Bell, G.: The VRML 2.0 annotated reference manual. Addison-Wesley, Reading, MA, USA (1997)
13. Reiners, D.: OpenSG: A Scene Graph System for Flexible and Efficient Realtime Rendering for Virtual and Augmented Reality Applications. PhD thesis, Technische Universität Darmstadt (2002)
14. Badler, N.I., Phillips, C.B., Webber, B.L.: Simulating Humans: Computer Graphics Animation and Control. Oxford University Press, New York, NY, USA (1992)
15. Park, K., Cho, Y., Krishnaprasad, N., Scharver, C., Lewis, M., Leigh, J., Johnson, A.: CAVERNsoft G2: A toolkit for high performance tele-immersive collaboration. In: VRST, Seoul, Korea, ACM Press (2000) 8–15
16. Lewis, G.J., Hassan, S.M., Alexandrov, V.N., Dove, M.T., Calleja, M.: Multicast application sharing tool - facilitating the eminerals virtual organization. In: International Conference on Computational Science, Springer Verlag (2005) 359–366

# Using Haptics to Improve Immersion in Virtual Environments

Priscilla Ramsamy, Adrian Haffegee, Ronan Jamieson, and Vassil Alexandrov

Centre for Advanced Computing and Emerging Technologies,
The University of Reading, Reading, RG6 6AY,
United Kingdom
sir04pr@reading.ac.uk

**Abstract.** Current immersive Virtual Reality (VR) system strategies do not fully support dynamic Human Computer Interaction (HCI) and since there is a growing need for better immersion, due consideration should be given to integrate additional modalities for improved HCI. While feedback in Virtual Environments (VE) is predominantly provided to the user through the visual and auditory channels, additional modalities such as haptics can increase the sense of presence and efficiency in VE simulations. Haptic interfaces can enhance the VE interaction by enabling users to "touch" and "feel" virtual objects that are simulated in the environment. This paper examines the reasons behind its integration based on the limitations of present immersive projection system.

**Keywords:** Virtual Reality, Haptics, Tactile feedback, Force feedback.

## 1 Introduction

There are three main categories of virtual reality systems whereby each implementation is ranked by the sense of immersion or the degree of presence it provides to the user. While moving from desktop systems (non-immersive) through semi-immersive (power-walls) to finally, a fully immersive system (e.g. the CAVE [20]) end users are provided with a much richer and engaging experience [2].

Virtual reality permits the user to 'step-into' a computer generated world or a Virtual Environment by immersing the user in the synthetic world. Considering that virtual worlds are completely conceived and created by the user, it is under their complete control and is therefore not confined to the laws of physics [3]. Human interaction capabilities are restricted if VEs deprive the users of sensorial cues that they experience in the real world; not only do we rely on our visual and auditory cues but we also depend on information that is conveyed through touch and force. These haptic cues complement the usual visual (graphics) and sound feedback modalities used in current VR simulations [1]. As such, providing greater realism to users could be achieved by integrating these sensory cues during the manipulation and interaction of virtual objects in VEs. A virtual object is a representation of an object in the virtual environment.

V.N. Alexandrov et al. (Eds.): ICCS 2006, Part II, LNCS 3992, pp. 603–609, 2006.

## 2 Haptics and Limitations of Our Current System

Haptic is a term derived from the Greek "haptesthai" which means "to come in contact with" by providing the sense of touch with both tactile (cutaneos) and kinaesthetic (proprioception) feedback [4]. Tactile, or touch feedback describes the sensations felt by the skin. Tactile feedback permits users to feel things such as the texture of surfaces, temperature and vibration. Kinaesthetic force feedback on the other hand enables us to recognise a force that is applied to our body by the help of sensoric cells located at the end of the tendons or between the muscle strands. This permits us to measure the weight of grasped virtual objects.

Haptics can be further broken down into three different areas.

1. **Human haptics**: Understanding the roles played by the mechanical, sensory, motor and cognitive subsystem of the human haptic system.
2. **Machine haptics**: Constructing haptic hardware device interfaces to replace or enhance the human touch.
3. **Computer haptics**: Considers the techniques to generate and display touch and feel sensations to users by using a force reflecting device. It considers both the behaviour of virtual objects as well as the rendering algorithm for the time displays.

Input devices such as computer keyboard, mice and wands can convey the user's commands to the computer but are unable to provide a natural sense of touch and feel to the user. It is crucial to understand the nature of touch interaction, since we react to our surroundings based on what and how we perceive things [5]. A study conducted by researchers at Texas University to identify the effects of the visual component of haptics proved to show, that by adding sensory cues end-users were able to identify hidden objects. The tests carried out consisted of introducing a 3-dimensional object that could be made invisible [6]. By separating the haptic and the visual components the user had to rely on the sense of touch to identify the object in question. The results obtained made it possible to quantify the effectiveness of using in this case, point-based haptic devices and was considered to be beneficial for applications for visually impaired users. The user's sensorial experience in a virtual environment is a fusion of information received through his perception channels. Based on the above observations and on[1][7][8][9] there is great potential to improve VR simulations. By adding haptic enabled interfaces in our current immersive projection system at Reading University it is hoped that by providing tactile and force feedback human performance and efficiency can be enhanced if the realism of the VE is closer to the physical world. The additional information from haptic feedback makes certain tasks such as manipulation much easier compared to that of the traditional computer interface [5][11]. The next section provides an overview of haptic rendering techniques.

## 3 Haptic Rendering Algorithms

The haptic algorithm consists of 2 steps:

1. A *collision detection* step that determines if the user's pointer is colliding with a virtual object. Information about the extent of penetrations and where collisions have occurred is determined during this step.
2. A *collision response* step that computes the interaction force between the user's pointer and virtual objects when a collision is detected. The return values are normally force and torque vectors that are applied to the haptic interface. If there is a collision, the contact information is sent to the collision response algorithm which generates the relevant tactile feedback based on the force model used.

There are several collision detection algorithms and each one of them possesses certain advantages in different situations. The choice of the algorithm will depend on the speed of calculation to detect the collision and also on the haptic interface used. For the purpose of our implementation we will consider using the friction cone algorithm with the associated face transition algorithms discussed in [12].

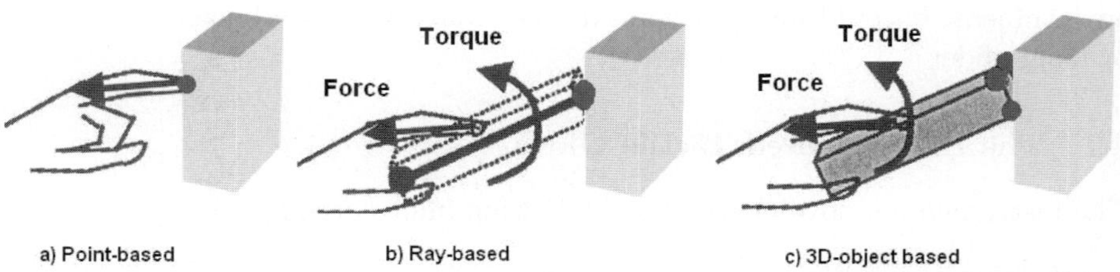


**Fig. 1.** Interaction methods [5]

Haptic rendering with force display also depends on the techniques used to model the probing object. Figure 1 provides models of the probing object.

- The simplest one is the point-based model Figure 1a where the probe is modelled as a point. Here only the tip of the end-effector is used to explore and manipulate objects. Checks are made to verify whether the tip of the end-effector is inside the virtual object and if so, the penetration depth is calculated based on the difference between the current position of the probe and the surface contact point.
- Ray-based interaction Figure 1b is where the probe is represented as a line segment. In this case the entire length of the line segment as well as its tip is used to explore and manipulate objects.
- The probe can be modelled as a 3D-object and is made up of points, line segments and polygons as shown in Figure 1c. Collisions between vertices of its geometry and the objects of the scene are checked every time the probe is moved. This is the most realistic technique but is also the most computationally expensive one [5][14][15][16].

The type of interaction method used in simulations greatly depends on the needs and complexity of the application. Moving from point based, ray-based to 3D-object we increase the accuracy but while so doing we would require intensive computation to detect collisions [13][17]. We therefore have to consider the trade offs between realism of the VEs and computational costs. For our implementation we will consider the ray-based technique. The next section covers some of current haptic device operations that have been implemented.

## 4 Current Feedback Device Operations

The device operations are methods that are concerned with setting and getting state. The device will require initialisation before use and will be initialised by setting the values to zero. In our implementation we have a instrumented tactile glove and a grounded haptic mobile platform both are currently being developed at Reading University. The tactile glove can replace or be used in conjuction with the traditional wand to provide a more realistic and intuitive way of manipulating virtual objects. The haptic mobile platform will be replacing the traditional joystick and will permit the user to navigate within the virtual environment. It could mimic several devices ranging from a bicycle to a trolley or hang glider.

### 4.1 The Instrumented Tactile Glove

The instrumented glove provides the following information:

- Position of the hand
- Finger angles - provides the information on whether the hand is clenched or open. To be noted that in our implementation of the haptic glove only three fingers (thumb, index and the middle finger) are actively being used and calibrated.

The device state and other information will be retrieved by using the haptic glove functions. The functions return -1 on failure indicating a hardware or network problem and 0 on success. The existing functions can retrieve the hand orientation and finger angles. Based on these values and on a collision detection algorithm we can render the appropriate feedbacks. If collision is detected with a manipulatable object the person would be able to interact with and manipulate the object. While the grasp aperture (the distance between thumb and fingers) is less than a critical distance (Gc) the object will be attached to the glove. If the grasp aperture is greater than Gc then we apply object physics to the object. Gc could either be set to a fixed value or vary depending on the size of the object being grasped. In order to produce the tactile cues on contact with virtual objects we need to generate some kind of reaction to the user's skin. While the skin responds to several distributed physical quantities such as high-frequency vibrations, pressure distribution and thermal properties in our first attempt, we will incorporate a vibro-tactile actuator in the centre of the palm.

The actuator will vibrate informing the user when he/she has made first contact with the object upon which the user can then manipulate the virtual object by picking it up (clenching the hand).

### 4.2 Grounded Haptic Mobile Platform

The grounded haptic mobile platform provides information on the following:

- The forward and backward velocity,
- The platform turn rates
- The amount of inertia, damping and friction - this would be dynamic depending on the force applied by the end-user onto the mobile platform handle.

The haptic enabled mobile platform functions will have similar error checks to that of the tactile glove. The class functions can retrieve the force applied, the displacement forward or backwards and the angle of rotation and can set the relevant constraints on the grounded mobile platform. If the mobile platform comes in contact with a surface in order to imitate what is experienced in real life, it should not be able to rotate or move forward. The programmer should also be able to query the forces applied as well as being able to define and set new forces to the object (in this case the mobile platform).

## 5 Future Work

We are currently in the early stages of the project and the main focus has been on the development and integration of the haptic hardware interface. Our future work will concentrate on the refinement of the current haptic device operations. We could enhance and provide more functionalities to the instrumented glove by increasing the number of vibrotactile stimulators (incorporating them onto each finger and not only in the palm) where, each stimulator can be individually programmed to vary the strength of touch sensation. Complex tactile feedback patterns can be produced by using a combination of stimulators to generate simple sensations such as pulses or sustained vibration. This would bring an enhanced realistic mode of interaction in immersive VEs whereby the user can pick up and maniputate for instance objects in a the virtual museum, it can also be used during the simulation of complex surgeries [18] by rendering accurate tactile information and by providing a natural and more intuitive way of control and manipulation.

Development of our own haptic library to provide collision detection or the integration of existing haptic rendering software subsystems will also be considered. Future investigation is required as to the possibility of using open source haptic software frameworks that are available for haptic programming and control of haptic force feedback devices. Other issues that can be considered is the fact that haptic rendering complex objects in virtual environments is computationally intensive and therefore due consideration should be given to the reduction of computational cost. We should have the ability to control the Level of Detail (LoD) of objects in VEs by substituting fine models with coarser ones to reduce the computation cost [19].

## 6 Conclusion

While several architectures have been reviewed it has been found that fewer fully immersive systems make use and integrate haptic feedback devices. By integrating sensorial cues in our immersive projection VE system, end-users would be provided with a more intuitive and efficient way of manipulating and interacting with virtual objects. The interface has been successfully developed and permits both the tactile and force feedback devices to integrate with our immersive system, and is being used in ongoing application development.

## Acknowledgements

I would like to thank Dr William Harwin and those at ISRG for their help.

## References

1. Burdea, G.C. Haptics issues in virtual environments. In Computer Graphics International,Proceedings. (2000)
2. P.J. Costello, Health and Safety Issues associated with Virtual Reality - A Review of Current Literature. Advisory Group on Computer Graphics (AGOCG) Technical Reports, July (1997)
3. K.P. Beier, Virtual Reality: A Short Introduction. http://www-vrl.umich. www-vrl.umich.edu/intro/
4. M. Mokhtari, F. Bernier, F. Lemieux, H. Martel, J.M. Schwartz, D. Laurendeau and A. Branzan-Albu. Virtual Environment and Sensori-Motor Activities: Haptic, Audition and Olfaction. WSCG POSTERS proceedings Plzen, Czech Republic., Vol.12, No.1-3, February, (2004) 2-6
5. M.A. Srinivasan and C. Basdogan, "Haptics in Virtual Environments: Taxonomy, Research Status, and Challenges," Computers and Graphics, Special Issue on Haptic Displays in Virtual Environments, Vol. 21, No. 4, (1997)
6. Eric Acosta, Bharti Temkin, "Touch&Tell: A game-based tool for learning to use the PHANToM," Seventh PHANToM Users Group Workshop, October (2002) 26-29
7. F. P. Brooks, M. Ouh-Young, J. J. Batter, and P. J. Kilpatrick. Project GROPE – Haptic Displays for Scientific Visualization. In Proc. ACM SIGGRAPH, Dallas, TX, Aug (1990) 177–185
8. Michael Dinsmore, Noshir Langrana, Grigore Burdea, Jumoke Ladeji. "Virtual Reality Training Simulation for Palpation of Subsurface Tumors," vrais, Virtual Reality Annual International Symposium, (1997) 54
9. Burdea, G., G. Patounakis, V. Popescu, & R. E. Weiss. Virtual Reality Training for the Diagnosis of Prostate Cancer. In IEEE International Symposium on Virtual Reality and Applications, Atlanta, Georgia, March (1998) 190-197
10. Burdea G. Haptic Interfaces for Virtual Reality. Proceedings of International Workshop on Virtual prototyping, Laval, France, May (1999) 87-96
11. Andrew G. Fischer, Judy M. Vance, "Implementing Haptic Feedback in a Projection Screen Virtual Environment," Seventh PHANToM Users Group Workshop, October (2002) 26-29

workspace of the interface, Iwata [2] suggested the principal of cancellation: while one platform will follow one foot during the swing phase, the other platform will move back the other foot. However, there are no general algorithms for navigation of locomotion interface using robotic manipulators. Therefore, it is necessary to develop a generalized navigation control algorithm that allows a user to interact with virtual environments while walking over various terrains using 6-dof programmable devices. In this paper, control and navigation algorithms that can allow continuous walking over various terrains using the proposed interface are developed.

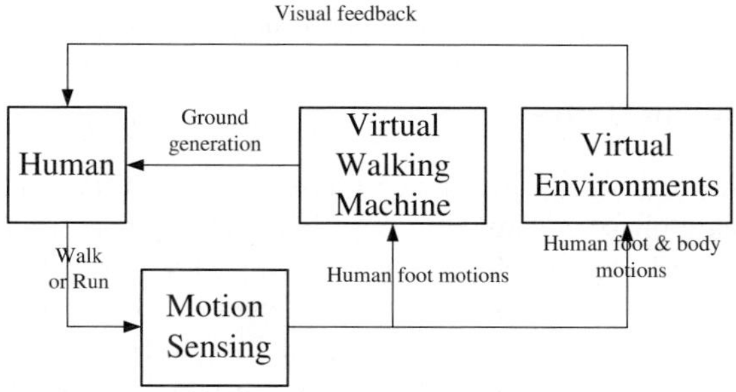


**Fig. 1.** Navigation using a virtual walking machine

## 2 Locomotion Interface

### 2.1 Overview

The suggested locomotion interface is composed of two planar devices on which two footpad devices are mounted. The planar device [4] is planar parallel manipulator that is composed of a platform and three limbs, each of which has three serial revolute joints (RRR) with the actuated first joint. The 3-dof footpad device [5] is composed of platform, and two limbs. The footpad device can generate pitch, roll, and heave motions at the platform.

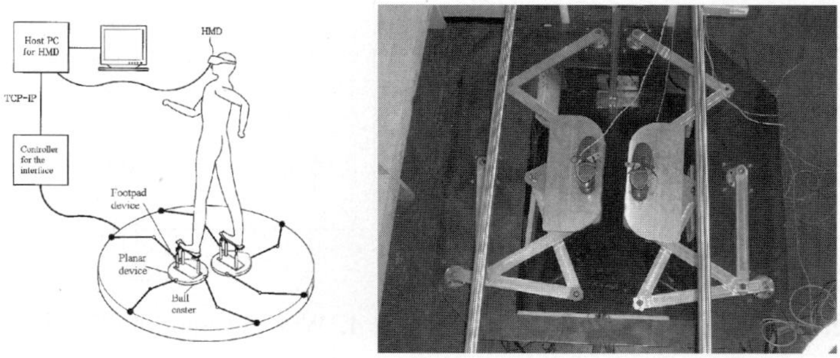


**Fig. 2.** The locomotion Interface with programmable foot platforms

Fig. 2 shows the structure of the proposed locomotion interface which is based on thorough understanding of the human gait. The user standing on the interface can walk and interact with the virtual environment while wearing a Head Mount Display (HMD), or watching big display screen. The position and orientation of a human foot can be measured using a Polhemus 3D magnetic tracker, which is tightly connected to the shoe so that it should precisely trace the foot motion without delay.

# 3 Walking Control Algorithm

The locomotion interface control system should enable a user to walk smoothly and continuously in a confined area. Thus, the control algorithm should be designed to keep the position of the human at a neutral position during walking.

## 3.1 New Cancellation Method

For a single normal gait cycle, the stance phase accounts for approximately 60 percent, while the swing phase accounts for approximately 40 percent. It should be noted however that a double support phase exists during which both limbs are in contact with the ground. During this phase, the body's center of gravity is at its lowest position. These double supports happen during initial 10% and final 10% of stance phase. Therefore, we suggest new cancellations method, in which the walking motions consider double stance phase. Thus, each platform will follow the magnetic tracker attached to a foot during swing phase when human foot is moving forward without contacting any object, while the other platform will move back during single stance phase when only one foot is in contact with ground. If two feet are in contact with the platforms, the two platforms will keep their current positions. The transitions between swing and stance phase are detected by using switch sensor system exerted by the human foot.

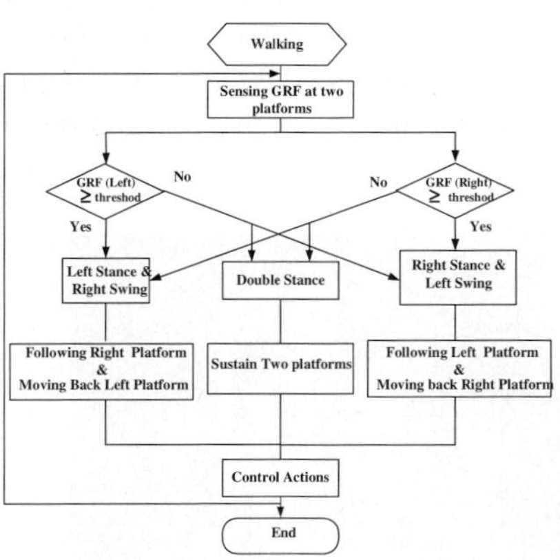


Fig. 3. New cancellation method

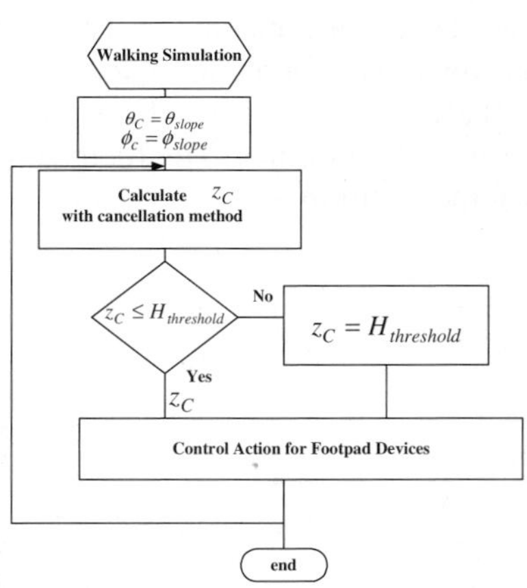


Fig. 4. Walking algorithm for spatial motions

Fig. 3 shows the block diagram of the proposed cancellation method. The proposed cancellation algorithm can allow a user to stop and start naturally according to user's intentions because of the added double support phases. Therefore, this algorithm will allow more natural walking on any programmable locomotion interfaces, satisfying normal gait conditions.

The proposed cancellation method is basically designed for natural walking with constant speed. However, if there is a velocity change, a human foot may be in the swing phase before or after the backward movement to the same positions with average velocities of previous swing phase during single stance phase. Due to this reason, home positioning errors generated by velocity change are inevitable. Therefore, by moving center positions of the two platforms to home positions without changing the relative positions between the two platforms, the neutral positions can be maintained although a user changes walking velocities.

### 3.2 Spatial Motions

For spatial motions of the locomotion interface, the footpad device will be used to generate various terrains such as stairs and slopes. In order to simulate stairs, the platform of the footpad device should have zero angles since the stairs have no slope. Therefore, cancellation method about planar motions can also be applied to lift motion control. If the $z_C$ is higher than the ground height $H_{threshold}$ , the command lift motion $z_C$ of the footpad device will be $H_{threshold}$ to keep stairs surface. This algorithm for spatial motions is shown in Fig. 4. For slope surface generation, if the pitch angles of the footpad device have constant values and the roll angle is zero, the surface will be uphill or downhill slopes. If ground has up-slope, the pitch angle should be positive and the ground height should be increased as human foot proceeds in forward direction as shown in Fig. 5. Therefore, to sustain the contact of the human foot with the ground at slope surfaces, the ground height threshold $H_{thereshold}$ should be computed in equation (1), while the same walking scheme for planar motions and lift motions will be applied to retreat the human foot back for continuous walking.

$$\theta_C = \theta_{slope} \, , \; \phi_C = \phi_{slope} \, , \; H_{threshold} = H_{ground} + y_C \tan(\theta) \tag{1}$$

where $\theta_C$ and $\phi_C$ are the desired pitch and roll angles of the footpad device, and $y_C$ is the back-and-forth desired control command of a human foot. This walking control algorithm, therefore, will sustain continuous walking over various terrains with the 6-dof locomotion interface in a limited area.

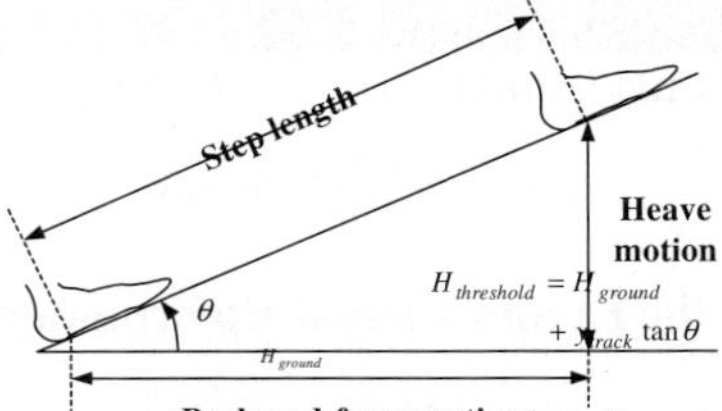


**Fig. 5.** The walking surface generation on slopes

# 4 Navigation Using the Locomotion Interface

## 4.1 Virtual Environments

Virtual terrains (see Fig. 6) have been developed by using OpenGL API (Application Program Interface) based on Microsoft Visual C++ programs for window environment. The viewpoint is selected as the first person viewpoint so that operator may see virtual environment and feel more realistic. Virtual environments are displayed with large screen to guarantee safety of the subject during the locomotion interactions. Even though HMD (Head Mounted Display) can provide full sight of virtual environment, the wearing of the HMD may prevent the user from coping with dangerous situations during walking interactions on the locomotion interface.

The virtual environment consists of three components:3D object modeler for buildings and terrains, scene manager and communication manager component. The 3D object modeler component provides geometric modeling functionality for virtual environment through polygon extrusion of 2D profile in the 2D map and 3D CAD model. The 3D model can be created for 2D map by extruding 2D profile geometry with *b* and *m*. In simple features specification of the Open GIS Consortium (OGC)[6] , the Well-Known Binary (WKB) representation for geometry provides a portable representation of a geometry value as a contiguous stream of bytes. Also, WKB representation is used to store geometry information as in OGC simple features specification.

The scene manager component manages the scene graph for real-time rendering according to the VWM requests. The communication manager component manages the communication between virtual walking machine and virtual environment via remote procedure call (RPC) to interact the VWM user. The interface using the interface definition language (IDL) for RPC is defined to communicate between the VWM and virtual environment. The proposed RPC interface provides virtual environment switching and interaction functions.

(a) Upright navigation (b) Slope navigation

**Fig. 6.** Virtual terrains

## 4.2 Interaction Between the LI and Virtual Environments

There are two types of RPC methods for interaction: position-based interaction and action-based interaction. The *position-based interaction* performs the navigation in virtual environments according to the position changes in local coordinate system of the VWM, while the action-based interaction performs relative motion at current virtual environments according to the velocity changes of the VWM. Since there are

many virtual environments with various terrains, the *action-based interaction* is easy to control the virtual environments in that it performs the navigation according to the VWM user actions rather than the position changes. Therefore, the walking velocity of the user will be directly translated to VR actions for navigation. The update rate for the VWM velocity is selected to be 20Hz, which is enough to show smooth walking actions. Then, the user walking velocity is estimated as;
Single Right Stance Phase:

$$V_{walking} = V_{avg,L}^{sw} \tag{2}$$

Single Left Stance Phase:

$$V_{walking} = V_{avg,R}^{sw} \tag{3}$$

Double Stance Phase:

$$V_{walking} = 0 \tag{4}$$

where $V_{avg,i}^{sw}$ is the average velocity during the swing phase and the $V_{walking}$ is the estimated walking velocity of a user on the locomotion interface. Since human foot moves forward only during swing phase, the average velocity of the human walking can be estimated as average velocity during only swing phase motion. Therefore, after the each foot moves forward and when the foot contacts on the platform, the walking velocity will be updated. During double limb stance, since the foot will not move forward, the velocity should be set to zero. Finally, the functions of the RPC interface shown in Fig.7 are utilized for each interaction mode.

The functions ***Forward*** *(speed)* and ***Backward*** *(speed)* for back-and forth motions will move the current camera position to the next camera position towards the direction vector of the current camera. The *speed* will determine differential displacement during sampling time. Similarly, the functions ***UpStair****(speed)* and ***DownStair****(speed)* will change the camera position of virtual environments for z direction. The functions ***Look-Left*** *(angle),* ***LookRight*** *(angle),* ***LookUp*** *(angle),*and ***LookDown*** *(angle)* are defined to change directions of the camera, which will be connected to HMD(Head Mounted Display) to reflect the change in user's view in virtual environments.

```
// switch various virtual environments such as floor, sloop, turning and stairway.
void SetEnvMode (int mode);

// control position-based interaction.
void SetVWMParam (VWMPosition point, VWMOrientation orientation);
void SetVWMCameraParams (VWMPosition focal, VWMPosition point, VWMOrientation orientation);

// control action-based interaction.
void Forward (float speed);
void Backward (float speed);
void LookLeft (float angle);
void LookRight (float angle);
void LookUp (float angle);
void LookDown (float angle);
void UpStair (float speed, float ratio);
void DownStair (float speed, float ratio);
```

**Fig. 7.** RPC interface for interaction between the VWM and virtual environments

### 4.3 Evaluations

For the preliminary interface evaluation, five subjects among our laboratory students participated in walking over the designed locomotion interface device. The subjects who have no experience of the walking on the interface were instructed how to walk over the walking interface. Since there is velocity limitation (maximum 1.2m/s) for the locomotion interface, they were requested to walk with normal speeds that they generally walk. Even though the number of subjects is small, it is useful for discussing the overall performance of the walking interface. Initially, most of them were afraid of the waking on the interface but after trial walking, they were able to walk without anxiety. Fig. 8 (a) shows the scores of the each item (safety, reality, and overall) for upright walking with respect to subjects A…E. After interactions with the walking device, they scored each item. Most of them are satisfied with the walking control algorithm since the algorithm has been developed based on the real human gait motions. But, they indicated that for moving back motions during stance phase, they felt the body inertia of moving back motions due to the neutral home positioning during double stance phase. For overall evaluations, subjects generally were satisfied with the reality of the walking with the interface device. For slope walking on the interface shown in Fig. 8 (b), the subjects were more positive for real walking in that there was not much of the body inertia for backward motions, especially for the back-and-forth motions. Since real bodies were moving up and moving down for walking interaction on slope grounds, they felt that their walking motions were more similar to real walking. For turning motions [7-8], since the two platforms may have possibility to collide each other when the platform follows the human foot, the only passive motions are allowed.

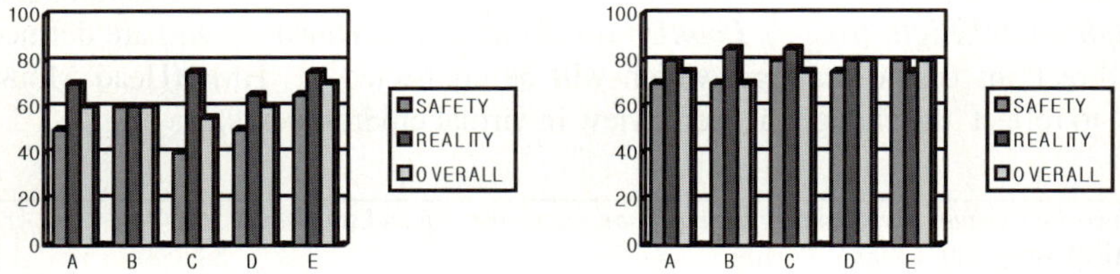


(a) upright walking (b) slope walking

**Fig. 8.** Evaluation results

With preliminary evaluations, some important understandings of the developed locomotion interface are addressed. The walking control and navigation algorithm was operating well with the proposed novel programmable locomotion interface. The walking interactions with the upright and slope grounds were successful on the locomotion interface.

## 5 Conclusions[1] and Future Work

This paper proposes novel navigation system that can induce user's real walking and generate realistic visual feedback during navigation, using robotic manipulators. For realistic visual feedback, the virtual environment is designed with three components; 3D object modeler for buildings and terrains, scene manager and communication manager component. The suggested navigation system can allow a user to explore into various virtual terrains with real walking and realistic visual feedback. As future works, the various haptic effects such as soft and hard grounds, or slippage will be simulated by changing the impedance parameters. Also, comparison between VWM and joystick navigation will be performed. After enhancing the control performances and the safety of the VWM, the final goal is to let user wear the HMD for full immersion in the virtual navigation with natural walking.

## References


1. Hollerbach, M.,"Locomotion interfaces," in: *Handbook of Virtual Environments Technology,* K.M. Stanney, ed., Lawrence Erlbaum Associates, Inc. (2002) 239-254.
2. Iwata, H., Yano, H., and Nakaizumi, F., Gait Master: a versatile locomotion interface for uneven virtual terrain, Prof. of IEEE Virtual Reality (2001) 131 - 137.
3. Schmidt, H., Sorowka, D., Hesse, R., and Bernhardt, R., Design of a Robotic Walking Simulator for Neurological Rehabilitation, IEEE/RSJ Int. Conf. On Intelligent Robots and Systems (2002) 1487-1492.
4. Yoon, J. and Ryu, J., The Development of the 3-DOF Planar Parallel Robot (RRR Type) for Omni-directional Locomotion Interface, 3rd IFAC Symposium on Mechatronic Systems, Sept.6-8, Sydney, Australia, 2004.
5. Yoon, J. and Ryu, J., A New Family of 4- Dof Hybrid Parallel Mechanisms with Two Platforms and Its Application to a Footpad Device, Journal of Robotic Systems, 22:(5) (2005) 287-298.
6. Open GIS Consortium, OpenGIS Simple Feature Specification for OLE/COM, OpenGIS Implementation Specifications, Revision 1.1, 1999.
7. Bouguila, L., Ishii, M., and Sato, M., Realizing a New Step-in-place Locomotion Interface for Virtual Environment with Large Display System. Proceedings of the Workshop on Virtual Environments (2002) 197-207.
8. http://www.virtusphere.com.


[1] Research reported here was supported by grant (No. R01-2002-000-00254-0) from the Basic Research Program of the Korea Science & Engineering Foundation and by the Korea Research Foundation Grant funded by the Korean Government(MOEHRD)" (KRF-2005-005-J09902).

# Efficient and Practical Algorithms for Deducing the History of Recombination in Populations
## (Keynote Talk Abstract)

Dan Gusfield

University of California, Davis
gusfield@cs.ucdavis.edu

A phylogenetic network (or Ancestral Recombination Graph) is a generalization of a tree, allowing structural properties that are not tree-like. With the growth of genomic and population data, much of which does not fit ideal tree models, and the increasing appreciation of the genomic role of such phenomena as recombination (crossing-over and gene-conversion), recurrent and back mutation, horizontal gene transfer, and mobile genetic elements, there is greater need to understand the algorithmics and combinatorics of phylogenetic networks.

In this talk I will survey a range of our recent results on phylogenetic networks with recombination, and show applications of these results to association mapping; finding recombination hotspots in genotype sequences; imputing the values of missing haplotype data; determining the extent of recombination in the evolution of LPL sequences; distinguishing the role of crossing-over from gene-conversion in Arabidopsis; and characterizing some aspects of the haplotypes produced by the program PHASE.

I will discuss the fundamental problem of constructing a phylogenetic network for a given set of binary (SNP) sequences derived from a known or unknown ancestral sequence, when each site in the sequence can mutate at most once in the network (the infinite sites model in population genetics), but recombination between sequences is allowed. The goal is to find a phylogenetic network that generates the given set of sequences, minimizing the number of recombination events used in the network. When all the "recombination cycles" are disjoint (which is likely with low recombination rates), we have developed efficient provably correct algorithms that find a network minimizing the number of recombinations, and have proven that the optimal solution is "essentially unique". I will also mention a network decomposition theory that shows the extent that these results can be generalized to arbitrary networks. For general phylogenetic networks (when the cycles are not constrained) we have developed algorithms that are efficient in practice and that empirically obtain close upper and lower bounds on the number of recombinations needed. In real data and simulations, these practical (heuristic) computations often produce bounds that match, demonstrating that a minimum recombination solution has been found. In small-size data we can guarantee that an optimal solution will been found by running an exponential-time algorithm to completion. For small-size data, we can also (provably) sample uniformly from the set of optimal solutions, and can also determine the phase of genotypic data so as to minimize the number of recombinations needed to derive

V.N. Alexandrov et al. (Eds.): ICCS 2006, Part II, LNCS 3992, pp. 618–619, 2006.

the resulting haplotypes. More recently, we have extended the lower and upper bound algorithms to incorporate gene-conversion as an allowed operation. Those new algorithms allow us to investigate and distinguish the role of gene-conversion from single-crossover recombination.

Various parts of this work are joint work with Satish Eddhu, Chuck Langley, Dean Hickerson, Yun Song, Yufeng Wu, V. Bansal, and Zhihong Ding. Support for this work was provided by NSF grant IIS-0513910. All the papers and associated software can be accesses at wwwcsif/cs.ucdavis.edu/~gusfield.

## References

1. Gusfield, D., Eddhu, S., Langley, C.: Optimal, efficient reconstruction of phylogenetic networks with constrained recombination. Journal of Bioinformatics and Computational Biology **2**(1) (2004) 173–213
2. Gusfield, D., Eddhu, S., Langley, C.: The fine structure of galls in phylogenetic networks. INFORMS J. on Computing, Special issue on Computational Biology **16**(4) (2004) 459–469
3. Gusfield, D.: Optimal, efficient reconstruction of root-unknown phylogenetic networks with constrained recombination. J. Computer and Systems Sciences, Special issue on Computational Biology **70** (2005) 381–398
4. Gusfield, D., Bansal, V.: A fundamental decomposition theory for phylogenetic networks and incompatible characters. In: Proc. 9th Annual International Conference on Computational Biology. Volume 3500 of Lecture Notes in Bioinformatics. (2005) 217–232
5. Gusfield, D., Hickerson, D., Eddhu, S.: A fundamental, efficiently computed lower bound on the number of recombinations needed in a phylogenetic history. Discrete Applied Math, Special issue on Computational Biology (to appear)
6. Song, Y., Wu, Y., Gusfield, D.: Efficient computation of close lower and upper bounds on the minimum number of needed recombinations in the evolution of biological sequences. Bioinformatics **21**(Supplement 1, Proceedings of the ISMB 2005 Conference) (2005) 413–422
7. Song, Y., Gusfield, D., Ding, Z., Langley, C., Wu, Y.: Algorithms to distinguish the role of gene-conversion from single-crossover recombination in the derivation of snp sequences in populations. In: Proc. 10th Annual International Conference on Computational Biology. Volume 3909 of Lecture Notes in Bioinformatics. (2006) 231–245 (to appear)

# Chordal Graphs in Computational Biology - New Insights and Applications

### (Keynote Talk Abstract)

Teresa M. Przytycka

NCBI/NLM/NIH
przytyck@ncbi.nlm.nih.gov

Recent advances in experimental techniques resulted in accumulation of vast amounts of information, which is often represented by various types of biological networks. Therefore it is not surprising that increasingly more complex graph-theoretical tools are developed to analyze such biological networks and extract biologically meaningful patterns. In this talk I will describe the research done in my group directed towards computational analysis of biological networks.

Graph theoretical representation of biological relationships provides insight into the evolution of biological systems and their function. For example, in order to focus on the properties of multi domain proteins and the relationships between them, we introduced and studied graph theoretical representation of multidomain proteins called domain overlap graph. In the domain overlap graph, the vertices are protein domains and two domains are connected by an edge if there is a protein that contains both domains. We demonstrate how properties of this graph such as chordality and the Helly property can indicate various evolutionary mechanisms [1].

The concept of domain overlap graph can be seen as an example of a more general construction, the construction of character overlap graphs. We show that character overlap graphs for characters that are appropriate to use in parsimony methods are characterized by significant under-representation of holes, and thus are relatively close to chordal graphs. This characterization explains success in constructing evolutionary trees using parsimony method for some characters (e.g. protein domains) and lack of such success for other characters (e.g. introns). In the latter case, the understanding of mathematical obstacles to applying the parsimony method in a direct way has lead us to a new algorithm that is able to bypass these obstacles [2].

A major challenge in systems biology is to understand the intricate network of interacting molecules. The complexity in biological systems arises not only from various individual protein molecules but also from their organization into systems with numerous interacting partners forming protein complexes and functional modules. We focus on the analysis of protein-protein interaction networks directed towards recovering temporal relation and overlaps between functional groups. We developed a graph-theoretical framework, based on properties of chordal graphs and cographs [3]. We apply our approach to delineate pheromone signaling pathway from the high throughput protein-protein interaction network.

V.N. Alexandrov et al. (Eds.): ICCS 2006, Part II, LNCS 3992, pp. 620–621, 2006.

## Acknowledgement

This work was supported by the intramural research program of the National Institutes of Health, National Library of Medicine.

## References

1. Przytycka, T.M., Davis, G., Song, N., Durand, D.: Graph Theoretical Insights into Multidomain Protein Evolution. Proc. RECOMB 2005, Lecture Notes in Computer Science 3500 (2005) 311–325.
2. Przytycka, T.M.: An important connection between network motifs and parsimony models. Proc. RECOMB 2006, to appear (2006).
3. Zotenko, E., Guimaraes, K.S., Jothi, R., Przytycka, T.M.: Decomposition of Overlapping Protein Complexes: A Graph Theoretical Method for Analyzing Static and Dynamic Protein Associations. Proc. 1st RECOMB Satellite Meeting on Systems Biology, to appear (2006).

# Exemplar Longest Common Subsequence

Paola Bonizzoni[1], Gianluca Della Vedova[2], Riccardo Dondi[1], Guillaume Fertin[3], and Stéphane Vialette[4]

[1] Dipartimento di Informatica, Sistemistica e Comunicazione, Università degli Studi di Milano-Bicocca Milano - Italy
[2] Dipartimento di Statistica, Università degli Studi di Milano-Bicocca Milano - Italy
[3] LINA - FRE CNRS 2729 Université de Nantes, Nantes Cedex 3, France
[4] LRI - UMR CNRS 8623 Faculté des Sciences d'Orsay, Université Paris-Sud Bât 490, Orsay Cedex, France

**Abstract.** In the paper we investigate the computational and approximation complexity of the Exemplar Longest Common Subsequence of a set of sequences (ELCS problem), a generalization of the Longest Common Subsequence problem, where the input sequences are over the union of two disjoint sets of symbols, a set of mandatory symbols and a set of optional symbols. We show that different versions of the problem are **APX**-hard even for instances with two sequences. Moreover, we show that the related problem of determining the existence of a feasible solution of the Exemplar Longest Common Subsequence of two sequences is **NP**-hard. On the positive side, efficient algorithms for the ELCS problem over instances of two sequences where each mandatory symbol can appear totally at most three times or the number of mandatory symbols is bounded by a constant are given.

## 1 Introduction

Algorithmic studies in comparative genomics have produced powerful tools for the analysis of genomic data which has been successfully applied in several contexts, from gene functional annotation to phylogenomics and whole genome comparison. A main goal in this research field is to explain differences in gene order in two (or more) genomes in terms of a limited number of rearrangement operations.

When there are no duplicates in the considered genomes, the computation of the similarity measure is usually polynomial-time solvable, *e.g.*, number of breakpoints, reversal distance for signed genomes, number of conserved intervals, number of common intervals, maximum adjacency disruption, summed adjacency disruption. However, aside a few exceptions, several copies of the same gene or several highly homologous genes are usually scattered across the genome, and hence it is major problem to handle those duplicates when computing the similarity between two genomes. One approach to overcoming this difficulty is based on the concept of *exemplar*: for each genome, an exemplar sequence is constructed by deleting all but one occurrence of each gene family. Another approach is based on *matching*: in this two-step procedure, the two genomes are

V.N. Alexandrov et al. (Eds.): ICCS 2006, Part II, LNCS 3992, pp. 622–629, 2006.

first made *balanced* (the number of occurrences of each gene from the same family must be the same in both genomes) by removing a minimum number of genes and next a one-to-one correspondence (among genes of each family) between genes of the genomes is computed.

Unfortunately, in the presence of duplicates, most similarity measures turn out to be **NP**-hard to compute for both the exemplar and the matching models, so that we generally have to rely on approximation algorithms or heuristic approaches. We discuss here one such general purpose heuristic approach (the EXEMPLAR LCS problem) which is basically a constrained string alignment problem. The basic idea of the general framework we propose here is based on the observation that, for most similarity measures and for both the exemplar and the matching models, specific common subsequences may correspond to highly conserved sets of genes. This suggests the following greedy heuristic algorithm: find a common subsequence of significant length – but compact enough – between the two genomes, replace in the two genomes the substring that contains the common subsequence (the substring that starts at the first character of the common subsequence and ends at the last character of the common subsequence) by a new letter and continue in a similar way.

At each iteration of this simple heuristic algorithm, one however has to be cautious in how to choose the common subsequence, as bad choices may have a disastrous impact for the rest of the algorithm. Let us take the exemplar model as a very simple explanatory example, and suppose that we are searching for a common subsequence between two precise substrings of the two genomes. For one, if one gene family has occurrences elsewhere in the two genomes, then taking or not one occurrence of this particular gene family in the common subsequence is thus not based on necessity but on the length of the obtained solution. For another, if there do not exist any other occurrences of one gene family except one in the two considered substrings, definitively one has to take this occurrence in the common subsequence (observe that in this case the obtained common subsequence may not be the longest one). This simple example suggests to consider a LCS-like problem that deals with two types of letters (*mandatory* and *optional* symbols) to allow greater flexibility in the searching process.

In this paper we will formally define such framework with a simple combinatorial problem that generalizes the well-known LCS problem and we will study its computational and approximation complexity. We show that different versions of the problem are **APX**-hard even for instances with two sequences and that even determining if a feasible solution exists or not is **NP**-hard. On a positive side the hardness of the problem can be limited in some cases, in fact we show that it is possible to determine efficiently a feasible solution, provided that each symbol appears at most three times totally in the input sequence. Finally we give a polynomial-time algorithm in case the number of mandatory symbols is a constant.

## 2 The Problems

The LONGEST COMMON SUBSEQUENCE problem (shortly LCS) is a well-known problem in Computational Biology. Let $s = s[1], s[2], \ldots, s[m]$ and $t = t[1], t[2], \ldots, t[l]$ be two sequences, $s$ is a subsequence of $t$ if for some $j_1 < j_2 < \ldots < j_m$ $s[h] = t[j_h]$. Let $s_1$, $s_2$ be two sequences, a *longest common subsequence* of $s_1$ and $s_2$ is a sequence $s$ of maximum length, such that $s$ is a subsequence of both $s_1$ and $s_2$. Let $S$ be a set of sequences, then a *longest common subsequence* of $S$ is a longest possible sequence $s$ that is a subsequence of each sequence in $S$.

**Table 1.** Versions of EXEMPLAR LCS

| Problem name | Occurrences mandatory symbols | Occurrences optional symbols |
|---|---|---|
| ELCS$(1, \leq 1)$ | exactly 1 | at most 1 |
| ELCS(1) | exactly 1 | unrestricted |
| ELCS$(\geq 1, \leq 1)$ | at least 1 | at most 1 |
| ELCS$(\geq 1)$ | at least 1 | unrestricted |

Given a set of sequences $S$, the LCS problem asks for a *longest common subsequences* of $S$. The complexity of LCS problem has been deeply studied in the past. In [7] it is shown that the problem is **NP**-hard even for sequences over an alphabet of size 2. However, when the instance of the problem consists of a fixed number of sequences, the LCS can be solved in polynomial time via dynamic programming algorithms [5, 4]. The input of the EXEMPLAR LCS problem (ELCS) consists of a set $S$ of sequences over alphabet $A_o \cup A_m$, $A_o \cap A_m = \emptyset$, where $A_o$ is the set of *optional* symbols and $A_m$ is the set of *mandatory* symbols. The output of the problem is a longest common subsequence of all sequences in $S$ and containing all mandatory symbols.

Given an instance $S$ of ELCS, by *exemplar common subsequence* we mean a feasible solution of ELCS over $S$. It is possible to define different versions of the problem, according to the number of occurrences of each symbol in the solution, as represented in Table 1. In this paper we will deal with such different versions of ELCS. First notice that ELCS(1) and ELCS$(\geq 1)$ are generalizations of the LCS problem, where no mandatory symbols are present. Therefore all the hardness results for LCS apply to ELCS(1) and ELCS$(\geq 1)$. Moreover, we will show that the above problems are hard also on instances of only two sequences (while the LCS problem can be solved in polynomial time for any fixed number of sequences). When dealing with the restriction of ELCS containing only a fixed number of sequences, we will denote such restriction prefixing the problem name with the number of sequences, e.g. 2-ELCS$(1, \leq 1)$ is the restriction of ELCS$(1, \leq 1)$ to instances of two sequences.

## 3 The Results

**Theorem 1.** *The 2-ELCS$(1, \leq 1)$ problem is **APX**-hard even when each symbol appears at most twice in each input sequence.*

*Proof.* We prove the theorem describing an L-reduction from MAX INDEPENDENT SET on Cubic Graph to 2-ELCS$(1, \leq 1)$, since the latter problem is known to be **APX**-hard[1]. Let $G = (V, E)$ be a cubic graph. Then for each vertex $v_i$ there are three edges $e_1(v_i)$, $e_2(v_i)$, $e_3(v_i)$ incident on it. In the reduction each vertex $v_i$ is associated with a symbol $v_i$ of $A_o$ and a symbol $x_i$ in $A_m$. Each edge is associated with a distinct symbol of $A_m$. Define a *block* associated with a vertex $v_i$, as a string consisting of a vertex symbol $v_i$, the symbols associated with edges incident to $v_i$ in $G$ and the symbols $x_i$. There are two possible blocks associated with $v_i$, one contained in $s_1$ and defined as $b_1(v_i) = v_i e_1(v_i) e_2(v_i) e_3(v_i) x_i$, the second contained in $s_2$ and defined as $b_2(v_i) = e_1(v_i) e_2(v_i) e_3(v_i) v_i x_i$. The instance of 2-ELCS$(1, \leq 1)$ consists of the following two sequences: $s_1 = b_1(v_1) b_1(v_2) \cdots b_1(v_n)$, $s_2 = b_2(v_1) b_2(v_2) \cdots b_2(v_n)$. Observe that since each symbol $x_i$ is mandatory and it occurs only in blocks $b_1(v_i)$, $b_2(v_i)$ of $s_1$, $s_2$ respectively, any symbol $x_i$ in a feasible solution of 2-ELCS$(1, \leq 1)$ over $s_1$ and $s_2$ must be taken from $b_1(v_i)$ and $b_2(v_i)$. It follows that if $v_i$ is in an exemplar common subsequence, then the exemplar common subsequence does not contain any symbol of $e_1(v_i) e_2(v_i) e_3(v_i)$ of $b_1(v_i)$ and $b_2(v_i)$. Let $s$ be a feasible solution of 2-ELCS$(1, \leq 1)$ over $s_1$, $s_2$, then $s$ consists of $f_1 x_1 \ldots f_i x_i \ldots f_n x_n$, where each $f_i$ is either $v_i$ or a subsequence of $e_1(v_i) e_2(v_i) e_3(v_i)$. Observe that each edge symbol is mandatory, which means that it must appear exactly once in a common subsequence. Moreover, an edge symbol encoding edge $(v_i, v_j)$ appears in blocks $b_1(v_i)$ and $b_1(v_j)$ of $s_1$ and in blocks $b_2(v_i)$ and $b_2(v_j)$ of $s_2$. Thus a common subsequence takes such edge symbol either from $b_1(v_i)$ and $b_2(v_i)$ or from $b_1(v_j)$ and $b_2(v_j)$.

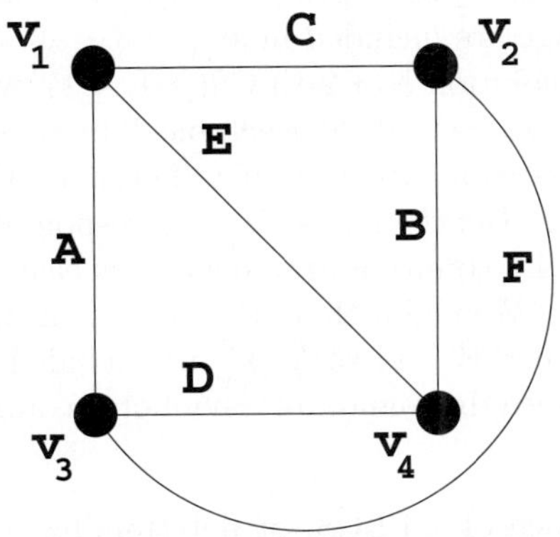


$s_1 = v_1 CAEx_1 v_2 CFBx2 v_3 AFDx_3 v_4 EBDx_4$
$s_2 = CAEv_1 x_1 CFBv_2 x2AFDv_3 x_3 EBDv_4 x_4$

**Fig. 1.** Reducing the graph $K_4$

Let $I$ be the set of vertices appearing in $s$, we will show that $I$ is an independent set of $G$. Assume that symbols $v_i, v_j \in I$. Then $(v_i, v_j)$ is not an edge of $G$, otherwise $s$ in $f_i$ and $f_j$ contains symbols $v_i$ and $v_j$ respectively. It follows that the edge symbol associated with $(v_i, v_j)$, that can appear only in $f_i$ and $f_j$, is not contained in $s$. Since each edge symbol is mandatory, it must appear in any feasible solution of 2-ELCS$(1, \leq 1)$, which is a contradiction. Observe that the length of a feasible solution of 2-ELCS$(1, \leq 1)$ over $s_1$, $s_2$ is $|V| + |E| + |I|$, where $I$ is an independent set of $G$. On the other side, let $I$ be an independent set of $G$, it is easy to compute a feasible solution of 2-ELCS$(1, \leq 1)$ over $s_1$, $s_2$ of size $|V| + |E| + |I|$, retaining in the exemplar common subsequence only the symbols associated with vertices in $I$. □

A similar proof can be given also for 2-ELCS$(\geq 1, \leq 1)$.

**Theorem 2.** *The 2-ELCS($\geq 1, \leq 1$) problem is* ***APX****-hard even when each symbol appears at most twice in each input sequence.*

*Proof.* The reduction is similar to the previous one, but for each vertex $v_i$ of the graph, we have four symbols $v_i^a v_i^b v_i^c v_i^d$ and the blocks $b_1(v_i)$ and $b_2(v_i)$ associated with $v_i$ in sequences $s_1$ and $s_2$ respectively are defined as follows: $b_1(v_i) = v_i^a v_i^b v_i^c v_i^d e_1(v_i)e_2(v_i)e_3(v_i)x_i$; $b_2(v_i) = e_1(v_i)e_2(v_i)e_3(v_i)v_i^a v_i^b v_i^c v_i^d x_i$.

Again the symbols $x_i$ are mandatory. Since each $x_i$ appears in blocks $b_1(v_i)$ and $b_2(v_i)$ of $s_1$ and $s_2$ respectively, it follows that any symbol $x_i$ in an exemplar common subsequence must be taken from the blocks of $s_1$, $s_2$ associated with $v_i$, that is $b_1(v_i)$ and $b_2(v_i)$. Since each mandatory edge symbols appears twice in each input sequence, it must appear once or twice in a common subsequence.

Clearly if sequence $v_i^a v_i^b v_i^c v_i^d$ is in a feasible solution of 2-ELCS($\geq 1, \leq 1$) over $s_1$, $s_2$, then this solution does not contain occurrence of symbols of sequence $e_1(v_1)e_2(v_1)e_3(v_1)$ in $b_1(v_i)$ and $b_2(v_i)$. This means that a feasible solution $s$ of 2-ELCS($\geq 1, \leq 1$) over $s_1$, $s_2$ consists of $g_1 x_1 \ldots g_i x_i \ldots g_n x_n$, where each $g_i$ is either a subsequence of $v_i^a v_i^b v_i^c v_i^d$ or a subsequence of $e_1(v_i)e_2(v_i)e_3(v_i)$. Observe that each edge symbol is mandatory, which means that it must appear exactly once in an exemplar common subsequence. Thus an exemplar common subsequence takes each edge symbol from one of the two blocks where it appears.

Let $I$ be an independent set of $G$, then $s = g_1 x_1 \ldots g_i x_i \ldots g_n x_n$, where each $g_i = v_i^a v_i^b v_i^c v_i^d$ if $v_i \in I$ and $g_i = e_1(v_i)e_2(v_i)e_3(v_i)$ otherwise. It is immediate to note that $s$ is a common subsequence of $s_1$ and $s_2$ of length $|V| + 3|V| + |I|$ and that all mandatory symbols encoding an edge are included in $s$.

Assume now that there exists a feasible solution $s$ of 2-ELCS($\geq 1, \leq 1$) over $s_1$, $s_2$ with length $|V| + 3|V| + |I|$. We can assume that, for each block in $s_1$, $s2$, either $v_i^a v_i^b v_i^c v_i^d$ or $e_1(v_i)e_2(v_i)e_3(v_i)$ appears as a substring of $s$. Let $Y$ be the set of blocks for which $v_i^a v_i^b v_i^c v_i^d$ is part of $s$. Hence the vertices corresponding to $Y$ are an independent set of $G$. By a trivial counting argument, it is easy to show that for $|I|$ blocks $s$ includes $v_i^a v_i^b v_i^c v_i^d$. We claim that such blocks encode an independent set. W.l.o.g. assume that $v_1^a v_1^b v_1^c v_1^d$ and $v_2^a v_2^b v_2^c v_2^d$ are included in $s$, then there is no edge $(v_1, v_2)$ in $G$, otherwise the mandatory symbol encoding such edge would not be in $s$. □

A related problem is that, given an instance of 2-ELCS, of determining if a feasible solution exists. In what follows we will consider a general version of the 2-ELCS problem. Notice that both reductions described above hold for instances that are known to admit a feasible solution, therefore they are not sufficient for dealing with the problem. Observe that we can assume that both input sequences consist only of mandatory symbols, since only mandatory symbols are relevant for the existence of a solution.

**Theorem 3.** *The problem of determining if a feasible solution exists for an instance of 2-ELCS where each mandatory symbol appears totally at most three times in the input sequences, can be solved in polynomial time.*

*Proof.* We prove the theorem reducing an instance of 2-ELCS where each mandatory symbol appears totally at most three times in the input sequences to an instance of 2SAT. Notice that 2SAT can be solved in polynomial time [2].

For each symbol $s$, let $occ_1(s)$ (respectively $occ_2(s)$) be the set of positions of the input sequence $s_1$ (resp. $s_2$) where the $s$ appears. Clearly both $occ_1(s)$ and $occ_2(s)$ are not empty and $|occ_1(s)|+|occ_2(s)| \leq 3$. For each symbol $s$ there are at most two pairs in $occ_1(s) \times occ_2(s)$, for otherwise $|occ_1(s)| + |occ_2(s)| > 3$, let us associate with each of such pairs a variable $x_{s,i}$, where $i \in \{1, 2\}$ if there are two pairs in $occ_1(s) \times occ_2(s)$ and $i = 1$ if there is only one pair in $occ_1(s) \times occ_2(s)$.

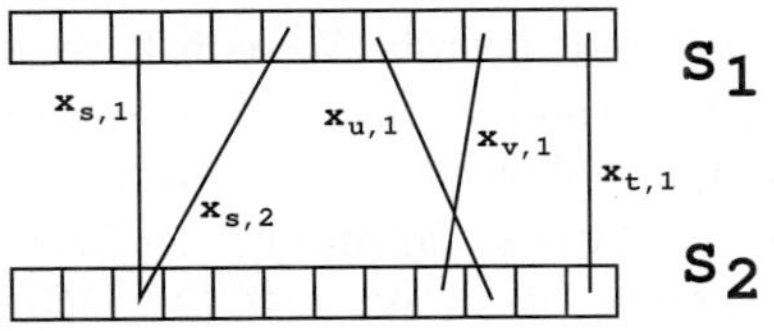


**Fig. 2.** Reducing 2-ELCS to 2SAT

Graphically the possible variables are represented in Fig. 2 with a line connecting two identical symbols belonging to different sequences. The case $|occ_1(s)| + |occ_2(s)| = 3$ is represented by the two leftmost lines and the variables $x_{s,1}$, $x_{s,2}$, while the case $|occ_1(s)| + |occ_2(s)| = 2$ is represented by the rightmost line and the variable $x_{t,1}$. Each truth assignment to the variables can be viewed as picking the lines corresponding to true variables. Let $C$ be the set of clauses of the instance of 2SAT that we are constructing. For each pair $x_{s,1}$, $x_{s,2}$ of variables, the clauses $\neg x_{s,1} \vee \neg x_{s,2}$ and $x_{s,1} \vee x_{s,2}$ are added to $C$. Moreover, for each symbol $s$ such that there is only one pair in $occ_1(s) \times occ_2(s)$, add the clause $x_{s,1}$ to $C$ (this corresponds to forcing the variable $x_{s,1}$ to be true). The fact that all these clauses are satisfied in any feasible solution of 2SAT, corresponds to pick exactly one of the lines associated with each symbol. Two lines (or two variables) are called *crossing* if they cross in the drawing built as in Fig. 2. More formally, notice that each variable $x_{s,i}$ is associated with an occurrence of $s$ in $s_1$ (denoted as $s_1(s, i)$) and one occurrence of $s$ in $s_2$ (denoted as $s_2(s, i)$). A pair $x_{s,i}, x_{t,j}$ of variables is crossing if in $s_1$ the symbol $s_1(s, i)$ precedes $s_1(t, j)$ and in $s_2$ the symbol $s_2(s, i)$ does not precede $s_2(t, j)$ or, symmetrically, if in $s_1$ the symbol $s_1(s, i)$ does not precede $s_1(t, j)$ and in $s_2$ the symbol $s_2(s, i)$ precedes $s_2(t, j)$. For each pair $x_{s,i}, x_{t,j}$ of crossing variables, the clause $\neg x_{s,i} \vee \neg x_{t,j}$ is added to $C$. We can prove that the original instance of 2-ELCS has a feasible solution if and only if the instance of 2SAT is satisfiable, that is there is a truth assignment for all variables such that all clauses in $C$ are evaluated true. Assume that there is a feasible solution $z$ of the instance of 2-ELCS then, for each symbol $s$, we pick the lines connecting the symbols retained in $z$. By definition of common subsequence there cannot be two crossing lines, and exactly one of the lines associated with each symbol must be picked as $|occ_1(s)|+|occ_2(s)| \leq 3$, therefore we have constructed a feasible solution of 2SAT. Conversely given a truth assignment for all variables that satisfies all clauses in $C$, it is immediate to note that there are not two crossing lines, and that there is exactly one line for each symbol, therefore it is immediate to construct a feasible solution of 2-ELCS that contains all symbols. □

Notice that the above result holds for all the restrictions of 2-ELCS considered here, as no symbol appears twice in both input sequences, therefore it can appear at most once in any solution. We will show in the following theorem that slightly relaxing this constraint makes the problem **NP**-hard.

**Theorem 4.** *The problem of determining if a feasible solution exists for an instance of 2-ELCS where each mandatory symbol appears at most three times in each input sequence, is* ***NP****-hard.*

*Proof.* We will prove the theorem reducing 3SAT to 2-ELCS, with a reduction very similar to the one shown before. Let $C = \{C_1, \ldots, C_k\}$ be a set of clauses, each one consisting of at most three (possibly negated) literals. We construct an instance of 2-ELCS associating a block with each variable. The block of $s_1$ associated with variable $x_i$ is defined as the symbol $x_i$, followed by the sequence of clauses containing $x_i$, followed by the sequence of clauses containing $\neg x_i$, where in each sequence the clauses are ordered according to the index in $\{C_1, \ldots, C_k\}$. In $s_2$ the block associated with variable $x_i$ is defined as the symbol $x_i$, followed by the sequence of clauses containing $\neg x_i$, followed by the sequence of clauses containing $x_i$ (again the clauses are ordered according to the index in $\{C_1, \ldots, C_k\}$). Both sequences $s_1$ and $s_2$ consist of the sequence of all blocks associated with the variables of the original instance of 3SAT. All symbols are mandatory, also notice that each symbol appear at most three times in each sequence as each clause contains at most three literals.

Each symbol $x_i$ appears exactly once in each sequence, hence there is no ambiguity on which occurrence is retained in any exemplar common subsequence. Consequently each symbol retained must correspond to occurrences taken from the same block. Inside the block associated with $x_i$, retaining the clauses where $x_i$ appears as a positive literal is mutually exclusive with retaining the clauses where $x_i$ appears as a negative literal, by definition of exemplar common subsequence. The first case corresponds to setting $x_i$ to true, while the second case corresponds to setting $x_i$ to false. In both case the clauses retained are satisfied by the assignment of variables $x_i$. It is immediate to note that any feasible solution must contain all clauses, therefore we have computed a truth assignment of the variables that satisfies all clauses in $C$, completing the proof. □

The above results have a definitive consequence on the approximability of the 2-ELCS problem where each mandatory symbol appears at most three times in both input sequences, as they rule out any possible polynomial-time approximation algorithm. Since the problem can be extended to instances consisting of a set of sequences, it is interesting to know if the above results can be made stronger. In fact, the well-known inapproximability results in [6] for the LCS problem, immediately apply also to the ELCS($\geq 1$) problem, since ELCS($\geq 1$) is more general than LCS. A closer inspection of their proofs shows that their results also apply to all versions of ELCS, as the optimal solutions in their reductions contain at most one occurrence of each symbol, excluding any $O(n^{1-\epsilon})$ ratio polynomial-time approximation algorithm unless **ZPP**=**NP**, even if no mandatory symbol is allowed and all symbols appear at most twice in each sequence.

# 4 Restricting the Problem

In this section we consider the restriction of 2-ELCS(1) where the number of mandatory symbols is at most a constant. We describe a polynomial time algorithm for this restriction based on two phases: the first step consists of guessing the exact ordering of all mandatory symbols in the optimal solution, the second step basically fills in the gaps between each pair of mandatory symbol. Since each mandatory symbol appears exactly once in a feasible solution, the correct ordering of the mandatory symbol is a permutation of $A_m$, which can be computed in constant time, since $|A_m|$ is a constant.

Let $s$ be a permutation of mandatory symbol, the second phase consists of computing a longest common subsequence $s^*$ of $\{s_1, s_2\}$ such that $s$ is a subsequence of $s^*$. Let us denote by $s[i]$ the $i$-th character of the sequence $s$ and by $s[i \dots j]$ the substring of $s$ starting with $s[i]$ and ending with $s[j]$. The recurrence equation for $\mathrm{EL}[i, j, k]$, that is the length of an optimal solution over $s_1[1 \dots i]$, $s_2[1 \dots j]$ that is a supersequence of the sequence $s[1] \cdots s[k]$ is:

$$\mathrm{EL}[i,j,k] = \max \begin{cases} \mathrm{EL}[i-1, j-1, k] + 1 \text{ if } s_1[i] = s_2[j],\ s_1[i] \in A_o \\ \mathrm{EL}[i-1, j-1, k-1] + 1 \text{ if } s_1[i] = s_2[j] = s[k] \\ \mathrm{EL}[i-1, j, k], \mathrm{EL}[i, j-1, k] \text{ if } s_1[i] = s_2[j] \neq s[k],\ s_1[i] \in A_m \\ \mathrm{EL}[i-1, j, k], \mathrm{EL}[i, j-1, k] \text{ if } s_1[i] \neq s_2[j] \end{cases}$$

The boundary conditions are $\mathrm{EL}[0, j, 0] = 0$ and $\mathrm{EL}[i, 0, 0] = 0$ for $0 \le i \le |s_1|$ and $0 \le j \le |s_2|$. The value of the optimal solution can be read in $\mathrm{EL}[|s_1|, |s_2|, |s|]$. Once the matrix EL has been completely filled in the actual optimal subsequence can be constructed with standard backtracking techniques [3]. The recurrence equation described above can be easily modified for the 2-ELCS($\geq 1$), by removing the requirement $s_1[i] \in A_o$ in the first condition of the equation.

## References


1. P. Alimonti and V. Kann. Some APX-completeness results for cubic graphs. *Theoretical Computer Science*, 237(1–2):123–134, 2000.
2. B. Aspvall, M. F. Plass, and R. E. Tarjan. A linear-time algorithm for testing the truth of certain quantified boolean formulas. *Information Processing Letters*, 8(3):121–123, 1979.
3. T. H. Cormen, C. E. Leiserson, R. L. Rivest, and C. Stein. *Introduction to Algorithms.* MIT Press, 2nd edition, 2001.
4. K. Hakata and H. Imai. The longest common subsequence problem for small alphabet size between many strings. In *Proc. 3rd International Symp. on Algorithms and Computation (ISAAC)*, pages 469–478, 1992.
5. W. Hsu and M. Du. New algorithms for the LCS problem. *Journal of Computer and System Sciences*, 19:133–152, 1984.
6. T. Jiang and M. Li. On the approximation of shortest common supersequences and longest common subsequences. *SIAM Journal on Computing*, 24(5):1122–1139, 1995.
7. D. Maier. The complexity of some problems on subsequences and supersequences. *Journal of the ACM*, 25:322–336, 1978.

# Synonymous Codon Substitution Matrices*

Adrian Schneider, Gaston H. Gonnet, and Gina M. Cannarozzi

Computational Biology Research Group,
Institute for Computational Science, ETH Zürich,
Universitätstrasse 6, 8092 Zürich, Switzerland
cgina@inf.ethz.ch

**Abstract.** Observing differences between DNA or protein sequences and estimating the true amount of substitutions from them is a prominent problem in molecular evolution as many analyses are based on distance measures between biological sequences. Since the relationship between the observed and the actual amount of mutations is very complex, more than four decades of research have been spent to improve molecular distance measures. In this article we present a method called SynPAM which can be used to estimate the amount of synonymous change between sequences of coding DNA. The method is novel in that it is based on an empirical model of codon evolution and that it uses a maximum-likelihood formalism to measure synonymous change in terms of codon substitutions, while reducing the need for assumptions about DNA evolution to an absolute minimum. We compared the SynPAM method with two established methods for measuring synonymous sequence divergence. Our results suggest that this new method not only shows less variance, but is also able to capture weaker phylogenetic signals than the other methods.

## 1 Introduction

Measures of synonymous substitution are fundamental for many analyses in molecular evolution such as in the calculation of dN/dS to provide information about the degree of selection operating on homologous sequences, in the construction of phylogenetic trees, and for molecular dating. Measuring synonymous change between coding DNA sequences requires a model of evolution at the codon or nucleotide level. Because DNA evolution has properties such as unequal transition (purine-purine or pyrimidine-pyrimidine mutation) and transversion (purine to pyrimidine and vice versa) rates and unequal nucleotide and codon frequencies, modeling of DNA evolution is not trivial. The first methods to estimate the amount of synonymous mutations were introduced more than two decades ago [1, 2] and have since then continuously been improved [3, 4, 5, 6]. Usually, a $4 \times 4$ or a $64 \times 64$ Markovian model is employed with varying numbers of rate constants and/or nucleotide or codon frequencies. These models are used to estimate the real number of substitutions from the observed number of

* This work was supported by the intramural research program of the National Institutes of Health, National Library of Medicine.

V.N. Alexandrov et al. (Eds.): ICCS 2006, Part II, LNCS 3992, pp. 630–637, 2006.

substitutions. The models differ in the number of variables needed to describe transitions between nucleotides, the equilibrium frequencies and the definition of 'synonymous site' used.

Two measures of synonymous sequence divergence are the dS metric that estimates the ratio of synonymous substitutions per synonymous site and the NED or TREx metric, based on mutations at the third position of two-fold redundant codons at conserved amino acid sites. The most commonly used implementation of dS is found in the *PAML* software package [7] and is based on the maximum-likelihood method by Yang and co-workers [3, 8]. NED was developed by Benner and co-workers [5, 6] to model the decay of the two-fold redundant amino acids at conserved sites with a first-order rate equation. This method has the advantage that it is very simple and can easily be applied in any molecular analysis. The use of only two-fold codons makes it independent of all the complications introduced by transition/transversion rate differences or mutations at the first base positions. However, the drawback is that there are often not enough of this specific type of mutation to yield statistically significant results.

The SynPAM method presented here employs a 64 × 64 Markov model with transition rates measured from empirical data to estimate the amount of synonymous change between two sequences of coding DNA using maximum-likelihood. SynPAM is different from the other methods in two aspects: Instead of a parameterized model of codon evolution, an empirically determined matrix [9] is used to assign likelihood scores to all the synonymous mutations in an alignment. Maximum-likelihood scoring matrices in the tradition of Dayhoff [10] are then employed to estimate the synonymous distance of the two sequences. Additionally, the SynPAM method directly measures synonymous change in the form of codon substitutions. This makes it unnecessary to identify the sites that are synonymous (which always introduces additional uncertainty).

## 2 Methods

**Synonymous Substitution Matrix.** The synonymous substitution probabilities are derived from an empirical codon matrix $M$ where every entry $M_{a,b}$ is the probability that codon $b$ mutates to codon $a$ [9]. Starting with a codon substitution matrix for a particular evolutionary distance, the probabilities of all nonsynonymous substitutions are set to 0. The remaining probabilities are then rescaled such that all possible codon substitutions between codons coding for a given amino acid sum to 1:

$$M^*_{a,b} = \begin{cases} \frac{M_{a,b}}{\sum_{x \in \text{syn}(b)} M_{x,b}} & \text{if } a \in \text{syn}(b) \\ 0 & \text{otherwise} \end{cases} \tag{1}$$

with syn(x) being the set of all codons that are synonymous to $x$.

From the the empirical codon substitution matrix, a wide range of matrices approximating different evolutionary distances are extrapolated through matrix exponentiation [11], thereby providing substitution matrices representing different evolutionary distances that can all be converted to synonymous substitution matrices. It is important to state that the exponentiation has to be executed on the full mutation matrix and only then the transformation described above and the computation of the scores should be performed. The reason is that the reduced substitution matrix no longer describes the full Markovian process of codon evolution. E.g. some codons have a larger chance to undergo nonsynonymous mutations than others, but this information would be lost, if only the synonymous matrices were exponentiated. In addition, by including all elements of the substitution matrix until the time of reduction, all alternative pathways, back mutations and multiple hits are taken into consideration.

Analogous to the definition of 1 PAM as the amount of evolution in which 1 percent of the amino acids are expected to undergo mutations, 1 SynPAM is defined as the amount of evolution in which 1 percent of the synonymous positions are subject to a codon substitution.

The substitution matrix $M(t)$ (and derived from it also $M^*(t)$) represents $t$ times the distance of 1 SynPAM and is computed as $M(1)^t$, where $M(1)$ is the full substitution matrix corresponding to 1 SynPAM (which was derived from the original $M$ in an iterative process).

**Likelihood Scores.** In almost the same way that log-likelihood scores are computed for amino acid scoring matrices, they can be calculated for the synonymous substitution matrices. In addition to the codon frequencies, $\pi$, also the synonymous frequencies, $\pi^*$, are needed, which is the codon distribution for given amino acids. This means that the sum of all $\pi_x^*$ which encode the same amino acid is 1. The scores are computed from the ratio of the probabilities of two synonymous codons having mutated from a common ancestor during time $t$ compared to being paired by random chance (which corresponds to $t = \infty$):

$$S_{i,j}(t) = 10 \log_{10} \frac{\pi_j \cdot M_{i,j}^*(t)}{\pi_i^* \cdot \pi_j} = 10 \log_{10} \frac{M_{i,j}^*(t)}{\pi_i^*} \tag{2}$$

Synonymous codon scores are only defined for synonymous substitutions. Because amino acid altering mutations should have no influence on the synonymous scores, all of these mutations are assigned a score of 0.

Given a codon-wise alignment of two coding DNA sequences and a SynPAM matrix for a given distance $t$, the synonymous score of the entire alignment is obtained by adding up the scores, $S_{i,j}$, for all synonymous codon positions in the alignment. The alignment is scored with a range of SynPAM matrices, the highest-scoring matrix can be determined and the distance of this matrix is chosen as the SynPAM distance of the alignment. This corresponds to a maximum-likelihood estimation of the synonymous distance.

**Implementations of NED and dS.** The dS values were computed using the *codeml* program from the *PAML* software package [7]. Version 3.14 of the program was used with nine free parameters used to account for codon frequencies (F3x4). Using other models of codon frequencies did not appreciably change the results. Since only pairwise sequence comparisons were done, the parameters were set as follows: runmode=−2 (pairwise comparisons) and the $d_N/d_S$ ratio was not allowed to vary between sites (NSsites = 0). The other parameters have no influence when pairwise comparison is chosen.

The NED method was implemented in *Darwin* [12]. Given a codon-wise alignment, the fraction $f2$ of conserved codons for all amino acids encoded by two codons is computed. Then the fraction of conserved codons is modeled as an exponential decay to equilibrium via the equation: $NED = -ln\frac{f_2-b}{1-b}$ where $b$ is the codon bias [6] and was originally taken from the Codon Usage Database [13]. Here, the equilibrium fraction of conserved codons was calculated from the codon frequencies from the codon substitution matrix and was found to be .5136.

**Databases.** The complete genome databases are from *ENSEMBL* [14] (*H. sapiens*, *P. troglodytes*, *C. familiaris*, *B. taurus*, *G. gallus*, *X. tropicalis*, *B. rerio*, *T. rubripes*, *T. nigroviridis*, *C. intestinalis*, *D. melanogaster*, *A. gambiae*, *A. mellifera*, *C. elegans* and *C. briggsae*).

## 3 Results and Discussion

Distance estimation methods can be assessed either using simulations or real data. Simulations have the advantage that many parameters can be chosen and a detailed analysis of the method can be performed. However, the use of simulations is not appropriate in this case because distance measures are based on a model of evolution while at the same time such a model is needed to create simulated data. Therefore simulations can not be a fair way of comparing distance measures.

The drawback with real data is that the true distance between sequences cannot be known. But at least two criteria can be used when comparing different distance measures: the variance of the estimates and the range in which they yield valid and reliable results.

An unbiased estimation of the variance can be found using each method to measure the distances between many pairs of genes with the same divergence time and then calculating the variance of these different estimates. Sets of gene pairs that fulfill this criteria of equal divergence times are the sets of orthologs between two species. Here, we used the metazoan set from the OMA orthologs project [15], a large-scale cross-comparison of complete genomes with the goal of systematically identifying groups of orthologous proteins.

### 3.1 Variance of the Estimates

The variation of the different time estimates between the different methods was compared. Because the time metrics of the different methods are on different scales, the variances can not be compared directly. Instead, the ratio of the standard deviation to the average value can be used, which is called the coefficient of variance (CV).

It is clear that since DNA mutations are a stochastic process, different sequence pairs are likely to have different amounts of synonymous mutations. Also, selection pressure on synonymous mutations can not be ruled out [16, 17]. Therefore, although we are looking at orthologous sequences between the same two

**Table 1.** Ratio of the standard deviation to mean for date estimates between human and animals using different dating methods

| | # values | SynPAM | NED | dS |
|---|---|---|---|---|
| *Canis familiaris* | 13235 | 0.29 | 0.32 | 0.38 |
| *Mus musculus* | 13316 | 0.23 | 0.28 | 0.30 |
| *Gallus gallus* | 8515 | 0.26 | 0.32 | 0.62 |
| *Xenopus tropicalis* | 8288 | 0.26 | 0.30 | 1.61 |
| *Brachydanio rerio* | 5845 | 0.26 | 0.33 | 1.07 |
| *Ciona intestinalis* | 1696 | 0.59 | 0.32 | 0.56 |
| *Drosophila melanogaster* | 1643 | 1.95 | 0.38 | 0.44 |
| *Caenorhabditis elegans* | 1038 | 1.60 | 0.33 | 0.48 |

species, it is in the nature of these sequence pairs to already have a certain amount of variation in the true sequence divergence. But when this sequence divergence is measured, an additional random component is added because the models of evolution and the ways to estimate the divergence can never precisely reflect the real pattern of molecular evolution.

Since in this comparison the same set of sequence pairs was used for all the tested methods, the inherent variance due to these stochastic (and possibly selection) effects are exactly the same for all methods. The only difference is how much additional noise is introduced by the model of evolution and the measurement technique used.

Table 1 displays these values for distance estimates of human against some other metazoans. For each pair, the coding DNA was aligned using local dynamic programming [18, 19] with the codon substitution matrices. Only alignments with at least 100 synonymous positions were considered in order to gain some significance in the results. For $f2$ values lower or equal to the codon bias $b$, no NED estimate can be computed. Alignments with this problem were discarded for all methods, therefore the comparison is done on the same set of alignments for all three methods. In addition, for each method, the 10% of alignments with the highest distance estimate were discarded to exclude obvious outliers. The number of sequence pairs used is also shown in Table 1 and decreases with increasing distance because there are less orthologs between distant species. Also, as the distance increases, the synonymous mutations approach their equilibrium values and produce more invalid results.

Table 1 shows that the SynPAM method consistently causes the least amount of variation for species up to the human-fish divergence. Since SynPAM is the estimator with the least amount of variance, it is the preferred method to estimate distances within this time range.

It is remarkable that the CV for dS clearly decreases after the human-frog distance and that it becomes very small for the longer distances. In order to explain this unexpected behavior, we plotted the histograms for all the pairwise values between any two species. A selection of those are shown in Figure 1. There, the explanation of this artifact can be seen: Gene pairs, where the synonymous substitutions are close to saturation, are assigned a seemingly arbitrary dS value around 65. Since this high amount of substitutions per site is impossible to

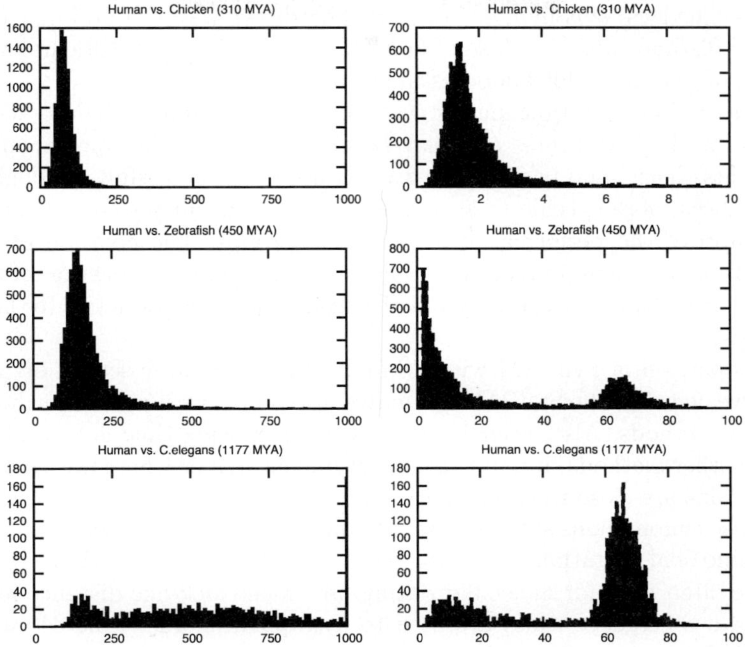


**Fig. 1.** Histograms for SynPAM (left) and dS (right) of values for the alignments of orthologs between human and other metazoans. There are 9130 alignments of human with with chicken, 7647 with zebrafish and 2512 with *C. elegans*.

estimate with confidence, this is clearly a limitation in the calculation of dS. Already at the primate-fish distance, a significant number of dS values are in this nonsense range, while the SynPAM distribution still forms a sharp curve.

For distances larger than primate-fish (i.e. approximately 450 million years), all methods tend to suffer from saturation effects and the single gene estimates become very unreliable. In the SynPAM histograms, many values of 1000 can be found, which corresponds to the highest matrix used and means that the synonymous mutations have reached an equilibrium state.

## 4 Conclusions

A new method called SynPAM for the estimation of the amount of synonymous change between two codon sequences has been introduced which is unique in its use of empirical codon transition probabilities combined with a maximum likelihood formalism in the tradition of the Dayhoff scoring matrices. This method has the property that it is works well without estimating mutational parameters. Due to the Markovian model of evolution, multiple hits, alternative pathways and back mutations are not only included in the model, but are also based on empirical observations. Since the distance is expressed in the form of expected number of synonymous codon substitutions, no individual nucleotide sites have to be assigned as 'synonymous' or not. The direct way of obtaining the synonymous distance makes it faster than other methods while still employing the strength of maximum likelihood estimation.

As the codon mutation matrix was built from vertebrate DNA, its usable range is for the vertebrates and may be extendable to other metazoans. Tests with plants, yeasts and bacteria have so far not been successful, although mutation matrices more specific to specific species sets are under consideration and could improve the performance for these subgroups. The creation of all new matrices for only pairs of species would allow for the incorporation of species-specific substitution patterns without the need to adjust parameters to the actual sequences.

A comparison of SynPAM with dS and NED using large sets of orthologous genes revealed that the SynPAM estimates have lower variances than those from the other methods. Also, their behavior with increasing time is favorable compared to other methods since it captures also weak signals where the synonymous substitutions are close to saturation.

Because synonymous substiutions are under less functional selection pressure than amino acid mutations, they are expected to happen more regularly and are therefore often used for molecular dating, in which clock-like distance measures and known divergence times are used to extrapolate actual time estimates for unknown divergences. Since the SynPAM method presented here for measuring synonymous divergence shows advantages over previous such distance metrics, we propose the use of SynPAM also for dating purposes in the realm of vertebrates and are currently investigating this topic in more detail.

## References

1. Miyata, T., Yasunaga, T.: Molecular evolution of mRNA: a method for estimating evolutionary rates of synonymous and amino acid substitutions from homologous nucleotide sequences and its application. J. Mol. Evol. **16** (1980) 23 – 36
2. Perler, F., Efstratiadis, A., Lomedico, P., Gilbert, W., Kolodner, R., Dodgson, J.: The evolution of genes: the chicken preproinsulin gene. Cell **20**(2) (1980) 555–566
3. Goldman, N., Yang, Z.: A codon-based model of nucleotide substitution for protein-coding DNA sequences. Mol. Biol. Evol. **11**(5) (1994) 725–736
4. Yang, Z., Nielsen, R., Goldman, N., Pedersen, A.M.K.: Codon-substitution models for heterogeneous selection pressure at amino acid sites. Genetics **155** (2000) 432–449
5. Benner, S.A.: Interpretive proteomics– finding biological meaning in genome and proteome databases. Advances in Enzyme Regulation **43** (2003) 271–359
6. Caraco, M.D.: Neutral Evolutionary Distance: A New Dating Tool and its Applications. PhD thesis, ETH Zürich, Zürich, Switzerland (2002)
7. Yang, Z.: Paml: A program package for phylogenetic analysis by maximum likelihood. CABIOS **13** (1997) 555 – 556
8. Yang, Z., Nielsen, R.: Estimating synonymous and nonsynonymous substitution rates under realistic evolutionary models. Mol. Biol. Evol. **17**(1) (2000) 32 – 43
9. Schneider, A., Cannarozzi, G.M., Gonnet, G.H.: Empirical codon substitution matrix. BMC Bioinformatics **6**(134) (2005)

10. Dayhoff, M.O., Schwartz, R.M., Orcutt, B.C.: A model for evolutionary change in proteins. In Dayhoff, M.O., ed.: Atlas of Protein Sequence and Structure. Volume 5. National Biomedical Research Foundation (1978) 345–352
11. Cox, D., Miller, H.: The Theory of Stochastic Processes. Chapman and Hall, London (1965)
12. Gonnet, G.H., Hallett, M.T., Korostensky, C., Bernardin, L.: Darwin v. 2.0: An interpreted computer language for the biosciences. Bioinformatics **16**(2) (2000) 101–103
13. Nakamura, Y., Gojobori, T., Ikemura, T.: Codon usage tabulated from the international DNA sequence database. Nucleic Acids Res. **28** (2000) 292
14. Hubbard, T., Andrews, D., Caccamo, M., Cameron, G., Chen, Y., Clamp, M., Clarke, L., Coates, G., Cox, T., Cunningham, F., Curwen, V., Cutts, T., Down, T., Durbin, R., Fernandez-Suarez, X.M., Gilbert, J., Hammond, M., Herrero, J., Hotz, H., Howe, K., Iyer, V., Jekosch, K., Kahari, A., Kasprzyk, A., Keefe, D., Keenan, S., Kokocinsci, F., London, D., Longden, I., McVicker, G., Melsopp, C., Meidl, P., Potter, S., Proctor, G., Rae, M., Rios, D., Schuster, M., Searle, S., Severin, J., Slater, G., Smedley, D., Smith, J., Spooner, W., Stabenau, A., Stalker, J., Storey, R., Trevanion, S., Ureta-Vidal, A., Vogel, J., White, S., Woodwark, C., Birney, E.: Ensembl 2005. Nucleic Acids Res. **33**(suppl-1) (2005) D447–D453
15. Dessimoz, C., Cannarozzi, G., Gil, M., Margadant, D., Roth, A., Schneider, A., Gonnet, G.: OMA, a comprehensive, automated project for the identification of orthologs from complete genome data: Introduction and first achievements. In McLysath, A., Huson, D.H., eds.: RECOMB 2005 Workshop on Comparative Genomics. Volume LNBI 3678 of Lecture Notes in Bioinformatics., Springer-Verlag (2005) 61 – 72
16. Bielawski, J.P., Dunn, K.A., Yang, Z.: Rates of nucleotide substitution and mammalian nuclear gene evolution: Approximate and maximum-likelihood methods lead to different conclusions. Genetics **156** (2000) 1299–1308
17. Dunn, K.A., Bielawski, J.P., Yang, Z.: Substitution rates in drosophila nuclear genes: Implications for translational selection. Genetics **157** (2001) 295–305
18. Waterman, M.S., Smith, T.F., Beyer, W.A.: Some biological sequence metrics. Advances in Mathematics **20** (1976) 367–387
19. Gotoh, O.: An improved algorithm for matching biological sequences. J. Mol. Biol. **162** (1982) 705–708

# SEPA: Approximate Non-subjective Empirical $p$-Value Estimation for Nucleotide Sequence Alignment

Ofer Gill and Bud Mishra

Courant Institute of Mathematical Sciences,
New York University, 251 Mercer Street, New York NY 10012, USA,
gill@cs.nyu.edu,
http://bioinformatics.nyu.edu/~gill/index.shtml

**Abstract.** In the bioinformatics literature, pairwise sequence alignment methods appear with many variations and diverse applications. With this abundance, comes not only an emphasis on speed and memory efficiency, but also a need for assigning confidence to the computed alignments through $p$-value estimation, especially for important segment pairs within an alignment. This paper examines an empirical technique, called SEPA, for approximate $p$-value estimation based on statistically large number of observations over randomly generated sequences. Our empirical studies show that the technique remains effective in identifying biological correlations even in sequen es of low similarities and large expected gaps, and the experimental results shown here point to many interesting insights and features.

## 1 Introduction

In the field of comparative genomics, an emphasis is placed on its functional genomics aspects. Most often we align two or more sequences, because we expect that the important areas selected from that alignment will point to a significant common biological function, even when we realize that there can be no absolute guarantee of this. In order to draw our attention very quickly to the most pertinent similar subsequences, it is necessary to compare the important areas of alignments and rank them in order of their relevance. For instance, by comparing alignments in related sequences to those of unrelated sequences with no common biological function, we may derive, for any alignment, the probability that its important areas occur by mere coincidence. This probability measure is also known as a $p$-value, and low $p$-values relate to high relevance rank.

Many $p$-value estimation techniques have been suggested and examined previously, for instance, Karlin-Altschul [4] and Siegmund-Yakir [7], but none have proven completely satisfactory. In this paper, we focus on using empirical results to improve the $p$-value approximation in case of alignments of noncoding nucleotide sequences of lengths varying from .5 Kb to 12 Kb, with expected large gaps and low similarities. These alignments are often computed with the complex but biologically faithful model involving piecewise-linear gap penalty

V.N. Alexandrov et al. (Eds.): ICCS 2006, Part II, LNCS 3992, pp. 638–645, 2006.

functions as in PLAINS [2]; nonetheless, other techniques such as LAGAN and EMBOSS have also proven effective. We demonstrate the effectiveness of a $p$-value approximation technique called SEPA (Segment Evaluator for Pairwise Alignments) as it selects and scores important segments pairs. Furthermore, for random sequences, we also empirically characterize how various alignment statistics, such as the segment pair lengths, scores, and magnitudes, distribute as a function of sequence lengths. From this analysis, the parameters for a $p$-value approximation are estimated, and used to demonstrate the method of sensitivity in distinguishing important homologies from unimportant chance occurrences of subalignments within sequences. Furthermore, SEPA is non-subjective, since it can easily be applied to any alignment tool. We will illustrate this advantage by using it to compare the results of PLAINS with LAGAN and EMBOSS. Because of these strengths and despite its empirical foundation, SEPA fulfills a practical computational need by speeding up the core search processes in comparative genomics.

## 2 Overview

We introduce some notations as follows: Assume the sequences to be aligned are $X$ and $Y$, and their respective lengths are $m$ and $n$, where $m \geq n$. Let $X_u$ and $Y_v$ denote respectively the $u^{\text{th}}$ character of $X$ and the $v^{\text{th}}$ character of $Y$, where $1 \leq u \leq m$ and $1 \leq v \leq n$.

Let us suppose that aligning $X$ and $Y$ with some arbitrary alignment tool produces an alignment $A$ of length $a$, where $m \leq a \leq m + n$. We will represent an alignment $A$ as follows: For each $i$, $A[i]$ denotes the $i^{\text{th}}$ position in alignment $A$, and it is represented as a pair of index coordinates $(u, v)$ taken from $X$ and $Y$, and this corresponds to $X_u$ and $Y_v$ being aligned to each other at position $i$ in $A$ if $u > 0$ and $v > 0$, or one of $X_u$ or $Y_v$ being aligned against a gap if either $v \leq 0$ or $u \leq 0$.

Next, let $A[i:j]$ denote the portion of alignment $A[i], A[i+1], \ldots, A[j]$. We will refer to $A[i:j]$ as a *strip* or *segment pair* from position $i$ to position $j$.

Let $ww(i)$ denote the penalty for a gap of length $i$. $ww(\cdot)$ can be any arbitrary function, but for this paper, we will assume it is a $d$-part piecewise-linear function where each successive slope is smaller than the previous one. A more specific version of this score-function is where $d = 1$, which is the affine function used in the Smith-Watermann algorithm.

Also, let $S(i, j)$ denote the score for strip $A[i:j]$ where the score is computed in the following way: $m_a$ is a score for each match, $m_s$ is the penalty for each mismatch, and $ww(\cdot)$ is used to penalize the gaps. To compute $S(i, j)$ from $A[i:j]$, each match and mismatch within it is added or deducted from the score individually, while each region of $X$ against a gap and $Y$ against a gap is penalized as a whole using $ww(\cdot)$ based on the length of that region.[1]

[1] For this paper, the $m_a$ reward is 1, the $m_s$ penalty is 0.346445, and the $ww(i)$ penalty is the piecewise-linear approximation of $1.01742 + 1.015724 \ln(i + 1)$. This selection empirically provides a good numerical contrast in scores between strips of high homology, and strips of lower homology.

Suppose we have a scheme that marks $r$ non-overlapping strips as important. Suppose that the endpoints for these strips are denoted as $(i_1, j_1), \ldots, (i_r, j_r)$. For each $k$, we wish to measure in some way how strip $A[i_k : j_k]$ provides a meaningful correlation between $X$ and $Y$. One common mathematical approach is to, given a certain null hypothesis, compute the $p$-value of $Pr(x \geq s)$ where $s = S(i_k, j_k)$. This $p$-value is known as the coincidental probability of obtaining a strip with score at least $s$. For this paper, we will assume the null-hypothesis is the behavior of important strips taken from pairwise-aligning randomly generated DNA sequences. Also, if the total scores of all strips is $t = \Sigma_{k=1}^{r} S(i_k, j_k)$, then $\zeta = Pr(x \geq t, y \leq r)$, the probability of obtaining at least a total score of $t$ using at most $r$ strips.

One should note that coincidental probabilities of the segments (both $p$-values and $\zeta$) are dictated by the scheme used to determine the segments as important. One scheme might deem strip $A[i : j]$ as important, but SEPA might not, and instead SEPA may consider a possibly overlapping strip $A[i' : j']$ as important. As a result, the formula for the $p$-values and $\zeta$ value could differ from one scheme to the other. For instance, in the method used to obtain important segments mentioned in Karlin-Altschul [4], $Pr(x \geq s) = 1 - exp(Kmne^{-\lambda s})$ holds. However, as argued later in this paper, the way SEPA obtains the segments from an alignment $A$ leads us to approximate the $p$-value as $Pr(x \geq s) = \frac{K}{\lambda} e^{-\lambda s}$.

## 2.1 Obtaining High-Scoring Strips from an Alignment

Given an alignment $A$ produced from sequences $X$ and $Y$, we produce important strips as follows: Given fixed constants $W$ and $\omega$, and $\rho$ (where $W$ is an integer, and $\omega$ and $\rho$ are real numbers in the range $[0, 1]$), let $W$ denote the window size to be used, $\omega$ denote the value used to prevent portions of $A$ of lowest match percentage from becoming considered as important strips, and $\rho$ denote the value used to filter away areas of $A$ that have too low of a $p$-value. We obtain our segment pairs in the following steps:

(1) For all $i$ from 1 to $a-(W-1)$, we compute $p_a(i)$, the percentage of entries in $A[i : i+W-1]$ where a match has occurred. Let $\mu$ and $\sigma$ denote the mean and standard deviation of our $p_a(\cdot)$ values. Next, for each $i$, we mark[2] $p_a(i)$ values as "special" if they exceed a threshold value of $\mu + \omega\sigma$. Hence, we filter away $A[i : i + W - 1]$ if it fails to meet this threshold value.

(2) For each $u$ and $u'$ (with $u \leq u'$), if $p_a(u)$, $p_a(u + 1)$, ..., $p_a(u')$ are all marked as "special", but $p_a(u - 1)$ and $p_a(u' + 1)$ are not, then we consider the strip $A[u : u' + W - 1]$ as important (i.e., we consider as important the strip starting the leftmost entry repsented by $p_a(u)$, up till the rightmost entry represented by $p_a(u')$).

(3) For each strip $A[i : j]$ deemed important, we trim it so that it starts and ends at a position in the alignment where a match occurred. Thus, if $i'$ is the smallest value such that $i' \geq i$ and $A[i']$ is a match position, and $j'$ is the largest

[2] The choice of using $\mu + \omega\sigma$ as the cutoff value instead of a fixed constant gives us the flexibility of catching important regions in the two sequences, regardless of how homologous they are to each other.

value such that $j' \leq j$ and $A[j']$ is a match position, then we trim strip $A[i:j]$ into strip $A[i':j']$.

(4) Next, we merge together any important strips that overlap. Namely, if we have two strips $A[i:j]$ and $A[k:l]$ such that $i \leq k \leq j$, then we merge these strips into one larger strip $A[i:\max(j,l)]$.

(5) With all strips now representing non-overlapping regions, we then proceed to give each strip $A[i:j]$ its corresponding score $S(i,j)$, as well as its $p$-value. We delete $A[i:j]$ if its $p$-value exceeds $\rho$, since that indicates that $A[i:j]$ may be coincidental. We can optionally also collect other information at this point, such as the length of each strip.

(6) The $r$ strips kept at this step are considered the "good" ones. We now compute $t$, the sum of the scores of the these strips. Using this value, we can compute $\zeta$, coincidental probability for all $r$ strips obtained.

Note that these steps for SEPA are similar to that of [2], except that the calculation for each segment pair's coincidental probability differs. Based on empirical experimentation, setting $W = 50$, $\omega = 0.5$, and $\rho = 0.5$ yields segment pairs that are reasonably long, non-coincidental, and have significantly higher matches than the alignment "background". We reasoned that since our method of obtaining segment pairs differs from that of Karlin-Altschul, then the method for computing $p$-values for each segment pair cannot build upon their assumptions.

### 2.2 Methods: Analyzing Segment Pairs

In order to approximate an appropriate $p$-value estimation for SEPA, we analyzed segment pairs' behavior over our assumed null hypothesis of alignments for randomly generated nucleotide sequences. For length values ranging from 1000 bp to 8000 bp, we generated 25 random sequences. We also generated 25 random sequences of length 500 bp. For each combination of these length pairs, we ran all 625 possible pairwise alignments using PLAINS, and analyzed results using SEPA where $\rho = 1$ (to avoid excluding any segments because of low $p$-value). The results for mean length-to-score and mean segment scores are shown in Fig. 1. From this analysis, we infer that both are constant in terms of $m$ and $n$. We also analyzed the behavior of $r$ (the number of segment pairs observed) and $t$ (the total score of all the segments) over these random sequences, and found them to depend on $m$ and $n$.[3] See the full version of this paper[4] for figures that elaborate further.

Since the average ratio of segment lengths to score is almost uniform in these plots, it suggests that the gap penalty used to score the strips can be treated as if it is a differently-weighted mismatch. Also, note that the $p$-values computed with the model studied by Siegmund-Yakir[7] differs mildly from the model using

[3] The mean of $r$ is $\approx 10^3 + \ln(7.95 \times 10^{-10}mn + 1.54 \times 10^{-7}(m+n) + 1.01)$, the variance of $r$ is $\approx 10^3 + \ln(1.93 \times 10^{-10}mn + 1.97 \times 10^{-7}(m+n) + 1.00)$, the mean of $t$ is $\approx 10^5 + \ln(4.29 \times 10^{-10}mn + 1.33 \times 10^{-8}(m+n) + 1.00)$, and the deviation for $t$ is $\approx \max\{100, -5.54 \times 10^{-5}j \cdot d + 4.63 \times 10^{-1}j + 1.04 \times 10^{-2}d - 65.01\}$, where $j = \min(m,n)$, and $d = \|m-n\|$.

[4] http://bioinformatics.nyu.edu/~gill/index.shtml

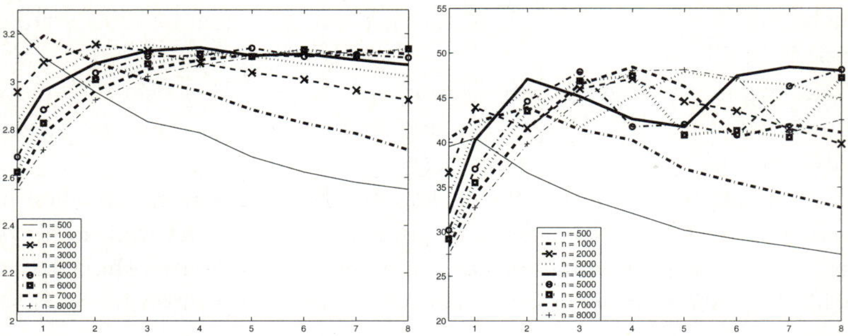


**Fig. 1.** Shown in the above graphs respectively are the mean length-to-score ratio and mean segment scores observed in the strips from aligning randomly generated DNA sequences. In these plots, a unique line is plotted corresponding to each value of $n$ in the thousand lengths ranging from 1000 to 8000, and $x$ represents the $m$ value divided by 1000, and $y$ represents the mean observed for that particular $m$ and $n$. These plots indicate that, for small $n$ values, the average length-to-score ratio and average score decrease with increasing $m$. However, asymptotically (for large $n$) the average length-to-score ratio and average segment scores stay roughly constant in terms of $m$ (at 3.1 : 1 and 45 respectively), and attempts at using Gumbel distributions failed to provide better approximations than this.

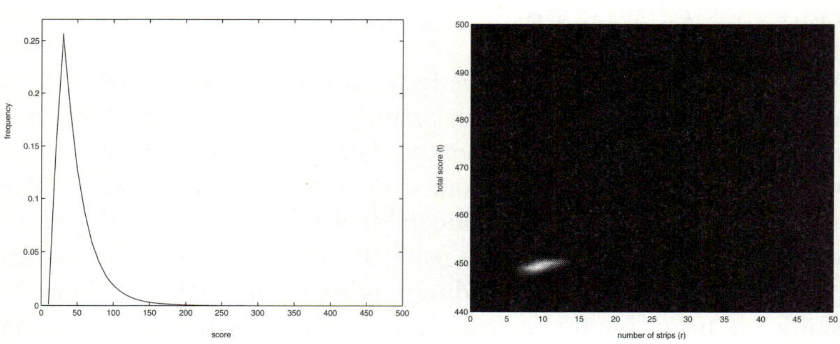


**Fig. 2.** On the left, is a plot of segment scores ($x$-axis) to frequency ($y$-axis) for randomly generated sequences using our assumption that segment score is length-independent. On the right, is a surface plot of observed frequency for number of segments $r$ and total score $t$ after adjusting both variables for average and variance behaviors, where lighter spots indicate higher frequencies. From the left graph, the tail of this plot for $x \geq 30$ indicates an exponential distribution of form $P(S = x) = Ke^{-\lambda x}$, where $K = 8.69 \times 10^{-2}$ and $\lambda = 3.26 \times 10^{-2}$. From the right graph, we observe that the majority of the data is concentrated in one area, and this area approximates to $e^c e^{-a_t T^2 + b_t T + c_t} e^{-a_r R^2 + b_r R + c_r}$, where $c = -183.90, a_t = 10.1, b_t = 9070, c_t = -2.04 \times 10^6, a_r = 0.241, b_r = 4.71, c_r = -27.5$.

the simplifying assumption that gaps are differently-weighted mismatches. For this reason, it is common for tools to ignore the effects of gaps in generating their $p$-values, much like BLAST, and in our case, SEPA as well. The left graph in fig. 2 plots segment scores to frequency, from which we derive our

**Table 1.** Shown here for PLAINS, EMBOSS, and LAGAN are the $r$, $t$, and $\zeta'$ values obtained from aligning genomic DNA sequences of lengths between 0.5 Kb and 12 Kb within human, mouse, dog, and fugu, where the pairs are biologically related and mainly noncoding DNA with expected large gaps and low homology regions. The conversion from $\zeta$ to $\zeta'$ was carried out for convenience in comparing lab results, where higher $\zeta'$ indicates results that are less coincidental. Also, note the loss of precision involved in reporting $\zeta'$ values. Hence, if for a paricular alignment, PLAINS and LAGAN receive $\zeta'$ values that differ by less than $1 \times 10^2$, then their $\zeta'$ values would "appear" equal in this table. Further information regarding the sequences used can be found in the full version of this paper at site `http://bioinformatics.nyu.edu/~gill/index.shtml`.

| Test Name | PLAINS | | | LAGAN | | | EMBOSS | | |
|---|---|---|---|---|---|---|---|---|---|
| | $t$ | $r$ | $\zeta'$ | $t$ | $r$ | $\zeta'$ | $t$ | $r$ | $\zeta'$ |
| fugu2r | 534.14 | 5 | 11.15 | 360.22 | 3 | 13.05 | 151.39 | 2 | 14.07 |
| HFortho1 | 734.82 | 7 | 10.94 | 349.33 | 4 | 14.18 | 374.35 | 5 | 13.05 |
| HFortho2 | 600.22 | 4 | 16.78 | 555.61 | 4 | 16.78 | 327.91 | 1 | 20.18 |
| HFortho3 | 637.52 | 7 | 14.53 | 259.44 | 3 | 19.05 | 409.99 | 5 | 16.71 |
| HFortho4 | 1004.97 | 10 | 21.74 | 529.16 | 5 | 0.00 | 367.86 | 4 | 0.00 |
| HFortho5 | 739.71 | 7 | 11.07 | 450.93 | 5 | 13.07 | 453.61 | 5 | 13.07 |
| human_mouse.1_1 | 676.29 | 10 | 8.46 | 52.36 | 1 | 18.29 | 186.98 | 2 | 17.00 |
| human_mouse.1_3 | 552.55 | 6 | 15.14 | 406.79 | 6 | 15.14 | 429.51 | 6 | 15.14 |
| human_mouse.3_9 | 1260.69 | 15 | 15.47 | 432.25 | 7 | 24.23 | 801.15 | 12 | 18.44 |
| human_mouse.4_3 | 262.19 | 3 | 15.44 | 74.91 | 1 | 17.79 | 176.83 | 2 | 16.59 |
| human_mouse.4_5 | 421.71 | 6 | 7.35 | 221.57 | 3 | 10.47 | 401.71 | 5 | 8.32 |
| human_mouse.7_11 | 594.32 | 8 | 9.06 | 164.10 | 2 | 15.44 | 476.71 | 7 | 9.99 |
| human_mouse.17_11 | 608.75 | 7 | 13.93 | 171.96 | 3 | 18.57 | 451.60 | 6 | 15.02 |
| human_dog.6_12 | 1284.79 | 14 | 13.88 | 548.19 | 7 | 21.23 | 394.04 | 6 | 22.44 |
| human_dog.7_16 | 1042.19 | 13 | 10.45 | 128.07 | 2 | 22.40 | 309.03 | 4 | 19.84 |

$p$-value approximation. Using it, we approximate that $P(x = s) = Ke^{-\lambda s}$, and our $p$-value of $P(x \geq s)$ is therefore:

$$P(x \geq s) = \int_s^\infty Ke^{-\lambda x}dx = \frac{K}{\lambda}e^{-\lambda s}$$

Further, notice that by this construction, $P(x \geq 30) = \frac{K}{\lambda}e^{-30\lambda} \approx 1$. We have designed our $p$-value estimation this way since strip scores below 30 are empirically observed to be unimportant.

Our next natural step, after obtaining $p$-values for each segment pair, is to provide a $p$-value estimate $\zeta$ for coincidental probability for the whole alignment, determined by the strips found. As mentioned earlier, we have learned that both $r$ and $t$ depend on sequence lengths $m$ and $n$. Hence, if $R$ and $T$ are supposed to be respectively the number of segment pairs and the total score of the segment pairs after adjusting for mean and variance based on sequence length, then the coincidental probability $\zeta = P(x \geq T, y \leq R)$. More specifically, $\zeta$ is the coincidental probability of seeing a total score of at least $T$ using at most $R$ segment pairs.

The right graph in figure 2 shows the distribution of $r$ and $t$ values observed from randomly generated sequences after adjusting for mean and variance. From it, our approximation of $T$ and $R$ for $P(x = T, y = R)$ gives us for $\zeta$ that[5]:

$$\zeta = P(x \geq T, y \leq R) = \int_T^\infty \int_0^R e^c e^{-a_t x^2 + b_t x + c_t} e^{-a_r y^2 + b_r y + c_r} dy dx =$$

$$= \frac{\pi e^{c + c_t + c_r + \frac{b_t^2}{4a_t} + \frac{b_r^2}{4a_r}}}{4\sqrt{a_t a_r}} \left(1 - Erf(\frac{-b_t + 2a_t T}{2\sqrt{a_t}})\right)\left(Erf(\frac{-b_r + 2a_r R}{2\sqrt{a_r}}) - Erf(\frac{-b_r}{2\sqrt{a_r}})\right)$$

[5] Note that $Erf(z) = \frac{2}{\sqrt{\pi}} \int_0^z e^{-x^2} dx$

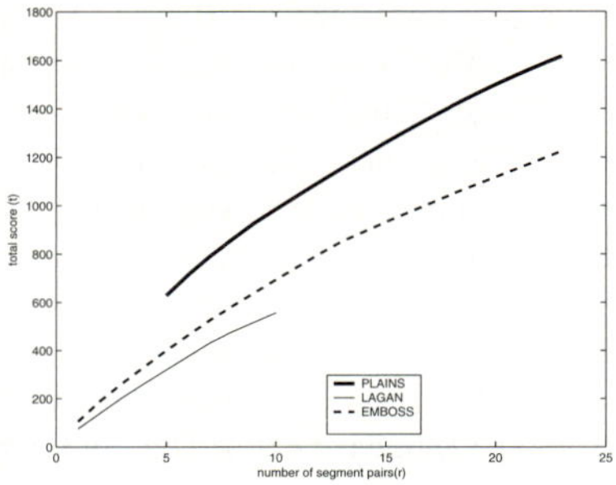


**Fig. 3.** In this figure, we observe the unadjusted $r$ and $t$ values produced by PLAINS, LAGAN, and EMBOSS from the human-mouse.3 – 9 experiment where we vary the $\rho$ variable used to filter our segment pairs. On each curve, we observed the $t$ and $r$ values of each tool when varying $\rho$ over various values from 0.1 till 0.9. Observe from table 1 that PLAINS performed poorly in terms of $\zeta'$ values for $\rho = 0.5$ for the human-mouse.3 – 9 experiments. The key to note here is that the $r$-to-$t$ ratio is almost uniform for all tools, but the $y$-intercept differs from one tool to the next, with PLAINS having the highest $y$-intercept. This means that, for any fixed $r$, PLAINS yields higher $t$ and hence better $\zeta'$ results. Many other experiments from table 1 have a similar plot to this one.

Table 1 shows a comparison of alignments for biologically related sequences in terms of unadjusted $r$ and $t$ values, and $\zeta'$ values, where $\rho = 0.5$, and $\zeta' = -\ln(\zeta)$. Note that PLAINS does not always yield the results of least coincidental probability in this table. This situation arises because the nature of PLAINS is to aggressively align as many regions as possible, thus leading to noticeably higher $r$ values than other tools, and hurts its $\zeta'$ values. In order to better understand how $r$ and $t$ vary for all the tools used, we chose to observe what happens when we vary $\rho$, and figure 3 elaborates in detail for one such experiment. From it, we see that, for any fixed $r$, the $r$ best segment pairs generated by PLAINS have smaller coincidental probabilities than the best $r$ segment pairs generated by other tools.

## 3 Conclusions, Future Work, and Acknowledgements

Our empirical analysis leads us to the conclusion that the SEPA-based $p$-value technique models coincidental probabilities much more accurately than the earlier technique employed in [2]. Furthermore, we note that aggressively incorporating too many segments into an alignment can lower the $\zeta'$ coincidental probability value, despite an apparent improvement in the total score, as illustrated by PLAINS. However, if we keep only the best $r$ segments from an alignment, the strength of PLAINS becomes obvious, since its $r$ segments are less coincidental than its competition.

However, in spite of the promising results from SEPA, there is still plenty of room for further improvements by using random portions of DNA from Human, Mouse, and Fugu instead of randomly generated DNA sequences. In that case, our concern shifts from the coincidental probability of a segment's score from aligning random DNA, to the coincidental probability of a segment's score from aligning unrelated random regions of organisms under comparison. Further extension includes development of better statistics that realistically captures the base-pair and coding/noncoding distributions within the sequences, as well as the effects of secondary and tertiary structures.

This project was sponsored by the Department of the Army Award # W81XWH-04-1-0307.

## References


1. Brudno, M., Do, C., Cooper, G., Kim, M.F., Davydov, E., Green, E.D., Sidow, A., Batzoglou, S.: LAGAN and Multi-LAGAN: efficient tools for large-scale multiple alignment of genomic DNA. Genome Research **13(4)** (2003) 721–731
2. Gill, O., Zhou, Y., Mishra, B.: Aligning Sequences with Non-Affine Gap Penalty: PLAINS Algorithm. Series in Mathematical Biology and Medicine **8** (2005). An unabridged version can be found at: `http://bioinformatics.nyu.edu/~gill/index.shtml`
3. Gu, X., Li, W.H.: The size distribution of insertions and deletions in human and rodent pseudogenes suggests the logarithmic gap penalty for sequence alignment. J. Mol. Evol. **40(4)** (1995) 464–473
4. Karlin, S., Altschul, S.F.: Methods for assessing the statistical significance of molecular sequence features by using general scoring schemes. Proc. Natl. Acad. Sci. USA **87** (1990) 2264–2268
5. Karlin, S., Altschul, S.F.: Applications and statistics for multiple high-scoring segments in molecular sequences. Proc. Natl. Acad. Sci. USA **90** (1993) 5873–5877
6. Rice, P., Longden, I., Bleasby, A.: EMBOSS: the European Molecular Biology Open Software Suite. Trends Genetics **Jun 16(6)** (2000) 276–277
7. Siegmund, D., Yakir, B.: Approximate $p$-Values for Local Sequence Alignments. The Annals of Statistics **28 (3)** (2000) 657–680
8. Smith, T.F., Waterman, M.S.: Identification of Common Molecular Subsequences. Journal of Molecular Biology **147** (1981) 195–197
9. States, D.J., Gish, W., Altschul, S.F.: Basic Local Alignment Search Tool. Journal of Molecular Biology **215** (1990) 403–410
10. Zhang, Z., Gerstein, M.: Patterns of nucleotide substitution, insertion and deletion in the human genome inferred from pseudogenes. Nucleic Acids Res. **31(18)** (2003) 5338-5348

# Multiple Sequence Alignment by Ant Colony Optimization and Divide-and-Conquer

Yixin Chen[1], Yi Pan[2], Juan Chen[3], Wei Liu[3], and Ling Chen[3]

[1] Department of Computer Science, Washington University in St. Louis, St. Louis, MO 63130, USA
chen@cse.wustl.edu
[2] Department of Computer Science, Georgia State University, Atlanta, GA 30303, USA
pan@cs.gsu.edu
[3] Department of Computer Science, Yangzhou University, Yangzhou 225009, China
lchen@yzcn.net

**Abstract.** Multiple sequence alignment is a common task in molecular biology and bioinformatics. Obtaining an accurate alignment of protein sequences is a difficult computational problem because many heuristic techniques cannot achieve optimality in a reasonable running time. A novel multiple sequence alignment algorithm based on ant colony optimization and divide-and-conquer technique is proposed. The algorithm divides a set of sequences into several subsections vertically by bisecting the sequences recursively using the ant colony optimization method. We also present two methods that adaptively adjust the parameters and update the pheromones to avoid local optimal traps. Experimental results show that the algorithm can achieve high quality solution and significantly reduce the running time.

## 1 Introduction

Multiple sequence alignment (MSA) is a common task in bioinformatics. It plays an essential role in detecting regions of significant similarity among a collection of primary sequences of nucleic acid or proteins. MSA is also used to help support the reconstruction of the phylogenetic trees, find the patterns of protein families, detect homology between new and existing sequences, and predict the secondary and tertiary structures of protein sequences.

Given a family $S = (S_1, ..., S_N)$ of *N* sequences, a multiple alignment of *S* is a new set of sequences $S' = (S_1', ..., S_N')$ so that all the strings in *S'* are of equal length and each $S_i'$ is generated from $S_i$ by inserting gaps. To evaluate the quality of the alignment, the *SP* (sum-of-pairs) function is the most popular scoring method. The goal of general multiple sequence alignment algorithms is to find out the alignment with the highest *SP* . MSA based on *SP* scores is an NP problem[1]. For practical reasons the dynamical programming method is only capable of aligning a maximum of a few sequences.

**Previous Methods.** There are numerous existing methods for MSA. The MSA program [2] can align up to ten closely related sequences. It is an implementation of the

V.N. Alexandrov et al. (Eds.): ICCS 2006, Part II, LNCS 3992, pp. 646 – 653, 2006.

Carrillo and Lipman algorithm [3] that identifies in advance the portion of the hyper-space not contributing to the solution and excludes it from computation. Stoye described a new divide and conquer algorithm DCA [4] sits on top of MSA and expands its capabilities. Recently, OMA, an iterative implementation of DCA [5] is proposed to speed up the DCA strategy and reduce memory requirements. ClustalW [6] is an efficient program based on the progressive algorithm by Feng and Doolittle [7]. Other progressive alignment methods, e.g. Dialign [8,9], assemble the alignment in a sequence-independent manner by combining segment pairs in an order dictated by their score, until every residue of every sequence has been incorporated in the alignment. Prrp proposed by Gotoh [10], AMPS algorithm [11], and Berger and Munsen's algorithm [12] use deterministic iterative methods, while SAGA [13], Gibbs sample algorithm for MSA [14], MSA algorithm based on simulated annealing [15] are stochastic methods. T-Coffee[16], MUSCLE[17], and PROBCONS[18] are integrative algorithms. Other approaches to MSA problem include Hidden Markov Model (HMM), an important method for sequence analysis [19]; POA, a novel MSA algorithm based on the partial order graph[13]; MAFFT, a MSA algorithm based on fast Fourier transform[20]; and the heuristic algorithm based on blocks [21].

Ant Colony Algorithm is a new evolution simulation algorithm proposed by M. Dorigo et al[22]. It successfully resolves the TSP problems by simulating the ants' food-hunting activities. It is recently used to solve the NP-complete problems of job-shop scheduling [23], quadratic assigning[24,25], and sequential ordering [26], and has shown exceptional performance in resolving complex optimization problems especially combinational optimization problem.

In this paper, we present an efficient algorithm for multiple sequence alignment based on ant colony algorithm and a divide-and-conquer strategy. The algorithm divides a set of sequences into several subsections vertically by bisecting the sequences recursively using the ant colony optimization method. The recursive procedure of bisecting the sequences ends when the length of all the sections is equal to one and hence the result of alignment is obtained. In the ant colony algorithm, we also present two methods that adaptively adjust the parameters and update the pheromones so as to avoid local convergence. Experimental results show that the algorithm can get high quality solution and reduce the running time.

## 2 Aligning Sequence Set by Divide and Conquer

Given a family $S = (S_1, ..., S_N)$ of $N$ sequences. Our algorithm first partitions the sequences into several subsets of segments vertically. Each sequence $S_i$ is divided at a suitable character $c_i$ near the midpoint. Here, $c_i$ is a position of a character in $S_i$ or a position between two characters where a gap should be inserted. Therefore we obtain two new families of shorter sequences with one family consist of the prefixes $S^P$ = ( $S_1^P(c_1),...,S_N^p(c_N)$ ) and one of suffixes $S^s$ = ( $S_1^s(c_1),...,S_N^s(c_N)$ ). Consequently, the algorithm partitions the prefix family and the suffix family in a recursive manner to reduce the original multiple sequences alignment problem into more alignment problems involving shorter sequences. The algorithm bisects the sequences using the ant colony optimization method. The recursive procedure of bisecting the sequences ends when the length of all the sections is equal to one and hence the result of alignment is obtained. The algorithm can be recursively described as follows:

```
Algorithm. D&C_Align(S,R)
Input•S=(S0,S1,..,SN-1): sequences to be partitioned; Out-
put•R: alignment of S;
begin
  If the maximal length of S0,S1,..,SN-1 is larger than 1
    Bisect S into Sp and  Ss using the ant algorithm;
    D&C_Align (Sp, Rp);  D&C_Align(Ss , Rs);
    R= Rp∪Rs;
  end if
end
```

## 3 Searching for the Cut-Off Points

In the proposed D&C_Align algorithm, we use the ant colony algorithm to iteratively bisect the sequence set $S$ at approximately optimum cut-off points. In the ant colony algorithm, each ant searches for a set of cutting points by starting from the midpoint of a sequence and moving on the other sequences to choose the matching characters. The searching result of the ant population is expressed by a 2-dimensional array $d$ where $d_{ij}$ denotes the position of cut-off of the $j$-th sequence in the $i$-th solution.

### 3.1 Bisecting the Sequence Set

Let $S_0,\ldots,S_{N-1}$ be $N$ sequences. An artificial ant starts from $S_0[m_0]$ which is a character randomly selected in the middle area of $S_0$, $m_0 \in [mid_0-\delta, mid_0+\delta]$, here $mid_0 = \left\lceil \frac{|S_0|}{2} \right\rceil$, $\delta$ is the search scope of the ants in $S_0$. The ant selects one character from or inserts a gap into the middle part of the sequences of $S_1,\ldots,S_{N-1}$ matching with $S_0[0]$. From the sequence $S_i$, $i$=1,2,…,$n$-1, the ant selects a character $S_i[j]$ by a probability determined by the matching score with $S_0[m_0]$, deviation of its location from the middle of $S_i$, $mid_i = \left\lceil \frac{|S_i|}{2} \right\rceil$, and pheromone on the logical edge between $S_i[j]$ and $S_0[m_0]$. In each step, the ant might select an empty character, which means a gap is inserted into the sequence in the alignment.

The other ants select their path in the same manner, but they start from different sequences. The $i$-th ant starts from $S_i$, and successively goes through $S_{i+1}$, $S_{i+2}$, …, $S_n$. When it reaches $S_n$, it continues going through $S_0$, $S_1$,…,$S_{i-1}$ to complete the path.

At the end of each iteration, the algorithm calculates the fitness of bisecting results of the ants. Then the pheromone on the logical edges is updated according to the fitness of the bisecting result passing through the edge. After a certain number of iterations, the solutions will converge to a near-optimal bisecting point.

### 3.2 Probability Function for Selecting the Characters

Let $P(k,l,n,m)$ be the probability for the ant starting from $S_k[l]$ to select the character $S_n[m]$ in $S_n$ . We define $P(k,l,n,m)$ as follows:

$$p(k,l,n,m)=\frac{a\times phe(k,l,n,m)+b\times mat(k,l,n,m)+c\times dev(k,l,n,m)}{\sum_{r=mid(n)-\delta}^{mid(n)+\delta}[a\times phe(k,l,n,r)+b\times mat(k,l,n,r)+c\times dev(k,l,n,r)]} \tag{1}$$

Here, $phe(k, l, n, m)$ is the pheromone on the logical edge between $S_k[l]$ and $S_n[m]$ , $mat(k, l, n, m)$ is the *SP* score of the characters the ant starting from $S_k[l]$ selected before reaching $S_n[m]$, $loc(k, l, n)$ is the start location in $S_n$ when the ant searches the character matching with $S_k[l]$ within $S_n$ , $dev(k, l, n, m)$ is the location deviation between $m$ and the the character which $S_k[l\text{-}1]$ has selected in $S_n$ , $a$,$b$,$c$ are the weights of pheromone, matching score and location deviation. $\delta$ is the range in $S_n$ for the ant selects the character matching with $S_k[l]$.

When the ant selects a character in $S_n$ to match with $S_k(l)$, it first calculates the selecting probabilities for all characters in a certain range of $S_n$. If $S_n(m)$ has the largest probability and $S_n(m)$ is equal to $S_k(l)$, then the ant selects $S_n(m)$. Otherwise, the ant selects the characters (include the gap) according to their selecting probabilities by the "roulette wheel" method. The characters in $S_n$ will have higher probability to be selected if it has higher pheromone on the logical edge connecting with $S_k[l]$, higher matching score with $S_k[l]$, and less deviation to $mid(n)$.

### 3.3 Updating the Pheromone

Suppose the $r$-th ant partitions the $S_i$ at position $c_i^r$ into two new families of shorter sequences, one family consisting of the prefixes $S^p = (S_1^P(c_1),\ldots,S_N^p(c_N))$ and one of suffixes $S^s = (S_1^s(c_1),\ldots,S_N^s(c_N))$ are obtained. To evaluate the bisecting result, the fitness of the $r$-th ant is defined as (2) based on our score estimating algorithm SE recently developed [27]:

$$fitness(r)=\sum_{i=1}^{N-1}\sum_{j=1}^{i-1}SE(S_i^p(c_i^r),S_j^p(c_j^r))+\sum_{i=1}^{N-1}\sum_{j=1}^{i-1}SE(S_i^s(c_i^r),S_j^s(c_j^r)) \tag{2}$$

After each ant completes a solution, the pheromone on its logical edges should be updated according the fitness of the solution. Let $evap1$ be the evaporation coefficient, $0\le evap1\le 1$, the value of $phe(k, l, n, m)$ is updated according to formula (3).

$$phe(k,l,n,m)=(1-evap1)\times phe(k,l,n,.m)+evap1\times\sum_{r\in set(k,k,n,m)}fitness(r) \tag{3}$$

Here $set(k,l,n,m)=\{r|\ c_k^r=l,\ c_n^r=m\}$ is the set of ants whose solution includes the logic edge between $S_k[l]$ and $S_n[m]$.

To avoid local convergence, the pheromone on the logical edges should be adjusted adaptively. In the ant algorithm, if the scores of the alignments in iteration are all less than the average score of the alignments in the $d$ iterations before, the pheromone on the logical edges should be adjusted. Here $d$ is a parameter which is adjustable according to the number of the iterations:

$$d = k_1 + \left\lceil \frac{k_2}{generation} \right\rceil \tag{4}$$

Here $k_1$, $k_2$ are constants, and *generation* is the number of current iteration. It is obvious that in the early iterations $d$ is assigned a greater value since the pheromone is evenly distributed. In the later iterations, the $d$ value becomes smaller since the pheromone might accumulate on a few logical edges. The rule of pheromone adjusting is: if the pheromone is more than a threshold, it should be evaporated according to a coefficient $evap2$ ($0.5 \le evap2 \le 1$) :

$$phe(k,l,n,m) = evap2 \times phe(k,l,n,m) \tag{5}$$

By reducing the pheromone on these logical edges, they have less probability to be chosen by the ants in the later iterations.

### 3.4 Adaptive Parameters *a*, *b* and *c*

The ants tend to select a path with more pheromone than other paths even if the difference is very slight. As a consequence, the paths the ants select would be concentrated on several paths where the edges have higher pheromone intention. This could result in local optima. To avoid local convergence, we adjust the parameters $a$, $b$, and $c$ in probability function (1) adaptively. In the early stages, we set the values of $b$ and $c$ be larger than $a$ so that the heuristic information of matching score and the location deviation have more importance. This helps the ants to search in a wide range to avoid local convergence in the early stages. In later iterations, we should make full use of the feedback information represented by the pheromone. Therefore, we increase $a$, and decrease $b$ and $c$ adaptively. Combined with the techniques of global and local pheromone updating, the algorithm can successfully speedup the convergence while maintaining the diversity of the solutions to avoid being trapped into local optimal solutions. The initial values of $a$, $b$, and $c$ are $s_a$, $s_b$, and $s_c$ respectively. At the end of each iteration, the algorithm adjusts $a$, $b$, and $c$ as follows:

$$a = a \times (1 + v_a) \qquad if\,(a > t_a) \;\; a = s_a \tag{6}$$

$$b = b \times (1 - v_b) \qquad if\,(b < t_b) \;\; b = s_b \tag{7}$$

$$c = c \times (1 - v_c) \qquad if\,(c < t_c) \;\; c = s_c \tag{8}$$

where $v_a$, $v_b$, $v_c$ are rates of update in the range of (0,1), and $t_a$, $t_b$, $t_c$ are upper/lower bounds for a, b, and c, respectively.

## 4 Experimental Result

We test our algorithm D&C_Align using the sequences randomly selected from benchmark database BAliBASE 2.0. As a comparison, we also test SAGA, a multiple sequence alignment program based on an evolutionary algorithm.

Experimental results show that algorithm D&C_Align improves the alignment accuracy for long sequences and requires less computational time than SAGA. The comparison of the running times of D&C_Align and SAGA, on sequences with different lengths is shown in Figure 1.

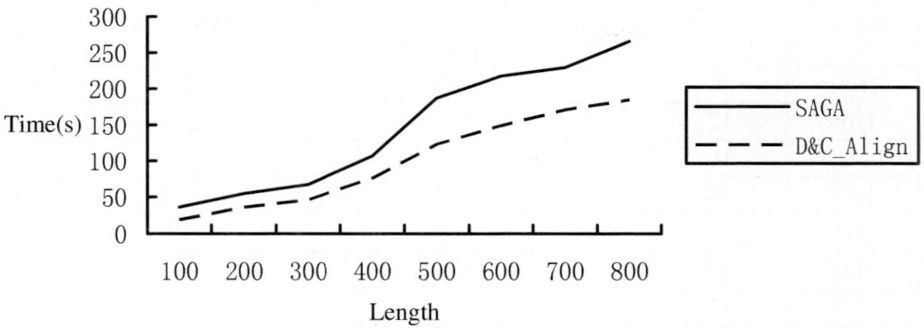


**Fig. 1.** Comparison of D&C_Align and SAGA on ten sequences

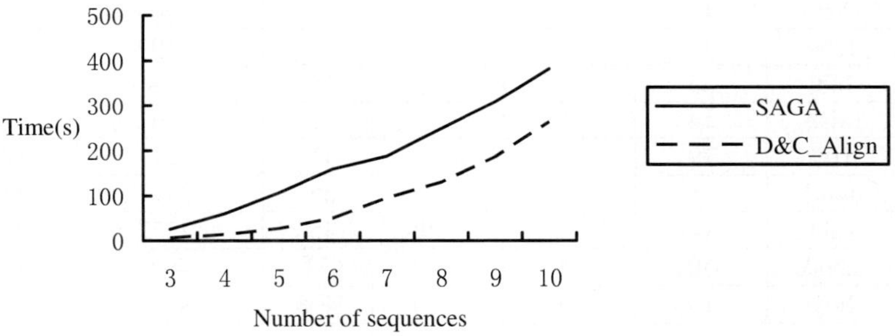


**Fig. 2.** Comparison of D&C_Align and SAGA of sequences with length of 1000

Figure 2 shows the comparison of the running times of D&C_Align and SAGA on sequences sets with different numbers of sequences, all with 1000 characters.

Comparing with SAGA, the running time of D&C_Align is much faster than that of SAGA. For example, for 4 sequences with 200 characters, the running time of SAGA is 200 to 400 seconds, while the running time of D&C_Align is 10 to 20 seconds.

Table 1 shows the experimental results of D&C_Align on the randomly selected sequences sets from the benchmark database BAliBASE 2.0. The last column in the table is the score of the reference alignment offered by BAliBASE which is very close to the theoretically optimal alignment.

In bioinformatics, if the proportion of similarity in a set of sequences is below 30%, most of the sequence alignment methods cannot find the correct alignment. Therefore, it is called the "twilight zone". The experiment results in Table1 show that D&C_Align can align not only sets of similar sequences but also those sets in the "twilight zone". It can improve the solution quality over the near-optimal reference alignments in a number of cases in Table 1.

**Table 1.** Experimental results of D&C_Align

| Name of sequences | Similarity | Number of sequences | Length of the longest sequence | Running time (s) | Score of D&C_align | Score of reference alignment |
|---|---|---|---|---|---|---|
| 1fmb | 49 | 4 | 106 | 5 | 823 | 828 |
| 1r69 | 12 | 4 | 78 | 3 | -274 | -289 |
| 1tvxA | 16 | 4 | 69 | 3 | -250 | -268 |
| 1ubi | 16 | 4 | 94 | 4 | -298 | -342 |
| 2trx | 15 | 4 | 102 | 4 | -301 | -293 |
| 1idy | 13 | 5 | 67 | 3 | -282 | -285 |
| 2fxb | 51 | 5 | 63 | 3 | 760 | 784 |
| 1krn | 45 | 5 | 82 | 5 | 749 | 814 |
| 1uky | 17 | 4 | 218 | 11 | -740 | -731 |
| 3grs | 17 | 4 | 237 | 12 | -778 | -768 |
| 1bavA | 15 | 5 | 199 | 18 | -1510 | -1508 |
| 1bbt3 | 14 | 5 | 192 | 18 | -1403 | -1395 |
| 1sbp | 19 | 5 | 263 | 26 | -1199 | -1193 |
| 1adj | 35 | 4 | 418 | 35 | 301 | 306 |
| 1ajsA | 14 | 4 | 387 | 31 | -1462 | -1458 |
| 1lvl | 19 | 4 | 449 | 41 | -1210 | -1214 |
| 1pamA | 18 | 5 | 572 | 66 | -3487 | -3443 |
| gal4 | 14 | 5 | 395 | 45 | -2257 | -2281 |
| laboA | 30 | 15 | 80 | 671 | 403 | 790 |
| lidy | 19 | 27 | 60 | 3498 | -8905 | -4701 |

## References


1. Wang L., Jiang T., On the complexity of multiple sequences alignment, Journal of Computational Biology, vol. 1(1994):337-348,
2. Lipman DJ, Altschul SF, Kececioglu JD: A tool for multiple sequence alignment. Proc.Natl. Acad. Sci. USA 86, (1989) 4412-4415.
3. Carrillo H, Lipman DJ: The multiple sequence alignment problem in biology.SIAM J. Appl. Math. Vol. 48, (1988)1073-1082.
4. Stoye J, Moulton V, Dress AW: DCA: an efficient implementation of the divide-andconquer approach to simultaneous multiple sequence alignment. Comput. Appl. Biosci. Vol. 13(6), (1997) 625-626.
5. Reinert K, Stoye J, Will T: An iterative method for faster sum-of-pair multiple sequence alignment. Bioinformatics , vol. 16(9), (2000) 808-814.
6. Thompson, JD, Higgins, DG and Gibson, TJ (1994) CLUSTAL W: improving the sensitivity of progressive multiple sequence alignment through sequence weighting, position specific gap penalties and weight matrix choice. Nucleic Acids Research,(1994), vol.22,No.22.4673-4680.
7. Feng D-F, Doolittle RF: Progressive sequence alignment as a prerequisite to correct phylogenetic trees. J. Mol. E, vol. 25, (1987)351-360
8. B. Morgenstern and T. Werner DIALIGN: Finding local similarities by multiple sequence alignment. Bioinformatics vil. 14, (1997)290-294.

9. Morgenstern B DIALIGN 2: improvement of the segment-to-segment approach to multiple sequence alignment. Bioinformatics, , vol.15, no. 3, (1999) 211-218.
10. Gotoh O: Significant Improvement in Accuracy of Multiple Protein Sequence Alignments by Iterative Refinements as Assessed by Reference to Structural Alignments. J. Mol. Biol. Vol.264, no.4, (1996)823-838.
11. Barton GJ, Sternberg MJE: A strategy for the rapid multiple alignment of protein sequences: confidence levels from tertiary structure comparisons. J. Mol. Biol. Vol. 198, (1987)327-337.
12. Berger MP, Munson PJ: A novel randomized iterative strategy for aligning multiple protein sequences. Comput. Appl. Biosci. Vol. 7, (1991)479-484.
13. Notredame C, Higgins DG: SAGA:sequence alignment by genetic algorithm. Nucleic Acids Res. Vol. 24, (1996)1515-1524.
14. Lawrence CE, Altschul SF, Boguski MS, Liu JS, Neuwald AF, Wootton JC: Detecting subtle sequence signals: a Gibbs sampling strategy for multiple alignment. Science vol. 262, (1993)208-214.
15. Kim J, Pramanik S, Chung MJ: Multiple Sequence Alignment using Simulated Annealing. Comp. Applic. Biosci. Vol. 10, no.4, (1994)419-426.
16. Notredame C, Higgins DG, Heringa J T-Coffee: A novel method for fast and accurate multiple sequence alignment. J Mol Biol. 2000 Sep 8;302(1):205-217
17. Edgar RC. MUSCLE: a multiple sequence alignment method with reduced time and space complexity. BMC Bioinformatics. 2004 Aug 19;5(1):113.
18. Do, CB, Brudno, M., and Batzoglou, S. ProbCons: Probabilistic.consistency-based multiple alignment of amino acid sequences. In. Proceedings of the Thirteenth National Conference on Artificial. Intelligence, (2004) 703–708.
19. A.Krogh. An introduction to hidden markov models for biological sequences. In: S. L. Salzberg, D. B. Searls and S. Kasif. Computational Methods in Molecular Biology. Elsevier, (1998) 45-63
20. Kazutaka Katoh , Kazuharu Misawa1 , Kei-ichi Kuma and Takashi Miyata MAFFT: a novel method for rapid multiple sequence alignment based on fast Fourier transform. Nucleic Acids Res.vol.30, no.14: (2002)3059-3066
21. A heuristic algorithm for multiple sequence alignment based on blocks. P.Zhao and Tao Jiang J. Combinatorial Optimization vol.5,no.1 (2001)95–115
22. Dorigo M., Maniezzo V., Colorni A. Ant system: Optimization by a colony of cooperating agents [J]. IEEE Transactions on Systems, Man, and Cybernetics-Part B, vol. 26, no.1, (1996) 29-41.
23. Gutjahr J.W. A Graph-based Ant System and its convergence. Future Generation Computer Systems, vol.16, no.8,(2000) 873-888.
24. Talbi E.G.,Roux O.,Fonlupt C.,Robillard D.Parallel Ant Colonies for the quadratic assignment problem. Future Generation Computer Systems, vol.17, no.4 (2001) 441-449.
25. Maniezzo V., Carbonaro A.. An ANTS heuristic for the frequency assignment problem, Future Generation Computer Systems, vol.16, no.8, (2000) 927-935.
26. Gambardella L.M., Dorigo M., Ant-Q: A reinforcement learning approach to the traveling salesman problem, Proceedings of the 11th International Conference on Evolutionary Computation, IEEE Press, (1996) 616-621
27. Yixin Chen, Yi Pan, Ling Chen, Juan Chen, Partitioned optimization algorithms for multiple sequence alignment, To appear in The Second IEEE Workshop on High Performance Computing in Medicine and Biology, Vienna, Austria

# COMBAT: Search Rapidly for Highly Similar Protein-Coding Sequences Using Bipartite Graph Matching

Bing Sun, Jacob T. Schwartz, Ofer H. Gill, and Bud Mishra

Courant Institute of Mathematical Sciences, New York University
251 Mercer Street, New York, NY 10012, USA
bingsun@cs.nyu.edu, jack@brainlink.com, gill@cs.nyu.edu, mishra@nyu.edu

**Abstract.** Comparing vertebrate genomes requires efficient cross-species sequence alignment programs. We describe COMBAT, a new mer-based method which can search rapidly for highly similar translated genomic sequences, with the stable-marriage algorithm with incomplete lists (SMI) as a filter scheme. We apply the COMBAT program to the comparative analysis of the human with the most recent bovine genome assemblies, and 84%$\sim$95% of the homologous blocks identified by this program are confirmed by BLASTZ.

## 1 Introduction

In the past decade many genome projects have produced complete genomes for increasingly many organisms. Since 1999 many tools have proven effective in aligning large genomic sequences of two closely related organisms. These include MUMmer [4], GLASS [1], AVID [2], DIALIGN [8], LAGAN [3], BLASTZ [9], BLAT [7], and etc. Common characteristics in many of these programs are: i) they assume the conserved regions of the sequences being aligned appear in the same order and orientation, as is particularly likely for closely related organisms; ii) they build tables of scores for matches and mismatches between amino acids or nucleotides incorporating penalties for insertions or deletions, and from these constructs obtain mathematically 'optimal' alignments; iii) many local alignment programs search for exact or spaced exact matches, and then extend the local similarities in both directions in passes directed by specified scoring functions.

However, certain shortcomings limit the use of many of these programs. First, genomic order and orientation need not be conserved between species of interest. Secondly, the scoring matrix (eg. a PAM or a BLOSUM matrix) most appropriate for aligning a set of sequences should be determined by the level of relatedness of sequences. Hence the percentage of similarity between two genomes has to be preestimated to choose a proper scoring matrix. Also, the fact that the rate of evolution varies across the genome makes it impractical to pick a universal scoring matrix or a set of gap costs [5]. Finally, by using the "match and extend" strategy many local alignment algorithms pay a steep cost in extending short matches in both directions.

V.N. Alexandrov et al. (Eds.): ICCS 2006, Part II, LNCS 3992, pp. 654–661, 2006.

This paper describes a novel local alignment algorithm, called COMBAT (Clean Ordered Mer-Based Alignment Tool), which addresses the above challenges by implementing the following two critical stages: i) generating an index of all overlapping K-mers in translated genomic sequences, where the index represents the blocks to which a K-mer belongs and is used to search efficiently for homologous blocks. ii) using the *SMI* algorithm to find the optimal one-to-one mapping from a list of multiple local mappings and so form a global matching map. COMBAT makes no assumption of gene order and orientation, does not utilize any sophisticated scoring matrix, and does not have the expensive "extend" stage of many local alignment programs. The COMBAT algorithm is fully described in the next section.

## 2 Method for Pairwise Genomic Comparison

The goal of COMBAT is to identify protein-encoding regions in genomic sequences using a genome comparison approach. Let us suppose the two genomes being compared are called genome $A$ and genome $B$. We define some of the terms used below, and present the parameters involved in Table 1.

J-interval: a continuous genomic sequence of length $J$. Adjacent J-intervals are spaced $J/2$ bases apart. The J-interval index is simply called J-index.

Partner Interval Pair (PIP): an instance of PIP $(a, b)$ consists of a J-interval $a$ in genome $A$ and a J-interval $b$ in genome $B$ if there are more than $T$ K-mers shared by both $a$ and $b$.

**Table 1.** The involved parameters in COMBAT program

| | |
|---|---|
| $J$ | The length of a J-interval |
| $K$ | The K-mer size |
| $T$ | The minimum number of common K-mers required in any PIP |
| $S$ | The actual number of common K-mers in a PIP |
| $E, F$ | The chaining filtering criterion requires that there must be at least $F$ PIPs, each no further than $E$ intervals from each other. |

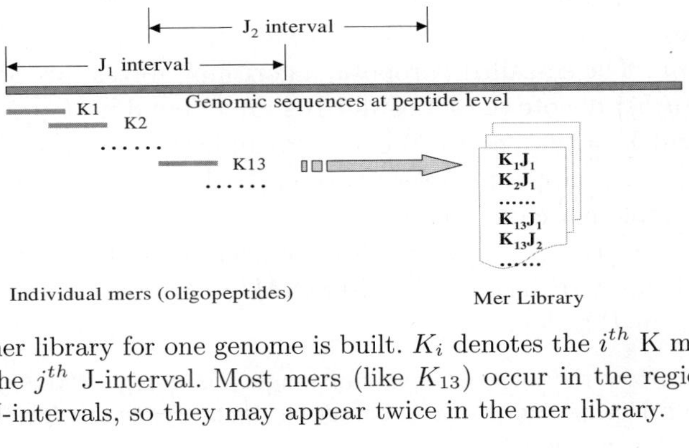


**Fig. 1.** How mer library for one genome is built. $K_i$ denotes the $i^{th}$ K mer. $J_j$ denotes the index of the $j^{th}$ J-interval. Most mers (like $K_{13}$) occur in the region covered by two adjacent J-intervals, so they may appear twice in the mer library.

The COMBAT algorithm comprises the following steps:

***Step 1: Build Clean Ordered Mer Libraries***
First, we translate genomic sequences of genome $A$ and $B$ in all three frames over both forward and reverse orientations. After choosing a mer-size $K$ we generate overlapping K-mers starting at every base position, ignoring mers in repeats annotated by RepeatMasker. We cover the considered genome with J-intervals. The "representation of position" that we attach to each K-mer is the index of each J-interval to which it belongs, shown in Figure 1. We keep only one copy of duplicate K-mers in each J-interval from the mer library. This makes all the K-mers unique in every interval. Next we sort the mer library by the mer sequences. Such mer libraries for genome $A$ and genome $B$ are built separately.

***Step 2: Search For Common Mers***
Next we scan the clean ordered mer libraries prepared in the manner just described to compute offsets between the pairs of matching mers found. When the offset $d_{ij} = A_i - B_j$ exists, it is the J-index difference between the $i$-th mer occurring in genome $A$ and the matching mer in genome $B$. It is easy to recover the J-index of genome $B$ using $d_{ij}$. We then sort this list of mers/offset pairs by their offsets and their J-indexes on genome $A$. For each J-index of genome $A$ in this list we count the number of K-mers that have the same offsets. We keep only those intervals as PIPs whose number of common K-mers is beyond the threshold $T$ for the next step.

***Step 3: Find One-to-One Correspondence***
As a natural result of genome duplications one region of one genome might match several regions of another genome. Usually the single best, orthologous match for each conserved region gains most biologists's attention. BLASTZ uses the axtBest program to produce the best alignments [9]. In this paper we first present the application of the stable marriage ($SM$) problem in large-scale genome comparison as an alignment filter[1]. The well-know $SM$ problem was introduced by Gale and Shapley [6]. The problem can be stated as below: given two finite equal-sized sets of players, called *men* and *women* ($m_i \in men, w_j \in women$), where each $m_i/w_j$ ranks $w_j/m_i$ in strict order forming his/her preference list, find a one-to-one stable matching $M$ between the two sexes. $M$ is "stable" if there is no two couples $(m, w)$ and $(m', w')$ in $M$ such that $m$ prefers $w'$ to $w$ and $w'$ prefers $m$ to $m'$. The so-called "proposal algorithm" solves this problem.

Let $P = \{(a, b)\}$ denote the set of all PIPs $(a, b)$ found in step 2, $X = \{a \mid \exists b : (a, b) \in P\}$, and $Y = \{b \mid \exists a : (a, b) \in P\}$. $P$ can be viewed as a bipartite graph which is a multiple mapping between two sets of J-indexes. We wish to find $M$, a one-to-one stable matching, from $P$. Since normally some $a$ in $X$ matches to a true subset of $Y$, this $SM$ problem becomes the relaxed version — the Stable Marriage Problem with Incomplete Lists ($SMI$). We form a preference list for each J-interval in $P$ as follows:

[1] One might borrow the idea of maximum weight matching ($MWM$) for this task. The $MWM$ solver maximizes the cumulative similarity, thus might not give single best matches for individual regions.

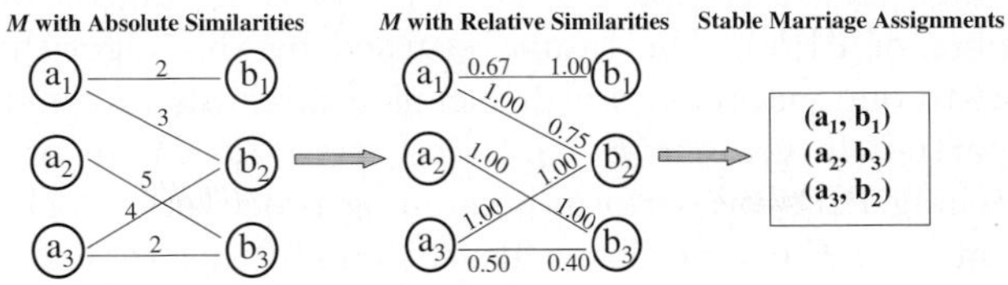


**Fig. 2.** An example of the stable marriage problem procedure. First compute relative similarities from absolute similarities in the bipartite graph, then follow the *SMI* algorithm to find the stable marriage assignments. $a_1 \sim a_4/b_1 \sim b_4$ denote the J-indexes on genome $A/B$. In multiple mapping $M$ with absolute similarities the numbers on the edges show the number of K-mers shared by the partner intervals. The numbers associated with an edge in the middle panel are the relative similarities for a pair of partner intervals.

1. A measure of absolute interval similarity $S$ is calculated, and $S = \{S_{(a,b)} \mid (a,b) \in P\}$, where $S_{(a,b)}$ denotes the number of K-mers shared by a PIP (a,b).
2. Relative similarities are computed subsequently as fractions of the absolute similarities of the best match partner for any J-interval in $P$. Then each J-interval $j$ ranks its match partners in strict order of their relative similarities to $j$, forming $j'$s preference list.

In the example in Figure 2, $b_2$ is the best match for $a_1$, so we set $R_{(a_1,b_2)} = 1.00$. The relative similarity for the other match partner of $a_1$ is computed as a fraction of $S_{(a_1,b_2)}$. Thus, $R_{(a_1,b_1)} = \frac{S_{(a_1,b_1)}}{S_{(a_1,b_2)}} = \frac{2}{3} \approx 0.67$. Relative similarities are asymmetric. Under the marriage interpretation, this means that any two match partners like each other to the different extent. We modify the proposal algorithm and explain the *SMI* algorithm used by COMBAT as follows:

```
1. X={a},Y={b},M={}. Every a and b has an ordered preference list.
2. WHILE X is not empty, LOOP
3.   choose an interval a from X
4.   b=the first interval on a's list(If have ties, randomly choose one)
5.   IF a is not on b's preference list, THEN
6.        delete b from a's list;
7.        IF a's preference list is empty, THEN
8.           delete a from X; goto line 2
9.        ELSE goto line 4
10.  ELSE
11.       IF (x, b) is in M for some x in X, THEN
12.          remove (x, b) from M; add x to X;
13.       add (a, b) to M
14.       FOR each successor x (x ranks after a) in b's list, LOOP
15.          delete x from b's list, and b from x's list;
16.       END LOOP
17. END LOOP
18. RETURN M
```

This *SMI* algorithm's complexity is $O(n^2)$ in time, and is linear in space ($n$ is the number of PIPs). The result returned by this algorithm is a list of incomplete one-to-one mapping, which means J-intervals in genome $A$ map to at most one partner in genome $B$, and vice versa. Lastly, in order to remove randomly matching PIPs we perform a chaining procedure which requires that there must be at least $F$ partner intervals, each no further than $E$ intervals from each other. This step is not necessary if we choose strict values of $J$ and $K$.

## 3 Results for the Human/Cow Genome Comparison

We have applied COMBAT to Human Assembly (hg17, May 2004) and Cow Assembly (bosTau1, Sep. 2004, BCM HGSC Btau_1.0), both from the UCSC Genome Bioinformatics Site. As an example illustrating our results, we take chromosome I from hg17 and the first 33,000 cow scaffolds, and align them by COMBAT. These two sequences are approximately 250MB in size. Let us call the first sequence *chr1*, and the second sequence *cow1*. The resulting alignment maps using different configurations are shown for positive strands in Figure 3. Figures 3-(1),(2),(4), and (5) are the results produced by COMBAT, with each plus sign representing the index coordinates of a pair of matching intervals found by COMBAT. Figures 3-(3) and (6) are the matches produced by BLASTZ, filtered by the axtBest program [9] (downloaded from the UCSC Genome Bioinformatics Site and transformed to fit our J-intervals context), with each dot representing the index coordinates of the starting positions of two matched regions. The BLASTZ result is transformed twice according to two values of $J$ used.

The chaining criterion used by COMBAT turns out to be relatively insensitive to the value of $E$ used (see Figure 3-(1) and 3-(2)). To evaluate COMBAT, we have tested the appearance of every matching pair of intervals found by COMBAT in the BLASTZ result (transformed by the same $J$ used by COMBAT)[2]. In Figure 3-(1), 95% of the 625 partner interval pairs found by COMBAT are true positives. In the other direction, out of 8,389 matching regions in the BLASTZ result, 7% are confirmed by COMBAT. In Figure 3-(4), there are 84% true positives out of 1235 PIPs, and they cover 11% of the BLASTZ result. In Figure 3-(5), there are 85% true positives out of 971 PIPs, and they cover 9% of the BLASTZ result. This high specificity indicates a promising wide use of COMBAT. The low coverage is not surprising because only highly similar protein-coding regions are expected to be found.

The computational core of the COMBAT algorithm was implemented as a C++ program and all experiments were performed on NYU Bioinformatics Group's cluster of Pentium IV machines with 3 GB memory running RedHat

[2] Consider a pair of matching J-interval ($a$, $b$) in COMBAT result as a true positive case if there exists a pair of matching regions ($x$, $y$) ($a$ and $x$ in genome $A$, $b$ and $y$ in genome $B$) in BLASTZ result and one of the following conditions is satisfied: 1) $a$ is contained in $x$ and $b$ is contained in $y$; 2) $x$ is contained in $a$ and $y$ is contained in $b$; 3) the starting positions of $a/b$ is within $J$ bases of those of $x/y$, respectively; 4) the ending positions of $a/b$ is within $J$ bases of those of $x/y$, respectively.

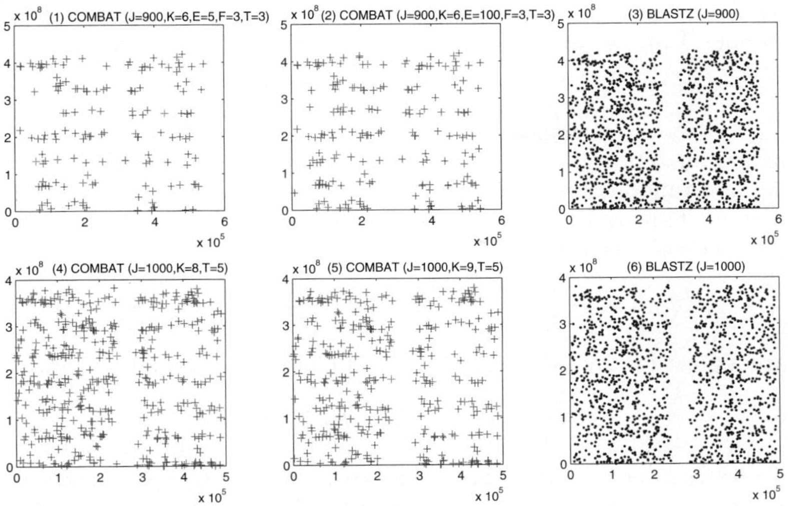


**Fig. 3.** Alignment maps on positive strands between *chr1* and *cow1*, with the X-axis showing the J-indexes along the *chr1* sequence, and the Y-axis showing those along the *cow1* sequence. (1),(2),(4), and (5) are the results produced by COMBAT; (3) and (6) are the transformed results produced by the BLASTZ. (4) and (5) are done without using the chaining procedure.

Linux 7.3. To compare 0.25 Gb of human sequence against 0.24 Gb of cow sequence ($\sim 1/10$ of total genomes) and produce the one-to-one mapping list of highly similar regions, it took 23 CPU hours under the configuration shown in Figure 3-(1), and took 2 CPU hours under the configuration shown in Figure 3-(4). For the sake of performance comparison, we point to the published report of BLASTZ taking 481 days of CPU time to align 2.8 Gb of human sequence against 2.5 Gb of mouse sequence on a cluster of 1024 833-Mhz Pentium III [9].

## 4 Error Estimation

Consider two random J-intervals $a$ in genome $A$ and $b$ in genome $B$ (each of length $J$ over an alphabet of 20 amino acids and 1 stop codon) . For the sake of simplicity, we will consider these intervals in one orientation only. Let $P_k$ denote the probability that there is a common K-mer at any position. Assuming that letters occur at any given position with equal probability and independency, we get $P_k = 1/(21)^K$. Let the positive-valued random variable $w$ denote the number of common K-mers in $a$ and $b$. We can show that $w$ follows a Poisson distribution with parameter $\lambda_w = J^2 P_k$. The expectation of a new random variable $\binom{w}{i}$ can be estimated by considering all possible $\binom{J}{i}$ subsets of K-mers from $a$ and counting the probability of each such subset having exact matches with $i$ K-mers in $b$.

$$E\left[\binom{w}{i}\right] = \binom{J}{i}(JP_k)((J-1)P_k)\cdots((J-i+1)P_k) \approx \frac{J^{(i)}}{i!}\left(\frac{J}{21^K}\right)^i \approx \frac{\left(J^2/21^K\right)^i}{i!} \tag{1}$$

Using Brun's sieve, the probability that two randomly selected J-intervals from genome $A$ and genome $B$ have exactly $m$ K-mers in common is:

$$Pr[w = m] = e^{-(J^2/21^K)} \frac{(J^2/21^K)^m}{m!} \tag{2}$$

Using parameters of this Poisson distribution, we can choose a lower threshold such that two random J-intervals are unlikely (with probability $> 1 - \epsilon$) to have more than $\theta_w$ K-mers in common. Using Chebychev's inequality, we see that a conservative choice would be:

$$\theta_w = \mu_w + \frac{\sigma_w}{\sqrt{\epsilon}}, \qquad \text{where} \quad \mu_w = \frac{J^2}{21^K}, \ \sigma_w = \frac{J}{21^{K/2}} \tag{3}$$

As argued earlier, by using the one-tailed Chebychev bound, we have:

$$Pr(w > \theta_w) = Pr(w - \mu_w > \frac{\sigma_w}{\sqrt{\epsilon}}) < \epsilon \tag{4}$$

By choosing a very small value of $\epsilon$ (for example, $\epsilon \approx O(1/G)$, where $G$ is the genome size), we could make the probability of false positive adequately small.

**Table 2.** Exemplary choices of parameters given $G$ and $s$ when $\epsilon = 1/G$. The $\theta$ here has the same meaning of the $T$ parameter in Table 1. Since $\epsilon$ is extremely small here, the suggested range of $\theta$ is very conservative.

| $s = 0.8$ | | $s = 0.6$ | |
|---|---|---|---|
| $G = 10^9$ | $G = 10^6$ | $G = 10^9$ | $G = 10^6$ |
| $J = 1000, K = 8$ | $J = 1000, K = 6$ | $J = 1000, K = 9$ | $J = 1000, K = 6$ |
| $162 < \theta < 200$ | $108 < \theta < 200$ | $35 < \theta < 150$ | $108 < \theta < 150$ |

In the other direction, let $s$ be a desired similarity value, in the sense that COMBAT must almost always find pairs $a$ and $b$, whenever they have a similarity value of $s$ or higher. The number of observed K-mers shared by $a$ and $b$ can be viewed as a random variable $v$: $B(|a \cap b|, s)$ which has a Binomial distribution with mean $\mu = |a \cap b|s$ and variance $\sigma^2 = |a \cap b|s(1 - s)$. Using the Chernoff bound, we can choose an upper threshold of $|a \cap b|s/2 > Js/4$ to guarantee a probability of success larger than $(1 - \epsilon)$, if $J$ is sufficiently large, i.e., $Js > 16 \ln(1/\epsilon)$. Assuming $\epsilon = 1/G$, and $16 \ln(G)/s < J \ll G$, we will need to satisfy the following inequality:

$$\frac{J^2}{21^K} + J\sqrt{\frac{G}{21^K}} < \theta < Js/4 \quad \text{or} \quad \frac{J}{21^K} + \sqrt{\frac{G}{21^K}} < \theta' < s/4 \tag{5}$$

Since $G$ and $s$ are determined by the genomes, we need only to choose $K$ and $J$. Table 2 shows some exemplary choices of parameters. Note that since our estimations are rather conservative, we found that, in practice, COMBAT performs quite well even for suboptimal choices of parameters.

## 5 Summary and Acknowledgements

To get adequate speed when performing comparison at the scale of whole genomes many high-speed alignment programs have a fast search stage that uses a heuristic to identify regions likely to be homologous. Providing a way of indexing sequences is a key to an efficient search stage. COMBAT indexes both genomic sequences. By using the index of intervals instead of genomic positions we have been able to decrease by J-fold the size of the index for a vertebrate genome, and make it practical to run on a single CPU machine. We show that COMBAT is capable of rapidly finding matching regions across vertebrate species working in translated mode. Then a detailed alignment can be easily retrieved by using the standard alignment algorithms [Smith-Waterman,1970; Needleman-Wunsch,1981]. Therefore, the complex large-scale genome comparison problem is simplified by COMBAT. We also solve the problem of finding a one-to-one mapping in a multiple mapping list by using the $SMI$ algorithm.

Since COMBAT looks for exact K-mers matches, it cannot find regions of relatively low similarity. However, the basic COMBAT scheme can be varied to increase its sensitivity. For example, we can generate K-mers consisting of $n$ exactly matching submers $K_1 \sim K_n$ with $g$ number of bases between them ($g \in [0, \alpha]$, where $\alpha$ is a threshold). This scheme makes it possible to find inexact K-mer matches with gaps or mismatches, and will be experimented in the future.

This project was sponsored by the Department of the Army Award Number W81XWH-04-1-0307.

## References

1. Pachter, L., Mesirov, J.P., Berger, B., Batzoglou, S. and Lander., E.S. Human and mouse gene structure: Comparative analysis and application to exon prediction. *Genome Res.*, pages 950–958, 2000.
2. Bray, N., Dubchak, I., and Pachter, L. Avid: A global alignment program. *Genome Res.*, pages 97–102, 2003.
3. Brudno, M., Do, C.B., Cooper, G.M., Kim, M.F., and Davydov, E.D. Lagan and multi-lagan: Efficient tools for large-scale multiple alignment of genomic dna. *Genome Res.*, pages 721–731, 2003.
4. Delcher, A.L., Kasif, S., Fleischmann, R.D., Peterson, J., White, O., and Salzberg, S.L. Alignment of whole genomes. *Nucleic Acids Res.*, pages 2369–2376, 1999.
5. Frazer, K.A., Elnitski, L., Church, D.M., Dubchak, I., and Hardison, R.C. Cross-species sequence comparisons: A review of methods and available resources. *Genome Res.*, pages 1–12, 2003.
6. Gale, D., and Shapley, L.S. College admissions and the stability of marriage. *Am. Math. Monthly*, pages 9–15, 1962.
7. Kent, W.J. Blat - the blast-like alignment tool. *Genome Res.*, (4):656–664, 2002.
8. Morgenstern, B., Rinner, O., Abdeddaïm, S., Haase, D., Mayer, K.F.X., Dress, A.W.M., and Mewes, H.W. Exon discovery by genomic sequence alignment. *Bioinformatics*, (6):777–787, 2002.
9. Schwartz, S., Kent, W.J., Smit, A., Zhang, Z., et al. Human-mouse alignments with blastz. *Genome Res.*, page 103–107, 2003.

# Missing Values Estimation in Microarray Data with Partial Least Squares Regression*

Kun Yang[1], Jianzhong Li[1], and Chaokun Wang[1,2]

[1] Department of Computer Science and Engineering, Harbin Institute of Technology, Harbin, China
[2] School of Software, Tsinghua University, Beijing, China

**Abstract.** Microarray data usually contain missing values, thus estimating these missing values is an important preprocessing step. This paper proposes an estimation method of missing values based on Partial Least Squares (PLS) regression. The method is feasible for microarray data, because of the characteristics of PLS regression. We compared our method with three methods, including ROWaverage, KNNimpute and LLSimpute, on different data and various missing probabilities. The experimental results show that the proposed method is accurate and robust for estimating missing values.

## 1 Introduction

Microarray technology can be used to detect the expression levels of thousands of genes under a variety of conditions. Microarrays have been successfully applied in many studies over a broad range of biological processes [1, 2, 3, 4]. Despite the popular usage of microarray, there are frequently missing values in microarray data. The missing-value phenomenon can occur for various reasons, including insufficient resolution, slide scratches, dust or hybridization error.

However, many multivariate analyses, such as Support vector machines (SVMs) [5], principal component analysis (PCA) [6] and singular value decomposition (SVD) [7], have difficulty to be applied straightforwardly to the data with missing values. One solution to the missing values problem is to repeat the experiments. But this strategy is often unfeasible for economic reasons and the limitations of available biological material. Thus, it is an important preprocessing step to estimate these missing values.

There are some approaches for estimating the missing values. A simple and common used method is to fill missing values by Zero (ZEROimpute) or by the row(or gene)/ column(or sample) average (ROWaverage) [4]. Furthermore, some advanced estimation methods have been introduced. The SVD-based method (SVDimpute) and weighted k-nearest neighbors (KNNimpute) have been proposed by Troyanskaya *et al* [8]. Recently, several methods including Bayesian

* This work was supported by the 863 Research Plan of China under Grant No. 2004AA231071 and the NSF of China under Grant No. 60533110.

V.N. Alexandrov et al. (Eds.): ICCS 2006, Part II, LNCS 3992, pp. 662–669, 2006.

PCA (BPCA) [9], least squares imputation (LSimpute) [10] and local least squares imputation (LLSimpute) [11] have been introduced.

Partial least squares(PLS) regression is a novel multivariate data analysis method and popularly used in the field of chemometrics. PLS regression has many advantages, which ordinary multiple linear regression does not have, such as avoiding the harmful effects in modeling due to the collinearity of explanatory variables and regressing when the number of observation is less than the number of explanatory variables, etc[12, 13, 14].

This paper presents a specialized missing values estimation method based on Partial Least Squares (PLS) regression. This method is referred as PLSimpute, which uses PLS regression to construct the prediction equation between the target gene with missing values and the similar genes, and then estimates the missing values. The estimation accuracy of our method is compared with that of the wildly used ROWaverage, KNNimpute and recent LLSimpute by introducing artificial missing values. In addition, the normalized root mean squared error (NRMSE) is used to quantitatively measure the estimation accuracy.

The remainder of the paper is organized as follows. Section 2 presents a brief description the missing values problem, while the method PLSimpute is detailed in Section 3. Section 4 provids the experimental results and discussion. Section 5 contains the conclusion.

## 2 Preliminaries

Throughout the paper, microarray gene expression data is represented as an $n \times p$ matrix $A$, where $n$ is the number of samples (observations) and $p$ is the number of genes. The rows correspond to samples, the columns correspond to the genes, and element $A_{i,j}$ is the expression value of gene $j$ in sample $i$. The $i$-th row vector of $A$, noted by $s_i$, and the $j$-th column vector, denoted by $g_j$, are called the expression profile of the $i$-th sample and the $j$-th gene, respectively.

Assume the target gene $g_1$ has one missing value in sample 1. For $2 \leq j \leq p$, the gene expression vector $g_j$ is noted by $(w_j, x_j^T)^T$, and let $g_1 = (a, y^T)^T$. Then the matrix A can be denoted by

$$\mathbf{A} = \begin{pmatrix} a \; w_2 \; \dots \; w_p \\ y \; x_2 \; \dots \; x_p \end{pmatrix}.$$

Firstly we should find others genes, which have a value in sample 1, highly similar to $g_1$, based on the expression profile from sample 2 to sample $n$. Then, the values of those similar genes in sample 1 are used to predict the value of $a$. That is, in order to estimate the value of $a$, we should find the genes which have expression vectors $x_j$ similar to $y$, and use the corresponding $w_j$ to recover $a$. After finding the first $k$ similar gene expression vectors (denoted as $x_j, 1 \leq j \leq k$), vector $y$ can be represented as a linear combination

$$y = b_1 x_1 + b_2 x_2 + \ldots + b_k x_k + \varepsilon, \qquad (1)$$

where $b_i$ are the coefficients of the linear combination. Accordingly, the missing value $a$ can be predicted by

$$\hat{a} = b_1 w_1 + b_2 w_2 + \ldots + b_k w_k. \tag{2}$$

The equation(1) is a multiple linear regression equation between $y$ and $x_1, \ldots, x_k$, and these parameters can be estimated based on least squares principle.

It is essential for good estimation of the parameters in a multiple linear regression equation that the number of variables is a part of the number of observations. When there are much more variables than the observations in presence, the original least squares method can not be applied straightforwardly. But in microarray data, the number of samples is much smaller than the number of genes. Thus, ordinary multiple regression is only feasible when few genes is included in the regression model. This may lost some useful information. Furthermore, the collinearity of the varibles $x_i$ will degrade the prediction ability of equation(1).

## 3 Method

Partial Least Squares (PLS) regression is considered especially useful for constructing prediction equations when there are many explanatory variables and comparatively little samples. PLS can overcome the collinearity problem of the explanatory variables. Our method based on Partial Least Squares regression consists of two main steps. The first step is to select $k$ highly similar genes. In the second step, univariate partial least squares regression is used to form prediction equation, then the prediction equation is used for estimating the missing values.

### 3.1 Selecting genes

To predict the missing values in gene $g_j$ in $A$, the top $k$ similar genes for $g_j$ are selected. The similarity is measured by the Euclidean distance or the absolute Pearson correlation coefficients. In the process of finding similar genes, the components, corresponding to missing values, of the target gene $g_j$ are ignored in computation. After computing the similarity between $g_j$ and these candidate genes, top $k$ highly similar genes are selected.

### 3.2 Partial Least Squares Regression

To describe the PLS regression, some notations are introduced here. Let $X = (x_1, \ldots, x_p)$ be an $N \times p$ matrix of $N$ samples and $p$ genes, where column vector $x_i$ is $N$-dimensional, corresponding to the expression profile of the $i$-th similar gene on $N$ samples. Also, let $Y$ be the expression profile of the target gene on $N$ samples. In the situation of regression, $Y$ is the response variable while $x_i (1 \leq i \leq p)$ are the explanatory variables. Additionally, we should note that $p$ can be much larger than $N$.

The main idea of PLS regression is to construct new explanatory variables (called *components* or *factors*) that capture most of the information in the $X$, where each component is a linear combination of $x_i$, while reducing the dimensionality of the regression problem by using fewer components than the number of variables $x_i$ to predict $Y$.

The objective criterion of forming components is to sequentially maximize the covariance between the response variable $Y$ and the component $t$, which is a linear combination of the columns in $X$, i.e. $t = Xw$. Thus the objective can be summarized as the following formula

$$w_k = arg \underset{\|w\|=1}{\mathrm{Max}} \, \mathrm{cov}^2(\mathrm{Xw}, \mathrm{Y})$$

$$\text{S. T.:} \quad w^T S w_j = 0 \text{ for all } \quad 1 \leq j < k \tag{3}$$

where $S = X^T X$. The maximum number of components is the rank of $X$. The details of PLS algorithm are given in Table1. According the algorithm of PLS, we can use equation $Y \approx X \cdot \alpha = X \cdot \sum_{k=1}^{d} \gamma_k \cdot \prod_{j=1}^{k-1} (I - w_j \cdot p_j^T) \cdot w_k$ to predict the missing values. For some theoretical aspects and properties of PLS, please see [12, 13, 14].

**Table 1.** The Algorithm of PLS

| | |
|---|---|
| 1 | For $k = 1$ To $d$ |
| 2 | $w_k = X_{k-1}^T \cdot Y_{k-1} / \|X_{k-1}^T \cdot Y_{k-1}\|$ ; |
| 3 | $t_k = X_{k-1} \cdot w_k$; |
| 4 | $p_k = X_{k-1} \cdot t_k / \|t_k\|^2$; |
| 5 | $\gamma_k = Y_{k-1} \cdot t_k / \|t_k\|^2$; |
| 6 | $Residual: X_k = X_{k-1} - t_k \cdot p_k^T$; (with $X_0 = X$); |
| 7 | $Residual: Y_k = Y_{k-1} - \gamma_k \cdot t_k$; (with $Y_0 = Y$); |
| 8 | End For |

The last problem is to decide the number of components in the regression model. Helland[12] reported that partial least squares regression often needs few components to give its optimal prediction. So we can determine the number of components easily at empirical results. Furthermore, another method for this problem is to use cross validation procedure[15]. In this procedure, the optimal number of components is found at the minimal sum of squared errors of prediction.

## 4 Results and Discussion

In this section, we compare the performance of our method with three missing values estimation methods, including ROWaverage, KNNimpute and LLSimpute. Two microarray data have been used in our experiments, including both time series and no-time series data. The first data (*Sp*) is attained from 784 cell-cycle-regulated genes in 14 experiments, which was studied by Spellman *et al* [16]. After deleting gene row with missing values, this data has 474 genes

**Table 2.** Probabilty $\rho$ and the corresponding expected number of genes with and without missing values on two data

| | Colon | | | | Sp | | | |
|---|---|---|---|---|---|---|---|---|
| $\rho$ | 0.005 | 0.01 | 0.02 | 0.03 | 0.005 | 0.01 | 0.02 | 0.03 |
| $E_0$ | 1792 | 1603 | 1283 | 1023 | 442 | 412 | 358 | 309 |
| $p - E_0$ | 208 | 397 | 717 | 977 | 32 | 62 | 116 | 165 |

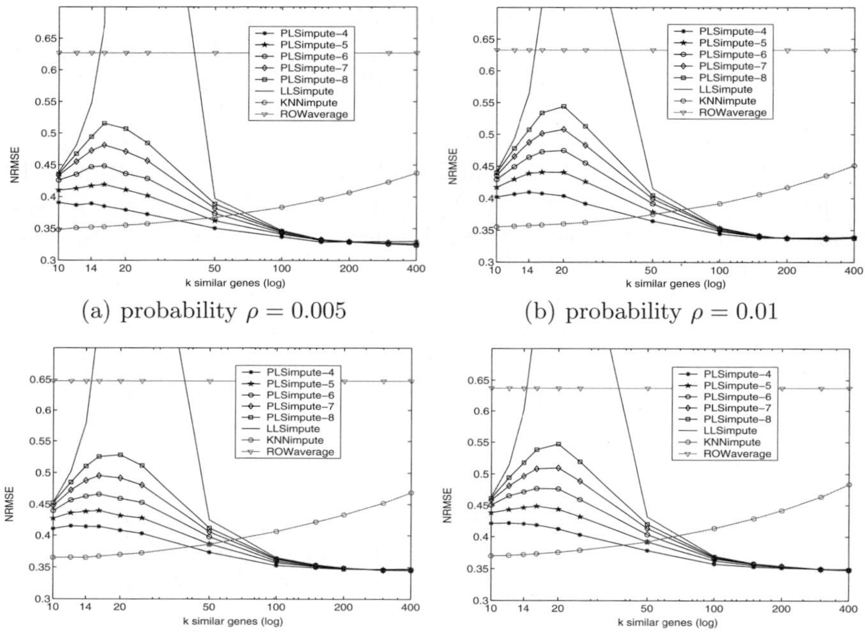


(a) probability $\rho = 0.005$ (b) probability $\rho = 0.01$

(c) probability $\rho = 0.02$ (d) probability $\rho = 0.03$

**Fig. 1.** NRMSEs of various methods on the Colon data over a wide range of similar genes ($k$) used in estimation

and 14 experiments (i.e. samples), and is the same data used in LLSimpute [11]. The second data (*Colon*) consists of 22 normal tissues samples and 2000 human genes from the colon data analyzed initially by Alon *et al*[17].

To evaluate the performance of different methods, artificial missing entries are created from a complete expression matrix (i.e. without missing values) according to the method of Ouyang *et al*[18]. That is, each entry in the complete matrix is randomly and independently treated as missing value with a probability $\rho$. If a probability $\rho$ and a complete matrix $A$ with $N$ samples and $p$ genes are given, then the expected number of genes with $k$ missing values in the artificial data can be calculated by function(4).

$$E_k = p \cdot C_N^k \cdot \rho^k \cdot (1-\rho)^{N-k}. \tag{4}$$

Table 2 shows the different probabilities $\rho$ used in experiments on both data and the corresponding expected numbers of genes with and without missing values.

For the missing entries are artificial, the performance of each method is measured by normalized root mean squared error(NRMSE).

$$NRMSE = \sqrt{\text{mean}[(\text{y}_{\text{guess}} - \text{y}_{\text{answer}})^2]} / \text{std}[\text{y}_{\text{answer}}] \tag{5}$$

where the mean and the SD are calculated over all missing entries in the whole matrix.

The experimental method for each complete data is: randomly create a artificial missing matrix with probability $\rho$, use all methods to estimate these missing values, calculate the NRMSE of each method by the values of estimation and that of the origin. This procedure is repeated 10 times. And the similarity is measured by the Euclidean distance. The results of all methods are the average NRMSEs on 10 randomly created data with an exact probability $\rho$. Because Troganskaya *et al* [8] reported that KNNimpute was insensitive to the exact value of $k$ in the range 10-20 and the best results were in this range, different values of $k$ (i.e. 10, 12, 14, 16, 20) are always used in every experiment. Furthermore, in order to investigate the effect of the number of components used in PLS regression on the performance of PLSimpute, the results attained from several fixed

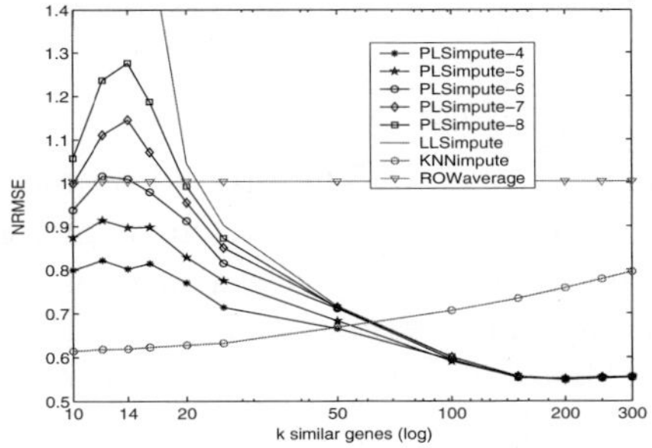


(a) probability $\rho = 0.005$

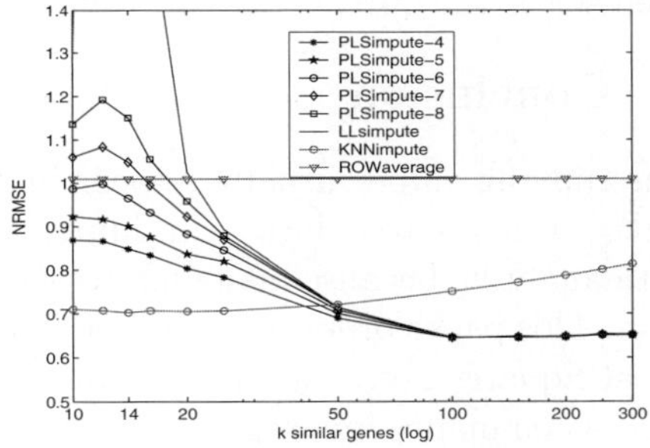


(b) probability $\rho = 0.01$

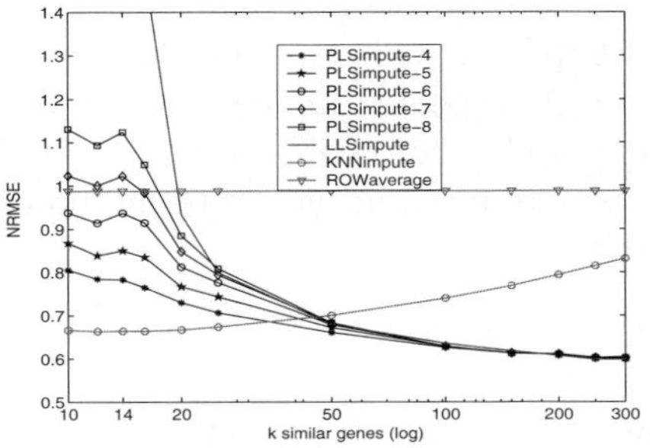


(c) probability $\rho = 0.02$

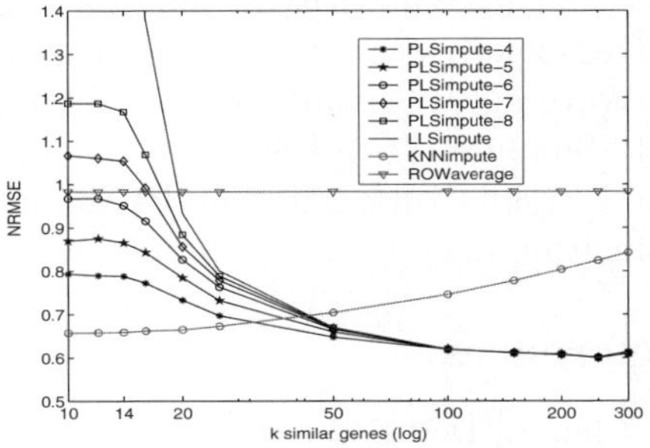


(d) probability $\rho = 0.03$

**Fig. 2.** NRMSEs of various methods on the SP data over a wide range of similar genes ($k$) used in estimation numbers of components (i.e. 4,5,6,7,8) are reported in all experiments. Figure 1 and 2 show the experimental results on both data. We observe the following:

For various missing probabilities $\rho$ on both datasets, PLSimpute shows the lowest estimation error, while LLSimpute shows less accurate results than PLSimpute. The performance of ROWaverage is the worst one and the method of KNNimpute is the medium. This result is the same as previous studies [8, 11].

We can see that PLSimpute just needs few components used in PLS regression to give a good estimation. This result is coincident with the result reported by Helland[12].

Moreover, the estimation errors of PLSimpute with different numbers of components are always less than LLSimpute when various numbers of similar genes are used in estimation. Additionally, When the components used in PLS regression increase, the performance of PLSimpute comes near to LLSimpute. It turns out that finding an empirical result of the feasible number of components in PLSimpute is relatively easy, despite the optimal number of components is possibly difficult to be selected. Anyway, we can determine the number of components used in PLSimpute by cross validation procedure [15].

The same as KNNimpute and LLSimpute, the number of similar genes (parameter $k$) is important to select for a high performance of PLSimpute, while the number of principal axes (eigenvectors) is important for both BPCA and SVDimpute. We can determine the parameter $k$ by a heuristic method [11]: treating some no-missing elements as missing values, and making a estimation, then select the special value of $k$ in respect to the best result of estimation.

## 5 Conclusion

The missing values usually appear in microarray data. The process of missing values in microarry data is an important preprocessing step in gene expression data analysis, because many analysis methods require complete gene expression data. This paper proposes a method of missing values estimation based on Partial Least Squares regression (PLSimpute) in Microarray data. The method can be used when many similar genes and comparatively little experiments are included in the estimation procedure. Furthermore, the method can reduce the harmful effect of the collinearity of the similar genes used for prediction.

Our experimental results show that PLSimpute gives better performance than ROWaverage, KNNimpute and LLSimpute. As reported in the previous study of LLSimpute [11], LLSimpute has a better ability for estimation than BPCA. These results turn out that PLSimpute is a robust and accurate method for estimating missing values in microarray data.

## References

1. Chu, S., DeRisi, J., *et al*: The transcriptioal program of sporulation in budding yeast. Science **278** (1998) 680–686
2. Alon, U., Barkai, N., *et al*: Broad patterns of gene expression revealed by clustering analysis of tumor and normal colon tissues probed by oligonucleotid arrays. Proc. Natl. Acad. Sci. USA **96** (1999) 6745–6750

3. Golub, T.R., Slonim, D.K., *et al*: Molecular classification of cancer: class discovery and class prediction by expression monitoring. Science **286** (1999) 531–537
4. Alizadeh, A.A., Eisen, M.B., *et al*: Distinct types of diffuse large B-cell lymphoma identified by gene expression profiling. Nature **403** (2000) 503–511
5. Vapnik, V. : The Nature of Statistical Learning Theory. Springer-Verlag. New York (1995)
6. Raychaudhuri, S., Stuart, J.M. and Altman, R.: Principal components analysis to summarize microarray experiments: application to sporulation time series. Pac. Symp. Biocomput. (2000) 455–466
7. Alter, O., Brown, P.O. and Botstein, D. : Singular value decomposition for genome-wide expression data processing and modeling. Proc. Natl Acad. Sci. USA **97** (2000) 10101–10106
8. Troyanskaya, O., Cantor, M. *et al* : Missing value estimation methods for DNA microarray. Bioinformatics **17** (2001) 520–525
9. Oba, S., Sato, M., *et al*: A Bayesian missing value estimation method for gene expression profile data. Bioinformatics **19** (2003) 2088–2096
10. Bø, T.H., Dysvik, B. and Jonassen, I. : LSimpute: accurate estimation of missing values in microarray data with least squares methods. Nucleic Acids Res. **32(e34)** (2004)
11. Kim, H., Golub, G.H. and Park, H. : Missing value estimation for DNA microarray gene expression data: local least squares imputation. Bioinformatics **21** (2005) 187–198
12. Helland, I.S. : On the structure of partial least squares regression. Commun. Stat. -Simul. Comput. **17** (1988) 581–607
13. Garthwaite, P.H. : An interpretation of partial least squares. J. Am. Stat. Assoc. **89** (1994) 122-127
14. Wang, H.: Partial Least-squares Regression — Method and Applications. National Defence Industry Press. China (1999)
15. Stone, M. : Cross-validatory choice and assessment of statistical predictions. J. Roy. Statist. Soc. **36** (1974) 111–133
16. Spellman, P.T., Sherlock, G., *et al*: Comprehensive identification of cell cycle-regulated genes of the yeast Saccharomyces cerevisiae by microarray hybridization. Mol. Biol. Cell **9** (1998) 3273–3297
17. Alon, U., *et al*: Broad patterns of gene expression revealed by clustering analysis of tumor and normal colon tissues probed by ologonucleotide arrays. Proc. Natl Acad. Sci. USA, **96** (1999) 6745–6750
18. Ouyang, M., Welsh, W.J. and Georgopoulos P.: Gaussian mixture clustering and imputation of microarray data. Bioinformatics. **20** (2004) 917–923

# Boost Feature Subset Selection: A New Gene Selection Algorithm for Microarray Dataset

Xian Xu and Aidong Zhang

State University of New York at Buffalo, Buffalo, NY 14260, USA
{xianxu, azhang}@cse.buffalo.edu

**Abstract.** Gene selection is usually the crucial first step in microarray data analysis. One class of typical approaches is to calculate some discriminative scores using data associated with a single gene. Such discriminative scores are then sorted and top ranked genes are selected for further analysis. However, such an approach will result in redundant gene set since it ignores the complex relationships between genes. Recent researches in feature subset selection began to tackle this problem by limiting the correlations of the selected feature set. In this paper, we propose a novel general framework BFSS: Boost Feature Subset Selection to improve the performance of single-gene based discriminative scores using bootstrapping techniques. Features are selected from dynamically adjusted bootstraps of the training dataset. We tested our algorithm on three well-known publicly available microarray data sets in the bioinformatics community. Encouraging results are reported in this paper.

## 1 Introduction

Recent technological advances in large scale DNA profiling enable researchers to simultaneously monitor the expression levels of thousands of genes or ESTs [1, 6, 10]. This provides unique opportunities to uncouple the relationship between disease phenotypes and their biochemical causations. As a first step, researchers have attempted to classify disease using such molecular evidences [1, 6, 10].

Microarray technology also raises new challenges for data analyzing algorithms because of the uniqueness of the resulting microarray dataset. In a typical microarray study, the genes or ESTs being monitored number from thousands to tens of thousands, while the number of different tissue samples is much smaller ranging from tens to hundreds. This results in the situation where the number of features (or genes) well outnumbers the number of observations. The term "peaking phenomenon" is coined in the machine learning and pattern recognition community, referring to the phenomenon that inclusion of excessive features may actually degrade the performance of a classifier if the number of training examples used to build the classifier is relatively small compared to the number of features [8]. Gene selection is commonly performed before sample classification is even attempted to alleviate the above stated problem. A smaller subset of informative genes is also a good start point of further biological investigations. The broadly used gene selection algorithms on microarray data sets share a common workflow: 1. some single-gene based discriminative score is selected; 2. genes are ranked based on such discriminative score; and 3. top scored genes are then selected

V.N. Alexandrov et al. (Eds.): ICCS 2006, Part II, LNCS 3992, pp. 670–677, 2006.

for further investigation. We term this class of algorithms single-gene based gene selection. Although relatively simple, various single-gene based gene selection algorithms have been proposed and demonstrated to be effective for improving sample classification accuracy. Some of them are statistical tests (t-test, F-test) [3], non-parametric tests like TNoM [2], mutual information [13], S2N ratio (signal to noise ratio) [6] etc.

However, the assumption of independence between genes oversimplifies the complex relationship between genes. Genes are well known to interact with each other through gene regulative networks. As a matter of fact, the common assumption of popular cluster analysis on microarray data sets [9] is that co-regulated genes have similar expression profiles. Several of recent researches on feature subset selection especially gene selection [7, 11, 12, 15] explicitly took into consideration the correlations among features (genes) by limiting redundancy in resulting feature (gene) set. We showed in earlier research that the concept of virtual gene (correlations between genes) [14] could help improve gene selection.

In this work, we propose a novel meta-algorithm for boosting the performance of single-gene based gene selection algorithms. In our framework, genes are selected not from original training samples, but from its bootstraps. The probability table of sampling for the different training samples is dynamically adapted based on how previous selected genes behave. Our proposed framework is a general purpose meta-algorithm. Instead of tying with some fixed single-gene based discriminative scores, our algorithm accepts as input most if not all single-gene based gene selection algorithms and produces a better gene selection algorithm. It is worth mentioning that our algorithm is closely related to bagging [4] and boosting [5] proposed for ensemble classifier design.

The rest of this paper is organized as follows. Our proposed algorithm is discussed in Section 2 in detail with an illustrating example. Extensive experimental results on three publicly available microarray data sets are reported in Section 3. In Section 4, we conclude this paper and give directions of our future work.

## 2 BFSS: Boost Feature Subset Selection

In this section we formulate our boost feature subset selection algorithm (BFSS). Notation used throughout this paper is introduced in the first subsection. An illustrating example is given in subsection two. Detailed algorithm is given in the third subsection.

### 2.1 Notation

Let $\mathcal{R}$ be the set of real numbers and $\mathcal{N}$ be the set of natural numbers including 0. Let $\mathcal{G} = \{g_1, g_2, \cdots, g_n\}$ be the set of all genes that are used in one study, $\mathcal{S} = \{s_1, s_2, \cdots, s_m\}$ be the set of all experiments performed, $\mathcal{L} = \{l_1, l_2, \cdots, l_l\}$ be the set of sample class labels of interest. We assume $\mathcal{G}, \mathcal{S}, \mathcal{L}$ are fixed for any given study. Let $n = |\mathcal{G}|$ be the total number of genes, $m = |\mathcal{S}|$ be the total number of experiments and $l = |\mathcal{L}|$ be the total number of class labels. A microarray gene expression dataset used in our study can be defined as $\mathcal{E} = (\mathcal{G}, \mathcal{S}, \mathcal{L}, L, E)$, where $L$ is a function $\mathcal{S} \to \mathcal{L}$ such that for $s \in \mathcal{S}$, $L(s) \in \mathcal{L}$ is the class label for sample $s$; $E$ is a function $\mathcal{G} \times \mathcal{S} \to \mathcal{R}$. For $g \in \mathcal{G}$ and $s \in \mathcal{S}$, $E(g, s)$ is the expression level of gene $g$ in experiment $s$. In the bioinformatics community, the function $E$ is normally presented as a two dimensional array of real numbers.

Sometimes we need to treat the set of samples $\mathcal{S}$ as a multiset (or bag). In this case we refer to the set of samples $\mathcal{S}$ as $\mathcal{S}^{\mathcal{M}_1} = (\mathcal{S}, \mathcal{M}_1)$, where $\mathcal{M}_1$ is a function that is always 1 ($\mathcal{M}_1(s) = 1, s \in \mathcal{S}$). A multiset is a set that allows duplication. In the case of $\mathcal{S}^{\mathcal{M}_1}$, $\mathcal{S}$ is the underlying set and $\mathcal{M}_1$ is the multiplicity function for elements in the underlying set. More generally we refer to multiset $\mathcal{S}^{\mathcal{M}} = (\mathcal{S}, \mathcal{M})$ as the **bootstrap sample set**, where $\mathcal{M}$ is an arbitrary function $\mathcal{S} \to \mathcal{N}$. We will discuss **bootstrap sample set** in detail later in this section.

For simplicity of presentation, we use a subscripting scheme to refer to elements in $\mathcal{E}$. Let $\mathcal{E}(G, \mathcal{S}^{\mathcal{M}}) = (G, \mathcal{S}^{\mathcal{M}}, L, E)$ where $G \subseteq \mathcal{G}$ and $\mathcal{S}^{\mathcal{M}}$ is a **bootstrap sample set**. We further use $L(\mathcal{S}^{\mathcal{M}})$ to denote the set of class labels for the set of experiments $\mathcal{S}^{\mathcal{M}}$.

## 2.2 Motivating Example

The most obvious drawback of single-gene based gene selection algorithms is the fact that those algorithms ignore the relationships that exist between genes. There is no guarantee that the combination of two "good" features will necessarily produce a "better" classifier. As an illustrating example, please refer to Figure 1, in which the expression levels of three genes across 100 samples are plotted. Samples are labeled using two class labels: either caner (grey background) or normal (white background).

The first two genes, gene 1 and gene 2 behave similarly. In majority of the samples (samples 1 to 40 and samples 61 to 100, or 80% of samples), the expression levels of gene 1 or gene 2 can be used to predict sample class labels effectively. Actually the expression levels of these two genes are generally higher in cancer samples than in normal ones. However, the expression levels of these two genes in samples 41 to 60 (20% of samples) are more mixed across cancer/normal class distinction.

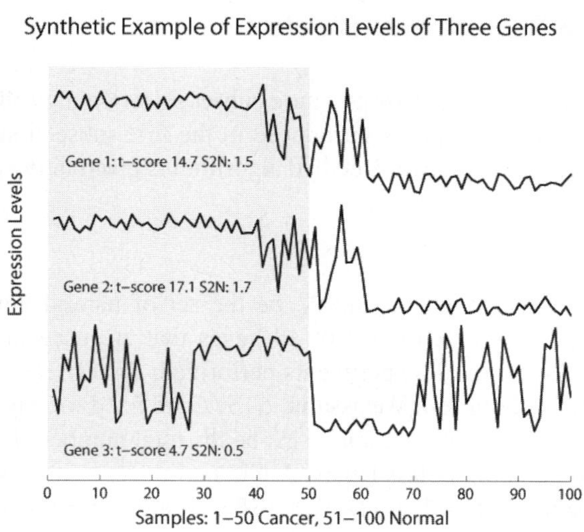


**Fig. 1.** Illustrating example of redundancy in selected gene set

Gene 1 and gene 2 score high in term of t-score and S2N scores as shown in Figure 1. Gene 3 is obviously a less capable predictor when considered alone. Clear trend exists in the expression levels of gene 3 in samples 31 to sample 70 (40% of samples). However it varies across cancer/normal labels in the rest samples (60%). Gene 3 scores much lower than gene 1 and gene 2 in term of t-score and S2N as expected. Using t-score and S2N, we can rank these three genes based on their salience in predicting cancer/normal class labels as: gene 2 > gene 1 > gene 3.

However, t-score and S2N do not consider the fact that gene 1 and gene 2 behave similarly. They both work well in samples 1 to 40 and samples 61 to 100. They both share more difficult samples, namely samples 41 to 60. If two genes out of the three are to be selected for further data analysis, would it be wise to use both gene 1 and gene 2, as suggested by their relatively high single-gene based discriminative score (t-score and S2N) ranking? This is the very question we address in this paper. We empirically show that it is not the case. Choosing gene 2 and gene 3 might be a better idea as gene 3 "covers" the more difficult samples where gene 2 fails.

### 2.3 BFSS: Boost Feature Subset Selection

In this section, we elaborate our new algorithm. First we will define some concepts and then describe our BFSS algorithm.

A **bootstrap sample set** $\mathcal{S}^{\mathcal{M}} = (\mathcal{S}, \mathcal{M})$ is a multiset of samples randomly drawn with replacement from the original set of samples $\mathcal{S}$. $\mathcal{M}(s), s \in \mathcal{S}$ is the multiplicity of item $s$. As a result, the same sample $s \in \mathcal{S}$ can appear more than once or does not appear at all in $\mathcal{S}^{\mathcal{M}}$. The cardinality of $\mathcal{S}^{\mathcal{M}}$ is denoted by $m_b$. The sampling probability of each sample in $\mathcal{S}$ is determined by a probability table $p(s)$ where $s \in \mathcal{S}$.

**Definition 1. A bootstrap sample set** $\mathcal{S}^{\mathcal{M}} = (\mathcal{S}, \mathcal{M})$ *of size* $m_b$ *is a multiset of samples resulting from random sampling from* $\mathcal{S}$ *with replacement. The probability of each sample* $s \in \mathcal{S}$ *being sampled is* $p(s)$. $m_b = \sum_{s \in \mathcal{S}} \mathcal{M}(s)$.

**Definition 2.** *A* ***bootstrap*** $\mathcal{B}$ *of a training dataset* $\mathcal{E} = (\mathcal{G}, \mathcal{S}, L, E)$ *using bootstrap sample set* $\mathcal{S}^{\mathcal{M}}$ *is a dataset defined as*

$$\mathcal{B} = (\mathcal{G}, \mathcal{S}^{\mathcal{M}}, L, E)$$

**Definition 3.** *The* **worst set of samples** $\mathcal{S}_{worst}$ *of size* $\delta$ *with respect to bootstrap dataset* $\mathcal{E}(g, \mathcal{S}^{\mathcal{M}})$ *and a single-gene based scoring function* $F$ *is defined as a multiset:*

$$\underset{S \subseteq \mathcal{S}^{\mathcal{M}} \textbf{ and } |S| = \delta}{argmax} \quad (F(\mathcal{E}(g, \mathcal{S}^{\mathcal{M}} - S)))$$

*Here* $\mathcal{S}^{\mathcal{M}} - S$ *means a set by removing* $S$ *from* $\mathcal{S}^{\mathcal{M}}$. *We also call* $\mathcal{S}^{\mathcal{M}} - \mathcal{S}_{worst}$ *the* **best set of samples**.

A **bootstrap** of training set is defined using the definition of **bootstrap sample set**. A bootstrap $\mathcal{B}$ is a new dataset defined using the four-tuple notation we used to define microarray dataset. The only difference is that the second element is a bootstrap sample

**Algorithm 1.** *WorstSampleSet* : Calculate the **worst set of samples** using a greedy algorithm

**Require:** $\mathcal{E} = (\{g\}, S, L, E)$, $F$ as a single-gene based discriminative score
**Ensure:** $S'$ is the **worst set of samples** with respect to $\mathcal{E}$ and $F$

```
S', S_0 to be empty sets
for all s ∈ S do
    S_1 ← S − {s}
    calculate F(E({g}, S_1)), add score to S_0
end for
sort S_0, add samples s corresponding to top δ scores in S_0 to S'
return S'
```

**Algorithm 2.** *BFSS* : Boost Feature Subset Selection

**Require:** $\mathcal{E} = (G, S, L, E)$; $n'$ as the number of genes to be selected; $F$ as a single-gene based discriminative score
**Ensure:** $G'$ as the selected gene set by BFSS using $F$.

```
Initialize p(s) to be 1/m (m is the number of samples in E). Set G' as an empty set.
E' ← E
for |G'| < n' do
    generate bootstrap sample set S^M and bootstrap B of training set E' by random sampling
    using probabilities p(s)
    calculate score F on bootstrap B, refer to this score as F', keep track of the best score so
    far
    add top ranked gene g based on F' to G'
    find worst δ samples S_worst based on B(g, S^M) using Algorithm 1
    reduce (p(s) where s ∈ S^M − S_worst by a factor of ε and normalize p(s) so that it represents
    a distribution
    remove g from E'
end for
return G'
```

set. $\mathcal{B}$ shares the same set of genes $\mathcal{G}$, same sample class label mapping $L$, and same expression levels mapping $E$ with $\mathcal{E}$.

Given a bootstrap $\mathcal{B}$, a gene $g$ and a score function $F$, the **worst set of sample** of size $\delta$ is a set of samples such that by removing them from the dataset $\mathcal{B}$, best $F$ score for gene $g$ is achieved. We refer to all other samples in $\mathcal{S}^{\mathcal{M}}$ other than those in the **worst set of samples** the **best set of samples**. Both the **worst set of samples** and the **best set of samples** are multisets.

By definition of the **worst set of samples** with respect to gene $g$ and score $F$, it is exponentially hard to find such set of samples since the power set of samples needs to be examined. We employ a simply greedy algorithm as described in Algorithm 1. For each sample $s \in \mathcal{S}^{\mathcal{M}}$, scores $F$ for gene $g$ on each sample set $\mathcal{S}^{\mathcal{M}} - \{s\}$ are computed. Such scores are ranked and the samples corresponding to the best $\delta$ scores are treated as the **worst set of samples**. $\delta$ is one parameter of our algorithm. However, as shown later in this section, fixed value of $\delta$ is used for all datasets we tested with good results.

Boost feature subset selection algorithm (BFSS) is shown in Algorithm 2. After some initialization, the algorithm first generates a bootstrap $\mathcal{B}$ of training set $\mathcal{E}$, where $i$ iteration counter. This involves the generation of **bootstrap sample set** of size $m_b$ and then the **bootstrap** itself. This involves random sampling with replacement from $S$ using probability table $p(s)$.

After **bootstrap** $\mathcal{B}$ of a training set is generated, the $F$ score is then calculated for each gene in $\mathcal{B}$. Best $F$ score so far is kept during the computation, so is the gene associated with it. In the next step, the gene with best $F$ score for current bootstrap $B$ is selected and added to the selected gene set. Based on the selected gene, BFSS then identifies the **worst set of samples** with respect to the currently selected gene and the single-gene based scoring function $F$ using Algorithm 1. The probability table $p(s)$ for generating bootstraps is modified by reducing the probabilities for the **best set of samples** by a constant factor. The probability of those good samples being selected in subsequent analysis is thus reduced, focusing BFSS onto those samples that previously selected genes would not perform well. The currently selected gene is then marked as selected and not considered further by the BFSS algorithm. BFSS repeats this process until $n'$ genes are selected.

There are three parameters $m_b$ (size of **bootstrap sample set**), $\delta$ (size of **worst set of samples**) and $\varepsilon$ (the degradation factor of sampling probability) used in our algorithm. We experimentally chose $\delta$ to be 0.96 of the number of training samples in a dataset and $\varepsilon$ to be 0.96. We set $m_b$ to be twice the number of samples in the training set so that a **bootstrap** is more representative. After fixing these three parameters, there is virtually no more need of tweaking our BFSS algorithm. We used these same parameters for all the three data sets we experimented with and achieved good performance on all of them. This indicates BFSS's good property of requiring little tuning for different data sets. We omit complexity analysis of our algorithm in this paper due to space constraint.

## 3 Experiments

We performed extensive experiments on three publicly available microarray datasets: Colon Cancer [1], Leukemia [6] and multi-class cancer [10]. Data preprocessing is the same as described in [14]. Performance of classifiers is used as a measure of the performance of feature subset selection algorithms. In order not to be biased on which classifiers we use, three very different general purpose classifiers are used: DLD (diagonal linear discriminant), KNN (k-nearest neighbor, k=3) and SVM (support vector machine, with radial kernel). A 2 fold cross-validation procedure is used to estimate classification performance. We systematically examine the performance of different gene subset selection algorithms using these three classifiers with different number of genes selected. The number of genes selected ranges from 2 to 100 in the increment of 2. All experiments are repeated 100 times to get more accurate results.

This paper focuses on improving performance of single-feature based feature subset selection algorithms. Two simple widely used algorithms are used for testing: t-score and S2N. We plug these two algorithms into our BFSS algorithm and refer to the resulting new algorithms as "boost t-score" and "boost S2N" respectively. We also report classification performance without using gene selection algorithms as a comparison.

**Table 1.** Performance(%) of single-gene based methods and their boosted versions

| *Num of genes* | *FSS method* | *Colon Cancer* | | | *Leukemia* | | | *Multi-class* | | |
|---|---|---|---|---|---|---|---|---|---|---|
| | | *KNN* | *DLD* | *SVM* | *KNN* | *DLD* | *SVM* | *KNN* | *DLD* | *SVM* |
| N/A | No FSS | 75.9 | 64.7 | 71.3 | 93.1 | 65.3 | 92.0 | 85.6 | 77.2 | 83.4 |
| 20 | t-score | 81.58 | 80.10 | 81.35 | 92.75 | 94.97 | 94.25 | 78.34 | 78.13 | 79.92 |
| | Boost t-score | 82.81 | 81.35 | 82.87 | 93.67 | 94.69 | 94.47 | 78.86 | 78.51 | 80.47 |
| | S2N | 81.23 | 79.94 | 81.30 | 93.33 | 95.19 | 94.56 | 78.56 | 78.26 | 80.71 |
| | Boost S2N | 82.90 | 81.71 | 82.16 | 94.11 | 95.11 | 94.94 | 78.85 | 78.37 | 81.14 |
| 50 | t-score | 81.52 | 78.10 | 81.81 | 94.14 | 95.47 | 95.19 | 81.57 | 78.65 | 82.13 |
| | Boost t-score | 83.87 | 81.00 | 83.71 | 94.47 | 95.64 | 94.94 | 81.93 | 78.96 | 82.62 |
| | S2N | 81.84 | 77.39 | 82.65 | 94.31 | 95.77 | 95.44 | 81.73 | 78.84 | 82.46 |
| | Boost S2N | 83.32 | 81.55 | 83.19 | 94.94 | 95.86 | 95.00 | 81.83 | 79.14 | 82.72 |
| 100 | t-score | 82.13 | 76.65 | 81.74 | 95.11 | 95.81 | 94.36 | 83.08 | 78.97 | 83.30 |
| | Boost t-score | 84.39 | 80.39 | 83.65 | 95.75 | 96.22 | 95.22 | 83.21 | 79.41 | 83.73 |
| | S2N | 81.52 | 76.39 | 81.45 | 95.42 | 96.00 | 95.61 | 83.29 | 79.19 | 83.21 |
| | Boost S2N | 83.84 | 80.19 | 83.68 | 95.75 | 96.17 | 95.81 | 83.09 | 79.50 | 83.63 |

Table 1 shows part of our experimental result. Boosted version of single gene based discriminative score yields higher classification accuracy in most cases in our experiments. Significant improvement is observed in colon cancer data set, where performance improved as much as 4%. Less yet consistent performance gain is observed in the other two data sets we tested. For Leukemia data set, the average performance is already in mid 90 percent, where the space for improvement is limited. For multi-class data set we tested, classification performance of KNN classifier is actually better than any of the gene selection algorithms in our experiment. Considering there are more than 16000 genes in this dataset, this seems indicating selecting top 100 genes may not be enough. Multi-class data set is also more heterogeneous since the cancer tissues came from 14 common cancer types. It is reasonable to expect more genes are needed to characterize such vast different phenotypes.

## 4 Conclusion and Future Work

In this paper, we presented a novel general feature subset selection framework to improve the performance of single-gene based discriminative scores. In our approach, genes are selected from bootstraps of training set instead of training set itself. The sampling probability is dynamically adapted based on the performance of previously selected genes on different bootstrap samples. Extensive experiments were performed on three publicly available microarray datasets. According to our experiments, boosted versions of those single-gene based discriminative scores perform consistently better in most cases and in many cases boosted versions perform considerably better than the original scores. A nice feature of our approach is that most if not all single-gene based discriminative scores can be plugged into our system and the resulted BFSS feature selectors are expected to perform better than the original scores according to our experiments. Our approach is also independent of the classifier used.

Although not targeted as feature selection algorithm for ensemble classifiers, our algorithm may work well with such classifiers nonetheless. Better performance could be achieved by diversifying the feature set selected for each member classifier. Since our BFSS algorithm is based on bootstraps of the training set and sampling probability is dynamically adapted, diversity of the selected features is already built in. It is also interesting to examine the behavior of BFSS in conjunction with these ensemble classifiers, especially these bootstrap based ensemble classifiers (bagging and boosting). Furthermore, since BFSS and bootstrap based ensemble classifiers are all based on the bootstrapping concept, it is possible that they can be combined into a uniform framework. We are currently researching into these directions.

## References


1. U. Alon, N. Barkai, D.A. Notterman, K. Gish, S. Ybarra, D. Mack, and A.J. Levine. Broad patterns of gene expression revealed by clustering analysis of tumor and normal colon tissue probed by oligonucleotide arrays. *Proc. Natl. Acad. Sci. U.S.A.*, 96(12):6745–50, 1999.
2. A. Ben-Dor, L. Bruhn, N. Friedman, I. Nachman, M. Schummer, and Z. Yakhini. Tissue classification with gene expression profiles. volume 7, pages 559–83, 2000.
3. T.H. Bø and I. Jonassen. New feature subset selection procedures for classification of expression profiles. *Genome Biology*, 3(4):research0017.1–0017.11, 2002.
4. Leo Breiman. Bagging predictors. *Machine Learning*, 1996.
5. Yoav Freund and Robert E. Schapire. Experiments with a new boosting algorithm. In *in Proc. ICML 1996*, 1996.
6. T. R. Golub et al. Molecular classifications of cancer: Class discovery and class prediction by gene expression monitoring. *Science*, 286(5439):531–7, 1999.
7. J. Jaeger, R. Sengupta, and W. L. Ruzzo. Improved gene selection for classification of microarrays. In *Proc. PSB*, 2003.
8. Anil K. Jain, Robert P.W. Duin, and Jianchang Mao. Statistical pattern recognition: A review. *IEEE Transactions on pattern analysis and machine intelligence*, 22(1):4–37, 2000.
9. D. Jiang, C. Tang, and A. Zhang. Cluster analysis for gene expression data: A survey. *IEEE Transactions on Knowledge and Data Engineering*, 16(11):1370–1386, 2004.
10. S. Ramaswamy, P. Tamayo, R. Rifkin, S Mukherjee, C.H. Yeang, M. Angelo, C. Ladd, M. Reich, E. Latulippe, J.P. Mesirov, T. Poggio, W. Gerald, M. Loda, E. S. Lander, and T.R. Golub. Multiclass cancer diagnosis using tumor gene expression signatures. *PNAS*, 98(26):15149–15154, 2001.
11. Yuhang Wang, Fillia S. Makedon, James C. Ford, and Justin Pearlman. Hykgene: a hybrid approach for selecting marker genes for phenotype classification using microarray gene expression data. *Bioinformatics*, 21(8):1530–1537, 2005.
12. Y. Wu and A. Zhang. Feature selection for classifying high-dimensional numerical data. In *IEEE Conference on Computer Vision and Pattern Recognition 2004*, volume 2, pages 251–258, 2004.
13. E. P. Xing, M. I. Jordan, and R. M. Karp. Feature selection for high-dimensional genomic microarray data. In *Proc. 18th International Conf. on Machine Learning*, pages 601–608. Morgan Kaufmann, San Francisco, CA, 2001.
14. Xian Xu and Aidong Zhang. Virtual gene: Using correlations between genes to select informative genes on microarray datasets. *LNCS Transactions on Computational Systems Biology II, LNBI 3680*, pages 138–152, 2005.
15. L. Yu and H. Liu. Redundancy based feature selection for microarray data. In *Proc. of SIGKDD*, 2004.

# A Hybrid Feature Selection Approach for Microarray Gene Expression Data

Feng Tan, Xuezheng Fu, Hao Wang, Yanqing Zhang, and Anu Bourgeois

Department of Computer Science, Georgia State University, Atlanta, GA 30303, USA
{ftan, xfu1, hwang10}@student.gsu.edu,
{yzhang, anu}@cs.gsu.edu

**Abstract.** Due to the huge number of genes and comparatively small number of samples from microarray gene expression data, accurate classification of diseases becomes challenging. Feature selection techniques can improve the classification accuracy by removing irrelevant and redundant genes. However, the performance of different feature selection algorithms based on different theoretic arguments varies even when they are applied to the same data set. In this paper, we propose a hybrid approach to combine useful outcomes from different feature selection methods through a genetic algorithm. The experimental results demonstrate that our approach can achieve better classification accuracy with a smaller gene subset than each individual feature selection algorithm does.

## 1 Introduction

As one of the important recent breakthroughs in the field of experimental molecular biology, microarray technology allows scientists to monitor changes in the expression levels of genes in response to changes in environmental conditions or in healthy versus diseased cells. It increases the possibility of disease classification and diagnosis at the gene expression level. However, due to a huge number of genes (features) and comparatively small number of samples, biologists are often frustrated by microarray gene expression data when they are trying to search for meaning from it. It is important for diagnostic tests to be able to remove redundant and irrelevant genes and find a subset of discriminative genes. Selecting informative genes is crucial to improving the diagnostic/classification accuracy [2, 7].

Many methods have been proposed for feature selection. Among them, feature ranking techniques are particularly attractive because of their simplicity, scalability, and good empirical success. A set of top ranked features can be selected for data analysis or building a classifier. Some feature ranking methods use information-theoretic functions, such as $\chi^2$-statistics [4], T-statistics [2], MIT correlation [5], information gain [1], and entropy-based measure [6]. Some other approaches utilize Support Vector Machines (SVMs) for feature ranking/selection [7, 8].

However, different feature selection methods result in different gene rankings or gene subsets. It is hard to decide which feature selection algorithm is best fit for a data set because the performance of an algorithm varies with different data sets. Our work

V.N. Alexandrov et al. (Eds.): ICCS 2006, Part II, LNCS 3992, pp. 678 – 685, 2006.

in this paper is to effectively combine information from various feature selection methods for more reliable classification. We propose a hybrid approach that combines valuable information from multiple feature selection methods through a genetic algorithm (GA). To evaluate our method, several feature selection algorithms, that is, entropy-based [6], T-statistics, and SVM-RFE (Recursive Feature Elimination) [7] are used to provide candidate features for the GA. We test our approach on two microarray data sets (Colon Cancer and Prostate Cancer). Experimental results show that our approach is effective and efficient in finding subsets of informative genes for reliable classification.

The rest of the paper is organized as follows. Section 2 describes our hybrid feature selection method. Section 3 introduces three existing feature selection algorithms used in later experiments. Section 4 compares our method with three existing feature selection algorithms on two data sets. Conclusion is drawn in Section 5.

## 2 Method

The idea of our hybrid approach (shown in Fig.1.) is to absorb useful information from different feature selection algorithms to find better feature subsets that can have smaller size or better classification performance than those individual algorithms. We use a genetic algorithm (GA) to fuse multiple feature selection criteria to accomplish this goal.

**Gene/Feature Pool.** The feature pool is a collection of candidate features to be selected by the genetic algorithm to find a feature subset. Instead of using all features from the original data, sets of features selected by multiple feature selection algorithms are input to the pool.

**Representation of Feature Subset.** Each feature subset (an individual) is encoded by $n$-bit binary vectors. The bits with value 1 in a vector represent the corresponding features being selected, while the bits with value 0 mean the opposite.

**Induction Algorithm.** The genetic algorithm is independent of the inductive learning algorithm used by the classifier. Multiple induction algorithms, such as Naïve Bayes, artificial neural network, and decision trees can be flexibly incorporated into our method. In this paper, we use SVM classifier [9] in our experiments.

**Fitness Function.** Our genetic algorithm is designed to maximize classification accuracy of the chosen classifier. So the accuracy obtained from the induction algorithm is used to evaluate each individual in a population.

**Genetic Operators**

(1) Selection: Roulette wheel selection is used to probabilistically select individuals from a population for later breeding.
(2) Crossover: We use single-point crossover operator. The crossover point $i$ is chosen at random.
(3) Mutation: Each individual has a probability $p_m$ to mutate. We randomly choose a number of $n$ bits to be flipped in every mutation stage.

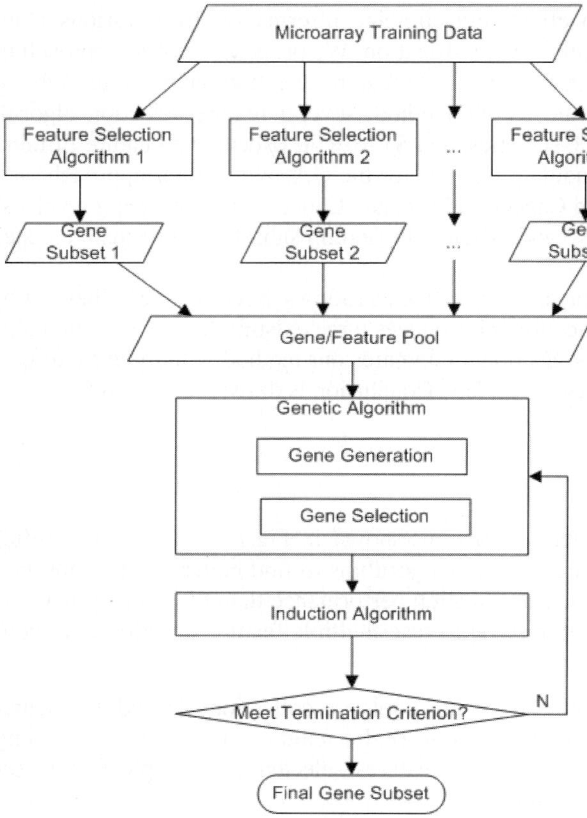


**Fig. 1.** Hybrid feature selection with genetic algorithm

## 3 Feature Selection Methods

To evaluate our method, three feature selection algorithms (i.e. entropy-based, T-statistics and SVM-RFE) are applied to microarray data sets. Then a number of top-ranked features from each algorithm are selected to form the feature pool for the GA. The idea of entropy-based method [6] is that removing an irrelevant feature would reduce the entropy more than that for a relevant feature. The algorithm ranks the features in descending order of relevance by finding the descending order of the entropy after removing each feature one at a time. T-statistics is a classical feature selection approach [2]. Each sample is labeled with {1, -1}. For each feature $f_j$, the mean $\mu_j^1$ (resp. $\mu_j^{-1}$) and standard deviation $\delta_j^1$ (resp. $\delta_j^{-1}$) are calculated using only the samples labeled 1 (resp. -1). Then a score T($f_j$) can be obtained by the equation defined in [2]. When making a selection, those features with the highest scores are considered as the most discriminatory features. Guyon et al. [7] proposed SVM-RFE, a backward feature elimination algorithm that removes one "worst" gene (i.e., the one changed the objective/cost function $J$ least after being removed) at one time. $w_i^2$ is taken as the ranking criterion in SVM-RFE.

# 4 Experiments

Leave-One-Out (LOO) cross-validation is used in Experiment 1 to obtain an estimate of the classification accuracy. Since a testing data set is available in Experiment 2, five-fold cross-validation is used to estimate the training accuracy. We use SVM with linear kernel as the classifier in both experiments. However, it is flexible to apply multiple induction algorithms to our hybrid method. Our experiments are implemented in a PC with Pentium 4(2.64MHz) and 512M RAM. All algorithms are coded in C++ and Matlab R14.

## 4.1 Experiment 1

Colon cancer data set [13] contains the expression of the 2000 genes across the 62 tissues/samples (40 tumor and 22 normal) collected from colon-cancer patients.

The experimental results of entropy, T-statistics, and SVM-RFE on Colon cancer data are presented in Table 1. We can see that in general, SVM-RFE gives the best performance among the three.

**Table 1.** LOO accuracy of entropy-based, t-statistics, and SVM_RFE on colon cancer data

| Top Features | Entropy (%) | T-statistics (%) | SVM-RFE (%) |
|---|---|---|---|
| 2 | 65.52 | 84.48 | 75.86 |
| 4 | 65.52 | 86.21 | 89.66 |
| 8 | 65.52 | 86.21 | 96.55 |
| 16 | 65.52 | 87.93 | 98.28 |
| 32 | 65.52 | 89.66 | 96.55 |
| 64 | 60.34 | 89.66 | 94.83 |
| 128 | 68.97 | 89.66 | 93.10 |
| 256 | 81.03 | 87.93 | 91.38 |
| 512 | 84.48 | 82.76 | 86.21 |
| 1024 | 82.76 | 81.03 | 84.48 |
| 2000 | 81.03 | 81.03 | 79.31 |

**Table 2.** Top-20 features from entropy-based, t-statistics, and SVM-RFE on colon cancer data

| Feature Selection Algorithms | Top-20 Features |
|---|---|
| Entropy-based | 169,1451,1430,1538,1660,375,1277,1150,445,1697,761,1170,825,609,1055,603, 1882,1910,1341,808 |
| T-statistics | 493,765,377,1423,249,245,267,66,**14**,1772,625,822,622,411,137,1674,1771,111, 1582,513 |
| SVM-RFE | 350,164,**14**,1378,43,1976,1325,353,44,16,250,175,159,115,458,24,988,47,33,30 |

**Table 3.** GA experiments on colon cancer data

| Gene/Feature Pool | | | Feature Subsets Selected by GA | LOO Acc. (%) |
|---|---|---|---|---|
| Entropy-based | T-statistics | SVM-RFE | | |
| Top 2 (**169**, 1451) | Top 2 (**493,765**) | Top 2 (**350,164**) | 5 (164,169,350,493,765) | 89.66 |
| Top 2 (169, **1451**) | Top 4 (493,765,377,**1423**) | Top 4 (**350,164,14, 1378**) | 6 (14,164,350,1378,1423,1451) | 98.28 |
| Top 4 (169, **1451,1430, 1538**) | Top 4 (493,**765, 377,1423**) | Top 4 (**350,164,14, 1378**) | 10 (14,164,350,377,765,1378, 1423,1430,1451, 1538) | 96.56 |
| Top 4 (169,**1451,1430,1538**) | Top 8 (**493,765,**377,1423,249,**245,267, 66**) | Top 8 (**350,164,14,**1378, **43,**1976,**1325,**353) | 12 (14,43,66,164,245,267,350,493,765,1325,1430,1451) | 98.28 |

The numbers in bold are the genes selected by the GA from the feature pool.

The Top-20 features given by the three algorithms on Colon data set are presented in Table 2. Only one gene/feature with column ID 14 is shared by T-statistics and SVM-RFE and entropy-based method has no common feature with the other two algorithms in Top-20 features. Besides the top 20 features, we notice that the ranks of features are very different in the three algorithms.

Table 3 shows the experimental results of applying our algorithm to Colon cancer data. To show the robustness of GA approach, we test several feature pools each of which contains a different number of top-ranked features chosen from the three methods. Our feature pools are relatively small because we assume a smaller informative gene subset (e.g., no more than 20 genes) is usually preferred for data analysis for a given accuracy due to the cost of performing the necessary clinical test and analysis. Features finally selected by the genetic algorithm are highlighted. In this experiment, the genetic algorithm is terminated with a maximum of 10 iterations. The size of the population is 30. It is possible to achieve better results if more iterations or a larger population size are allowed. As Table 3 shows, the genetic algorithm is capable of selecting a subset of 6 genes and achieves 98.28% accuracy while SVM-RFE needs 16 genes to get the same accuracy. In another case, our approach finds a subset of 12 genes from a different gene pool, which also reaches 98.28% accuracy.

### 4.2 Experiment 2

We further test the T-statistics, SVM-RFE and our approach on Prostate cancer data set [14]. Entropy-based algorithm is not used on this data set for the reason of time consumption. This data set consists of training data and testing data. The training data contain 52 prostate tumor samples and 50 non-tumor (normal) prostate samples with 12600 genes. The testing data contain 34 samples (25 tumor and 9 normal samples) obtained from a different experiment. In this experiment, feature selection algorithms first select features from training data. Then SVM classifies the testing data based on these selected features.

**Table 4.** Training and testing accuracy of t-statistic and SVM-RFE on prostate cancer data

| Features | Training Accuracy (%) | | Testing Accuracy (%) | |
|---|---|---|---|---|
| | T-statistics | SVM-RFE | T-statistics | SVM-RFE |
| 2 | 76.47 | 84.31 | 97.06 | 73.53 |
| 4 | 78.43 | 86.27 | 97.06 | 70.59 |
| 8 | 86.27 | 96.08 | 88.24 | 73.53 |
| 16 | 83.33 | 100.00 | 88.24 | 85.29 |
| 32 | 89.22 | 100.00 | 88.24 | 94.12 |
| 64 | 90.20 | 100.00 | 76.47 | 91.18 |
| 128 | 91.18 | 99.02 | 91.18 | 91.18 |
| 256 | 93.14 | 95.10 | 82.35 | 91.18 |
| 512 | 93.14 | 95.10 | 82.35 | 91.18 |
| 1024 | 91.18 | 94.12 | 85.29 | 94.12 |
| 2048 | 91.18 | 93.14 | 88.24 | 94.12 |
| 4096 | 89.22 | 92.16 | 94.12 | 94.12 |
| 8192 | 90.20 | 91.18 | 97.06 | 94.12 |
| 12600 | 89.22 | 91.18 | 97.06 | 94.12 |

**Table 5.** Top-20 features from t-statistics and SVM-RFE on prostate cancer data

| Feature Selection Algorithms | Top-20 Features |
|---|---|
| T-statistics | 6185,10138,3879,7520,4365,9050,**205**,5654,3649,**12153**,3794,9172, 9850,8136,7768,5462,12148,9034,4833,8965 |
| SVM-RFE | 10234,**12153**,8594,9728,11730,**205**,11091,10484,12495,49,12505,10 694,1674,7079,2515,11942,8058,8658,8603,7826 |

Table 4. presents the results of T-statistics and SVM-RFE on Prostate cancer data. SVM-RFE again performs better than T-statistics in most of cases. However, its testing accuracy is much lower than that of T-statistics when using top 2 and top 4 features. Among the Top-20 features ranked by the two methods, there are only two common genes as shown in Table 5.

The results of our algorithm on the Prostate cancer data are presented in Table 6. Since this data set is relatively large with 12600 features, we terminate the genetic algorithm with 5 iterations. We adopt five-fold validation for training accuracy. For all of the three feature pools in Table 6, our approach can obtain the best testing accuracy (94.12%) with much smaller feature subsets compared to SVM-RFE. However, the testing accuracy is not as good as the one achieved by Top-2 or Top-4 genes selected by T-statistics. In the first case, our approach selects a subset of 3 genes which can achieve 93.14% training accuracy, while SVM-RFE and T-Statistics cannot even reach the same training accuracy with 4 genes. In the second case, a subset of 4 genes selected by our GA achieves 95.10% training accuracy, which is higher than SVM-RFE and T-statistics with the same size of gene subset. In the last case, we select a subset of 8 genes, which achieves the same training accuracy as SVM-RFE does, but with higher testing accuracy.

**Table 6.** GA experiments on prostate cancer data

| Gene/Feature Pool | | Feature Subsets Selected by GA | Training Acc. (%) | Testing Acc. (%) |
|---|---|---|---|---|
| T-statistics | SVM-RFE | | | |
| Top2 (**6185 10138**) | Top2 (**10234**,12153) | 3 (6185,10138,10234) | 93.14 | 94.12 |
| Top4 (**6185**,10138, **3879**,7520) | Top 4 (**10234**,12153,**8594**, 9728) | 4 (3879,6185,8594, 10234) | 95.10 | 94.12 |
| Top8 (6185,10138, 3879,7520,4365 9050, **205**,5654) | Top 8 (**10234,12153, 8594,9728,11730,205,11091**,**10484**) | 8 (205,8594,9728, 10234,10484,11091,11730,12153) | 96.08 | 94.12 |

The numbers in bold are the genes selected by the GA from the feature pool.

## 5 Conclusions

In this paper, we present a hybrid feature selection approach for microarray gene expression data. We use a genetic algorithm to combine valuable outcomes of multiple feature selection algorithms to try to find subsets of informative genes. We compare our method with entropy-based, T-statistics, and SVM-RFE on two data sets. Our method is able to robustly find a small-sized feature set for reliable classification. The experimental results show that our method is efficient for microarray gene selection and classification. In our experiments, we randomly choose top features to form the feature pool. However, it can be seen that top features cannot guarantee better classification. How to choose features from the outcomes of various algorithms is an important problem to be solved. The genetic algorithm can also be improved. We leave it for future work.

## References

1. Liu,Y., A comparative study on feature selection methods for drug discovery, Journal of Chemical Information and Computer Sciences 44(5): 1823-1828,2004
2. Liu, H., J. Li, L. Wong, A comparative study on feature selection and classification methods using gene expression profiles and proteomic pattern, Genomic Informatics, 13, 51-60, 2002
3. J. Weston et al., Feature selection for SVMs, Advances in Neural Information Processing Systems 13 , 2000
4. H. Liu, R. Setiono, Chi2: Feature Selection and Discretization of Numeric Attributes, Proc. IEEE 7$^{th}$ International Conference on Tools with artificial Intelligence, 338-391, 1995
5. Golub, T.R. et al., Molecular classification of cancer: Class discovery and class prediction by gene expression monitoring, Science, 286, 1999
6. Dash, M., Liu, H., Handling Large Unsupervised Data via Dimensionality Reduction, ACM SIGMOD Workshop on Research Issues in Data Mining and Knowledge Discovery 1999

7. Guyon, I., Weston, J., Barnhill, S., Vapnik, V., Gene Selection for Cancer Classification using Support Vector Machines, Machine Learning, Vol. 46, No. 1-3, pp. 389-422,2002
8. Weston, J., S. Mukherjee, O. Chapelle, M. Pontil, T. Poggio and V. Vapnik: Feature Selection for SVMs. Advances in Neural Information Processing Systems 13, 2000
9. Burges, C.J.C., A tutorial on support vector machines for pattern recognition, Data mining and Knowledge Discovery, 2(2), 121-167, 1998
10. LeCun, Y., J.S. Denker, S.A.Solla, Optimum Brain Damage, Advances in Neural Information Processing Systems II, D.S.Touretzky, Ed. Mateo, CA: Morgan Kaufmann, 1990
11. W. S. Noble, Support vector machine applications in computational biology, Kernel Methods in Computational Biology. B. Schoelkopf, KTsuda and J.-P. Vert, ed. MIT Press, 71-92, 2004
12. B. Schölkopf, I. Guyon, and J. Weston, Statistical Learning and Kernel Methods in Bioinformatics, Artificial Intelligence and Heuristic Methods in Bioinformatics 183, (Eds.) P. Frasconi und R. Shamir, IOS Press, Amsterdam, The Netherlands,1-21, 2003
13. Alon, U., et al. Broad Patterns of Gene Expression Revealed by Clustering Analysis of Tumor a Normal Colon Tissues Probed by Oligonucleotide Arrays, PNAS, 96:6745-6750, 1999
14. Dinesh Singh, et al. Gene Expression Correlates of Clinical Prostate Cancer Behavior. Cancer Cell, 1:203-209, March, 2002

# A Self-supervised Learning Framework for Classifying Microarray Gene Expression Data

Yijuan Lu[1], Qi Tian[1], Feng Liu[2], Maribel Sanchez[3], and Yufeng Wang[3]

[1] Department of Computer Science, University of Texas at San Antonio, TX, USA
{lyijuan, qitian}@cs.utsa.edu
[2] Department of Pharmacology, University of Texas Health Science Center at San Antonio, TX, USA
liuf@uthsca.edu
[3] Department of Biology, University of Texas at San Antonio, TX, USA
msanchez@lonestar.utsa.edu, yufeng.wang@utsa.edu

**Abstract.** It is important to develop computational methods that can effectively resolve two intrinsic problems in microarray data: high dimensionality and small sample size. In this paper, we propose a self-supervised learning framework for classifying microarray gene expression data using Kernel Discriminant-EM (KDEM) algorithm. This framework applies self-supervised learning techniques in an optimal nonlinear discriminating subspace. It efficiently utilizes a large set of unlabeled data to compensate for the insufficiency of a small set of labeled data and it extends linear algorithm in DEM to kernel algorithm to handle nonlinearly separable data in a lower dimensional space. Extensive experiments on the *Plasmodium falciparum* expression profiles show the promising performance of the approach.

## 1 Introduction

High throughput microarray technology enables the large-scale characterization of gene expression profiles rapidly. Genes that are involved in correlated functions tend to yield similar expression patterns in microarray hybridization experiments.

To date, various machine learning methods have been applied to analyze microarray data to capture specific expression patterns. Despite that some of the methods have achieved useful classification results, two problems still plague efforts to analyze high throughput microarray data: (i) the high dimensionality of data, and (ii) the relatively small sample size.

As *high dimensionality* obscures the details in the data and *small sample* size precludes the influence of solidly supported conclusions, an approach that is relatively unaffected by these problems will allow us to get more from less.

The small sample problem can be alleviated by self-supervised learning techniques, which take a hybrid of labeled and unlabeled data to train classifiers. These techniques normally assume that only a fraction of the data is labeled with ground truth, but still take advantage of the entire dataset to generate a good classifier. They make the assumption that unlabeled data contains information about the joint data distribution over features, which can be used to help supervised learning. This

V.N. Alexandrov et al. (Eds.): ICCS 2006, Part II, LNCS 3992, pp. 686 – 693, 2006.

learning paradigm could be viewed as an integration of supervised learning and unsupervised learning.

The problem of high dimensionality can be alleviated by discriminant analysis. It is to find a mapping such that the data are clustered in the reduced feature space, in which the probabilistic structure could be simplified and captured by simpler model assumption, e.g., Gaussian mixtures.

Discriminant-EM (DEM) [1] effectively combines self-supervised learning and discriminant analysis for content-based image retrieval. It applies self-supervised learning techniques in a lower dimensional space projected by discriminant analysis. The basic idea is to learn discriminating features and a classifier simultaneously by inserting a multi-class linear discriminating step in the standard expectation-maximization (EM) [2] iteration loop. However, since its discriminating step is linear, DEM has an obvious drawback in handling data that are not linearly separable.

In this paper, we generalize the DEM algorithm from a linear setting to a nonlinear one, and propose a self-supervised learning framework for microarray analysis using Kernel Discriminant-EM (KDEM). KDEM transforms the original data space **X** to a higher dimensional kernel *"feature space" F* and then projects the transformed data to a lower dimensional discriminating subspace, such that nonlinear discriminating features could be identified and training data could be better classified in a nonlinear feature subspace. Extensive experiments are performed on the microarray dataset of malaria parasite *Plasmodium falciparum* for classification of specific functional classes.

The rest of the paper is organized as follows. In Section 2, we present kernel DEM algorithm. In Section 3, we apply KDEM and other algorithms to gene classification and use KDEM to identify putative genes of specific functional classes. Finally conclusions and future work are given in Section 4.

## 2 Kernel Discriminant-EM Algorithm

Kernel DEM (KDEM) is a generalization of DEM [1] in which instead of a simple linear transformation to project the data into discriminant subspaces, the data is first projected nonlinearly into a high dimensional feature space $F$ where the data are better linearly separated. After that, the original Multiple Discriminant Analysis (MDA) [2] algorithm is applied in a kernel feature space $F$, which is related to the original space by a non-linear mapping $\phi : x \rightarrow \phi(x)$. To avoid working with the mapped data explicitly, the kernel function computes a dot product in a feature space F: $k(x,z) = (\phi(x)^T \cdot \phi(z))$. Formulating MDA using only dot products of the form $\phi_i^T \cdot \phi_j$, the reproducing kernel matrix can be substituted into the formulation and eliminate the need for direct nonlinear transformation.

Using superscript $^{\phi}$ to denote quantities in the new space, we have the objective function of MDA in the following form:

$$W_{opt} = \arg\max_{W} \frac{|W^T S_B^{\phi} W|}{|W^T S_W^{\phi} W|} \tag{1}$$

$$S_B^{\phi} = \sum_{j=1}^{C} N_j \cdot (m_j^{\phi} - m^{\phi})(m_j^{\phi} - m^{\phi})^T \tag{2}$$

$$S_W^{\phi} = \sum_{j=1}^{C} \sum_{i=1}^{N_j} (\phi(\mathbf{x}_i^{(j)}) - m_j^{\phi})(\phi(\mathbf{x}_i^{(j)}) - m_j^{\phi})^T \tag{3}$$

with $S_B$ and $S_W$ are between-class and within-class scatter matrices, $m^{\phi} = \frac{1}{N}\sum_{k=1}^{N}\phi(\mathbf{x}_k)$, $m_j^{\phi} = \frac{1}{N_j}\sum_{k=1}^{N_j}\phi(\mathbf{x}_k)$, where $j = 1,\ldots,C$, and $N$ is the total number of samples.

In general, there is no other way to express the solution $W_{opt} \in F$, either because $F$ is too high or infinite dimension, or because we do not even know the actual *feature space* connected to a certain kernel. Hence the goal of kernel multiple discriminant analysis (KMDA) [3] is to find

$$\mathbf{A}_{opt} = \arg\max_{A} \frac{|\mathbf{A}^T K_B \mathbf{A}|}{|\mathbf{A}^T K_W \mathbf{A}|} \tag{4}$$

Where $A = [\vec{\alpha}_1, \cdots, \vec{\alpha}_{C-1}]$, $K_B$ and $K_W$ are $N \times N$ matrices which require only kernel computations on the training samples [3].

Kernel DEM can be initialized by selecting all labeled data as kernel vectors, and training a weak classifier based on only labeled samples. Then, the three steps of Kernel DEM are iterated until some convergence criterion is satisfied:

- E-step: set $\hat{Z}^{(k+1)} = E[Z \mid D; \hat{\Theta}^{(k)}]$
- D-step: set $\mathbf{A}_{opt}^{k+1} = \arg\max_{A} \frac{|\mathbf{A}^T K_B \mathbf{A}|}{|\mathbf{A}^T K_W \mathbf{A}|}$, and project a data point $\mathbf{x}$ to a linear subspace of feature space $F$.
- M-Step: set $\hat{\Theta}^{(k+1)} = \arg\max_{\Theta} p(\Theta \mid D; \hat{Z}^{(k+1)})$

The same notation is used as in [1]. The E-step gives probabilistic labels to unlabeled data, which are then used by the D-step to separate the data.

## 3 Experiments and Analysis

### 3.1 Dataset

The microarray dataset used in this study is a time-course expression profile of malaria parasite *Plasmodium falciparum*, during the 48-hour red blood cell cycle [4]. The original data are downloadable from http://malaria.ucsf.edu/SupplementalData.php, which include the profiles of 46 consecutive time points excluding 23-hour and 29-hour during which synchronized samples were not available. After standard quality control filtering and normalization, a complete dataset consisted of signals for 7091 oligonucleotides corresponding to over 4000 Open Reading Frames (ORFs) [5]. Note that the spots with array features were recorded as empty and thus were not included for further study if the sum of median intensities is smaller than the local background plus two times its standard deviation.

In the original paper [4], 14 functional classes of proteins were shown to exhibit distinct developmental profiles by Fourier Transform, including components involved in genetic information flow, metabolic pathways, cellular regulatory networks, organellar activities, and parasite-specific activities. Because the number of genes in class 5 and 7 were too small to train, we combined class 3 with class 7 to form a large group given that they are naturally consequential in metabolic pathways, and combine class 4 with class 5 given that they both represent nucleotide synthetic pathways. Finally, the entire ground truth dataset included the expression of 472 annotated genes in a total of 12 functional classes (Table 1).

### 3.2 Experiments

In a well-cited microarray classification study [6], SVM, decision tree and multi-layer perceptrons (MLP) etc. have been investigated, and SVM, especially SVM with radial basis (RBF) kernel significantly outperformed the other algorithms in the functional classification. Therefore we were focused on the comparison of KDEM with SVM using the same radial basis kernel (RBF) functions. RBF functions used are $K(\mathbf{X},\mathbf{Y}) = \exp(-\|\mathbf{X}-\mathbf{Y}\|^2/2\alpha^2)$. In our testing, $\alpha$ was set to be a widely used value, the median of the Euclidean distances from each positive example to the nearest negative example [6].

We performed a two-class classification with positive genes from one functional class and the negative genes from the remaining classes. Hence, each gene can be classified in one of the four ways: true positive (TP), true negative (TN), false positive (FP) and false negative (FN), according to the ground truth and classifier results. The malaria dataset is an imbalanced dataset, in which the number of negative genes is much larger than the number of positive genes. For example, for class *Mitochondrial*, the number of positive instances was only 16 whereas the number of negative instances reached 456. In this case, we chose to use *f*_measure=2·(recall×precision)/(recall+precision) to measure the overall performance of each classifier. Besides, for each class, we randomly selected 2/3 positive genes and 2/3 negative genes as training set and the remaining gene data as testing set for classification. This procedure was repeated for 30 times. Finally, we obtained the average values of *f*_measure of 30 rounds for each class.

Table 1 shows the *f*_measure for three different classifiers on the twelve functional classes. From this table, we can clearly see that: (i) KDEM outperformed SVM for total of eleven classes. SVM failed for most classes with small size and yielded zero *f*_measure. It is because given small sample size, SVM could hardly find sufficient labeled data to train classifiers well. By contrast, KDEM releases the pain of small sample size problem by incorporating a large number of unlabeled data. (ii) When the sample size was large, for example, for class 2, KDEM also performed at least as good as SVM. (iii) Compared to DEM, KDEM also achieved superior performance on most classes except for class 5. This shows that KDEM, provided good kernel functions, has a better capacity than DEM to separate linearly non-separable data. For example, for the functional class *Proteasome*, the *f*_measure of KDEM was 87.4, whereas DEM could only achieved 28.8. For class 5, KDEM doesn't show significant advantages over DEM, which is probably due to that the data are more likely linearly separable. We will further investigate this to prove our assumption. (iv) Figure 1 shows the performance of KDEM, DEM and SVM as the size of training samples drops from 2/3 to 1/5 of the total samples. It is clear that the performance of KDEM is good and stable while the performance of DEM and SVM declines with smaller training samples.

Overall, the superior classification results of KDEM over other methods demonstrate its promise for classifying microarray gene expression data.

**Table 1.** Comparison of *f*_measure for KDEM, DEM, and SVM on twelve classes

| Class_ID | Functional Class | Number | SVM-rbf | DEM | KDEM-rbf |
|---|---|---|---|---|---|
| 1 | Transcription | 23 | 0.0 | 16.0 | **30.0** |
| 2 | Cytoplasmic Translation | 149 | **87.5** | 79.7 | 87.2 |
| 3 | Glycolysis pathway and TCA cycle | 23 | 1.33 | 17.6 | **35.6** |
| 4 | nucleotide synthesis | 21 | 0.0 | 22.0 | **23.6** |
| 5 | DNA replication | 36 | 17.6 | **59.9** | 58.4 |
| 6 | Proteasome | 35 | 70.9 | 28.8 | **87.4** |
| 7 | Plastid genome | 18 | 57.9 | 67.3 | **81.3** |
| 8 | Merozoite invasion | 80 | 84.1 | 80.7 | **86.5** |
| 9 | Actin myosin motors | 13 | 0.0 | 32.7 | **35.3** |
| 10 | Early ring transcripts | 31 | 91.3 | 90.6 | **91.4** |
| 11 | Mitochondrial | 16 | 0.0 | 27.3 | **35.5** |
| 12 | Organellar Translation | 27 | 0.0 | 26.2 | **42.4** |

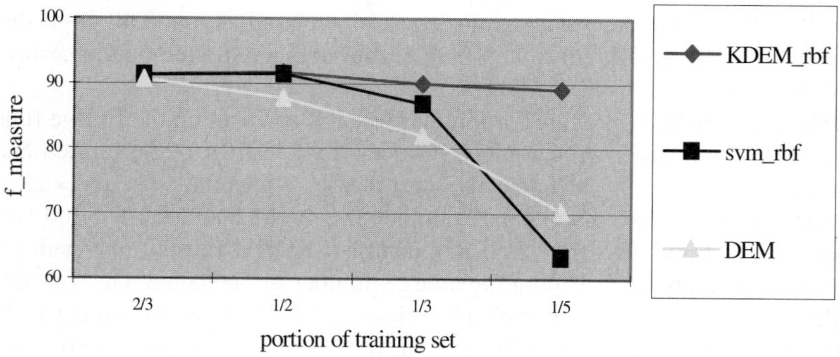


**Fig. 1.** Comparison of *f*_measure for KDEM, DEM, and SVM on *Mitochondrial* class with different sizes of training set

**Table 2.** Representative co-expressed genes of specific functional classes. The classification is based on their expression profiles during erythrocytic developmental cycle in malaria parasite.

| Class | Oligo_ID | Gene_ID | Annotation | Prob(%) |
|---|---|---|---|---|
| Transcription | f22770_1 | PFC0805w | DNA-directed RNA pol II | 96.4 |
| | opfc0750 | | | 77.9 |
| | m44300_14 | PF13_0152 | sir2 homologue | 99.6 |
| | f21506_2 | MAL8P1.131 | Gas41 homologue | 51.3 |
| | M33088_1 | MAL13P1.213 | transcription activator | 98.2 |
| Translation | c430 | PFC0635c | TIF E4 | 89.8 |
| | f26262_1 | PF07_0117 | eukaryotic TIF2 α | 55.4 |
| | j346_4 | PF10_0103 | eukaryotic TIF2, β | 96.3 |
| | j353_17 | PF10_0136 | Initiation factor 2 subunit | 76.9 |
| | ks142_1 | | | 92.1 |
| | popfj52810 | | | 88.9 |
| | opfl0097 | PFL2430c | eukaryotic TIF2b | 63.5 |
| | a3310_7 | PFA0495c | elongation factor | 69.5 |
| | c578 | PFC0870w | elongation factor 1 | 97.9 |
| | f64345_2 | PFL1590c | elongation factor g | 85.0 |
| | f41218_2 | MAL7P1.20 | peptide chain release factor | 50.9 |
| | opff72453 | MAL6P1.210 | nascent polypeptide associated complex alpha c | 92.8 |
| DNA replication | D17715_47 | PFD0475c | replication factor a protein | 99.9 |
| | D12635_36 | PFD0950w | ran binding protein 1 | 94.5 |
| | F64125_2 | PFE0520c | topoisomerase I | 51.2 |
| | F16271_1 | PF07_0105 | exonuclease I | 99.9 |
| | F16210_1 | MAL7P1.145 | DNA mismatch repair protein pms1 homologue | 79.6 |
| | oPFG0045 | | | 95.1 |
| | F57777_1 | MAL6P1.125 | DNA polymerase epsilon | 99.9 |
| Nucleotide synthesis | m38941_10 | PF13_0349 | diphosphate kinase b | 67.2 |
| Glycolysis TCA | j21_14 | PF10_0363 | pyruvate kinase | 89.7 |
| | ks152_12 | PF11_0157 | glycerol-3-phosphate dehydrogenase (GPDH) | 89.7 |
| | L2_270 | PFL0780w | GPDH | 81.7 |
| | Z_5_70 | PF13_0141 | L-lactate dehydrogenase | 99.9 |
| | Z_5_80 | | | 99.2 |
| | Z_5_90 | | | 99.9 |
| | m16243_2 | PF13_0269 | glycerol kinase | 99.9 |
| | N132_136 | PF14_0341 | glucose-6-phosphate isomerase | 54.1 |
| | E714_14 | PFE0225w | 3-methyl-2-oxobutanoate dehydrogenase | 97.5 |
| | J158_3 | PF10_0218 | citrate synthase | 97.3 |
| | PFBLOB0009 | PF10_0334 | succinate dehydrogenase | 91.8 |
| Proteasome | D6287_29 | PFD0165w | ubiquitin-specific protease | 74.6 |
| | D23156_23 | PFD0680c | Ubiquitin terminal hydrolase a | 99.9 |
| | Z_7_90 | MAL8P1.142 | proteasome β-subunit | 93.3 |
| | oPFL0014 | PFL2345c | tat-binding protein homolog | 99.9 |

### 3.3 Putative Genes of Specific Functional Classes Identified by KDEM

After validation of our algorithms on a small set of genes (472) with ground truth from *P. falciparum* microarry dataset, we applied KDEM to the rest of unknown 3776 putative malaria genes to classify six specific functional categories.

Table 2 shows several representative genes that were predicted to belong to six functional classes. Their potential functionality is confirmed by independent predictions based on Gene Ontology [8], demonstrating that self-supervised learning is a powerful expression classification method. Such classification could shed light on novel network components and interactions:

(1) Transcription, translation, and DNA replication machineries are complex networks that involve fine regulations of DNA (RNA)-protein and protein-protein interactions. For instance, besides essential enzymes (DNA-directed RNA polymerase complex), transcriptional factors such as Gas41 and Sir2 homolog and transcriptional activators may participate in the regulation of transcription (Table 2). The promoter regions of these regulators are yet to be discovered.

(2) Glycolysis/TCA cycle and Nucleotide (DNA or RNA) synthesis exemplify metabolic networks which involve protein-metabolite interactions. For example, the presence of a cascade of co-expressed enzymes, including glucose-6-phosphate isomerase, glycerol-3-phosphate dehydrogenase, pyruvate kinase, lactate dehydrogenase (Table 2), not only suggests that malaria parasite possesses conserved key components in carbohydrate metabolism, but also portrays the various co-factors and metabolites that are involved in the activity of each enzyme.

(3) Proteasome is a tightly-wrapped complex of threonine proteases and regulatory proteins that mediate protein-protein interactions in cell cycle control and stress response. In previous work [9], we predicted a number threonine proteases and ubiquitin hydrolases, sketching the core elements of malarial proteasome. A concerted regulation pattern revealed by this study is consistent with the postulation of an essential ATP-dependent ubiquitin-proteasome pathway, which was inferred from the results of inhibition assays [10].

## 4 Discussions

Kernel Discriminant-EM (DEM) proposed a framework to address the small sample problem and the high dimensionality problem by applying self-supervised learning in an optimal nonlinear discriminant subspace. The proposed algorithm is applied on gene classification on the *Plasmodium falciparum* dataset, and KDEM outperforms linear DEM and SVM in the extensive tests.

The insights provided by self-supervised learning on transcriptomic data into the dynamics of gene networks could shed light on as yet unrecognized network interactions [11]. A significant roadblock on the use of genomic data to better understand infectious diseases is our inability to assign gene functionality. Malaria parasite *Plasmodium falciparum* appears among the most problematic: 60% of the open reading frames are annotated as "hypothetical" [5]. Our study may provide an effective means to circumvent this problem. By identifying co-expressed genes in developmental cycle, it also helps us to identify what could conceivably be network modules. Any network module could contain a range of proteins and regulatory elements [11]. The key components of these modules may have stringent functional constraint and hence are conserved across species [12]. Subtracting these known from the modules, the remaining "hypothetical" in transcriptomic maps represent lineage-specific gaps in gene networks. The ability to assign a "hypothetical" gene to a specific network module opens an opportunity toward a tempo-specific functional characterization, because for a parasite with multiple hosts (human and mosquito) and a dynamic life cycle, "when and where" to initial wet-lab experiments is of critical importance. This network view should allow us to locate choke points in the parasite - potential vulnerabilities that could result in new malarial control strategies.

## Acknowledgements

This work is support in part by San Antonio Life Science Institute (SALSI) and ARO grant W911NF-05-1-0404 to Q. Tian, and San Antonio Area Foundation, NIH RCMI grant 2G12RR013646-06A1, and UTSA Faculty Research Award to Y. Wang.

## References

1. Wu, Y., Tian, Q., and Huang, T. S.: Discriminant EM algorithm with application to image retrieval, *Proc. of IEEE Conf. Computer Vision and Pattern Recognition* (2000)
2. Duda, R. O., Hart, P. E., and Stork, D. G.: 2nd *Pattern Classification*, John Wiley & Sons, Inc. (2001)
3. Schölkopf, B. and Smola, A. J.: *Learning with Kernels*. Mass: MIT Press. (2002)
4. Bozdech, Z. Llinas M., Pulliam, B. L., Wong, E. D., Zhu, J. DeRisi, J. L.: The transcriptome of the intraerythrocytic development cycle of plasmodium falciparum. *Plos Biology*, (2003)
5. 1(1): 1-16
6. Gardner, M.J., N. Hall, E. Fung, O. White, M. Berriman, R.W. Hyman, J.M. et al.: Genome sequence of the human malaria parasite *Plasmodium falciparum. Nature,* (2002) 419: 498-511
7. Brown, M.P., Grundy, W.N., Lin, D., et al.: Knowledge-based analysis of microarray gene expression data by using support vector machines, *Proc. Natl, Acad. Sci.* USA, (2000), 97(1): 262-267
8. The Gene Ontology Consortium, "Gene Ontology: tool for the unification of biology," *Nature Genet.* (2000) Vol. 25, 25-29
9. Wu, Y., Wang, X., Liu, X., and Wang Y.: Data-mining approaches reveal hidden families of proteases in the genome of malaria parasite, *Genome Res.* (2003) 13:601-616
10. 9.Gantt S.M., Myung J.M., Briones M.R., Li W.D., Corey E.J., Omura S., Nussenzweig V., Sinnis, P.: Proteasome inhibitors block development of Plasmodium spp., *Antimicrob Agents Chemother.* (1998) 42:2731-2738
11. Kitano, H.: Systems biology: A brief overview. *Science,* (2002) 295: 1662-1664
12. Bowers, P.M., Cokus, S.J., Eisenberg, D., Yeates, T.O.: Use of logic relationships to decipher protein network organization, *Science.* (2004) 306:2246-2249

# Pooling Evidence to Identify Cell Cycle–Regulated Genes

Gaolin Zheng[1], Tom Milledge[1], E. Olusegun George[2], and Giri Narasimhan[1]

[1] Bioinformatics Research Group (BioRG), School of Computer Science,
Florida International University, Miami, Florida, 33199, USA
{gzhen001, tmille01, giri}@cis.fiu.edu
[2] Department of Mathematical Sciences, University of Memphis, Memphis, TN 38152, USA
eogeorge@memphis.edu

**Abstract.** Most of the biological studies have embraced statistical approaches to make inferences. It is common to have several independent experiments to test the same null hypothesis. The goal of research on pooling evidence is to combine the results of these tests to ask if there is evidence from the collection of studies to reject the null hypothesis. In this study, we evaluated four different pooling techniques (Fisher, Logit, Stouffer and Liptak) to combine the evidence from independent microarray experiments in order to identify cell cycle-regulated genes. We were able to identify a better set of cell cycle-regulated genes using the pooling techniques based on our benchmark study on budding yeast (*Saccharomyces cerevisiae*). Our gene ontology study on time series data of both the budding yeast and the fission yeast (*Schizosaccharomyces pombe*) showed that the GO terms that are related to cell cycle are significantly enriched in the cell cycle-regulated genes identified using pooling techniques.

## 1 Introduction

In bioinformatics analyses, the process of making inferences and extracting knowledge from data is complicated by the fact that the data may arrive from diverse sources. These varied sources of data may be heterogeneous or homogeneous. For example, in order to predict potential transcription factor binding sites, one may use heterogeneous sources of data such as gene expression data (to infer co-expressed genes), ChIP-chip data (to obtain information on binding sites for known transcription factors), upstream promoter sequence data (to infer shared and over-expressed sequence motifs), text literature (to use information from reported biological experiments on known transcription factors and their binding sites), sequence data on other related organisms (to infer conserved regions in the genome sequence and to get supporting evidence for the predictions), gene ontology information (to identify functionally-related genes), and much more. Even for a focused study (with homogeneous sources of data) that attempts to identify which set of genes show significant change in expression in a tissue sample from a lung cancer patient relative to that from a normal patient, it is possible that the data may be collected from a wide variety of individuals and the experiments may be performed by different research groups under varying conditions. While a host of bioinformatics methods exist for making

V.N. Alexandrov et al. (Eds.): ICCS 2006, Part II, LNCS 3992, pp. 694 – 701, 2006.

inferences from a single data set (along with providing statistical significance for these inferences), a significant challenge is posed when inferences have to be made based on multiple sources of data. It is non-trivial to integrate different sources of data (with different levels of variability and reliability) to arrive at one coherent set of conclusions.

Several procedures for pooling of such evidence in meaningful ways exist in the statistical literature [1-4]. The earliest work on implementing pooling strategies in bioinformatics is to be found in the work of Bailey and Gribskov, where they developed an algorithm called QFAST for the sequence homology search problem [5]. Their algorithm used a variant of Fisher's inverse Chi-square method without reference to the Chi-square distribution for computational efficiency [5]. In this study we investigate and evaluate a variety of statistical techniques to combine or pool together evidence from many homogeneous sources to arrive at meaningful conclusions. To evaluate these techniques, we revisit the familiar problem of using gene expression data to identify cell cycle-regulated genes in two related organisms: budding yeast, *Saccharomyces cerevisiae*, and fission yeast *Schizosaccharomyces pombe*. We consider existing tools to identify cell cycle-regulated genes and investigate the change in their performance after they are applied to three different yeast data sets generated by Spellman et al. and the evidence is pooled.

The cell-division cycle is fundamental to the proliferation of all organisms, and knowledge of its regulation helps in understanding various diseases [6]. Periodic gene expression seems to be a universal feature of cell cycle regulation. In 1998, the first genome-wide DNA microarray studies were conducted in the budding yeast, *S. cerevisiae*, to reveal a large number of cell cycle-regulated genes that peaked only once per cell cycle [7, 8]. More recently, microarray experiments have been conducted to study the cell cycle regulation in the fission yeast *S. pombe*, which is a distant relative of budding yeast [6, 9]. Both *S. cerevisiae* and *S. pombe* are useful organisms for the study of the cell division cycle because they have many well-characterized cell division cycle mutants [10]. In particular, the budding yeast data sets have driven the development of various computational methods for identifying periodically expressed genes [11-15]. Some of these methods even provided statistical significance (in terms of *P*-values) for their predictions [14, 15].

In order to demonstrate the efficacy of pooling, we considered two existing tools for identifying cell cycle-regulated genes [14, 15]. Although several other tools exist for this task, they were not appropriate for our study since they did not return *P*-values for their predictions for the individual experiments.

## 2 Methods

### 2.1 Identifying Cell Cycle-Regulated Genes

We adopted Fisher's *g*-test to test the periodicity of a gene [14]. A *P*-value was obtained for each time course data [14]. The lower the *P*-value, the more periodic is the gene. We also tested the periodicity of a gene using a permutation-based method [15].

### 2.2 Combining Evidence Using *P*-Values

Given several independent sets of samples along with the corresponding *P*-values for a certain property, the idea of combining the *P*-values to obtain a unified measure of significance has a long history. The existing methods are Fisher's inverse Chi-square method [1], George and Mudholkar's logit method [2], Stouffer's overall z method [4], and the Liptak-Stouffer [3] method.

In bioinformatics, combining inferences has been applied to sequence homology search, protein classification and scoring motifs [5, 16]. Their method is a variant of the Fisher's inverse Chi-square method. A more recent application can be found in the work of Oliva et al. on the fission yeast *Schizosaccharomyces pombe*, where they used Stouffer's method to obtain a single *P*-value for each gene from the three *P*-values inferred from the two elutriation experiments and the cdc25 block-release experiment in order to identify oscillating transcripts [9].

With the goal of evaluating the various *P*-value pooling techniques, we picked the problem of identifying cell cycle-regulated genes in budding yeast and fission yeast. The null hypothesis was that there is no periodic component in the time course data for the gene in question. The first advantage of choosing this problem was the ready availability of several independent cell cycle microarray experiments (alpha, cdc15 and cdc28) to test whether they collectively reject the null hypothesis. The second and more important reason for using the budding yeast data set is that de Lichtenberg et al. had also designed benchmark data to evaluate the results of the predictions [15]. The techniques we evaluate will provide us with *P*-values (for each gene) supporting the claim that the gene in question is cell cycle-regulated. The benchmark data sets allow us to measure the ability of different methods to identify the cell cycle-regulated genes. Below we describe the pooling techniques that were evaluated in this work. Since benchmark data is not readily available for the fission yeast, we were unable to evaluate the data sets in the same way that the budding yeast data set was. We identified the periodically expressed genes in the fission yeast using the pooling methods and the results were evaluated for functional enrichment.

#### 2.2.1 The Fisher Method

The Fisher procedure [1] for pooling the *P*-values is given by $F = \sum_{i=1}^{N} -2\ln(P_i)$ where N is the number of tests to be combined. Under the null hypothesis, the $P_i$'s can be assumed to have a uniform distribution in the range of [0,1]. Thus under this condition F has a $\chi^2$-distribution with 2N degrees of freedom. The *P*-value associated with the F-score is given by $P(X \geq F)$, where $X \sim x^2(2N)$. Although Fisher's method is the most commonly used method for combining information across multiple test of the same null hypothesis, one significant drawback of Fisher's method is that it is asymmetrically sensitive to small *P*-values compared to large *P*-values.

### 2.2.2 The Logit Method

A logit procedure is given by $L = -\sum_{i=1}^{N} (P_i/(1-P_i))$ [2]. Under the null hypothesis, $L$ has the distribution of a convolution of N logistic random variables. The distribution of $L$ can be approximated as $L \sim \pi\sqrt{\frac{N(5N+2)}{3(5N+4)}}t_{5N+4}$. The $P$-values associated with $L$ is obtained as $P\left(X > \frac{L}{\pi}\sqrt{\frac{3(5N+4)}{N(5N+2)}}\right)$, where, $X$ is a student-t random variable with 5N+4 degrees of freedom.

### 2.2.3 Stouffer's Z-Transform Method

The Z-transform $Z_i = \Phi^{-1}(P_i)$, converts the $P$-values, $P_i$, from each of the N independent tests into standard norm deviates $Z_i$. The sum of the $Z_i$'s, divided by the square root of the number of tests, N, has a standard normal distribution if the common null hypothesis is true [4]. Thus, $Z_s = \sum_{i=1}^{N} Z_i \Big/ \sqrt{N}$ can be compared to the standard normal distribution to provide a test of the cumulative evidence on the common null hypothesis [4].

### 2.2.4 Liptak-Stouffer's Weighted Z-Method

Liptak generalized Stouffer's Z-transform method by giving different weights to each study according to their power [3]. The scoring is given by $Z_w = \sum_{i=1}^{N} W_i Z_i \Big/ \sqrt{\sum_{i=1}^{N} W_i^2}$. There are different ways to choose the weights. Usually sample size or the inverse of the sample error variance is chosen as the weight for each test. It is debatable whether the weighted Z-method is appropriate since the $P$-values are already weighted by the sample size.

## 2.3 Gene Expression Data and Benchmark Data

We used three budding yeast cell cycle gene expression data. These three gene expression data set were obtained from samples of budding yeast cultures synchronized by three different methods: α-factor arrest, and arrest of two different temperature-sensitive mutants (cdc15 and cdc28) [8].

Fission yeast cell cycle data were obtained from *S. pombe* Functional Genomics website [http://www.sanger.ac.uk/PostGenomics/S_pombe/projects/cellcycle/]. Three elutriation and three cdc25 block-release cell cycle data were used in this study. The goal of using the fission yeast data was to be able to compare the results obtained from the budding and the fission yeast. Plus, fission yeast data sets are very new and substantial[9].

The performance of the *P*-value pooling methods were evaluated by measuring their abilities to identify genes from two of the three benchmark sets provided by de Lichtenberg et al. [15]. Benchmark data sets B1 and B2 were used in this comparative study. B1 consisted of a total of 113 genes previously identified as periodically expressed in small-scale experiments [15]. B2 consisted of genes whose promoters were bound by at least one of the nine known cell cycle transcription factors in the Chromatin IP studies [15]. After removing 50 genes that belonged to B1, B2 consisted of 352 genes which were deemed to be cell cycle-regulated. Benchmark data B3 was not used because it was deemed unreliable by the authors [15].

## 3 Results and Discussion

Summarizing our benchmarking experiments on *S. cerevisiae*, we have two methods to identify cell cycle-regulated genes from a single data set (Fisher's *g*-test and permutation method), three different gene expression data sets for *S. cerevisiae* (alpha, cdc15 and cdc28), four methods to pool *P*-values when any given method is used to identify the periodic genes from the three data sets, and two different benchmark sets of known cell cycle-regulated genes (B1 and B2 for *S. cerevisiae*). For *S. pombe*, we have six independent gene expression data sets and no readily available benchmark sets of cell cycle-regulated genes. Finally, for each of the experiments, the lists of significantly periodic genes obtained from the budding and the fission yeast data sets were queried against Gene Ontology database to obtain a list of significantly enriched GO terms.

### 3.1 Benchmark Results for *S. cerevisiae*

A set of *P*-values were obtained for each data set using either Fisher's *g*-test or permutation method. Each gene was associated with a *P*-value obtained from the individual data sets. The *P*-values were combined using the four *P*-value pooling methods mentioned above for *S. cerevisiae*. For each method, the genes were ranked based on the corresponding *P*-values. Lower *P*-values were ranked higher. Figure 1 shows the performance of Fisher's *g*-test and permuation method on each of three individual microarray expression data sets for *S. cerevisiae* along with that of the four *P*-value pooling techniques to combine the evidence from the three experimental data sets.

### 3.2 Function Enrichment of Periodically Expressed Genes

To assess the ability of pooling technique to identify cell cycle-regulated genes, gene ontology information was used to evaluate whether the identified cell cycle- regulated genes have significant enrichment of one or more terms that are related to cell cycle. This was done using GOMiner [17]. A Fisher's exact test is used to test if a GO term is enriched or depleted. For the budding yeast, three *P*-values were generated for each gene based on the three experiments using Fisher's *g*-test [14]. A single *P*-value was obtained for each gene using the logit pooling procedure. (Only the logit method was used in this analysis, since the four pooling methods were comparable in performance, with the logit method being marginally better). To correct the *P*-values for multiple testing, the method of Benjamini and Hochberg [18] was used to control the false discovery rate (FDR). The FDR method controls the expected portion of false positives at a given rate q. A list of 1686 genes was identified as cell cycle-coupled

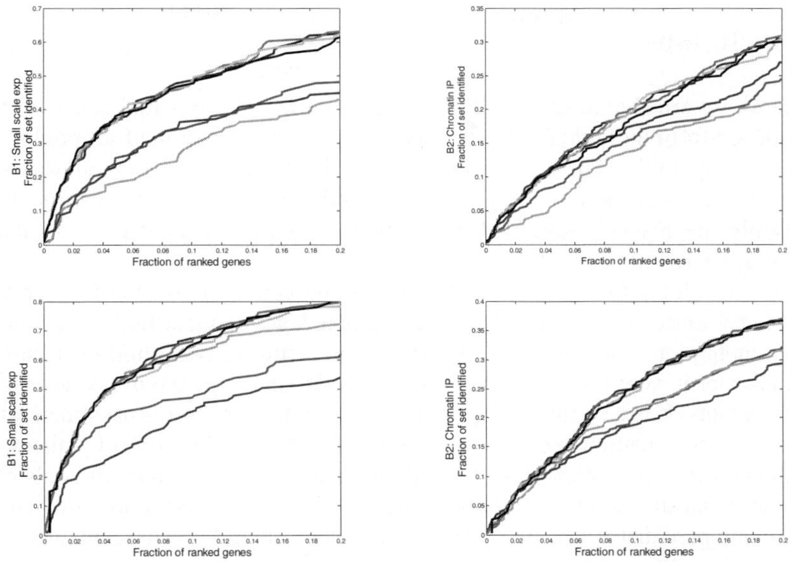


**Fig. 1.** Comparison of the three synchronization methods and the four *P*-value pooling methods. The fraction of the benchmark set that is identified is plotted as a function of the fraction of all the 5550 ranked genes. The different methods are colored as follows: alpha only (red), cdc15 only (green), cdc28 only (dark red), Fisher (blue), logit (magenta), Stouffer (cyan), and Liptak-stouffer (black). The graph on the left is based on the benchmark evaluation gene set B1, while the graph on the right is based on the gene set B2. Top row: results from Fisher's *g*-test. Bottom row: results from permutation method.

**Table 1.** Functional enrichment of periodically expressed genes identified by pooling the evidences with George and Mudholkar's logit method

| Organism | Enriched GO Term (total) | Number of periodic genes in GO term | Significance of Enrichment |
|---|---|---|---|
| Budding yeast | M-phase (114) | 62 | 0.0062 |
| | Cytokinesis (82) | 43 | 0.0429 |
| | Cytokinesis, contractile ring formation (10) | 9 | 0.0028 |
| | Cytokinesis, formation of actomyosin apparatus (10) | 9 | 0.0028 |
| | Chromosome condensation (11) | 9 | 0.0095 |
| | G1/S transition of mitotic cell cycle (20) | 13 | 0.0358 |
| Fission yeast | M-phase (125) | 50 | 0.0155 |
| | Cytokinesis (86) | 33 | 0.0761 |
| | Cytokinesis, completion of separation (5) | 4 | 0.0333 |
| | Cell separation during cytokinesis (5) | 4 | 0.0333 |
| | Chromosomal condensation (9) | 6 | 0.0281 |
| | G1/S transition of mitotic cell cycle (40) | 20 | 0.0079 |
| | G2/M transition of mitotic cell cycle (30) | 17 | 0.0027 |

for the budding yeast when q is set to 0.05. In a similar fashion, a list of 1315 genes was found to be cell cycle-regulated. These periodic genes were fed to GOMiner to find enriched GO terms. Table 1 shows the functional enrichment results.

## 4 Conclusions

In general, we are able to identify a better set of cell cycle-regulated genes by pooling the evidence from individual experiments. The performance of the four pooling methods was comparable. The performance of the Liptak-Stouffer method was slightly inferior to the other three pooling methods, while that of logit was marginally superior. While the Fisher's inverse Chi-square for pooling is quite popular, the other methods appeared to be less known.

In our functional enrichment studies, we found cell cycle-related GO terms were significantly enriched for the cell cycle-regulated genes identified using the Logit pooling method for both organisms. More interestingly, two children terms under cytokinesis term were more significantly enriched than the parent cytokinesis term for both organisms. It is also interesting to note that while there is an enrichment of genes from the G1/S transition process of the mitotic cell cycle in both organisms, genes from the G2/M transition process were significant only in *S. pombe*. It is also important to note that the study of data sets from two related organisms provided an opportunity to compare differences in significant genes for the cell cycle regulation.

We compared four different *P*-value pooling techniques and showed that they perform better than methods that use only a single data set. There are numerous potential applications for *P*-value pooling methods in the field of Bioinformatics. They are useful in any application where one has access to different data sets and statistical significance for the predictions.

## Acknowledgement

Research of GN and EOG was supported in part by NIH Grant P01 DA15027-01.

## References

1. Fisher, R.A., Statistical Methods for Research Workers. 14th ed. 1932, Edinburgh: Oliver and Boyd.
2. George, E.O. and G.S. Mudholkar, On the Convolution of Logistic Random Variables. Metrika, 1983. 30: p. 1-14.
3. Liptak, T., On the combination of independent tests. Magyar Tud. Akad. Mat. Kutato Int. Kozl, 1958. 3: p. 171-197.
4. Stouffer, S.A., E.A. Suchman, L.C. Devinney, S.A. Star, and R.M.J. Williams, The American Soldier, Vol. 1: Adjustment during army life. 1949, Princeton: Princeton University Press.
5. Bailey, T.L. and M. Gribskov, Combining evidence using p-values: application to sequence homology searches. Bioinformatics, 1998. 14(1): p. 48-54.

6. Rustici, G., J. Mata, K. Kivinen, P. Lio, C.J. Penkett, G. Burns, J. Hayles, A. Brazma, P. Nurse, and J. Bahler, Periodic gene expression program of the fission yeast cell cycle. 2004. 36(8): p. 809-817.
7. Cho, R.J., M. Campbell, E. Winzeler, L. Steinmetz, A. Conway, L. Wodicka, T. Wolfsberg, A. Gabrielian, D. Landsman, D. Lockhart, and R. Davis, A genome-wide transcriptional analysis of the mitotic cell cycle. Molecular Cell, 1998. 2: p. 65-73.
8. Spellman, P.T., G. Sherlock, M.Q. Zhang, V.R. Iyer, K. Anders, M.B. Eisen, P.O. Brown, D. Botstein, and B. Futcher, Comprehensive Identification of Cell Cycle-regulated Genes of the Yeast Saccharomyces cerevisiae by Microarray Hybridization. Mol. Biol. Cell, 1998. 9(12): p. 3273-3297.
9. Oliva, A., A. Rosebrock, F. Ferrezuelo, S. Pyne, H. Chen, S. Skiena, B. Futcher, and J. Leatherwood, The Cell Cycle Regulated Genes of Schizosaccharomyces pombe. PLoS Biology, 2005. 3(7): p. e225.
10. Hartwell, L., J. Culotti, and B. Reid, Genetic control of the cell-division cycle in yeast. I. Detection of mutants. Proc Nat Acad Sci, 1970. 66: p. 352–359.
11. Johansson, D., P. Lindgren, and A. Berglund, A multivariate approach applied to microarray data for identification of genes with cell cycle-coupled transcription. Bioinformatics, 2003. 19(4): p. 467-473.
12. Zhao, L.P., R. Prentice, and L. Breeden, Statistical modeling of large microarray data sets to identify stimulus-response profiles. Proc. Natl. Acad. Sci., 2001. 98: p. 5631-5636.
13. Lu, X., W. Zhang, Z.S. Qin, K.E. Kwast, and J.S. Liu, Statistical resynchronization and Bayesian detection of periodically expressed genes. Nucl. Acids Res., 2004. 32(2): p. 447-455.
14. Wichert, S., K. Fokianos, and K. Strimmer, Identifying periodically expressed transcripts in microarray time series data. Bioinformatics, 2004. 20(1): p. 5-20.
15. de Lichtenberg, U., L.J. Jensen, A. Fausboll, T.S. Jensen, P. Bork, and S. Brunak, Comparison of computational methods for the identification of cell cycle-regulated genes. Bioinformatics, 2005. 21(7): p. 1164-1171.
16. Bailey, T.L. and W.N. Grundy. Classifying proteins by family using the product of correlated p-values. in Proceedings of the Third international conference on computational molecular biology (RECOMB99). 1999.
17. Zeeberg, B.R., W. Feng, G. Wang, M.D. Wang, A.T. Fojo, M. Sunshine, S. Narasimhan, D.W. Kane, W.C. Reinhold, S. Lababidi, K.J. Bussey, J. Riss, J.C. Barrett, and J.N. Weinstein, GoMiner: A Resource for Biological Interpretation of Genomic and Proteomic Data. Genome Biology, 2003. 4(4): p. R28.
18. Benjamini, Y. and Y. Hochberg, Controlling the false discovery rate: a practical and powerful approach to multiple testing. Journal of the Royal Statistical Society, Series B, 1995. 57: p. 289-300.

# Discovering Sequence-Structure Patterns in Proteins with Variable Secondary Structure

Tom Milledge, Gaolin Zheng, and Giri Narasimhan

Bioinformatics Research Group (BioRG), School of Computer Science,
Florida International University, Miami, Florida, 33199, USA
{tmille01, gzhen001, giri}@cs.fiu.edu

**Abstract.** Proteins that share a similar function often exhibit conserved sequence patterns. Sequence patterns help to classify proteins into families where the exact function may or may not be known. Research has shown that these domain signatures often exhibit specific three-dimensional structures. We have previously shown that sequence patterns combined with structural information, in general, have superior discrimination ability than those derived without structural information. However in some cases, divergent backbone configurations and/or variable secondary structure in otherwise well-aligned proteins make identification of conserved regions of sequence and structure problematic. In this paper, we describe improvements in our method of designing biologically meaningful sequence-structure patterns (SSPs) starting from a seed sequence pattern from any of the existing sequence pattern databases. Improved pattern precision is achieved by including conserved residues from coil regions that are not readily apparent from examination of multiple sequence alignments alone. Pattern recall is improved by systematically comparing the structure of all known true family members and to include all the allowable variations in the pattern residues.

## 1 Introduction

Databases such as PROSITE [1] [2], eMOTIF [3] [4], PRINTS [5], and SPAT [6] have been created as repositories for sequence patterns that describe and distinguish functional and structural domains in proteins. Kasuya et al. [7] systematically investigated the three-dimensional structures of protein fragments whose sequences contain a specific PROSITE pattern. They observed that in a large number of cases, the three-dimensional conformations of the residues from the PROSITE pattern were nearly identical in all the true positives (i.e., proteins belonging to the family and containing the sequence pattern). The main drawback with the approach followed by existing databases to generate sequence patterns is that they base their computations on multiple sequence alignments, which are often inaccurate, especially when the sequences exhibit considerable diversity. We have previously described a method [8] which uses both sequence and structure information to construct patterns consisting of a sequence component (a "PROSITE-style" regular expression pattern) and a structure component (a structure template). In our method, sequence-structure patterns (SSPs) are generated by starting from "seed" sequence patterns from PROSITE, eMOTIF, PRINTS, or other sources, and improving them using a novel method that alternates between sequence and structure alignment of proteins, while using the knowledge of

V.N. Alexandrov et al. (Eds.): ICCS 2006, Part II, LNCS 3992, pp. 702 – 709, 2006.

substitution groups [9]; protein structures are obtained from the Protein Data Bank (PDB) [10]. We say that a protein has a sequence match with the SSP if it contains the sequence component of the SSP. The SSPs are evaluated with regard to their specificity (TP/(TP+FP)) and sensitivity (TP/(TP+FN)), where TP is the set of true positive sequence matches, while FP and FN are the sets of false positive and false negative sequence matches with respect to their membership in a SCOP (Structural Classification of Proteins) protein family. The SCOP database is a comprehensive classification of all proteins of known structure [11]. The basic classification unit in SCOP is the domain, a unit of the protein that is either observed isolated in nature or in more than one context in different multi-domain proteins. A related database is the ASTRAL Compendium, which provides sequences and structures for all domains filtered according to percentage sequence similarity [5]. The ASTRAL 40% database (version 1.65) contains a subset of proteins from the PDB database with less than 40% sequence identity to each other, and this database will be referred to as ASTRAL40 in this paper. We also refer to ASTRAL95 and ASTRAL100 (or full PDB) to refer to the corresponding databases with 95% and 100% sequence identity respectively. For our purposes, the ASTRAL40 database was used to generate SSPs for families that were well represented in the PDB and the ASTRAL95 database was used to generate SSPs for protein families with fewer PDB examples. In both cases, the ASTRAL100 database was used for testing.

## 2 SSP Method Improvements

Fig. 1 gives a brief description of our algorithm for generating SSPs and is reproduced from our earlier work. It takes as input a "seed" PROSITE-style pattern along with a training set database (in our case, we use ASTRAL40 unless there are not enough structures in it, in which case we use ASTRAL95). It produces as output a SSP, which is a pair <P, T>, where P is a sequence pattern, and T is a structure template for the sequence pattern. As mentioned earlier, it also produces sequence and structure alignments of proteins with this SSP. We first identified the PROSITE patterns with the highest number of hits in the ASTRAL40 database. The algorithm was then experimentally tested on these protein families. In the case where an SSP for a SCOP family already had a corresponding PROSITE sequence pattern, the PROSITE pattern was improved about 90% of the time with respect to the SSP sequence pattern. This represents an average improvement of specificity of +27.3% and an average improvement of sensitivity of +16.2%. Although patterns generated by this method were shown to have higher precision and recall values than comparable patterns in PROSITE, problems were encountered when variations in protein backbone configuration among the true members of a protein family did not allow structurally conserved residues to be detected in the corresponding sequence alignment. In this paper, we describe methods for improving the SSP discovery method in cases where variability in the protein backbone and/or secondary structure obscures instances of residue conservation.

### 2.1 Zinc Finger Example

The function of the C2H2 (SCOP family G.37.1.1) zinc finger proteins is zinc-dependent DNA or RNA binding, where the first pair of zinc coordinating residues

**SSP ALGORITHM**

**Input:** **(a)** A database of protein structures, and associated protein sequences, **N**,
**(b)** A PROSITE-style sequence pattern , **P.**

**Output:** **(a)** Sequence-structure pattern <**P'**, **T**>,
**(b)** Structure alignment S of proteins with pattern **P'**, and
**(c)** Sequence alignment **Q** of proteins with pattern **P'**.

1. Search for pattern **P** in database N to generate a list of candidate proteins **C**.
2. Pick a "cluster" **L** of proteins from **C** that belong to the same SCOP family.
3. Create a structure alignment **S** for **L** using the residues of pattern **P**.
4. Extract sequence alignment **Q** from structure alignment **S**.
5. Identify all positions in sequence alignment **Q** that have residues from a substitution group.
6. If stopping condition is not satisfied, then create a new structure alignment S for L using the positions identified in Step 6. Then go to Step 5.
7. Construct a PROSITE-style sequence-structure pattern **P'** and template T from the positions in **Q**.
8. Iterate the whole process if new candidates from database N are matched.

**Fig. 1.** SSP Algorithm

are cysteines and the second pair are histidines. The PROSITE pattern for this family, PS00028, is **C-x(2,4)-C-x(3)-[LIVMFYWC]-x(8)-H-x(3,5)-H** (Fig. 2a). The pattern generated by the SSP algorithm, SSP91022, is **[AHITFVY]-x(1)-C-x(2,5)-C-x(8,12)-[RIMFYL]-x(2)-H-x(3,5)-H** (Fig. 2b). Only the fully conserved zinc-coordinating cysteine and histidine residues are common to both patterns**.** In the PROSITE pattern, the first histidine is required to follow exactly 13 positions after the second cysteine. In the ASTRAL40, the C2H2 protein 1rmd (shown in Fig. 2c in alignment with 1a1i) has a gap between these two residues of eleven positions and 2gli (Fig. 2d) has a gap of 15. In addition to the C2H2 residues of the PROSITE pattern, the zinc finger SSP output by the SSP algorithm (SSP91022) includes a partially aliphatic position two residues before the first cysteine and another partially aliphatic position three residues before the first histidine. The first of these locations is usually occupied by an aromatic residue (phenylalanine or tyrosine) 70% of the time. The exceptions of alanine, histidine, isoleucine, threonine, and valine appear to form a closed set as no other residues are observed at this position. The second of these locations is occupied by a long-chain aliphatic group consisting of [FILMRY]. Of these residues, leucine is found in 55% of the ASTRAL40 proteins at this position. Valine is the only aliphatic residue not seen here and arginine is the only non-aliphatic residue found to be tolerated at this position.

With the adjustment made for the gap length between the second cysteine and first histidine as described above, the SSP of **[AHITFVY]-x-C-x(2,5)-C-x(8,12)-[RIMFYL]-x(2)-H-x(3,5)-H** generated by the original method was found to match all 102 known C2H2 proteins in the ASTRAL100, SCOP family G.37.1.1, thus providing for 100% recall with respect to the ASTRAL100 database. However, due to this increased variable gap length, the precision of SSP91022 is 71.8% compared to the precision of 80.5% for PROSITE pattern PS00028. In order to increase the precision

carbon atoms of these residues to be structurally aligned. In Fig. 3a, the region between the arrows shows a backbone variation of the domain around the second cysteine residues. Although the sidechains of both cysteines are clearly aligned, the sequence alignment produced from this structure alignment does not have these two cysteines in the same position. In Fig. 3b, the arrows indicate two Phenylalanine residues that are separated by three positions in the sequence alignment, but nevertheless occupy the same position in the structure of the zinc finger domain. Recognition of these offset residues as legitimately conserved elements of the domain allows the creation of the new SSP (SSP95022): **[AIHTYFV]-x(1)-C-x(2,5)-C-x(1,6)-[IYF]-x(5,7)-[RFYMIL]-x(2)-H-x(3,5)-H**. This pattern has a precision of 96.2% and a recall of 98.0% with respect to the ASTRAL100, which represents a significant improvement over the 80.5% precision and 93.1% recall of the PROSITE zinc finger signature.

## 2.2 Cytochrome C Example

In the case of the cytochrome c family of proteins, the original SSP method identified a number of positions in the domain that showed a high degree of conservation in sequence and structure in addition to those in the short PROSITE pattern. However, the resulting SSP (SSP91008) did not match all known cytochrome c proteins in the ASTRAL100 (see Table 1) and, unlike in the zinc finger example above, the pattern region was bordered by a region of variable secondary structure. In particular, a long and variable length coil between the first and second alpha helices of the domain interposed itself between the pattern residues. In some cases this coil folded upon itself to form a beta strand. Since the backbone of this coil region did not display a high degree of conformation among the cytochrome c proteins in the multiple structural alignment, conserved residues in this region did not, in general, share the same positions in the corresponding multiple sequence alignment.

**Table 1.** Cytochrome c (SCOP Family A.3.1.1) sequence patterns. Pattern residue position **[ILPVY]** is added to SSP91008 to improve the pattern precision. Other, less conserved, positions are removed to improve recall to 100% with respect to the ASTRAL100.

| Accn. num. | Pattern | Precision (TP/TP+FP) | Recall (TP/TP+FN) |
|---|---|---|---|
| PS00190 | C-{CPWHF}-{CPWR}-C-H-{CFYW}. | 41.7% (196/470) | 100% (196/196) |
| SSP91001 | [PSGA]-x(2,3)-[FELKIV]-[MAYFV]-x(2,13)-C-x(2)-C-H-x(41,90)-[KRTEDNGA]-[ESFATLDV]-[MKVILA]-[EKTGNA]-[AFNHYW]-[MTIVL]. | 94.6% (175/185) | 89.3% (175/196) |
| SSP91008 | [FELKIV]-[MAYFV]-x(2,13)-C-x(2)-C-H-x(3,19)-**[ILPVY]**-x(31,77)-[EFTALV]-x(2)-[AKFHYW]-[MTYIVL]. | 65.3% (196/300) | 100% (196/196) |

However upon visual inspection of the van der Waals radii of the original SSP, a higher level of correspondence between sequence and structure became apparent. The sidechains of the SSP pattern residues formed a scaffold around the cytochrome C binding between the first and third alpha helices. For each cytochrome c structure, this scaffold formed a contiguous section of the domain tertiary structure with the exception of a "hole" adjoining the coil region (Fig. 4a-c). By identifying the residues whose sidechains filled this space, a new SSP (SSP91001) was created (Fig. 4d-f). Like the original PROSITE pattern, the recall is complete with respect to the ASTRAL100. And like the original SSP, the precision of SSP91001 was significantly improved over the shorter PROSITE pattern (see Table 1).

## 3 Discussion and Conclusion

As we described previously, the addition of information on conserved residue types gained from multiple structure alignment greatly improves the accuracy of multiple sequence alignment and that new information derived from the improved sequence alignment can likewise improve structure alignment such that both sequence and structure alignments can be improved simultaneously in an iterated manner. Although the precision and recall of sequence patterns derived from the SSP method are usually quite good for protein domains with a well-ordered secondary structure, pattern discovery is more challenging for proteins with variable loop regions where residue type may be well conserved, but nevertheless the "position" of the residue is not well defined. In some cases we have discovered ways to extract this "hidden regularity", as we describe in the examples above. The key concept is determining when a residue type within a structure is playing the same role in all members of the protein family. Typically the protein backbone will be offset among proteins in the structure alignment, however the residue sidechain at the offset position will be rotated such that all family members have a residue of that type in the same location in the structure. Occasionally a conserved residue type will exist at one of several nearby positions in the alignment where the sidechain will also demonstrate this form of "structural compensation". Using the methods we described above, the identification of these variable positions and incorporating the residue types in the protein family signatures can greatly increase their performance.

## Acknowledgement

This research was supported in part by NIH grant P01 DA15027-01.

## References

1. Falquet, L., M. Pagni, P. Bucher, N. Hulo, C.J.A. Sigrist, K. Hofmann, and A. Bairoch, *The PROSITE database, its status in 2002.* Nucl. Acids. Res., 2002. **30**(1): p. 235-238.
2. Sigrist, C.J., L. Cerutti, N. Hulo, A. Gattiker, L. Falquet, M. Pagni, A. Bairoch, and P. Bucher, *PROSITE: a documented database using patterns and profiles as motif descriptors.* Brief Bioinform, 2002. **3**(3): p. 265-74.

3. Nevill-Manning, C.G., T.D. Wu, and D.L. Brutlag, *Highly specific protein sequence motifs for genome analysis.* Proc Natl Acad Sci U S A, 1998. **95**(11): p. 5865-71.
4. Huang, J.Y. and D.L. Brutlag, *The EMOTIF database.* Nucl. Acids. Res., 2001. **29**(1): p. 202-204.
5. Attwood, T.K., *The PRINTS database: a resource for identification of protein families.* Brief Bioinform, 2002. **3**(3): p. 252-63.
6. Hart, R., A. Royyuru, G. Stolovitzky, and A. Califano, *Systematic and Fully Automatic Identification of Protein Sequence Patterns.* J. Comput. Biol., 2000. **7**((3/4)): p. 585-600.
7. Kasuya, A. and J.M. Thornton, *Three-dimensional structure analysis of PROSITE patterns1.* Journal of Molecular Biology, 1999. **286**(5): p. 1673-1691.
8. Milledge, T., S. Khuri, X. Wei, C. Yang, G. Zheng, and G. Narasimhan, *Sequence-Structure Patterns: Discovery and Applications.* 6th Atlantic Symposium on Computational Biology and Genome Informatics (CBG), 2005: p. 1282-1285.
9. Wu, T.D. and D.L. Brutlag. *Discovering Empirically Conserved Amino Acid Substitution Groups in Databases of Protein Families.* in *ISMB-96.* 1996.
10. Berman, H.M., J. Westbrook, Z. Feng, G. Gilliland, T.N. Bhat, H. Weissig, I.N. Shindyalov, and P.E. Bourne, *The Protein Data Bank.* Nucleic Acids Res, 2000. **28**(1): p. 235-42.
11. Brenner, S.E., C. Chothia, T.J.P. Hubbard, and A.G. Murzin, *Understanding protein structure: Using SCOP for fold interpretation.* Methods in Enzymology, 1996: p. 635-643.

# Clustering Support Vector Machines and Its Application to Local Protein Tertiary Structure Prediction

Jieyue He[1], Wei Zhong[2], Robert Harrison[2,3,4] Phang C. Tai[3], and Yi Pan[2,*]

[1] Department of Computer Science
Southeast University, Nanjing 210096, China
jieyuehe@seu.edu.cn
[2] Department of Computer Science,
[3] Department of Biology
Georgia State University, Atlanta, GA 30303-4110, USA
[4] GCC Distinguished Cancer Scholar
pan@cs.gsu.edu

**Abstract.** Support Vector Machines (SVMs) are new generation of machine learning techniques and have shown strong generalization capability for many data mining tasks. SVMs can handle nonlinear classification by implicitly mapping input samples from the input feature space into another high dimensional feature space with a nonlinear kernel function. However, SVMs are not favorable for huge datasets with over millions of samples. Granular computing decomposes information in the form of some aggregates and solves the targeted problems in each granule. Therefore, we propose a novel computational model called Clustering Support Vector Machines (CSVMs) to deal with the complex classification problems for huge datasets. Taking advantage of both theory of granular computing and advanced statistical learning methodology, CSVMs are built specifically for each information granule partitioned intelligently by the clustering algorithm. This feature makes learning tasks for each CSVMs more specific and simpler. Moreover, CSVMs built particularly for each granule can be easily parallelized so that CSVMs can be used to handle huge datasets efficiently. The CSVMs model is used for predicting local protein tertiary structure. Compared with the conventional clustering method, the prediction accuracy for local protein tertiary structure has been improved noticeably when the new CSVM model is used. The encouraging experimental results indicate that our new computational model opens a new way to solve the complex classification for huge datasets.

## 1 Introduction

Support Vector Machines (SVMs) [18] are new generation of machine learning techniques and have been successfully applied to a wide variety of application domains [8] including bioinformatics [15]. SVMs are searching the optimal separating hyperplane by solving a convex quadratic programming (QP). The typical running time for

* Corresponding author.

V.N. Alexandrov et al. (Eds.): ICCS 2006, Part II, LNCS 3992, pp. 710 – 717, 2006.

the convex quadratic programming is $\Omega$ ($m^2$) for the training set with m samples. The convex quadratic programming is NP-complete in the worst case [19]. Therefore, SVMs are not favorable for a large dataset [7]. In the local protein tertiary structure prediction, our dataset contains half millions samples. According to [23], it would take years to train SVMs on a dataset containing one million records.

Many algorithms and implementation techniques have been developed to enhance SVMs in order to increase their training performance with large data sets. The most well-known techniques include chunking [16], Osuna's decomposition method [13], and Sequential Minimal Optimization (SMO)[14]. The success of these methods depends on dividing the original Quadratic Programming (QP) problem into a series of smaller computational problems in order to reduce the size of each QP problem. Although these algorithms accelerate the training process, these algorithms do not scale well with the size of the training data.

The second class of algorithms tries to speed up the training process by reducing the number of training data. Since some data points such as the support vectors are more important to determine the optimal solution, these algorithms provide SVMs with high quality data points during the training process. Random Selection [1,3], Bagging [17], and clustering analysis [2, 9, 23] are representatives of these algorithms. These algorithms are highly scalable for the large dataset while the performance of training depends greatly on the selection of training samples.

Unlike those two classes of methods to deal with huge dataset training, we propose a new computational model called Clustering Support Vector Machines Model (CSVMs Model) by combining the theory of granular computing and principles of the statistical learning algorithms. The difference between the CSVMs model and clustering analysis [2, 9, 23] is that all the training dataset is kept during the training process. Using the principle of granular computing, the CSVMs model is able to divide a complex data-mining problem into a series of smaller and computationally simpler problems [22].

Han and Baker have used the conventional clustering algorithm to predict local protein structure [5, 6]. In their work, the K-means clustering algorithm is essential to understand how protein sequences correspond to local protein tertiary structures. Since K-means clustering algorithm may not reveal the nonlinear sequence-to-structure relationship effectively, many sequence segments in one cluster are weakly mapped to the representative structure of their assigned clusters. Consequently, the structure prediction quality for these clusters is not satisfactory.

In order to overcome the problems of the conventional clustering algorithm, the CSVMs model is used to predict local protein structure in this study. In this new computational model, one SVM is built particularly for each information granule defined by sequence clusters created by the clustering algorithm. CSVMs are modeled to learn the nonlinear relationship between protein sequences and their structures in each cluster. SVM is not favorable for large amount of training data samples. However, CSVMs can be easily parallelized to speed up the modeling process. After gaining the knowledge about the sequence to structure relationship, CSVMs are used to predict the local tertiary structure for protein sequence segments. Local tertiary structure is represented by distance matrices, torsion angles and secondary structures for backbone α-carbon atoms of protein sequence segments. In this study, the performance of the CSVM model and the conventional clustering algorithm are compared based on accuracy for local protein structure prediction.

The paper is organized as follows. Section 2 describes a new computational model called Clustering Support Vector Machines Model. Section 3 explains the experimental setup and result analysis. Finally, the conclusion and the future work are presented in section 4.

## 2 Clustering Support Vector Machines Model (CSVM Model)

Model construction and performance evaluation are two major tasks for Clustering Support Vector Machines Model (CSVMs Model). In the model construction phase, the whole sequence space is granulated into a series of information granule and each CSVM is built specifically for each information granule. In the performance evaluation phase, the established CSVM model is used to predict local tertiary structures for sequence segments.

### 2.1 Granulating the Whole Sequence Space into Clusters in the Model Construction Phase

Granular computing decomposes information in the form of some aggregates and solves the targeted problems in each granule [21]. Fuzzy sets, probabilistic sets, decision trees, clusters and association rules are some of granulation methods under the framework of granular computing [22]. Since K-means clustering is computationally efficient for large datasets [10], K-means clustering is chosen as the granulation method in our study. With the K-means clustering algorithm, data samples with similar characteristics can be grouped together. As a result, the whole sample space is partitioned into subspaces intelligently and the complex data mining work is mapped into a series of computationally tractable tasks. Different number of initial clusters were tried and based on these results, 800 clusters were chosen. 800 clusters are relatively suitable for the K-means clustering algorithm in our application.

### 2.2 Building Clustering Support Vector Machines (CSVMs) for Each Cluster in the Model Construction Phase

Because the distribution patterns for frequency profiles in clusters are diverse, the functionality of CSVMs is customized for each cluster belonging to different cluster groups. The definition of different cluster groups is introduced in the section explaining the experimental setup. The CSVMs for clusters belonging to the bad cluster group are designed to select potentially reliable prediction since a large percentage of prediction by the bad cluster group is not accurate. The CSVMs for clusters belonging to the good cluster group are designed to filter out potentially unreliable prediction since the large percentage of prediction by the good cluster group is highly reliable.

Since CSVM solves the binary classification problem, samples in a cluster must be labeled as positive or negative before the training process. In each cluster, positive samples are defined as those samples whose structure deviation from the representative structure of this cluster is within a given threshold. Negative samples are similarly defined. Positive samples have the potential to be closely mapped to the representative structure of the specified cluster and negative samples may not correspond to the representative structure of the specified cluster closely. Each CSVM is trained specifically for one cluster.

## 2.3 Local Protein Tertiary Structure Prediction by CSVMs Model

Local protein tertiary structure prediction by CSVMs is based on the prediction method from the clustering algorithm. At first, the sequence segments whose structures to be predicted are assigned to a specific cluster in the cluster group by the clustering algorithm. The clustering algorithm and cluster membership assignment can be found in [25, 26]. Then CSVM modeled for this specific cluster is used to identify how close this sequence segment is nonlinearly correlated to the representative structure of this cluster. If the sequence segment is predicted as the positive sample by CSVM, this segment has the potential to be closely mapped to the representative structure for this cluster. Consequently, the representative local tertiary structure of this cluster can be safely assigned to this sequence segment. The method to decide the representative local tertiary structure of each cluster can be found in [25, 26]. If the sequence segment is predicted as the negative sample by CSVMs, this segment does not closely correspond to the local tertiary structure for this cluster. The structure of this segment cannot be reliably predicted. The assigned cluster will not be considered in the next iteration. In the next iteration, the cluster membership function is used to select the next cluster from the remaining clusters of the cluster group. The previous procedure will be repeated until one SVM modeled for the selected cluster predict the given sequence segment as positive and structure prediction is performed. The complete prediction process is shown in figure 1. CSVMs are used to reclassify sequence segments, which are misclassified by the conventional clustering algorithms.

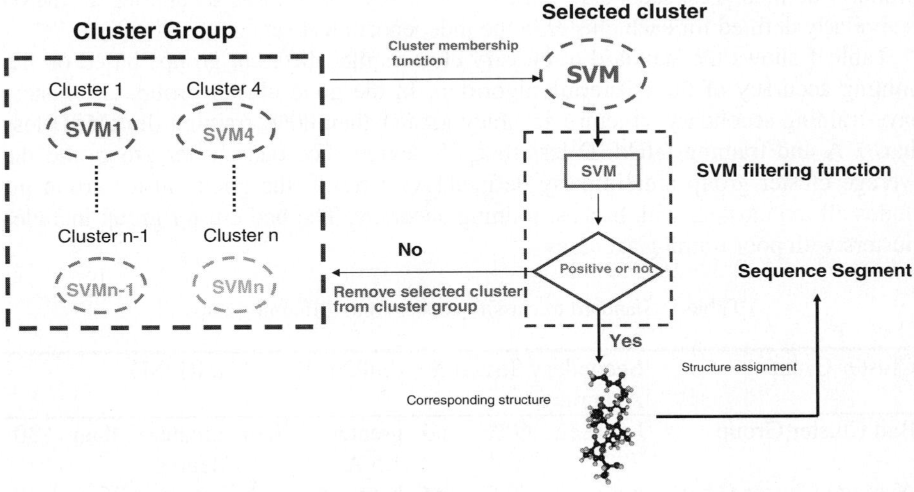


**Fig. 1.** Local Protein Structure Prediction by CSVMs Algorithm

# 3 Experimental Evaluation

## 3.1 Training Set and Independent Test Set

The training dataset used in our work includes 2090 protein sequences obtained from the Protein Sequence-Culling Server (PISCES)[20]. The training set is utilized to construct the CSVM model. 200 protein sequences from the recent release of PISCES are included into the independent test set. The test set is used in the performance evaluation phase. The structures of protein sequences in the training set and the testing set are available from Protein Data Bank (PDB) [4]. Any two sequences in the training set and the test set share less than 25% similarity.

## 3.2 Experimental Setup

Structure prediction accuracy for sequence segments in terms of secondary structure accuracy, Distance Matrix Root Mean Square Deviation (dmRMSD) and Torsion angle RMSD (taRMSD) is calculated to evaluate the performance of the conventional clustering algorithm and our new computational model. The definition for representative structures of clusters was introduced in [25, 26]. We use the formula in [11, 12, 24] to calculate secondary structure accuracy, dmRMSD, and taRMSD.

During the prediction process, structures of sequence segments are first predicted by clusters with the high training accuracy. If the structures of sequence segments cannot be predicted by clusters with high training accuracy, clusters with the lower training accuracy will be used for structure prediction.

Training secondary structure accuracy for a given cluster is the average training accuracy of sequence segments in the training set predicated by this cluster. Training dmRMSD of a given cluster is the average training dmRMSD of sequence segments in the training set predicated by this cluster. Training taRMSD of a given cluster is similarly defined. Test secondary structure accuracy, test dmRMSD and test taRMSD is similarly defined for each cluster in the independent test set.

Table 1 shows the standard to classify clusters into different groups based on the training accuracy of the clustering algorithm. In the good cluster group, all clusters have training secondary structure accuracy greater than 80%, training dmRMSD less than 1 Å and training taRMSD less than 25 degree. The bad cluster group and the average cluster group are similarly defined. As a result, the good cluster group includes all the clusters with highest training accuracy. The bad cluster group includes clusters with poor training accuracy.

**Table 1.** Standard to classify clusters into different groups

| Cluster Group | Secondary Structure Accuracy | dmRMSD | taRMSD |
|---|---|---|---|
| Bad Cluster Group | between 60% and 70% | greater than 1.5 Å | greater than 30 degree |
| Average Cluster Group | between 70% and 80% | between 1 Å and 1.5 Å | between 25 and 30 degree |
| Good Cluster Group | greater than 80% | Less than 1 Å | less than 25 degree |

As described in [25, 26], only combined information of secondary structure, torsion angle and distance matrix can represent protein structure precisely. In order to rigorously evaluate the prediction quality for these algorithms, we designed accuracy criterion for each cluster. Accuracy criteria for one cluster is the percentage of sequence segments with secondary structure accuracy greater than 70%, dmRMSD less than 1.5 Å and taRMSD less than 30 degree in the test set for this cluster. This accuracy criterion for one cluster reflects the percentage of sequence segments with the acceptable level of structure prediction accuracy.

### 3.3 Result and Analysis

By the experiment, we get the average accuracy, prediction and recall of SVMs for different cluster groups in Table 2. Average SVM accuracy for different cluster groups is over 80% in Table 2. This indicates that the generalization power for CSVMs is strong enough to recognize the complicated pattern of the sequence-to-structure relationship for each cluster. Table 3 compares the prediction accuracy between the conventional clustering algorithm and the CSVMs model using the definition for the accuracy criterion defined in the section 3.2.

**Table 2.** Comparison of the average accuracy, precision and recall of SVMs for different cluster groups

| Cluster Group | Average Accuracy | Average Precision | Average Recall |
|---|---|---|---|
| Bad Cluster Group | 75% | 74% | 77% |
| Average Cluster Group | 82% | 84% | 95% |
| Good Cluster Group | 85% | 86% | 96% |

**Table 3.** Comparison of the accuracy between the conventional clustering algorithm and the CSVMs model for different cluster groups

| Cluster Group | Conversional Clustering Algorithm | CSVMs Model |
|---|---|---|
| Bad Cluster Group | 57.13% | 62.34% |
| Average Cluster Group | 68.76% | 70.43% |
| Good Cluster Group | 78.95% | 79.34% |

Table 3 shows that the prediction accuracy for local protein tertiary structure has improved by average 2% when the CSVMs model is applied. Since K-means clustering may introduce noisy and irreverent information into each cluster, the machine learning techniques are required to identify the strength of the sequence-to-structure relationship for each sequence segment. After learning the relationship between sequence distribution patterns and representative structures of each cluster, CSVMs can filter out potentially unreliable prediction and selects potentially reliable prediction for each cluster. Our experimental results indicate that building CSVMs for each granule respectively can increase effectiveness and efficiency of data mining algorithms.

## 4 Conclusion

SVM is not efficient for very large datasets due to the high training time complexity. To solve this problem, a new model called Clustering Support Vector Machines (CSVMs) is proposed. The special characteristics of CSVMs convert a complex classification problem into multiple smaller computational problems so that learning tasks for each CSVM are more specific and efficient. Each CSVM can concentrate on highly related samples in each cluster without being distracted by noisy data from other clusters. Because of data partitioning, the training tasks for each CSVM are parallelized. The parallel training process makes the data-mining task for very large datasets possible. The satisfactory experimental results for local protein structure prediction show that our computational model opens a new approach for solving the complex classification problem for huge datasets.

Further improvement for the CSVMs model will be made in the future work. Under the framework of granular computing, there are many granulation methods such as fuzzy sets, probabilistic sets, decision trees, clusters and association rules. In the future work, the more effective granulation method need be studied.

## References

1. Agarwal, D. K.: Shrinkage estimator generalizations of proximal support vector machines. in Proc.of the 8th ACM SIGKDD international conference of knowledge Discovery and data mining, Edmonton, Canada (2002)
2. Award, M., Khan, L. Bastani, F. and Yen, I.: An Effective Support Vector Machines(SVMs)Performance Using Hierarchical Clustering. in Proc. of the 16th IEEE International Conference on Tools with Artificial Intelligence(ICTAI 2004).
3. Balcazar, J. L., Dai, Y. and Watanabe, O.: Provably Fast Training Algorithms for Support Vector Machines. in Proc.of the 1stIEEE International Conference on Data mining, IEEE Computer Society (2001) 43-50
4. Berman, H. M., Westbrook, J. and Bourne, P. E. :The protein data bank. Nucleic Acids Research, Vol. 28, (2000) 235-24
5. Bystroff, C. and Baker,D.:Prediction of local structure in proteins using a library of sequence-structure motifs. J. Mol. Biol, vol. 281, (1998) 565-77
6. Bystroff, C., Thorsson, V. and Baker, D.: HMMSTR: A hidden markov model for local sequence-structure correlations in proteins. J. Mol. Biol, vol. 301, (2000) 173-90
7. Chang, C. C and Lin, C. J.: Training nu-support vector classifiers: Theory and algorithms. Neural Computations, vol. 13, (2001) 2119-2147
8. Cristianini, N., Shawe-Taylor,J.: An Introduction to Support Vector Machines and other Kernel-based Learning Methods. Cambridge University Press, Cambridge, UK (2000)
9. Daniael, B. and Cao, D.: Training Support Vector Machines Using Adaptive Clustering. in Proc. of SIAM International Conference on Data Mining 2004,Lake Buena Vista, FL, USA.
10. Gupta, S. K., Rao, K. S. and Bhatnagar, V.:K-means clustering algorithm for categorical attributes. Data Warehousing and Knowledge Discovery DaWaK-99, Florence, Italy (1999) 203-208

11. Hu, H., Pan, Y., Harrsion, R. and Tai, P. C.: Improved protein secondary structure prediction using support vector machine with a new encoding scheme and advanced tertiary classifier. IEEE Transactions on NanoBioscience, Vol. 2, (2004) 265-271
12. Kolodny, R. and Linial, N.:Approximate protein structural alignment in polynomial time. Proc Natl. Acad. Sci., Vol. 101, (2004)12201-12206
13. Osuna, E. Freund, R. and Girosi, F.: An improved training algorithm for support vector machines. In Proc. Of IEEE Workshop on Neural Networks for Signal Processing,Pages (1997)276-285
14. Platt, J.: Fast training of support vector machines using sequential minimal optimization. In advances in Kerenel Methods-Support Vector Learning, (1999) 185-208
15. Schoelkopf, B., Tsuda, K. and Vert, J. P.: Kernel Methods in Computational Biology. MIT Press, (2004)71-92
16. Scholkopf, B., Burges, C. and Smola, A. (eds); Advances in Kernel Methods-Support Vector Learning. MIT Press,Cambridge,MA(1999)
17. Valentini, G. and Dietterich,T. G.: Low Bias Bagged Support vector Machines. in Proc. of the 20th International Conference on Machine Learning ICML 2003,Washington D.C.USA,(2003)752-759
18. Vapnik, V. : Statistical Learning Theory. John Wiley&Sons, Inc., New York (1998)
19. Vavasis,S.A.: Nonlinear Optimization: Complexity Issues. New York: Oxford Science.(1991).
20. Wang, G. and Dunbrack, RL. Jr.:PISCES: a protein sequence-culling server. Bioinformatics, vol. 19, no. 12, (2003) 1589-1591
21. Yao,Y. Y: Granular Computing. Computer Science (Ji Suan Ji Ke Xue), Proceedings of The 4th Chinese National Conference on Rough Sets and Soft Computing, Vol. 31 (2004) 1-5
22. Yao,Y. Y.: Perspectives of Granular Computing. to appear in 2005 IEEE Conference on Granular Computing.
23. Yu, H., Yang, J. and Han, J.: Classifying Large Data sets Using SVMs with Hierarchical Clusters. in Proc. Of the 9th ACM SIGKDD 2003, Washington DC, USA (2003)
24. Zagrovic B. and Pande, V. S.:How does averaging affect protein structure comparison on the ensemble level? Biophysical Journal, Vol. 87, (2004) 2240-2246
25. Zhong, W., Altun, G., Harrison, R., Tai, P. C. and Pan, Y.: Mining Protein Sequence Motifs Representing Common 3D Structures. Poster Paper of IEEE Computational Systems Bioinformatics (CSB2005), Stanford University (2005)
26. Zhong, W., Altun, G., Harrison, R., Tai, P. C., and Pan, Y.: Improved K-means Clustering Algorithm for Exploring Local Protein Sequence Motifs Representing Common Structural Property. IEEE Transactions on NanoBioscience, Vol. 4, (2005) 255-65

# Extracting Protein-Protein Interactions from the Literature Using the Hidden Vector State Model

Deyu Zhou, Yulan He, and Chee Keong Kwoh

School of Computer Engineering, Nanyang Technological University,
Nanyang Avenue, 639798, Singapore
{zhou0063, asylhe, asckkwoh}@ntu.edu.sg

**Abstract.** In the field of bioinformatics in solving biological problems, the huge amount of knowledge is often locked in textual documents such as scientific publications. Hence there is an increasing focus on extracting information from this vast amount of scientific literature. In this paper, we present an information extraction system which employs a semantic parser using the Hidden Vector State (HVS) model for protein-protein interactions. Unlike other hierarchical parsing models which require fully annotated treebank data for training, the HVS model can be trained using only lightly annotated data whilst simultaneously retaining sufficient ability to capture the hierarchical structure needed to robustly extract task domain semantics. When applied in extracting protein-protein interactions information from medical literature, we found that it performed better than other established statistical methods and achieved 47.9% and 72.8% in recall and precision respectively.

## 1 Introduction

Understanding protein functions and how they interact gives researchers a deeper insight into understanding of living cell as a complex machine, disease process and provides target for effective drug designs. To date, many databases, such as PDB [1], Swiss-Prot [2] and BIND [3], have been built to store various types of information for protein. However, data in these databases were mainly hand-curated to ensure their correctness and thus limited speed in transferring textual information into searchable structure data. As of to date, vast knowledge of protein-protein interactions are still locked in the full-text journals. As a result, automatically extracting information about protein-protein interactions is crucial to meet the demand of the researchers.

Existing approaches can be broadly categorized into two types, based on simple pattern matching, or employing parsing methods. Approaches using pattern matching [4, 5, 6] rely on a set of predefined patterns or rules to extract protein-protein interactions. For example, Ono's method [5] manually defines some rules and patterns which are augmented with additional restrictions based on syntactic categories and word forms to give better matching precision. It achieves high performance with a recall rate of 85% and precision rate of 84% for Saccharomyces cerevisiae (yeast) and Escherichia coli. Another method [6] tries to

V.N. Alexandrov et al. (Eds.): ICCS 2006, Part II, LNCS 3992, pp. 718–725, 2006.

use dynamic programming to automatically discover patterns which describe protein-protein interactions. Their results give precision of 80.5% and recall of 80.0%. It is however not feasible in practical applications as it requires heavy manual processing to define patterns when shifting to another domain.

Parsing based methods employ either full or shallow parsing. Unlike word-based pattern matchers, shallow parsers [7, 8] break sentences into none overlapping phases. They extract local dependencies among phases without reconstructing the structure of an entire sentence. The precision and recall rates reported for shallow parsing approaches are estimated at 50-80% and 30-70%, respectively.

Systems based on full-sentence parsing [9, 10, 11] deal with the structure of an entire sentence and therefore are potentially more accurate. Yakushiji [9] defines grammars for biomedical domain and uses a general full parser to extract interaction events. However, no recall or precision value using this approach was reported. Another full parser-based approach uses the context-free grammar to extract protein interaction information with a recall rate of 63.9% and a precision rate of 70.2% [10]. The major drawback of the aforementioned methods is that they may require complete redesign of the grammar in order to be tuned to different domains.

In this paper, we describe a statistical method using the hidden vector state model (HVS) to automatically extract protein-protein interactions from biomedical literature. The HVS model has been successfully used to discover semantic information in spoken utterances [12]. However, it is not straightforward to extend the usage of the HVS model to the biomedical literature domain. One major reason is that spoken utterances are normally simple and short. Thus, unlike written documents, there are normally no complex syntactic structures in spoken utterances. It therefore poses a challenge on how to effectively and efficiently extract semantic information from much more complicated written documents.

The rest of the paper is organized as follows. Section 2 briefly describes the HVS model and how it can be used to extract protein-protein interactions from the biomedical literature. Section 3 presents the overall structure of the extraction system. Experimental results are discussed in section 4. Finally, section 5 concludes the paper.

## 2 The Hidden Vector State Model

The Hidden Vector State (HVS) model [12] is a discrete Hidden Markov Model (HMM) in which each HMM state represents the state of a push-down automaton with a finite stack size. This is illustrated in Fig. 1 which shows the sequence of HVS stack states corresponding to the given parse tree. State transitions are factored into separate stack pop and push operations constrained to give a tractable search space. The result is a model which is complex enough to capture hierarchical structure but which can be trained automatically from only lightly annotated data.

The HVS model computes a hierarchical parse tree for each word string $W$, and then extracts semantic concepts $C$ from this tree. Each semantic concept

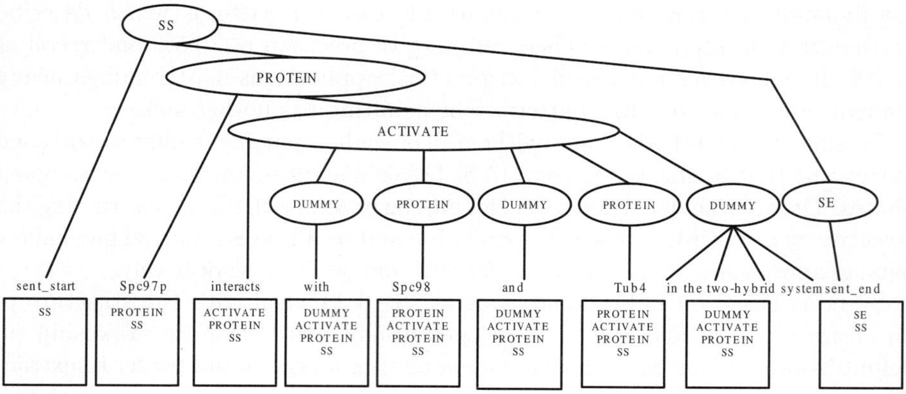


**Fig. 1.** Example of a parse tree and its vector state equivalent

consists of a name-value pair where the name is a dotted list of primitive semantic concept labels. For example, the top part of Fig. 1 shows a typical semantic parse tree and the semantic concepts extracted from this parse would be in equation 1

$$\begin{array}{l}\texttt{PROTEIN=Spc97}\\\texttt{PROTEIN.ACTIVATE=interacts}\\\texttt{PROTEIN.ACTIVATE.PROTEIN=Spc98}\\\texttt{PROTEIN.ACTIVATE.PROTEIN=Tub4}\end{array} \tag{1}$$

In the HVS-based semantic parser, conventional grammar rules are replaced by three probability tables. Let each state at time $t$ be denoted by a vector of $D_t$ semantic concept labels (tags) $c_t = [c_t[1], c_t[2], ..c_t[D_t]]$ where $c_t[1]$ is the preterminal concept label and $c_t[D_t]$ is the root concept label (SS in Fig. 1). Given a word sequence $W$, concept vector sequence $\mathbf{C}$ and a sequence of stack pop operations $N$, the joint probability of $P(W, \mathbf{C}, N)$ can be decomposed as

$$P(W, \mathbf{C}, N) = \prod_{t=1}^{T} P(n_t|c_{t-1})P(c_t[1]|c_t[2 \cdots D_t])P(w_t|c_t) \tag{2}$$

where $n_t$ is the vector stack shift operation and takes values in the range $0, \cdots, D_{t-1}$, and $c_t[1] = c_{w_t}$ is the new pre-terminal semantic label assigned to word $w_t$ at word position $t$.

Thus, the HVS model consists of three types of probabilistic move, each move being determined by a discrete probability table:

1. popping semantic labels off the stack - $P(n|c)$;
2. pushing a pre-terminal semantic label onto the stack - $P(c[1]|c[2 \cdots D])$;
3. generating the next word - $P(w|c)$.

Each of these tables are estimated in training using an EM algorithm and then used to compute parse trees at run-time using Viterbi decoding. In training, each word string $W$ is marked with the set of semantic concepts $C$ that it contains. For example, if the sentence shown in Fig. 1 was in the training set, then it would be marked with the four semantic concepts given in equation 1. For each word $w_k$ of each training utterance $W$, EM training uses the forward-backward algorithm to compute the probability of the model being in stack state $c$ when $w_k$ is processed. Without any constraints, the set of possible stack states would be intractably large. However, in the HVS model this problem can be avoided by pruning out all states which are inconsistent with the semantic concepts associated with $W$. The details of how this is done are given in [12].

## 3 System Overview

The extraction system works as follows. At the beginning, abstracts are retrieved from MEDLINE and split into sentences. Protein names and other biological terms are then identified based on a pre-constructed biological term dictionary. After that, each sentence is parsed by the HVS semantic parser. Finally, information about protein-protein interactions is extracted from the tagged sentences using a set of manually-defined simple rules. An example of the procedure is illustrated in Fig. 2.

The details of each step are described below.

1. *Identification of protein names, other biological terms and interaction keywords.*
   To extract information on protein-protein interactions from literature, protein names need to be first identified, which still remains as a challenging problem. In our system, protein names are identified based on a dictionary of manually constructed biological term. In addition, a category/keyword dictionary for identifying terms describing interactions has also been built

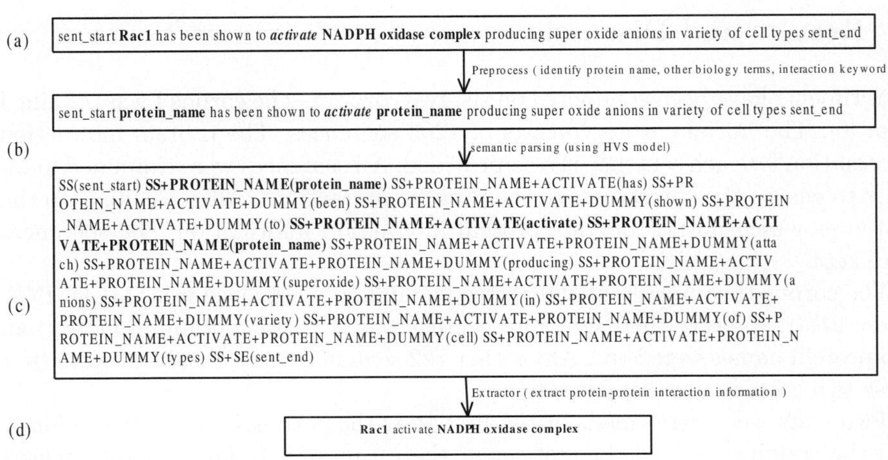


**Fig. 2.** An example of a procedure for information extraction using the HVS model

based on [10]. All identified biological terms and interaction keywords are then replaced with their respective category labels as can be seen in Fig. 2(b). By doing so, the vocabulary size of the training corpus can be reduced and the data sparseness problem would be alleviated.

2. *Parsing sentences using the HVS model.*
A sentence which contains at least two proteins identified by Step 1 is then parsed with the HVS model. Before doing so, the HVS model needs to be trained using a lightly annotated training corpus. An annotation example is shown below.

| | |
|---|---|
| Sentence: | CUL-1 was found to interact with SKR-1, SKR-2, SKR-3, SKR-7, SKR-8 and SKR-10 in yeast two-hybrid system |
| Annotation: | PROTEIN_NAME ( ACTIVATE ( PROTEIN_NAME ) ) |

It can be seen that unlike fully-annotated treebank data, no explicit semantic tag/word pairs are given. Only the abstract annotations are provided to guide the EM training of the HVS model [12].

3. *Extraction of protein-protein interactions.*
Given the HVS parsing result as shown in Fig. 2(c), the protein-protein interactions can be easily extracted follows the rules below:
   - ignore the semantic tag if its preterminal tag is DUMMY;
   - if the semantic tag is of the form SS+PROTEIN_NAME+REL+PROTEIN _NAME, SS+REL+PROTEIN_NAME+PROTEIN_NAME, and so on, REL can be any of the category names describing the interactions such as "activate", "inhibit" etc, extract the corresponding protein name, then search backwards or forward for the interaction keyword and the other protein name.

Based on the rules described above, the protein-protein interactions can be easily extracted as shown in Fig. 2(d).

## 4 Results and Discussion

Experiments have been conducted on the two corpora. The corpus I was obtained from [6]. The initial corpus consists of 1203 sentences. The protein interaction information for each sentence is also provided. All sentences were examined manually to ensure the correctness of the protein interactions. After cleaning up the sentences which do not provide protein interaction information, 800 sentences were kept.

The corpus II comprises of 300 abstracts randomly retrieved from MEDLINE. These abstracts were then split into sentences and those containing more than two protein names were kept. Altogether 722 sentences were obtained. Note that these two corpora are disjoint sets.

Two tests were performed. In the first test, the corpus I was split randomly into the training set and the test set at the ration of 9:1. The test set consists of 80 sentences and the remaining 720 sentences were used as the training set. The experiments were conducted three times with different training and test

**Table 1.** Results evaluated in sentence level on Corpus I

| Experiment | TP | TP+TN | Recall(%) |
|---|---|---|---|
| 1 | 59 | 80 | 73.8 |
| 2 | 65 | 80 | 81.3 |
| 3 | 63 | 80 | 78.8 |

**Table 2.** Results evaluated in interaction level on corpus I

| Experiment | TP | TP+TN | FP | Recall(%) | Precision(%) | F-Score(%) |
|---|---|---|---|---|---|---|
| 1 | 55 | 138 | 18 | 39.9 | 75.3 | 52.1 |
| 2 | 64 | 130 | 29 | 49.2 | 68.8 | 57.4 |
| 3 | 67 | 140 | 25 | 47.9 | 72.8 | 57.8 |

**Table 3.** Comparsion of Results on corpus I and corpus II

| Experiment | F-Score(%) |
|---|---|
| corpus I | 56.0 |
| corpus II | 50.4 |

data each round. The average processing speed on Itanium-1 model Linux server equipped with 733Mhz processor and 4 GB RAM was 0.23s per sentence.

Table 1 shows the recall values evaluated at the sentence level. True Positive (TP) is the number of sentences which contain at least one correctly extracted protein interaction. (TP+TN) is the total number of sentences which contain protein-protein interactions. As corpus I does not have negative examples, this value is always 80. It can be seen from Table 1 that the best possible recall value that can be achieved is 81.3%.

Table 2 shows the evaluation results measured in the interaction level. TP is the number of correctly extracted interactions. (TP+TN) is the number of all interactions in the test set and (TP+NP) is the number of all extracted interactions. F-score is computed using the formula below:

$$\text{F-score} = \frac{2 \cdot \text{recall} \cdot \text{precision}}{\text{recall} + \text{precision}} \tag{3}$$

In the second test, the HVS model is trained on corpus I and tested on corpus II. Table 3 gives comparison of the results between on corpus I and on corpus II. The value on corpus I is average of results based on Table 2. It was observed that a F-score of 50.4% was achieved when tested on a general corpus randomly extracted from MEDLINE, which is impossible to those systems based on predefined semantic grammar rules. For example, MedScan [13] can only successfully parse and generate semantic structures for about 34% sentences randomly picked from MEDLINE. The recall rate of MedScan was found to be 21% [13]. This demonstrated the robustness of the HVS model.

Generally, it is difficult to compare our method with other existing systems fairly, because there is neither an accurate task definition on processing the

MEDLINE abstracts nor a standard benchmark dataset. Since the corpus I data used in our experiments came from [6], it would be interested to see how our system performed compared to the method based on pattern matching proposed in [6]. If simply comparing the respective recall and precision rates, our method is less efficient. However, by examining the experimental results more carefully, we have the following findings:

1. The method proposed in [6] employed a part-of-speech (POS) tagger to preprocess the data. Some tags such as adjective, determiner and so on were removed. Since some interactions can be defined by adjectives, it therefore inevitably affected the system performance.
   For example, The sentence "The class II proteins are expressed constitutively on B-cells and EBV-transformed B-cells, and are inducible by IFN-gamma on a wide variety of cell types." provides an protein-protein interaction as shown by the underlying text. However, in [6], adjectives such as "inducible" were excluded and the system thus failed to extract the above interaction. On the contrary, our system was able to give the correct result.
2. Our system is able to generate reasonable results on a general domain as illustrated in the experiments on the corpus II, whilst the method proposed in [6] did not provide any results in this aspect.

## 5 Conclusions

In this paper, we have presented a system using the HVS model to automatically extract information on protein-protein interactions from text sources. The system is able to give reasonable performance measured in recall and precision. We have also shown the robustness of the system as it can be used in any general biomedical domain. Our results may provide a useful supplement to manually created resources in established public databases.

In future work we will work on the enhancement of the HVS model in order to improve the extraction accuracy. We will also study the adaptation issue of the HVS model and see how the model could give better performance by providing a small amount of adaptation data when the HVS model trained on one particular protein domain is used in another protein domain.

## Acknowledgements

The authors would like to thank Minlie Huang of Tsinghua University, China, for providing the corpus I data.

## References

1. H.M. Berman, J. Westbrook, Z. Feng, G. Gilliland, T.N. Bhat, H. Weissig, I.N. Shindyalov, and P.E. Bourne. The protein data bank. *Nucleic Acids Research*, pages 235–242, 2000.

2. B. Boeckmann, A. Bairoch, R. Apweiler, M. Blatter, A. Estreicher, E. Gasteiger, M.J. Martin, K. Michoud, C. ODonovan, and I. Phan. The swiss-prot protein knowledgebase and its supplement trembl in 2003. *Nucleic Acids Research*, pages 365–370, 2003.
3. GD. Bader, D. Betel, and CW. Hogue. Bind: the biomolecular interaction network database. *Nucleic Acids Research*, 31(1):248–250, 2003.
4. J. Thomas, D. Milward, C. Ouzounis, and S. Pulman. Automatic extraction of protein interactions from scientific abstracts. In *Proceedings of the Pacific Symposium on Biocomputing*, pages 541–552, Hawaii, U.S.A, 2000.
5. Toshihide Ono, Haretsugu Hishigaki, Akira Tanigam, and Toshihisa Takagi. Automated extraction of information on protein-protein interactions from the biological literature. *Bioinformatics*, 17(2):155–161, 2001.
6. Minlie Huang, Xiaoyan Zhu, and Yu Hao. Discovering patterns to extract protein-protein interactions from full text. *Bioinformatics*, 20(18):3604–3612, 2004.
7. Mark Craven and Johan Kumlien. Constructing biological knowledge bases by extracting information from text sources. In *Proceedings of the Seventh International Conference on Intelligent Systems for Molecular Biology*, pages 77–86, Heidelberg, Germany, 1999. AAAI Press.
8. J. Pustejovsky, J. Castano, J. Zhang, M. Kotecki, and B. Cochran. Robust relational parsing over biomedical literature: Extracting inhibit relations. In *Proceedings of the Pacific Symposium on Biocomputing.*, pages 362–373, Hawaii, U.S.A, 2002.
9. A. Yakushiji, Y. Tateisi, Y. Miyao, and J. Tsujii. Event extraction from biomedical papers using a full parser. In *Proceedings of the Pacific Symposium on Biocomputing*, volume 6, pages 408–419, 2001.
10. Joshua M. Temkin and Mark R. Gilder. Extraction of protein interaction information from unstructured text using a context-free grammar. *Bioinformatics*, 19(16):2046–2053, 2003.
11. Shengyang Tang and Chee Keong Kwoh. Cytokine information system and pathway visualization. *International Joint Conference of InCoB, AASBi and KSBI (BIOINFO2005).*, 2005.
12. Yulan He and Steve Young. Semantic processing using the hidden vector state model. *Computer Speech and Language*, 19(1):85–106, 2005.
13. S. Novichkova, S. Egorov, and N. Daraselia. Medscan, a natural language processing engine for medline abstracts. *Bioinformatics*, 19(13):1699–1706, 2003.

# A Multilevel Approach to Identify Functional Modules in a Yeast Protein-Protein Interaction Network

S. Oliveira and S. C. Seok

Department of Computer Science, 14 MLH,
University of Iowa, Iowa City IA 52242, USA
{oliveira, sseok}@cs.uiowa.edu

**Abstract.** Identifying functional modules is believed to reveal most cellular processes. There have been many computational approaches to investigate the underlying biological structures [9, 4, 10, 6]. A spectral clustering method plays a critical role identifying functional modules in a yeast protein-protein network in [6, 4]. We present an unweighted-graph version of a multilevel spectral algorithm which more accurately identifies protein complexes with less computational time.

## 1 Introduction

Most cellular processes are carried out by groups of proteins. Identifying functional modules in protein-protein networks is considered as one of the most important and challenging research topic in computational systems biology. There has been a lot of recent computational approaches to disclose the underlying biological structures [9, 4, 10, 6].

Pothen et al. [6] propose a two-level architecture for a yeast proteomic network. They construct a smaller network from a protein-protein interaction network by removing proteins which interact with too many or too few proteins. A clustering algorithm is applied to this residual network. Validation of clusters is performed by comparing the clustering result with a protein complex database, called MIPS. A spectral clustering method plays a critical role for identifying functional modules in the protein-protein network in their research.

We successfully applied a multilevel spectral algorithm to cluster a group of documents in [16] using similarity matrices which are mostly dense with entries between 0 and 1. Like large-scale networks, the vertex connectivities of proteomic networks follow a scale-free power-law distribution. That is, the proteomic network consists of a small number of high degree nodes and a majority of low degree nodes. However, there are no edge weights. In this paper, we present an unweighted-graph version of a multilevel spectral algorithm in [16] which identifies more protein complexes with less computation time.

Multilevel algorithms have a long history, mostly for PDE in numerical analysis but also for graph partitioning such as in METIS [12]. Recently multilevel schemes have been applied to graph clustering [16, 11]. Multilevel algorithms

V.N. Alexandrov et al. (Eds.): ICCS 2006, Part II, LNCS 3992, pp. 726–733, 2006.

conventionally consists of three main steps: coarsening, partitioning and decoarsening. Multilevel clustering algorithms, like multilevel partitioning algorithms, mostly try to improve existing clustering algorithms using good coarsening or matching algorithms. Multilevel algorithms not only improve qualities of the clustering but also significantly reduce computational time.

The most well-known algorithms include *random-edge matching* (REM) and *heavy-edge* matching (HEM) [13]. And there are more recent matching algorithms like LAM [1]. These algorithms are mainly designed for weighted graphs. We also compared two different coarsening algorithms for weighted graphs in [16]. We have shown that when nodes are merged in order from the heaviest edge weight we can expect better results. We call this Sorted Matching (SM), because each edge weight represents how close two nodes are. The unweighted or uniform weight graph cases can not use any of these edge-oriented methods directly because all edge weights are initially all 1. But after one level of coarsening some edges may represent more than one edge (up to 4). That is, we have groups of different edge weights after coarsening when we define the weight of an edge as the sum of edge weights combined into it. So after one level of coarsening we are able to use Sorted Matching (SM) for unweighted graph coarsening: merge nodes with highest edge weight. But there are many nodes with the same edge weight. Thus we give the higher priority to the edge with fewer combining node weights, which is defined as the number of all nodes included in the supernode. The maximum edge with $1/w(n_i) + 1/w(n_j)$ is taken as a tie-break rule, where $w(n_i)$ and $w(n_j)$ are the node weights (the number of nodes) of supernodes $n_i$ and $n_j$. Let us call this algorithm Heavy-Edge-Small-Node (HESN). We show that HESN algorithm outperforms a randomly matching algorithm and a matching algorithm which focuses on maximizing the number of nodes collapsed.

We also introduce self-edge weightings for unweighted graphs. The spectral clustering algorithm in this research uses similarities between vertices. Note that in unweighted graphs all edges have the same edge weights, that is the same similarities. In a weighted graph each node has the highest similarity with itself. To cause the same effect we add self edges to each node with weight equal to the degree of the node. This weighting process contributes to our clustering algorithm, especially for the refining step.

Further applications of our clustering algorithm may include other complex networks such as genetic networks, the World Wide Web, citation networks, biological networks and social networks [15].

## 2 Background on Multilevel Approach and Proteomic Networks

Let $G = (V, E)$ be a graph with vertex set $V$ and set of undirected edges $E$. One of the most commonly used data structures for Graphs are matrices. Matrix representation is very useful to store weights for edges and vertices. We can also use a lot of well known computational techniques from Linear Algebra. In our matrix representations $S = (s_{ij})$, diagonal entries $s_{ii}$ store the weights of vertices and off-diagonal entries $s_{ij}$ represent edge weights.

### 2.1 Multilevel Algorithms

The basic concept of multilevel clustering algorithms is that when we have a big graph $G = (V, E)$ to partition, we construct a smaller graph $\bar{G} = (\bar{V}, \bar{E})$ whose vertices are groups of vertices from $G$. We can apply a clustering method to this smaller graph, and transfer the partition to the original graph. This idea is very useful because smaller matrices or graphs require much less time to cluster. The process of constructing the smaller matrix is called coarsening, and the reverse process is called decoarsening.

**Coarsening** and decoarsening steps are implemented by multiplying a special coarsening matrix $C$ by the graph matrix $S$. Each column of $C$ has 1's for vertices to merge and 0's for the rest. The way merging or matching is accomplished is explained in section 3.1. Then a series of $S_0, S_1, \cdots, S_l$ are recursively constructed using $C_1, \cdots, C_l$ in the form of $S_i = C_i' * S_{i-1} * C_i$ with $i = 1, \cdots, l$. The coarsest matrix is used to get the initial partition $Cut$.

**Partitioning** algorithms fall into two categories: direct partitioning and recursive bipartitioning. One recursive bipartitioning algorithm which has been successfully applied to identify functional modules is Divisive MinMaxCut algorithm [4, 6]. Divisive MinMaxCut algorithm repeatedly performs two main steps. One is selecting a cluster to split and the other is applying the two-way MinMaxCut algorithm. The two-way MinMaxCut algorithms tries to find a pair of disjoint subsets $(A, B)$ of $V$ which minimizes the objective function

$$J_{MMC} = \frac{s(A,B)}{s(A,A)} + \frac{s(A,B)}{s(B,B)} = \frac{s(A,\bar{A})}{s(A,A)} + \frac{s(B,\bar{B})}{s(B,B)}, \tag{1}$$

where $s(A, B) = \sum_{i \in A, j \in B} s_{ij}$. It is well known [3] that the optimal solution of (1) is the eigenvector $q_2$ associated with the second smallest eigenvalue of the system $(D - S)q = \lambda Dq$, where $D = diag(d_1, d_2, \cdots, d_n)$ and $d_i = \sum_j s_{ij}$.

The optimum value of (1) is called the cohesion of the cluster and is an indicator which shows how closely vertices of the current cluster are related [3].

Divisive MinMaxCut algorithm recursively chooses a cluster which has the least cohesion value, until we have a predefined number of clusters or when all current clusters have cohesion values greater than a threshold.

**Decoarsening** is the way back to the original graph. The partition from the coarsest level is mapped into finer levels by multiplying a proper coarsening matrix $C$. Then a Kernighan-Lin (KL) type refinement algorithm [14] is applied to improve the quality at each level. KL starts with an initial partition; it iteratively searches for nodes from each cluster of the graph if swapping of a node to one of the other clusters leads to a better partition. For each node, there would be more than one cluster to give smaller objective function values than the current cut. So the node moves to the cluster that gives the biggest improvement. The iteration terminates when it does not find any node to improve the partition.

### 2.2 Functional Modules in Protein-Protein Interaction Networks

Proteomic networks have two important features [2]. One is that the degree distribution function $P(k)$ follows a power law (and so is considered a scale-free network). That is, most vertices have low degrees and a few are highly connected. The other feature is the *small world* property which is also known as *six degrees of separation*. This means the diameter of the graph is small compared with the size.

**DIP** is a protein interaction database. The protein-protein interactions can be detected by high-throughput experiments. However, many of them (as high as 50%) are false positive. Database of Interacting Proteins (DIP: http://dip.doe-mbi.ucla.edu) provides a protein-protein interaction dataset in the budding yeast, Saccharomyces cerevisiae, that is experimentally determined and cross-referenced to the major sequence databases (SWISS-PROT, GenBank, PIR). They also provide a smaller dataset called CORE which contains the pairs of interacting proteins identified that were validated according to the criteria described in Deane at al.[5].

Pothen et al. presented a two-level architecture on this CORE dataset. The network has 2610 proteins and 6236 interactions. Their idea is that removing high degree proteins (called hub proteins) and low degree proteins (low-shell proteins) from the network before clustering leads to better partitioning and then the removed nodes can be added to the partitioning. The residual network after removing hub proteins and low shell proteins has 499 proteins and 1229 interactions.

Instead of using the small network (CORE dataset), we use the DIP network which has 4931 nodes 17471 edges. We construct a residual network after removing nodes that have degree 20 or more and 3 or less from the original DIP dataset. We do this in similar way to Pothen et al. [6]. The residual network has 1078 nodes and 2778 edges.

**MIPS** is the Munich information center for protein sequence information on the molecular structure and functional networks and data of yeast for comparative analysis. The currently annotated functional modules for yeast have 6451 proteins in all with 4022 id known proteins and 2429 id unknown proteins. 800 proteins are common to the residual network of DIP protein-protein network.

Many functional modules in the yeast are found in a hierarchy of as many as four levels. That is, some proteins are related to only one cellular process and some proteins work in a group and groups of them are involved in another cellular process. For example, proteins in the Anaphase promoting complex (id:60) are not involved in any other cellular process. However, proteins in the Dynactin complex (id:140.30.30.30) cooperate with proteins in Kinesin-related motorproteins (id:140.30.30.10) and Dynein-complex motorproteins (id:140.30.30.20) to make bigger, complex, Tubulin-associated motorproteins. Similarly, all proteins in complexes whose id start with 140 are included in highest complex Cytoskeleton (id:140). Let us call these smallest functional units leaf modules.

## 3 Coarsening Unweighted Graphs and Computational Experiments

We first present three different matching algorithms for unweighted graphs. A computational comparison of them follows. And then we show how multilevel approach improves the current clustering algorithms. We use two measures to validate our computational results. One is the triplet, $\kappa$. The first item of $\kappa$ is the total number of edges inside clusters and the second is number of edges between clusters. The last item is the maximum number of edges between any two clusters. This measure does not exactly show how well our network is clustered. However, this measure provides a some insight. The bigger the first number and the smaller the second and third, the more cohesive clusters a network has. The other measure we use is the number of nodes exactly clustered. Each cluster partitioned from the DIP residual network is compared with all leaf functional modules of MIPS. We define $\tau$ as the sum of maximum number of correctly matched proteins of each cluster.

One property of $\tau$ is that it increases as the number of clusters increases because the sizes of supernodes decrease.

### 3.1 Coarsening Algorithms

A matching in a graph is a set of edges in which no two of them are incident on the same node. A matching is maximal when any edge in the graph that is not in the matching has at least one of its endpoints matched. Some algorithms aim to match as many nodes as possible [17] and some aim to maximize the sum of all edge weights [8]. These algorithms are too time intensive [1] and designed for weighted graphs. Our new matching algorithm tries to find a compromise.

The simplest matching for unweighted graphs is random matching. One node is randomly visited and one of unmatched node is randomly chosen to be merged with the node (RVRM). A drawback is that the nodes with low degrees have higher chance to be left unmerged than high degree nodes. In order to avoid this problem we can pick the lowest degree node among unmerged nodes and choose one of the unmerged nodes randomly to merge (LVRM). Thus this algorithm tends to merge more nodes than RVRM.

We define the weights of edges as follows. The edge weights are all 1's to start with, but become the sum of the number of edges combined in a matching step. Similarly, a node weight is defined as the total number of nodes merged into it. We developed a matching algorithm which uses the edge weights and the node weights. After one level of coarsening with RVRM or LVRM, some edges may represent more than one edge (up to 4). That is, we have groups of different edge weights after coarsening. So we now can adapt the Sorted Matching (SM) idea for the unweighted graph coarsening: merge nodes with highest edge weight. But there are too many nodes with the same edge weight. Then we give the higher priority to the edge with lower combined node weights, like LVRM. We take the edge with maximum $1/w(n_i) + 1/w(n_j)$ as a tie-break rule, where $w(n_i)$ and

$w(n_j)$ are the node weights, that is, the number of nodes, of supernodes $n_i$ and $n_j$. We call this matching Heavy-Edge-Small-Node (HESN).

### 3.2 Comparison of Coarsening Algorithms

Table 1 shows the comparison of three coarsening algorithms without partitioning and refinement. The total numbers of correctly matched proteins are listed after various levels of coarsening. The first row for each algorithm has the sizes of the graphs (no. of nodes) at the various levels of our multilevel scheme. LVRM results in the smallest graph in coarsest level than other two algorithms. But HESN generates more cohesive clusters as the second row of each coarseningon this table shows. That is, there are more edges inside clusters and less edges crossing them. Moreover, more proteins match MIPS dataset in the third rows by HESN.

**Table 1.** Comparison of different coarsening algorithms with different levels. First row of each algorithm has sizes of graphs. Second rows have the number edges inside, between and maximum between edges ($\kappa$) . Third rows have the number correctly grouped nodes ($\tau$).

| level | 3 | 4 | 5 | |
|---|---|---|---|---|
| RVRM | 607-340-185 | 612-346-192-105 | 616-339-184-100-53 | size of graph |
| | 1036-1742-17 | 1216-1562-22 | 1283-1495-41 | $\kappa$ |
| | 459 | 346 | 218 | $\tau$ |
| LVRM | 573-296-151 | 573-302-157-79 | 571-302-156-79-40 | size of graph |
| | 1062-1716-18 | 1226-1552-17 | 1254-1524-23 | $\kappa$ |
| | 419 | 282 | 182 | $\tau$ |
| HESN | 601-333-182 | 601-333-182-102 | 601-333-182-102-56 | size of graph |
| | 1351-1427-37 | 1602-1176-11 | 1749-1029-8 | $\kappa$ |
| | 514 | 397 | 295 | $\tau$ |

### 3.3 Various Levels and Identifying Functional Modules

One issue when we use this spectral clustering algorithm is that (1) may have 0 for denominators because all diagonal entries of $S$ are 0. Then the objective function values can be computed without including the clusters whose inner similarity $s(A, A)$ is 0. We tried to add weights on diagonal entries of $S$ with 1's and with degrees. To compare these three algorithms, we generated two groups of clusters without multilevel algorithm. Table 2 shows that we expect better results when we add degrees of nodes into diagonal entries.

Table 3 shows the effect our multilevel algorithm (ML) on finding functional modules. We considered the ML algorithms with levels 0 through 3 and use HESN and 60 clusters in all experiments for this table. As shown in this table, the number of correctly clustered proteins ($\tau$) increases and the edge connectivities ($\kappa$) become better as the number of levels increases. The last row of the table shows the total timing and the third row shows the timings for the 3 parts

**Table 2.** Comparison of three different weighting methods on diagonal entries of $S$ with 40 and 60 clusters. The entries show the number of correctly clustered proteins ($\tau$)and the edge information (triplet $\kappa$) in parenthesis.

| no. of clusters | 40 | 60 |
|---|---|---|
| no weighting | 151 (1296 1482 209) | 224 (1198 1580 86) |
| Adding 1 | 192 (1596 1182 50) | 226 (1477 1301 37) |
| Adding degree | 201 (1652 1126 50) | 234 (1490 1288 46) |

**Table 3.** Multilevel algorithm results with different number of levels. Timings are measured in seconds.

| level | 0 | 1 | 2 | 3 |
|---|---|---|---|---|
| no. of nodes | 601 | 333 | 182 | 102 |
| edge info ($\kappa$) | 1490 1288 46 | 1769 1009 31 | 1788 990 20 | 1834 944 40 |
| no. of correct ($\tau$) | 234 | 300 | 312 | 329 |
| Time | 0/261.1/53.4 | 0.3/69.3/112.3 | 0.4/7.6/106.3 | 0.4/1.7/117.9 |
| Total time | 314.5 | 181.9 | 114.3 | 120.0 |

of ML (coarsening, partitioning and decoarsening). The total time consumed decreases up to 3 levels and then starts increasing. Particularly the time for partitioning clearly decreases although the time for coarsening increases with the number of levels. The refining step takes up most of time for ML with 3 levels: 117.9 out of 120 seconds. This is because KL refinement algorithm takes $O(N^2)$ complexity. So we can improve the timings even further using more efficient refining algorithms such as Fiduccia-Mattheyses linear time heuristic [7].

## 4 Conclusion

We presented a multilevel algorithm for unweighted graphs which represent protein-protein interactions. This research focuses on matching groups of proteins which are more likely to be part of the same functional modules. These groups are considered as single nodes so the graph with them is much smaller than the original graph. We showed that this multilevel approach favors not only less computation time but also more accurate groupings of proteins.

## References


1. R. Diekmann B. Monien, R. Preis. Quality matching and local improvement for multilevel graph-partitioning. *Parallel Comput.*, 26(12):1609–1634, 2000.
2. S. Bornholdt and H.G. Schuster, editors. *Handbook of Graphs and Networks.* Wiley VCH, 2003.
3. R. Meraz S. Holbrook C. Ding, X. He. A unified representation for multi-protein complex data for modeling protein interaction networks. *Proteins: Structure, Function, and Bioinformatics*, 57:99–108, 2004.

4. Richard F. Meraz Stephen R. Holbrook Chris Ding, Xiaofeng He. A unified representation of multiprotein complex data for modeling interaction networks. *Proteins: Structure, Function, and Bioinformatics*, 57(1):99–108, 2004.
5. Xenarios I Eisenberg D. Deane CM, Salwinski L. Protein interactions: two methods for assessment of the reliability of high throughput observations. *Mol Cell Proteomics.*, 1(5):349–56, May 2002.
6. A.Pothen E.Ramadan, C.Osgood. The architecture of a proteomic network in the yeast. International Workshop on Distributed Data Mining in Life Science (LifeDDM05), 2005.
7. C.M. Fiduccia and R.M. Mattheyses. A linear time heuristic for improving network partitions. pages 175–181. 19th IEEE Design Automation Conference, 1982.
8. H. N. Gabow. Data structures for weighted matching and nearest common ancestors with linking. In *SODA*, pages 434–443, 1990.
9. C.W. Hogue G.D. Bader. An automated method for finding molecular complexes in large protein interaction networks. *BMC Bioinformatics*, 4(1), 2003.
10. C. Ding Y. Zhang V. Kumar S. Holbrook H. Xiong, X. He. Identification of functional modules in protein complexes via hyperclique pattern discovery. Pacific Symposium on Biocomputing (PSB 2005).
11. B. Kulis I. Dhillon, Y. Guan. A fast kernel-based multilevel algorithm for graph clustering. The eleventh ACM SIGKDD international conference on Knowledge discovery in data mining, 2005.
12. G. Karypis and V. Kumar. *MeTis: Unstrctured Graph Partitioning and Sparse Matrix Ordering System, Version 2.0*, 1995.
13. G. Karypis and V. Kumar. A fast and high quality multilevel scheme for partitioning irregular graphs. Technical Report 95-035, Department of Computer Science, University of Minnesota, Minneapolis, MN, 1998.
14. B.W. Kernighan and S. Lin. An efficient heuristic procedure for partitioning graphs. *The Bell System Technical Journal*, 1970.
15. M. E. J. Newman. Properties of highly clustered networks. *Physics Review*, 57:99–108, 2003.
16. S. Oliveira and S.C. Seok. A multi-level approach for document clustering. *Lecture Notes in Computer Science*, 3514:204–211, Jan 2005.
17. V. Vazirani. A theory of alternating paths and blossoms for proving correctness of the (sqrt{$E$}) general graph maximum matching algorithm. *Combinatorica*, 14(1):71–109, 1994.

# Towards Detecting Protein Complexes from Protein Interaction Data

Pengjun Pei and Aidong Zhang*

Department of Computer Science and Engineering,
State University of New York at Buffalo,
Buffalo NY 14260, USA

**Abstract.** High-throughput methods for detecting protein-protein interactions (PPI) have given researchers an initial global picture of protein interactions on a genomic scale. These interactions connect proteins into a large protein interaction network (PIN). However, both the size of the data sets and the noise in the data pose big challenges in effectively analyzing the data. In this paper, we investigate the problem of protein complex detection, i.e., finding biologically meaningful subsets of proteins, from the noisy protein interaction data. We identify the difficulties and propose a "seed-refine" approach, including a novel subgraph quality measure, an appropriate heuristics for finding good seeds and a novel subgraph refinement method. Our method considers the properties of protein complexes and the noisy interaction data. Experiments show the effectiveness of our method.

## 1 Introduction

Proteins must interact with other molecular units to execute their function. Discovering proteins that interact in a cell is key to elucidate its functional networks. Recent advances in biotechnology have made it possible to detect protein interactions on a global scale [10, 17, 7, 9]. We can construct a protein interaction network (PIN) [7] from existing protein-protein interaction data by connecting each pair of vertices (proteins) involved in an interaction.

Proteins are likely to form closely-coupled protein complexes as functional units to participate in a certain biological process. Detecting the protein complexes from protein interaction network will help find the building modules of the protein network. These complexes can be roughly considered as dense subgraphs of the protein interaction network. However, protein complexes are likely to overlap and the interaction data are very noisy. Therefore, intelligent methods are in great demand to effectively detect protein complexes.

This paper identifies the difficulties of this problem and proposes some favorable properties of the methods for this purpose. Then we propose a 'seed-refine'

* This research was partially supported by National Science Foundation Grants DBI-0234895, IIS-0308001 and National Institutes of Health Grant 1 P20 GM067650-01A1.

V.N. Alexandrov et al. (Eds.): ICCS 2006, Part II, LNCS 3992, pp. 734–741, 2006.

approach, including a novel subgraph quality measure, a heuristic to find good seeds, and a novel method to control subgraph overlapping. Experiments show the effectiveness of our method. Finally, we conclude the paper and propose some future work.

## 2 Challenges in Protein Complex Detection

Though the problem of detecting protein complexes from interaction data shares some commonality with clustering problem, there are some additional difficulties:

- Protein complexes may overlap with each other. Therefore, the traditional paradigm for clustering of putting each protein into one single cluster [14] does not suit our problem well. Instead, we would prefer finding 'dense' subgraphs.
- Protein complexes generally correspond to small but dense subgraphs. Divisive hierarchical clustering approaches like [4] are more useful in finding large protein clusters representing biological processes. For small dense subgraphs, we would prefer using extensive local search and optimization.
- Protein interaction data are very noisy. How to define the quality of a subgraph in the presence of noisy edges is a non-trivial task. Normal quality measurement of subgraphs either considers only the worst connected vertices or the averaged overall density.

To sum up, we need a seed-refine approach for protein complex detection. Specifically, we need to generate some promising seed subgraphs, followed by refining these seed subgraphs based on some quality measure. We can stop the refining process when either the quality is smaller than a predefined value, or preferably, the quality achieves a local maximum. The latter is much more preferred because it will not require a predefined threshold.

## 3 A Seed-Refine Approach for Finding Protein Complexes

Throughout the paper, we use an unweighted, undirected graph $G = (V, E)$ to represent the protein interaction network where $V$ represents the set of vertices (proteins) and $E$ represents the set of edges (interactions). An *induced subgraph* is a subset of the vertices of the graph together with all the edges of the graph between the vertices of this subset. As we only consider an induced subgraph of the original graph, we abbreviate the term and simply call it a *subgraph*. We denote the set of neighbors of a vertex $v$ in graph $G$ as $N(v) = \{u|(u, v) \in E\}$.

### 3.1 Subgraph Quality Definition

As observed in [13, 8], topology features can provide some insight on the biological significance of the interactions. For proteins that interact with a lot of other proteins, the biological significance of these interactions might be questionable. Therefore, the degree of a vertex should be taken into consideration when evaluating the strength of an edge. Correspondingly, the quality of a subgraph $G' = (V', E')$ is related to not only $|V'|$ and number of inside links $|E'|$ but also outside links $|E_{out} = \{(u,v)|u \in V', v \notin V'\}|$. Previous quality definitions like density $(Density = 2 * |E'|/(|V'| * (|V'| - 1)))$ [15], k-core [1,2] and cliques disregard outside links.

Meanwhile, we would prefer a subgraph in which each vertex contributes similarly to the quality of the subgraph so that every vertex is likely to be an authentic part. Comparatively, the density considers only the average quality. Cliques and 'k-core' define the subgraph quality by the worst-connected vertex. Theses definitions are too stringent and therefore will miss a lot of potentially biologically meaningful subgraphs.

We first define the quality of a vertex in a subgraph, denoted as $Q(v, G')$: For a vertex $v \in V'$, the number of edges within the subgraph $G' = (V', E')$ is $|N(v) \cap V'|$. Under the null hypothesis that the set of neighbors of $v$ is chosen randomly from the vertices of the graph, i.e., $V$, the probability of observing at least $|N(v) \cap V'|$ neighbors within the subgraph $G'$ can be expressed as:

$$PV_{v,G'} = \sum_{i=|N(v)\cap V'|}^{\min(|N(v)|,|V'|)} \binom{|N(v)|}{i} \times \binom{|V| - |N(v)|}{|V'| - i} / \binom{|V|}{|V'|}.$$

We define $Q(v, G')$ as the minus log of this probability, i.e., $Q(v, G') = -\log(PV_{v,G'})$.

Then we seek to combine $Q(v, G')$ values for all $v \in V'$. Since we prefer a subgraph in which each vertex contributes similarly to the quality of the subgraph, we treat the logs of $Q(v, G')$ for all $v \in V'$ as a random sample from a Normally distributed population with mean $\mu$. [1] We estimate the sample mean, denoted as $\bar{x}$ and variance, denoted as $s^2$: $\bar{x} = \frac{\sum_{v\in V'} \log(Q(v,G'))}{|V'|}$, $s^2 = \frac{\sum_{v\in V'}(\log Q(v,G') - \bar{x})^2}{|V'|-1}$. Then the sampling distribution of $\frac{\bar{x}-\mu}{\sqrt{s^2/|V'|}}$ follows Student's t distribution with $|V'| - 1$ degrees of freedom [3]. We use the lower boundary of the 95% confidence interval of the population mean as our subgraph quality measure:

$$Q(G') = e^{\bar{x} - t_{|V'|-1} * \sqrt{s^2/|V'|}},$$

where $t_{|V'|-1}$ is the cut-off value of the t distribution for 95% confidence interval with $|V'| - 1$ degrees of freedom. This quality measure gives a boundary of the underlying population mean, and therefore, is a statistically meaningful combination of the quality of each vertex.

### 3.2 The Seed-Refine Algorithm

Our algorithm iteratively finds an initial **seed graph** centered on an edge $(u, v)$, denoted as $G^{(0)}_{(u,v)}$ and refines it until no quality improvement can be achieved.

[1] The log transformation of the quality values is used to stabilize variance and thus to make the sample satisfy the Normal distribution requirement.

We call this optimized subgraph a **refined subgraph**, denoted as $G_{(u,v)}$. The set of refined subgraphs, denoted as $GS$, represents our predicted complexes. We use $Visited$ to represent the set of edges that have been covered by refined subgraphs. The "seed-refine" algorithm is illustrated in Algorithm 1 in Figure 1.

After we get $GS$, a simple postprocessing can be applied to filter out those subgraphs with quality less than a threshold.

### 3.3 Finding Seed Subgraphs

We define two layer seeds: we use an edge $(u,v)$ not in previously refined subgraphs as the **seeding edge** and find corresponding **seeding vertices**, denoted as $SV_{(u,v)}$: $SV_{(u,v)} = \{w|(u,w) \in E \setminus Visited, (v,w) \in E \setminus Visited\}$.

The seed subgraph is the subgraph induced by $SV_{(u,v)} \cup \{u,v\}$. This definition guarantees that those edges in $Visited$ can not be used as part of the seed graph. Since each of the seeding vertices is connected to both vertices of the seeding edge, the seed subgraph can be regarded as **centered on** the seeding edge $(u,v)$. Therefore, we denote it as $G^{(0)}_{(u,v)}$. Given all candidate seed subgraphs with at least 3 vertices, we choose the most promising one: the one with the largest number of vertices. If there is a tie, we choose the one with highest quality, i.e., we define the function $isMorePromising(G1, G2) = true$ iff

$$(|V(G1)| > |V(G2)|) \textbf{ or } ((|V(G1)| = |V(G2)|) \textbf{ and } (Q(G1) > Q(G2))).$$

The subroutine is described in Algorithm 2 in Figure 1.

### 3.4 Refining Subgraphs

Given a seed subgraph, the subroutine $refineSubGraph$ tries all possible actions of adding one vertex to or removing one vertex from the subgraph and takes the action that achieves highest quality improvement. In this process, we require that (i) the subgraph contains the seeding edge $(u,v)$, and (ii) the subgraph remains connected. This process is repeated until no quality improvement action can be found. The pseudocode is listed in Algorithm 3 in Figure 1.

### 3.5 Analysis of the Algorithm

Since the subgraph quality improves monotonously in the $refineSubGraph$ subroutine, the refinement recursion can end. Also, since we increase the set $Visited$ after finding one refined subgraph, the algorithm will end after visiting all edges.

For computational complexity, notice that we are finding dense but *small* subgraphs and we require that the subgraph always contains the seeding edge. Therefore the refined subgraph will be still close to the seeding edge, suggesting that the $refineSubGraph$ recursion will not take too much time. Also, considering the sparsity of the PIN, computational time is not a serious issue here.

Notice that our objective is to predict a 'reasonable' number of protein complexes without excessive overlapping. The control of subgraph overlapping is

**Algorithm 1.** *SRA*: Seed-Refine Algorithm for Protein Complex Detection

**Input:** $G = (V, E)$: protein interaction network
**Output:** $GS$: the set of predicted protein complexes
1. Initialization: $GS \leftarrow \phi$, $Visited \leftarrow \phi$
2. $G^{(0)}_{(u,v)} \leftarrow findSeedSubgraph(G, Visited)$
3. **while** $G^{(0)}_{(u,v)} \neq empty$ **do**
4. $\quad G_{(u,v)} \leftarrow refineSubGraph(G^{(0)}_{(u,v)})$
5. $\quad GS \leftarrow GS \cup \{G_{(u,v)}\}$
6. $\quad Visited \leftarrow Visited \cup E(G_{(u,v)})$
7. $\quad G^{(0)}_{(u,v)} \leftarrow findSeedSubgraph(G, Visited)$
8. **end while**
9. **return** $GS$

**Algorithm 2.** *findSeedSubgraph*: Find a Seed Subgraph

**Input:** $G = (V, E)$: protein interaction network, $Visited$
**Output:** $G^{(0)}_{(u,v)}$: a seed subgraph centered on $(u, v)$
1. Initialization: $G^{(0)}_{(u,v)} \leftarrow empty$, $Candidate \leftarrow E \setminus Visited$
2. **for all** edge $(i, j) \in Candidate$ **do**
3. $\quad$ Construct the seed subgraph centered on $(i, j)$: $G^{(0)}_{(i,j)}$
4. $\quad$ **if** $|V(G^{(0)}_{(i,j)})| \geq 3$ **and** $isMorePromising(G^{(0)}_{(i,j)}, G^{(0)}_{(u,v)})$ **then**
5. $\quad\quad G^{(0)}_{(u,v)} \leftarrow G^{(0)}_{(i,j)}$
6. $\quad$ **end if**
7. **end for**
8. **return** $G^{(0)}_{(u,v)}$

**Algorithm 3.** *refineSubGraph*: Refine a subgraph

**Input:** $G^{(i)}_{(u,v)}$: a subgraph centered on $(u, v)$
**Output:** $G_{(u,v)}$: a refined subgraph centered on $(u, v)$
1. Generate graphs $\{G'_{(u,v)}\}$ by adding a vertex to/deleting a vertex from $G^{(i)}_{(u,v)}$
2. **for all** graph $G'_{(u,v)}$ **do**
3. $\quad G^{(i+1)}_{(u,v)} \leftarrow \underset{G'_{(u,v)}}{\operatorname{argmax}}(Q(G'_{(u,v)}))$
4. **end for**
5. **if** $Q(G^{(i+1)}_{(u,v)}) > Q(G^{(i)}_{(u,v)})$ **then**
6. $\quad$ **return** $refineSubGraph(G^{(i+1)}_{(u,v)})$
7. **else**
8. $\quad$ **return** $G^{(i)}_{(u,v)}$
9. **end if**

**Fig. 1.** A seed-refine algorithm for protein complex detection achieved by the two layer seeds: the seeding edge is used as the center of the seed subgraph. It is fixed in each refinement iteration to prevent the subgraph from being attracted towards another dense area far from the original seed. The seeding vertices give the preliminary shape of the seed subgraph for further refinement. After finding one refined subgraph, we prevent the edges in the subgraph from existing in later seed subgraphs. Therefore, the next seed subgraph tends to be a bit far from already discovered refined subgraphs. Also notice that we do not prohibit the inclusion of edges in previously refined subgraphs in the refinement process. This gives the possibility of overlapping subgraphs. Our choice of $isMorePromsing$ function is more likely to select larger subgraphs as seeds and thus less likely to branch out into dense regions of the graph that have already been discovered. To conclude, we design an algorithm that allows outputting overlapping subgraphs but methodologically makes it possible only when there is strong evidence to do so.

## 4 Experiments

Due to space limit, we only report the performance of our algorithm for two data sets: *PreHTMS* includes all yeast interactions except high-throughput mass spectrometry studies. *HTP* data set includes purely large scale studies.

To assess our predictions, we use the curated protein complexes in MIPS [11] (including 267 complexes with at least two proteins) and manually curated Gavin complexes [7] (including 221 complexes) as the ground truth. Similar to [2], for a predicted complex $G' = (V', E')$, we find the best-match complex in the ground truth complex set, denoted as $GT' = (GTV', GTE')$ and use $MatchRatio = \frac{|V' \cap GTV'|}{|V'|} * \frac{|V' \cap GTV'|}{|GTV'|}$ to evaluate the match. We consider a predicted complex *matches* a ground truth complex if $MatchRatio > 0.2$. For a total number of $N$ predicted complexes and $M$ ground truth complexes, suppose $CN$ predicted complexes match $CM$ ground truth complexes, we define $precision = \frac{CN}{N}$ and $recall = \frac{CM}{M}$.[2] We report the precision and recall of our method in Figure 2. For reference, we also list the result from [2].

Figure 2(a) shows that our method outperforms MCODE method. Also notice that our quality threshold is used in the postprocessing step and therefore, we only need to run the main algorithm once and can choose different thresholds on the unfiltered results afterwards. Though MCODE scores each vertex only once, its complex-finding subroutine still needs to run for different parameter choices.

Comparing our results in Figure 2 (a) and (b), we notice that the performance of the PreHTMS data set is higher than that of HTP data set when assessed by MIPS protein complexes. This is because PreHTMS includes more reliable interactions. However, HTP data set has higher performance when assessed by Gavin protein complexes. This is because the HTP data itself includes the interaction data inferred from the Gavin raw purifications.

[2] Our definition of the precision equals the specificity in [2]. However, their sensitivity is defined as $\frac{CN}{CN+M-CM}$. In our experiments, the difference has little effect on the final performance comparison or parameter selection.

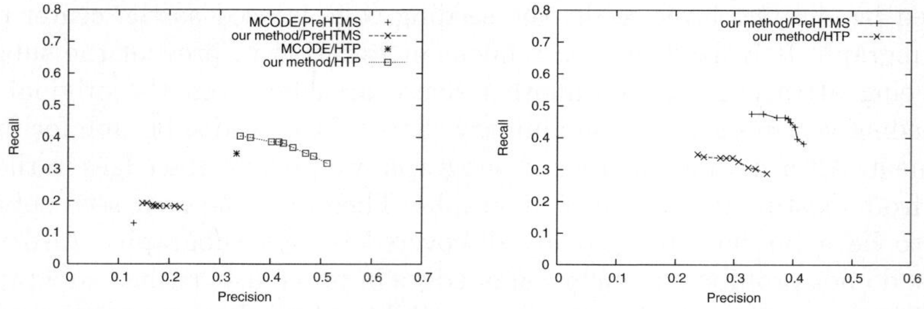


(a) Assessed with Gavin complexes (b) Assessed with MIPS complexes

**Fig. 2.** Algorithm Performance. PreHTMS (9049 interactions among 4325 proteins) includes several large scale studies [10, 17, 16, 12, 5, 6] and the small scale studies in DIP [18]. HTP includes 12243 interactions among 4554 proteins from [10, 17, 5, 6, 9, 7, 16] . We use the spoke model [1] to extract binary interactions from the raw purifications in [9, 7]. For our approach, we choose the postprocessing quality threshold over the range from 6 to 10 in 0.5 increments and report the results. In (a), we use Gavin protein complexes for the assessment. We list reported results from [2] for MCODE using its most optimized parameter settings. In (b), we use MIPS protein complexes for the assessment. Since we use a later version of MIPS complexes, the performance of our method and MCODE is not directly comparable, thus we omit the MCODE result.

**Table 1.** STE5-MAPK complex correctly predicted by our method

| Protein | Neighbors in the Subgraph | Total Degree | $PV_{v,G'}$ Score |
|---|---|---|---|
| FUS3 | STE5, STE7, STE11 | 12 | 14.9 |
| KSS1 | STE5, STE7, STE11 | 10 | 15.5 |
| STE5 | FUS3, KSS1, STE7, STE11 | 11 | 21.2 |
| STE7 | FUS3, KSS1, STE5 | 5 | 18.0 |
| STE11 | FUS3, KSS1, STE5 | 13 | 14.6 |

Table 1 gives the details of a correctly predicted protein complex using PreHTMS data set. The $PV_{v,G'}$ scores for these five vertices in the subgraph, as listed in the table, are all very high and similar. The final quality of the subgraph is 13.7. This subgraph corresponds to the 'STE5-MAPK complex' in MIPS.

## 5 Conclusion and Future Work

In this paper, we have investigated the problem of finding protein complexes from protein interaction network and proposed a novel method. Experiments have shown the effectiveness of our method.

Similar to [2], our method can be used in a *directed mode* to find the complex that a specified protein is part of. This directed mode enables researchers to focus on the proteins of interest. Our method for generating seeds and subgraph refinement may also alleviate the problem of seed subgraph branching out into denser regions in the graph in [2]. We plan to investigate this usage of the algorithm in the future.

## References


1. G. D. Bader and C. W. Hogue. Analyzing yeast protein-protein interaction data obtained from different sources. *Nat Biotechnol*, 20:991–997, 2002.
2. G. D. Bader and C. W. Hogue. An automated method for finding molecular complexes in large protein interaction networks. *BMC Bioinformatics*, 4:2, 2003.
3. M. Bland. *An Introduction to Medical Statistics.* Oxford University Press, USA, 2000.
4. C. Ding et al. A unified representation of multiprotein complex data for modeling interaction networks. *Proteins*, 57:99–108, 2004.
5. B. L. Drees et al. A protein interaction map for cell polarity development. *J Cell Biol*, 154:549–571, 2001.
6. M. Fromont-Racine et al. Genome-wide protein interaction screens reveal functional networks involving sm-like proteins. *Yeast*, 17:95–110, 2000.
7. A. C. Gavin et al. Functional organization of the yeast proteome by systematic analysis of protein complexes. *Nature*, 415:141–147, 2002.
8. D. S. Goldberg and F. P. Roth. Assessing experimentally derived interactions in a small world. *Proc. Natl. Acad. Sci. USA*, 100:4372–4376, 2003.
9. Y. Ho et al. Systematic identification of protein complexes in saccharomyces cerevisiae by mass spectrometry. *Nature*, 415:180–183, 2002.
10. T. Ito et al. Toward a protein-protein interaction map of the budding yeast: A comprehensive system to examine two-hybrid interactions in all possible combinations between the yeast proteins. *Proc. Natl. Acad. Sci. USA*, 93(3):1143–7, 2000.
11. H. W. Mewes et al. Mips: a database for genomes and protein sequences. *Nucleic Acids Res*, 30:31–34, 2002.
12. J. R. Newman, E. Wolf, and P. S. Kim. A computationally directed screen identifying interacting coiled coils from saccharomyces cerevisiae. *Proc Natl Acad Sci U S A*, 97:13203–13208, 2000.
13. R. Saito et al. Interaction generality, a measurement to assess the reliability of a protein-protein interaction. *Nucleic Acids Res*, 30:1163–1168, 2002.
14. M. P. Samanta and S. Liang. Predicting protein functions from redundancies in large-scale protein interaction networks. *Proc Natl Acad Sci U S A*, 100:12579–12583, 2003.
15. V. Spirin and L. A. Mirny. Protein complexes and functional modules in molecular networks. *Proc Natl Acad Sci U S A.*, 100:12123–12128, 2003.
16. A. H. Tong et al. A combined experimental and computational strategy to define protein interaction networks for peptide recognition modules. *Science*, 295:321–324, 2002.
17. P. Uetz et al. A comprehensive analysis of protein-protein interactions in saccharomyces cerevisiae. *Nature*, 403:623–627, 2000.
18. I. Xenarios et al. Dip, the database of interacting proteins: a research tool for studying cellular networks of protein interactions. *Nucleic Acids Res*, 30:303–305, 2002.

# High-Throughput SNP Genotyping by SBE/SBH*

Ion I. Măndoiu and Claudia Prăjescu

Computer Science & Engineering Department, University of Connecticut,
371 Fairfield Rd., Unit 2155, Storrs, CT 06269-2155, USA
{ion.mandoiu, claudia.prajescu}@uconn.edu

**Abstract.** Despite much progress over the past decade, current Single Nucleotide Polymorphism (SNP) genotyping technologies still offer an insufficient degree of multiplexing when required to handle user-selected sets of SNPs. In this paper we propose a new genotyping assay architecture combining multiplexed solution-phase single-base extension (SBE) reactions with sequencing by hybridization (SBH) using universal DNA arrays such as all $k$-mer arrays. Our contributions include a study of multiplexing algorithms for SBE/SBH genotyping assays and preliminary experimental results showing the achievable multiplexing rates. Simulation results on datasets both randomly generated and extracted from the NCBI dbSNP database suggest that the SBE/SBH architecture provides a flexible and cost-effective alternative to genotyping assays currently used in the industry, enabling genotyping of up to hundreds of thousands of user-specified SNPs per assay.

## 1 Introduction

After the completion of the Human Genome Project genomics research is now focusing on the study of DNA variations that occur between individuals, seeking to understand how these variations confer susceptibility to common diseases such as diabetes or cancer. The most common form of genomic variation are the so called *single nucleotide polymorphisms* (SNPs), i.e., the presence of different DNA nucleotides, or *alleles*, at certain chromosomal locations. Determining the identity of alleles present in a DNA sample at a given set of SNP loci is called *SNP genotyping*. Despite much progress over the past decade, current SNP genotyping technologies still offer an insufficient degree of multiplexing when required to handle user-selected sets of SNPs.

In this paper we propose a new genotyping assay architecture combining multiplexed solution-phase single-base extension (SBE) reactions with sequencing by hybridization (SBH) using universal DNA arrays such as all $k$-mer arrays. SNP genotyping using SBE/SBH assays requires the following steps (see Figure 1): (1) Synthesizing primers complementing the genomic sequence immediately preceding SNPs of interest; (2) Hybridizing primers with the genomic

* Work supported in part by NSF CAREER Award IIS-0546457 and a Faculty Large Research Grant from the University of Connecticut Research Foundation.

V.N. Alexandrov et al. (Eds.): ICCS 2006, Part II, LNCS 3992, pp. 742–749, 2006.

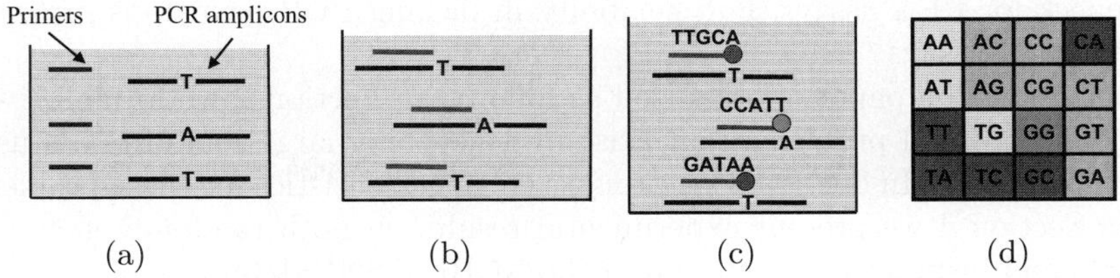


**Fig. 1.** SBE/SBH assay: (a) Primers complementing genomic sequence upstream of each SNP locus are mixed in solution with the genomic DNA sample. (b) Temperature is lowered allowing primers to hybridize to the genomic DNA. (c) Polymerase enzyme and dideoxynucleotides labeled with 4 different fluorescent dyes are added to the solution, causing each primer to be extended by a nucleotide complementing the SNP allele. (d) Extended primers are hybridized to a universal DNA array (an all $k$-mer array for $k$=2 is shown). SNP genotypes are determined by analyzing the resulting hybridization pattern.

DNA; (3) Extending each primer by a single base using polymerase enzyme and dideoxynucleotides labeled with 4 different fluorescent dyes; and finally (4) Hybridizing extended primers to a universal DNA array and determining the identity of the bases that extend each primer by hybridization pattern analysis.

Although both SBE and SBH are well-established techniques, their combination in the context of SNP genotyping has not been explored thus far. The most closely related genotyping assay is the generic Polymerase Extension Assay (PEA) recently proposed in [1]. In PEA, short amplicons containing the SNPs of interest are hybridized to an all $k$-mers array of *primers* that are subsequently extended via single-base extension reactions. Hence, in PEA the SBE reactions take place on solid support, similar to *arrayed primer extension* (APEX) assays which use SNP specific primers spotted on the array [2].

As the SBH multiplexing technique of [3], the SBE/SBH assay leads to high array probe utilization since we hybridize to the array a large number of short extended primers. However, the main power of the method lies in the fact that the sequences of the labeled oligonucleotides hybridized to the array are a priori known (up to the identity of extending nucleotides). While genotyping with SBE/SBH assays uses similar general principles as the PEA assays proposed in [1], there are also significant differences. A major advantage of SBE/SBH is the much shorter length of extended primers compared to that of PCR amplicons used in PEA. A second advantage is that *all* probes hybridizing to an extended primer are informative in SBE/SBH assays, regardless of array probe length; in contrast, only probes hybridizing with a substring containing the SNP site are informative in PEA assays. As shown by the experimental results in Section 4 these advantages translate into an increase by orders of magnitude in multiplexing rate compared to the results reported in [1]. We further note that PEA's effectiveness crucially depends on the ability to amplify very short genomic fragments spanning the SNP loci of interest. This limits the achievable degree of multiplexing in PCR amplification, making PCR amplification the main bottleneck for PEA assays. Full flexibility in picking PCR primers is preserved in SBE/SBH assays.

The rest of the paper is organized as follows. In Section 2 we formalize several computational problems that arise in genotyping large sets of SNPs using SBE/SBH assays. In Section 3 we propose efficient heuristics for these problems, and in Section 4 we present experimental results on both randomly generated datasets and instances extracted from the NCBI dbSNP database.

## 2 Problem Formulations and Complexity

A set of SNP loci can be unambiguously genotyped by SBE/SBH if every combination of SNP genotypes yields a different hybridization pattern (defined as the vector of dye colors observed at each array probe). To formalize the requirements of unambiguous genotyping, let us first consider a simplified SBE/SBH assay consisting of four parallel *single-color* SBE/SBH reactions, one for each possible SNP allele. Under this scenario, only one type of dideoxynucleotide is added to each SBE reaction, corresponding to the Watson-Crick complement of the tested SNP allele. Therefore, a primer is extended in such a reaction if the tested allele is present at the SNP locus probed by the primer, and is left un-extended otherwise.

Let $\mathcal{P}$ be the set of primers used in a single-color SBE/SBH reaction involving dideoxynucleotide $e \in \{\text{A,C,G,T}\}$. From the resulting hybridization pattern we must be able to infer for every $p \in \mathcal{P}$ whether or not $p$ was extended by $e$. The extension of $p$ by $e$ will result in a fluorescent signal at all array probes that hybridize with $pe$. However, some of these probes can give a fluorescent signal even when $p$ is not extended by $e$, due to hybridization to other extended primers. Since in the worst case *all* other primers are extended, it must be the case that at least one of the probes that hybridize to $pe$ does not hybridize to any other extended primer.

Formally, let $X \subset \{A,C,G,T\}^*$ be the set of array probes. For every string $y \in \{A,C,G,T\}^*$, let the *spectrum of $y$ in $X$*, denoted $Spec_X(y)$, be the set of probes of $X$ that hybridize with $y$. Under the assumption of perfect hybridization, $Spec_X(y)$ consists of those probes of $X$ that are reverse Watson-Crick complements of substrings of $y$. Then, a set of primers $\mathcal{P}$ is said to be *decodable* with respect to extension $e$ if and only if, for every $p \in \mathcal{P}$,

$$Spec_X(pe) \setminus \bigcup_{p' \in \mathcal{P} \setminus \{p\}} Spec_X(p'e) \neq \emptyset \tag{1}$$

Decoding constraints (1) can be directly extended to 4-color SBE/SBH experiments, in which each type of extending base is labeled by a different fluorescent dye. As before, let $\mathcal{P}$ be the set of primers, and, for each primer $p \in \mathcal{P}$, let $E_p \subseteq \{A,C,G,T\}$ be the set of possible extensions of $p$, i.e., Watson-Crick complements of corresponding SNP alleles. If we assume that any combination of dyes can be detected at an array probe location, unambiguous decoding is guaranteed if, for every $p \in \mathcal{P}$ and every extending nucleotide $e \in E_p$,

$$Spec_X(pe) \setminus \bigcup_{p' \in \mathcal{P}\setminus\{p\}, e \in E_{p'}} Spec_X(p'e) \neq \emptyset \quad (2)$$

In the following, we refine (2) to improve practical reliability of SBE/SBH assays. More precisely, we impose additional constraints on the set of probes considered to be *informative* for each SNP allele. First, to enable reliable genotyping of genomic samples that contain SNP alleles at very different concentrations (as a result of uneven efficiency in the PCR amplification step or of pooling DNA from different individuals), we require that a probe that is informative for a certain SNP locus must not hybridize to primers corresponding to different SNP loci, *regardless of their extension.* Second, since recent studies by Naef et al. [4] suggest that fluorescent dyes can significantly interfere with oligonucleotide hybridization on solid support, possibly destabilizing hybridization to a complementary probe on the array, in this paper we use a conservative approach and require that each probe that is informative for a certain SNP allele must hybridize to a strict substring of the corresponding primer. On the other hand, informative probes are still required not to hybridize with any other extended primer, even if such hybridizations involve fluorescently labeled nucleotides. Finally, we introduce a *decoding redundancy* parameter $r \geq 1$, and require that each SNP have at least $r$ informative probes. Such a redundancy constraint facilitates reliable genotype calling in the presence of hybridization errors. Clearly, the larger the value of $r$, the more hybridization errors that can be tolerated. If a simple majority voting scheme is used for making allele calls, the assay can tolerate up to $\lfloor r/2 \rfloor$ hybridization errors involving the $r$ informative probes of each SNP.

The refined set of constraints is captured by the following definition, where, for every primer $p \in \{A, C, G, T\}^*$ and set of extensions $E \subseteq \{A, C, G, T\}$, we let $Spec_X(p, E) = \bigcup_{e \in E} Spec_X(pe)$.

**Definition 1.** *A set of primers $\mathcal{P}$ is said to be* strongly $r$-decodable *with respect to extension sets $E_p$, $p \in \mathcal{P}$, if and only if, for every $p \in \mathcal{P}$,*

$$\left| Spec_X(p) \setminus \bigcup_{p' \in \mathcal{P}\setminus\{p\}} Spec_X(p', E_{p'}) \right| \geq r \quad (3)$$

Note that testing whether or not a given set of primers is strongly $r$-decodable can be easily accomplished in time linear in the total length of the primers.

For each SNP locus there are typically two different SBE primers that can be used for genotyping (one from each strand). As shown in [5] for the case of SNP genotyping using tag arrays, exploiting this degree of freedom significantly increases achievable multiplexing rates. Therefore, we next extend our Definition 1 to capture this degree of freedom. Let $P_i$ be the *pool of primers* that can be used to genotype the SNP at locus $i$. Similarly to Definition 1, we have:

**Definition 2.** *A set of primer pools $\mathcal{P} = \{P_1, \ldots, P_n\}$ is said to be* strongly $r$-decodable *if and only if there is a primer $p_i$ in each pool $P_i$ such that $\{p_1, \ldots, p_n\}$ is strongly $r$-decodable with respect to extension sets $E_{p_i}$, $i = 1, \ldots, n$.*

Primers $p_i$ in Definition 2 are called the *representative primers* of the pools in $\mathcal{P}$, respectively.

Genotyping a large set of SNPs will, in general, require more than one SBE/SBH assay. This rises the problem of partitioning a given set of SNPs into the smallest number of subsets that can each be genotyped using a single SBE/SBH assay, which is formulated as follows:

**Minimum Pool Partitioning Problem (MPPP):** *Given primer pools $\mathcal{P} = \{P_1, \ldots, P_n\}$, associated extension sets $E_p$, $p \in \cup_{i=1}^n P_i$, probe set $X$, and redundancy $r$, find a partitioning of $\mathcal{P}$ into the minimum number of strongly $r$-decodable subsets.*

A natural strategy for solving MPPP, similar to the well-known greedy algorithm for the set cover problem, is to find a maximum strongly $r$-decodable subset of pools, remove it from $\mathcal{P}$, and then repeat the procedure until no more pools are left in $\mathcal{P}$. This greedy strategy for solving MPPP has been shown to empirically outperform other algorithms for solving the similar partitioning problem for PEA assays [1]. In the case of SBE/SBH, the optimization involved in the main step of the greedy strategy is formalized as follows:

**Maximum $r$-Decodable Pool Subset Problem (MDPSP):** *Given primer pools $\mathcal{P} = \{P_1, \ldots, P_n\}$, associated extension sets $E_p$, $p \in \cup_{i=1}^n P_i$, probe set $X$, and redundancy $r$, find a strongly $r$-decodable subset $\mathcal{P}' \subseteq \mathcal{P}$ of maximum size.*

**Theorem 1.** *MDPSP is NP-hard, even when restricted to instances with $r = 1$ and $|P| = 1$ for every $P \in \mathcal{P}$.*

Theorem 1 is proved by reduction from the *Maximum Induced Matching* (MIM) problem in bipartite graphs (see [6] for details). Since the reduction preserves the size of the optimal solution, it follows that any hardness of approximation result for the latter problem also holds for MDPSP. From the hardness result in [7] we get:

**Theorem 2.** *It is NP-hard to approximate MDPSP within a factor of 6600/6659, even when restricted to instances with $r = 1$ and $|P| = 1$ for every $P \in \mathcal{P}$.*

## 3 Algorithms

In this section we describe three heuristic approaches to MDPSP. The first one is a naive greedy algorithm that sequentially evaluates the primers in arbitrary order. The algorithm picks a primer $p$ to be the representative of pool $P \in \mathcal{P}$ if $p$ together with the representatives already picked satisfy condition (3). The other two algorithms are inspired by the Min-Greedy algorithm in [7], which approximates MIM in $d$-regular graphs within a factor of $d - 1$. For the MIM problem, the Min-Greedy algorithm picks at each step a vertex $u$ of minimum degree and a vertex $v$, which is a minimum degree neighbor of $u$. All the neighbors of $u$ and $v$ are deleted and the edge $(u, v)$ is added to the induced matching. The algorithm stops when the graph becomes empty.

Each instance of MDPSP can be represented as a bipartite *hybridization graph* $G = ((\bigcup_{i=1}^{n} P_i) \cup X, E)$, with the left side containing all primers in the given pools and the right side containing the array probes, i.e., $X$. There is an edge between primer $p$ and probe $x \in X$ iff $x \in Spec_X(p, E_p)$. As discussed in Section 2, we distinguish between the hybridizations that involve the extending nucleotides and those that do not. Thus, for every primer $p$, we let $N^+(p) = Spec_X(p)$ and $N^-(p) = Spec_X(p, E_p) \setminus Spec_X(p)$. Similarly, for each probe $x \in X$, we let $N^+(x) = \{p | \ x \in N^+(p)\}$ and $N^-(x) = \{p | \ x \in N^-(p)\}$.

We considered two versions of the Min-Greedy algorithm when run on the bipartite hybridization graph, depending on the side from which the minimum degree vertex is picked. In the first version, referred to as MinPrimerGreedy, we pick first a minimum degree node from the primers side, while in the second version, referred to as MinProbeGreedy, we pick first a minimum degree node from the probes side. Thus, MinPrimerGreedy greedy picks at each step a minimum degree primer $p$ and pairs it with a minimum degree probe $x \in N^+(p)$. MinProbeGreedy greedy, selects at each step a minimum degree probe $x$ and pairs it with a minimum degree primer $p$ in $N^+(x)$. In both algorithms, all neighbors of $p$ and $x$ and their incident edges are removed from $G$. Also, at each step, the algorithms remove all vertices $u$, for which $N^+(u) = \emptyset$. These deletions ensure that the primers $p$ selected at each step satisfy condition (3). Both algorithms stop when the graph becomes empty.

As described so far, the MinPrimerGreedy and MinProbeGreedy algorithms work when each pool contains only one primer and when the redundancy is 1. We extended the two variants to handle pools of size greater than 1 by simply removing from the graph all primers $p' \in P \setminus \{p\}$ when picking primer $p$ from pool $P$. If the redundancy $r$ is greater than 1, then whenever we pick a primer $p$, we also pick it's $r$ probe neighbors from $N^+(p)$ with the smallest degrees (breaking ties arbitrarily). The primer neighbors of all these $r$ probes will then be deleted from the graph. Moreover, the algorithm maintains the invariant that $|N^+(p)| \geq r$ for every primer $p$ and $|N^+(x)| \geq 1$ for every probe $x$ by removing primers/probes for which the degree decreases below these bounds. Full pseudocode and efficient implementation details for proposed algorithms are available in [6].

## 4 Experimental Results

We performed experiments with two types of array probe sets. First, we used probe sets containing all $k$-mers, for $k$ between 8 and 10. All $k$-mer arrays are well studied in the context of sequencing by hybridization. However, a major drawback of all $k$-mer arrays is that the $k$-mers have a wide range of melting temperatures, making it difficult to ensure reliable hybridization results. For short oligonucleotides, a good approximation of the melting temperature is obtained using the simple 2-4 rule of Wallace [8], according to which the melting temperature of a probe is approximately twice the number of A and T bases, plus four times the number of C and G bases. The second type of arrays that we

**Table 1.** Number of SBE/SBH assays needed to cover 90 − 95% of extracted reference SNPs using SBE primers of length 20

| Chr | # | # | # 10-mer arrays | | | | | | # 13-token arrays | | | | | |
|---|---|---|---|---|---|---|---|---|---|---|---|---|---|---|
| ID | Ref. | Extracted | r=1 | | r=2 | | r=5 | | r=1 | | r=2 | | r=5 | |
| | SNPs | Pools | 90% | 95% | 90% | 95% | 90% | 95% | 90% | 95% | 90% | 95% | 90% | 95% |
| 1 | 786058 | 736850 | 5 | 7 | 8 | 11 | 15 | 24 | 10 | 14 | 17 | 23 | 39 | 56 |
| 2 | 758368 | 704415 | 5 | 6 | 7 | 9 | 14 | 18 | 9 | 12 | 14 | 18 | 32 | 42 |
| 3 | 647918 | 587531 | 5 | 6 | 7 | 8 | 13 | 16 | 8 | 10 | 12 | 15 | 26 | 35 |
| 4 | 690063 | 646534 | 5 | 6 | 7 | 9 | 14 | 17 | 8 | 10 | 12 | 15 | 26 | 34 |
| 5 | 590891 | 550794 | 5 | 6 | 6 | 8 | 12 | 16 | 7 | 10 | 12 | 15 | 26 | 34 |
| 6 | 791255 | 742894 | 10 | 20 | 14 | 29 | 30 | 54 | 15 | 29 | 23 | 38 | 49 | 73 |
| 7 | 666932 | 629089 | 6 | 9 | 8 | 12 | 16 | 25 | 10 | 15 | 16 | 22 | 36 | 48 |
| 8 | 488654 | 456856 | 4 | 5 | 5 | 7 | 10 | 12 | 7 | 8 | 10 | 13 | 22 | 29 |
| 9 | 465325 | 441627 | 4 | 6 | 6 | 8 | 11 | 17 | 7 | 10 | 11 | 16 | 26 | 36 |
| 10 | 512165 | 480614 | 4 | 6 | 6 | 8 | 11 | 16 | 8 | 10 | 12 | 16 | 27 | 38 |
| 11 | 505641 | 476379 | 4 | 6 | 6 | 8 | 11 | 15 | 8 | 10 | 12 | 15 | 26 | 35 |
| 12 | 474310 | 443988 | 4 | 6 | 6 | 8 | 11 | 18 | 7 | 10 | 11 | 15 | 25 | 36 |
| 13 | 371187 | 347921 | 3 | 4 | 5 | 6 | 9 | 11 | 5 | 7 | 8 | 10 | 16 | 22 |
| 14 | 292173 | 271130 | 3 | 4 | 4 | 5 | 7 | 10 | 5 | 7 | 8 | 10 | 16 | 23 |
| 15 | 277543 | 258094 | 3 | 4 | 4 | 5 | 7 | 11 | 5 | 7 | 8 | 10 | 17 | 24 |
| 16 | 306530 | 288652 | 4 | 6 | 5 | 9 | 9 | 18 | 7 | 10 | 11 | 15 | 25 | 35 |
| 17 | 269887 | 249563 | 3 | 5 | 4 | 8 | 9 | 18 | 7 | 10 | 11 | 15 | 25 | 37 |
| 18 | 268582 | 250594 | 3 | 3 | 4 | 5 | 7 | 9 | 4 | 6 | 6 | 8 | 14 | 18 |
| 19 | 212057 | 199221 | 4 | 6 | 5 | 9 | 11 | 21 | 8 | 11 | 12 | 17 | 29 | 43 |
| 20 | 292248 | 262567 | 3 | 4 | 4 | 5 | 7 | 11 | 6 | 8 | 9 | 12 | 20 | 27 |
| 21 | 148798 | 138825 | 2 | 3 | 3 | 3 | 5 | 6 | 3 | 4 | 5 | 6 | 10 | 13 |
| 22 | 175939 | 164632 | 3 | 4 | 3 | 6 | 6 | 13 | 6 | 8 | 9 | 12 | 21 | 29 |
| X | 380246 | 362778 | 4 | 6 | 6 | 8 | 10 | 15 | 6 | 9 | 9 | 13 | 19 | 26 |
| Y | 50725 | 49372 | 2 | 2 | 2 | 2 | 3 | 3 | 2 | 2 | 2 | 3 | 4 | 5 |

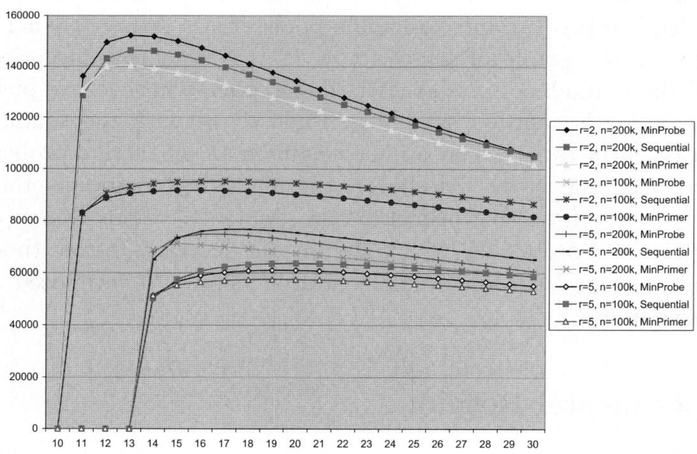


**Fig. 2.** Size of the strongly $r$-decodable pool subset computed by the three MDPSP algorithms as a function of primer length, for random instances with $n = 100 - 200k$ pools of 2 primers and all 10-mer arrays (averages over 10 test cases)

considered are all $c$-token arrays. For a given integer $c$, a DNA string is called a $c$-token if it has a weight $c$ or more and all its proper suffixes have weight strictly less than $c$, where the *weight* of a DNA string is defined as the number of A and

T bases plus twice the number of C and G bases. Since the weight of a $c$-token is either $c$ or $c + 1$, it follows that the 2-4 rule computed melting temperature of $c$-tokens varies in a range of 4°C.

The results of a comprehensive set of experiments comparing the three proposed MDPSP algorithms on both synthetic and genomic datasets are reported in [6]. In Table 1 we report the number of SBE/SBH assays required to cover 90%, respectively 95%, of a total of over 9 million 2-primer pools extracted from the NCBI dbSNP database build 125. We disregarded reference SNPs for which two non-degenerate SBE primers of length 20 could not be determined from the genomic sequence. The results are obtained with a simple MPPP algorithm which iteratively finds maximum $r$-decodable pool subsets using the sequential greedy algorithm.

Further improvements in the multiplexing rate can be achieved by optimizing the length of SBE primers (see Figure 2). Notice that constraints (3) imply a minimum length for SBE primers. Increasing the primer length beyond this minimum primer length is at first beneficial, since this increases the number of array probes that hybridize with the primer. However, if primer length increases too much, a larger number of array probes become non-specific, and the multiplexing rate starts to decline.

## References


1. Sharan, R., Gramm, J., Yakhini, Z., Ben-Dor, A.: Multiplexing schemes for generic SNP genotyping assays. Journal of Computational Biology **12**(5) (2005) 514–533
2. Tonisson, N., Kurg, A., Lohmussaar, E., Metspalu, A.: Arrayed primer extension on the DNA chip - method and application. In Schena, M., ed.: Microarray Biochip Technology, Eaton Publishing (2000) 247–263
3. Hubbell, E.: Multiplex sequencing by hybridization. Journal of Computational Biology **8**(2) (2001) 141–149
4. Naef, F., Magnasco, M.: Solving the riddle of the bright mismatches: Labeling and effective binding in oligonucleotide arrays. In: Physical Review E. Volume 68. (2003) 11906–11910
5. Măndoiu, I., Prăjescu, C., Trincă, D.: Improved tag set design and multiplexing algorithms for universal arrays. LNCS Transactions on Computational Systems Biology **II**(LNBI 3680) (2005) 124–137
6. Măndoiu, I., Prăjescu, C.: High-throughput SNP genotyping by SBE/SBH. ACM Computing Research Repository, cs.DS/0512052 (2005)
7. Duckworth, W., Manlove, D., Zito, M.: On the approximability of the maximum induced matching problem. In: Journal of Discrete Algorithms. Volume 3. (2005) 79–91
8. Wallace, R., Shaffer, J., Murphy, R., Bonner, J., Hirose, T., Itakura, K.: Hybridization of synthetic oligodeoxyribonucleotides to phi chi 174 DNA: the effect of single base pair mismatch. Nucleic Acids Res. **6**(11) (1979) 6353–6357

# Tag SNP Selection Based on Multivariate Linear Regression

Jingwu He* and Alex Zelikovsky**

Department of Computer Science, Georgia State University, GA 30303, Atlanta
{jingwu, alexz}@cs.gsu.edu

**Abstract.** The search for the association between complex diseases and single nucleotide polymorphisms (SNPs) or haplotypes has been recently received great attention. For these studies, it is essential to use a small subset of informative SNPs (tag SNPs) accurately representing the rest of the SNPs. Tagging can achieve budget savings by genotyping only a limited number of SNPs and computationally inferring all other SNPs and compaction of extremely long SNP sequences (obtained, e.g., from Affimetrix Map Array) for further fine genotype analysis. Tagging should first choose tags from the SNPs under consideration and then knowing the values of chosen tag SNPs predict (or statistically cover) the non-tag SNPs. In this paper we propose a new SNP prediction method based on rounding of multivariate linear regression (MLR) analysis in sigma-restricted coding. When predicting a non-tag SNP, the MLR method accumulates information about all tag SNPs resulting in significantly higher prediction accuracy with the same number of tags than for the previously known tagging methods. We also show that the tag selection strongly depends on how the chosen tags will be used – advantage of one tag set over another can only be considered with respect to a certain prediction method. Two simple universal tag selection methods have been applied: a (faster) stepwise and a (slower) local-minimization tag selection algorithms. An extensive experimental study on various datasets including 6 regions from HapMap shows that the MLR prediction combined with stepwise tag selection uses significantly fewer tags (e.g., up to two times less tags to reach 90% prediction accuracy) than the state-of-art methods of Halperin et al. [8] for genotypes and Halldorsson et al. [7] for haplotypes, respectively. Our stepwise tagging matches the quality of while being faster than STAMPA [8]. The code is publicly available at http://alla.cs.gsu.edu/~software.

## 1 Introduction

The search for the association between complex diseases and single nucleotide polymorphisms (SNPs) or haplotypes has been recently received great attention.

* Partially supported by GSU Molecular Basis of Disease Fellowship.
** Partially supported by NIH Award 1 P20 GM065762-01A1 and US CRDF Award #MOM2-3049-CS-03.

V.N. Alexandrov et al. (Eds.): ICCS 2006, Part II, LNCS 3992, pp. 750–757, 2006.

For these studies, it is essential to use a small subset of informative SNPs (tag SNPs) accurately representing the rest of the SNPs. Tagging can achieve budget savings by genotyping only a limited number of SNPs and computationally inferring all other SNPs and compaction of extremely long SNP sequences (obtained, e.g., from Affimetrix Map Array) for further fine genotype analysis.

Tagging methods have been previously explored in statistical and pattern recognition community as well as optimization community. In statistics, tags are required to *statistically cover* individual (non-tagged) SNPs or haplotypes (sets of SNPs), where the quality of statistical covering is usually measured by correlation, e.g., find minimum number of tags such that for any non-tag SNP there exists a highly correlated (squared correlation $R^2 > .8$) tag SNP [2, 4]. In the optimization community, the number of tags is usually minimized subject to upper bounds on *prediction error* measured in leave-one-out cross-validation experiments [7, 8].

The generic tagging problem can be formulated as follows: Given a sample $S$ of a population $P$ of *individuals* (either haplotypes or genotypes) on $m$ SNPs, find positions of $k$ ($k < m$) tag SNPs such that one can predict (or statistically cover) an entire *individual* (haplotype or genotype) from its restriction onto the $k$ tag SNPs. This tagging problem formulation implicitly relies on a certain prediction or statistical covering method. The corresponding **SNP prediction problem** is formulated as follows: Given the values of $k$ tags of the individual $x$ with unknown SNP $s$ and $n$ individuals with $k$ tag SNP and known value of SNP $s$, find the value of $s$ in $x$. Respectively, each non-tag SNP must be statistically covered by a tag SNP or multiple tag SNPs as proposed below.

In this paper we propose a new SNP prediction based on rounding of multivariate linear regression (MLR) analysis in sigma-restricted coding. The MLR method predicts the non-tag SNP $s$ so that the predicted $s$ is the closest to its projection on the span of vectors corresponding to tag SNPs. Respectively, the MLR method statistically covers $s$ by its projection. The MLR method accumulates information about *all* tag SNPs resulting in significantly higher prediction accuracy with the same number of tags. In contrast, the previous SNP prediction methods rely either on a single SNP (see, e.g., [2]), a pair of tag SNPs [8], or small number of tag SNPs from the block with limited haplotype diversity [14]. Our previous linear-algebraic tag selection and SNP prediction [9, 10] picks linear independent tag SNPs but cannot easily handle bounds on *prediction error* or number of tags. We show that the tag selection strongly depends on how the chosen tags will be used – advantage of one tag set over another can only be considered with respect to a certain prediction method. We suggest two simple universal methods for tag selection based on known prediction method: a (faster) stepwise and a (slower) local-minimization tag selection algorithms.

An extensive experimental study on various datasets including 6 regions from HapMap [1] shows that the MLR prediction combined with stepwise tag selection uses significantly fewer tags (e.g., up to two times less tags to reach 90% prediction accuracy) than the state-of-art methods of Halperin et al. [8] for genotypes and Halldorsson et al. [7] for haplotypes, respectively. Our stepwise tagging matches the quality of while being faster than STAMPA [8]. When applied to statistical covering, our method also uses fewer tags than the state-of-the-art method of Carlson et al. [2] especially when high correlation ($r^2 > .8$) is required.

The rest of the paper is organized as follows: Section 2 describes our MLR SNP prediction algorithm. Section 3 shows how to separate tag selection problem from SNP prediction or statistical covering and describes the general approach for tag selection based on the way how tags are used. Section 4 presents an experimental results and discussions.

## 2 The MLR SNP Prediction Algorithm

Usually, a genotype is represented by a vector with coordinates 0,1, or 2, where 0 represents the homozygous site with major allele, 1 represents the homozygous site with minor allele, and 2 represents the heterozygous site. Respectively, each haplotype's coordinate is 0 or 1, where 0 represents the major allele and 1 represents the minor allele. The sample population $S$ together with the tag-restricted individual $x$ are represented as a matrix $M$. The matrix $M$ has $n + 1$ rows corresponding to $n$ sample individuals and the individual $x$ and $k + 1$ columns corresponding to $k$ tag SNPs and a single non-tag SNP $s$. All values in $M$ are known except the value of $s$ in $x$. In case of haplotypes, there are only two possible resolutions of $s$, namely, $s_0$ and $s_1$ with the unknown SNP value equal to 0 or 1, respectively. For genotypes, there are 3 possible resolutions $s_0$, $s_1$, and $s_2$ corresponding to SNP values 0, 1, or 2, respectively. The SNP prediction method should chose correct resolution of $s$.

The proposed MLR SNP prediction method considers all possible resolutions of $s$ together with the set of tag SNPs $T$ as the vectors in $(n + 1)$-dimensional Euclidean space. It assumes that the most probable resolution of $s$ should be the "closest" to $T$. The distance between resolution of $s$ and $T$ is measured between $s$ and its projection on the vector space $span(T)$, the span of the set of tag SNPs $T$ (see Figure 1).

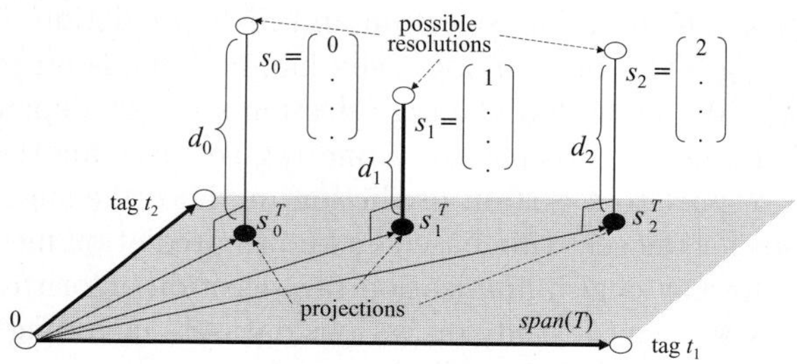


**Fig. 1.** MLR SNP Prediction Algorithm. Three possible resolutions $s_0$,$s_1$, and $s_2$ of $s$ are projected on the span of tag SNPs (a dark plane). The unknown SNP value is predicted 1 since the distance between $s_1$ and its projection $s_1^T$ is the shorter than for $s_0$ and $s_2$.

Computationally, the distance between a resolution $s_i$ and $T$ is measured as $dist(T, s_i) = |T \cdot (T^t \cdot T)^{-1} \cdot T^t \cdot s_i - s_i|$. The runtime of the MLR SNP prediction algorithm is $O(kn^2)$. In general, there are $m-k$ non-tag SNPs in each individual but the matrix $T \cdot (T^t \cdot T)^{-1} \cdot T^t$ is the same for all these non-tag SNPs and should be computed only once. Thus, the total runtime for predicting a complete individual is $O(kn(n+m))$.

Instead of $\{0, 1, 2\}$ notation, we use sigma-encoding $\{-1, 1, 0\}$ of SNP values proposed in [10]: 0's are replaced by $-1$'s and 2's are replaced with 0's. Our experiments show that MLR with $\{-1, 0, 1\}$ notations needs up to 30% less tags than with the original $\{0, 2, 1\}$ notations for reaching same prediction accuracy.

## 3 Tag SNP Selection Methods

Assuming self-similarity of data, one can expect that an algorithm predicting with high accuracy SNPs of an unknown individual will also predict with high accuracy SNPs of the sampled individual. Then, we expect that the better prediction algorithm will have fewer errors when predicting SNPs in the sample $S$. This expectation allows us to find tags using prediction algorithm as follows: We can check each $k$-tuple of tags and choose the $k$-tuple with the minimal number of errors in predicting the non-tag SNPs in the sampled individuals. Even though the sample elements are completely typed, prediction algorithms can make still errors because the number of SNPs may be not sufficient to distinguish any two sampled individuals. Thus, tag SNP selection based on prediction is reduced to the following problem:

**Tag SNP Selection for Prediction.** *Given a prediction algorithm $A_k$ and a sample $S$, find $k$ tags such that the prediction error $e$ of $A_k$ averaged over all SNPs in $S$ (including tags) is minimized.*

Similarly, a *statistical covering criteria* has as an input a set of $k$ column-tags $(t_1, \cdots, t_k)$ and a single SNP $s$ on the sample $S$. $A_k$ checks if the set of tags has a statistically significant correlation with $s$. For example, in Carlson et al. [2] $A_k$ checks if the $R^2$ between $s$ and $t_i$'s is higher than a certain threshold (e.g., 0.8). A similar criteria is defined in Stram et al. [13]. In this paper, we suggest to compute $A_k$ as a correlation between $s$ and $s^i$ predicted using MLR algorithm. The statistical covering version of the tag SNP selection can be formulated as follows:

**Tag SNP Selection for Statistical covering.** *Given a statistical covering criteria $A_k$ and a sample $S$, find $k$ tags such that the number of statistically covered SNPs (including tags) according to $A_k$ is maximized.*

In general, these problems are computationally difficult and the runtime of an exact algorithm may become prohibitively slow. Below we propose two universal heuristics which can be applied to an arbitrary prediction algorithm or statistical covering criteria $A_k$.

The *Stepwise Tagging algorithm* (STA) starts with the best tag $t_0$, i.e., tag that minimizes error when predicting with $A_k$ all other tags. Then STA finds such tag $t_1$ which would be the best extension of $\{t_0\}$ and continue adding best tags until reaching the set of tags of the given size $k$. STA produces *hereditary* set of tags, i.e., the chosen $k$ tags contain the chosen $k-1$ tags. This hereditary property may be useful in case if the set of tags can be extended. The runtime of STA is $O(knmT)$, where $T$ is the runtime of the prediction algorithm. Note that for statistical covering, STA is equivalent to the greedy algorithm used by Carlson et al. [2].

The *Local-Minimization Tag Selection algorithm* (LMT) is more accurately searching for a better set of tags among much larger possibilities. LMT starts with the $k$ tags produced by STA and then iteratively replaces each single tag with the best possible choice while not changing other tags. Such replacements will be continued until no significant improvement in the prediction quality (i.e., by more than given amount of $\epsilon\%$) can be achieved. The runtime of LMT is $O(knmT\epsilon^{-1})$ since the number of iterations cannot exceed $\frac{100}{\epsilon}$.

## 4 Experimental Results

We first list the test datasets, then compare the MLR prediction method combined with STA and LMT. Next comparison of prediction accuracy and statistical covering is given as well comparison of MLR with Idselect. Finally, we give the comparison results of MLR with the state-of-art genotype tagging of STAMPA [8], and haplotype tagging of Haldorsson et al. [7] and He et al. [10].

The following datasets are used to measure the quality of our algorithms. Currently, our algorithms cannot tolerate missing data. Following Halperin et al.[8], we use GERBIL [6] to phase the genotypes and then combine the resulting two haplotypes to recover any missing data.

**Three ENCODE regions.** Three regions (ENm013, ENr112, ENr113) from 30 CEPH family trios obtained from HapMap ENCODE Project [1]. The number of SNPs genotyped in each region is 361, 412 and 515 respectively.

**Two gene regions.** Two gene regions STEAP and TRPM8 from 30 CEPH family trios obtained from HapMap [1]. The number of SNPs genotyped in each gene region is 23 and 102 SNPs.

**Chromosome 5q31.** The data set collected by Daly et al. [5] derived from the 616 kilobase region of human Chromosome $5q31$ from 129 family trios.

**LPL & Chromosome 21.** The Clark et al. [3] data set consists of the haplotypes of 71 individuals typed over 88 SNPs in the human lipoprotein lipase (LPL) gene. The Chromosome 21 data set consists the first 1,000 of 24,047 SNPs typed over 20 haploid copies of human Chromosome 21 [11].

We apply leave-one-out cross-validation to evaluate the quality of the solution given by the tag SNP selection and prediction methods. One by one, each individual is removed from the sample. Then, tag SNPs are selected using only the remaining individuals. The "left out" individual is reconstructed based on its tag SNPs and the remaining individuals in the sample. The prediction accuracy is percent of correctly predicted SNPs in all individuals.

**Table 1.** The number of tag SNPs sufficient for MLR/STA to reach prediction accuracy between 80% and 99%. The *'s indicate cases when MLR/LMT needs one tag less than MLR/STA.

| | prediction accuracy, % | | | | | | | | | | | |
|---|---|---|---|---|---|---|---|---|---|---|---|---|
| Datasets | 80 | 85 | 90 | 91 | 92 | 93 | 94 | 95 | 96 | 97 | 98 | 99 |
| ENm013 | 2 | 3 | 6 | 6 | 7 | 8 | 9 | 9 | 11 | 15 | 22 | 254 |
| ENr112 | 6 | 9 | 14 | 16 | 18 | 20 | 24 | 33 | 63 | 95 | 126 | 187 |
| ENr113 | 4 | 5 | 10 | 11 | 13 | 15 | 18 | 40 | 55 | 80 | 104 | 200 |
| STEAP | 1 | 1 | 1 | 2 | 2 | 2 | 2 | 2 | 3* | 3 | 4* | 4 |
| TRPM8 | 1 | 2 | 4 | 5 | 5 | 6 | 7 | 8 | 10 | 15 | 15 | 24 |
| 5q31 | 1 | 2 | 5 | 7 | 7 | 9 | 13 | 16 | 21 | 31 | 41 | 55 |

**Table 2.** The number of tag SNPs used by MLR/STA and STAMPA to achieve 80% and 90% prediction accuracy in leave-one-out tests for 6 datasets

| Acc. | Algorithm | ENm013 | ENr112 | ENr113 | STEAP | TRPM8 | 5q31 |
|---|---|---|---|---|---|---|---|
| 80% | MLR | 2 | 6 | 4 | 1 | 1 | 1 |
| | STAMPA | 5 | 9 | 11 | 2 | 3 | 2 |
| 90% | MLR | 6 | 14 | 10 | 1 | 4 | 5 |
| | STAMPA | 12 | 17 | 18 | 2 | 6 | 6 |

Table 1 presents the results of leave-one-out experiments on the 6 genotype datasets. STA combined with MLR (MLR/STA) has almost the same quality as LMT combined with MLR (MLR/LMT) while being much faster. For example, when we perform the test on TRPM8 dataset, MLR/STA needs 4 seconds to reach 95% prediction accuracy while MLR/LMT needs 51 seconds.[1]

Table 2 compares the prediction accuracy of MLR/STA and STAMPA (Halperin et al. [8]). MLR/STA requires fewer tag SNPs and faster than STAMPA. For example, for TRPM8 data, MLR/STA in 6 seconds selects 12 tags to reach 96.6% accuracy, while STAMPA in 34 seconds selects 12 tags to reach only 94.2% accuracy. Since the MLR prediction algorithm can be applied to haplotypes, we compare it with the methods of Halldorson et al. [7] and He et al. [10] in leave-one-out tests on the LPL & Chromosome 21 datasets. Figure 2 shows that the MLR/STA considerably improves over the both methods.

For maximizing statistical covering, each (non-tag) SNP-column $s$ is predicted with the MLR prediction algorithm. We say that the SNP-column $s$ is counted as *statistically covered* if squared correlation $R^2$ between the predicted SNP-column $s'$ and given SNP-column $s$ is at least 0.8. In Table 3, the first two rows show the correlation of prediction accuracy and number of statistically covered SNPs. The third row shows that it is slightly better to use the correct objective (i.e., statistical covering) rather than prediction accuracy in order to maximize the number of statistically covered SNPs. Table 4 shows that MLR/STA uses on average 30% fewer tags than IdSelect [2] for statistical covering all SNPs.

[1] All experiments are performed on a computer with Intel Pentium 4, 3.06Ghz processor and 2 GB of RAM.

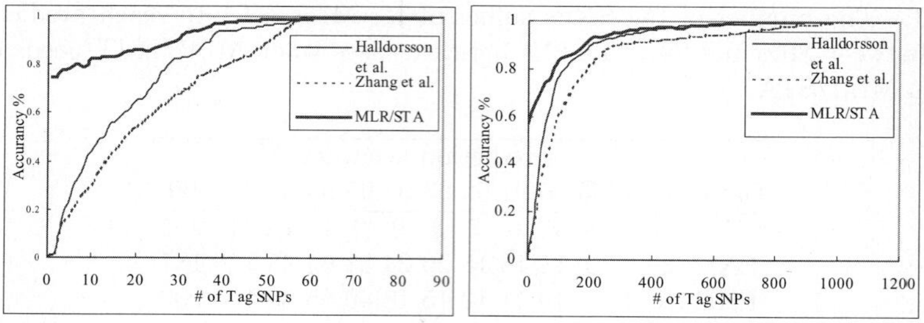


**Fig. 2.** Comparison of three haplotype prediction methods: Haldorsson et al. [7], He et al. [10], and MLR/STA. The x-axis shows the number of SNPs typed, and the y-axis shows the fraction of SNPs correctly imputed in a leave-one-out experiment. (A) Results from the LPL data set. (B) Results from the first 1000 SNPs of Chromosome 21 data set.

**Table 3.** The quality of MLR/STA on Daly et al. [5] data with two different tagging objectives over different number of tag SNPs

| objective of tagging | | number of tag SNPs 0 | 1 | 2 | 4 | 6 | 8 | 10 |
|---|---|---|---|---|---|---|---|---|
| SNP prediction | prediction accuracy, % | 61.54 | 81.35 | 83.94 | 88.65 | 91.11 | 92.96 | 93.89 |
| SNP prediction | # of SNPs covered | 0 | 10 | 16 | 36 | 47 | 53 | 59 |
| statistical covering | # of SNPs covered | 0 | 11 | 24 | 38 | 50 | 54 | 61 |

**Table 4.** The number of tag SNPs for statistical covering of all SNPs required by three methods: MLR/STA with prediction objective, MLR/STA with statistical covering objective, and IdSelect [2]

| Algorithm | ENm013 | ENr112 | ENr113 | STEAP | TRPM8 | 5q31 |
|---|---|---|---|---|---|---|
| MLR (prediction) | 56 | 82 | 106 | 13 | 46 | 44 |
| MLR (statistical covering) | 51 | 71 | 85 | 11 | 41 | 41 |
| IdSelect | 71 | 122 | 132 | 16 | 53 | 51 |

## 5 Conclusions

The paper gives a new SNP prediction method based on multivariate linear regression and suggests two universal tag selection methods. The MLR method combined with stepwise tag selection uses significantly fewer tags (e.g., up to two times less tags to reach 90% prediction accuracy) than the state-of-the-art methods of Halperin et al. [8] for genotypes and Halldorsson et al. [7] for haplotypes, respectively. Our stepwise tagging matches the quality of while being faster than STAMPA [8]. When applied to statistical covering, our method also uses fewer tags than the state-of-the-art method of Carlson et al. [2] especially when high correlation ($R^2 > .8$) is required.

## References


1. http://www.hapmap.org
2. Carlson, C.S., Eberle, M.A., Rieder, M.J., Yi, Q., Kruglyak, L. and Nickerson, D.A. (2004) 'Selecting a maximally informative set of single-nucleotide polymorphisms for association analyses using linkage disequilibrium', *American Journal of Human Genetics*, Vol. 74, No. 1, pp. 106–120.
3. Clark, A., Weiss, K., Nickerson, D., Taylor, S., Buchanan, A., Stengard, J., Salomaa, V., Vartiainen, E., Perola, M., Boerwinkle, E. (1998) 'Haplotype structure and population genetic inferences from nucleotide-sequence variation in human lipoprotein lipase', *American Journal of Human Genetics*, Vol. 63, pp. 595–612.
4. Chapman, J.M., Cooper, J.D., Todd, J.A. and Clayton, D.G. (2003). 'Detecting disease associations due to linkage disequilibrium using haplotype tags: a class of tests and the determinants of statistical power', Human Heredity, Vol. 56, pp. 18–31.
5. Daly, M., Rioux, J., Schaffner, S., Hudson, T. and Lander, E. (2001) 'High resolution haplotype structure in the human genome', *Nature Genetics*, Vol. 29, pp. 229–232.
6. Kimmel, G and Shamir R. (2004) 'GERBIL: Genotype resolution and block identification using likelihood', *PNAS* vol. 102, pp 158–162.
7. Halldorsson, B.V., Bafna, V., Lippert, R., Schwartz, R., de la Vega, F.M., Clark, A.G. and Istrail, S. (2004) 'Optimal haplotype block-free selection of tagging SNPs for genome-wide association studies', *Genome Research* Vol. 14, pp. 1633–1640.
8. Halperin, E., Kimmel, G. and Shamir, R. (2005) 'Tag SNP Selection in Genotype Data for Maximizing SNP Prediciton Accuracy', *Bioinformatics* 21:i195-i203;.
9. He, J. and Zelikovsky, A. (2004) 'Linear Reduction Methods for Tag SNP Selection', *Proceedings of the International Conference of the IEEE Engineering in Medicine and Biology*, pp. 2840–2843.
10. He, J. and Zelikovsky, A. (2005) 'Linear Reduction Method for Predictive and Informative Tag SNP Selection', *International Journal Bioinformtics Research and Applications*, Vol. 3, pp. 249-260.
11. Patil, N., Berno, A., Hinds, D., Barrett, W., Doshi, J., Hacker, C., Kautzer, C., Lee, D., Marjoribanks, C., McDonough, D., Nguyen, B., Norris, M., Sheehan, J., Shen, N., Stern, D., Stokowski, R., Thomas, D., Trulson, M., Vyas, K., Frazer, K., Fodor, S. and Cox, D. (2001) 'Blocks of limited haplotype diversity revealed by high-resolution scanning of human chromosome', *Science*, Vol. 294, pp. 1719–1723.
12. StatSoft, Inc. (1999). Electronic Statistics Textbook. Tulsa, OK: StatSoft. WEB: http://www.statsoft.com/textbook/stathome.html.
13. Stram, D., Haiman, C., Hirschhorn, J., Altshuler, D., Kolonel, L., Henderson, B. and Pike, M. (2003). 'Choosing haplotype-tagging SNPs based on unphased genotype data using as preliminary sample of unrelated subjects with an example from the multiethnic cohort study', *Human Heredity*, Vol. 55, pp. 27–36.
14. Zhang, K., Qin, Z., Liu, J., Chen, T., Waterman, M., and Sun, F. (2004) 'Haplotype block partitioning and tag SNP selection using genotype data and their applications to association studies', *Genome Research*, Vol. 14, pp. 908–916.

# Minimum Multicolored Subgraph Problem in Multiplex PCR Primer Set Selection and Population Haplotyping

M.T. Hajiaghayi[1], K. Jain[2], L.C. Lau[3], I.I. Măndoiu[4], A. Russell[4], and V.V. Vazirani[5]

[1] Laboratory for Computer Science, MIT
hajiagha@mit.edu
[2] Microsoft Research
kamalj@microsoft.com
[3] Department of Computer Science, University of Toronto
chi@cs.toronto.edu
[4] CSE Department, University of Connecticut
{ion, acr}@cse.uconn.edu
[5] College of Computing, Georgia Institute of Technology
vazirani@cc.gatech.edu

**Abstract.** In this paper we consider the minimum weight multicolored subgraph problem (MWMCSP), which is a common generalization of minimum cost multiplex PCR primer set selection and maximum likelihood population haplotyping. In this problem one is given an undirected graph $G$ with non-negative vertex weights and a color function that assigns to each edge one or more of $n$ given colors, and the goal is to find a minimum weight set of vertices inducing edges of all $n$ colors. We obtain improved approximation algorithms and hardness results for MWMCSP and its variant in which the goal is to find a minimum number of vertices inducing edges of at least $k$ colors for a given integer $k \leq n$.

## 1 Introduction

In this paper we consider the following *minimum weight multicolored subgraph problem* (MWMCSP): given an undirected graph $G$ with non-negative vertex weights and a color function that assigns to each edge one or more of $n$ given colors, find a minimum weight set of vertices of $G$ inducing edges of all $n$ colors. We also consider the generalization of MWMCSP in which one seeks a minimum weight set of vertices inducing edges of at least $k$ colors for a given integer $k \leq n$, referred to as the *minimum weight $k$-colored subgraph problem* (MW$k$CSP), and the unweighted versions of MWMCSP and MW$k$CSP, denoted MMCSP and M$k$CSP, respectively. As detailed below, MWMCSP and its variants model two important bioinformatics problems: minimum cost multiplex PCR primer set selection and maximum likelihood population haplotyping.

V.N. Alexandrov et al. (Eds.): ICCS 2006, Part II, LNCS 3992, pp. 758–766, 2006.

### 1.1 Primer Set Selection for DNA Amplification by PCR

A critical step in many high-throughput genomic assays is the cost-effective amplification of DNA sequences containing loci of interest via biochemical reactions such as the *Polymerase Chain Reaction* (PCR). In its basic form, PCR requires a pair of short single-stranded DNA sequences, referred to as *PCR primers*, flanking the amplification locus on the two strands of the template. In *multiplex PCR*, multiple genomic loci are amplified simultaneously (and a primer may simultaneously participate in multiple amplifications). In addition to constraints on individual primer properties that affect reaction efficiency, such as primer melting temperature and lack of secondary structure, multiplex PCR primer set selection must ensure various *pairwise* compatibility constraints between selected primers. Since the efficiency of PCR amplification falls off exponentially as the length of the amplification product increases, an important practical constraint is that the two primer sites defining a product must be within a certain maximum distance $L$ of each other. In applications such as spotted microarray synthesis [1] a further pairwise compatibility constraint is the requirement of unique amplification: for every desired amplification locus there should be a pair of primers that amplifies a DNA fragment surrounding it but no other fragment.

Subject to these constraints, one would like to minimize the total cost of the primer set required to amplify the $n$ given loci. As noted by Fernandes and Skiena [1], primer selection problem subject to pairwise compatibility constraints can be easily reduced to M(W)MCSP: each candidate primer becomes a graph vertex and each pair of primers that feasibly amplifies a desired locus becomes an edge colored by the respective locus number. More generally, the problem of selecting the minimum size/cost set of primers required to amplify at least $k$ of the $n$ loci reduces to M(W)$k$CSP; this problem arises when several multiplex reactions are required to amplify the given loci.

### 1.2 Maximum Likelihood Population Haplotyping

The most common form of genomic variation between individuals, is the presence of different DNA nucleotides, or *alleles*, at certain chromosomal locations, commonly referred to as *single nucleotide polymorphisms* (SNPs). For diploid organisms such as humans, the combinations of SNP alleles in the maternal and paternal chromosomes of an individual are referred to as the individual's *haplotypes*. Finding the haplotypes in human populations is an important step in determining the genetic basis of complex diseases [2].

With current technologies, it is prohibitively expensive to directly determine the haplotypes of an individual, but it is possible to obtain rather easily the conflated SNP information in the so called *genotype*. The *population haplotyping problem* (PHP) seeks to infer the set of haplotypes explaining the genotypes observed in a large population. Formally, a haplotype is represented as a 0/1 vector – e.g., by representing the most frequent SNP allele as a 0 and the alternate allele as a 1 – while a genotype is a 0/1/2 vector, where 0 (1) means that both chromosomes contain the respective SNP allele and 2 means that the two chromosomes contain different SNP alleles. We say that a set $\mathcal{H}$ of haplotypes *explains* a given set $\mathcal{G}$ of genotypes if, for every $g \in \mathcal{G}$, there exist $h, h' \in \mathcal{H}$ with $h + h' = g$, where $h + h'$ is the vector whose $i$-th component is equal to 2 when $h_i \neq h'_i$, and to the common value of $h_i$ and $h'_i$ when $h_i = h'_i$.

Several optimization objectives have been considered for PHP and the related *genotype phasing* problem, which seeks a pair of haplotypes explaining each of the given genotypes – see, e.g., [3, 4, 5] for recent surveys. In the *maximum likelihood* approach to PHP, one assumes an a priori probability $p_h$ for every possible haplotype $h$ (inferred, e.g., from genotype frequencies [6]), and seeks the *most likely* set $\mathcal{H}$ of haplotypes explaining the observed genotypes, where the likelihood of a set $\mathcal{H}$ is given by $L(\mathcal{H}) = \prod_{h \in \mathcal{H}} p_h$. In the special case when all a priori haplotype probabilities are equal, likelihood maximization recovers the *maximum parsimony* approach to PHP [7, 8], in which one seeks the *smallest* set $\mathcal{H}$ of haplotypes explaining $\mathcal{G}$.

The maximum likelihood PHP can be reduced to MWMCSP by associating a vertex of weight $-\log p_h$ to each candidate haplotype $h$, and adding an edge $(h, h')$ colored by $h+h'$ whenever $h+h'$ is one of the given genotypes. Maximum parsimony PHP reduces to MMCSP in a similar way. Notice that in resulting M(W)MCSP instances each edge is assigned at most one color (in fact, color classes form a matching in the underlying graph). This property is no longer true for the more general versions of PHP in which the input contains missing data, i.e., when the input consists of *partial genotypes* which are vectors over the alphabet $\{0, 1, 2, *\}$, and the goal is to resolve each "$*$" symbols into a 0, 1, or a 2, and find a most likely/smallest set of haplotypes that explain the resolved genotypes. We also remark that the reductions of PHP to M(W)MCSP are not polynomial, as the number of haplotypes compatible with the given genotypes may be exponential. Nevertheless, in practice the reductions yield instances of manageable size [7].

### 1.3 Previous Work

Gusfield [7] proposed an (exponential size) integer program formulation for the maximum parsimony PHP. He reports that the commercial integer programming solver CPLEX finds optimal solutions in practical running time for instances with up to 150 genotypes and up to 100 SNPs. For the same problem, Wang and Xu [9] proposed a greedy heuristic and an exact branch and bound algorithm. Lancia et al. [8] proved that maximum parsimony PHP is APX-hard, and gave two straightforward algorithms with approximation factors of $\sqrt{n}$ and $q$, where $n$ is the number of genotypes and $q$ is the maximum number of haplotype pairs compatible with a genotype. These results immediately imply APX-hardness of M(W)MCSP and M(W)$k$CSP (even when only one color can be assigned to each edge), and can be shown to yield approximation factors of $\sqrt{n}$ and $\mathfrak{m}$ for MMCSP with one color per edge and MWMCSP, respectively, where $\mathfrak{m}$ is the maximum size of a color class (i.e., the maximum number of edges sharing the same color).

Brown and Harrower [10] and Lancia et al. [11] independently proposed polynomial size integer programs for maximum parsimony PHP. Although these formulations are more compact than the one proposed by Gusfield [7], experimental results in [10] indicate that they often take longer to solve for instances of practical interest, even when augmented with sophisticated sets of valid constraints. This may be explained by the fact that there is no known integrality gap for the formulations in [11] and [10], whereas the results in Section 4, imply an integrality gap of $O(\sqrt{q}\log n)$ for Gusfield's formulation. The formulations in [10] and [11] do not seem to extend to the maximum likelihood PHP problem.

Fernandes and Skiena [1] studied MMCSP with at most one color per edge in the context of multi-use primer selection for synthesis of spotted microarrays. They gave practical greedy and densest-subgraph based heuristics for the problem and proved, by a direct reduction from set cover, that even this special case of M$k$CSP cannot be approximated within a factor better than $(1-o(1))\ln n - o(1)$, where $n$ is the number of colors. Konwar et al. [12] introduced a string-pair covering formulation for multiplex PCR primer set selection when there are only amplification length constraints, and proved that in this special case a modification of the classical greedy algorithm for set cover gives an approximation factor of $1 + \ln(nL)$, where $L$ is the upperbound on the amplification length. The algorithm in [12] cannot enforce arbitrary pairwise compatibility constraints, such as ensuring amplification uniqueness.

Very recently, Hassin and Segev [13] showed that a suitable adaptation of the greedy set cover algorithm yields an approximation factor of $O(\sqrt{n}\log n)$ for the MMCSP, and Huang et al. [14] gave a factor $O(\log n)$ approximation algorithm for maximum parsimony PHP based on semidefinite programming.

### 1.4 Our Results and Techniques

In this paper we give several approximation algorithms and hardness results for MWMCSP and its variants. Unlike approximation factors in [8, 13, 14], our results hold for the weighted version of the problem and do not require the assumption that edges belong to a single color class. Our contributions are as follows:

- First, in Section 2, we present a $\sqrt{k(1+\ln\Delta)}$ approximation algorithm for M$k$CSP using an algorithm of Slavik [15] for the partial set cover problem. Here $\Delta$ is the maximum number of colors assigned to an edge.
- Then, in Section 3, we present evidence of potential polynomial inapproximability for M$k$CSP problem by showing a novel reduction from the densest $k$-subgraph maximization problem to our minimization problem. We believe that our approach can serve as a general technique to reduce hardness from other budgeted graph-theoretic maximization problems to the corresponding minimization problems.
- Finally, in Section 4, we give an $O(\sqrt{\mathfrak{m}}\log n)$ approximation algorithm for MWMCSP, where $\mathfrak{m}$ is the maximum size of a color class and $n$ is the number of colors. For PCR primer set selection with arbitrary pairwise compatibility

constraints in addition to amplification length constraints, $\mathfrak{m} = O(L^2)$. Thus, reduction to MWMCSP gives an approximation factor of $O(L \log n)$. For maximum likelihood PHP, $\mathfrak{m} = O(2^t)$, where $t$ is the maximum number of 2's in a genotype. Thus, our algorithm yields an approximation factor of $O(2^{t/2}\sqrt{\log n})$ in this case. Our approximation algorithm for MWMCSP is based on LP-rounding, and we show that the approximation factor is almost tight by showing a matching (up to the logarithmic factor) integrality gap for the underlying linear program.

## 2 Approximation Algorithm for M$k$CSP

Notice that arbitrarily picking a set of $k$ color classes and an arbitrary edge from each color class yields a factor $O(\sqrt{k\Delta})$ approximation for M$k$CSP, where $\Delta$ denotes the maximum number of colors that can be assigned to an edge. The following theorem gives an improved approximation algorithm.

**Theorem 1.** *There exists an approximation algorithm with factor* $\sqrt{2kH(\Delta)} = O(\sqrt{k(1+\ln \Delta)})$ *for MkCSP, where* $H(\Delta) = 1 + \frac{1}{2} + \ldots + \frac{1}{\Delta}$.

*Proof.* The algorithm is as follows. Let $X$ be the set of selected vertices; initially empty. While the number of colors covered is less than $k$, we choose an edge with maximum number of uncovered colors and add both of its endpoints to $X$ (if they are not already in $X$). Let $i$ be the number of edges that we choose in this process, clearly $i \leq k$. We know that $|X| \leq 2i$. On the other hand, by a result of Slavik [15], we know that the above greedy algorithm for the partial set cover problem, i.e., finding the minimum number of sets to cover at least $k$ elements, is an $H(\Delta)$ approximation algorithm. This means that the minimum number of edges needed to cover at least $k$ colors is at least $i/H(\Delta)$. It is easy to see that, in order to induce at least $i/H(\Delta)$ edges, the optimum M$k$CSP solution should pick at least $\sqrt{2i/H(\Delta)}$ vertices. The approximation factor follows immediately by using this lower bound.

**Remark.** For the case when $k = n$ and $\Delta = 1$, i.e., for MMCSP with one color per edge, the above algorithm corresponds to the $\sqrt{k}$-approximation algorithm of [8]. It is also worth mentioning that using the approximation algorithm of Gandhi, Khuller and Srinivasan [16] for partial set cover in the proof of Theorem 1, we can obtain an $\sqrt{2k\mathfrak{m}}$ approximation algorithm for M$k$CSP, where $\mathfrak{m}$ is the maximum number of edges sharing the same color. The reduction established in next section suggests that the approximation factor for M$k$CSP cannot be easily improved.

## 3 Hardness Result for M$k$CSP

In this section, we show an interesting relation between the approximability of M$k$CSP and that of the densest $k$-subgraph problem. Formally, we show that if there is a polynomial time $f$-approximation algorithm $\mathcal{A}$ for M$k$CSP, then there is a polynomial time $2f^2$-approximation algorithm for the densest $k$-subgraph problem. Given a graph $G$ and a parameter $k$, the densest $k$-subgraph problem is to find a set of $k$ vertices with maximum number of induced edges. The densest $k$-subgraph problem is well-studied in the literature [17, 18]. The best known approximation factor for the densest $k$-subgraph problem is $O(\min\{n^\delta, n/k\})$ for some $\delta < 1/3$ and improvement seems to be hard [19, 18]. The connection between M$k$CSP and the densest $k$-subgraph problem suggests that significant improvements in the approximation ratio for M$k$CSP would require substantially new ideas.

**Theorem 2.** *If there is a polynomial time $f$-approximation algorithm $\mathcal{A}$ for M$k$CSP, then there is a polynomial time $2f^2$-approximation algorithm for the densest $k$-subgraph problem.*

*Proof.* Given a graph $G$ with $m$ edges, we would like to find a set of $k$ vertices with maximum number of edges in the subgraph induced by this set. We assign to each edge of $G$ a different color and use $\mathcal{A}$ to find the approximate solutions for M$k$CSP on the resulting graph. Suppose $l$ is the maximum color coverage requirement for which $\mathcal{A}$ outputs a solution $Y$ with at most $k$ vertices. That is, there are $l$ colors assigned to the subgraph induced by $Y$, and the approximate solution returned by $\mathcal{A}$ when $l+1$ colors are required to be covered contains at least $k+1$ vertices. Let the optimal solution to the densest $k$-subgraph problem contain *opt* edges. We shall prove that $opt \le 2f^2 l$ and thus $Y$ is a solution to the densest $k$-subgraph problem which is within a factor of $\frac{1}{2f^2}$ to the optimal solution.

By our choice of $l$ and the fact that $\mathcal{A}$ is an $f$-approximation algorithm, any $\lfloor \frac{k}{f} \rfloor$ vertices of $G$ can induce at most $l$ colors. Consider a subset $X$ with $k$ vertices. The total number of colors induced by all possible subsets of $\lfloor \frac{k}{f} \rfloor$ elements of $X$ is at most $\binom{k}{\lfloor \frac{k}{f} \rfloor} l$. Notice that each edge is counted exactly $\binom{k-2}{\lfloor \frac{k}{f} \rfloor - 2}$ times. So, the total number of edges in $X$ is at most

$$\frac{\binom{k}{\lfloor \frac{k}{f} \rfloor} l}{\binom{k-2}{\lfloor \frac{k}{f} \rfloor - 2}} = \frac{k(k-1)}{\lfloor \frac{k}{f} \rfloor \lfloor (\frac{k}{f} - 1) \rfloor} l \le f^2 l \frac{k(k-1)}{(k-f)(k-2f)} < 2f^2 l.$$

The last inequality holds since we can assume without loss of generality that $k > 4f^2$ (otherwise, any connected subgraph on $k$ vertices is a $2f^2$-approximation), and also that $k$ is a constant such that $\frac{k(k-1)}{(k-f)(k-2f)} < 2$. Since $X$ is an arbitrary set with $k$ vertices, $opt \le 2f^2 l$ and this completes the proof.

## 4 LP-Rounding Based Approximation for MWMCSP

Let $G = (V, E)$ be the input graph and $\mathcal{X} = (\chi_1, \ldots, \chi_n)$ be the family of nonempty "color classes" of edges (without loss of generality we assume that $\bigcup_i \chi_i = E$). We use the following integer program formulation of MWMCSP:

$$\min \sum_v w_v x_v, \text{ subject to } \quad \begin{array}{l} \forall \chi \in \mathcal{X}, \sum_{e \in \chi} y_e \geq 1 \ , \\ \forall v \in V, \forall \chi \in \mathcal{X}, \sum_{v \in e \in \chi} y_e \leq x_v \ , \\ \forall e \in E, y_e \geq 0, \forall v \in V, x_v \geq 0 \ . \end{array}$$

Here the $x_v$ and $y_e$ are variables associated with the vertices and edges of the graph, and the $w_v$ denote the positive weights given in the problem instance. Our formulation is related to that introduced by Gusfield [7] for maximum parsimony PHP. Gusfield's formulation lacks weights, and replaces our second set of constraints by the simpler requirement that $y_e \leq x_v$ for every edge $e$ incident to a vertex $v$. The two sets of constraints are identical for MMCSP instances obtained by reduction from maximum parsimony PHP, since color classes are independent sets of edges in this case. However, using the stronger set of constraints is essential in establishing our approximation guarantee for arbitrary M(W)MCSP instances; a simple example shows that in this case the integrality gap with Gusfield's constraints is $\Omega(\mathfrak{m})$. Due to space constraints we omit the proof of the following two theorems:

**Theorem 3.** *There is an LP-rounding approximation algorithm for MWMCSP with approximation factor of* $O(\sqrt{\mathfrak{m} \log |\mathcal{X}|})$, *where* $\mathfrak{m} = \max_{\chi \in \mathcal{X}} \mid \chi \mid$.

**Theorem 4.** *The linear relaxation of the above MWMCSP integer program has an integrality gap of* $\Omega(\sqrt{\mathfrak{m}})$.

Theorem 4 suggests that the linear relaxation of the MWMCSP integer program may have limited value in achieving approximation results beyond the $\sqrt{\mathfrak{m}}$ threshold. It is worth mentioning that the integrality gap in Theorem 4 holds for Gusfield's maximum parsimony PHP formulation [7] as well. As mentioned in Subsection 1.2, in this case, the graph is more restricted, that is, each vertex is a 0/1 vector and each edge between vertices $h$ and $h'$ has a unique color $h+h'$ (which is a 0/1/2 vector). Still we can construct such a restricted graph which shows the integrality gap is the same as that of Theorem 4.

## 5 Conclusions

In this paper we have proposed the first non-trivial approximation and inapproximability results for the MWMCSP problem and several of its variants capturing important applications in computational biology. Interesting open problems include closing the gap between approximation guarantees and inapproximability results for MWMCSP, and obtaining non-trivial approximations for MW$k$CSP (an approximation factor of $k$ is obtained, e.g., by picking the lightest edge of each color.) An important constraint on the primers for multiplex PCR not modeled by MWMCSP is that they shouldn't cross-hybridize. This motivates studying the variant of MWMCSP in which certain edges are marked as "forbidden", and the goal is to find a minimum multicolored induced subgraph with no forbidden edges.

## Acknowledgments

IIM's work was supported in part by NSF CAREER Award IIS-0546457 and a Faculty Large Research Grant from the University of Connecticut Research Foundation.

## References

1. Fernandes, R., Skiena, S.: Microarray synthesis through multiple-use PCR primer design. Bioinformatics **18** (2002) S128–S135
2. Clark, A.: The role of haplotypes in candidate gene studies. Genet. Epid. **27** (2004) 321–333
3. Bonizzoni, P., Vedova, G.D., Dondi, R., Li, J.: The haplotyping problem: An overview of computational models and solutions. Journal of Computer Science and Technology **18** (2003) 675–688
4. Halldorsson, B., Bafna, V., Edwards, N., Lippert, R., Yooseph, S., Istrail, S.: A survey of computational methods for determining haplotypes. In: Proc. of the DIMACS/RECOMB Satellite Workshop on Computational Methods for SNPs and Haplotype Inference (2004) 26–47
5. Niu, T.: Algorithms for inferring haplotypes. Genet. Epid. **27** (2004) 334–347
6. Halperin, E., Hazan, E.: HAPLOFREQ - estimating haplotype frequencies efficiently. In: Proc. 9th Annual International Conference on Research in Computational Molecular Biology (2005) 553–568
7. Gusfield, D.: Haplotyping by pure parsimony. In: Proc. 14th Annual Symp. on Combinatorial Pattern Matching (2003) 144–155
8. Lancia, G., Pinotti, C., Rizzi, R.: Haplotyping populations: complexity and approximations. Technical Report DIT-02-0080, University of Trento (2002)
9. Wang, L., Xu, Y.: Haplotype inference by maximum parsimony. Bioinformatics **19** (2003) 1773–1780
10. Brown, D., Harrower, I.: A New Integer Programming Formulation for the Pure Parsimony Problem in Haplotype Analysis. In: Proc. 4th International Workshop on Algorithms in Bioinformatics (2004) 254–265
11. Lancia, G., Pinotti, M., Rizzi, R.: Haplotyping populations by pure parsimony: Complexity of exact and approximation algorithms. INFORMS Journal on Computing **16** (2004) 348–359
12. Konwar, K., Măndoiu, I., Russell, A., Shvartsman, A.: Improved algorithms for multiplex PCR primer set selection with amplification length constraints. In: Proc. 3rd Asia-Pacific Bioinformatics Conference (2005) 41–50
13. Hassin, R., Segev, D.: The set cover with pairs problem. In: Proc. 25th Annual Conference on Foundations of Software Technology and Theoretical Computer Science (2005) 164–176
14. Huang, Y.T., Chao, K.M., Chen, T.: An approximation algorithm for haplotype inference by maximum parsimony. Journal of Computational Biology **12** (2005) 1261–1274
15. Slavik, P.: Improved performance of the greedy algorithm for partial cover. Information Processing Letters **64** (1997) 251–254
16. Gandhi, R., Khuller, S., Srinivasan, A.: Approximation algorithms for partial covering problems. Journal of Algorithms **53** (2004) 55-84

17. Feige, U., Kortsarz, G., Peleg, D.: The dense $k$-subgraph problem. Algorithmica **29**(3) (2001) 410–421
18. Khot, S.: Ruling out PTAS for graph min-bisection, densest subgraph and bipartite clique. In: Proc. 45th Annual IEEE Symposium on Foundations of Computer Science (2004) 136-145
19. Feige, U.: Relations between average case complexity and approximation complexity. In: Proc. 34th Annual ACM Symposium on Theory of Computing (2002) 534–543

# Phasing of 2-SNP Genotypes Based on Non-random Mating Model

Dumitru Brinza* and Alexander Zelikovsky**

Department of Computer Science, Georgia State University, Atlanta, GA 30303
{dima, alexz}@cs.gsu.edu

**Abstract.** Emerging microarray technologies allow genotyping of long genome sequences resulting in huge amount of data. A key challenge is to provide an accurate phasing of very long single nucleotide polymorphism (SNP) sequences. In this paper we explore phasing of genotypes with 2 SNPs adjusted to the non-random mating model and then apply it to the haplotype inference of complete genotypes using maximum spanning trees. The runtime of the algorithm is $O(nm(n+m))$, where $n$ and $m$ are the number of genotypes and SNPs, respectively. The proposed phasing algorithm (2SNP) can be used for comparatively accurate phasing of large number of very long genome sequences. On datasets across 79 regions from HapMap[7] 2SNP is several orders of magnitude faster than GERBIL and PHASE while matching them in quality measured by the number of correctly phased genotypes, single-site and switching errors. For example, 2SNP requires 41 s on Pentium 4 2Ghz processor to phase 30 genotypes with 1381 SNPs (ENm010.7p15:2 data from HapMap) versus GERBIL and PHASE requiring more than a week of runtime and admitting no less errors than 2SNP. 2SNP software is publicly available at http://alla.cs.gsu.edu/~software/2SNP.

## 1 Introduction

The difference between individual DNA sequences mostly occurs at a single-base site, in which more than one nucleic acid or gap is observed across the population. Such variations are called single nucleotide polymorphisms (SNPs). The number of sufficiently frequent SNPs in the human population is estimated to be around 10 million [11]. For complex diseases caused by more than a single gene it is important to identify a set of alleles inherited together. Identification of haplotypes, the sequences of alleles in contiguous SNP sites along a chromosomal region, is a central challenge of the International HapMap project [7]. The number of simultaneously typed SNPs for association and linkage studies is reaching 250,000 for SNP Mapping Arrays (Affymetrix).

Diploid organisms, like human, have two near-identical copies of each chromosome. Most experimental techniques for determining SNPs do not provide

* Partially supported by GSU Molecular Basis of Disease Fellowship.
** Partially supported by NIH Award 1 P20 GM065762-01A1 and US CRDF Award MOM2-3049-CS-03.

V.N. Alexandrov et al. (Eds.): ICCS 2006, Part II, LNCS 3992, pp. 767–774, 2006.

the haplotype information separately for each of the two chromosomes. Instead, they generate for each site an unordered pair of allele readings, one from each copy of the chromosome,which is called a genotype.

The input to the phasing problem consists of $n$ genotype vectors each with $m$ coordinates corresponding to SNPs.The phasing problem asks for explaining each genotype with two haplotypes corresponding to chromosomes. In general as well as in common biological setting, there are $2^{k-1}$ of possible haplotype pairs for the same input genotype with $k$ heterozygous sites.

Computational inferring of haplotypes from the genotypes (or *phasing*) has been initiated by Clark[2] who proposed a parsimony-based approach. It has been shown later that the likelihood based expectation-maximization (EM) is more accurate [13]. Markov chain Bayesian haplotype reconstruction methods have been used in PHASE [14] and HAPLOTYPER [12]. A combinatorial model based on the perfect phylogeny tree assumption was suggested in [5]. HAP [6] exploits perfect phylogeny model and block structure showing good performance on real genotypes with low error rates. Recently, GERBIL [10] has combined block identification and phasing steps for reliable phasing of long genotypes.

In this paper we first explore phasing of genotypes with 2 SNPs which have ambiguity when the both sites are heterozygous. Then there are two possible phasing and the phasing problem is reduced to inferring their frequencies. Given phasing solution for 2-SNP genotypes, complete haplotypes for a given genotype can be inferred based on the maximum spanning tree of a complete graph with vertices corresponding to heterozygous sites and edge weights given by inferred 2-SNP frequencies. In tests on real datasets across 79 different genomic regions from HapMap [7] 2SNP was several orders of magnitude faster than GERBIL and PHASE while matching them in quality. We also found that HAPLOTYPER is almost as fast as 2SNP and close in quality, but currently does not handle populations with more than 500 SNPs. A brief description of 2SNP software can be found in the application note [1].

The rest of the paper is organized as follows. The next section describes proposed 2-SNP genotypes phasing. Section 3 describes the phasing of long genotypes using 2-SNP genotypes phasing. Section 4 describes tested datasets and defines phasing accuracy measures and compares the quality and runtime of PHASE, GERBIL, HAPLOTYPER and 2SNP algorithms.

## 2 Phasing of 2-SNP Genotypes

In this section we first formally introduce the phasing problem and suggest a LD-based formula for the expected frequencies of cis- or trans-phasing of 2-SNP genotypes. We conclude with adjusting of expected haplotype frequencies to deviation from the random mating model.

The input to the phasing problem consists of $n$ genotype vectors each with $m$ coordinates corresponding to SNPs. SNP values belong to $\{0, 1, 2, ?\}$, where 0's and 1's denote homozygous sites with major allele and minor allele, respectively; 2's stand for heterozygous sites, and ?'s denote missed SNP values. Phasing replaces each genotype vector by two haplotype vectors with SNP values in $\{0,1\}$ such that any genotype 0-SNP (resp. 1-SNP or 2-SNP) is replaced with two haplotype 0-SNPs (resp. two 1-SNPs or 0-SNP and 1-SNP). A 2-SNP genotype 22 can be cis-phased, i.e., represented as 00 and 11 haplotypes, or trans-phased, i.e. represented as 01 and 10 haplotypes.

**Certainty of cis- or trans- phasing.**It is natural for the certainty of cis- or trans- phasing of homozygous SNPs $i$ and $j$ to take in account odds ratio of phasing being cis- or trans-, which is $\lambda = \frac{F_{00}\times F_{11}}{F_{01}\times F_{10}}$ where $F_{00}, F_{01}, F_{10}, F_{11}$ are unknown true frequencies of haplotypes with the first and the second binary index denoting alleles of the $i$-th and $j$-th SNP, respectively. In our experiments, we have noticed that the modified odds ratio $\lambda' = \frac{F_{00}+F_{11}}{F_{01}+F_{10}}$ better describes real cis- or trans-phasing. We measure LD (linkage disequilibrium) between endpoints by the ratio of the modified odds ratio $\lambda'$ over the expected value of $\lambda'$ computed in assumption of no linkage between endpoints, $LD_{ij} = \frac{\lambda'}{exp(\lambda')}$. Finally, it has been observed[12, 13, 10] the higher LD between pairs of closer SNPs. In order to discard falsely encountered LD between non-linked SNPs which are far apart, we divide LD by the square of the distance between the SNPs obtaining $c_{ij} = \frac{\log LD_{ij}}{(i-j)^2}$. The complete formula for the certainty of cis- or trans-phasing of two homozygous SNPs $i$ and $j$ is logarithm of linkage disequilibrium for the cis-/trans- odds ratio divided by squared distance between corresponding SNPs.

$$c_{ij} = \log\left(\frac{n + (F_{00}F_{11} - F_{01}F_{10})/(F_{01} + F_{01})}{n - (F_{00}F_{11} - F_{01}F_{10})/(F_{00} + F_{11})}\right) / (i-j)^2 \quad (1)$$

where $n$ is number of input genotypes, and $F_{00}, F_{01}, F_{10}, F_{11}$ are frequencies of haplotypes with the first and the second binary index denoting alleles of the $i$-th and $j$-th SNP, respectively. Haplotype frequencies are computed based on all genotype frequencies except 22. For 22 genotypes, the haplotype frequencies are chosen to fit best Hardy-Weinberg equilibrium adjusted to observed deviation in single-site genotype distribution.

**Adjusting observed frequencies to deviations from the random mating model.** The certainty formula (1) cannot be directly used in phasing since the participating haplotype frequencies are *true* frequencies and, therefore, are unknown. We have access only to *observed* haplotype frequencies which can be extracted from all types of 2-SNP genotypes except 22, genotypes heterozygous in the both SNPs.

The distribution of cis- and trans-phasing of 22-genotypes can be adjusted to the unknown mating model as follows. Let $C_{22}$ and $P_{22}$ denote unknown numbers of tarns- and cis- phasings, then $C_{22} + P_{22} = G_{22}$, where $G_{22}$ is the observed number of 22-genotypes. Then the adjusted odds ratio $\lambda'$ can be expressed as $\lambda' = \frac{F^*_{00}+F^*_{11}+P_{22}}{F^*_{01}+F^*_{10}+C_{22}}$ where $F^*_{ij}$ denote haplotype frequencies observed from all genotypes except 22. The best-fit values of $C_{22}$ and $P_{22}$ should minimize the sum of differences between expected and observed genotype frequencies. It is easy to compute the expected genotype frequencies for the random mating model, which unfortunately can significantly deviate from the real mating. Unknown for pairs of SNPs, such deviation can be accurately measured for individual SNPs. In our algorithm, we assume that the deviation for pairs of SNPs is similar to the deviation observed for the corresponding single SNP. Therefore, the expected 2-SNP frequencies are adjusted proportionally to the observed single-site deviation.

## 3 Phasing of Complete Genotypes

Below we describe phasing of long genotype using certainty of cis- or trans-phasing between any pairs of 2's computed in the previous section. We then explain how we resolve missing data recovery and conclude with runtime analysis of the 2SNP algorithm.

**Genotype Graph.** For each genotype $g$, 2SNP constructs a *genotype graph*, which is a complete graph with vertices corresponding to 2's (i.e., heterozygous sites) of $g$. The weight of the edge between heterozygous sites $i$ and $j$ represents the certainty (formula 1) in that $i$ and $j$ are cis- or trans-phased. The maximum spanning tree of the genotype graph uniquely determines the phasing of the corresponding genotype since it gives cis-/trans- phasing for any two 2's. Obviously, if for any pair of 2's we know if they are cis- or trans-phased, then the entire phasing is known. Note that [10] have applied the same construction for preliminary estimation of haplotype frequencies rather than phasing *per se*. Therefore, for the edge weight, they have chosen LD-based formula over probabilities of *full* ($i$-$j$)-haplotypes given by maximum-likelihood solution. Instead, edge weights in 2SNP do not account for SNPs between $i$ and $j$.

**Missing Data Recovery.** Missing data (?'s) are recovered after phasing of 2's. For each haplotype $h$ we find the closest (w.r.t. Hamming distance) haplotype(s) $h'$ and recover ?'s in $h$ with the corresponding values from $h'$.

**Runtime 2SNP Algorithm.** The runtime of 2SNP algorithm has two bottlenecks. The first is computing of the observed haplotype frequencies for each pair of SNPs, which takes $O(nm^2)$ since we have $n$ genotypes each with $m$ SNPs. The second is recovering of the missing data, which needs $O(n^2m)$ runtime since it results in computing all pairwise Hamming distances between $2n$ haplotypes each with $m$ SNPs. As a result, the total runtime of the algorithm is $O(nm(n+m))$, where $n$ and $m$ are the number of genotypes and SNPs, respectively.

## 4 Results

In this section we first describe the datasets and quality measures. Then, we compare our 2SNP method with PHASE-2.1.1[14], HAPLOTYPER[12] and GERBIL[10].

**Data Sets.** The comparison of phasing methods were performed on 46 real datasets from 79 different genomic regions and on 4 simulated datasets. All real datasets represent family trios – the computationally inferred offspring haplotypes for offspring have been compared with haplotypes inferred from parental genotypes.

**Table 1.** Mean single-site, individual, and switching errors with 95% confidence intervals and runtimes of PHASE, GERBIL, HAPLOTYPER (HTYPER), and 2SNP on real datasets

| Data | Measure | PHASE | GERBIL | HTYPER | 2SNP |
|---|---|---|---|---|---|
| 5q31, Daly et al.[3] | single-site | 1.9 ± 0.3 | 1.8 ± 0.3 | 2.2 ± 0.4 | 1.5 ± 0.3 |
| # genotypes = 129 | individual | 25.9 ± 4.4 | 30.4 ± 5.2 | 29.2 ± 4.9 | 25.8 ± 4.5 |
| # SNPs = 103 | switching | $3.5^{*}$ ± 0.1 | 3.3 ± 0.1 | 4.0 ± 0.1 | 3.0 ± 0.1 |
| | runtime | $1.4 \times 10^4$ | $1.0 \times 10^2$ | $1.2 \times 10^2$ | $2.0 \times 10^0$ |
| Gabriel et al.[4] | single-site | 2.7 ± 0.6 | 3.3 ± 0.7 | 4.4 ± 1.0 | 3.6 ± 0.8 |
| average on 51bk | individual | 26.6 ± 5.5 | 32.2 ± 6.7 | 34.7 ± 6.9 | 32.9 ± 7.0 |
| # genotypes = 29 | switching | 13.8 ± 2.9 | 15.5 ± 3.2 | 23.0 ± 3.3 | 16.1 ± 3.4 |
| # SNPs = 50 | runtime | $5.6 \times 10^2$ | $8.0 \times 10^0$ | $8.0 \times 10^0$ | $8.0 \times 10^{-2}$ |
| ENm010.7p15:2[7] | single-site | 5.8 ± 0.4 | 6.1 ± 0.3 | – | 5.6 ± 0.2 |
| CEU | individual | 50.0 ± 2.2 | 50.0 ± 2.5 | – | 50.0 ± 2.7 |
| # genotypes = 30 | switching | 13.6 ± 0.2 | 11.9 ± 0.1 | – | 10.9 ± 0.2 |
| # SNPs = 1381 | runtime | $1.1 \times 10^6$ | $3.7 \times 10^5$ | – | $4.1 \times 10^1$ |
| ENm010.7p15:2[7] | single-site | 0.4 ± 0.1 | 0.5 ± 0.1 | 0.6 ± 0.1 | 0.4 ± 0.1 |
| CEU-nonrs | individual | 40.0 ± 2.1 | 36.6 ± 1.1 | 43.3 ± 0.7 | 36.6 ± 1.8 |
| # genotypes = 30 | switching | 19.2 ± 0.5 | 21.0 ± 0.6 | 28.7 ± 0.7 | 16.8 ± 0.6 |
| # SNPs = 459 | runtime | $1.3 \times 10^4$ | $5.0 \times 10^1$ | $1.0 \times 10^1$ | $4.0 \times 10^0$ |
| ENm010.7p15:2[7] | single-site | 5.7 ± 0.7 | 5.7 ± 0.5 | 6.9 ± 1.1 | 5.5 ± 0.4 |
| YRI | individual | 50.0 ± 1.1 | 50.0 ± 1.3 | 50.0 ± 0.4 | 50.0 ± 1.4 |
| # genotypes = 30 | switching | 24.3 ± 0.2 | 23.1 ± 0.3 | 32.6 ± 0.2 | 22.7 ± 0.3 |
| # SNPs = 371 | runtime | $4.5 \times 10^4$ | $1.3 \times 10^3$ | $4.9 \times 10^1$ | $3.0 \times 10^0$ |
| ENr123.12q:12[7] | single-site | 3.3 ± 0.1 | 2.4 ± 0.2 | – | 2.8 ± 0.1 |
| CEU-nonrs | individual | 39.7 ± 1.4 | 46.6 ± 3.0 | – | 36.6 ± 1.4 |
| # genotypes = 30 | switching | 2.2 ± 0.1 | 1.0 ± 0.1 | – | 2.0 ± 0.1 |
| # SNPs = 1026 | runtime | $2.2 \times 10^5$ | $3.3 \times 10^5$ | – | $3.7 \times 10^1$ |
| ENm013.7q21:13[7] | single-site | 2.5 ± 0.2 | 2.3 ± 0.1 | – | 2.4 ± 0.2 |
| YRI | individual | 41.6 ± 2.2 | 38.3 ± 1.8 | – | 36.6 ± 1.5 |
| # genotypes = 30 | switching | 5.1 ± 0.2 | 3.7 ± 0.1 | – | 4.6 ± 0.2 |
| # SNPs = 758 | runtime | $9.9 \times 10^4$ | $2.2 \times 10^4$ | – | $1.6 \times 10^1$ |
| ENr113.4q26[7] | single-site | 3.8 ± 0.1 | 4.8 ± 0.3 | – | 4.6 ± 0.4 |
| CEU | individual | 41.6 ± 2.8 | 43.3 ± 3.2 | – | 43.3 ± 3.2 |
| # genotypes = 30 | switching | 2.3 ± 0.1 | 1.5 ± 0.1 | – | 1.6 ± 0.1 |
| # SNPs = 1017 | runtime | $5.0 \times 10^5$ | $6.1 \times 10^5$ | – | $3.9 \times 10^1$ |
| ENr113.4q26[7] | single-site | 5.5 ± 0.2 | 6.8 ± 0.3 | – | 6.5 ± 0.1 |
| YRI | individual | 50.0 ± 2.7 | 50.0 ± 2.8 | – | 50.0 ± 2.6 |
| # genotypes = 30 | switching | 6.0 ± 0.1 | 5.5 ± 0.1 | – | 6.0 ± 0.1 |
| # SNPs = 885 | runtime | $2.1 \times 10^5$ | $1.3 \times 10^5$ | – | $2.5 \times 10^1$ |
| ENr112.2p16:3[7] | single-site | 6.3 ± 0.2 | 5.1 ± 0.1 | – | 5.3 ± 0.1 |
| YRI | individual | 50.0 ± 2.2 | 48.0 ± 2.8 | – | 50.0 ± 2.9 |
| # genotypes = 30 | switching | 8.0 ± 0.2 | 3.9 ± 0.1 | – | 4.6 ± 0.1 |
| # SNPs = 1090 | runtime | $5.6 \times 10^5$ | $3.3 \times 10^5$ | – | $4.2 \times 10^1$ |
| **ENCODE project[7]** | **single-site** | **3.4 ± 0.1** | **3.5 ± 0.1** | – | **3.4 ± 0.1** |
| **average over** | **individuals** | **43.4 ± 1.9** | **44.9 ± 2.0** | – | **43.6 ± 1.9** |
| **40 datasets** | **switching** | **10.2 ± 1.8** | **10.1 ± 1.9** | – | **9.7 ± 1.7** |
| | **runtime** | $3.0 \times 10^5$ | $1.7 \times 10^5$ | – | $2.7 \times 10^1$ |

**Table 1.** (*continued*)

| Data | Measure | PHASE | GERBIL | HTYPER | 2SNP |
|---|---|---|---|---|---|
| TRPM8[7] | single-site | $2.4 \pm 0.1$ | $3.3 \pm 0.2$ | $2.1 \pm 0.1$ | $2.6 \pm 0.2$ |
| CEU | individual | $35.0 \pm 2.3$ | $35.0 \pm 1.7$ | $31.7 \pm 1.8$ | $28.0 \pm 5.6$ |
| # genotypes = 30 | switching | $2.8 \pm 0.1$ | $6.5 \pm 0.2$ | $2.3 \pm 0.1$ | $1.7 \pm 0.1$ |
| # SNPs = 315 | runtime | $8.8 \times 10^3$ | $1.4 \times 10^3$ | $2.3 \times 10^1$ | $3.0 \times 10^0$ |
| TRPM8[7] | single-site | $2.8 \pm 0.2$ | $4.4 \pm 0.4$ | $4.7 \pm 0.5$ | $3.6 \pm 0.4$ |
| YRI | individual | $36.6 \pm 1.8$ | $48.3 \pm 2.2$ | $43.3 \pm 2.3$ | $43.3 \pm 2.9$ |
| # genotypes = 30 | switching | $5.0 \pm 0.2$ | $6.8 \pm 0.2$ | $11.1 \pm 0.2$ | $6.4 \pm 0.2$ |
| # SNPs = 290 | runtime | $9.7 \times 10^3$ | $9.8 \times 10^2$ | $2.7 \times 10^1$ | $3.0 \times 10^0$ |
| STEAP[7] | single-site | $0.3 \pm 0.1$ | $0.6 \pm 0.2$ | $0.5 \pm 0.1$ | $0.6 \pm 0.2$ |
| YRI | individual | $6.6 \pm 0.6$ | $13.3 \pm 0.7$ | $11.6 \pm 0.7$ | $13.3 \pm 0.6$ |
| # genotypes = 30 | switching | $5.2 \pm 0.6$ | $7.9 \pm 0.4$ | $5.9 \pm 0.5$ | $7.8 \pm 0.6$ |
| # SNPs = 52 | runtime | $1.2 \times 10^2$ | $2.0 \times 10^0$ | $4.0 \times 10^0$ | $1.0 \times 10^{-3}$ |
| STEAP[7] | single-site | $0.6 \pm 0.1$ | $1.0 \pm 0.2$ | $1.1 \pm 0.2$ | $1.0 \pm 0.1$ |
| CEU | individual | $6.6 \pm 0.7$ | $10.0 \pm 0.9$ | $11.7 \pm 1.0$ | $8.3 \pm 1.2$ |
| # genotypes = 30 | switching | $7.3 \pm 0.9$ | $11.1 \pm 1.0$ | $11.6 \pm 1.0$ | $11.5 \pm 1.1$ |
| # SNPs = 60 | runtime | $1.3 \times 10^2$ | $3.0 \times 10^0$ | $5.0 \times 10^0$ | $1.0 \times 10^{-3}$ |
| STEAP[7] | single-site | $1.3 \pm 0.1$ | $2.1 \pm 0.2$ | $2.0 \pm 0.2$ | $1.9 \pm 0.2$ |
| YRI+CEU | individuals | $9.2 \pm 0.9$ | $15.8 \pm 1.3$ | $15.0 \pm 1.2$ | $15.8 \pm 1.2$ |
| # genotypes = 60 | switching | $15.1 \pm 1.0$ | $24.6 \pm 1.0$ | $22.2 \pm 0.9$ | $22.9 \pm 1.0$ |
| # SNPs = 49 | runtime | $3.2 \times 10^2$ | $5.0 \times 10^0$ | $1.2 \times 10^1$ | $1.0 \times 10^{-3}$ |
| TRPM8[7] | single-site | $4.0 \pm 0.9$ | $5.6 \pm 1.1$ | $6.5 \pm 0.9$ | $5.2 \pm 1.0$ |
| YRI+CEU | individuals | $43.3 \pm 1.2$ | $47.5 \pm 1.7$ | $47.5 \pm 1.6$ | $46.6 \pm 1.7$ |
| # genotypes = 60 | switching | $18.7 \pm 0.3$ | $28.6 \pm 0.3$ | $36.1 \pm 0.3$ | $27.7 \pm 2.5$ |
| # SNPs = 231 | runtime | $3.7 \times 10^4$ | $4.5 \times 10^2$ | $4.7 \times 10^1$ | $2.0 \times 10^0$ |

- *Chromosome* $5q31$*:* 129 genotypes with 103 SNPs derived from the 616 KB region of human Chromosome $5q31$ [3].
- *Yoruba population (D):* 30 genotypes with SNPs from 51 various genomic regions, with number of SNPs per region ranging from 13 to 114 [4].
- *HapMap datasets:* 30 genotypes of Utah residents and Yoruba residents available on HapMap by December 2005. The number of SNPs varies from 52 to 1381 across 40 regions including ENm010, ENm013, ENr112, ENr113 and ENr123 spanning 500 KB regions of chromosome bands 7p15:2, 7q21:13, 2p16:3, 4q26 and 12q12 respectively, and two regions spanning the gene STEAP and TRPM8 plus 10 KB upstream and downstream. Two more datasets with 60 genotypes each were obtained by mixing two populations YRI+CEU where SNPs with non-zero frequency in the both populations are kept.
- *Random matching* $5q31$*:* 128 genotypes each with 89 SNPs from $5q31$ cytokine gene cluster generated by random matching from 64 haplotypes of 32 West African reported by [9].
- *MS-simulated data:* 258 populations have been generated by MS[8] haplotype generator (using recombination rates 0,4 and 16). From each population of 100 haplotypes with 103 SNPs we have randomly chosen one haplotype and generated 129 genotypes by random matching.

**Table 2.** Mean single-site, individual, and switching errors with 95% confidence intervals and runtimes of PHASE, GERBIL, HAPLOTYPER (HTYPER), and 2SNP on simulated datasets

| Data | Measure | PHASE | GERBIL | HTYPER | 2SNP |
|---|---|---|---|---|---|
| random mating | single-site | $3.9 \pm 0.1$ | $8.7 \pm 0.7$ | $2.8 \pm 0.5$ | $9.1 \pm 0.5$ |
| Hull et al.[9] | individual | $25.0 \pm 4.1$ | $47.2 \pm 6.9$ | $20.9 \pm 3.1$ | $46.4 \pm 6.8$ |
| # genotypes = 128 | switching | $4.1 \pm 0.1$ | $12.1 \pm 0.1$ | $4.9 \pm 0.1$ | $11.8 \pm 0.1$ |
| # SNPs = 89 | runtime | $3.9 \times 10^4$ | $4.7 \times 10^1$ | $2.4 \times 10^1$ | $2.0 \times 10^0$ |
| MS[8] | single-site | $0.1 \pm 0.1$ | $0.1 \pm 0.1$ | $0.5 \pm 0.1$ | $0.2 \pm 0.1$ |
| recomb rate=0 | individual | $1.6 \pm 0.1$ | $2.7 \pm 0.2$ | $5.7 \pm 0.3$ | $4.6 \pm 0.3$ |
| # genotypes = 100 | switching | $0.7 \pm 0.1$ | $1.2 \pm 0.1$ | $3.3 \pm 0.3$ | $2.1 \pm 0.2$ |
| # SNPs = 103 | runtime | $1.5 \times 10^2$ | $6.0 \times 10^0$ | $3.0 \times 10^0$ | $1.0 \times 10^0$ |
| MS[8] | single-site | $0.2 \pm 0.1$ | $0.3 \pm 0.1$ | $0.3 \pm 0.1$ | $0.5 \pm 0.2$ |
| recomb rate=4 | individual | $2.0 \pm 0.1$ | $3.66 \pm 0.2$ | $6.5 \pm 0.2$ | $6.6 \pm 0.3$ |
| # genotypes = 100 | switching | $0.6 \pm 0.1$ | $1.4 \pm 0.2$ | $2.8 \pm 0.2$ | $3.4 \pm 0.4$ |
| #SNPs = 103 | runtime | $1.2 \times 10^2$ | $6.0 \times 10^0$ | $3.0 \times 10^0$ | $1.0 \times 10^0$ |
| MS[8] | single-site | $0.4 \pm 0.1$ | $0.5 \pm 0.1$ | $0.4 \pm 0.1$ | $1.0 \pm 0.1$ |
| recomb rate=16 | individual | $6.7 \pm 0.5$ | $6.3 \pm 0.3$ | $5.3 \pm 0.3$ | $9.8 \pm 0.4$ |
| # genotypes = 100 | switching | $1.7 \pm 0.2$ | $2.1 \pm 0.2$ | $2.6 \pm 0.2$ | $3.2 \pm 0.2$ |
| # SNPs = 103 | runtime | $1.3 \times 10^2$ | $6.0 \times 10^0$ | $3.0 \times 10^0$ | $1.0 \times 10^0$ |

**Error measures.** A *single-site error*[15] is the percent of erroneous SNPs among all SNPs in phased haplotypes. An *individual error*[12] is the percent of genotypes phased with at least one error among all genotypes. A *switching error*[10] is the percent of switches (among all possible switches) between inferred haplotypes necessary to obtain a true haplotype. For each dataset we bootstraped phasing result 100 times, and for each bootstrap sample we computed an error. The 95% confidence interval for the error mean was computed based on the 100 error values.

**Comparison of phasing methods.** Table 1 shows performance of four phasing methods on real datasets, and Table 2 shows the performance for simulated datasets. All runs were performed on computer with Intel Pentium 4, 2.0Ghz processor and 2 Gigabytes of Random Access Memory. HAPLOTYPER was run with 20 rounds and the 2.1.1 version of PHASE was run with the default parameters. The $*$-marked switching error is 2.9% for the earlier version PHASE-2.0.2. The dashes in HAPLOTYPER columns correspond to the cases when it does not output valid phasing.

The tables show that 2SNP is several orders of magnitude faster than two other phasing methods handling large datasets in a matter of seconds. The reported mean errors with the respective 95% confidence intervals show that GERBIL, PHASE, and 2SNP have the same accuracy for real data (Chromosome 5q31, Yoruba(D), HapMap datests). On the other hand, 2SNP and GERBIL are considerably outperformed by PHASE and HAPLOTYPER on some simulated data (Random matching 5q31). Poor performance of 2SNP can be caused by the absence of deviation from the Hardy-Weinberg equilibrium observed on real data. For the mixed populations one can see deterioration of all phasing methods.

In conclusion, we have presented a new extremely fast and simultaneously highly accurate phasing algorithm 2SNP based on 2-SNP haplotypes. We hope that it will be very useful for high-throughput genotype data processing, e.g., SNP Mapping Arrays (Affymetrix). We are going to extend our method by applying 3-SNP haplotype analysis.

## References


1. Brinza, D. and Zelikovsky, A. (2006) 2SNP: Scalable Phasing Based on 2-SNP Haplotypes. *Bioinformatics*, **22(3)**, 371–374.
2. Clark, A. (1990) Inference of haplotypes from PCR-amplified samples of diploid populations. *Mol Biol Evol.*, **7**, 111–122.
3. Daly, M., Rioux, J., Schaffner, S., Hudson, T. and Lander, E. (2001) High resolution haplotype structure in the human genome. *Nat Genet.*, **29**, 229–232.
4. Gabriel, G., Schaffner, S., Nguyen, H., Moore, J., Roy, J., Blumenstiel, B., Higgins, J., et al. (2002) The structure of haplotype blocks in the human genome. *Science*, **296**, 2225–2229.
5. Gusfield, D. (2003) Haplotype inference by pure parsimony. *Proc. Symp. on Comb. Pattern Matching*, LNCS **2676**, 144–155.
6. Halperin, E. and Eskin, E. (2004) Haplotype Reconstruction from Genotype Data using Imperfect Phylogeny. *Bioinformatics*, **20**, 1842–1849.
7. International HapMap Consortium. (2003) The International HapMap Project. *Nature*, **426**, 789–796, *http://www.hapmap.org*.
8. Hudson, R. (1990) 'Gene genealogies and the coalescent process', *Oxford Survey of Evolutionary Biology*, **7**, 1–44.
9. Hull, J., Rowlands, K., Lockhart, E., Sharland, M., Moore, C., Hanchard, N., Kwiatkowski, D.P. (2004) Haplotype mapping of the bronchiolitis susceptibility locus near IL8. *Am J Hum Genet.*, **114**, 272-279
10. Kimmel, G. and Shamir, R. (2005) GERBIL: Genotype resolution and block identification using likelihood. *Proc Natl Acad Sci.*, **102**, 158–162.
11. Kruglyak, L. and Nickerson, D. A. (2001) Variation is the spice of life. *Nat Genet.*, **27**, 234-236.
12. Niu, T., Qin, Z., Xu, X. and Liu, J.S. (2002) Bayesian haplotype inference for multiple linked single-nucleotide polymorphisms. *Am J Hum Genet.*, **70**, 157–169.
13. Niu T. (2004) Algorithms for inferring haplotypes. *Genet Epidemiol.*,**27(4)**, 334–47.
14. Stephens, M., Smith, N. and Donnelly, P. (2001) A new statistical method for haplotype reconstruction from population data. *Am J Hum Genet.*, **68**, 978–989.
15. Stephens, M., and Donnelly, P. (2003) 'A Comparison of Bayesian Methods for Haplotype Reconstruction from Population Genotype Data', *Am. J. Human Genetics*, 73:1162-1169.

# Event Models for Tumor Classification with SAGE Gene Expression Data

Xin Jin[1], Anbang Xu[1], Guoxing Zhao[2,3], Jixin Ma[3], and Rongfang Bie[1,*]

[1] College of Information Science and Technology,
Beijing Normal University, Beijing 100875, P.R. China
[2] School of Mathematical Sciences, Beijing Normal University, Beijing 100875, P.R. China
[3] School of Computing and Mathematical Science,
University of Greenwich, London SE18 6PF, U.K
xinjin796@126.com, anbangxu@mail.bnu.edu.cn,
G.Zhao@gre.ac.uk, j.ma@gre.ac.uk,
rfbie@bnu.edu.cn

**Abstract.** Serial Analysis of Gene Expression (SAGE) is a relatively new method for monitoring gene expression levels and is expected to contribute significantly to the progress in cancer treatment by enabling a precise and early diagnosis. A promising application of SAGE gene expression data is classification of tumors. In this paper, we build three event models (the multivariate Bernoulli model, the multinomial model and the normalized multinomial model) for SAGE data classification. Both binary classification and multicategory classification are investigated. Experiments on two SAGE datasets show that the multivariate Bernoulli model performs well with small feature sizes, but the multinomial performs better at large feature sizes, while the normalized multinomial performs well with medium feature sizes. The multinomial achieves the highest overall accuracy.

## 1 Introduction

Tumor classification from gene expression data is meaningful because there is high correlation between certain classes of cancers and respective gene expression levels. Traditional cancer classification methods are based on *microarray* gene expression data [9,13], which cannot tell us the expression levels of unknown genes (they may be useful for cancer prediction).

SAGE is a recently developed method which can monitor both known and unknown gene expression levels [2,5]. SAGE identifies a short mRNA tag from each individual transcript and concatenates them into long DNA molecules, which are then sequenced. By counting these tags one can estimate the gene expression levels in a cell. The information gained from performing this technique on a tissue sample is called a SAGE library.

In this study, we build three different event models for SAGE data. The multivariate Bernoulli model captures the information of which tags are found in a library, but

* Corresponding author.

V.N. Alexandrov et al. (Eds.): ICCS 2006, Part II, LNCS 3992, pp. 775 – 782, 2006.

not the number of times each tag appears. The multinomial model captures the information about how many times a tag appears in each library. The normalized multinomial model not only captures tag frequency but also takes into account library length.

The remainder of this paper is organized as follows: We first describe the gene selection method in Section 2. Then, in Section 3, we present the three event models for SAGE data. In Section 4 we evaluate the performance of the three models. Conclusions are presented in Section 5.

## 2 Feature Selection

Gene selection, or feature selection, can reduce over-fitting to the training samples and to speed up the classification process for gene expression data [3].

In our study, we use information gain to rank and select *useful* genes for SAGE data classification. Information gain is the expected entropy reduction caused by partitioning the data according to an attribute [7].

Let $S$ be a set of SAGE samples, $E$(S) be the entropy of $S$, the information gain $I(S;t)$ of a tag (or feature) $t$ is defined as:

$$I(S,t) = E(S) - \sum_{v \in V(t)} \frac{|S_v|}{|S|} E(S_v), \tag{1}$$

where $|S|$ is the cardinality of $S$, $S_v$ is the subset of $S$ for which tag has value $v$, and $V(t)$ is set of all possible value of tag $t$.

## 3 Three Models for SAGE Data Classification

Suppose that SAGE data is generated by a mixture model parameterized by $\theta$. Since the true parameters $\theta$ of the mixture model are not known, we need to estimate the parameters $\theta'$ from labeled training libraries. Taking into account that one SAGE library only belongs to one class (type of cancer), for a new library $l$ we search for a class $c_i$ that maximizes the posterior probability $P(c_i| l;\theta')$, by applying Bayes rule [11,12,6]. Note that $P(l|\theta')$ is the same for all classes, thus $l$ can be classified by computing:

$$c_l = \arg\max_{c_i \in C} P(c_i | \theta') P(l | c_i; \theta') \tag{2}$$

The estimation of $P(l|c_i;\theta)$ depends on the choice of generative models which are described below.

### 3.1 Multivariate Bernoulli Event Model

Under the multivariate Bernoulli event model, a SAGE library $l_j \in L$ ($L$ is the library set, $|L|$ is the number of tags appearing in the set) is generated by a series of $|L|$ Bernoulli experiments, one for each tag. The outcome of each experiment determines whether the corresponding tag will be included in the library. Thus a SAGE library is a binary vector over the space of tags. Dimension $k$ of the vector for library $l_j$ is written $B_{ik}$, and is either 0 or 1, indicating whether tag $t_k$ occurs in the library. We can also represent each library as $|l|+1$ variables $V_0, V_1,\ldots, V_{|l|}$. $V_0$ takes values in $\{c_1,\ldots,c_M\}$ and presents the class of the library. $V_1,\ldots, V_l$ take values in $\{0,1\}$ and represent whether the particular tags appear in the library. Fig.1 presents this model as a graphical Markov model (or Bayesian network).

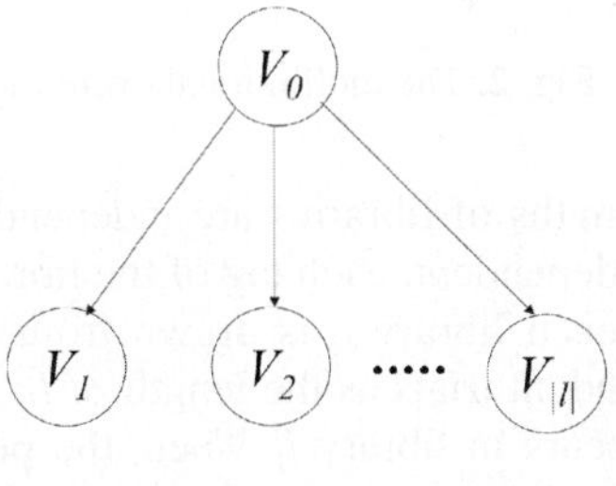


**Fig. 1.** The multivariate Bernoulli event model

We assume that the $|L|$ trials are independent on each other, that is, the probability of each tag occurring in a library is independent of the occurrence of other tags in a library. Thus we can estimate the probability of a library given a class from the probabilities of the tags given the class,

$$P(l_j \mid c_i;\theta) = \prod_{k=1}^{|L|} (B_{jk} P(t_k \mid c_i;\theta) + (1 - B_{jk})(1 - P(t_k \mid c_i;\theta))) \quad (3)$$

The multivariate Bernoulli event model does not capture the number of times each tag occurs, however, it includes the non-occurrence probability of tags that do not appear in the library but appear in the whole library set.

The parameters $\theta_{t_k|c_i} = P(t_k \mid c_i;\theta)$ of the mixture component $c_i$ can be estimated as the fraction of training documents in $c_i$ that contain $t_k$:

$$\theta'_{t_k|c_i} = P(t_k \mid c_i;\theta') = \frac{\sum_{i=1}^{N} B_{jk} P(c_i \mid l_j)}{\sum_{i=1}^{N} P(c_i \mid l_j)} \quad (4)$$

### 3.2 Multinomial Event Model

Multinomial approach to modeling tag frequency is to treat tags for a length $|l_j|$ (the number of tags of the library) SAGE library $l_j$ as resulting from $|l_j|$ drawing events on a $d$-valued multinomial variable $V$. Each library can then be represented by a set of random variables $V_0, V_1,\ldots, V_{|l|}$. $V_0$ takes values in $\{c_1,\ldots,c_M\}$ and presents the class of the library. $V_1,\ldots, V_l$ take values in $\{0,1,\ldots\}$ and represent the number of occurrences of particular tags in the library. Fig. 2 presents a graphical Markov model representation where a single node represents a SAGE library's vector of tag counts. The multinomial model captures tag frequency information in SAGE libraries.

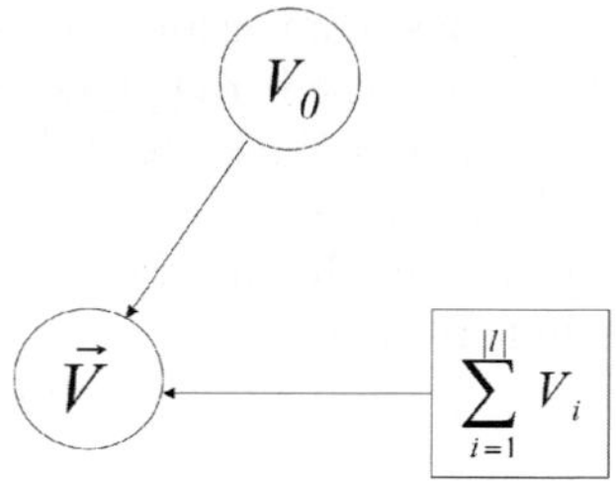


**Fig. 2.** The multinomial event model

We assume that the lengths of libraries are independent of class. We also assume that the draws on $V$ are independent, each tag of the library is generated independently from every other. Thus, each library $l_j$ is drawn from a multinomial distribution of tags with as many independent trials as the length of $l_j$. If $N_{jk}$ denotes the count of the number of times tag $t_k$ occurs in library $l_j$. Then, the probability of a library given a class is the multinomial distribution:

$$P(l_j \mid c_i;\theta) = P(|l_j|)\,|l_j|!\prod_k \frac{P(t_k \mid c_i;\theta)^{N_{jk}}}{N_{jk}!} \tag{5}$$

where $k$= 1, 2,…, $T$, $T$ is the number of tags in library $l_j$.

The parameters of the generative component for each class $c_i$ are the probabilities for each tag, written $\theta_{t_k|c_i} = P(t_k \mid c_i;\theta)$, where $\sum_k \theta_{t_k|c_i} = 1$. It can be estimated from the training libraries:

$$\theta'_{t_k|c_i} = P(t_k \mid c_i;\theta') = \frac{\sum_j N_{jk} P(c_i \mid l_j)}{\sum_s \sum_j N_{js} P(c_i \mid l_j)} \tag{6}$$

where $s$=1,…, $T$, $T$ is the total number of tags in library $l_j$.

### 3.3 Normalized Multinomial Event Model

Normalized multinomial event model for SAGE data not only capture tag frequency but also take into account library length. We normalize tag counts by transforming the tag frequencies according to

$$N'_{jk} = \frac{N_{jk}}{\sqrt{\sum_{s=1}^{T} (N_{js})^2}} \tag{7}$$

yielding a length 1 tag frequency vector for each library.

Normalization is common within the gene expression data clustering community because the probability of assigning a sample to a cluster is estimated by calculating distance across samples; in such a case it is not meaningful to use some of the distance functions such as the Euclidean distance to compare the libraries to each other, since their tag frequencies are not on the same scale [1,5]. For classification, however, because comparisons are made across classes, and not across libraries, the benefit of such normalization is subtler, especially as the multinomial model accounts for length very naturally [8,10]. In experiments we find that normalized multinomial model performs better than standard multinomial for *breast* SAGE library set (classify tumor types), but the other way round for *brain* libraries (classification between tumor and normal tissue).

## 4 Experiments

We experiment on classification of SAGE data under three event models. We run 10 random splits with 50% training data and 50% testing data per class. The number of selected genes varies from 10 to all for each experiment. The performance results are measured by the common used *accuracy*, the percentage of correct predictions on the test sets.

### 4.1 Data and Preprocessing

The experiments are based on two SAGE data sets, the raw data is available on the NCBI SAGE website [4]. One problem with the raw SAGE data is that many tags in each library are expected to contain sequencing errors, and since these errors result in noise and increase the dimensionality of the data, error removal is a must need. Within one library, some tags have a frequency of 1; these unique tags are either sequencing errors or representations of very low expression level genes. In our experiment, we just remove these single frequency tags to filter out data noise.

*Brain dataset*: The dataset is based on 52 Hs (human sapiens) SAGE brain libraries. These libraries are made of samples from human brain and fall in to four categories: Astrocytoma (11 libraries), Ependymoma (9 libraries), Glioblastoma (8 libraries) and Medulloblastoma (24 libraries). There are 64558 tags (after noise removal) in the dataset. We used the dataset for multicategory classification experiments. We also used the biggest two categories, Astrocytoma and Medulloblastoma, for binary classification.

*Breast dataset*: The dataset is based on 26 Hs (human sapiens) SAGE breast libraries. These libraries are made of samples from human breast and fall in to two classes: Normal (10 libraries) and Cancer (16 libraries). There are 36087 tags (after noise removal) in the dataset. We used this dataset for binary classification experiments.

### 4.2 Results

Fig.3 shows the performance results of three event models for binary classification on the human *brain* dataset. The multinomial model achieves a maximum of 98.3% accuracy with the 3000 top-ranked genes; the multivariate Bernoulli model is best with 96.1% accuracy at 500 features; the normalized multinomial model reaches a maximum of 94.1% accuracy at 1000 selected genes.

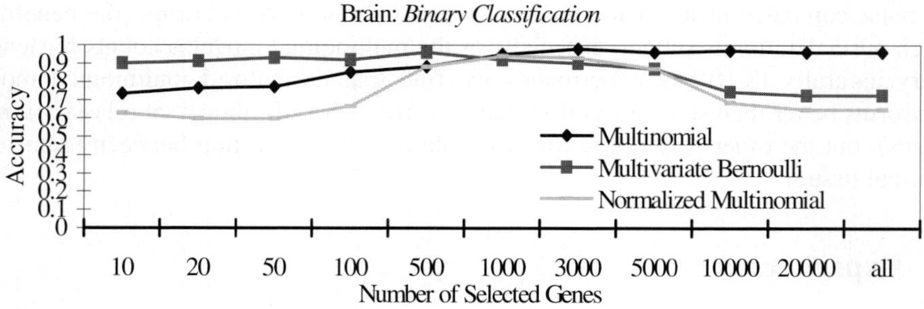


**Fig. 3.** Three event models for *brain* SAGE libraries on different feature sizes: *binary classification* (astrocytoma or medulloblastoma)

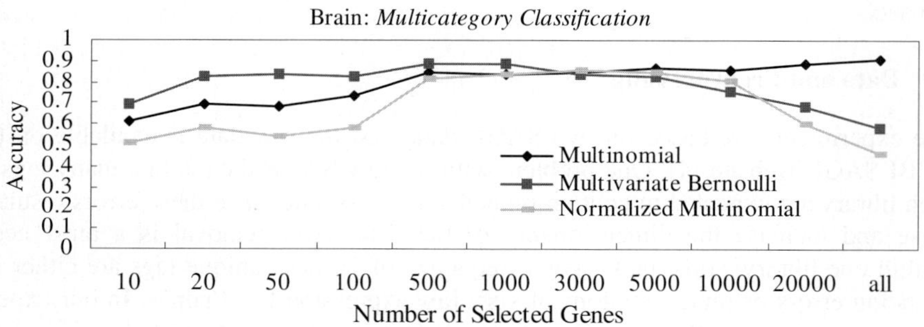


**Fig. 4.** Three event models for *brain* SAGE libraries on different feature sizes: *multicategory classification*

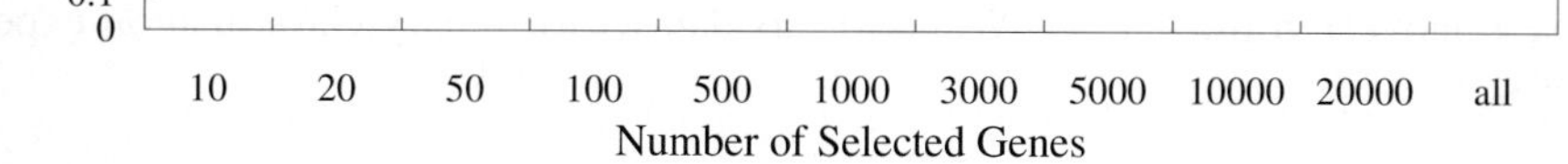


**Fig. 5.** Three event models for *breast* SAGE libraries on different feature sizes: *binary classification* (tumor or normal)

Fig.4 shows the results of multicategory classification on the human *brain* libraries. The multinomial event model reaches a maximum of 90.4% accuracy when all genes are involved in the classifier. The multivariate Bernoulli model achieves a maximum of 88.5% accuracy at both 500 and 1000 features, while the normalized multinomial 85.3% at 3000 genes.

From Fig.3 and 4 we can see that for *brain* libraries, the three models do best when feature selection is used. The multivariate Bernoulli performs best with a small feature size and the multinomial performs best at a larger feature size, while normalized multinomial model do best at a moderate feature size. The multinomial model achieves higher accuracy than the multivariate Bernoulli model and the normalized multinomial model.

For classification on the human *breast* libraries, as shown in Fig.5, the best performance is achieved by the normalized multinomial event model with a maximum of 89.8% at both 1000 and 5000 features. The standard multinomial model reaches a maximum of 87.2% accuracy at 5000 features, while the multivariate Bernoulli model reaches is best with 77.0% accuracy at 100 features. Like the results for *brain* libraries, the multivariate Bernoulli performs best with a few features, the multinomial performs best with a larger feature size and normalized multinomial model do best with a moderate amount of selected genes. However, for *breast* libraries, the normalized multinomial model performs better than the multinomial and the multivariate Bernoulli model. In ongoing work we are exploring the reasons that the *breast* libraries show results different from the *brain* libraries. Note that for breast libraries the classification is done between tumor and normal samples, while for brain libraries it is between tumor types.

## 5 Conclusions

In this paper, we build and evaluate three event models (a multivariate Bernoulli model, a multinomial model and a normalized multinomial model) for SAGE data classification. We used information gain ranking function for gene selection.

In empirical results on publicly available SAGE libraries, we find that the multivariate Bernoulli performs well with a few features, the normalized multinomial model do well with a medium amount of selected genes and the un-normalized multinomial performs well with a larger feature size. The normalized multinomial model performs better than the multinomial for tumor-normal classification of *breast* SAGE libraries, but the multinomial is better for tumor-tumor *brain* libraries classification (both binary and multicategory). The multinomial achieves the highest overall accuracy.

## Acknowlegments

This work was supported by the National Science Foundation of China under the Grant No. 60273015 and No. 10001006.

## References

1. Raymond T. Ng, Jörg Sander, Monica C. Sleumer: Hierarchical Cluster Analysis of SAGE Data for Cancer Profiling. BIOKDD, 65-72 (2001)
2. Velculescu V. E., Zhang L., Vogelstein B., Kinzler K.W.: Serial Analysis of Gene Expression. Science, Vol. 270, Oct 20 484-487 (1995)
3. I. Guyon, J. Weston, S. Barnhill, V. Vapnik: Gene Selection for Cancer Classification Using Support Vector Machines. Machine Learning, 46(1/3) 389–422 (2002)

4. NCBI SAGE: http://www.ncbi.nlm.nih.gov/SAGE (2005)
5. J. Sander, R.T. Ng, M.C. Sleumer, M. Saint Yuen, S.J. Jones: A Methodology for Analyzing SAGE Libraries for Cancer Profiling. ACM Transactions on Information Systems, 23(1) 35-60 (2005)
6. Andrew McCallum, Kamal Nigam: A Comparison of Event Models for Naive Bayes Text Classification. In Proceedings of AAAI-98 Workshop on Learning for Text Categorization, 41–48. AAAI Press (1998)
7. Cover, Thomas: Elements of Information Theory. Wiley & Sons (1991)
8. Jason D. Rennie, Lawrence Shih, Jaime Teevan, David R. Karger: Tackling the Poor Assumptions of Naive Bayes Text Classifiers. Twentieth International Conference on Machine Learning. August 22 (2003)
9. Marcel Dettling: BagBoosting for Tumor Classification with Gene Expression Data. Bioinformatics, Vol. 20, No. 18, 3583-3593 (2004)
10. Lewis, D. D. (1998). Naive (Bayes) at forty: The Independence Assumption in Information Retrieval. Proceedings of ECML98 (1998)
11. J. Hilden: Statistical Diagnosis Based on Conditional Independence Does not Require It. Computational Methods in Biology and Medicine, 14(4) 429–435 (1984)
12. J. Hellerstein, Jayram Thathachar, I. Rish: Recognizing End-user Transactions in Performance Management. In Proceedings of AAAI-2000, Austin, Texas, 596–602 (2000)
13. Helman P, Veroff R, Atlas, SR., Willman CL: A Bayesian Network Classification Methodology for Gene Expression Data. Journal of Computational Biology, 11(4) 581-615 (2004)

# Genomes Containing Duplicates Are Hard to Compare*
## (Extended Abstract)

Cedric Chauve[1], Guillaume Fertin[2], Romeo Rizzi[3], and Stéphane Vialette[4]

[1] LaCIM, CGL, Département d'Informatique, Université du Québec À Montréal
CP 8888, Succ. Centre-Ville, H3C 3P8, Montréal (QC) - Canada
chauve@lacim.uqam.ca

[2] Laboratoire d'Informatique de Nantes-Atlantique (LINA), FRE CNRS 2729
Université de Nantes, 2 rue de la Houssinière, 44322 Nantes Cedex 3 - France
fertin@lina.univ-nantes.fr

[3] Dipartimento di Matematica e Informatica - Università di Udine - Italy
Romeo.Rizzi@dimi.uniud.it

[4] Laboratoire de Recherche en Informatique (LRI), UMR CNRS 8623
Faculté des Sciences d'Orsay - Université Paris-Sud, 91405 Orsay - France
vialette@lri.fr

**Abstract.** In this paper, we are interested in the algorithmic complexity of computing (dis)similarity measures between two genomes when they contain duplicated genes. In that case, there are usually two main ways to compute a given (dis)similarity measure $M$ between two genomes $G_1$ and $G_2$: the first model, that we will call the *matching model*, consists in computing a one-to-one correspondence between genes of $G_1$ and genes of $G_2$, in such a way that $M$ is optimized in the resulting permutation. The second model, called the *exemplar model*, consists in keeping in $G_1$ (resp. $G_2$) exactly *one* copy of each gene, thus deleting all the other copies, in such a way that $M$ is optimized in the resulting permutation. We present here different results concerning the algorithmic complexity of computing three different similarity measures (number of common intervals, MAD number and SAD number) in those two models, basically showing that the problem becomes **NP**-complete for each of them as soon as genomes contain duplicates. In the case of MAD and SAD, we actually prove that, under both models, both MAD and SAD problems are APX-hard.

## 1 Introduction

A classical method to compute phylogenies based on gene orders in whole genomes is to first compute a pairwise distance matrix, obtained by comparing every pair of genomes in a considered dataset. Note that the pairwise distance between two gene orders $G_1$ and $G_2$ is often deduced from a similarity or dissimilarity measure, say $M$, between $G_1$ and $G_2$. Gene orders are usually represented by

* Work partially supported by the French-Italian Galileo Project PAI 08484VH and the French-Québec 60th CPCFQ.

V.N. Alexandrov et al. (Eds.): ICCS 2006, Part II, LNCS 3992, pp. 783–790, 2006.

signed *sequences* on the alphabet of *gene families*, where every element in a genome is a *gene* (each gene being represented either as a letter or an integer). Any gene belongs to a unique gene family, and two genes that have the same label, regardless of the sign, belong to the same gene family. A family $f$ of genes is said to be *balanced* between two genomes if the number of occurrences of genes of $f$ is the same in both genomes. Two genomes $G_1$ and $G_2$ are consequently said to be *balanced* if all families of genes in $G_1$ and $G_2$ are balanced. In this paper we consider only balanced pairs of genomes, for reasons that we describe later in this section. When there are no duplicates in a balanced pair of genomes, it can be represented by a signed permutation, and the computation of the measure $M$ is polynomial-time solvable in several models that have been used in practice (e.g. number of breakpoints, reversal distance, number of conserved intervals, number of common intervals, MAD and SAD numbers among others). However, aside of a few exceptions, like small genomes of viruses and organelles, duplicated genes and other genomic markers are common in genomes, due to several evolutionary mechanisms, like gene duplication or whole genomes duplications [7]. Hence, it is an important problem to be able to handle those duplicates when measuring the distance or (dis)similarity between pairs of genomes. The main approach is to transform the two signed sequences $G_1$ and $G_2$ in a signed permutation, that allows to use one of the models defined in this case. This has been done in two different ways, called the *matching* model and the *exemplar* model. In the *matching* model, we want to end up in a situation where the pair of genomes is balanced. For this, for each family $f$ of genes, we establish a maximal one-to-one correspondence between the genes of $f$ in $G_1$ and the genes of $f$ in $G_2$, before deleting the genes of $f$ which have not been paired. The result is a complete matching between $G_1$ and $G_2$ that induces a signed permutation. The goal is to compute the matching that induces a signed permutation that is optimal, among all signed permutation given by all possible matchings, for the chosen measure $M$. In the *exemplar* model, introduced by Sankoff [6], the idea is to keep exactly *one copy* of each family in each genome, that can be seen as the ancestral copy of all occurrences of the corresponding gene family. Consequently, we end up here again with a signed permutation and the goal is to compute the exemplar copy of each gene family that induces a signed permutation that is optimal for $M$. Let $occ(G, g)$ denote the maximum number of occurrences of a gene $g$ in genome $G$ (regardless of the signs), and let $occ(G)$ be the maximum of $occ(G, g)$ over all genes $g$ in $G$. Note that if $occ(G_1) = 1$ (that is, if $G_1$ contains no duplicates), then for any genome $G_2$, both the matching and the exemplar models coincide.

In this paper, we present new results on the algorithmic complexity of computing different (dis)similarity measures, for balanced[1] pairs of genomes that

[1] Note that comparing real genomes often lead to unbalanced pairs of signed sequences and requires either models of measure that accounts for the insertion and deletion of gene and gene segments [1] or methods to balance a pair of genomes [4]. However, for all models we are aware of, insertion/deletions are not necessary for a balanced pair of genomes, and then the complexity of the balanced problem is sufficient to give the complexity of the more general problem.

contain duplicates. Several measures have already been studied like the number of breakpoints [3, 1], the number of reversals [3, 4] and the number of conserved intervals [2]. Basically, it has been shown that, for each of the above mentioned measures, the problem is **NP**-complete as soon as duplicates are present in genomes, in both the matching and exemplar approaches ; some inapproximability results are also given in some cases [9]. Here, we follow the same line and we describe results for three other similarity measures, namely number of common intervals, Maximum Adjacency Disruption number (MAD) and Summed Adjacency Disruption number (SAD), which will be defined in Section 2. In order to simplify notations, and since none of those three measures depends on the signs given to the genes (unlike measures like breakpoints, conserved intervals and reversals), we consider only *unsigned genomes*. We focus in Section 3 on the problem of computing the number of common intervals in genomes containing duplicates, and show that the problem is **NP**-complete in both the matching and exemplar models. In Sections 4 and 5, we prove that, under both models, both the MAD and SAD problems are APX-hard when genomes contain duplicates. Due to space constraints, no formal proof will be given in this extended abstract ; they will appear in the journal version of the paper.

## 2 Preliminaries

In this section, we define the three similarity measures we are interested in. As mentioned before, each of those measures asks that the genomes are balanced, and that a one-to-one correspondence exists between any gene of $G_1$ and a gene of $G_2$. Hence we will often conveniently rename genome $G_1$ into the identity permutation on $n$ genes, $Id_n$ (that is, $1\ 2\ 3 \ldots n$) and $G_2$ can be recomputed accordingly into a new permutation. We now define those three measures.

*Number of common intervals*: a *common* interval between $G_1$ and $G_2$ is a substring of $G_1$ for which the exact same content can be found in a substring of $G_2$. For example, let $G_1 = Id_5$ and $G_2 = 1\ 5\ 3\ 4\ 2$, then the interval $[3, 5]$ of $G_1$ is a common interval.

*Maximum Adjacency Disruption Number (MAD)*: this notion has been recently introduced by Sankoff and Haque [8], where a genome is represented by a string of integers. This number, say $\mathcal{M}$, is defined as the maximum between two values $M_{1,2}$ and $M_{2,1}$, where $M_{1,2}$ (resp. $M_{2,1}$) is the maximum difference between two consecutive genes (i.e., integers) in $G_2$ (resp. $G_1$) , supposing that $G_1 = Id_n$ (resp. that $G_2 = Id_n$) and that $G_2$ (resp. $G_1$) has been renamed accordingly. We need to compute both $M_{1,2}$ and $M_{2,1}$ in order to restore symmetry, since those two measures might differ.

*Summed Adjacency Disruption Number (SAD)*: this notion has also been introduced by Sankoff and Haque [8], and can be seen as a global variant of the MAD number. Similarly to the previous case, suppose $G_i = Id_n$ and $G_j = g_1^j\ g_2^j \ldots g_n^j$ has been renamed accordingly ($i \neq j \in \{1, 2\}$). The Summed Adjacency Disruption number is then defined as $\mathcal{S} = \sum_{k=1}^{n-1} |g_k^1 - g_{k+1}^1| + \sum_{k=1}^{n-1} |g_k^2 - g_{k+1}^2|$. In other words, we sum the differences between consecutive genes, and we do that in both "directions" to avoid asymmetry.

Note that the two last measures are actually *dissimilarity* measures, which means that the goal is to minimize them, while the first is a similarity measure that we wish to maximize.

## 3 Number of Common Intervals

Let EComI (resp. MComI) denote the problem of computing the maximum number of common intervals in the exemplar (resp. matching) model. We show that both EComI and MComI are **NP**-complete, even for restricted instances. The reduction we give below is valid for both models, since it shows NP-completeness in the case $occ(G_1) = 1$. However, in order to simplify notations, we will mention here only the exemplar model (i.e., the EComI problem). The reduction is from VertexCover. Starting from any instance of VertexCover (that is, a graph $G = (V, E)$ with $V = \{v_1, v_2 \dots v_n\}$ and $E = \{e_1, e_2 \dots e_m\}$), we will first describe a polynomial-time construction of two genomes $G_1$ and $G_2$ such that $occ(G_1) = 1$ and $occ(G_2) = 2$. We first describe $G_1$: $G_1 = b_1, b_2 \dots b_m, x, a_1, C_1, a_2, C_2 \dots a_n, C_n, y, b_{m+n}, b_{m+n-1} \dots b_{m+1}$

The $a_i$s, the $b_i$s, $x$ and $y$ are genes, while $C_i$s are sequences of genes. They are defined as follows:

- for any $1 \leq i \leq n$, $a_i = 2(i-1)m + i$ ;
- for any $1 \leq i \leq n$, $C_i = (a_i + 1), (a_i + 2) \dots (a_i + 2m)$ ;
- for any $1 \leq i \leq n + m$, $b_i = a_n + 2m + i$ ;
- $x = b_{n+m} + 1$ ;
- $y = b_{n+m} + 2$.

It can be seen that no gene appears more than once in $G_1$, thus $occ(G_1) = 1$. Now we describe the construction of $G_2$:
$G_2 = y, a_1, D'_1, b_{m+1}, a_2, D'_2, b_{m+2} \dots b_{m+n-1}, a_n, D'_n, b_{m+n}, x$

The duplicated genes in $G_2$ are $b_1, b_2 \dots b_n$, and are spread within the $D'_i$s. Moreover, each $b_i$, $1 \leq i \leq n$ will appear only twice in $G_2$. We now describe the contents of $D'_i$, $1 \leq i \leq n$. Basically, $D'_i$ is constructed in two steps: (1) we first construct, for each $i$, a sequence of genes $D_i$, which is a specific shuffle of the contents of $C_i = (a_i + 1), (a_i + 2) \dots (a_i + 2m)$. More precisely, let min $= a_i + 1$ and max $= a_i + 2m$ ; then $D_i = (a_i + 3), (a_i + 5) \dots (a_i + 2m - 3), (a_i + 2m - 1)$, min, max, $(a_i + 2), (a_i + 4) \dots (a_i + 2m - 4), (a_i + 2m - 2)$ ; (2) for any $1 \leq i \leq n$, we obtain $D'_i$ by adding some $b_j$s ($1 \leq j \leq m$) into $D_i$, accordingly to the initial graph $G$ we are given. More precisely, for any edge $e_j$ that is incident to a vertex $v_i$ in $G$, we add the gene $b_j$ between the $j$-th and the $(j+1)$-th gene of $D_i$. This process gives us the $D'_i$s. Note that no two $b_j$s ($1 \leq j \leq m$) can appear contiguously in a $D'_i$, and that no $D'_i$ starts or ends with a $b_j$ (all $D'_i$s start and end with a gene that only appears in $C_i$ in $G_1$).

**Lemma 1.** *Let $G$ be a graph and $G_1$ and $G_2$ be the two genomes obtained by the construction described above. $G$ admits a Vertex Cover $VC$ such that $|VC| \leq k$ iff there exists an exemplar genome $G_2^E$ obtained from $G_2$ having at least $\mathcal{I} = 2nm + 4n + m + 3 - 2k$ common intervals.*

As a direct consequence of Lemma 1, we conclude that the ECOMI problem is **NP**-complete. Moreover, as mentioned before, the proof and the result are also valid for the MCOMI problem, since our construction implies $occ(G_1) = 1$. We thus have the following theorem.

**Theorem 1.** *The* ECOMI *and* MCOMI *problems are both* **NP***-complete, even when $occ(G_1) = 1$ and $occ(G_2) = 2$.*

We also consider, for the matching model, instances for which the constraints do not rely on the maximum number of duplicates per family, but on the number of families that contain duplicates. We have the following result.

**Theorem 2.** *The* MCOMI *problem is* **NP***-complete, even when $f(G_1) = f(G_2) = 1$, where $f(G)$ denotes the number of different families of genes that contain duplicates in $G$.*

## 4 Maximum Adjacency Disruption (MAD)

Let EMAD (resp. MMAD) denote the problem of computing the minimum MAD number of in the exemplar (resp. matching) model. In this section, we prove inapproximability results for both the EMAD and MMAD problems. More precisely, we show that for no $\varepsilon > 0$, EMAD (resp. MMAD) admits a $(2 - \varepsilon)$-approximation algorithm, unless P=NP. This inapproximability result does not rely on the PCP theorem. We will also remark however, how, reconsidering the reduction proposed in view of APX-hardness results based on the PCP theorem, one can replace the constant 2 above with a strictly bigger constant. The proof is split into two: we first study the complexity of a restricted form of SAT, which we call UNIFORM-SAT, and in particular we show that it is **NP**-complete. Next, we show that a $(2-\varepsilon)$-approximation algorithm for EMAD (resp. MMAD), for some $\varepsilon > 0$, would imply the existence of a polynomial-time algorithm for UNIFORM-SAT. Finally, we obtain the inapproximability result for EMAD (resp. MMAD).

In the following, 3SAT will denote the restriction of SAT for which each clause contains at most 3 literals. We introduce a restricted form of 3SAT called UNIFORM-SAT, as follows: an instance $\langle X, \mathcal{C} \rangle$ of 3SAT is an instance of UNIFORM-SAT when the following two conditions are met: (i) for each clause $C \in \mathcal{C}$, either all literals occurring in $C$ are positive occurrences of variables from $X$ or all literals occurring in $C$ are negated occurrences of variables from $X$ and (ii) for each variable $x \in X$, $x$ has at most 3 positive and at most 2 negated occurrences within $\mathcal{C}$. A 3SAT formula $F = \bigwedge_{C \in \mathcal{C}} C$ is called 3-*bounded* if no variable has more than 3 occurrences within $C$ and is called $(2, 2)$-*bounded* if no variable has more than 2 positive occurrences and no more than 2 negated occurrences within $C$. The following two facts are known: (1) the decision problem 3SAT is **NP**-complete even when restricted to 3-bounded formulas and (2) the optimization problem MAX-3SAT is APX-hard even when restricted to 3-bounded formulas [5]. Since both problems admit a trivial self-reduction in case a variable has only positive (or only negated) occurrences, then the following two facts also hold: (1) 3SAT is **NP**-complete even when restricted to $(2,2)$-bounded formulas and (2) MAX-3SAT is APX-hard even when restricted to $(2,2)$-bounded formulas. Notice that, of the above two results, only the second is related to the PCP-theorem.

**Theorem 3.** *Deciding whether a given* UNIFORM-SAT *formula is satisfiable is* **NP***-complete.*

Theorem 3 here above does not need the PCP theorem and is all what is required in the following for proving that, for no $\varepsilon > 0$, EMAD (resp. MMAD) admits a $(2-\varepsilon)$-approximation algorithm, unless P=NP. With dependence on PCP, we have the following result, which, besides being of independent interest, can be used to show that the right constant for the approximability of EMAD (resp. MMAD) is not 2.

**Theorem 4.** *Given a* UNIFORM-SAT *formula, the problem of finding a truth assignment maximizing the number of satisfied clauses is APX-hard.*

We now prove that both the EMAD and MMAD problems are APX-hard. The result holds for both problems, since we prove it in the case where $occ(G_1) = 1$, where they coincide. The result rests on a reduction form UNIFORM-SAT. Assume we are given an instance $\langle X, \mathcal{C} \rangle$ of UNIFORM-SAT, where $X = \{x_1, x_2, \ldots, x_n\}$. Here, $\mathcal{C}$ can be partitioned into the family $\mathcal{P} = \{P_1, P_2, \ldots, P_{m_p}\}$ of clauses comprising only positive literals and the family $\mathcal{N} = \{N_1, N_2, \ldots, N_{m_n}\}$ of clauses comprising only negated literals. Let $M_\varepsilon$ be a sufficiently big positive integer that we will fix later in order to force our conclusions. We propose to compare two genomes $G_1$ and $G_2$. Here, $G_1$ is the simple (that is, without repetitions) genome $G_1$ of length $L_1 = 2\,M_\varepsilon + m_p + m_n + n - 1$ defined as follows: $G_1 = 1\ 2\ 3 \ldots L_1$. A gene at position $i$ in $G_1$ with $i \leq m_p$ or $i \geq L_1 - m_n + 1$ is called a $*$*-gene*. Genome $G_2$ has length $L_2 = 2\,M_\varepsilon + 6\,n - 1$, and conforms to the following pattern, where we have found it convenient and pertinent to spot out the displacement of the $*$-genes within genome $G_2$.

$$G_2 = m_p + 1, \ldots, m_p + M_\varepsilon, *, *, *, *, *, m_p + M_\varepsilon + 1, *, *, *, *, *, m_p + M_\varepsilon + 2, \ldots$$
$$\ldots, *, *, *, *, *, m_p{+}M_\varepsilon{+}n, m_p{+}M_\varepsilon{+}n{+}1, m_p{+}M_\varepsilon{+}n{+}2, \ldots, m_p{+}2\,M_\varepsilon{+}n{-}1$$

We will specify later the precise identity of the $*$-genes within genome $G_2$. For now, notice that in $G_2$ we have precisely $n$ runs of 5 consecutive $*$-genes. We put these runs into $(1,1)$-correspondence with the $n$ variables in $X$, so that the $i$-th run corresponds to variable $x_i$, for $i = 1, 2, \ldots, n$. For each $i = 1, 2, \ldots, n$, let $\mathcal{P}_i$ and $\mathcal{N}_i$ be the lists of index sets of the clauses from $\mathcal{P}$ and $\mathcal{N}$ which contain variable $x_i$. E.g., if $x_i$ appears in $P_3$, in $P_7$, and in $N_2$, then $\mathcal{P}_i = (3, 7)$, whereas

$\mathcal{N}_i = (2)$. Notice that the lengths of the lists $\mathcal{P}_i$ and $\mathcal{N}_i$ are at most 3, and 2, respectively. ¿From the list $\mathcal{P}_i$ we obtain a list $\mathcal{P}'_i$ of length precisely 3 by possibly iterating the last element in $\mathcal{P}_i$ the required number of times (that is, $3 - |\mathcal{P}_i|$ times). A list $\mathcal{N}'_i$ of length precisely 2 is similarly obtained from list $\mathcal{N}_i$. Now, for each $i = 1, 2, \ldots, n$, the $i$-th run of 5 consecutive $*$-genes consists in the following 5 characters: $(*, *, *, *, *) \rightarrow (\mathcal{P}'_i[1], \mathcal{P}'_i[2], \mathcal{P}'_i[3], L_1 - m_n + \mathcal{N}'_i[1], L_1 - m_n + \mathcal{N}'_i[2])$

The above reduction leads us to the following result.

**Theorem 5.** *For no* $\varepsilon > 0$, EMAD *(resp.* MMAD*) admits a* $(2 - \varepsilon)$*-approximation algorithm, unless P=NP.*

*Remark 1.* There actually exists a constant $c > 2$ such that EMAD (resp. MMAD) admits no $c$-approximation algorithm unless P=NP. We can get to this stronger conclusion if in the proof of Theorem 5 here above we apply Theorem 4 instead of Theorem 3.

## 5 Summed Adjacency Disruption (SAD)

Let ESAD (resp. MSAD) denote the problem of computing the minimum SAD number of in the exemplar (resp. matching) model. In this section, we prove that both problems ESAD and MSAD, expressed on two genomes $G_1$ and $G_2$ such that $|G_1| \leq |G_2|$ cannot be better than $\log(|G_1|)$ approximated. This result holds for both the exemplar and the matching models, since we prove it in the case where $occ(G_1) = 1$, for which the two problems coincide. The inapproximability of ESAD (resp. MSAD) is obtained starting from the inapproximability of SetCover. This result will hence depend on the PCP theorem, but will deliver stronger SetCover-like inapproximability thresholds than for the EMAD ans MMAD problems discussed in the previous section.

Let $\langle V, \mathcal{S} \rangle$ be an instance of SetCover, where $V = \{1, 2, \ldots, n\}$, and $\mathcal{S} = \{S_1, S_2, \ldots, S_m\}$ is a family of subsets of $V$. We can assume $n$ is even, say $n = 2k$, and each set $S_i$ contains precisely $k = \frac{n}{2}$ elements, say $s^i_1, s^i_2, \ldots, s^i_k$. The well-known inapproximability results for SetCover hold also under these assumptions, since we can think of enlarging a ground set $V$, originally on $k$ elements, by adding a set $V'$ of $k$ new elements, adding $V'$ to $\mathcal{S}$, and enlarging the other sets in $\mathcal{S}$ with elements from $V'$ until their size rises up to $k$. Let $M = m^3 n^3$ play the role of a sufficiently big positive integer. We propose to compare two genomes $G_1$ and $G_2$. Here, $G_1$ is the simple genome $G_1$ of length $L_1 = M + n + m$ defined as follows: $G_1 = 1\ 2\ 3 \ldots L_1$. Genome $G_2$ has length $L_2 = M + m(k+1)$, and is constructed as follows:

$$G_2 = n{+}1, n{+}2, \ldots, n{+}M, s^1_1, s^1_2, \ldots, s^1_k, n{+}M{+}1, s^2_1, s^2_2, \ldots, s^2_k, n{+}M{+}2, \ldots$$
$$\ldots, s^{m-1}_1, s^{m-1}_2, \ldots, s^{m-1}_k, n + M + m - 1, s^m_1, s^m_2, \ldots, s^m_k, n + M + m$$

The above reduction leads us to the following result.

**Theorem 6.** *There exists a constant* $c > 0$ *such that* ESAD *(resp.* MSAD*) admits no* $(c \log |G_1|)$*-approximation algorithm, unless P=NP.*

## 6 Conclusion

In this paper, we have investigated the algorithmic complexity of the problem of computing similarity measures between genomes, in the case where genomes contain duplicates. This has been done for three measures: common intervals, MAD and SAD. We have shown that the three problems are **NP**-complete, for both the exemplar and matching variants. Moreover, we have provided APX-hardness results concerning MAD and SAD. Our results basically show that as soon as duplicates are present, the problems become hard, even in very restricted instances. Moreover, as can be seen, no APX-hardness result is known concerning common intervals ; we are currently investigating those questions.

## References

1. G. Blin, C. Chauve, and G. Fertin. The breakpoint distance for signed sequences. In *1st Int. Conference on Algorithms and Computational Methods for Biochemical and Evolutionary Networks, CompBioNets 2004*, volume 3 of *Texts in Algorithms*, pages 3–16. KCL Publications, 2004.
2. G. Blin and R. Rizzi. Conserved interval distance computation between non-trivial genomes. In *11th Int. Comp. and Combinatorics Conference (COCOON'05)*, volume 3595 of *LNCS*, pages 22–31, 2005.
3. D. Bryant. The complexity of calculating exemplar distances. In D. Sankoff and J. Nadeau, editors, *Comparative Genomics: Empirical and Analytical Approaches to Gene Order Dynamics, Map Alignment, and the Evolution of Gene Families*, pages 207–212. Kluwer Acad. Pub., 2000.
4. X. Chen, J. Zheng, Z. Fu, P. Nan, Y. Zhong, S. Lonardi, and T. Jiang. Assignment of orthologous genes via genome rearrangement. *IEEE/ACM Trans. on Comp. Biology and Bioinformatics*, 2(4):302–315, 2005.
5. M.R. Garey and D.S. Johnson. *Computers and Intractability: a guide to the theory of NP-completeness.* W.H. Freeman, San Franciso, 1979.
6. D. Sankoff. Genome rearrangement with gene families. *Bioinformatics*, 15(11):909–917, 1999.
7. D. Sankoff. Gene and genome duplication. *Curr. Opin. Genet. Dev.*, 11(6):681–684, 2001.
8. D. Sankoff and L. Haque. Power boosts for cluster tests. In *Comparative Genomics, RECOMB 2005 International Workshop, RCG 2005*, volume 3678 of *LNBI*, pages 121–130. Springer, 2005.
9. N. Cam Thach. Algorithms for calculating exemplar distances. Honours Year Project Report, National University of Singapore, 2005.

# Rearrangement of Noisy Genomes

Chunfang Zheng[1] and David Sankoff[2]

[1] Department of Biology,
University of Ottawa, Canada K1N 6N5
[2] Department of Mathematics and Statistics,
University of Ottawa, Canada K1N 6N5
{czhen033, sankoff}@uottawa.ca

**Abstract.** Measures of distance between genomic maps are inflated by high levels of noise due to incorrectly resolved paralogy and error at the mapping, sequencing and alignment levels. Comparison is also hampered by lack of information on gene orientation and lack gene order. We suggest a suite of algorithms for genome rearrangement analysis in the presence of noise and incomplete information, and test its robustness as noise levels increase.

## 1 Introduction

The comparison of genomic maps is susceptible to high levels of noise that threaten the accuracy and significance of the results. This noise is due to error at the mapping, sequencing and alignment levels, and to incorrect assignment of orthology due to gene, segment or genome duplication. Comparison is also sometimes hampered by lack of gene orientation information and poor resolution of gene order [7]. We suggest an approach to genome rearrangement analysis in the presence of noise and estimate its robustness as simulated noise levels increase.

Our method, designed particularly for unsigned genes (unknown DNA strand) in multichromosomal contexts, is focused on the combinatorial search for the optimal set of compatible "pre-strips". A pre-strip is a set of two or more genes, not necessarily contiguous, on a single chromosome in one genome matched with genes in the same order (or reversed) on any single chromosome in the other genome, i.e., a non-trivial common subsequence of the two chromosomes (one of them possibly reversed). Any set of compatible pre-strips, when all other genes have been deleted from the genomes, becomes a set of "strips"; a strip [5, 4] is a pre-strip with all terms contiguous on the chromosome.

The compilation of pre-strips for our analysis proceeds by an initial identification of all maximal, or inextensible, pre-strips, i.e., pre-strips not contained in another pre-strip, followed by the addition of certain non-maximal pre-strips contained in each maximal pre-strip. We then define a compatibility relation among all these pre-strips, namely whether or nor they can coexist in the same chromosome. This is submitted to a search for a maximum-weight clique (MWC), using the algorithm in [6], where the weights are just the number of genes in the

V.N. Alexandrov et al. (Eds.): ICCS 2006, Part II, LNCS 3992, pp. 791–798, 2006.

pre-strip. Though of course this algorithm may not execute in polynomial time, it is relatively efficient for dense graphs [1], such as the one representing pre-strips for multichromosomal genomes: here, a given pre-strip will not usually involve the same two chromosomes as another pre-strip, and hence will necessarily be compatible with it.

Once the maximal compatible set of pre-strips *cum* strips is output from the MWC routine, these can be signed by the method in [4], and then submitted to standard rearrangement (GR) algorithms [3, 8, 9].

A large proportion of singletons, i.e., genes not in any pre-strip and hence not considered in the MWC analysis, are nonetheless usually compatible with the output of the MWC and can be reincorporated into the solution, but this must be done judiciously, depending on the origins of the original data.

We test our method on genomes with 100 genes partitioned among four chromosomes, subjecting them to a fixed series of random translocations and inversions before the addition of various levels of noise. Under conditions of moderate noise, the method can assign almost all genes to strips and recover the rearrangements. With increasing noise levels, the distance calculation remains surprisingly stable, though the original rearrangements can no longer be inferred.

In the succeeding sections, we will describe each step as follows:

i) Compile set of all maximal (inextensible) pre-strips. Augment set by adding certain non-maximal pre-strips (Section 2).
ii) Apply MWC [6], with weight = number of genes in a strip (Section 3).
iii) Assign gene orientation according to strip analysis [5, 4] (Section 4).
iv) Input into standard genome rearrangement (GR) algorithm (Section 5).
v) Restoration of singletons (Section 6).
vi) Simulations (Section 7).

Steps i - iii are exact for the problem of maximum weight strip analysis when singletons are disregarded. Step i is NP-hard, but is tractable for a fixed number of pre-strips, and is very rapid for realistic data. Step ii is NP-hard and is the rate-limiting step in our analysis. Step iii takes polynomial time and is linear if there are no two-term strips in the output of Step ii. Step iv is polynomial. There are several options for Step v. We present exact solutions for three of them.

## 2 The Pre-strips

Because mapping and other errors are likely to result in singleton genes in discordant genomic contexts in the two species being compared, our strategy is to rely entirely on strips to decide which data are erroneous and which are meaningful. But we have to start with structures that are more general than strips in two ways. Intervening error singletons may disrupt a strip, destroying contiguity, or a gene that should be in a strip may be itself erroneously mapped out of the strip, leaving its left and right neighbours adjacent in one of the genomes.

The common order (or its reverse) of the terms of a strip in the two genomes is the only aspect of its structure we have access to. Thus we search for common

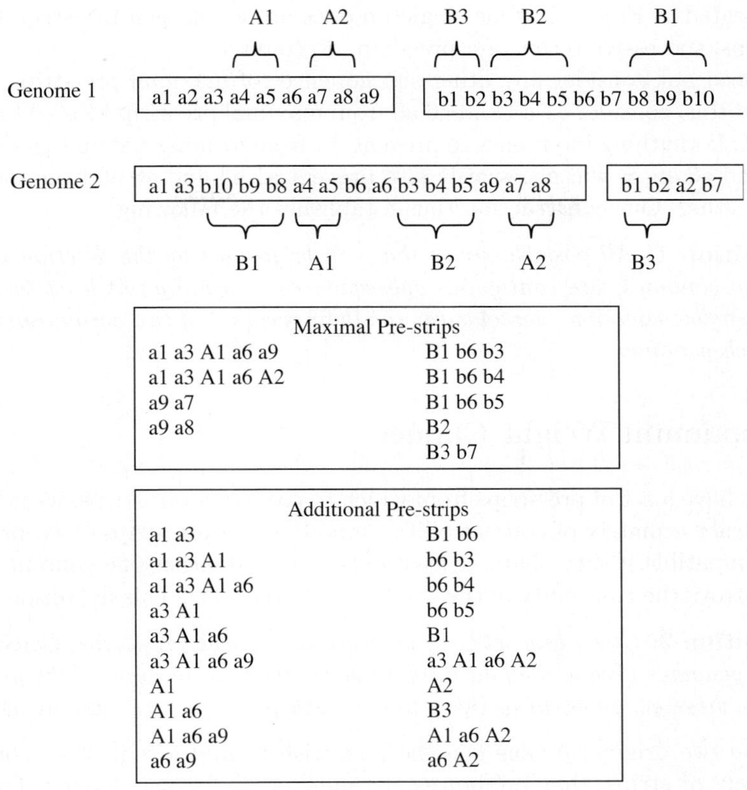


**Fig. 1.** Example of maximal pre-strips and additional pre-strips to be tested for compatibility. Genomes 1 and 2 both contain two chromosomes. Groups of contiguous genes in both genomes labeled with capital letters. All maximal pre-strips containing A1 intersect with all those containing B1, and are hence incompatible with them. Similarly, the three maximal pre-strips containing a9 are incompatible with each other.

subsequences in the two genomes as potential strips – pre-strips – relying on the subsequent analyses to eliminate the singletons and thus reveal the underlying strips. As a first step then we identify all maximal pre-strips by finding all inextensible common subsequences (and reversed order common subsequences) of the genes in all pairs of chromosomes, one chromosome from each genome. This is easily programmed to run in polynomial time, as long as the number of maximal pre-strips is a fixed parameter, or even when the number of such pre-strips is allowed to grow polynomially as a function of the number of genes.

Our goal is to find strips, i.e., to reduce the pre-strips to strips by eliminating extraneous genes from the chromosome that interrupt the contiguity of pre-strip genes. As shown in Figure 1, two maximal pre-strips may intersect, so that both cannot be present in a solution. However all or part of one and part of the other may be present, so that some non-maximal pre-strips must be considered in our search strategy. It suffices to use all contiguous fragments of a maximal pre-strip, as illustrated in Figure 1. (One singleton does not count as a pre-strip, but two singletons, successive terms of a pre-strip, do count.)

We need not consider any other subsequences of maximal pre-strips. To understand this, consider a1 a3 and a6 a9 from maximal pre-strip a1 a3 A1 a6 a9 in Figure 1. If anything intervenes to prevent A1 from forming a strip together with a1 a3 and a6 a9, it will necessarily also prevent a1 a3 and a6 a9 from forming a strip together. Our construction thus establishes the following:

**Proposition 1:** *All possible strips that can be formed by the deletion of genes from two genomes, are contiguous subsequences, containing at least two genes, of inextensible common subsequences (or their reverse) of two chromosomes, one from each genome.*

## 3 Maximum Weight Cliques

Once we have a set of pre-strips from which to construct our strips, we next need to construct a matrix of compatibilities among these pre-strips. Two pre-strips are incompatible if they share at least one gene or if either one contains a gene that destroys the contiguity of the genes in the other one. This definition entails:

**Proposition 2:** *Given any set $C$ of pairwise compatible pre-strips. Consider the reduced genomes produced by deleting all genes that are in none of the pre-strips in $C$. In these genomes all of the genes in each pre-strip in $C$ appear as strip.*

From the two original $n$-gene genomes, we wish to find a reduction, composed completely of strips, that minimizes the number $d$ of genes deleted. From the compatibility matrix we construct a graph $G$ with the pre-strips as the vertices and with compatible vertices joined by an edge.

**Proposition 3:** *The solution $C$ of the maximum weighted clique problem on $G$, where the weight $w(i)$ on a pre-strip $i$ is simply the number of genes it contains, induces a reduction of the original genomes so that they are composed completely of strips and so that the number of genes deleted is minimized.*

*Proof:* Let $W = \sum_{i \in C} w(i)$. Then the statement follows from Propositions 1 and 2 and

$$d = n - W. \tag{1}$$

Kumlander's algorithm [6] is based on a heuristic vertex colouring of sets of independent vertices, followed by two sorts of pruning of the clique search tree, one based on colour classes and the other a backtrack search. Empirically, it was demonstrated to work better than competing algorithms on denser graphs. In the case of genome comparison, we would expect $G$ to become dense, in the sense of [1], as the number of chromosomes increases, as explained in Section 1.

In a preliminary implementation of this algorithm, it worked well with hundreds of pre-strips but ran into computing time problems with thousands of vertices.

For the example in Figure 1, the algorithm gives the solution in Figure 2.

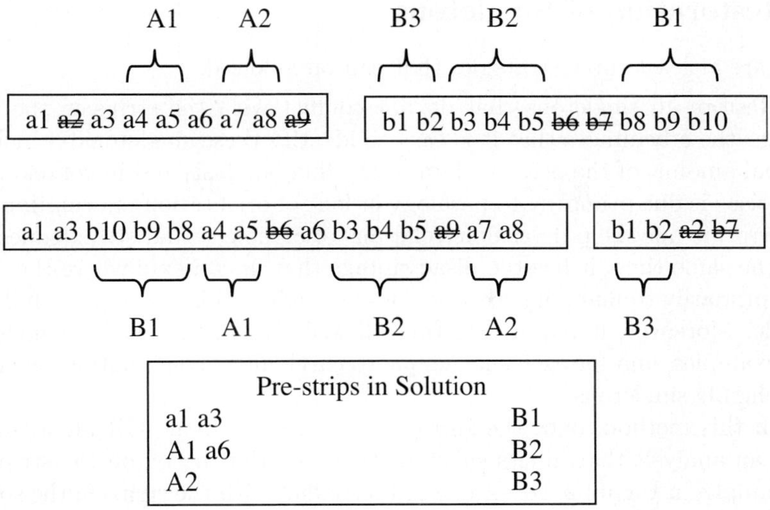


**Fig. 2.** MWC solution of problem in Figure 1. Rejected singletons a2, a9, b6 and b7 crossed out. Weight =15 (out of 19 genes), number of discarded singletons $d = 4$.

## 4 Sorting Unsigned Genomes

The problem of adding signs to an unsigned genome so as to a achieve a minimal reversal distance to the identity permutation $1, \cdots, n$ is equivalent to the problem of sorting an unsigned permutation, known to be NP-hard [2].

However, as conjectured in [5] and proved in [4], for all segments of the permutation consisting of three or more consecutive integers (strips) in increasing order, plus signs can be given to all these integers, and for all decreasing strips, such as B1 in the example of Figures 1 and 2, minus signs can be given, and this assignment is consistent with a solution. In [4], it is also shown how to give signs to 2-strips in polynomial time. The algorithm these authors develop is exponential only in $s$, the number of singletons, and is polynomial if $s$ is $O(\log n)$. In our case, however, $s = 0$, so sign assignment can be done in polynomial time.

## 5 Genome Rearrangement

To try to infer the rearrangements we used to generate the original data, we constructed all the cycles in the breakpoint graph of the two genomes [3, 9]. A summary statistic is provided by $r = \frac{2(b-c)}{b}$, the reuse rate, where $b$ is the number of breakpoints in the comparison of the two genomes, and $c$ is the number of cycles in their breakpoint graph. The addition of noise generally causes $r$ to increase by adding proportionately more breakpoints than cycles and increasing the sizes of cycles. In the example, $b = 5, c = 1, d = 4$ and $r = 1.6$.

## 6 Restoration of Singletons

There are at least three strategies that can be followed.

1) Restore all singletons that do not conflict with the strips in the MWC. This has the advantages that it is easy to identify these singletons, it includes a maximal amount of the original data in the final analysis and in contexts where the "noise" is due primarily to paralogy or lack of orientation information rather than error, it may help determine relations of orthology and strandedness.

At the same time, it has the disadvantage that in contexts where the "noise" is due primarily to mapping error, it does not reduce this noise to the full extent possible. Moreover, it requires testing all MWCs, which may be numerous in some examples, and for each such clique to check all $2^s$ combinations of signs on the $s$ eligible singletons.

With this method, once the singletons are restored, the GR algorithm may return an analysis that differs substantially from that based on the strips only. For example, in Figure 2, genes a2 and b6 conflict with the strips in the solution, and hence cannot be restored to the genomes. At the other extreme, the singleton b7 may be restored without materially affecting the rearrangement inference of three inversions and one translocation. The singleton a9 does not conflict with any strips in the solution, but its inclusion in the GR arrangement changes the solution, requiring an additional inversion. This is a disadvantage only if a9 represents erroneous information. If its singleton status is due to genomic processes, lack of information, or simply insufficient data, the GR analysis may give a more accurate result when such singletons are included.

2) Restore all singletons that do not substantially change the analysis based on the strips only, such as b7. This has the advantage that it only includes items that are maximally compatible with the strip-based analysis, so it should be more effective in excluding noise, while helping resolve lack of orientation. It does increase the amount of the original data included in the final analysis, but this may represent only a small proportion of the singletons. A clear difficulty with this approach is that it requires some operationalization of the notion that two GR analysis based on two different data sets, one subsumed in the other, are substantially the same. For example, we could ask that some optimal GR scenario for the larger set of genes, reduced by omitting all genes absent from the smaller set, also be optimal for the smaller set, taking into account rearrangements that disappear, such as inversions containing genes only in the larger set.

3) Restore no singletons. While this reduces noise maximally, it has the clear disadvantage that it takes into account none of the potentially informative singletons, especially when the noise is not simply mapping error. However, for our artificial (simulated) data, this is the most appropriate method.

We have implemented a program that permits any of these three options, but we present here simulations that test only the final option, generating data that are randomly noisy, without any attempt to model biological or methodological sources of error. Note that a global solution, treating singletons as if they were strips would require a different weighting system for MWC to avoid a trivial solution, and would also have to confront the NP-hardness of GR in this context.

## 7 Simulations

Prior to our simulations, we divided 100 genes approximately equally among four chromosomes, subjected them to four translocations and 13 inversions, with randomly chosen breakpoints.

For each simulation trial, we add noise by randomly choosing one gene and moving it to a new, random, position in the genome. This models errors such as the mistaken identification of a gene in one genome to be orthologous to one in the other, when it is only a paralog of the true ortholog. We repeat this process, eventually moving a total of 50 genes in a trial. We do 50 such trials in all.

In each trial, we apply our method after each of the 50 random changes and calculate the number of genes in the MWC solution and the distance between the genomes. On the left of Figure 3 we show how the distance remains remarkably stable, despite the loss of over 40 % of the genome to noise. More telling, we calculate the reuse rate (2 x distance / number of breakpoints), a measure of how random one genome is with respect to the other, and thus how much doubt should be thrown on the genome rearrangement analysis. The right side of Figure 3 shows a steady increase in reuse, eventually approaching 2.0, the maximum value, indicative of complete randomness. Although the distance remains more or less constant, the rearrangements inferred between the two genomes change drastically as noise is added. Thus while our method seems to infer genome distance in a robust way, the details of the rearrangement scenario are obscured by increased noise levels.

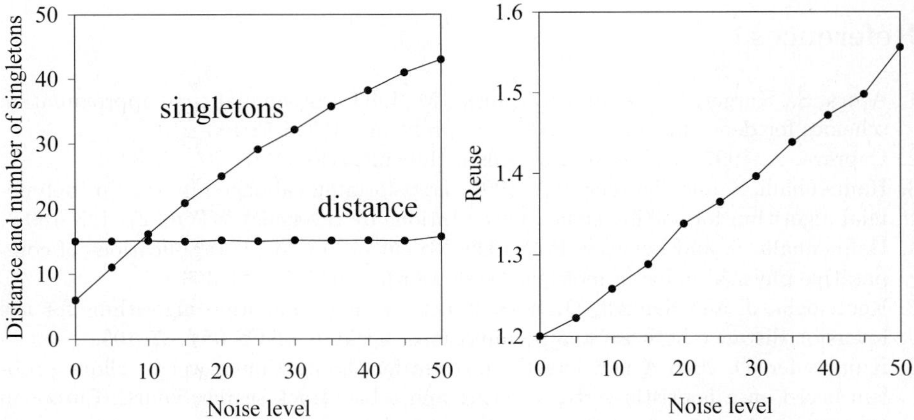


**Fig. 3.** Left: stability of genome distance (inversions plus translocations) despite the loss of large numbers of singletons. Right: increase in reuse as a result of noise.

## 8 Conclusions

The main contribution of this work is the conversion of the maximal weight strip problem to the MWC problem, based on the induced elimination of as few genes as possible from the genomes being compared. Our method recovers the underlying rearrangements while the noise level is low, but is degraded by higher levels. In our simulations, the total genomic distance, though not the actual rearrangements, remain fairly constant despite high levels of noise.

Though our pre-strip computation is not worst-case polynomial-time, in real problems it is not the bottleneck. Nevertheless, we are now implementing a polynomial time algorithm for producing only 2- and 3- strips. Applying the MWC algorithm to these small strips, followed by piecing all overlapping ones together, gives the same solution as the present method.

The MWC itself which is the bottleneck. Since our compatibility graph is dense, methods such as those introduced in [1] might speed up the MWC search.

It might be thought that with the advent of genome sequencing, comparative mapping without strandedness would become obsolete. In fact, the trend towards low-coverage sequencing without finishing, leaving many gene order ambiguities, makes it likely that physical and genetic mapping methods will continue to predominate, aside from very few model organisms.

## Acknowledgements

Research supported in part by grants from the NSERC (Canada). DS holds the Canada Research Chair in Mathematical Genomics and is a Fellow of the Evolutionary Biology Program of the Canadian Institute for Advanced Research.

## References

1. Arora, S., Karger, D. and and Karpinski, M. 1999. Polynomial time approximation schemes for dense instances of NP-hard problems. JCCS 58:193–210.
2. Caprara, A. 1997. Sorting by reversals is difficult. RECOMB '97. 75–83.
3. Hannenhalli, S. and Pevzner, P.A. 1995. Transforming cabbage into turnip (polynomial algorithm for sorting signed permutations by reversals). STOC 27. 178–189.
4. Hannenhalli, S. and Pevzner, P.A. 1996. To cut or not to cut (applications of comparative physical maps in molecular evolution). SODA 7. 304–313.
5. Kececioglu, J. and Sankoff, D. 1993. Exact and approximation algorithms for the inversion distance between two permutations. CPM 4, LNCS 684. 87–105.
6. Kumlander, D. 2005 A new exact algorithm for the maximum-weight clique problem based on a heuristic vertex-coloring and a backtrack search. Fourth European Congress of Mathematics. Poster and manuscript.
7. Sankoff, D., Lenert, A. and Zheng, C. 2005. Reversals of fortune. RECOMB 2005 Ws on Comparative Genomics, LNBI 3678. 131–141.
8. Tesler, G. 2002. Efficient algorithms for multichromosomal genome rearrangements. JCSS 65:587–609.
9. Yancopoulos, S., Attie, O. and Friedberg, R. 2005. Efficient sorting of genomic permutations by translocation, inversion and block interchange. Bioinformatics 21: 3340 – 3346.

# Simple Reconstruction of Binary Near-Perfect Phylogenetic Trees*

Srinath Sridhar[1], Kedar Dhamdhere[2], Guy E. Blelloch[1], Eran Halperin[3], R. Ravi[4], and Russell Schwartz[5]

[1] Computer Science Dept, CMU
srinath@cs.cmu.edu
[2] Google Inc, Mountain View, CA
kedar.dhamdhere@gmail.com
[3] ICSI, University of California, Berkeley
heran@icsi.berkeley.edu
[4] Tepper School of Business, CMU
ravi@cmu.edu
[5] Department of Biological Sciences, CMU
russells@andrew.cmu.edu

**Abstract.** We consider the problem of reconstructing near-perfect phylogenetic trees using binary character states (referred to as BNPP). A perfect phylogeny assumes that every character mutates at most once in the evolutionary tree, yielding an algorithm for binary character states that is computationally efficient but not robust to imperfections in real data. A near-perfect phylogeny relaxes the perfect phylogeny assumption by allowing at most a constant number $q$ of additional mutations. In this paper, we develop an algorithm for constructing optimal phylogenies and provide empirical evidence of its performance. The algorithm runs in time $O((72\kappa)^q nm + nm^2)$ where $n$ is the number of taxa, $m$ is the number of characters and $\kappa$ is the number of characters that share four gametes with some other character. This is fixed parameter tractable when $q$ and $\kappa$ are constants and significantly improves on the previous asymptotic bounds by reducing the exponent to $q$. Furthermore, the complexity of the previous work makes it impractical and in fact no known implementation of it exists. We implement our algorithm and demonstrate it on a selection of real data sets, showing that it substantially outperforms its worst-case bounds and yields far superior results to a commonly used heuristic method in at least one case. Our results therefore describe the first practical phylogenetic tree reconstruction algorithm that finds guaranteed optimal solutions while being easily implemented and computationally feasible for data sets of biologically meaningful size and complexity.

## 1 Introduction

Reconstruction of evolutionary trees is a classical computational biology problem [9]. In the maximum parsimony (MP) model of this problem one seeks the

* Supported in part by NSF grant CCR-0105548 and ITR grant CCR-0122581(The ALADDIN project)

V.N. Alexandrov et al. (Eds.): ICCS 2006, Part II, LNCS 3992, pp. 799–806, 2006.

smallest tree to explain a set of observed organisms. Parsimony is a particularly appropriate metric for trees representing short time scales, which makes it a good choice for inferring evolutionary relationships among individuals within a single species or a few closely related species. The intraspecific phylogeny problem has become especially important in studies of human genetics now that large-scale genotyping and the availability of complete human genome sequences have made it possible to identify millions of single nucleotide polymorphisms (SNPs) [18], sites at which a single DNA base takes on two common variants.

Minimizing the length of a phylogeny is the problem of finding the most parsimonious tree, a well known NP-complete problem [7]. Researchers have thus focused on either sophisticated heuristics or solving optimally for special cases (e.g. fixed parameter variants [1, 3, 13]). Previous attempts at such solutions for the general parsimony problem have only produced theoretical results, yielding algorithms too complicated for practical implementation. A large number of related work has been published but it is impossible to mention all of them here.

Fernandez-Baca and Lagergren recently considered the problem of reconstructing optimal near-perfect phylogenies [6], which assume that the size of the optimal phylogeny is at most $q$ larger than that of a perfect phylogeny for the same input size. They developed an algorithm to find the most parsimonious tree in time $nm^{O(q)}2^{O(q^2s^2)}$, where $s$ is the number of states per character, $n$ is the number of taxa and $m$ is the number of characters. This bound may be impractical for sizes of $m$ to be expected from SNP data, even for moderate $q$. Given the importance of SNP data, it would therefore be valuable to develop methods able to handle large $m$ for the special case of $s = 2$, a problem we call Binary Near Perfect Phylogenetic tree reconstruction (BNPP).

**Our Work:** Here we present theoretical and practical results on the optimal solution of the BNPP problem. We completely describe and analyze an intuitive algorithm for the BNPP problem that has running time $O((72\kappa)^q nm + nm^2)$, where $\kappa$ is the number of characters that violate the *four gamete* condition, a test of perfectness of a data set explained below. Since $\kappa \leq m$ this result significantly improves the prior running time by removing the big-oh from the exponent. Furthermore, the complexity of the previous work would make practical implementation daunting; to our knowledge no implementation of it has ever been attempted. Our results thus describe the first practical phylogenetic tree reconstruction algorithm that finds guaranteed optimal solutions while being easily implemented and computationally feasible for data sets of biologically meaningful size and complexity. We implement our algorithm and demonstrate it on a selection of real mitochondrial, Y-chromosome and bacterial data sets, showing that it substantially outperforms its worst-case bounds and yields far superior results to a commonly used heuristic method in at least one case.

## 2 Preliminaries

A phylogenetic tree $T$ is called *perfect* if for all states $s$ and characters $c$, all taxa having state $s$ at character $c$ lie in a connected component of the phylogeny.

Since the problem of reconstructing a perfect phylogeny is NP-complete [2, 17], Gusfield considered an important special case when the number of states is bounded by 2, called the binary perfect phylogeny problem (BPP). He showed that the BPP problem can be solved in linear time [8]. The problem we consider is an extension called the binary *near* perfect phylogeny reconstruction (BNPP).

In defining formal models for parsimony-based phylogeny construction, we borrow definitions and notations from Fernandez-Baca and Lagergren [6]. The input to the BNPP problem is an $n \times m$ matrix $I$ where rows $R$ represent *taxa* and are strings over states. The columns $C$ are referred to as *characters.* Thus, every taxon $r \in \{0,1\}^m$. In a *phylogenetic tree*, or *phylogeny*, each vertex $v$ corresponds to a taxon and has an associated label $l(v) \in \{0,1\}^m$.

**Definition 1.** *A* phylogeny *for a set of $n$ taxa $R$ is a tree $T(V,E)$ with the following properties:*

1. *if a taxon $r \in R$ then $r \in l(V(T))$*
2. *for all $(u,v) \in E(T)$, $H(l(u),l(v)) = 1$ where $H$ is the Hamming distance*

**Definition 2.** *For a phylogeny $T$:*

- *length(T) = $|E(T)|$*
- *penalty(T) = length(T) $- m$*
- *vertex $v$ of $T$ is* terminal *if $l(v) \in R$ and* Steiner *otherwise.*

**The BNPP problem:** Given an integer $q$ and an $n \times m$ input matrix $I$, where each row(taxon) $r \in \{0,1\}^m$, find a phylogeny $T$ such that *length(T)* is minimized or declare `NIL` if all phylogenies have penalty larger than $q$. The problem is equivalent to finding the minimum Steiner tree on a hyper-cube if the optimal tree is at most $q$ larger than the number of dimensions or declaring `NIL` otherwise. The problem is fundamental and therefore expected to have diverse applications besides phylogenies.

**Definition 3.** *We define the following additional* notations*:*

- *$r[i] \in \{0,1\}$: the state in character $i$ of taxa $r$*
- *$\mu(e) : E(T) \to C$: the character corresponding to edge $e = (u,v)$ with the property $l(u)[\mu(e)] \neq l(v)[\mu(e)]$*

We say that an edge $e$ *mutates* character $c'$ if $\mu(e) = c'$. We will use the following well known definition and lemma on phylogenies:

**Definition 4.** *The* set of gametes *$G_{i,j}$ for characters $i,j$ is defined as: $G_{i,j} = \{(k,l) | \exists r \in R, r[i] = k, r[j] = l\}$. Two characters $i,j \in C$ contain (all) four gametes when $|G_{i,j}| = 4$.*

**Lemma 1.** *[8] The most parsimonious phylogeny for input $I$ is not perfect if and only if $I$ contains the four-gamete property.*

**Input Assumptions:** If no pair of characters in input $I$ contains the four-gamete property, we can use Gusfield's elegant algorithm [8] to reconstruct a perfect phylogeny. We assume that the all zeros taxa is present in the input. If not, using our freedom of labeling, we convert the data so that it contains the same information with the all zeros taxa (see section 2.2 of Eskin et al [4] for details). We now remove any character that contains only one state. Such characters do not mutate in the whole phylogeny and are therefore useless in any phylogeny reconstruction. We now repeat the following preprocessing step. For every pair of characters $c', c''$ if $|G_{c',c''}| = 2$, we (arbitrarily) remove character $c''$. After preprocessing, we have the following lemma:

**Lemma 2.** *For every pair of characters $c', c''$, $|G_{c',c''}| \geq 3$.*

We will assume that the above lemma holds on the input matrix for the rest of the paper. Note that such characters $c', c''$ are identical (after possibly relabeling one character) and are usually referred to as non-informative. It is not hard to show that this preprocessing step does not change the correctness or running time of our algorithm.

**Conflict Graph $G$:** The *conflict graph $G$*, introduced by Gusfield et al. [11], is used to represent the imperfectness of the input in a graph. Each vertex $v \in V(G)$ of the graph represents a character $c(v) \in C$. An edge $(u, v)$ is added if and only if all the four gametes are present in $c(u)$ and $c(v)$. Let $VC$ be any minimum vertex cover of $G$. Damaschke [3] showed that the minimum number of characters that needs to be removed to support a perfect phylogeny is the minimum vertex cover of the conflict graph. Therefore $|VC|$ is a lower bound on $\texttt{penalty}(T_{opt})$ and this is often useful in practice. We now introduce new definitions that will be used to decompose a phylogeny:

**Definition 5.** *For any phylogeny $T$ and set of characters $C' \subseteq C$:*

- *a* super node *is a maximal connected subtree $T'$ of $T$ s.t. for all edges $e \in T', \mu(e) \notin C'$*
- *the* skeleton *of $T$, $s(T, C')$, is the tree that results when all super nodes are contracted to a vertex. The vertex set of $s(T, C')$ is the set of super nodes. For all edges $e \in s(T, C')$, $\mu(e) \in C'$.*

**Definition 6.** *A* tag *$t(u) \in \{0, 1\}^m$ of super node $u$ in $s(T, C')$ has the property that $t(u)[c'] = l(v)[c']$ for all $c' \in C'$, vertices $v \in u$; $t[u][i] = 0$ for all $i \notin C'$.*

Throughout this paper, w.l.o.g. we will deal with phylogenies and skeletons that are rooted at the all zeros taxa and tag respectively. Furthermore, the skeletons used in this work themselves form a perfect phylogeny in the sense that no character mutates more than once in the skeleton. Note that in such skeletons, tag $t(u)[i] = 1$ i.f.f. character $i$ mutates exactly once in the path from the root to $u$. Figure 3(a) shows an example of a skeleton of a phylogeny. We will use the term *sub-phylogeny* to refer to a subtree of a phylogeny.

## 3 Algorithm Description

Throughout the analysis, we fix an optimal phylogeny $T_{opt}$ and show that our algorithm finds it. We assume that both $T_{opt}$ and its skeleton is rooted at the

**function buildNPP** ( **binary matrix** $I$, **integer** $q$ )

1. let $G(V,E)$ be the conflict graph of $I$
2. let $V_{nis} \subseteq V$ be the set of non-isolated vertices
3. for all $M \in 2^{c(V_{nis})}$, $|M| \leq q$
   (a) construct rooted perfect phylogeny $\mathtt{PP}(V_{PP}, E_{PP})$ on characters $C \setminus M$
   (b) define $\lambda : R \mapsto V_{PP}$ s.t. $\lambda(r) = u$ i.f.f. for all $i \in C \setminus M$, $r[i] = t(u)[i]$
   (c) $T_f :=$ **linkTrees** (PP)
   (d) if $\mathtt{penalty}(T_f) \leq q$ then return $T_f$
4. return NIL

**Fig. 1.** Pseudo-code to find the skeleton

**function linkTrees** ( **skeleton** $Sk(V_s, E_s)$ )

1. let $S := \mathtt{root}(Sk)$
2. let $R_S := \{s \in R | \lambda(s) = S\}$
3. for all children $S_i$ of $S$
   (a) let $Sk_i$ be subtree of $Sk$ rooted at $S_i$
   (b) $(r_i, c_i) :=$ **linkTrees**$(Sk_i)$
4. let $\mathtt{cost} := \sum_i c_i$
5. for all $i$, let $l_i := \mu(S, c_i)$
6. for all $i$, define $p_i \in \{0,1\}^m$ s.t. $p_i[l_i] \neq r_i[l_i]$ and for all $j \neq l_i$, $p_i[j] = r_i[j]$
7. let $\tau := R_S \cup (\cup_i \{p_i\})$
8. let $D \subseteq C$ be the set of characters where taxa in $\tau$ differ
9. guess root taxa of $S$, $r_S \in \{0,1\}^m$ s.t. $\forall i \in C \setminus D, \forall u \in \tau,\ r_S[i] = u[i]$
10. let $c_S$ be the size of the optimal Steiner tree of $\tau \cup \{r_S\}$
11. return $(r_S, \mathtt{cost} + c_S)$

**Fig. 2.** Pseudo-code to construct and link imperfect phylogenies

all zeros label and tag respectively. The high level idea of our algorithm is to first guess the characters that mutate more than once in $T_{opt}$. The algorithm then finds a perfect phylogeny on the remaining characters. Finally, it adds back the imperfect components by solving a Steiner tree problem. The algorithm is divided into two functions: `buildNPP` and `linkTrees` and the pseudo-code is provided in Figures 1 and 2.

Function `buildNPP` starts by determining the set of characters $c(V_{nis})$ that corresponds to the non-isolated vertices of the conflict graph in Step 2. From set $c(V_{nis})$, the algorithm then selects by brute-force the set of characters $M$ that mutate more than once in $T_{opt}$. Only characters corresponding to non-isolated vertices can mutate more than once in any optimal phylogeny (a simple proof follows from Buneman graphs [16]). Since all characters of $C \setminus M$ mutate exactly once, the algorithm constructs a perfect phylogeny on this character set using Gusfield's linear time algorithm [8]. The perfect phylogeny is unique because of Lemma 2. Note that $PP$ is the skeleton $s(T_{opt}, C \setminus M)$. Since the tags of the skeleton are unique, the algorithm can now determine the super node where every taxon resides as defined by function $\lambda$ in Step 3b. This rooted skeleton $PP$ is then passed into function `linkTrees` to complete the phylogeny.

Function `linkTrees` takes a rooted skeleton $Sk$ (sub-skeleton of $PP$) as argument and returns a tuple $(r, c)$. The goal of function `linkTrees` is to convert skeleton $Sk$ into a phylogeny for the taxa that reside in $Sk$ by adding edges that mutate $M$. Notice that using function $\lambda$, we know the set of taxa that reside in skeleton $Sk$. The phylogeny for $Sk$ is built bottom-up by first solving the phylogenies on the sub-skeleton rooted at children super nodes of $Sk$. Tuple $(r, c)$ returned by function call to `linkTrees`$(Sk)$ represents the cost $c$ of the optimal phylogeny when the label of the root vertex in the root super node of $Sk$ is $r$. Let $S =$ `root`$(Sk)$ represent the root super node of skeleton $Sk$. $R_S$ is the set of input taxa that map to super node $S$ under function $\lambda$. Let its children super nodes be $S_1, S_2, \ldots$. Assume that recursive calls to `linkTrees`$(S_i)$ return $(r_i, c_i)$. Notice that the parents of the set of roots $r_i$ all reside in super node $S$. The parents of $r_i$ are denoted by $p_i$ and are identical to $r_i$ except in the character that mutates in the edge connecting $S_i$ to $S$. Set $\tau$ is the union of $p_i$ and $R_S$, and forms the set of vertices inferred to be in $S$. Set $D$ is the set of characters on which the labels of $\tau$ differ i.e. for all $i \in D, \exists r_1, r_2 \in \tau, r_1[i] \neq r_2[i]$. In Step 9, we guess the root $r_S$ of super node $S$. This guess is 'correct' if it is identical to the label of the root vertex of $S$ in $T_{opt}$. Notice that we are only guessing $|D|$ bits of $r_S$. Corollary 1 of Lemma 3 along with optimality requires that the label of the root vertex of $T_{opt}$ is identical to $\tau$ in all the characters $C \setminus D$:

**Lemma 3.** *There exists an optimal phylogeny $T_{opt}$ that does not contain any degree 2 Steiner roots in any super node.*

*Proof.* Figure 3(b) shows how to transform a phylogeny that violates the property into one that doesn't. Root `10` is degree 2 Steiner and is moved into parent supernode as `01`. Since `10` was Steiner, the transformed tree contains all input.

**Corollary 1.** *In $T_{opt}$, the LCA of the set $\tau$ is the root of super node $S$.*

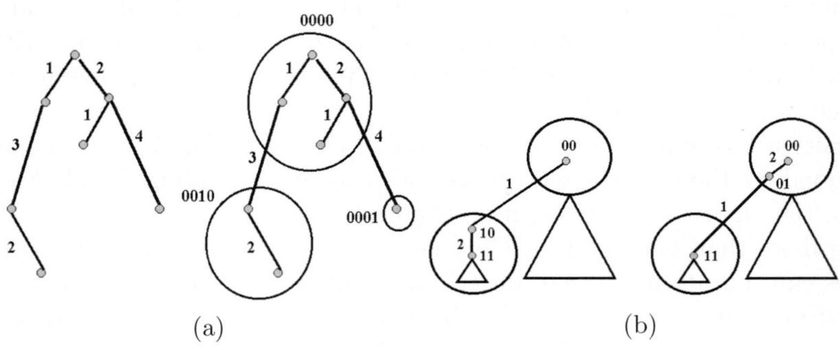


**Fig. 3.** (a) Phylogeny $T$ and skeleton $s(T, C')$, $C' = \{3, 4\}$. Edges are labeled with characters that mutate $\mu$ and super nodes with tags $t$. (b) Transform to remove a degree 2 Steiner root from a super node. Note: the size of the phylogeny is unchanged.

| Input size (Rows × Cols) and Desc | Penalty of opt(q) | Parsimony Score | Remarks and *total* run-time of our algorithm Intel P4 2.4Ghz, 1G RAM |
|---|---|---|---|
| 24 × 1041 mtDNA genus Pan [19] | 2 | 63 | `pars` program of `phylip` [5](default parameters) gives parsimony score 252; 0.59 secs |
| 15 × 98 chr Y, genus Pan [19] | 1 | 99 | identical to original paper which uses branch-and-bound; 0.33 secs |
| 17 × 1510 Bacterial DNA sequence [14] | 7 | 96 | 0.47 secs |
| 150 × 49, HapMap chr Y, 4 ethnic groups [12] | 1 | 16 | 0.3 secs |

**Fig. 4.**

In step 10, the algorithm finds the cost of the optimum Steiner tree for the terminal set of taxa $\tau \cup \{r_S\}$. We use Dreyfus-Wagner recursion [15] to compute this minimum Steiner tree. The function now returns $r_S$ along with the cost of the phylogeny rooted in $S$ which is obtained by adding the cost of the optimum Steiner tree in $S$ to the cost of the phylogenies rooted at $c_i$. The following Lemma bounds the running time of our algorithm and completes the analysis:

**Lemma 4.** *The algorithm described above runs in time* $O((18\kappa)^q nm + nm^2)$ *and solves the BNPP problem with probability at least* $2^{-2q}$*. The algorithm can be easily derandomized to run in time* $O((72\kappa)^q nm + nm^2)$*.*

*Proof.* The probability of a correct guess at Step 9 in function `linkTrees` is exactly $2^{-|D|}$. Notice that the Steiner tree in super node $S$ has at least $|D|$ edges. Since $\texttt{penalty}(T_{opt}) \leq q$, we know that there are at most $2q$ edges that can be added in all of the recursive calls to `linkTrees`. Therefore, the probability that all guesses at Step 9 are correct is at least $2^{-2q}$. The time to construct the optimum Steiner tree in step 10 is $O(3^{|\tau|}2^{|D|})$. Assuming that all guesses are correct, the total time spent in Step 10 over all recursive calls is $O(3^{2q}2^q)$. Therefore, the overall running time of the randomized algorithm is $O((18\kappa)^q nm + nm^2)$. To implement the randomized algorithm, since we do not know if the guesses are correct, we can simply run the algorithm for the above time, and if we do not have a solution, then we restart. *Although presented as a randomized algorithm for ease of exposition, it is not hard to see that the algorithm can be derandomized by exploring all possible roots at Step 9.* The derandomized algorithm has total running time $O((72\kappa)^q nm + nm^2)$.

## 4 Experiments and Conclusion

We tested the derandomized algorithm using non-recombining DNA sequences. In such sequences, the most likely explanation for a pair of characters exhibitingall four gametes is recurrent mutations. The results are summarized in Figure 4.

**Conclusion:** We have presented an algorithm for inferring optimal near-perfect binary phylogenies that improves the running time of the previous method. This problem is of considerable practical interest for phylogeny reconstruction from SNP data. In practice, we find that the algorithm significantly outperforms its worst case running time. Our algorithm is easily implemented unlike previous theoretical algorithms. At the same time, the algorithm returns guaranteed optimal solution unlike popular fast heuristics such as `pars`.

## References

1. R. Agarwala and D. Fernandez-Baca. A Polynomial-Time Algorithm for the Perfect Phylogeny Problem when the Number of Character States is Fixed. In: *SIAM Journal on Computing*, 23 (1994).
2. H. Bodlaender, M. Fellows and T. Warnow. Two Strikes Against Perfect Phylogeny. In proc *ICALP*, (1992).
3. P. Damaschke. Parameterized Enumeration, Transversals, and Imperfect Phylogeny Reconstruction. In proc *IWPEC*, (2004).
4. E. Eskin, E. Halperin and R. M. Karp. Efficient Reconstruction of Haplotype Structure via Perfect Phylogeny. In *JBCB* 2003.
5. J. Felsenstein. PHYLIP version 3.6. Distributed by the author. Department of Genome Sciences, University of Washington, Seattle. (2005).
6. D. Fernandez-Baca and J. Lagergren. A Polynomial-Time Algorithm for Near-Perfect Phylogeny. In: *SIAM Journal on Computing*, 32 (2003).
7. L. R. Foulds and R. L. Graham. The Steiner problem in Phylogeny is NP-complete. In: *Advances in Applied Mathematics* (3), (1982).
8. D. Gusfield. Efficient Algorithms for Inferring Evolutionary Trees. In: *Networks*, 21 (1991).
9. D. Gusfield. Algorithms on Strings, Trees and Sequences. *Cambridge University Press*, (1999).
10. D. Gusfield and V. Bansal. A Fundamental Decomposition Theory for Phylogenetic Networks and Incompatible Characters. In proc: *RECOMB* (2005).
11. D. Gusfield, S. Eddhu and C. Langley. Efficient Reconstruction of Phylogenetic Networks with Constrained Recombination. In Proc *IEEE CSB* (2003).
12. The International HapMap Consortium. The International HapMap Project. *Nature* 426 (2003).
13. S. Kannan and T. Warnow. A Fast Algorithm for the Computation and Enumeration of Perfect Phylogenies. In *SIAM Journal on Computing*, 26 (1997).
14. M. Merimaa, M. Liivak, E. Heinaru, J. Truu and A. Heinaru. Functional co-adaption of phenol hydroxylase and catechol 2,3-dioxygenase genes in bacteria possessing different phenol and p-cresol degradation pathways. *unpublished*
15. H. J. Promel and A. Steger. The Steiner Tree Problem: A Tour Through Graphs Algorithms and Complexity. *Vieweg Verlag* (2002).
16. C. Semple and M. Steel. Phylogenetics. *Oxford University Press* (2003).
17. M. A. Steel. The Complexity of Reconstructing Trees from Qualitative Characters and Subtrees. In *J. Classification*, 9 (1992).
18. S. T. Sherry, M. H. Ward, M. Kholodov, J. Baker, L. Pham, E. Smigielski, and K. Sirotkin. dbSNP: The NCBI Database of Genetic Variation. In *Nucleic Acids Research*, 29 (2001).
19. A. C. Stone, R. C. Griffiths, S. L. Zegura, and M. F. Hammer. High levels of Y-chromosome nucleotide diversity in the genus Pan. In *Proceedings of the National Academy of Sciences* (2002).

# Reconstructing Ancestor-Descendant Lineages from Serially-Sampled Data: A Comparison Study

Patricia Buendia[1], Timothy M. Collins[2], and Giri Narasimhan[1]

[1] Bioinformatics Research Group (BioRG),
School of Computing and Information Science,
Florida International University, Miami, FL 33199, USA
giri@cis.fiu.edu
[2] Department of Biological Sciences,
Florida International University, Miami, FL 33199, USA

**Abstract.** The recent accumulation of serially-sampled viral sequences in public databases attests to the need for development of algorithms that infer phylogenetic relationships among such data with the goal of elucidating patterns and processes of viral evolution. Phylogenetic methods are typically applied to contemporaneous taxa, and result in the taxa being placed at the tips or leaves of the tree. In a serial sampling scenario an evolutionary framework may offer a more meaningful alternative in which the rise, persistence, and extinction of different viral lineages is readily observable. Recently, algorithms have been developed to study such data. We evaluate the performance of 5 different methods in correctly inferring ancestor-descendant relationships by using empirical and simulated sequence data. Our results suggest that for inferring ancestor-descendant relationships among serially-sampled taxa, the MinPD program is an accurate and efficient method, and that traditional ML methods, while marginally more accurate, are far less efficient.

## 1 Introduction

The modeling of viral evolution can be greatly improved through the study of samples isolated at different periods in time which can lead to a better understanding of the diseases caused by pathogens such as HIV-1, human Influenza A and Hepatitis C. Understanding the within-host evolution of pathogens also has implications for the development of new therapies. An increasing amount of data from rapidly evolving viral organisms sampled serially from a single host is now available in the public databases. Unlike contemporaneous data, serially-sampled data contains taxa ancestral to other taxa, and may be placed at internal nodes of an evolutionary framework (see Fig. 1) providing a more specific hypothesis of evolutionary relationships.

Holmes et al. investigated the evolution of the V3 region of the HIV envelope gene by analyzing sequences of plasma viral RNA donated over a seven-year period by a single patient [1] and created the following "evolutionary framework" (Fig. 1), stating that *"unlike most molecular phylogenies, real ancestors may be present in the data and the framework expresses the postulated ancestor-descendent relationships."* [1].

V.N. Alexandrov et al. (Eds.): ICCS 2006, Part II, LNCS 3992, pp. 807 – 814, 2006.

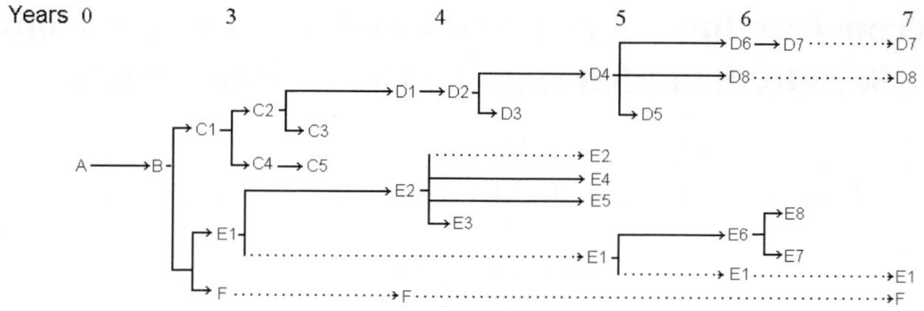


**Fig. 1.** Evolutionary Framework relating 24 different amino acid sequences found in the V3 loop. Redrawn in rectangular format from Holmes et al. Fig. 2 [1]. Time scale is given along the top. Dashed lines indicate identical sequences.

Hillis et al. pioneered the use of known molecular phylogenies, producing a known T7 phage phylogeny in the laboratory [2]. Cunningham et al. extended this work by serially propagating six bifurcating lineages of bacteriophage T7 according to the protocol of Hillis et al. resulting in a data set with known phylogeny [3]. We used Cunningham's experimental evolution data in our analysis as well as the Holmes viral data set and will henceforth refer to the HIV data as the Holmes92 data set, and the T7 phage sequences as the Cunningham97 data set.

More recently, several researchers have attempted to adapt existing phylogenetic methods to analyze serially-sampled data (see Methods section). The goal of our study is to compare five phylogenetic methods and assess how well they capture ancestor/descendant relationships in lineages of serially-sampled sequence data. The methods were tested on the two published phylogenies described above and on simulated data. The coalescent method was used to generate the simulated data sets. Since in practice only a small fraction of the total number of sequences representing a lineage is sampled, we incorporated a random sampling step into our simulation strategy.

Although it has been well documented that recombination is an important process in retroviral evolution [4, 5], there is only one existing method (MinPD) that directly addresses the study of serially-sampled data in the presence of recombination [6]. In this study however, MinPD's recombination detection feature was turned off.

## 2 Methods

In the last five years, several methods have been devised to study serially-sampled data, many of which are variants of phylogenetic methods for contemporaneous taxa. Three of the methods–sUPGMA, TipDate, and BEAST [7-9]–assume a molecular clock, i.e., assume a constant rate of evolution. They are further constrained by the traditional tree style of handling contemporaneous data. We are not including BEAST in this comparison study as it outputs a distribution of trees, while the computation of our performance score requires a single tree with branch lengths. Moreover, when choosing the tree with the highest likelihood, BEAST performs comparable to TipDate, but its computation time is at least twice as long as TipDate (see also Table 2).

Although traditional phylogenetic methods typically assume taxa are contemporaneous, it is possible to modify these methods to allow taxa to be designated as ancestors. Therefore, we also chose fastDNAml, an efficient implementation of the traditional ML method [10, 11]. Below we provide a detailed description of each method.

**Sequential-linking algorithm.** (SeqLink) We chose to implement (in C language) the NJ version of the algorithm (as described in [12, 13]). The algorithm is based on the evolutionary framework published by Holmes et al. [1] and is based on two assumptions:

1. The sequence from time point n with the minimum distance to some sequence in sampling period n+1 is the ancestor of "all" sequences from sampling period n+1.
2. The ancestor of a sequence was sampled at the previous time period.

Ties are broken by using additional criteria involving NJ branch lengths. Distances were measured using the JC69 distance, as other distance measures decreased the accuracy of the algorithm.

**MinPD.** The distance-based MinPD method attempts to improve on the performance of the previous algorithm, SeqLink, by avoiding the strong assumptions noted above. MinPD calculates pairwise distances using the Tamura-Nei 93 method with gamma rate heterogeneity and finds a closest ancestor among all preceding sampling time periods by searching the distance matrix for minimum distances.

**TipDate.** (Version 1.2) was designed to compute Maximum Likelihood (ML) estimates of the mutation rate from a set of non-contemporaneous input sequences (dated tips) assuming a molecular clock and a known tree topology [8]. As tree input to TipDate we used the topology estimated by the fastDNAml method. TipDate recomputes the tree branch lengths to fit the molecular clock assumption.

**sUPGMA.** This is a distance-based program modified from the UPGMA method, which by definition assumes a constant rate of evolution [7]. The program was implemented as a command line script using the JAVA PAL 1.4 package available from the URL: http://bioweb.pasteur.fr/docs/PAL/. We used the rate estimated by TipDate as input for sUPGMA to analyze the data sets created with the clock model of evolution. To ensure a better performance for sUPGMA for the non-clock data sets, we settled for a mutation rate of 0.004.

**Maximum Likelihood.** (fastDNAml) As mentioned earlier, fastDNAml is an efficient implementation of the ML method chosen to examine how the best traditional phylogenetic methods perform with non-contemporaneous data [10, 11]. FastDNAml was used with default settings (HKY85, $T_s/T_v$ of 2, and empirical base frequencies).

### 2.1 Evaluation Measures and Tools

We performed comprehensive experimentation to compare the five methods mentioned above for their ability to correctly infer ancestor-descendant lineages. By studying the placement of ancestor sequences in the trees of empirical and simulated data sets, we could observe that ancestors are often assigned to very short branches, and therefore devised a score based on branch lengths. This measure is referred to as the *Performance Score* and is based on the percentage of correctly inferred relationships. For a given taxon, the closest ancestral relative is defined as the closest sequence from some previous sampling period corresponding to either the most recent sampled ancestor or to the closest sampled relative of an unsampled most recent ancestor.

As some methods output a phylogenetic tree without explicitly inferring ancestral relationships we created a program, Nwk2Ances, which reads in a phylogenetic tree in Newick format, and returns for each sequence the "closest" sequence from any previous sampling period. Given a phylogenetic tree with inferred lengths, Nwk2Ances uses an additive metric to search for the minimum path between a sampled sequence and a sampled ancestral sequence, where the path length is the sum of branch lengths along the path.

# 3 Results

## 3.1 Inferring Ancestor-Descendant Relationships: Empirical Data

The 31 T7 page Cunningham97 sequences were 2733 nucleotides long [3], a much more robust data set compared to the 89 short sequences (each of length 35 aa) used by Holmes et al. [1]. Other features, such as the presence of parallel evolution and the skewing of the mutational bias and the number of invariable sites by the mutagen, nitrosoguanidine, used in the propagation of the T7 page, presented additional challenges to phylogenetic reconstruction methods, especially those based on an assumption of a clocklike rate of evolution. Table 1 shows the performance scores for all five programs on the two empirical data sets. SeqLink recovered most of the Holmes92 relationships. It fared poorly with the Cunningham97 data set. The poor performance of the algorithm with the T7 page phylogeny may be due to the strong assumptions of the algorithm. MinPD recovered 100% of the Cunningham97 lineage relationships. As for the Holmes92 framework, the one notable difference was with sequence E8 sampled in year 6, for which MinPD chose sequence B (from year 3) as being a closer representative of its ancestral lineage over sequence E1 as postulated by Holmes et al. [1]. FastDNAml and the clock-based methods, TipDate and sUPGMA performed better on the Cunningham97 data set. Nwk2Ances was applied to the output trees of the ML and clock-based programs to calculate the performance score.

**Table 1.** Performance scores for empirical data

| Programs | Performance Scores | | |
|---|---|---|---|
| | Holmes92 | Cunningham97 | Average |
| *MinPD* | 95.65% | 100.00% | 97.83% |
| *fastDNAml* | 65.22% | 96.43% | 80.82% |
| *TipDate* | 69.57% | 92.86% | 81.21% |
| *sUPGMA PAL+TD rate* | 69.57% | 75.00% | 72.29% |
| *Seq-Link* | 78.26% | 10.71% | 44.49% |

## 3.2 Inferring Ancestor-Descendant Relationships: Simulated Data

A large number of DNA sequences were generated using the coalescent model of evolution [14]. These sequences were provided as input for the five programs under consideration. As before, evaluation was based on the Performance Score measure.

### 3.2.1 Generated Data Sets

The program Treevolve v1.3.2 was modified to return the sampled sequences from the internal nodes and the genealogy of only the sampled sequences [14]. The twister randomization function of SeqGen 1.2.7 was also added. Our modified version of Treevolve performed the following steps:

1. Generate the random tree with different combinations of tree generation parameters: Mutation rate, Recombination Rate, Clock, Number of Leaves (See Fig. 2).
2. Assign all nodes to sampling periods and randomly sample sequences from sampling periods using specified sampling parameters: Sample Size, Start of Sampling, Number of Periods (Fig. 2).
3. Output smaller tree containing only sampled sequences and linking nodes.

Twelve sets of 100 replicates each were generated for different parameter combinations. Parameters were selected based on information from published studies [15].

### 3.2.2 Simulation Results

Results were analyzed with the standard statistical software package, SPSS 13, by running 2-way ANOVAs on the program performance scores using one of the simulation or sampling parameters as a second variable. As variances were large and mostly overlapping, Post Hoc analysis (Bonferroni, $p<0.05$) was used to determine which differences in means were significant.

The data did not meet the assumptions of a normal distribution or homogeneous variances, suggesting an arcsine-sqrt transformation of the performance score variable. However, it is known that ANOVA is robust to violations of these assumptions as long as the data consists of large and roughly equal-sized samples [16]. Furthermore, boxplots (or normality plots) of the arcsine-sqrt score did not show any marked skewness, and the interaction and Post Hoc results remained identical with or without the transformation. Here we report the untransformed performance scores and show the graphs for the interaction effects that were found to be significant (see Fig. 2).

*Clock:* Among all our experiments, the only case when a clock method, TipDate, outperformed MinPD and ML, was when the data sets were generated assuming the clock-like model of evolution. Even for these data sets, the difference was within a margin of 4-6% (see Fig. 2) and was significant from MinPD's performance scores, but not from fastDNAML scores. The effect size, measured by the partial eta-squared value of 0.23, was the largest among all the interactions.
*Recombination Rate:* Picking one recombinant donor (parental sequence) out of two or more (a partial match) counted as a correct prediction. This scoring choice explains why the performances were not affected by the higher recombination rates, but also suggests that the programs placed one of the recombinant donors closer to the reference sequence about as frequently as they did with ancestors of non-recombinant sequences. Effect Size was only 1%. (MinPD's recombination detection feature was turned off).
*Mutation Rate:* MinPD and fastDNAml displayed improved performance at a mutation rate of $10^{-5}$, while the clock methods achieved their best performance at a faster rate of $10^{-4}$, which resulted in highly divergent sequences. At a slower rate of $2\times10^{-6}$ (at $10^{-6}$ fastDNAml would severely slow down) all of the methods had difficulty inferring ancestor-descendant relationships. The effect size was small at 3%.

*Leaves:* More leaves imply a larger tree and therefore, a smaller sampling rate. The performance scores declined slightly for the larger trees. The partial eta-squared value was just 0.013. The performance scores of fastDNAml for the 50K leaves data sets were significantly better than that of the other programs, including MinPD.
*First Sampling Period (Start):* Earlier initiation of sampling resulted in sampling of sequences closer to the root. The plots confirmed that inferring relationships is more difficult when sampling starts after sequences have diverged and separated into different lineages. The interaction effect size was 9%.

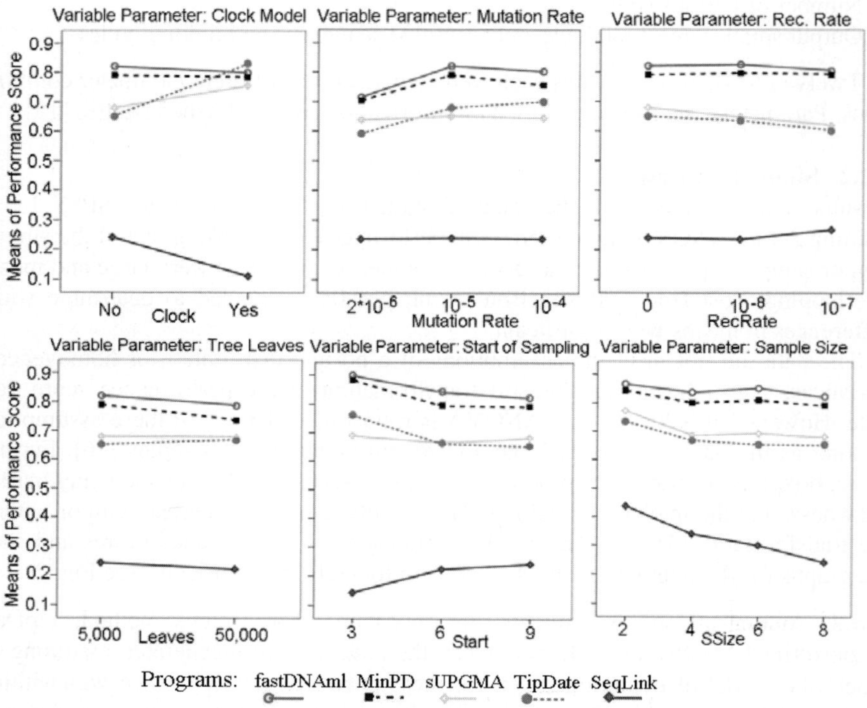


**Fig. 2.** Graphs showing dependence of the algorithm's performance scores on the different parameters. The analyses were based on experiments with 100 replicates generated with a modified version of Treevolve1.3.2. All parameters were fixed except the ones described in the horizontal x-axis. Wherever unspecified, parameters were fixed as follows: Mutation rate: $10^{-5}$; Start: 9; Periods: 6; Clock: No; Sample size: 8; Recombination rate: 0; Leaves: 5000, Sequence Length:1000, Model HKY, $T_s/T_v$:4, Gamma:0.5.

*Sample Size (SSize):* The interaction effect was very low (1%). Only the decline in performance (-5%) from size 2 to 4 for the top three programs was deemed significant by the ANOVA analysis. The lower error rates for smaller sample sizes are attributed to the smaller pool of ancestors that the programs end up choosing from.

## 4 Discussions

The investigation of ancestor-descendant lineages from serially-sampled viral sequences data will provide new insights into evolutionary patterns and potential development of the viral population in a host. We studied five different phylogenetic methods on two significantly different sets of viral sequence data: the Holmes92 [1] and the Cunningham97 data set [3], as well as on simulated data. We carried out simulations to better understand the effect of sampling and quality of data sets on the performances of these five methods. The results show that for inferring ancestor-descendant relationships, the distance method MinPD outperforms methods specifically designed to analyze serially-sampled sequence data. An important characteristic of MinPD is that it does not require an explicit assumption of a molecular clock.

The only method that performed consistently better (but rarely was the difference statistically significant) than MinPD is the ML-based fastDNAml, which was designed for contemporaneous data. Due to the similarity in performance of MinPD and fastDNAml, they were often grouped together in the same homogeneous subset by the Tukey range test procedure. MinPD is, however, several orders of magnitude faster than fastDNAml. Table 2 shows the average computation time for the five methods.

**Table 2.** Comparison of computation times for all five methods on 10 runs with 80 sequences of length 1000 on a Pentium 4, 2.40 GHz CPU with 512MB of RAM running Windows XP

| Program | Average Time | Standard Deviation |
|---|---|---|
| sUPGMA | <1 second | 0 |
| Tipdate | 67 minutes | 52.44 |
| fastDNAml | 38 minutes | 4.47 |
| MinPD | <1 second | 0 |
| SeqLink | <1 second | 0 |

The results of our simulation studies showed that certain sampling practices can adversely affect the outcome of the programs. Sampling from a larger tree increased the error rate by about 5%. Sampling early on, before the sequences have considerably diverged, produces better-defined phylogenetic relationships. In contrast, different sample sizes did not significantly affect program performance. The presence of a high rate of recombination did not increase the error rate when partial matches were counted towards the score. There is a decrease in performance for faster or slower mutation rates. As expected the two clock-based methods performed well with data generated by a clock model.

MinPD had a better performance than fastDNAml with the two empirical data sets in this study, it is possible however, that our simulation sampling procedure may have provided a slight advantage to a traditional method like fastDNAml as most sampling started later, i.e., closer to the leaves of the final time line. By starting to sample later we tried to emulate the scenario in which sampling from a patient starts in the later stages of infection when the virus has had ample time to evolve and diverge into different lineages. More studies with empirical data and simulated data generated under new combinations of parameters may provide a better understanding into the effects of data set composition on program performance.

## Acknowledgements

P.B. was supported by MBRS-RISE Fellowship (NIH/NIGMS R25GM61347). G.N. was supported in part by NIH Grant P01 DA15027-01.

## References

1. Holmes, E.C., et al.: Convergent and divergent sequence evolution in the surface envelope glycoprotein of human immunodeficiency virus type 1 within a single infected patient. Proc. Natl. Acad. Sci. USA.(89) 1992 p. 4835-4839
2. Hillis, D.M., et al.: Experimental phylogenetics: generation of a known phylogeny. Science. 255.(5044) 1992 p. 589-592
3. Cunningham, C.W., et al.: Parallel molecular evolution of deletions and nonsense mutations in bacteriophage T7. Mol Biol Evol. **14.**(1) 1997 p. 113-6
4. Robertson, D., et al.: Recombination in HIV-1. Nature. **374.** 1995 p. 124-6
5. Mikkelsen, J.G. and F.S. Pedersen: Genetic reassortment and patch-repair by recombination in retroviruses. Journal of Biomedical Sciences. **7.** 2000 p. 77-99
6. Buendia, P. and G. Narasimhan. MinPD: Distance-based Phylogenetic Analysis and Recombination Detection of Serially-Sampled HIV Quasispecies. In IEEE Computational Systems Bioinformatics Conference. Stanford, CA. (2005)
7. Drummond, A. and A.G. Rodrigo: Reconstructing genealogies of serial samples under the assumption of a molecular clock using serial-sample UPGMA (sUPGMA). Molecular Biology and Evolution. **17.** 2000 p. 1807-1815
8. Rambaut, A.: Estimating the rate of molecular evolution: Incorporating non-contemporaneous sequences into maximum likelihood phylogenies. Bioinformatics. **16.** 2000 p. 395-399
9. Drummond, A. and A. Rambaut 2003. BEAST v1.0. http://evolve.zoo.ox.ac.uk/beast/
10. Olsen, G.J., et al.: fastDNAml: A tool for construction of phylogenetic trees of DNA sequences using maximum likelihood. Comput. Appl. Biosci. **10.** 1994 p. 41-48
11. Felsenstein, J.: Evolutionary trees from DNA sequences: A maximum likelihood approach. J. Mol. Evol. **17.** 1981 p. 368-376
12. Ogishima, S., F. Ren, and H. Tanaka: Reconstruction and Analysis of Within-host Longitudinal HIV-1 Evolution by a Distance-based Sequential-linking Algorithm. Chem-Bio Informatics Journal. **1(2).** 2001 p. 73-83
13. Ren, F., S. Ogishima, and H. Tanaka: Longitudinal phylogenetic tree of within-host viral evolution from noncontemporaneous samples: a distance-based sequential-linking method. Gene. **317(1-2).** 2003 p. 89-95
14. Grassly, N., P. Harvey, and E. Holmes: Population dynamics of HIV-1 inferred from gene sequences. Genetics. **151.** 1999 p. 427-438
15. Shankarappa, R., et al.: Consistent Viral Evolutionary Changes Associated with the Progression of HIV 1 Infection. Journal of Virology. **73.**(12) 1999 p. 10489-10502
16. Glass, G., P. Peckham, and J.R. Sanders: Consequences of failure to meet assumptions underlying the fixed effects analysis of variance and covariance. Review of Educational Research. **42.** 1972 p. 237-288

# Robustness of Greedy Type Minimum Evolution Algorithms

Takeya Shigezumi

c/o Prof. O. Watanabe,
Dept. of Mathematical and Computing Sciences,
Tokyo Institute of Technology
Takeya.Shigezumi@is.titech.ac.jp

**Abstract.** For a phylogeny reconstruction problem, Desper and Gascuel [2] proposed Greedy Minimum Evolution algorithm (in short, GME) and Balanced Minimum Evolution algorithm (in short, BME). Both of them are faster than the current major algorithm, Neighbor Joining (in short, NJ); however, less accurate when an input distance matrix has errors. In this paper, we prove that BME has the same optimal robustness to such errors as NJ but GME does not. Precisely, we prove that if the maximum distance error is less than a half of the minimum edge length of the target tree, then BME reconstruct it correctly. On the other hand, there is some distance matrix such that maximum distance error is less than $\frac{2}{\sqrt{n}}$ of the minimum edge length of the target tree, for which GME fails to reconstruct the target tree.

## 1 Introduction

A phylogeny reconstruction problem is to determine an evolutionary tree representing relationships between a set of species from a distance matrix expressing the "closeness" of each pair of species. Recently, the phylogeny reconstruction is not only used for biology, but also for clustering various kinds of data, such as languages[10], music, etc.[11] To state the phylogeny construction problem formally, we need some notations. For any $n$, we use $L = \{1, \dots, n\}$ to denote a set of $n$ species. A distance matrix $D = (d_{ij})$ over $L$ , for each pair $i$ and $j$ in $L$, a distance $d_{ij} \geq 0$. Note that $d_{ij} = d_{ji}$ and that $d_{ii} = 0$. An evolutionary tree over $L$ is a tree such that its set of leaves is $L$ and all internal vertices are of degree three. For any tree $T$ and for any $i$ and $j$ in $L$, let $d_{ij}^T$ denote the length of the unique path between $i$ and $j$, i.e., the sum of the length of edges on the path. Now the phylogeny construction problem is stated as follows.

**Instance:** A distance matrix $D = (d_{ij})$ over $L = \{1, \dots, n\}$.
**Question:** Find a tree $T^*$ minimizing the following square error:

$$\sum_{i<j} \left| d_{ij} - d_{ij}^{T^*} \right|^2$$

V.N. Alexandrov et al. (Eds.): ICCS 2006, Part II, LNCS 3992, pp. 815–821, 2006.

Day [8] showed that this problem is NP-hard in general. It is, however, possible to "reconstruct" a tree in polynomial time if a distance matrix is sufficiently close enough to the one for a given tree. Note that for any tree $T$, this $T$ itself is the unique solution for the phylogeny construction problem on the distance matrix $D^T = (d^T_{ij})$. Several polynomial time algorithms have been proposed for reconstructing the original tree $T$ from $D^T$. Among them, Neighbor-Joining[4], its variants BIONJ [5] and WEIGHBOR[6], and Fitch-Margoliash as implemented by Felsenstein [9] are well known. NJ and its variants run in $O(n^3)$ for reconstructing a tree of $n$ leaves. Though polynomial, this running time is not sufficient enough for constructing a large evolutionary trees, and several attempts have been made to improve the efficiency .

Recently, Desper and Gascuel [2] introduced Greedy Minimum Evolution (GME) algorithm based on the ordinary least squares (OLS) branch length estimation and Balanced Minimum Evolution (BME) algorithm based on a branch length estimation scheme of Pauplin [3]. BME runs in $O(n^2\text{diam}(T))$ where $\text{diam}(T)$ is a diameter of the target tree, and GME runs in $O(n^2)$, which is optimal considering that the size of the input distance matrix is $O(n^2)$. Moreover, BME has an important property that the output tree has no edge with negative length [1]. Since edges with negative length are biologically meaningless, this property is also an advantage of BME.

In practice, a distance matrix has errors, that is, some distance $\delta_{ij}$ differs from the original distance $d^T_{ij}$. For measuring such errors, Atteson [7] introduced the notation of "$l_\infty$ radius." We say that a distance matrix $\Delta = (\delta_{ij})$ has *$l_\infty$ radius* $\alpha$ to an original distance $D^T = (d^T_{ij})$ if

$$\|D^T - \Delta\|_\infty = \max_{i<j} |d^T_{ij} - \delta_{ij}| < \alpha \min_{e \in E(T)} \{l(e)\},$$

where $E(T)$ is the set of edges of $T$ and $l(e)$ denotes the length of an edge $e$. An algorithm has *$l_\infty$ radius $\alpha$* if it yields the original $T$ for any distance matrix of $l_\infty$ radius $\alpha$ to $D^T$. Note that no algorithm has $l_\infty$ radius larger than $\frac{1}{2}$. Because if we allow errors larger than $\frac{1}{2}l(e)$, there exists one distance matrix made from two different trees with such errors. Atteson showed that NJ has $l_\infty$ radius $\frac{1}{2}$.

Desper and Gascuel showed the relationship between BME length estimation and weighted least squares branch length estimation [1] as a theoretical foundation of BME. However, there is still no theoretical analysis about $l_\infty$ radius for GME and BME. We analyze BME and GME, and obtain the following two theorems, which claim that BME has $l_\infty$ radius $\frac{1}{2}$ but GME does not.

**Theorem 1.** *BME has $l_\infty$ radius $\frac{1}{2}$, i.e., if the input distance matrix $\Delta$ satisfies*

$$\left\| D^{T^*} - \Delta \right\|_\infty < \frac{1}{2} \min_{e \in E(T^*)} \{l(e)\}$$

*BME outputs $T^*$ .*

**Theorem 2.** *Let $\alpha_G$ be an $l_\infty$ radius of GME. Then,*

$$\alpha_G < \frac{2}{\sqrt{n}}.$$

According to these theorems, BME has the same robustness as NJ, and it is faster than NJ. On the other hand, GME is faster than BME, it is less accurate in some cases. Our proof is obtained by a precise analysis of the proof of Desper and Gascuel [1] for the correctness of BME on error free distance matrix. Desper and Gascuel also proposed Balanced Nearest Neighborhood Interchange (BNNI) algorithm to improve a phylogeny tree into minimal one. However, according to our theorem, we do not need to use BNNI when BME is used for a distance matrix $\Delta$ with sufficiently small $\|D^{T^*} - \Delta\|_\infty$ error.

## 2 Preliminaries

We call $T^*$ the target tree being studied with an edge set $E(T^*)$ and edge weight function $l : E(T^*) \to \mathbb{R}$. Let $l^{T^*}_{\min} = \min_{e \in E(T^*)}\{l(e)\}$ and $\Delta = (\delta_{ij})$ be an estimate of the evolutionary distance between leaves $i$ and $j$. Let $E = (\varepsilon_{ij})$ be an error matrix given by following equation:

$$\varepsilon_{ij} = \delta_{ij} - d^{T^*}_{ij},$$

and $\varepsilon_0$ be the maximum error such that $\varepsilon_0 = \max_{i,j} |\varepsilon_{ij}|$. For a given tree $T$, we say $P^T_{ij}$ be a number of edges on the path connecting $i$ and $j$ on the tree $T$, and say $r(A)$ for any subtrees $A$ of $T$ be a root vertex of the subtree. Let $P^T_{AB}$ be a number of edges from $r(A)$ to $r(B)$ for any disjoint subtrees $A$ and $B$ . We will treat a subtree $A$ as a set of leaves for a simpler notation.

Pauplin [3] showed BME principle to estimate the edge length by modifying the OLS analytical formulae. Under this principle , the tree length is estimated as $\hat{l}(T, \Delta) = \sum_{i,j} 2^{1-P^T_{ij}} \delta_{ij}$. Consider the tree $T$ in Figure 1, where $k$ is inserted between subtrees $C$ and $A \cup B$, and assume that we have the size $\hat{l}(T, \Delta)$ of this new tree. If we know the size of $T$, we can quickly calculate the size of $T'$ using the following equation:

$$\hat{l}(T', \Delta) = \hat{l}(T, \Delta) + \frac{1}{4}\left\{\left(\delta^T_{AC} + \delta^T_{kB}\right) - \left(\delta^T_{AB} + \delta^T_{kC}\right)\right\}. \tag{1}$$

In this formula, $\delta^T_{AB}$ is defined as follows. If $A$ and $B$ each contains only one leaf, denoted as $a$ and $b$, respectively, then $\delta^T_{AB} = \delta_{ab}$, while if one of the two, say B, is made of two subtrees denoted as $B_1$ and $B_2$, then $\delta^T_{AB} = \frac{1}{2}\left(\delta^T_{AB_1} + \delta^T_{AB_2}\right)$.

Under the OLS length estimation principle, (1) is represented as follows:

$$\begin{aligned}\hat{l}(T', \Delta) - \hat{l}(T, \Delta) = \frac{1}{2}\big[&(\lambda - \lambda')\left(\Delta_{k|A} + \Delta_{B|C}\right)\\ &+(\lambda' - 1)\left(\Delta_{A|B} + \Delta_{k|C}\right) + (1 - \lambda)\left(\Delta_{A|C} + \Delta_{k|B}\right)\big]\end{aligned} \tag{2}$$

where $\lambda = \frac{(|A|+|B||C|)}{(|A|+|B|)(|C|+1)}$ , $\lambda' = \frac{(|A|+|B||C|)}{(|A|+|C|)(|B|+1)}$ and $\Delta_{A|B} = \frac{1}{|A||B|}\sum_{i\in A, j\in B} \delta_{ij}$.

We here introduce BME algorithm by Desper and Gascuel [2]. This algorithm starts with the tree of three leaves and adds the leaf to minimize the estimated tree length.

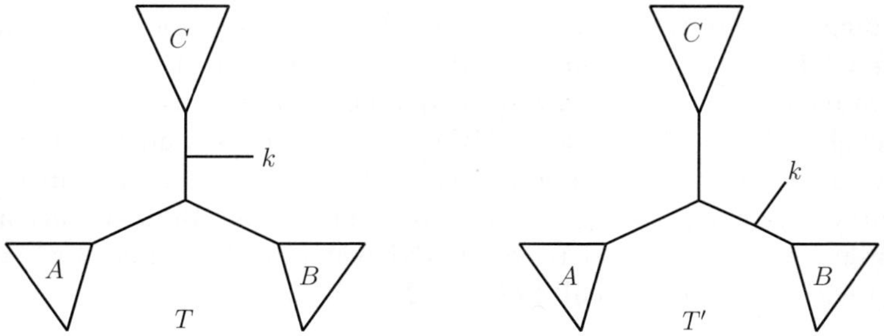


**Fig. 1.** Positions the leaf $k$ is inserted to

BME algorithm is as follows:

- For $k = 3$, initialize the matrix of balanced average distances between distant-2 subtrees and the array counting the number of leaves par subtree. Form $T_3$ with leaf set $\{1, 2, 3\}$.
- For $k = 4$ to $n$,
  1. Compute all $\delta_{kS}^{T_{k-1}}$ balanced average distances;
  2. Starting from an initial edge $e_0$ of $T_{k-1}$, set $f(e_0) = 0$, and recursively search each edge $e$ to obtain $f(e)$ from (1).
  3. Select the best edge by minimizing $f$, insert $k$ on the edge to form $T_k$, and update the balanced average distance between every pair of distant-2 subtrees as well as the number of leaves per subtree.
- Return $T_n$.

The time complexity of BME is $O(n^2 \mathrm{diam}(T^*))$. GME is almost same but using (2) and average distance $\Delta_{A|B}$ instead of the equation 1 and $\delta_{AB}^T$. The time complexity of GME is $O(n^2)$.

## 3 Robustness of BME

BME algorithm chooses the position of the leaf $k$ step by step. Let $T_k^*$ be a tree over $\{1, \ldots, k\}$ induced from the target tree $T^*$. The proof of Theorem 1 is by the induction of $k$. In $k = 3$, there is only one topology $T_3^*$. Let us consider the $k$-th round. By the induction hypothesis, BME reconstructs the target tree topology $T_{k-1}^*$. BME chooses $T_k$ minimizing $\hat{l}(T_k, \Delta_k)$, where $\Delta_k$ is a distance matrix on $\{1, 2, \ldots, k\}$ induced from $\Delta$.

For any $T_k' \neq T_k^*$, we prove the following inequality holds if $\varepsilon_0 < \frac{1}{2} l_{\min}^{T^*}$.

$$\hat{l}(T_k', \Delta_k) - \hat{l}(T_k^*, \Delta_k)$$
$$= \sum_{ij} 2^{1-P_{ij}^{T_k'}} \varepsilon_{ij} - \sum_{ij} 2^{1-P_{ij}^{T_k^*}} \varepsilon_{ij} + \hat{l}(T_k', D_k^{T^*}) - \hat{l}(T_k^*, D_k^{T^*}) > 0.$$

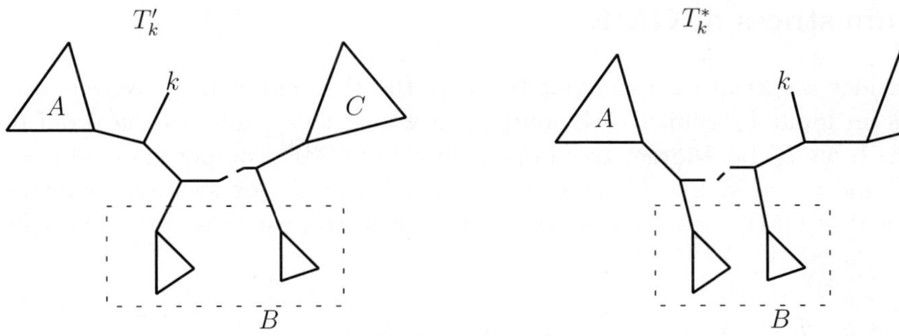


**Fig. 2.** $T_k^*$ and $T_k'$

We call the value of the first two terms of the last equation the error difference and the value of the last two terms the tree difference.

We here provide two lemmas about the error difference and tree difference. Their precise proofs are shown in [12]. Let $T_k'$ and $T_k^*$ , and name subtrees $A, B$ and $C$ of $T_{k-1}$ as shown in Figure 2. Let $p = P_{AC}^{T_{k-1}}$, i.e., number of edges on the path between the root of $A$ and the root of $C$ of $T_{k-1}$.

**Error Difference**

**Lemma 1**

$$\sum_{ij} 2^{1-P_{ij}^{T_k'}} \varepsilon_{ij} - \sum_{ij} 2^{1-P_{ij}^{T_k^*}} \varepsilon_{ij}$$
$$\geq -3\varepsilon_0 + 2 \cdot 2^{-p}\varepsilon_0 + \begin{cases} 3 \cdot 2^{-\frac{p}{2}}\varepsilon_0 & (p: \mathit{even}) \\ 2 \cdot 2^{-\lfloor \frac{p}{2} \rfloor}\varepsilon_0 & (p: \mathit{odd}) \end{cases}$$

**Tree Difference**

**Lemma 2**

$$\hat{l}(T_k', D_k^{T^*}) - \hat{l}(T_k^*, D_k^{T^*}) \geq l_{\min}^{T_k^*} \left(p - 2 + 2 \cdot 2^{-p}\right).$$

**Proof of the Theorem 1**

Now, we obtained that for any topology $T_k' \neq T_k^*$, the difference of the estimated tree length is

$$\hat{l}(T_k', \Delta_k) - \hat{l}(T_k^*, \Delta_k)$$
$$\geq l_{\min}^{T_k^*} \left(p - 2 + 2 \cdot 2^{-p}\right) - 3\varepsilon_0 + 2 \cdot 2^{-p}\varepsilon_0 + \begin{cases} 3 \cdot 2^{-\frac{p}{2}}\varepsilon_0 & (p: \text{even}) \\ 2 \cdot 2^{-\lfloor \frac{p}{2} \rfloor}\varepsilon_0 & (p: \text{odd}) \end{cases}$$

Under the condition $\varepsilon_0 = \left\| D^{T^*} - \Delta \right\|_\infty < \frac{1}{2} l_{\min}^{T^*} \leq \frac{1}{2} l_{\min}^{T_k^*}$, the estimated difference is larger than 0 for any $p \geq 2$. Note that if $p = 1$, then $T_k^* = T_k'$. ∎

# 4 Robustness of GME

We consider a special case of input to prove the theorem 2. If $\varepsilon_0$ exceeds $\alpha_G$, there is an input for which GME outputs a wrong tree. Same as the proof of theorem 1, let $T_k^*$ be a target tree in $k$-th round of GME, composed of subtrees $A, B, C$ and a leaf $k$. Let $T_k'$ be a tree obtained from $T_k^*$ by swapping subtree $A$ and leaf $k$. GME uses the following equation to compare the estimated tree lengths:

$$\hat{l}(T_k', \Delta_k) - \hat{l}(T_k^*, \Delta_k) = \frac{1}{2}\left[(\lambda - \lambda')\left(\Delta_{k|A} + \Delta_{B|C}\right) \right. \\ \left. +(\lambda' - 1)\left(\Delta_{A|B} + \Delta_{k|C}\right) + (1 - \lambda)\left(\Delta_{A|C} + \Delta_{k|B}\right)\right],$$

where $\lambda = \frac{(|A|+|B||C|)}{(|A|+|B|)(|C|+1)}$ and $\lambda' = \frac{(|A|+|B||C|)}{(|A|+|C|)(|B|+1)}$.

We obtained the following lemma by the same way as the theorem 1.

**Lemma 3.** *Let $e$ be an edge on $T_k^*$ separating $A$ and $B$ from $C$ and $k$.*

$$\hat{l}(T_k', \Delta_k) - \hat{l}(T_k^*, \Delta_k) = (1 - \lambda')l^{T_k^*}(e) - \begin{cases} 2(1-\lambda')\varepsilon_0 & (|B| \geq |C|) \\ 2(1-\lambda)\varepsilon_0 & (|B| < |C|) \end{cases}$$

**Proof.** See [12]. ∎

If $|B| \geq |C$ and $\varepsilon_0 < \frac{1}{2}l^{T_k^*}(e)$, GME chooses $T_k^*$. However, for $A, B, C$ and $\varepsilon_0$ satisfying $|B| < |C|$ and $\varepsilon_0 > \frac{(1-\lambda')}{2(1-\lambda)}l^{T_k^*}(e)$, GME chooses $T_k'$. Consider when the leaf $k$ is the last leaf. in this case, FASTNNI after GME cannot correct this fault.

Let $F(a, b, c)$ be:

$$F(a,b,c) = \frac{(1-\lambda')}{2(1-\lambda)} = \frac{(a+b)(c+1)(ab+c)}{(a+c)(b+1)(ac+b)}$$

where $a, b$ and $c$ are the cardinality of $A, B$ and $C$ respectively. By the definition of $\alpha_G$, if $\varepsilon_0 < \alpha_G l_{\min}^{T^*}$ then GME succeeds to output $T^*$, we obtained $\alpha_G \leq \min_{b<c, a+b+c+1=n} F(a,b,c)$. We observed that $F(a, b, c)$ is minimized when $a = \sqrt{n}, b = \sqrt{n} - 4$ and $c = n - 2\sqrt{n} + 3$. Now the theorem is obtained:

$$\alpha_G \leq \min_{\substack{0<a, 0<b<c, \\ a+b+c+1=n}} F(a,b,c) \leq F(\sqrt{n}, \sqrt{n}-4, n-2\sqrt{n}+3) \\ = \frac{2}{\sqrt{n}} - \frac{2}{n} - \frac{5}{n\sqrt{n}} + \frac{12}{n^2} + \ldots < \frac{2}{\sqrt{n}}.$$

# 5 Concluding Remarks

We proved BME has $l_\infty$ radius $\frac{1}{2}$ and GME has $l_\infty$ radius less than $\frac{2}{\sqrt{n}}$. Desper and Gascuel presented BNNI and FASTNNI to modify an initial tree to the minimal one. However, if $\varepsilon_0 < \frac{1}{2}\min_{e \in E(T^*)} l(e)$, we does not need to use BNNI. If $\varepsilon_0 > \frac{1-\lambda'}{2(1-\lambda)}\min_{e \in E(T^*)} l(e)$, and GME fails to put the last leaf, FASTNNI cannot correct such fault.

## References


1. R. Desper, O. Gascuel. Theoretical Foundation of the Balanced Minimum Evolution Method of Phylogenetic Inference and Its Relationship to Weighted Least-Squares Tree Fitting. *J. of Mol. Biol. Evol.* 2004 21(3): 587–598.
2. R. Desper, R. Gascuel. Fast and accurate phylogeny reconstruction algorithms based on the minimum evolution principle. *J. Comp. Biol.* 2002 9: 687–705.
3. Y. Pauplin. Direct calculation of a tree length using a distance matrix. *J Mol Evol.* 2000 51(1): 41–7.
4. N. Saitou, M. Nei. The Neighbor-Joining Method: a New Method for Reconstructing Phylogenetic Trees. *Mol Biol Evol.* 1987 4(4): 406–25.
5. O. Gascuel. BIONJ: an Improved Version of the NJ Algorithm Based on a Simple Model of Sequence Data. *Mol Biol Evol.* 1997 14(7):685–95.
6. Bruno WJ, Socci ND, Halpern AL. Weighted Neighbor Joining: a Likelihood-Based Approach to Distance-Based Phylogeny Reconstruction. *Mol Biol Evol.* 2000 17(1):189–97.
7. K. Atteson. The Performance of Neighbor-Joining Algorithms of Phylogeny Reconstruction. *Proceedings of the Third Annual International Conference on Computing and Combinatorics, Lecture Notes In Computer Science Vol. 1276* 1997 :101 – 110.
8. W. Day, D. Sankoff. Computational Complexity of Inferring Phylogenies by Compatibility. *Systematic Zoology,*1986 35(2):224–229.
9. J. Felsenstein. An Alternating Least Squares Approach to Inferring Phylogenies from Pairwise Distances *Systematic Biology,*1997 46(1):101–111.
10. D. Benedetto, E. Caglioti and V. Loreto. Language Trees and Zipping. *Physical Review Letters* 2002 88(4):048702-1–04872-4.
11. R. Cilibrasi, P. Vitányi. Clustering by Compression. *IEEE Transactions on Information Theory,* 2005 51(4):1523–1545.
12. T. Shigezumi. Robustness of Greedy Type Minimum Evolution Algorithms. *Dept. of Math. and Comp. Sciences Tokyo Institute of Technology Research Reports (Series C),* C-218, 2005. `http://www.is.titech.ac.jp/research/research-report/C/C-218.ps.gz`

# Space and Time Efficient Algorithms for Planted Motif Search*

Jaime Davila, Sudha Balla, and Sanguthevar Rajasekaran

CSE Department at University of Connecticut, Storrs
{jdavila, ballasudha, rajasek}@engr.uconn.edu

**Abstract.** We consider the $(l, d)$ Planted Motif Search Problem, a problem that arises from the need to find transcription factor-binding sites in genomic information. We propose the algorithms PMSi and PMSP which are based on ideas considered in PMS1 [10]. These algorithms are exact, make use of less space than the known exact algorithms such as PMS and are able to tackle instances with large values of $d$. In particular algorithm PMSP is able to solve the challenge instance $(17, 6)$, which has not reported solved before in the literature.

## 1 Introduction

The Planted Motif Search Problem arises from the need to find *transcription factor-binding sites* in genomic information and has been studied extensively in the biocomputing literature –see [11] for a literature survey.

The problem can be defined in the following formal way.

**Definition 1.** *Given a string $s$ with $|s| = m$ and a string $x$ with $|x| = l$ with $l < m$. We say $x \lhd_l s$ if $x$ is a subsequence of $s$. Equivalently we say that $x$ is an $l$-mer of $s$.*

**Definition 2.** *Given a set of strings $\{s_i\}_{i=1}^n$ over an alphabet $\Sigma$, with $|s_i| = m$ and $l$, $d$ with $0 \leq d < l < m$ we define the $(l, d)$* motif search problem *as that of finding a string $x$ with $|x| = l$ such that $s_i$ has an $l$-mer $x_i$ with*

$$d_H(x, x_i) = d \text{ for } i = 1, \ldots, n.$$

*We will call $x$ a* motif.

This problem is known to be NP-complete [6] and a PTAS exists for variants of the problem known as the Common Approximate Substring and Common Approximate String [7], [1]. However the high degree in the polynomial complexity of the PTAS makes it of little practical use.

Numerous algorithms have been implemented in order to solve instances of this problem. Among them we have Random Projection [2], MITRA [4], Winnower [8], Pattern Branching [9], Hybrid Sample and Pattern Driven Approaches [12], PMS1 [10], CENSUS [5] and Voting [3].

* This research was supported in part by the NSF Grant ITR-0326155.

V.N. Alexandrov et al. (Eds.): ICCS 2006, Part II, LNCS 3992, pp. 822–829, 2006.

Out of these algorithms, CENSUS, MITRA, PMS1 and Voting are in the exact category. The last two algorithms are able to work in reasonable time for practical instances where $m = 600$ and $n = 20$. In [2] the notion of a *challenging instance* was defined, as one where if the strings are selected at random, the expected number of $(l, d)$ motifs is greater than 1. We have that $(9, 2)$, $(11, 3)$, $(13, 4)$, $(15, 5)$ and $(17, 6)$ are challenging instances.

PMS1 is able to solve $(9, 2)$,$(11, 3)$ and $(13, 4)$ in less than a couple of minutes, and Voting is able to solve these instances and $(15, 5)$ –the last one in 22 min–. However as $d$ becomes larger the higher memory requirements get.

In this paper we propose the algorithms PMSi and PMSP that build upon ideas similar to the ones in PMS1, but are able to achieve better time results and less use of memory for practical as well as challenging instances. In particular we notice that PMSP is able to handle challenging instances like $(l, d) = (15, 5)$ and $(17, 6)$. No other exact algorithm has been reported to work on $(17, 6)$.

In section 2 we describe the algorithm PMS1 and our improvements PMSi and PMSP and prove some complexity bounds on them.

In section 3 we describe some of the experimental results we have obtained, and compare the results with the already existing exact algorithms.

## 2 Improved Algorithms Based on PMS1

PMS1 is a simple exact algorithm introduced in [10] which works as follows. It considers each $l$-mer in each input sequence and for each such $l$-mer $q$ it constructs a list of neighbors (i.e., $l$-mers) that are at a distance of $d$ from $q$. Neighbor lists of the input sequences are then intersected using radix sort to identify the planted motif. This algorithm works well in practice for values of $d \leq 3$ however as $d$ increases the memory requirement tends to be large.

We propose two algorithms PMSi and PMSP that improve on the memory requirements of PMS1 at the cost of possibly more computation time. However in practical instances we show how they perform better than PMS1 and are able to tackle instances which could not be solved by PMS1.

### 2.1 PMS1

Before we describe PMS1 [10] we consider the following definitions.

**Definition 3.** *For any string $x$, with $|x| = l$:*

1. *Let $B(x, d) := \{y : |y| = l \text{ and } d_H(y, x) = d\}$.*
2. *Let $\mathcal{N}(l, d) := |B(x, d)|$.*

**Definition 4.** *Given $s$, with $|s| = m$ and $0 \leq d < l \leq m$ let $L_s := \bigcup_{x \triangleleft_l s} B(x, d)$.*

PMS1 works by doing the following simple steps.

1. Build $L_i$ for $i = 1, \ldots, n$.
2. The set of $(l, d)$ motifs will be $M := \bigcap_{i=1}^{n} L_i$.

In order to do the intersections of step 2 in an efficient way, we keep the sets $L_i$ sorted in lexicographical order.

**Theorem 1.** *PMS1 can be implemented in $O(mn\frac{l}{w}\mathcal{N}(l,d))$ time and $O(m\frac{l}{w}\mathcal{N}(l,d))$ space, where $w$ is the word length of the computer.*

### 2.2 PMSi

PMSi works by generating $L_{s_{2i-1}} \cap L_{s_{2i}}$ for $i = 1, \ldots, \frac{n}{2}$ and then taking the intersection of these lists. The advantage is that $L_{s_{2i-1}} \cap L_{s_{2i}} \subset L_{s_{2i}}$, so it uses less memory than PMS1. In practice, the size of these sets can be substantially smaller than the size of $L_{s_{2i}}$.

In a more formal way we have the following.

**Definition 5.** *Let $1 \leq i \leq \lceil\frac{n}{2}\rceil$ we define $\bar{L}_i := L_{s_{2i-1}} \cap L_{s_{2i}}$.*

**Algorithm.** PMSi

1. Let $M := \bar{L}_1$.
2. Sort $M$ by lexicographical order using radix sort.
3. For $i = 2, \ldots, \frac{n}{2}$
   (a) Construct $\bar{L}_i$.
   (b) Sort $\bar{L}_i$ by lexicographical order using radix sort.
   (c) Construct $M \cap \bar{L}_i$ by merging $\bar{L}_i$ with $M$ and considering the $l$-mers which appear twice.
4. Output $M$.

In order to generate $\bar{L}_i$ we need to introduce the following definitions and results.

**Definition 6.** *Given $x$, with $|x| = l$, $s$ with $|s| = m$ and $0 \leq h \leq l$, we define $\mathcal{L}_{i,d}(x) := \{y \triangleleft_l s_i : d_H(x,y) \leq d\}$.*

**Lemma 1.** $\bar{L}_i = \bigcup_{x \triangleleft_l s_{2i-1}} \left( \bigcup_{y \triangleleft_l s_{2i}} B(x,d) \cap B(y,d) \right).$

*Proof.* It follows by using De Morgan's theorem.

**Lemma 2.** $\bar{L}_i = \bigcup_{x \triangleleft_l s_{2i-1}} \left( \bigcup_{y \in \mathcal{L}_{2i,2d}(x)} B(x,d) \cap B(y,d) \right).$

*Proof.* It follows from fact that if $d_H(x,y) > 2d$ then $B(x,d) \cap B(y,d) = \emptyset$ and Lemma 1.

Taking into account Lemma 2 we define the procedure GenInter($i$) which outputs $\bar{L}_i$.

**Procedure.** GenInter($i$)

1. Let $M = \emptyset$.
2. For each $l$-mer $x$ of $s_{2i-1}$:
   (a) Let $N := \mathcal{L}_{2i,2d}(x)$.
   (b) For each $l$-mer $y \in B(x, d)$
      i. For each $x' \in N$, if $d_H(y, x') = d$ add $y$ to $M$.
3. Output $M$.

It is useful when generating $\mathcal{L}_{2i,2d}(x)$, to order the elements of it in ascending order of distance towards $x$, in this way we calculate the distance against the $l$-mers which are closer, which have greater probability of being in the intersection.

It should also be noted that we represent every $l$-mers as a sequence of integers, as in [10]. In step 2(b)i the distance is calculated by using an array in which the distances of the sequences represented by one integers are cached.

Taking these into account, the following two results are straightforward.

**Theorem 2.** *PMSi can be implemented in* $O(nm^2 + \mathcal{S}\frac{l}{w}\mathcal{N}(l,d))$ *time and in* $O(\max_{i=1,\dots,\frac{n}{2}} |\bar{L}_i|)$ *space, where* $\mathcal{S} = \sum_{i=1}^{\frac{n}{2}} \sum_{x \triangleleft_l s_{2i-1}} |\mathcal{L}_{2i,2d}(x)|$.

**Corollary 1.** *PMSi can be implemented in* $O(nm^2\frac{l}{w}\mathcal{N}(l,d))$ *time using* $O(m\frac{l}{w}\mathcal{N}(l,d))$ *space.*

## 2.3 PMSP

PMSP follows the following simple idea. For every $l$-mer $x$ in $s_1$ it generates the set of neighbors of $x$ and tries to guess if an $l$-mer $y$ in that neighborhood is a motif by checking whether there are $l$-mers in $s_i$ for $i = 2, \dots, n$ that are at distance $d$ from it. This algorithm has been given in [10]. But we modify this algorithm in a critical way. The key observation is to notice that we do not need to evaluate the distance in all $l$-mers of $s_i$, but rather in the $l$-mers of $s_i$ which are at distance $\leq 2d$ from $x$.

This can be said in a formal way in the straighforward lemmas 3 and 4.

**Lemma 3.** *The set of motifs* $M$ *can be written as*

$$M = \bigcup_{x \triangleleft s_1} \left( B(x,d) \cap \bigcap_{i=2}^{n} L_i \right).$$

*Proof.* Follows by definition of $L_i$ and $M$.

**Lemma 4.** *Let $x \lhd_l s_1$ and $x' \in B(x,d)$. We have that $x' \in \bigcap_{i=2}^{n} L_i$ iff*

$$\forall i = 2, \ldots, n : \exists y_i \in \mathcal{N}(x, s_i, 2d) : d_H(x', y_i) = d.$$

*Proof.* Follows by lemma 3 and the use of lemma 2 iteratively.

Taking these considerations into account, we can write PMSP in the following way.

**Algorithm.** PMSP

1. Let $M = \emptyset$.
2. For each $x \lhd_l s_1$:
   (a) Construct $N(i) = \mathcal{L}_{i,2d}(x)$ for $i = 2, \ldots, n$.
   (b) For each $x' \in B(x,d)$:
      i. Check if for **every** $i$ ($2 \leq i \leq n$) there exists a $y_i \in N(i)$ such that $d_H(x', y_i) = d$.
      ii. In the affirmative add $x'$ to $M$.
3. Output $M$.

In PMSP we are using $O(m^2 n)$ space due to the fact that we need to calculate the distance from all $l$-mers in $s_1$ to all $l$-mers in $s_i$ for $i = 2, \ldots, n$. This can be done in $O(m^2 n)$ time and space by using the strategy described in [8]. Hence the following results hold.

**Theorem 3.** *PMSP can be implemented in $O(nm^2 + \frac{l}{w}\mathcal{N}(l,d)\mathcal{S})$ time and in $O(nm^2)$ space, where $\mathcal{S} = \sum_{i=1}^{\frac{n}{2}} \sum_{x \lhd_l s_{2i-1}} |\mathcal{L}_{2i,2d}(x)|$.*

**Corollary 2.** *PMSP can be implemented in $O(nm^2 \frac{l}{w} N(l,d))$ time and in $O(nm^2)$ space.*

## 2.4 Complexity Bounds in the Expected Case

According to corollaries 1 and 2, we have that the newly designed algorithms are slower than PMS1 by a factor of $m$ in the worst case. In this section we will argue than in the expected case the behavior is better.

If we assume that the set of strings $s_i$ is constructed by picking each character from every sequence with equal probability we can prove sharper estimates in the expected case. In doing so we follow the approach used in [2], which basically assumes that the probabilty of occurence of two different $l$-mers at different positions is independent from each other –which is not the case when the $l$-mers are in close proximity–. Furthermore we assume $\Sigma = \{A, C, G, T\}$.

The following results can be found in [2].

**Lemma 5.** *For a fixed* $0 \le d \le l$ *we can estimate the probability of two random* $l$*-mers being at a Hamming distance of* $\le d$ *as*

$$p_d := \sum_{i=0}^{d} \binom{l}{i} \left(\frac{3}{4}\right)^i \left(\frac{1}{4}\right)^{l-i}.$$

**Lemma 6.** *The expected number of* $(l, d)$ *motifs for a set of strings* $s_i$ *with* $i = 1, \ldots, n$, $|s_i| = m$ *and such that* $s_i$ *is picked at random using the previous model can be estimated by*

$$E(l, d, n) = 4^l \left(1 - (1 - p_d)^{m-l+1}\right)^n.$$

By applying the previous lemma we can estimate in a better way the time and space complexity in the expected case.

**Lemma 7.** *The expected value of* $\mathcal{S}$ *can be estimated as* $p_{2d} m^2 \frac{n}{2}$.

*Proof.* It follows from the definition of $\mathcal{L}_{i,d}(x)$ and $\mathcal{S}$, linearity of expectation and Lemma 5.

**Lemma 8.** *The expected value of* $|\bar{L}_i|$ *for* $i = 1, \ldots, \frac{n}{2}$ *can be estimated as* $E(l, d, 2) = 4^l \left(1 - (1 - p_d)^{m-l+1}\right)^2$.

*Proof.* It follows from Lemma 6.

**Theorem 4.** *The expected time complexity of PMSi is* $O(p_{2d} \frac{l}{w} n m^2 \mathcal{N}(l, d))$ *and the expected space complexity is* $O(\frac{l}{w} E(l, d, 2))$. *The expected time complexity of PMSP is* $O(p_{2d} \frac{l}{w} n m^2 \mathcal{N}(l, d))$.

One consequence of Theorem 4 is that the proposed algorithms will be better if $p_{2d} < \frac{1}{m}$. Also notice that they imply that if we fix $d$, in the expected case we are going to get better results in PMSi and PMSP if we increase the value of $l$. Notice that it also implies the worst time and space complexities are attained for values of $l$ and $d$ such that $E(l, d, n)$ is greater than 1.

## 3 Experimental Results

As described before we follow the experimental setting described in [8] and [2]. That is, we fix $n = 20$, $m = 600$ and consider different values of $l$ and $d$. In this model the strings are generated uniformly at random, and an $l$-mer is planted in random positions of these strings, mutating it in exactly $d$ places.

We implemented versions of PMS1, PMSi and PMSP in C, and run our programs on a Pentium4 2.40 GHz machine with a core memory size of 1GB.

Our version of PMS1 is coded in C as well –the original version of PMS1 is implemented in Java–. Our version of PMS1 is slower than the original one but it is a good framework to test our results.

**Table 1.** Comparison of PMS1 and PMSi for $d = 3$

| Algorithm | $l = 13$ Time (sec) | Mem (%) | $l = 14$ Time(sec) | Mem (%) | $l = 15$ Time(sec) | Mem (%) |
|---|---|---|---|---|---|---|
| PMSP | 3 | 2.8% | 2 | 2.8 % | 1.5 | 2.8% |
| PMSi | 28 | 1.2% | 15 | < 1 % | 10 | < 1% |
| PMS1 | 70 | 20.0% | 79 | 21.0% | 94 | 22.0% |

**Table 2.** Comparison of PMS2, PMSi, PMSP for $d = 4$

| Algorithm | $l = 13$ Time (min) | Mem (%) | $l = 14$ Time(min) | Mem (%) | $l = 15$ Time(min) | Mem (%) |
|---|---|---|---|---|---|---|
| PMSP | 2:12 | 2.8% | 1:36 | 2.8% | 1:34 | 2.8% |
| PMSi | 28:30 | 35% | 17:54 | 20% | 10:23 | 12% |
| PMS2 | 3:48 | 95% | 3:46 | 95% | 3:37 | 95% |

**Table 3.** Time Comparison In Challenging Problems Instances

| Algorithm | (9,2) | (11,3) | (13,4) | (15,5) | (17,6) |
|---|---|---|---|---|---|
| PMSP | 0.6s | 6.9s | 152s | 35m | 12h |
| Voting | 0.4s | 8.6s | 108s | 22m | – |
| PMSi | 4.6s | 111.0s | 18m | – | – |
| PMS1 | 3.0s | 44.9s | – | – | – |

Table 1 shows the running time in seconds as well as the memory usage in percentage for our implementations of PMS1, PMSi and PMSP for $d = 3$ and $l = 13, 14, 15$. When we compare PMSi with PMS1 we get a speedup between 5 and 8 times and between 10 and 20 times less memory. In the case of PMSP we get a speedup between 20 and 50 times and the amount of memory used is very low. Let us also notice that the speedup in time as well as in memory use gets higher for higher values of $l$ which is consistent with Theorem 4.

Table 2 shows the running time and memory usage of PMSi and PMSP for instances that PMS1 could not solve because of large memory requirements. We have to notice that already existing extensions of PMS1 such as PMS2 are able to handle the case of $d = 4$ as reported in [10]. In this table we see that PMSi takes more time than PMS2 and for bigger values of $l$ it gets faster and uses less memory. PMSP outperforms the two previous algorithms, using significantly less memory and the time it expends decreases as the value of $l$ increases.

In Table 3 we get to see the behavior of PMSP, PMSi, PMS1 and Voting Algorithm in different challenging instances. A '–' in the table means that the algorithm uses too much memory in that instance or its time is not reported. PMSi takes 3 times as much time as PMS1 in some cases, but it uses less memory, which allows it to solve the (13, 4) instance which cannot be solved by PMS1. PMSP clearly outperforms PMSi and PMS1 and uses less memory which allows it to solve bigger challenging instances. PMSP behaves well when compared with Voting, sometimes performing better and sometimes worse by no more than twice in time. However the main advantage of PMSP is its low use of memory which allows it to solve the $(17, 6)$ instance, which was not reported solved before.

## 4 Conclusions

We presented algorithms PMSi and PMSP that are based upon ideas used in PMS1. These algorithms are more space efficient than PMS1 and improve the running time of PMS1 in several cases. PMSP clearly outperforms PMS1 and PMSi in challenging instances and compares well with already existing exact algorithms while making use of significantly less memory. This property allows it to solve the challenging instance $(17, 6)$ which was not reported solved before.

## References


1. Balla, S., Davila, J., Rajasekaran, S.: Approximation Algorithms for the Primer Selection, Planted Motif Search and Related Problems. In: Approximation Algorithms and Metaheuristics. (To Appear)
2. Buhler, J., Tompa, M.: Finding motifs using random projections. Journal of Computational Biology **9**(2) (2002) 225–242
3. Chin, F.Y., Leung, H.C.: Voting algorithms for discovering long motifs. In: APBC. (2005)
4. Eskin, E., Pevzner, P.A.: Finding composite regulatory patterns in dna sequences. In: ISMB. (2002) 354–363
5. Evans, P., Smith, A.: Toward optimal motif enumeration. In: WADS. (2003)
6. Evans, P., Smith, A., Wareham, H.T.: On the complexity of finding common approximate substrings. TCS: Theoretical Computer Science **306** (2003)
7. Li, M., Ma, B., Wang, L.: On the closest string and substring problems. Journal of the ACM **49**(2) (2002) 157–171
8. Pevzner, P.A., Sze, S.H.: Combinatorial approaches to finding subtle signals in DNA sequences. In: ISMB. (2000) 269–278
9. Price, A.L., Ramabhadran, S., Pevzner, P.A.: Finding subtle motifs by branching from sample strings. In: ECCB. (2003) 149–155
10. Rajasekaran, S.: Exact algorithms for planted motif problems. Journal of Computational Biology **12**(8) (2005) 1117–1128
11. Rajasekaran, S.: Motif Search Algorithms. In: Handbook of Computational Molecular Biology. CRC Press (2005)
12. Sze, S.H., Lu, S., Chen, J.: Integrating sample-driven and pattern-driven approaches in motif finding. In: WABI. (2004)

# Predictability of Rules in HIV-1 Protease Cleavage Site Analysis

Hyeoncheol Kim[1], Tae-Sun Yoon[1], Yiying Zhang[2], Anupam Dikshit[2], and Su-Shing Chen[2]

[1] Dept. of Computer Science Education, Korea University, Seoul, 136-701, Korea
hkim@comedu.korea.ac.kr
[2] Computer & Info. Sciences and Engineering, University of Florida, Gainesville, FL. 32611, U.S.A.
suchen@cise.ufl.edu

**Abstract.** Symbolic rules play an important role in HIV-1 protease cleavage site prediction. Recently, some studies have done on extraction of the prediction rules with some success. In this paper, we demonstrated a decompositional approach for rule extraction from nonlinear neural networks. We also compared the prediction rules to the ones extracted by other approaches and methods. Empirical experiments are also shown.

## 1 Introduction

HIV-1 protease is an enzyme in the AIDS virus that is essential to its replication. Understanding HIV-1 protease cleavage site specificity is very desirable, because efficiently cleaved subtrates are also excellent templates for synthesis of tightly binding chemically modified inhibitors. HIV-1 protease inhibitor drugs are small molecules that bind to the active site in HIV-1 protease and stay there, so that the normal functioning of the enzyme is prevented.

Recently, many machine learning methods have been applied to the problem of HIV-1 protease cleavage site prediction [2, 3, 8, 9, 12, 7]. Neural networks have been successful for the cleavage site prediction yielding a prediction accuracy of 90-92%, which is superior than decision tree C5.0 of 85.5% and similar to Gaussian SVM [2, 8, 3]. Linear classifiers such as a perceptron and linear SVM were also used yielding accuracy of 88-92% while MLP produced accuracy of 86-92% according to size of training dataset [9].

Narayanan *et al.* addressed importance of symbolic interpretation in HIV-1 protease cleavage site prediction, and they demonstrated symbolic rule generation using decision trees [8]. Rules can guide future in vitro experiments which take into account various patterns of residues and identify the most important positions to mutate in the n-residue regions to test for cleavability [8, 9]. Even though their experiments showed that neural networks outperformed decision trees, they used a decision tree to generate symbolic rules because MLP is hard to interprete. Rögnvaldsson *et al.* used a linear model such as a simple perceptron or LSVM to generate rules because it is more easily interpreted than a nonlinear

V.N. Alexandrov et al. (Eds.): ICCS 2006, Part II, LNCS 3992, pp. 830–837, 2006.

one such as MLP. Rögnvaldsson *et al.* used a so-called exhaustive search method to generate rules from simple linear perceptron, in which they generated all possible candidate combinations and tested each one with a trained perceptron that is used as a function. However, this method is not practical for nonlinear model (i.e., MLP).

In this paper, we applied a decompositional approach to extract symbolic rules from trained nonlinear neural networks (i.e., MLP), which is computationally practical and safe. Also we compared the extracted rules with others from various methods in terms of their predictability.

## 2 Rule Discovery Using Neural Networks

As addressed in [8], machine learning algorithm needs to take into account combined influence of attributes on classification. Algorithms such as MLP do so by adjusting all the connection weights for each instance to ensure more accurate classification, while decision tree looks for just one attribute (i.e., the most important residue position) at each decision point in the tree on which to split the samples. However the knowledge learned by MLP can not be converted easily into symbolic rules.

In neural networks, since the knowledge learned through training is distributed to connection weights that are interconnected nonlinearly, it is very hard to interpret the black box into human-understandable symbolic forms. In recent years, there have been studies on rule extraction from feed-forward neural networks [1, 4, 5, 6, 10, 11]. The extracted rules provide human users with the capability to explain how the patterns are classified and may provide better insights about the domain.

There are two approaches of rule extraction from neural networks(i.e., MLP): pedagogical (black-box) approach and decompositional (white-box) approach. Pedagogical approaches take a trained neural network as a black-box and observe input patterns and their corresponding output patterns. Generic methods of the approaches take into account all possible combinations of an input vector which are then presented to the trained neural network for their validity evaluation. For our cleavage site prediction problem, domain size would be $20^8$ and the number of all possible candidate rules would be $21^8$. Also there are some complicated efficiency issues on how to evaluate a candidate rule and how to generate maximally general and accurate rules. Obviously, pedagogical approaches are not effective when the size of the neural network increases. In order to overcome this limitation, decompositional approaches apply heuristically guided search to the process of rule extraction.

Decompositional approaches to rule extraction from a trained neural network (i.e., MLP) involves the following phases:

1. Intermediate rules are extracted at the level of individual units within the network (i.e., hidden and output nodes).
2. The intermediate rules from each unit are aggregated to explain the logical mapping between the input and output nodes of the network.

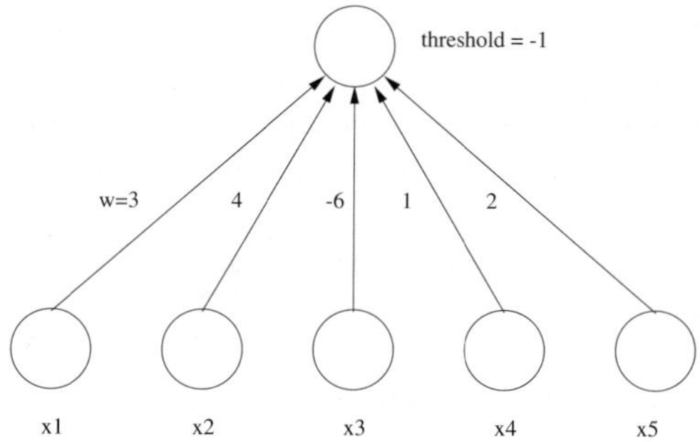


**Fig. 1.** An individual rule with five incoming connection weights and a threshold

At the first phase, it searches valid and maximally general (i.e., size of a combination is as small as possible) combinations by finding the combinations of weights of a node that will make it active. It is illustrated in figure 1 in which a non-input unit with 5 binary inputs, their corresponding incoming connection weights and a threshold is given. For example, a combination $(x_1, x_2)$ is a valid rule because its weight-sum (i.e., $3 + 4 = 7$) is always greater than the threshold (i.e., $-1$) regardless of the values of other incoming units.

However the process in phase 1 is inherently combinatorial. For $n$ binary inputs, the number of combinations is $3^n$ (i.e., exponential complexity) which is computationally very expensive as the $n$ grows. Thus, heuristics have been proposed to reduce the search space. We used Kim's OAS algorithm for our experiments, which orders incoming weights by their contributions to node activity and then searches valid and maximally-general rules using tree-based search structure in $O(n \log n)$ most cases and less than $O(2^{n-1})$ worst case [6]. Our implementation of the OAS algorithm uses two parameters to control the quantity and quality of the rules extracted: $k$ and *CF threshold*. At each node, it generates $k$ rules with the highest CF values among all the rules whose CF values are greater than the *CF threshold*. The CF value is defined as the output value of sigmoid function in the node. *CF threshold* is normally set to 0.8 for positive rules.

Decompositional approaches take each node in the network as a concept unit and the concepts in the concept node are logically aggregated with other concepts from other nodes. Thus, the approaches using the parameters can generate general rules excluding many other trivial rules. Therefore the rules extracted by this approach can be different from the rules by black-box approaches.

## 3 Experimental Results

### 3.1 Datasets

Cai and Chou [2] published a dataset with 362 peptides in 1998, which has been used as the standard dataset in many publications of this research domain [8, 9, 12]. The data set includes 362 examples with 114 positives (i.e., cleaved) and

**Table 1.** HIV-1 protease cleavage site datasets. *Default cleavability* represents probability of the cleaved of a default rule (i.e., "*Whatever it is, it is always cleaved*").

| Dataset | Cleaved | Uncleaved | Total | Default cleavability |
|---|---|---|---|---|
| Cai *et al.* | 114 | 248 | 362 | 31.49% |
| New dataset | 281 | 111 | 392 | 71.68% |
| Combined | 395 | 359 | 754 | 52.52% |

248 negatives (i.e., non-cleaved). Recently, You *et al.* collected 384 more peptides to create a 746-peptide dataset [12]. Since the new dataset is not available to public domain currently, we created our own new dataset in a similar way by collecting and refining peptide sequences from published literatures containing experimental data on oligopeptide sequences that have been exposed to the HIV-1 protease [12]. We collected 392 new examples and thus created a 754-peptide dataset without duplication. In this article, we call the datasets as a 362-dataset, a 392-dataset and a 754-dataset, respectively. Table 1 shows their distributions.

A sequence in the datasets is composed of 8 positions and each position represents one of 20 different amino acids. A peptide is denoted by $p_4p_3p_2p_1p_1'p_2'p_3'p_4'$ where each $p$ represents one of 20 amino acids. The scissile bond is located between positions $p_1$ and $p_1'$. For our experiments, we use 9 attributes for each octapeptide: location 1 though location 8 and class attribute (i.e., cleavagability). Location 1 corresponds to $p_4$ and location 8 to $p_4'$. We also used orthonormal encoding scheme that maps the sequence to a sparse orthonormal representation. Each amino acid $p_i$ is then represented by a 20 bit vector with only one bit set to one and the rest 19 bits to zero. This places an octamer in one of the corners of a $20 * 8 = 160$ dimensional hypercube.

Like other cleavage site prediction domains, the HIV-1 protease cleavage domain has a problem of dataset size. That is, the number of available sequences is too small compared to domain size: $362 \ll 2^8 \ll 2^{160}$. And there is no evidence that the 362 or 754 examples are representative for the domain. Thus, a question comes to us that classifiers created from just 362 or 754 examples can really classify any previously unknown instance that is picked from the space of $20^8$ instances.

### 3.2 Rule Extraction and Comparison

In our experiments, we first extract rules using different approaches and methods, and then compare quality of the rules with their predictability. The predictability of a classifier means how well a classifier created by a small number of examples can predict other instances previously unknown.

We used the 362-dataset as a training set, and the 392-dataset as a test set. Since the rules published in other papers have been created on the 362-dataset [8, 9, 7], we used the same training dataset to generate our rules and used the new 392-dataset for testing all the rules together.

**Table 2.** The rules extracted by See5 [8]. Positions @1-4 represent the left hand sequence in front of cleavage point and positions @5-8 the right hand sequence. TP for true positives and FP for false positives. Rule 1, 2, 3 are generated by normal mode and rule 4 is generated by *boosting* mode.

| | DT Rules on 362-dataset | P/FP on 362-dataset (362/248) | P/FP on 392-dataset (392/111) | P/FP on combined set (754/359) |
|---|---|---|---|---|
| 1 | F@4, cleavage | 35/5 | 42/4 | 77/9 |
| 2 | L@4, cleavage | 38/9 | 47/4 | 85/13 |
| 3 | Y@4 and P@5, cleavage | 32/5 | 44/14 | 76/19 |
| 4 | E@6, cleavage | 38/6 | 17/0 | 55/6 |

Narayanan, *et al.* [8] tried a decision tree C5.0 to extract prediction rules for HIV cleavage site, which are listed in table 2. Position 4 and 5 are considered to be important attributes by the decision tree. A decision tree searches the most informative attribute first and divides examples by the attributes, and then do the same with the subgroup of the examples. The most informative attribute is determined by information gains of attributes, and the information gain is obtained by amino acid's frequency in the examples. Thus, the extracted rules are attribute (i.e., position)-oriented and frequency-based. Frequency-based (or probability-based) mining normally requires a training set to be large enough and/or training examples to be high domain-representative. With the 362-dataset whose quantity and quality of examples are not good enough, quality of the rules extracted directly from the examples might not be guaranteed.

We also tried association rule mining which is another frequency(or probability)-based rule mining method. While decision trees search for attributes (i.e., positions) one by one to generate rules, association rule mining searches for a set of attributes that are associated each other. The association is defined by frequency of co-occurrence of the attributes in a set. The frequency-based association rules are shown in table 3. The frequency-based method is statistically meaningful when the size of data samples are large enough. Since the size HIV data set is not large enough, the generated rules does not cover all of meaningful patterns in the domain as seen in table 3.

A feed-forward neural network trained by backpropagation is not a frequent-based attribute-by-attribute method. A perceptron uses gradient descent-based learning for linear classification, and multi-layered perceptron is a set of perceptrons that are networked in a layered architecture and thus it provides nonlinear separability by combining linear separable perceptrons. Rögnvaldsson *et al.*

**Table 3.** Association rules

| | Assoc. Rules on a 362-dataset | support count | confidence | correlation |
|---|---|---|---|---|
| 1 | P@5, cleavage | 60 | 80.0% (60/12) | 2.54 |
| 2 | E@6, cleavage | 38 | 84.2% (38/6) | 2.67 |

**Table 4.** Rules extracted from a simple linear perceptron trained on a 362-dataset [9]. 6 other rules containing E@6 are subsumed to the rule1.

| | SP Rules on 362-dataset | P/FP on 362-dataset | P/FP on 392-dataset | P/FP on combined set |
|---|---|---|---|---|
| 1 | E@6, cleavage | 38/6 | 17/0 | 55/6 |
| 2 | F@4 and P@5, cleavage | 17/0 | 3/0 | 20/0 |
| 3 | F@4 and Q@6, cleavage | 14/0 | 8/0 | 22/0 |
| 4 | F@4 and V@3, cleavage | 0/0 | 20/1 | 20/1 |
| 5 | L@4 and Q@6, cleavage | 3/0 | 1/0 | 4/0 |

used a exhaustive rule search method which is a generic black box approach to generate rules from a trained simple perceptron which is a linear model [9, 12]. In an exhaustive rule search, all possible combinations of an input vector are taken into account in the process of rule generation. For each combination (i.e., candidate rule), it is presented to the trained neural network to get its validity. Even for 2-position rules only (i.e., size of antecedent of a rule is 2), there are ${}^{8}C_2 * 20 * 20 = 11,200$ candidate rules to test. They generated all the cases and ranked them by their probabilities obtained from a trained linear perceptron. Table 4 lists the 2-position rules. They showed top 10 rules [9], but we generalized 6 rules containing "E@6" in their antecedents into a 1-position rule (i.e., rule1 in table 4). Since the rules in table 4 are top-10 rules with the highest probabilities among 11,200 combinations, rules with very high accuracy are first selected.

**Table 5.** The rules extracted by a neural network trained on a 362-dataset. CF threshold at each hidden unit was set to 0.6 for the top table and 0.8 for the bottom table. The lower the CF threshold is, the more general rules are generated.

With CF threshold = 0.6

| | NN Rules on a 362-dataset | P/FP on 362-dataset | P/FP on 392-dataset | P/FP on combined set |
|---|---|---|---|---|
| 1 | P@5, cleavage | 60/12 | 66/28 | 126/40 |
| 2 | L@4, cleavage | 38/9 | 47/4 | 85/13 |
| 3 | E@6, cleavage | 38/6 | 17/0 | 55/6 |
| 4 | N@3, cleavage | 78/31 | 96/30 | 174/61 |
| 5 | V@3, cleavage | 33/13 | 138/13 | 171/26 |
| 6 | F@4, cleavage | 35/5 | 42/4 | 77/9 |
| 7 | A@6, cleavage | 28/22 | 83/3 | 111/25 |

With CF threshold = 0.8

| | NN Rules on a 362-dataset | P/FP on 362-dataset | P/FP on 392-dataset | P/FP on combined set |
|---|---|---|---|---|
| 1 | E@6, cleavage | 38/6 | 17/0 | 55/6 |
| 2 | F@4 and P@5, cleavage | 17/0 | 3/0 | 20/0 |
| 3 | N@3 and F@4, cleavage | 17/1 | 13/1 | 30/2 |
| 4 | N@3 and A@1, cleavage | 8/1 | 1/1 | 9/1 |
| 5 | N@3 and A@6, cleavage | 2/1 | 2/0 | 4/1 |
| 6 | V@3 and A@6, cleavage | 4/2 | 56/1 | 60/3 |

However, the exhaustive method is not practical at all in terms of its computational complexity, and it is also very difficult to generate rules from non-linear classifiers such as MLP. Thus, it is not practical to use a black-box approach to generate rules from a non-linear model such as MLP.

We used a decompositional approach to extract rules from MLP neural networks with 8 hidden units. The extracted rules from the MLP with two different *CF threshold* parameter values are listed in table 5. We could extract more general rules from neural networks. Notice that the rules "*A@6*" and "*V@3 and A@6*" show low accuracy on training set but very high prediction accuracy on test set. Frequency-based methods such as decision trees can not generate this kind of rules because the methods are more dependent on training examples. The rule"*E@6*" was discovered by any methods, and the rule "*F@4 and P@5*" which is already well known by domain experts was discovered by neural network methods.

## 4 Conclusion and Discussion

Narayanan *et al.* [8] used a decision tree to generate symbolic rules for HIV-1 protease cleavage site prediction because they could not interpret neural network knowledge in symbolic format even though neural networks outperform decision trees. Because of the small number of examples available, probability (or frequency)-based mining (i.e., decision trees and association rule mining) might be misleading if the examples are not high domain-representative. Rögnvaldsson *et al.* [9, 12] used linear perceptron to generate symbolic rules because nonlinear multilayered perceptron (i.e., MLP) is hard to be interpreted. They had to show that linear model performs as good as the nonlinear model before they used the linear model to generate rules. However, the exhaustive method they used is not computationally practical and have some efficiency issues on rule generation. Also we don't have any clear evidence that the HIV-1 protease cleavage site prediction problem is naturally linear problem.

In this paper, we applied a decompositional approach to extract symbolic rules from a trained nonlinear neural network(i.e., MLP), and compared the rules to the ones extracted by other methods. Experimental results show that a neural network works better especially when only small number of training examples are available. The extracted rules from the MLP seems to have good generalization capability and thus predictability.

## References

1. Robert Andrews, Joachim Diederich, and Alan B. Tickle. Survey and critique of techniques for extracting rules from trained artificial neural networks. *Knowledge-Based Systems*, 8(6):373–389, 1995.
2. Yu-Dong Cai and Kuo-Chen Chou. Artificial neural network model for predicting HIV protease cleavage sites in protein. *Advances in Engineering Software*, 29:119–128, 1998.

3. Yu-Dong Cai, X.J. Liu, X.B. Xu, and Kuo-Chen Chou. Support vector machines for predicting hiv protease cleavage sites in protein. *journal of Computational Chemistry*, 23:267–274, 2002.
4. LiMin Fu. *Neural Networks in Computer Intelligence.* McGraw Hill, Inc., 1994.
5. LiMin Fu. Rule generation from neural networks. *IEEE Transactions on Systems, Man, and Cybernetics*, 24(8):1114–1124, 1994.
6. Hyeoncheol Kim. Computationally efficient heuristics for if-then rule extraction from feed-forward neural networks. *Lecture Notes in Artificial Intelligence*, 1967:170–182, 2000.
7. Alessandra Lumini and Nanni Loris. Machine learning for hiv-1 protease cleavage site prediction. *Proceedings of Artificial Intelligence and Application (AIA2005)*, 2005.
8. Ajit Narayanan, Xikun Wu, and Z. Rong Yang. Mining viral protease data to extract cleavage knowledge. *Bioinformatics*, 18(Suppl.1):S5–S13, 2002.
9. Thorstein Rögnvaldsson and Liwen You. Why neural networks should not be used for HIV-1 protease cleavage site prediction. *Bioinformatics*, 20(11):1702–1709, July 2004.
10. Rudy Setino and Huan Liu. Understanding neural networks via rule extraction. *Proceedings of the 14th International Conference on Neural Networks*, (1):480–485, 1995.
11. Ismali A. Taha and Joydeep Ghosh. Symbolic interpretation of artificial neural networks. *IEEE Transactions on Knowledge and Data Engineering*, 11(3):443–463, 1999.
12. Liwen You, Daniel Garwicz, and Thorstein Rögnvaldsson. Comprehensive bioinformatic analysis of the specificity of human immunodeficiency virus type 1 protease. *Journal of Virology*, 79(19):12477–12486, Oct 2005.

# Statistical Feature Selection from Chaos Game Representation for Promoter Recognition

Orawan Tinnungwattana and Chidchanok Lursinsap

Advanced Virtual and Intelligent Computing (AVIC) Center
Department of Mathematics, Chulalongkorn University
Bangkok 10330, Thailand
orakik@yahoo.com, lchidcha@chula.ac.th

**Abstract.** The accuracy of promoter recognition depends upon not only the appropriate representation of the promoter sequence but also the essential features of the sequence. These two important issues are addressed in this paper. Firstly, a promoter sequence is captured in form of a Chaos Game Representation (CGR). Then, based on the concept of Mahalanobis distance, a new statistical feature extraction is introduced to select a set of the most significant pixels from the CGR. The recognition is performed by a supervised neural network. This proposed technique achieved 100% accuracy when it is tested with the E.coli promoter sequences using a leave-one-out method. Our approach also outperforms other techniques.

## 1 Introduction

Currently, promoter prediction is one of the most challenging problems in bioinformatics studied by many researchers. A promoter is the region of DNA sequence which is usually located on the upstream of a transcription start site (TSS). If the position of promoter is known then the starting position of the coding region, which will be translated into the protein sequence, can be correctly located. In this paper we focus on recognition of E.coli promoter.

E.coli promoter contains two binding sites are the -35 and -10 hexamer boxes upstream of the TSS (position +1). The consensus sequences, i.e. the prototype sequences are composed of the most frequently occurring nucleotides at each position, for the -35 box is TTGACA and -10 box is TATAAT. However, very few of existing E.coli promoters exactly contains the two consensus sequences [1]. The region between two binding sites is called the *spacer*. Figure 1 shows the structure of promoter with two binding sites and two spacers. The distance of spacer between the -35 box and the -10 box varies from 15 to 21 bases. Another spacer between the -10 box and the transcription start site varies from 3 to 11 bases. Since the distances and positions of two binding sites are uncertain, recognizing the correct location of the promoter is rather difficult. Therefore, these features are not appropriate for promoter prediction. In this paper we do not try to find these regions.

Several algorithms have been proposed for E.coli promoter recognition, including artificial neural network (ANN) [2, 3, 4, 5, 6], hidden markov models (HMM)

V.N. Alexandrov et al. (Eds.): ICCS 2006, Part II, LNCS 3992, pp. 838–845, 2006.

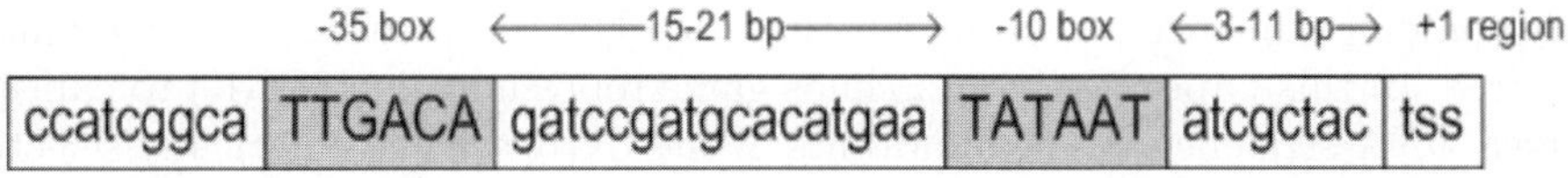


**Fig. 1.** The structure of E.coli promoter with two binding sites are known as the -35 and the -10 hexamer boxes and two spacers

[7], weight matrices [8], as well as graph-based induction method [9]. Many algorithms used hybrid systems which integrated neural network (NN) and other statistical decision techniques. Ma et al. [1] combined expectation maximization (EM) algorithms with NN. The EM algorithm is used for locating the -35 and -10 binding sites. Then, the features in each training promoter are chosen according to their information content and fed to an ANN for promoter recognition. Matsuyama and Kawamura [10] proposed Independent Component Analysis (ICA) algorithm including a position-dependent conversion based on symbol frequencies. Huang and Wang [11] proposed a hybrid learning system to calculate the distribution of oligo-nucleotides statistics as position weight matrices and fed as inputs to the KBANN, SVM and multilayer feed-forward neural network for discriminate promoters and non-promoters. Their result is better than other promoter prediction methods [3, 9, 11, 20, 21] with 97.2% accuracy.

The accuracy of promoter prediction is based on two factors, i.e. the representation of the given DNA sequence and the essential features of the sequence. The goal of this paper concerns both issues of above solutions that can provide a distinct classification between the promoter and non-promoter sequences. A chaos game representation (CGR) is adopted for transforming a DNA sequence having promoters and non-promoters into an image. The essential features of the CGR is selected by applying the concept of Mahalanobis distance. Then, the recognition is performed by a supervised neural network. Our method compared with [11] using same dataset, the result shown 100% accuracy. The rest of the paper is organized as follows. Section 2 presents our proposed methods that are composed of CGR for representing DNA sequences, feature selection and structure of neural network used in this problem. Section 3 shows the dataset that used in this paper, explain the evaluation methods and the results. The conclusion is in Section 4.

## 2 Proposed Methods

A DNA sequence are composed of four different nucleotides, adenine ($A$), thymine ($T$), cytosine ($C$) and guanine ($G$). The important features of each sequence must be extracted in order to classify it as in the promoter class of in non-promoter class. Here, the classification is achieved by using a supervised neural network. An important issue in applying neural networks to classify promoter and non-promoter sequences is how to extract the important features representing a given DNA sequence [12]. The key concept is based on the observation that all features of promoter sequences must be distinguishable from all features of non-promoter sequences. This implies that each DNA sequence must be uniquely captured. In this paper, we used a chaos game representation (CGR) to capture a given DNA sequence. Some quadrants of the CGR are systematically selected as important features of the DNA sequence using the concept of Mahalanobis distance. The details of CGR and feature selection are discussed in the next two following subsections.

## 2.1 Chaos Game Representation (CGR)

Chaos Game Representation (CGR) is a method for uniquely representing DNA sequence patterns in forms of an image [13, 14, 15, 16]. The value of each pixel in the image is based on iterated function systems of fractal theory mapping a discrete sequence of symbols onto a continuous space. Initially, the image is divided into four quadrants which each of them represents one of the four possible nucleotides (A, C, G and T). CGR position is calculated by moving a pointer to a half way between the previous point and the corner of current symbol in the sequence. Let $S = (s_1 s_2 \ldots s_n)$ be a DNA sequence, either having promoters or having no promoters, where $s_i \in \{A, C, G, T\}$ and $l$ denotes the length of $S$. Initially, quadrants whose corners are at coordinates $(0,0)$, $(0,1)$, $(1,1)$, and $(1,0)$ are assigned to nucleotides $A$, $C$, $G$, and $T$, respectively. Each $s_i$ is mapped to the appropriate quadrant corresponding to its nucleotide symbol. The position of $s_i$, denoted by $p_i$, is defined by the following steps.

$$\begin{aligned} p_0 &= (0.5, 0.5) \\ p_i &= p_{i-1} + 0.5(x_i - p_{i-1}), \quad i = 1, 2, \ldots, n \end{aligned} \tag{1}$$

$x_i$ is the coordinate variable of nucleotide $s_i$. The value of $x_i$ is set as follows. Position $p_i$ is denoted by a dot. The corresponding CGR image can be viewed as

| $s_i$ | $x_i$ (coordinate) |
|---|---|
| A | $(0,0)$ |
| C | $(0,1)$ |
| G | $(1,1)$ |
| T | $(1,0)$ |

an image of distributed dots. This image is, then, partitioned by grids into a set of square entries of equal size. The number of square entries is equal to $2^n \times 2^n$. For example, suppose a considered DNA sequence is the following. The value $l$ for this sequence is equal to 18.

ATAACATCGTGTCGGACT

This sequence is captured by a CGR as shown in Figure 2(a) and the CGR is partitioned into a set of square entries of equal size. Suppose $n$ is set to 2. The number of entries is equal to $2^2 \times 2^2$ or 16 and the partitioned CGR is shown un Figure 2(b).

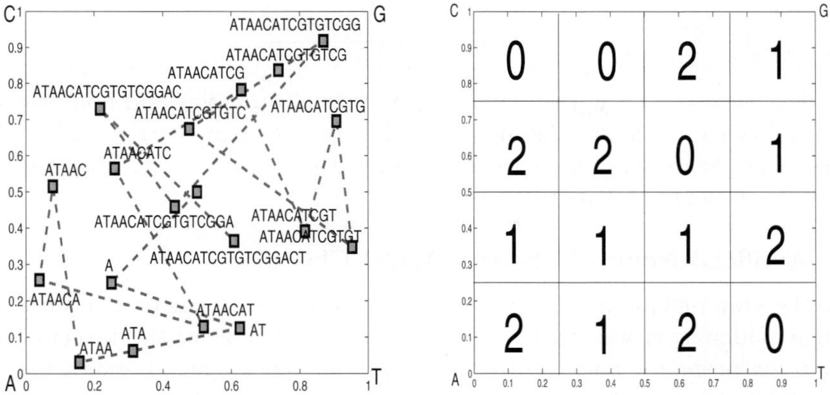


(a) (b)

**Fig. 2.** The CGR of sequence GATCCGATGCACATGAA. (a) The CGR after mapping each mucleotide to its corresponding position. (b) The CGR after being partitioned into 16 equal square entries.

## 2.2 Feature Selection of DNA Sequences

The number of dots in each square entry is counted and used as a feature. Thus, there are $2^n \times 2^n$ features for each DNA sequence of length $l$. This number can make a neural training non-convergent. To overcome this problem, only relevant square entries or features must be selected from the CGR image and used as the training input. Feature selection is an effective technique in dealing with dimensionality reduction [17]. Our selection is based on the concept of Mahalanobis distance. All DNA sequences having promoters are assigned to class 1 denoted by set $S_1$ and those having no promoters are assigned to class 0 denoted by set $S_0$. All sequences in both $S_1$ and $S_0$ are represented by a set of CGR images. Let $f^{(c)}_{i,j,k}$, $c \in \{0, 1\}$, be the value of square entry $(i, j)$ of the $k^{th}$ CGR image for class $c$. A distance measure between promoter class and non-promoter class at entry $(i, j)$ is defined as follows.

$$D_{i,j} = \frac{(m_1^2 - m_0^2)}{\delta_1^2 - \delta_0^2} \tag{2}$$

$$m_0 = \frac{\sum_{k \in S_0} f^{(0)}_{i,j,k}}{|S_0|} \tag{3}$$

$$m_1 = \frac{\sum_{k \in S_1} f^{1}_{i,j,k}}{|S_{(1)}|} \tag{4}$$

$$\delta_0 = \sqrt{\frac{\sum_{k \in S_0} (f^{(0)}_{i,j,k} - m_0)^2}{|S_0| - 1}} \tag{5}$$

$$\delta_1 = \sqrt{\frac{\sum_{k \in S_1} (f^{(1)}_{i,j,k} - m_1)^2}{|S_1| - 1}} \qquad (6)$$

The high value of $D_{i,j}$ implies that the features of promoter and non-promoter at this entry can be highly distinguished. The first $d$ square entries with highest values of $D_{i,j}$'s are selected as the essential features for training. The value of $d$ in our experiment is set to 16

### 2.3 Artificial Neural Network (ANN) Classifier

From the proposed method, a feed-forward backpropagation neural network with a single hidden layer was used. There are one output unit for two target classes, class 1 for promoter and class 0 for non-promoter. Figure 3 shows the network architecture composed of input, hidden and output layers. In this paper, the ANN implementation was achieved using Stuttgart Neural Network Simulator (SNNS) 4.2 which is freely downloadable at http://www-ra.informatik.uni-tuebingen.de/SNNS/. At the beginning of each simulation, the weights were initialized with random values. The training of the network was carried out using error back-propagation with a sum square error function. In this study, there are 16 input units, 10 hidden units and one output unit.

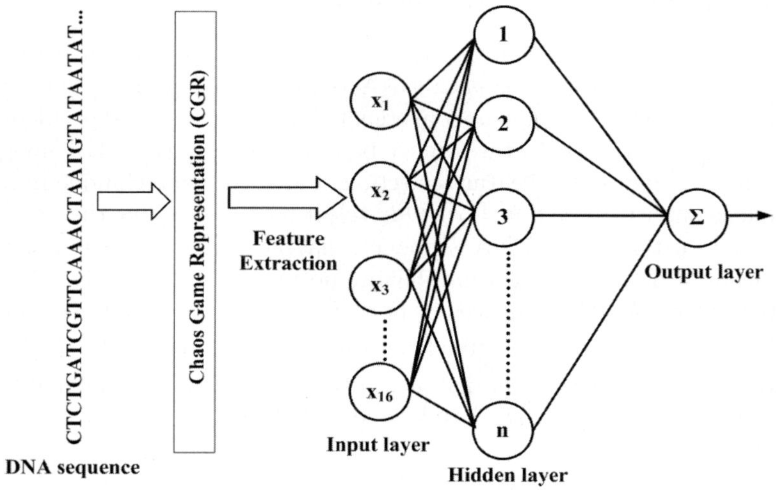


**Fig. 3.** The structure of neural network used in this problem

## 3 Experiments and Results

### 3.1 Data

In this paper, the data sets were taken from the UCI Machine Learning Repository [18]. It consists of 106 sequences including 53 promoter sequences and 53 non-promoter sequences. All sequences are 57 bp long. Each promoter sequence contains 49 bp upstream and 7 bp downstream of the TSS.

### 3.2 Performance Evaluation

Prediction performance was determined by measuring the precision, specificity (sp) and sensitivity (sn). These are defined as Eqs. (7)-(9), respectively

$$Precision = \frac{TP + TN}{TP + TN + FP + FN} \tag{7}$$

$$Specificity = \frac{TP}{TP + FP} \tag{8}$$

$$Sensitivity = \frac{TP}{TP + FN} \tag{9}$$

where $TP, TN, FP, FN$ are the number of true positives, true negatives, false positives and false negatives, respectively. A *true positive* is a promoter sequence that was also classified as a promoter sequence. A *false positive* is a non-promoter sequence that was misclassified as a promoter sequence.

### 3.3 Results

In this paper, our method was compared with Huang and Wang [11]. They used leave-one-out method for evaluating the performance and comparing the result with several other promoter prediction systems. We used the same data sets and the same evaluation method as theirs. Table 1 shows the number of errors compared with other methods including our method. The first number denotes the number of error patterns and the second one is the total number of testing patterns. Table 2 we compared our method with [11] by using precision, specificity and sensitivity criteria. It is clear that our method performs better than the others.

**Table 1.** Number of errors on several methods

| Methods | Errors | Methods | Errors |
|---|---|---|---|
| ID3[19] | 19/106 | BP | 8/106 |
| C4.5[20] | 18/106 | NTSS[21] | 7/106 |
| GBI[9] | 16/106 | HM-layer | 6/106 |
| KNN | 13/106 | KBANN | 4/106 |
| O'Neill[3] | 12/106 | HSVM[11] | 3/106 |
| FTSS[21] | 9/106 | Our method | 0/106 |

**Table 2.** Performances evaluation using precision, specificity and sensitivity

| | HSVM[11] | Our method |
|---|---|---|
| Precision | 97.2% | 100% |
| Specificity | 96.2% | 100% |
| Sensitivity | 98.1% | 100% |

## 4 Conclusion

In this paper, we proposed a new method for selecting all essential features in recognizing E.coli promoters in DNA sequences captured by Chaos Game Representation. These good features are used for training and classifying the given promoter sequences. Each selected feature is determined by using a distance based on the concept of Mahalanobis distance. We do not consider some well-known patterns around TSS, such as TATAAT-box and TTGACA-box, which were previously used by many researchers. The experimental results indicated that our proposed method perform better than the others'. The future research will focus on applying this technique to predict promoter in different species.

## References

1. Ma, Q., et al.: DNA Sequence Classification via an Expectation Maximization Algorithm and Neural Networks: A Case Study. IEEE Trans. Systems, Man and Cybernetics, Part-C: Applications and Reviews. **31** (2001) 468–475
2. Mahadevan, I., Ghosh, I.: Analysis of E.coli promoter structures using neural networks. Nucleic Acids Research. **22(11)** (1994) 2158–2165
3. O'Neill, M.C.: Escherichia coli promoters: neural networks develop distinct descriptions in learning to search for promoters of different spacing classes. Nucleic Acids Research. **20** (1992) 3471–3477
4. Pedersen, A.G., Engelbrecht, J.: Investigations of Escherichia coli promoter sequences with artificial neural network: New signals discovered upstream of the transcriptional startpoint. In: Proceedings of ISMB95. (1995)
5. Horton, P.B., Kanehisa, M.: An assessment of neural network and statistical approaches for prediction of E.coli promoter sites. Nucleic Acids Research. **20** (1992) 4331–4338
6. Demeler, B., Zhou, G.W.: Neural network optimization of E.coli promoter prediction. Nucleic Acids Research. **19** (1991) 1593–1599
7. Pedersen, A.G., et.al.: Characterization of prokaryotic and eukaryotic promoters using hidden markov models. In: Proceedings of ISMB98. (1998)
8. Bucher, P.: Weight matrix descriptions of four eukaryotic RNA polymerase II promoter elements derived from 502 unrelated promoter sequences. J.Mol.Biol. **212** (1990) 563–578
9. Matsuda, T., Motoda, H., Washio, T.: Graph based induction and its applications. Advanced Engineering Informatics. **16** (2002) 135–143
10. Matsuyama, Y., Kawamura, R.: Promoter Recognition for E.coli DNA Segments by Independent Component Analysis. In: Proceedings of CSB2004. (2004) 686–691
11. Huang, Y.F., Wang, C.M.: Integration of Knowledge-Discovery and Artificial-Intelligence Approaches for Promoter Recognition in DNA Sequences. In: Proceedings of ICITA'05. **1** (2005) 259–264
12. Hirsh, H. and Noordewier, M.: Using Background Knowledge to Improve Inductive Learning of DNA sequences. In: Proceeding of the Tenth Annual Conference on Artificial Intelligence for Applications, San Antonio, TX, (1994) 351–357
13. Jeffrey, H.J.: Chaos game representation of gene structure. Nucleic Acids Research. **18(8)** (1990) 2163–2170

14. Goldman, N.: Nucleotide, dinucleotide and trinucleotide frequencies explain patterns observed in chaos game representations of DNA sequences. Nucleic Acids Research. **21(10)** (1993) 2487–2491
15. Almeida, J.S., et al.: Analysis of genomic sequences by Chaos Game Representation. Bioinformatics. **17(5)** (2001) 429–437
16. Deschavanne, P.J., et al.: Genomic Signature: Characterization and Classification of Species Assessed by Chaos Game Representation of Sequences. Mol. Biol. Evol. **16(10)** (1999) 1391–1399
17. Dash, M. and Liu, H.: Consistency-based search in feature selection. Elsevier **151** (2003) 155–176
18. Blake, C.L., Merz, C.J.: http://www.ics.uci.edu/ mlearn/mlrepository.html. UCI Repository of machine learning databases. (1998)
19. Quinlan, J.R.: Induction of decision trees. Machine Learning. **1** (1986) 81–106
20. Quinlan, J.R.: C4.5: programs for machine learning. Morgan Kaufmann, Los Altos, CA. (1993)
21. Huang, J.-W., Yang, C.-B., Tseng, K.-T.: Promoter Prediction in DNA Sequences. In: Proceedings of National Computer Symposium, Workshop on Algorithm and Computation Theory, Taichung, Taiwan. (2003)

# Blue Matter: Strong Scaling of Molecular Dynamics on Blue Gene/L

Blake G. Fitch[1], Aleksandr Rayshubskiy[1] Maria Eleftheriou[1], T.J. Christopher Ward[2], Mark Giampapa[1], Yuri Zhestkov[1], Michael C. Pitman[1], Frank Suits[1], Alan Grossfield[1], Jed Pitera[3], William Swope[3], Ruhong Zhou[1], Scott Feller[4], and Robert S. Germain[1]

[1] IBM Thomas J. Watson Research Center, 1101 Kitchawan Road/Route 134, Yorktown Heights, NY 10598, USA
[2] IBM Hursley Park, Hursley, Hursley SO212JN, United Kingdom
[3] IBM Almaden Research Center, 650 Harry Road, San Jose, CA
[4] Department of Chemistry, Wabash College, Crawfordsville, Indiana 47933

**Abstract.** This paper presents strong scaling performance data for the Blue Matter molecular dynamics framework using a novel n-body spatial decomposition and a collective communications technique implemented on both MPI and low level hardware interfaces. Using Blue Matter on Blue Gene/L, we have measured scalability through 16,384 nodes with measured time per time-step of under 2.3 milliseconds for a 43,222 atom protein/lipid system. This is equivalent to a simulation rate of over 76 nanoseconds per day and represents an unprecedented time-to-solution for biomolecular simulation as well as continued speed-up to fewer than three atoms per node. On a smaller, solvated lipid system with 13,758 atoms, we have achieved continued speedups through fewer than one atom per node and less than 2 milliseconds/time-step. On a 92,224 atom system, we have achieved floating point performance of over 1.8 TeraFlops/second on 16,384 nodes. Strong scaling of fixed-size classical molecular dynamics of biological systems to large numbers of nodes is necessary to extend the simulation time to the scale required to make contact with experimental data and derive biologically relevant insights.

## 1 Introduction

Blue Matter [1, 2, 3] is a molecular simulation framework and application developed to support the scientific goals of IBM's Blue Gene project [4], to serve as a platform for research into application programming patterns for massively parallel architectures, and to explore ways to exploit hardware features of the Blue Gene/L architecture. A major design goal for Blue Matter has been to achieve strong scaling of molecular dynamics for moderately sized systems (10,000 – 100,000 particles) to node counts corresponding to ratios of atoms per node of order one. This supports one of the aims of the scientific component of the project, to carry out simulations on a scale that allows meaningful comparisons with experimental data. Results on a 43,222 atom protein/lipid system obtained from early production use of prototype Blue Gene/L hardware were recently

V.N. Alexandrov et al. (Eds.): ICCS 2006, Part II, LNCS 3992, pp. 846–854, 2006.

published in the Journal of the American Chemical Society [5], another paper on the same system in in press and larger scale studies including microsecond-scale simulations of solvated protein and membrane protein systems are currently underway.

The use of Blue Gene hardware to advance our understanding of biologically important processes has been an integral part of the Blue Gene mission from the beginning of the project [4]. As part of that strategy, we started an application effort to support the scientific goals of the project and to also act as a concrete test-bed for research into application development for massively parallel machines. The Blue Matter development effort has tracked the evolution of the machine architecture with the goal of exploiting hardware facilities on the target machine and enabling a systematic exploration of parallel decompositions for molecular dynamics. The most recent results of this exploration are described in this paper. As part of our efforts to exploit the current Blue Gene/L machine architecture [6] we have explored the following: (1) Use of the global collective network (2) Machine topology effects (3d-torus) [7] (3) Low level interfaces vs. MPI (4)Use of both processors on the BG/L compute chip.

The Blue Matter architecture requires infrastructure to support extensive regression and validation because of the aggressive and experimental nature of the computational platform we are targeting and because of its support for multiple force fields (the models and their parameters used for classical molecular simulation). The two main requirements are that the force field parameters be properly implemented, and that the integrator correctly measures the forces on each atom and makes the appropriate update of position and velocity for each time step. The JACS publication [5] was based on a 118ns NVE simulation of a membrane-bound protein, and the total and kinetic energy drift over that long period of simulation was negligible, indicating that there was consistently correct bookkeeping and integration of all the interaction forces.

## 2 Parallelization Strategies and Challenges

Classical molecular dynamics uses a model of the interactions between particles as the basis for a numerical integration of the equations of motion of the n-body system. In the case of biomolecular simulation, the existence of molecules with well-separated partial charges means that long range electrostatic interactions must be treated properly or unphysical behavior can be observed [8]. This issue is most commonly addressed through the use of periodic boundary conditions and the Ewald [9] or related mesh [10] techniques. Use of these techniques involves partitioning the computation of the long range forces into a real-space component that is short-ranged and a reciprocal space component. In the case of the mesh techniques, the reciprocal space component involves a convolution, implemented using three-dimensional FFTs, of the "meshed" charge distribution.

One of our goals for the Blue Matter framework was to allow a systematic exploration of parallelization strategies, progressing from the relatively straightforward to the more complex. Our starting point was a version of the "replicated data" [11] approach that leveraged the Blue Gene/L hardware collective network (to globalize positions) as well as the torus (to perform a global force reduction) [2]. This allowed us to get out onto hardware very quickly and allowed us to create the testing and validation infrastructure that has been used from very early on the Blue Matter effort. While the replicated data approach makes load balancing straightforward, its scalability is limited by the performance of the floating point "all reduce" collective used for the forces.

In the current phase of this exploration, we use a minimal communication radius decomposition, a class of spatial decompositions which enables a very fine-grained decomposition of interactions and effective load balancing across a large number of nodes[3]. Our requirements included the ability to load balance based on pair interactions and the maintenance of locality for the real-space portion of the calculation so that a "natural" domain decomposition of the simulation volume onto the 3D torus layout of BG/L would minimize contention on the links. Currently, we assign the interaction between two particles to the node containing the mid-point of the pair of interacting particles although any node in the intersection of the broadcast spheres of the pair could carry out this computation and other algorithmic choices for the assignment are possible.

Another approach that combines a spatial decomposition with some of the advantages of the interaction decomposition invented by Plimpton and Hendrickson [11] is taken by the NAMD code[12] which has enabled this code to scale to over 1500 processors[13] prior to the advent of Blue Gene/L and more recently to 8192 nodes on BG/L[14]. Recently, two additional approaches that combine spatial and interaction decompositions have been proposed [15, 16], but neither outlines a detailed strategy for dealing with load imbalance and to our knowledge, no published performance results on biomolecular systems using either technique are available at this time.

Our decomposition requires many-to-many personalized communication operations [17] that are not efficiently represented by collective operations within the MPI standard. These operations entail each node concurrently originating a broadcast of positions to a local neighborhood and a corresponding concurrent reduction of computed forces back to the originating node. Given a three-dimensional simulation domain and a three-dimensional torus interconnect, communication locality on the machine can be achieved via a "natural" mapping of the simulation domain onto the machine.

Although MPI only allows a task to participate in one collective operation at a time, it is possible to construct equivalent function within the standard in several ways:

1. Sequentially invoking MPI broadcast/reduce collectives with common members (collectives involving disjoint task groups can proceed concurrently).
2. Using `ISEND/IRECV` to implement the same communication function in a non-serialized fashion.
3. Use of `ALLTOALLV` on `MPI_COMM_WORLD` with many node pairs transferring no data to implement the same communication function while avoiding MPI internal message-handling overhead. This can be achieved by using a lower

overhead messaging protocol within the implementation of `ALLTOALLV` rather than just using a set of `ISEND/IRECV` calls.

We have implemented the second and third options and have found that as a result of optimizations of the MPI collectives for BG/L [20], the third option gives superior performance. Even so, the realized performance on MPI does not yet reflect the full capabilities of the hardware. We have implemented the collective operations required by Blue Matter via the low-level System Programming Interface (SPI) of the Blue Gene/L Advanced Diagnostics Environment [21]. This is the environment used by the BG/L hardware team to test and validate hardware performance. Results comparing the performance of one communication-intensive kernel, the 3D FFT, on both the MPI and SPI communication layers have been presented previously[22] and measurements in the context of Blue Matter of the time taken by the FFT as well by the neighborhood broadcast and reduce are provided in Table 2.

**Table 1.** Details about the systems benchmarked with Blue Matter. Runs on SOPE and Rhodopsin were made with the velocity Verlet integrator [18] while the ApoA1 runs used RESPA[19]. All used the a particle mesh technique (P3ME or PME) to handle long range electrostatic interactions and were constant particle number, volume, and energy (NVE) simulations.

| System | Total Atoms | Cutoff/Switch (Å) | P3ME Mesh |
|---|---|---|---|
| SOPE | 13,758 | 9.0/1.0 | $128^3$ |
| Rhodopsin | 43,222 | 9.0/1.0 | $128^3$, $64 \times 128^2$ |
| ApoA1 | 92,224 | 10.0/2.0 | $128^3$ |

## 3 Performance Results

We present detailed performance scaling results for three molecular systems whose sizes span the range from 10,000 to 100,000 atoms in Table 2 for both MPI and BG/L ADE SPI communications protocols. Two of these systems, a small solvated lipid bilayer system, SOPE, and a solvated membrane protein system, Rhodopsin, have been used in published scientific work by the Blue Matter science team[5, 23, 24]. The third, the 92,224 atom ApoA1 solvated lipoprotein system, is used as a benchmark by the NAMD package[13]. Additional details about these systems are provided in Table 2.

A summary plot of the strong scaling behavior of Blue Matter is shown in Figure 1, in which the computational rate in time-steps per hour is plotted as a function of atoms per node. This scaling plot shows that use of the BG/L ADE SPI communications interfaces allows continued performance gains to values of atoms per node well below those achievable using MPI. The MPI implementation on Blue Gene/L[20] is quite good as evidenced by the results achieved on the 3D-FFT[22], but the scalability of Blue Matter using MPI appears to be limited by the performance of the neighborhood broadcast and reduce collectives discussed above as can be seen in Table 2.

**Table 2.** Tabulated performance data for the various molecular systems described in this paper using Blue Matter on BG/L. The total time per time-step and selected components are provided for both MPI and BG/L ADE SPI implementations. All of these data were taken in a dual core mode in which k-space and real-space operations are carried out on separate cores and used a complex to complex single precision 3D-FFT. The SOPE and Rhodopsin benchmarks used the velocity Verlet integrator and carried out the FFT-based P3ME operations on every time-step while the ApoA1 benchmark used RESPA[19] and only carried out P3ME on every fourth time-step. Performance is reported for two different processor mesh geometries at 4096 nodes because communication-intensive operations can be sensitive to the aspect ratio of the processor mesh.

| Nodes | | | | Time/time-step (milliseconds) | | | | | | | | Atoms/Node |
|---|---|---|---|---|---|---|---|---|---|---|---|---|
| | | | | Total | | FFT | | Bcast | | Reduce | | |
| Total | $P_x$ | $P_y$ | $P_z$ | MPI | SPI | MPI | SPI | MPI | SPI | MPI | SPI | |
| 512 | 8 | 8 | 8 | 7.46 | 6.81 | 2.17 | 1.88 | 0.55 | 0.35 | 0.44 | 0.35 | 26.8 |
| 1024 | 8 | 8 | 16 | 5.24 | 4.29 | 1.33 | 1.21 | 0.63 | 0.30 | 0.52 | 0.29 | 13.4 |
| 2048 | 8 | 16 | 16 | 4.66 | 2.80 | 0.93 | 0.77 | 0.87 | 0.24 | 0.85 | 0.22 | 6.7 |
| 2048 | 16 | 8 | 16 | 4.55 | 2.81 | 0.91 | 0.70 | 0.87 | 0.24 | 0.77 | 0.23 | 6.7 |
| 4096 | 16 | 16 | 16 | 5.07 | 1.94 | 0.71 | 0.46 | 1.39 | 0.22 | 1.30 | 0.21 | 3.3 |
| 4096 | 8 | 32 | 16 | 5.61 | 2.56 | 1.00 | 0.78 | 1.38 | 0.24 | 1.32 | 0.23 | 3.3 |
| 8192 | 16 | 32 | 16 | 7.31 | 1.88 | 0.84 | 0.52 | 2.48 | 0.22 | 2.20 | 0.20 | 1.6 |
| 16384 | 32 | 32 | 16 | 12.51 | 1.84 | 0.95 | 0.50 | 4.88 | 0.22 | 4.29 | 0.20 | 0.8 |

(a) SOPE (13,758 atoms)

| Nodes | | | | Time/time-step (milliseconds) | | | | | | | | Atoms/Node |
|---|---|---|---|---|---|---|---|---|---|---|---|---|
| | | | | Total | | FFT | | Bcast | | Reduce | | |
| Total | $P_x$ | $P_y$ | $P_z$ | MPI | SPI | MPI | SPI | MPI | SPI | MPI | SPI | |
| 512 | 8 | 8 | 8 | 16.77 | 16.82 | 1.27 | 1.97 | 0.76 | 0.47 | 0.53 | 0.50 | 84.4 |
| 1024 | 8 | 8 | 16 | 9.42 | 9.49 | 0.91 | 1.28 | 0.77 | 0.38 | 0.58 | 0.38 | 42.2 |
| 2048 | 8 | 16 | 16 | 6.45 | 5.57 | 0.73 | 0.81 | 0.94 | 0.34 | 0.77 | 0.33 | 21.1 |
| 2048 | 16 | 8 | 16 | 6.40 | 5.61 | 0.70 | 0.76 | 0.94 | 0.34 | 0.75 | 0.33 | 21.1 |
| 4096 | 8 | 32 | 16 | 5.82 | 3.54 | 0.78 | 0.81 | 1.44 | 0.28 | 1.23 | 0.25 | 10.5 |
| 4096 | 16 | 16 | 16 | 5.56 | 3.47 | 0.59 | 0.50 | 1.42 | 0.28 | 1.15 | 0.27 | 10.5 |
| 8192 | 16 | 32 | 16 | 7.17 | 2.50 | 0.75 | 0.54 | 2.46 | 0.24 | 2.04 | 0.22 | 5.2 |
| 16384 | 32 | 32 | 16 | 12.88 | 2.27 | 0.96 | 0.52 | 5.29 | 0.25 | 4.52 | 0.24 | 2.6 |

(b) Rhodopsin (43,222 atoms)

| Nodes | | | | Time/time-step (milliseconds) | | | | | | | | Atoms/Node |
|---|---|---|---|---|---|---|---|---|---|---|---|---|
| | | | | Total | | FFT | | Bcast | | Reduce | | |
| Total | $P_x$ | $P_y$ | $P_z$ | MPI | SPI | MPI | SPI | MPI | SPI | MPI | SPI | |
| 512 | 8 | 8 | 8 | 35.51 | 36.30 | 0.59 | 0.53 | 1.04 | 1.07 | 0.67 | 0.97 | 180 |
| 1024 | 8 | 8 | 16 | 19.28 | 19.17 | 0.36 | 0.33 | 1.02 | 0.73 | 0.71 | 0.75 | 90 |
| 2048 | 16 | 8 | 16 | 11.34 | 10.70 | 0.25 | 0.19 | 1.04 | 0.59 | 0.84 | 0.55 | 45 |
| 4096 | 16 | 16 | 16 | 7.53 | 5.96 | 0.19 | 0.13 | 1.37 | 0.50 | 1.22 | 0.47 | 22.5 |
| 4096 | 8 | 32 | 16 | 8.57 | 6.31 | 0.26 | 0.20 | 1.67 | 0.50 | 1.53 | 0.46 | 22.5 |
| 8192 | 16 | 32 | 16 | 7.32 | 3.67 | 0.21 | 0.14 | 2.21 | 0.43 | 2.15 | 0.38 | 11.3 |
| 16384 | 32 | 32 | 16 | 11.82 | 2.50 | 0.24 | 0.13 | 4.83 | 0.39 | 4.79 | 0.34 | 5.6 |

(c) ApoA1 (92,422 atoms)

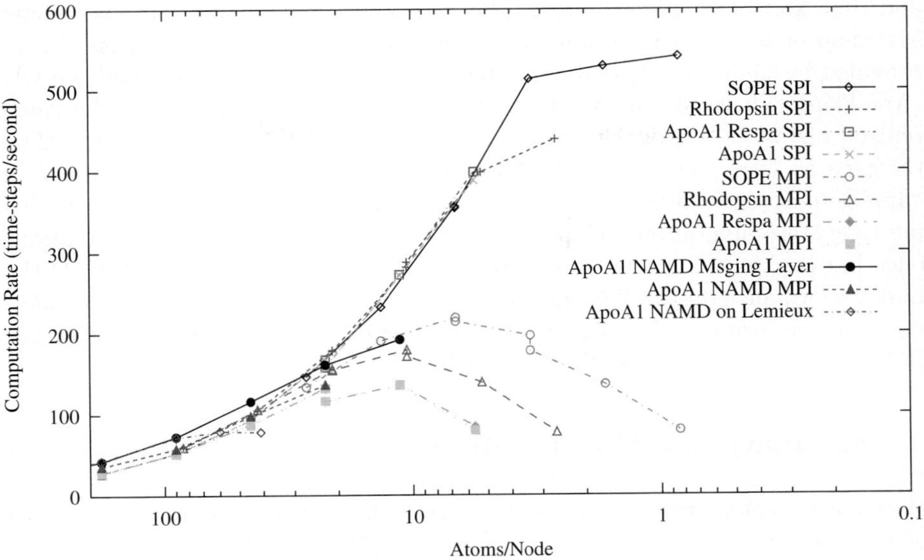


**Fig. 1.** This figure compares the strong scaling characteristics of the Blue Matter MPI and BG/L ADE SPI-based implementations for the molecular systems described in Table 2. The simulation parameters employed for the SOPE and Rhodopsin systems were those used in production, specifically a Verlet integrator with P3ME being computed on every time-step. For ApoA1, we attempted to match the parameters used by the NAMD benchmark as closely as possible, but we were forced to use a larger, $128 \times 128 \times 128$, mesh for our FFT because our current implementation is restricted to dimensions that are powers of two. This plot also includes data reported for NAMD on Blue Gene/L [14] and for the Lemieux Alpha system at the Pittsburgh Supercomputing Center [13]. Because benchmarking data for NAMD was not available for the SOPE and Rhodopsin system, the only comparison possible was on the 92,224 atom ApoA1 system. The NAMD benchmarks used a smaller FFT mesh size ($108 \times 108 \times 80$) than that used by Blue Matter ($128^3$) and on the Lemieux system used a specially tuned version of the Charm++ library written to the Elan communication library provided by Quadrics. On Blue Gene/L, NAMD results are shown for both a specially tuned version of the Charm++ library using a custom messaging layer and for an MPI-based implementation of Charm++[14].

**Table 3.** Floating point performance of Blue Matter on the ApoA1 system using multiple time-stepping (P3ME every four time steps) derived from measurements using floating point performance counters in the BG/L chip

| **Node Count** | 512 | 1024 | 2048 | 4096 | 8192 | 16384 |
|---|---|---|---|---|---|---|
| **GFLOP/sec.** | 127 | 241 | 433 | 774 | 1258 | 1846 |

As a way to place Blue Matter running on Blue Gene/L in context, Figure 1 also shows published results using the NAMD package [12] on the Lemieux system at the Pittsburgh Supercomputing Center [13] and on Blue Gene/L [14]

and Blue Matter on Blue Gene/L. The results on Lemieux were obtained using a version of the Charm++ library written to the Elan communication library provided by Quadrics. The systems benchmarked by Blue Matter and NAMD (ApoA) are identical, and we have made every effort to use either the same (cut-off distances) or higher cost (FFT mesh size) parameters in the Blue Matter runs as were used in the NAMD study to get as close as possible to an "apples-to-apples" comparison. Also, our comparison was made using a multiple time step integration technique that only carried out the P3ME operations once in every four time steps because this was the mode that gave NAMD the best performance on the PSC Lemieux system. Table 3 gives the realized floating point performance for Blue Matter running on the ApoA1 system using the BG/L ADE SPI communication protocols.

## 4 Summary and Conclusions

We have presented strong scaling data on biomolecular systems spanning a range of sizes using two communications protocols (MPI and the BG/L ADE SPI interface) on Blue Gene/L and we have compared our performance results on the largest system, ApoA1 with published values obtained using the NAMD code on Blue Gene/L and on the PSC Lemieux system. Using Blue Matter on BG/L with communications via the BG/L ADE SPI interface, we have achieved under 2.3 milliseconds per time-step on 16,384 nodes for a 43,222 atom protein/lipid system. The continued speed-up through values of less than three atoms/node on this system and to under one atom/node on the smaller SOPE system is the first time that this level of strong scaling has been obtained with classical biomolecular simulation.

The improvement in performance over the MPI baseline obtained through use of the SPI communications interface shows the advantages that can be realized through use of application-aware communications collectives that fully leverage the available hardware capabilities.

The time-to-solution measured for the 43,222 atom rhodopsin system on 16,384 nodes corresponds to over 76 nanoseconds of simulation time per day or a microsecond of simulation in only two weeks. This capability enables studies of biologically relevant systems on time-scales that were previously impractical. Scientific results using Blue Matter on prototype BG/L hardware have already been published and additional scientific studies are underway.

Work is currently underway to explore further optimizations of the 3D-FFT, such as implementing a real FFT to reduce communication data volume below that of the current single precision complex FFT implementation and to continue to improve the performance of neighborhood broadcast and reduce operations. We are also continuing to refine our load balancing techniques and are working with the compiler team to improve the floating point efficiency of the Blue Matter code.

## References


1. Fitch, B., Germain, R., Mendell, M., Pitera, J., Pitman, M., Rayshubskiy, A., Sham, Y., Suits, F., Swope, W., Ward, T., Zhestkov, Y., Zhou, R.: Blue Matter, an application framework for molecular simulation on Blue Gene. Journal of Parallel and Distributed Computing **63** (2003) 759–773
2. Germain, R., Zhestkov, Y., Eleftheriou, M., Rayshubskiy, A., Suits, F., Ward, T., Fitch, B.: Early performance data on the Blue Matter molecular simulation framework. IBM Journal of Research and Development **49**(2/3) (2005) 447–456
3. Germain, R.S., Fitch, B., Rayshubskiy, A., Eleftheriou, M., Pitman, M.C., Suits, F., Giampapa, M., Ward, T.C.: Blue Matter on Blue Gene/L: massively parallel computation for biomolecular simulation. In: CODES+ISSS '05: Proceedings of the 3rd IEEE/ACM/IFIP international conference on Hardware/software codesign and system synthesis, New York, NY, USA, ACM Press (2005) 207–212
4. Allen, F., et al.: Blue Gene: a vision for protein science using a petaflop supercomputer. IBM Systems Journal **40**(2) (2001) 310–327
5. Pitman, M.C., Grossfield, A., Suits, F., Feller, S.E.: Role of cholesterol and polyunsaturated chains in lipid-protein interactions: Molecular dynamics simulation of rhodopsin in a realistic membrane environment. Journal of the American Chemical Society **127**(13) (2005) 4576–4577
6. Gara, A., et al.: Overview of the Blue Gene/L system architecture. IBM Journal of Research and Development **49**(2/3) (2005) 195–212
7. Adiga, N., et al.: Blue Gene/L torus interconnection network. IBM Journal of Research and Development **49**(2/3) (2005) 265–276
8. Bader, J., Chandler, D.: Computer simulation study of the mean forces between ferrous and ferric ions in water. The Journal of Physical Chemistry **96**(15) (1992)
9. De Leeuw, S., Perram, J., Smith, E.: Simulation of electrostatic systems in periodic boundary conditions I. lattice sums and dielectric constants. Proc. Roy. Soc. Lond. A **373** (1980) 27–56 and references therein.
10. Deserno, M., Holm, C.: How to mesh up ewald sums. i. a theoretical and numerical comparison of various particle mesh routines. J. Chem. Phys. **109**(18) (1998) 7678–7693
11. Plimpton, S., Hendrickson, B.: A new parallel method for molecular dynamics simulation of macromolecular systems. Journal of Computational Chemistry **17**(3) (1996) 326–337
12. Kale, L., Skeel, R., Bhandarkar, M., Brunner, R., Gursoy, A., Krawetz, N., Phillips, J., Shinozaki, A., Varadarajan, K., Schulten, K.: NAMD2: Greater scalability for parallel molecular dynamics. Journal of Computational Physics **151** (1999) 283–312
13. Phillips, J., Zheng, G., Kumar, S., Kale, L.: NAMD: biomolecular simulation on thousands of processors. In: Supercomputing 2002 Proceedings. (2002) `http://www.sc2002.org/paperpdfs/pap.pap277.pdf`.
14. Kumar, S., Huang, C., Almasi, G., Kale, L.V.: Achieving strong scaling with NAMD on Blue Gene/l. 20th IEEE International Parallel and Distributed Processing Symposium, IEEE (2006) http://charm.cs.uiuc.edu/papers/NAMDIDPDS06.pdf.
15. Snir, M.: A note on n-body computations with cutoffs. Theory of Computing Systems **37** (2004) 295–318 DOI: 10.1007/s00224-003-1071-0.
16. Shaw, D.E.: A fast, scalable method for the parallel evaluation of distance-limited pairwise particle interactions. Journal of Computational Chemistry **26**(13) (2005) 1318–1328

17. Kale, L., Kumar, S., Varadarajan, K.: A framework for collective personalized communication. In: Parallel and Distributed Processing Symposium, 2003. Proceedings. International, IEEE (2003) http://dx.doi.org/10.1109/IPDPS.2003.1213166.
18. Swope, W., Andersen, H., Berens, P., Wilson, K.: A computer simulation method for the calculation of equilibrium constants for the formation of physical clusters of molecules: Application to small water clusters. Journal of Chemical Physics **76** (1982) 637–649
19. Tuckerman, M., Berne, B., Martyna, G.: Reversible multiple time scale molecular dynamics. J. Chem. Phys. **97**(3) (1992) 1990–2001
20. Almasi, G., et al.: Design and implementation of message-passing services for the Blue Gene/L supercomputer. IBM Journal of Research and Development **49**(2/3) (2005) 393–406
21. Giampapa, M., et al.: Blue Gene/L advanced diagnostics environment. IBM Journal of Research and Development **49**(2/3) (2005) 319–332
22. Eleftheriou, M., Fitch, B., Rayshubskiy, A., Ward, T., Germain, R.: Performance measurements of the 3d FFT on the Blue Gene/L supercomputer. In Cunha, J., Medeiros, P., eds.: Euro-Par 2005 Parallel Processing: 11th International Euro-Par Conference, Lisbon, Portugal, August 30-September2, 2005. Volume 3648 of Lecture Notes in Computer Science., Springer-Verlag (2005) 795–803
23. Suits, F., Pitman, M.C., Feller, S.E.: Molecular dynamics investigation of the structural properties of phosphatidylethanolamine lipid bilayers. Journal of Chemical Physics **122**(24) (2005)
24. Pitman, M.C., Suits, F., Gawrisch, K., Feller, S.E.: Molecular dynamics investigation of dynamical properties of phosphatidylethanolamine lipid bilayers. Journal of Chemical Physics **122**(24) (2005)

# DigitalTree: A Tool for Displaying Biological Data in Tree Structure

Robin Kramer[1], Victor Olman[2], Ying Xu[2], and Dong Xu[1]

[1]Computer Science Department and Christopher S. Bond Life Sciences Center, University of Missouri-Columbia, Columbia MO 65211-2060, USA
kramer@mizzou.edu, xudong@missouri.edu
[2]Department of Biochemistry and Molecular Biology and Institute of Bioinformatics, University of Georgia-Athens, Athens, GA 30602-7229, USA
olman@csbl.bmb.uga.edu, xyn@bmb.uga.edu

**Abstract.** We developed a computational tool, DigitalTree, for displaying a two-dimensional weighted tree of biological data. Existing methods, Radial Layout, Hyperbolic Layout, Phyllotactic Placement, and Force-Directing were integrated into DigitalTree. DigitalTree preserves the distances between nodes in the tree and the trees topology. The tool has been applied in analyzing microarray data and phylogenetic trees. DigitalTree is available for download at `http://digbio.missouri.edu/~digtree`.

## 1 Introduction

Weighted trees, representing biological data, have been widely used. Phylogenetic trees can represent biological species and genes in a tree structure. Biological data can be organized in a Minimum Spanning Tree (MST) for clustering [9]. Most of these biological data are complex and have high dimensions. Visualizing the tree structure for the biological data on a two-dimensional (2D) plane is useful in terms of understanding the relationships among the vertices in the tree. Various techniques have been developed for displaying trees or graphs, including Force-Directed[4], Radial Layout[7], Hyperbolic Layout[3], Phyllotactic Layout[1], and Information Visualization [5]. Although some distortion in the weight relationship among vertices is expected due to the projection from high dimension to 2D, existing methods typically do not preserve the ratios among edges of the displayed tree. In addition, these tools generally lack utilities for handling standardized biological data, such as the widely used phylogenetic tree format.

We developed an algorithm for laying out general weighted, both rooted and unrooted, trees on a 2D plane. This algorithm was implemented as a tool for displaying specific biological data on a personal computer. A weighted tree contains a set of weighted edges such that no vertex can be its own ancestor. A layout is a set of abstract coordinates that can be displayed, with only simple transformations (rotation, translation, scaling, and shearing) being performed. Computer visualization preprocesses for the eye and as such adheres to aesthetic

V.N. Alexandrov et al. (Eds.): ICCS 2006, Part II, LNCS 3992, pp. 855–862, 2006.

principles. Conflicting aesthetic principles like spreading every vertex away from each other vertex and preventing edge crossings are balanced by DigitalTree, while maintaining the integrity and quantity of information being displayed.

## 2 Algorithm and Methods

In this section, we will discuss various techniques that we used for tree display, including choosing a root, calculating angles, placing vertices, optimizing layouts, adjusting angles, and choosing orderings of the children. The layout includes two parts: (1) placing an initial 2D layout based on certain angles of children around their parent vertex, while ratios among the edge weights are preserved; (2) optimizing the placement through an objective function using the steepest descent optimization.

### 2.1 Choosing a Root

From a tree $T = \{V, E\}$, with the set $V$ of vertices with parents $P(v)$ or children $C(v)$ and a set $E$ of edges with weights $w(e)$ where $e \in E, v \in V$, DigitalTree first chooses a root that balances the tree. For this purpose, we define a path $\gamma$ between two vertices $i$ and $j$ as $\gamma_{i,j}$ containing edges $e \in \gamma_{i,j}$. A path's length is the sum of the weights of edges on the path. The root quality of the $i$-th vertex is calculated as follows.

$$\Delta_i = \sum_{j \neq i} \left\{ \sum_{e \in \gamma_{i,j}} w(e) \right\}$$

Root is a vertex with index, such that

$$i_{root} = arg \left\{ \min_i \Delta_i \right\}.$$

### 2.2 Calculating Angles

Once the root is determined, all descendants are placed on a 2D plane at angles relative to the first child placed, with the lengths of edges proportional to corresponding edge weights. Then the first grandchild of each child is placed relative to the boundary between the child and its closest sibling, and the rest of the children for a vertex are placed relative to the previously placed sibling, and so on and so forth. Although the placement algorithm for the edges is different, the angles themselves are chosen in a similar manner to the algorithm in Nicheworks [7].

For a sub-branch of $T$ rooted at a vertex $v$, $L(v)$ is the number of vertices that are leaves. Note that $L(v)$ for a parent vertex is equal to the sum of the $L(v)$ values of its children. Therefore, $L(v)$ of the root has as many leaves as the entire tree, and a leaf has $L(v) = 1$, i.e.,

$$L(v) = \begin{cases} 1 & \text{if } v \text{ a leaf,} \\ \sum_{u \in C(v)} L(u) & \text{otherwise.} \end{cases}$$

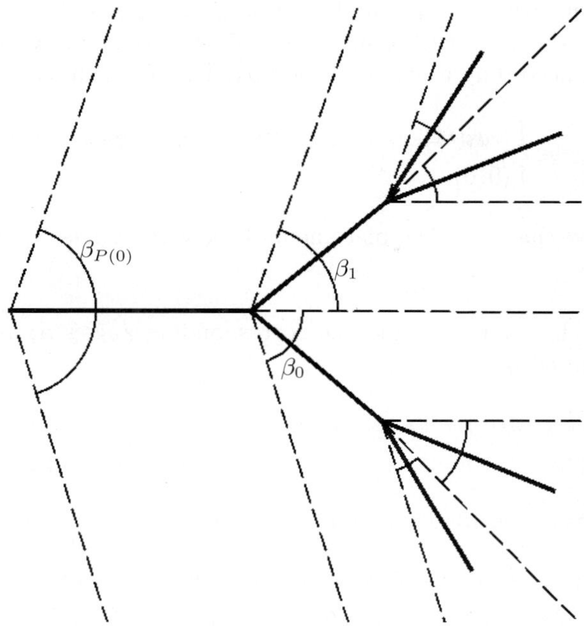


**Fig. 1.** A drawing of the edges and boundaries for a tree layout. The solid straight lines represent the edges of the tree. The dashed lines extending from the vertex represent the boundaries on the portion of the drawing surface which is allotted for drawing an edge and its descendant edges. The $\beta$s represent the angles between the two dashed lines.

As shown in Figure 1, we define the boundaries between the siblings and the parents, relative to the first boundary of the first child of the root. The children of a vertex are ordered $v = 0, 1 \ldots v_n$. The angle $\beta_v$ at each child vertex is proportional to its parent angle $\beta_{P(v)}$ and its leaf count $L(v)$, and inversely proportional to the parent leaf count $L(P(v))$. In the case of the root vertex the angle is $2\pi$, i.e.,

$$\beta_v = \begin{cases} 2\pi & \text{if } v \text{ is the root,} \\ \beta_{P(v)} \left( \frac{L(v)}{L(P(v))} \right) & \text{otherwise.} \end{cases} \tag{1}$$

We align the first child of the root relative to the horizontal axis of the 2D plane. Then the angle of a boundary relative to the horizontal axis is

$$\theta_v = \begin{cases} 0 & \text{for the first child of the root,} \\ \theta_{P(v)} & \text{for the first child of a parent,} \\ \theta_{(v-1)} + \beta_{(v-1)} & \text{for every consecutive child with the same parent.} \end{cases} \tag{2}$$

Then the initial layout of the angles $\alpha_v^{(0)}$ between an edge and the horizontal axis can be calculated as

$$\alpha_v^{(0)} = \frac{\beta_v}{2} + \theta_v .$$

As a result, for each vertex $v$ and its vertex $(x_v, y_v) = V(\alpha_v)$, we place the coordinates of the root vertex at $(0,0)$, and the coordinates for every other vertex $(x_v, y_v)$ according to $\alpha_v$, with preserved edge weight $\omega_v$, i.e.,

$$V(\alpha_v) = \begin{cases} (\cos(\alpha_v)\omega_v + x_{P(v)}, \sin(\alpha_v)\omega_v + y_{P(v)}) & v \neq \text{root}, \\ (0,0) & v = \text{root}. \end{cases}$$

We can prove that an initial placement of the vertices has no crossing between edges.

**proof:** For a child vertex $v$ and the corresponding angles $\alpha_v$ and $\theta_v$, an edge can be represented by

$$\left\{tcos(\alpha_v) + x_{P(v)}, tsin(\alpha_v) + y_{P(v)}\right\}, 0 < t < \omega_v, \tag{3}$$

$$\theta_v < \alpha_v < \theta_{v+1}. \tag{4}$$

The boundaries of the edge can be represented by

$$\left\{tcos(\theta_v) + x_{P(v)}, tsin(\theta_v) + y_{P(v)}\right\}, 0 \leq t, \tag{5}$$

$$\left\{tcos(\theta_{v+1}) + x_{P(v)}, tsin(\theta_{v+1}) + y_{P(v)}\right\}, 0 \leq t. \tag{6}$$

By Equations (3-6) the above edge will not intersect the boundaries of its siblings or its parent given Equation (2):

$$\theta_{P(v)} \leq \theta_v, \tag{7}$$

$$\theta_{v+1} \leq \theta_{P(v)+1}. \tag{8}$$

with the boundaries of the parent of this edge represented as

$$\left\{tcos(\theta_{P(v)}) + x_{P(P(v))}, tsin(\theta_{P(v)}) + y_{P(P(v))}\right\}, 0 \leq t, \tag{9}$$

$$\left\{tcos(\theta_{P(v)+1}) + x_{P(P(v))}, tsin(\theta_{P(P(v))+1}) + y_{P(P(v))}\right\}, 0 \leq t. \tag{10}$$

Therefore, given that a vertex $(x_{P(v)}, y_{P(v)})$ is chosen within its own bounding area according to Equations (9,10), the edges of its children cannot intersect their own bounding areas.

## 2.3 Optimizing the Layout

Based on force-directed methods for creating MST layouts [4], the initial set of angles are refined. The coordinates of vertices are then evaluated with an objective function $f(x_v, y_v)$ given by:

$$f(x_v, y_v) = \sum_{i=1}^{n} \frac{1}{1 + e^{-(x_i - x_v)^2 - (y_i - y_v)^2}}.$$

where $n$ is the total number of vertices on the tree. The set of angles $\alpha$ and coordinates $(x_v, y_v) = V(\alpha_v)$ can be found through minimizing $f(x_v, y_v)$, i.e.,

$$\alpha = arg\left\{\min_{\alpha}\sum_{i} f(V(\alpha_i))\right\}.$$

A step size of $\delta$ equal to one half of the smallest $\beta$ between any of the boundaries was chosen for calculating the next set of angles $\alpha' = \alpha + \delta(-1)^n$. Each vertex was updated as its gradient was calculated (rather than after all the gradients had been calculated). This way each edge responds to the updated position of the previous edge, and will not accidentally try to occupy the same location. Also, as one branch is updated all descendants are updated by that same amount. Hence, the gradient algorithm would be less likely to make edge crossings. The next $\alpha$ was calculated according to

$$\alpha_v^{(l+1)} = \delta\left\{f(x_v', y_v') - f(x_v, y_v)\right\} + \alpha_v^{(l)},$$

where $l$ is the optimization iteration step.

To take advantage of the space introduced by displacing the child vertex from the parent, and to facilitate the gradient optimization, the angle $\beta_i$ is enlarged proportional to the distance from the parent $\omega_i$, i.e.,

$$\beta_v' = \beta_v + \frac{c\omega_v}{\log(n)},$$

where $\beta_v$ is from Equation (1), $n$ is the number of vertices in the tree, and $c$ is a constant, which was empirically found $c \approx 0.1$ for good results. The adjustment improves initial appearance and convergence time for both large and small trees.

### 2.4 Ordering the Children: Phyllotaxis

Phyllotaxis is the placement of new leaves, branches, petals, etc. based on ordering of the children for improving the quality of the tree layout. The idea of Phyllotaxis came from plants. Although plants with limited surface areas are uncertain about the final shape of their branches, plants know that earlier branches are likely to be larger with more sub-branches than later branches. The Golden Ratio $\phi \approx 1.618$ is used to derive the Golden Angle $\chi = 2\pi\frac{1}{1+\phi}$ for placing the new petal of a flower, i.e., $\rho_{(i+1)}$ relative to the last $\rho_i$ satisfies the phylotactic angles ($\rho_i = \rho_{(i+1)} + \chi$). Analogously, the final shape of each DigitalTree branch is uncertain, and placing the longer branches away from each other, takes better advantage of space. In particular, at a vertex in tree with $N$ children and $N$ phyllotactic angles, we have

$$\left\{c_k : k = (1 \ldots N), \beta_k > \beta_{(k+1)}\right\},$$
$$\left\{\rho_i : \rho_i \in (0, 2\pi), i = (1 \ldots N), \rho_i < \rho_{(i+1)}\right\}.$$

The phyllotactic arcs are $\left\{\hat{\rho}_i = \frac{1}{2}\left(\rho_{(i-1)} - \rho_{(i+1)}\right), i_0 = i_N\right\}$. With the ordering $\left\{\hat{\rho}_k : \hat{\rho}_k > \hat{\rho}_{(k+1)}\right\}$, the order of the children $c_k$ is mapped to the phyllotactic order $c_i$. This ensures that the longest branches are placed where the most space in the phyllotaxis would be.

### 2.5 Complexity Analysis

Picking the root vertex is $O(n^2)$, and is a small portion of the time involved in calculating the layout. Breadth first traversals can be used to calculate both $\{\theta, \beta, V(\alpha^{(0)})\}$ and $V(\alpha^{(l)})$. These are linear in the number of vertices in the data set $O(n)$. At each gradient step and with $m$ gradient steps, the objective function is evaluated $n$ times. The objective function for general unweighted force-directed methods is $O(n)$ [6], because there is only one vertex moving at a time. Hence, with the average number of descendants from a vertex approximated by $O(log(n))$, the objective function is on average $O(n \log(n))$ as only the vertex and its descendants need to measure its distance from each other. This makes DigitalTree's total computational complexity $O(mn^2 log(n))$.

## 3 Implementation and Features

The algorithm was implemented in the C programming language for speed, while the graphical user interface (GUI) was implemented in Java which called the C code through native interfaces. The algorithm makes use of breadth first and depth first searches in a number of places, these are all coded in while loops for efficiency, rather than with recursive function calls. DigitalTree can handle more than 10,000 vertices including optimization and display in 10 minutes on a typical

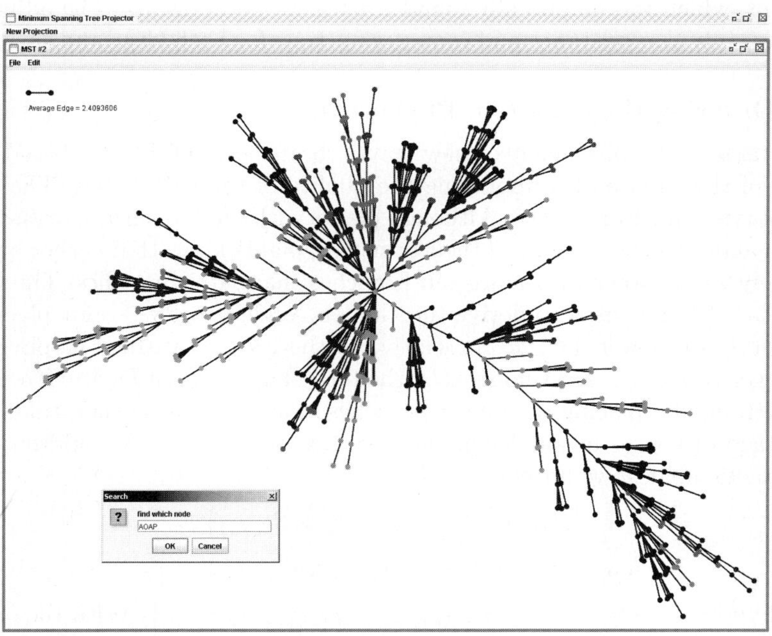


**Fig. 2.** A screen shot of DigitalTree interface

desktop PC with 1 GB memory. There is a DigitalTree file format output from the C package, which the GUI reads and writes. This format stores the parent-child relationships, weights, vertex numbers, vertex names, and coordinates and colors of the vertices.

DigitalTree is a standalone package and it is also incorporated into EXCAVATOR [8], a MST-based clustering package. To produce even better displays, there are several editing features, which allow the user to rotate, zoom, re-center the tree (see Figure 2). DigitalTree also allows the user to move branches of the tree. DigitalTree can parse a number of input formats, in particular the most common phylogenetic tree file format, Newick. The Newick file parser was written using spirit++. The outputs of common phylogenetic tree construction tools Phylip, ClustalW and PAUP* can be used. In Phylip there is no need to change formats. In ClustalW, the result can be saved into the Phylip format. In PAUP*, one can use the command (with the file extension .dnd):

```
savetrees file=BestMLtree.phy brlens format=phylip;
```

## 4 Discussion

In summary, we developed a package DigitalTree for displaying a weighted tree on a 2D plane with preserving the ratios among edges. It incorporates and improves various existing algorithms. DigitalTree can handle large-scale biological data and support various existing formats (e.g., the Newick and EXCAVATOR file formats). It facilitates pattern discovery and visualization with interactive features, such as editable vertices, a scale key, and vertex search.

DigitalTree provides a nice layout for weighted tree structures. The layout is especially useful when the underlying measure of distance behaves according to the properties of norms, such as Euclidean distance or Manhattan distance, because inferences about vertices which are connected through other vertices can be made. DigitalTree is also useful for displaying trees with correlation coefficients and phylogenetic distances, which do not satisfy the properties of norms. With correlation coefficients it is useful to recognize which vertices are highly correlated with many other vertices. With phylogenetic trees it is especially useful to see what the evolutionary distance might be.

DigitalTree fills a unique niche which has not been covered by previous layout algorithms. Force directed methods with random layouts, performing well on cyclic graph structures, converge with edge crossings on large tree structures. While edge crossings are admissible or necessary in large cyclic graph structures, they are not in tree structures. Although Rradial Layouts are linear in the number of nodes for tree structures and have no edge crossings, they do not take full advantage of the screen space. Hyperbolic Radial Layouts provide a user with a method to zoom in on parts of the graph at the expense of distorting the relative sizes of the tree. DigitalTree provides a method which preserves the ratios among the lengths of edges, while minimizing edge crossings and optimizing screen usage.

## Acknowledgments

This project is supported by a National Science Foundation grant NSF/ITRIIS-0407204. RK was also supported by a short-term fellowship sponsored by National Library of Medicine, National Institute of Health during the summer of 2005. VO and YX's work is also supported in part by the U.S. Department of Energys Genomes to Life program under project Carbon Sequestration in *Synechococcus* sp: From Molecular Machines to Hierarchical Modeling, National Science Foundation Grants NSF/DBI-0354771, and a "distinguished scholar" grant from Georgia Cancer Coalition.


## References


1. Carpendale, S., Agarawala, A.: PhylloTrees: Harnessing Nature's Phyllotactic Patterns for Tree Layout. InfoVis. (2004).
2. Chen, C., Morris, S.: Visualizing Evolving Networks: Minimum Spanning Trees versus Pathfinder Networks. InfoVis. (2003) 9.
3. Lamping, J., Rao, R., Pirolli: A focus+context technique based on hyperbolic geometry for visualizing large hierarchies. Conference on Human Factors in Computing Systems. (1995) 401-408.
4. Noel, S., Chu, C.H., Raghavan, V.V.: Visualization of document co-citation counts. In Proceedings of the Intl. Conf. on Data Mining. (2002) 691-696.
5. Plaisant, C.: The challenge of information visualization evaluation. In Proceedings of the Working Conference on Advanced Visual interfaces. (2004) 109-116.
6. Pulo, K.J.: Recursive space decompositions in force-directed graph drawing algorithms. In Australian Symposium on Information Visualisation. 9 (2001) 95-102.
7. Wills, G.J.: NicheWorks-Interactive Visualization of Very Large Graphs. Journal of Computational and Graphical Statistics. 8 No. 2 (1999) 190-212.
8. Xu, D., Olman, V., Wang, L., Xu, Y.: EXCAVATOR: a computer program for gene expression data analysis. Nucleic Acid Research. 31 (2003) 5582-5589.
9. Xu, Y., Olman, V., Xu, D.: Clustering Gene Expression Data Using a Graph-Theoretic Approach: An Application of Minimum Spanning Trees. Bioinformatics. 18 (2002) 526-535.

# HiSP: A Probabilistic Data Mining Technique for Protein Classification

Luiz Merschmann* and Alexandre Plastino**

Departamento de Ciência da Computação
Universidade Federal Fluminense
Niterói, Brazil
{lmerschmann, plastino}@ic.uff.br
http://www.ic.uff.br

**Abstract.** In this work, we propose a new computational technique to solve the protein classification problem. The goal is to predict the functional family of novel protein sequences based on their motif composition. In order to improve the results obtained with other known approaches, we propose a new data mining technique for protein classification based on Bayes' theorem, called Highest Subset Probability (HiSP). To evaluate our proposal, datasets extracted from Prosite, a curated protein family database, are used as experimental datasets. The computational results have shown that the proposed method outperforms other known methods for all tested datasets and looks very promising for problems with characteristics similar to the problem addressed here.

## 1 Introduction

Proteins are complex organic macromolecules made up of amino acids. They are fundamental components of all living cells including many substances, such as enzymes, structural elements, and antibodies, that are directly related with the functioning of an organism. Hence, the knowledge of the proteins' functions is very important. Until recently, the function of the proteins could be identified only by time-consuming and expensive experiments. However, in the post-genomic era, with the huge amount of available sources of information, new challenges arise in protein function characterization. Moreover, computer-based methods to assist in this process are becoming increasingly important.

Various protein sequence databases are readily available and can be used in the task of assigning proteins to functional families. This is a typical data mining task, and therefore, this problem can be viewed as a classification problem. Hence, the need of an efficient method to solve this problem is evident. In this work, we propose a new data mining method, based on motif analysis, for the protein classification problem.

Since the protein function is closely related with the occurrence of motifs in its sequence, the motif composition has been used for the function prediction of

* Work sponsored by CNPq PhD scholarship.
** Work sponsored by CNPq research grant 300879/00-8.

V.N. Alexandrov et al. (Eds.): ICCS 2006, Part II, LNCS 3992, pp. 863–870, 2006.

proteins [1]. The difficulty in this task arises because many proteins share one or more motifs with proteins that belong to different functional families.

The data mining method proposed in this work uses a database which stores information about previously studied proteins (i.e., proteins grouped into families according to the functions they perform) to predict the functional family of novel proteins. Some motif databases have been developed, such as Prosite, Prints, Blocks, Pfam, among others [2].

The datasets used in this study, named genbase10, genbase18, genbase28, and genprosite, were extracted from Prosite database [3]. Each documentation entry in this database represents a protein class and is associated with a function. The entry PDOC00343, for example, corresponds to class "Kinesin motor domain signature and profile". In Prosite database, every documentation entry is coded by a PDOCxxxxx number. Each protein class is associated with pattern(s) and, in some cases, also with profile(s).

Pattern is a short conserved amino acid sequence which is part of a region known to be important for the biological function of a protein or which includes biologically significant residue(s), while profile is a table (also called weight matrix) built on alignment of multiple protein sequences. Unlike patterns, profiles seek to characterize a protein family over its entire length [4]. Patterns and profiles are represented by PSxxxxx codes. For simplicity, in this paper, both pattern and profile will be referred as motifs.

The classification problem we address in this work can be stated as follows. Given a dataset of proteins, called training dataset, where each protein is characterized by their motifs and functional family (also referred as protein class), the goal is to assign newly discovered protein sequences to one of the protein families represented in the training dataset.

### 1.1 Contributions

In this work, we propose a computational technique to solve the protein classification problem[1]. The main motivation was to improve results presented in previous works for large datasets. The method proposed in [6] reached only 41.4% of accuracy for a database containing approximately 40000 proteins in 1100 different classes.

The new data mining technique we propose is inspired in Bayes' theorem [7]. The main characteristic of the proposal is the protein classification based on the evaluation of its subsets of motifs. We compare the results of our approach with those obtained by other methods proposed in literature. In addition, accuracy and speed performance of our approach are presented using all proteins present in Prosite database (release 18.0).

The remaining of this paper is structured as follows. Section 2 gives an overview of related works. Section 3 presents our approach for the protein classification problem. The experiments and results are discussed in Section 4. Finally, in Section 5, the conclusions and some future work directions are presented.

[1] This proposal was preliminarily and briefly explored in [5].

## 2 Related Works

Reference [8] adopted the decision tree method for assigning protein sequences to functional families based on their motif composition. The datasets used in the experiments were extracted from Prosite database. The experimental results showed that the obtained decision tree classifiers presented a good performance.

Reference [9] presented a preprocessing software tool, called GenMiner, which is capable of processing three important protein databases and transforming their data into a suitable format for the Weka data mining tool [10] and MS SQL Analysis Manager 2000 [11]. GenMiner was developed in Java to communicate with the Weka data mining module. Weka was chosen because it provides a variety of data mining algorithms and its open sourceness facilitates the interconnection with GenMiner. The decision tree technique was used for mining protein data and experiments produced interesting results that confirmed the system's capability of efficiently discovering properties of novel proteins.

Reference [6] proposed a data mining approach for motif-based classification of proteins. In this work, finite state automata were used to induce protein classification rules, which were applied to the test datasets in the experiments. The form of the extracted rules is $X \rightarrow Y$. The left part is a set of motifs, while the right part is a set of protein families. Every extracted classification rule had three qualitative attributes: support, confidence and interest. In the performed experiment, only simple classification rules of the form "$Motif \rightarrow Class$" were considered. The extracted rules were applied in descending order, following their interest attribute. A conducted comparative experiment showed that the achieved results outperformed those obtained in [9].

Reference [1] adopted the decision tree technique for assigning protein sequences to functional families. A training dataset of peptidase sequences (extracted from Merops protease database [12]), represented by their motifs composition, was used to automatically construct decision trees. Every protein sequence in the training dataset was labeled with the corresponding Merops functional family. The conducted experiments aimed at comparing the decision trees constructed using motifs generated by a tool called Meme with trees constructed using expert annotated Prosite motifs. The classification results showed that decision trees built using a motif-based representation of protein sequences constructed using Meme outperformed decision trees constructed using Prosite.

## 3 The Proposed Method

Based on the experience obtained when exploring this protein classification problem, we noted the promising idea of classifying a protein by discovering subsets of its motifs that strongly characterize some class. In this way, the classifier assigns the protein to the class which is better characterized by one of the subsets of its motifs.

Consider the protein P22216 contained in genbase18, which is composed by the following set of motifs M={PS50006, PS50011, PS50321, PS50322} and belongs to class PDOC50006. We can derive from this dataset that all proteins which contain the subset of motifs S={PS50006, PS50321} are associated with the class PDOC50006, that is, 100% of such proteins belongs to PDOC50006. This is a strong probability. On the other hand, the entire set of motifs M is never associated with the class PDOC50006. We believe that the search for subsets associated to high class probabilities can lead to an appropriate classifier technique.

Therefore, we propose the following approach to predict the class of an unknown protein $X$, based on the exploration of subsets of its characteristics, called Highest Subset Probability (HiSP) classifier.

Let $M_X$ be the set of motifs present in the protein $X$. Analogous to naive Bayes classifiers [7], supposing that there are $m$ classes, $\{C_1, C_2, \ldots, C_m\}$, $X$ is assigned to class $C_k$, $1 \leq k \leq m$, if and only if

$$P(C_k|X) > P(C_j|X) \quad \text{for all } j,\ 1 \leq j \leq m, \quad j \neq k. \tag{1}$$

In a different way from the naive Bayes classifier, we calculate $P(C_i|X)$, $1 \leq i \leq m$, as follows:

$$P(C_i|X) = max\{P(C_i|t)\} \quad \text{for all } t \subseteq M_X,\ t \neq \emptyset, \tag{2}$$

where

$$P(C_i|t) = \frac{P(C_i \wedge t)}{P(t)}. \tag{3}$$

Now, one needs to calculate $P(C_i \wedge t)$ and $P(t)$. $P(C_i \wedge t)$ stands for the probability of a protein pertaining to $C_i$ and having the subset $t$. $P(t)$ is the probability of the subset $t$ occurring. They are estimated from the training dataset in the following way:

$$P(C_i \wedge t) = \frac{F_{C_i t}}{N}, \tag{4}$$

where $F_{C_i t}$ and $N$ are the number of samples (proteins) of class $C_i$ having the subset $t$, and the total number of training samples, respectively. And

$$P(t) = \frac{F_t}{N}, \tag{5}$$

where $F_t$ is the number of samples having the subset $t$.

Nevertheless, this method can incur expensive computational costs when the number of motifs per protein is large. However, in the case of the problem addressed in this work, the application of this method is feasible, once in the Prosite database (release 18.0), for example, the average number of motifs per protein is 1.56 with a standard deviation of 0.90.

The pseudo-code for HiSP is presented in Fig. 1. Let $C = \{C_1, C_2, \ldots, C_m\}$ be the set of classes, and $TrainingProteins = \{a_1, a_2, \ldots, a_z\}$ be the set of training proteins, each one labeled with a class $C_i \in C$. $M_x$ is the set of motifs present in the protein $x$. The test protein $b_j$ will be assigned to a class $C_k \in C$ using the `CLASSIFIER` procedure. In line 1, the variables *bestClass* and *biggestProb* are set to initial values. For each subset of motifs present in the test protein

```
procedure CLASSIFIER(C, TrainingProteins, b_j)
1: bestClass ← NO CLASS; biggestProb ← 0;
2: for each t ⊆ M_{b_j} (such as t ≠ ∅) do
3:    F_t ← 0;
4:    F_t[i] ← 0; ∀i = 1, ..., m
5:    for each a_r ∈ TrainingProteins do
6:       if (t ⊆ M_{a_r}) then
7:          F_t[s] ← F_t[s] + 1, where C_s is the class of the protein a_r;
8:          F_t ← F_t + 1;
9:       end if
10:   end for
11: end for
12: for each C_i ∈ C do
13:   for each t ⊆ M_{b_j} (such as t ≠ ∅) do
14:      P(C_i|t) ← F_t[i]/F_t;
15:      if (P(C_i|t) > biggestProb) then
16:         bestClass ← C_i;
17:         biggestProb ← P(C_i|t);
18:      end if
19:   end for
20: end for
21: Return (bestClass);
end.
```

**Fig. 1.** Pseudo-code for HiSP

(line 2), $F_t$ (frequency of the subset $t$) and the vector $F_t[i]$ (frequency of training samples of class $C_i$ having the subset $t$) are initialized (lines 3 and 4), and the training dataset is scanned in order to compute them (lines 5 to 10). In lines 12 to 20, $P(C_i|t)$ is calculated for each class $C_i \in C$ (as described in Equation (3)). In line 15, if $P(C_i|t) = biggestProb$, then the following criterion is adopted for solving this tie: if $P(C_i|t) = P(C_j|t')$ and $F_t[i] > F_{t'}[j]$, where $t'$ is the subset associated with $biggestProb$, then $C_i$ takes precedence over $C_j$. Finally, the protein classification is returned in line 21.

## 4 Evaluating the Proposed Method

In this section, we compare the results of our approach with those obtained by the other two classifiers: the automata-based technique [6], and decision tree method [9].

For the classification problem addressed in this work, the samples correspond to the proteins, each protein is composed by a set of motifs, and the classes are associated with the protein functional families.

Three different datasets are used in the comparative tests. The datasets named genbase10, genbase18, and genbase28 contain 662, 2233, and 2934 proteins belonging to 10, 18, and 28 classes, respectively. These datasets were also used in

[6]. However, such datasets correspond to a subset of proteins and classes present in Prosite database. Therefore, in order to evaluate the proposed method considering a more realistic scenario, we also conducted tests using a dataset, called genprosite, which contains all proteins and classes from Prosite database (release 18.0), i.e., 75384 proteins in 1200 different classes.

Although $k$-fold cross-validation [13] is the most accepted method for estimating classifier accuracy in current machine learning and data mining researches, for comparison purpose, we adopted the same technique used in [6], called holdout [13]. According to this approach, the dataset is randomly divided into two disjoint datasets, a training dataset and a test dataset. Then, the training dataset is used by the classifier in the classification task, whose performance is estimated using the test dataset.

The accuracy ($ACC$) was used to measure the classification performance in [6]. Again, for comparison purpose, we also adopted it, which is given by

$$ACC = \frac{\text{number of proteins correctly classified}}{\text{total number of proteins in the dataset}}. \tag{6}$$

In the same way performed in [6], we repeat ten times the holdout method. We randomly generate ten tests, each one containing a training and a test dataset. Each test dataset was constituted by 10% of the total number of proteins. The accuracy was obtained by calculating the average of the values obtained from the ten tests.

Prosite database contains some proteins associated with more than one functional family. Therefore, the datasets used in the tests include proteins belonging to more than one class. In this case, the protein is repeatedly represented in the dataset so that each protein is associated with an unique class. This format of protein representation in the datasets has a drawback in the evaluation of classifier performance. For example, suppose that the test dataset has a protein appearing $k$ times, each one with a different class. Since the classification result of this protein will be unique (independently of the number of times that it is represented in the test dataset), considering only the $k$ repetitions of this protein, the classification accuracy will be at most $1/k$. Therefore, for the test datasets that contain protein(s) belonging to more than one class, the accuracy measure can not achieve 100%. Then, the maximum accuracy possible for each test dataset (Upper Bound column), as well as the comparative results for each classification method are shown in Table 1. The datasets are listed in the first column. Each other column shows the accuracy (average of ten tests) for each method. The parenthesis in HiSP column contain the standard deviation for each average.

The experiments evidence that the proposed approach reached encouraging classification performance. The results presented in Table 1 show that HiSP outperformed the methods applied in [6] and [9] for all tested datasets. It is worth noting that HiSP reached high accuracy on a more realistic scenario, i.e., for genprosite database.

Although genprosite dataset has not been used in the experiments conducted in [6], a dataset containing all proteins and classes from a previous Prosite

**Table 1.** Accuracy comparison

| Datasets | Approach in [6] | Approach in [9] | HiSP | Upper Bound |
|---|---|---|---|---|
| genbase10 | 93.71 | 80.13 | **99.85** (0.48) | 100 |
| genbase18 | 94.31 | 84.27 | **97.57** (1.12) | 99.91 |
| genbase28 | 75.98 | 74.49 | **83.24** (2.18) | 99.45 |
| genprosite | — | — | **85.71** (0.49) | 98.55 |

database version was considered in their performance evaluation. Such dataset was composed by approximately 40000 proteins in 1100 different classes. The accuracy was 41.4%. HiSP reached a performance more than twice better (85.71%) for a dataset containing 75384 proteins in 1200 different classes.

As mentioned in Section 3, HiSP can incur expensive computational costs for proteins with a large number of motifs. Thus, we decided to measure the CPU time spent to classify one instance. The experiments were carried out on an AMD XP 2600+ PC, with 512 Mbytes of RAM.

For genprosite dataset, the average classification CPU time per instance of HiSP was 0.6 seconds. The good behavior of HiSP is due to the small number of motifs per proteins on average. However, the speed performance of HiSP is directly influenced by the number of motifs per protein in a test dataset, since the amount of subsets of motifs evaluated in the classification process depends on the number of motifs of each protein.

In Prosite database (release 18.0), each protein is constituted by at most ten motifs, and the average number of motifs per protein is 1.56 with a standard deviation of 0.9. For genprosite dataset, which contains all proteins from Prosite database, when a protein contains two motifs, HiSP spent, on average, 0.3 seconds to classify it. On the other hand, for proteins containing ten motifs (the worst case in Prosite database), HiSP spent, on average, 108 seconds to classify them. However, we may observe that 98% of the proteins in genprosite dataset contain at most five motifs. And, for five motifs, the HiSP strategy takes on average 3.2 seconds to classify a protein. So, we can conclude that the proposed method works very appropriately for the problem addressed in this work.

## 5 Conclusions

We have presented a computational technique to classify protein sequences based on their motif composition. In order to improve the results presented by automata-based technique [6] and decision tree method [9], we proposed a new data mining technique, called HiSP. Its main idea is the evaluation of subsets of attribute values to classify an instance. The computational experiments have shown that HiSP outperformed the results presented in [6] and [9] for all tested datasets. Also, the run time evaluation showed that, on average, HiSP presented a good behavior. Therefore, we can conclude that the idea of classifying samples according to the class which is better characterized by one of the subsets of their attribute values looks very promising for problems with characteristics similar to the one addressed in this work.

In future works, we intend to evaluate this classification approach considering different kinds of applications and databases.

## Acknowledgments

The authors are grateful to research group of Bioinformatics Laboratory at the National Laboratory for Computing Sciences (LNCC, Brazil), especially to Cláudia Barros Monteiro Vitorello for all help and suggestions. Also, the authors would like to thank Fotis Psomopoulos for his willingness to share the datasets with them and to answer their numerous questions about his paper.

## References

1. Wang, X., Schroeder, D., Dobbs, D., Honavar, V.: Automated data-driven discovery of motif-based protein function classifiers. Information Sci. **155** (2003) 1–18
2. Henikoff, S., Henikoff, J.G.: Protein family databases. In: Encyclopedia of life sciences. Macmillan Publishers Ltd, Nature Publishing Group (2001) http://www.els.net.
3. Falquet, L., Pagni, M., Bucher, P., Hulo, N., Sigrist, C.J., Hofmann, K., Bairoch, A.: The prosite database, its status in 2002. Nucleic Acids Res. **30** (2002) 235–238
4. Sigrist, C., Cerutti, L., Hulo, N., Gattiker, A., Falquet, L., Pagni, M., Bairoch, A., Bucher, P.: Prosite: a documented database using patterns and profiles as motif descriptors. Brief Bioinformatics **3** (2002) 265–274
5. Merschmann, L., Plastino, A.: A bayesian approach for protein classification. In: Proc. of the $21^{st}$ Annual ACM Symposium on Applied Computing, Dijon, France (2006) (to appear as a short paper).
6. Psomopoulos, F., Diplaris, S., Mitkas, P.A.: A finite state automata based technique for protein classification rules induction. In: Proc. of the $2^{nd}$ European Workshop on Data Mining and Text Mining in Bioinf., Pisa, Italy (2004) 54–60
7. Duda, R., Hart, P.: Pattern Classification and Scene Analysis. John Wiley & Sons, New York (1973)
8. Wang, D., Wang, X., Honavar, V., Dobbs, D.L.: Data-driven generation of decision trees for motif-based assignment of protein sequences to functional families. In: Proc. of the Atlantic Symposium on Computational Biology, Genome Information Systems & Technology, North Carolina, USA (2001)
9. Hatzidamianos, G., Diplaris, S., Athanasiadis, I., Mitkas, P.A.: GenMiner: A data mining tool for protein analysis. In: Proc. of the $9^{th}$ Panhellenic Conference On Informatics, Thessaloniki, Greece (2003) 346–360
10. Witten, I.H., Frank, E.: Data Mining: Practical Machine Learning Tools and Techniques with Java Implementations. Morgan Kaufmann (1999)
11. Seidman, C.: Data Mining with Microsoft SQL Server. Microsoft Press, Redmond, WA (2000)
12. Rawlings, N.D., Barret, A.J.: Merops: The peptidase database. Nucleic Acids Res. **28** (2002) 323–325
13. Han, J., Kamber, M.: Data Mining: Concepts and Techniques. Morgan Kaufmann Publishers, New York (2001)

# Cross-Ontological Analytics: Combining Associative and Hierarchical Relations in the Gene Ontologies to Assess Gene Product Similarity

C. Posse, A. Sanfilippo, B. Gopalan, R. Riensche,
N. Beagley, and B. Baddeley

Pacific Northwest National Laboratory
902 Battelle Blvd, Richland, PO Box 999, WA 99352, USA
{Christian.Posse, Antonio.Sanfilippo, Banu.Gopalan,
Rick.Riensche, Nathaniel.Beagley, Bob.Baddeley}@pnl.gov

**Abstract.** Gene and gene product similarity is a fundamental diagnostic measure in analyzing biological data and constructing predictive models for functional genomics. With the rising influence of the gene ontologies, two complementary approaches have emerged where the similarity between two genes/gene products is obtained by comparing gene ontology (GO) annotations associated with the gene/gene products. One approach captures GO-based similarity in terms of hierarchical relations within each gene ontology. The other approach identifies GO-based similarity in terms of associative relations across the three gene ontologies. We propose a novel methodology where the two approaches can be merged with ensuing benefits in coverage and accuracy.

## 1 Introduction

The analysis of microarray data has become a mainstay of biological knowledge discovery, with a consequent explosive growth in the amount of DNA, RNA, and protein sequences and expression values to be interpreted. Gene and gene product similarity is a primary method that biologists use to carry out this interpretation.

When genes and gene products are represented by their sequences, gene and gene product similarity is usually performed using word-based heuristics to select maximal segment pair alignments similar to those implemented in BLAST [1] and FASTA [12] to approximate the Smith-Waterman algorithm [17]. Once gene or protein sequences are mapped into higher level representations, other similarity methods become available that provide additional insights to sequence similarity measures. For example, [13] demonstrate how gene product similarity can be performed in terms of PFAM domains (http://www.sanger.ac.uk/Software/Pfam) and [14] show how a computational analysis of scientific literature can be leveraged to define and recognize gene expression clusters.

With the rising influence of the gene ontologies (http://www.geneontology.org), new approaches have emerged where the similarity between genes or gene products is obtained by comparing gene ontology (GO) code annotations associated with them. The gene ontologies provide three orthogonal networks of functional genomic concepts structured in terms of semantic relationships such as inheritance and

V.N. Alexandrov et al. (Eds.): ICCS 2006, Part II, LNCS 3992, pp. 871 – 878, 2006.

meronymy, which encode biological process (BP), molecular function (MF) and cellular component (CC) properties of gene and gene products. The use of GO code annotations in establishing gene and gene product similarity provides significant added functionality to methods such as BLAST and FASTA as it explicitly relates genes and gene products in terms of participation in the same/similar biological processes, presence in the same/similar cellular components and expression of the same/similar molecular functions.

Two main GO-based similarity approaches have emerged so far. One approach assesses GO code similarity in terms of shared *hierarchical* relations *within* each gene ontology (BP, MF, or CC) [2], [5], [9], [10]. For example, the relative semantic closeness of two biological processes would be determined by the informational specificity of the most immediate parent that the two biological processes share in the BP ontology. The other approach establishes GO code similarity by leveraging *associative* relations *across* the three gene ontologies [3]. Such associative relations make predictions such as which cellular component is most likely to be the location of a given biological process and which molecular function is most likely to be involved in a given biological process.

While these two approaches are fully complementary, no effort has been made so far to combine them. One major difficulty in carrying out such a combination resides in the heterogeneity of the two measures, one based on a hierarchical assessment and the other on an associative one. The aim of this paper is to present a methodology (XOA for *Cross-Ontological Analytics*) where this difficulty can be solved so that the two approaches can be integrated, with ensuing benefits in coverage and accuracy.

## 2 Background

GO-based similarity methods that focus on measuring intra-ontological relations have adopted the information theoretic treatment of semantic similarity developed in Natural Language Processing–see [4] for an extensive survey. An example of such an approach is given by [15], who defines semantic similarity between two concept nodes *c1 c2* in a semantic graph as the information content of the least common superordinate (*lcs*) of *c1* and *c2*, as shown in (1) where *p(c)* is the probability of encountering instances of *c* in a specific corpus and *-log p(c)* indicates the information content of *c*.

$$sim(c1,c2) = -log\ p(lcs(c1,c2)) \quad (1)$$

[6] provide a refinement of Resnik's measure by factoring in the distance from each concept to the least common superordinate, as shown in (2).

$$sim(c1,c2) = -[2*log\ p(lcs(c1,c2))-(log\ p(c1)+log\ p(c2))] \quad (2)$$

[8] introduces a slight modification to Jiang's and Conrath's measure:

$$sim(c1,c2) = 2*log\ p(lcs(c1,c2))/(log\ p(c1)+log\ p(c2)) \quad (3)$$

The information theoretic approach is very well suited to assess GO code similarity since each gene ontology is formalized as a directed acyclic graph, In addition, the GO database includes numerous curated GO annotations which can be used to calculate the information content of each GO code with high reliability. Evaluations of this methodology have yielded promising results. [9] and [10] demonstrate that there is strong correlation between GO-based similarity judgments for human proteins and similarity judgments obtained through BLAST searches for the same proteins. [2] show that there is a strong connection between the degree of GO-based similarity and the expression correlation of gene products.

As [3] remark, the main problem with the information theoretic approach to GO code similarity is that it does not take into account associative relations across the gene ontologies. For example, the two GO codes 0050909 (*sensory perception of taste*) and 0008527 (*taste receptor activity*) belong to different gene ontologies (BP and MF) but are undeniably very closely related. The information theoretic approach would simply miss associations of this kind as it is not designed to capture inter-ontological relations.

[3] propose to recover associative relations across the gene ontologies using a variety of statistical techniques which estimate the similarity of two GO codes intra-ontologically in terms of the distribution of the gene product annotations associated with the two GO codes in the GO database. One such technique is an adaptation of the vector space model frequently used in Information Retrieval [16], where each GO code is represented as a vector of gene-based features weighted according to their distribution in the GO annotation database, and the similarity between two GO codes is computed as the cosine of the vectors for the two codes.

The ability to measure associative relations across the gene ontologies can significantly augment the functionality of the information theoretic approach so as to provide a more comprehensive assessment of gene and gene product similarity. However, in spite of their complementarity, the two GO code similarity measures are not easily integrated. This is because the two measures are obtained through different methods, express distinct senses of similarity (i.e. intra- and inter-ontological) and are thus incomparable. To date, this remains a challenge which is yet to be addressed.

## 3 Cross-Ontological Analytics (XOA)

Our basic approach in developing a GO-based similarity algorithm capable of combining intra- and inter-ontological relations consists in "translating" each associative relation across the gene ontologies into a hierarchical relation within a single ontology. More precisely, let *c1* (*c2*) denote a GO code in the gene ontology *O1* (*O2*), the XOA similarity between *c1* and *c2* is defined as follows:

$$XOA(c1, c2) = max\{sim(c1, c3)*cos(c2, c3), sim(c2, c4)*cos(c1, c4)\} \quad (4)$$

where *cos(ci,cj)* denotes the cosine associative measure proposed by [3], *sim(ci,cj)* denotes any of the three intra-ontological semantic similarities defined in the previous section, see (1)-(3), and the maximum is taken over all GO codes *c3* in *O1* and *c4* in *O2*. Note that if *O1* and *O2* refer to the same ontology, then:

$$XOA(c1, c2) = sim(c1, c2). \quad (5)$$

This shows that the XOA approach extends intra-ontological similarities to the inter-ontological context. Conversely, *c3* (*c4*) can be seen as a surrogate for *c2* (*c1*) in the same ontology as *c1* (*c2*). The inter-ontological semantic similarity between *c1* (*c2*) and *c2* (*c1*) can be then estimated from the intra-ontological semantic similarity between *c1* (*c2*) and *c3* (*c4*) and the associative relation between *c3* (*c4*) and *c2* (*c1*).

The XOA similarity between genes/gene products derives from the XOA similarity between GO codes. Let *GP1* and *GP2* be two genes/gene products. Let *c11*, *c12*,..., *c1n* denote the set of GO codes associated with *GP1* and *c21*, *c22*,...., *c2m* the set of GO codes associated with *GP2*. The XOA similarity between *GP1* and *GP2* is given as shown in (6), where *i=1*,..., *n* and *j=1*,...,*m*.

$$XOA(GP1, GP2) = max\{XOA(c1i\,, c2j)\} \tag{6}$$

Note that [10] use the average instead of the maximum. However, genes/gene products with many GO annotations have higher chance of exhibiting a few lower XOA similarities between GO codes. This automatically lowers the similarity between genes/gene products. In our experiments, using the maximum provided more consistent results than the average. [2] also find that the average does not always meaningfully estimate similarity and plan on investigating the maximum approach. The maximum approach is also consistent with the first information content based semantic similarity proposed by [15] where the similarity between two words is defined as the maximum of the similarities between all concept pairs emerging from difference senses for the two words.

## 4 Evaluation

To evaluate the XOA approach, we follow the benchmarking study carried out by [10], which compares intra-ontological semantic similarities with sequence similarities as measured with BLAST. In doing so, we use current data from the GO database. This has several advantages. First, the replication of the above benchmarking study three years later lets us assess the effects of the changes of the gene ontologies and reference databases on the findings unearthed by [10]. Second, it allows us to improve the benchmarking analysis. Finally, it allows us to compare our XOA results with the intra-ontological semantic similarities obtained by [10].

All results reported are from analyses carried out on the May 2005 release of GO database available from http://www.godatabase.org/dev. A large set of protein pairs with varying degrees of sequence similarities was obtained by

1) performing BLAST searches with expect value of 100 over the set of approximately 7000 SWISS-PROT human proteins for which GO codes were available, and
2) retaining proteins from BLAST results for which GO codes were available. Only those codes with "Traceable Author Statement" tags were used.

The full set of BLAST searches was completed in less than 30 minutes using ScalaBLAST [11], a powerful parallel BLAST engine. This process generated a set of 255,529 protein pairs. Table 1 summarizes the distribution of GO codes in the benchmarking set. It reveals that XOA provides coverage for around 15% more protein pairs than the combination of intra-ontological semantic similarities applied separately to the BP, MF and CC ontologies. When compared to any single gene ontology the XOA coverage is at least twice as large.

**Table 1.** Distribution of TAS GO codes in benchmarking set

| GO Code Distribution | Number of protein pairs |
|---|---|
| Each protein in the pair has at least one GO code | 255,529 |
| Each protein in the pair has at least one BP GO code | 128,921 |
| Each protein in the pair has at least one MF GO code | 112,027 |
| Each protein in the pair has at least one CC GO code | 94,677 |
| Both proteins in the pair have at least one GO code in the same ontology (BP, MF or CC) | 216,714 |
| Both proteins in the pair have only GO codes in different ontologies | 38,788 |

[9] and [10] compare semantic and sequential similarities by calculating the correlation coefficient between the intra-ontological semantic similarities and the logarithm of the BLAST bit score LBS for each of the gene ontologies. Table 2 provides these correlations based on our more recent data. In addition, it shows the correlation between XOA and LBS.

**Table 2.** Correlation coefficients between LBS and semantic similarities. Correlation coefficients were calculated from the subset of proteins given in Table 1.

| Gene Ontology | Resnik | Lin | Jiang and Conrath |
|---|---|---|---|
| Molecular Function | 0.555 | 0.516 | 0.468 |
| Biological Process | 0.394 | 0.394 | 0.378 |
| Cellular Component | 0.318 | 0.345 | 0.320 |
| XOA | 0.412 | 0.362 | 0.270 |

We observe that despite the major changes in the gene ontologies in the last 3 years and the enrichment of the SWISS-PROT database these correlations are very similar to those produced by [10] in 2003 for the intra-ontological semantic similarities. This is confirmed in Figure 1, which shows the functional relationships between LBS and the semantic similarities. Furthermore, Table 2 agrees with [10] findings that Resnik-based semantic similarities on one hand and the Molecular Function ontology on the other end tend to produce higher correlations. A novel finding is that XOA correlations are comparable to those obtained with intra-ontological semantic similarities though they are based on at least 100% more protein pairs than any single ontology comparison.

These results however should be taken with some caution. Correlation coefficients can only measure linear relationship between two variables. If taking the logarithm of the bit score straightens the relationship between the Resnik similarity measure and the bit score, this is not the case for the other similarity measures. One can show that, for the Lin and Jiang & Conrath measures, higher order transformations of the bit score, such as $a+b*LBS+c*LBS^2$ with a, b and c being some constants, are required in order to get a linear relationship. Rather than pursuing this direction we note that what we are truly interested in measuring is a *monotonic* relationship between the bit score

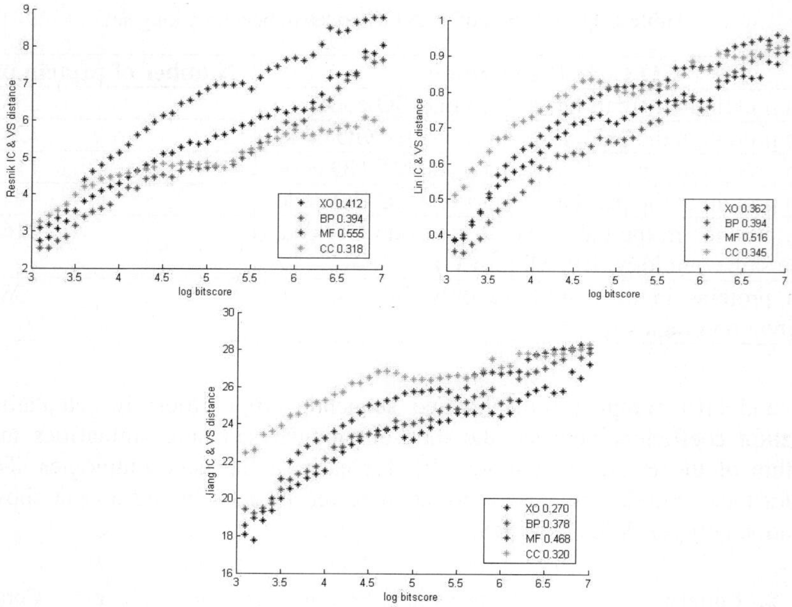


**Fig. 1.** Comparing sequence and semantic similarity. BLAST search were performed for each SWISS-PROT human protein, and matches were analyzed for semantic similarity with the search protein. Following [10], intervals were taken along the x-axis and log[bit score] and semantic similarity values were averaged.

**Table 3.** Spearman rank order correlation coefficients between bit score and semantic similarities. Correlation coefficients were calculated from the subset of proteins given in Table 1.

| Gene Ontology | Resnik | Lin | Jiang and Conrath |
|---|---|---|---|
| Molecular Function | 0.575 | 0.544 | 0.516 |
| Biological Process | 0.389 | 0.417 | 0.423 |
| Cellular Component | 0.343 | 0.387 | 0.382 |
| XOA | 0.405 | 0.393 | 0.368 |

and the semantic similarity. A well-known measure of monotonicity is Spearman rank order correlation coefficient [7]. Table 3 reports this coefficient for the configuration set in Table 2.

One can see that correlations tend to increase using the Spearman rank order correlation coefficient. Even more interesting is the fact that each similarity measure yields best correlation results when applied to a different ontology: Resnik for MF, Jiang and Conrath for BP and Lin for CC. This is not too surprising. Because Resnik measure rests only on the closest parent node of the node pair, it will tend to produce higher values in well developed ontologies, such as MF, where nodes tend to be more specific. By normalizing the Resnik measure with information content of the node pair, the Lin and Jiang & Conrath measures prefer sparser ontologies with fewer child nodes. Because Jiang and Conrath measure normalizes via a difference and Lin measure via a ratio, the former favors more uniform ontologies while the latter prefers ontologies more developed in regions of higher specificity within the ontology.

Another issue with the results in Tables 2 and 3 is that they do not reflect the significant difference in coverage across the different methods. XOA correlations rely on all protein pairs for which there are GO codes available. By contrast, MF- and BP-based correlations use only half of them and CC-based correlations use only a third. Though the three gene ontologies are orthogonal, they can be subsumed under a single root level term whose information content is 0. Conceptually, this root node allows us to compute the Resnik, Lin and Jiang & Conrath semantic similarities for nodes in different gene ontologies. In practice, this always results in a zero value for the Resnik and Lin similarities and very small values for the Jiang and Conrath similarity. Using this mechanism, we computed Spearman rank order correlations between the bit score and the various semantic similarities using all protein pairs in our benchmarking set. Table 4 summarizes these values. All correlations decrease except for the XOA ones which remain unchanged and become the strongest ones.

**Table 4.** Spearman rank order correlation coefficients between bit score and semantic similarities. Correlation coefficients were calculated using the entire benchmarking set of 255,529 protein pairs.

| Gene Ontology | Resnik | Lin | Jiang and Conrath |
|---|---|---|---|
| Molecular Function | 0.307 | 0.301 | 0.296 |
| Biological Process | 0.195 | 0.202 | 0.203 |
| Cellular Component | 0.229 | 0.234 | 0.233 |
| XOA | 0.405 | 0.393 | 0.368 |

## 5 Conclusions and Future Work

Our XOA approach provides a more comprehensive approach to the use of GO annotations for the assessment of gene and product similarity than any of the methods previously proposed. Due to its ability to yield a greater number of GO code comparisons, XOA accounted for 15% more human protein pairs than the combined three intra-ontological semantic similarity approaches (see Table 1), with a consequent improvement of the overall correlation with BLAST results (see Table 4).

Moving forward, we are planning the use of textual evidence extracted from relevant biomedical literature to improve inter-ontological GO-based similarities. We also intend to use manually curated ground truth datasets such as the PFAM database (http://www.sanger.ac.uk/Software/Pfam/) in addition to BLAST gene (product) similarity search results, to carry out a more meaningful evaluation of XOA. Finally, we have started to engage XOA in the analysis of gene expression data to aid the identification of biological pathways and the detection of biomarkers. These applications are yielding promising results and will constitute the ultimate testing ground for XOA.

## References


1. Altschul, S.F., T. L. Madden, A. A. Schaffer, J. Zhang, Z. Anang, W. Miller and D.J. Lipman (1997) Gapped BLAST and PSI-BLST: a new generation of protein database search programs. *Nucl. Acids Res.* 25:3389-3402.
2. Azuaje F., H. Wang and O. Bodenreider (2005) Ontology-driven similarity approaches to supporting gene functional assessment. In *Proceedings of the ISMB'2005 SIG meeting on Bio-ontologies* 2005, pages 9-10.
3. Bodenreider, O., M. Aubry and A. Burgun (2005) Non-lexical approaches to identifying associative relations in the gene ontology. In *Proceedings of Pacific Symposium on Biocomputing*, pages 104-115.
4. Budanitsky, A. (1999) Lexical semantic relatedness and its application in natural language processing. Technical report CSRG-390, Department of Computer Science, University of Toronto.
5. Couto, F. M., M. J. Silva and P. Coutinho (2003) Implementation of a functional semantic similarity measure between gene-products. *Technical Report*, Department of Informatics, University of Lisbon, http://www.di.fc.ul.pt/tech-reports/03-29.pdf.
6. Jiang J. and D. Conrath (1997) Semantic similarity based on corpus statistics and lexical taxonomy. In *Proceedings of International Conference on Research in Computational Linguistics*, Taiwan.
7. Kendall M. G. (1949) Rank and product-moment correlation. *Biometrika* 36: 177-193.
8. Lin, D. (1998) An information-theoretic definition of similarity. In *Proceedings of the 15th International Conference on Machine Learning*, Madison, WI.
9. Lord P.W., R.D. Stevens, A. Brass, and C.A.Goble (2002) Investigating semantic similarity measures across the Gene Ontology: the relationship between sequence and annotation. *Bioinformatics* 19(10):1275-1283.
10. Lord P.W., R.D. Stevens, A. Brass, and C.A.Goble (2003) Semantic similarity measures as tools for exploring the Gene Ontology. In *Proceedings of Pacific Symposium on Biocomputing*, pages 601-612.
11. Oehmen C.S. and J.N. Nieplocha (2006) ScalaBLAST: A scalable implementation of BLAST for high performance data-intensive bioinformatics analysis. *IEEE Transactions on Parallel and Distributed Systems*, in press.
12. Pearson, W. R. and D. J. Lipman (1988) Improved tools for biological sequence analysis. In *Proceedings of the National Academy of Sciences* 85:2444-2448.
13. Popescu, M., J. M. Keller and J.A. Mitchell (2005) Gene ontology automatic annotation using a domain based gene product similarity measure. In *Proceedings of 14th IEEE International Conference on Fuzzy Systems*, Reno, Nevada, May 21-25, pages 108-111.
14. Raychaudhuri, S., J. T. Chang, F. Imam and R. B. Altman (2003) The computational analysis of scientific literature to define and recognize gene expression clusters. *Nucleic Acids Research* 31(15):4553-4560.
15. Resnik, P. (1995) Using information content to evaluate semantic similarity. In *Proceedings of the 14th International Joint Conference on Artificial Intelligence*, pages 448–453, Montreal.
16. Salton, G., A. Wong and C. S. Yang (1975) A Vector space model for automatic indexing, *CACM* 18(11):613-620.
17. Smith, T. and M. S. Waterman (1981) Identification of common molecular subsequences. *J. Mol. Biol.* 147:195-197.

# A GO-Based Method for Assessing the Biological Plausibility of Regulatory Hypotheses

Jonas Gamalielsson, Patric Nilsson, and Björn Olsson

Systems Biology Group, University of Skövde,
Box 407, 541 28 Skövde, Sweden
jonas.gamalielsson@his.se, patric.nilsson@his.se,
bjorn.olsson@his.se

**Abstract.** Many algorithms have been proposed for deriving regulatory networks from microarray gene expression data. The performance of such algorithms is often measured by how well the resulting network can recreate the gene expression data that it was derived from. However, this kind of performance does not necessarily mean that the regulatory hypotheses in the network are biologically plausible. We therefore propose a method for assessing the biological plausibility of regulatory hypotheses using prior knowledge in the form of regulatory pathway databases and Gene Ontology-based annotation of gene products. A set of templates is derived by generalising from known interactions to typical properties of interacting gene product pairs. By searching for matches in this set of templates, the plausibility of regulatory hypotheses can be assessed. We evaluate to what degree the collection of templates can separate true from false positive interactions, and we illustrate the practical use of the method by applying it to an example network reconstruction problem.

## 1 Introduction

It is highly desirable to be able to derive causal gene regulatory networks using gene expression data. This is known as reverse engineering of genetic networks[1]. Time series expression data is often used, and the task of the reverse engineering algorithm is to find a set of activation rules that fits these data. Methods for reverse engineering of genetic networks that have been proposed include techniques such as boolean [2, 3], neural [4, 5] and Bayesian networks [6, 7, 8]. In most cases, however, many different reverse engineered networks are consistent with the observed data, but we can expect only a few of these networks to be biologically plausible. A drawback of reverse engineering methods which are based solely on fit to the data is that they do not provide any way of distinguishing between biologically plausible and implausible networks. Therefore, we here propose a method for assessing the biological plausibility of regulatory hypotheses using prior biological knowledge in the form of Gene Ontology (GO) annotation of gene products and examples of regulatory interactions from pathway databases. Using GO, we derive templates encoding general knowledge by generalising from examples of known interactions to typical properties of interacting gene product pairs. By matching regulatory hypotheses to such templates, their plausibility

V.N. Alexandrov et al. (Eds.): ICCS 2006, Part II, LNCS 3992, pp. 879–886, 2006.

can be assessed. The assumption is that if the properties of the gene products in a regulatory hypothesis are similar to those of gene products which are already known to interact, this increases the plausibility of the hypothesis.

Gene Ontology [9] is a structured vocabulary of molecular biology. It contains three different sub-ontologies covering the molecular functions, biological processes and cellular components of gene products, and is structured as a directed acyclic graph showing how terms are related to each other, using inheritance (IS-A) and aggregation (PART-OF). Using these relations, abstraction hierarchies of terms with different specificity are created. A gene product can be associated with several terms in each sub-ontology. One important use of an ontology is the calculation of semantic similarity between two terms. Different information theoretic measures have been proposed and applied for this purpose [10, 11, 12] and ideas from these measures are used in our method.

The concept of templates is appealing for identifying hypothetical regulatory relations that are similar to known regulatory relations. Templates have previously been used in the context of association rule discovery and expression data in [13], where a template-language was designed that allows users to group, filter and inspect a large number of rules produced by association rule discovery algorithms. Of particular interest for our work are rule templates $F1 \rightarrow F2$ which detect rules where the antecedent part contains any of the genes in a predefined functional group $F1$ and the consequent part contains a gene from another predefined group $F2$.

The basic idea behind the method proposed here is to use what we know about regulation in documented pathways, generalise this knowledge, and apply it for assessment of the plausibility of regulatory hypotheses. The GO molecular function classification of the gene products participating in regulatory relations in pathways is used to derive templates. The templates encode the types of gene products known to be involved in a particular type of regulatory relation.

## 2 Methods and Algorithms

The three main steps of the method are: (1) GO term probability calculation, (2) binary relation extraction, and (3) template derivation.

**Step 1: GO term probability calculation.** The GO term probability calculation is done as specified in [14] using annotations of the *S. cerevisiae* Genome Database (SGD) available at the GO website (http://www.geneontology.org, September 2005). The term probability calculation is defined as follows, where $D_A$ is a GO annotation database for gene products (in this case SGD):

- For each gene product $GP_i \in D_A$:
  - Increment a counter $CNT_j$ for each GO term $GT_j$ in the annotation of $GP_i$ and increment the counter of each ancestor term of $GT_j$.
- For each GO term $GT_k$:
  - Calculate the term probability $p(GT_k) = \frac{CNT_k}{N_A}$, where $N_A$ is the total number of annotations in $D_A$.

The term probabilities indicate how specific, or common, different GO terms are, and they are used as a measure of specificity of relation templates. The specificity increases with decreasing term probability.

**Step 2: Binary relation extraction.** Each relation between protein complexes is decomposed into a set of binary relations between pairs of gene products. These binary relations are subsequently used to derive templates. In the following, $D_P$ is a regulatory pathway database, $C_i$ and $C_j$ refer to complexes containing sets of gene products, and $rel$ is a specific relation type, e.g. "expression":

- For each complex relation $C_i[rel]C_j \in D_P$:
  - Create $|C_i| \cdot |C_j|$ binary relations $GP_k[rel]GP_l$ by combining each gene product in $C_i$ with each gene product in $C_j$, and add all new relations to a set $MOD$ of model relations.

**Step 3: Template derivation.** Templates representing generalised knowledge of regulatory relations are derived. For each template a GO-score $GS$ is calculated, based on how common the two GO terms appearing in the template are in the annotation database. This is formalised as the average information content of the template terms, using standard information theoretic principles. Templates are derived by the following procedure, where $GID$ represents a GO identifier:

- For each binary relation $GP_k[rel]GP_l \in MOD$:
  - Create templates $GID_i[rel]GID_j$ where $GID_i \in S_k$=$\{GP_k$ terms with ancestors$\}$ and $GID_j \in S_l$=$\{GP_l$ terms with ancestors$\}$, yielding $|S_k| \cdot |S_l|$ templates. Add these templates to a set $T$.
- For each template $GID_i[rel]GID_j \in T$:
  $GS = -log_2((p(GID_i) + p(GID_j))/2)$.

Using this information-theoretic definition of $GS$ we get $GS \in [0, -log_2(1/N_A)]$, where $N_A$ is the total number of annotations in $D_A$. In order to facilitate interpretation of $GS$ values and to make it easier to set a threshold for which hypotheses to consider plausible we also calculate an "expect" value $E$ for each hypothesis $H_i$ according to:

$$E(H_i) = |H| \cdot \frac{|\{t : t \in T \wedge GS(t) \geq GS(H_i)\}|}{|T|} \tag{1}$$

where $H$ is the hypothesis set and $E(H_i)$ the expected number of template hits for $H_i$ with $GS(t) \geq GS(H_i)$ if templates are randomly drawn from $T$.

We refer to templates based only on annotation level terms (terms directly associated with the gene products in $D_A$) as "basic templates", while templates containing GO terms of higher abstraction levels are referred to as "variant templates". When initially testing the approach using the *S. cerevisiae* cell-cycle pathway relations from KEGG as input, the method derived 68 basic templates, all with GO-scores in the interval [4.84,11.29], and 1236 variant templates with GO-scores in the interval [0,10.97]. 54% of these variant templates had a $GS$ lower than the lower bound for the basic templates.

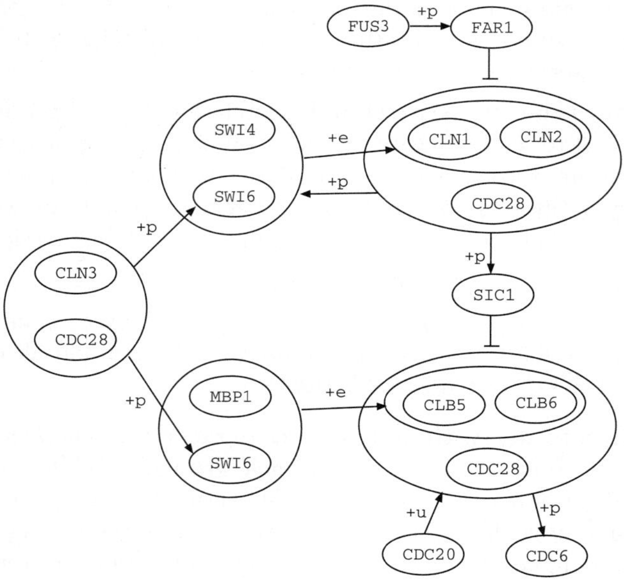


**Fig. 1.** Part of the cell-cycle regulatory network of *S. cerevisiae* (adapted from [7])

As an example to illustrate the process of template derivation, we here consider the complex-to-complex relation {SWI4,SWI6} [+e] {CLN1,CLN2} in the *S. cerevisiae* cell cycle pathway (figure 1). This relation between gene product complexes is decomposed into the four binary relations "SWI4 [+e] CLN1", "SWI4 [+e] CLN2", "SWI6 [+e] CLN1" and "SWI6 [+e] CLN2". Each of these are used for template derivation. When "SWI4 [+e] CLN1" is used, each of the terms in the GO subgraph of SWI4 in figure 2 are combined with those for CLN1 to create templates. The basic template in this case would be "GO:0003700 [+e] GO:0016538", with a GO-score of $-log_2((0.010+0.003)/2)) = 7.27$. Variant templates are for example "GO:0003700 [+e] GO:0019887" and "GO:0030528 [+e] GO:0019207", with GO-scores $-log_2((0.010+0.005)/2)) = 7.06$ and $-log_2((0.052 +0.005)/2))=5.13$, respectively. With six terms for SWI4 and five terms for CLN1, there is a total of 30 templates for this relation (one basic and 29 variant templates).

The derived set of templates is used in the assessment of hypothetical regulatory relations derived using e.g. reverse engineering algorithms. A hypothetical relation (or hypothesis for short) is defined as a directed relation between two gene products. There can be different relation types, e.g. "expression" and "phosphorylation". Each hypothesis $GPh_k[rel]GPh_l$ from a set $H$ of regulatory hypotheses is assessed by the templates using the following procedure:

– For each template $GID_i[rel]GID_j \in T$:
  - Report template match if $GID_i \in \{GPh_k$ terms with ancestors$\}$ and $GID_j \in \{GPh_l$ terms with ancestors$\}$.

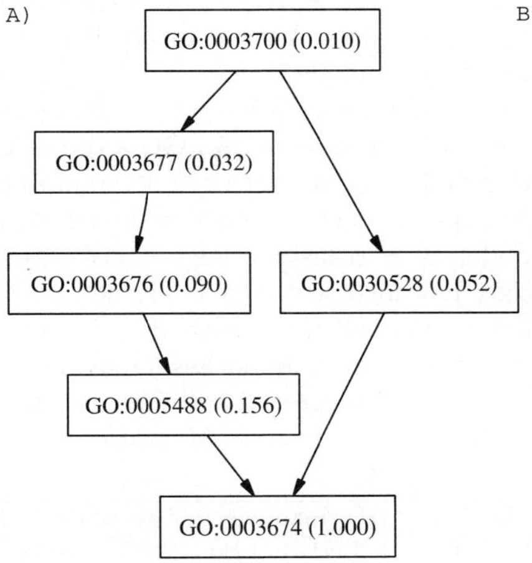


**Fig. 2.** GO subgraphs for SWI4 (A) and CLN1 (B). GO terms are represented by nodes with probabilities in brackets. Directed edges represent IS-A relationships. A) SWI4: GO:0003700 = transcription factor activity, 0003677 = DNA binding, 0003676 = nucleic acid binding, 0005488 = binding, 0003674 = molecular function, 0030528 = transcription regulator activity. B) CLN1: 0016538 = cyclin-dependent kinase regulator activity, 0019887 = protein kinase regulator activity, 0019207 = kinase regulator activity, 0030234 = enzyme regulator activity.

This procedure results in a list of templates conforming to the hypothesis, ranked in GO-score order. Subsequently, all hypotheses in $H$ can be ranked in GO-score order according to the highest-scoring template of each hypothesis.

## 3 Results

This section presents experiments done to assess the performance and to show the practical use of the proposed method for plausibility assessment. Out of the 12 types of regulatory relations supported by KEGG, we selected "activation", "inhibition", "expression", "phosphorylation" and "dephosphorylation". The only relation representing transcriptional regulation is "expression", while the other four represent post-translational regulatory mechanisms.

In the evaluation, we study sets of hypothetical relations derived by applying a dynamic Bayesian network technique to gene expression data measuring the expression of genes during the *S. cerevisiae* cell cycle. Hypotheses derived in [15] are shown together with their best-scoring templates in table 1. Each hypothesis in the table is classified into one of four different categories of conformance with respect to the target network in figure 1: correct, transitive, misdirected and incorrect [7]. The hypothetical relations are limited to cover only the part of the cell cycle shown in figure 1. Relations from the whole *S. cerevisiae* cell cycle pathway were used to derive the template set.

**Table 1.** Regulatory hypotheses and best matching templates. *Hypothesis*: a directed hypothetical relation between two gene products, derived in ref. [15]. *Template*: GO term identifier for the left hand side of the template, followed by type of relation in square brackets (E = "expression", P = "phosphorylation", I = "inhibition"), and GO term identifier for the right hand side of the template. *Sc*: GO-score of the template. *E*: the expected number of hits with a GO-score greater than or equal to the matching template's GO-score (see eq. 1). *Cl*: hypothesis class, coded as in ref. [7] (C = correct edge with respect to the known model, T = transitive edge which skips exactly one gene product, M = misdirected edge, I = incorrect edge). The left half of the table shows hypotheses with $E < 1$ and the right half those with $E \geq 1$. Note that all five correct hypotheses have $E < 1$. In addition, correct hypotheses had the lowest average $E$, followed by transitive hypotheses, while mis-directed and incorrect hypotheses had substantially higher average $E$ ($C$: 0.62, $T$: 0.96, $I$: 3.07, $M$: 3.48).

| Hypothesis | Template | Sc | E | Cl | Hypothesis | Template | Sc | E | Cl |
|---|---|---|---|---|---|---|---|---|---|
| FUS3→SIC1 | 4707[P]19210 | 10.7 | 0.07 | I | CLN1→CLB6 | 19887[I]16538 | 7.97 | 1.29 | T |
| CDC28→CDC6 | 4693[P]3689 | 9.97 | 0.18 | C | CLB6→CLN2 | 19887[I]16538 | 7.97 | 1.29 | I |
| CLN1→SIC1 | 16538[P]19210 | 9.04 | 0.47 | C | CLN1→CLN3 | 19887[I]16538 | 7.97 | 1.29 | I |
| SIC1→CLN2 | 19210[I]16538 | 9.04 | 0.47 | M | CLB6→CLN1 | 19887[I]16538 | 7.97 | 1.29 | I |
| CLN2→FAR1 | 16538[P]19210 | 9.04 | 0.47 | M | CLN2→SWI4 | 16538[P]3700 | 7.24 | 2.36 | M |
| CLB5→CDC6 | 16538[P]3689 | 8.97 | 0.52 | C | FUS3→SWI4 | 4674[P]3700 | 6.57 | 3.59 | I |
| CLN3→CDC6 | 16538[P]3689 | 8.97 | 0.52 | I | CDC28→FUS3 | 4674[E]4674 | 6.51 | 3.68 | I |
| FAR1→FAR1 | 4861[I]19887 | 8.59 | 0.59 | I | CDC20→FAR1 | 30234[P]19210 | 6.38 | 4.03 | I |
| FAR1→SIC1 | 19887[P]19210 | 8.53 | 0.63 | T | CDC20→CLN2 | 30234[I]16538 | 6.23 | 4.60 | I |
| FUS3→CLN3 | 4707[P]19887 | 8.53 | 0.63 | I | CDC6→CLB6 | 3677[E]16538 | 5.82 | 7.07 | M |
| CLB5→CDC28 | 19887[I]4693 | 8.33 | 0.96 | C | CDC6→CLB5 | 3677[E]16538 | 5.82 | 7.07 | M |
| CDC28→CLB5 | 4693[P]19207 | 8.24 | 0.99 | C | SWI6→SWI6 | 30528[E]5515 | 4.26 | 12.9 | I |

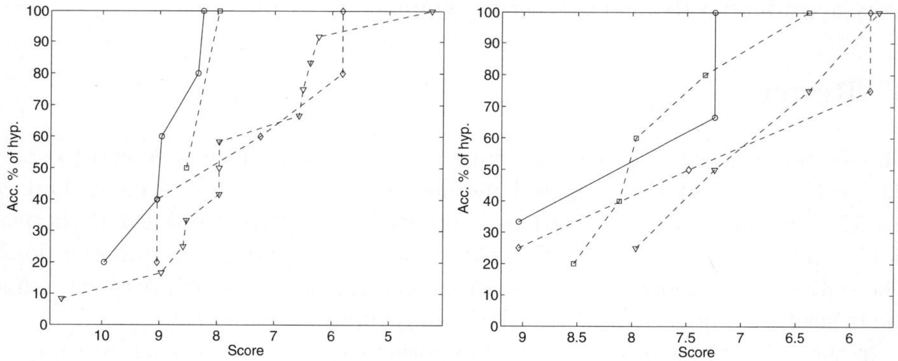


**Fig. 3.** Percentage of accumulated hypotheses as a function of GO-score for different hypothesis classes for the relations derived in ref. [15] (left) and ref.[7] (right). Circles = correct, squares = transitive, diamonds = misdirected, triangles = incorrect.

As can be seen in table 1, all correct hypotheses have $E < 1$, and the average $E$ of the class of correct hypotheses is higher than for any of the other classes. This suggests that the approach is useful for reducing the number of false positives from a set of derived regulatory hypotheses by setting an $E$-value threshold for their plausibility. The results are visualised in figure 3, showing the accumulation of hypotheses from different classes as a function of GO-score. It can be noted that the curve representing the class of "correct" hypotheses reaches 100% before any other class, while the class "incorrect" is the last to reach this level. This suggests that templates and their GO-scores are useful in the evaluation of hypothesis sets derived from data mining. For another hypothesis set, derived in [7], the diagram at the right in figure 3 shows that the results are similar.

## 4 Discussion

We proposed a method based on the generalisation of knowledge of regulatory pathways for assessing the biological plausibility of hypotheses derived during regulatory network reconstruction. Our results demonstrate that the method is able to filter out a large proportion of implausible hypotheses, thus improving the specificity of the regulatory network reconstruction process.

Gene products in some hypotheses are identical or very similar to gene products in relations used for template derivation, with respect to GO functional annotation. This means that the hypotheses are highly feasible, according to our current general knowledge, while at the same time current specific knowledge would suggest that they are incorrect. It is possible that a future more fine-grained GO annotation will allow us to better discriminate between such hypotheses. It is also possible that future experiments show that some of these hypothetical relations actually exist. A way to get higher accuracy for our method could be to incorporate other sources of biological knowledge, such as transcription factor binding site information or protein interactions.

Our method decomposes relations between gene product complexes into atomic binary relations between individual gene products. This leads to certain problems, as can be observed e.g. for the CLN3-CDC28 complex in figure 1. In order to form this complex, CLN3 will bind to the CDC28 kinase and the complex as a unit will phosphorylate SWI6. Hence, it is not entirely true that CLN3 and CDC28 individually phosphorylate SWI6 as our method suggests. We plan to extend our method so that complex relations can be handled without being decomposed into binary relations.

Our method automatically identifies the biologically most plausible hypotheses by a measure based on GO term specificity, but it would also be beneficial to have the top scoring hypotheses assessed by a domain expert in order to reduce the number of false positives.

## References

1. D'haeseleer, P., Liang, S., Somogyi, R.: Genetic network inference: from co-expression clustering to reverse engineering. Bioinformatics **16** (2000) 707–726
2. Liang, S., Fuhrman, S., Somogyi, R.: REVEAL - a general reverse engineering algorithm for inference of genetic network architectures. Pac Symp Biocomput **3** (1998) 18–29

3. Akutsu, T., Miyano, S., Kuhara, S.: Identification of genetic networks from a small number of gene expression patterns under the Boolean network model. Pac Symp Biocomput **4** (1999) 17–28
4. D'haeseleer, P., Wen, X., Fuhrman, S., Somogyi, R.: Linear modeling of mRNA expression levels during CNS development and injury. Pac Symp Biocomput **4** (1999) 41–52
5. Weaver, D. C., Workman, C. T., Stormo, G. D.: Modeling regulatory networks with weight matrices. Pac Symp Biocomput **4** (1999) 112–123
6. Friedman, N., Linial, M., Nachman, I., Pe'er, D.: Using Bayesian networks to analyze expression data. RECOMB '00: Proceedings of the 4th Annual International Conference on Computational Molecular Biology (2000) 127–135
7. Kim, S. Y., Imoto, S., Miyano, S.: Inferring gene networks from time series microarray data using dynamic Bayesian networks. Briefings in Bioinformatics **4** (2003) 228–235
8. Husmeier, D.: Sensitivity and specificity of inferring genetic regulatory interactions from microarray experiments with dynamic Bayesian networks. Bioinformatics **19** (2003) 2271–2282
9. Ashburner, M., Ball, C. A., Blake, J. A., Botstein, D., Butler, H., Davis, A. P., Dolinski, K., Dwight, S. S., Eppig, J. T., Harris, M. A., Hill, D. P., Issel-Tarver, L., Kasarskis, A., Lewis, S., Matese, J. C., Richardson, J. E., Ringwald, M., Rubin, G. M. and Sherlock, G.: Gene Ontology: tool for the unification of biology. Nature Genetics **25** (2000) 25–29
10. Lin, D.: An information-theoretic definition of similarity. Proceedings of the 15th International Conference on Machine Learning (1998) 296–304
11. Jiang, J. J., Conrath, D. W.: Semantic similarity based on corpus statistics and lexical taxonomy. Proceedings of the 10th International Conference on Research on Computational Linguistics, ROCLING X (1997) 19–33
12. Resnik, P.: Semantic similarity in a taxonomy: An information-based measure and its application to problems of ambiguity in natural language. Journal of Artificial Intelligence Research **11** (1999) 95–130
13. Tuzhilin, A., Adomavicius, G.: Handling very large numbers of association rules in the analysis of microarray data. Proceedings of the 8th ACM SIGKDD International Conference on Data Mining and Knowledge Discovery (2002) 396–404
14. Lord, P. W., Stevens, R. D., Brass, A. and Goble, C. A.: Investigating semantic similarity measures across the Gene Ontology: the relationship between sequence and annotation. Bioinformatics **19** (2003) 1275–1283
15. Sigursteinsdottir, G.: Learning gene interactions from gene expression data using dynamic Bayesian networks. Master's thesis, HS-IKI-MD-04-202, Univ. of Skövde (2004)

# Delays in Biological Regulatory Networks (BRN)

Jamil Ahmad[1], Adrien Richard[2], Gilles Bernot[2],
Jean-Paul Comet[2], and Olivier Roux[1]

[1] IRCCyN UMR CNRS 6597
BP 92101, 1 rue de la Noë, 44321 Nantes Cedex 3, France
[2] Programme épigénomique and IBISC, Université d'Evry
Tour Evry 2, Boulevard François Mitterrand, 91025 Evry cedex, France

**Abstract.** In this article, we propose a refinement of the modeling of genetic regulatory networks based on the approach of René Thomas. The notion of delays of activation/inhibition are added in order to specify which variable is faster affected by a change of its regulators. The formalism of linear hybrid automata is well suited to allow such refinement. We then use HyTech for two purposes: (1) to find automatically all paths from a specified initial state to another one and (2) to synthesize constraints on the delay parameters in order to follow any specific path.

## 1 Introduction to Biological Regulatory Networks

Biologists often represent their knowledge on a biological system in terms of graphs[1]. Biological regulatory networks (BRN) represent interactions between biological entities which can be genes or their products, proteins. For example, genetic regulatory networks are graphs where vertices represent genes or regulatory products (e.g. RNA, proteins) and edges represent interactions between them. These interactions are further directed (regulators are distinct from targets) and signed (+ for activation, − for inhibition).

It is now clear for researchers that the semantics of a biological regulatory system and more generally an interaction system, is encoded in the dynamics of the system and not only in the entities of this system. Biologists often use the previously described regulatory graphs as a basis for generating dynamical models using either continuous representation or discrete ones.

- In differential models the activity of each gene is represented by a concentration of the associated RNA or proteins $x_i$, and the evolutions of all concentrations $x = (x_i)_{i \in [1,n]}$ obey a differential equation system $dx/dt = f(x)$. Observation leads biologists to consider only highly non-linear models with some strong threshold effects. The derivation of the dynamics from the interaction graph is not trivial even if the type of each interaction is known, because a lot of parameters have to be inferred, and a tiny modification of a parameter can lead to a strong change in the dynamics.
- In discrete models, the threshold effects are highlighted and allow modellers to discretize the concentrations. The first approach has been based on drastic discretization since all genes can be either *on* (present) or *off* (absent) [2].

V.N. Alexandrov et al. (Eds.): ICCS 2006, Part II, LNCS 3992, pp. 887–894, 2006.

This boolean model has been generalized into a multi-valued model [3, 4], in which logical identification of all steady states [5, 6] becomes possible. The dynamics of these networks are based on abstraction of continuous-time switching networks which are a special type of hybrid systems as studied in, for example, control theory. Such continuous-time switching networks have been used to model dynamics in, for example, the sporulation network of Bacillus subtilis [7]. The derivation of the dynamics from the interaction graph remains difficult even if the number of possible models is now finite. Since the formalism consists essentially in the discretization of the continuous differential equation system, the state space is divided into set of domains representing the symbolic qualitative states of the network. The transitions between the different states depend on logical parameters that play the role of limits of the solutions of the differential equation system of each domain in the continuous space. These limits are sometimes called attractors or targets [8].

The modeling activities then focus on the determination of parameters of the model which lead to a dynamic coherent with the specification (formal translation of experimental facts). Formal verification is not possible in the general framework of differential equation systems. In [9] authors focus on a particular discrete model and use model checking in order to verify if the temporal properties are satisfied. Bernot et al. [8] proposed to consider all possible parameterizations, to generate all possible dynamics, to call for each of them a model checker for verification and to select only models which lead to a dynamic coherent with the specification. The enormous number of models limits this brute force approach.

One can also notice that the transition systems obtained in the formalism of R. Thomas [2] or of H. de Jong [9] are not deterministic: they abstract all possible continuous trajectories but they introduce some traces which do not correspond to continuous ones. This is due to a complete and total abstraction of time. To overcome this point Adélaïde and Sutre [10] showed that under some conditions of equality of degradation constants, this abstraction can lead to a dynamic which does not present the same drawback.

In this article we propose to take into account activation and inhibition delays in the formalism of R. Thomas following [11] where delays have been introduced to study traces closer to the experiment facts. After having briefly presented R. Thomas modelling in section 2, we introduce in section 3 the refinement based on delays. The example presented in section 4 allows us to present an algorithm for searching paths between two specified states (section 5). Finally we show how this algorithm can be helpful for parameter synthesis (section 6). The section 7 is devoted to conclusion.

## 2 Modeling of R. Thomas

In a directed graph $G = (V, A)$, we note $G^-(v)$ and $G^+(v)$ the set of predecessors and successors of a node $v \in V$ respectively.

**Definition 1.** *A biological regulatory network, or BRN for short, is a tuple* $\mathcal{G} = (V, A, l, s, t, K)$ *where*

- $(V, A)$ *is a directed graph denoted by* $G$,
- $l$ *is a function from* $V$ *to* $\mathbb{N}$ *with* $l(v) > 0$ *if* $G^+(v) \neq \{\}$,
- $s$ *is a function from* $A$ *to* $\{+, -\}$,
- $t$ *is a function from* $A$ *to* $\mathbb{N}$ *such that* $\{t(u, v) | v \in G^+(u)\} = \{1, ..., l(u)\}$,
- $K = \{K_v | v \in V\}$ *is a set of maps: for each* $v \in V$, $K_v$ *is a function from* $2^{G^-(v)}$ *to* $\{0, ..., l(v)\}$ *such that* $K_v(\omega) \leq K_v(\omega')$ *for all* $\omega \subseteq \omega' \subseteq G^-(v)$.

The map $l$ describes the domain of each variable $v$: if $l(v) = k$, the abstract concentration on $v$ holds its value in $[0, 1...k]$. Similarly, the map $s$ represents the sign of the regulation (+ for an activation , - for an inhibition).

$t(u, v)$ is the threshold of the regulation: the regulation takes place iff the abstract concentration of $u$ is above $t(u, v)$, in such a case the regulation is said active. The condition on these thresholds states that each abstract level of $u$ plays a role in the set of regulations of $v$. For all $x \in [0...l(u) - 1[$, the set of active regulations of $u$, when the expression level of $u$ is $x$, differs from the set when the expression level is $x + 1$.

Finally, the map $K_v$ allows us to define what is the effect of a set of regulators on the specific target $v$. If this set is $\omega \subseteq G^-(v)$, then, the target $v$ is subject to a set of regulations which makes it to evolve towards a particular level $K_v(\omega)$.

**Definition 2 (States).** *A state* $\mu$ *of a BRN* $\mathcal{G} = (V, A, l, s, t, K)$ *is a function from* $V$ *to* $\mathbb{N}$ *such that* $\mu(v) \in \{0, ..., l(v)\}$ *for all variable* $v \in V$. *We denote* $E^{\mathcal{G}}$ *the set of states of* $\mathcal{G}$.

When $\mu(u) > t(u, v)$ and $s(u, v) = +$, we say that $u$ is a resource of $v$ since the activation takes place. Similarly when $\mu(u) < t(u, v)$ and $s(u, v) = -$, $u$ is also a resource of $v$ since the inhibition does not take place (the absence of the inhibition is treated as an activation).

**Definition 3 (Resource function).** *Let* $\mathcal{G} = (V, A, l, s, t, K)$ *be a BRN. For each* $v \in V$ *we define the resource function* $\omega_v : E^{\mathcal{G}} \to 2^{G^-(v)}$ *by:*

$$\omega_v(\mu) = \{u \in G^-(v) \mid (\mu(u) \geq t(u, v) \text{ and } s(u, v) = +) \text{ or } (\mu(u) < t(u, v) \text{ and } s(u, v) = -)\}.$$

As said before, at the state $\mu$, $K_v(\omega_v(\mu))$ gives the level towards which the variable $v$ tends to evolve. We consider three cases, (1) if $\mu(v) < K_v(\omega_v(\mu))$ then $v$ can increase by one unit, (2) if $\mu(v) > K_v(\omega_v(\mu))$ then $v$ can decrease by one unit and (3) if $\mu(v) = K_v(\omega_v(\mu))$ then $v$ can not evolve. The state graph of BRN represents the set of the states that a BRN can adopt with transitions between them deduced from these rules.

**Definition 4 (State graph).** *Let* $\mathcal{G} = (V, A, b, s, t, K)$ *be a BRN. The state graph of* $\mathcal{G}$ *is a directed graph* $\mathbf{G} = (E^{\mathcal{G}}, T)$ *with* $(\mu, \mu') \in T$ *if there exists* $v \in V$ *such that :*
$K_v(\omega_v(\mu)) \neq \mu(v)$ *and* $\mu'(v) = \mu(v) + \alpha_v(\mu)$ *and* $\mu(u) = \mu'(u), \forall u \in V \backslash \{v\}$ *where* $\alpha_v(\mu) = +1$ *if* $K_v(\omega_v(\mu)) > \mu(v)$ *and* $-1$ *otherwise.*

## 3 Refinement Based on Delays

In the semantics that we use, a state can have several successors, each of them corresponding to the evolution of the expression level of a unique gene (the dynamic is asynchronous). One way to overcome this indeterminism is to use timing constraints in terms of parameters with a path algorithm (section 5). These parameters represent time delays for the discrete changes in expression levels of a gene. To be more precise, when an order of activation/inhibition arrives, the biological machinery starts to increase/decrease the corresponding protein concentration, but this action takes time. These times are abstracted by these time delays. We use two types of time delays $d_v^+(x)$ and $d_v^-(x)$ to represent the time required to change the expression level of a gene $v$ from an abstract level $x$ to $x+1$ and from the level $x$ to $x-1$ respectively as shown in figure 1.

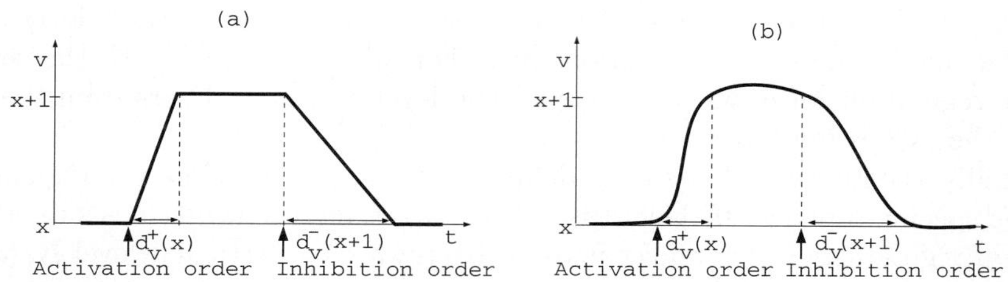


**Fig. 1.** Approximate evolution (a) of the actual evolution (b) of a gene's expression

## 4 Example of a BRN

Consider three genes $a, b$ and $c$ which interact according to the graph of figure 2-(a). From this BRN, we obtain the table of resources (figure 2-(c)) and state graph as shown in figure 2-(b).

In the table of figure 2, $\omega_v$ is the set of resources of a gene $v \in \{a, b, c\}$, that is the set of biological entities which helps variable $v$ to increase. The evolution of the expression level of a gene $v$ depends on its resources $\omega_v$. Table of figure 2-(c)

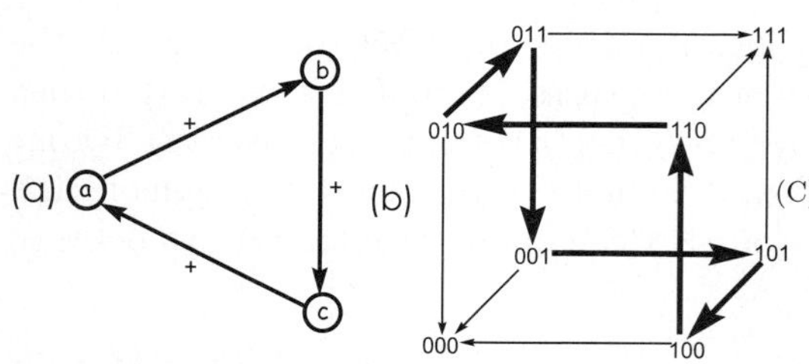


(C)

| | $a$ | $b$ | $c$ | $\omega_a$ | $\omega_b$ | $\omega_c$ | $K_a(\omega_a)$ | $K_b(\omega_b)$ | $K_c(\omega_c)$ |
|---|---|---|---|---|---|---|---|---|---|
| $\mu_0$ | 0 | 0 | 0 | $\{\}$ | $\{\}$ | $\{\}$ | 0 | 0 | 0 |
| $\mu_1$ | 0 | 0 | 1 | $\{c\}$ | $\{\}$ | $\{\}$ | 1 | 0 | 0 |
| $\mu_2$ | 0 | 1 | 0 | $\{\}$ | $\{\}$ | $\{b\}$ | 0 | 0 | 1 |
| $\mu_3$ | 0 | 1 | 1 | $\{c\}$ | $\{\}$ | $\{b\}$ | 1 | 0 | 1 |
| $\mu_4$ | 1 | 0 | 0 | $\{\}$ | $\{a\}$ | $\{\}$ | 0 | 1 | 0 |
| $\mu_5$ | 1 | 0 | 1 | $\{c\}$ | $\{a\}$ | $\{\}$ | 1 | 1 | 0 |
| $\mu_6$ | 1 | 1 | 0 | $\{\}$ | $\{a\}$ | $\{b\}$ | 0 | 1 | 1 |
| $\mu_7$ | 1 | 1 | 1 | $\{c\}$ | $\{a\}$ | $\{b\}$ | 1 | 1 | 1 |

**Fig. 2.** BRN (a) and its state graph (b) according to the attractor parameters: $K_a(\{\}) = K_b(\{\}) = K_c(\{\}) = 0$ *and* $K_a(\{c\}) = K_b(\{a\}) = K_c(\{b\}) = 1$.(c) Associated table of resources

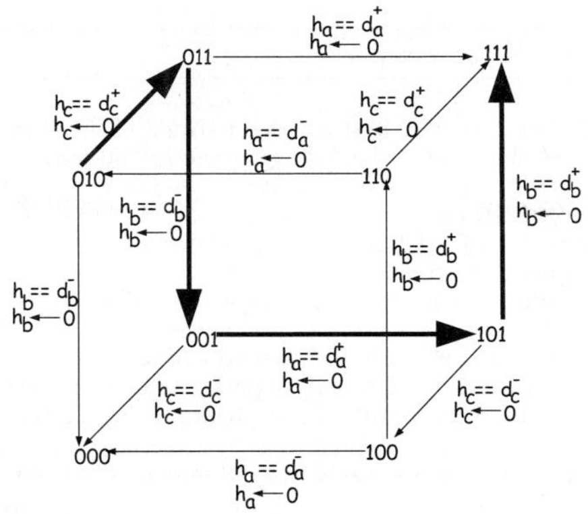


**Fig. 3.** Transition system for the BRN of figure 2 along with guard conditions

helps the reader to reconstruct the state graph representing the dynamics of the BRN. The edges in the graph give the possible transitions between states which can occur with time. The circuit (010, 011, 001, 101, 100, 110, 010) represents an unstable circuit in the dynamics and the two states not involved in the circuit are the only two stable states.

We associate a clock $h_v$, to each variable $v \in \{a, b, c\}$. All the clocks of the system increase continuously and simultaneously[1]. The guard $h_v == d_v^\alpha$, where $\alpha \in \{+, -\}$ gives the condition for the transition. The clock of a variable is set to zero when this variable changes its abstract level. In figure 3 all the transitions are labeled with guards and clock initializations.

## 5 Searching Paths Between Two States

The analysis of the hybrid refinement of the BRN is performed by using a linear hybrid model checker HyTech [12]. The delays are defined as parameters whose values are unknown. We associate timing constraints with qualitative states of a BRN in terms of time delay parameters. Our path algorithm finds all the possible paths between two qualitative states of a regulatory network. We have implemented this algorithm in HyTech and obtained the exact path between any two qualitative states of a BRN by applying timing constraints in terms of delay parameters.

Algorithm 1 is the pseudocode of the HyTech implementation. `pre` and `post` operators returns respectively the predecessors and successors of a state including the state itself. The difficulty of the algorithm lies in the fact that it converts the breadth-first search (induced by the `post` operator) into the depth-first search of a path. The algorithm consists of two main loops. In the outer loop the algorithm exhaustively searches the `final_state` from the `initial_state` and accumulates the accessed states in a set named `states_accumulated`. When

[1] It is possible to go from $x$ to $x'$ via the increasing (resp. decreading) of variable $v$ if the time delay $d_v^+$ (resp. $d_v^-$) is achieved by the clock associated to $v$.

**Algorithm 1.** Finds paths between two states

```
Path(initial_state, final_state)
states_accessed :=initial_state; // The first accessed states is the initial state
reached := initial_state; // The first visited state is the initial state
path := initial_state; // Path is set to initial state
states_accumulated := initial_state;
// while exist accessible states of a BRN from initial state
while not_empty(states_accessed) do
   // The set of accessed states is now the successors of the previously accessed states
   states_accessed := post(states_accessed) - states_accessed;
   path := states_accessed - path; // Find new states of a path
   states_accumulated := states_accumulated ⋃ path; // Set that accumulates the states
   states_accumulated := states_accumulated - initial_state; // Remove the initial state from
   the set
   states_accessed := states_accessed - reached; // Remove all previously visited states from set
   of accessed states
   reached :=reached ⋃ states_accessed; // The previously visited states will now be the accessed
   states
   // Check if final state is accessed
   if not_empty(path ⋂ final_state) then
      states_accessed1:= final_state;
      reached1 := final_state;
      path1:=final_state;
      //The nested loop starts here
      while not_empty(states_accessed1) do
         // The set of accessed states is now the predecessors of the previously
         // accessed states in a set of accumulated states
         states_accessed1 := (pre(states_accessed1)-states_accessed1) ⋂ states_accumulated;
         path1:=states_accessed1 - path1;
         path_states := path_states ⋃ path1; // Set that accumulates the states of one discovered
         path
         states_accessed1 := states_accessed1 - reached1;
         reached1:= reached1 ⋃ states_accessed1;
      end while
      print_path(path_states); // To print the states of a path
   end if
end while
```

the algorithm finds the `final_state` then it starts the nested loop and begins backward search from `final_state` and takes the intersection of each accessed states with the set `states_accumulated`. If the intersection is not empty then the algorithm gives the intersection as a state of the path which is accumulated in a set `path_states`. Finally the algorithm invokes the procedure `print_path(path_states)` to print the states of a path in proper order.

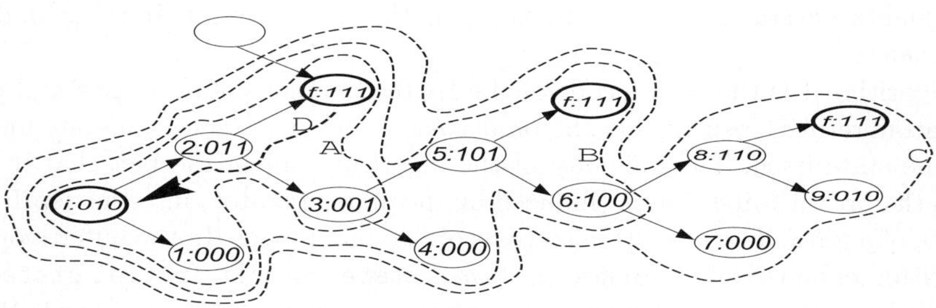


**Fig. 4.** How the algorithm finds the final states from initial state and vice versa. The empty state shows the accessibility of final state through other path.

The dashed lines `(A)`, `(B)` and `(C)` of figure 4 represent the successive sets of accumulated states when the algorithm finds the final state `f` during the outer loop. The inner loop is used for backward search and the dashed arrow `(D)` shows this search for the first path. Algorithm 1 finds the three paths between states (0,1,0) and (1,1,1) in the example of figure 3.

## 6 Parameters Synthesis

The delay parameters used for the increasing and decreasing of the expression level of a gene $v$ can be synthesized in HyTech to form timing constraints for each transition that takes place in paths between two states. The conjunction of constraints along any sequence of transitions gives the synthesized parameters constraint for the given path.

- For the transition (0,1,0) → (0,1,1): $d_c^+ \leq d_b^-$
- For the transition (0,1,1) → (0,0,1): $d_b^- \leq d_a^+$
- For the transition (0,0,1) → (1,0,1): $d_c^+ + d_a^+ \leq d_a^- + d_b^- + d_c^-$
- For the transition (1,0,1) → (1,1,1): $d_a^+ + d_b^+ \leq d_c^- \wedge d_a^+ + d_c^+ \leq d_a^- + d_b^-$

Now if we desire to find only the path shown by the bold line in figure 3, then we use the timing constraints which are synthesized[2] by HyTech for the transitions in one path starting from a state where $h_a = h_b = h_c = 0$ (see below). Thus, we draw an equivalence between the path $(0,1,0) \rightarrow (0,1,1) \rightarrow (0,0,1) \rightarrow (1,0,1) \rightarrow (1,1,1)$ and the region described by the conjunction of constraints: $(d_c^+ \leq d_b^-) \wedge (d_b^- \leq d_a^+) \wedge (d_c^+ + d_a^+ \leq d_a^- + d_b^- + d_c^-) \wedge (d_a^+ + d_b^+ \leq d_c^-) \wedge (d_a^+ + d_c^+ \leq d_a^- + d_b^-)$.

## 7 Conclusion

We propose in this paper a refinement based on delays for BRN modelling. The introduction of delays allows one to distinguish paths from one state to another one. This refinement reintroduces time in the abstraction of R. Thomas, and this way is different from the refinement of [13] and [10] which split the state space by partitioning the domains of the state space. The present work describes how the introduction of time can be helpful for modeling such networks, allowing the modeller to verify temporal properties. It is now crucial to confront this modelling with real experiments. Our experience in modeling in a multidisciplinary context will help to initiate biological modeling with delays.

Our future work is threefold: (1) formal representation of parameters constraints, (2) parameter synthesis: we have to check when a path is equivalent to an empty region and what this really means, and (3) cycles: we have to check if there are cycles in the state graph of a BRN.

[2] The interested readers can contact the authors for the HyTech codes and results.

## References


1. de Jong, H.: Modeling and simulation of genetic regulatory systems: a literature review. J. Comput. Biol. **9**(1) (2002) 67–103.
2. Thomas, R.: Logical analysis of systems comprising feedback loops. J. Theor. Biol. **73**(4) (1978) 631–56
3. Snoussi, E.: Qualitative dynamics of a piecewise-linear differential equations : a discrete mapping approach. DSS **4** (1989) 189–207
4. Thomas, R.: Regulatory networks seen as asynchronous automata : A logical description. J. theor. Biol. **153** (1991) 1–23
5. Snoussi, E., Thomas, R.: Logical identification of all steady states : the concept of feedback loop caracteristic states. Bull. Math. Biol. **55**(5) (1993) 973–991
6. Devloo, V., Hansen, P., Labbé, M.: Identification of all steady states in large networks by logical analysis. Bull. Math. Biol. **65**(6) (2003) 1025–51
7. de Jong, H., Geiselmann, J., Batt, G., Hernandez, C., Page, M.: Qualitative simulation of the initiation of sporulation in *Bacillus subtilis*. Bulletin of Mathematical Biology **66**(2) (2004) 261–299
8. Bernot, G., Comet, J.P., Richard, A., Guespin, J.: Application of formal methods to biological regulatory networks: Extending Thomas' asynchronous logical approach with temporal logic. Journal of Theoretical Biology **229**(3) (2004) 339–347
9. de Jong, H., Geiselmann, J., Hernandez, C., Page, M.: Genetic network analyzer: qualitative simulation of genetic regulatory networks. Bioinformatics **19**(3) (2003) 336–44.
10. Adélaïde, M., Sutre, G.: Parametric analysis and abstraction of genetic regulatory networks. In: Proc. 2nd Workshop on Concurrent Models in Molecular Biology (BioCONCUR'04), London, UK, Aug. 2004. Electronic Notes in Theor. Comp. Sci., Elsevier (2004)
11. Thomas, R., Kaufman, M.: Multistationarity, the basis of cell differentiation and memory. Chaos **11** (2001) 180–195
12. Henzinger, T.A., Ho, P.H., Wong-Toi, H.: HYTECH: A model checker for hybrid systems. International Journal on Software Tools for Technology Transfer **1**(1–2) (1997) 110–122
13. Batt, G., Ropers, D., de Jong, H., Geiselmann, J., Page, M., Schneider, D.: Qualitative analysis and verification of hybrid models of genetic regulatory networks: Nutritional stress response in *Escherichia coli*. In Morari, M., Thiele, L., eds.: Eighth International Workshop on Hybrid Systems: Computation and Control, HSCC 2005. Volume 3414 of Lecture Notes in Computer Science., Springer (2005) 134–150

# Phase Transitions in Gene Knockdown Networks of Transitive RNAi

Shibin Qiu and Terran Lane

Dept. of Computer Science, University of New Mexico, Albuquerque, NM, 87131
{sqiu, terran}@cs.unm.edu

**Abstract.** Since gene silencing by RNA interference (RNAi) has been observed when inexact matches exist in siRNA-mRNA binding, the number of mismatched nucleotides allowed by nature becomes an important quantity in characterizing RNAi specificity. We use scale-free graphs to model the knockdown interactions among different genes and estimate the allowable flexibility by examining transitive RNAi, which amplifies siRNA and cyclically silences targets. Simulation results in *S. pombe* indicate that continually increasing the number of mismatches risks transcriptome-wide knockdown and eventually turns RNAi from defensive to self-destructive. At the phase transition, the number of mismatches indicates a critical value beyond which tRNAi would cause an organism instable. This critical value suggests an upper limit of no more than 6 nt mismatches in the binding.

## 1 Introduction

RNAi becomes unspecific when close or exact identity between siRNA and non-target mRNA produces "off-target" degradation. Empirical studies have demonstrated that such off-target effects occur in the presence of single and multiple nucleotide mismatches between siRNA and off-target mRNA [1, 2, 3, 4]. It is conceivable that an siRNA-mRNA binding that is too permissive might knock down a substantial portion of genes in an organism and cause loss of function of too many genes, which would in turn make an organism unstable. Since such instability is not observed, there must be an upper limit on the flexibility allowed by nature. To date, it is unclear how much flexibility is actually allowed in this binding function. Here we estimate the allowable mismatches by examining the knockdown interactions of transitive RNAi using a network model.

In transitive RNAi (tRNAi), gene knockdown is mediated by ~21 nucleotide (nt) siRNAs produced by enzyme Dicer from an initiating dsRNA. These siRNAs also serve as primers for RNA-directed RNA polymerase (RdRP) to generate elongated dsRNA from the target mRNA, which are cleaved again by Dicer to produce secondary siRNAs, yielding degradation of secondary targets [5]. tRNAi (or post-transcriptional gene silencing, PTGS) has been observed in plants, fungi, flies, and worms [5, 6, 7]. tRNAi is highly efficient for gene silencing and has been termed as "nature's own PCR machine" due to its siRNA amplification [6]. But

V.N. Alexandrov et al. (Eds.): ICCS 2006, Part II, LNCS 3992, pp. 895–903, 2006.

siRNA proliferation in tRNAi also increases the chances of undesirable knockdowns, especially when the siRNA-mRNA binding is not stringent.

In *D. melanogaster*, where tRNAi has been observed [6], mismatches of 3 or 4 nt between siRNA and mRNA were tolerated [1]. Although tRNAi has not been confirmed in mammals, mutations in siRNA of 3 to 4 nt were allowed [2]. Moreover, in some cases siRNAs with 9 nt mismatches have been shown to directly silence non-target genes in cultured human cells [3]. In RNAi experiments, mismatches can also be intentionally introduced during siRNA synthesis to improve silencing efficacy, as RISC prefers lower binding energy at one of the termini of siRNA [8]. However, excessive flexibility is dangerous since it may silence too many unintended targets and, more seriously, may cause instability.

Since tRNAi may cause chain reactions, and its effects are difficult to fully observe [5, 9], it is expensive, if feasible, to experimentally investigate tRNAi at a genomic scale by silencing each gene as primary target. However, a computational simulation is able to model the knockdown interactions at a transcriptomic scale. In this work, we construct *gene knockdown networks* based on principles and parameters observed in biological experiments to investigate the impact of mismatches on genome stability.

We find that the knockdown network became dense when less stringent RNAi bindings were employed in our simulation and a giant component (a large group of connected genes) emerged. In a knockdown network, genes connected to a common node can be silenced by a single primary siRNA and the giant component can be silenced with only a few primary siRNAs. This phase transition with respect to binding permissiveness will turn RNAi from defensive into self-destructive and imply a loss of genomic stability. Such instability is not observed in nature, implying that the biological siRNA-mRNA binding function does not support a large number of mismatches in general. This phase transition suggests a critical value for the permitted flexibility of RNAi binding. Due to evolutionary conservations, the results can be used as a references for allowable mismatches in organisms not having tRNAi.

## 2 Network Model of Transitive RNAi

In this section, we describe the biological mechanism of tRNAi and the construction of our network model.

### 2.1 Transitive RNAi

We can outline dsRNA-mediated tRNAi in the following steps [5].

1. A dsRNA is introduced into the cell to initiate gene silencing;
2. Dicer cleaves the dsRNA into siRNAs of length 21-25 nt;
3. RISC unwinds the siRNA to help target the appropriate mRNA;
4. RISC-siRNA complex recognizes the primary target mRNA;
5. RISC cleaves the siRNA/mRNA hybrid and the primary target is degraded;
6. RdRP produces elongated dsRNA from the target site on the primary target;

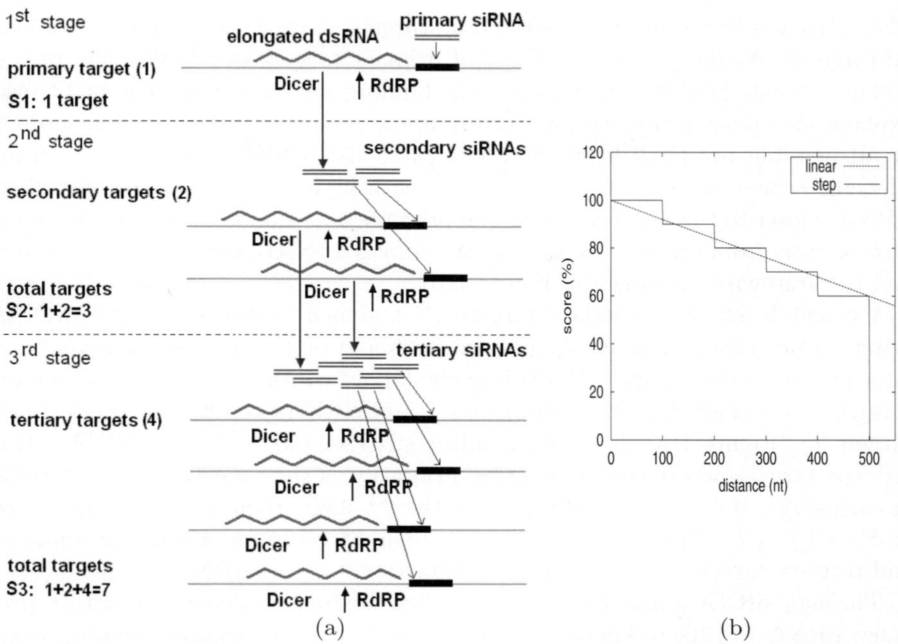


**Fig. 1.** Transitive RNAi. (a) First three stages of tRNAi, silencing 7 targets with a single primary siRNA. (b) Scoring schemes for secondary siRNAs.

7. Dicer cleaves the elongated dsRNA into secondary siRNAs;
8. RISC destroys the secondary targets using the secondary siRNAs. Now the secondary targets become primary targets, and tRNAi cycles to step 3.

In addition to degrading the primary target mRNA as in ordinary RNAi, the siRNA hybridized with its complimentary target can be used by RdRP to produce an elongated dsRNA upstream of the target site(step 6). Dicer then generates more siRNAs, the secondary siRNAs, using the elongated dsRNA (step 7). These secondary siRNAs associate with RISC which knocks down secondary targets (step 8). The secondary siRNAs are in turn elongated to generate tertiary siRNAs, which silence tertiary targets. Since tRNAi is cyclic [5], it may cause a chain reaction of gene knockdown [5]. Figure 1 (a) shows tRNAi of the first three stages, where seven targets are silenced with a single primary siRNA.

Even though G-U wobbles and bulges may also ber permitted in siRNA-target binding [2], mismatches contribute the most to off-targeting effects [10]. We used contiguous mismatches, because they are frequently employed experimentally and are computationally easy to implement [1, 2, 3].

## 2.2 Simulation of Transitive RNAi

We randomly pick an siRNA from a gene $g_i$, $i = 1, 2, ..., N$, where $N$ is the number of genes in the transcriptome of an organism. And find its target genes. We introduce variable mismatches in the siRNA-mRNA binding as a parameter.

This siRNA is the primary siRNA and its targets are primary targets (including off-targets). We denote the collection of primary targets as $T_i^1$. We then pick a 500 nt elongated dsRNA upstream of the binding site from each gene in $T_i^1$ (see explanations later in this section). Overlapping secondary siRNAs are generated at all possible positions from these elongated dsRNAs. We assign differential silencing scores to the secondary siRNAs, with higher scores being assigned to siRNAs closer to the binding site on the primary target and lower scores to those farther away from the binding site [5]. We employ two scoring schemes, i.e. linear and the step-wise, as shown in Figure 1(b).

We search for the secondary targets $T_i^2$ matched by the secondary siRNAs using a filter based on a silencing score threshold of $f_s$=90% (see explanations later in this section). Since tRNAi is cyclic [5], we continue the target search by using the secondary targets to produce tertiary siRNAs and finding the tertiary targets $T_i^3$. Figure 1(a) shows 4 secondary siRNAs and 7 tertiary siRNAs. The induced targets, when gene $g_i$ is aimed primarily, is the collection of the targets at each stage. If we observe tRNAi up to the $t^{th}$ stage, then the total targets are in $S_i^t = \bigcup_{j=1}^{t} T_i^j$. For example, if $t = 3$, then $S_i^3$ contains primary, secondary, and tertiary targets when $g_i$ is the primary target of the siRNA.

Through siRNA amplification and binding flexibility, silencing $g_i$ with a primary siRNA will likely knock down other genes. The knockdown relationships between genes can be represented by a graph, as described below. As every gene in $S_i^t$ will be degraded as a consequence of a primary siRNA when $g_i$ is intended to be silenced, we add an edge from $g_i$ to each gene in $S_i^t$. We conduct a target search for each gene of an organism and construct a directed graph $G_t$ as the knockdown network of $t$-stage tRNAi. The graph $G_t$ is parameterized by $m$, length of mismatch allowed by the siRNA-target binding, and $t$, the number of stages. We then study the statistics of $G_t$ and compute the largest connected component of $G_t$. The phase transition of $G-t$ is used to estimate the allowable mismatches in the binding. We average the statistics over 100 networks, each of which is initiated by a random siRNA primarily aiming each gene.

To simulate the general principles of tRNAi, we ignore some of the subtle variations reported in the literature since they do not yield significant differences. Particularly, we need to point the following specifics in our computation.

1. The length of the elongated region is restricted to 500 nt, which is more consistent with the model proposed by Sijen et al. [5], although Lipardi and colleagues have proposed that the full length of the target mRNA be used [6].

2. We use the region upstream of the binding site to generate elongated dsRNA, although some experiments suggest both upstream and downstream produce secondary siRNAs [11, 12].

3. If an siRNA matches multiple sites on a target, all upstream regions are used for elongated dsRNA for the next stage. This is consistent with the extension (ligation) model of Lipardi and colleagues [6].

4. We do not allow self-cycles: once a gene has been included as a target in $S^{t-1}$ covering stage $t-1$, it will not be included in $S^t$.

5. Though tRNAi may contain infinitely many stages, we only consider up to 9 stages. This consideration is based on the threshold model [5, 13], which suggests that an abundance of target mRNA is required to maintain tRNAi. Since siRNA amplification leads to acceleration in RNA degradation, which reduces the mRNA level dramatically, the possibility of very long chain is unlikely.

6. Although every siRNA in the elongated region of 500 nt is possibly functional, and is scored according to the model of Sijen [5] (Figure 1(b)), we used a cutoff threshold of $f_s$=90%. We make this simplification since we found that lowering $f_s$ to increase the length of effective region did not have a significant effect on the graphs. However, adjusting $f_s$ provides flexibility in our simulation.

### 2.3 Algorithms and Graph Statistics

We use string kernels and binary search trees for efficient searches [4]. For a random graph, the relative largest component size $\rho$ is the number of nodes in the largest connected cluster divided by the total number of nodes in the graph. When $\rho$ is greater than a certain level, the connected genes form a giant component [14]. To ensure the largest component is indeed a giant component, we use $\rho \geq 90\%$ as the criterion. When a graph increases its density to eventually contain a giant component as $m$ gets larger, it is said to go through a phase transition. To quantify the density of a graph, we also measure the average distance between any two reachable vertices in the graph. We use breadth first-search algorithms to calculate the component sizes.

## 3 Results

Since evidence of tRNAi has been observed in *S. pombe* [11], we use its 5401 cDNA sequences (obtained from Sanger Institute `ftp://ftp.sanger.ac.uk`) to test our network model. Due to large amounts of genomic scale searches and graphical computations, data of this small organism is tractable. Results are shown below.

### 3.1 Degree Distributions

We evaluated degree distributions in the gene knockdown networks of *S. pombe* using 21 nt siRNAs, simulating tRNAi for three stages, $t = 3$, and allowing for two mismatch lengths, $m = 4$ and 6. Figure 2 (a) shows the log-log plot of the cumulative frequencies $P_{cum}(k) = \sum_{d \geq k} P(d)$, where $P(d)$ is the frequency of vertices having $d$ edges. As indicated, the degrees roughly follow power distributions. If expressed as $P(k) \sim k^{-\gamma}$, then $\gamma = 3.28$ and 3.59 for the two curves (statistically significant, p-value $< 10^{-3}$) . Since a power law distribution is not modal, the connectivity of a knockdown network is characterized by a continuum hierarchy with most of the genes connecting to a small number of genes, and a small number of genes connecting to a large number of genes. A power law distribution also suggests that the knockdown network is scale free [14].

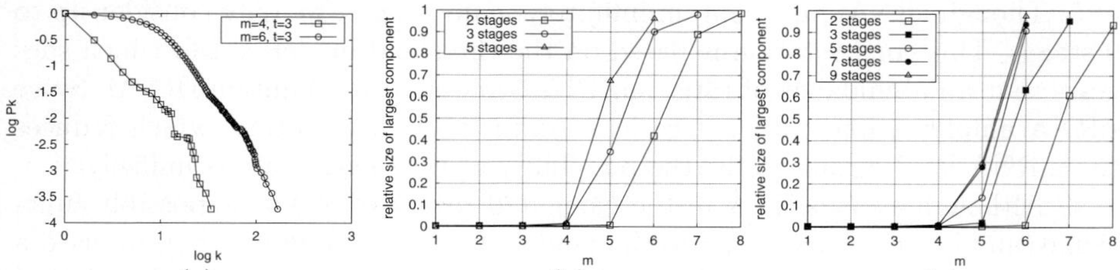


(a) (b) (c)

**Fig. 2.** Knockdown network properties. (a) Log-log plots of degree distributions. X-axis is the degree in logarithmic scale; Y-axis is the cumulative frequency $P_{cum}(k) = \sum_{d \geq k} P(d)$ in logarithmic scale. (b) Phase transitions: $m$, on X-axis, is the number of mismatches in siRNA-mRNA bindings. Y-axis is $\rho$, the largest component size divided by the number of genes. (c) Convergence of phase transitions up to $t = 9$ stages.

### 3.2 Phase Transitions

We computed the relative size $\rho$ of the largest component of the knockdown networks. Figure 2 (b) shows $\rho$ for transitive RNAi covering two, three, and five stages in the chain reaction, when increasing flexibility was allowed by siRNA-target bindings ($m$=1-8) using 21-mers as siRNA. As Figure 2 (b) suggests, when siRNA-target binding was stringent ($m \leq 4$ nt), the graphs were sparse and the largest components were small ($\rho < 1\%$). The average distance between any two vertices in the graph was about 12 hops. But the graphs experienced very sharp phase transitions to have giant components when increasing flexibility was allowed ($m \geq 5$). If we consider 5 transitive stages, a 6 nt mismatch yielded a giant component containing 95% of the nodes in the network. In such a dense network, the average distance was about 3 hops. Figure 2 (b) also indicates that more stages increased the chance of phase transitions when $m$ was small.

In the knockdown network, genes connected to a common gene can be knocked down by a single primary siRNA. And the giant component can be knocked down with only a few primary siRNAs, since the average distance is short. This phase transition will turn RNAi from defensive into self-destructive and suggests a loss of stability in an organism. Therefore, mismatches of 6 nt or longer would cause instability in an organism, and thus are not biologically plausible.

Though the degree distributions, $P(k)$, both followed a power law before and after the phase transition, they had slightly different slopes. $\gamma = 3.28$ for the graph corresponding to $m = 4$ before the phase transition, and $\gamma = 3.59$ for $m = 6$ after the phase transition. The network having a giant component also had a larger maximum degree and average degree.

### 3.3 Convergence of Phase Transitions

Previously, we examined phase transitions of tRNAi up to 5 stages. To further investigate the effect of stages, we simulated 9 stages and display the results in Figure 2 (c). As indicated in the figure, phase transition patterns of 7 and 9 stages did not differ much from that of 5 stages and the patterns of stages 7 and 9 were almost indistinguishable. The curves tend to converge with stages. Even though we did not simulate more stages than 10 due to computational intensity of the simulation, we expect a convergence of phase transition w.r.t stages for even more stages. Our results were solely derived from sequence data in the transcriptome and did not consider mRNA concentration. Nonetheless, they are consistent with the threshold model of tRNAi [5, 13].

# 4 Discussions and Conclusions

We modelled gene knockdown interactions as scale free graphs, which demonstrated that increasing the flexibility of siRNA-target binding increased the density of the graph. Statistics indicated that giant components existed if mismatches of more than 6 nt were allowed. This phase transition would turn RNAi from defensive to self-destructive. Therefore, the critical mismatch allowed in RNAi, especially in tRNAi, is unlikely to be higher than 6 nt.

The power-law-like degree distribution of the knockdown network in *S. pombe* implies that the knockdown network is scale free, suggesting that larger genomes exhibit similar patterns of knockdown interactions. Therefore, our results on *S. pombe* may be generalizable to other organisms. In particular, excessive mismatches in tRNAi will cause instability and there must be an upper limit to the allowable flexibility in all organisms. Since *S. pombe* has a small transcriptome, we are not sure whether the limits in other species are the same, or shorter due to higher chance of knockdown interactions.

Our results about the convergence of phase transition were solely derived from transcript sequence and did not consider mRNA concentration. Nonetheless, they are consistent with the threshold model of tRNAi [5, 13]. The threshold model suggests that an abundance of target mRNA is required to maintain tRNAi. Since siRNA amplification leads to acceleration in RNA degradation, which in turn reduces the mRNA level dramatically, the existence of very long chains are unlikely. Our results implies that sequence usage and concentration are consistent and somewhat related.

When we simulated tRNAi, we made conservative assumptions about the length of elongated dsRNA, only considered upstream of the target site, and limited the number of stages. If the assumptions are violated in nature, the graphs would be denser and the permissive flexibility even more stringent. Therefore our estimation is an optimistic upper bound. The similarity model used to characterize siRNA-target binding only considered contiguous mismatches. If we relax the contiguity restriction, the graph might get denser and the critical mismatch might decrease slightly. If we consider variable silencing potencies of different siRNAs, some siRNAs may be non-functional and the graph would be somewhat sparser and the critical mismatch might be a little longer. But if siRNAs are uniformly picked, then the knockdown networks are still scale-free. If both upstream and downstream sequences were used for elongated dsRNA, then the graph would get denser and the critical mismatch might decrease slightly.

Additionally, dsRNA may induce DNA methylation, which degrades more targets at transcription level and causes smaller critical mismatches [12].

Although we only studied tRNAi, high degree of evolutionary conservations of the RNAi mechanism implies that similar behaviors may exist in organisms not having tRNAi, such as *H. sapiens.* Furthermore, it is possible for larger mammalian genomes to have smaller critical mismatches. Therefore, our estimations are more consistent with the results in [1, 2]. However, mismatches longer than 6 nt, such as 9 nt in [3], are generally infrequent, and would probably cause extensive undesirable knockdowns and even instabilities.

## Acknowledgement

This work is supported by NIH grant P20RR18754 from the Institutional Development Award Program of the National Center for Research Resource. We thank Coenraad M. Adema for helpful suggestions.

## References

1. Elbashir, S.M., Martinez, J., Patkaniowska, A., Lendeckel, W., Tuschl, T.: Functional atonamy of siRNA for mediating efficient RNAi in *Drosophila melanogaster* embryo lysate. The EMBO Journal **20** (2001) 6877–6888
2. Saxena, S., Jonsson, Z.O., Dutta, A.: Small RNAs with imperfect match to endogenous mRNA repress translation. Journal of Biological Chemistry **278** (2003) 44312–44319
3. Jackson, A., Bartz, S., Schelter, J., Kobayashi, S., Burchard, J., Mao, M., Li, B., Cavet, G., Linsley, P.S.: Expression profiling reveals off-target gene regulation by RNAi. Nature Biotechnology **21** (2003) 635–637
4. Qiu, S., Adema, C.M., Lane, T.: A computational study of off-target effects of RNA interference. Nucleic Acids Research **33** (2005) 1834–1847
5. Sijen, T., Fleenor, J., Simmer, F., Thijssen, K., Parrish, S., Timmons, L., Plasterk, R., Fire, A.: On the role of RNA amplification in dsRNA-triggered gene silencing. Cell **107** (2001) 465–476
6. Lipardi, C., Wei, Q., Paterson, B.: RNAi as random degradative PCR: siRNA primers convert mRNA into dsRNA that are degraded to generate new siRNA. Cell **107** (2001) 297–307
7. Dalmay, T., Hamilton, A., Rudd, S., Angell, S., Baulcombe, D.: An RNA-dependent RNA polymerase gene in *Arabidopsis* is required for posttranscriptional gene silencing mediated by a transgene but not by a virus. Cell **101** (2000) 543–553
8. Ui-Tei, K., Naito, Y., Takahashi, F., Haraguchi, T., Ohki-Hamazaki, H., et al.: Guidelines for the selection of highly effective siRNA sequences for mammalian and chick RNA interference. Nucleic Acids Research **32** (2004) 936–948
9. Dillin, A.: The specifics of small interfering RNA specificity. Proc. Natl. Acad. Sci. USA **100** (2003) 6289–6291
10. Qiu, S., Lane, T.: String kernels of imperfect matches for off-target detection in RNA interferance. In Sunderam, V., et al., eds.: Proc. 1st IWBRA and 5th ICCS. LNCS vol. 3515, Springer-Verlag (2005) 894–902
11. Allshire, R.: RNAi and heterochromatina hushed-up affair. Science **297** (2002) 1818–1819

12. Vaistij, F., Jones, L., Baulcombe, D.: Spreading of RNA targeting and DNA methylation in RNA silencing requires transcription of the target gene and a putative RNA-dependent RNA polymerase. The Plant Cell **14** (2002) 857–867
13. Metzlaff, M., O'Dell, M., Cluster, P., Flavell, R.: RNA-mediated RNA degradation and chalcone synthase A silencing in Petunia. Cell **88** (1997) 845–854
14. Barabási, A.L., Albert, R.: Emergence of scaling in random networks. Science **286** (1999) 509–512

# Compile-Time Energy Optimization for Parallel Applications in On-Chip Multiprocessors*

Juan Chen[1], Huizhan Yi[1], Xuejun Yang[1], and Liang Qian[2]

[1] School of Computer, National University of Defense Technology, Changsha 410073, P.R.China
{juanchen, huizhanyi, xjyang}@nudt.edu.cn
[2] Interdisciplinary Research Centre in Materials, University of Birmingham, UK
liangqian99@gmail.com

**Abstract.** Energy consumption is becoming one of the key optimization objects in on-chip multiprocessor. Minimizing the energy consumption without parallel performance loss is concerned. In this paper, we focus on a DVFS-enabled on-chip multiprocessor architecture, which allows dynamically adjusting each processor's voltage/frequency or shut down unused processors so to obtain energy savings. A detailed analytical model is provided and validated by experiments. Experimental results show energy saving can be up to 10.34% without performance loss.

## 1 Introduction

Many embedded systems using rechargeable batteries are strongly constrained by power, such as satellite systems, hand-held devices, and solar powered systems. Low-power technology is critical for these systems. In our article, we focus on parallel compilation technique that optimizes the energy consumption of parallel applications.

We adopt a DVFS-enabled (dynamical voltage/frequency scaling-enabled) on-chip multiprocessor architecture to support this low-power parallel compilation technique. Such a DVFS-enabled on-chip multiprocessor architecture meets current high-performance and low-power demands through two main techniques: *on-chip multi-processing* and *dynamical voltage/frequency scaling (DVFS)*. *On-chip multi-processing* exploits the high-level and coarse-grain parallelism in parallel programs.

*DVFS* is mainly driven by the demand of energy reduction during parallel program execution. Current DVFS-enabled processors include Transmeta's Crusoe [1] and Intel's XScale [2]. They adjust processor voltage/frequency in terms of execution load imbalance. Considering the high-performance and low-power demands, we expect these two techniques (on-chip multi-processing and *DVFS*) to be co-exist in the future architectures.

Considering such an architecture, we explore load imbalance existing in parallel applications to adjust multiprocessors' voltage/frequency. For a given parallel application, there exist serial regions and parallel regions. In the serial execution process, only one of the processors runs and other processors are idle, which causes serious load imbalance.

* This work was supported by the National High Technology Development 863 Program of China under Grant No. 2002AA1Z2101 and No. 2004AA1Z2210.

V.N. Alexandrov et al. (Eds.): ICCS 2006, Part II, LNCS 3992, pp. 904 – 911, 2006.

We can shut down idle processors during the serial executions for energy savings. In parallel execution processes, load imbalance still cannot be fully avoided even if program has been well parallelized. During this parallel region execution process, the parallel execution time depends on the processor with the largest workload and other light-load processors must wait this processor. At this time, reducing these light-load processors frequencies will save energy consumption without performance loss.

We must note that during the transition from one serial region to next parallel region, it takes some extra energy consumption to bring a processor from the power-down state to the active state. And during the transition from one parallel region to next parallel region, each processor's voltage/frequency maybe needs re-adjusting, among which switching overheads must be considered.

We develop an analytical model to make a research on such an energy optimization approach for parallel applications, and then do the experiments to validate our model. Experimental results show energy saving can be up to 10.34% without performance loss.

I. Kadayif also did the related work. In [3], he found loop nests usually need fewer processors than the maximum and his strategy determines each loop nest to use different number of processors. In [4], he provided an integer linear programming method to solve this problem. In his later work [5], a new energy reduction method is provided. Each processor voltage/frequency is set to different level for energy savings. Our previous energy optimization approach [6] combines shutting-down technique and *DVFS* technique. We exploited the workload imbalance from serial regions and parallel regions and validated our approach through simulation. The major difference between our current work and [6] is that we build a detailed analytical model instead of simulation experiments. And our analytical model avoids the performance loss while the approach in [6] allows some performance loss.

The rest of the paper is organized as follows: Section 2 illustrates our analytical model for energy optimization. Section 3 gives the detailed experimental results and evaluations. Section 4 makes the conclusions.

## 2 Analytical Model for Energy Optimization

This analytical model is based on a DVFS-enabled on-chip multiprocessor architecture, which consists of eight DVFS-enabled processors. Each processor can change its voltage from 0.9V to 2.05V and change its frequency from 200 MHz to 1 GHz. Each processor has its own L1 cache and L2 cache and they share an off-chip DRAM memory. All the inter-processor communications occur through reading from and writing into this shared off-chip DRAM memory. Multithread parallel applications are executed on this architecture.

Two main energy optimization methods are: *DVFS* and *shutting down*. We allow each processor can continuously adjust its voltage/frequency for energy savings. Switching overheads are taken into accounts carefully. Besides, the unused processor can be placed into a power-down state [7][8]. When the processor/cache is placed into power-down state, the energy consumption is reduced significantly. *Power reduction factor* defines the proportion of the power-down state power dissipation to the active state power dissipation. The re-synchronization penalty is incurred when a processor/cache pair is activated from power-down state. We assume 50-cycle re-synchronization cycle and $10-\mu J$ energy cost.

Our energy optimization problem is stated as follows: given a parallel application with $N_s$ serial regions, $N_p$ loop regions and a DVFS-enabled on-chip multiprocessor with P processors, find the optimal voltage and frequency modes for each regions, then the whole energy consumption is optimized without performance loss.
Model parameters are shown in Table 1, where $1 \le i \le N_s$ and $1 \le j \le N_p$ .

**Table 1.** Model parameters

| Name | Meanings |
|---|---|
| $ts_i$ | the execution time for the *i-th* serial region |
| $IC_i^s$ | the instruction cycles for the *i-th* serial region (profiling data) |
| $ES_i$ | the energy consumption for the *i-th* serial region |
| $t_{syn}$ | the re-synchronization time penalty (from power-down to active state) |
| $f_{\max}$ | the maximum clock frequency used during the active period |
| $P_{acti}(f)$ | the active power dissipation with variable frequency *f* |
| $P_{stat}$ | the power dissipation when the processor is shut down |
| $E_{syn}$ | the energy consumption for one re-synchronization |
| $tps_{j,k}$ | the execution time of the *k-th* processor for the *j-th* parallel region |
| $tp_j$ | the longest execution time among the *P* processors for the *j-th* parallel region |
| $IC_{j,k}^{ps}$ | instruction cycles for the *k-th* processor during the *j-th* parallel region (profiling data) |
| $EP_j$ | the energy consumption for *j-th* parallel region |
| $t_{switch}(V_{j-1,k}, V_{j,k})$ | the execution time switching cost from voltage $V_{j-1,k}$ to $V_{j,k}$ |
| $f_{j,k}$ | the optimal frequency of the *k-th* processor for the *j-th* parallel region |
| $V_{j,k}$ | the optimal voltage of the *k-th* processor for the *j-th* parallel region |
| $E_{switch}(V_{j-1,k}, V_{j,k})$ | the energy switching cost in switching form voltage $V_{j-1,k}$ to $V_{j,k}$ |

We use normalized clock frequency $f^*$ and normalized operating voltage $V^*$ instead of *f* and *V*, where $f^* = f / f_{\max}, V^* = V / V_{\max}$ . Then,

$$V^* = k_1 + k_2 \cdot f^* \tag{1}$$

where the constants $k_1 = V_{th} / V_{\max}$ and $k_2 = 1 - k_1$ . This simple relationship closely matches industrial data [9]. From today's technology, $k_1$ is approximately 0.3, and $k_2$ equals to 0.7.

According to [10], power dissipation can be calculated by[1] formula (2), where *A* is the switching activity factor, *C* is the load capacitance, *V* is the supply voltage and *f* is the clock frequency.

$$P = ACV^2 f \tag{2}$$

Then the energy consumption consumed by *P* processors for the $N_s$ serial regions and $N_p$ parallel regions at the maximum frequency is:

$$E_{ori} = \sum_{i=1}^{N_s} (P \cdot ts_i \cdot ACV_{\max}^2 f_{\max}) + \sum_{j=1}^{N_p} (P \cdot tp_j \cdot ACV_{\max}^2 f_{\max}) \tag{3}$$

[1] For 130 nm technology, static power dissipation due to leakage current has been ignored.

For all serial regions, we shut down all unused ($P$-1) processors. For all parallel regions, we use the optimal frequency/voltage values ( $f_{j,k}$ and $V_{j,k}$ ) instead of the maximum ones ( $f_{\max}$ and $V_{\max}$ ). After that, the total energy consumption is :

$$E_{opt} = \sum_{i=1}^{N_s} (ts_i \cdot ACV_{\max}^2 f_{\max} + (P-1) \cdot ts_i \cdot P_{stat}) + \sum_{j=1}^{N_p} \sum_{k=1}^{P} (tp_j \cdot ACV_{j,k}^2 f_{j,k}) \quad (4)$$

where

$$f_{j,k} = f_{j,k}^* \cdot f_{\max} \quad k = 1,..., P,\ j = 1,..., N_p \quad (5)$$

$$V_{j,k} = V_{j,k}^* \cdot V_{\max} \quad (6)$$

$$f_{j,k}^* = \frac{tps_{j,k}}{tp_j} \quad (7)$$

$$V_{j,k}^* = k_1 + k_2 \cdot f_{j,k}^* \quad (8)$$

where $ts_i, tps_{j,k}$ is calculated by $IC \cdot CPI \cdot f^{-1}$. So we require profiling data $IC_i^s$ and $IC_{j,k}^{ps}$. Assume $\beta$ represents power reduction factor, then

$$P_{stat} = \beta \cdot ACV_{\max}^2 \cdot f_{\max} \quad (9)$$

we assume $\beta = 0.1$, $V_{\max} = 2.05V$, $f_{\max} = 1GHz$.

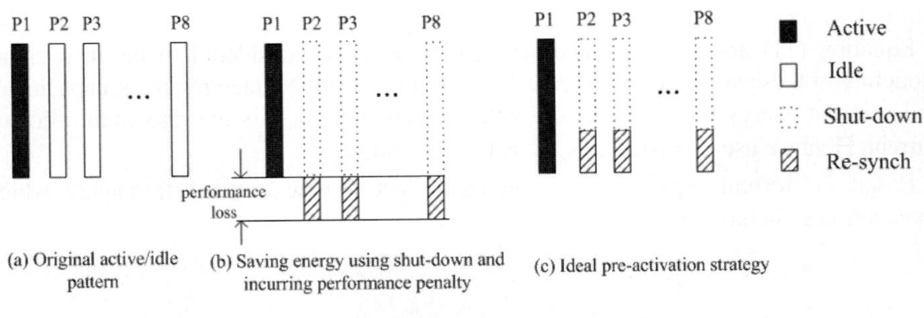


**Fig. 1.** Illustration for ideal pre-activation strategy for serial regions

In fact, there exist two kinds of switching overhead: one is re-synchronization penalty; the other is voltage/frequency switching overhead. A processor activation call brings a processor from the power-down state to the active state. In common, performance loss will exist as shown in Fig. 1(b), we use an ideal pre-activation strategy to avoid the performance loss as Fig. 1(c) shows.

Although performance loss can be avoided by ideal pre-activation strategy, re-synchronization energy overhead is inevitable. So, the total energy consumption is:

$$E^*_{opt} = \sum_{i=1}^{N_s}(ts_i \cdot ACV^2_{\max} f_{\max} + (P-1)\cdot((ts_i - t_{syn})\cdot P_{stat} + E_{syn})) + \sum_{j=1}^{N_p}\sum_{k=1}^{P}(tp_j \cdot ACV^2_{j,k} f_{j,k}) \quad (10)$$

where $E_{syn} = 10 \,\µ J$ , $t_{syn} = 50 cycles$ .

Another switching overhead is from DVS, which is illustrated in Fig. 2. Fig. 2(a) illustrates the normal activities of all the processor at the maximum frequency/voltage, where idleness is inevitable due to imbalance. DVFS can eliminate idleness, but switching time yields performance loss as Fig. 2(b) shows. In Fig. 2(c), optimal voltage/frequency for each processor is re-calculated for no performance loss.

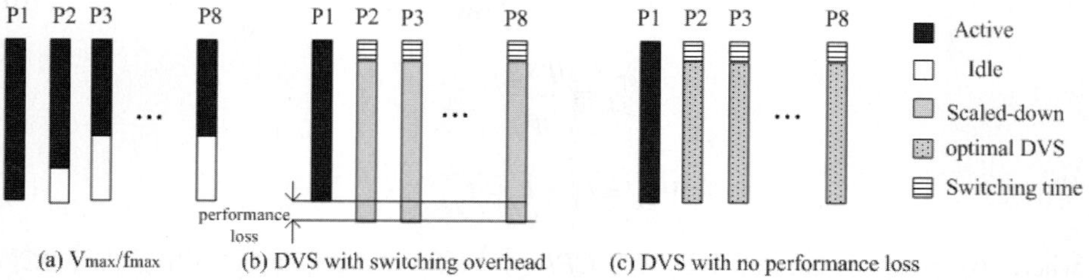


**Fig. 2.** Illustration for voltage/frequency switching overheads

Switching energy and switching time are calculated by:

$$E_{switch}(V_{j-1,k}, V_{j,k}) = (1-u)\cdot c \cdot | V^2_{j-1,k} - V^2_{j,k} | \quad j = 2,\ldots,N_p \quad (11)$$

$$t_{switch}(V_{j-1,k}, V_{j,k}) = \frac{2c}{I_{MAX}} \cdot | V_{j-1,k} - V_{j,k} | \quad j = 2,\ldots,N_p \quad (12)$$

Equation (11) and (12) are taken from [11], and are considered to be an accurate modeling of these switching overhead. The variable $c$ is the voltage regulator capacitance and $u$ is the energy-efficiency of the voltage regulator. $I_{MAX}$ is the maximum allowed current. Here we use $c = 10\mu f$ , $I_{MAX} = 1A$ , $u = 90\%$ .

Based on formula (5), (7), we can easily get a new optimal frequency which considering switching time:

$$f_{j,k} = \frac{tps_{j,k}}{tp_j - t_{switch}(V_{j-1,k}, V_{j,k})} \cdot f_{\max} \quad (13)$$

Based on formula (13), we can deduce the optimal frequency $f_{j,k}$ using formula (12), (6), (8) and (5). And $V_{j,k}$ can be calculated easily.

$$f_{j,k} = \begin{cases} \dfrac{tps_{j,k}}{tp_j} \cdot f_{\max} & j = 1 \\[2ex] \dfrac{tps_{j,k} \cdot f_{\max}}{tp_j - \dfrac{2c \cdot k_2 \cdot V_{\max}}{I_{MAX} \cdot f_{\max}} | f_{j-1,k} - f_{j,k} |} & j = 2,\ldots,N_p \end{cases} \quad k = 1,\ldots,P \quad (14)$$

Considering voltage/frequency switching overhead, the total energy consumption can be optimized into:

$$E_{opt}^{*} = \sum_{i=1}^{N_s} (ts_i \cdot ACV_{\max}^2 f_{\max} + (P-1)\cdot((ts_i - t_{syn})\cdot P_{stat} + E_{syn})) + \sum_{j=1}^{N_p}\sum_{k=1}^{P}(tp_j \cdot ACV_{j,k}^2 f_{j,k}) + \sum_{j=1}^{N_p-1}\sum_{k=1}^{P}(1-u)\cdot c \cdot | V_{j,k}^2 - V_{j+1,k}^2 | \quad (15)$$

where $V_{j,k} / f_{j,k}$ is achieved through the previous formulas.

According to the above analyses, we can optimize the energy consumption without performance loss as long as $ts_i > t_{syn}$.

# 3 Experiments

The whole compiler implementation framework consists of three phases: profiling phase, frequency/voltage scaling phase and optimization phase. In profiling phase, some necessary data are profiled, and then the optimal voltage/frequency for each region is calculated based on the above formulas. In the second phase, some compiler-directed commands are inserted into the codes to shut down the unused processors and adjust voltage/frequency at the right points. In the third phase, processor reuse optimization is applied. We assume the processor assignment for each parallel region is done independently of the other ones. As a result of this, as we move from one parallel region to another one, the same frequency perhaps is set for the different processor, which causes some unnecessary overhead for changing processor. We reuse as many processors as possible to reduce this overhead. For example, we assume frequency level $f_{j,k}$ is set for processor *k* during the *j-th* parallel region. If the next adjoined region is also a parallel region and we still need to assign $f_{j,k}$ to one processor, we still use processor *k*. It is necessary to re-schedule load assignment for each processor. Finally, the processors are reused as much as possible and unnecessary energy consumption is avoided.

We use a representative array-intensive program—sparse matrix-vector product (SMVP)—to validate the effective of our energy optimization approach. SMVP calculates the product of a sparse matrix and a vector. In common, this sparse matrix is big and only a few non-zero elements. Due to the number of non-zero elements in all rows are quite different, load imbalance always occurs [6].

The mp_simplesim simulator [12] combined with energy analytical model is used to validate our energy optimization model. We implemented multithread SMVP

**Table 2.** Descriptions for the sparse matrices

| Sparse Matrix | Size | Descriptions |
|---|---|---|
| orsirr1 | 1030 | Oil reservoir simulation for a 21*21*5 irregular grid |
| orsirr2 | 886 | Oil reservoir simulation for a 21*21*5 irregular grid |
| sherman2 | 1080 | A black oil, impes simulation, 35*11*13 grid |
| sherman3 | 5005 | A fully implicit black oil simulator, 16*23*3 grid |

**Table 3.** Profiling cycles for each serial region and parallel region

| Sparse Matrix | Cycles for serial region | Cycles for the 1st parallel region / Cycles for the 2nd parallel region | | | | | | | |
|---|---|---|---|---|---|---|---|---|---|
| | | P1 | P2 | P3 | P4 | P5 | P6 | P7 | P8 |
| orsirr1 | 75620 | 531381/ 531680 | 531743/ 531916 | 540119/ 540320 | 531736/ 531993 | 531295/ 531531 | 539931/ 539572 | 531449/ 531580 | 539697/ 539661 |
| orsirr2 | 90160 | 393208/ 393354 | 393557/ 393492 | 400844/ 400983 | 393620/ 393688 | 393480/ 393688 | 400383/ 401101 | 393571/ 393457 | 400499/ 400526 |
| sherman2 | 72590 | 580267/ 583660 | 588839/ 592406 | 580181/ 583581 | 588839/ 592078 | 580181/ 583140 | 588839/ 591924 | 580454/ 583175 | 588905/ 589542 |
| sherman3 | 72600 | 12502099/ 12546222 | 12543121/ 12547473 | 12544017/ 12508724 | 12544815/ 12548586 | 12547937/ 12547802 | 12511428/ 12549846 | 12548979/ 12552057 | 12546366/ 12551027 |

**Table 4.** DVS results for each parallel region

| Sparse Matrix | The optimal voltage/frequency during two successive parallel regions | | | | | | | | |
|---|---|---|---|---|---|---|---|---|---|
| | | P1 | P2 | P3 | P4 | P5 | P6 | P7 | P8 |
| orsirr1 | 1st parallel region | 2.0268V/ 983.82MHz | 2.0277V/ 984.49MHz | 2.05V/1GHz | 2.0277V/ 984.48MHz | 2.0266V/ 983.66MHz | 2.0495V/ 999.65MHz | 2.027V/ 983.95MHz | 2.0489V/ 999.22MHz |
| | 2nd parallel region | 2.0271V/ 984.02MHz | 2.0277V/ 984.45MHz | 2.05V/1GHz | 2.0279V/ 984.59MHz | 2.0267V/ 983.74MHz | 2.0481V/ 998.67MHz | 2.0268V/ 983.83MHz | 2.0483V/ 998.8MHz |
| orsirr2 | 1st parallel region | 2.0227V/ 980.95MHz | 2.0239V/ 981.82MHz | 2.05V/1GHz | 2.0241V/ 981.98MHz | 2.0236V/ 981.63MHz | 2.0483V/ 998.85MHz | 2.024V/ 981.86MHz | 2.0488V/ 999.14MHz |
| | 2nd parallel region | 2.0223V/ 980.7MHz | 2.0229V/ 981.08MHz | 2.0496V/ 999.73MHz | 2.0235V/ 981.55MHz | 2.0235V/ 981.53MHz | 2.0501V/ 1GHz | 2.0227V/ 981MHz | 2.048V/ 998.6MHz |
| sherman2 | 1st parallel region | 2.029V/ 985.33MHz | 2.0498V/ 999.89MHz | 2.0287V/ 985.19MHz | 2.0498V/ 999.89MHz | 2.0287V/ 985.19MHz | 2.0498V/ 999.89MHz | 2.0294V/ 985.65MHz | 2.05V/1GHz |
| | 2nd parallel region | 2.0288V/ 985.24MHz | 2.05V/1GHz | 2.0286V/ 985.11MHz | 2.0492V/ 999.47MHz | 2.0276V/ 984.4MHz | 2.0489V/ 999.22MHz | 2.0277V/ 984.47MHz | 2.0434V/ 995.39MHz |
| sherman3 | 1st parallel region | 2.0446V/ 996.26MHz | 2.0493V/ 999.53MHz | 2.0494V/ 999.6MHz | 2.0495V/ 999.67MHz | 2.0499V/ 999.92MHz | 2.0457V/ 997.01MHz | 2.05V/1GHz | 2.0497V/ 999.79MHz |
| | 2nd parallel region | 2.0493V/ 999.54MHz | 2.0495V/ 999.64MHz | 2.0451V/ 996.55MHz | 2.0496V/ 999.72MHz | 2.0495V/ 999.66MHz | 2.0498V/ 999.83MHz | 2.05V/1GHz | 2.0499V/ 999.92MHz |

**Table 5.** Energy saving results

| Sparse Matrix | Serial energy before DVFS (J) | Parallel energy before DVFS (J) | Total energy before DVFS (J) | Serial energy after DVFS (J) | Parallel energy after DVFS (J) | Total energy after DVFS (J) | Serial energy savings | Parallel energy savings | Energy savings |
|---|---|---|---|---|---|---|---|---|---|
| orsirr1 | 0.0047 | 0.0667 | 0.0714 | 0.0011 | 0.0651 | 0.0662 | 76.60% | 2.40% | 7.28% |
| orsirr2 | 0.0056 | 0.0495 | 0.0551 | 0.0013 | 0.0481 | 0.0494 | 76.79% | 2.83% | 10.34% |
| sherman2 | 0.0045 | 0.0730 | 0.0775 | 0.0010 | 0.0716 | 0.0726 | 77.78% | 1.92% | 6.32% |
| sherman3 | 0.0045 | 1.5502 | 1.5547 | 0.0010 | 1.5471 | 1.5481 | 77.78% | 0.20% | 0.42% |

running on mp_simplesim. Four different sparse matrices are used: orsirr1, orsirr2, sherman2, and sherman3 from Harwell-Boeing sparse matrix set [13][14]. Table 2 provides the sparse matrix descriptions. Fig. 4 in [6] gives the non-zero element distributions of each row for these four sparse matrices.

We choose the same SMVP parallel program for four sparse matrices, which consists of one serial region followed by two successive parallel regions. The serial region implements the initialization; sparse matrix is divided into two parts and two parallel regions complete sparse matrix-vector product. Table 3 shows the detailed profiling data for these regions. And Table 4 shows the detailed DVS results.

The detailed energy saving result is showed in Table 5. From statistics results, the total energy saving can be up to 10.34%. For each input matrix set, serial regions provides much more energy saving than parallel regions. That is because SMVP application has been well parallelized, and the opportunity for DVFS is little.

## 4 Conclusions

We provide a DVFS-enabled on-chip multiprocessor architecture, which allows dynamically adjusting each processor voltage/frequency or shut down some processors to obtain energy savings. Our energy optimization approach adjusts each processor's voltage/frequency in terms of load imbalance. We use pre-activation strategy to effectively avoid performance loss. Besides, voltage/frequency switching costs are also considered. Experimental results show our energy optimization approach is successful in reducing the energy consumption of the parallel applications. The maximum energy saving can be up to 10.34%.

## References


[1] http://www.transmeta.com/crusoe_docs/tm5900_databook_040204.pdf.

[2] Intel XScale Technology. http://www.intel.com/design/intelxscale/.

[3] I. Kadayif, M. Kandemir and M. Karakoy. An Energy Saving Strategy Based on Adaptive Loop Parallelization. In the *Proceedings of the 39th Design Automation Conference (DAC 2002)*, New Orleans, LA, USA, June 10-14, 2002. p.195-200.

[4] I. Kadayif, M. Kandemir and U. Sezer. An Integer Linear Programming Based Approach for Parallelizing Applications in On-Chip Multiprocessors. In the *Proceedings of the 39th Design Automation Conference (DAC 2002)*, New Orleans, LA, USA, June 10-14, 2002. p. 703-708.

[5] I. Kadayif, M. Kandemir, N. Vijaykrishnan, M. J. Irwin and I. Kolcu. Exploiting Processor Workload Heterogeneity for Reducing Energy Consumption in Chip Multiprocessors. In the *Proceedings of DATE'04*. pp.1158-1163. Feb 2004.

[6] Juan Chen, Yong Dong, Xue-jun Yang, Dan Wu. A Compiler-Directed Energy Saving Strategy for Parallelizing Applications in On-Chip Multiprocessors. In the *Proceedings of Fourth International Symposium on Parallel and Distributed Computing*. France, July 4-6 2005. IEEE Computer Society Press, pp. 147-154.

[7] A. Chandrakasan, W. J. Bowhill, and F. Fox. Design of High-Performance Microprocessor Circuits. IEEE Press, 2001.

[8] W. Zhang, N. Vijaykrishna, M. Kandemir, M. J. Irwin, D. Duare, and Y. Tsai. Exploiting VLIW schedule slacks for dynamic and leakage energy reduction. In the *Proceedings of the 34th Annual International Symposium on Microarchitecture*, Austin, TX, December 2001.

[9] K. Nowka et al. A 0.9V to 1.95V Dynamic Voltage-Scalable and Frequency-Scalable 32b PowerPC Processors. In the *Proceedings of International Solid-State Circuits Conference (ISSCC)*, IEEE Press, 2002, pp. 340-341.

[10] J. L. Hennessy and D. A. Patterson. Computer Architecture: A Quantitative Approach. Elsevier Science Pte Ltd., third edition, 2003.

[11] T. Burd and R. Brodersen. Design issues for dynamic voltage scaling. In the *Proceedings of International Symposium on Low Power Electronics and Design (ISLPED-00)*, June 2000.

[12] Naraig Manjikian. Multiprocessor Enhancements of the SimpleScalar Tool Set. In ACM Computer Architecture News, Vol.29 No. 1, March 2001, pp. 8-15.

[13] I. S. Duff, R. G. Grimes and J. G. Lewis: User's Guide for the Harwell-Boeing Sparse Matrix Collection, Tech. Report TR-PA-92-96, CERFACS, Toulouse Cedex, France, Oct. 1992.

[14] http://www.cise.ufl.edu/research/sparse/HBformat/HB/

# Using SBASCO to Solve Reaction-Diffusion Equations in Two-Dimensional Irregular Domains

Manuel Díaz, Sergio Romero, Bartolomé Rubio,
Enrique Soler, and José M. Troya

Department of Languages and Computer Science, University of Málaga, 29071 SPAIN
{mdr, sromero, tolo, esc, troya}@lcc.uma.es

**Abstract.** The SBASCO programming environment provides the developer of parallel and distributed applications with high-level programming capabilities. This is achieved as a result of the combination of two technologies: algorithmic skeletons and software components. This paper is a case study on the use of SBASCO. Specifically, we present a scientific application to study the propagation of reaction waves in two-dimensional irregular domains which can be divided into overlapping rectangular regions. Domain decomposition techniques are used to solve a system of two non-linear reaction-diffusion equations. The structure of the application is established by means of a high-level skeleton, which captures all the communication and synchronization details that take place in parallel component interaction, thereby releasing the programmer from coding them. In addition, the use of software components facilitates the development process and allows the creation of more flexible and adaptable software.

## 1 Introduction

Domain decomposition techniques have received great attention especially for the numerical solution of partial differential equations (PDEs) [8]. In our context, the term "domain decomposition" means the separation of the physical domain into different regions and so, many of these techniques accept a parallelization that allows them to tackle large-scale and realistic engineering problems [7].

This paper presents a practical case study using SBASCO in the development of a parallel scientific application in order to obtain the numerical solution of a reaction-diffusion problem. The problem is modelled as a system of two time-dependent, nonlinearly coupled PDEs, and is solved by means of domain decomposition methods. Two-dimensional domains that exhibit an irregular geometry with re-entrant corners are considered. Overlapping domain decomposition techniques used in this paper are based on Schwarz's methods which, at the differential level, use the solution in one subdomain to update the Dirichlet data of the other, the convergence rate being influenced by the overlapping length [5].

The application combines two different parallelism levels. On the one hand, the solution in the different subdomains is computed in parallel by assigning

V.N. Alexandrov et al. (Eds.): ICCS 2006, Part II, LNCS 3992, pp. 912–919, 2006.

one task per subdomain. On the other hand, the numerical method each task implements, which is basically a procedure for solving a large linear system of algebraic equations, may be parallel itself, e.g. data parallel red-black Gauss-Seidel relaxation is used in this work. Moreover, the updating of the interior boundaries of adjacent subdomains, which takes place in every iteration, involves the communication and synchronization among either sequential or parallel tasks. These communications depend on the data distribution belonging to the participant tasks. The development of such applications can be tedious and error-prone when based on traditional high-performance solutions such as HPF/MPI, C/PVM, etc. and so, we encourage the use of languages and tools that offer a higher degree of abstraction.

SBASCO (Skeleton-Based Scientific Components) [2] is a new programming environment focused on the efficient development of parallel and distributed numerical applications, also integrating two different technologies: algorithmic skeletons [1] and software components [4]. This unified approach provides interesting features in terms of interoperability, high-level programmability, compositionality, and code reusability.

The *multiblock* skeleton defined in SBASCO captures the pattern of parallel computation and communication that takes place in the proposed application. This skeleton allows the establishment of the application structure in an elegant and declarative way, and also abstracts the programmer from most of the low-level aspects of parallelism exploitation such as the creation, communication and synchronization of tasks. They are addressed by the runtime support which implements the skeleton so that the programmer can focus on writing the scientific code, which is encapsulated into software components.

This paper is structured as follows. The next section outlines the main characteristics of SBASCO. Section 3 introduces the physical problem taken into consideration. The design of the application, including some implementation details are described in section 4. Experimental results are shown in section 5. Finally, some conclusions are outlined.

## 2 SBASCO Overview

The skeleton-based composition language of SBASCO is used for both scientific components and application construction. The internal structure of a component can be established by means of the application of a fixed set of skeletons. So, the interaction of the different tasks integrating the components is expressed in a high-level and declarative way, according to static and predictable patterns. The following is a brief description of the skeletons provided:

- The *multiblock* skeleton is focussed on the solution of multi-block and domain decomposition-based problems, which form an important kind of problem in the high-performance computing area.
- The *farm* skeleton improves a task throughput as different data sets can be computed in parallel on different sets of processors.

- Problem solutions that have a communication pattern based on array interchange can be defined and solved easily by using the *pipeline* skeleton, which pipelines sequences of tasks.

Scientific component interfaces are described by means of two different views. The *application view* contains information related to data types of component input/output. This view is used by the programmer in order to develop his/her applications by means of the composition language. The *configuration view* extends the application view with information about input and output data distribution, processor layout and component internal structure (in terms of the skeleton composition scheme).

The knowledge at the component interface level of data distribution and processor layout allows the system to obtain an efficient implementation of the communication scheme among components, which follows a "data flow" style by means of a typical `put_data/get_data` scheme.

The implementation of SBASCO is based on the extension of the message passing interface, namely MPI-2 [3]. The system exploits the functions for the creation and management of processes, as well as the mechanisms for connecting and communicating parallel MPI applications. The component model is supported by a compiler and libraries to facilitate the programming task.

This section has summarized the main features of SBASCO. A more detailed explanation can be found in [2].

## 3 Problem Formulation and Discretization

Physical phenomena involving heat and mass transfer, combustion, etc. are characterized by reaction-diffusion equations with non-linear source terms. Here, we consider the following set of two time-dependent, nonlinearly coupled PDEs:

$$\frac{\partial U}{\partial t} = \frac{\partial^2 U}{\partial x^2} + \frac{\partial^2 U}{\partial y^2} + S(U), \tag{1}$$

where

$$U = (u, v)^T, \qquad S = (-uv, uv - \lambda v)^T, \tag{2}$$

$u$ and $v$ represent the concentration of a reactant and the temperature, respectively, $u = 1$ and $v = 0$ on the external boundaries, $t$ is time, $x$ and $y$ denote Cartesian coordinates, $\lambda$ is a constant (in this paper, $\lambda = 0.5$), and the superscript $T$ denotes transpose. Eq. (1) has been previously studied in [6], where a comparison of several numerical techniques for tackling domain decomposition problems in irregular domains is presented.

Eq. (1) was discretized by means of an implicit, linearized, $\theta$-method in an equally spaced grid where the non-linear term $S_{i,j}^{n+1}$ was approximated by means of its Taylor polynomial of first degree around $(t^n, x_i, y_j)$ to obtain the following system of linear algebraic equations:

$$\begin{aligned}\frac{\Delta U_{i,j}}{k} &= \frac{1}{\Delta x^2}\left[\theta\delta_x^2 \Delta U_{i,j} + \delta_x^2 U_{i,j}^n\right] + \frac{1}{\Delta y^2}\left[\theta\delta_y^2 \Delta U_{i,j} + \delta_y^2 U_{i,j}^n\right] \\ &\quad + S_{i,j}^n + \theta J_{i,j}^n \Delta U_{i,j},\end{aligned} \tag{3}$$

where

$$\Delta U_{i,j} = U_{i,j}^{n+1} - U_{i,j}^{n}, \quad S_{i,j}^{n} = S(U_{i,j}^{n}), \quad J_{i,j}^{n} = \frac{\partial S}{\partial U}(t^n, x_i, y_j)$$
$$\delta_x^2 U_{i,j} = U_{i+1,j} - 2U_{i,j} + U_{i-1,j}, \quad \delta_y^2 U_{i,j} = U_{i,j+1} - 2U_{i,j} + U_{i,j-1}, \quad (4)$$

$i$ and $j$ denote $x_i$ and $y_j$, respectively, $t^n$ denotes the $n$th time level, $k$ is the time step, $\Delta x$ and $\Delta y$ represent the grid spacing in the $x-$ and $y-$directions, respectively, and $0 < \theta \leq 1$ is the implicitness parameter. In this paper, $\theta = 0.5$, i.e., second-order accurate finite difference methods are employed.

## 4 Application Design and Implementation

In order to apply domain decomposition, the geometry of the problem being considered is divided, as shown in Fig. 1, into several overlapping regions, i.e. $\Omega_1$, $\Omega_2$ and $\Omega_3$. The part of the boundary of $\Omega_i$ that is interior to $\Omega_j$ is denoted by $\Gamma_{i,j}$. In the Dirichlet method considered here, the current solution in one subdomain is used to update the boundary data in the adjacent subdomains. Then, the interior points are recalculated by solving the corresponding linear system of algebraic equations. This iterative procedure is repeated until convergence. The following algorithm gives an abstract description of the global application.

```
1.  Set initial conditions                          (on each subdomain)
2.  For time_step = 1..MAX_TIME_STEPS Do
3.     Repeat
4.        Update boundaries                         (on each subdomain)
5.        Solve system of linear equations          (on each subdomain)
6.     Until convergence on ALL subdomains
```

This algorithm involves up to three major nested loops: the `For` loop in line 2, the `Repeat` loop in line 3 and the linear system solver in line 5 which, due

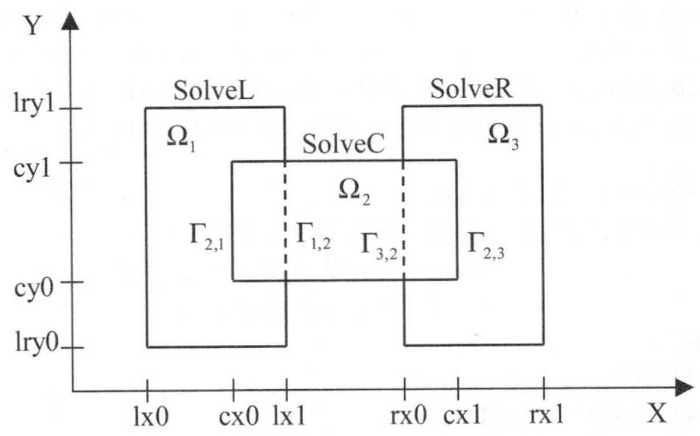


**Fig. 1.** Original domain decomposed into overlapping regions

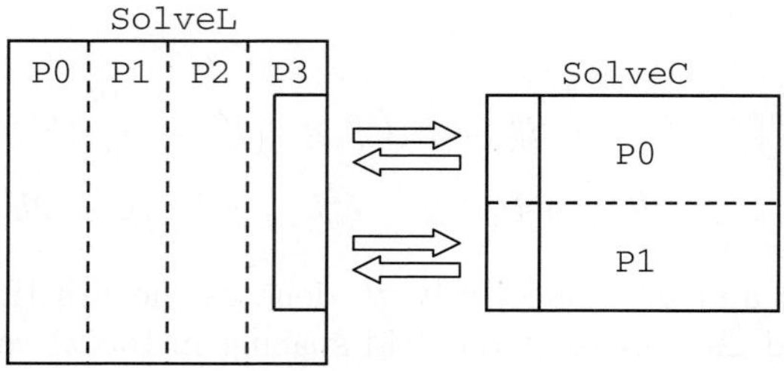


**Fig. 2.** Boundaries updating involves parallel communications

to the large number of unknowns per domain, is usually based on an iterative procedure, instead of a direct method. Regarding the convergence criterion in line 6, this is satisfied when boundary values are not significantly modified in two iterations ($\|\Gamma_{i,j}^{k} - \Gamma_{i,j}^{k-1}\| < 10^{-10}$ for every pair of adjacent subdomains).

A way to execute the above algorithm in parallel is to associate a task with each one of the regions. In our approach, these tasks are encapsulated into software components which comprise the application so that, we have instances of components that run in parallel to solve the problem. Furthermore, the scientific code implemented in components can be sequential or parallel as well. For example, we use parallel red-black Gauss-Seidel relaxation as a linear equations solver and so, data parallel components are considered.

The updating of the boundaries and the convergence criterion (lines 4 and 6 respectively) require the synchronization and communication of components. Data distribution and processor layout are the key elements that influence the implementation of an efficient communication scheme, as is shown in Fig. 2. The component on the left, called `SolveL`, is executed on four processors its domain being distributed by columns. The component on the right, called `SolveC`, runs on two processors having data distributed by rows. When boundaries are being updated, processor `P3` of `SolveL` needs to communicate with processors `P0` and `P1` of `SolveC`. It is important to remark that the system manages these communications automatically, according to the high-level description of boundaries, data distribution and processor layout that the programmer has provided.

The SBASCO solution uses a *multiblock* skeleton, which is oriented to these types of domain-decomposition problems, in order to structure the application and establish the interaction scheme among the components.

```
PROGRAM ReactionDiffusion
complex, DOMAIN2D :: left/lx0,lry0,lx1,lry1/,
                     center/cx0,cy0,cx1,cy1/,
                     right/rx0,lry0,rx1,lry1/
STRUCTURE
  MULTIBLOCK
    SolveL(left) ON PROCS(4),
    SolveC(center) ON PROCS(2),
    SolveR(right) ON PROCS(4)
```

```
WITH BORDERS
  left(lx1,cy0,lx1,cy1)<-center(_),
  center(cx0,cy0,cx0,cy1)<-left(_),
  /* ... */
```

The program declares three domains based on the geometry shown in Fig. 1. The `complex` datatype was chosen because every domain point needs to save a pair of values, i.e. $u$ and $v$. The `MULTIBLOCK` definition includes the selected components, the number of processors assigned to each one, and the boundaries definition. The expression `left(lx1,cy0,lx1,cy1)<-center(_)` indicates that the zone of `left` delimited by points `(lx1,cy0)` and `(lx1,cy1)` will be updated (in every iteration) by the values belonging to the zone of `center` delimited by the same points. No data distribution is declared at the composition level, as this information will be obtained from the configuration view of the different components involved. Distribution types are similar to those used in HPF. The following shows the configuration view of `SolveC`:

```
CONFIGURATION INTERFACE SolveC
  complex, INOUT, DOMAIN2D :: my_domain,
  DISTRIBUTE my_domain(BLOCK,*)
```

By using the high-level *mutiblock* skeleton, the programmer is abstracted from details such as the synchronization and communication among parallel components or the execution pattern implemented in the domain decomposition method. So, he/she is only concerned with the programming of the initial conditions, the desired convergence criterion and the linear system solver. As a result, the complexity of the application development decreases significantly.

The programming of the components is carried out in an object-oriented style using C++ with MPI, together with a class library provided by SBASCO. The software architecture is comprised of a group of MPI applications that implements the parallel components being executed on disjoint sets of processors. The creation and interconnection of scientific components are carried out by means of the process management mechanisms defined in MPI-2. Prior to the application execution, the components exchange information related to data distribution of arguments and processor layout so that message passing can be performed efficiently when communication takes place.

## 5 Experimental Results

The presented application has been used to solve Eq.(1) on a domain with reentrant corners, which is decomposed into three overlapping regions as described in Fig. 1. The problem is subjected to the following initial conditions:

$$u(x, y, 0) = 1, \quad v(x, y, 0) = e^{-((x-x_{ig})^2+(y-y_{ig})^2)}, \tag{5}$$

where $(x_{ig}, y_{ig})$ is located at the center of $\Omega_1$ and denotes the ignition location. These conditions establish a peak of temperature in the left subdomain needed to start the reaction process.

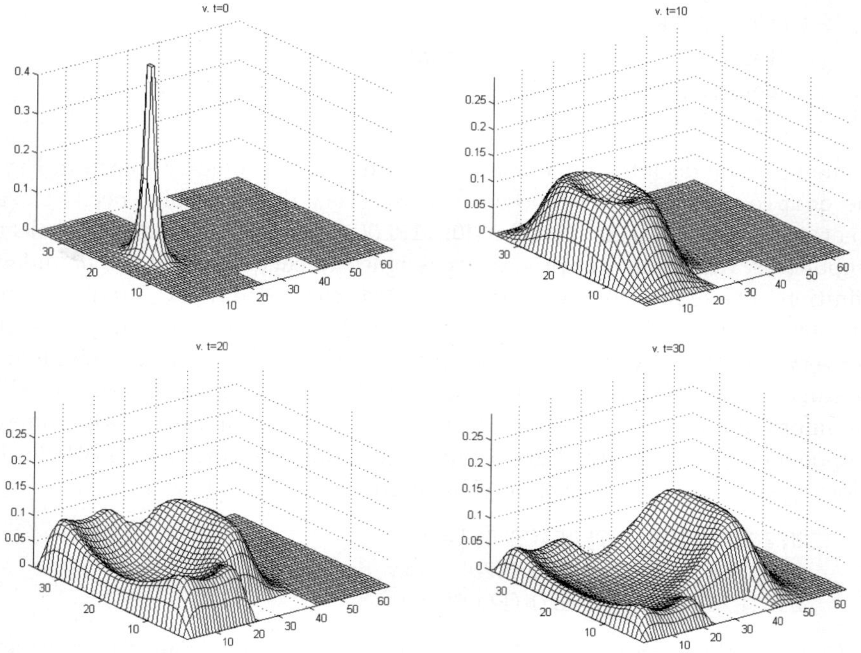


**Fig. 3.** Spatial distribution of nondimensional temperature, $v$, at $t = 0, 10, 20$ and 30

**Table 1.** Execution time, in seconds, for the reaction-diffusion problem. Values in brackets represent speed-up.

| Problem size | Subdomain size | Sequential | Number of processors | | | |
|---|---|---|---|---|---|---|
| | | | 3 | 6 | 9 | 12 |
| 240x440 | 240x160 | 54.45 | 30.22 | 19.72 | 14.71 | 13.12 |
| | | | (1.80) | (2.76) | (3.70) | (4.15) |
| 480x880 | 480x320 | 189.62 | 109.54 | 65.80 | 48.67 | 42.78 |
| | | | (1.73) | (2.88) | (3.89) | (4.43) |
| 960x1760 | 960x640 | 805.23 | 548.18 | 329.16 | 195.09 | 173.44 |
| | | | (1.46) | (2.44) | (4.12) | (4.64) |

The solution is characterized by a reaction front that propagates from the ignition point to the rest of the domain. The concentration of reactant, $u$, decreases as the reaction progresses, whereas the temperature, $v$, exhibits the behavior shown in Fig. 3.

Table 1 summarizes the computational time, in seconds, for different combinations of grid sizes and processors. Each problem instance is composed of three same-sized subdomains which are overlapped. Column one denotes the size of the problem as the area of the smallest rectangle that contains the complete domain. The number of processors is equally distributed among the three subdomains. In addition, we provide a sequential case in which all components are executed on a single processor. Experiments were carried out in a cluster of Pentium 4, 2.66GHz, 1GB RAM Linux workstations interconnected with a 1Gb/s Myrinet network and running LAM/MPI as message passing interface implementation. As the results show, the high-level programming approach used does not result in loss of performance when the number of processors is increased.

## 6 Conclusions

This paper has shown a practical utilization of modern high-level programming techniques in scientific computing. The presented application solves a system of two reaction-diffusion equations by means of domain decomposition methods on irregular domains that exhibit re-entrant corners. Based on the high-level parallel programming environment SBASCO, the approach takes advantage of using algorithmic skeletons and software components, two technologies that elevate the abstraction degree of the programming model, as a consequence, making the development of the parallel application easier, as well as improving the software evolution and maintenance. Some experimental results have shown the suitability and efficiency of the proposal.

## References


1. Cole, M., Algorithmic Skeletons: Structured Management of Parallel Computation. MIT Press, 1989.
2. Díaz, M., Rubio, B., Soler, E., Troya, J.M., SBASCO: Skeleton-Based Scientific Components, in "Proceedings of the 12th Euromicro Conference on Parallel, Distributed and Network-based Processing (PDP 2004)", pp. 318–324, IEEE Computer Society, A Coruña, Spain, 2004.
3. Gropp, W., Huss-Lederman, S., Lumsdaine, A., Lusk, E., Nitzberg, B., Saphir, W., Snir, M., MPI: The Complete Reference, volume 2–The MPI-2 Extensions. MIT Press, 1998.
4. Heineman, G.T., Council, W.T., Component-Based Software Engineering: Putting the Pieces Together. Addision Wesley, 2001.
5. Lions, P.L., On the Schwarz Alternating Method III, in "Proceedings of the 3rd International Symposium on Domain Decomposition Methods for Partial Differential Equations", pp. 202–223, Chain et al. eds, Philadelphia, SIAM, USA, 1989.
6. Ramos, J.I., Soler, E., Domain Decomposition Techniques for Reaction-Diffusion Equations in Two-Dimensional Regions with Re-entrant Corners, *Applied Mathematics and Computation*, **118**, 2-3 (2001), pp. 189–221.
7. Smith, B., Bjørstad, P., Gropp, W., Domain Decomposition. Parallel Multilevel Methods for Elliptic P.D.E.'e. Cambridge University Press, 1996.
8. Quarteroni, A., Valli, A., Domain Decomposition for Partial Differential Equations. Oxford Science Publications, 1999.

# Higher Order Flattening*

Roman Leshchinskiy**, Manuel M. T. Chakravarty, and Gabriele Keller

University of New South Wales
School of Computer Science and Engineering
Programming Languages and Systems
{rl, chak, keller}@cse.unsw.edu.au

**Abstract.** We extend the *flattening transformation*, which turns nested into flat data parallelism, to the full higher-order case, including lambda abstractions and data parallel arrays of functions. Our central observation is that flattening needs to transform the closures used to represent functional values. Thus, we use closure conversion before flattening and introduce *array closures* to represent arrays of functional values.

## 1 Introduction

Nested data parallelism [2] enables the concise specification of irregular parallel computations involving sparse data structures (e.g., trees and sparse matrixes) and comes with a simple language-based cost model. Following [5], we add nested data parallelism to Haskell by including a type of *parallel arrays*, denoted $[:\alpha:]$, such that we can use variants of all special list syntax and *Prelude* list operations that only involve finite lists. To distinguish arrays from lists, we add colons to the square brackets (as in $[:1, 2, 3:]$) and append the suffix *P* to function names (e.g., *mapP* $(+1)$ $[:1, 2, 3:]$ adds 1 to each element of $[:1, 2, 3:]$). The main difference between lists and arrays is that the latter have a *parallel evaluation semantics;* i.e., all elements are evaluated as soon as one is demanded.

Given an array of arrays, $xs :: [:[:Float:]:]$, we may evaluate the sum of each subarray by *mapP sumP xs*. As *sumP* itself is a parallel operation, we overall have a nested parallel computation, where *lengthP xs* parallel summations are performed in parallel. A nested array, such as *xs*, generally constitutes an irregular structure as the subarrays may be of varying length; hence it goes beyond Fortran's notion of an array. Nevertheless, we want to avoid a pointer-based representation, where the outer array is an array of pointers to the subarrays, and prefer a flat representation, which stores the elements of all subarrays in one contiguous block of memory and keeps the information about subarray boundaries in a separate structure called a *segment descriptor*. For example, $[:[:1, 2:], [::], [:3, 4, 5:]:]$ would be represented by the pair $([:2, 0, 3:], [:1, 2, 3, 4, 5:])$ of segment descriptor and flat value array. This improves locality of reference, and hence, increases cache utilisation as well as decreases communication on a distributed-memory parallel machine.

* Partly funded by the Australian Research Council under grant number DP0211203.
** The first author was, in part, at the Technische Universität Berlin, Germany.

V.N. Alexandrov et al. (Eds.): ICCS 2006, Part II, LNCS 3992, pp. 920–928, 2006.

*Flattening* is a program transformation that turns nested data structures into flat representations (of the form just described) and vectorises the code that operates on these nested structures [1]. For example, *mapP sumP xs* turns into $sumP^{\uparrow}\ xs$, where $sumP^{\uparrow}$ is the *lifted* variant of *sumP* that simultaneously computes the sum of all segments of a flat representation of a nested array. In previous work [3], we showed how flat data representations and vectorised code can be derived for functional programs operating on nested arrays of arbitrary recursive product-sum types. However, we could not flatten arrays of functions nor vectorise partial function applications. This effectively restricted the approach to first-order programs and precluded a seamless integration of nested data parallelism into Haskell or any other functional language. The present paper closes this gap. We show how to extend flattening to the higher-order case by introducing an explicit notion of closures, which leads to the concepts of *array closures* and *closure vectorisation.* Consequently, we need to apply *closure conversion* [7] to a program before flattening.

In summary, our main contributions are threefold:

– a flat representation of arrays of functions and closures (Sect. 3);
– a method to combine and concatenate arrays of functions (Sect. 4); and
– a new approach to flattening nested *mapP*s (Sect. 5).

Before addressing these technical contributions, Sect. 2 explains the shortcomings of previous approaches to flattening with respect to arrays of functions and partial applications. Space constraints restrict us to an illustration of the core ideas by example in this paper; more details are available in [6].

## 2 Why Flattening Higher-Order Functions Is Hard?

The flattening transformation comprises (1) a *flattened array representation* and (2) code *vectorisation.* The flattened array representation turns arrays containing arbitrary tree structures into tree structures that contain arrays of *unboxed primitive* types.[1] As an example, an array of rose trees storing floats in their leaves is flattened to a list of arrays of floats. The length of the list corresponds to the height of the tallest rose tree, and each list element (i.e., array) contains the floats of one level of the original rose trees.

Code vectorisation pairs every function $f$ with a version $f^{\uparrow}$ *lifted* into vector space—think of $f^{\uparrow}$ as the data-parallel version of $f$. Code vectorisation also replaces all computations of the form *mapP* $f$ by $f^{\uparrow}$. It turns out that code vectorisation transforms a program exactly such that it can directly operate on a flattened representation of its array data.

Now for some examples. The code of the identity function $id = \lambda n.n$ remains the same after lifting, but where *id*'s type is $\alpha \rightarrow \alpha$, that of $id^{\uparrow}$ is $[:\alpha:] \rightarrow [:\alpha:]$. The increment function $inc = \lambda n.1 + n$ is slightly more interesting. Assuming its type is $Int \rightarrow Int$, the type of of $inc^{\uparrow}$ is $[:Int:] \rightarrow [:Int:]$. The code of $inc^{\uparrow}$ is

[1] Arrays of unboxed primitive types are C- or Fortran-style arrays that are essentially represented as chunks of memory full of integer and floating-point values.

$\lambda ns.(\mathit{replicateP}\ (\mathit{lengthP}\ ns)\ 1) +^{\uparrow} ns$ , where $\mathit{replicateP}\ n\ e\ =\ [{:}e,\ \ldots,\ e{:}]$ (i.e., $e$ repeated $n$ times) and $(+^{\uparrow})$ denotes the pairwise addition of two arrays.

Slightly more involved is the treatment of sum types. In Haskell, we have

$$\begin{array}{ll} \mathbf{data}\ \mathit{Either}\ \alpha\ \beta & = \mathit{Left}\ \alpha \mid \mathit{Right}\ \beta \\ \mathit{either} & :: (\alpha \to \gamma) \to (\beta \to \gamma) \to \mathit{Either}\ \alpha\ \beta \to \gamma \\ \mathit{either}\ f\ g\ x & = \mathbf{case}\ x\ \mathbf{of}\ \{\mathit{Left}\ x' \to f\ x'; \mathit{Right}\ x' \to g\ x'\} \end{array}$$

The flat representation of an array of sums $[{:}\mathit{Either}\ \alpha\ \beta{:}]$ is a product of a *selector* and two arrays containing the values of the two alternatives $([{:}\mathit{Bool}{:}], [{:}\alpha{:}], [{:}\beta{:}])$; e.g., $[{:}\mathit{Left}\ 5,\ \mathit{Right}\ 4,\ \mathit{Left}\ 2{:}]$ becomes $([{:}\mathit{True},\ \mathit{False},\ \mathit{True}{:}],\ [{:}5,\ 2{:}],\ [{:}4{:}])$.

Let us combine these two examples into a larger one:

$$\begin{array}{lll} \mathit{foo} & :: & \mathit{Either}\ \mathit{Int}\ \mathit{Int} \to \mathit{Int} \\ \mathit{foo}\ x & = & \mathit{either}\ \mathit{inc}\ \mathit{id}\ x \end{array}$$

We lift this into vector space as $\mathit{foo}^{\uparrow}\ xs\ =\ \mathit{either}^{\uparrow}\ \mathit{inc}^{\uparrow}\ \mathit{id}^{\uparrow}\ xs$ with

$$\mathit{either}^{\uparrow}\ f\ g\ (\mathit{sel},\ \mathit{left},\ \mathit{right})\ =\ \mathit{combineP}\ \mathit{sel}\ (f\ \mathit{left})\ (g\ \mathit{right})$$

The array primitive $\mathit{combineP} :: [{:}\mathit{Bool}{:}] \to [{:}\alpha{:}] \to [{:}\alpha{:}] \to [{:}\alpha{:}]$ merges its two $[{:}\alpha{:}]$ arguments, arranging the elements in the order determined by the selector of type $[{:}\mathit{Bool}{:}]$. For example, if we apply *foo* to $[{:}\mathit{Left}\ 5,\ \mathit{Right}\ 4,\ \mathit{Left}\ 2{:}]$, *combineP* will be invoked as $\mathit{combineP}\ [{:}\mathit{True},\ \mathit{False},\ \mathit{True}{:}]\ [{:}6,\ 3{:}]\ [{:}4{:}]$, resulting in $[{:}6,\ 4,\ 3{:}]$.

In its full glory, the story is slightly more complicated as we need to replace a function $f$ by a pair of its original and lifted version $(f, f^{\uparrow})$ and adjust all function applications by adding appropriate projections. However, these details are secondary for demonstrating the core difficulty in flattening higher-order functions; please refer to [3, 4] for a more detailed introduction to flattening.

Now, let us turn to the core of the problem that we solve in this paper. We vary the example function *foo* to get

$$\begin{array}{lll} \mathit{bar} & :: & (\mathit{Either}\ \mathit{Int}\ \mathit{Int},\ \mathit{Int}) \to \mathit{Int} \\ \mathit{bar}\ (x,\ y) & = & \mathit{either}\ ((+)\ y)\ \mathit{id}\ x \end{array}$$

The first argument to *either*, namely $(+)\ y$, is a partial application of addition to the new argument $y$, replacing the previously added constant 1 with the variable $y$. It might seem as if we can lift *bar* in essentially the same way as *foo*:

$$\mathit{bar}^{\uparrow}\ (xs,\ ys)\ =\ \mathit{either}^{\uparrow}\ ((+^{\uparrow})\ ys)\ \mathit{id}^{\uparrow}\ xs$$

Unfortunately, we will see that this gets us into trouble quickly. We flatten an array of products $[{:}(\alpha,\ \beta){:}]$ as a product of arrays $([{:}\alpha{:}],\ [{:}\beta{:}])$. For example, $[{:}(\mathit{Left}\ 5,\ 1),\ (\mathit{Right}\ 4,\ 2),\ (\mathit{Left}\ 2,\ 7){:}]$ is represented by $(([{:}\mathit{True},\ \mathit{False},\ \mathit{True}{:}],\ [{:}5,\ 2{:}],\ [{:}4{:}]),\ [{:}1,\ 2,\ 7{:}])$. We can now calculate as follows:

$$\begin{array}{l} \quad \mathit{bar}^{\uparrow}(([{:}\mathit{True}, \mathit{False}, \mathit{True}{:}], [{:}5, 2{:}], [{:}4{:}]), [{:}1, 2, 7{:}]) \\ = \{\text{Unfolding } \mathit{bar}^{\uparrow}\} \\ \quad \mathit{either}^{\uparrow}\ ((+^{\uparrow})\ [{:}1, 2, 7{:}])\ \mathit{id}^{\uparrow}\ ([{:}\mathit{True}, \mathit{False}, \mathit{True}{:}], [{:}5, 2{:}], [{:}4{:}]) \\ = \{\text{Unfolding } \mathit{either}^{\uparrow}\} \\ \quad \mathit{combineP}\ [{:}\mathit{True}, \mathit{False}, \mathit{True}{:}]\ ((+^{\uparrow})\ [{:}1, 2, 7{:}]\ [{:}5, 2{:}])\ (\mathit{id}^{\uparrow}\ [{:}4{:}]) \end{array}$$

Now we have a bogus subexpression: $(+^{\uparrow})$ $[:1,\ 2,\ 7:]$ $[:5,\ 2:]$. The length of the two arguments to $(+^{\uparrow})$ does not match! This is not only a matter of choosing an appropriate prefix of the longer array (as with Haskell's *zipWith*). We would need to evaluate $(+^{\uparrow})$ $[:1,\ 7:]$ $[:5,\ 2:]$ to achieve the correct result; i.e., drop the middle element of the first array. Why does the first argument have to be $[:1,\ 7:]$? Because the selector $[:\mathit{True},\ \mathit{False},\ \mathit{True}:]$ determines that only the first and third element may be used by the first argument of *either*. Unfortunately, the partial application $(+^{\uparrow})$ $[:1,\ 2,\ 7:]$ drags the entire array $[:1,\ 2,\ 7:]$ into the body of *either*, completely disregarding the selector, instead of narrowing the array according to the selector, as in $\mathit{packP}\ [:\mathit{True},\ \mathit{False},\ \mathit{True}:]\ [:1,\ 2,\ 7:]\ =\ [:1,\ 7:]$.

Similar problems arise from local function definitions with free variables; e.g., if we replace $(+)\ y$ in *bar* by $\lambda x'.y\ +\ x'$, we arrive at the same situation as above. As we will see next, key to solving these problems is an appropriate flattened representation of arrays of functions.

## 3 Flattening Closures

The crucial observation of last section was that $\mathit{either}^{\uparrow}$ needs to *packP* its functional arguments, or rather the arrays embedded in these arguments. Hence, we need to represent function values in a form that permits array operations (such as *packP*) to inspect these values and modify them. The standard method to separate code from data, and thus make data embedded in function values explicit, is *closure conversion* [7]. It replaces all function values by *closures*, which are pairs of a closed binary function and its *environment*. The latter is a value for the first argument of the binary function and contains all the data items embedded in the function value represented by the closure.

With flattening, each closure contains two binary functions: the standard version and its lifted counterpart—we'll discuss later why this is crucial to our approach. Thus, we denote closures as expressions $\langle\!\langle (f, f^{\uparrow}),\ e\rangle\!\rangle$ of type $\alpha \Rightarrow \beta$ if $f :: (\gamma, \alpha) \rightarrow \beta$, $f^{\uparrow} :: ([:\gamma:],\ [:\alpha:]) \rightarrow [:\beta:]$ and $e :: \gamma$; e.g., we represent $(+)\ 1$ by the closure $\langle\!\langle (\lambda(v_1, v_2).v_1\ +\ v_2,\ \lambda(vs_1, vs_2).vs_1 +^{\uparrow} vs_2),\ 1\rangle\!\rangle$, or just $\langle\!\langle ((\underline{+}),\ (\underline{+}^{\uparrow})),\ 1\rangle\!\rangle$, where $\underline{f} :: (\alpha,\ \beta) \rightarrow \gamma$ is the uncurried version of $f :: \alpha \rightarrow \beta \rightarrow \gamma$. The application of a closure of type $\alpha \Rightarrow \beta$ to a value $x :: \alpha$ is defined as $\langle\!\langle (f, f^{\uparrow}),\ e\rangle\!\rangle \dagger x = f\ (e,\ x)$.

To solve our problem with lifting partial applications, we need a suitable representation of partial applications of lifted functions, such as $(+^{\uparrow})$ $[:1,\ 2,\ 7:]$. Is a closure of the form $\langle\!\langle ((\underline{+}^{\uparrow}),\ (\underline{+}^{\uparrow\uparrow})),\ [:1,\ 2,\ 7:]\rangle\!\rangle$, where $(\underline{+}^{\uparrow\uparrow})$ is addition lifted twice, sufficient? Unfortunately, no! In particular, it prevents typing closure converted, flattened programs. After all, the type of *packP* is $[:\mathit{Bool}:] \rightarrow [:\alpha:] \rightarrow [:\alpha:]$; so, it's second argument should be an array. Hence, if we want to apply *packP* to a closure value, that value should better be the flattened representation of an array type. Not surprisingly, it turns out that packable closures are the flattened representation of arrays of functions. They differ slightly from vanilla closures, as their environment is always an array. In other words, an *array closure* has the form $\langle\!\langle\!:(f,\ f^{\uparrow}),\ es\!:\!\rangle\!\rangle$ of type $\alpha \Rrightarrow \beta$ if $f :: (\gamma, \alpha) \rightarrow \beta$, $f^{\uparrow} :: ([:\gamma:], [:\alpha:]) \rightarrow [:\beta:]$, and $es :: [:\gamma:]$. The type $\alpha \Rrightarrow \beta$ represents arrays of standard closures $[:\alpha \Rightarrow \beta:]$.

For example, we represent both the partial application $(+^{\uparrow})$ [:1, 2, 7:] and the local function $\lambda xs'.ys +^{\uparrow} xs'$, where $ys =$ [:1, 2, 7:], by the array closure $\langle\!\langle{:}((\underline{+}), (\underline{+}^{\uparrow})),\ [{:}1, 2, 7{:}]{:}\rangle\!\rangle$. The closure's environment fixes the length of arrays acceptable as the second argument to $(\underline{+}^{\uparrow})$. We define the application of an array closure of type $\alpha \Rrightarrow \beta$ to a value $xs :: [{:}\alpha{:}]$ as $\langle\!\langle{:}(f,\ f^{\uparrow}),\ es{:}\rangle\!\rangle \ddagger xs \ = \ f^{\uparrow}\ (es,\ xs)$.

As we will see in the rest of the paper, this representation directly supports the full range of array operations. For instance, standard closures can be replicated to array closures simply by replicating the environment:

$$\mathit{replicateP}\ n\ \langle\!\langle (f, f^{\uparrow}),\ e\rangle\!\rangle \ = \ \langle\!\langle{:}(f, f^{\uparrow}),\ \mathit{replicateP}\ n\ e{:}\rangle\!\rangle$$

Conversely, we obtain a standard closure from an array closure by indexing:

$$\langle\!\langle{:}(f, f^{\uparrow}), es{:}\rangle\!\rangle\ !{:}\ n \ = \ \langle\!\langle (f, f^{\uparrow}),\ es\ !{:}\ n\rangle\!\rangle \qquad \text{— !: is indexing of parallel arrays}$$

NB: To convert between standard closures and array closures, it is crucial that both closure types include the standard and the lifted version of the function.

Pivotal to our running example is the ability to *packP* array closures:

$$\mathit{packP}\ bs\ \langle\!\langle{:}(f, f^{\uparrow}),\ es{:}\rangle\!\rangle \ = \ \langle\!\langle{:}(f, f^{\uparrow}),\ \mathit{packP}\ bs\ es{:}\rangle\!\rangle$$

More generally, an array closure $\langle\!\langle{:}(f, f^{\uparrow}),\ es{:}\rangle\!\rangle$ represents $\mathit{mapP}\ (\mathit{curry}\ f)\ es$ and, therefore, satisfies the usual parametric *map* laws [8]. Thus, a wide range of polymorphic array operations, including permutations, can be implemented analogously to *packP* by propagating the operation to the environment.

Now, let us return to the problematic function *bar* of the previous section. This time, we replace the two function arguments to $\mathit{either}^{\uparrow}$ by array closures:

$$\begin{aligned} \mathit{bar}^{\uparrow}\ (xs,\ ys) &= \mathit{either}^{\uparrow}\ \langle\!\langle{:}((\underline{+}),\ (\underline{+}^{\uparrow})),\ ys{:}\rangle\!\rangle\ \langle\!\langle{:}(\mathit{id}_0,\ \mathit{id}_0^{\uparrow}),\ [{:}(),\ (),\ (){:}]{:}\rangle\!\rangle\ xs \\ \textbf{where}\ \ \mathit{id}_0 &= \lambda(v_1, v_2).v_2 \\ \mathit{id}_0^{\uparrow} &= \lambda(vs_1, vs_2).vs_2 \end{aligned}$$

and re-define $\mathit{either}^{\uparrow}$ such that it packs its first two arguments:

$$\begin{aligned} &\mathit{either}^{\uparrow}\ f\ g\ (\mathit{sel},\ \mathit{left},\ \mathit{right}) = \\ &\qquad \mathit{combineP}\ \mathit{sel}\ ((\mathit{packP}\ \mathit{sel}\ f) \ddagger \mathit{left})\ ((\mathit{packP}\ (\mathit{mapP}\ \mathit{not}\ \mathit{sel})\ g) \ddagger \mathit{right}) \end{aligned}$$

Now, we can redo the calculation for the *bar* example:

$$\begin{aligned} &\quad \mathit{bar}^{\uparrow}(([{:}\mathit{True}, \mathit{False}, \mathit{True}{:}], [{:}5, 2{:}], [{:}4{:}]), [{:}1, 2, 7{:}]) \\ &= \{\text{Unfolding } \mathit{bar}^{\uparrow}\} \\ &\quad \mathit{either}^{\uparrow}\ \langle\!\langle{:}((\underline{+}), (\underline{+}^{\uparrow})), [{:}1, 2, 7{:}]{:}\rangle\!\rangle\ \langle\!\langle{:}(\mathit{id}_0, \mathit{id}_0^{\uparrow}), [{:}(), (), (){:}]{:}\rangle\!\rangle \\ &\qquad ([{:}\mathit{True}, \mathit{False}, \mathit{True}{:}], [{:}5, 2{:}], [{:}4{:}]) \\ &= \{\text{Unfolding } \mathit{either}^{\uparrow}\} \\ &\quad \mathit{combineP}\ [{:}\mathit{True}, \mathit{False}, \mathit{True}{:}] \\ &\qquad ((\mathit{packP}\ [{:}\mathit{True}, \mathit{False}, \mathit{True}{:}]\ \langle\!\langle{:}((\underline{+}), (\underline{+}^{\uparrow})), [{:}1, 2, 7{:}]{:}\rangle\!\rangle) \ddagger [{:}5, 2{:}]) \\ &\qquad ((\mathit{packP}\ [{:}\mathit{False}, \mathit{True}, \mathit{False}{:}]\ \langle\!\langle{:}(\mathit{id}_0, \mathit{id}_0^{\uparrow}), [{:}(), (), (){:}]{:}\rangle\!\rangle) \ddagger [{:}4{:}]) \\ &= \{\text{Applying } \mathit{packP}\} \\ &\quad \mathit{combineP}\ [{:}\mathit{True}, \mathit{False}, \mathit{True}{:}] \\ &\qquad \langle\!\langle{:}((\underline{+}), (\underline{+}^{\uparrow})), [{:}1, 7{:}]{:}\rangle\!\rangle \ddagger [{:}5, 2{:}])\ ((\langle\!\langle{:}(\mathit{id}_0, \mathit{id}_0^{\uparrow}), [{:}(){:}]{:}\rangle\!\rangle \ddagger [{:}4{:}]) \end{aligned}$$

$= \{\text{Unfolding } \ddagger\}$
$\quad combineP\ [{:}\mathit{True}, \mathit{False}, \mathit{True}{:}]\ ((\underline{+}^{\uparrow})\ ([{:}1, 7{:}], [{:}5, 2{:}]))\ (id_0^{\uparrow}\ ([{:}(){:}], [{:}4{:}]))$
$= \{\text{Applying } \underline{+}^{\uparrow} \text{ and } id_0^{\uparrow}\}$
$\quad combineP\ [{:}\mathit{True}, \mathit{False}, \mathit{True}{:}]\ [{:}6, 9{:}]\ [{:}4{:}]$
$= \{\text{Applying } combineP\}$
$\quad [{:}6, 4, 9{:}]$

In summary, to flatten higher-order programs, we first apply *closure conversion* to make closures explicit, hence admitting the manipulation of closures as first-class values. All remaining function values (which now only occur inside closures) are supercombinators; i.e, closed functions, which also only use supercombinators inside. Moreover, we use array closures as the representation of arrays of functional values, so that we can perform standard array operations, such as *packP*, on lifted partial applications and arrays of functions.

## 4 Combining Closures

Data parallelism dictates that an array closure $\langle\!\langle{:}(f, f^{\uparrow}),\ e{:}\rangle\!\rangle$ represents the partial application of a single function to multiple arguments. Many array operations, such as *replicateP*, *mapP*, and *packP*, maintain this property, but unfortunately some don't. For instance, consider

$$combineP\ [{:}\mathit{True}, \mathit{False}, \mathit{True}{:}]\ \langle\!\langle{:}((\underline{+}), (\underline{+}^{\uparrow})), [{:}1, 2{:}]{:}\rangle\!\rangle\ \langle\!\langle{:}((\underline{*}), (\underline{*}^{\uparrow})), [{:}3{:}]{:}\rangle\!\rangle$$

The result is of type $\mathit{Int} \Rrightarrow \mathit{Int}$ and represents the partial applications of both addition and multiplication. In a data parallel environment, can we express this as a partial application of a *single* function?

The following observation suggests a solution: Elementwise application of a combined closure to an argument array is equivalent to splitting the argument array, applying the two closures individually, and combining the results:

$\quad (combineP\ [{:}\mathit{True}, \mathit{False}, \mathit{True}{:}]\ \langle\!\langle{:}((+), (+^{\uparrow})), [{:}1, 2{:}]{:}\rangle\!\rangle\ \langle\!\langle{:}((*), (*^{\uparrow})), [{:}3{:}]{:}\rangle\!\rangle)$
$\quad \ddagger\ [{:}4, 5, 6{:}]$
$= \{\text{Splitting the argument array}\}$
$\quad combineP\ [{:}\mathit{True}, \mathit{False}, \mathit{True}{:}]$
$\quad\quad (\langle\!\langle{:}((+), (+^{\uparrow})), [{:}1, 2{:}]{:}\rangle\!\rangle\ \ddagger\ [{:}4, 6{:}])\ (\langle\!\langle{:}((*), (*^{\uparrow})), [{:}3{:}]{:}\rangle\!\rangle\ \ddagger\ [{:}5{:}])$
$= \{\text{Merging the results}\}$
$\quad [{:}5, 15, 8{:}]$

This scheme is easily encoded by a closure $\langle\!\langle{:}(capp, capp^{\uparrow}), es{:}\rangle\!\rangle$ whose environment *es* contains the two original closures along with the selector. The function $capp^{\uparrow}$, defined below, uses the selector to split the argument and combines the result of applying the two closures.

Crucially, this approach fits nicely with the rest of the flattening transformation, as the environment can be directly represented by a parallel array of type $[{:}\mathit{Either}\ (\mathit{Int} \Rightarrow \mathit{Int})\ (\mathit{Int} \Rightarrow \mathit{Int}){:}]$. As described in Section 2, this is flattened to the product $([{:}\mathit{Bool}{:}], \mathit{Int} \Rrightarrow \mathit{Int}, \mathit{Int} \Rrightarrow \mathit{Int})$ which stores precisely the selector and the two closures and is constructed by *combineP* as follows:

$$combineP\ bs\ fs\ gs = \langle\!\langle{:}(capp, capp^{\uparrow}), (bs, fs, gs){:}\rangle\!\rangle$$

where we define the split/combine procedure as follows

$$\begin{array}{l} capp^{\uparrow} :: (([{:}Bool{:}], \alpha \Rightarrow \beta, \alpha \Rightarrow \beta), [{:}\alpha{:}]) \rightarrow [{:}\beta{:}] \\ capp^{\uparrow}\ ((bs, fs, gs), xs) = \\ \quad combineP\ bs\ (fs \ddagger packP\ bs\ xs)\ (gs \ddagger packP\ (mapP\ not\ bs)\ xs) \end{array}$$

Our definition of $capp^{\uparrow}$ sequentialises the two applications of *fs* and *gs*. This is a natural consequence of data parallelism, as *fs* and *gs* will generally be entirely different computations. This corresponds closely to $either^{\uparrow}$ from Section 3.

The result of *combineP* is a regular array closure and, thus, naturally supports all array operations. For instance, packing clearly has the desired semantics by propagating the operation into the environment. Moreover, indexing extracts a single element from the environment by, depending on the index, selecting it from either of the two closures. Depending on which closure is selected, the result will be embedded into a *Left* or *Right* constructor. Applying the resulting closure to an argument invokes *capp* which is implemented as

$$\begin{array}{lcl} capp & :: & (Either\ (\alpha \Rightarrow \beta)\ (\alpha \Rightarrow \beta),\ \alpha) \rightarrow \beta \\ capp\ (h,\ x) & = & \textbf{case}\ h\ \textbf{of}\ \{Left\ f \rightarrow f \dagger x;\ Right\ g \rightarrow g \dagger x\} \end{array}$$

In fact, $capp^{\uparrow}$ is, as expected, precisely the lifted version of *capp*; i.e., we can derive $capp^{\uparrow}$ from *capp* by vectorisation.

This idea can be extended to other joining operations, such as concatenation, which is a special case of combining, but we only have space to discuss *combineP*.

## 5 Nesting Closures

The purpose of the flattening transformation is the elimination of nested parallelism, of which the archetypal example is the nested application of *mapP*, as in $mapP\ (mapP\ (1+))\ [{:}[{:}1,2{:}],[{:}{:}],[{:}3,4,5{:}]{:}]$, which increments every integer element while preserving the array's nesting structure. As discussed in Section 1, the nested array of our example is implemented by the flattened representation $([{:}2,0,3{:}],[{:}1,2,3,4,5{:}])$, where the the segment descriptor $[{:}2,0,3{:}]$ encodes the nesting structure. In the flat representation, we can increment all integer elements without affecting the nesting structure by simply applying the non-nested $mapP\ (+1)$ to the second component of the representation. So, we have

$$mapP\ (mapP\ f)\ (segd,\ xs) = (segd,\ mapP\ f\ xs)$$

In the following, we discuss how to realise this with array closures.

After closure conversion, *mapP* expects a closure as its first argument, which we simply replicate to get an array closure that we can apply as follows:

$$\begin{array}{lcl} mapP & :: & (\alpha \Rightarrow \beta) \rightarrow [{:}\alpha{:}] \rightarrow [{:}\beta{:}] \\ mapP\ c\ xs & = & replicateP\ (lengthP\ xs)\ c \ddagger xs \end{array}$$

The partial application $mapP\ (1+)$ in our nested example ultimately evaluates to $\langle\!\langle(\underline{mapP}, \underline{mapP^{\uparrow}}), \langle\!\langle((\underline{+}), (\underline{+^{\uparrow}})), 1\rangle\!\rangle\rangle\!\rangle$, which we map over the nested array:

$$mapP\ (\langle\!\langle(\underline{mapP}, \underline{mapP^{\uparrow}}), \langle\!\langle((\underline{+}), (\underline{+^{\uparrow}})), 1\rangle\!\rangle\rangle\!\rangle)\ ([{:}2,0,3{:}],[{:}1,2,3,4,5{:}])$$

$$
\begin{aligned}
&= \{\text{Unfolding } \mathit{mapP}\}\\
&\quad \mathit{replicateP}\ 3 \langle\!\langle(\underline{\mathit{mapP}}, \underline{\mathit{mapP}^{\uparrow}}), \langle\!\langle((\underline{+}), (\underline{+}^{\uparrow})), 1\rangle\!\rangle\rangle\!\rangle \ddagger ([{:}2, 0, 3{:}], [{:}1, 2, 3, 4, 5{:}])\\
&= \{\text{Unfolding } \mathit{replicateP} \text{ twice}\}\\
&\quad \langle\!\langle{:}(\underline{\mathit{mapP}}, \underline{\mathit{mapP}^{\uparrow}}), \langle\!\langle{:}((\underline{+}), (\underline{+}^{\uparrow})), [{:}1, 1, 1{:}]{:}\rangle\!\rangle{:}\rangle\!\rangle \ddagger ([{:}2, 0, 3{:}], [{:}1, 2, 3, 4, 5{:}])\\
&= \{\text{Unfolding } \ddagger\}\\
&\quad \underline{\mathit{mapP}^{\uparrow}}\ (\langle\!\langle{:}((\underline{+}), (\underline{+}^{\uparrow})), [{:}1, 1, 1{:}]{:}\rangle\!\rangle, ([{:}2, 0, 3{:}], [{:}1, 2, 3, 4, 5{:}]))
\end{aligned}
$$

Not surprisingly, the two nested applications of *mapP* are resolved to a single application of $\mathit{mapP}^{\uparrow}$. The implementation of the latter can be derived by lifting the definition of *mapP*; here, we present a slightly simplified version:

$$
\begin{aligned}
&\mathit{mapP}^{\uparrow} :: (\alpha \Rrightarrow \beta) \to ([{:}\mathit{Int}{:}], [{:}\alpha{:}]) \to ([{:}\mathit{Int}{:}], [{:}\beta{:}])\\
&\mathit{mapP}^{\uparrow}\ \langle\!\langle{:}(f, f^{\uparrow}), \mathit{es}{:}\rangle\!\rangle\ (\mathit{segd}, \mathit{xs}) = (\mathit{segd}, \langle\!\langle{:}(f, f^{\uparrow}), \mathit{expandP}\ \mathit{segd}\ \mathit{es}{:}\rangle\!\rangle \ddagger \mathit{xs})\\
&\quad \textbf{where}\\
&\qquad \mathit{expandP}\ [{:}n_1, \ldots, n_k{:}]\ [{:}x_1, \ldots, x_k{:}] = [{:}\underbrace{x_1, \ldots, x_1}_{n_1 \text{ times}}, \ldots, \underbrace{x_k, \ldots, x_k}_{n_k \text{ times}}{:}]
\end{aligned}
$$

Conceptually, the mapped array closure contains one element for each subarray of the nested array. Hence, we blow the array closure up, such that each element is replicated as often as the length of the corresponding subarray dictates—this is the job of *expandP*. Afterwards, we can simply apply the expanded array closure to the representation of the nested array. The nesting structure remains invariant under this operation. Returning to our example, we now have:

$$
\begin{aligned}
&\quad \mathit{mapP}^{\uparrow}\ \langle\!\langle{:}((\underline{+}), (\underline{+}^{\uparrow})), [{:}1, 1, 1{:}]{:}\rangle\!\rangle\ ([{:}2, 0, 3{:}], [{:}1, 2, 3, 4, 5{:}])\\
&= \{\text{Unfolding } \mathit{mapP}^{\uparrow}\}\\
&\quad ([{:}2, 0, 3{:}],\ \langle\!\langle{:}((\underline{+}), (\underline{+}^{\uparrow})), [{:}1, 1, 1, 1, 1{:}]{:}\rangle\!\rangle \ddagger [{:}1, 2, 3, 4, 5{:}])\\
&= \{\text{Unfolding } \ddagger\}\\
&\quad ([{:}2, 0, 3{:}],\ [{:}2, 3, 4, 5, 6{:}])
\end{aligned}
$$

As expected, the result is the flat representation of $[{:}[{:}2, 3{:}], [{:}{:}], [{:}4, 5, 6{:}]{:}]$; moreover, it has been computed in one parallel step.

## References

1. G. E. Blelloch. *Vector Models for Data-Parallel Computing.* MIT Press, 1990.
2. G. E. Blelloch. Programming parallel algorithms. *CACM*, 39(3):85–97, 1996.
3. M. M. T. Chakravarty and G. Keller. More types for nested data parallel programming. In P. Wadler, editor, *5th ACM SIGPLAN Intl. Conf. on Functional Programming (ICFP'00)*, pages 94–105. ACM Press, 2000.
4. M. M. T. Chakravarty and G. Keller. An approach to fast arrays in Haskell. In J. Jeuring and S. Peyton Jones, editors, *Lecture notes for The Summer School on Advanced Functional Programming '02*, LNCS 2638, 2003.
5. M. M. T. Chakravarty, G. Keller, R. Lechtchinsky, and W. Pfannenstiel. Nepal—Nested data parallelism in Haskell. In R. Sakellariou, J. Keane, J. R. Gurd, and L. Freeman, editors, *Euro-Par 2001: Parallel Processing*, number 2150, pages 524–534. Springer-Verlag, 2001.

6. R. Leshchinskiy. *Higher-Order Nested Data Parallelism: Semantics and Implementation.* PhD thesis, Technische Universität Berlin, 2005.
7. Y. Minamide, G. Morrisett, and R. Harper. Typed closure conversion. In *POPL '96: Proc. of the 23rd ACM SIGPLAN-SIGACT Sym. on Principles of Programming Languages*, pages 271–283. ACM Press, 1996.
8. P. Wadler. Theorems for free! In *FPCA'89: Proceedings 4th International Conference on Functional Programming Languages and Computer Architecture*, pages 347–359, New York, 1989. ACM Press.

# Combining Measurement and Stochastic Modelling to Enhance Scheduling Decisions for a Parallel Mean Value Analysis Algorithm

Gagarine Yaikhom, Murray Cole, and Stephen Gilmore

School of Informatics, The University of Edinburgh,
James Clerk Maxwell Building, Mayfield Road, Edinburgh EH9 3JZ, UK
G.Yaikhom@sms.ed.ac.uk,{mic, stg}@inf.ed.ac.uk
http://groups.inf.ed.ac.uk/enhance/

**Abstract.** In this paper we apply the high-level modelling language PEPA to the performance analysis of a parallel program with a pipeline skeleton which computes the Mean Value Analysis (MVA) algorithm for queueing networks.

## 1 Introduction

From the algorithmic and performance perspectives, Grid systems are characterised by the dynamically heterogeneous nature of the resources at the programmer's disposal. This property adds a new dimension to the already challenging parallel programming task. The Enhance project [1, 2] aims to simplify the effective programming of such systems by exploiting and synthesising results from two underlying research programmes. Stochastic process algebras such as PEPA [3] are already used to model the behaviour of concurrent systems in which some aspects of behaviour are not precisely predictable. Meanwhile, skeletal parallel programming [4] recognises that many real applications draw their structure from a range of well known solution paradigms and seeks to make it easy for an application developer to tailor such a paradigm to a specific problem.

The choice of a particular skeleton carries with it considerable information about implied scheduling dependencies. In response to changing workload on servers, or unplanned unavailability due to software or hardware faults, the application can be restructured to use an alternative implementation skeleton or can simply reevaluate the parameters to the skeleton which is currently in use. Such resilience to operational faults is not typically found in low-level parallel programming approaches and is one of the strengths of a structured approach to parallel programming. Furthermore, the use of skeletons allows the programmer to provide explicit information about the future interaction structure of the application, which would be difficult or impossible to derive statically from an equivalent unstructured program source. By modelling skeletons with PEPA, and thereby being able to address aspects of uncertainty which are inherent to Grid computing, we aim to underpin systems which can make better scheduling and dynamic rescheduling decisions than less sophisticated approaches.

V.N. Alexandrov et al. (Eds.): ICCS 2006, Part II, LNCS 3992, pp. 929–936, 2006.

The ability to reconfigure and redeploy an application on-the-fly can be used even more effectively if the future load on the available servers can be estimated at least tolerably well. The Network Weather Service (NWS) [5] provides us with these estimates. Analogously with meteorological climate prediction, a short-range forecast is made on the basis of a record of recent activity. Perfect forecasting of future load is, of course, not achievable but in the experience of the users of the NWS, its predictions turn out to be useful more often than they are misleading and so scheduling decisions based on NWS information can be favourable ones.

In previous papers [1, 2], we have reported on development of our *pipeline* and *deal* skeletons, on their associated PEPA models, and upon the construction of AMoGeT, a modelling tool which automatically generates PEPA models for the pipelines (optionally with embedded deals), given information on the performance of the customising software components, and on resource performance obtained dynamically from NWS [5]. These PEPA models are solved using procedures of numerical linear algebra implemented in the PEPA Workbench [6].

We summarise the main technical ideas of the work here to make the present paper self-contained. Briefly, the PEPA process algebra is a timed process algebra where the duration of activities is modelled by exponentially-distributed random variables. Processes can choose between alternatives, and can be composed in parallel. Processes in a composition can be required to synchronise on shared activities. For example, a producer can enter a job into a pipeline only if the first stage is willing to accept it, and a stage in the pipeline can pass on a completed job only if a consumer is ready to receive it. Both of these are synchronisations. The slower partner in the synchronisation determines the rate of the joint activity.

Timed activities, choice, parallel composition and synchronisations are the modelling tools offered by the PEPA language. The analysis procedures are implemented in the PEPA Workbench software tool. We use PEPA models which denote finite-state stochastic (in fact, Markovian) processes. A PEPA model submitted to the Workbench will be compiled into a Continuous-Time Markov Chain (CTMC) and the long run ("steady state") probability distribution over all of the states of the model is computed using an iterative numerical procedure. From this can be calculated performance metrics such as the throughput of the pipeline, which is the basis of the comparison which allows us to rank scheduling decisions.

This paper presents the first serious application of our technology. We have taken an existing MPI program [7] for the Mean Value Analysis problem, refactored it as an instance of our pipeline skeleton (replacing the explicit MPI with our pipeline skeleton) and used AMoGeT to deploy the resulting program across a subset of the available processors of our local Beowulf cluster, while deliberately skewing the performance characteristics of the processors with a variety of artificial loads. Section 2 reviews the MVA problem and its pipelined parallelisation. Section 3 describes our experimental plan, and the expected results. Section 4 describes and discusses the actual results.

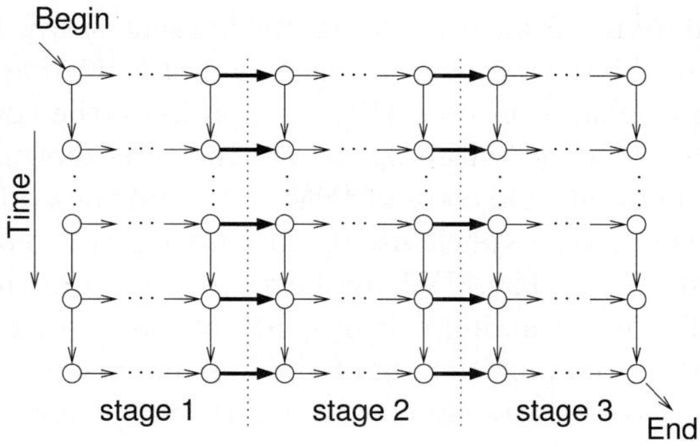


(a)

```
long int m[100]; double a[100];
void main (void) {
    int i;
    for (i = 0; i < 100; i++) {
        a[i] = (double) random();
        m[i] = random();
    }
    for (i = 0; i < 100; i++) {
        a[i] = a[100 − i] ∗ m[i];
        if (i == 99) i = 0;
    }
}
```

(b)

**Fig. 1.** (a) Pipeline implementation of the MVA algorithm, (b) load generation

## 2 The Mean Value Analysis Algorithm

The application under investigation is a numerical procedure which is used in the quantitative analysis of queueing networks. A queueing network is used to describe a system as a collection of service centres visited by customers attempting to fulfil their service requirements. The customers under study may very often be jobs executing on a variety of servers. The objective is to determine meaningful performance metrics about the system such as average queue length and utilisation of servers. This information can be used to rectify performance problems which stem from the design of the network. Typically this will involve network re-structuring or the introduction of additional servers in order to decrease mean queue length and increase utilisation.

Three distinct types of analysis can be applied to queueing networks, depending on their structure and the service discipline used at service centres. Firstly, closed-form analytic solutions exist for open queueing networks for a limited class of service disciplines if customers do not re-enter the network after their service is complete. Secondly, numerical algorithms may be applied to open or closed product-form networks with Poisson arrivals and exponentially-distributed service times with the possibility for customers to loop back and re-enter the system. Finally, Monte Carlo discrete-event simulation may be applied in other cases. These three types of analysis have increasing evaluation cost. We are considering the second type here.

Numerical algorithms compute the probability of a network of $M$ service centres being in a state $\mathbf{n} = (n_1, n_2, \ldots, n_M)$ using a descriptor of the form

$$\Pr(\mathbf{n}) = \frac{1}{G} \prod_{i=1}^{M} f_i(n_i) \qquad (1)$$

where $G$ is a normalisation constant for the probability values and $f$ is a function which depends upon the type of the service centre [8].

The Mean Value Analysis (MVA) algorithm [9] computes performance metrics such as utilisation and average queue length for such queueing networks using increasing customer populations, from 0 to $N$. In the present study we base our skeletal program on an existing parallel implementation of MVA due to Gennaro and King [7]. Their implementation uses MPI to coordinate the computation and has the structure of a pipeline, with task dependencies as shown in Figure 1(a). We have taken the computational cores of their code, and embedded them in a skeleton program, replacing the explicit use of MPI with a call to our pipeline skeleton (shown in Figure 2(a)). The MVA application is novel in that the number of stages in the pipeline is not an implicit property of the algorithm, but is dictated by the granularity of the partitioning of the implicit task graph (Figure 1(a)). As illustrated, we have arbitrarily chosen a three stage version. Changing the granularity is trivial (redefining a constant) in both original MPI and skeletal versions of the code.

## 3 Experimental Plan

Our immediate goal is to demonstrate experimentally that AMoGeT can use its PEPA pipeline model, informed by dynamically gathered resource performance information, to choose a good allocation of pipeline stages to processors. To this end, we have chosen to work with a three stage version of the MVA pipeline, and a five processor subset of a Beowulf cluster. These numbers are small enough to allow us to ask AMoGeT to generate and evaluate all possible schedules (which three processors to use, and in which order). As the implied search space grows exponentially, exhaustive search will eventually become prohibitive. Pruning the search space is an orthogonal challenge. The work reported here demonstrates that AMoGeT provides a sound objective function upon which to base heuristics.

Our experiments are grouped into sets of six runs. In each set, we artificially create situations in which zero, one, two, three, four and five of the available processors are seeded with heavy computational loads. Without loss of generality, and to assist the reader, we choose processors with lowest index (or, rank) when loading processors. For example, to load three processors we choose processors 1, 2, and 3. Since our Beowulf currently has a "free-for-all" allocation policy, there may be other sporadic communication and computation loads placed by other users.

For each run, in each set, we ask AMoGeT to gather information and predict which three of the five available processors will provide the best and worst overall application performance (remembering that in a pipeline, this choice should be affected by both computation and communication resource availability). We then run (in sequence) these two schedules, noting the execution times on each of the selected processors.

For each set of experiments, we present a total of 36 measured run times. These are grouped into six pairs of three measurements (each measurement corresponding to a pipeline stage). Each pair corresponds to experiments with a fixed number of "bad" processors, grouped left to right in the graph from zero bad processors to five bad processors. Within each pair, to the left we find the real run times on the three processors in the predicted best case, and to the right the real run times on the three processors in the predicted worst case. For example, in Figure 2(b), the second pair of runs (indexed (2,5,4) (5,1,3)) indicate that it was predicted that using processors 2, 5 and 4 would give the best performance and that using processors 5, 1 and 3 would give the worst performance. The height of the bars gives the actual run time on each of these processors in the corresponding run, and the figure of 63.29% at the top indicates the improvement from worst run to best run.

We present results for two such sets of experiments. In the first set, conducted on a cluster with low background interference from other users, our "bad" processors were artificially loaded to have around 33% of the availability of the "good" processors. In the second set, conducted on a cluster with higher background interference from other users, the "bad" processors' availability is improved to around 66%.

An intuitive assessment of the situation, based on conventional wisdom about pipeline performance, might suggest the following outcomes:

- we should choose three "good" processors, when these are available;
- the gap between best and worst case schedules should be widest in those situations for which it is possible to choose all "good" processors in the best case, and at least one "bad" processor in the worst case (which for our set-up occurs only in the "one bad processor" and "two bad processors" runs);
- both of the preceding observations may be tempered for situations in which the performance of computationally "good" processors is hampered by excessively bad communications performance to intended neighbours.

## 4 Experimental Outcomes

The experiments were conducted on two separate Beowulf clusters (16 nodes and 64 nodes) with RedHat Linux Fedora Core 3 release, version 2.6.12. We used LAM-MPI implementation of the MPI standard, version 7.1.1. For extracting CPU availability per node, and inter-node TCP communication latency, we used the Network Weather Service (NWS), version 2.13.

The number of classes in the queueing network is set to three, with each class having a population of 1024. The number of resources in the queueing network is set to two resource queues. For the artificial load simulation, we use the program shown in Figure 1(b), adapted from [10, page 171].

The experimental results are shown in Figure 2. The results show that the mapping suggested by our approach produces satisfactory improvements, as expected. The results for the cluster with low background interference and CPU availability reduced to 33%, show improvements of up to 63% when stages can be mapped to unloaded nodes. This choice is reflected in the corresponding mapping. When more than two nodes are loaded, our model still chooses the best possible mapping, producing an improvement of around 10%. The results for the cluster with more background interference and CPU availability reduced to 66%, show improvements of up to 29%. Because of the sporadic interference, the

(a)

```
void first_stage (void) {
    do { /* Perform computations and update queue lengths. */
        if (node_idx[0] == end) PipeGive (tmp_qlen, dtype, 1); /* Send queue lengths. */
    } while (update_node_idx()); /* Continue if not done. */ }
void last_stage (void) {
    do { if (node_idx[0] == start) PipeTake (tmp_qlen, dtype, 1); /* Receive queue lengths. */
        /* Perform computations and updates. */ } while (update_node_idx());
    /* Generate output. */ }
void inter_stage (void) {
    do {/* Receive, compute, update and send queue lengths. */} while (update_node_idx()); }
int main (int argc, char *argv[]) {
    SkelLibInit(); /* Initialise skeleton library. */
    for (i = 0; i < NSTAGES+1; i++) {spreads[i] = SPLOCAL; types[i] = dtype;} /* Data map. */
    for (i = 0; i < NSTAGES; i++) amodes[i] = SKEL; /* Stage type. */
    mystage = myrank(); /* Process to stage mapping. */
    stages[0] = first_stage; stages[NSTAGES-1] = last_stage; /* Assign stage function. */
    for (i = 1; i < NSTAGES - 1; i++) stages[i] = inter_stage; /* Assign stage function. */
    Pipeline(NSTAGES, amodes, stages, mystage, BUF, spreads, types,
            NULL, 0, 0, NULL, 0, &outmul, 0, mycomm()); /* Execute pipeline. */
    /* Output results, and finalise. */
}
```

(b)

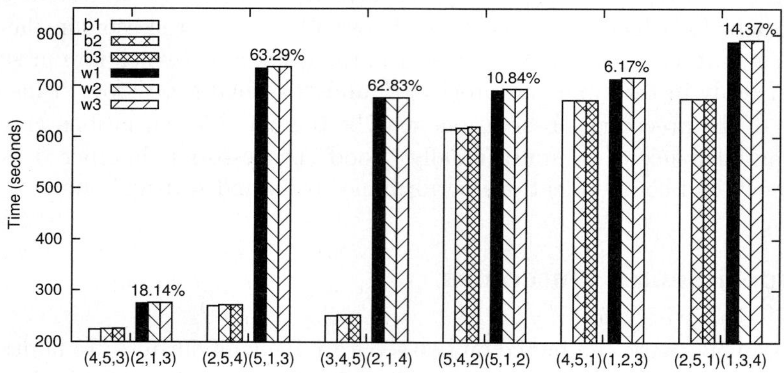


(c)

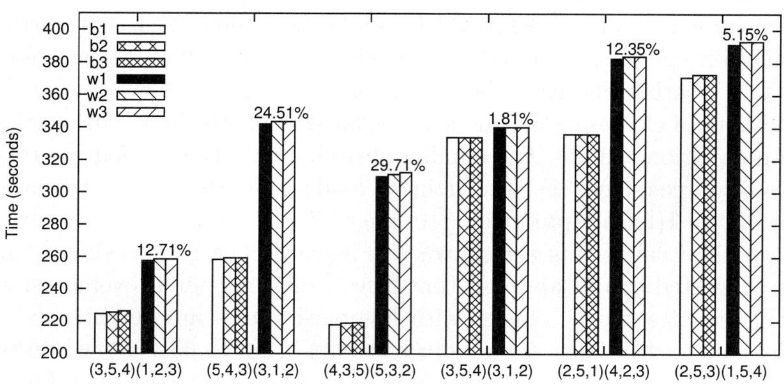


**Fig. 2.** (a) Implementation source code (simplified), (b) 33 percent availability (16 node cluster with low background interference), (c) 66 percent availability (64 node cluster with higher background interference). $b_i$ and $w_i$ respectively correspond to the best and worst stage $i$ mapping.

performance improvements display some variation, giving an average improvement of about 8% when all nodes are unloaded, or one or all of the stages are mapped to loaded nodes. Considering the average execution time of 40 seconds to generate and solve the pipeline PEPA models for the 60 possible mappings, our approach proves to be pragmatic and advantageous.

In contrast to our use of computational load generation, we made no attempt to impose artificial communication load. Nevertheless, detailed inspection of the NWS readings (not reproduced here) reveals that real imbalances in communication load arose from the background on a number of occasions, sometimes with interesting effects on predicted best and worst schedules. For example, in Figure 2(c), in the two loaded processor case (these being processors 1 and 2, as explained above), there is an unusual spike in predicted latency between processor 5 and processor 3. The predicted worst case schedule (5,3,2) chooses to exploit this weakness, even at the expense of not choosing the weak processor 1. Similarly, in the (rightmost) "all" loaded case, there is a very high spike in the latency between processors 1 and 5. Again this is exploited in the worst case and avoided in the best, although the difference in actual run times between these cases is smaller than might have been expected. It may be that this variation is attributable to the inevitable distinction between predicted latencies, as used in scheduling, and actual latencies experienced at run time.

## 5 Conclusions

In this paper we have applied performance analysis methods based on stochastic process algebra to an exemplar of the class of structured parallel programs with a pipeline configuration of tasks. Our exemplar is the parallel implementation of the MVA algorithm for queueing networks. Our methods are general ones: we are using MVA simply as an example pipeline-structured application. Thus we have not made alterations to our methods to take advantage of specific features of MVA which are not usually found in a pipeline-structured parallel program.

By basing our performance predictions on measurements of network latency and compute load we are able to predict ahead-of-run those scheduling decisions which will lead to the shortest run-time and those which lead to the longest run-time. In the present study we did not find a single instance where the performance predictions computed by the PEPA Workbench were misleading: predicted best cases always significantly outperformed predicted worst cases in practice. Our analysis is fully automatic and runs without user intervention. The analysis cost is modest. This is essential for maintaining the freshness of the measurement data and so that the scheduling overhead does not impede the progress of the main computation. The relative speed of the analysis points towards its suitability for use in on-line scheduling decisions.

In the present paper our dynamic scheduling is performed *before run*. Our eventual plan is to allow dynamic rescheduling *within a run* so that downstream computations benefit from measurements obtained during the upstream computation.

## 6 Related Work

The combination of stochastic modelling, structured parallelism and dynamically derived performance monitoring used to inform our scheduling approach is, we believe, unique. Previous skeletal and similarly structured systems have used analytical performance models to guide scheduling on homogenous resources [11, 12, 13], and information gathered from NWS to guide scheduling in heterogenous contexts [14].

*Acknowledgements.* The authors thank Jane Hillston for helpful discussions and Horacio Gonzalez-Velez for helpful suggestions on the usage of NWS. The Enhance project is funded by the UK EPSRC grant GR/S21717/01.

## References

1. Benoit, A., Cole, M., Gilmore, S., Hillston, J.: Evaluating the performance of skeleton-based high level parallel programs. In: ICCS 2004: Proceedings of the 4th International Conference on Computational Science, Kraków, Poland. Volume 3038 of LNCS., Springer-Verlag (2004) 289–296
2. Benoit, A., Cole, M., Gilmore, S., Hillston, J.: Scheduling skeleton-based grid applications using PEPA and NWS. The Computer Journal 48(3):369-378, 2005. **48** (2005) 369–378
3. Hillston, J.: A Compositional Approach to Performance Modelling. CUP (1996)
4. Cole, M.: Algorithmic Skeletons: Structured Management of Parallel Computation. MIT Press & Pitman (1989)
5. Wolski, R., Spring, N., Hayes, J.: The Network Weather Service: a distributed resource performance forecasting service for metacomputing. Future Generation Computer Systems **15** (1999) 757–768
6. Gilmore, S., Hillston, J.: The PEPA Workbench: A Tool to Support a Process Algebra-based Approach to Performance Modelling. In: Proceedings of the Seventh International Conference on Modelling Techniques and Tools for Computer Performance Evaluation. Volume 794 of LNCS., Vienna (1994) 353–368
7. Gennaro, C., King, P.: Parallelising the Mean Value Analysis Algorithm. Transactions of The Society for Computer Systems International **16** (1999) 16–22
8. Baskett, F., Chandy, K., Muntz, R., Palacios, F.: Open, Closed, and Mixed Networks of Queues with Different Classes of Customers. JACM **22** (1975) 248–260
9. Reiser, M., Lavenberg, S.: Mean Value Analysis of Closed Multichain Queueing Networks. JACM **27** (1980) 313–322
10. Gunther, N.J.: Analyzing Computer System Performance with Perl::PDQ. Springer (2005)
11. Scaife, N., Horiguchi, S., Michaelson, G., Bristow, P.: A parallel SML compiler based on algorithmic skeletons. Journal of Functional Programming **15** (2005) 615–650
12. Morajko, A., Cesar, E., Caymes-Scutari, P., Margalef, T., Sorribes, J., Luque, E.: Automatic Tuning of Master Worker Applications. In: Proceedings of Euro-Par 05. Volume 3648 of LNCS., Springer-Verlag (2005) 95–103
13. Cosmo, R.D., Li, Z., Pelagatti, S., Weis, P.: Skeletal parallel programming with ocamlp3l 2.0. In: To appear in Parallel Processing Letters. (2006)
14. Gonzalez-Velez, H.: An Adaptive Skeletal Task Farm for Grids. In: Proceedings of Euro-Par 05. Volume 3648 of LNCS., Springer-Verlag (2005) 401–410

# Joint Structured/Unstructured Parallelism Exploitation in `muskel`*

M. Danelutto[1,4] and P. Dazzi[2,3,4]

[1] Dept. Computer Science, University of Pisa, Italy
[2] ISTI/CNR, Pisa, Italy
[3] IMT – Institute for Advanced Studies, Lucca, Italy
[4] CoreGRID Institute on Programming model

**Abstract.** Structured parallel programming promises to raise the level of abstraction perceived by programmers when implementing parallel applications. In the meanwhile, however, it restricts the freedom of programmers to implement arbitrary parallelism exploitation patterns. In this work we discuss a data flow implementation methodology for skeleton based structured parallel programming environments that easily integrates arbitrary, user-defined parallelism exploitation patterns while preserving most of the benefits typical of structured parallel programming models.

## 1 Introduction

Structured parallel programming models provide the user/programmers with native high level parallelism exploitation patterns that can be instantiated, possibly in a nested way, to implement a wide range of applications [6, 10, 3, 2]. In particular, those programming models do not allow programmers to program parallel applications at the "assembly level", i.e. by directly interacting with the distributed execution environment via communication or shared memory access primitives and/or via explicit scheduling and code mapping. Rather, the high-level native, parametric parallelism exploitation patterns provided encapsulate and abstract from all these parallelism exploitation related details. As an example, to implement an embarrassingly parallel application processing all the data items in an input stream or file, the programmers simply instantiates a "task farm" skeleton by providing the code necessary to (sequentially) process each input task item. The system, either a compiler and run time tool based implementation or the library based one, takes care of devising the proper distributed resources to be used, to schedule proper tasks on the resources and to distribute input tasks and gather output results according to the process mapping used. In contrast, when using a traditional system, the programmers have usually to explicitly program code for distributing and scheduling the processes on the available resources and for moving input and output data between the

* This work has been partially supported by Italian national FIRB project no. RBNE01KNFP GRID.it and by the FP6 Network of Excellence CoreGRID funded by the European Commission (Contract IST-2002-004265).

V.N. Alexandrov et al. (Eds.): ICCS 2006, Part II, LNCS 3992, pp. 937–944, 2006.

involved processing elements. The cost of this appealing high-level way of dealing with parallel programs is paid in terms of programming freedom. The programmer is normally not allowed to use arbitrary parallelism exploitation patterns, but he must only use the ones provided by the system, usually including all those reusable patterns that happen to have efficient distributed implementations available. This is mainly aimed at avoiding the possibly for the programmers to write code that can potentially impairs the efficiency of the implementation provided for the available, native parallel patterns. This is a well-known problem. Cole recognized its importance in his "manifesto" paper [6]. In this work we discuss a methodology that can be used to implement mixed parallel programming environments providing the user with both structured and unstructured ways of expressing parallelism while preserving most of the benefits typical of structured parallel programming models. The methodology is based on data flow. Structured parallel exploitation patterns are implemented translating them into data flow graphs executed by a scalable, efficient, distributed macro data flow interpreter (the term *macro* data flow refers to the fact that the computation of a single data flow instruction can be huge computations). Unstructured parallelism exploitation can be achieved by explicitly programming data flow graphs. The system provides suitable ways to interact with the data flows graphs derived from structured pattern compilation in such a way that mixed structured and unstructured parallelism exploitation patterns can be used within the same application.

## 2 Data Flow Implementation of Structured Programming Environments

Muskel is a full Java skeleton programming environment derived from Lithium [2]. Currently, it only provides the stream parallel skeletons of Lithium, namely stateless task farm and pipeline. These skeletons can be arbitrary nested, to program pipelines with farm stages, as an example, and they process a single stream of input tasks to produce a single stream of output tasks. Muskel implements skeletons exploiting data flow technology and Java RMI facilities. The programmer using muskel can express parallel computations simply using the provided `Pipeline` and `Farm` classes. As an example, to express a parallel computation structured as a two-stage pipeline with a farm in each one of the stages, the user should write a code such as the one of Figure 1 left. `f` and `g` are two classes implementing the `Skeleton` interface, i.e. supplying a `compute` method with the signature `Object compute(Object t)` computing `f` and `g`, respectively. In order to execute the program, the programmer first sets up a `Manager` object, tells it which is the program to execute, which is the performance contract required (in this case, the parallelism degree required for the execution), the filename for the input file hosting the input stream items and the filename of the output file that will eventually host the result stream. Then he can ask parallel program execution simply issuing an `eval` call to the manager. When the call terminates, the output file has been produced.

```
...
Skeleton main =
    new Pipeline(new Farm(f),
                 new Farm(g));
Manager manager = new Manager();
manager.setProgram(main);
manager.setContract(new ParDegree(10));
manager.setInputStream(inFName);
manager.setOutputStream(outFName);
manager.eval();
...
```

(a) Sample muskel code

```
DFGraph gh = manager.newGraph(main);
int n = gh.getMaxInstrId() + 1;
int gid = gh.getGraphId();
// MDFi parmas: instrId, graphID, inTokNo, outTok
gh.addDFi(new MDFi(n+1,gid,f1,1,
        new outToken(n+2,1), new outToken(n+4,1)));
gh.addDFi(new MDFi(n+2,gid,g1,1, new outToken(n+3,1)));
gh.addDFi(new MDFi(n+3,gid,g2,1, new outToken(n+4,2)));
gh.addDFi(new MDFi(n+4,gid,h1,2,
        new outToken(skg.getInputTokenIds())));
manager.setStartInstr(n+1);
```

(b) Adding custom MDF graph to skeleton MDF graph

**Fig. 1.** Sample muskel code (left) and custom/standard MDF graph integration (right)

Actually, the `eval` method execution happens in steps. First, the application manager looks for available processing elements using a simplified, multicast based peer2peer discovery protocol, and recruits the required remote processing elements. Each remote processing element runs a data flow interpreter, actually. Then the skeleton program (the `main` of the example) is compiled into a macro data flow graph (actually capitalizing on normal form results shown in [1, 2]) and a thread is forked for each one of the remote processing elements recruited. Then the input file is read. For each task item, an instance of the macro data flow graph is created and the task item token is stored in the proper place (initial data flow instruction(s)). The graph is placed in the task pool, the repository for data flow instructions to be executed. Each thread looks for a fireable instruction in the task pool and delivers it for execution to the associated remote data flow interpreter. The remote interpreter is initialized by being sent the serialized code of the data flow instructions once and forall before the computation actually starts. Once the remote interpreter terminates the execution of the data flow instruction, the thread either stores the result token in the proper "next" data flow instruction(s) in the task pool, or it directly writes the result to the output file. Currently, the task pool is a centralized one, associated with the centralized manager. We are currently investigating the possibility to distribute both task pool and manager, in such a way this bottleneck will eventually disappear. The `manager` takes care of ensuring that the performance contract is satisfied. If a remote node "disappears" (e.g. due to a network failure, or to the node failure/shutdown), the manager looks for another node and starts dispatching data flow instructions to the new node instead [9]. As the manager is a centralized entity, if it fails, the whole computation fails. However, the manager is usually run on the user machine, which is assumed to be safer than the remote nodes recruited as remote interpreter instances.

The policies implemented by the muskel managers are *best effort*. The muskel library tries to do its best to accomplish user requests. In case it is not possible to completely satisfy the user requests, the library accomplishes to establish the closest configuration to the one implicitly specified by the user with the performance contract. In the example above, the library tries to recruit 10 remote interpreters. In case only $n < 10$ remote interpreters are found, the parallelism degree is set exactly to $n$. In the worst case, that is if no remote interpreter is found, the computation is performed sequentially, on the local processing element.

In the current version of the muskel prototype, the only performance contract actually implemented is the `ParDegree` one, asking for the usage of a constant number of remote interpreters in the execution of the program. The prototype has been thought to support at least another kind of contract: the `ServiceTime` one. This contract can be used to specify the maximum amount of time expected between the delivery of two program result tokens. Thus with a line code such as `manager.setContract(new ServiceTime(500))`, the user may ask to deliver one result every half a second (time is in millisecs, as usual in Java). We do not enter in more detail in the implementation of the distributed data flow interpreter here. The interested reader can refer to [8, 9]. Instead, we will try to give a better insight into the compilation of skeleton code into data flow graphs.

A muskel parallel skeleton code is described by the grammar:

$$\mathsf{P} ::= \mathsf{seq}(\mathit{className}) \mid \mathsf{pipe}(\mathsf{P},\mathsf{P}) \mid \mathsf{farm}(\mathsf{P})$$

where the `classNames` refer to classes implementing the `Skeleton` interface, and a macro data flow instruction is a tuple: $\langle id, gid, opcode, \mathcal{I}^n, \mathcal{O}^k\rangle$ where $id$ is the instruction identifier, $gid$ is the graph identifier (both are either integers or the special $NoId$ identifier), $opcode$ is the name of the `Skeleton` class providing the code to compute the instruction (i.e. computing the output tokens out of the input ones) and $\mathcal{I}$ and $\mathcal{O}$ are the input tokens and the output token destinations, respectively. An input token is a pair $\langle value, presenceBit\rangle$ and an output token destination is a pair $\langle destInstructionId, destTokenNumber\rangle$. With this assumptions, a data flow instruction such as $\langle a, b, \mathtt{f}, \langle\langle 123, \mathtt{true}\rangle, \langle \mathtt{null}, \mathtt{false}\rangle\rangle, \langle\langle i, j\rangle\rangle$ is the instruction with identifier $a$ belonging to the graph with identifier $b$. It has two input tokens, one present (the integer 123) and one not present yet. It is not fireable, as one token is missing. When the missing token will be delivered to this instruction, either coming from the input stream or from another instruction, the instruction becomes fireable. To be computed, the two tokens must be given to the `compute` method of the `f` class. The method computes a single result that will be delivered to the instruction with identifier $i$ in the same graph, in the position corresponding to input token number $j$. The process compiling the skeleton program into the data flow graph can therefore be more formally described as follows. We define a pre-compile function $PC[]$ as

$$\begin{aligned}
&PC[\mathsf{seq}(\mathtt{f})]_{gid} = \lambda i.\{\langle newId(), gid, \mathtt{f}, \langle\langle \mathtt{null}, \mathtt{false}\rangle\rangle, \langle\langle i, NoId\rangle\rangle\rangle\}\\
&PC[\mathsf{farm}(\mathsf{P})]_{gid} = C[\mathsf{P}]_{gid}\\
&PC[\mathsf{pipe}(\mathsf{P}_1,\mathsf{P}_2)]_{gid} = \lambda i.\{C[\mathsf{P}_1]_{gid}(getId(C[\mathsf{P}_2]_{gid})), C[\mathsf{P}_2]_{gid}(i)\}
\end{aligned}$$

where $\lambda x.T$ it the usual function representation ($(\lambda x.T)(y) = T|_{x=y}$) and $getID()$ is the function returning the $id$ of the first instruction in its argument graph, that is, the one assuming to receive the input token from outside the graph, and a compile function $C[]$ such as

$C[\mathsf{P}] = PC[\mathsf{P}]_{newGid()}(\texttt{NoId})$

where $newId()$ and $newGid()$ are state full functions returning a fresh (i.e. unused) instruction and graph identifier, respectively. The compile function returns therefore a graph, with a fresh graph identifier, hosting all the data flow instructions relative to the skeleton program. The result tokens are identified as those whose destination is `NoId`. As an example, the compilation of the `main` program pipe(farm(seq(f)), farm(seq(g))) produces the data flow graph
$\{\langle 1, 1, \texttt{f}, \langle\langle \texttt{null}, \texttt{false} \rangle\rangle, \langle\langle 2, 1 \rangle\rangle\rangle \ , \langle 2, 1, \texttt{g}, \langle\langle \texttt{null}, \texttt{false} \rangle\rangle, \langle\langle \texttt{NoId}, \texttt{NoId} \rangle\rangle\rangle\}$
(assuming that identifiers and token positions start from 1).

When the application manager is told to actually compute the program, via an `eval()` method call, the input file stream is read looking for tasks to be computed. Each task found is used to replace the data field of the lower id data flow instruction in a new $C[\mathsf{P}]$ graph. In the example above, this results in the generation of a set of independent graphs such as:
$\{\langle 1, i, \texttt{f}, \langle\langle \texttt{task}_i, \texttt{true} \rangle\rangle, \langle\langle 2, 1 \rangle\rangle\rangle \ , \langle 2, i, \texttt{g}, \langle\langle \texttt{null}, \texttt{false} \rangle\rangle, \langle\langle \texttt{NoId}, \texttt{NoId} \rangle\rangle\rangle\}$
for all the tasks ranging from $task_1$ to $task_n$.

All the resulting instructions are put in the task pool of the distributed interpreter in such a way that the control threads taking care of "feeding" the remote data flow interpreter instances can start fetching the fireable instructions. The output tokens generated by instructions with destination tag equal to `NoId` are directly delivered to the output file stream by the threads receiving them from the remote interpreter instances. Those with a non-`NoId` flag are delivered to the proper instructions in the task pool that will eventually become fireable.

# 3 Joint Structured/Unstructured Parallelism Exploitation

We now take into account how unstructured parallelism exploitation mechanisms are provided to the muskel user. First, we provide the programmer with primitives to access the fundamental parameters of the data flow graph generated out of the compilation of a skeleton program. We provide methods to deliver data to and retrieve data from this graph. We provide the programmers with the ability to instantiate a new graph $C[\mathsf{P}]$ in the task pool by providing the input task token and to redirect the output token of $C[\mathsf{P}]$ to an arbitrary data flow instruction in the pool. This possibility is given through a new specific method of the `Manager` class, a `MdfGraph newGraph(Skeleton p)` method returning a `MdfGraph` handle for the new data flow graph generated out of the skeleton program `p`, and by two methods of the `MdfGraph` class, one (`IdPair getInputTokenId()`) returning the id identifying the first instruction (to be used to redirect a data token to this instruction) and the other one (`setOutputTokenId(int instrId, int graphId)`) writing the destination *id*s in the output token of the graph, used to redirect the output of macro data flow instruction to first instruction of the skeleton program. Second, we provide the programmer with direct access to the definition of data flow graphs, in such a way he can describe his particular parallelism exploitation patterns that cannot be efficiently implemented with the

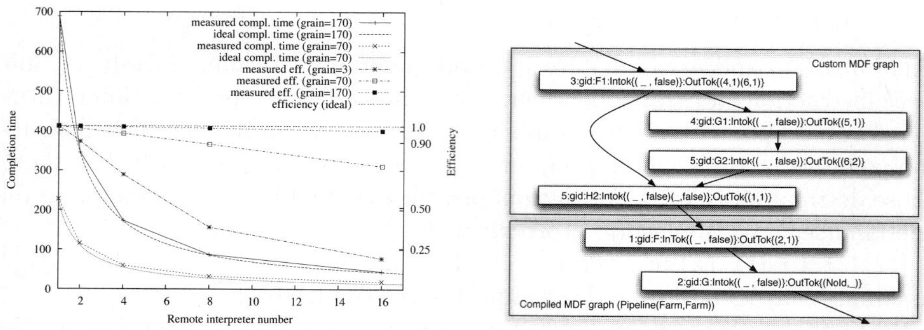


**Fig. 2.** Mixed sample MDF graph (right), scalability and effect of computation grain (left)

available skeletons. In particular, the programmer can instantiate new data flow instructions using the `MDFi` class constructors and accessory methods.

These two mechanisms can be jointly used to program all those parts of the application that cannot be easily and efficiently implemented using the skeletons subsystem. Let us suppose we want to pre-process the input tasks in such a way that for each task $t_i$ a new task $t'_i = h_1(f_1(t_i), g_2(g_1(f_1(t_i))))$ is produced. This computation cannot be programmed using the stream parallel skeletons currently provided by muskel. Then we want to process the preprocessed tasks through the `main` two-stage pipeline described above, in order to produce the final result. In this case the programmer can set up a new graph using the code shown in Figure 1 right. The resulting MDF graph eventually executed by the muskel interpreter is outlined in Figure 2 (right), where the parts derived from the `main` and the custom parts provided by the user are outlined.

Making good usage of these mechanisms, the programmer can arrange to express computations with arbitrary mixes of arbitrary data flow graphs and graphs coming from the compilation of structured, stream parallel skeleton computations. The execution of the resulting data flow graph is supported by the muskel distributed data flow interpreter as the execution of any other data flow graph derived from the compilation of a skeleton program.

## 4 Experimental Results

A muskel prototype implementation supporting the extensions described in this work is currently being tested. The muskel interpreter engine has been left basically unchanged, whereas the part supporting parallelism exploitation pattern programming has been changed to support linking of custom MDF graphs to the code produced by the compiler out of plain muskel skeleton trees. Figure 2 summarizes the typical performance results of the enhanced interpreter. We run several synthetic programs using the custom MDF graph features introduced in muskel. We designed the programs in such a way the MDF instructions appearing in the graph had a precise "average grain" (i.e. average ration among the time spent computing the instruction at the remote interpreter and the time spent communicating data to and from the remote interpreter). Each run processed 1K input tasks. When grain is small, muskel do not scale, even using a very small number of remote interpreter instances. When the grain is high enough (about 200 times the time spent in communications actually spent in computation of MDF instructions) the efficiency is definitely close to the ideal one. Despite the shown data refer to synthetic computations, actual computations (e.g. image processing ones) achieved very similar results. This because the automatic load balancing mechanism implemented in the muskel distributed interpreter through auto scheduling perfectly optimized the execution of variable grain MDF instructions. All the experiments have been performed on a Linux (kernel 2.4.22) RLX Pentium III blade architecture, with Fast Ethernet interconnection among the blades, equipped with Java 1.4.1_01 run time.

## 5 Related Work

Macro data flow implementation for algoritmical skeleton programming environment was introduced by the authors in late '90 [8] and then has been used in other contexts related to skeleton programming environments [13]. Cole suggested in [6] that "we must construct our systems to allow the integration of skeletal and ad-hoc parallelism in a well defined way", and that structured parallel programming environments should "accommodate diversity", that is "we must be careful to draw a balance between our desire for abstract simplicity and the pragmatic need for flexibility". Actually, his eSkel [4, 7] MPI skeleton library addresses these problems by allowing programmers to program their own peculiar MPI code within each process in the skeleton tree. Programmers can ask to have a stage of a pipeline or a worker in a farm running on $k$ processors. Then, the programmer may use the $k$ processes communicator returned by the library for the stage/worker to implement its own parallel pipeline stage/worker process. As far as we know, this is the only attempt to integrate ad hoc, unstructured parallelism exploitation in a structured parallel programming environment. The implementation of eSkel, however, is based on process templates, rather than on data flow.

Other skeleton libraries, such as Muesli [10, 12], provide programmers with a quite large flexibility in skeleton programming following a different approach. They provide a number of data parallel data structures along with elementary, collective data parallel operations that can be arbitrary nested to get more and more complex data parallel skeletons. However, this flexibility is restricted to the data parallel part, and it is anyway limited by the available collective operations.

COPS [11] is a design pattern based parallel programming environment written in Java and targeting symmetric multiprocessors. In COPS, programmers are allowed to program their own parallel design patterns (skeletons) by interacting with the intermediate implementation level [5]. Again, this environment does not use data flow technology but implements design patterns using proper process network templates.

## 6 Conclusions

We have shown how the existing data flow implementation of muskel can be extended to allow programmers to mix arbitrary parallelism exploitation patterns and structured ones in the same parallel application. We discussed preliminary performance results showing that the approach is feasible and efficient. The current version of muskel does not allow completely arbitrary custom graph and skeleton mix. As an example, it still does not support the usage of arbitrary MDF graphs as parameters of the skeletons. We are currently working to provide also this possiblity, however. Once completed, the new version of the muskel parallel Java library that will eventually be released under GPL on the muskel web site `www.di.unipi.it/∼marcod/muskel`.

## References

1. M. Aldinucci and M. Danelutto. Stream parallel skeleton optimisations. In *Proc. of the IASTED International Conference Parallel and Distributed Computing and Systems*, pages 955–962. IASTED/ACTA Press, November 1999. Boston, USA.
2. M. Aldinucci, M. Danelutto, and P. Teti. An advanced environment supporting structured parallel programming in Java. *FGCS*, 19(5):611–626, 2003.
3. B. Bacci, M. Danelutto, S. Pelagatti, and M. Vanneschi. SkIE: a heterogeneous environment for HPC applications. *Parallel Computing*, 25:1827–1852, Dec. 1999.
4. A. Benoit, M. Cole, J. Hillston, and S. Gilmore. Flexible Skeletal Programming with eSkel. In *Proceedings of EuroPar 2005*, LNCS. Springer Verlag, Sept. 2005.
5. S. Bromling. Generalising Pattern-based Parallel Programming Systems. In *Proceedings of Parco 2001*. Imperial College Press, 2002.
6. M. Cole. Bringing Skeletons out of the Closet: A Pragmatic Manifesto for Skeletal Parallel Programming. *Parallel Computing*, 30(3):389–406, 2004.
7. M. Cole and A. Benoit. The Edinburgh Skeleton Library home page, 2005. http://homepages.inf.ed.ac.uk/abenoit1/eSkel/.
8. M. Danelutto. Dynamic Run Time Support for Skeletons. In *Proc. of the International Conference ParCo99*, Parallel Computing Fundamentals & Applications, pages 460–467. Imperial College Press, 1999.
9. M. Danelutto. QoS in parallel programming through application managers. In *Proceedings of the 13th Euromicro Conference on Parallel, Distributed and Network-based processing*. IEEE, 2005. Lugano (CH).
10. H. Kuchen. A skeleton library. In *Euro-Par 2002, Parallel Processing*, number 2400 in LNCS, pages 620–629. Springer Verlag, August 2002.
11. S. McDonald, D. Szafron, J. Schaeffer, and S. Bromling. Generating Parallel Program Frameworks from Parallel Design Patterns. In *Euro-Par 2000 Parallel Processing*, LNCS, No. 1900, pages 95–105. Springer Verlag, August/September 2000.
12. Muesli home, 2005. www.wi.uni-muenster.de/PI/forschung/Skeletons/ index.php
13. J. Serot and D. Ginhac. Skeletons for parallel image processing: an overview of the SKIPPER project. *Parallel computing*, 28(12):1685–1708, 2002.

# Co-Array Collectives: Refined Semantics for Co-Array Fortran

Matthew J. Sottile[1], Craig E Rasmussen, and Richard L. Graham

Los Alamos National Laboratory,
Los Alamos NM 87545, USA
{matt, rasmussn, rlgraham}@lanl.gov

**Abstract.** Co-array notation provides a compact syntax for programming parallel programs. Co-array Fortran (CAF) introduced and implements this notation, and CAF is currently proposed as an extension to the Fortran language standard. We believe that co-array notation requires a revised semantic definition beyond that specified by CAF for both pragmatic reasons within Fortran and to make the notation attractive for incorporation into other programming languages. The revised semantics make the language model easier to understand and reduces the potential for programmer error. Furthermore, these revised semantics allow CAF to be extended to capture collective operations in co-array notation.

## 1 Introduction

Co-Array Fortran (CAF) is a simple, parallel extension to Fortran [3]. For CAF, a single program is replicated `p` number of times in SPMD fashion (Single Program, Multiple Data). Each copy of the program (referred to as an *image*) executes asynchronously with its own set of data objects. The co-array notation extends the normal array syntax with an additional array dimension called the *co-dimension*. The local portion of a co-array is denoted with the normal parentheses while the co-dimension is specified with square brackets.

For example, if `a` and `b` both have local rank 1, local size 100, and a co-rank 1, they are declared as

```
integer :: a(100)[*], b(100)[*]
```

The co-dimension is declared to be assumed size, *i.e.* `[*]`, as the co-size depends on the number of participating images (a compile time or run-time parameter).

Indexing in the co-dimension is the same as in the local dimension. For example, given constants `i` and `j` that are identical on all images, `a(1)[i] = b(1)[j]` states that the first element of the array `b` on image `j` is to be copied to the first element of the array `a` on image `i`. Because this assignment, as written, would be executed on every image, it is actually illegal (a non-conforming program) because concurrent writes into the same storage location are not allowed.

To emphasis this point, consider the simple assignment, `a(1)[i] = 3`. Without additional context, all images will execute this statement and cause the

V.N. Alexandrov et al. (Eds.): ICCS 2006, Part II, LNCS 3992, pp. 945–952, 2006.

constant 3 to be stored in the first element of `a` on image `i`. As this is a concurrent write and therefore illegal, the programmer must embed the statement within other control structures to ensure that it adheres to the rules established for CAF. One must explicitly declare *which* image is to cause the store to occur, e.g.,

```
if (this_image() == j) a(1)[i] = 3
sync_all()
```

If image `j` is not equal to `i`, image `j` *puts* the value 3 to image `i`.

We suggest that this is an unfortunate definition for the result of the CAF assignment operator. Assuredly all images must execute an assignment statement (unless restricted by an if statement), but there is no reason all images must *cause* data to be transferred to image $i$. Certainly in this case, the value of `a(1)[i]` will be 3 after the `sync_all` is executed, no matter which image caused the store to occur. (The intrinsic function `sync_all` is a barrier that all images must reach before any image may continue execution.)

In particular, this execution model causes problems if array sections are allowed in the co-dimension (co-sections). For example, the assignment statement `a(1)[:] = 3` would require that *all* images store 3 into co-memory on *all* images. Indeed, although co-sections were allowed in the original Reid and Numrich definition [3], they were removed in a recent CAF specification [4]. One reason they were removed, is that they are not present in CAF implementations and limiting co-subscripts to scalars simplifies the language specification [5]. The goal of this paper is to examine the ramifications of permitting co-sections, with the hope that they can be restored to the language.

Specifically, we propose that one-sided *get* semantics be adopted in all assignment statements involving co-array variables, making the image that "owns" the memory the only one allowed to initiate a write operation to it. This simplifies the language specification and automatically satisfies the requirement that concurrent stores to the same memory location should not occur. In this paper we examine the consequences of this change. The rules defining the result of a co-variable assignment statement are described in Section 3 and then applied to collective operations in Section 4. Section 5 contains two programming examples employing these changes and Section 6 summarizes the paper.

## 2 Refined CAF Assignment Operators

In this section we propose a set of semantic rules for co-array notation that define the result of a co-variable assignment statement. These rules are intended to provide the compiler with as much freedom as reasonable to generate efficient parallel code by removing the burden of explicit delegation of message passing initiation from the user. Like CAF, as defined in [4], these rules are such that the compiler is able to make decisions solely on the basis of an analysis of the program as a sequential entity, without requiring cross image analysis to perform messaging optimizations. In addition, they allow the use of array sections in the co-dimension.

The Fortran grammar defines an assignment statement as *variable* = *expr* [2]. The first rule describes the result of a co-array expression: For any co-array expression `b(:)[begin:end:stride]` executed on any image `i`, the result will be as if a temporary rank 2 array `tmp` were created on `i` and initialized as,

```
do k = begin, end, stride
  do j = 1, nrows
    tmp(j,k) = b(j)[k]
  end do
end do
```

This rule generalizes to other co-array ranks. In general, the rank of the temporary array will be equal to the total rank (the sum of the local rank and the co-rank) of the co-array. Note that the elements of `tmp` are obtained using get semantics.

The second rule defines the result of an assignment statement if the variable is a co-array and is sectioned (using subscript triplet notation) in the co-dimension. We define this as a parallel assignment statement. Consider `a(:)[begin:end:stride] = tmp(:)`, where `tmp` is a local array and may be a temporary array resulting from the evaluation of a co-array expression, as defined above (note the change in the rank of the temporary). The results of the assignment statement, when executed on all images, is

```
do i = begin, end, stride
  if (this_image() == i) then
    a(:)[i] = tmp(:)
  end if
end do
```

This rule also generalizes to other co-array ranks. In general, the local rank of the co-array variable must be equal to the rank of the array temporary and their shapes must conform.

In addition, we propose that a *put* assignment operator be provided. This would allow third party transfers, not allowed with get semantics. While not intending to provide guidance for the syntax of a put assignment, the following,

```
if (this_image() == 2) then
  a(1)[1] .recvs. b(1)[3]
end if
```

could be adopted to cause image two to copy memory from image three to image one. The same code segment with standard assignment syntax,

```
if (this_image() == 2) then
  a(1)[1] = b(1)[3]
end if
```

results in no change to the state of the program because of the implied `if (this_image() == 1)` with the standard assignment operator.

These are the only two changes proposed for CAF: 1. Get semantics as embodied in the second rule above; and 2. a put assignment statement. Nothing has been removed from the language. Only a clear distinction between the syntax for the two assignment statements, thus removing the overloading of the assignment operator. Furthermore, this removes ambiguity with respect to the meaning (get vs. put) of co-array notation when it appears on different sides of the = operator. With serial execution this distinction is unimportant. However, this is not true for parallel execution (multiple images). The put assignment (with no implied if statement) can lead to unnecessary and repetitious memory transfer, *.e.g.,* `a(1)[1] = 3`, and concurrent stores. Therefore, it is suggested that the distinguishing syntax is helpful to the programmer.

## 3 Collective Assignment Statements

Given the rules specified in Section 2, we can now enumerate the outcome of the 16 possible combinations (of array subscript versus array section) that can occur in the assignment statement `a(i)[j] = b(k)[l]`. This is done in order to insure that the results of any statement containing a co-section, *i.e.* `[:]`, is well understood.

The examples are shown in two columns: the left-hand column is written using array section notation in both the local and the co-dimensions, while the right-hand column is written using it only for the local dimension. The statements all assume that `me = this_image()`, `p = num_images()`, and that the local size of the arrays equals `p`.

Simple scalar and array assignment are well understood, as can be seen below:

```
a(1)[1] = b(3)[3]    ;    if (me == 1) a(1) = b(3)[3]
a(:)[1] = b(3)[3]    ;    if (me == 1) a(:) = b(3)[3]
a(:)[1] = b(:)[3]    ;    if (me == 1) a(:) = b(:)[3]
```

Likewise are the scalar and array broadcasts:

```
a(1)[:] = b(3)[3]    ;    a(1) = b(3)[3]
a(:)[:] = b(3)[3]    ;    a(:) = b(3)[3]
a(:)[:] = b(:)[3]    ;    a(:) = b(:)[3]
```

Broadcast statements execute on all images, with result that the same data are stored in the local portion of `a` on each image. Strictly speaking, these are not broadcasts, as the assignment has get semantics, though the implementation may be a broadcast by recognizing the array section in the co-dimension.

The gather collectives read data from all images and store the results either on one image (gather) or on all images (allgather).

```
a(:)[1] = b(3)[:]    ;    if (me == 1) then  ! gather
                            do i = 1, p; a(i) = b(3)[i]; end do
                          end if
a(:)[:] = b(3)[:]    ;    do i = 1, p;  a(i) = b(3)[i];  end do
```

There is no collective assignment statement corresponding to a scatter operation. It would have to be expressed explicitly:

```
a(1)[me] = b(me)[3]
```

There are several statements (of the 16 combinations) that are well defined but result in illegal assignments:

```
a(1)[1] = b(:)[3]   ;   if (me == 1) a(1) = b(:)[3]
a(1)[1] = b(3)[:]   ;   if (me == 1) a(1) = b(3)[:]
a(1)[1] = b(:)[:]   ;   if (me == 1) a(1) = b(:)[:]
a(:)[1] = b(:)[:]   ;   if (me == 1) a(:) = b(:)[:]
a(1)[:] = b(:)[3]   ;   a(1) = b(:)[3]
a(1)[:] = b(3)[:]   ;   do i = 1, p;  a(1) = b(3)[i];  end do
a(1)[:] = b(:)[:]   ;   do i = 1, p;  a(1) = b(:)[i];  end do
a(:)[:] = b(:)[:]   ;   do i = 1, p;  a(:) = b(:)[i];  end do
```

All of these statements could be made legal by increasing the rank of `a` by one, or by a reduction of the right-hand side, *.e.g.*, `a(1)[:] = sum(b(:)[3])`.

## 4 Examples

Two algorithms are implemented using co-arrays. The first example requires a high degree of synchronization while the second does not. These examples were chosen to compare and contrast the differences between the current CAF definition of the assignment operator, with that of our proposed definition.

### 4.1 Maximum Value of a Co-Array

The original Reid and Numrich Co-Array Fortran paper [3] contains several examples. One in particular, an algorithm that finds the maximum value of a co-array, was chosen because it highlights many important points. It should be noted that this example is inefficient and therefore, the actual CAF intrinsic function `CO_MAXVAL` is preferred.

Consider the following subroutine that will return the maximum value of co-array `a` in the co-scalar `great`:

```
subroutine greatest(a, great)  ! find maximum value of a(:)[*]
  real, intent(in)  :: a(:)[*]
  real, intent(out) :: great[*]
  great[:] = maxval(a(:))       ! all calculate local maxima
  sync_all()                    ! wait for all to calc local max
  great[1] = maxval(great[:])   ! gather to 1, then find maximum
  sync_all()                    ! wait for 1 to calc global max
  great[:] = great[1]           ! all collect the results
  sync_all()                    ! wait for all to finish
end subroutine greatest
```

The algorithm first computes local maxima on all local portions of a. It then gathers the local maxima to image one, where the global maximum is calculated. This global maximum is then copied by the other images.

Note the three synchronization calls in the example. The second is required because all images (other than one) must wait for the global maximum to be calculated before getting a copy of it. The code section containing the outer sync_all calls could be replaced by:

```
sync_all()                      ! wait for all to finish
great[1] = maxval(great[:])     ! gather to 1, then find maximum
if (this_image() == 1) then
  great[:] .recvs. great[1]     ! broadcast the results to all
end if
sync_all()                      ! wait for all to finish
```

This implementation using puts requires one less synchronization call because image one broadcasts the global maximum, only after it has calculated it. This example points out subtle differences between put and get semantics. Different control flow and synchronization patterns may be required, as demonstrated.

The relevant code segment implementing greatest using the original CAF execution model is shown below:

```
real :: work(num_images())
great = maxval(a(:))              ! all calculate local maxima
sync_all()                        ! wait for all to calc local max
if (this_image(great) == 1) then
  do i = 1, num_images()          ! gather local maxima
    work(i) = great[i]
  end do
  great = maxval(work)
  do i = 1, num_images()
    great[i] = great              ! broadcast the results to all
  end do
end if
sync_all()                        ! all wait for image 1 to finish
```

Note the lack of explicit, temporary arrays (work), if statements, and do loops in the first implementation. We suggest that the lack of these constructs (in the first implementation) makes the code more readable and their overuse can get in the way of program understanding. The first implementation is expressed at a higher level of abstraction; the programmer instructs the compiler regarding *what* should be done, without unnecessarily informing the compiler *how* it should be accomplished. This is left up to the compiler to decide.

### 4.2 Producer/Consumer

The second example we consider is that of an agent producing information that another consumes. This example was chosen to examine implementation of an algorithm that is more loosely coupled than the first, requiring no algorithmic synchronization.

A genetic algorithm (GA) [1] implementation could be run in parallel by instructing each image to work on an entirely separate and distinct population, with no communication between images. This would be as if a real population were evolving on an island, separated by other populations of the same species. A population is a set of individual genomes that exchange genetic information during the creation (breeding) of a new generation. We define a population as an integer co-array `pop` with both a local and a co-rank of one, `integer :: pop(pop_size)[*]`.

For better results, it is desirable to exchange genetic information between separate populations. The algorithm does not require synchronization between images because the timing of the information exchange is not crucial. For each population and at each generation, we choose to replace one individual chosen randomly with another individual, also chosen randomly from the global population:

```
do n = 1, num_generations
  call new_generation(pop) ! create new generation, local memory
  i = random_image();  k = random_index();  l = random_index()
  pop(k) = pop(l)[i]  ! (*) replace local individual
end do
```

The implementation of this GA is identical, whether using the original or our revised CAF assignment operator definition.

Note that while no synchronization is required by the algorithm, two `sync_all` calls would be necessary around the statement `(*)` to ensure that no remote reads can occur while the local population is being modified. This becomes clear when the elements representing individuals in the `pop` array are not atomic with respect to local memory stores operations. Global synchronization is an expensive operation. Thus, this example points out the need for a CAF language construct to provide mutually exclusive access to non-atomic data structures. To do this, however, consideration must be given to performance aspects, because such a language construct could harm the performance of all remote read operations.

## 5 Conclusions

We have proposed an alternative definition of the assignment operator for Co-Array Fortran. This new definition was required to allow collective operations to be expressed using co-array section notation while ensuring that the language specification does not imply that excessive and redundant messages would result. Our alternative definition specifies that assignment statements involving cooperating images have one-sided, get semantics. Get semantics allow the use of collective assignment statements without causing inefficiencies to arise, as unneeded data is transferred, or causing illegal, concurrent writes to memory. Put semantics are allowed only via a different syntax. The proposed changes provide the following advantages:

- Array-section notation may be specified in either the local or the co-dimensions, thus unifying its usage.
- `If` statements denying concurrent writes are not needed. A programmer is not able to *express* concurrent writes to a single address in the language, unless the separate `.recvs.` syntax is used.
- The programmer is able to express collectives without explicit loops.
- Explicit temporaries can be eliminated when expressing collective all-reduce and gather-scatter operations.
- The programmer does not need to be artificially aware of memory affinity, as may occur with the explicit use of `if` blocks to restrict execution to a single image; get semantics allows a programmer to think of the co-dimension as a memory address, rather than an image index.
- While get semantics is specified as the pattern to which a programmer designs, the compiler is free to choose the optimal method for data transfer, provided that these semantics hold for the program.
- Collective notation allows the programmer to explicitly specify which images are expected to participate within an expression or assignment statement.

Perhaps most importantly, the new Co-Array Fortran semantics leave room for future extensions to the language in ways that the original semantics do not. For example, suppose CAF could be extended to allow a generalized vector subscript notation (or generalized section notation), that would select co-array elements from the *combined* local and co-dimensions.

## Acknowledgments

Los Alamos National Laboratory is operated by the University of California for the National Nuclear Security Administration of the United States Department of Energy under contract W-7405-ENG-36, LA-UR-05-9203.

## References

1. J. H. Holland. *Adaptation in natural and artificial systems.* The University of Michigan Press, Ann Arbor, MI, 1975.
2. American National Standards Institute. Information technology – programming languages – Fortran – part 1: Base language. Technical Report ISO/IEC 1539-1:2004, 2004.
3. Robert W. Numrich and John K. Reid. Co-Array Fortran for parallel programming. *ACM Fortran Forum*, 17(2):1–31, 1998. http://www.co-array.org.
4. Robert W. Numrich and John K. Reid. Co-arrays in the next Fortran standard. Technical Report ISO/IEC JTC/SC22/WG5 N1642, May 2005.
5. John K. Reid, Personal communication, Delft, May 2005.

# An Approach to Buffer Management in Java HPC Messaging

Mark Baker, Bryan Carpenter, and Aamir Shafi

Distributed Systems Group, University of Portsmouth

**Abstract.** One of the most challenging aspects to designing a Java messaging system for HPC is the intermediate buffering layer. The lower and higher levels of the messaging software use this buffering layer to write and read messages. The Java New I/O package adds the concept of direct buffers, which—coupled with a memory management algorithm—opens the possibility of efficiently implementing this buffering layer. In this paper, we present our buffering strategy, which is developed to support efficient communications and derived datatypes in MPJ Express—our implementation of the Java MPI API. We evaluate the performance of our buffering layer and demonstrate the usefulness of direct byte buffers.

## 1 Introduction

The challenges of making parallel hardware usable have, over the years, stimulated the introduction of many novel languages, language extensions, and programming tools. Lately, though, practical parallel computing has mostly adopted conventional (sequential) languages, with programs developed in relatively conventional programming environments usually supplemented by libraries like MPI that support parallel programming. This is largely a matter of economics: creating entirely novel development environments matching the standards programmers expect today is expensive, and contemporary parallel architectures predominately use commodity microprocessors that can best be exploited by off-the-shelf compilers.

This argues that if we want to "raise the level" of parallel programming, one practical approach is to move towards advanced commodity languages. Compared with C or Fortran, the advantages of the Java programming language include higher-level programming concepts, improved compile-time and run-time checking, and as a result, faster problem detection and debugging. Also, it supports multi-threading and provides simple primitives like `wait()` and `notify()` that can be used to synchronize access to shared resources. Recent JDKs (Java Development Kits) provide greater functionality in this area, including semaphores and atomic variables. In addition, Java's automatic garbage collection, when exploited carefully, relieves the programmer of many of the pitfalls of lower-level languages.

We have developed MPJ Express (MPJE) [6], a thread-safe implementation of Java MPI API. A challenging aspect of implementing Java HPC messaging

V.N. Alexandrov et al. (Eds.): ICCS 2006, Part II, LNCS 3992, pp. 953–960, 2006.

software is providing an efficient intermediate buffering layer. The low-level communication devices and higher levels of the messaging software use this buffering layer to write and read messages. The heterogeneity of these low-level communication devices poses additional design challenges. To appreciate this fully, assume that the user of a messaging library sends ten elements of an integer array. The C programming language can retrieve the memory address of this array and pass it to the underlying communication device. If the communication device is based on TCP, it can then pass this address to the socket's write method. For proprietary networks, like Myrinet [7], this memory region can be registered for Direct Memory Access (DMA) transfers, or copied to a DMA capable part of memory and sent using low level Myrinet communication methods. Until quite recently doing this kind of thing in Java was hard.

JDK 1.4 introduced Java New I/O (NIO) [8]. In NIO, read and write methods on files and sockets (for example) are mediated through a family of buffer classes handled specially by the Java Virtual Machine (JVM). The underlying `ByteBuffer` class essentially implements an array of bytes, but in such a way that the storage can be outside the JVM heap (so called *direct* byte buffers).

So now if a user of a Java messaging system sends an array of ten integers, they can be copied to a `ByteBuffer`, which is used as an argument to the `SocketChannel` write method. For proprietary networks like Myrinet, NIO provides a viable option because it is now possible to get memory addresses of direct byte buffers, which can be used to register memory regions for DMA transfers. Using direct buffers may eliminate the overhead [9] incurred by additional copying when using Java Native Interface (JNI) [4]. On the other hand, it may be preferable to create a native buffer using the JNI. These native buffers can be useful for a native MPI or a proprietary network device.

We are convinced that NIO provides essential ingredients [2] of an efficient messaging system via non-blocking I/O and direct buffers.

Based on these factors, we have designed an extensible buffering layer that allows various implementations based on different storage mediums like direct or indirect `ByteBuffers`, byte arrays, or memory allocated in the native C code. The higher levels of MPJE use the buffering layer through an interface. This implies that functionality is not tightly coupled to the storage medium. The motivation behind developing different implementations of buffers is to achieve optimal performance for lower level communication devices. Our buffering strategy uses a pooling mechanism to avoid creating a buffer instance for each communication method. The creation time of these buffers can affect overall communication time, especially for large messages. Our current implementation is based on Knuth's buddy algorithm [5], but it is possible to use other pooling techniques.

The main contribution of this paper is the design and implementation of our buffering layer for HPC supported by two different pooling mechanisms. In addition, we have evaluated the performance of these two pooling mechanisms. We show that one of them is faster with a smaller memory footprint. Also, we demonstrate the usefulness of direct byte buffers in Java messaging systems.

The remainder of this paper is organized as follows. Section 2 discusses related work. The strategy itself with an explanation of our memory management algorithms is described in section 3. In section 4, we present performance evaluation of our buffering strategies. Section 5 concludes the paper outlining future research work.

## 2 Related Work

The most popular Java messaging system is mpiJava [1], which uses a JNI wrapper to the underlying native C MPI library. Being a wrapper library, mpiJava does not use a clearly distinguished buffering layer. After packing a message onto a contiguous buffer, a reference to this buffer is passed to the native C library. But in achieving this, additional copying may be required between the JVM and the native C library. This overhead is especially noticeable for large messages.

Javia [3] is a Java interface to the Virtual Interface Architecture (VIA). An implementation of Javia exposes communication buffers used by the VI architecture to Java applications. These communication buffers are created outside the Java heap and can be registered for DMA transfers. This buffering technique makes it possible to achieve performance within 1% of the raw hardware.

An effort similar to Javia is JAGUAR [9]. This uses compiled-code transformations to map certain Java bytecodes to short, in-lined machine code segments. These two projects, JAGUAR and Javia were the motivating factors to introduce the concept of direct buffers in the NIO package. The design of our buffering layer is based on direct byte buffers. In essence, we are applying the experiences gained by JAGUAR and Javia to design a general and efficient buffering layer that can be used for pure Java and proprietary devices in Java messaging systems alike.

## 3 The Buffering Layer in MPJE

In this section, we discuss our approach to designing and implementing an efficient buffering layer supported by a pooling mechanism. The self-contained API developed as a result is called the MPJ Buffering (`mpjbuf`) API. The functionality provided includes packing and unpacking of user data.

An mpjbuf buffer object contains two data storage structures. The first is a static buffer, in which the underlying storage primitive is an implementation of the `RawBuffer` interface. The implementation of static buffer called `NIOBuffer` uses direct or indirect `ByteBuffers`. The second is a dynamic buffer where a byte array is the storage primitive. The size of the static buffer is predefined, and can contain only primitive datatypes. The dynamic buffer is used to store serialized Java objects, where it is not possible to determine the length of the serialized objects beforehand. The class structure of our package is shown in Figure 1.

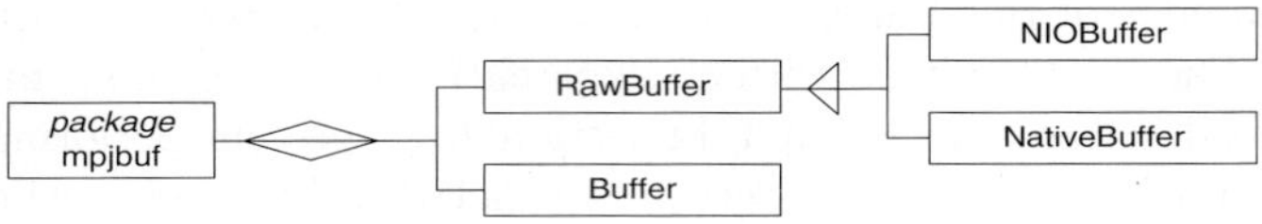


**Fig. 1.** Primary Buffering Classes in `mpjbuf`

### 3.1 Memory Management

We have implemented our own application level memory management mechanism based on a buddy allocation scheme [5]. The motivation is to avoid creating an instance of a buffer (`mpjbuf.Buffer`) for every communication operations like `Send()` or `Recv()`, which may dominate the total communication cost, especially for large messages. We can make efficient use of resources by pooling buffers for future reuse instead of letting the garbage collector reclaim the buffers and create them all over again. The functionality provided by the buffering API is exported to the users through a `BufferFactory`.

In the MPJ buffering API it is possible to plug in different implementations of buffer pooling. A particular strategy can be specified during the initialisation of `mpjbuf.BufferFactory`. Each implementation can use different data structures like trees or doubly linked lists. In the current implementation, the primary storage buffer for mpjbuf is an instance of `mpjbuf.NIOBuffer`. Each `mpjbuf.NIOBuffer` has an instance of `ByteBuffer` associated with it. The pooling strategy boils down to reusing `ByteBuffers` encapsulated in `NIOBuffer`.

Our implementation strategies are able to create smaller thread-safe `ByteBuffers` from the initial `ByteBuffer` associated with the region. We achieve this by using `ByteBuffer.slice()` for creating new byte buffer whose contents are a shared sub sequence of original buffers contents. In the sub-sections to follow, we discuss two implementations of memory management techniques.

In a buddy algorithm, the region of available storage is conceptually divided into blocks of different levels, hierarchically nested in a binary tree. A free block at level $n$ can be split into two blocks of level $n-1$, half the size. These sibling blocks are called buddies. To allocate a number of bytes $s$, a free block is found and recursively divided into buddies until a block at level $\lceil \log_2(s) \rceil$ is produced. When a block is freed, one checks to see if its buddy is free. If so, buddies are merged (recursively) to consolidate free memory.

**The First Pooling Strategy.** Our first implementation (called Buddy1 below) is developed with the aim of keeping a small memory footprint of the application. This is possible because a buffer only needs to know its offset in order to find its buddy. This offset can be stored at the start of the allocated memory chunk.

Figure 2 outlines the implementation details of our first pooling strategy. `FreeList` is a list of `BufferLists`, which contains buffers at different levels. Here, level refers to the different sizes of buffer available. If a buffer is of size $s$, then its corresponding level will be $\lceil \log_2(s) \rceil$. Initially, there is no region associated with `FreeList`. An initial chunk of memory of size $M$ is allocated. At this point,

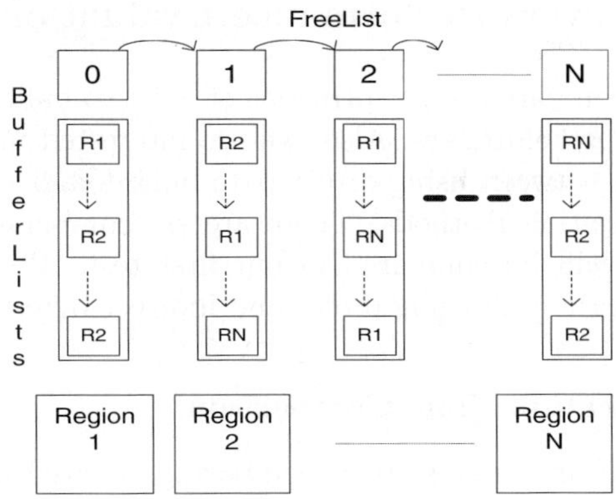


**Fig. 2.** The First Implementation of Buffer Pooling

BufferLists are created starting from 0 to $\log_2(M)$. When buddies are merged, a buffer is added to the BufferList at the higher level and the buffer itself and its buddy are removed from the BufferList at the lower level. Conversely, when a buffer is divided to form a pair of buddies, a newly created buffer and its buddy is added to the BufferList at the lower level while removing a buffer that is divided from the higher level BufferList. An interesting aspect of this implementation is that FreeList and BufferLists are independent of a region and these lists grow as new regions are created to match user requests.

**The Second Pooling Strategy.** Our second implementation (called Buddy2 below) stores higher-level buffer abstractions (NIOBuffer) in BufferLists. Unlike the first strategy, each region has its own FreeList and has a pointer to the next region as shown in Figure 3. While finding an appropriate buffer for a user, this implementation starts sequentially starting from the first region until it finds the requested buffer or creates a new region. We expect some overhead associated with this sequential search. Another downside for this implementation is a bigger memory footprint.

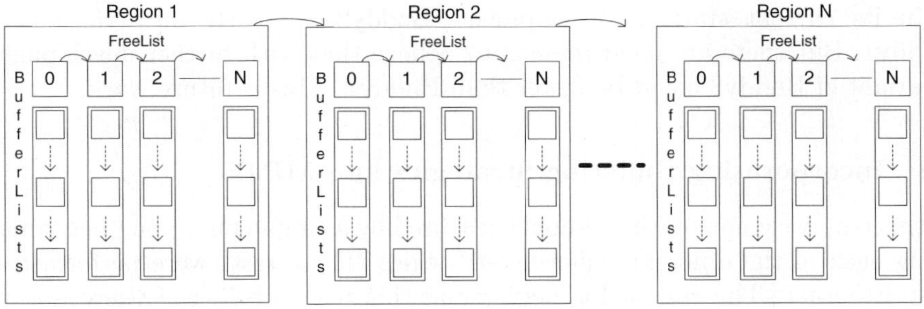


**Fig. 3.** The Second Implementation of Buffer Pooling

# 4 Buffering Layers Performance Evaluation

In this section, we compare the performance of our two buffering strategies with direct allocation of `ByteBuffers`. Also, we are interested in exploring the performance difference between using direct byte buffers and indirect byte buffers in MPJE communication methods. There are six combinations of our buffering strategies that will be compared in our first test—Buddy1, Buddy2, and simple-minded allocation all using direct and indirect byte buffers.

## 4.1 Simple Allocation Time Comparison

In our first test, we are interested in comparing isolated allocation times for a buffer for our six allocation approaches. Only one buffer is allocated at one time throughout the tests. This means that after measuring allocation time for a buffer, it is de-allocated in the case of our buddy schemes (forcing buddies to merge into original region chunk of 8 Mbytes before the next allocation occurs), or the reference is freed in the case of straightforward `ByteBuffer` allocation.

Figure 4 shows a comparison of allocation times. The first thing to note is, that all the buddy-based schemes are dramatically better than relying on the JVMs management of `ByteBuffer`. This essentially means that without a buffer pooling mechanism, creation of intermediate buffers for sending or receiving messages in a Java messaging system can have detrimental effect on the performance. Results are averaged over many repeats, and the overhead of garbage collection cycles will be included in the results in an averaged sense; this is a fair representation of what will happen in a real application. In a general way we attribute the dramatic increase in average allocation time for large `ByteBuffers` as due to forcing proportionately many garbage collection cycles. All the buddy variants (by design) avoid this overhead. The allocation times for buddy based schemes decrease for larger buffer sizes because less time is spent in traversing the data structures to find an appropriately sized buffer. The size of the initial region is 8 Mbytes—resulting in the least allocation time for this buffer size. The best strategy in almost all cases is Buddy1 using direct buffers.

Qualitative measurements of memory footprint suggest the current implementation of Buddy2 also has about a 20% bigger footprint because of the extra objects stored.

In its current state of development, Buddy2 is clearly outperformed by Buddy1. But there are good reasons to believe that with further development, a variant of Buddy2 could be faster than Buddy1. This is future work.

## 4.2 Incorporating Buffering Strategies into MPJE

In this test, we compare throughput measured by a simple ping-pong benchmark using each of the different buffering strategies. These tests were performed on Fast Ethernet. The reason for performing this test is to see if there are any performance benefits for using direct `ByteBuffers`. From the source-code of the NIO package, it appears that the JVM maintains a pool of direct ByteBuffers

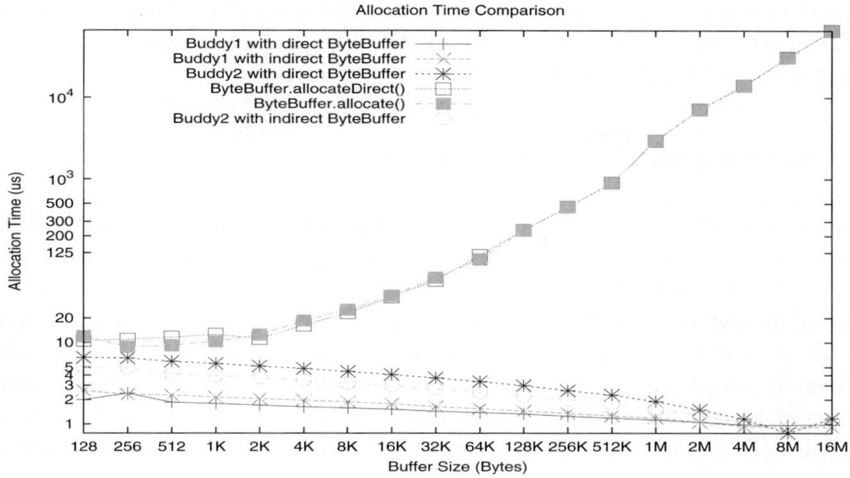


**Fig. 4.** Allocation Time Comparison

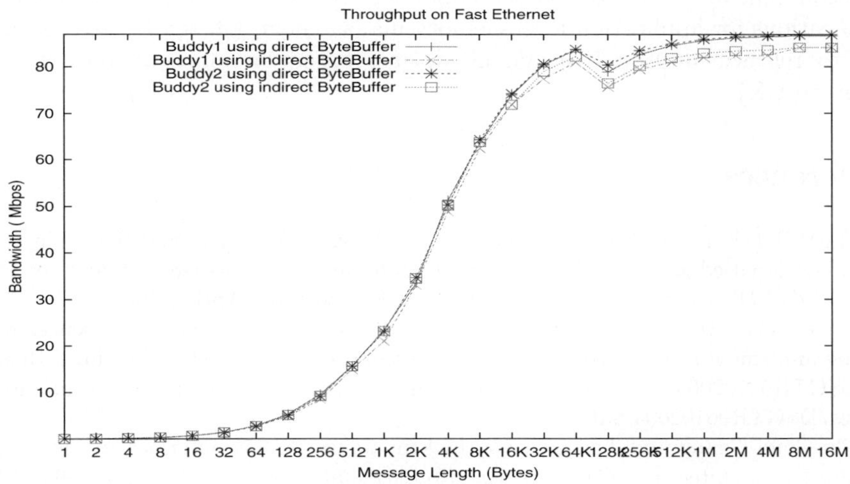


**Fig. 5.** Throughput Comparison

for internal purposes. These buffers are used for reading and writing messages into the socket. A user provides an argument to `SocketChannel`s write or read method. If this buffer is direct, it is used for writing or reading messages. If this buffer is indirect, a direct byte buffer is acquired from direct byte buffer pool and the message is copied first before writing or reading it into the socket. Thus, we expect to see an overhead of this additional copying for indirect buffers.

Figure 5 shows that MPJE achieves maximum throughput when using direct buffer in combination with either of the buddy implementations. We expect to see this performance overhead related to indirect buffers to be more significant for faster networks like Gigabit Ethernet and Myrinet. The drop in throughput at 128Kbytes message size is because of the change in communication protocol from eager send to rendezvous.

## 5 Conclusions and Future Work

In this paper, we have discussed the design and implementation of our buffering layer, which uses our own implementation of buddy algorithm for buffer pooling. For a Java messaging system, it is useful to rely on an application level memory management technique instead of relying on JVM's garbage collector because constant creation and destruction of buffers can be a costly operation. We benchmarked our two pooling mechanisms against each other using combinations of direct and indirect byte buffers. We found that one of the pooling strategies (Buddy1) is faster than the other with a smaller memory footprint. Also, we demonstrated the performance gain of using direct byte buffers.

We released a beta version of our software in early September 2005. This release contains our buffering API with the two implementations of buddy allocation scheme. This API is self-contained and can be used by other Java applications for application level memory management. Currently, we are working to release additional messaging devices based on mpiJava and Myrinet eXpress(MX).

## References

1. Mark Baker, Bryan Carpenter, Geoffrey Fox, Sung Hoon Ko, and Sang Lim. An object-oriented Java interface to MPI. In *International Workshop on Java for Parallel and Distributed Computing*, San Juan, Puerto Rico, April 1999.
2. Mark Baker, Hong Ong, and Aamir Shafi. A status report: Early experiences with the implementation of a message passing system using Java NIO. Technical Report DSGTR06102004, DSG, October 2004. http://dsg.port.ac.uk/projects/mpj/docs/res/DSGTR06102004.pdf.
3. Chi-Chao Chang and Thorsten von Eicken. Javia: A Java interface to the virtual interface architecture. *Concurrency - Practice and Experience*, 12(7):573–593, 2000.
4. The Java Native Interface Specifications. http://java.sun.com/j2se/1.3/docs/guide/jni.
5. Donald Knuth. *The Art of Computer Programming: Fundamental Algorithms*. Addison Wesley, Reading, Massachusetts, USA, 1973.
6. MPJ Express. http://dsg.port.ac.uk/projects/mpj.
7. Myricom, The MX (Myrinet eXpress) library. http://www.myri.com.
8. The Java New I/O Specifications. http://java.sun.com/j2se/1.4.2/docs/guide/nio.
9. Matt Welsh and David Culler. Jaguar: enabling efficient communication and I/O in Java. *Concurrency: Practice and Experience*, 12(7):519–538, 2000.

# A Low Complexity and Robust Frequency Offset Estimation Algorithm for OFDM-Based WLAN Systems*

Sanghun Kim[1], Seokho Yoon[1], Hyoung-Kee Choi[1], and Sun Yong Kim[2]

[1] School of Information and Communication Engineering, Sungkyunkwan University, 300 Chunchun-dong, Jangan-gu, Suwon, Kyunggi-do, 440-746, Korea
{ksh7150, syoon, hkchoi}@ece.skku.ac.kr
[2] Department of Electronics Engineering, Konkuk University, 1 Hwayang-dong, Gwangjin-gu, Seoul 143-701, Korea
kimsy@konkuk.ac.kr

**Abstract.** This paper addresses the frequency offset estimation problem in the presence of the timing error for OFDM-based WLAN systems. When the timing error exists, the correlation value used for the frequency offset estimation could be reduced significantly due to the timing error, resulting in considerable degradation in estimation performance. In this paper, using the coherence phase bandwidth (CPB) and a threshold, a novel frequency offset estimation algorithm is proposed and based on which, an efficient timing error estimation algorithm is also proposed for the re-estimation of the frequency offset.

## 1 Introduction

Due to its high spectral efficiency and immunity to multipath fading and impulsive noise, orthogonal frequency division multiplexing (OFDM) technology has been widely adopted as a modulation scheme in wireless local area network (WLAN) standards, such as institute of electrical and electronics engineers (IEEE) 802.11a, high performance local area network type 2 (HiperLAN/2), and mobile multimedia access communication (MMAC), [1]. The OFDM technology, however, is very sensitive to frequency offset caused by Doppler effect and the mismatch of the oscillators in the transmitter and receiver [2]. Therefore, exact estimation of the frequency offset is one of the important technical issues in OFDM-based systems including WLAN [3].

Although a number of frequency offset estimation algorithms have been proposed [4]-[7], most of the investigations have been discussed under the assumption of no timing error. Practically, it is very difficult to determine the OFDM symbol arrival time at the receiver prior to the frequency offset estimation, because the signal-to-noise ratio (SNR) before the frequency offset estimation is

* This research was supported by grant No. R01-2004-000-10690-0 from the Basic Research Program of the Korea Science & Engineering Foundation. Dr. Yoon is the corresponding author.

V.N. Alexandrov et al. (Eds.): ICCS 2006, Part II, LNCS 3992, pp. 961–968, 2006.

very low. Therefore, it is not reasonable to assume that the timing error does not exist during the frequency offset estimation process. An efficient frequency offset estimation algorithm that is robust to the timing error was presented by Bang [8]. Bang's algorithm has good performance even though a timing error exists; however, the complexity in implementation rapidly increases as the frequency offset range increases.

In this paper, we propose a novel frequency offset estimation algorithm using the coherence phase bandwidth (CPB) and a threshold. The proposed algorithm has similar performance to that of Bang' algorithm in the presence of the timing error, yet has even lower computational complexity when compared with Bang's algorithm. We also address the problem of estimating the timing error, using the OFDM symbol compensated by the proposed frequency offset estimation algorithm.

## 2 The Effect of Symbol Timing Error on Frequency Offset Estimation

An OFDM symbol is generated by the inverse fast Fourier transform (IFFT) and expressed as, for $n = 0, 1, 2, \cdots, N-1$,

$$s_n = \frac{1}{\sqrt{N}} \sum_{l=0}^{N-1} S_l e^{j2\pi nl/N}, \tag{1}$$

where $S_l$ is a phase shift keying (PSK) or quadrature amplitude modulation (QAM) data symbol transmitted through the $l$th subcarrier and $N$ is the size of the IFFT. In the presence of the timing error, the received signal $r_n$ can be expressed as

$$r_n = s_{n-\delta} e^{j2\pi\epsilon(n-\delta)/N} + w_n, \tag{2}$$

where $\epsilon$ and $\delta$ represent the frequency offset and the timing error, which are normalized to the subcarrier spacing $1/N$, respectively and $w_n$ is the zero-mean complex additive white Gaussian noise (AWGN).

To recover the data symbol, the received signal is first demodulated using the fast Fourier transform (FFT) operation. The $k$th FFT output $R_k$ is then given by, for $k = 0, 1, 2, \cdots, N-1$,

$$R_k = S_{k-\epsilon} e^{-j2\pi\delta(k-\epsilon)/N} + W_k, \tag{3}$$

where $W_k$ is the FFT output of $w_n$. We consider the frequency offset estimation using the preamble, as in [4]-[8]. Thus, an estimate $\hat{\epsilon}$ of the frequency offset $\epsilon$ can be obtained as

$$\hat{\epsilon} = \arg\max_d \left\{ \left| \sum_{k=0}^{N-1} Z_k^* R_{(k+d)_N} \right| \right\}. \tag{4}$$

where $Z_k$ is the known preamble, $*$ is the complex conjugate operator, $d$ is the amount of cyclic shift, and $(\cdot)_N$ is the modulo-$N$ operator.

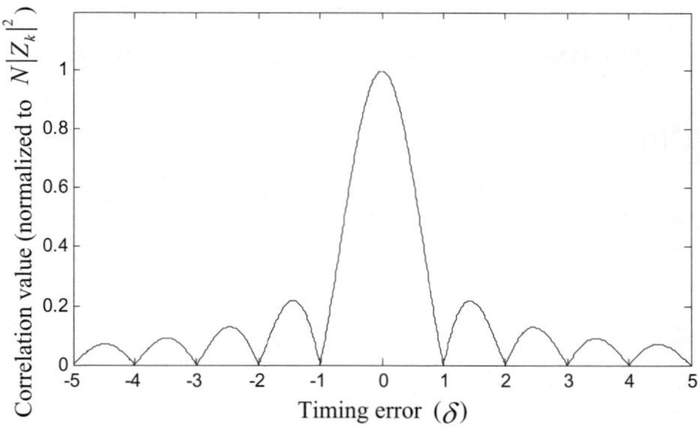


**Fig. 1.** The correlation value plotted as a function of timing error $\delta$ when $d = \epsilon$ and $N = 1024$.

When $d = \epsilon$ (i.e., when $\epsilon$ is correctly estimated), the correlation value (normalized to $N\left|Z_k\right|^2$) in (4) becomes $\left|e^{\frac{-j\pi\delta(N-1)}{N}}\frac{\sin(\pi\delta)}{\sin(\pi\delta/N)}\right|$ and is plotted as a function of the timing error $\delta$, as shown in Fig. 1. From Fig. 1, we can clearly observe that the correlation value used in the frequency offset estimation is very sensitive to the variation of the timing error, which implies that the correlation value could be reduced significantly due to the timing error even if the frequency offset is correctly estimated, resulting in considerable degradation in estimation performance (note that a large correlation value when $d = \epsilon$, is essential for the detection of the correct frequency offset estimate). This is the very effect of the timing error on the frequency offset estimation.

## 3 Proposed Algorithm

In Bang's algorithm, the correlation is performed in parallel for all possible values of $d$. As the spacing between the consecutive values of $d$ becomes smaller, therefore, the performance becomes more robust to the timing error variation; however, the computational complexity increases rapidly. In this paper, we propose an efficient algorithm that has similar performance, but less complexity for a given value of $d$, when compared with Bang's algorithm.

### 3.1 Frequency Offset Estimation

To alleviate the effect of the timing error on the frequency offset estimation, we consider the CPB within which the correlation value increases monotonically. The CPB can be expressed as [8]

$$\mathrm{CPB} = \frac{N}{2\delta^t}, \tag{5}$$

where $\delta^t$ is a maximum tolerated value of the timing error $\delta$. Using (4) and (5), we can calculate the correlation value $C$ for the frequency offset estimation as

$$C = \sum_{m=0}^{K-1} \left| \sum_{k=0}^{\mathrm{CPB}-1} Z^*_{k+m\mathrm{CPB}} R_{(k+m\mathrm{CPB}+d)_N} \right| , \tag{6}$$

where $K = N/\mathrm{CPB}$.

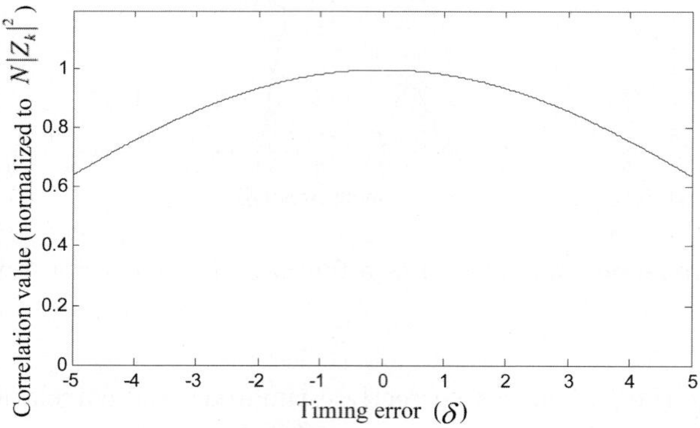


**Fig. 2.** The correlation value plotted as a function of the timing error $\delta$ when $d = \epsilon$, $\delta^t = 5$, and $N = 1024$.

Fig. 2 shows the correlation value plotted as a function of the timing error $\delta$ when $d = \epsilon$, $\delta^t = 5$, and $N = 1024$. It can be observed that the correlation value is relatively constant under the timing error variation. In [8], the correlation value $C$ is calculated in parallel for all possible values of $d$ and $\hat{\epsilon}$ is chosen as the $d$ corresponding to the maximum correlation value. In the proposed algorithm, on the other hand, the correlation value $C$ is first calculated for an initial $d$ and then compared with a given threshold. If the correlation value exceeds the threshold, the $d$ is decided to be the correct estimate of $\epsilon$. Otherwise, the FFT output is cyclically shifted by $d$ and the above procedure is repeated. The operation of the proposed algorithm is described in Fig. 3. The threshold $\eta$ is obtained as follows.

From (6), correlation value $\alpha$ over one CPB block is

$$\alpha = \left| \sum_{k=0}^{\mathrm{CPB}-1} Z^*_k R_{(k+d)_N} \right| . \tag{7}$$

If we assume $d = \epsilon$ and ignore AWGN,

$$\alpha = |Z_k|^2 \left| \sum_{k=0}^{\mathrm{CPB}-1} e^{-j2\pi\delta k/N} \right| . \tag{8}$$

It is easy to see that $\alpha$ has its minimum value $\alpha_{\min}$ when $\delta = \delta^t$. Thus,

$$\alpha_{\min} = |Z_k|^2 \left| 1 - j\cot\left(\frac{\pi\delta^t}{N}\right) \right| \tag{9}$$

and the minimum value $C_{\min}$ of $C$ can be obtained as

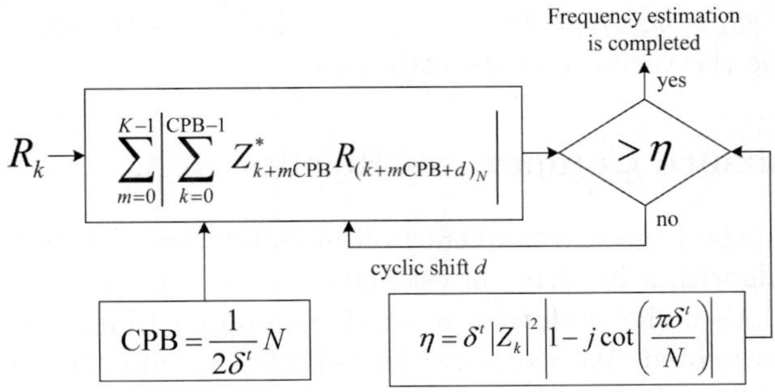


**Fig. 3.** Block diagram of the proposed frequency offset estimation algorithm

$$C_{\min} = 2\delta^t |Z_k|^2 \left| 1 - j\cot\left(\frac{\pi\delta^t}{N}\right) \right| . \tag{10}$$

Finally, the threshold $\eta$ is chosen such that

$$\eta = C_{\min}/2 = \delta^t |Z_k|^2 \left| 1 - j\cot\left(\frac{\pi\delta^t}{N}\right) \right| . \tag{11}$$

As a threshold becomes smaller, the detection probability for the correct estimate of $\epsilon$ and the false alarm probability increase, and vice versa. Thus, the threshold chosen as one half of $C_{\min}$ is one of the reasonable choices in view of the detection and false alarm probabilities, which decide the overall system performance.

### 3.2 Timing Error Estimation

The frequency offset is often time varying and, thus, the re-estimation for the frequency offset is required. In that case, it would be very useful if an estimate of the timing error is available, although the proposed algorithm is robust to the timing error. So, we address the problem of estimating the timing error, using the FFT output (i.e., OFDM symbol in the frequency domain) compensated by the proposed frequency offset estimation algorithm.

Assuming the frequency offset is perfectly estimated, we have the $k$th FFT output $R_k$ such that

$$R_k = H_k Z_k e^{j2\pi\delta k/N} . \tag{12}$$

where $H_k$ represents the frequency response of the channel. The argument $\Phi_k$ of $Z_k^* R_k$ is then given by

$$\Phi_k = \angle\left(Z_k^* R_k\right) = 2\pi\delta\frac{k}{N} + \angle\left(H_k\right) , \tag{13}$$

where $\angle$ denotes the argument operator of a complex number. As a result, if the channel does not change significantly during one OFDM symbol duration, the timing error can be estimated as follows.

$$\hat{\delta} = \frac{N}{2\pi} E\left\{\Delta_k\right\} , \quad \text{for } k = 1, 2, \cdots, N-1, \tag{14}$$

where $|\Delta_k| = |\Phi_k - \Phi_{k-1}| \leq 2\pi\delta^t/N$ and $E\{\cdot\}$ denotes the expectation and is taken to reduce the variance of the estimator.

## 4 Performance Comparison Results

In this section, the proposed frequency offset estimation algorithm is compared with Bang's algorithm in terms of estimation accuracy and complexity in implementation. Also, the performance of the proposed timing error estimation algorithm is presented. We consider the following parameters: $N = 1024$ and 2048, $\epsilon \in [0, 500]$, and $\delta^t = 16$. Two channel models are used: AWGN and multipath channels, the parameters of which are shown in Table 1. The time delays and amplitude attenuations of the second, third, and fourth paths are relative to those of the first path. For reference, the conventional frequency offset algorithm proposed by Nogami [4], which uses the assumption of no timing error, is also included in the comparison results.

**Table 1.** The channel models

| Path | First | Second | Third | Fourth |
|---|---|---|---|---|
| Time delay (sample) | 0 | 5 | 10 | 15 |
| Amplitude attenuation (dB) | 0 | 4 | 8 | 12 |
| Phase | Distributed uniformly over $[0, 2\pi)$ | | | |
| AWGN | 10 dB | | | |

Table 2 shows the complexity of Nogami's, Bang's, and the proposed algorithms. We can see from Table 2, for $L \gg 1$, where $L$ denotes the frequency offset range, the proposed algorithm has about a half computational complexity when compared with Nogami's and Bang's algorithms. Another important result is that the proposed algorithm does not require any memory for correlation value, unlike Nogami's and Bang's algorithms.

**Table 2.** Complexity comparison for Nogami's, Bang's, and the proposed algorithms

| Algorithm | Number of complex multiplication | Number of comparison operation | Size of memory for correlation value |
|---|---|---|---|
| Nogami's Algorithm | $LN$ | $L-1$ | $L$ |
| Bang's Algorithm | $LN$ | $L-1$ | $L$ |
| Proposed Algorithm | $\frac{L+1}{2}N$ | $\frac{L+1}{2}$ | - |

Tables 3 and 4 show the frequency offset estimation accuracy of Nogami's, Bang's, and the proposed algorithms in AWGN and multipath channels, respectively. From the tables, it is seen that Bang's and the proposed algorithms are quite robust to the timing error and dramatically outperform Nogami's algorithm.

**Table 3.** Accuracy comparison of the frequency offset estimation algorithms in AWGN channel

| Timing error (sample) | Accuracy (%) | | | | | |
|---|---|---|---|---|---|---|
| | Number of subcarrier ($N$): 1024 | | | Number of subcarrier ($N$): 2048 | | |
| | **Nogami's Algorithm** | **Bang's Algorithm** | **Proposed Algorithm** | **Nogami's Algorithm** | **Bang's Algorithm** | **Proposed Algorithm** |
| **0** | 100 | 100 | 100 | 100 | 100 | 100 |
| **1** | 0 | 100 | 100 | 0 | 100 | 100 |
| **2** | 0 | 100 | 100 | 0 | 100 | 100 |
| **5** | 0 | 100 | 100 | 0 | 100 | 100 |

**Table 4.** Accuracy comparison of the frequency offset estimation algorithms in multipath channel

| Timing error (sample) | Accuracy (%) | | | | | |
|---|---|---|---|---|---|---|
| | Number of subcarrier ($N$): 1024 | | | Number of subcarrier ($N$): 2048 | | |
| | **Nogami's Algorithm** | **Bang's Algorithm** | **Proposed Algorithm** | **Nogami's Algorithm** | **Bang's Algorithm** | **Proposed Algorithm** |
| **0** | 100 | 100 | 100 | 100 | 100 | 100 |
| **1** | 0 | 100 | 100 | 0 | 100 | 100 |
| **2** | 0 | 100 | 100 | 0 | 100 | 100 |
| **5** | 0 | 100 | 100 | 0 | 100 | 100 |

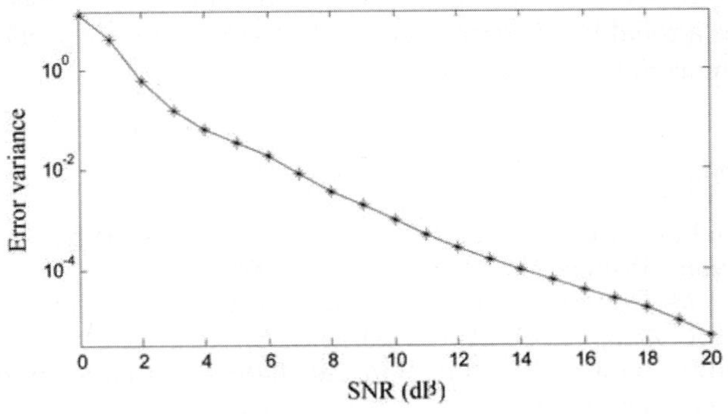


**Fig. 4.** Error variance of the proposed timing error estimation algorithm in AWGN channel

Figs. 4 and 5 show the error variance of the proposed timing error estimation algorithm as a function of SNR in AWGN and multipath channels, respectively. From the figures, we can observe that the error variance decreases rapidly as the SNR increases: specifically, the error variance becomes about $10^{-3}$ at SNR=10 dB, which is acceptable in general wireless applications and, thus, the proposed estimator should be able to be employed for the re-estimation of the frequency offset.

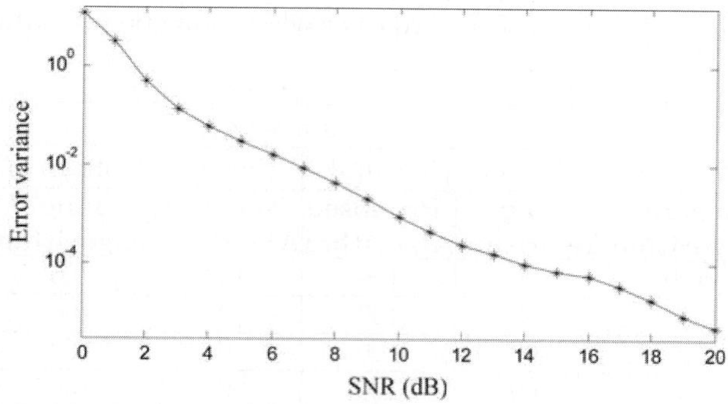


**Fig. 5.** Error variance of the proposed timing error estimation algorithm in multipath channel

## 5 Conclusion

We have first investigated the effect of the timing error on the frequency offset estimation in OFDM-based WLAN systems. Using the CPB and a threshold, then, we have proposed a low complexity and robust frequency offset estimation algorithm in the presence of the timing error. Based on which, an efficient timing error estimation algorithm has also been proposed for the re-estimation of the frequency offset. The comparison results have shown that the proposed frequency offset estimation algorithm dramatically outperforms Nogami's algorithm and performs similarly to Bang's algorithm in the presence of the timing error, with about a half complexity. It has also been shown that the proposed timing error estimator gives a reliable estimate of the timing error and, thus, should be useful for the re-estimation of the frequency offset.

## References

1. N. Prasad and A. Prasad, *WLAN Systems and Wireless IP for Next Generation Communications.* Boston, MA: Artech House, 2002.
2. T. Pollet and M. Peeters, "Synchronization with DMT modulation," *IEEE Commun. Mag.*, vol. 37, pp. 80-86, Apr. 1999.
3. R. V. Nee and R. Prasad, *OFDM for Wireless Multimedia Communications.* London, England: Artech House, 2000.
4. H. Nogami and T. Nagashima, "A frequency and timing period acquisition technique for OFDM systems," *in Proc. IEEE PIRMC*, Toronto, Canada, pp. 1010-1015, Sep. 1995.
5. B. Y. Prasetyo, F. Said, and A. H. Aghvami, "Fast burst synchronisation technique for OFDM-WLAN systems," *IEE Proceedings: Commun.*, vol. 147, pp. 292-298, Oct. 2000.
6. J. Li, G. Liu, and G. B. Giannakis, "Carrier frequency offset estimation for OFDM-based WLANs," *IEEE Signal Process. Lett.*, vol. 8, pp. 80-82, Mar. 2001.
7. K. Fazel and S. Kaiser, *Multi-Carrier and Spread Spectrum Systems.* West Sussex, England: John Wiley and Sons, 2003.
8. K. Bang, N. Cho, H. Jun, K. Kim, H. Park, and D. Hong, "A coarse frequency offset estimation in an OFDM system using the concept of the coherence phase bandwidth," *IEEE Trans. Commun.*, vol. 49, pp. 1320-1324, Aug. 2001.

# Simplified Signal Detection for BLAST Architecture with ML and DFE Detectors

Myung-Sun Baek, Byung-Su Kang, So-Young Yeo,
Young-Hwan You, and Hyoung-Kyu Song

uT Communication Research Institute, Sejong University, Seoul, Korea
sabman@sju.ac.kr, saturnus@sju.ac.kr, yeossoh@sju.ac.kr,
yhyou@sejong.ac.kr, songhk@sejong.ac.kr

**Abstract.** In this paper, a simplified signal detection technique is proposed for V-BLAST systems. In this proposed scheme, since the performance of V-BLAST system depends on the first sub-stream detection capability, $V$ probable streams are detected according to the first detected sub-stream of DFE detector and most probable stream is selected by likelihood test. And to reduce the computational complexity of DFE detector, a simple DFE detection scheme is also proposed. It has been shown that the proposed technique can detect the transmitted data more accurately than conventional DFE decoding scheme, and has very lower complexity than ML detector.

## 1 Introduction

Recently, there has been increasing interest in providing high quality and efficient broadband services over wireless and mobile links. One powerful performance enhancing technique, which holds the promise of offering significant increase in the system information capacity, is multiple input multiple output (MIMO) antenna technique [1]-[3]. Although current communication architectures such as the Vertical Bell Laboratories layered space-time (V-BLAST) have demonstrated very high spectral efficiencies, they have not reached the available capacities. In particular, there is a wide gap between the performance obtained by original detection algorithm of V-BLAST and optimum performance algorithms [4]. In general V-BLAST detection scheme which is called decision feedback equalization (DFE) or successive interference cancelation (SIC), since the performance depends on initial layer detection capability [5], the accurate detection of first layer is very important to increase an overall system performance.

In this paper, for more accurate detection of initial layer, simplified ML detection technique with DFE detection is proposed. At the same time, to reduce the computational quantity of the DFE detector we present simple DFE scheme with Cholesky factorization.

This paper is organized as follows. After describing the system model, general detection and proposed DFE detection is provided. Next, the proposed simplified signal detection algorithm is addressed. Simulation results and conclusions are finally showed.

V.N. Alexandrov et al. (Eds.): ICCS 2006, Part II, LNCS 3992, pp. 969–976, 2006.

## 2 System Description

We consider $M$ transmit antennas and $N \geq M$ receive antennas. The data is demultiplexed in $M$ data layers of equal length. These layers are mapped into certain modulation symbols. The layers are transmitted over the $M$ transmit antennas simultaneously.

Let $\mathbf{x} = [x_1\ x_2\ \cdots\ x_M]$ denote the $M \times 1$ vector of transmit symbols, then the corresponding $N \times 1$ receive signal vector is given by

$$\mathbf{r} = \mathbf{Hx} + \mathbf{n} \tag{1}$$

where $\mathbf{n}$ is zero-mean Gaussian noise with variance $\sigma_n^2$ and $\mathbf{H} = [\mathbf{h}_1\ \mathbf{h}_2\ \cdots\ \mathbf{h}_N]^T$ is an i.i.d random complex vector of multipath channel with each element of $\mathbf{h}_n = [h_{n1}\ h_{n2}\ \cdots\ h_{nM}]$.

## 3 ML and DFE Detector

### 3.1 General ML and V-BLAST Detector

Under the assumption of channel state information (CSI) perfectly known by the receiver, ML detection and decoding correspond to choosing the codeword $\mathbf{x}$ which determines the symbol combination with the smallest distance metric as a decision value as follows

$$\hat{\mathbf{x}}_{ML} = \underset{\hat{\mathbf{x}}}{\mathbf{argmin}} ||\mathbf{r} - \mathbf{H} \cdot \hat{\mathbf{x}}||. \tag{2}$$

Theoretically, ML detection would be the optimum way of recovering the transmitted data at the receiver. But as the computational effort is of order $L^M$ in $L$-QAM system, ML detection is not feasible for real time implementations [5]. Therefore, suboptimum detection schemes are generally used.

The representative and general detection scheme is ordered SIC (OSIC) detection scheme [2]. In the OSIC detector, the received signal vector $\mathbf{r}$ is multiplied by filter matrix $\mathbf{G}$ which is the Moore-Penrose pseudo-inverse denoted by $(\cdot)^\dagger$ of the channel matrix [2]. The zero-forcing (ZF) Moore-Penrose pseudo-inverse matrix is

$$\mathbf{G}_{ZF} = \mathbf{H}^\dagger = (\mathbf{H}^H\mathbf{H})^{-1}\mathbf{H}^H. \tag{3}$$

And in the MMSE detector, minimizing the mean squared error (MSE) between the actually transmitted symbols and the output of a linear detector leads to the filter matrix

$$\mathbf{G}_{MMSE} = (\mathbf{H}^H\mathbf{H} + \sigma_n^2)^{-1}\mathbf{H}^H. \tag{4}$$

With the definition of a $(M+N) \times M$ extended channel matrix $\underline{\mathbf{H}}$ through

$$\underline{\mathbf{H}} = \begin{bmatrix} \mathbf{H} \\ \sigma_n \mathbf{I}_M \end{bmatrix}, \tag{5}$$

the MMSE filter given by Eqn. (4) can be rewritten as follows [5]

$$\mathbf{G}_{MMSE} = (\underline{\mathbf{H}}^H \underline{\mathbf{H}})^{-1} \underline{\mathbf{H}}^H. \tag{6}$$

Assume that sub-stream $i$ yields the smallest estimation error or, equivalently, the largest signal-to-noise ratio (SNR) after linear nulling of the interference. It can be concluded that this sub-stream is associated with the row $\mathbf{g}^{(i)}$ of $\mathbf{G}$ that has minimum Euclidean norm, because this vector brings out the smallest noise enhancement. So, during the first step of the algorithm, only the decision static

$$\hat{x}_i = \mathbf{g}^{(i)}\mathbf{r} = \mathbf{g}^{(i)}(\mathbf{Hx} + \mathbf{n}) = x_i + \eta_i \tag{7}$$

with the effective noise $\eta_i = \mathbf{g}^{(i)}\mathbf{n}$ is used to find an estimate $\hat{x}_i$ for the transmit signal $x_i$.

The interference caused by this signal is then subtracted from the receive signal vector $\mathbf{r}$ and the $i$-th column is removed from the channel matrix, leading to a new system with only $M-1$ transmit antennas. This procedure consisting of nulling and cancelling is repeated for the reduced system until all signals are detected [2], [5]. In this procedure, the pseudo-inverse matrix calculation of the channel matrix is required in every layer detection. For example, the V-BLAST system with 8 transmit antennas requires 8 times pseudo-inverse matrix calculations. Therefore, as the number of transmit antenna is increased, the decoding complexity is highly increased. To reduce this high complexity, the decomposition theory is applied in V-BLAST decoding.

### 3.2 Proposed DFE Detector with Cholesky Factorization

In this subsection, the sorted MMSE DFE detector with Cholesky factorization is proposed.

Let $\mathbf{R} = \mathbf{H}^H\mathbf{H}$ denote the cascade of $\mathbf{H}$; since $\mathbf{R}$ is Hermitan, it possesses a unique Cholesky factorization of the form $\mathbf{R} = \mathbf{D}^H\mathbf{D}$, where $\mathbf{D}$ is lower triangular with real diagonal elements. Multiplying the received signal $\mathbf{r}$ in Eqn. (1) by the unitary matrix $\mathbf{D}^{-H}\mathbf{H}^H$ yields [5]

$$\mathbf{z} = \mathbf{D}^{-H}\mathbf{H}^H\mathbf{r} = \mathbf{Dx} + \boldsymbol{\eta} \tag{8}$$

where $\boldsymbol{\eta} = \mathbf{D}^{-H}\mathbf{H}^H\mathbf{n}$.

The whole proposed DFE algorithm based on Cholsky factoriztion is described as follows.

*Step 1.* The column of the channel matrix is sorted according to MMSE criterion in Eqn. (4). In this step, the MMSE criterion calculates $||\mathbf{G}_i||^2$ in Eqn (4) for $i = 1, 2, \cdots, M$, sorts them from the largest to the smallest and saves the sorted column index in sequence $\mathbf{k}$, $\mathbf{k} = \{k_1, k_2, \cdots, k_M\}$. Finally rearrange the columns of $\mathbf{H}$ according to the sorted index sequence $\mathbf{k}$

$$\mathbf{H}_{sort} = [\mathbf{H}_{k_1}\mathbf{H}_{k_2} \cdots \mathbf{H}_{k_M}] \tag{9}$$

where $\mathbf{H}_{k_i}$ is $[H_{1,k_i}\ H_{2,k_i}\ \cdots\ H_{N,k_i}]^T$.

*Step 2.* The rearranged channel matrix $\mathbf{H}_{sort}$ in Eqn. (9) is extended to $\underline{\mathbf{H}}_{sort}$ and Cholesky factorization is executed as follows

$$\underline{\mathbf{H}}_{sort}^{H}\underline{\mathbf{H}}_{sort} = \underline{\mathbf{R}}_{sort} = \underline{\mathbf{D}}^{H}\underline{\mathbf{D}}. \quad (10)$$

*Step 3.* In extended $\underline{\mathbf{D}}$ of Eqn. (10), the required part $\underline{\mathbf{D}}'$ which is $M$ by $M$ matrix is picked and signal detection is executed as follows

$$\begin{aligned}\mathbf{z}' &= (\underline{\mathbf{D}}')^{-H}\mathbf{H}_{sort}^{H}\mathbf{r} = \underline{\mathbf{D}}'\mathbf{x} + \boldsymbol{\eta}' \\ &= \begin{bmatrix} d_{1,1} & 0 & \cdots & 0 \\ d_{2,1} & d_{2,2} & \cdots & 0 \\ \vdots & \vdots & \ddots & \vdots \\ d_{M,1} & d_{M,2} & \cdots & d_{M,M} \end{bmatrix} \begin{bmatrix} x_{k_1} \\ x_{k_2} \\ \vdots \\ x_{k_M} \end{bmatrix} + \begin{bmatrix} \eta_1 \\ \eta_2 \\ \vdots \\ \eta_M \end{bmatrix}\end{aligned} \quad (11)$$

*step 4.* The detected signal $\hat{\mathbf{x}} = (\hat{x}_{k_1}\hat{x}_{k_2}\cdots\hat{x}_{k_M})$ is rearranged according to the order of transmit antenna by using index sequence $\mathbf{k}$.

In this process, the pseudo-inverse matrix calculation and sorting procedure of channel matrix are required only once.

## 4 Simplified Signal Detection

The proposed ML technique consists of three steps. In first step, $V$ sub-streams are detected, which are first outputs of DFE detector. Next, $V$ streams are detected according to the detected $V$ sub-streams in first step. In this step, the DFE decoding process is executed $V$ times. In final step, most probable stream among $V$ streams in second step is selected by likelihood test. The whole algorithm is described as follows.

*Step 1.* Detecting $V$ sub-streams at the first detection order of sorted DFE detector: The first detected sub-stream of Eqn. (11) can be presented as follows

$$\tilde{x}_{k_1} = z_1'/d_{1,1} = (d_{1,1}x_{k_1} + \eta_1)/d_{1,1}, \quad (12)$$

and $V$ probable sub-streams are determined. It can be expressed as follows

$$\hat{\mathbf{x}}_{k_1} = Q^{(V)}(\tilde{x}_{k_1}), \quad (13)$$

where $Q^{(V)}(k)$ is decision function which determines $V$ symbols of $L$-QAM or $L$-PSK system ($V \in L$) by checking the Euclidean distance from $k$ to each symbol, and $\hat{\mathbf{x}}_{k_1} = \{\hat{x}_{k_1}^{(1)}, \ldots, \hat{x}_{k_1}^{(v)}, \ldots, \hat{x}_{k_1}^{(V)}\}$ where $\hat{x}_{k_1}^{(v)}$ is $v$-th estimated symbol whose Euclidean distance from $\tilde{x}_{k_1}$ is $v$-th. Fig. 1 shows the process of $Q^{(V)}(\tilde{\mathbf{x}}_{k_1})$ when $V = 4$.

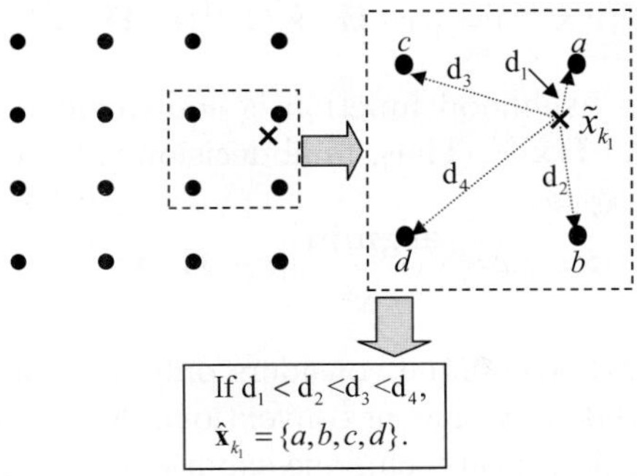


**Fig. 1.** Process of $Q^{(V)}(\tilde{\mathbf{x}}_{k_1})$ when $V = 4$

*Step 2.* Determining $V$ sub-stream by using sorted DFE detector according to detected $V$ sub-streams of step 1:

$$\begin{aligned}
&\hat{x}_{k_2}^{(1)} = Q[(z_2 - d_{2,1}\hat{x}_{k_1}^{(1)})/d_{2,2}], \cdots \\
&\hat{x}_{k_M}^{(1)} = Q[(z_M - \textstyle\sum_{i=1}^{M-1} d_{M,i}\hat{x}_{k_i}^{(1)})/d_{M,M}], \\
&\vdots \\
&\hat{x}_{k_2}^{(v)} = Q[(z_2 - d_{2,1}\hat{x}_{k_1}^{(v)})/d_{2,2}], \cdots \\
&\hat{x}_{k_M}^{(v)} = Q[(z_M - \textstyle\sum_{i=1}^{M-1} d_{M,i}\hat{x}_{k_i}^{(v)})/d_{M,M}], \\
&\vdots \\
&\hat{x}_{k_2}^{(V)} = Q[(z_2 - d_{2,1}\hat{x}_{k_1}^{(V)})/d_{2,2}], \cdots \\
&\hat{x}_{k_M}^{(V)} = Q[(z_M - \textstyle\sum_{i=1}^{M-1} d_{M,i}\hat{x}_{k_i}^{(V)})/d_{M,M}].
\end{aligned} \tag{14}$$

In this step, the DFE decoding process is executed $V$ times.
*Step 3.* Selecting the most probable stream among $V$ streams in step 2: In this step, final stream maximizing likelihood is selected among $V$ streams from second step.

To minimize the probability of error for a particular stream, the receiver should select $\hat{\mathbf{x}}^{(v)} = [\hat{x}_{k_1}^{(v)}, \hat{x}_{k_2}^{(v)}, \ldots, \hat{x}_{k_M}^{(v)}]$ that maximizes the a posteriori log-probability (LP)

$$\begin{aligned}
LP(\hat{\mathbf{x}}^{(v)}|\mathbf{r}) &= logp(\hat{\mathbf{x}}^{(v)}|\mathbf{r}) \\
&= logp(\hat{\mathbf{x}}^{(v)}) + logp(\mathbf{r}|\hat{\mathbf{x}}^{(v)}) - logp(\mathbf{r}) \\
&= LP(\hat{\mathbf{x}}^{(v)}) + LP(\mathbf{r}|\hat{\mathbf{x}}^{(v)}) - LP(\mathbf{r})
\end{aligned} \tag{15}$$

where the first term corresponds to a prior information on $\hat{\mathbf{x}}^{(v)}$, the second term corresponds to the extrinsic information (or likelihood function), and the third term is just a normalization factor [7].

Since consideration of a prior information on $\hat{\mathbf{x}}^{(v)}$ is not required, the receiver needs to evaluate the second term, the likelihood function. The corresponding likelihood can be expressed as

$$LP(\mathbf{r}|\hat{\mathbf{x}}^{(v)}) = [(\mathbf{r} \cdot \mathbf{H} \cdot \hat{\mathbf{x}}^{(v)})^H (\mathbf{r} \cdot \mathbf{H} \cdot \hat{\mathbf{x}}^{(v)})]. \quad (16)$$

Note that maximizing likelihood function is equivalent to minimize Euclidean distance between $\mathbf{r}$ and $\mathbf{H}\hat{\mathbf{x}}^{(v)}$. Thus, final decision value can be obtained as [8]

$$\hat{\mathbf{x}}_{final} = \underset{\hat{\mathbf{x}}^{(v)}}{\mathbf{argmin}} ||\mathbf{r} - \mathbf{H} \cdot \hat{\mathbf{x}}^{(v)}||. \quad (17)$$

Since this simplified ML technique considers only $V$ candidate streams, much lower complexity is need than that of conventional ML detection scheme.

The introduced ML detection technique can use any other DFE detector such as OSIC and QR decomposition as well as the Proposed DFE detector.

## 5 Simulation Results and Discussions

To evaluate the performance of the proposed technique, V-BLAST system with QPSK modulation is considered. It is assumed that the receiver acquires the perfect CSI.

Fig. 2 shows the bit error rate (BER) performance of the system with $M = N = 4$. It can be seen that the proposed sorted ZF and MMSE DFE detectors with Cholesky factorization achieve the similar performance as the ZF and MMSE SIC schemes, with significantly lower complexity.

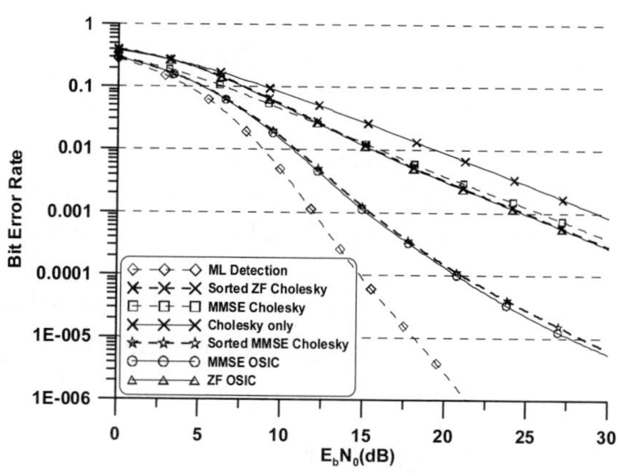


**Fig. 2.** BER performance comparison for $M = N = 4$

Fig. 3 shows the BER performance of the proposed technique with $M = N = 3$ according to the number of $V$. As expected, the more the number of $V$ is adopted, the better BER performance is acquired. In the case of $V = 4$, the performance of proposed method can be dramatically improved due to reduction of error propagation.

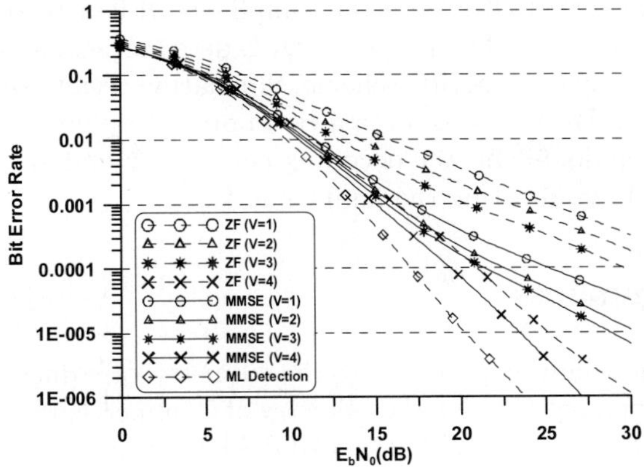


**Fig. 3.** BER performance of proposed technique with both ZF and MMSE detections according to the number of $V$

Fig. 4 illustrates the effect of the number of transmit antennas on the BER performance of proposed ML with V=4 and conventional SIC. In this figure, it can be easily observed that the performance of the proposed technique is better than that of classical VBLAST detection and as the number of transmit antennas is increased, the performance differences are larger due to the efficient diminution of noise enhancement.

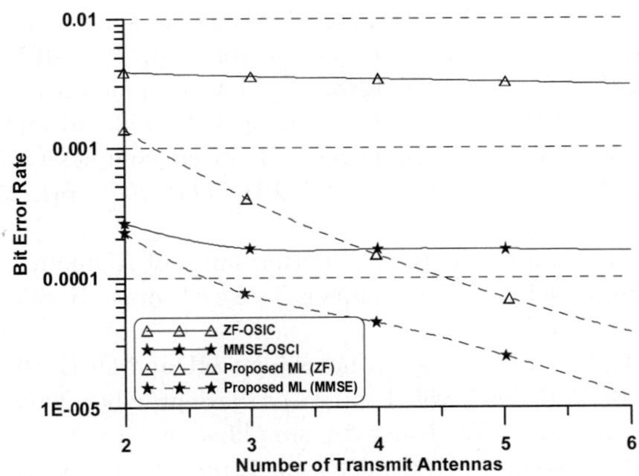


**Fig. 4.** BER performance according to the number of transmitting antennas for proposed ML ($V = 4$) and conventional OSIC

## 6 Conclusions

Since the performance of V-BLAST system is limited by the first detected sub-stream due to error propagation, we propose the simplified ML detection technique which can detect the first sub-stream more accurately. And to reduce the complexity of sorted DFE detector, simplified DFE detector with Cholesky factorization is presented. The proposed DFE detector achieves the similar performance as the conventional SIC scheme. Simulation results show that the performance of the V-BLAST system can be improved by adopting the proposed ML detection technique, and the decoding complexity and system performance can be controlled by adjusting the number of $V$.

## Acknowledgements

This work is financially supported by the Ministry of Education and Human Resources Development (MOE), the Ministry of Commerce, Industry and Energy (MOCIE) and the Ministry of Labor (MOLAB) through the fostering project of the Lab of Excellency and is supported by MIC Frontier R&D Program in KOREA.

## References

1. G. J. Foschini, "Layered space-time architecture for wireless communications in a fading environment when using multi-element antennas," *Bell Labs Technical Journal,* vol. 1, no. 2, pp. 41-59, Autumn 1996.
2. P. W. Wolniansky, G. J. Foschini, G. D. Golden, and R. A. Valenzuela, "V-BLAST: An architecture for achieving very high data rate over rich-scattering wireless channels," *Proc. of ISSSE'98,* pp.295-300, Italy, September 1998.
3. W. Y. Tao, R. S. Cheng, and K. B. Letaief, "Adaptive space-time coding system in fading channels," *Proc. of VTC 2001 Spring,* vol. 1, pp.103 - 107, May 2001.
4. A. Bhargave, R. Figueiredo, and T. Eltoft, "A detection algorithm for the V-BLAST system ," *Proc. of GLOBECOM '01,* vol. 1, pp.494 - 498, November 2001.
5. R. Bohnke, D. Wudden, K. Kammeyer, "Reduced complexity MMSE detection for BLKAST architectures," *Proc. of GLOBECOM 2003,* pp.2258-2262, September 2003.
6. A. Vielmon, Y. G. Li and J. R. Barry, "Performance of Alamouti transmit diversity over time-varying rayleigh-fading channels," *IEEE Trans. Wireless Comm.*, vol. 3, pp. 1369-1373, September 2004.
7. J. Zhao, X. Xu, C. Wu, and K. Ng, "A novel joint ML and DFE scheme for V-BLAST system," *Proc. WCNM 2005,* vol. 1, pp.472-475, September 2005.
8. H. Sung, K. B. L and J. W. Kang "A simplified maximum likelihood detection scheme for MIMO systems," *VTC 2003*, vol. 1, pp. 419-423, Octorber 2003.

# Scenario Decomposition Based Analysis of Next Generation Mobile Services

Dongchun Shin[1], Jinbae Kim[2], Seungwan Ryu[1,⋆], Donsung Oh[3], Joowan Lee[1], and Minhyung Kang[1]

[1]Department of Information Systems, Chung-Ang University, Korea
[2] Department of Computer&Information, Anyang Technical College, Anyang, Korea
[3] Mobile Telecommunications Research Division, ETRI, Taejon, Korea
{dcshin, rush2384, expertz}@cau.ac.kr, jbkim@ianyang.ac.kr, dsoh@etri.re.kr

**Abstract.** The next-generation mobile services have not been studied actively yet because there is no clear definition what the next generation mobile communication and its services will be. In this paper, we propose a service analysis framework for eliciting key service functionalities and core technologies using given various future mobile service scenarios. The proposed analysis framework, called *scenario based service analysis process (2SAP)*, consists of 4 sequential analysis steps, a scenario transformation, a scenario decomposition, an analysis of functions for each situation, and an elicitation of service functionalities.

## 1 Introduction

It is expected that the next generation mobile communication system will be a service-driven developed system differs from the technology-driven development approach of the second and the third generation mobile communication systems [1, 2]. Therefore, in the service-driven system development process for next generation system, various possible future mobile communication service scenarios are generated first. Then core technologies and network elements required to realize those service scenarios are elicited through scenario based analysis. Therefore, we believe that it is imperative to have a systematic scenario analysis method in such service-driven system development approach.

In this paper, we propose a service analysis framework for eliciting key service functionalities and core technologies when various future mobile service scenarios are given. The proposed analysis framework, called *scenario-based service analysis process (2SAP)*, consists of 4 sequential analysis steps, a scenario refinement, a scenario decomposition, an analysis of function for each situation, and an elicitation of service functionalities. Then the refined scenarios are divided into several sets of situations based on several criteria. Each situation is also divided into several scenes to elicit service functionalities that are required to realize such scenes. Finally, various basic core technologies that are required to implement each service functionality are obtained.

⋆ Prof. S. Ryu is the corresponding author.

V.N. Alexandrov et al. (Eds.): ICCS 2006, Part II, LNCS 3992, pp. 977–984, 2006.

This paper is organized as follows. In next section, concept of scenario is introduced followed by generic process of scenario based system design approaches. In section 3, we propose the scenario based service analysis process (2SAP) for next generation mobile services. Finally, we conclude this paper with giving further study issues.

## 2 Scenarios and Scenario Based System Design

### 2.1 Scenarios

In designing systems we need better means to describe contexts of user activities. A possible approach is explicit envisioning and documentation of typical and significant user activities early and continuingly in the development process. Such descriptions are often called *scenarios* that can support reasoning about situations of use, even before those situations are actually created [3]. A scenario is defined as a set of situations having common characteristics generated while using a system. In short, scenarios are stories about people and their activities. A scenario should contain basic characteristics such as agent (or users), proper objectives to use, and operational conditions [4]. In other words, scenarios could highlight goals, required procedures and role of users and subsystems by describing behavior of the system and activities of users. Such behaviors and activities include what users try to do with the system, what procedures are required or not adopted, and to be carried out to achieve, and what interpretations users make of what happens to them.

### 2.2 Scenario Based System Design

The basic idea behind scenario-based methods is that descriptions of technologies in use are essential in discussing and analyzing how those technologies could be used for reshaping of their activities. A scenario is used to analyze system requirement because of following properties [3].

- Scenarios are concrete but rough. On the one hand, a scenario could be a concrete design proposal that can be evaluated and improved by the designer. On the other hand, it is rough enough for designers to change it easily and to design details of the system later.
- Scenarios maintain an orientation to users and their needs. Since scenarios are work-oriented design objects that describes systems in terms of the work that users will try to do when they use those systems, it makes the design work remain with the needs and concerns of users. Therefore, scenarios can be made even more effective as work-oriented design objects when users are directly involved in creating them.
- A narrative scenario is written to evoke an image of people doing things, pursuing goals, using technology in support of these goals. The story enables readers to empathize with the people in the situation, which in turn leads to questions about motivations, intentions, reactions, and satisfaction.

The concrete and work-oriented nature of scenarios gives effective representation means for human-centered design activities, particularly when these activities include participation of end-users. Because of the fore-mentioned properties of scenarios, we can get various benefits described below from the scenario-based requirement analysis in designing a system.

- It gives useful information by decomposing complex problems into small sub-problems systematically. In addition, since useful information can be obtained from various problems independently, it is possible to synthesize requirements from technologies extracted from scenarios.
- By analyzing coherency of scenarios, it is possible to discover distinct objects and operation conditions that may give effective ways to avoid conflict among end users.
- It provides basic verification tools via coherency analysis and completeness of well-defined scenarios.
- It provides effective and rapid communication ways about system usage and other concerning matters between all person involved in system development work.

### 2.3 A generic Scenario Based Design Approach

The scenario-based design is a design approach that describes future system usage at the beginning of system development process. Similar to user-centric design approaches, the scenario based design approach focuses on how people use the target system to perform their works and activities not on detailed functionalities. The scenario based design approach consists of three sequential steps, the analysis step, the design step and the prototyping and evaluation step.

- In the analysis step, general understanding of the current problem status is acquired from the problem scenario and related claims. The problem scenario describes participants (called actors) involved in meaningful activities, and the related claims and characteristics of the current status that may result significant effects to actors. The problem scenario is transformed and refined repeatedly by means of repetitive design attempts. Then a root concept that supports detailed analysis and refinement of the claims by describing essential aspects of initial vision is derived from the initial requirement analysis. Several important elements that constitute the root concept are a high-level vision, basic rationale, and stakeholder group.
- The design step is consists of three activity design stages, the activity design stage, the information design stage and the reciprocal action design stage.
  - The activity design stage describes how the activities are improved or how does it changed by applying usable technologies. In this stage, new activities are generated by analyzing current meaningful activities. Then, activity scenario and related claims are obtained in this stage.

  - Information design stage describes how work information is organized and/or created while users are using the system. To achieve this purpose, special treatment of beginners or some who has no experience to use the system is required. In this stage, information scenarios and related claims are obtained.
  - The last stage of design step is the reciprocal action design stage. This stage will take concrete shape when detailed exchanged items between users and the system are specified and listed.
- Usage list is generated from scenarios and claims. Usage list can be divided into several detailed operations corresponding to representative operation situation. Then target usage objective is assigned to each detailed operation. Design scenarios provide realistic work situations, whereas claims make it possible to analyze those scenarios in terms of expected usage objectives. When a prototype has sufficient functionalities, detailed operations are evaluated and compared with target objectives.

Figure 1 shows a generic scenario-based design framework.

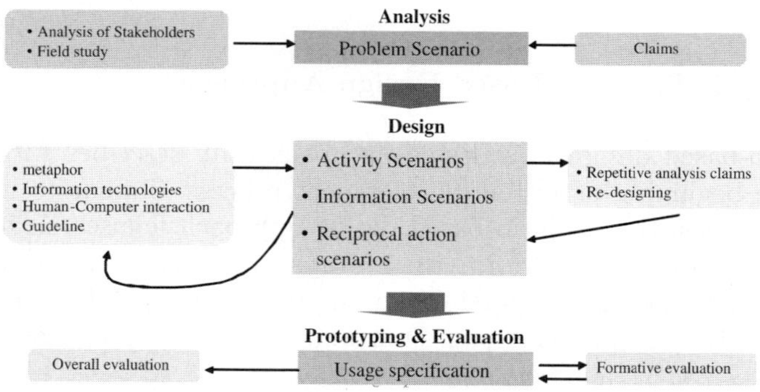


**Fig. 1.** A generic scenario-based design framework

## 3 Scenario Based Service Analysis Process

In this section we propose the scenario based analysis method for next generation mobile services, called the *scenario-based service analysis process (2SAP)*. The proposed analysis method consists of four sequential steps, *the scenario refinement process*, *the scenario decomposition process*, *the functional analysis process for each situation*, and *the service functionalities elicitation process*. In the first process, various crude narrative scenarios are refined or better analysis. In the second process, based on several criteria, each scenario is decompose into a set of situations having common characteristics generated while using a system. In the third process, each situation obtained in previous process is also divided into several scenes to elicit service functionalities that are required to realize such scenes. Finally, various basic core technologies that are required to implement each service functionality are obtained in the last process.

In this paper, initial and refined scenarios are described in natural language, and then expressed by scenario trees in the analysis step. Finally it is modelled as a *use case diagram* in both of functionality analysis and core technologies and service functionalities elicitation processes.

### 3.1 Scenario Refinement

In order to obtain core technologies and service functionalities of next generation mobile communication systems, various crude next generation mobile service scenarios are generated. These scenarios are written as a narrative story with lots of modifiers such as adjectives and adverbs. Although the "story" type crude service scenario is easy to understand, it is hard to get core technologies and service functionalities via analyzing such service scenarios. Thus in the scenario refinement process, initially generated story type scenario is refined to a concise one containing essential and necessary contents. In particular, all modifiers exists in a scenario are removed because they are used to make the scenario elegant and understandable for people. Then, the refined scenario is transformed into an enumerative type containing list of different situations. A possible initial story type service scenario and a refined and transformed enumerative scenario are shown in figure 2 and 3 respectively.

Tom living in LA, CA. is getting up in the morning with listening beautiful morning call music coming from his mobile phone. Today he is leaving to Toronto, Canada to meet a researcher work at Nortel Research Center to get useful information about network gaming business. He gets weather information about Toronto using his mobile phone. However, thanks to the ubiquitous computing and communication environment equipped in his house, he sees scenery of Toronto via display devices located in the wall and gets weather forecast via his audio system. Then various sensing devices having context information collection capabilities checks his physical status, and report his physical status to his mobile phone which is the gateway of the personal area network around him. Then the collected context information is transmitted to Dr. John who is Tom's private doctor. Then Tom gets chatting request from Dr. John for discussing today's physical treatment. After having good time with Dr. John via chatting service on his mobile phone, prescription data is sent to a pharmacy. The medicine will be delivered to the airport by the pharmacy making it according to prescription. Before leaving home, he downloads Marry's photo on his mobile phone immediately to see it in Toronto. Tom starts up his car using his mobile phone. After arriving airport, he ...........

**Fig. 2.** An example narrative story type scenario

### 3.2 Scenario Decomposition

In previous process a scenario is refined and transformed into an enumerative type having list of different situations. Although it is possible to analyze all the listed situations together, a better and efficient approach is to analyze it after decomposing the whole scenario into a number of small sub-problems. Therefore,

1. Tome get alarming service from his mobile phone.
2. Tom gets weather information about place which he is planning to visit.
3. Physical status of him is sent to his private doctor, and discuss with him via chatting service. Prescription is sent to pharmacy and medicine will be delivered to the airport.
4. Download photos of his daughter on his mobile phone.
5. .....

**Fig. 3.** An example of refined and transformed scenario into an enumerative one

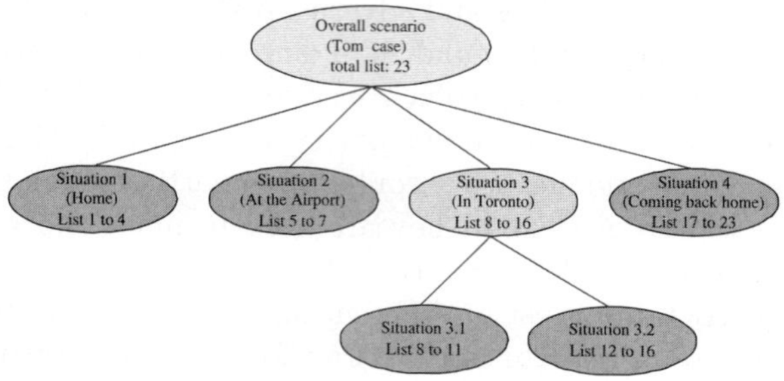


**Fig. 4.** An example of scenario tree representation based on decomposition and grouping approach

in 2SAP methods, the refined scenario is decomposed into a number of groups of situations based on coherency of them using grouping criteria.

The fundamental grouping criteria are location of a user, devices, and time. For example, in most cases, a new service is initiated when location and/or devices are changed. In addition, when, time especially date, is changed, it is highly probable to begin a new situation. However, this grouping task should be conducted carefully by taking circumstances of such changes on location, time and devices. When there are groups having larger numbers of list than other groups, they are decomposed into small sub-groups by applying the secondary decomposition criteria such as the number of actors and objectives of each situation. Figure 4 shows an example of scenario decomposition and grouping process.

### 3.3 Functional Analysis of Each Situation

In order to get service functionalities from situations described at the end nodes of the scenario tree, it is required to decompose each situation into smaller units called *scene*. A scene is defined as a target service moment that we are interested in. A scene is thus corresponds service functionality that is essential basis service technology felt directly by users [2]. The decomposition criteria used for getting several scenes from a situation are the initiation and/or releasing moment of services for a user gets it through his(or her) five senses. By means of intensive analysis of these scenes, we can obtain functionalities that make these scenes possible. We first establish a pool of possible service functionalities required to realize next generation mobile services using our previous research result [2]. Detailed functional analysis procedures for each situation are described below.

- Establish a pool of service functionalities. In this case, the pool can be established by assistant of expert groups or investigation of related technologies.
- derives scenes from a situation, and obtains service functionalities required to realize each scene using the pool of service functionalities.
- Gets a list of required functionalities for each scene.
- Finally, represents each scene (or a situation) as a use case diagram which describes relations between a user and service functions.

Table 1 gives example lists of scenes and required functionalities for the situation 1. Figure 5 also shows an example use case diagram of the situation 1.

**Table 1.** Example lists of scenes and required functionalities for the situation 1

| Scenes | Description | Functionalities |
|---|---|---|
| Scene 1 | Gets alarming service via mobile phone | • Mobile phone alarming |
| Scene 2 | Checking weather of the place to visit | • web browsing |
| Scene 3 | Health check & data transferring to a doctor | • Remote Control<br>• data messaging & file transfer |
| Scene 4 | Chatting request from a doctor | • Chatting / Messenger |
| Scene 5 | prescription is delivered to a pharmacist | • File transfer |
| Scene 5 | Download daughter's photos | • File transfer |

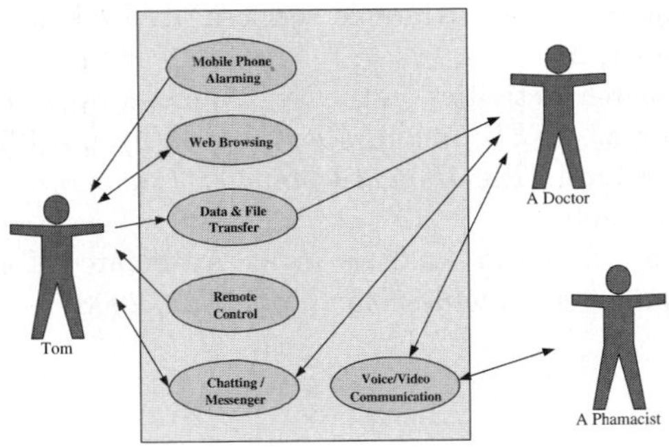


**Fig. 5.** An example use case diagram of the situation 1

### 3.4 Elicitation of Service Technologies

Several types of core technologies such as platform technologies, mobile terminal technologies, and service application technologies are necessary to implement service functionalities. Service technologies elicitation procedure is described as follow

- Establish a pool of service technologies with assistant of expert groups or investigation of related reference books or technical documents.
- Derives service technologies for each service functionalities using use case diagrams obtained in previous process. If there is a service technology that is required in realizing a service functionality but does not exist in the pool, creates a new service technology and makes a detailed technological list.
- Gets a list of required service technologies for each service functionalities.
- Finally, for each service functionality, draw a use case diagram describes relations between a user and service technologies.

## 4 Conclusion and Further Study Issues

In this article, we propose a service analysis framework for eliciting key service functionalities and core technologies when various future mobile service scenarios are given. The proposed analysis framework, called *scenario-based service analysis process (2SAP)*, consists of 4 sequential analysis steps, a scenario refinement, a scenario decomposition, an analysis of function for each situation, and an elicitation of service functionalities. We believe that our research activities and results on next generation mobile services, especially the 2SAP framework, will play important role in developing core mobile communication technologies as well as next generation ubiquitous mobile applications and services.

## References

1. S. Arbanowski et al., "I-centric Communications: Personalization, Ambient Awareness, and Adaptability for Future Mobile Services," *IEEE Comm. Magazine*, 42(10), pp. 63–69, September, 2004.
2. S. Ryu et al., "Research Activities on the Next Generation Mobile Communications and Services in Korea," *IEEE Comm. Magazine*, 43(9), pp. 122–131, 2005.
3. M. Rosson and J. Caroll, *The Human-Computer Interaction Handbook*, Lawrence Erlbaum Associates, 2002.
4. H. Zhu and L. Jin, "Scenario Analysis in an Automated Tool for Requirements Engineering," *Requirement Engineering*, 5, pp. 2–22, 2000.

# A Power Saving Scheme for Integrated WLAN and Cellular Networks*

SuKyoung Lee

Dept. of Computer Science, Yonsei University, Seoul, Korea
sklee@cs.yonsei.ac.kr.

**Abstract.** For the integration of WLAN and cellular networks, there have been recently several efforts to provide an optimized system design. However, it is noted from the existing schemes that in idle state, WLAN interface is supposed to be periodically turned on to receive periodic beacons from access points with cellular interface turned on at the same time for the paging from base station, resulting in significant power consumption. Therefore, in order to save the power consumption, when WLAN interface goes to idle state, we propose to turn off the WLAN interface without any periodic wake-up while at the same time, the existing paging of cellular networks are utilized. It is confirmed via both analytical modeling and simulation that the proposed scheme achieves better performance as compared to techniques currently adopted in WLAN in terms of power consumption.

## 1 Introduction

Most recently, a significant number of carriers are migrating towards heterogeneous wireless networking where Wireless Local Area Networks (WLANs) based on IEEE 802.11 standards and 3G (third-generation) cellular networks co-exist, in order to offer Internet access to end users with better Quality of Service (QoS). However, cost-effective integration of heterogeneous wireless networks is not yet accomplished, so that there have been many research activities having focused on this integration [1]-[3].

It is a challenge to maintain mobile terminals' active connections as they move across over different types of wireless networks (that is known as "vertical handoff"), while minimizing power consumption [1]. There have been several efforts to connect mobile device equipped with multiple interfaces, to the most optimal one among the access networks in terms of network performance. In [1], new performance metrics are introduced for providing seamless mobility. The authors of [2] proposed an end-to-end mobility management system that reduces ping-pong effect by obtaining the condition of different networks. Various network layer based inter-network handover techniques are addressed and their performance is investigated in [3].

* This work was supported by the Korea Research Foundation Grant funded by the Korean Government(MOEHRD) (KRF-204-2005-D00029).

V.N. Alexandrov et al. (Eds.): ICCS 2006, Part II, LNCS 3992, pp. 985–991, 2006.

Here, it is worthy of notice that a large portion of the power consumption in wireless interface is idle power. In most of the existing vertical handoff management schemes [1]-[3], MN (Mobile Node) must turn on both WLAN and cellular interfaces at once even in the idle state with power save mode to receive periodic beacons from Access Points (APs) while receiving paging messages from Base Station (BS), resulting in significant power consumption.

Therefore, in this paper, in order to save power consumption of MN in the integrated WLAN and cellular networks, when WLAN interface goes to idle state, we propose to turn off the WLAN interface without any periodic wake-up while at the same time, the existing Paging Channel (PCH) of cellular networks is utilized to turn on the WLAN interface for incoming data. Noting that WLAN is primarily used for accessing Internet where mainly downloading relatively long-lived multimedia data service and the WLAN interface is forced to be in the idle state for most of the time, it is beneficial to turn off WLAN interface for the idle period.

It is confirmed via both analytical modeling and simulation results that the proposed power saving scheme achieves better performance as compared to techniques currently adopted in WLAN in terms of the total power consumption.

## 2 Power-Efficient Interface Switching

When connected to a WLAN, WLAN interface card is usually in idle mode for around 70% of the overall time [4]. In idle state, the power consumption level for WLAN interface is about 10 times as great as that of cellular interface [5]. Our scheme targets on saving this power consumption, where WLAN interface is turned off without any periodic wake-up during idle period, which is called "inactive" state in this paper. Herein, WLAN interface has following states:

- Communication state: WLAN interface sends or/and receives data.
- Non-communication state: WLAN interface goes to this state when the data session completes.
  1. Typical WLAN: Idle state; 2. Proposed system: Inactive state

In our system model, in idle state, the cellular interface is assumed to listen continuously to the PCH to detect messages directed at APs in its cell as well as itself. The PCH of cellular network is utilized also to provide the information about APs in its current cell– Serving GPRS Support Node (SGSN) has an acquisition of IP address and Service Set Identifier (SSID) of all APs in its coverage where some 802.11 gateways serving the APs have a direct link to the SGSN, at the stage of configuration where BSs are initialized.

In this paper, we focus on downlink traffic since it is envisioned that 4th generation wireless system's traffic pattern will be highly asymmetrical, with 50/1 ratio or more favoring the downlink. Basing our system on 3GPP system, note that Radio Network Controller (RNC) is responsible for controlling user connections between a user and the core network with buffers for different users [6]. Thus, for downlink transmission, BS notifies MN when the number of packets

**Table 1.** Detailed procedure of the proposed power saving scheme for downlink traffic for integrated WLAN and cellular networks

**(1)** For downlink transmission, data traffic comes into a per-user-buffer at RNC when MN's WLAN interface is in inactive state. **(2)** Once the number of packets in the buffer reaches to threshold $n$, the BS notifies the MN about the existence of downlink data via its PCH. **(3)** On receiving the notification, WLAN interface is turned on (i.e. goes to communication state) and WLAN_IF_READY message is sent to corresponding 802.11 gateway while receiving the incoming data through cellular network. **(4)** Upon receiving the WLAN_IF_READY message, the 802.11 gateway sends VERTICAL_HANDOFF_COMPLETE message including the SSID and MAC address of the AP selected for the MN. **(5)** Once the SGSN receives the VERTICAL_HANDOFF_COMPLETE message, it starts to forward the remaining data to 802.11 gateway instead of to the BS. **(6)** The 802.11 gateway transmits the remaining data received from the SGSN to the MN.

in a per-user-buffer at RNC [6] for incoming data reaches a certain threshold $n$ (usually, $\leq$ maximum buffer size) so that the MN should not consume its power due to frequent turn-on and off actions which might occur if MN should be awakened upon receiving a packet from cellular network. The proposed power saving scheme for downlink traffic is described at length in Table 1.

## 3 Analytical Model of Power Consumption

Typically, users' packets are separated into buffers at RNC [6] while BS simultaneously serves a number of users so that BS should provide the opportunity for a scheduler to select the optimal user to transmit to at any instant [7]. To investigate the performance of the proposed scheme, we now develop an analytical model for heterogeneous networks treating the per-user-buffer at RNC with the traffic common in WLAN system, also considering the power on/off state of WLAN interface.

Unlike a cellular network, as far as traffic patterns are concerned, WLAN system is characterized as its traffic has an on/off behavior [8]. For example, for a web page transfer, a MN alternates between on period, $t_{\text{on}}$ and off period, $t_{\text{off}}$ during which a set of web pages is used to be downloaded as an application session and there is no traffic due to user's thinking time, respectively. The probabilities of a MN being in $t_{\text{on}}$ and of a MN being in $t_{\text{off}}$ are given by $p_{t_{\text{on}}} = \frac{\alpha}{\alpha+\beta}$ and $p_{t_{\text{off}}} = \frac{\beta}{\alpha+\beta}$, respectively. Both on and off periods are assumed to have an exponential distribution with the mean, $\beta^{-1}$ and $\alpha^{-1}$, respectively. During $t_{\text{on}}$ period, we assume that traffic arrives with a Poisson distribution the mean of which is $\lambda$. It is also assumed that each mobile user has only one TCP session active at a time from cellular network due to the lack of multi-tasking ability in current operating system of mobile device for cellular networks.

Let $N$ be the maximum number of packets allowed in a per-user-buffer at RNC. If we set the threshold $n$ to the buffer size, $N$, the arrival process to each

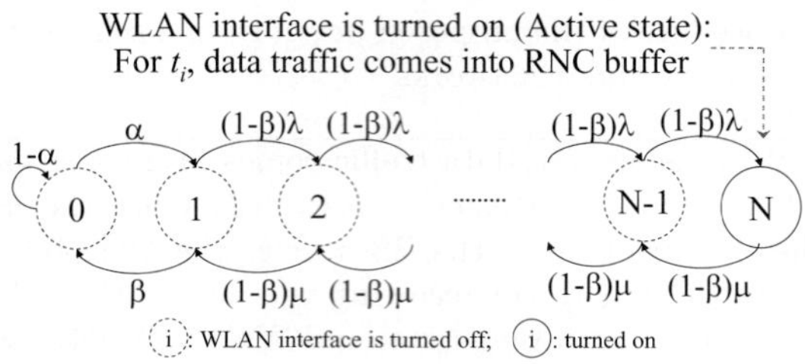


**Fig. 1.** State transition diagram for $n = N$

buffer of RNC can be modeled as an Interrupted Bernoulli Process (IBP). In the first place, to analyze the buffer under our proposed scheme with $n = N$, we note that during $t_{\text{off}}$ period, the buffer content must be zero. When the state of the buffer at RNC first makes a transition to $t_{\text{on}}$ state, for each subsequent transition to the same $t_{\text{on}}$ state, a packet arrives to the buffer with mean $\lambda$ while at the same time, the content of the buffer is transmitted to the MN through cellular interface with mean rate $\mu$ until the BS wakes up the corresponding WLAN interface by its PCH or the traffic flow ends without turning on WLAN interface due to the transient characteristic of the current traffic. Therefore, we are able to construct a Markov chain model for per-user-buffer at RNC as shown in Fig 1. If we denote $p_i$ as the steady-state probability that the buffer contains $i$ packets, then the steady-state probabilities are given by

$$p_0 = p_{t_{\text{off}}} \tag{1}$$

$$\alpha p_0 = \beta p_1 \tag{2}$$

$$\alpha p_0 + (1-\beta)\mu p_2 = (1-\beta)\lambda p_1 + \beta p_1 \tag{3}$$

$$\lambda p_{i-1} + \mu p_{i+1} = (\lambda + \mu) p_i \qquad 2 \le i \le N-2 \tag{4}$$

Let $x$ and $t_i$ denote the time elapsed from the moment when the WLAN interface is turned on and the time taken to initialize WLAN, respectively. Since $p_0 = \frac{\beta}{\alpha+\beta}$, from Eqs. 1-4, the rate at which WLAN interface is turned on from inactive state is given by

$$\begin{aligned} p_{on}^{(N)} &= (1-\beta)\lambda p_{N-1}(1-(1-\beta)\mu p_N)u(x) \\ &= \frac{\alpha(1-\beta)\lambda\rho^{N-2}}{\alpha+\beta}\left(1 - \frac{\alpha(1-\beta)\lambda\rho^{N-2}}{\alpha+\beta}\right)u(x) \end{aligned} \tag{5}$$

where $u(x) = \begin{cases} 1 & t_i - x > 0 \\ 0 & t_i - x \le 0 \end{cases}$ and $\rho = \frac{\lambda}{\mu}$. Thus, it can be easily known from the case of $n = N$ that the rate at which WLAN interface is turned on with $n = k$ becomes $p_{on}^{(k)} = (1-\beta)\lambda p_{k-1}(1-(1-\beta)\mu p_k)u(x) = \frac{\alpha(1-\beta)\lambda\rho^{k-2}}{\alpha+\beta}(1 - \frac{\alpha(1-\beta)\lambda\rho^{k-2}}{\alpha+\beta})u(x)$.

For the proposed scheme with $n = 1$ where WLAN interface is turned on from inactive state once the first packet of traffic flow comes to the buffer at RNC, the rate at which WLAN interface is turned on from inactive state is expressed as

$$p_{on}^{(1)} = \alpha p_0 = \frac{\alpha\beta}{\alpha+\beta}. \quad (6)$$

For the case of typical WLAN interface with power save mode, the sleeping interface must wake up periodically to receive regular beacons coming from the AP, which identify whether idle interface has data buffered at the AP and waiting for delivery. For idle mode, let $w$ and $s$ denote the wake-up rate for receiving beacons from AP and the sleep rate after receiving beacons from AP, respectively, both of which are fixed. Then, for idle state with power save mode, we have the rate at which WLAN interface is turned on as followings:

$$\begin{aligned} p_{on}' &= (\alpha + (1-\alpha)\frac{w}{w+s})p_0 \\ &= \frac{(w+s\alpha)\beta}{(w+s)(\alpha+\beta)}. \end{aligned} \quad (7)$$

Now we can compute the average power consumed for non-communication state during unit time $t$. Let $C_i$ and $C_d$ be the power consumed for wake-up to receive beacon/data and the baseline power consumption for doze state of idle period, respectively. Let $C_{ia}$ be the power consumption due to the wake up from inactive state to receive incoming data which will be vertically handed off from cellular network. For our proposed scheme, there is no baseline power consumption since MN goes to inactive state in non-communication state. Then, for non-communication state, the average power-consumption during time $t$ is

$$PW_{nc} = \begin{cases} (C_i P_{on}' + C_d p_0)t & \text{for typical WLAN} \\ C_{ia} P_{on}^{(k)} t & \text{for PCP with } n = k \end{cases} \quad (8)$$

where in general, $C_{ia}$ is known to be greater than $C_i$, so that $C_i$ and $C_{ia}$ are set to 0.68 W and 1 W, respectively while $C_d$ is assumed to have the value of 0.06 W in this study, where the power consumption values are obtained from [5].

## 4 Performance Evaluation

To obtain some simulation results as well as numerical results, we assume that during on period, the downlink transmission rate to MN is 80 Kbps while $t_{\text{on}}$ is 120 and 360 sec. It is also assumed that $t_i$ is 1 sec. The buffer size, $L_b$ and the packet size are set to is 20 Kbytes and 1000 bytes, respectively. For typical WLAN interface with power save mode, beacon period is set to 100 ms.

In Figs. 2 (a) and (b), the average power consumed by WLAN interface for non-communication state, $PW_{nc}$ is plotted as a function of off period ranging from 30 to 360 sec for $t_{on}$=120 and 360 sec, respectively when $R$=50 Kbps which denotes the data rate to BS during $t_{\text{on}}$. The graphs in Fig. 2 show that $PW_{nc}$ obtained with typical WLAN system is higher than the proposed system, where when $n = N$, the power consumption under the proposed scheme is lower than when $n = 1$ over all the ranges of off period. We also know that in terms of power consumption, the above observed performance improvement is valid, irrespective of whether the active period is 120 sec or 360 sec.

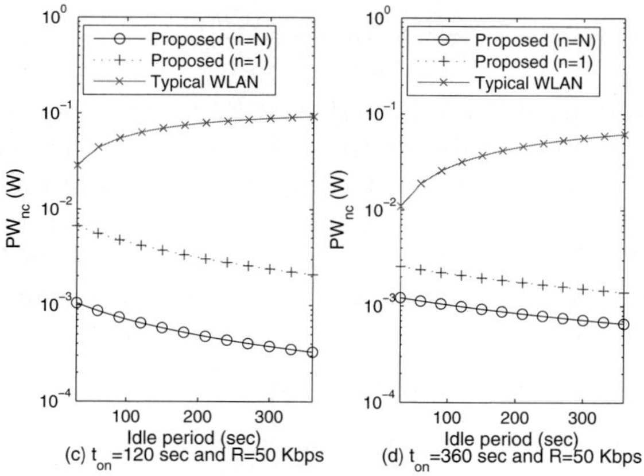


**Fig. 2.** Numerical power consumption versus off period

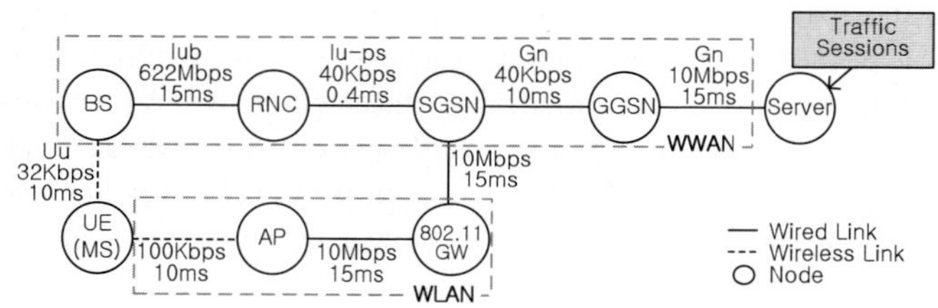


**Fig. 3.** Network topology

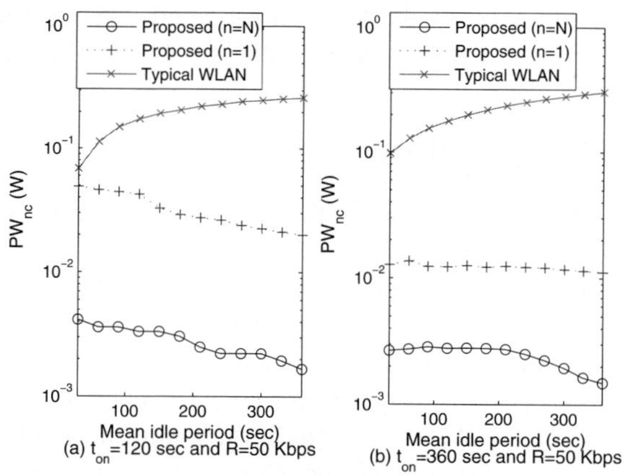


**Fig. 4.** Simulated power consumption versus varying mean idle period

Through simulation, we also compare the performance of the proposed scheme with typical WLAN system with periodic wake-up in idle state, in terms of power consumption for non-communication state. For this comparison, a simulation environment is created by extending the UMTS Terrestrial Radio Access Network (UTRAN) support modules [9]. The simulation topology is depicted in Fig. 3.

Fig. 4 plots the power consumed by the WLAN interface for non-communication state versus idle period ranging from 30 to 360 sec for $R$=50 Kbps when $t_{\mathrm{on}}$ is 120 and 360 sec. We observe that the power consumption behavior patterns shown in Fig. 4 are aligned along with the numerical results in Fig. 2. From the graphs in Fig. 4, it is also observed that the proposed scheme with $n = N$ works better with regard to power consumption, compared to that with $n = 1$ when the active period is smaller.

## 5 Conclusion

In this paper, we propose a power saving scheme for integrated WLAN and cellular networks, that utilizes the existing paging of cellular networks in order to turn off the WLAN interface during idle time. We aimed to save the power consumption resulting from periodic wake-up during the idle time.

Performance results for the proposed system are derived from the analysis as well as the simulation with regard to power consumption. The numerical and simulation results show that the power consumed in non-communication state for our scheme is lower than typical WLAN system because for the idle time, the power consumption resulting from the periodic wake-up is removed.

## References


1. J. McNair and F. Zhu, "Vertical Handoffs in Fourth-Generation Multinetwork Environments", *IEEE Wireless Commun. Mag.*, vol.11, no.3, pp.8-15, Jun. 2004.
2. C. Guo, Z. Guo, Q. Zhang and W. Zhu, "A Seamless and Proactive End-to-End Mobility Solution for Roaming across Heterogeneous Wireless Networks", *IEEE JSAC*, vol.22, no.5, pp.834-848, Jun. 2004.
3. R. Chakravorty, P. Vidales, K. Subramanian, I. Pratt, and J. Crowcroft, "Performance Issues with Vertical Handovers - Experiences from GPRS Cellular and WLAN Hot-spots Integration", *IEEE Proceedings Percom: Pervasive Computing and Communications*, 2004
4. Ram Chary, et al, "Power Management Technologies for WLAN enabled Handheld Devices", Intel Developer Forum, Fall 2003
5. K. Mahmud, M. Inoue, H. Murakami, M. Hasegawa, and H. Morikawa, "Energy Consumption Measurement of Wireless Interfaces in Multi-Service User Terminals ofr Heterogeneous Wireless Networks", *IEICE Trans. Commun.*, vol.E88-B, no.3, Mar. 2005.
6. 3GPP TS 25.322. "Radio Link Control (RLC) Protocol Specification ", version 6.2, Dec. 2004, Available via *http://www.3gpp.org*.
7. R.G. Mukhtar, S.V. Hanly, and L. Andrew, "Efficient Internet Traffic Delivery over Wireless Networks", *IEEE Commun. Mag.*, vol. 41, no. 21, pp. 46-53, Dec. 2003.
8. V.D. Park and M.S. Corson, "Models for an IP-based Cellular Network", *http://www.ieee802.org/20/Contribs/ C802.20-03-13r1.pdf*.
9. Enhanced UMTS Radio Access Network Extensions for ns-2 (ERUANE), Available via *http://www.ti-wmc.nl/eurane/*

# The Optimal Subchannel and Bit Allocation Problem for OFDM

Taehyung Park[1,⋆] and Sungbin Im[2]

[1] Department of Industrial and Information Systems Engineering,
Soongsil University, Seoul, Korea
tpark@ssu.ac.kr
[2] School of Electrical Engineering, Soongsil University, Seoul, Korea
sbi@nuri.net

**Abstract.** The advantages of the orthogonal frequency division multiplexing (OFDM) are high spectral efficiency, resiliency to RF interference, and lower multi-path distortion. To further utilize vast channel capacity of the multiuser OFDM, one has to find the efficient adaptive subchannel and bit allocation among users. In this paper, we propose an 0-1 integer programming model formulating the optimal subchannel and bit allocation problem of the multiuser OFDM. We provide an efficient Lagrangian dual formulation with fast solution approach. Simulation results are provided to show the effectiveness of the 0-1 integer programming model. MATLAB simulation on a system employing M-ary quardarature amplitude modulation (MQAM) assuming a frequency-selective channel consisting of five independent Rayleigh multipaths are carried with the optimal subchannel and bit allocation solution generated by 0-1 integer programming model.

## 1 Introduction

In this paper, we consider the optimal subchannel and bit allocation problem in the multiuser OFDM (Orthogonal Frequency Division Multiplexing) network. Multicarrier modulation technique is utilized for reducing intersymbol interference. Assuming that the transmitter knows the instantaneous channel transfer functions of all users, significant performance improvement is achieved if adaptive modulation is used with OFDM. Specially, subchannels with large channel gains carry more bits/OFDM symbol, while subchannels with deep fade carry one or zero bits/symbol.

Previous research on adaptive subchannel assignment and bit allocation problem are references [1], [3], [4], [5], and [7]. These research can be divided into two groups; In [1], [3], and [7], the problem is defined as minimizing the total channel power with the constrains of satisfying all user's data transmission demand and exclusive channel assignment for each user; In [4] and [5], the problem is maximizing the minimum of user's throughput subject to the power constraints. In

⋆ Both authors are supported by the research fund from Soongsil University, Seoul, Korea.

V.N. Alexandrov et al. (Eds.): ICCS 2006, Part II, LNCS 3992, pp. 992–995, 2006.

[1] and [7], the optimization problem is mainly formulated as a convex nonlinear programming while our model explicitly incorporates the integrality of the bit loading and channel assignment.

The remainder of this paper is organized as follows. In the next section, we introduce the assumptions underlying our model and formulate the adaptive subchannel assignment and bit loading problem into an equivalent 0-1 integer linear programming problem. In Section 3, we compare our solution with a static assignment approaches via Monte Carlo simulation.

## 2 Problem Formulation

We assume that there are $m$ users and the $i$th user has a data rate of $R_i$ bits per OFDM symbol. In the transmitter, the serial data from the $m$ users are fed into subchannel and bit allocation blocks which allocate bits from different users to different subchannels. We assume that there are $n$ subchannels. The instantaneous channel information on all the subchannels of all the users is assumed to be available at the transmitter. It is also assumed that the subchannel and bit allocation information is sent to each user by a separate channel. The complex symbols at the output of the modulators are transformed into the time domain samples by inverse fast Fourier transform. Cyclic extension of the time domain samples, the guard interval, is then added to ensure orthogonality among the subchannels. The transmit signal is then passed through different frequency selective fading channels to different users.

To formulate the problem into an optimization problem, we introduce the decision variables $x_{ij}$ and $y_{ij}$. $x_{ij}$ denotes the number of bits/symbol of the $i$th user that are assigned to the $j$th subchannel and has values in $C = \{0, 1, \ldots, M\}$. 0-1 variable $y_{ij}$ is one if user $i$ is assigned to subchannel $j$ and zero, otherwise. Since each user's data transmission requirement must be satisfied, we have $\sum_j x_{ij} = R_i \ \forall i$. Also, each subchannel cannot be shared by more than one user, we have $\sum_i y_{ij} \le 1 \ \forall j$. Since the allocated number of bits $x_{ij} = 0$ if user $i$ is not assigned to subchannel $j$, and if $y_{ij} = 1$, $x_{ij}$ can be positive, we have $x_{ij} \le M y_{ij} \ \forall i, j$.

The transmission power allocated to user $i$'s subchannel $j$ is assumed to be a convex nonlinear function $f_i(x_{ij})/\alpha_{ij}^2$ where $f_i(x_{ij})$ denotes the required power supporting $x_{ij}$ bits/symbol at a given bit per error(BER) when the channel gain is equal to unity, and $\alpha_{ij}$ the instantaneous channel gain that is assumed to be known to the transmitter. The specific form of $f_i(x_{ij})$ used is found on page 1752 in [7].

Instead of using a general integer variable $x_{ij}$, we can represent $x_{ij}$ in terms of 0-1 varibles $x_{ijk}$ with constraints $x_{ij} = \sum_{k \in C} k x_{ijk} \ \forall i, j$, $\sum_k x_{ijk} = 1 \ \forall i, j$, and $x_{ijk}$ integer $\forall i, j, k$. Since $f_i(x_{ij}) \in \{0, f_i(1), \ldots, f_i(M)\}$, the transmission power term $f_i(x_{ij})/\alpha_{ij}^2$ has value $c_{ijk} = f_i(k)/\alpha_{i,j}^2 \ \forall i, j$ and $k \in \{0, 1, 2, \ldots, M\}$. Now the optimization problem becomes the following 0-1 integer linear programming problem.

$$\begin{aligned}
\textbf{OFDM:} \quad \text{Minimize} \quad & \textstyle\sum_i \sum_j \sum_k c_{ijk} x_{ijk} && \text{(1a)}\\
\text{subject to} \quad & \textstyle\sum_j \sum_k k x_{ijk} = R_i \quad \forall i && \text{(1b)}\\
& \textstyle\sum_k x_{ijk} \le y_{ij} \quad \forall i, j && \text{(1c)}\\
& \textstyle\sum_i y_{ij} \le 1 \quad \forall j && \text{(1d)}\\
& x_{ijk} \ge 0 \quad \forall i, j, k && \text{(1e)}\\
& y_{ij} \ge 0 \quad \forall i, j && \text{(1f)}\\
& x_{ijk} \text{ integer} \quad \forall i, j, k && \text{(1g)}\\
& y_{ij} \text{ integer} \quad \forall i, j. && \text{(1h)}
\end{aligned}$$

Note that if $m = 1$, problem (1) becomes a Generalized Upper Bounded (GUB) constrained knapsack problem and [2] contains a specialized algorithm for the linear programming relaxation of formulation (1) when $m = 1$.

If we discard the integrality constraints upon $x_{ijk}$ and $y_{ij}$ from the formulation (1) and dualize the constraints (1b) employing dual variables $\mu = (\mu_i)$, the Lagrangian subproblem has

$$\sum_i \sum_j \sum_k (c_{ijk} + k\mu_i) x_{ijk} - \sum_i \mu_i R_i$$

as the objective function and has constraints (1c) - (1f). Since the Lagrangian subproblem is separable with respect to $j$, we have the following Lagrangian dual problem LD.

$$\textbf{LD:} \ \text{Maximize}_\mu \quad \theta(\mu) = \sum_j \theta^j(\mu) - \sum_i \mu_i R_i$$

where $\theta^j(\mu) \equiv \text{Minimum}\{\sum_i \sum_k (c_{ijk} + k\mu_i) x_{ijk} | \sum_k x_{ijk} \le y_{ij} \ \forall i; \sum_i y_{ij} \le 1; x_{ijk} \ge 0 \ \forall i, k; y_{ij} \ge 0 \ \forall i\}$.

It can be easily proved that $\theta^j(\mu) = \min\ \{\min_{i,k}\{c_{ijk} + k\mu_i\}, 0\}$, so we have $\theta(\mu) = \sum_j \min\ \{\min_{i,k}\{c_{ijk} + k\mu_i\}, 0\} - \sum_i \mu_i R_i$. The Lagrangian dual LD is nondifferentiable concave maximization problem and the subgradient maximization approach such as the Variable Target Value Method(VTVM) [6] can be applied.

## 3 Simulation Result and Summary

In this section, we present simulation results to show the performance of the proposed algorithm. In the simulations presented here, the wireless channel is modeled as a frequency-selective channel consisting of five independent Rayleigh multipaths with an exponential power delay profile, which are implemented using the MATLAB functions. We consider a system that employs M-ary quadrature amplitude modulation (MQAM) with $C = \{0, 2, 4, 6\}$. That is, 4-QAM, 16-QAM, and 64-QAM are used for carrying 2, 4, and 6 bits/symbol.

In order to evaluate the performance of the proposed algorithm, we carried out 500 independent experiments for each $E_b/N_0$. In the experiment, 128 subcarriers are used along with a total transmission rate of 512 bits/symbol over

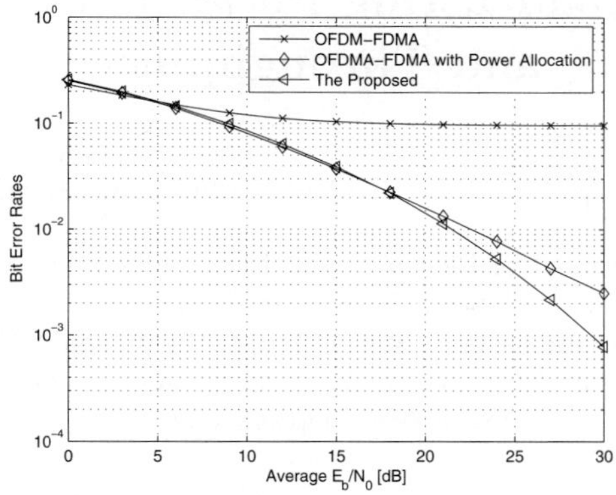


**Fig. 1.** Bit error rates versus $E_b/N_0$ for the three schemes

all users. The results are depicted in Fig. 1. For the purpose of comparison, the BER curves of two different schemes are provided in Fig. 1. One is "OFDM-FDMA"(frequency division multiple access), which distributes the same number of bits to each subcarrier, (4 bits per subcarrier in this experiment), while "OFDM-FDMA with Power Allocation" is to employ the adaptive power allocation algorithm with OFDM-FDMA. As shown in Fig. 1, the curve of OFDM-FDMA shows the error floor effect due to the frequency-selective channel. The proposed algorithm is superior to the other two methods, especially, for high $E_b/N_0$.

## References

1. Aggarwal, A., Meng, T.H.: A convex interior-point method for optimal OFDM PAR reduction. IEEE International Conference on Communications (ICC) (2005)
2. Johnson E.L., Padberg M.W.: A note on the knapsack problem with special ordered sets. Oper. Res. Lett. **1** (1981) 18–22
3. Kim, I., Lee, H.L., Kim, B., Lee, Y.H.: On the use of linear programming for dynamic subchannel and bit allocation in multiuser OFDM. GLOBECOM '01, **6** (2001) 3648–3652
4. Rhee, W., Cioffi, J.M.: Increasing in capacity of multiuser OFDM system using dynamic subchannel allocation. Proceedings of the IEEE VTC (2000)
5. Shen, Z., Andrews, J.G., Evans, B.L.: Optimal Power Allocation in Multiuser OFDM Systems. Proc. IEEE Global Comm. Conference **1** (2003) 337–341
6. Sherali, H.D., Choi, G., Tuncbilek, C.H.: A variable target value method for non-differentiable optimization. Oper. Res. Lett. **26** (2000) 1–8
7. Wong, C.Y., Cheng, R.S., Letaief, K.B., Murch, R.D.: Multiuser OFDM with adaptive subcarrier, bit, and power allocation. IEEE Journal on Selected Areas in Communications **17** (1999) 1747–1758

# Bluetooth Broadcasting Performance: Reliability and Throughput*

Kaan Dogan[1], Guray Gurel[1], A. Kerim Kamci[1], and Ibrahim Korpeoglu[2]

[1] Department of Electrical and Electronics Engineering,
Bilkent University, Turkey
{kaan, guray, kamci}@ee.bilkent.edu.tr
[2] Department of Computer Engineering, Bilkent University, Turkey
{korpe}@cs.bilkent.edu.tr

**Abstract.** This paper studies the performance of Bluetooth broadcasting scheme. The transmission of a Bluetooth broadcast packet is repeated several times to increase the reliability of broadcast. We have analyzed the effects of baseband ACL packet type preference, on the broadcast performance in terms of reliability and effective throughput, over different channel characteristics (i.e. bit error rate). As the result of our analysis, we determined the optimal packet type and re-transmission count combinations that can provide the highest effective throughput values for various practical BER ranges. These results can be used at Bluetooth baseband layer to dynamically adapt to varying channel conditions and to achieve a good broadcast performance.

## 1 Introduction

Bluetooth (BT) standard [1][2] includes a reliable broadcasting mechanism for applications that require data packets to be sent to more than one node in a BT piconet. Broadcasting is more energy and bandwidth efficient than separately unicasting the data to several nodes.

As broadcast packets are not acknowledged, reliability in BT broadcasting is achieved by simply repeating the transmission of the same packet $N_{BC}$ times. This causes, however, a tradeoff to be made between reliability and efficient channel utilization (high effective throughput).

In a BT piconet, the master node learns slaves' desired repetition count, $N_{BC}$, and announces the new value using the LMP_quality_of_service and LMP_quality_ of_service_req messages [2]. The master node can also determine an appropriate packet type to be used in broadcasting, since each packet type has different maximum size and error protection capability (different FEC coding). Broadcasting uses ACL (asynchronous connectionless link) packets and there are several types of ACL packets [2][3]. Hence it is possible for the master node to select the best combination for the packet type and repetition count in order to make a good tradeoff between reliability and throughput.

* This work is supported in part by The Scientific and Technical Research Council of Turkey (TUBITAK) under projects EEEAG-103E014 and EEEAG-104E028.

V.N. Alexandrov et al. (Eds.): ICCS 2006, Part II, LNCS 3992, pp. 996–999, 2006.

In this work, we have investigated the optimum values for packet types and retransmission counts in terms of their effects on channel utilization under different bit error rates with a custom simulator. As far as we know this is the first paper that analyzes the BT broadcasting scheme in terms of throughput and reliability. In the next section we describe our methodology and simulation model. Then we provide our results and discussions.

## 2 Methodology and Simulation Model

A master node can acquire feedback about the channel condition and about the efficiency of broadcasting using the BT link manager messages. After broadcasting a certain number of packets to slaves, the master can poll the slaves so that the slaves can inform the master about the $N_{BC}$ values they want the master to use. The desired $N_{BC}$ value can be computed by a slave based on the quality of the reception and this value corresponds to a BER range. In order to ensure a minimum reliability level for all slaves, largest $N_{BC}$ feedback should be taken into account. Then, the master can choose the most appropriate packet type and repetition count based on these offline calculations and the desired $N_{BC}$ feedback from the slaves. This can be achieved without modifying the broadcasting mechanism defined in the BT specification [2].

A BT baseband packet consists of mainly three parts: an access code, a packet header, and the payload. The header is protected by 1/3 FEC with whitening, providing an aggregate code distance of $d_0$=6, so the code is capable of correcting all single and detecting all quadruple errors [4]. The access code is assumed to be protected by pseudo-random noise along with the expurgated code, which provides large Hamming distances ($d_{min} = 14$). The trailer is a 4 bit pattern of either 1010 or 0101, and is used for DC compensation. Thus, the errors on access code and trailer are going to be ignored in the simulator [5].

DH packet payloads do not employ any protection scheme and error detection is achieved using CRC bits added to the end of the packets. Payloads of DM packets are protected by 2/3 FEC (10/15 shortened Hamming Code), i.e. it can correct all single bit and detect all double bit errors in a 15 bit block.

The analytical approach discussed in [6] clearly reflects the BER and Packet Loss Rate (PLR) relationship. For packet types without FEC protection, the packet loss rate $p$ is related to $BER$ and payload size $s$ (in bits) with the formula

$$p = 1 - (1 - BER)^s \tag{1}$$

For packets with 2/3 FEC protection, 15 bits are used to encode 10 bits of data, and this can correct one bit in every 15 bits. So the packet loss rate $p$ becomes:

$$p = 1 - ((1 - BER)^{15} + 15 \times BER \times (1 - BER)^{14})^{s/15} \tag{2}$$

In our simulations, when the packet type is specified, our simulator calculates PLR from BER and payload size. During the simulations, number of packets lost and the amount of data successfully transported are recorded. These values are then used to calculate effective throughput (goodput) and reliability. Effective throughput is expected to decrease when the $N_{BC}$ value increases, since several packets occupying the channel contain the same data. Using large packets increases throughput, on the other hand using FEC decreases the effective throughput due to redundant bits inserted into packets. We define reliability as the ratio of correctly received packets to the packets transmitted. Various applications may require different degrees of reliability in terms of correct delivery of baseband broadcast packets to slaves from a master. Increasing $N_{BC}$ or using FEC protection increases reliability. In contrast enlarging packet payload, increases PLR and therefore reduces reliability.

We performed simulations to find out the repetition counts and packet types that will yield the maximum throughput for a given reliability range.

## 3 Simulation Results

We simulated the BT broadcast mechanism for various practical RF channel BER values. For each BER value that is fixed, we have evaluated the performance of each of the BT baseband packet types in terms of reliability and throughput. The 95% confidence interval of the simulation results are very tight and they are not discernible from the graphs. In the presented results, we have only depicted confidence mean values that are obtained from batches of 10 independent simulation runs in which the master sends 5 MBs worth of broadcast data.

Performance of packet types under various BER values is closely related to each packet's vulnerability to error, which depends on packet payload size and the FEC scheme employed. In order to present the effect of $N_{BC}$ on reliability more clearly, we have simulated the broadcasting scenario for a fixed BER of $10^{-3}$. Figure 1 shows that relatively small values of $N_{BC}$ is sufficient for DM packets to provide reliable transmission. On the other hand, lack of FEC protection causes DH packets to exhibit poor performance. Finally, using packets with larger payloads reduces reliability, whereas increasing $N_{BC}$ yields more reliable broadcasts. When a desired level of reliability is specified, it is possible to find optimal packet types and $N_{BC}$ values that satisfy this constraint while providing maximum achievable throughput. Figure 2 show the maximum throughput obtained by ACL packet types under reliability constraints of 99.9%. As BER is increased, $N_{BC}$ value is increased to satisfy reliability constraint and consequently maximum throughput decreases. The packet type that exhibits the best throughput performance changes over the practical BER range. Optimal repetition count and packet selection for 99.9% reliability is tabulated in Table 1.

Having received the desired $N_{BC}$ values from the slaves, piconet master can find the corresponding BER according to the packet type being used in broadcast. The master can look up for optimal packet type and $N_{BC}$ value in Table 1 using this BER value.

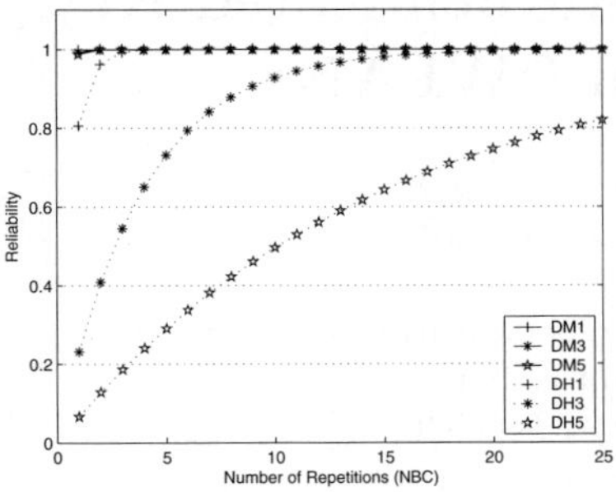


**Fig. 1.** Reliability vs $N_{BC}$ for BER of $10^{-3}$

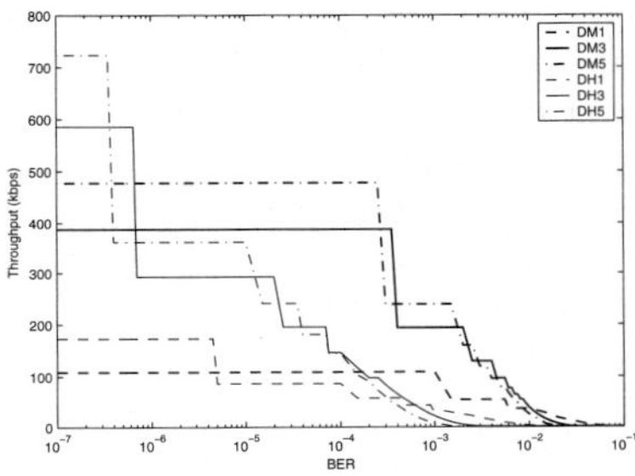


**Fig. 2.** Max. throughput for 99.9%

**Table 1.** Optimum packet types & $N_{BC}$ values for 99.9% reliability in all BER range

| Bit Error Rate | Packet Type | $N_{BC}$ | T (kbps) |
|---|---|---|---|
| [BER <7.5x10$^{-7}$] | DH5 | 1 | 723.00 |
| [7.5x10$^{-7}$, 10$^{-6}$] | DH3 | 1 | 585.60 |
| [10$^{-6}$, 10$^{-4}$] | DM5 | 1 | 477.80 |
| [10$^{-4}$, 5x10$^{-4}$] | DM3 | 1 | 387.30 |
| [5x10$^{-4}$, 5.5x10$^{-4}$] | DM5 | 2 | 238.90 |
| [5.5x10$^{-4}$, 2.5x10$^{-3}$] | DM3 | 2 | 193.50 |
| [2.5x10$^{-3}$, 3x10$^{-3}$] | DM5 | 3 | 159.30 |
| [3x10$^{-3}$, 5x10$^{-3}$] | DM3 | 3 | 129.00 |
| [5x 10$^{-3}$, 6.5x10$^{-3}$] | DM3 | 4 | 96.80 |
| [6.5x10$^{-3}$, 7x10$^{-3}$] | DM3 | 5 | 77.40 |
| [7x10$^{-3}$, 8x10$^{-3}$] | DM3 | 6 | 64.50 |
| [8x10$^{-3}$, 9.5x10$^{-3}$] | DM3 | 7 | 48.40 |
| [9.5x10$^{-3}$, 10$^{-2}$] | DM3 | 8 | 45.00 |
| [10$^{-2}$, 1.5x10$^{-2}$] | DM1 | 4 | 27.20 |
| [1.5x10$^{-2}$, 2x10$^{-2}$] | DM1 | 5 | 21.76 |
| [2x10$^{-2}$, 2.5x10$^{-2}$] | DM1 | 7 | 15.54 |
| [2.5x10$^{-2}$, 3x10$^{-2}$] | DM1 | 9 | 12.10 |

## 4 Conclusion

In this paper, we analyze the Bluetooth broadcasting scheme and provide results that can be used to improve broadcasting throughput and reliability. We used previously covered methods of reliability calculations in Bluetooth to come up with a novel method to create an adaptive packet type and repetition count ($N_{BC}$) selection scheme. We observed that our scheme can successfully improve the performance of current Bluetooth broadcasting scheme (in terms of throughput and reliability) for various radio channel conditions.

## References

1. Haarsten, J., Naghshineh, M., Inouye, J., Joeressen, O., Allen, W.: Bluetooth: Vision, Goals and Architecture, Mobile Computing and Communications Review, ACM SIGMOBILE, Vol. 2, No. 4 (1998)
2. Bluetooth Core Specification Version 1.1, Bluetooth SIG (1999)
3. Ronai, M.A.: Packet Level Simulation of Bluetooth Physical Layer, M.S. Thesis, Budapest University of Technology and Economics (2001)
4. Proakis, J., Salehi, M.: Communication Systems Engineering, Prentice Hall (2002)
5. Krassi, B.: Reliability of Bluetooth, 12th Conf. on Extreme Robotics, RTC, St. Petersburg (2001)
6. Chen, L-J.,Kapoor,R., Sanadidi,M.Y.,Gerla,M.:Enhancing Bluetooth TCP Throughput via Link Layer Packet Adaptation, In: Proceedings of IEEE ICC, France (2004)

# An Optimized Two Factor Authenticated Key Exchange Protocol in PWLANs

Eun-Jun Yoon and Kee-Young Yoo*

Department of Computer Engineering, Kyungpook National University, Daegu 702-701, Republic of Korea
Tel.: +82-53-950-5553; Fax: +82-53-957-4846
ejyoon@infosec.knu.ac.kr, yook@knu.ac.kr

**Abstract.** Recently, Park and Park proposed a new two factor authenticated key exchange (PP-TAKE) protocol that can be applied to low-power PDAs in Public Wireless LANs (PWLANs). The current paper proposes an efficient TAKE protocol based upon the PP-TAKE protocol. The computational cost of the proposed TAKE protocol is less than that of the PP-TAKE protocol, as is the number of steps needed to communicate is one fewer in which needs only three steps.

**Keywords:** Authentication, Wireless security, Key exchange, Optimized.

## 1 Introduction

Two factor authentication refers to the authentication of an entity by identifying (1) what a user remembers (the password) and (2) what the user has (a token or wireless terminal) in an integrated fashion. Two factor authentication requires a token as the second factor and a token-reading input device (a token reader). In generally, a token might be a smart card, a USB (Universal Serial Bus)-based smart key, or a wireless device. If a wireless device or USB-based smart key is used as a token in a PWLAN environment [1], no token reader is required. As the token stores within it the user's secret key or certificate data, however, it should be stored in a safe module that provides a certain level of tamper resistance.

In 2004, Park and Park [2] proposed a new mutual authentication and key establishment (PP-TAKE) protocol that can be applied to low-power PDAs in Public Wireless LANs (PWLANs), by using two factor authentication and precomputation. The PP-TAKE protocol provides mutual authentication, identity privacy, and half forward-secrecy. Also, the complex computations that the client must perform include only one symmetric key encryption and five hash functions during the protocol runtime. The number of communication steps in the PP-TAKE protocol, however, has clearly increased. A protocol has two important efficiency metrics - the number of steps and the number of rounds. One step involves the sending of data items from one party to a single destination at one time. A round includes all independent steps that can be sent and received

* Corresponding author.

V.N. Alexandrov et al. (Eds.): ICCS 2006, Part II, LNCS 3992, pp. 1000–1007, 2006.

in parallel. Thus, a participant can simultaneously send different messages to different destinations in a single round, and so multiple participants can send messages in a single round [3][4]. Accordingly, many protocols can be executed in less time by rearranging and sending the messages in parallel; such protocols are round efficient versions. The current paper proposes a modified PP-TAKE protocol in which the number of communication steps in the protocol is reduced by one and the computational cost is low. The computational cost of the proposed TAKE protocol is less than that of the PP-TAKE protocol, and the number of steps regarding communication is one fewer, only three steps are needed. Therefore, the proposed TAKE protocol is more efficient than the PP-TAKE protocol, which can be applied to low-power PDAs in Public Wireless LANs (PWLANs).

This paper is organized as follows: In Section 2, we briefly review the PP-TAKE Protocol. In Section 3, we propose an optimized TAKE protocol. In Sections 4 and 5, we analyze the security and efficiency of our proposed TAKE protocol, respectively. Finally, the conclusion is given in Section 6.

## 2 The Review of the PP-TAKE Protocol

This section briefly reviews Park-Park's TAKE (PP-TAKE) protocol. Some of the notations used in this paper are defined as follows:

- $A$, $B$ : client (supplicant) and authentication server ($AS$)
- $p$ : password
- $t$ : symmetric key used in symmetric key encryption
- $ID_A$ : client $A$'s identifier
- $E_K\{\}$, $D_K\{\}$ : encrypt and decrypt with symmetric key $K$
- $H()$ : a cryptographic hash function
- $sk_A$ : session key generated by client $A$
- $p$, $q$ : a large prime number and prime divisor of $(p-1)$
- $g$ : an element of order $q$ in $Z_p^*$
- $b$, $g^b$ : static private key and public key of $B$
- $\oplus$: a bitwise exclusive or operation

The TAKE protocol is based on DH (Diffie-Hellman) key agreement and can be modified to work in an arbitrary finite group. In terms of phases, the TAKE protocol includes enrollment phase, a precomputation phase and a real-execution phase. Fig. 1 shows the real-execution phase of the PP-TAKE protocol. For simplicity, the operator mod$p$ will be omitted.

**The Enrollment Phase:** The client and the Authentication Server ($AS$) determine and share the password $\langle \pi \rangle$ and symmetric key $\langle t \rangle$, that are used for symmetrical algorithms such as 3DES or Rijndael. The $AS$ selects a random number $\langle b \rangle$ from numbers ranging between $[1\ (q-1)]$ as its private key to a specific client, and the selected number is stored in a secure database. The client is informed of $AS$'s public key $\langle g^b \rangle$ and domain parameters $\langle p, q, g \rangle$. The client stores the symmetric key $\langle t \rangle$ in a secure token. Since the $AS$'s public key $\langle g^b \rangle$ and domain parameters $\langle p, q, g \rangle$ can be made public, they do not have to be stored in a secure location.

| Client $A$ $(\langle\pi\rangle, \langle t\rangle)$ | | Authentication Server $B$ $(\langle\pi\rangle, \langle t\rangle, \langle b\rangle)$ |
|---|---|---|
| Precomputation $\begin{bmatrix} x \in_R Z_q \\ g^x, c = g^{bx} \end{bmatrix}$ | | |
| | $\xrightarrow{H(ID_A, g^b)}$ | |
| $f = H(\pi, t, ID_A)$ | $\xleftarrow{r}$ | $r \in_R Z_q$ |
| $e = E_f\{g^x\}$ | | |
| $sk_A = H(c, g^x, r, ID_A)$ | | |
| $M_A = H(sk_A, \pi, t, g^b)$ | $\xrightarrow{e, M_A}$ | $f = H(\pi, t, ID_A)$ |
| | | $g^x = D_f\{e\}$ |
| | | $c = g^{xb}$ |
| | | $sk_B = H(c, g^x, r, ID_A)$ |
| | | $M_A \stackrel{?}{=} H(sk_B, \pi, t, g^b)$ |
| $M_B \stackrel{?}{=} H(sk_A, \pi, t, ID_A)$ | $\xleftarrow{M_B}$ | $M_B = H(sk_B, \pi, t, ID_A)$ |

Session key $sk_A = sk_B = H(c, g^x, r, ID_A)$

**Fig. 1.** The PP-TAKE protocol

**The Precomputation Phase:** Precomputation is performed off-line prior to the execution of the protocol. It reduces time and the computational load during the protocol execution. The client's wireless device performs precomputation during idle time or at the time of power on. In order to be more specific, a random number $\langle x\rangle$ is selected from $[1 \sim (q-1)]$, and then $g^x$ and $c = g^{bx}$ are calculated off-line prior to the protocol execution.

**The Real-Execution Phase:** The real-execution phase performs mutual entity authentication and session key establishment and it consists of the following four steps:

Step 1. $A \rightarrow B$: $H(ID_A, g^b)$

In order to connect to PWLANs service, client $(A)$ sends to $AS(B)$ its identifier $ID_A$ and $AS$'s public key $\langle g^b\rangle$, which has been hashed into $H(ID_A, g^b)$. If the client identifier uses a NAI (Network Access $ID$) to support global roaming and accounting (ex: userid@realm.com), the user name portion and the $g^b$ are hashed into $H(userid, g^b)$ and the realm portion are sent together as well.

Step 2. $B \rightarrow A$: $r$

Upon receipt of $H(ID_A, g^b)$, $B$ extracts $\langle H(ID_A, g^b)\rangle$, $\langle ID_A\rangle$, $\langle\pi\rangle$, $\langle t\rangle$, $\langle b\rangle$ from its database. $B$ selects a random number $r \in_R Z_q^*$ and sends it to $A$.

Step 3. $A \rightarrow B$: $e, M_A$

Upon receipt of $\langle r\rangle$, $A$ computes $f = H(r, \pi, t)$ and $e = E_f\{g^x\}$, using $f$ as a symmetric key for the encryption of $g^x$. $A$ then computes the session key $sk_A = H(c, g^x, r, ID_A)$ and generates $M_A = H(sk_A, \pi, t, g^b)$. $A$ sends $\langle e\rangle$ and $\langle M_A\rangle$ to $B$.

Step 4. $B \rightarrow A$: $M_B$

Upon reception of $\langle e\rangle$ and $\langle M_A\rangle$, $B$ computes $f = H(r, \pi, t)$ and $g^x = D_f\{e\}$, using $\langle f\rangle$ as a symmetric key for the decryption of $\langle e\rangle$. Then

$B$ computes $c = g^{xb}$ and $sk_B = H(c, g^x, r, ID_A)$ and confirms if $\langle M_A \rangle$, received from $A$ and $H(sk_B, \pi, t, g^b)$ computed by $B$ are identical. If they are identical, $A$'s authentication is successful and $B$ accepts $\langle M_A \rangle$. $B$ then computes $M_B = H(sk_B, \pi, t, ID_A)$ and sends it to $A$. $A$ checks to see if $\langle M_B \rangle$, that has been received from $B$ and $H(sk_A, \pi, t, ID_A)$, as computed by $A$, are identical. If they are identical, $B$'s authentication is successful and $A$ accepts $\langle M_B \rangle$. If $A$ and $B$ accept $\langle M_B \rangle$ and $\langle M_A \rangle$ respectively, mutual authentication is successful.

## 3 Optimized TAKE Protocol

This section proposes an optimized TAKE protocol. An enrollment phase and a precomputation phase are equal to the PP-TAKE protocol. Fig. 2 shows a real-execution phase of the proposed optimized TAKE protocol. Unlike the PP-TAKE protocol, the proposed protocol does not need a symmetric encryption/decryption operation. The proposed protocol uses bitwise exclusive or ($\oplus$) operation for the protection of the precomputed value $g^x$. The proposed real-execution phase requires only three steps. They are as follows:

**The Real-Execution Phase:** The real-execution phase performs mutual entity authentication and session key establishment and it consists of the following steps:

Step 1. $A \rightarrow B$: $H(ID_A, g^b), e$
In order to connect to the PWLANs service, client ($A$) computes $f = H(\pi, t, ID_A)$ and $e = f \oplus g^x$, and sends $H(ID_A, g^b)$ and $e$ to $AS(B)$. If the client identifier uses a NAI (Network Access $ID$) to support global roaming and accounting (ex: userid@realm.com), the user name portion and the $g^b$ hashed into $H(userid, g^b)$ and the realm portion are sent together as well.

Step 2. $B \rightarrow A$: $M_B, r$
Upon receipt of $H(ID_A, g^b)$ and $e$, $B$ extracts $\langle H(ID_A, g^b) \rangle$, $\langle ID_A \rangle$, $\langle \pi \rangle$, $\langle t \rangle$, $\langle b \rangle$ from its database. $B$ selects a random number $r \in_R Z_q^*$ and computes $f = H(\pi, t, ID_A)$ and $g^x = e \oplus f$. $B$ then computes $c = g^{xb}$, $sk_B = H(c, g^x, r, ID_A)$ and $M_B = H(sk_B, \pi, t, ID_A)$, and sends $M_B$ and $r$ to $A$.

Step 3. $A \rightarrow B$: $M_A$
Upon receipt of $M_B$ and $r$, $A$ then computes session key $sk_A = H(c, g^x, r, ID_A)$ and checks to see if $\langle M_B \rangle$, that has been received from $B$ and $H(sk_A, \pi, t, ID_A)$, as computed by $A$, are identical. If they are identical, $B$'s authentication is successful and $A$ accepts $\langle M_B \rangle$. $A$ then generates $M_A = H(sk_A, \pi, t, g^b)$ and sends it to $B$. Upon reception of $\langle M_A \rangle$, $B$ confirms if $\langle M_A \rangle$ received from $A$ and $H(sk_B, \pi, t, g^b)$, as computed by $B$, are identical. If they are identical, $A$'s authentication is successful and $B$ accepts $\langle M_A \rangle$. If $A$ and $B$ accept $\langle M_B \rangle$ and $\langle M_A \rangle$ respectively, mutual authentication is successful.

| Client $A$ $(\langle\pi\rangle, \langle t\rangle)$ | | Authentication Server $B$ $(\langle\pi\rangle, \langle t\rangle, \langle b\rangle)$ |
|---|---|---|
| Precomputation $\begin{bmatrix} x \in_R Z_q \\ g^x, c = g^{bx} \end{bmatrix}$ | | |
| $f = H(\pi, t, ID_A)$ <br> $e = f \oplus g^x$ | $\xrightarrow{H(ID_A, g^b), e}$ | $r \in_R Z_q$ <br> $f = H(\pi, t, ID_A)$ <br> $g^x = e \oplus f$ <br> $c = g^{xb}$ <br> $sk_B = H(c, g^x, r, ID_A)$ <br> $M_B = H(sk_B, \pi, t, ID_A)$ |
| $sk_A = H(c, g^x, r, ID_A)$ <br> $M_B \stackrel{?}{=} H(sk_A, \pi, t, ID_A)$ <br> $M_A = H(sk_A, \pi, t, g^b)$ | $\xleftarrow{M_B, r}$ <br> $\xrightarrow{M_A}$ | $M_A \stackrel{?}{=} H(sk_B, \pi, t, g^b)$ |

Session key $sk_A = sk_B = H(c, g^x, r, ID_A)$

**Fig. 2.** The optimized TAKE protocol

After Step 3, $A$ and $B$ have had it validated that the common secret session key $sk_A = sk_B = H(c, g^x, r, ID_A)$.

## 4 Security Analysis

This section analyzes the security of the proposed TAKE protocol. Here, the following security properties must be considered in the proposed [2]: identity protection, explicit mutual authentication, session key establishment, forward-secrecy, resistance to an off-line dictionary attack, key confirmation, and non-repudiation. The followings are used to analyze the security properties in the proposed protocol.

(1) *The proposed TAKE protocol provides identity protection:* To ensure the privacy of personal communication, it is necessary to protect a client's identity from passive attacks such as eavesdropping. Also, identity protection is particularly useful for the client to whom a dynamic IP address is allocated by the DHCP. In Step 1 of the proposed TAKE protocol, upon receiving an $ID$ request from the $AP$, the client sends $H(ID_A, g^b)$ instead of its real identity $ID_A$, to prevent passive attackers such as eavesdroppers, from knowing the client's identity. The $AS$, however, needs to be able to match the pseudonym of the client to its real identity $ID_A$.

(2) *The proposed TAKE protocol provides explicit mutual authentication:* Since attackers can launch Man-in-the-Middle (MitM) attacks by installing a rouge AP or a rouge radio NIC between the client and the Authentication Server (AS), explicit mutual authentication between the client and the network is necessary to prevent MitM attacks. The client must know the password $\langle\pi\rangle$, the symmetric key $\langle t\rangle$, the session key $\langle sk_A\rangle$, and the $AS$'s public key $\langle g^b\rangle$ in order to compute the $M_A$ for client authentication, while the $AS$ must know the password $\langle\pi\rangle$, the symmetric key $\langle t\rangle$, the session key $\langle sk_B\rangle$, and its private key $\langle b\rangle$ in order to compute the $M_B$ for $AS$ authentication.

The session key is explicitly authenticated by mutual confirmation values $M_A$ and $M_B$, respectively. Therefore, the proposed TAKE protocol provides explicit mutual authentication.

(3) *The proposed TAKE protocol provides session key establishment:* The session key must be established between the client and the $AS$, so that it can support the dynamic WEP key. For a random challenge, random numbers $\langle x \rangle$ and $\langle r \rangle$, which are separately generated by each entity, are different every time. Therefore, the established session keys $\langle sk_A \rangle$ and $\langle sk_B \rangle$ are freshness and randomness.

(4) *The proposed TAKE protocol can provide full forward-secrecy:* Forward-secrecy should be provided to ensure that attackers cannot compute previous session keys from the sessions which were eavesdropped on previously, even when a long-term secret keying material of the entity participating in the protocol has been revealed. If $\langle ID_A \rangle$, $\langle \pi \rangle$, $\langle t \rangle$, and $\langle g^b \rangle$, possessed by a client, are all exposed to an attacker, the attacker may learn about $g^x$ by computing $e \oplus f$. The value $c = g^{xb}$, however, is hard to compute due to the DDH (Decision Diffie-Hellman) problem [5]. Therefore, forward-secrecy on the client's side is ensured. On the other hand, if $\langle b \rangle$, $\langle \pi \rangle$, and $\langle t \rangle$, which are stored on the $AS$ side, and the $\langle ID_A \rangle$ of a client are all exposed to an attacker, the attacker can compute a session key. Therefore, forward-secrecy on the $AS$ side is not provided. In order to provide full forward secrecy, the Diffie-Hellman key exchange algorithm can be used to compute the session key. In this approach, we let $c = g^{xr}$, where $x$ and $r$ are the random exponents chosen by the client and $AS$ separately, and $sk_A = sk_B = H(c, g^x, r, ID_A, g^b)$. Then, the protocol can provide full forward secrecy. The client, however, is required to perform, in addition, one exponentiation operation in a real-execution phase of the protocol as a trade-off.

(5) *The proposed TAKE protocol can resist an off-line dictionary attack:* It should be guaranteed that secret information (passwords and keys) shared between a client and the $AS$ is resistant to an off-line dictionary attack. As the symmetric key $\langle t \rangle$ with high entropy, the password $\langle \pi \rangle$, and a random number $\langle r \rangle$ are hashed into $\langle f = H(r, \pi, t) \rangle$ and are used as a key for the protection of $g^x$, therefore, off-line dictionary attacks stand little chance of success. In other words, an attacker cannot help but guess the password $\langle \pi \rangle$, symmetric key $\langle t \rangle$ and random value $\langle g^x \rangle$ at the same time.

(6) *The proposed TAKE protocol provides key confirmation:* It should be confirmed to a legitimate user participating in the protocol that he or she actually shares a common secret session key with an entity with which communication is intended. The proposed TAKE protocol includes the session key in $M_A$ and $M_B$, in order to confirm the keys.

(7) *The proposed TAKE protocol can provide non-repudiation:* Fraudulent clients must know that the accounting is correct but that illegal access is not paid for. The proposed TAKE protocol doesn't employ a digital signature, so it doesn't support non-repudiation which provides proof of the integrity and origin of data. Using the two factor authentication, however, makes it more

difficult for fraudulent clients to deny the use of the PWLANs service than using the single factor authentication.

## 5 Efficiency Analysis

The computational costs of the PP-TAKE protocol and the proposed TAKE protocol, in the precomputation and real-execution phases, are summarized in Table 1. The followings are used to analyze the efficiency properties in the proposed protocol.

(1) *The proposed TAKE protocol uses a low computational load:* In general, a protocol requires a low computational load that can be borne by even low-power devices such as PDAs and precomputation, in order to minimize on-line computation operations. In the precomputation phase of the PP-TAKE protocol and the proposed TAKE protocol, the client is required to perform two exponentiation operations. In a real-execution phase, PP-TAKE requires a one-time exponentiation operation, a one-time symmetric encryption operation, a one-time symmetric decryption operation and a nine-time hash operation, but the proposed TAKE protocol does not need symmetric encryption/decryption operations. It requires a total of one exponentiation operation, nine hash operations and two bitwise exclusive-or operations. On the client side of the PP-TAKE protocol, the computational load is one symmetric key encryption, and five hashes, but the computational load of the proposed TAKE protocol is one bitwise exclusive-or, and four hashes.

(2) *The proposed TAKE protocol uses a minimum number of message exchanges:* In terms of network resource efficiency and network delay, it is advantageous to have as few communication rounds as possible. Therefore, the number of messages that are exchanged between client and $AS$ should be kept to a minimum. The PP-TAKE protocol requires four steps in order to perform mutual authentication and key establishment, while the proposed TAKE protocol requires only three steps.

(3) *The proposed TAKE protocol uses a minimum communication bandwidth:* The protocol message should be as short as possible. Among the five messages, three are hash output bits, one is random number bits and the other is the bitwise exclusive or encryption output bit of $g^x$.

**Table 1.** A comparisons of the computational costs

| | PP-TAKE Protocol | | Proposed TAKE Protocol | |
|---|---|---|---|---|
| | Client | Server | Client | Server |
| Precomputation Phase | 2Exp | · | 2Exp | · |
| Real-Execution Phase | 1Sym + 5Hash | 1Exp + 1Sym + 4Hash | 5Hash + 1Xor | 1Exp + 4Hash + 1Xor |
| # of Steps | 4 Steps | | 3 Steps | |

Exp: Exponentiation operation; Sym: Symmetric key encryption/decryption operation; Hash: Cryptographic hash operation; Xor: Bitwise exclusive-or (⊕) operation.

As shown in the efficiency properties (1)-(3), it is obvious that our protocol is more efficient than the PP-TAKE protocol for both the client and server, respectively.

## 6 Conclusions

The current paper proposed an efficient TAKE protocol, which is based upon the PP-TAKE protocol. The computational cost of the proposed TAKE protocol is less than that of the PP-TAKE protocol, and the number of steps in the communication process is one fewer that what is normally needed (only three steps are required). Furthermore, the security requirements of the proposed TAKE protocol are the same as the original PP-TAKE protocol, as described in literature [2]. Therefore, the proposed TAKE protocol is more efficient than the PP-TAKE protocol.

## Acknowledgements

This research was supported by the MIC (Ministry of Information and Communication), Korea, under the ITRC (Information Technology Research Center) support program supervised by the IITA (Institute of Information Technology Assessment).

## References

1. IEEE.: Standard for local and metropolitan area networks-Port based network access control. IEEE Std 802.1x. (June 2001)
2. Park, Y.M., Park, S.K.: Two factor authenticated key exchange (TAKE) protocol in public wireless LANs. IEICE Trans. Commun., Vol. E87-B, No. 5. (May 2004) 1382-1385
3. Gong, L.: Efficient network authentication protocols: lower bounds and implementations. Distrib Comput. Vol. 9. No. 3. (1995) 131-145
4. Lee, T.F., Hwang, T., Lin, C.L.: Enhanced three-party encrypted key exchange without server public keys. Computers & Security. Vol. 23. (2004) 571-577
5. Boneh, D.: The decision Diffie-Hellman problem. Proc. Third Algorithmic Number Theory Symposium. (1998) 48-63

# Adaptive Clustering with Virtual Subnets Support in Ad Hoc Networks

Tzu-Chiang Chiang[1,2], Ming-Hui Tsai[3], and Yueh-Min Huang[1]

[1] Department of Engineering Science, National Cheng-Kung University, Taiwan, ROC
[2] Department of Information Management, Hisng-Kuo University of Management, Taiwan
[3] National Tainan First Senior High School, Taiwan, R.O.C
n9892105@mail.ncku.edu.tw, colux@tnfsh.tn.edu.tw, huang@mail.ncku.edu.tw

**Abstract.** This paper concerns how the virtual subnet mechanism is engaged with the hierarchical architecture in ad hoc networks. The employment of virtual subnets performed at the Data Link Layer can prevent the reveal of broadcast storm and further improve the efficiency of traffic between members of specific group in ad hoc networks. A convenient method is proposed to select crucial backbone nodes which will furnish the ability of filtering frames in virtual subnet against the unnecessary flooding, termed CVA (Clustering with Virtual subnets support Algorithm). The chosen nodes include two different types, station nodes and port nodes, which act functionally as the switch and the ports of the switch to filter frames, respectively. We evaluate CVA with the metrics of average nodes in the backbone and duration of the protocols. The simulation result manifests the proper performance while supporting the virtual subnets in ad hoc networks.

## 1 Introduction

A self-organizing and self-configuring multihop wireless network is called a mobile ad hoc network (MANET) where the network structure changes dynamically due to the mobility of nodes. Nevertheless, a flat ad hoc network applied in such circumstance has poor scalability. Clustering is regarded as the familiar solution to provide the flat network with a hierarchical structure [5]. There are at least three benefits [2]. First, it can increase the system capacity by facilitating the spatial reuse of resources. Second, it can use bandwidth efficiently by spreading restrictedly the routing information in the set of nodes ( clusterheads / clustergateways ). Last, it can make an ad hoc network appear smaller and more stable in this hierarchical view. Oppositely, it also must suffer the cost of clustering, for instance, extra control message for clustering, computation load of nodes, re-clustering due to nodal movements, and so on.

Most of the backbone formation algorithms have the following two phases: the first phase (clustering) is to find a set of nodes as the *dominators* (or clusterheads) and the second phase is to find a set of nodes, called *connector*, to connect these dominators to form the connected dominating set. Such the connected set, also referred to *spine*, plays a very important role in routing, broadcasting and connectivity

V.N. Alexandrov et al. (Eds.): ICCS 2006, Part II, LNCS 3992, pp. 1008 – 1015, 2006.

management at layer 3 in wireless ad hoc networks [6, 7, and 9]. It integrates the clustering hierarchy to form a communication hierarchical structure that is practical in providing desirable properties such as minimizing communication overhead and increasing the probability of aggregating redundant data [1].

We aim to select the adaptive nodes in this hierarchical structure to form the virtual backbone and integrate virtual subnet filtering mechanism into an ad hoc network. Virtual subnets (IEEE 802.1Q) [17] can be used to effectively control broadcast traffic at layer 2 in wired circumstance. Broadcast traffic from servers and clients in a particular virtual subnet is propagated only on those switch ports connected to clients belonging to the same virtual subnet. Hence, we exercise the virtual backbone to establish virtual subnets as switches do in wireless circumstance. Some nodes of virtual backbone act functionally as the physical ports of switches for filtering frames.

The paper is organized as follows. In Section 2 we give a quite intensive overview of distributed clustering algorithm, backbone formation and virtual subnet architecture. Section 3 presents our convenient backbone formation algorithm with the ease of virtual subnet membership maintenance. In Section 4 we describe in details our simulation environment and we evaluate and comment on the performance of the protocols with respect to the selected metrics of interest. Finally, Section 5 concludes the paper.

## 2 Related Works

### 2.1 Clustering

Clustering for ad hoc networks has been investigated for over twenty years. Many clustering algorithm had been proposed including [8, 10, and 13]. Clusterhead selection and backbone formation, although differently from the methods used in the mentioned solutions, are also the two fundamental steps in [11]. Moreover, regarding the backbone formation, most of the proposed protocols end up generating a clustering and a corresponding backbone whose nodes form a dominating set of nodes [1], [3]. Clusterheads are joined through gateway nodes to form a backbone that is a *connected dominating set* (CDS) [6, 7, 14, and 16]. For instance, [12] is a simple distributed procedure for creating the desired CDS efficiently but tends to create a very large CDS. Trade-off between message complexity and size of CDS exists evidently. A different approach to create a small CDS is to compute a *maximal independent set* (MIS) first and connect it up [4, 14, and 15].

### 2.2 Virtual Subnet

In wired networks, a switch is a device responsible for multiple functions such as flooding, sending frames, and filtering. Layer 2 switching is *shardware* based, which means it uses the MAC address from the host's NIC cards to filter the frames. The layer 2 switches check the frames' hardware addresses instead of the Network layer (layer 3) header information before deciding to either forward or drop them. There are two basic functions at layer 2 switch.

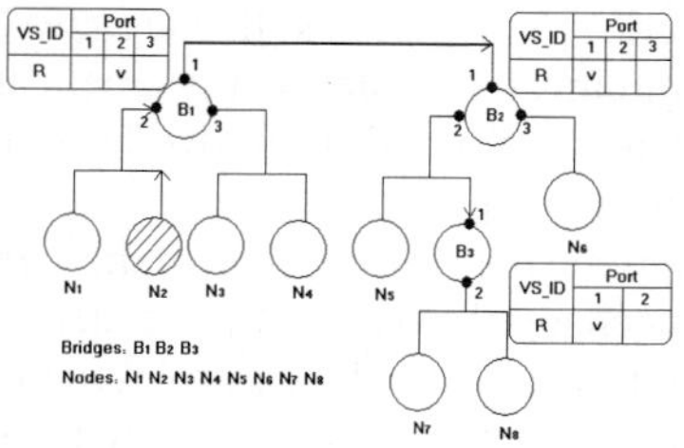


**Fig. 1.** (left Fig) N2 declares its initiation with virtual subnet R

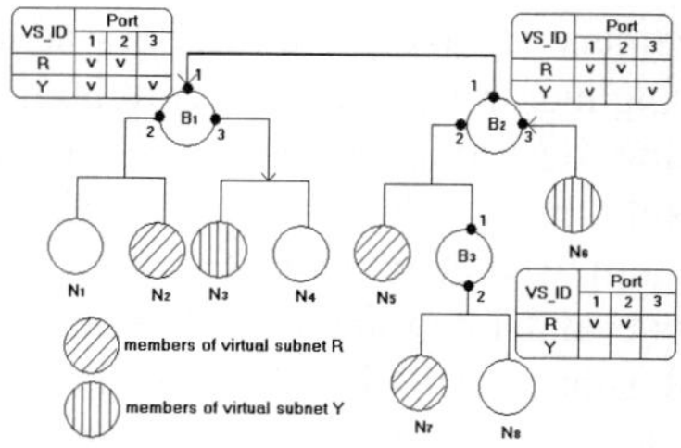


**Fig. 2.** (right Fig) N6 sends frames to its members of virtual subnet Y

(1) Address learning: they can remember the source hardware address of each frame received on an interface and enter this information into a MAC database.
(2) Forward/filter decisions: when a frame is received on an interface, they look at the destination hardware address and find the exit interface in the MAC database.

The detailed functionality of the virtual subnet refer to the standard of IEEE 802.1Q. Only frames that are destined for addresses outside the virtual subnet need to proceed to switch ports for forwarding. We illustrate the concept of virtual subnets with the both Figures 1 and 2.

Whiles N2 declares its initiation into virtual subnet R, the B1 (a switch) updates its filtering database at port 2 to maintain the virtual subnet membership. This declaration is then forwarded and excites the update of the filtering database in B2 and B3 as the same way. See Figure 1. Figure 2 also illustrated, N2, N5, N7 and N3, N6 are the members of virtual subnet R and virtual subnet Y, respectively. If N6 sends a packet to its members of virtual subnet Y, both port 2 of B2 and B1 do not relay this packet according to the filtering database.

## 3 The Virtual Subnet Cluster Architecture

Our concerns are how to automatically segment these mobile nodes into the correct virtual subnets and how to prevent the unnecessary broadcasts from interfering with the other subnets. In wired networks, however, virtual subnet membership is typically defined by the physical Layer 2 switch port to which a node/device is connected. In wireless mobile networks, all nodes act as Layer 3 router but are not lightly tied to a physical port of router.

In order to implement the filtering functionality of virtual subnets, some nodes chosen by the underlying backbone formation algorithm must be able to maintain the virtual subnet filtering tables and act as the switch ports. The chosen nodes include two different types, station nodes and port nodes, which act functionally as the switch and the ports of the switch, respectively. Therefore, we first present the clustering and backbone formation for supporting virtual subnets (Clustering with Virtual subnets support Algorithm, termed as CVA), that is to say, the selection of station and port nodes after clustering. As for the implementation of virtual subnet functionality within the hierarchical structure and the maintenance of virtual subnet memberships, they are then presented as the following.

### 3.1 Clustering and Backbone Formation

We aim to select the adaptive nodes to form the virtual backbone and act as a virtual subnet filtering devices no matter which clustering algorithm is adopted. The main reason that we adopt DCA here to partition networks is its advantages in terms of low communication complexity. During clustering and consequent backbone formation, there are three types of backbone nodes:

(1) Participant nodes are candidates for station/port nodes whether their *order* is greater than two. (Order, i.e. how many clusters a node can access one hop away)
(2) Port nodes that filter the subnet frames depending on its filtering database.
(3) Station nodes that just forward the virtual subnets frames without filtering functionality or act as port nodes in the specific condition described later.

The chosen nodes would be station nodes or port nodes depending on their physical connection status. For avoiding the performance degradation by the excess filtering verification occurring at an oversized backbone, how many nodes should be determined as station/port nodes in this hierarchical clustering architecture will affect significantly the performance of the whole networks especially throughput.

A node will be set to "participant" node while it can connect more than two clusters. Note that the order of each node could be lightly and rapidly counted after clustering. After sending/receiving the broadcasts to/from neighbors, a node can claim itself as a station node only if it has the maximum order more than all its neighbors have or the same maximum order among its neighbors but with the minimum ID. And this station node then broadcast its claim to neighbors in order to set adjacent participant nodes to port nodes. It is worth to mention that if a port node really attached to a station node and received another station node claim, it would transform its state into station. In addition, some station nodes act as port nodes while they can connect not only station/port nodes but cluster members close to them. All parallel connected station nodes are regarded as a virtual switch/router to support virtual subnet by its joint port nodes.

Consider the topology snapshot and cluster architecture in Figure 3 and Figure 4, respectively. There are six clusters in an ad hoc network and all of the participant nodes are marked (which possibly but not certainly be selected as station/port nodes later, indicated as ◇). Determination whether a node is a station/port node from those participant nodes is following our backbone formation algorithm. As shown in Figure 5, station nodes are indicated as ◎ and port nodes as □.

First, node 4, 6, and 20 have transformed as station nodes. Participant nodes close to them such as node 2, 3, 5, 9, 14, and 16 should transform their state into "port". Note that the node 5 and node 16, both of the nodes receiving the dissemination from their adjacent station nodes (node 4, 20 and node 6, 20, respectively) transform their original state "port" into "station". Despite that node 18 have been previously set to participant node, it have not ever transformed as station/port node. It is just a cluster member affiliated to its clusterhead (node 3).After the backbone formation, there are five station nodes and four port nodes in the system, which are {4, 5, 6, 16, 20} and {2, 3, 9, 14} respectively. All the parallel connected station nodes are regarded as a virtual switch to support virtual subnet by its joint port nodes, as shown in Figure 6.

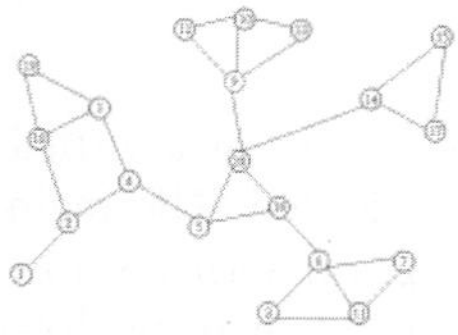

**Fig. 3.** Topology

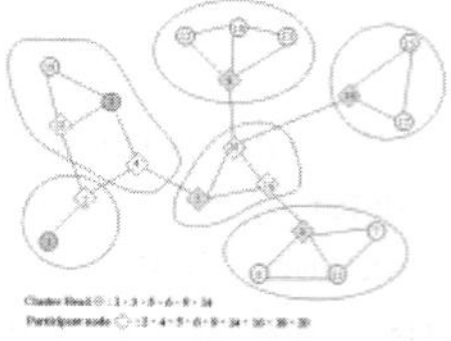

**Fig. 4.** DCA clustering

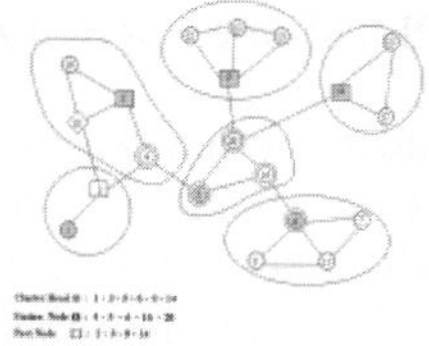

**Fig. 5.** Backbone formation

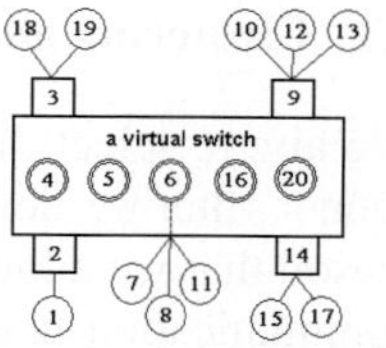


**Fig. 6.** Parallel connected station nodes are regarded as a virtual switch

## 3.2 Virtual Subnet Membership Maintenance

Virtual subnets configured by using Media Access Control addresses (or node ID) can recognize nodes even that has changed their affiliation with original virtual subnet. The virtual subnet membership maintenance should automatically reconfigure when a node transforms its affiliation into an appropriate virtual subnet without the need of changing the node's MAC address. The communication among virtual subnets in our system will be forwarded without any filtering functionality by some station node (for example, node 4, 5, 16, and 20 in Figure 6) whose virtual subnet filtering table (VSFT) is null. On the contrary, the others of station nodes (for example, node 6 in Figure 6) acting as port nodes are not always permeable for all virtual subnet frames depending on their VSFTs. In addition, it is necessary for all port nodes (for example, node 2, 3, 9, and 14 in Figure 6) to maintain the virtual subnets membership. There are two basic kind of virtual subnet management to be interpreted:

### 3.2.1 Join in a Virtual Subnet

While a node declares initiation into a virtual subnet or transforms its original affiliation into a new certain virtual subnet, it has to send the virtual subnet join message to it's accompany clusterhead first in consequence of the hierarchical organization. Its clusterhead forwards (broadcasts) the message to the adjacent backbone node to maintain the virtual subnet membership. If the clusterhead is station/port (backbone) node, the forwarding would not be processed. The backbone node updates subsequently its VSFT to enable the certain virtual subnet frames to pass through from outside in accordance with the received join message. The virtual subnet join message contains the following fields:

$$< node\_id,\ virtual\_subnet\_id,\ sequence\_\# >$$

For instance, while node 8 disseminates its initiation into a virtual subnet R (VS_R), this affiliation information will be send to its clusterhead node 6 first and this station node subsequently update its virtual subnet filtering table (VSFT) to enable the filtering function, refer to Figure 6 and Table 1. In addition, associated with virtual subnet ID entries is a timer, called the membership expiration timer. The purpose of this timer is to purge the stale virtual subnet membership. Each time a frame tagged to VS_R is forwarded through this node and the time out for the associated membership entry is reset to the current time plus a constant expiration time parameter $\tau$ (i.e., called as *ageing time* in IEEE 802.1Q).

**Table 1.** Virtual Subnet Filtering Table of node 6

| Virtual Subnet Filtering Table | |
|---|---|
| VID | Membership Expiration Time (s) |
| R | *current time* * + τ |

### 3.2.2 Depart from a Virtual Subnet

There is no specific message which is disseminated from a node to depart from a virtual subnet. The stale entries of VSFT in station/port nodes are automatically removed depending on their individual ageing time. As mentioned above, whenever a virtual subnet frame passes through the station/port nodes, the timer of those corresponding entries is extended. The dynamically ageing out of filtering entries ensures that nodes that have been affiliated with a different virtual subnet will not be permanently prevented from receiving frames.

For example, the VSFT of node 14 is shown as Table 2. It indicates that frames tagged for both of virtual subnet R and B can be forwarded bidirectionally through this node. Whenever node 14 receives a frame tagged for virtual subnet R, it checks the filtering entry first to determine whether the entry is expired. If not, the frame is successfully forwarded and the ageing time is reset. Oppositely if the entry is expired, the frame tagged for VS_B is filtering out and the entry is removed form the VSFT.

## 4 Simulation Analysis

The adaptive clustering with virtual subnets support architecture has been evaluated using the VINT project network simulator (ns2) [18]. We also advanced C++ Partition Toolkit (CCPT) to determine the nodes affiliation with virtual subnets in a clustering ad hoc network. Our simulations refer to static scenarios in which *n* wireless nodes with maximum transmission radius of 30 meters and the number of nodes *n* has been assigned the values between 20 and 200 with an interval of 20. All nodes are initially placed randomly within a predefined 200m x 200m grid area. In particular, we have considered the following metrics:

(1) The average percentage of nodes in the backbone.
(2) The average time (duration) needed by the protocol to complete the clustering and backbone formation.

**Backbone Sizes**

We choose the algorithms of Wan et al [11], Wu and Li [17], DCA [16], and our CVA to evaluate the backbone size. A smaller backbone improves network performance not only with respect to control overhead but also on account of a smaller number of nodes needed to involve in the data transmission. The average percentage of nodes in the backbones is shown in Figure 8. All the curves present that the percentage of nodes in the backbone decreases as *n* increases: as the higher density of network, the cluster size grows into larger scale and the selection of connectors become ampler. Our protocol curve is close to the DCA's as the result of clustering algorithm.

**Protocol Duration**

It show the result of the average time needed by the above four mentioned protocols to complete their operations in Figure 9. The more complex of protocol is, the longer duration of backbone formation is needed. This is due to the higher complexity, the raised number of information that each node needs to exchange with its neighbors, and the increasing possibility of communication collision. On the other side, the clusterheads selection of DCA is performed based on a generic weight associated to a node, so the operations require a node to wait for all its neighbors with maximum weight to communicate their role before it can decide its own. Our protocol show reasonably good performance, requiring a time which ranges from 1.44s when $n = 40$ to 5.53s when $n = 200$.

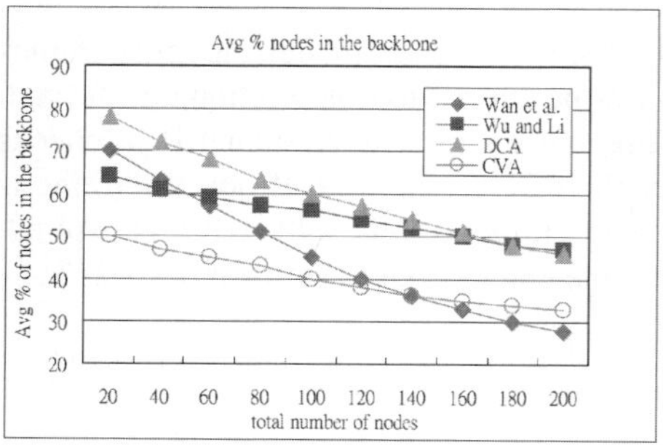


**Fig. 7.** Percentage of backbone nodes

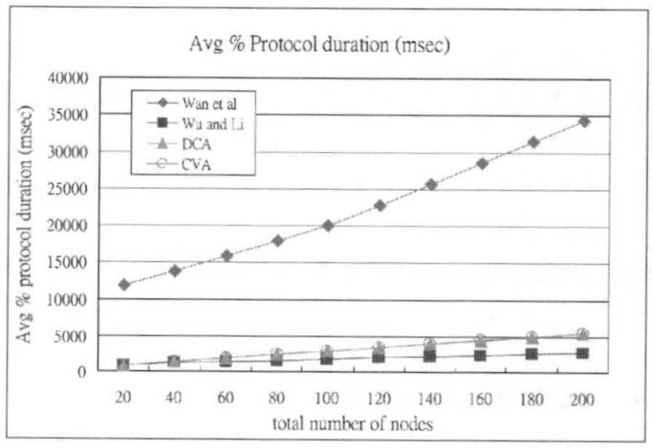


**Fig. 8.** Duration of the protocols

## 5 Conclusions

This paper represented an ingenious algorithm of backbone formation which is elaborately fitting for adopting the mechanism of virtual subnet membership maintenance, termed as CVA. Our protocol partitions an ad hoc network into physical clusters on it physical topology and partitions the network into logically virtual subnets on independent nodes' affiliation. The communication among the same virtual subnet will decrease the interference with irrelative clusters and can significantly improve the networks' performance. Topics for further research include the investigation of backbone nodes maintenance in presence of mobility.

## References

1. S. Basagni, M. Mastrogiovanni and C. Petrioli, "A Performance Comparison of Protocols for Clustering and Backbone Formation in Large Scale Ad Hoc Networks," the 1st IEEE International Conference on Mobile Ad-hoc and Sensor Systems, October 24-27, 2004, Fort Lauderdale, Florida, USA.
2. Jane Y. Yu and Peter H.J. Chong, "A Survey of Clustering Schemes for Mobile Ad Hoc Networks", IEEE Communications Surveys & Tutorials First Quarter 2005, Volume 7, No.1, pp. 32-48.

3. B. Liang and Z. Haas, "Virtual backbone generation and maintenance in ad hoc network mobility management," in Proceedings of the 19th IEEE Infocom, vol. 3, Tel Aviv, Israel, March 26–30 2000, pp. 1293–1302.
4. I. Stojmenovic, M. Seddigh, and J. Zunic, "Dominating sets and neighbor elimination based broadcasting algorithms in wireless networks," IEEE Transactions on Parallel and Distributed Systems, vol. 13, no. 1, pp. 14-25, 2002.
5. Y.M. Huang and Tzu-Chinag Chiang, "An Efficient Encoding Tree for Location-Aware Multicast Routing Protocol in Wireless Ad Hoc Networks", Journal of Internet Technology,Vol. 6, No. 1, 2005.
6. B. Das and V. Bhargavan, "Routing in ad-hoc networks using minimum connected dominating sets," IEEE International Conference on Communications (ICC '97), June 1997.
7. B. Dss, E. Sivakumar, and V. Bhargavan, "Routing in ad-hoc networks using a virtual backbone," Proceedings of the 6th International Conference on Computer Communications and Networks(IC3N '97), pages 1 - 20, Sept. 1997.
8. C. C. Chiang, H. K. Wu, W. Liu and M. Gerla, "Routing in Clustered Multihop, Mobile Wireless Networks with Fading Channel," Proc. IEEE Singapore international Conference on Networks (SICON97), 1997, pp. 197-211
9. Ming-Hui Tsai, Tzu-Chinag Chiang and Y.M. Huang, "On Scalability and Mobility Management of Hierarchical Large-Scale Ad Hoc Networks", Lecture Notes in Computer Science,Volume 3823 ,pp. 714 - 723,12/2005.
10. C. R. Lin, M. Gerla, "Adaptive Clustering for Mobile Wireless Networks," IEEE Jour. Selected Areas in Communications, Sept. 1997, pp. 1265-1275.
11. P.-J. Wan, K. M. Alzoubi, and O. Frieder, "Distributed construction of connected dominating sets in wireless ad hoc networks," in proceedings of INFOCOM 2002.
12. J. Wu and H. Li, "On calculating connected dominating sets for efficient routing in ad hoc wireless networks," Telecommunication Systems, Special Issue on Mobile Computing and Wireless Networks, vol. 18, no. 1/3, pp. 13–36, September 2001.
13. F. Garcia Nocetti, J. Solano Gonzales, and I. Stojmenovic, "Connectivity based k -hop clustering in wireless networks," TelecommunicationSystems, vol. 22, no. 1–4, pp. 205–220, 200
14. D. Dubhashi, A. Mei, A. Panconesi, J. Radhakrishnan, and A. Srinivasan, "Fast distributed algorithms for (weakly) connected dominating sets and linear-size skeletons," in Proceedings of the 14th Annual ACM-SIAM Symposium on Discrete Algorithms (SODA), Baltimore, MD, 12–14 January 2003, pp. 717–724.
15. S. Guha and S. Khuller, "Approximation algorithms for connected dominating sets," Algorithmica, vol. 20, no. 4, pp. 374–387, April 1998.
16. I. Chlamtac and A. Farag´o, "A new approach to the design and analysis of peer-to-peer mobile networks," Wireless Networks, vol. 5, no. 3, pp. 149–156, May 1999.
17. IEEE draft standard for virtual bridged local area networks, 802.1Q/D11 July.30, 1998.
18. The VINT Project, The ns Manual. http://www.isi.edu/nsnam/ns/,2002.

# A Service Management Architecture for NEMO in IPv4 and IPv6 Networks*

Jin Ho Kim, Choong Seon Hong**, Dae Sun Kim

Department of Computer Engineering, Kyung Hee University,
Seocheon, Giheung, Yongin, Gyeonggi, 449-701, Korea
jhkim@networking.khu.ac.kr, cshong@khu.ac.kr, dskim@networking.khu.ac.kr

**Abstract.** The current design of Mobile IPv4 and Mobile IPv6 protocols doesn't provide the compatibility between networks of different versions. A Network Mobility(NEMO) protocol also provides the limited mobility in IPv6 networks. However, we can easily expect that the current IPv4 network will coexist with IPv6 network in the near future internet environment. So, in this paper, we propose the architecture of mobility between networks of different versions which is supported the NEMO basic protocol. This new mechanism can manage the mobility through a Tunnel Agent(TA). And the TA maintains a NEMO tunnel when a Mobile Router(MR) moves from IPv6 to IPv4 network.

## 1 Introduction

With the rapid development of the radio communication technology, users tend to exchange all kinds of data through wireless internet. It increases the desire that users want to access the internet from anywhere, at anytime and the connection is expected to be seamless. As the solutions for these requirements, Internet Engineering Task Force Working Group(IETF WG) made an effort to standardize the method to support mobility of terminals. In particular, the specification of the Mobile IPv6[1] has been standardized as RFC 3775. Recently, the support for mobility of network such as buses, airplanes, cars and trains as well as that of terminal unit is discussed actively. Discussion about supporting network mobility method is progressing vigorously in IETF NEMO WG[2]. First, NEMO WG established the extended Network Mobility(NEMO) basic protocol[3] from the Mobile IPv6 in RFC 3963. It can offer the seamless connectivity of internet for terminals in mobile network. Furthermore, they have plans to standardize the extended NEMO such as route optimization and multihoming.

The most distinguished feature of network mobility is that the mobile router, not each node, is wholly responsible for maintaining the mobility. In other words, whenever the point of internet connection is changed, only one signal of the MR can replace several signals processed by each terminal. So, the signaling overhead of network is decreasing. Also, usual IPv6 appliances which don't support IP mobility can still use the internet even if the internet connection point is changed. But, if the introduction of IPv6 can't provide interoperability with

* This work was supported by MIC and ITRC Project.

** corresponding author.

V.N. Alexandrov et al. (Eds.): ICCS 2006, Part II, LNCS 3992, pp. 1016–1023, 2006.

existing IPv4 environment, it isn't useful. Therefore, conversion technologies between IPv4 and IPv6 are essential. IETF Network Generation Transition Working Group(NGTrans WG) had already established transition technologies of dual stack, tunneling, translator for fixed hosts as its standard. But, mobility support isn't considered in these mechanisms.

If session connectivity is supported, mobile nodes and MNNs(Mobile Network Nodes)[4] can move from IPv6 to IPv4 network. But, the session connectivity for supporting mobility in the protocols such as Mobile IPv6 and NEMO isn't guaranteed because it isn't compatible with the networks of different versions. Therefore, in this paper, we propose the architecture that guarantees to retain session connectivity of the NEMO tunnel by the Tunnel Agent(TA) in case where mobile network moves to the IPv4 network.

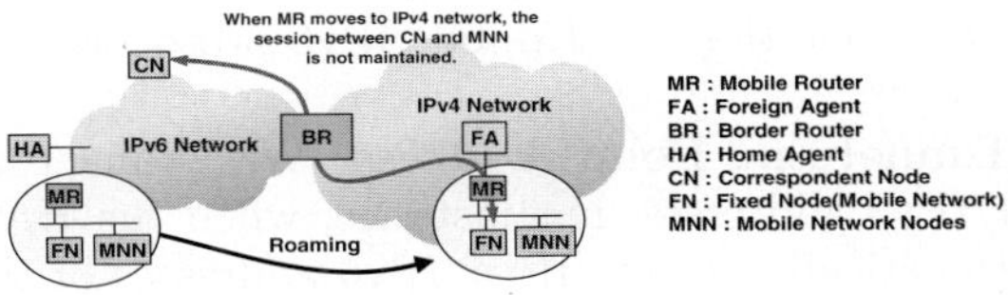


**Fig. 1.** Roaming scenario of Mobile Network

This paper is organized as follows. Section 2 introduces some related works such as dual stack, Doors mechanisms and IPv4 Care-of Address(CoA) Registration scheme. In section 3, we describe the operation of the TA and the MR. Then we show the result of performance evaluations in section 4. Finally, we conclude in section 5.

## 2 Related Works

The simplest mechanism for transition between Mobile IPv4[5] and Mobile IPv6 is a dual stack structure[6]. But, a designer must implement the dual stack of the Mobile Node(MN) supporting each version of two mobility protocols. When the terminal moves to other links, it needs to send signal to the Home Agent(HA)[1][3][5] of IPv4 and IPv6 networks to keep connectivity. For example, when terminals move to the network of another version, they have the burden which must process all Registration Request[5] and Registration Reply[5] messages in Mobile IPv4, Binding Update(BU)[1] and Binding Acknowledgement(BA)[1] messages in Mobile IPv6.

Doors mechanism[7] suggests using a 'Doors Router' between IPv4 and IPv6 networks. In this mechanism, when an MR moves to a Private IPv4 network as well as a Public IPv4 network, sessions of each network are maintained. But, when the MR roams to the IPv4 network, MNNs carry tunneled packet of IPv4-IPv6-IPv6 headers to an HA. Then tunnel overhead increases additionally.

IPv4 CoA Registration mechanism[8] is introduced to reduce tunnel overhead compared with tunnel in the Doors mechanism. When the MR moves to IPv4 network, it adds an option including the IPv4 CoA in the BU message. And the

MR registers the IPv4 CoA of the MR to the HA which carries packets with a reduced IP header, IPv4-IPv6, to the MNNs. So, the tunnel overhead decreases.

## 3 Proposed Architecture

This paper proposes an architecture for supporting mobility through a TA when an MR including MNNs move to an IPv4 network. That is, as the TA relays tunnel end point of IPv6 and IPv4 networks, a NEMO tunnel is kept to work. Basically, the TA and the MR are dual stack(IPv4/IPv6) structure, and the MR must include the minimal Mobile IPv4 protocol function which can process an Agent Advertisement[5] message to compose an IPv4 CoA in the IPv4 network environment. In other words, the signaling overhead of the proposed architecture decreases as compared to the other proposed scheme using the dual stack which has all functions of Mobile IPv4 and Mobile IPv6 protocols.

**Obtaining the Tunnel End Point Address.** We assume that an HA already knows TA's IPv4/IPv6 addresses. For instance, when the HA gets tunnel end point address automatically, it gets IPv4/IPv6 addresses of the TA in current network using DSTM[9]. But, if the HA cannot use the DSTM, it can use configured address. Figure 2 is a proposed message format of Home Agent Address Discovery Request/Reply[3] messages which has a "V" flag and a Tunnel Agent Information(TAI) option additionally. If the MR moves to the IPv4 network, and wants to maintain its session from the TA, it sets the "V" flag and sends the Home Agent Address Discovery Request message by an Anycast address. Then, the HA knows the tunnel end point address and responds the Home Agent Address Discovery Reply message which includes the TAI option with the "V" flag to the MR. At this time, the TAI option must include the TA's IPv4/IPv6 addresses necessarily, and the MR must store this information.

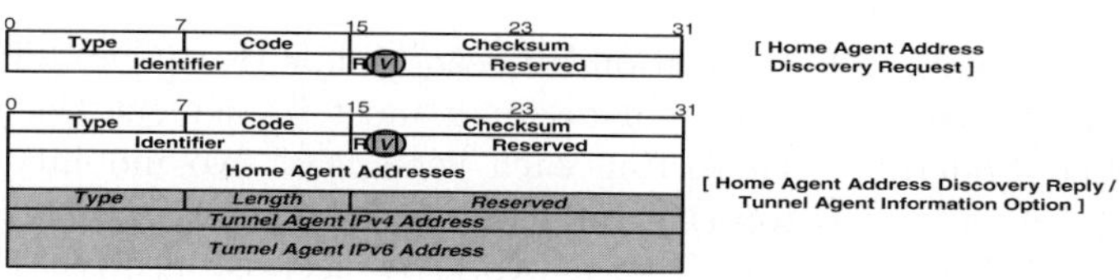


**Fig. 2.** Home Agent Address Discovery Request/Reply and Tunnel Agent Information(TAI) option message format

**Movement Detection of IPv4 Networks and Home Registration.** If we use the mechanism defined in the NEMO basic protocol standard, mobility of the MNNs in the MR which roams only in the IPv6 network can be supported without any problem. If the MR already knows the TA's address through Home Agent Address Discovery mechanism[3], mobility can be supported in the IPv4 network. This paper assumes that the MR has dual stack and has the minimum functionality to compose the IPv4 CoA for detecting movement in the IPv4

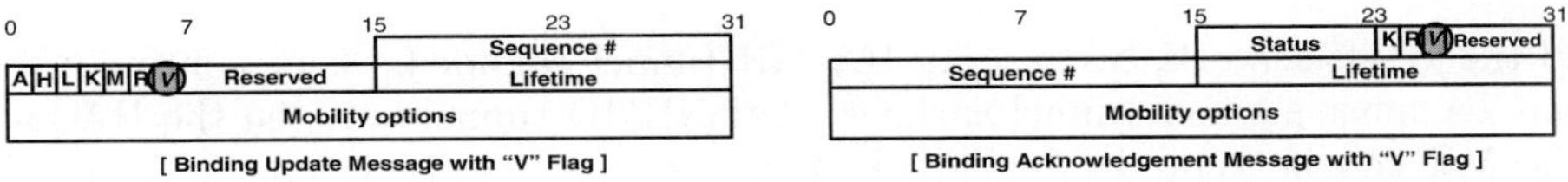


**Fig. 3.** Binding Update/Acknowledgement messages format

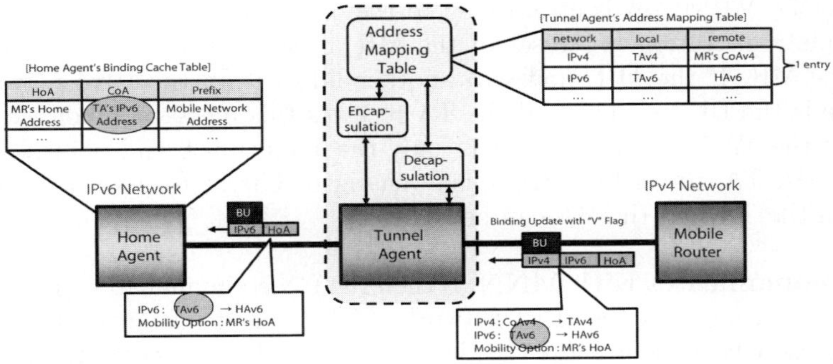


**Fig. 4.** Home Registration process and Tunnel Agent architecture

network. That is, the MR can detect movement in the IPv4 network by receiving the Agent Advertisement message and compose the IPv4 CoA.

If the MR detects its movement in the IPv4 network, then the MR sends the BU message with the "V" flag to the HA in order to register the information. Figure 3 shows BU and BA messages format having the "V" flag.

Figure 4 shows the address mapping table of the TA, the HA's binding cache table architecture and the BU message format which the MR sends to the HA. The MR composes the source address of the BU message as own CoA in the NEMO basic protocol. But, in this paper, when the MR moves to the IPv4 network, it sends the BU message to the HA, its source address is decided by the tunnel end point's IPv6 address which is known by the MR already. And the MR sends the encapsulated the IPv4 header to the TA. In other words, an IPv4 protocol number in the BU header field must be 41 to inform that it is the IPv6 packet, source address is the MR's IPv4 CoA, and destination address is the TA's IPv4 address. After the TA receives the BU message, it decapsulates the IPv4 header and forward it to the HA. The HA receives the BU from the TA registers the MR's CoA moving to the IPv4. Therefore, as the source address of the BU message is the TA's IPv6 address, the HA makes the MR's IPv6 CoA to the TA's IPv6 address and updates binding cache[3]. The HA sends all packets to the TA while the MR moves to the IPv4 network. The reason is that the CoA which is mapped to an MR's home address[3] in binding cache had the TA's IPv6 address.

**NEMO Tunnel Setup.** If BU and BA messages are exchanged successfully between the HA and the TA, the NEMO tunnel is established. In the NEMO basic protocol, when the MNNs communicate with a Correspondent Node(CN), packets are always transmitted through an HA-MR tunnel[3]. Therefore, it is the most important that the NEMO tunnel is maintained. In case the MR moves to the IPv4 network, because an HA-MR tunnel cannot be kept continuously, the TA must manage tunnel cache so that NEMO tunnel between the HA and the MR can be kept. In dual stack structure and Doors mechanism, three IP headers, IPv4-IPv6-IPv6 are created. But, in this paper, the MR establishes an efficient a TA-MR(IPv6 in IPv4) tunnel considering handoff to other links in the IPv4 network. The HA makes a IPv6-IPv6 tunnel between the HA and the TA because the MR's CoA is established to the TA's address. And the TA finishes the home registration processes of the MR after creating a IPv4-IPv6 tunnel with the MR. If the MR moves to another link within the IPv4 network, the MR sends the BU message and the TA receives this message. Then we need to only set the TA-MR tunnel. That is, only the TA's tunnel end point is changed and the HA-TA tunnel(IPv6-IPv6) isn't changed. Therefore, additional signaling between the TA and the HA can be reduced.

**CN Communicates with MNN.** When the CN communicates with the MR's MNNs which move to the IPv4 network, packets are transmitted like figure 5. First, if the CN sends a packet to the MNNs, the HA intercepts the packet because the destination address is an MR's mobile network prefix[4]. After searching the binding cache entry, the HA forwards the packet to the TA using the HA-TA tunnel because registration of the MR's IPv6 CoA to the TA's IPv6 address. If the TA receives the packet from the HA, it transmits the packet to the MR's IPv4 CoA which is the tunnel end point of the TA-MR tunnel. Simultaneity, the TA searches mapping table and translates IPv6 and IPv4 headers. This process uses methods such as header translation mechanism of NAT-PT[11] and SIIT[12]. If the MR receives all packets through the TA-MR tunnel it decapsulates a packet of them for removing its IPv4 header. Finally, the MR forwards the packet to the MNNs. On the contrary, the packet transmitted from the MNNs to the CN is the format that the source address is changed to the destination address as figure 5 shows.

**IPv4 CoA De-registration.** If the MR returns from the IPv4 network to the IPv6 network, it sends the BU message of which the "V" flag isn't sets to the TA. After that, the TA must release the TA-MR tunnel and remove the tunnel cache. And the TA forwards the BU message to the HA. If the HA receives the BU for de-registration, it updates the CoA with actual the MR's CoA instead

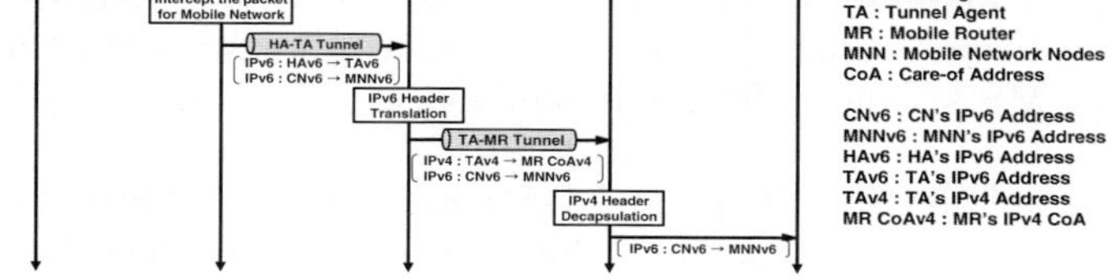


**Fig. 5.** Communication process flowchart between CN and MNN

of the TA's address in the binding cache. The handoff in IPv6 network is similar to the operation defined in the NEMO basic protocol standard. In other words, communication between the MNNs and the CN uses the HA-MR tunnel.

## 4 Performance Measures and Analysis

We can show that the proposed architecture is efficient by measuring the handoff signaling and the packet delay. Figure 6 is the network topology for measurement of the delay. Figure 6(a) shows one scenario of the MR's handoff in an IPv4 network after the node roams from the IPv6 to the IPv4 network. And figure 6(b) shows another scenario that MNNs which move to the IPv4 network communicates with the network which includes an HA and a CN in other IPv6 networks. The following figure 6(c) and equation is referred from [12]. Figure 6(c) shows parameters that indicate meaning and value of each symbol for measuring the performance.

**Handoff Signaling Delay.** As the MR goes away from the HA in the IPv4 network, an HD measures the value of handoff delay. In other words, we calculate the handoff delay as the hop count is increasing in the IPv4 network(The hop count of the IPv6 network isn't changed). The delay of handoff signaling in the two mechanisms is shown in figure 6(a):

- In Doors mechanism, the MR should communicate BU/BA messages with the own the HA directly when it handoffs in the IPv4 network.

$$HD = 2(\frac{L_{FA-MR} \cdot H_{4-M}}{BW_{wired}}) + \frac{(L_{TA-HA} + L_{HA-TA}) \cdot H_{6-H}}{BW_{wired}} + 2(\frac{L_{FA-MR}}{BW_{wireless}}) + 4L_{wired} + 2L_{wireless} \quad (1)$$

- Because a TA handles BU/BA messages in the handoff of the proposed architecture, it is more efficient than the Doors mechanism.

$$HD = 2(\frac{L_{FA-MR} \cdot H_{4-M}}{BW_{wired}}) + 2(\frac{L_{FA-MR}}{BW_{wireless}}) + 2L_{wired} + 2L_{wireless} \quad (2)$$

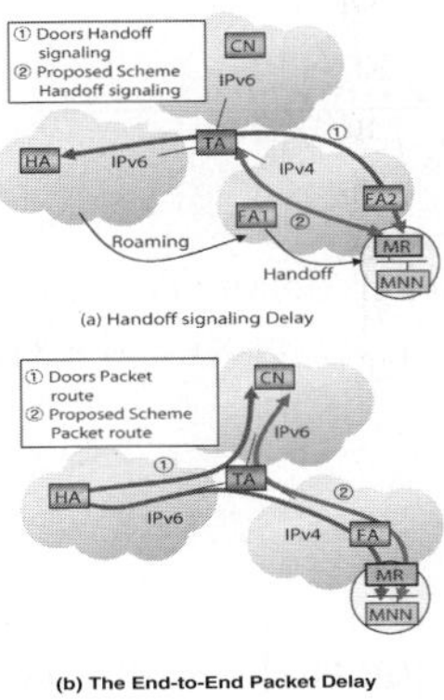


| Symbol | Meaning | Value |
|---|---|---|
| $L_{CN-TA}$ | IP header + payload length of the packet from CN to TA | |
| $L_{TA-HA}$ | IP header + payload length of the packet from TA to HA | |
| $L_{HA-TA}$ | IP header + payload length of the packet from HA to TA | |
| $L_{TA-FA}$ | IP header + payload length of the packet from TA to FA | |
| $L_{FA-MR}$ | IP header + payload length of the packet from FA to MR | |
| $L_{MR-MNN}$ | IP header + payload length of the packet from MR to MNN | |
| $BW_{wired}$ | Bandwidth of wired links | 100 Mb/s |
| $BW_{wireless}$ | Bandwidth of wireless links | 11 Mb/s |
| $L_{wired}$ | Propagation delay + link layer delay of wired links | 0.5 ms |
| $L_{wireless}$ | Propagation delay + link layer delay of wireless links | 2 ms |
| $H_{6-H}$ | Number of hops belong to HA's IPv6 Networks | |
| $H_{6-C}$ | Number of hops belong to CN's IPv6 Networks | |
| $H_{4-M}$ | Number of hops belong to MR's IPv4 Networks | |

(c) Parameters for measuring performance

**Fig. 6.** Network topology, scenario and parameters for delay evaluation

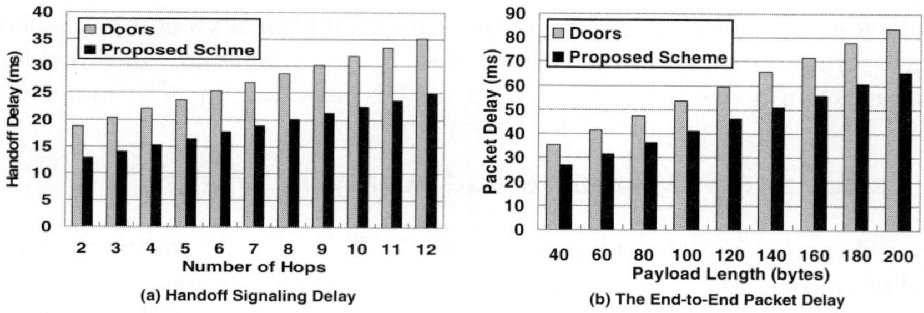


**Fig. 7.** Performance measures and analysis

Figure 7(a) shows the result of handoff signaling delay. It means that the proposed architecture provides fast handoff when the MR moves within the IPv4 network.

**End-to-End Packet Delay.** A PD measures the values of the packet delay when the CN communicates with the MNNs in the MR which roam from the IPv6 to the IPv4 network. We calculated the values of the packet delay by increasing the payload length. The delay of end-to-end packet in the two mechanisms is the shown in figure 6(b):

- In Doors mechanism, the packet delay of the IPv6 network which includes the HA is occurred because the MNNs communicates with the CN through the HA.

$$PD = \frac{L_{CN-TA} \cdot H_{6-C}}{BW_{wired}} + \frac{(L_{TA-HA} + L_{HA-TA}) \cdot H_{6-H}}{BW_{wired}} + \frac{L_{TA-FA} \cdot H_{4-M}}{BW_{wired}} + \frac{L_{FA-MR} + L_{MR-MNN}}{BW_{wireless}} + 4L_{wired} + 2L_{wireless} \quad (3)$$

- The proposed architecture forwards packets to the MR directly through a TA-MR tunnel without relaying. So we can reduce the packet delay in the IPv6 network which includes the HA because the packet cannot be carried to the HA.

$$PD = \frac{L_{CN-TA} \cdot H_{6-C}}{BW_{wired}} + \frac{L_{TA-FA} \cdot H_{4-M}}{BW_{wired}} + \frac{L_{FA-MR} + L_{MR-MNN}}{BW_{wireless}} + 2L_{wired} + 2L_{wireless} \quad (4)$$

Figure 7(b) shows the result of the end-to-end packet delay. We can see that packet delay decreases as the packet length increases in the proposed architecture compared with Doors mechanism.

## 5 Conclusion and Future Work

In this paper, we proposed an efficient method to support the seamless mobility to IPv4 network through a TA in NEMO protocol supporting the mobility only in IPv6 network. We reduced the signaling overhead by minimizing the Mobile IPv4 protocol stack which compared with dual stack mechanism. When a CN carries packets to the IPv4 network, the tunnel overhead is decreased by using IPv4-IPv6 headers through a TA-MR tunnel. Also, we decreased additional binding message between HAs by registering to the TA. Results of Performance measurement show that the proposed mechanism minimizes the handoff and the packet delay. The TA is used with 6to4 Router, DSTM-TEP, ISTAP and NAT-PT. So, The TA supports transitions to mobile routers and mobile nodes as well as static hosts between IPv4 and IPv6 networks. In future work, we will implement and verify the proposed architecture of this paper.

## References

1. D. Johnson, C. Perkins, J. Arkko, "Mobility Support in IPv6", IETF RFC 3775, June 2004.
2. IETF NEMO Working Group, "http://ietf.org/html.charters/nemo-charter.html".
3. V. Devarapalli, R. Wakikawa, A. Petrescu, P. Thubert, "Network Mobility (NEMO) Basic Support Protocol", IETF RFC 3963, January 2005.
4. T. Ernst, H-Y. Lach, "Network Mobility Support Terminology", draft-ietf-nemo-terminology-04, October 2005.
5. C. Perkins, "IP Mobility Support for IPv4", IETF RFC 3220, January 2002.
6. H. Soliman, G. Tsirtsis, "Dual Stack Mobile IPv64", draft-soliman-v4v6-mipv4-02, July 2005.
7. P. Thubert, P. Molteni, P. Wetterwald, "IPv4 traversal for MIPv6 based Mobile Routers", draft-thubert-nemo-ipv4-traversal-01, May 2003.
8. R. Wakikawa, C. Williams, K. Uehara, J. Murai, "IPv4 Care-of Address Registration", draft-wakikawa-nemo-v4tunnel-01, February 2005.
9. J. Bound, L. Toutain, JL. Richier, "Dual Stack IPv6 Dominant Transition Mechanism (DSTM)", draft-bound-dstm-exp-04, October 2005.
10. G. Tsirtsis, P. Srisuresh, "Network Address Translation Protocol Translation (NAT-PT)", IETF RFC 2766, February 2000.
11. E. Nordmark, "Stateless IP/ICMP Translation Algorithm (SIIT)", IETF RFC 2765, February 2000.
12. N. Banerjee, "Mobility Support in Wireless Internet", IEEE Wireless Commun., vol.10, no.5, pp.54-61, October 2003.

# Overlapped Detection Via Approximate Entropy Estimation Against Flooding Attack in Mobile Sensor Networks*

Mihui Kim and Kijoon Chae

Department of Computer Science and Engineering, Ewha Womans University, Korea
mihui@ewhain.net, kjchae@ewha.ac.kr

**Abstract.** To achieve security in sensor networks, it is important to be able to defend against flooding attack recently considered as an extremely threatening attack. In this paper, we propose a flooding attack detection method as the first defense step, through approximate entropy estimation reflecting resource constraints of sensors. Each detector performs both basic estimation for its own region and overlapped estimation for its own and neighbor regions, against the mobility of attack node. Also, in order to enhance the accuracy of detection even in the various deployments of attack agents, we deploy hierarchically detectors according to network topology. This detector by entropy estimation is simplified by only multiplication calculation instead of logarithm, in addition to providing higher estimation precision of entropy compared to the conventional entropy estimation. Our simulation results indicate that this hierarchical defense is a feasible method, being especially promising for accurate decision through overlapped detection even in frequent handoffs of mobile attack agents.

## 1 Introduction

Wireless sensor networks (WSNs) are rapidly growing in their importance and relevance to both the research community and the public at large. Their promising applications are expected to help protect and monitor critical military, environmental, safety-critical, or domestic infrastructures and resources. To date, researches for security in sensor networks have laid disproportionate emphasis on basic security mechanisms, such as authentication or key management. Due to wireless characteristic or easy physical compromise of sensor nodes, these basic security services are indispensable. However, defense against possible attacks is also essential in order not to negate much of the promise sensor networks hold, because attacks can still be performed even if the network communication provides confidentiality and authenticity. Especially, threat against flooding attack, one of Denial of Service (DoS) attacks considered as the most threatening attacks in the wired network is not an exception even on sensor networks. In addition to, flooding attacks may permit real world damage to the health and safety of people, exceeding simple damages such as loss of sensory data or energy exhaustion of nodes. Although the operations of

* This research was supported by the MIC, Korea, under the ITRC(Information Technology Research Center) support program supervised by the IITA.

V.N. Alexandrov et al. (Eds.): ICCS 2006, Part II, LNCS 3992, pp. 1024 – 1032, 2006.

sensors are simplified due to their low cost and overhead, the vulnerabilities of each layer on sensors exist for different flooding attacks, like jamming, exhaustion, or flooding on transport layer. A.D. Wood et al. [1] presented the basic defenses like spread-spectrum or authentication against these attacks, but only these straightforward defenses are not sufficient for protecting the availabilities of sensor networks.

As one of defense mechanisms against attack on sensor networks, Serdar Sancak et al. [2] proposed two rough detection methods against spam attacks that generate dummy packets to deplete the energy of relaying nodes. Even though the second method using the frequencies of messages sent by the sensors in the same region is a little similar with our detection using entropy for distribution estimation of messages, a sink in [2] should count packets per each sensor, that is not scalable for many units, and they roughly mentioned that this detection method regarded a sensor node that generated $\delta$ times more packets as an anti-node. Anthony D.Wood et al. [3] designed a defense method against jamming attack on link layer by unwanted and disruptive radio propagation. Heuristically, it detects the attack if communication channel drops below a certain threshold, and then maps collaboratively a jammed region by modifying MAC, sending a brief, high-priority, un-acknowledge broadcast message. Fan Ye et al. [4] proposed SEF (Statistical En-route Filtering) in order to detect and drop false reports through validating by multiple keyed message codes (MACs).

Moreover, there has been an increasing interest to study and build the mobile nodes as well as static nodes in wireless sensor networks [5]. The mobile nodes in the network will enhance its capabilities, for example, they could be used to physically collect and transport data or to recharge and repair the static nodes in the network. However, in the view of security service, the mobility of nodes makes the solving problems more complicate and difficult. There are three types of mobility: *node mobility*, *sink mobility*, and *event mobility* [6]. First case is a general mobility of mobile sensor network, and is important to reorganize itself frequently enough to be able to function correctly. Second mobility is a special case of node mobility, for example, a human user requested information via a PDA while walking in an intelligent building. Last mobility is a case that the cause of events or objects to be tracked can be mobile. We assume the first general node mobility whether sensor nodes attached to livestock or human are mobile or nodes themselves can be mobile.

The structures of sensor networks are various according to their application, the flat structure by random deployment and the hierarchical structure centering nodes on a sink are representative. In this paper, we will explain our detection method based on the hierarchical sensor network consisting of sink (SINK), aggregate nodes (AN), and sensor nodes (SN). However, this detection might be easily applied to flat sensor network considered as a specific case of hierarchical network, one-layer network.

In this paper, we focus on flooding DoS attacks on mobile sensor environment. As the first proactive defense, we propose a flooding detection through practical entropy estimation. Entropy is a representative algorithm used for decision tree, one of the most popular classification algorithms in current use in data mining and machine learning, and has frequently used for differentiating the abnormal situation from the normal on the wired networks [7]. However, the original entropy calculation by logarithm could be a burden to limited ability of nodes. Thus, we pursue three goals for flooding attack detection on mobile sensor networks like following.

- **High detection accuracy** and **scalability** through distributed detection with a few samples for various deployments of attack agents
- **Detection possibility** through overlapped detection even at frequent handoffs of mobile attack agents
- **Low processing** considering the limited ability of nodes

This paper is divided into five sections. In Section 2, we explain the basic of used approximate entropy method, as background. We introduce in Section 3 the proposed detection in detail. Next, we simulate our mechanism and explain our analysis of the simulation results, and a brief conclusion is finally presented.

## 2 Approximate Entropy Estimator for a Discrete Random Variable

To discriminate between attack and normal, we use the approximate entropy estimation by key information attached to all messages, based on our key mechanism [8]. In [8], the key management nodes calculate a key using key information provided by the holder of the key. Thus, we can imagine that if several compromised nodes would generate lots of DoS traffic, the distribution for used keys might be changed. If our key mechanism would not be used, this approximate entropy estimator could be used with the other information of message such as source identification, location information, or sensory data themselves for distinguishing unusual events. We pursued to simplify the entropy estimation considering limited capacity of nodes and result in accurate estimation, thus we used an approximate method for Bayesian entropy estimation [9].

Entropy estimation is one of statistical approaches used for distribution comparison where the measurements involved are discrete values. $M$-ary entropy function $H$ for a discrete random variable X which takes $M$ sorts of values $\{a_1,\ldots,a_M\}$ with probabilities $\{r_1,\ldots,r_M\}$ is $-\sum_{i=1}^{M} r_i \log_2 r_i$ . Thus, in conventional entropy estimation, occurrence probabilities $r$ of values $a$ are estimated by the maximal likelihood estimation method, i.e., $\check{r} = n/N$, in which values $a$ are assumed to be included respectively for n $\equiv (n_1,\ldots,n_M)$ in the observed sample set of size $N$. However, unfortunately, the conventional entropy estimator is not optimum in the meaning of least square error [9].

Thus, we use basically the approximated Bayesian entropy estimator [9] as shown in equation (1), instead of the conventional estimator. In the equation (1), the previous work of [9] revealed that $s_i$=0.5 ($i$=1,...,$M$), as center, and $K$=20 were sufficient from the viewpoint of approximation precision, and numerical experiments of [9] demonstrated that the proposed entropy estimation improved the estimation precision of entropy remarkably in comparison to conventional entropy estimation.

Equation (1) is seen a little complicatedly for our sensor network, but the only $n_i$ is changed in the entropy computation of each sample, and our simplified estimator will be explained in detail at next section.

$$\hat{H}_{Bayes} = \sum_{i=1}^{M} s_i - \frac{1}{N+M}\sum_{i=1}^{M}(1+\log s_i)(n_i+1) - \sum_{k=2}^{K}\frac{1}{k(k-1)}\sum_{j=0}^{k}\binom{k}{j}\frac{(N+M-1)!}{(N+M+j-1)!}\cdot(-1)^j\sum_{i=1}^{M} s_i^{-j+1}\frac{(n_i+j)!}{n_i!} \quad (1)$$

# 3 Overlapped Detecting Approach

## 3.1 Basic Concept

Our detection based on entropy estimation could use the source identification or location information if it is attached to message, sensory data that could be regularly changed as time, like temperature or light, or key information attached to message. We explain our detection based on our key management on hierarchical mobile sensor networks. However, this detection might be easily applied to flat sensor network considered as a specific case of hierarchical network, one-layer network.

The key idea of detection is that detector nodes such as SINK, aggregator, or cluster head, compute entropy on a sample of consecutive messages generated in each region. That is, at the previous work [8], sensor nodes or aggregators attach the information for a key used for authentication or encryption to messages, and the attached information consists of {key space number, key information}. Because every key is randomly distributed, the distribution for used key spaces would have a specific form normally. Thus, as a flooding attack detection method, we use entropy estimation on the key space number for a key used at messages. Respective entropy values of the normal case and the attack case that floods lots of messages, are very different [7].

For each detector node, simplifying equation (1), we result in equation (2) as our estimator for attack detection. $M$ can be the total number of key spaces in the key pool, or the subset number of key spaces. Value α and β are constant by $N$ and $M$ that are fixed when this estimator is used ($\alpha=0.5\cdot M$, $\beta=N+M-1$). Also, $(-2)^j$ ($j$=1,...,20) is made to constants over temporary memory. Thus, the only $n_i$ changes in the entropy computation of each sample, and moreover any $n_i$ ($i$=1,...,$M$) are not changed if we calculate the entropy through a sliding window of fixed width $\delta$ (<$N$) for consecutive messages as shown in figure 1. In conclusion, this estimator for flooding DoS detection is calculated with only multiplication operation.

$$\tilde{H}_{Bayes} = \alpha - \sum_{k=2}^{20} \frac{1}{k(k-1)} \sum_{j=0}^{k} \binom{k}{j} \frac{\beta!}{(\beta+j)!} \cdot (-2)^j / 2 \sum_{i=1}^{M} \frac{(n_i+j)!}{n_i!} \tag{2}$$

For speedy detection, detectors use $\delta$ sliding window for arrival packets. That is, each detection node updates each $n_i$ with the new arriving $\delta (1 \le \delta \le N)$ packets like figure 1, recalculates $\tilde{H}'_{Bayes}$, and compares it with the previous $\tilde{H}_{Bayes}$. If the gap between them is higher than a threshold, which could be chosen with the average and standard deviation of normally estimated entropy, the detection node notifies it to other detection node on higher layer (detection nodes toward a sink). If the sink judges the attack with estimation results of aggregators and itself, it performs follow-up measures in order to localize the damage as soon as possible. For the value of $\delta$, a trade-off between the speed of detection and the calculation overhead of entropy exists. Therefore, this value could be determined as the capacity of detection nodes. Also, for the value of $N$, the sample size for approximate entropy calculation, there is a trade-off between the accuracy of detection and the calculation overhead of entropy,

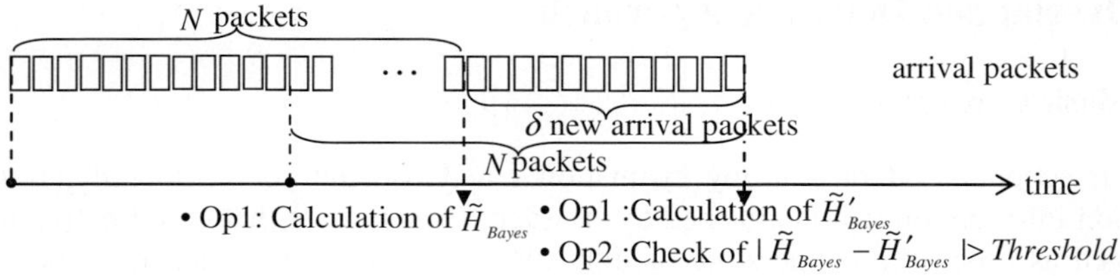


**Fig. 1.** Calculation and decision process on detection nodes

thus it also should be decided as the general generation speed of messages and the capacity of each detection node.

### 3.2 Overlapped Detection on Hierarchical Mobile Sensor Network

We assume that the subset of ANs and SINK perform two approximate entropy calculations for detection, a basic detection for its own management region and a overlapped detection for the region including a detection node's own region and its neighbors. Also, this subset of ANs could be changed as the control of SINK considering their residual energy and their covering region. Even if an AN does not perform the detection, the neighbor detection node can cover the region of the AN through overlapped detection. The reason why ANs perform the basic detection is to localize the damage by fast detection, and overlapped detection is for detecting moving attack agents. Also, SINK could synthesize the detection decision with estimation results of each region, aginst the various deployments of attack nodes.

Overlapped detection is that a detector calculates the approximate entropy for attack detection with the total traffic from sensor nodes of both its own region and neighbor regions, from where mobile sensor nodes could be moved to its own region. The goal of overlapped detection is to minimize the decrease of the detection performance by relatively reduction of attack traffic affecting a region though the mobility of attack node. The worst case could be that the handoff of an attack node would occurs frequently, thus it might be not disclosed by detectors. That is, the mobility of an attack node makes even the severe attack reduce to the stealthy attack of a region. If the mobility is considered, the centralized detection is better than the distributed method, but it is not scale to many sensor nodes.

Neighbor regions for overlapped detection are different according to network structure. For example, in the case of hexagonal grid-based network, there are 6 neighbor regions per a detector, and in the case of hierarchical sensor network, they are variable according to the partition of each layer. Neighbor regions of A3 in figure 2 are regions of SINK/A2/A4/A9/A10, and the information for neighbor regions on the same layer, such as A2/A4, could be conveyed as the combined information of higher detector, SINK, for overlapped detection, because the higher detector also periodically received the information of A2 and A4 regions.

We assume that the detectors are loosely time-synchronized, thus they periodically send their $n_i$ to detectors of neighbor regions through a high-priority message, and calculates the approximate entropy for the summation of $n_i$. If the detector does not receive $n_i$ from some regions, it requests the $n_i$ to them again. If the detector suspects the attack through overlapped detection process, it notifies the detection fact to neighbor detectors for the follow-up measure. Also, if the detector judges the attack through basic detection process, it sends its own $n_i$ to neighbor detectors for proactive detection against mobile attack agents.

## 4 Evaluation

To analyze the performance of our detection, we configured the simulation network and scenario like figure 2. This network consists of a SINK, 12 ANs covering respectively 50 SNs, and 650 SNs. The SINK has 5 key spaces, including especially a distributing key space 5 for newly deployed nodes, each AN at the first and second layer has the information for 3 and 2 key spaces respectively, and each SN has 3 randomly selected key sets coming under key spaces of higher AN. Even though the original key pre-distribution is performed randomly with the same probability for each key space, we intentionally distributes the key sets with probability (0.1, 0.2, 0.3, 0.1, 0.3) for each key space. However, normally each SN generates the sensory data by the same probability with randomly selected key among its own key sets, thus the intended traffic generation probability for each key space is (0.1, 0.2, 0.3, 0.1, 0.3).

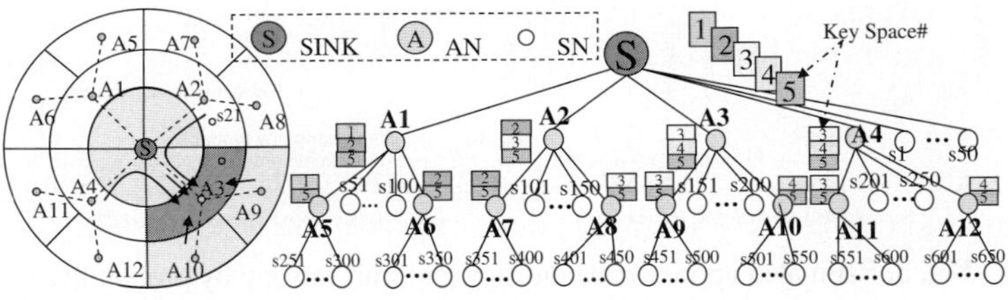


(a) A sample network for simulation (b) Simplified model including SNs for (a)

**Fig. 2.** A sample network for simulation

### 4.1 Detection Accuracy and Scalability

At first, in order to estimate the possibility of our DoS detection based on approximate entropy, we intent to mount DoS attack during the specific period. As a result, we get a variation of estimated approximate entropy on SINK and true entropy by probability as shown in figure 3. The x-axis of this figure is estimate round as time goes on. For the calculation of approximate entropy $\tilde{H}_{Bayes}$, we use $s_i$=0.5 and $K$=20 as recommended in [9], $M$=5 due to 5 key spaces and $N$=100 or $N$=500. At each estimate round, the calculation of $\tilde{H}_{Bayes}$ is performed with the new 100 or 500 packets, that is $\delta = N$ as the worst case in the view of detection speed. During estimate period from 101 to 120, we generate DoS attack traffic from SN s51, s52, s53 and s54 comprising 10% or 30%(*Pattack* in figure 3) of all packets. Thus, the true entropy by probability (0.1, 0.2, 0.3, 0.1, 0.3) is 1.504788 as shown in figure 3, and approximate entropy is also around the true entropy normally. But, during the simulated DoS attack period, the most approximate entropies increase considerably enough to easily detect the attack. Especially the bigger sample size $N$ and the severer attack exist, the more conspicuous entropy difference between normal and attack is.

This estimation for DoS detection is distributed to ANs as well as SINK. Also, as the more nodes under the charge of a management node are, the more capacity the management node has, each AN as a detection node may not be a big burden to the detection load which is proportional to the received traffic from nodes under its charge. In comparison with [2] that a SINK should count packets per each sensor, we support scalability for many sensor nodes, because our SINK or higher layer ANs as a detection node deal with the only aggregated messages, and estimation is distributed.

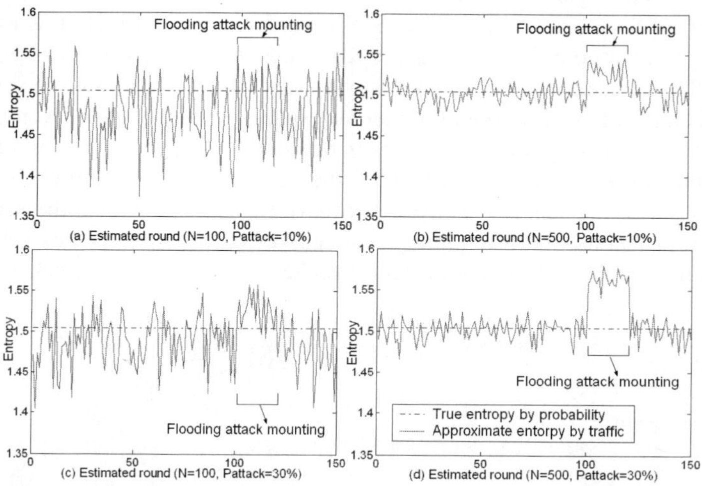


**Fig. 3.** Estimated approximate entropy and true entropy by probability

## 4.2 Detection Possibility on Frequent Handoffs

As the second simulation, we mount the DoS attack with 2 deployments of attack agents, concentrated attack and distributed attack. The attack agents in the former swarm around an AN, that are SN s51, s52, s53, and s54 of figure 2. In the latter case, they are distributed to different ANs, which are SN s51, s101, s151, and s201. Like the first simulation, we mount a DoS attack during estimate period from 101 to 120, and we estimate approximate entropy $\tilde{H}_{Bayes}$ for the DoS detection at SINK and AN1 respectively. Figure 4 shows the conspicuous increment of entropy at SINK in the concentrated attack case, but the entropy of AN1 is not changed nearly because the key spaces of 4 attack agents might cover all key spaces of AN1. On the other hand, in the distributed attack case of figure 4, the entropy increment at AN1 is more remarkable than at SINK because the attack traffic from only one attack agent makes the skew distribution of used key space number of AN1. Moreover, the uniformed distribution of attack agents in the view of SINK causes the less increment of entropy than their concentrated distribution. In conclusion, against various deployments of attack agents on hierarchical sensor networks, the synthetic decision needs for the accurate DoS detection.

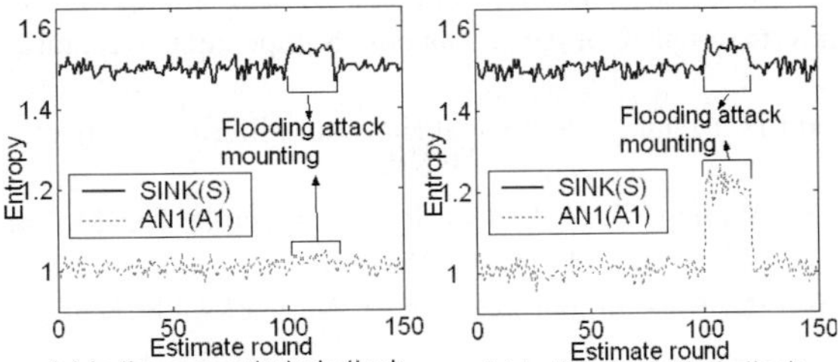


(a) In the concentrated attack (b) In the distributed attack

**Fig. 4.** Estimated entropy in two attack deployments

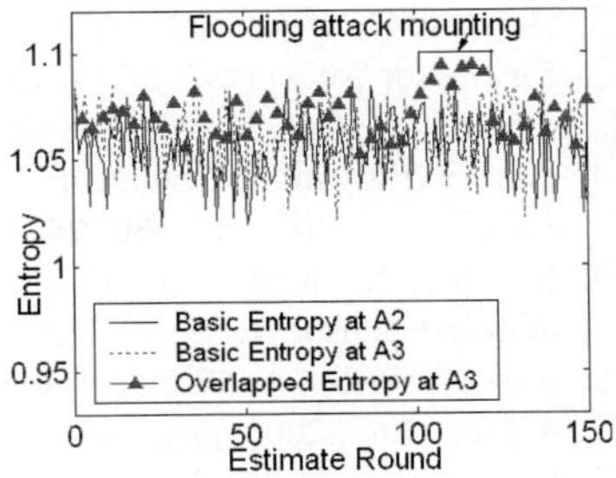


**Fig. 5.** Basic and overlapped entropy on A2 and A3 in attack of mobile nodes

As the last simulation result, figure 5 shows that basic and overlapped entropy in attack case of mobile agents. The mobile attack agents, from SN s101 to s104 mount stealthy attack during estimate period from 101 to 120, comprising 15% attack traffic of all packets, and they wander over the regions of A2 and A3. In result, basic entropy on each region does not change due to less attack traffic by mobility, but overlapped entropy at A3 indicates the difference between normal and attack. Thus, the overlapped detection is important especially for the stealthy attack of mobile agents.

### 4.3 Low Processing

Our estimator is calculated with only multiplication, instead of logarithm operation of traditional entropy estimator. In the view of operation complexity, the multiplication is much simpler and needs less cycle on the CPU board than logarithm. It could diminish the burden for detection load to each detection node.

## 5 Conclusion

In this paper, we have suggested a detection method against flooding attack on mobile sensor networks using the approximate entropy estimation. This approximate entropy estimator is calculated with only multiplication operation instead of logarithm, thus it fits for nodes with limited resources. we localized the detection function to each subgroup of sensors as resources is permissible, and SINK synthesizes localized reports to detect the attacks even at various deployments of attack agents. Moreover, the detector estimates the overlapped entropy for mobile attack agents. We have also simulated our mechanism, in order to analyze the possibility of detection via the entropy estimation and the accuracy of detection even at handoffs of mobile attack agents. Finally, we obtained a prominent result even though the attack agents are concentrated or distributed, or they move.

## References

1. Anthony Wood and John A. Stankovic, "Denial of Service in Sensor Networks," IEEE Computer, 35(10), pp.54-62, October 2002.
2. Serdar Sancak, Erdal Cayirci, Vedat Coskun and Albert Levi, "Sensor wars: Detecting and defending against spam attacks in wireless sensor networks," ICC 2004 - IEEE International Conference on Communications, vol. 27, no. 1, pp.3668-3672, June 2004.

3. Anthony D. Wood, John A. Stankovic and Sang H. Son, "Jammed-Area Mapping Service for Sensor Networks," 24th IEEE International Real-Time Systems Symposium, December 2003.
4. Fan Ye, Haiyun Luo, Songwu Lu and Lixia Zhang, "Statistical En-route Filtering of Injected False Data," IEEE Infocom'04, March 2004.
5. Schmidt, S., Krahn, H., Fischer, S., Watjen, D., "A Security Architecture for Mobile Wireless Sensor Networks," ESAS'04, LNCS3313 (2005), pp.166-177.
6. Karl, H., Wilig, A. "Protocols and Architectures for Wireless Sensor Networks," John Wiley & Sons, Ltd., 2005, pp.62-63.
7. Laura Feinstein, et al., "Statistical Approaches to DDoS AttackDetection and Response," Proc. of the DARPA Information Survivability Conference and Exposition, 2003.
8. Mihui Kim and Kijoon Chae, "A Hierarchical Key Management based on Multiple Regression in Mobile Sensor Networks," Submitted to a Conference.
9. Yasunari Yokota, "An Approximate Method for Bayesian Entropy Estimation for a Discrete Random Variable," Proc. of the 26th IEEE EMBS, September 2004.

# Implementation of Next Generation Mobile Service: The Context-Aware Follow-Me Service

Jungsook Bae[1], Seungwan Ryu[2,*], JaeYong Lee[3], and ByungChul Kim[3]

[1]Mobile Telecom. Research Division, ETRI, Taejon, Korea
[2]Department of Information Systems, Chung-Ang University, Korea
[3]School of Computer & Communications Engineering,
Chungnam National University, Korea
jsbae@etri.re.kr, rush2384@cau.ac.kr,
{jylee, bckim}@datacom.cnu.ac.kr

**Abstract.** With the advent of mobile communication technologies, next generation mobile service is evolving toward *user-centric* services, in that each user can get a personalized mobile service based on his/her needs. In this paper, we propose the *context-aware follow me (CAF)* service as a prototype next generation user-centric mobile service. The CAF service is designed to provide optimized service based on various context information such as user's location and preference, capability of devices located around a user and characteristics of services in use. We also present an example CAF service scenario, an overall system architecture, a test-bed system and a sample CAF service flows.

## 1 Introduction

The next-generation (NG) mobile communication is expected to be a system capable to provide a variety of multimedia services such as multimedia moving pictures, movies, and television broadcasts with high-speed and high quality via the converged network of wireless and wired infrastructures [1]. In particular, with the advent of next generation mobile communication technologies such as a high-speed data rate, converged network infrastructure, appearance of new digital devices having embedded sensing capabilities and appearance of next generation mobile terminal having higher computing capabilities, advanced mobile services will be realized over whole spheres of our everyday life. Therefore, in accordance with above technological advances, next generation mobile service is evolving toward *user-centric* services, in that each user can get a personalized mobile service based on his/her needs [2].

In order to provide the user-centric mobile services, a best suitable service should be generated and provided to a user based on his/her current situation by collecting context information around him without restrictions on his location or time. In other words, to provide various user-centric mobile services, next generation mobile services should be able to support various advanced service

* Corresponding author.

V.N. Alexandrov et al. (Eds.): ICCS 2006, Part II, LNCS 3992, pp. 1033–1040, 2006.

functions such as *context-awareness* for providing a user best suitable services based on context information around him, *personalization* for providing optimized services based on user's preference and needs, *service mobility* for providing seamless services without restrictions on location, time, device and network, and *service adaptation* for providing a best suitable type of service adaptively to the change of situation around a user [3].

In this paper, we propose the *context-aware follow me (CAF)* service as a prototype next generation user-centric mobile service that can supports above functions. The CAF service is designed to provide optimized service based on various context information such as user's location and preference, capability of devices located around a user and characteristics of services in use. In this service, users can enjoy the seamless call service and the video on demand (VOD) service anywhere and anytime via best suitable device.

This paper is consisted as follows. In next section, concepts, paradigms and key features of next generation mobile service are introduced. In section 3, design concepts and features of user-centric CAF services are proposed with example service scenarios and an overall system architecture. In section 4, a prototype test-bed system and example service flow of the CAF service is presented. Finally we conclude this paper.

## 2 Next Generation Mobile Services

With the progress of mobile communication technologies, mobile services are evolving to provide a variety of advanced services to guarantee people happiness of life [2]. The key feature of such evolutional trend of next generation mobile service is the concept of *user-centric service* which means that a best suitable service is provided for a user by analyzing user's needs and situational information. The concept of user-centric service is considered as a kernel paradigm of next generation mobile services and its enabling technologies are currently exploited by next generation mobile service research organizations such as Wireless World Research Forum (WWRF) [3] and Wireless World Initiative (WWI) [4]. In particular, after setting up *"I"* as a subject of mobile service, WWRF is trying to establish the concept of *I-centric* service and its enabling technologies with focusing on service requirements of "I" to support personalization, ambient awareness and adaptability. In the I-centric service environment, a personalized mobile service is provided to a user after service selection and adaptation process based on user's service requirements and context information.

In order to provide user-centric services, a best suitable service should be provided to a user after service selection and adaptation processes based on his/her current situation by collecting context information around him without restrictions on his location or time. The user-centric service should be able to provide various service features in terms of cost, data rate, maturity and application areas as follow [2].

- Low cost service: diverse and convenient services at low price
- High speed service: transmission of large amount of multimedia data with high data rate
- Context-Aware service: a best suitable service based on context information and situation of a user.
- Seamless service: service capable to overcome user's movement and/or changes of network environments
- Converged service: various services provided by connecting or combining mobile services with other services such as broadcasting and other information technologies.
- Personalized service: a tailored service to the user's expectation in accordance with diverse demand and desire and personalized life style.
- Virtual reality service: services that are true to nature provided by various user friendly interfacing methods.
- Safety service: services for keeping security and privacy.

In order to fulfill user-centric services having above service features, functions of context-awareness, personalization, service mobility and adaptation should be supported.

# 3 The Context Aware Follow-Me Service

## 3.1 Concepts and Features

A number of context-aware computing and communication concepts and applications have been proposed by research organizations and companies [5, 6]. However, most of such context-aware services were not suggested as mobile services in next generation [7]. For example, [6] introduces recent research and development activities on context-aware communications with focusing on how changes on individual context initiate and facilitate people's interaction with one another.

In this paper, we propose the *context aware follow-me (CAF)* service developed as a prototype next generation user-centric context-aware mobile service. CAF service is designed for providing optimal services using best suitable devices based on user's location and preference, capability of available devices and characteristics of services in use.

Key features of the CAF service are

- Context awareness: inference of current service availability using collected context information such as user's location, device availability and capability of devices exist in close proximity of a user.
- Service mobility: seamless transfer of on-going service from a device to other selected device based on service environment where a user is located and user's device preference.
- Service personalization: provisioning of service environment user prefer using context information such as user profile, user's preferences, and other environmental information.

| |
|---|
| Tom is driving a car on the way home. He takes French lesson using the mobile foreign language lesson service. Since he is driving, he takes lesson only in audio sound mode. After arriving home, Tom gets out of the car, but the lesson is continued via his mobile terminal with giving audio and video mode services simultaneously. As soon as he enters living room the lesson service is shifted to a TV. After finishing French lesson, Tom begins taking a bath. When he is in bath room, Jain calls him, and the call is forwarded to a smart mirror equipped in bath room only with voice mode. When the bath is over, the call is shifted to his mobile terminal with giving voice and video mode simultaneously. |

**Fig. 1.** An example CAF service scenario

- Service adaptation: adaptation of service data format and quality according to capability of the selected device and network.

A possible service scenario of the CAF service is given below.

### 3.2 System Architecture

As a prototype service of next generation context-aware service, the CAF service is provided within a local domain via interfacing with other remote domains. Each domain is consisting of several elements such as a domain server, a session control server, a location server, location management systems, a media adaptation server and smart devices. Functions of each system elements are.

- **A domain server** identifies and authorizes a user, collects context information of a user, gathers information related to other devices, and controls session mobility. Examples are the home server, the office server, the car server.
- **A session management server** manages session connections or hands over a session connection to a selected intelligent device.
- **A location management system** detects the user's current location.
- **A location server** identifies user's current location.
- **A media adaptation server** trans-codes media to be suitable to the selected intelligent device and, vise versa.
- **A smart device** has communication capability with system elements in domain and multimedia service support capability.
- **A mobile terminal** can support multi radio access technologies such as CDMA, WLAN and WPAN.

The overall system architecture for the CAF service is given in figure 2.

Generic operational rules and sequences of the CAF service under the given architecture are as follow.

1. A user enters a new domain such as a car, a house or an office.
2. A (local) domain server collects device related information device such as location, device capability and availability periodically using the device discovery protocol.

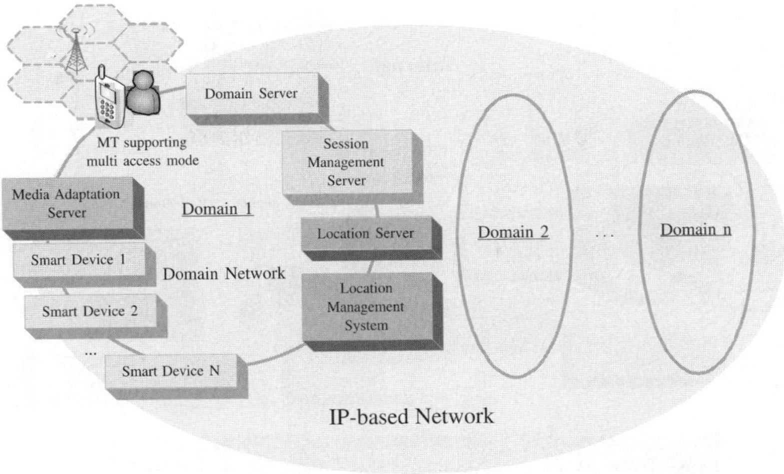


**Fig. 2.** An overall system architecture for CAF service

3. The domain server identifies and authorizes the user.
4. Location management systems and a location server keep track of location of the authorized user.
5. The domain server collects users current context information such as user location, using service, user profile and user preferences.
6. When the user is under calling with someone or getting media streaming service, the domain server selects the best suitable device based on users current location, users preferences, device information and service characteristics.
7. The domain server manages session control server to transfer the user's current session connections to the selected device
8. The domain server manages the media adaptation server which transforms media code to be suitable for the selected device.

### 3.3 The Test-Bed System for CAF Service

In order to investigate feasibility of the CAF service, especially for the example CAF scenario presented in figure 1, we developed a test-bed system. As shown in figure 3, the test-bed system consists of two domains, a car domain and a home domain, in addition to a mobile terminal that can roam those two domains. The car domain is consisting of a car server, location systems, a location server, a media adaptation server and a car A/V system which has communication and computing capability. All devices in the car domain are assumed to be connected via short-range communications. In this test-bed, we use WLAN for communications among devices. The home domain is consisting of a home server, location systems, a location server, a media adaptation server and smart devices such as an intelligent TV, audio and smart mirror are assumed to have communication and computing capability. In addition, such devices are also assumed to be connected via WLAN similar to the devices in the car domain. The home domain

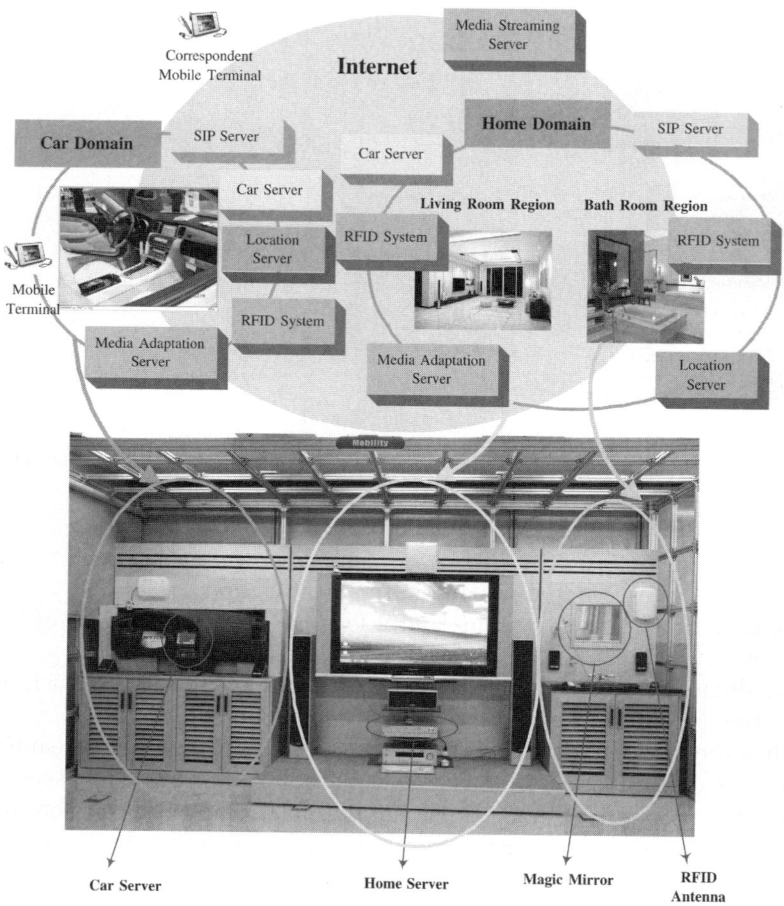


**Fig. 3.** Test bed system for CAF service

is divided into two sub-regions, the living room region and the bath room region. For tracking and recognition of user location, a passive type RFID tag is attached on a user, and 915Mhz RFID readers are installed in several parts of each domain. In this test-bed, the service mobility across a range of devices in various domains is supported by the session initiation protocol (SIP).

### 3.4 Service Flow in Test-Bed System

Because of page limitation, we give readers a sample CAF service flows for a specific situation instead of giving a complete set of CAF service flow for the example CAF service scenario shown in figure 1. In detail, we choose a specific situation when the user gets off the car and enters home, and show how the CAF service is provided in this situation. We describe service flows among service elements such as a mobile terminal (MT), a home server (HS), a location server (LS), a media adaptation server (MAS), a media streaming server (MSS), a

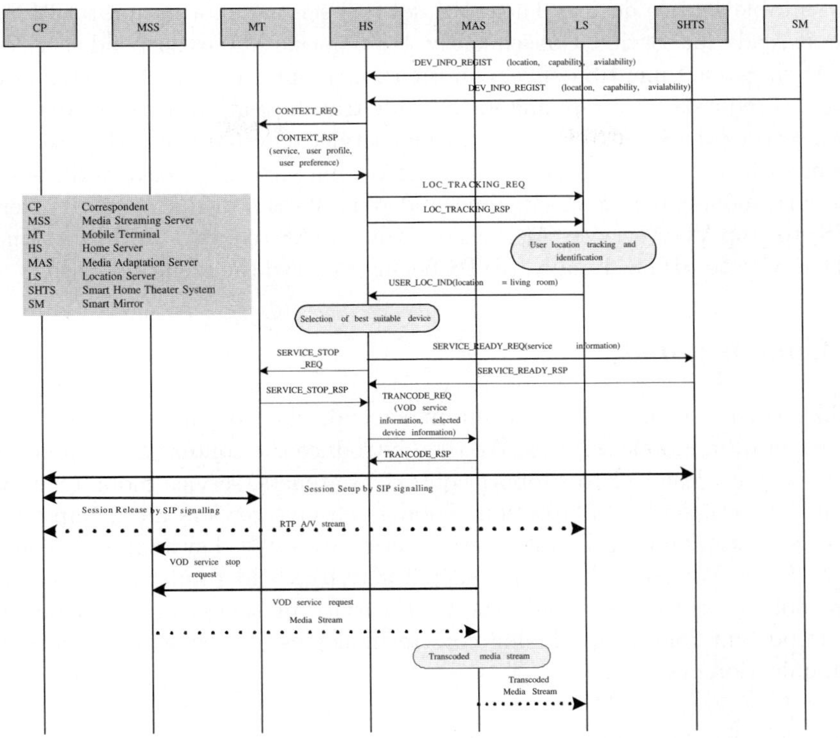


**Fig. 4.** CAF Service Flow in test bed system

smart home theater system (SHTS), and a correspondent (CP) who is a person connected with the user's terminal through SIP protocol. In the chosen situation, we assume that the user enters a home domain while using both of VoIP call service and the VOD service via his mobile terminal. Figure 4 describes how service flows are implemented among service elements in the test-bed system to maintain the CAF service seamlessly in this situation.

Each domain server collects device related information such as device location, device capability and availability from service devices exist in the same domain. When a new user is identified, a domain server (the home server (HS) in this case) requests context information of the identified user stored in the former domain server (the mobile terminal (MT) of the user or the car server in this case). Upon receiving such request from HS, the mobile terminal delivers context information such as services in use, user profile and preferences, and other context information. At the same time, HS requests location tracking information of the identified user to the location server (LS). Since LS detects the user enters living room, it informs the current location and location tracking information of the user to HS. Then, based on collected context information such as user's current location, user's profile and preference, device information and service information, HS selects the smart home theater system (SHTS) as the best suitable service device. Then, HS delivers service information to SHTS to make it ready for service. Consequently, HS requests MT to suspend services in use. At the same time, HS delivers information about service and selected device as well as requests for streaming media code transformation to the media adaptation server (MAS) to give SHTS optimal service environment. SHTS sets up a session with correspondent (CP) via SIP signaling, and MT releases the session with correspondent by SIP signaling and requests the media streaming server (MSS) to stop VOD service. At the same time, MAS requests MSS to resume of VOD service to SHTS. Finally, SHTS begins to serve the identified user.

## 4 Conclusion

In this paper, we investigate evolutional trend, concepts and key features of next generation mobile services. We then introduce the context-aware follow me (CAF) service designed to support context awareness, service mobility, service personalization, service adaptation. Finally, we give readers an example CAF service scenario, an overall system architecture, a test-bed system and a sample service flows. We hope that our research activities and results on next generation mobile services, especially the CAF service presented in this paper, will play important role in developing core technologies of next generation mobile communication systems.

## References

1. S. Ryu and D. Oh and G. Shin and K. Han and S. Hwang and S. Park: "Research Activities on the Next Generation Mobile Communications and Services in Korea," *IEEE Communicaions Magazine*, 43(9), pp. 122–131, September, 2005.
2. Next Generation Mobile Communication Forum: "Rainbow Vision," `http://www.ngmcforum.org/ngmc2/main.html`.
3. S. Arbanowski and P. Ballon and K. David and O. Droegehorn and H. Eertink and W. Kellerer and H. Kranenburg and K. Raatikainen and R. Popescu-Zeletin: "I-centric Communications: Personalization, Ambient Awareness, and Adaptability for Future Mobile Services," *IEEE Communicaions Magazine*, 42(10), pp. 63–69, September, 2004.
4. Wireless world initiative: "Mobi Life: Life goes on," `http://www.wireless-world-initiative.org/`.
5. G. Chen and D. Kotza: "A survey of context-aware mobile computing research," Technical Report, TR2000-381, Department of Computer Science, Dartmouth College, May, 2000.
6. B. Schilit and D. Hilbert and J. Trevor: "Context-Aware Communication," *IEEE Wireless Communicaions Magazine*, 9(5), pp. 2–10, October, 2002.
7. J. Kim and J. Bae and S. Yoon and J. Ha and G. Sihn: "Evolution of Context-Aware Services in Next Generation Mobile Communication," *WG contributions of 13th WWRF Meeting*, March, 2005.

# Multi-piconet Formation to Increase Channel Utilization in IEEE 802.15.3 High-Rate WPAN*

Ssang-Bong Jung[1], Soon-Bin Yim[1], Tae-Jin Lee[1],
Sun-Do June[2], Hyeon-Seok Lee[2], Tai-Gil Kwon[2], and Jin-Woong Cho[2]

[1] School of Information and Communication Engineering, Sungkyunkwan University, Korea
[2] Wireless Network Research Center, Korea Electronics Technology Institute, Korea
{jssbong, sbyim, tjlee}@ece.skku.ac.kr
{jsd, hslee75, tgkwon, chojw}@keti.re.kr

**Abstract.** IEEE 802.15.3 high-rate Wireless Personal Area Networks (WPANs) is a short range communication technology, in which devices (DEVs) of a piconet communicate with a Piconet Coordinator (PNC). An extension of the communication area in WPAN via formation of a multi-piconet is one of the most challenging issues. Although a concept of multi-piconet has been defined in the IEEE 802.15.3 WPAN standard, a detailed formation mechanism of multi-piconet is not presented. In this paper, we propose efficient multi-piconet formation algorithms, Piconet Coordinator To Device (PCTD) and Device To Piconet Coordinator (DTPC). We evaluate the performance of the algorithms via simulations. The proposed algorithms are shown to extend the communication area with one to three levels and to provide sufficient Channel Time Allocations (CTAs) for subsidiary piconets in IEEE 802.15.3 WPAN.

## 1 Introduction

Recently, we have witnessed a noticeable increase of personal devices. The devices are often required to interact wirelessly to make information accessible and to exchange data without physical cables. Wireless Personal Area Networks (WPANs) can connect various personal devices within a Personal Operating Space (POS). IEEE 802.15.1 (Bluetooth) [1] and IEEE 802.15.4 (Zigbee) [3] supports the transmission rate of less than 1Mbps, which is not appropriate for high-rate WPAN applications. IEEE 802.15.3 [2] [4] is an emerging high-rate WPAN technology, which enables portable multimedia devices to connect and communicate wirelessly. IEEE 802.15.3 WPAN operates in the unlicensed 2.4GHz ISM band with the data rate of up to 55Mbps and supports the ad-hoc mode allowing a number of devices to communicate with one another in a peer-to-peer mode. A piconet in IEEE 802.15.3 consists of one Piconet Coordinator (PNC) and more than one device (DEV). In IEEE 802.15.3 WPAN, a multi-piconet can be formed to extend a single piconet. A multi-piconet consists of a parent piconet and its child piconets. The child PNC is a member of the parent piconet and also plays a role of the PNC in the child piconet. A child piconet is dependent on the parent piconet [5].

* This research was supported by Electro-0580 funded by the Ministry of Information and Communication (MIC) in South Korea. Dr. Tae-Jin Lee is the corresponding author.

V.N. Alexandrov et al. (Eds.): ICCS 2006, Part II, LNCS 3992, pp. 1041–1049, 2006.

In the Bluetooth specification, a piconet consists of one master device and up to seven slave devices. A piconet can be extended in the form of a scatternet. A scatternet is a collection of multiple piconets and each piconet includes bridge devices to interconnect piconets. The formation of scatternets is essential to break the limit of a small piconet and allows interaction among Bluetooth piconets. There have been algorithms for scatternet formation, i.e., a distributed scatternet formation protocol [6], a formation mechanism to support device mobility [7], a Tree Scatternet Formation (TSF) algorithm [8], and the scatternet-route structure [9].

In the IEEE 802.15.3 specification, efficient multi-piconet formation protocols and Channel Time Allocation (CTA) mechanisms are not defined although basic frame structure for the multi-piconet extension is defined. A multi-piconet can extend the communication area of a single piconet. It is very important to form an interconnected network combining separate piconets and to provide new services to users. Therefore, in this paper, we propose an efficient multi-piconet formation scheme and evaluate its performance. In this paper, we present two multi-piconet formation algorithms, Piconet Coordinator-To-Device (PCTD) and Device-To-Piconet Coordinator (DTPC). In the PCTD multi-piconet formation algorithm, the formation of a multi-piconet is initiated by a PNC. The PNC allocates private CTAs to DEVs and the DEVs create child piconets to communicate one another in the allocated private CTAs. In the DTPC multi-piconet formation algorithm, the formation of a multi-piconet is initiated by some DEVs of a piconet and the PNC of the piconet allocates private CTAs to the DEVs. The initiating DEVs report the remote scan information to the PNC. The DEVs then create child picונts and communicate one another in the allocated private CTAs. An efficient private CTA allocation algorithm for PCTD and DTPC is also proposed. We conduct simulations to evaluate its performance. Simulation results validate the desired features of the PCTD and DTPC algorithms.

The rest of the paper is organized as follows. Section 2 gives an overview of the IEEE 802.15.3 High-rate WPAN. In Section 3, we describe our proposed PCTD and DTPC multi-piconet formation algorithms. Section 4 provides the performance analysis of the proposed algorithms. Finally, Section 5 concludes the paper.

## 2 IEEE 802.15.3 High-Rate WPAN Overview

In IEEE 802.15.3, a piconet consists of a PNC and more than one DEV. The PNC provides the basic timing for the piconet with beacons and it allows independent devices to communicate one another. The standard utilizes the 2.4GHz ISM unlicensed frequency band and two sets of channels are assigned for operation. The first set is the high-density mode which allocates 4 channels while the second one is the IEEE 802.11b co-existence mode which allocates 3 channels. Since the two outer channels of the sets overlap, there are total of five channels.

In IEEE 802.15.3 WPAN, the channel time is divided into superframes, and each superframe begins with a beacon. Fig. 1 illustrates the MAC superframe structure. A superframe has three parts, i.e., beacon, Contention Access Period (CAP), and Channel Time Allocation Period (CTAP). The beacon is used to carry control and channel time allocation information to the entire piconet, and it is generated and broadcast by the

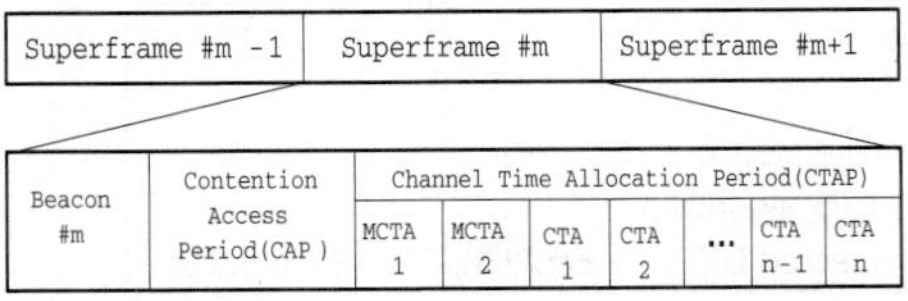


**Fig. 1.** Superframe structure of IEEE 802.15.3

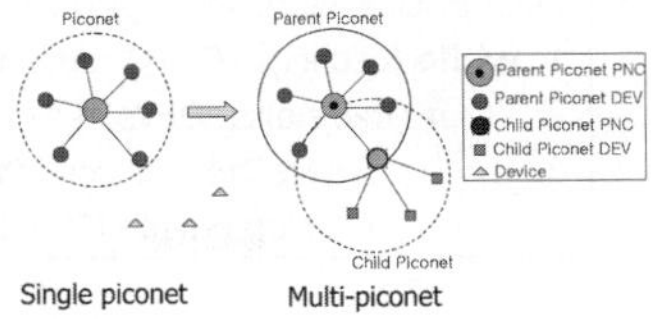


**Fig. 2.** Single piconet and multi-piconet

PNC. The CAP is used to exchange association request and response command, channel time request and response command, and asynchronous data. Carrier Sense Multiple Access/Collision Avoidance (CSMA/CA) is used for medium access control in the CAP. The CTAP is composed of channel time allocations (CTAs) and management CTAs (MCTAs), which makes TDMA-like medium access.

IEEE 802.15.3 WPAN allows a DEV to request the formation of a subsidiary piconet. There are subsidiary piconets of two types. One is the child piconet and the other one is the neighbor piconet. The original piconet is referred to as the parent piconet. The neighbor piconet can not communicate with the parent piconet, but the child piconet can communicate with the parent piconet. We call a multi-piconet as a larger piconet consisting of a parent piconet and its child piconets. Fig. 2 shows a single piconet and a multi-piconet. A parent piconet and its child piconets operate under the same channel. A child PNC requests private CTAs to the parent PNC and the parent PNC allocates private CTAs to the child PNC. A child piconet can then communicate in the private CTA period.

## 3 Proposed Multi-piconet Formation Algorithms

In this section, we present our proposed two multi-piconet formation algorithms to extend limited piconet communication area. One is the PCTD multi-piconet formation algorithm and the other one is the DTPC multi-piconet formation algorithm.

### 3.1 PCTD Multi-piconet Formation Algorithm

In this section, we describe the PCTD multi-piconet formation algorithm. In the PCTD algorithm, a PNC starts to form a multi-piconet, empowers some DEVs to be child piconet PNCs and allocates private CTAs to the DEVs. The DEVs which are assigned the private CTAs have the PNC capability. The PCTD multi-piconet formation procedure is shown in Fig. 3.

If a parent PNC has its available capacity to distribute and there exist DEVs in the current piconet, the PNC selects some DEVs as child PNCs by the child PNCs selection (CPS) algorithm shown in Fig. 4. Then, the PNC allocates private CTAs to the selected DEVs by Eqs. (1)∼(5). The selected DEVs make piconets, i.e., child piconets. Child piconets can now communicate in the allocated private CTAs. The mechanism of selecting some DEVs as child PNCs works as follows. If the number of DEVs in a piconet is less than the minimum number of DEVs (min_DEVs), we select all the DEVs.

```
PROCEDURE PCTD Multi-piconet Formation
while (capacity (CTAs) remains)
    if (there exist DEVs in the piconet)
        Select DEVs as child PNCs by CPS algorithm
        Allocate private CTAs to the selected DEVs by Eqs. (1), (2), (3), (4), (5)
```

**Fig. 3.** PCTD multi-piconet formation algorithm

```
PROCEDURE Child PNCs Selection (CPS) Algorithm
// n: number of DEVs in a current piconet, min_DEVs: minimum number of DEVs for child PNCs,
// p: probability of selecting DEVs for child PNCs, n_s: number of selected DEVs as child PNCs
if (n <= min_DEVs)
    n_s = n
else
    n_s = min_DEVs + ⌈n × p⌉
```

**Fig. 4.** Child PNCs Selection (CPS) algorithm

Otherwise, we randomly select some DEVs among the DEVs with the probability of $p$ while providing the minimum number of DEVs.

Allocation of private CTAs is done as follows based on the basic capacity for level $i$, $C^i_{capacity}$ and the allocation CTA remainder to one piconet for level $i$, $R^i_{CTA1}$. To explain the private CTA allocation algorithm for the PCTD multi-piconet formation algorithm, we define the following notations:

- $T_B$ : Beacon time
- $T_{CAP}$ : Channel access period
- $T_{CTA}$ : Channel time allocation
- $S$ : Maximum superframe size
- $N$ : Number of DEVs in the parent piconet
- $L$ : Number of levels
- $n_i$ : Number of DEVs in level $i$, $\quad i = 1, 2, \cdots, L$
- $m_i$ : Number of child piconets in level $i$, $\quad i = 1, 2, \cdots, L$
- $R^i_{CTA1}$ : CTA allocation remainder to one piconet in level $i$, $\quad i = 1, 2, \cdots, L$
- $C^i_{capacity}$ : Basic capacity for level $i$, $\quad i = 1, 2, \cdots, L$
- $C^0_{capacity}$ : Basic capacity for level 0 (parent piconet)
- $C^i_{max}$ : Maximum capacity for level $i$, $\quad i = 1, 2, \cdots, L$

The $C^0_{capacity}$ consists of a beacon time and a CAP, and the basic CTAs for the parent piconet DEVs and is computed as Eq. (1). We can calculate the basic capacity for level $i$, $C^i_{capacity}$ by Eq. (2).

$$C^0_{capacity} = (T_B + T_{CAP}) + (T_{CTA} \times N) \tag{1}$$

$$C^i_{capacity} = (T_B + T_{CAP}) \times m_i + (T_{CTA} \times n_i), \quad i = 1, 2, \cdots, L \tag{2}$$

```
PROCEDURE DTPC Multi-piconet Formation
while (capacity (CTAs) remains)
    if (there exist DEVs in the piconet)
        Select DEVs by CPS algorithm
        Request remote scan to the selected DEVs
        Collect remote scan responses from the DEVs
        if (there is reported remote scan information)
            Allocate private CTAs to the selected DEVs by Eqs. (1), (2), (6), (7), (8)
```

**Fig. 5.** DTPC multi-piconet formation algorithm

since $m_i$ child piconets require beacons and CAPs and $n_i$ DEVs in level $i$ are assigned $T_{CTA}$. The remaining capacity has to be allocated for the parent piconet and its child piconets. It the PCTD algorithm, the information on the next level is not known, the CTA allocation remainder to one child piconet for level 1 is found as follows:

$$R^{1}_{CTA1} = \frac{S - C^{0}_{capacity}}{m_1 + 1} \tag{3}$$

since $S - C^{0}_{capacity}$ is the remaining capacity for level 1 and $m_1$ is number of child piconets in level 1. Note that $C^{0}_{capacity}$ does not contain $n_1$, which is the information on the next level. The 1 in $(m_1 + 1)$ denotes the parent piconet. Similarly, we can find the CTA allocation remainder in level $i$, $R^{i}_{CTA1}$ as follows.

$$R^{i}_{CTA1} = \frac{R^{i\text{-}1}_{CTA1} \times m_{i\text{-}1} - C^{i\text{-}1}_{capacity}}{m_i + m_{i\text{-}1}}, \qquad i = 2, 3, \cdots, L. \tag{4}$$

Thus, the private CTAs for each child piconet is equal to $R^{i}_{CTA1}$ in the PCTD multi-piconet formation algorithm. The maximum capacity at level $i$, $C^{i}_{max}$ is given by

$$C^{i}_{max} = R^{i}_{CTA1} \times m_i, \qquad i = 1, 2, \cdots, L \tag{5}$$

from the fact that each child piconet of level $i$ is assigned $R^{i}_{CTA1}$ and there are $m_i$ child piconets.

### 3.2 DTPC Multi-piconet Formation Algorithm

In this section, we describe the DTPC multi-piconet formation algorithm. In the DTPC algorithm, a PNC requests remote scan to some selected DEVs. The selected DEVs respond remote scan information to the PNC. The PNC then allocates private CTAs to the DEVs. The DTPC multi-piconet formation procedure is shown in Fig. 5. First, DEVs make their own piconets or join already formed piconets. The PNC of a piconet requests remote scan to some selected DEVs and the selection is done by the CPS algorithm, and the DEVs respond remote scan information with the near piconets to the PNC. The DEVs then request private CTAs to the PNC to form child piconets. The PNC allocates private CTAs to the requesting DEVs by Eqs. (1), (2), (6), (7), (8) and the requesting DEVs make child piconets.

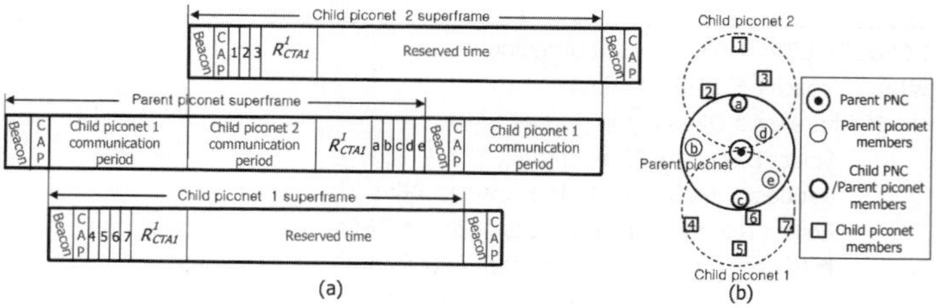


**Fig. 6.** Multi-piconet CTA allocation example (a - e: parent picinet DEVs, 1 - 7: child piconet DEVs), (a) superframe allocation to child piconets (b) a multi-piconet topology

The basic capacity at level 0 (parent piconet) $C^0_{capacity}$ and the basic capacity at level $i$ $C^i_{capacity}$ are the same as in Eqs. (1) and (2). The CTA allocation remainder at level 1, $R^1_{CTA1}$ and the CTA allocation remainder at level $i$, $R^i_{CTA1}$ are calculated as follows:

$$R^1_{CTA1} = \frac{S - (C^1_{capacity} + C^0_{capacity})}{m_1 + 1} \tag{6}$$

$$R^i_{CTA1} = \frac{R^{i\text{-}1}_{CTA1} \times m_{i\text{-}1} - C^i_{capacity}}{m_i + m_{i\text{-}1}}, \qquad i = 2, 3, \cdots, L \tag{7}$$

since $S$ - $(C^1_{capacity} + C^0_{capacity})$ is the remaining capacity for level 1 and $m_1$ is the number of child piconets in level 1. The 1 in $(m_1 + 1)$ denotes the parent piconet. In the PCTD algorithm the information on the next level is not known, we find $R^i_{CTA1}$ by Eqs. (3) and (4). However, in the DTPC algorithm, we can know the information on the next level. Therefore, $R^1_{CTA1}$ and $R^i_{CTA1}$ in DTPC algorithm is found as Eqs. (6) and (7).

The maximum capacity of level $i$, $C^i_{max}$is given by

$$C^i_{max} = R^i_{CTA1} \times m_i + C^i_{capacity}, \qquad i = 1, 2, \cdots, L \tag{8}$$

from the fact that each child piconet of level $i$ is assigned $R^i_{CTA1}$ and there are $m_i$ child piconet, and $C^i_{capacity}$ is the basic capacity of level $i$.

Fig. 6 (b) shows a multi-piconet with level $L$=1 and the number of child piconets $m_1$=2. Fig. 6 (a) shows the relationship of superframes for the topology. The superframe of a parent piconet consists of the beacon, CAP, the communication period for the child piconets, the local communication period for the parent piconet DEVs (a, b, c, d, e) and the CTA allocation remainder $R^1_{CTA1}$. The communication period of child piconet 1 is 4, 5, 6, 7 and the communication period of child piconet 2 is 1, 2, 3. The CTA allocation remainder ($R^1_{CTA1}$) is required to reserve communications between the parent piconet and child piconets, and to make child piconets of the next level.

# 4 Performance Evaluation

We evaluate the performance of the proposed two multi-piconet formation algorithms via simulations. The simulation parameters are summarized in Table 1. The DEVs are assumed to be located uniformly in a square area of 50m × 50m and 60m × 60m. And

**Table 1.** Simulation parameters

| Parameter | Value |
|---|---|
| Channel bandwidth | 11Mbps |
| Total number of DEVs | 10 ∼ 150 |
| CTA size | 2,048byte |
| CTA time ($T_{CTA}$) | 1.53ms |
| Beacon time ($T_B$) | 0.06ms |
| CAP time ($T_{CAP}$) | 1.00ms |
| Maximum Superframe size ($S$) | 65.535ms |
| Minimum number of DEVs ($min\_DEVs$) | 4 |
| Probability to select DEVs ($p$) | 0.2 |

the number of DEVs ranges from 10 to 150 in the area. The number of simulation runs is 500. The length of a MAC Service Data Unit (MSDU) is set to 2048 bytes, which equals to the unit CTA size [1]. We analyze the performance of the PCDT and the DTPC multi-piconet formation algorithm with respect to the number of formed multi-piconet levels and the remaining capacity.

We then analyze the performance for random networks. Fig. 7 and Fig. 8 present the formed multi-piconet levels by the proposed algorithms. The PCTD algorithm is shown to form multi-piconets with two to three levels, and the DTPC algorithm is shown to form multi-piconets with one to two levels. The proposed algorithms form more levels of a multi-piconet if the density of DEVs is high and the PCTD algorithm generates more levels than the DTPC algorithm does. With the proposed algorithms, one can extend the communication range two to three times. In particular, the communication range of a formed multi-piconet by the PCTD algorithm is extended up to 50m assuming the transmission range of a DEV is 10m. The PCTD algorithm exhibits higher efficiency than the DTPC algorithm. We define the efficiency as the ratio between the allocated capacity and the original total capacity. For example, in Fig. 7, if the number of DEVs is 70, the PCDT algorithm is 130% better than the DTPC algorithm in terms of the formation efficiency. And, if the number of DEVs is 110, the PCDT algorithm is 114 % better than the DTPC algorithm.

We also analyze the remaining capacity at the final level. Fig. 9 and Fig. 10 present the remaining multi-piconet capacity as the number of DEVs varies. If the number of DEVs increases, the remaining capacity decreases. Since the maximum superframe size is 65.535ms, multi-piconet formation is possible up to level 3. Thus, the more the number of DEVs, the higher the level of a multi-piconet becomes and the less the remaining capacity of the last level becomes.

In the PCTD algorithm, if the capacity remains to make child piconets, it can directly form new child piconets. However, in the DTPC algorithm, although the capacity remains to make child piconets, if the near piconet information as a result of remote scan does not exist, it can not make new child piconets. Thus, the remaining capacity at the last level in PCTD is smaller than that in DTPC as in Fig. 9 and Fig. 10. And the level of the PCTD algorithm is higher than that of the DTPC algorithm.

[1] In fact, physical and MAC header and preambles are appended as well.

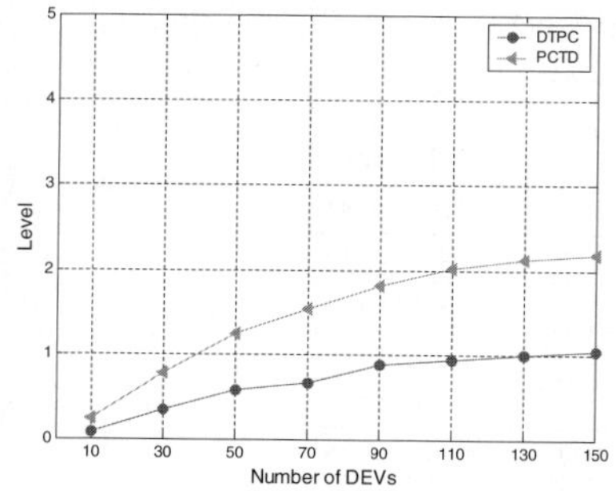


**Fig. 7.** Multi-piconet level as the number of DEVs (50m ×50m) varies

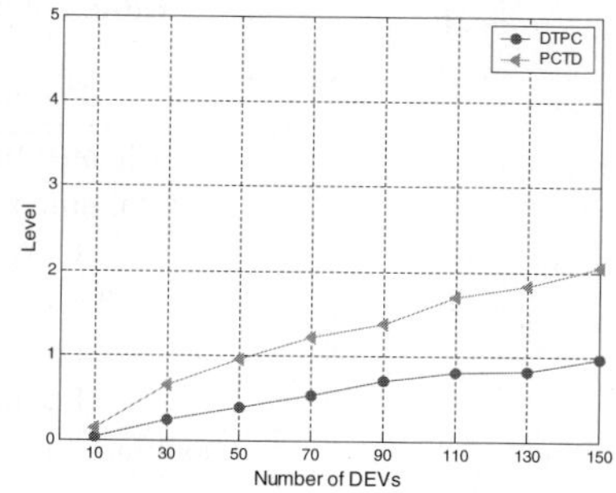


**Fig. 8.** Multi-piconet level as the number of DEVs (60m ×60m) varies

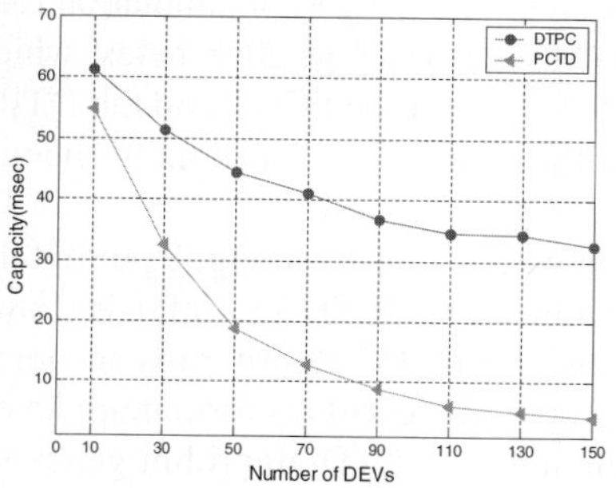


**Fig. 9.** Remaining multi-piconet capacity as the number of DEVs (50m ×50m) varies

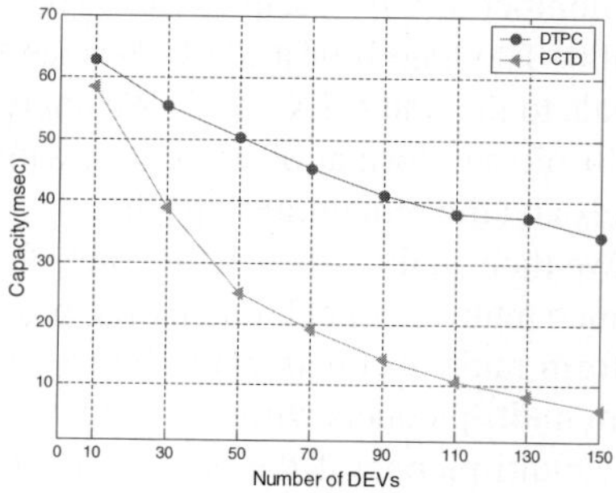


**Fig. 10.** Remaining multi-piconet capacity as the number of DEVs (60m ×60m) varies

## 5 Conclusion

In this paper, we have proposed the multi-piconet formation mechanisms with efficient private CTA allocation schemes to extend the piconet communication area of IEEE 802.15.3 WPAN. We evaluate the performance of the proposed algorithms via simulations. The results show the relationships of the level, the capacity and the number of DEVs for multi-piconet formation. The PCTD algorithm is shown to form multi-piconets with two to three levels, but the available capacity at the last level for child piconets is limited. The DTPC algorithm can form multi-piconets with one to two levels, and the available capacity at the last level for child piconets is sufficient. The proposed algorithms are able to extend the communication area and to generate large-scale WPAN service.

## References

1. IEEE, "Standards for Part 15.1: Wireless Medium Access Control (MAC) and Physical Layer (PHY) Specifications for Wireless Personal Area Networks (WPANs)," Jun. 2002.
2. IEEE, "Standard for Part 15.3: Wireless Medium Access Control (MAC) and Physical Layer (PHY) Specifications for High Rate Wireless Personal Area Networks (WPANs)," Sep. 2003.

3. IEEE, "Standard for Part 15.4: Wireless Medium Access Control (MAC) and Physical Layer (PHY) Specifications for Low-Rate Wireless Personal Area Networks (LR-WPANs)," Oct. 2003.
4. J. Karaoguz, "High-Rate Wireless Personal Area Networks," IEEE Communications Magazine, vol. 39, no. 12, pp. 96-102, Dec. 2001.
5. D. Trezentos, G. Froc, I. Moreau and X. Lagrange, "Algorithms for Ad-hoc Piconet Topology Initialization Evaluation for the IEEE 802.15.3 High Rate WPAN System," in Proc. of IEEE VTC, Vol. 5, pp. 3448-3452, Oct. 2003.
6. V. P. Verma and A. A. Chandak, "Distributed Bluetooth Scatternet Formation Algorithm," in Proc. of IEEE GLOBECOM, vol. 3, pp. 1274-1278, Dec. 2003.
7. C. H. Yang and Y. S. Chen, "Bluetooth Scatternet Formation for Supporting Device Mobility," in Proc. of IEEE ITRC, pp. 408-412, Jun. 2005.
8. G. Zaruba, S. Basagni, and I. Chlamtac, "Bluetooth-Scatternet Formation to Enable Bluetooth Based Ad Hoc Networks," in Proc. of IEEE ICC, vol. 1, pp. 273-277, Jun. 2001.
9. Y. Liu, M. Lee and T. Saadawi, "A Bluetooth Scatternet-Route Structure for Multihop Ad Hoc Networks," IEEE J. on Set. Areas in Communications, vol. 21, no. 2, pp. 229-239, Feb. 2003.

# Unencapsulated Mobile Multicast Routing for Next Generation Video Networks

Thomas C. Schmidt[1,2], Matthias Wählisch[2], Hans L. Cycon[3], Mark Palkow[3], and Henrik Regensburg[3]

[1] HAW Hamburg, Department Informatik, Berliner Tor 7, 20099 Hamburg, Germany
[2] FHTW Berlin, Hochschulrechenzentrum, Treskowallee 8, 10318 Berlin, Germany
[3] FHTW Berlin, FB I, Allee der Kosmonauten 20–22, 10315 Berlin, Germany
{t.schmidt, waehlisch}@ieee.org,
{hcycon, mpalkow,h.r}@fhtw-berlin.de

**Abstract.** Mobile multimedia communication will particularly rely on IP multicasting, as users commonly share frequency bands of limited capacities. This paper is dedicated to the problem of mobile Source Specific Multicast (SSM) senders. We propose extensions to multicast routing for transforming (morphing) previous delivery trees into optimal trees rooted at a relocated source. This extension scheme only requires basic signaling mechanisms, explicit joins and prunes. First evaluations grounded on real-world Internet topologies indicate network performance superior to traditional distribution schemes. As corresponding application we further on introduce a videoconferencing and communication software of distributed architecture. It is built as a simple, ready-to-use scheme for distributed presenting, recording and streaming multimedia content over next generation unicast or multicast networks.

## 1 Introduction

Mobile visual devices accepting synchronous or streaming media become more and more ubiquitous. It is a challenge for the next generation Internet and its recently released MIPv6 [1], to introduce voice and video conferencing over IP (VoIP/VCoIP) to mobiles, so that IP will be pervasive and prevalent across all digital communication devices. Conferencing parties will then soon request seamless real–time performance of a mobility aware group communication service, thereby attaining the simultaneous roles of mobile multicast listener and source. Intricate multicast routing procedures, though, are not easily extensible to comply with mobility requirements. Significant effort has been already invested in protocol designs for mobile multicast receivers. Only limited work has been dedicated to multicast source mobility, which poses the more delicate problem [2, 3].

Source Specific Multicast (SSM) [4], still in its design process, is considered a promising improvement of group distribution techniques. In contrast to Any Source Multicast (ASM), optimal source trees are constructed immediately from $(S, G)$ subscriptions at the client side, without utilizing network flooding or RendezVous Points. Source addresses are to be acquired by out of band channels,

V.N. Alexandrov et al. (Eds.): ICCS 2006, Part II, LNCS 3992, pp. 1050–1057, 2006.

which for conferencing is easily achieved by SIP. As a consequence, routing simplifies significantly, but invalidates with source addresses changing under mobility. SSM source mobility is known as an unsolved problem.

Addresses in Internet mobility carry the dual meaning of logical *and* topological identifiers. While MIPv6 operates dual addresses transparently at end points, SSM routing needs to account for logical subscription *and* topological forwarding. In the present paper we start from this observation and present an approach to SSM routing, which adapts to source mobility. This scheme operates fast, efficient without encapsulation and does not cause additional packet loss.

This paper is organized as follows: In section 2 we review the basic problems of SSM source mobility and related work. Section 3 introduces our new approach and presents first evaluations. The subsequent section 4 is dedicated to our group conferencing application, discussing selected deployment scenarios. Finally section 5 gives conclusions and an outlook.

## 2 The Mobile Multicast Source Problem

### 2.1 Problem Statement

Any next generation Internet support for multicast source mobility management is required to operate transparently w.r.t. the socket layer. Specific protocol operations or extensions are thus bound to a multicast aware MIPv6 stack and the Internet routing layer. Recalling the address duality problem, modified multicast routing protocols must be foreseen, as routing at the occurrence of source movement is required to transform any $(S, G)$ state into $(S', G)$, while listeners continue to receive multicast data streams. Hence any simple mobility solution such as the remote subscription approach of MIPv6 [1] loses its receivers and will no longer function in our context.

With SSM an additional address problem needs consideration: A multicast listener, willing to subscribe to an $(S, G)$ state, needs to account for the current location of the mobile source. Concurrently a general intricacy derives from the principle decoupling of multicast source and receivers: Any multicast source submits data to a group of unknown receivers and thus operates without feedback channel. Address updates on handovers of an SSM source have to proceed without means of the mobile source to inquire on properties of the delivery tree or the receivers.

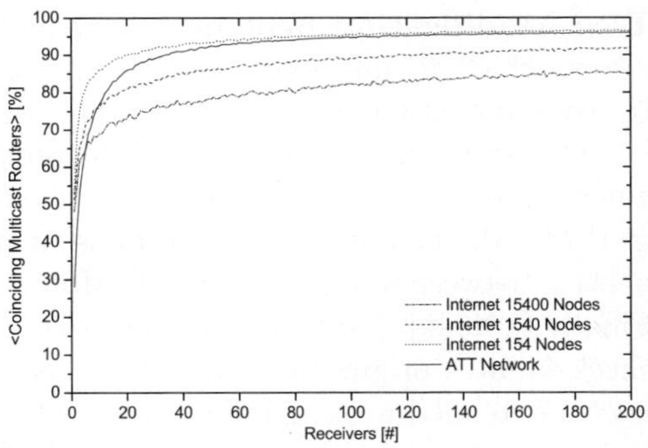


**Fig. 1.** Relative router coincidence in subsequent multicast trees

All of the above severely add complexity to a robust multicast mobility solution, which should converge to optimal routes and, for the sake of efficiency, should avoid data encapsulation. Bearing in mind characteristic applications, i.e. multimedia distribution, handover delays are to be considered critical. The distance of subsequent points of attachment, the 'step size' of the mobile, may serve as an appropriate measure of complexity. Figure 1 visualises the relative change of distribution trees as a function of receiver multiplicity for a medium step size of 5. It is interesting to note that even in large networks 75 to 85 % of multicast distributing routers remain unchanged under a mobility step. For details of the simulation we refer to section 3.

### 2.2 Related Work

Two principal approaches to SSM source mobility are presently around.

**Bi-directional Tunneling:** The MIPv6 standard proposes bi-directional tunneling through the home agent as a minimal multicast support for mobile senders and listeners. In this approach the mobile multicast source (MS) always uses its Home Address (HoA) for multicast operations.

**Inter–Tree Handovers:** Several authors propose to construct a completely new distribution tree after the movement of a mobile source. These schemes have to rely on client notification for initiating new trees. At the same time they need to preserve address transparency to the client. To account for the latter, Thaler [5] proposes to employ binding caches and to obtain source address transparency analogous to MIPv6 unicast communication. Initial session announcements and changes of source addresses are to be distributed periodically to clients via an additional multicast tree based at the home agent. Source–tree handovers are then activated on listener requests. Jelger and Noel [6] suggest handover improvements by employing anchor points within the source network, supporting a continuous data reception during client–initiated handovers. Even though it has not been applied to SSM, additional work is of relevance to this paper:

**Tree Modification Schemes:** Very little attention has been given to procedures, which modify existing distribution trees to continuously serve for data transmission of mobile sources. In the case of DVMRP routing, Chang and Yen [7] propose an algorithm to extend the root of a given delivery tree to incorporate a new source location in ASM. Focusing on interdomain mobile multicast routing in PIM-SM, the authors in [8] propose a tunnel–based backbone distribution of packets between newly introduced "Mobility-aware Rendezvous Points" (MRPs). Finally O'Neill [9] suggests a scheme to overcome reverse path forwarding (RPF) check failures originating from multicast source address changes, by introducing an extended routing information, which accompanies data in a Hop-by-Hop option header.

## 3 Tree Morphing: An Algorithm to Source Mobility

### 3.1 Introducing Routing Trees Adaptive to Mobility

In the present section we will briefly introduce our new concept of multicast routing, adaptive to source mobility. A mobile multicast source (MS) away from home will transmit *unencapsulated* data to a group using its HoA on the application layer and its current CoA on the Internet layer, just as unicast packets are transmitted by MIPv6. In extension to unicast routing, though, the entire Internet layer, i.e. routers included, will be aware of the permanent HoA. Maintaining address pairs in router states like in binding caches will enable all nodes to simultaneously identify $(HoA, G)$–based group membership and $(CoA, G)$–based tree topology.

When moving to a new point of attachment, the MS will alter its address from previous CoA (pCoA) to new CoA (nCoA) and eventually change from its previous Designated multicast Router (pDR) to a next Designated Router (nDR). Subsequent to handover it will immediately continue to deliver data along an extension of its previous source tree. Delivery is done by elongating the root of the previous tree from pDR to nDR (s. fig. 2). All routers along the path, located at root elongation or previous delivery tree, thereby will learn MS's new CoA and implement appropriate forwarding states.

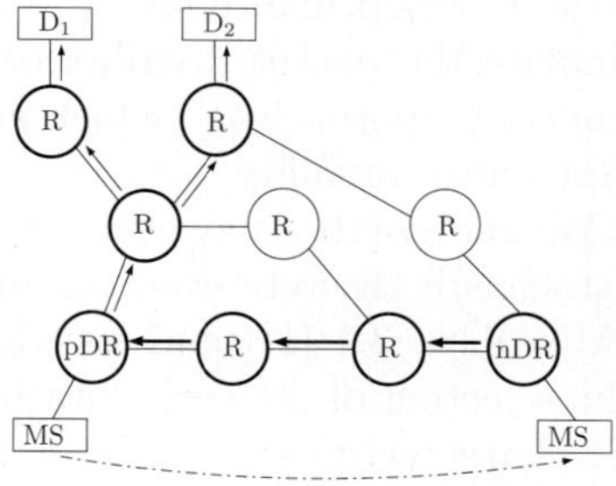


**Fig. 2.** Elongation of the Tree Root

Routers on this extended tree will use RPF checks to discover potential short cuts. Registering nCoA as source address, those routers, which receive the state update via the topologically incorrect interface, will submit a join in the direction of a new shortest path tree and prune the old tree membership, as soon as data arrives. All other routers will re-use those parts of the previous delivery tree, which coincide with the new shortest path tree. Only branches of the new shortest path tree, which have not previously been established, need to be constructed. In this way the previous shortest path tree will be morphed into a next shortest path tree as shown in figure 3. Note that this algorithm does not require data encapsulation at any stage.

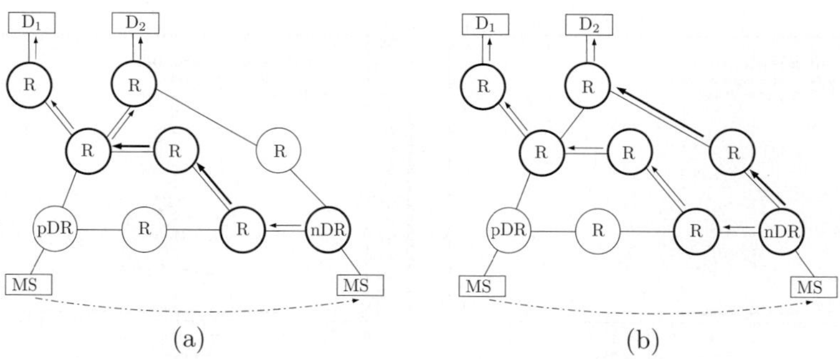


**Fig. 3.** Morphing States

### 3.2 Performance Evaluation & Simulation

Mobility initiated handovers may in general lead to packet loss and delay. The tree morphing multicast routing scheme will not produce any packet loss in addition to mobile IP handovers, as can be easily concluded from primary packet forwarding relying on unicast source routes. For a first evaluation measure we will subsequently concentrate on handover initiated packet delay as a result from initially suboptimal delivery trees. Based on real-world Internet topologies we simulate the packet distribution and compare our results to the bi-directional tunneling approach [1], which currently is the only stable mobility solution for SSM source mobility.

We analyze its delay effects within realistic Internet topologies. We performed a stochastic discrete event simulation based on the network simulator platform OMNeT++ 3.1 [11] and several real–world topologies of different dimensions. The selection of network data in our simulation must be considered critical, as key characteristics of multicast routing only make an impact in large networks, and as topological setup fixes a dominant part of the degrees of freedom in routing simulations.

We chose the ATT core network [12] as a large (154 nodes), densely meshed single provider example. For inter–provider data we extracted sub-samples of varying sizes from the "SCAN + Lucent" map [13]. Sample sizes, 154, 1.540 and 15.400 nodes, vary by two orders of magnitude. The delay excess relative to optimal routes has been calculated as characteristic performance measure under the assumption of homogeneous link delays. Extreme values, i.e. maximal delays at initial elongation phase and minimal after convergence, were evaluated for tree morphing (TM) as functions of the distance from $pDR$ to $nDR$. Comparisons are drawn with bi-directional tunneling (BT), which does not depend on designated router distances, but on HA position. The delay excess in BT as function of HA position does not converge to a characteristic value, but rather admits a broad distribution derived from scattering HA positions uniformly. It should be noted that these simulations concern delays for all three distribution trees in

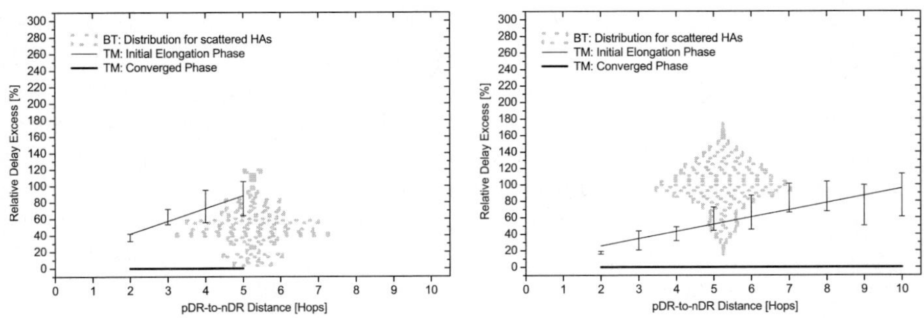


(a) ATT Core Network (b) Internet 15.400 Nodes

**Fig. 4.** Excess Delay of Optimal Routes: Comparison of BT and TM, Initial and Converged Phase, for Different Network Topologies presence and thus qualitatively cover the solutions discussed in section 2. Aside from signaling overhead, BT reflects the delays of [5], TM those of [6].

The results of our simulations are displayed in figure 4. $pDR$ to $nDR$ distances were chosen between 2 and 10, except for the ATT network, which exhibits a maximal edge router separation of 5. Error bars indicate the standard deviation of initial TM delay excess, as calculated from events differing in location of the mobile source. Delay excess distributions for scattered HAs in BT are laid underneath TM curves in grey dots. It can be observed that initially maximal delays of the tree morphing scheme tend to remain below the average of permanent BT packet retardation. Convergence of the TM then will lead to (relatively) undelayed packet delivery, which is never met in BT. Little dependence on network size becomes visible for TM — relative delays more strongly change with topologic characteristics. In a densely meshed provider network such as the ATT core, packet transitions are rapid and therefore initial delays from tree elongation account more dominantly for our relative measure. In the contrary it is interesting to note that delays from BT admit a systematic dependence on network size. The tree morphing even in its initially weakest phase exhibits fairly uniform performance, no matter how large the underlying network is (see [10] for a more detailed analysis).

## 4 A Distributed Video Communication System

### 4.1 The Basic Software

In this section we introduce a digital audio-visual conferencing system, realised as a server-less multipoint video conferencing software without MCU developed by the authors[14]. It has been designed in a peer-to-peer model as a lightweight Internet conferencing tool aimed at email-like friendliness of use. The system is built upon a fast H.264/MPEG-4 AVC standard conformal video codec implementation [15]. It is a Baseline profile implementation, optimized for real-time decoding and encoding by several accelerating measures like diamond shape motion search, MMX enhanced SAD motion estimation, fast mode selection and a fast subpel search strategy. There is also application-tailored fast wavelet-based video codec [16] used for higher available data rate. By controlling the coding parameters appropriately, the software permits scaling in bit rate from 24 to 1440 kbit/s on the fly. All streams can be transmitted by unicast as well as by multicast protocol. Audio streams are prioritized above video since audio communication is more sensitive to distortions in erroneous networks.

An application-sharing facility is included for collaboration and teleteaching. It enables participants to share or broadcast not only static documents, but also any selected dynamic PC actions like animations including mouse pointer movements. All audio/video (A/V) - streams including dynamic application sharing actions can be recorded on any site. This system is equally well suited to intranet and wireless video conferencing on a best effort basis, since the audio/video quality can be controlled to adapt the data stream to the available bandwidth.

The joined use of high bandwidth UDP traffic with TCP updates bound to real-time demands is known to suffer from distortions due to TCP traffic suppression. Application sharing in conferencing applications thus is endangered to encounter disruptions in the event of network congestion. For a service enhanced synchronous use of UDP media sessions and application sharing with reliable data transport requirements, we implemented end-to-end load balancing employing proprietary extensions to UDP, reliable (RUDP), we work on its packet identifiers to control application sharing data flows. On the occasion of a significantly many (e.g., 5) unacknowledged packets, we slow down video packet transmission to reserve required resources for real-time application updates. Audio communication remains undisturbed of load-balancing actions.

### 4.2 Application Scenarios

Based on the systems capabilities, various new scenarios for synchronous and asynchronous distributed learning and communication evolve:

(i) *Synchronous distributed learning scenario:* Teacher and students are connected by LAN or WLAN, establishing mutual connections. The teacher can send his PC presentations and applications to the students PCs. Students (outside or inside the lecture room) can participate active and/or passive by real-time audio/video with latencies (in LANs) well below 50 ms. Since all participants can send their presentations or applications via WLAN to a beamer in a conference room, this can be used as a "wireless" connected beamer which can present full video formats. All participants can also initiate cooperations in small groups, so that each student within the peer–to–peer network can share and work on any PC applications for collaboration. There is also a live streaming option.

(ii) *Asynchronous distributed learning scenario:* Each station can record all sessions. The recordings can be stored locally or made netwide accessible a by converting it e.g. into a MS streaming format and uploaded to an e-learning platform. Lecture room presentation or distributed group work are thus ready to be played back anywhere at any time by streaming video.

(iii) A *fast broadband conference version* of the system is in operation to connect two distant medical campuses in Berlin city by high resolution video. The presentations are digitized by a high performance XGA frame grabber. This video stream will be encoded with a very fast software wavelet-based codec.

(iv) The system has been used for *tests with mobile video over UMTS* transmissions within a convergence project carried out by Fraunhofer Research Institute Berlin and local companies. It aims to seamlessly interconnecting WLAN, UMTS 3Gbeyond (4G) and DVB services. Results from the T–Mobile network using a Multimedia Network Card show varying bandwidth results for the upstream direction. With TCP we observed in subsequent measurements of 30s intervals 26 – 59 kbps and with UDP 48 kbps (with 4 % loss) to 54 kbps (2 % loss). The downstream direction is less critical for our scenario. We received 211 kbps over TCP and 338 kbps without loss over UDP, also measured in 30 s intervals. Typical echo round trip delays were around 500 ms. These experiments show that mobile video over UMTS works in principle with reasonable quality.

## 5 Conclusions and Outlook

In this paper we discussed the mobility problem in multicast routing and presented and analysed a novel scheme to solve it by morphing a previous distribution tree into a new shortest path tree. Correspondingly a video conferencing and communication software was introduced ready for immediate use mobile Internet multicasting. In future work we will embed this application in SSM-capable extensions of SIP dialogs. Further on we will quantify and compare further characteristic measures of the multicast routing scheme.

## References

1. Johnson, D.B., Perkins, C., Arkko, J.: Mobility Support in IPv6. RFC 3775, (2004)
2. Romdhani, I., Kellil, M., Lach, H.Y., Bouabdallah, A., Bettahar, H.: IP Mobile Multicast: Challenges and Solutions. IEEE Comm. Surv. & Tut. **6** (2004) 18–41
3. Schmidt, T.C., Wählisch, M.: Multicast Mobility in MIPv6: Problem Statement. IRTF Internet Draft – work in progress 00, MobOpts (2005)
4. Holbrook, H., Cain, B.: Source-Specific Multicast for IP. Internet Draft – work in progress 07, IETF (2005)
5. Thaler, D.: Supporting Mobile SSM Sources for IPv6. Proc. IETF meeting (2001)
6. Jelger, C., Noel, T.: Supporting Mobile SSM sources for IPv6 (MSSMSv6). Internet Draft – work in progress (expired) 00, individual (2002)
7. Chang, R.S., Yen, Y.S.: A Multicast Routing Protocol with Dynamic Tree Adjustment for Mobile IPv6. Journ. Inform. Science and Engin. **20** (2004) 1109–1124
8. Romdhani, I., Kellil, M., Lach, H.Y., Bouabdallah, A., Bettahar, H.: Mobility-Aware Rendezvous Point for Mobile Multicast Sources. In et. al., P.L., ed.: Proceedings of WWIC 2004. LNCS **2957**, Berlin, Springer–Verlag (2004) 62–73
9. O'Neill, A.: Mobility Management and IP Multicast. Internet Draft – work in progress (expired) 01, IETF (2002)
10. Schmidt, T.C., Wählisch, M.: A First Performance Analysis of the Tree Morphing Approach to Source Mobility in SSM Routing. In Lorenz, P., Dini, P., eds.: Proceedings of the ICN'06, IEEE Press (2006) to appear.
11. Varga, A., et al.: The OMNeT++ simulation system. http://www.omnetpp.org
12. Heckmann, O., Piringer, M., Schmitt, J., Steinmetz, R.: On Realistic Network Topologies for Simulation. In: MoMeTools '03: Proceedings of the ACM SIGCOMM workshop on Models, methods and tools for reproducible network research, New York, NY, USA, ACM Press (2003) 28–32
13. Govindan, R., Tangmunarunkit, H.: Heuristics for internet map discovery. In: Proceedings IEEE INFOCOM 2000. Volume 3., Tel Aviv, Israel, IEEE Computer Society (2000) 1371–1380
14. : The daViKo homepage (2005) `http://www.daviko.com`
15. Ostermann, J., Bormans, J., List, P., Marpe, D., Narroschke, N., Pereira, F., Stockhammer, T., Wedi, T.: Video Coding with H.264/AVC: Tools, Performance and Complexity. IEEE Circuits and Systems Magazine **4** (2004) 7–28
16. Cycon, H.L., Palkow, M., Schmidt, T.C., Wählisch, M., Marpe, D.: A fast wavelet-based video codec and its application in an IP version 6-ready serverless videoconferencing system. Intern. J. of Wavelets, Multires. and Inf. Proc. **2** (2004) 165–171

# Channel Estimation of High Rate WPAN System with Diversity Technique

Byung-Su Kang, Myung-Sun Baek, Dong-Jun Cho, Young-Hwan You, and Hyoung-Kyu Song

uT Communication Research Institute
saturnus@sju.ac.kr, songhk@sejong.ac.kr

**Abstract.** A simple channel estimation technique for IEEE 802.15.3 HR WPAN system is proposed and assessed, which gives improved performance in multi-path channel. HR WPAN system can be designed for supporting multimedia traffic in the wireless home network. The demand for a high data rate and reliable technology will increase continuously. To satisfy the demand, we design a new architecture that applies the diversity technique to the HR WPAN system. The proposed scheme can estimate multi-path channel easily by simple operation using preamble, and also can be utilized in the situation of using one or two transmit antennas.

## 1 Introduction

There have been many communication systems for wireless home netwoke. The development of Wireless Personal Area Network(WPAN) is prompted by the needs for wireless connectivity of high data rate, low cost, low complexity and low power consumption. IEEE 802.15.3 standard was designed particularly for the High Rate WPAN(HR WPAN) which can support short-range communications among portable consumer electronic devices at high data rate up to 55Mbps[1].

In the future more and more homes will have multiple PCs and multiple connections to outdoor public telecommunication and broadcasting networks. So, high data rate wireless communication becomes more concerned for support of multimedia services. To satisfy these demand, we introduce the multiple-input multiple-output(MIMO) systems promise improved performance compared to conventional systems in Rayleigh fading environment[2]. By using two transmit antennas, we can achieve higher data rate and more reliable communication than using single antenna. High data rate and reliability can be obtained by space time block code(STBC) algorithm.

One of the most important condition to use MIMO technique is that the accurate channel information is needed. In HR WPAN system, the channel estimation uses preamble called constant amplitude zero auto correlation(CAZAC). We propose an efficient method that estimates multi-path channel in MIMO environment by using CAZAC. The proposed channel estimation algorithm takes advantage of the orthogonal property of CAZAC. The BER performances of the proposed scheme have been shown through the simulation results.

V.N. Alexandrov et al. (Eds.): ICCS 2006, Part II, LNCS 3992, pp. 1058–1064, 2006.

The paper is organized as follows. The system model is described in Section 2. In Section 3, we explain the channel estimation algorithm and the proposed architecture. Section 4 shows the simulation results and discussions. Finally, our conclusions are presented in Section 5.

## 2 System Model

### 2.1 Space-Time Coded WPAN System

In HR WPAN system, a set of CAZAC sequences is sent in order to allow the receiver to estimate the channel. Before the data is transmitted, the data source is encoded by QPSK, 16/32/64-QAM with 8-state TCM scheme or DQPSK modulation. TCM offers bandwidth and power advantages over other modulations. The symbol rate of IEEE 802.15.3 is 11Msymbol/sec and raw physical layer data rates supported are 11, 22, 33, 44, 55Mbits/sec (QPSK-TCM, DQPSK, 16/32/64-QAM-TCM, respectively)[9].

A space-time block code is defined by a $p \times n$ transmission matrix [6]-[7]. The entries of the transmission matrix are linear combinations of the variables $x_1, x_2, \cdots, x_k$ and their conjugates. The number of transmission antenna is $n$, and we usually use it to separate different codes from each other. Since $p$ time slots are used to transmit $k$ symbols, the rate $R$ of the code is $k/p$. We consider only the transmission matrices for one and two transmit antennas, because the proposed channel estimation algorithm is suitable for two transmit antennas.

For one transmit antenna, the transmission matrix $G_1$ is simply $G_1 = [x_1]$, for which the code rate is one. For two transmit antennas, there exists a rate one STBC defined by the transmission matrix $G_2$ given by [4]

$$\mathrm{G}_2 = \begin{bmatrix} x_1 & x_2 \\ -x_2^* & x_1^* \end{bmatrix}. \tag{1}$$

Suppose that there are $2^b$ signals in the constellation. At the first time slot, $2b$ bits arrive at the encoder and select two complex code symbol $x_1$ and $x_2$. These code symbols are transmitted simultaneously from antennas one and two, respectively. At the second time slot, code symbols $-x_2^*$ and $x_1^*$ are transmitted. Figure 1 represents the case of two-branch transmit diversity with one receive antenna[4].

### 2.2 WPAN Channel Model

A short range indoor propagation channel can be described by a Rayleigh fading model with exponentially decaying power profile [3]. The average root mean square(RMS) power of the taps can be obtained according to the RMS delay spread $T_{RMS}$. Then the coefficients $H(\tau)$ are selected according to

$$H(\tau) = G(0, \sigma_\tau^2/2) + jG(0, \sigma_\tau^2/2) \tag{2}$$

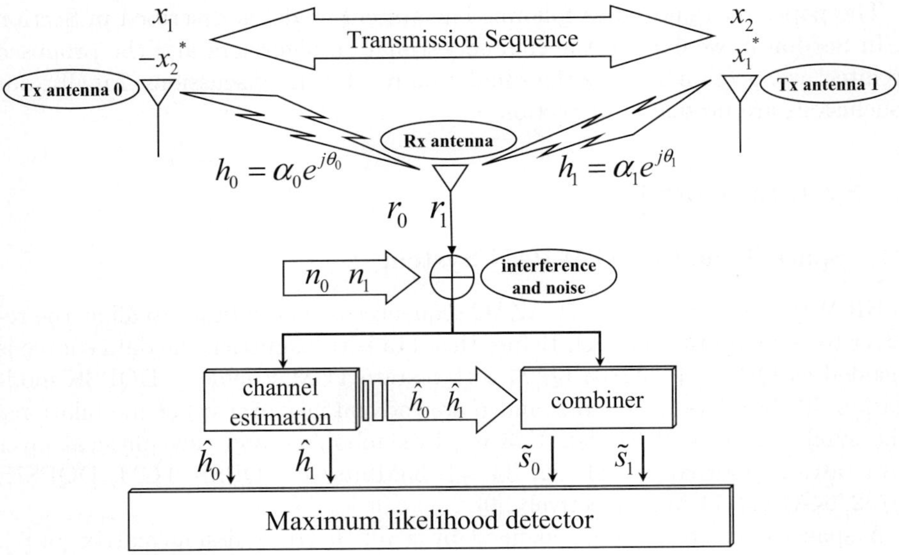


**Fig. 1.** Two-branch transmit diversity with one receiver

with

$$\sigma_\tau^2 = \sigma_0^2 \exp(-\tau T_s/T_{RMS}) \tag{3}$$

where $T_s$ is the sampling rate, $G(0, \sigma_\tau^2/2)$ is a zero mean Gaussian random variable with variance $\sigma_\tau^2/2$ and $\sigma_0^2 = 1 - \exp(-T_s/T_{RMS})$ is chosen so that the unit energy constraint $\sum_{\tau=0}^{L} \sigma_\tau^2 = 1$ for $L = 10T_{RMS}/T_s$ is satisfied where $L$ denotes the maximum number of paths. In practice, short range indoor channels usually induce a delay spread of 25-70ns at 2.4GHz.

## 3 The Proposed Channel Estimation Algorithm

According to the specification of the IEEE 802.15.3, the preamble contains 12 periods of a special sequence of $N = 16$ QPSK symbols called CAZAC sequence [9]. We apply two transmit antennas and $m$ receive antennas to IEEE 802.15.3 high rate WPAN systems. The transmitted signals from each antenna undergo multi-path fading independently. The expression for the received signal at the $j$-th receive antenna is as follows

$$y(n) = \sum_{i=1}^{2} \sum_{\tau=0}^{L-1} H_{ij}(\tau) P_{ij}(n-\tau) + w_j(n) \qquad j = 1, 2 \tag{4}$$

where $i$ is transmit antenna index, $L$ is the number of paths, $H_{ij}(\tau)$ is the channel impulse response(CIR) for antenna $i$ at $j$th receive antenna, $P_{ij}(n)$ is the detected CAZAC training sequence transmitted from antenna $i$ at $j$th receive antenna, and

$w_j(n)$ is complex additive white Gaussian noise (AWGN). At the receiver, we assume perfect frequency and timing synchronization. To separate multi-channel coefficients, we multiply the received signal by the conjugate of the shifted CAZAC sequence. The expression for the estimated CIR values is denoted by

$$\begin{aligned}\hat{H}_{ij}(l) &= 1/N\left(\sum_{i=1}^{2}\sum_{\tau=0}^{L-1} H_{ij}(\tau)P_{ij}(l-\tau)\right)\cdot P_i^*(l) \\ &\quad +\hat{w}_j(l) \\ &= H_{ij}(l)+\hat{w}_j(l) \qquad \text{for } 0<l\le L\end{aligned} \tag{5}$$

where $(\cdot)^*$ denotes the complex conjugation and $\hat{w}_j(l)$ is the AWGN at the $l$th path. Also, $P_i^*(l)$ denotes $l$ times shifted CAZAC sequence from antenna $i$ and $N$ is the CAZAC length. The first term in eq.(4) is always zero except that the transmitted CAZAC sequence exactly coincides with the conjugated version of transmitted CAZAC sequence at the receiver, which is called the orthogonality. When the two CAZACs from each transmit antenna do not guarantee the orthogonality, it is impossible to distinguish the two received signals because of breaking the autocorrelation property. So $P_2(n)$, which is $K$ times cyclic shifted version of $P_1(n)$, is transmitted from second antenna. If we choose $K = N/2$ and assume $K \ge L$, two CAZACs do not overlap at the receiver, without corrupting autocorrelation property. Based on this assumption, eq.(5) is satisfied. The mean square error (MSE) of the proposed estimator is obtained by

$$MSE = E\left[\sum_{j=1}^{m}\sum_{i=1}^{2}\sum_{l=0}^{L-1}\left(\hat{H}_{ij}(l)-H_{ij}(l)\right)^2\right] = E\left[\hat{w}_j(l)^2\right]. \tag{6}$$

Because the channel estimation error, $\hat{H}_i(l)-H_i(l)$, equals to the AWGN, we can conclude the MSE is same to the noise variance from eq.(6). Now, we consider imperfect channel estimation. For a given number of cyclic shift $K = N/2$, if $L > K$, some multipath signals of CAZAC sequences are overlapped, which results in corrupting orthogonality. In this case, the estimated channel for $i$, $k = 1, 2$ and $k \ne i$ can be expressed as

$$\hat{H}_{ij} = \begin{cases} H_{ij}(l)+H_{kj}(l+N/2)+\hat{w}_j(l) & 0<L-K \\ H_{ij}(l)+H_{kj}(l-N/2)+\hat{w}_j(l) & K<L \\ H_ij(l)+\hat{w}_j(l) & elsewhere \end{cases} \tag{7}$$

Eq.(7) shows that correct estimation is not possible because of the CIR estimates from the other antenna. Therefore, the additional channel estimation error, $H_{kj}(l+N/2)$ and $H_{kj}(l-N/2)$, is added to the noise variance. The equation for total MSE is as follows

$$\begin{aligned}MSE &= E\left[\sum_{k=1}^{2}\sum_{j=1}^{2}\sum_{l=0}^{L-K-1}\left(H_{kj}(l+N/2)\right)^2\right] \\ &\quad +E\left[\sum_{k=1}^{2}\sum_{j=1}^{2}\sum_{l=K}^{L-K-1}\left(H_{kj}(l-N/2)\right)^2\right] \\ &\quad +E\left[\hat{w}_j(l)^2\right]\end{aligned} \tag{8}$$

when $L > K$.

## 4 Simulation Results and Discussions

We illustrate the performance analysis of the proposed channel estimation algorithm in terms of MSE and BER.

Figure 2 shows the MSE performance in respect to RMS delay and $Eb/N0$. We simulated up to 100ns RMS delay, which means the number of paths is 10. When $Eb/N0$is increased by 10dB, the MSE is reduced by a factor of 0.1 where RMS delay increases up to 80ns. If the RMS delay is longer than 80ns, it can be known that the performance is degraded rapidly. This result is induced by exceeding the number of paths, $N/2$. Fortunately, since the RMS delay spread in an indoor wireless channel is less than 80ns[9], our scheme can be effectively applied to the IEEE 802.15.3 standard, without changing the training symbol.

Figure 3 displays the MSE of channel estimation versus $Eb/N0$. The MSE decreases linearly as $Eb/N0$ increases up to RMS delay of 80ns. However, the MSE converges to 0.7 approximately with RMS delay of 90ns. It is also observed from two figures that the analytic error curves according to Eq.(8) match the simulated error curves.

To estimate channel, we require only one CAZAC. The computational complexity for each CIR estimate is 16 multiplications, 16 additions, and 1 division. Also it is possible to apply the proposed algorithm to single transmit antenna. The algorithm can be easily implemented because of low complexity.

The BER performance comparison between the conventional WPAN system and the space-time coded WPAN system with two transmit antennas is illustrated in Figure 4. As shown in this figure the BER performance of space-time coded WPAN system outperforms that of the conventional WPAN system. Also

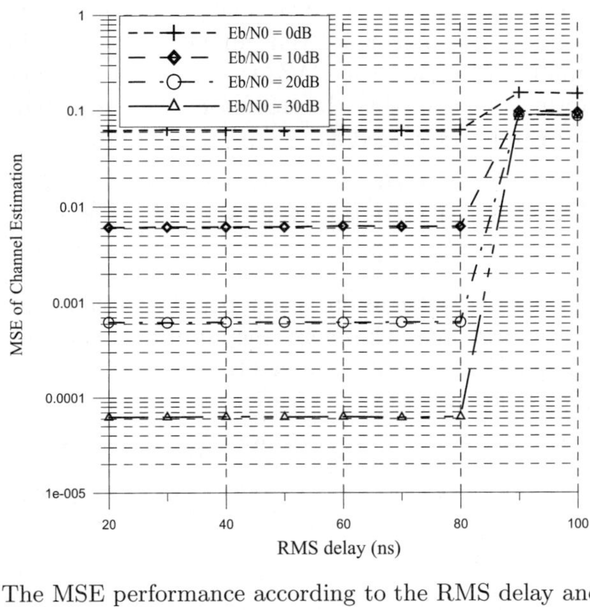


**Fig. 2.** The MSE performance according to the RMS delay and $Eb/N0$

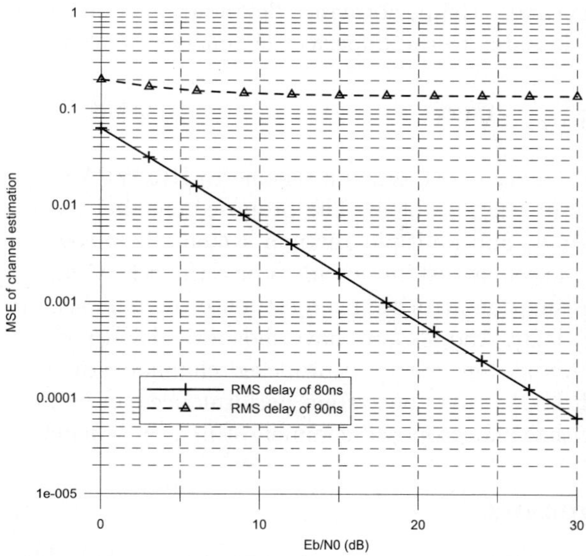


**Fig. 3.** The MSE performance of the proposed scheme employing $M = 2$

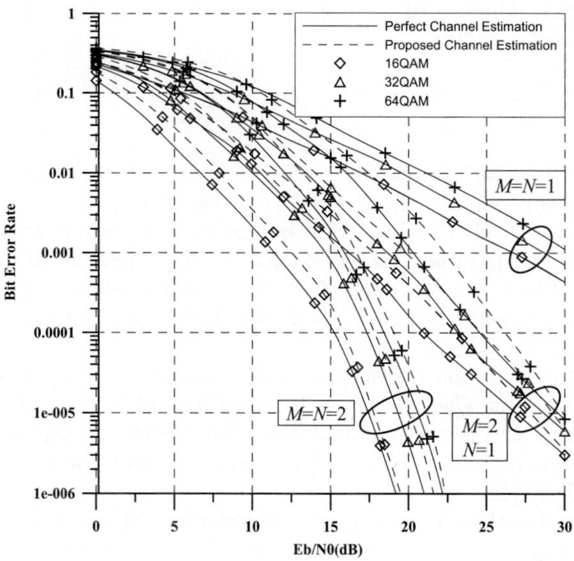


**Fig. 4.** BER performance of both perfect channel and channel estimation with STBC and conventional WPAN system

the more receive antennas are used, the better BER performance is obtained. The performance difference between the WPAN system with channel estimation and WPAN system with perfect estimation becomes similar for high SNR. In low

SNR, on the other hand the performance degradation of the channel estimation is observed, which is due to the imperfect channel estimation.

## 5 Conclusions

In this paper, we apply multi-antenna scheme to HR WPAN system and investigate channel estimation problem. The proposed channel estimation algorithm can operate in single and multiple antenna system. To recover the transmitted data in multiple antenna system, we introduce the STBC algorithm. The proposed channel estimation algorithm can estimate the multi-channel information up to 80ns RMS delay. And also, we have shown the BER performance of space-time coded WPAN system with the proposed channel estimation scheme. The simulation result shows that the system performance can be increased by using the STBC algorithm and proposed channel estimation scheme.

## Acknowledgement

This work is financially supported by the Ministry of Education and Human Resources Development (MOE), the Ministry of Commerce, Industry and Energy (MOCIE) and the Ministry of Labor (MOLAB) through the fostering project of the Lab of Excellency and is supported by MIC Frontier R&D Program in KOREA.

## References

1. J.Karaoguz, "High-Rate Wireless Wireless Personal Area Networks" *IEEE Communication Magazine*, vol. 39, no. 12, pp.96-102, Dec. 2001.
2. B. A. Bjerke and J. G. Proakis, "Multiple-antenna diversity techniques for transmission over fading channels" *IEEE WCNC'99*, Italy, pp.1038-1042, Sept. 1999.
3. Hashemi, H., Tholl, D., Morrison, G., "Statistical modeling of the indoor radio propogation channel-part I," *IEEE Vehicular Technology Conference*, vol.4, pp.338-342, May 1992.
4. S.M. Alamouti, "A simple transmit diversity scheme for wireless communications" *IEEE J. Select. Areas Commun.*, vol. 16, no. 8, pp. 1451-1458, Oct. 1998.
5. S.M. Alamouti, "Trellis-coded modulation and transmit diversity: design criteria and performance evaluation" *IEEE Universal Personal Communications.*, vol. 1, pp.703-707, Oct. 1998.
6. V. Tarokh, H. Jafarkhani, and A. R. Calderbank, "Space-time block codes from orthogonal designs," *IEEE Trans. On Info. Theory*, vol. 45, no. 5, pp.1456-1467, July 1999.
7. V. Tarokh, H. Jafarkhani, and A. R. Calderbank, "Space-time block coding for wireless communications: Performance results," *IEEE J. Select. Areas Commun.*, vol. 17, no.3, pp.461-471, March 1999.
8. G. J. Foschini Jr. and M. J. Gans "On limits of wireless communication in a fading environment when using multiple antennas" *Wireless Personal Commun.*, March 1998.
9. IEEE Std P802.15.3-2003, 2003.

# A Timestamp Tree-Based Cache Invalidation Report Scheme in Mobile Computing Environments*

Hakjoo Lee[1], Jonghyun Suh[2], Sungwon Jung[2], Sooyoung Lee[3], and Junguck Lee[3]

[1] DTV Lab. AIM Gr., LG Electronics, San 4-1, Bongcheon 7, Gwanak-gu, Seoul, Korea
or4nge@lge.com

[2] Department of Computer Science, Sogang University, Seoul, Korea
{joe77, jungsung}@sogang.ac.kr

[3] R&D Division, Advanced Power&IT Research Center, KATECH, Korea

**Abstract.** In mobile computing environments, cache invalidation techiniques are widely accepted. However, theses techniques require a large size of the invalidation report and show low cache utilization under high server update rate. We propose a new cache-level cache invalidation technique called TTCI(Timestamp Tree-based Cache Invalidation technique) to overcome the above problems. TTCI also supports selective tuning for a cache-level cache invalidation. We verify the effectiveness of our TTCI schemes through in-depth simulation experiments.

## 1 Introduction

In mobile computing environments, cache invalidation techiniques are widely accepted. But it is very difficult due to frequent disconnections and relatively unreliable wireless links. There have been many research efforts to develop efficient cache-level cache invalidation report schemes for mobile clients [3, 4]. However, under high update rate, invalidation report size increases proportionally and false invalidation will occur more often which causes low cache utilization. And none of the above schemes support selective tuning, which is very important for mobile clients to conserve the scarce battery resources. There have been other efforts like [2]. However, it lacks scalability and consumes more bandwidth including uplink. To cope with the above problems, we propose a new cache invalidation report scheme called Timestamp Tree-based Cache Invalidation(TTCI) without using any uplink. Also unlike group-based IR and BS-based IR schemes, TTCI doesn't require any priori knowledge.

* This research was supported in part by grant No.10016775 under the supervision of Korea Automotive Technology Institute from the next-generation new growth engine program of Ministry Of Commerce and Industry and Energy in Korea and in part by grant No.R01-2003-000-10197-0 from the Basic Research Program of the Korea Science&Engineering Foundation.

V.N. Alexandrov et al. (Eds.): ICCS 2006, Part II, LNCS 3992, pp. 1065–1068, 2006.

The rest of the paper is organized as follows. In Section 2 and 3, we describe our TTCI scheme and give performance analysis through simulation experiments. Finally, Section 4 provides concluding remarks.

## 2 Timestamp Tree-Based Cache Invalidation(TTCI)

### 2.1 TTCI Algorithm

In this subsection, we introduce the TTCI algorithm using an example. Basically, TTCI is composed of TT(Timestamp Tree) and IS(ID liSt). TT is a timestamp-based tree and works like an index helping clients to find a proper location of the updated information. IS is a list of the updated data IDs. Clients listen to the updated data from the position where TT points till the end of IS.

Suppose there are 20 updated data with $NT = 16$ non-identical timestamps, $FA = 3$ fanout of TT, and $NTS = 8$ the number of timestamps that TT uses. TTCI works in four steps as follows: First, build OLIST, a list of the updated data IDs sorted by timestamps in ascending order with no duplications, as shown in Table 1. Second, using timestamps from OLIST, build TT. A node in TT is composed of pointers($P_i$) and timestamps($TS_i$), like $<P_1, TS_1, P_2, TS_2, P_3, TS_3, ..., TS_{FA-1}, P_{FA}>$. This step is described in Algorithm 1. Third, build IS, which is a list of the updated data IDs in order of timestamps. We simply sort OLIST by timestamps. Finally, combine IS and TT. Pointers in the internal nodes are set pointing to another node and pointers in the leaf nodes are set pointing to a proper location in IS. After TTCI is built as shown in Figure 1, as an invalidation report, server broadcasts TT and IS.

### 2.2 Client-Side Algorithm

When a client receives an invalidation report, it checks whether the cache's most recent timestamp is within the history window. If it is not within the history window, the entire cache is dropped. Otherwise, the client starts the look-up process from the root node in TT to reach the updated data IDs. Finally, the client validates its cache with the updated information in IS.

In this example, let the last time it was validated be $T^C = 17$, the timestamp of the current TTCI be $T = 32$, invalidation broadcast window be $w = 2$, and the

**Table 1.** Sample Data and OLIST

(a) Updated Data with Timestamps

| Data ID | *d1* | *d2* | *d3* | *d4* | *d5* | *d6* | *d7* | *d8* | *d9* | *d10* | *d11* | *d12* | *d13* | *d14* | *d15* | *d16* | *d17* | *d18* | *d19* | *d20* |
|---|---|---|---|---|---|---|---|---|---|---|---|---|---|---|---|---|---|---|---|---|
| Timestamp | 24 | 16 | 10 | 6 | 22 | 18 | 26 | 32 | 2 | 20 | 14 | 30 | 8 | 4 | 12 | 28 | 8 | 14 | 24 | 20 |

(b) OLIST

| Timestamp | 2 | 4 | 6 | 8 | 10 | 12 | 14 | 16 | 18 | 20 | 22 | 24 | 26 | 28 | 30 | 32 |
|---|---|---|---|---|---|---|---|---|---|---|---|---|---|---|---|---|
| Data ID | *d9* | *d14* | *d4* | *d13*,*d17* | *d3* | *d15* | *d11*,*d18* | *d2* | *d6* | *d10*,*d26* | *d5* | *d1*,*d19* | *d7* | *d16* | *d12* | *d8* |

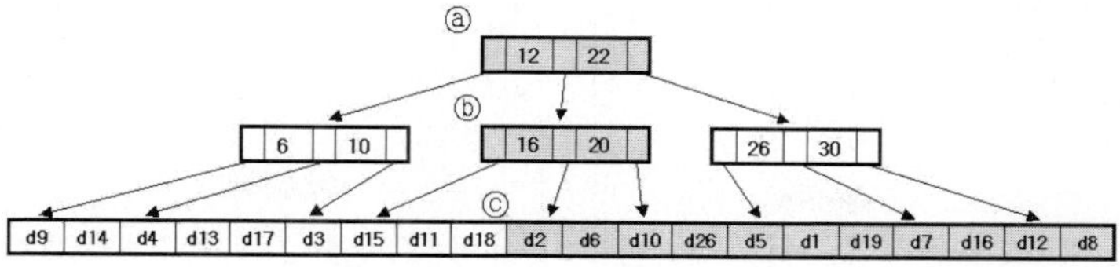


**Fig. 1.** TTCI

IR broadcast interval be $L = 5$. Then it is within the history window. Starting from root node $a$, the client reaches position $c$ in IS via node $b$. Figure 1 shows the TTCI for this example while the client uses only the shaded nodes.

## 3 Performance Analysis

We present analysis of our scheme by comparing to DRCI[3] and TS[1]. We address the invalidation report size as a performance metric, becuase it directly affects the consumption of scarce bandwidth and limited resources of mobile devices. The other is the cache usage ratio. Invalidation reports aim to maximize the cache utilization. Thus the number of valid data in cache after the validation reflects the effectiveness of an invalidation report. We define the cache usage ratio as the following equation 1, where $NumOfCachedData$ denote the entire size of the cache, and $NumOfValidData$ denote the number of currently cached data.

$$CacheUsageRatio = \frac{NumOfValidData}{NumOfCachedData} \times 100(\%) \tag{1}$$

As the update rate increases, the IR size of DRCI and TTCI increases because they should contain more updated information. In all ranges, TTCI outperforms DRCI with an average improvement of more than 25%. Additionally, we can predict that TTCI will also perform better on tuning time, because it uses smaller IRs and supports selective tuning. Cache usage ratio degrades with increasing update rate. However, even under 50% of the update rate, TTCI substantially outperforms DRCI and TS, proving that it is less sensitive to the update rate. Thus, as expected, TTCI efficiently copes with various update rate.

**Algorithm 1.** TTCI Algorithm Step 2: Building TT

$level = 1$; $gap = NT$; $DE = \lfloor \log_{FA} NTS \rfloor$;
**while** $level \leq DE$ **do**
  $node = 1$; $t_{max} = gap$; $t_{min} = 1$; $cluster = gap$;
  **while** $node \leq FA^{(level-1)}$ **do**
    $gap = \lceil \frac{t_{max} - t_{min}}{FA} \rceil$;
    **for** $i = 1$; $i < FA$; $i++$ **do**
      $TS_i = (gap \times i + t_{min})$th Timestamp in OLIST;
      $t_{min} = t_{max} + 1$; $t_{max} = t_{max} + cluster - 1$;
      **if** $(t_{max} > NT)$ **then**
        $t_{max} = NT$;
      **end if**
    **end for**
    $node++$;
  **end while**
  $level++$;
**end while**

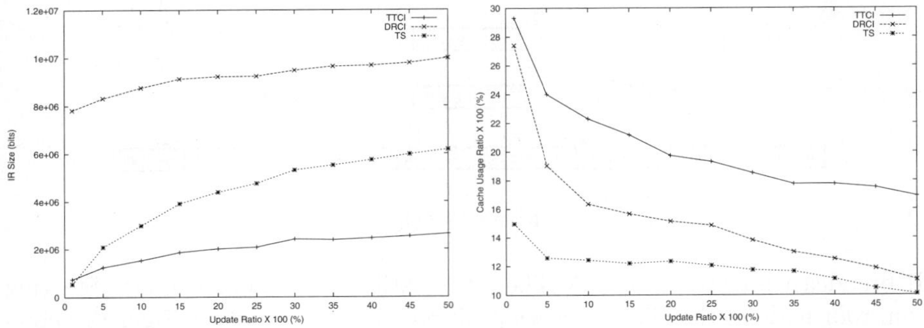


(a) Effect on the invalidation report size (b) Effect on the cache usage ratio

**Fig. 2.** Simualtion Results

# 4 Conclusion

In this paper, we have explored the problems of existing cache invalidation schemes. As the update rate increases, existing cache invalidation schemes suffer from a large invalidation report size and low cache utilizations. Furthermore, most cache-level cache invalidation schemes do not support selective tuning, which is important for the battery power conservation. We have proposed a new cache invalidation report scheme, called TTCI to cope with the above problems. TTCI dynamically organizes the updated data IDs into groups using a timestamp tree. As a result, the invalidation report size of TTCI becomes less sensitive to increasing update rate. TTCI also supports selective tuning while it performs cache-level invalidation. We have analyzed the performance of our TTCI scheme against existing schemes. We have verified the effectiveness of our TTCI scheme through in-depth simulation experiments.

# References

1. D. Barbara and T. Imielinski, "Sleepers and Workaholics: Caching in Mobile Distributed Environments",*Proc. ACM SIGMOD Int. Conf. on Management of Data*, pp. 1-12, 1994.
2. S-Y. Yi, W. Song, S. Jung, and S. Park, "A Cost Effective Cache Consistency Method for Mobile Clients in Wireless Environment",*DASFAA 2004*, pp. 908-915, 2004.
3. K-L. Tan, B.C. Ooi, "An Evaluation of Cache Invalidation Strategies in Wireless Environments",*IEEE Transactions on parallel and Distributed systems*, vol.12, no.8, pp. 789-807, 2001.
4. A. Elmagarmid, J. Jing, A. (Sumi) Helal, And C. Lee, "Scalable Cache Invalidation Algorithms for Mobile Data Access",*IEEE Transactions on knowledge and data engineering*, vol 15, no 6, pp. 1498-1511 2003.

# Clustering Versus Evenly Distributing Energy Dissipation in Wireless Sensor Routing for Prolonging Network Lifetime*

Guangyan Huang[1], Xiaowei Li[1], and Jing He[2]

[1] Advanced Test Technology Lab., Institute of Computing Technology, Chinese Academy of Sciences, Beijing, P. R.China 100080
{huanggy, lxw}@ict.ac.cn

[2] Chinese Academy of Sciences Research Center on Data Technology and Knowledge Economy, Beijing, P. R. China 100080
hejing@gucas.ac.cn

**Abstract.** A novel Cluster Heads (CH) choosing algorithm based on both Minimal Spanning Tree and Maximum Energy resource on sensors, named MSTME, is provided for prolonging lifetime of wireless sensor networks. MSTME can satisfy three principles of optimal CHs: to have the most energy resource among sensors in local clusters, to group approximately the same number of closer sensors into clusters, and to distribute evenly in the networks in terms of location. Simulation shows the network lifetime in MSTME excels its counterparts in two-hop and multi-hop wireless sensor networks.

## 1 Introduction

Recent advancements in wireless communications and electronics have made the development of low-cost, low-power smart wireless sensors available [1-3]. Tens or hundreds of sensors can be deployed in observation environment to sense physical parameters that are sent to the Base Station (BS) to provide useful knowledge for people [4-6]. Because sensors have limited battery resource and wireless communications consume large number of energy, energy efficiency of the communication is most important in the wireless sensor networks [7,8].

The energy consumed by wireless communications is related with the number of transmitting data and transmitting distance [9]. Optimal routes can achieve the minimal average transmitting distance. Furthermore, with only limited computing abilities, sensors can fuse data by compressing or getting rid of redundancy to reduce the number of data and then send the user-cared knowledge to BS [9]. Thus, sensors are clustered into several clusters and some sensors are chosen as Cluster Heads (CH) to do the fusing and middle-transferring jobs. However, different transmitting distances and different jobs of non-CH sensors and CHs bring different energy dissipation. Thus, all the sensors should work as a team and evenly distributing energy dissipation to prolong the whole network lifetime that is defined as the Time before the First sensor

* This paper was supported by the National Basic Research Program of China (No.2005CB321604 and No.2004CB720103), and the National Natural Science Foundation of China (No.90207002).

V.N. Alexandrov et al. (Eds.): ICCS 2006, Part II, LNCS 3992, pp. 1069 – 1072, 2006.

Node Dies (TFND). Also, the Ratio of TFND and the Time before All sensor Nodes Die (TAND), RTT, is used to evaluate the effectiveness of evenly distributing energy dissipation. Therefore, network lifetime is determined by both energy efficiency and effectiveness of evenly distributing energy dissipation.

Low Energy Adaptive Clustering Hierarchy (LEACH) [9-11] gives an elegant scheme for evenly distributing energy dissipation. Base Station Controlled Dynamic Clustering Protocol (BCDCP) [12] improves LEACH by reducing the transmitting distances and balancing the number of sensors in clusters [13]. LEACH and BCDCP show network lifetime can be improved further in four aspects: better closeness of sensors in each cluster (Closer), the same number of sensors in each cluster (Same Number), CHs with the most energy resource (Maximum Energy) and locations of CHs are distributed evenly (Even Location). LEACH and BCDCP do well in Maximum Energy and in both Closer and Same Number, respectively. This paper proposes a CH-choosing method, Minimal Spanning Tree (MST) and Maximum Energy resource on sensors determining CHs (MSTME), to do best in all the four aspects.

## 2 MSTME CH Choosing Algorithm

MSTME CH choosing algorithm is given in Table 1. The main goal is to choose expected number, $N_{CH}$, of CHs, which satisfy all of the four optimal principles. Then, non-CH sensors affiliate their closest CHs and thus clusters are formed. MSTME is reasonable for clustering while evenly distributing energy dissipation in wireless sensor networks. Firstly, MST connects all CH candidates with the minimal edges. Wherever the edge is broken, the nodes in the same split sub trees are closer. Because non-candidate sensors support the nearest CH candidates, CH candidates may delegate their supporters in terms of locations. Thus, sensors in the same clusters formed by MSTME are closer to each other. Also approximately the same number of supporters in each CH candidate subset ensures nearly the same number of sensors closest to the CH in this subset. At last, any nodes in MST are high-energy sensors. Therefore, MSTME synthetically does well in the four optimal principles.

Suppose BS knows the locations of all sensors and MSTME runs on BS. Maximum Energy principle is satisfied firstly, then CHs (solid dots) formed by MSTME shown in Fig. 1 (c) and (d) satisfy Closer, Same Number and Even Location better compared with LEACH and BCDCP in Fig. 1 (a) and (b). The number of CHs is greater, the MSTME method is better. However, optimal number of CHs is determined by network topology. Six CHs is optimal in two-hop networks according to [10], and nine CHs is optimal in multi-hop networks based on simulated annealing algorithm.

To simulate the network lifetime of MSTME, we implement an energy simulator of wireless sensor networks programmed in C/C++. We use a radio power model and related parameter values in [10] and a network shown in Fig. 1 with 100 nodes randomly deployed in 100m× 100m. Initial energy resource on each node, the package size of data message and BS position are 2J, 20kbits and (25m, 150m), respectively.

In two-hop networks, Fig. 2 (a) shows TFND in MSTME increases 3.2% of that in LEACH and also TAND of MSTME outperforms 3.6% that of LEACH. RTT of 96% in MSTME is nearly the same as RTT of 96.4% in LEACH. It means that MSTME gets better trade-offs on clusters with even numbers of closer sensors and CHs with more energy resources. In multi-hop networks, Fig. 2 (b) shows TFND in MSTME increases 6.4% of that in BCDCP and TAND in MSTME increases 5.6% of that in BCDCP. RTT of 98.7% in MSTME is nearly the same as RTT of 98% in BCDCP. MSTME performs better because it reduces average energy dissipation of the networks while keeps RTT approximately the same as both LEACH and BCDCP. In addition, it seems that MSTME performs better in Fig. 2 (b) than in Fig. 2 (a), because MSTME reduces average transmitting distance of CHs further in multi-hop networks.

**Table 1.** MSTME CH Choosing Algorithm

*Step 1*: Sensors with more energy resource than average level are selected into CH candidates set, *S*. Then a MST, *T*, is used to connect all the items in *S*. Also, supporters of a CH candidate *x* are those non-candidate sensors that are nearest to *x* among all CH candidates. Compute the number of supporters for each CH candidate including candidate itself.
*Step 2 (Initialization)*: Let the number of split edge *nSplit*=0, *T'*=*T* and *S'*=*S*.
*Step 3 (Loop)*: Find an edge to break *T'* into two sub MSTs of *T1* and *T2* and at the same time the nodes in *S'* are grouped into two subsets of *S1* and *S2* respectively with the nearest number of supporters in both subsets. Then let *nSplit*= *nSplit* +1.
*Step 4 (Termination test)*: If *nSplit*>=$N_{CH}$, go to Step 5. Otherwise, go on splitting *S1* and *S2* in turn. If the number of supporters in *S1* (or *S2*) is more than $N/N_{CH}$, then let *S'*=*S1* and *T'*=*T1* (or *S'*=*S2* and *T'*=*T2*). Go to Step 3.
*Step 5*: The CH candidate with the most energy resource in each subset is chosen as CH.

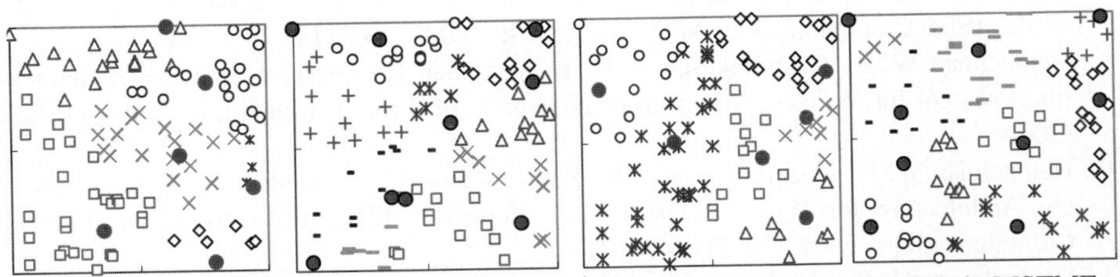

(a)6 CHs in LEACH (b)9 CHs in BCDCP (c)6 CHs in MSTME (d)9 CHs in MSTME

**Fig. 1.** Analysis of CH Distribution

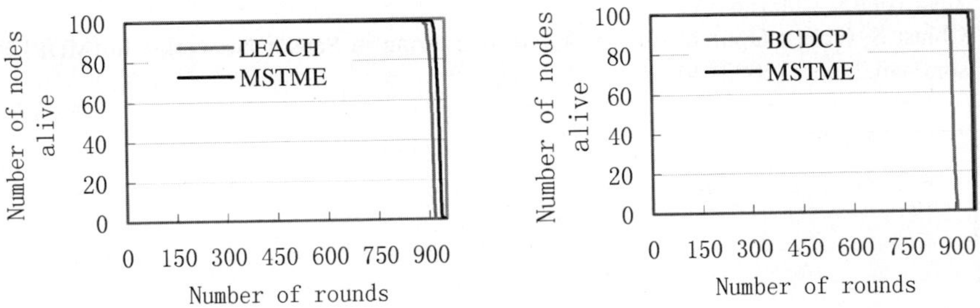


(a) Two-hop (b) Multi-hop

**Fig. 2.** Lifetime in Wireless Sensor Networks

## 3 Conclusions

MSTME is proposed to cluster sensors optimally in wireless sensor networks. It not only reduces the average transmitting distance but also distributes the energy dissipation evenly in the networks. Simulating of the network lifetime shows MSTME excels LEACH and BCDCP in two-hop and multi-hop networks respectively.

## References

1. Chandrakasan A., Amirtharajah R., Cho S.-H., Goodman J., Konduri G., Kulik J., Rabiner W., Wang A.: Design Considerations for Distributed Microsensor Networks. In: Proc. IEEE Custom Integrated Circuits Conference (1999) 279-286
2. Clare L., Pottie G., Agre J.: Self-Organizing Distributed Sensor Networks. In: Proc. SPIE Conf. Unattended Ground Sensor Technologies and Applications (1999) 229-237
3. Dong M., Yung K., Kaiser W.: Low Power Signal Processing Architectures for Network Microsensors. In: Proc. Int'l Symp. Low Power Electronics and Design (1997) 173-177
4. Estrin D., Govindan R., Heidemann J., Kumar S.: Next Century Challenges: Scalable Coordination in Sensor Networks. In: Proc. ACM/IEEE Mobicom (1999)
5. Kulik J., Rabiner W., Balakrishnan H.: Adaptive Protocols for Information Dissemination in Wireless Sensor Networks. In: Proc. ACM/IEEE Mobicom (1999)
6. Xu Y., Heidemann J., and Estrin D.: Geography-Informed Energy Conservation for Ad Hoc Routing. In: Proc. SIGMOBILE (2001)
7. Al-Karaki J. N., Kamal A. E.: Routing Techniques in Wireless Sensor Networks: A Survey. IEEE Wireless Communications, vol.11, (2004) 6-28
8. Akyildiz I. F., Weilian Su, Sankarasubramaniam Y., Cayirci E.: A Survey on Sensor Networks. IEEE Communications Magazine, vol. 40, (2002) 102-114
9. Heinzelman W. B., Chandrakasan A. P., Balakrishnan H.: Energy Efficient Communication Protocol for Wireless Microsensor Networks. In: Proc. Hawaii Int'l. Conf. Sys. Sci. (2000)
10. Heinzelman W. B., Chandrakasan A. P., Balakrishnan H.: An Application-Specific Protocol Architecture for Wireless Microsensor Networks. IEEE Transactions on Wireless Communications, vol.1, (2002) 660-670
11. Heinzelman W. B.: Application-Specific Protocol Architectures for Wireless Networks. Ph.D. Thesis, Massachusetts Institute of Technology (MIT), (2000)
12. Muruganathan S. D., Ma. D. C. F., Bhasin R. I., Fapojuwo A. O.: A Centralized Energy-Efficient Routing Protocol for Wireless Sensor Networks. IEEE Communications Magazine, vol.43, (2005) 8-13
13. Ghiasi S. (et al): Optimal Energy Aware Clustering in Sensor Networks. In: MDPI Sensors, vol.2, no.7 (2002) 40-50

# On the Effect of Heterogeneous Traffic Sources on the Network Availability for Wireless Sensor Grids

Ali Hammad Akbar[1], Ki-Hyung Kim[1,*], Shaokai Yu[2], and Won-Sik Yoon[3]

[1] Graduate School of Information and Communication
[2] Ajou University (Invited Faculty Programme)
[3] School of Electrical Engineering
Ajou University, Suwon, Korea, 443-749
{hammad, kkim86, wsyoon}@ajou.ac.kr,
yshaokai@yahoo.com

**Abstract.** Wireless sensor grids are poised for applications where pre-deployment information is highly available to optimize communication and computation costs. Using two widely known grid traversal schemes, we critique the effect of multiple services being rendered by a sensor network onto its availability. We derive an interesting result using Markovian model to appreciate the effect of load heterogeneity onto node and network lifetimes. Our mathematical model also analyzes synchronous sleep for its effect on energy conservation. We also propose a epicenter regulated, asynchronous sleep scheme which allows nodes to conserve energy yet relay data towards the sink effectively. The performance results show that the proposed scheme prolongs network availability in sensor grids.

## 1 Introduction

Wireless sensor networks are unique in the sense that they are supposed to work unattended. For their longevity, network designers propose schemes that reduce the computation and communication model. The fact that sensor networks have a growing application in telematics, home control, and ubiquitous services is a relief to sensor network community, wherein protocol designs are simple yet efficient.

In wireless sensor grids, pre-deployment information is much useful to the designers. They can trim heavy and cumbersome protocols of general purpose sensor networks to have simpler equivalents by exploiting location awareness.

One such kind of information is the type of traffic that flows over the sensor grid [1]. For non-real time services, traffic flow is highly adjustable. Vice versa, real time services' traffic flows have high variance due to *greedy* approach and resulting *contention*. Network availability as identified in [2] provides researchers with a baseline to diagnose and regulate node and network parameters for enhancing operational lifetime. In [3], the authors suggest a scheme for optimal transmission range distribution to balance out power consumption across sensor nodes as a linear programming

* Corresponding author.

V.N. Alexandrov et al. (Eds.): ICCS 2006, Part II, LNCS 3992, pp. 1073 – 1080, 2006.

problem. They propose a numerical model that gives upper bounds on energy inefficiencies due to energy balancing strategies. Likewise, in [4], Yang et al. give a dichotomy power control mechanism for prolonging network availability. They split the sensor network region of interest (ROI) into concentric rings. An algorithm then provides rules for adjusting transmission power such that all nodes dissipate energy evenly.

In this paper, we analyse sensor grids [5] for gauging the effect of traffic heterogeneity onto their longevity. In [6], we have already derived network availability expression of sensor grids for two widely known lattice path traversal that handle just a singular traffic type.

The contribution of this paper is to visualize network longevity from the application-centric viewpoint. We appreciate the innate nature of today's heterogeneous services i.e., real and non real time services. We also critique the notion that traffic generated sporadically versus a heavy network load has a multi-dimensional relationship with the longevity of a network. Furthermore, we propose a sleep schedule for sensor grids that is cognizant of the kind of traffic sensor network supports. Our algorithm allows real-time applications to make the choice of source-dependent sleep schedules or as spelled out by the sink. Non real-time applications conveniently allow the sink to announce sleep schedules. The performance results validate the strength of our proposal when combined with carrier sensing on the link layer. Our design considerations throughout the paper befit well with the state-of-the-art in silicon technology, such as sleep modes of sensor nodes, and in software e.g., TinyOS®.

## 2 Communication Model and Fundamental Assumptions

Fig. 1 illustrates a sensor grid that supports heterogeneous services viz real and non-real time services. For the sake of simplicity, the services are classified as both real time and non real time services. Following assumptions are made to formulate the model;

- Some sensor nodes generate real time data such as tracking information for military applications. It is typical example of a push model. Here traffic shaping is impossible.
- Other sensor nodes generate non real time data. A buffer, again shown in Fig. 1, gives an allowance to the network to shape the traffic in order to avoid collisions. First-in-first-out and drop tail are the queuing and scheduling mechanism enforced.
- Carrier sensing such as MACA is supported.
- Unicast communication is adopted. Link layer information such as remaining battery of one hop neighbours can determine the right of way for the next node to transmit, amongst the downstream neighbours.
- $h_{ij}$ is the initial energy of the sensor node $(i,j)$ at the reference time $t_0$, distributed across the network as $\eta_{ij}e^{-\eta_{ij}t}$ with mean $1/\eta_{ij}$.
- Sensing processes, both for real and non real-time applications are temporally independent We assume Poisson distribution with mean $\lambda_{ij}$ [7]. We further denote this variable with subscripts $r$ (real time) and $nr$ (non real-time) to distinguish between the two traffic types.

- Energy consumptions due to these activities are exponentially distributed with mean $1/\beta_{ij}$ and $1/\gamma_{ij}$ respectively.
- Sensor node $(i,j)$ consumes energy at a rate of $a_{ij}$ per second to remain awake.
- Sensor node $(i,j)$ consumes energy at a rate of $a'_{ij}$ per second during the time they sleep. While a node is sleeping, it does not perform relaying for other nodes' data. For the sake of simplicity in analytical modeling, we consider a sleeping node to behave the same as a dead node—especially in the context of routing. The dead node creates holes that result into network partitions.

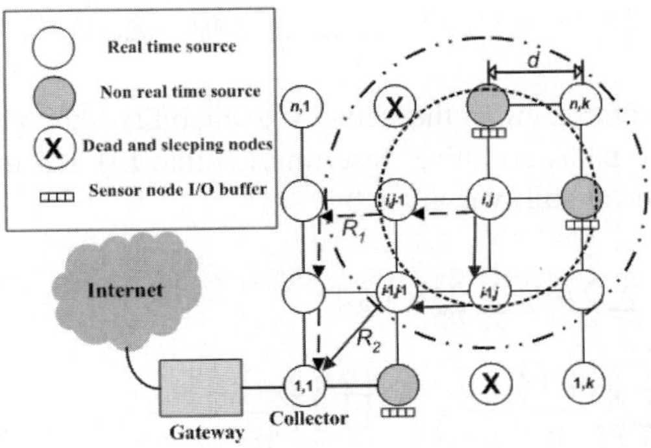


**Fig. 1.** A sensor grid with heterogeneous traffic sources

# 3 Numerical Analysis of Network Availability

We denote $y_{ij}$ as the total energy consumed by sensor node $(i,j)$ in sensing, relaying, during sleep, and in idle mode. Furthermore, $x_{ij}$ denotes energy consumption just being sensing and relaying data only. $F_t(x_{ij})$, $\forall ij=\{11, \ldots, 1k, 21, \ldots, 2k, \ldots n1, \ldots, nk\}$ is the joint probability density function (pdf) of all nodes at time $t$. Since Poisson processes are pure birth processes, the joint pdf of all the sensor nodes can be given by the differential-difference equation as

$$\frac{dF_t}{dt}\left(x_{ij}\right)=-\sum_{i=1}^{n}\sum_{j=1}^{k}\left(\lambda_{ij_r}+\lambda_{ij_{nr}}\right)F_t\left(x_{ij}\right)+\sum_{i=1}^{n}\sum_{j=1}^{k}\left(\lambda_{ij_r}+\lambda_{ij_{nr}}\right)\left(\beta_{ij_r}+\beta_{ij_{nr}}\right)$$

$$\times\prod_{l=1}^{i-1}\prod_{m=1}^{j-1}\left(\gamma_{lm_r}+\gamma_{lm_{nr}}\right)\int_{y_{11}=0}^{x_{11}}\cdots\int_{y_{ij}=0}^{x_{ij}}F_t\left(y_{11},\ldots,y_{ij},\ldots,x_{nk}\right)\times$$

$$\exp\left[-\left(\beta_{ij_r}+\beta_{ij_{nr}}\right)\left(x_{ij}-y_{ij}\right)-\sum_{l=1}^{i-1}\sum_{m=1}^{j-1}\left(\gamma_{lm_r}+\gamma_{lm_{nr}}\right)\left(x_{lm}-y_{lm}\right)\right]dy_{11}\ldots dy_{ij}\,. \tag{1}$$

Equation (1) reflects overall energy consumption for two kinds of traffic; real time and non real time. Considering (1) to be an initial value problem, we obtain $R_t(s_{ij})$ as Laplace transform of $F_t(x_{ij})$:

$$R_t(s_{ij})=\exp\left[-\left(\sum_{i=1}^{n}\sum_{j=1}^{k}(\lambda_{ij_r}+\lambda_{ij_{nr}})-\sum_{i=1}^{n}\sum_{j=1}^{k}(\lambda_{ij_r}+\lambda_{ij_{nr}})\frac{(\beta_{ij_r}+\beta_{ij_{nr}})}{s_{ij}+(\beta_{ij_r}+\beta_{ij_{nr}})}\prod_{l=1}^{(i-1)}\prod_{m=1}^{(j-1)}\frac{(\gamma_{lm_r}+\gamma_{lm_{nr}})}{s_{lm}+(\gamma_{lm_r}+\gamma_{lm_{nr}})}\right)\times t\right]. \quad (2)$$

In this paper, we analyze a synchronous sleep mode as identified in [8] for sensor nodes on detecting no activity. The total energy consumption of entire sensor network is now characterized by a single node's energy consumption, i.e., $y_{ij}=x_{ij}+a_{ij}t_1+a'_{ij}t_2$; $t_1$ and $t_2$ being the awake and sleep times respectively and that $t=t_1+t_2$. Let $Z_t(s_{ij})$ be the Laplace transform of $F_t(y_{ij})$.

$$Z_t(s_{ij})=R_t(s_{ij})\times\exp\left[\sum_{i=1}^{n}\sum_{j=1}^{k}a_{ij}s_{ij}t_1+\sum_{i=1}^{n}\sum_{j=1}^{k}a'_{ij}s_{ij}t_2\right]. \quad (3)$$

We define henceforth, $A_t$ as the network availability—the probability that all the nodes along all the paths are alive. Inserting (2) into (3) and manipulating the variables, the network availability is given by

$$A_t=\exp\left[\begin{array}{l}(-\sum_{i=1}^{n}\sum_{j=1}^{k}\eta_{ij}a_{ij}-\sum_{i=1}^{n}\sum_{j=1}^{k}(\lambda_{ij_r}+\lambda_{ij_{nr}})+\\ \sum_{i=1}^{n}\sum_{j=1}^{k}\frac{(\lambda_{ij_r}+\lambda_{ij_{nr}})(\beta_{ij_r}+\beta_{ij_{nr}})}{\eta_{ij}+(\beta_{ij_r}+\beta_{ij_{nr}})}\prod_{l=1}^{(i-1)}\prod_{m=1}^{(j-1)}\frac{(\gamma_{lm_r}+\gamma_{lm_{nr}})}{\eta_{lm}+(\gamma_{lm_r}+\gamma_{lm_{nr}})})\times t_1-(\sum_{i=1}^{n}\sum_{j=1}^{k}\eta_{ij}a'_{ij})\times t_2\end{array}\right]. \quad (4)$$

Considering two unicast data relaying models, i.e., Delannoy number-based and staircase lattice path traversals, we analyze the effect of traffic load and traffic heterogeneity on the network availability. If the transmission range is adjusted to $R_1$ as shown in Fig. 1, staircase lattice links (🡐🡓) are used en-route to relay data from the sensing node to the gateway, and assuming a square topology, (4) can be transformed as

$$A_{t_S}=\exp\left\{-\left[n^2\eta a+n^2(\lambda_r+\lambda_{nr})-\frac{(\lambda_r+\lambda_{nr})(\beta_r+\beta_{nr})}{(\eta+\beta_r+\beta_{nr})}\times D\right]\times t\right\} \quad (5)$$

where

$$D=\sum_{i=0}^{n-1}(1+i)\left(\frac{\gamma_r+\gamma_{nr}}{\gamma_r+\gamma_{nr}+\eta}\right)^i+\sum_{j=1}^{n-1}(n-j)\left(\frac{\gamma_r+\gamma_{nr}}{\gamma_r+\gamma_{nr}+\eta}\right)^{(n-1)+j}$$

Similarly, adjusting the power levels to transmission range of $R_2$, the data relaying activity turns out to be a different lattice path traversal (🡐🡓🡗), as given by Delannoy numbers [6]. The network availability of (4) is now given as

$$A_{t_D}=\exp\left\{-\left[n^2\eta a+n^2(\lambda_r+\lambda_{nr})-\frac{(\lambda_r+\lambda_{nr})(\beta_r+\beta_{nr})}{(\eta+\beta_r+\beta_{nr})}\left(\sum_{i=0}^{n-1}(2i+1)\left(\frac{\gamma_r+\gamma_{nr}}{\gamma_r+\gamma_{nr}+\eta}\right)^i\right)\right]\times t\right\} \quad (6)$$

The parameters in (5) and (6) are all assumed to be independent of $i$, $j$, $l$, $m$, and the traffic type, i.e., $a_{ij}=a$, $a'_{ij}=a'$, $\lambda_{ij}=\lambda$, $\beta_{ij}=\beta$, $\eta_{ij}=\eta$, $\eta_{lm}=\eta$, and $\gamma_{lm}=\gamma$. Numeric values are adopted from [9]: $a$ = 15 μJ/s, $a'$ = 1.5 μJ/s, $\lambda$ = 0.083 packets per second, $1/\eta$ = 12960 J, $1/\beta$ = 42.61 J, $1/\gamma_S$ = 140.87 J for staircase and $1/\gamma_D$ = 280 J for Delannoy. The number of nodes varies from 4 (or $n$ = 2) to 1600 (or $n$ = 40).

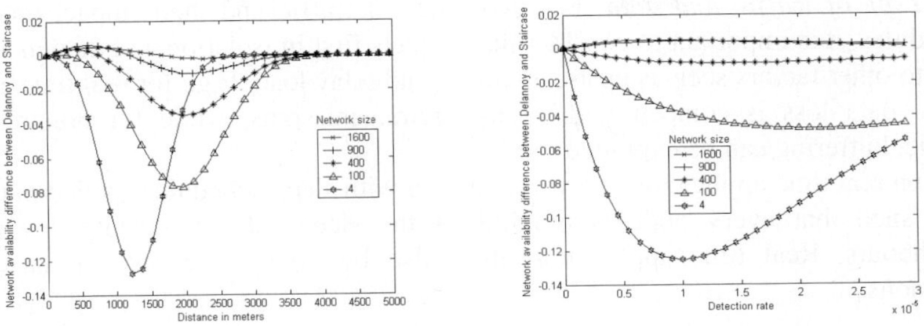


(a) Varying inter-node distance (b) Varying traffic of all sensor nodes

**Fig. 2.** Difference between network availabilities under topology size and traffic load variation

Fig. 2 (a) is the difference between (6) and (5) to show the effect of inter-node distance variation onto network availability. It was observed at $t$=1000000 seconds (11.57 days). The grid size was varied with a step size of 150 m. When the inter-node distance is increased for a fixed number of sensor nodes, staircase is a better choice for relaying until the network size becomes exceedingly large, for which Delannoy performs better.

Fig. 2 (b) is the difference between (6) and (5) to show the effect of traffic load and heterogeneity onto network availability. For medium to very large networks, e.g., for networks with nodes exceeding 1600, Delannoy offers better performance for increasing traffic load. For heterogeneous sources, Delannoy is poised to given a performance gain as for real-time traffic, diagonal paths are available for immediate data delivery. For delay-flexible traffic staircase equalises in performance.

## 4 Traffic Type-Cum-Load Aware Sleep Schedules for Sensor Grids

For the sake of simplicity in numerical modelling, we adopted synchronous sleep mode for the sensor grids. To capture the behaviour of sensor grid under asynchronous sleep schedule (sensing fidelity is relatively higher here) is NP-hard problem. Since such sleeping nodes do not participate in relaying activity [8], more sleeping nodes contribute to frequent path failures. Energy conservation through sleep mode is compromised by packet loss and retransmissions. In order to assign sleep schedules to sensor nodes, three considerations must be taken into account

I) ***Total number of paths available for a node to relay data***. It can be computed by the two relay models presented here. For Delannoy it is $\sum_{k=0}^{n} \frac{(n)!}{(n-k)!(k)!} \frac{(n+k)!}{(n)!(k)!}$. Likewise, for staircase it is $\frac{(2n-2)!}{(n-1)!(n-1)!}$.

II) ***The total relay load a node carries***. For a node *ij*, upto (*n*-*i*+1)(*n*-*j*+1)-1 nodes relay data through it. If the nodes adopt asynchronous sleep mode, and sleeping probability of each node is $P_s$, the minimum probability that data from node (*i*,*j*) will be delivered successfully is given by $P_D = (1 - P_s)^{[i+j-2]}(P_s)^{[i \times j-(i+j)-2]}$.
III) ***Type of traffic and data***. For two kinds of traffic and their impact onto sleep schedules, it is important to consider their effects first in isolation and then in association to other factors such as upstream nodes and relay load. E.g., for real time applications, data loss is compensated through retransmissions, while for non real time traffic, buffering can be exploited too.

Non real time applications however fit well with sleep schedules spelled out by the sink such that every node is mindful of the sleep schedule of its downstream neighbours. Real time applications may also be given a leeway to make local decisions.

### 4.1 Deterministic Sleep Mode Adoption

When a node selects a downstream relay agent, it registers with that agent through a *registration request*. This control packet is a notification of letting the downstream neighbour know of a prospective sender. Additional information such as `start time`, whether `real time` or `non real time` traffic, and `expected duration` may also be specified here. Registration process continues till the epicenter. The epicenter compiles and propagates a final record of all the applications for which it will relay the data. This compilation is subsequently used by the node to adjust its sleeping schedule as described through Fig 3.

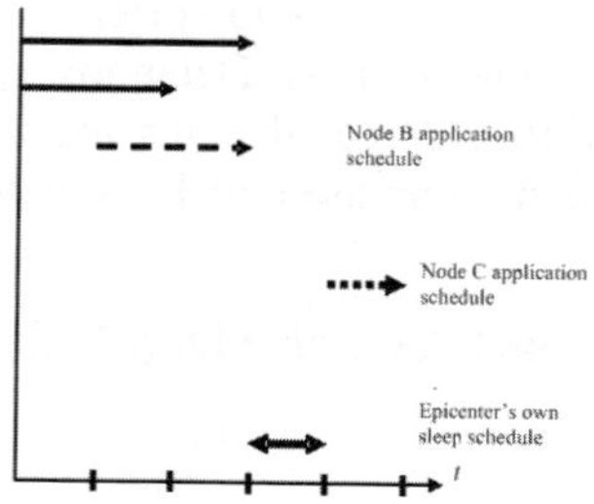


**Fig. 3.** Epicenter computes sleep schedule and shares with one hop neighbours

## 5 Performance Analysis

A simulator based on the system in section IV was developed in C++ to evaluate our scheme.

Fig. 4 (a) illustrates simulation results obtained for staircase lattice path traversal under the three scenarios as shown in the legend. The advantage using our scheme is clear. Nodes closer to the gateway either do not sleep or sleep less. Thus they do not loose any routing data, resulting into network-wide energy saving when compared with uniform sleeping scheme and no sleep at all.

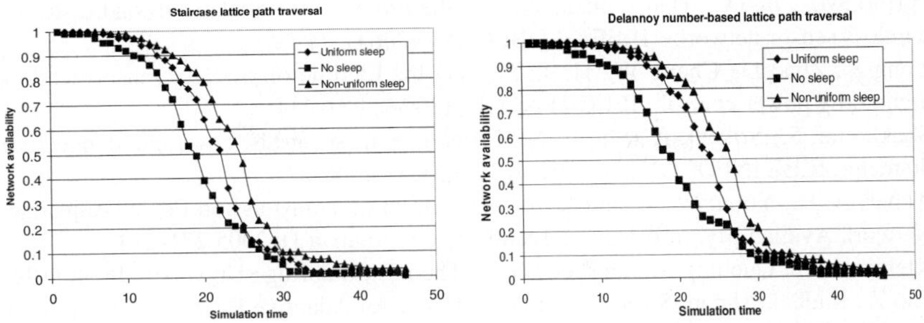


(a) Staircase lattice path traversal (b) Delannoy number-based path traversal

**Fig. 4.** Network availability comparison for sleep schemes

Fig. 4 (b) shows the comparison and simulation results asserting the usability of non-uniform sleep scheme in Delannoy number-based traversal. A fractional gain in network availability is observed compared to staircase traversal for all the three scenarios.

## 6 Conclusion

In this paper, we study wide range of parameters for wireless sensor grids. In particular, we focus on the relaying activity and traffic type for network longevity. We conclude that Delannoy performs better for real time applications by providing an additional path per hop. It also provides a gain in network availability due to reduction in hops for very large networks.

We propose a non-uniform, sleep scheme that is aware of the traffic load type of upstream neighbours. The paradox that nodes closer to the gateway sleep less yet enhance overall network connectivity and longevity is an important finding. The performance results show that the proposed scheme prolongs network availability effectively in sensor grids.

## Acknowledgement

This research is supported by the ubiquitous Autonomic Computing and Network Project, the Ministry of Information and Communication (MIC) 21st Century Frontier R&D Program in Korea.

## References

1. Akyildiz, I.F., Weilian S., Sankarasubramaniam, Y. Cayirci, E.: A survey on sensor networks. IEEE Communications Magazine, IEEE, vol. 40, (2002), 102 - 114
2. Hiroshi, S., Hiroya, M.: Performance Issues and Network Design for Sensor Networks. IEICE Transactions on Communications, vol. E87-B (2001), 294-301

3. Perillo, M., Zhao C., Heinzelman, W.: On the problem of unbalanced load distribution in wireless sensor networks. IEEE GLOBECOM, (2004), 74-79
4. Yang Z., Zhao D., Cheng W., He J.: Prolong the lifetime of wireless sensor network with dichotomy power control.: IEEE TENCON, (2004), 231-234
5. Shakkottai, S., Srikant, R., Shroff, N.: Unreliable sensor grids: coverage, connectivity and diameter. IEEE INFOCOM, vol. 51, no.12, (2002), 1448-53
6. Akbar, A. H., Yoon, W. S., and Kim, J. H.: Effect of Transmission Power Adjustments on Network Availability. Information Technology Journal, 4(3), 2005 271-273
7. Kleinrock, L.: Queuing Systems Volume I: Theory. Cambridge University Press (1997)
8. Tao Z., Radhakrishnan, S., Sarangan, V.: PMAC: An Adaptive Energy-Efficient MAC Protocol for Wireless Sensor Networks. IEEE Parallel and Distributed Processing Symposium, (2005), 65-72
9. Bhardwaj, M., Garnett, T., and Chandrakasan, A. P.: Upper Bounds on the Lifetime of Sensor Networks. in Proceedings of ICC, (2001), 785-790

# Selective Handover Technique on Multihomed Mobile Network Environment*

Kiyong Park, Sunyong Han**, and Jungwook Song
Computer and Information Communication Department,
Kon-kuk University, Seoul, Korea
kypark@cclab.konkuk.ac.kr, syhan@konkuk.ac.kr,
swoogi@konkuk.ac.kr

**Abstract.** Mobile Networks could have multiple wireless network interfaces due to various wireless technologies provided by many ISPs. Recently, those issues of the multiple connections, Multihomed Mobile Network, were highlighted in IETF NEMO WG. However, considering multiple connections has some tradeoffs between efficiency and fault tolerance, since the network must handle the status of all connections. Moreover, current NEMO basic support architecture does not have functions to handle those multiple connections when it encounters multiple new networks on new location. This article suggests a RA Message selection technique to support efficient Multihomed Mobile Network on NEMO basic support architecture even in case of the nested NEMO environment.

## 1 Introduction

Recently, to support Network Mobility [1], NEMO basic support architecture (NEMO BSA) [2], based on bi-directional tunnels [3] was selected as one of RFCs and its related techniques [4] are very similar to Mobile IP [5]. For these reasons, NEMO BSA has some critical problems such as route optimization problem called Pinball Route [6]. Meanwhile, the Mobile Router (MR) can receive several different Router Advertisement (RA) Messages in the new wireless network that the MR has moved. In the structure of NEMO BSA, because the MR does not have any functions to distinguish each of the RA messages except for Mobile Network Prefixes (MNP), the MR processes its handoff instantly whenever it receives a new RA message. In other words, if there are several RA messages that are being advertised, the MR might not know which RA message from each network is adequate and will handover whenever it detects different RA message. Hence, we could consider the NEMO Multihoming model [7] to support multiple connections on new network. However, previous NEMO BSA does not support multiple CoA registration on Home Agent [8]. In addition, one of the connections to a more deeply nested network could be of no use because of pinball route problem. Therefore, the MR must have some functions that select appropriate network and manage the connections.

In our paper, we are to propose a RA Message selection scheme for NEMO to prevent frequent handovers among new networks and to maintain an appropriate

---

* This research was supported by Seoul Research and Development program

** Corresponding author.

V.N. Alexandrov et al. (Eds.): ICCS 2006, Part II, LNCS 3992, pp. 1081 – 1088, 2006.

connection even in the nested-NEMO [9] case. Chapter 2 describes related researches and current problems; chapter 3 explains entire architecture with detailed extensions; chapter 4 shows simulation results. Finally, the conclusion was described in chapter 5.

## 2 Related Researches

### 2.1 Multihoming in NEMO and Nested NEMO Model

As said in [7], Mobile networks are typically connected by means of wireless and less reliable links; there could also be many nodes behind the MR. A loss of connectivity or a failure to connect to the Internet has a more significant impact than for a single node. Real-life scenarios as illustrated in [10] demonstrate that providing a permanent access to mobile networks such as vehicles typically require the use of several interfaces and technologies since the mobile network may be moving in distant geographical locations where different access technologies are provided and governed by distinct access control policies.

To support efficient handover in nested-NEMO model in terms of multihoming, previous research [11] considered the nested depth of mobile network. In this article, the nested depth of the mobile router is the most important factor of making a decision to select its parent-MR [9] in nested and NEMO multihoming model.

### 2.2 Pinball Route Problem on NEMO Basic Support Architecture

As we mentioned previously, there are bidirectional tunnels between a MR and its respective HA. These tunnels are used for forwarding data from CN [9] to MNN [9] behind the MR via the HA. Because all MNNs are using their addresses formed up from the home network prefix of the MR, the route to MNNs can not be on native route if the MR is not on the home network. Moreover, in assumption that the MR takes multiply nested-NEMO as its parent-NEMO, it causes a serious problem in the route path, called Pinball Route Problem. In this situation, the newly formed MR's CoA from its parent-MR is no more natively routable. Therefore n-times tunnel is indispensable for CN to reach MNNs under n-times nested-NEMO.

### 2.3 Summary of Current Problems

As previously mentioned, it can be supposed that NEMO multihoming model is applied to all MRs retaining the multiple wireless network interfaces with different technologies. Unfortunately, RA messages from one of MRs which has been nested and whose root-MR [9] has already lost its connection to the Internet can be advertised. We call this kind of MR 'Disabled MR' (D-MR). In this case, regarding only the depth of the nested-NEMO to determine the parent-MR among multiple parent-MRs is not sufficient. The worst case is that if the D-MR's signal strength is greater than the previous connection of a moving MR and each MR exchanges their RA messages simultaneously, the moving MR will form up its new connection with the D-MR's I-face [9] and the D-MR will form up its E-face [9] connection with the moving MR's I-face. We call this case 'Cross Registration in NEMO'. This is a critical problem on NEMO environment. Fig 1 shows these problems.

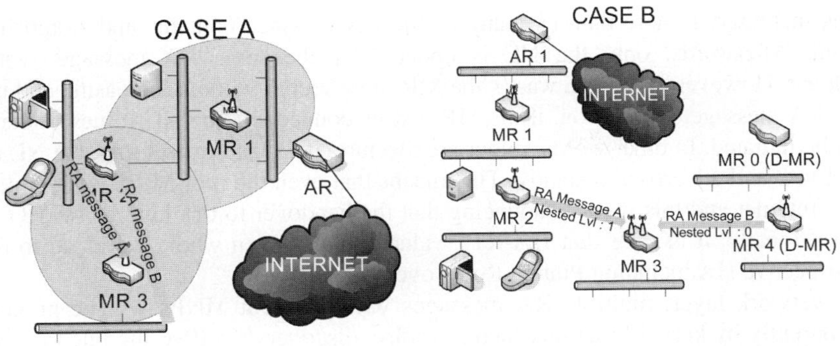


**Fig. 1.** Current Problems on Multihoming in nested Mobile Networking

Case A in Fig 1 shows 'Cross Registration' and Case B shows RA message comprehension using only the depth of nested-NEMO in RA message. However, in Case B, as you can see, MR3 can handover to MR4, which is D-MR because of lower depth value. Thus, that handover is of no use. As a result, more detailed information such as RTT value needs to be added for the sake of selecting appropriate RA message among multiple parent-MRs due to Pinball Route Problem.

In addition, the MR must register multiple CoAs to its respective HA. However, NEMO BSA does not have any capabilities to update the Binding List (BL) of the HA with multiple CoAs per HoA. Nevertheless, if there are multiple CoAs with different MNPs per HoA in the BL of the HA then it must monitor the status of all connections to CoAs due to some unexpected failures of the tunnels to those CoAs. These should cause huge workload on the HA.

## 3 Architecture

### 3.1 Basic Concepts

First of all, we must comprehend and separate appropriate one from various RA messages generated and advertised from parent-MRs. that means the MR in the intersection of the parent-MRs must select the best one for handover. Afterwards, the MR must select better one than previous selected or ignore recent one. To do so, we extended the nested depth of mobile network [11] as Nested Level (NL) and defined two more system values named Current Outgoing Link Status (COLS) and HA-RTT. The COLS implies whether the current connection to the Internet is available. In fact, current uplink status might already exist in each system. Therefore the system simply updates it. Finally, HA-RTT value between a MR and its respective HA is stored and updated periodically. And RA message is extended to store these three factors.

In the case of COLS, simply, if the uplink is working properly, COLS must be set to 1 or set to 0. In the case of NL, the original RA message generated from AR means 'not nested'. If the MR receives a RA message without any extension part for above three factors, it can send its RA message to its local network with NL '1' instead of '0' in [11] and sub-MRs [9] behind the MR can understand that they are located under 1-time nested-NEMO. If the MR uses one of the RA messages from one of nested-NEMO, the

NL is increased 1 more than as many as the RA message contains and stored in the system. Afterwards, only the NL is updated by the same RA messages without handover. However, if the NL was 0, the MR must keep 0 without increasing and use it in its RA message. In addition, if the MR lost its connection, the NL values of the MR must be initiated. In other words, in our architecture, if COLS was set to 0, the NL must be set to 0 and advertised instantly. This means that even the root-MR is not connected to the Internet and this is for announcing that the handover to this kind of NEMO is of no use. Finally, it is clear that HA-RTT value contains even whole workload to reach the respective HA including Pinball Route overload.

In network layer, multiple RA messages with different MNPs will be processed transparently by kernel functions such as 'ndisc_discovery' in IPv6 module in spite of multiple network interfaces which have different L2 technologies. For these reasons, using COLS, NL and HA-RTT are the main factors to decide the most appropriate RA message among different RA messages and some evaluation using these three factors was added in our scheme. It will be explained later. However, if the MR has only one E-face and it moved to the parent-MR which has 0 as NL or COLS by L2 process, it can not help turning into a D-MR.

### 3.2 Operation Overview

Fig. 2 shows how the RA messages that include newly defined three factors explained in previous chapter are sent to designated MR which is moving into new location.

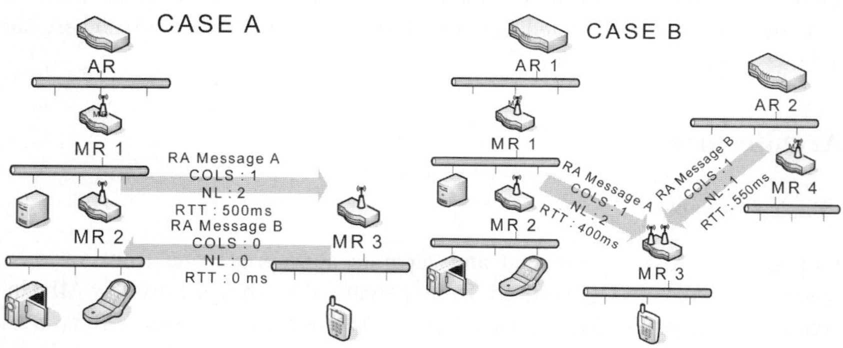


**Fig. 2.** Route Advertisement Message Flows

Case A is explaining that moving MR3 is approaching MR2 without its own connection to the Internet. So, MR3 is about to listen to the RA message from MR2 and MR2 is about to listen to it from MR3. Case B is that MR3 is approaching the range of the intersection between MR2 and MR4. RA message A was generated from MR2 and B is from MR4. Although, in the CASE B, we didn't draw the RA message generated from MR3, it is fully same RA message within CASE A.

First of all, in our architecture, MR will ignore all RA messages representing 0 as COLS. In such case, all MR would not start handover and it is clear that the NL is set to 0. Additionally, if the NL was set to 0 in the RA message in spite the COLS was set to 1, it means that the mobile network is nested and there is no point of attachment to the Internet even in the parent-MRs. Therefore, the MR that heard this kind of RA messages discards it. In CASE A, because the RA message B includes 0 as COLS, MR2 would not process to the RA message B but MR3 uses the RA message A from MR2. In addition, if the RA message does not contain NL, COLS and RTT fields, it means the RA message is from the AR. In this case, the MR3 must select the RA message without any complexity if it is not currently connected to another AR.

In CASE B, the MR3 is on the intersection of two networks which are advertising their own RA messages and we can imagine that the MR3 would process the RA message A first. After a moment, in this case, the MR3 will change the connection to the MR4 as soon as it receives the RA message B despite the handover to the MR2 because the evaluation using NL, COLS and RTT of the RA Message B is better than the value of the RA Message A. in the contrary, if MR3 received the RA message B first, the RA message A is discarded.

There is one more case that we didn't describe in the figure. Let's suppose a group of several NEMOs nested hierarchically and the root-MR has no connectivity to the Internet is moving and one (named nMR) of the MRs in that group, but not root-MR, is approaching the MR4 in CASE B in Fig 2. In this situation, the MR4 receives the RA message including 1 as COLS and 0 as NL from nMR. As we explained before, the MR4 would not accept the RA message because of the NL in spite of the COLS. In the contrary, the RA message B from MR4 affects the nMR and the group of nested NEMOs approaching the MR4 will be rearranged in nested status because the nMR becomes the root-MR in that group.

### 3.3 Extensions and Characteristics

As mentioned in the basic concept, we defined several new functions in NEMO BSA. Those extensions summarized below. First, the RA message extensions are:

(1) Nested Level Count Field (2 bytes)
- Extended part of Valid Options field to store the Nested Level

(2) RA Message Bit, named 'L'
- One of the reserved bits in RA message format
- If it was checked, COLS means 1 or 0

(3) RTT time between the parent-MR and its respective HA (4 bytes)

The most important extension is the functions of MR. We stated those below.

(1) Three system variables to store COLS, NL and RTT
(2) Parsing above three factors from the RA messages
(3) Periodically, the MR checks its COLS : 1 as 'on' or 0 as 'off', RTT and update periodically
- Additionally, if COLS system variable is set to 0, NL must be set to 0
(4) Store NL system value as 1 increased comparing to incoming RA message
(5) the NL value of 0 must be kept and stored in the MR and used in its RA message for its sub-MRs
(6) Ignore or accept the RA message by some evaluation function using COLS, NL and RTT
(7) If the same RA message received, only updates the NL from it.

As presented in the above, our mechanism eliminates the complexities of handoffs on multihomed-NEMO environment, in our method,

A. There are no handover storms and cross registrations between MRs
B. The HA need not to manage which CoA is the best one or unreachable, that means our method could minimize workloads of HA for the MR's multiple network interfaces
C. Supports Load Balance due to the appropriate parent-MR selection at that time
D. Our mechanism minimizes the delay caused by Pinball Route Problem in NEMO BSA

## 4 Performance Evaluation

In real world, because we couldn't make the test suit for huge nested and mobile networks, we used OMNet++ [12], one of the network simulators, to evaluate our model. We supposed three nested mobile networks which have different NL each other are forming up their intersection on the specific location geometrically and a moving MR, named oMR, located on the intersection. Finally, NEMO BSA and our architecture are compared in the point of the processing and delay time in communicating between CN and oMR.

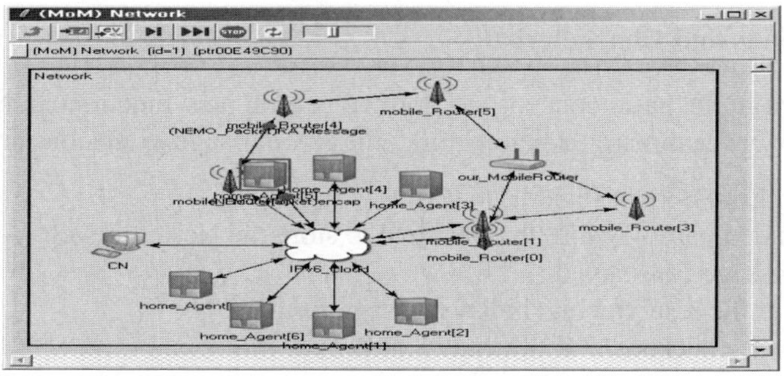


**Fig. 3.** The mobile networks to be simulated

In the Fig 3, MR3's NL is 2, MR5's NL is 3 and MR0's NL is 1. Each MR advertises its RA message with each of NL, COLS and RTT in each 1 second and the advertisement is started by MR5 first, MR3, MR0 and the oMR doesn't possess its own connectivity to the Internet. We calculated delay times of the packets from CN to oMR. We supposed 50~100ms delays to be routed in IPv6 Cloud, 10ms delays between sub-MR and parent-MR and each packet from CN to oMR is generated every 100ms. Fig 4 shows the result of our simulation and the total delay time from NEMO BSA takes much more than our architecture. In NEMO BSA, because oMR takes each RA message from MR3, MR5, MR0 by turn, the packet route is changed frequently. These make the packet delay unstable and taken more time. However, because our architecture uses only the most efficient RA message among them as Active CoA, thus the delay could be minimized.

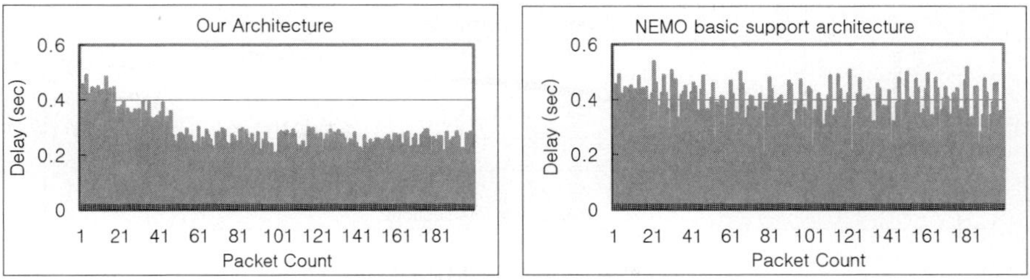


**Fig. 4.** Simulation results 1

Fig 5 shows the result of the case that only MR0 lost its connectivity to the Internet, that is Disabled MR, but advertising its RA to oMR. This evaluation was focused on the delay time whenever oMR sends its data to CN. In our architecture, MR0 was ignored due to COLS and NL, therefore, oMR choose MR3 as the parent-MR and sent the information to its HA. However, in NEMO BSA, because oMR selected all RA messages from the parent-NEMOs like we simulated the case before, Clear and the most significant point of this case was that oMR couldn't send its data to the CN through MR0. Moreover, since NEMO BSA is suffered from binding update storms, total cost for the communication must be greater than this result. Note that delay 0 means loss of packets. Finally, the results show that our method is more stable and efficient than the handoff processes of NEMO BSA for the communications.

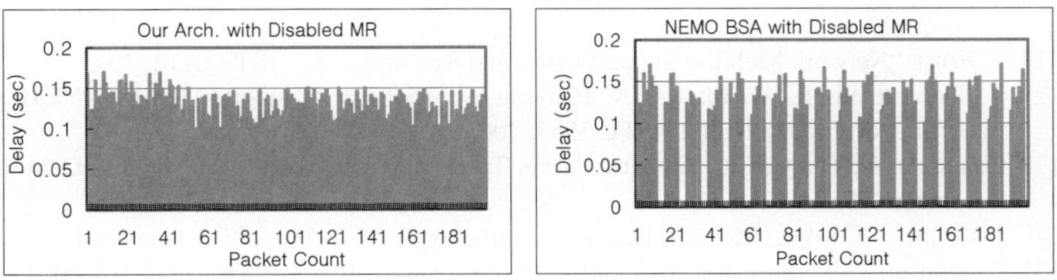


**Fig. 5.** Simulation results 2

Fig 6 shows how new MRs select their parent-MR among MR0, MR3 and MR5. in this situation, each MR calculates Workload with NL, RTT and 'weight value'. We evaluated like below. ('weight value' means tunnel process overhead and was temporarily supposed as 50)

$$\text{Evaluation Result} = \text{NL} * (\text{weight value}) + \text{RTT} \qquad (1)$$

These graphs mean each parent-MR's total Workload whenever new MR selects one as their parent-MR. As you can see, each Workload goes to around same value when new MRs are numerously increased. That means our technique is Load-balanceable.

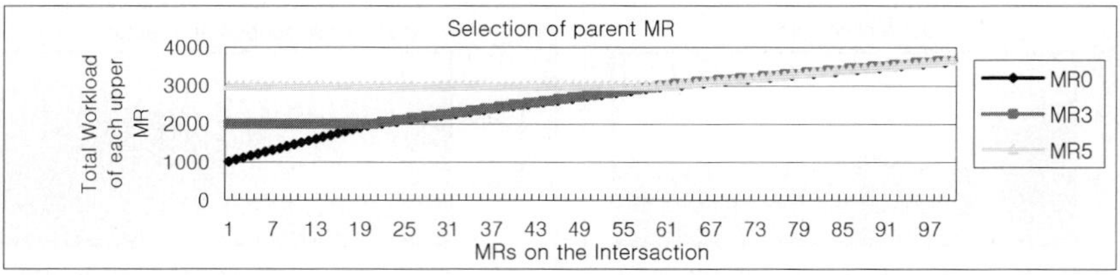


**Fig. 6.** Total Workload on MR0, MR3, MR5

# 5 Conclusion

In this paper, we examined and analyzed problems of NEMO multihoming model on NEMO BSA, and proposed new one that provides efficient and optimized RA message selection technique. With our architecture, we achieved optimal usage of multiple connections on NEMO multihoming model. In order to support our theory, we extended some functions of MR. Because this extension is based on NEMO BSA standard in terms of the functions and protocols; it does not result in much modification or loads to the network components. We also simulated some model to evaluate the performance of our approach. It showed that much lower delay for MNNs in a nested-NEMO to reach its correspondent nodes was taken, compared to NEMO BSA.

# References


[1] T. Ernst, “Network Mobility Support Goals and Requirements”, IETF Draft, 2005

[2] V. Devarapalli, R. Wakikawa, A. Petrescu, P. Thubert, “Network Mobility (NEMO) Basic Support Protocol”, IETF RFC 3963, 2005.

[3] Conta, A. and S. Deering, "Generic Packet Tunneling in IPv6 Specification", RFC 2473, December 1998.

[4] J. Manner, M. Kojo, “Mobility Related Terminology”, RFC3753, IETF, June 2004.

[5] D. Johnson, C. Perkins, Jari Arkko, “Mobility Support in IPv6”, IETF RFC 3775, 2004.

[6] Thubert, P., and Molteni, M., “Taxonomy of Route Optimization Models in the NEMO Context”, Internet Draft: draft-thubert-nemo-ro-taxonomy-00, Oct 2002

[7] C. Ng. “Analysis of Multihoming in Network Mobility Support”, IETF Draft, 2005

[8] Ernst, T. and J. Charbon, "Multihoming with NEMO Basic Support", Proceedings First International Conference on Mobile Computing and Ubiquitous Networking (ICMU), January 2004.

[9] T. Ernst, “Network Mobility Support Terminology”, IETF nemo Draft, Oct 2005

[10] Ernst, T., Montavont, N., Wakikawa, R., Paik, E., Ng, C., Kuladinithi, K., and T. Noel, "Goals and Benefits of Multihoming", draft-multihoming-generic-goals-and-benefits-02 (work in progress), October 2005

[11] Nicolas Montavont, “Multihoming in Nested Mobile Networking”, International Symposium on Applications and the Internet Workshops, 2004

[12] OMNeT++ version 3.1, Home Page http://www.omnetpp.org

# Collaborative Trust-Based Shortest Secure Path Discovery in Mobile Ad Hoc Networks*

Seungtak Oh, Chilgee Lee, and Hyunseung Choo

School of Information and Communication Engineering,
Sungkyunkwan University,
440-746, Suwon, Korea
+82-31-290-7145
choo@ece.skku.ac.kr

**Abstract.** As routes in MANET are composed of only mobile terminals based on a multi-hop mechanism, MANET may encounter severe situations where intermediate nodes disturb delivering packets adversely. Accordingly, the stability of routing protocols is very important for operating the ad hoc networks efficiently and securely. However finding a secure end-to-end path is not easy, since the nodes in MANET are characterized by uncertain trust relationship each other. In this paper, a comprehensive mechanism for discovering the most secure and shortest paths, is proposed. This proposed mechanism is based on the Dijkstra algorithm, and regards *distance weight* and *trust weight* highly. The metric value which can determine the routing path is to be extracted from the *distance value* and the *trust level*. According to the simulation results, even though the communication cost increases up to about 8%, the security level of the shortest path is improved significantly.

## 1 Introduction

Mobile Ad Hoc Network (MANET) is an infrastructureless network, consisting of a set of mobile nodes without any support of base stations or access points. A node in MANET is responsible for relaying packets to other nodes in addition to a terminal station. Then some packets transmitted from a source node can be delivered to a destination node by way of various intermediate nodes, under the store-and-forward function. Therefore, MANET can be easily applied to next generation applications such as telematics, USN and home networks, because of its flexibility and agility in establishment.

Unfortunately, in the presence of malicious nodes, MANET is highly vulnerable to attacks, due to dynamically changing topology, frequent appearance of nodes, and lack of centralized security infrastructure. An attacker can attempt to cause legitimate packets to be routed in dysfunctional ways or inject packets

* This research was supported by the Ministry of Information and Communication, Korea, under the Information Technology Research Center support program supervised by the Institute of Information Technology Assessment, IITA-2005-(C1090-0501-0019). Corresponding author: H. Choo.

V.N. Alexandrov et al. (Eds.): ICCS 2006, Part II, LNCS 3992, pp. 1089–1096, 2006.

into the network by an attack to consume valuable network resources. In the case of wormhole attack that is not easy to detect, an attacker receives packets in one part of the network and tunnels them to another part and replays them in the different part. Consequently, it is very important to realize robust and secure routing protocols.

There are basically two stems to solve these problems in MANET: the one is how to identify legitimate users, the other is how to establish secure paths. Many researches are currently interested in the user identification using certificates [4, 6, 7], but it is not easy to realize without any certificate authority(CA) infrastructure. And also several secure routing protocols have been developed so far, but it is very hard to find out the best one among these protocols because of its preliminary conditions such as sharing a secret key, requiring a certain infrastructure, and so on.

With regard to the insecurity of MANET environment, a further study on combining user identification and secure routing protocol will be intensely needed. By now, there have been some works on trust relationship so as to enhance the security level of the routing paths, but has not been any research upon how to initialize the *trust level* on each node and estimate the *trust level* of other nodes. In this paper, a solution is presented, discovering the shortest and most secure path by employing a new collaborative trust relationship mechanism. This mechanism considers the *distance weight* and the *trust weight*, and shows how to assign the initial *trust weight* and reckon the *trust level*. In simulation, it is confirmed that the proposed mechanism increases the communication cost up to about 8%, but the security level of the shortest path is improved significantly.

The remainder of this paper is organized as follows. The related works regarding secure routing protocols in MANET are presented in Section 2. In section 3, the solution is explained in detail. In Section 4, the results of simulation are discussed, and the effectiveness of the proposed mechanism is verified. Finally, Section 5 concludes this paper.

## 2 Related Works

Current efforts toward the design of a secure MANET environment have mainly focused on user identification by means of a certificate and representative secure routing protocols. SRP[1] relies on the availability of a Security Association($SA$), a shared key $K_{ST}$, between the source node($S$) and the destination node($T$), without intermediate nodes requiring cryptographic validation of control traffic. The success of this protocol depends on an efficient distribution of shared secret keys. S. Capkun and J. Hubaux introduce a realistic solution that enables the establishment of secret keys between all pairs of nodes through offline authority or fully self-organized mechanism[2]. This solution, however, has a potential risk to share a secret key with a malicious node, as the indirect user identification between a pair of nodes fully relies on the security association of its friend node. Another secure routing protocol, ARAN[3] requires the use of a trusted certificate server, and is composed of two distinct stages. The goal of the first stage is to enable the source node to verify whether the intended destination is reached or not. The second stage is primarily used for secure discovery of the shortest path. ARAN introduces authentication, message integrity and non-repudiation, but is not a suitable mechanism for an isolated MANET environment, due to the requirement for the certificate server. And SEAD routing protocol is proposed in [5]. It makes use of efficient one-way hash chains. The basic idea of SEAD is to authenticate the sequence number and the metric of a routing table update message using hash chains elements. An attacker cannot generate any value, in the hash chain of the node, which the node will use in the future as it is impossible to invert a one-way hash function. SEAD, however, cannot prevent the same distance attack where a node receives an advertisement for a particular sequence number and metric and then it readvertises the same sequence number and metric.

On the other hand, N. Pissinou, T. Ghosh and K. Makki introduce an extension of the AODV[8] routing protocol which is called T-AODV[9, 10]. When a node desires finding a path to another node, it initiates route discovery by broadcasting a Route Request packet(RREQ). The packet header contains a *trust level* field, in addition to the other fields in AODV RREQ. When an intermediate node receives the RREQ packet, it rebroadcasts the packet, after modifying the *trust level* field to include the *trust level* of the node transmitting it the RREQ. Every node verifies the rebroadcasted RREQ packet from its next node to verify transmitting of proper information. If the validation fails, it immediately broadcasts a warning message questioning the trustworthiness of that specific node. The final route selection is based upon the *trust level* metric. However, this is not sufficient in figuring out how to share the *trust level*, and calculate the *trust levels* set on each of the edges of a route path.

In [11], A. Pirzada and his group address a trust-based routing scheme. Whenever a node transmits data, it places its receiver in promiscuous mode. As soon as it hears its intermediate neighbor forwarding the packet, the sender node checks the integrity of the packet by verification process. If the integrity check passes, it confirms that the intermediate node has acted in a good manner and so its direct trust counter is increased. Similarly, if the integrity check fails or the intermediate node does not transmit the packet, its corresponding direct trust counter is decreased. In order to propagate trust information, each node appends the direct trust value of the preceding node from which it had received the RREQ packet before forwarding the packet. This mechanism can detect grey hole, black hole, and modification attacks, but has a critical flaw that most of nodes must always set its receiver in promiscuous mode, and then the information of other nodes can be easily disclosed.

## 3 Shortest Secure Path Discovery Mechanism

In Figure 1, simply applying the Dijkstra algorithm, which can search the shortest path, may undergo a serious problem such as wormhole attacked covertly by a malicious node M in MANET.

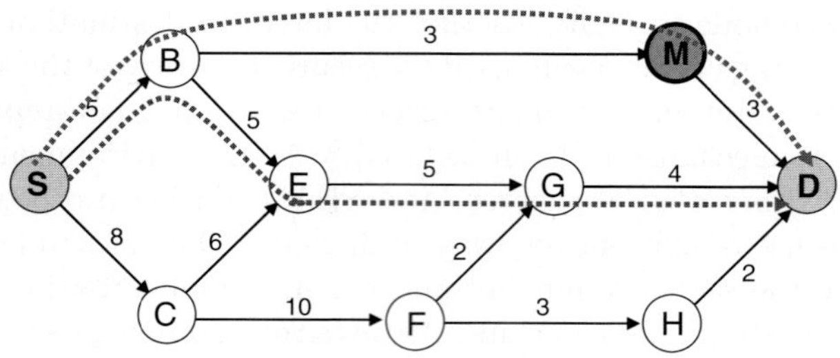


**Fig. 1.** Single-Source Shortest Paths

If the *trust level* is considered, the shortest path is established by following S → B → E → G → D, but if not, it is constructed by following S → B → M → D. Eventually, the shortest path which is composed of a malicious node can't afford a secure path absolutely. To overcome this problem, we propose a new mechanism, which can find an secure path by appending the *trust weight* to every edge among nodes, in addition to the *distance weight.* A graph *G(V,E)* represents a set of nodes and a set of edges in MANET respectively. Here a node means an element in V and an edge means a connection between two nodes containing the *distance weight* and the *trust weight.* Then the two edge elements are handled respectively. The running condition of the proposed mechanism requires that secure routing protocols, such as T-AODV, discovering the shortest path ought to have the *trust weight* information regarding the previous nodes in response packets.

### 3.1 *Trust Level*

The edge element $E$ may demonstrate the *trust weight* between a pair of nodes, and has a real number between 0 and 1. When a node attempts to connect to MANET in the proposed mechanism, initially, certificates have to be exchanged with each other through a point-to-point network, that is, secure side channel, following realistic mutual identification with neighbor nodes.

Otherwise, connection to MANET is impossible. Mutual authentication by certificates is not explained further in this paper.

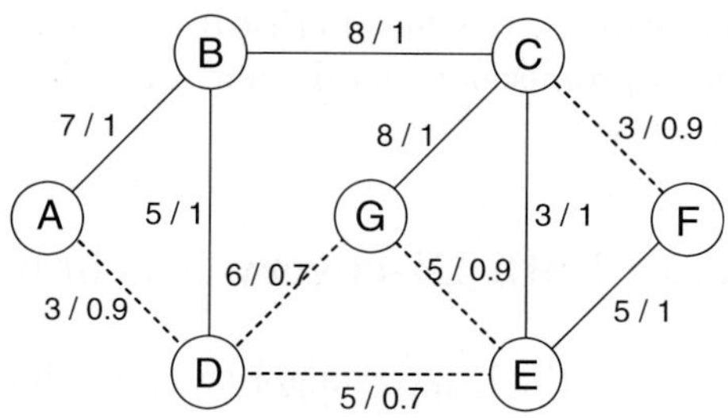


**Fig. 2.** Edges with Distance and Trust Weight

In Figure 2, all edges have two numbers that the first number means the *distance weight* and the second number the *trust weight.* And the solid line edges represent the situation that two nodes, facing each other, execute the mutual authentication directly, its *trust weight* is always 1. Otherwise, the dotted line edges mean that no mutual authentication occurs directly. Here *Trust(P)*, the *trust level* from a source node to a destination node, can be evaluated using the following equation.

$$Trust(P) = \begin{cases} tw, & \text{for } \eta = 1 \\ \frac{\min_{i=0}^{p}(tw_i)}{\ln(\eta+1)}, & \text{for } \eta > 1 \end{cases} \tag{1}$$

Here, $\eta$ represents for the number of hops. The *trust level* is always 1 on the condition that mutual authentications are executed offline and the number of hops is 1. Otherwise, the *trust level* depends on the minimum of the *trust weights* from a source node to a destination node and the number of hops. In Equation (1), the natural log, $\ln(\eta+1)$, is used when the number of hops is greater than 1, so that the *trust level* can be differentiated by the number of hops even though the same *trust level.* That is, the more the number of hops increases, the less the *trust level* decreases.

Furthermore, if several routing paths to a destination node are discovered, the path having the highest *trust level* is selected as the shortest secure path.

### 3.2 Routing Path Discovering Algorithm

In addition to *Trust(P)*, it is essential to aggregate all the *distance weights* of a path to find the shortest path. *Distance(P)*, the *distance value* from a source node to a destination node, can be defined as Equation (2).

$$Distance(P) = \sum_{i=1}^{p} dw(v_{i-1}, v_i) \tag{2}$$

$$Metric(P) = \frac{Distance(P)}{Trust(P)} \tag{3}$$

That is, $Distance(P)$ aggregates the *distance weight (dw)* from a source node to a destination node. The shortest routing path may be the path with the smallest distance value to the destination node. In addition to the distance, the proposed algorithm manages the *Trust(P)* and its weight. *Trust(P)* basically chooses the smallest *trust weight* of all the edges on a path. Eventually a new *Metric(P)*, which directs the shortest and secure routing path in Equation (3), is defined.

In Table 1, the proposed algorithm, regarding the critical functions, is described. The proposed algorithm uses the metric value in order to determine the most efficient path.

**Table 1.** Proposed Algorithm

```
Proposed(G,w,s)

Initialize-Single-Source(G,s)
d[s] ← 0
t[s] ← 1
m[s] ← 0
pred[s] ← NIL
Q ← V[G]
while Non-Empty(Q)
  do u ← Extract-Min(Q)
    S ← S ∪ {u}
    for each v ⊆ Adj[u]
      do Relax(u,v,dw,tw)
```

```
Relax(u,v,dw,tw)

if (m[u]+(dw(u,v)/tw(u,v)))<m[v]
  do d[v] ⊆ d[u]+dw(u,v)
    t[v] ⊆ min(t[u],tw(u,v))
/log(h[v]+1)
    m[v] ⊆ d[v]/t[v]
    Decrease-Key(Q,v,m[v])
    pred[v] ⊆ u
```

## 4 Performance Evaluation

The Dijkstra algorithm written in C++ language is used for simulation, and some functions are appended or modified to realize the proposed scheme. The modified program is executed a total of five times under such conditions, as the network consists of 130 nodes and 120 links. More detailed simulation environments are presented in Table 2. The *distance weight* is randomly generated from a minimum of 3 to a maximum of 52, and the *trust weight* from a minimum of 0.3 to a maximum of 1.

**Table 2.** Simulation Environments

| Times | Vertices | Edges | Paths |
|---|---|---|---|
| 1st | 130 | 503 | 124 |
| 2nd | 130 | 501 | 123 |
| 3rd | 130 | 500 | 126 |
| 4th | 130 | 503 | 127 |
| 5th | 130 | 503 | 128 |
| Total | 650 | 2,510 | 628 |

Figure 3 presents the characteristic of the number of paths in terms of the number of hops, which composes some of the simulation results. In Figure 3, we can find the phenomenon that the distribution curve of hops highly increases at approximately 4 hops of both algorithms, which have 628 paths equally, and that there is no path consisting of 8 hops or more, in the proposed algorithm. This is the reason that the proposed algorithm intends to discover the most secure path within a nearby boundary, because the more the number of hops increases, the less the *trust level* decreases.

Figure 4 shows the fact that distance average in the Gaussian distribution generally increases from 128.66 to 136.82, and then, compared with Dijkstra, the distance graph is shifted slightly toward the right. Therefore, it is expected that

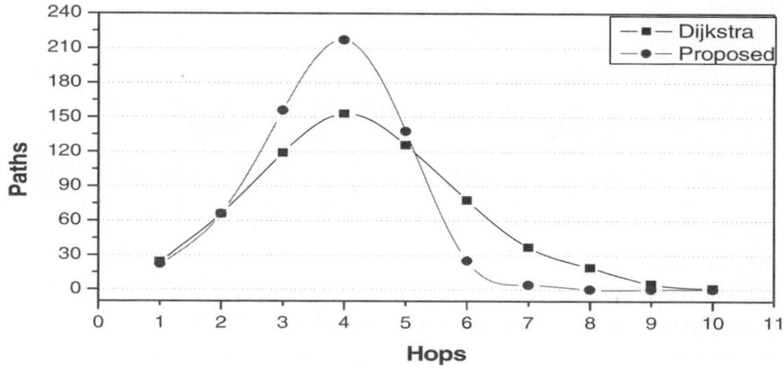


**Fig. 3.** No. of Paths in terms of the No. of Hops

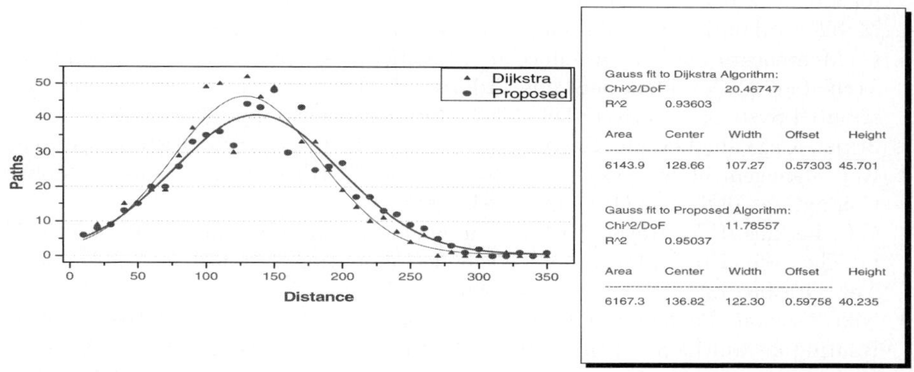


**Fig. 4.** Correlation of Distance and Paths

the increase in distance will bring the communication cost up to about 8%. However, it is possible to obtain more secure paths by adjusting the *threshold level* appropriately. For instance, if the *threshold level* is set to 0.4, only the path which its *trust level* is more than the *threshold level* is to be accepted as a secure one.

## 5 Conclusion

The proposed mechanism is located in the sublayer of secure routing protocols considering the *trust level*. In offering routing information from them, it is possible to select the most secure and shortest path from the unreliable MANET environment, by means of the ability to perceive the *trust levels* of opposite nodes. Even though the proposed mechanism works with the *trust weight* as well as the *distance weight*, it does not make any differences in a viewpoint of the number of hops, except for rising the communication cost slightly. Consequently, discovering the shortest secure paths in MANET by using the proposed mechanism, provides one of the best solutions, and basically prevents internal malicious nodes from attacking or disguising, such as in the case of a wormhole[4].

## References


1. P. Papadimitratos, Z.J. Haas, "Secure routing for mobile ad hoc networks", SCS Communication Networks and Distributed Systems Modeling and Simulation Conference(CNDS 2002), 2002.
2. Srdjan Capkun, Jean-Pierre Hubaux and Levente Buttyan, "Mobility Helps Peer-to-Peer Security", IEEE Transactions on Mobile Computing, VOL. 5, NO. 1, pp. 43-51, 2006.
3. K. Sanzgiri, B.Dahill, B.N. Levine, C. Shields, E.M. Belding-Royer, "A Secure Routing Protocol for Ad Hoc Networks", Proceedings of the 10 th IEEE International Conference on Network Protocols (ICNP02), 2002.
4. Karlof, David Wagner, "Secure routing in wireless sensor networks: attacks and countermeasures", Ad Hoc Networks 1, pp. 293-315, 2003.
5. Y.C. Hu, D.B. Johnson, A. Perrig, "SEAD: secure efficient distance vector routing for mobile wireless ad hoc networks", Proceedings of the 4th IEEE Workshop on Mobile Computing Systems and Applications(WMCSA 2002), pp. 3-13, 2002.
6. G. Montenegro and C. Castelluccia. "Statistically Unique and Cryptographically Verifiable(SUCV) Identifiers and Addresses", Proc. Symp. Networks and Distributed Systems Security(NDSS 2002), Internet Society, pp. 87-99, 2002.
7. Srdjan Capkun, Levente Buttyan, Jean-Pierre Hubaux, "Self-Organized Public-Key Management for Mobile Ad Hoc Networks", IEEE Transactions on Mobile Computing, VOL. 2, NO. 1, pp. 52-64, 2003.
8. C.E. Perkins, E.M. Royer, "Ad-hoc on-demand distance vector(AODV) routing", in: The Second IEEE Workshop on Mobile Computing Systems and Applications, New Orleans, LA, USA, February 1999.
9. Niki Pissinou, Tirthankar Ghosh, Kia Makki, "Collaborative Trust-Based Secure Routing in Multihop Ad Hoc Networks", Proceedings of Networking '04:Springer Verlag, Series:Lecture Notes in Computer Science, Vol.3042, pp. 1446-1451, Athens, Greece, May 9-14, 2004.
10. Tirthankar Ghosh, Niki Pissinou, Kia Makki, "Collaborative Trust-based Secure Routing Against Colluding Malicious Nodes in Multi-hop Ad Hoc Networks", Local Computer Networks, 2004. 29th Annual IEEE International Conference on, pp. 224-231, Nov 16-18, 2004.
11. Asad Amir Pirzada, Amitava Datta, Chris McDonald, "Trust-based routing for ad-hoc wireless networks", Networks, 2004.(ICON 2004). Proceedings. 12th IEEE International Conference on Volumn 1, pp. 326-330, Nov 16-19, 2004.

# An Efficient Neighbor Knowledge Based Broadcasting for Mobile Ad Hoc Networks*

Sung-Hee Lee and Young-Bae Ko

College of Information and Communication, Ajou University, Suwon, South Korea
{sunghee, youngko}@ajou.ac.kr

**Abstract.** Although a flooding is the most commonly used method for one-to-all communications, it often results in huge traffic in mobile ad hoc networks. Several optimization schemes have been proposed to alleviate this problem. However, none of them offers a complete solution and has their own limitations, such as, lack of reliability, high latency or excessive control packet overhead. In this paper, we propose the FONIAH (Flooding based on One-hop Neighbor Information and Adaptive Holding) scheme to overcome these limitations.

## 1 Introduction

In mobile ad hoc networks (MANET), a flooding is the simplest method to disseminate a certain message from a single node to everyone in a network. With the simplest blind flooding, any node is required to rebroadcast a receiving packet to its neighbors unless it has already received a duplicate packet. Although flooding is simple and easy to implement, it often results in a large number of network wide redundant packets. This situation is also referred in literature as the *broadcast storm* [1].

Many routing protocols and applications in MANET rely on flooding (or its variants) for a route discovery or data dissemination. Previous studies show that the reliability achieved by employing the blind flooding comes at the cost of overall network performance degradation [2]. There are several disadvantages associated with the blind flooding such as excessive redundant packets, heavy contentions and collisions [1]. To alleviate these problems, a considerable number of protocols have been presented with the same good of developing more efficient broadcasting schemes. In [3], the authors categorized these protocols into four families: Simple flooding (= blind flooding), Probability based flooding, Area based flooding, and Neighbor knowledge based flooding. *Probability based flooding* is defined to use a certain predetermined probability to decide whether each node rebroadcasts a receiving packet or not. *Area based flooding* requires a node to evaluate the additional

* This research is supported by the Ubiquitous Autonomic Computing and Network Project, the 21st Century Frontier R&D Program, and the ITRC (Information Technology Research Center) support program supervised by the IITA(Institute of Information Technology Assessment), the Ministry of Information and Communication(MIC) in Korea.

V.N. Alexandrov et al. (Eds.): ICCS 2006, Part II, LNCS 3992, pp. 1097 – 1100, 2006.

area covered by its rebroadcasting. If this additional coverage is less than some threshold value, the node will give up its rebroadcasting. *Neighbor knowledge based flooding* utilizes two hop neighbor information obtained via HELLO packets to reduce redundant broadcasting. Thus, this Neighbor knowledge method may allow a node receiving a packet to do rebroadcast, only when its rebroacasting results in any additional neighbors to be reached. As the authors in [3] argued, all these flooding approaches have their own strengths and weaknesses in terms of algorithm complexity, additional overhead, and so on.

In this paper, we take some hybrid approach combining Neighbor knowledge based flooding and Area based flooding, and propose a new algorithm called as *FONIAH (Flooding based on One-hop Neighbor Information and Adaptive Holding)*. The proposed FONIAH utilizes knowledge of 1-hop neighbors and the concept of "adaptive holding time," which is motivated from one of the area-based flooding protocol named GeoFlood [4]. In [4], the concept of "Packet holding time" is defined to allow a node receiving a packet to hold it for some amount of time before rebroadcasting. This holding time is incremental as the distance from the sender to the receiver decreases, so a node closer to the sender will become waiting longer. Although this concept can be useful to rapidly enlarge a coverage area, it may cause a high latency problem in a sparse network. Through the simulation study with ns-2, we have shown that the proposed scheme provides better efficiency without harming delivery effectiveness, compared to the existing schemes of either purely Neighbor knowledge based or Area based approaches.

## 2 The Proposed Scheme: FONIAH

We assume that each node knows its own geographical location, and periodically sends HELLO packets containing its location information. Upon receiving a HELLO packet, a node updates its neighbor table. Consequently, every node becomes to know geographical locations of all neighbors within one hop transmission radius.

In the proposed scheme, at first, a broadcasting initiator node looks up the neighbor table to find a farthest neighbor within one hop radius, and then calculates the distance from itself to the farthest neighbor. This distance is referred to as *Maximum Distance* ($d_{max}$), and it is embedded in the broadcast packet with the location of the sender node. $d_{max}$ value is utilized to calculate a packet holding time at the receiver node. This embedded information is updated by the successive next hop nodes.

On receiving a broadcast packet, a node waits for a certain holding time which is inversely proportional to the distance from the sender to itself. When a holding time expires, a node updates the value of $d_{max}$ and the location of the sender node in the packet, and then rebroadcasts it. We propose a simple linear function to obtain a packet holding time as follows:

$$HT_{cur} = HT_{max} - \frac{d_{cur} \times HT_{max}}{d_{max}} \quad (1)$$

$HT_{max}$ is the predetermined, constant maximum holding time. $d_{cur}$ is the distance from the previous hop node to the current node. It is calculated from the location of the previous hop node embedded in the receiving packet. $d_{max}$ is also embedded in the packet. A current holding time $HT_{cur}$ is a function of $d_{cur}$ and $d_{max}$. As a result, a current holding time is inversely proportional to the distance between the sender and receiver node. That is, the farthest node from the sender node rebroadcasts earlier than any others. It helps to enlarge a coverage area rapidly because outer nodes cover more new network space than inner nodes.

Since the maximum distance is fixed to the maximum transmission range in GeoFlood, a high latency problem occurs in a sparse network as we mentioned before. However, in our scheme, this value is not fixed and it is just supplied by the previous hop node via the packet. Therefore, unlike GeoFlood, the farthest neighbor always has zero holding time wherever it is placed. We refer this mechanism to the *Adaptive Holding Time*.

When a node receives a new broadcasting packet, it creates a *Mark Table* per packet. This mark table represents which neighbors have yet received the packet. Waiting for the packet holding time to expire, a node decides whether to rebroadcast a receiving packet or not, based on the following procedure. When a node receives a redundant packet during the holding time, it estimates which of its neighboring node has also received that packet. A receiver node calculates the distance from the packet sender to each own neighbor, and compares it with $d_{max}$ value embedded in the packet. If this distance is shorter or equal to $d_{max}$, this neighbor supposes to receive that packet. This process is repeated when a redundant packet arrives. Finally, before a holding time expires, if all neighbors in the mark table are marked – that is all neighbors have already received the packet, the deferred packet is dropped. Otherwise, that packet is rebroadcasted. A mark table is also removed when a holding time expires.

## 3 Performance Evaluation

We performed a simulation study using ns-2. In our simulation model, nodes are randomly deployed in $1000m \times 1000m$ square area and move according to a random movement pattern at maximum speed of $5m/s$ without any pause. A broadcast packet is initiated at every 1 second from a randomly selected node with fixed-size payload of 64 bytes which is equal to the size of typical RREQ message in AODV. We use the IEEE 802.11 MAC with a bandwidth of 2Mbps and a maximum transmission range of $250m$. In FONIAH, each node sends a HELLO packet at every 5 seconds. Total simulation time is 300 seconds and we repeat each scenario five times with different random seed numbers. We select $50ms$ of the maximum holding time for GeoFlood and FONIAH.

We evaluate our scheme using the following metrics. *Packet delivery ratio* is defined as a ratio of the number of nodes receiving the flooded packet to the number of nodes in the network. *Total packet overhead* is defined as the total size of all transmitted data and control packets. This metric reflects the total routing load involved in flooding and hence protocol efficiency. The end-to-end latency is measured as a difference in time from when the source node initiates flooding to when other nodes receives it. We report the *average latency* over all nodes.

Fig. 1(a) presents the delivery ratio as a function of the number of nodes. When the number of nodes is 20, the delivery ratio of three schemes is dropped up to 83%, because the network density is too sparse to support the network connectivity. In

Fig. 1(b), we observe that the total packet overhead in FONIAH is smaller than GeoFlood and Blind Flooding. The difference of the packet overhead becomes larger as the number of nodes increases. In Fig. 1(c), the average latency of our scheme is always smaller than GeoFlood. It represents that the adaptive holding time mechanism in FONIAH is more efficient than GeoFlood in the end-to-end packet latency. Especially, the difference between GeoFlood and FONIAH is larger in the small number of nodes, since our scheme alleviates the high-latency problem of GeoFlood in a sparse network. Even though the latency is still longer than simplest blind flooding, a reduced packet overhead is helpful to improve the network performance.

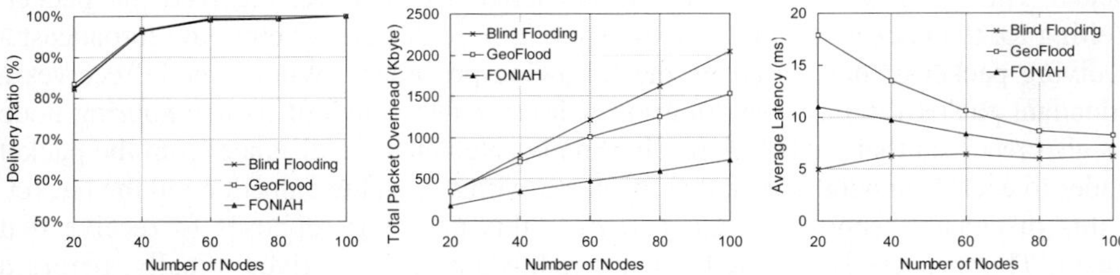


(a) Delivery Ratio (b) Total Packet Overhead (c) Average Latency

**Fig. 1.** Simulation results with respect to the number of nodes

# 4 Conclusion

The proposed FONIAH (Flooding based on One-hop Neighbor Information and Adaptive Holding) scheme provides a more efficient solution to alleviate the problem caused by blind flooding. Our scheme combines both Neighbor knowledge based flooding and Area based flooding to overcome the limitations in each approach. Simulation results shows that our scheme has low packet overhead, reasonable latency and high reliability. Future works would include some complexity analysis of the proposed algorithm and the study for further reducing its end-to-end latency.

# References

1. S.-Y. Ni, Y.-C. Tseng, Y.-S. Chen, and J.-P. Sheu, "The broadcast storm problem in a mobile ad hoc network," *in Proc. of MobiCom*, pages 151.162, 1999.
2. J. M. McCune, E. Shi, A. Perrig and M. K. Reiter, "Detection of Denial-of-Message Attacks on Sensor Network Broadcasts," in *Proceedings of IEEE Symposium on Security and Privacy*, pp. 64-78, 2005.
3. B. Williams, T. Camp "Comparison of Broadcasting Techniques for Mobile Ad Hoc Networks," *in Proc. of the third ACM international symposium on Mobile ad hoc networking & computiong*, pp.194-205, June, 2002
4. J. Arango, M. Degermark, A Efrat, S. Pink "An Efficient Flooding Algorithm for Mobile Ad-hoc Networks", *in Proc. of WiOpt*, 2004

# Maximum Lifetime Paths for the High Packet Delivery Ratio Using Fast Recovery in a Mobile Ad Hoc Network*

HyoJin Kim, SeungJae Han, and JooSeok Song

Department of Computer Science, Yonsei University, Seoul 120-749, Korea
hyojin@emerald.yonsei.ac.kr, sjhan@cs.yonsei.ac.kr,
jssong@emerald.yonsei.ac.kr

**Abstract.** Both movement of mobile nodes and path lifetime exhaustion cause to change the network topology in a mobile ad hoc network. Then, it is not easy to obtain the high packet delivery ratio. This paper proposes the Maximum Lifetime Paths (MLP) to transmit a large amount of data using nodes with long lifetime and using fast recovery. The proposed MLP establishes routes based on the shortest path and dynamically recovers routes using warning messages. The simulation experiment results show that MLP increases the packet delivery ratio and reduces the hop distance and the frequency of the route establishment.

## 1 Introduction

In a mobile ad hoc network, mobile nodes communicate with each others having the limited battery power without any infrastructure. Thus, both node movement and path lifetime exhaustion cause to change the network topology. Dynamic Source Routing (DSR) is one of routing protocols in a mobile ad hoc network. Although DSR can provide the packet salvaging and multiple routes, it usually uses only one route to transmit packets which we call to DSR1 [1]. However, the traditional routing protocols including DSR1 take only node movement into consideration. Thus, recently many researchers have studied about the path lifetime extension, for example, the Power-Aware Routing Technique (PARO) [2] and the Conditional Max-Min Battery Capacity Routing (CMMBCR) [3]. However, PARO and CMMBCR do not take fast recovery into consideration. Therefore, this paper proposes the Maximum Lifetime Paths (MLP) to transmit a large amount of data by nodes with high energies and fast recovery. The proposed MLP based on DSR establishes routes based on the minimum hops (i.e. the shortest path) and dynamically recovers routes using Route Warning (RWAR) message.

The rest of paper is as follows: Section 2 describes MLP, and Section 3 presents the performance analysis. Finally, Section 4 concludes the paper.

* This research was supported by the MIC(Ministry of Information and Communication), Korea, under the ITRC(Information Technology Research Center) support program supervised by the IITA(Institute of Information Technology Assessment).

V.N. Alexandrov et al. (Eds.): ICCS 2006, Part II, LNCS 3992, pp. 1101 – 1104, 2006.

## 2 Maximum Lifetime Paths (MLP)

Like DSR, MLP finds the shortest path from source to destination using Route Request (RREQ)-Route Reply (RREP) cycles. We choose nodes that have longer lifetime than the *Reference Point* (*RP*) to extend the path lifetime as the following steps:

**Step 1.** The source (*S*) calculates the appropriate number of nodes with the normalized node lifetime and finds the average value (*E*). *S* chooses *RP* as $E - \frac{MAX - E}{2}$ when *MAX* is the longest normalized node lifetime. Each node ($n_i$) including *S* compares the lifetime of the next node ($n_{i+1}$) with *RP*.

**Step 2.1** If the lifetime of $n_{i+1}$ is longer than *RP*, $n_i$ uses the link to $n_{i+1}$.

**Step 2.2** Otherwise, $n_i$ compares the lifetime of the following nodes such as $n_{i+2}$, $n_{i+3}$, and so forth on the primary route with *RP* and chooses a new alternative route to the chosen node.

**Step 2.2.1** If the alternative route is existed and the path lifetime is longer than *RP*, $n_i$ uses the route to forward packets. The path lifetime of the alternative route is determined by the node that has the minimum lifetime on the path. If the path lifetime of the alternative route is not longer than *RP*, $n_i$ chooses another route to the chosen node.

**Step 2.2.2** Otherwise, go to Step 2.1.

**Step 3.** *S* forwards packets using the established route. $n_i$ establishes another disjoint alternative route to $n_{i+2}$ or the following node and reserves the route if $n_i$ can.

When the lifetime of $n_{i+1}$ is shorter than *RP*, $n_{i+1}$ sends a RWAR to $n_i$. $n_i$ compares the path lifetime of the reserved route with *RP*. If the lifetime of the reserved route is longer than *RP*, $n_i$ uses the reserved route. If there is no reserved route in $n_i$ or the path lifetime of the reserved route is not longer than *RP*, $n_i$ chooses another route to $n_{i+2}$ or the following nodes such as $n_{i+3}$, $n_{i+4}$, and so on. Using the new alternative route, $n_i$ forwards the later packets. In this way, MLP protects the path lifetime exhaustion.

When $n_{i+1}$ moves out of the network during data transmissions, route maintenance is started by sending a Route Error (RERR) to $n_i$. $n_i$ forwards the later packets to the reserved alternative route immediately, establishes another disjoint route with the reserved route, and reserves the route. If there is no reserved route from $n_i$, $n_i$ forwards the RERR to *S*. In this way, MLP maintains routes against the node movement.

## 3 Performance Analysis

The packet delivery ratio (*PDR*) is defined as the ratio between the amount of packets sent by the source and received by the destination. We assume that packets are transmitted during time from $t_s$ to $t_d$ with the generating rate ($p$) and the receiving rate ($q$). When we assume that the number of nodes on the primary route in MLP is $l$, the path lifetime of the primary route ($f_1$) of MLP is as the following:

$$f_1 = \min\{i_0, i_1, i_2, i_3, \cdots, i_l\}. \tag{1}$$

When we denote $f_0$ to a failure time due to the node movement, a failure occurs in MLP as:

$$f_{MLP} = \min\{f_0, f_1\}. \tag{2}$$

Therefore, *PDR* of MLP ($PDR_{MLP}$) satisfies the following equation:

$$PDR_{MLP} = \frac{\int_{t_s}^{t_d}(q)\,dt}{\int_{t_s}^{t_d}(p)\,dt} = \frac{\int_{t_s}^{f_{MLP}}(q)\,dt + \int_{f_{MLP}}^{t_d}(q)\,dt}{\int_{t_s}^{t_d}(p)\,dt}. \tag{3}$$

We simulate 50 mobile nodes in a 1000 m by 1000 m area with 10 sources for 900 seconds focusing on *PDR* using Lucent 2 Mb/s WaveLAN 802.11 Lancard by ns-2 simulations [4]. Each node uses the random waypoint model and moves with a speed from 0 to 5 m/s. We change the pause time when every node is initially stopped. The Constant Bit Rate (CBR) sources send 512-byte User Datagram Protocol (UDP) packets with 5 packets/sec. The lifetime of each node is normalized from 0 to 1.

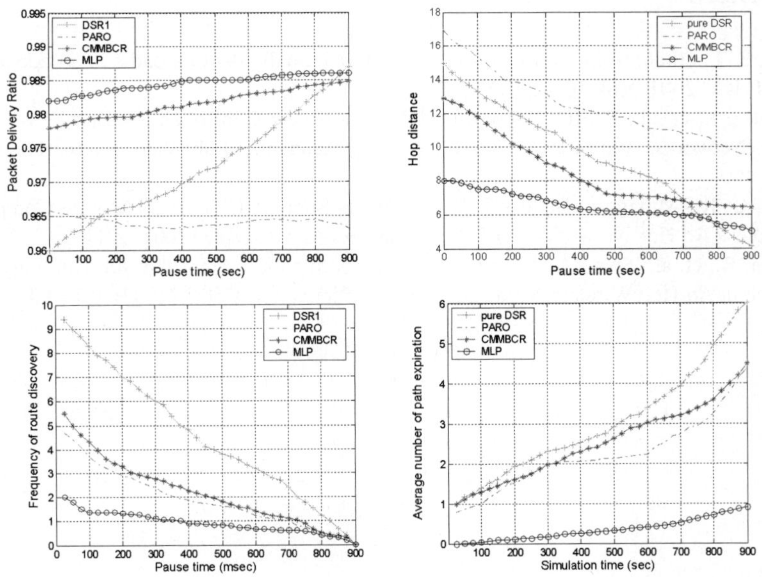


**Fig. 1.** (a) Packet delivery ratio vs. pause time (b) Hop distance vs. pause time (c) Frequency of route discovery vs. pause time (d) Average number of path expiration vs. simulation time

Fig. 1(a) shows that MLP has higher *PDR* than other protocols in high mobility. When mobile nodes are stopped, DSR1 has the highest *PDR* using the shortest path and PARO has the lowest *PDR* having additional hops for redirectors. The hop distance is defined as the number of hops that packets traverse using the shortest path before detecting a failure and the alternative route after detecting a failure. Fig. 1(b) presents that MLP has the shortest hop distance in high mobility. CMMBCR extends node lifetime but uses longer hops than MLP in high mobility because it does not provide fast recovery. In Fig. 1(c), MLP initiates the fewest route discoveries in high mobility, but DSR1 discovers routes frequently in high mobility because it does not reserve any alternative route. Both CMMBCR and PARO have to use route discovery in high mobility because they do not dynamically recover routes. Fig. 1(d) describes the average number of path expiration with varying simulation time. The path expiration is defined as the lifetime exhaustion of the path from source to destination. The paths of MLP are nearly expired contrary to DSR1.

## 4 Conclusion

This paper has proposed a new routing protocol for the high packet delivery ratio against both node movement and path lifetime exhaustion even though the proposed protocol has control overheads to establish local alternative routes dynamically. The simulation results show that the proposed protocol is more efficient than the existing protocols such as 1.26%, 2.08%, and 0.30% than DSR1, PARO, and CMMBCR.

## References


1. Lee, S.-J., et. al.: Split multipath routing with maximally disjoint paths in ad hoc networks. IEEE ICC 2001, Vol. 10. (2001) 3201-3205
2. Comez, J., et. al.: PARO: Supporting Dynamic Power Controlled Routing in Wireless Ad Hoc Networks. Wireless Networks. Kluwer Academic Publishers, Netherlands (2003) 443-460
3. Toh, C.-K.: Maximum Battery Life Routing to Support Ubiquitous Mobile Computing in Wireless Ad Hoc Networks. IEEE Communications Magazine (2001) 2-11
4. Chen, B., et. al.: Span: An Energy-Efficient Coordination Algorithm for Topology Maintenance in Ad Hoc Wireless Networks. Wireless Networks, Vol. 8 (2002) 481-494

# Author Index

Aalto, Samuli IV-420
Abawajy, J.H. IV-1015, IV-1071
Abbas, Ghulam I-802
Abbaspour, Amin III-578
Abbod, M.F. I-993
Abed, E.H. III-448
Abramson, David I-720
Absil, P.-A. I-210
Agarwal, Pankaj K. III-409
Aguado González, S. II-350
Aguilar-Saborit, Josep I-156
Ahmad, Jamil II-887
Ahn, Chunsoo I-952
Ahn, Hye-Young IV-894
Ahn, Hyun Gi I-989
Akbal, Ayhan IV-631, IV-638
Akbar, Ali Hammad II-1073
Akcelik, Volkan III-481
Akdim, B. I-372
Akioka, Sayaka I-242
Akritas, Alkiviadis II-486
Aksyonov, Konstantin A. III-879
Aldana-Montes, Jose F. III-936
Alemani, Davide II-70
Alexandrov, Vassil I-868, II-595, II-603, III-632, III-640
Alfonsi, Giancarlo I-465
Alípio, Pedro III-240
Al-khalifah, Ali I-868
Almeida, Francisco I-872
Alonso, J. III-313
Alonso, Pedro I-348
Altinakar, Mustafa Siddik II-58
Altintas, Ilkay III-69, III-920
Alves, Domingos I-1005, III-297
Amghar, T. I-603
Amorim, Ronan M. I-68
Anagnostou, Miltiades I-579, I-892
Anai, Hirokazu II-462
Ang, Chee-Wei IV-260
Angelova, Donka III-624
Anguita, M. II-518
Antaki, James IV-855
Aramudhan, M. IV-388
Argyrakis, Panos III-1048
Arora, Nitin I-16
Arrowsmith, J. Ramon III-920
Arslan, Ahmet II-247
Artoli, Abdel Monim II-78
Assous, Franck IV-623
Asthana, A. III-161
Atanassov, Emanouil III-616
Ausloos, Marcel III-1114
Avci, Mutlu IV-615
Avila, Andres I-856
Awan, Asad III-465

Babik, Marian III-980
Babuška, I. III-530
Bachman, Timothy IV-855
Baddeley, B. II-871
Bader, M. I-673
Bae, Jungsook II-1033
Bae, Sung Eun I-595
Baek, Myung-Sun II-969, II-1058
Bagheri, Ebrahim I-588
Bajaj, C. III-530
Bajuelos, António Leslie II-255
Baker, C.G. I-210
Baker, Mark II-953
Balas, Lale I-814
Balcan, Duygu III-1083
Baldridge, Kim III-69
Balicki, Jerzy III-863
Balik, Hasan H. IV-631, IV-638
Baliś, Bartosz II-542
Balla, Sudha II-822
Balogh, Zoltan III-980
Balos, Kazimierz IV-1039
Banaś, K. III-743
Bandini, Stefania III-289
Banicescu, Ioana II-430
Bao, Yukun I-728, IV-308, IV-517
Barabási, Albert-László III-417
Barbosa, Ciro B. I-68
Baronikhina, Anastasia A. III-879
Barrientos, Ricardo J. I-611
Barsky, Prof. Brian A. II-215

Bartlett, Roscoe A. IV-525
Bartocci, Ezio III-1012
Baru, Chaitan III-920
Bass, J. III-530
Battiato, Sebastiano II-334
Baumgartner, Gerald I-267
Bauschlicher Jr., C.W. III-128
Beagley, N. II-871
Becker, Michael F. I-443
Beezley, Jonathan D. III-522
Bell, Ron I-7
Belmonte Fernández, Ó. II-350
Benavent, Xaro III-13
Bergamaschi, L. IV-685
Bernabeu, Miguel O. I-348
Bernard, Julien IV-999
Bernhard, Fabrice IV-236
Bernholdt, David E. I-267
Bernier, J.L. II-518
Bernot, Gilles II-887
Bernreuther, Martin II-161
Bernstein, D.S. III-489
Bertozzi, Luigi IV-831
Berzins, M. II-147
Bhana, Ismail II-581
Bhattacharjee, Apurba K. I-387
Bianchini, Germán I-539
Bie, Rongfang II-775, IV-781
Biely, Christoly III-1067
Bieniasz, Sławomir III-759
Biros, George III-481
Biscay, R.J. I-132
Bishop, Marvin III-608
Blais, J.A.R. III-48
Blelloch, Guy E. II-799
Blower, J.D. III-996
Bonizzoni, Paola II-622
Borowski, Stefan I-196
Botana, Francisco II-470
Bourchtein, Andrei I-258
Bourchtein, Ludmila I-258
Bourgeois, Anu II-678
Branford, Simon III-632, III-640
Brendel, Ronny II-526
Breuer, Peter T. IV-765
Brill, Downey III-401
Brinza, Dumitru II-767
Briseid, Sverre IV-204
Brogan, David III-570
Brown, Martin IV-773
Browne, J.C. III-530
Brunst, Holger II-526
Bu, Jiajun I-449
Bubak, Marian II-542
Buendia, Patricia II-807
Buffle, Jacques II-70
Bunde, Armin III-1048
Bungartz, Hans-Joachim II-161
Burguillo-Rial, J.C. III-815
Bustos, Benjamin IV-196
Byun, Yanga I-276, I-284

Caballero-Gil, P. III-337
Cai, Guoyin I-292, I-876, III-1
Cai, Yang I-1001, IV-870
Caliari, M. IV-685
Calle, Eusebi IV-136
Camahort, Emilio II-287, II-310
Campos, Celso II-310
Campos, Fernando Otaviano I-68, I-76
Cannarozzi, Gina M. II-630
Cao, Chunxiang III-9
Cao, Jian III-948
Cao, Wuchun III-9
Cao, Yuanda IV-81
Čapkovič, František III-176
Cappello, Angelo IV-831
Carbonell, F. I-132
Cariño, Ricolindo L. II-430
Carnahan, Joseph III-570
Caron, David III-514
Carpenter, Bryan II-953
Carr, Nathan A. IV-228
Carvalho, Luis Alfredo V. de I-842
Carvalho, Paulo III-240
Castelló, Pascual II-263
Čepulkauskas, Algimantas II-407
Cetnarowicz, Krzysztof III-775, III-823, III-855
Chae, Kijoon II-1024
Chai, Zhilei I-1043
Chakravarty, Manuel M.T. II-920
Chambarel, André II-50
Chandrasekar, J. III-489
Chang, Chun-Hyon IV-280
Chang, F.K. III-456
Chang, Hao-Li IV-878
Chang, Kungyen I-226
Chang, Kyungbae IV-987
Chantzara, Maria I-579

Charoy, François III-976
Chatterjee, Abhijit III-77
Chaturvedi, Alok III-433
Chauve, Cedric II-783
Chen, Chun I-449
Chen, Guihai IV-404
Chen, Jianming IV-501
Chen, Juan II-646, II-904
Chen, Junliang IV-104
Chen, Lei IV-938
Chen, Ling II-646
Chen, Ming-Jen IV-184
Chen, Shudong III-1004
Chen, Su-Shing II-830
Chen, Yangzhou II-478
Chen, Yibing I-851
Chen, Yixin II-646
Chen, Yongqiang I-896
Chen, Yu-Sheng I-1026
Chen, Zhanglong I-1043
Chen, Zhengxin IV-476, IV-485
Cheng, Guang IV-144
Cheng, Haiying I-1047
Cheng, Jingde IV-797
Cheng, Junbo I-851
Cheng, Junxia I-851
Cheng, Ruixing IV-87
Cheng, Shiduan IV-128
Cheng, Wang-Cho IV-260
Chi, Hongmei IV-773
Chiang, Tzu-Chiang II-1008
Chinnasarn, Krisana I-403
Chinnasarn, Sirima I-403
Cho, Choongho IV-168
Cho, Dong-Jun II-1058
Cho, Han Wook IV-244
Cho, Insook III-1040
Cho, Jin-Woong II-1041
Cho, Keumwon III-972
Cho, KumWon IV-293
Cho, Kyu Bong II-587
Cho, Sung-Jin I-1067
Cho, Yongyun I-965, II-510
Cho, Yookun IV-946
Choi, Bumghi I-63
Choi, Hyoung-Kee II-961
Choi, Jaeyoung I-965, I-1059, III-972
Choi, Jeong-Yong IV-25
Choi, Jin-Ghoo IV-160
Choi, Jin-Hee IV-160, IV-172
Choi, Kee-Hyun III-895, III-899
Choi, Min-Hyung I-308, I-490
Choi, Seung-Hyuk I-969
Choi, Un-Sook I-1067
Chong, Kiwon IV-902
Choo, Hyunseung I-948, I-960, I-989, II-1089
Chopard, Bastien II-70, IV-653
Choppella, Venkatesh I-267
Chou, Chien-Lung I-900
Chover, Miguel II-263
Chow, Peter II-34
Chowdhury, A.K. III-161
Christov, N.D. I-697
Chu, Yuan-Sun IV-184
Chung, Min Young I-969, I-989
Chunguo, Wu I-547
Chuyi, Song I-547
Ciarlet Jr., Patrick IV-623
Cięciwa, Renata III-823
Cirak, Fehmi II-122
Ciszewski, Stanisław III-759
Čivilis, Alminas I-1034
Ciuffo, Leandro N. I-68
Clark, James S. III-409
Clarke, Bill I-218
Cocu, Adina I-172
Coen, Janice III-522
Cohen, Reuven III-1048
Cokuslu, Deniz I-571
Cole, Martin J. III-393
Cole, Murray II-929
Collier, R. III-727
Colling, D.J. III-956
Collins, Timothy M. II-807
Combes, P. IV-653
Comet, Jean-Paul II-887
Conner, Jeffery III-920
Cornford, Dan III-586
Corradini, Flavio III-1012
Cortés, Ana I-539
Costa-Montenegro, E. III-815
Cotoi, I. II-26
Cox, Simon J. III-928
Crosby, Christopher J. III-920
Crăciun, Marian Viorel I-172
Cuadrado, J.J. IV-789
Cui, Pingyuan II-478
Culebras, R. I-395
Curley, Martin I-4

Cycon, Hans L. II-1050
Czarnul, Pawel III-944
Czekierda, Łukasz III-940

Dagdeviren, Orhan I-571
da Silva, Fabrício A.B. I-1005, III-297
Dai, Wenchao I-1047
Dai, Yafei IV-412, IV-428
Dailyudenko, Victor F. I-846
Dal Negro, Marco III-264
Danelutto, M. II-937
Danilecki, Arkadiusz I-753
Darema, Frederica III-375
Darlington, John III-964
Das, Abhimanyu III-514
Das, A.K. III-161
Dauvergne, Benjamin IV-566
Davila, Jaime II-822
Dazzi, P. II-937
Degond, Pierre II-1
Deiterding, Ralf II-122
De la Cruz, H. I-132
de la Encina, Alberto II-207
Delaplace, Franck III-1056
Della Vedova, Gianluca II-622
Del Vecchio, David I-681
De Paoli, Serge IV-999
Demkowicz, L. III-530
Deng, Hui IV-17
Deussen, Oliver IV-196
Deville, Michel II-58
Dhamdhere, Kedar II-799
Dhariwal, Amit III-514
Díaz, Manuel II-912
Dickens, L.W. III-956
Dieci, Luca IV-677
Dieter, Bill I-226
Dietz, Hank I-226
Dikshit, Anupam II-830
Diller, K.R. III-530
Dimov, Ivan III-632, III-640
Ding, Koubao I-482
Ding, Shifei I-777
Ding, Wei IV-112, IV-120, IV-144
Dobnikar, Andrej III-345
Dogan, Kaan II-996
Dokken, Tor IV-204
Dondi, Riccardo II-622
Donetti, Luca III-1075
Dong, Hongbin III-216
Dong, Jin-xiang IV-839
Dong, Yabo IV-57
Dornaika, Fadi I-563
Dou, Wen-hua I-1030
Douglas, Craig C. III-393, III-522
Draganescu, Andrei III-481
Dreżewski, Rafał III-871, III-908
Drummond, Arielle IV-855
Duan, X. I-372
Dumitriu, Luminiţa I-172, II-199
Dyshlovenko, Pavel IV-599

Easterday, Scott IV-582
Efendiev, Yalchin III-393
Eleftheriou, Maria II-846
Ellis, Carla III-409
El Yacoubi, Samira III-360
Engelmann, Christian II-573
Ensan, Faezeh I-588
Eom, Young Ik IV-356
Erciyes, Kayhan I-571
Erlebacher, Gordon II-177
Erzan, Ayşe III-1083
Escrivá, Miguel II-287
Evans, Deidre W. IV-773
Ewing, Richard III-393

Falzon, Chantal T. III-82
Fan, Liangzhong II-367
Fan, Zhiping III-601
Fang, Liqun III-9
Fang, Wei III-847
Fangohr, Hans II-139
Fantozzi, Silvia IV-831
Farhat, C. III-456
Farinella, Giovanni Maria II-334
Farinelli, Simone IV-324
Fathy, M. I-744
Feixas, Miquel II-263
Felici, G. IV-460
Feller, Scott II-846
Feng, Dan I-1063, III-671, IV-396
Feng, Gang IV-645
Feng, Y. III-530
Feng, Yi IV-468
Fernández, A. III-313
Fernández de Vega, F. III-281
Fernández, J. II-518
Fernando, Terrence III-60
Ferrari, T. III-956

Fertin, Guillaume II-622, II-783
Fesehaye Kassa, Debessay IV-65
Fidanova, Stefka I-1009
Figueiredo, Renato III-546
Filatyev, Sergei III-433
Fisher, Randy I-226
Fitch, Blake G. II-846
Fjeldly, T.A. IV-607
Fladmark, Gunnar II-102
Flikkema, Paul G. III-409
Forestiero, Agostino IV-1047
Fort, H. III-313
Fortes, José III-546
Forth, Shaun A. IV-558
Fougère, Dominique II-50
Freundl, C. II-185
Fritz, Nicolas IV-200
Froelich, Wojciech III-839
Fu, Chong I-826
Fu, Karl I-1001
Fu, Qing-hua IV-878
Fu, Xuezheng II-678
Fuentes, D. III-530
Fujimoto, Richard II-41, III-425
Funika, Włodzimierz II-534, II-549
Furlan, Luciana B. I-1005
Fürlinger, Karl II-494
Fúster-Sabater, A. III-337

Gaaloul, Khaled III-976
Gagliardi, Henrique F. I-1005, III-297
Gaitán, Rafa II-287
Galante, M.A. IV-460
Galceran, Josep II-70
Gallivan, K.A. I-210
Gallos, Lazaros K. III-1048
Gálvez, A. II-414
Gamalielsson, Jonas II-879
Gan, Honghua I-204
Ganzha, Maria III-208
Gao, Lijun IV-501
Gao, Wenzhong II-430
Gao, Xiaodong III-601
Gao, Xiaoyang I-267
Gao, Zhigang IV-918
García, Víctor M. I-324
Garšva, Gintautas IV-364
Garzón, E.M. II-106
Gashkov, Igor I-912
Gasparo, Maria Grazia IV-677
Gastineau, Mickaël II-446
Gavrilenko, Vladimir I. III-89
Gay, David M. IV-525
Geist, Al II-573
Gelfand, Alan III-409
George, E. Olusegun II-694
Germain, Robert S. II-846
Gerndt, Michael II-494
Ghattas, Omar III-481
Giampapa, Mark II-846
Giannoutakis, Konstantinos M. I-506
Giering, Ralf IV-591
Gill, Ofer H. II-638, II-654
Gilmore, Stephen II-929
Girdzijauskas, Stasys IV-364
Głut, Barbara II-302
Godart, Claude III-976
Golubchik, Leana III-514
Gomes, André Severo Pereira III-97
Goncharova, Natalia V. III-879
Gong, Jian I-1022, IV-112, IV-120, IV-144
Gonnet, Gaston H. II-630
González, Daniel I-872
González, J. I-649
González, Luis III-305
González-Castaño, F.J. III-815
González de-la-Rosa, Juan-José I-316
Gopalan, B. II-871
Gopinath, K. III-679
Gore, Jay III-433
Górriz, J.M. I-234, I-316, I-356, I-395, I-649
Goscinski, A.M. IV-1015
Goualard, Frédéric I-332
Govaerts, W. II-391
Govindan, Ramesh III-514
Grabska, Ewa III-883
Graham, Richard L. II-945
Grama, Ananth III-465
Gravvanis, George A. I-506
Grivel, E. I-697
Grochowski, M. III-783
Grossfield, Alan II-846
Grunewaldt, Lars II-565
Grzech, Adam III-224
Grzesiak-Kopeć, Katarzyna III-883
Guan, Ximeng I-250
Guensler, R. III-425
Guibas, L.J. III-456

Guisado, J.L. III-281
Guitart, Jordi I-84
Guodong, Yuan I-864
Guo, Jianping I-292, I-876, III-1, III-9
Guo, Wu II-223
Gurel, Guray II-996
Gurov, Todor III-616
Gusfield, Dan II-618
Gutsev, G.L. III-128
Guzy, Krzysztof II-542

Habala, Ondrej III-980
Haffegee, Adrian II-595, II-603
Hagen, Trond Runar IV-204, IV-220
Hager, Svenja IV-340
Haines, K. III-996
Hajiaghayi, M.T. II-758
Halperin, Eran II-799
Ham, Eunmi IV-894
Han, Chang-Wook IV-862
Han, Joohyun I-965
Han, JungHyun III-40
Han, Kijun I-940, IV-180
Han, Kyungsook I-276, I-284
Han, SeungJae II-1101
Han, Sunyoung I-936, II-1081, IV-260
Han, Wook-Shin III-648
Handzlik, Piotr II-549
Harman, Mark IV-740
Harris, J. Clay III-393
Harrison, A.B. III-996
Harrison, Ken III-401
Harrison, Robert II-710
Harrison, Rodrigo III-1091
Hartell, Mark G. I-387
Hartono, Albert I-267
Hascoët, Laurent IV-566
Hassoun, Y. III-956
Havlin, Shlomo III-1048
Hayashi, Yukio III-1106
Hazle, J. III-530
He, Gaiyun I-822
He, Jieyue II-710
He, Jing II-1069, IV-509
He, Jingwu II-750
He, Yulan II-718
Hegeman, Kyle IV-228
Heo, Junyoung IV-946
Hermer-Vazquez, Linda III-546
Hernandez, Gonzalo I-856, III-1091
Hernández Encinas, L. II-438
Hicks, Rickey P. I-387
Hidalgo-Herrero, Mercedes II-207
Hill, Judith III-481
Hiller, Stefan IV-196
Hincapie, Doracelly I-920
Hiroaki, Deguchi II-490
Hirokazu, Hashiba II-490
Hluchy, Ladislav III-980
Honavar, Vasant G. III-440
Hong, Choong Seon II-1016
Hong, Jiman IV-946, IV-970, IV-991
Hong, Kwang-Seok IV-886
Hong, Min I-308, I-490
Hong, Tzung-Pei I-1026
Horie, Daisuke IV-797
Horng, Gwoboa I-1026
Hovland, Paul IV-574, IV-582
Howard, I.C. I-993
Hrach, Rudolf I-806
Hsiao, Chieh-Ming IV-757
Hsu, Chung-Chian IV-757
Hsu, Ji-Hsen II-295
Hu, Dewen I-689
Hu, Hongyu IV-81
Hu, Xiaoping I-689
Hu, Xiaoyong I-838
Hu, Yincui I-292, I-876, III-1
Huang, Changqin I-838
Huang, Guangyan II-1069
Huang, Houkuan III-216
Huang, Jin I-1051
Huang, Wayne IV-188
Huang, Wei I-728, IV-308, IV-493, IV-517
Huang, Xianglin I-997
Huang, Xin I-411
Huang, Y. III-554
Huang, Yan-Ping IV-757
Huang, Yueh-Min II-1008
Huerta, Joaquin II-310
Huh, Eui-Nam I-960
Humphrey, Marty I-681
Hunke, Elizabeth C. IV-533
Hunter, M. III-425
Huo, Mingxu I-482
Hurtado, Pablo I. III-1075
Hwang, Ho Seon I-944
Hwang, Ho-Yon IV-264
Hwang, In-Yong I-1018

Hwang, Jae-Hyun IV-172
Hwang, Tae Jin I-944
Hwang, Yoon-Hee I-1067

Iavernaro, Felice IV-724
Iglesias, A. II-383, II-414
Im, Sungbin II-992
Imielinska, Celina IV-822
Iñiguez, B. IV-607
Ipanaqué, R. II-383
Irwin, Mary Jane I-242
Iskandarani, Mohamed III-393
Iskra, K.A. III-281
Ismail, Suhaila IV-1071
Issakova, Marina I-928

Jackson, Steven Glenn II-422
Jaeger-Frank, Efrat III-920
Jain, K. II-758
Jakop, Yanto I-522
Jakubowska, Malgorzata I-498
Jaluria, Y. III-473
Jamieson, Ronan II-595, II-603
Janicki, Aleksander IV-301
Janik, Arkadiusz IV-1023
Janik, Pawel I-300
Jankowski, Robert III-56
Jarząb, Marcin IV-1039
Javaheri Javid, Mohammad Ali III-367
Jędrzejowicz, Piotr III-719
Jeon, Jun-Cheol III-329, IV-661
Jeon, Segil IV-272, IV-293
Jeong, Karpjoo IV-264, IV-293
Jeong, Taek Sang III-40
Jeong, Taikyeong T. I-443, I-761, III-105, III-113
Jeong, Yoon-Seok IV-280
Jeong, Yoon-Su I-908
Jeras, Iztok III-345
Jermann, Christophe I-332
Jhang, Seong Tae IV-979
Jhon, Chu Shik IV-979
Jia, Jinyuan II-342
Jia, Zupeng I-851
Jiang, Gangyi II-367
Jiang, Hong IV-396
Jiang, Peng I-794, IV-693
Jie, Min Seok I-108
Jimenez, J.C. I-132
Jiménez-Morales, F. III-281
Jin, Guang IV-57
Jin, Hai I-1051, IV-380, IV-1055
Jin, Shiyao I-769
Jin, Xiaogang I-140, III-1032
Jin, Xin II-775, IV-781
Jin, Yu Xuan IV-264
Jin, Yuehui IV-128
Jingqing, Jiang I-547
Johnson, Chris R. I-3, II-147, III-393
Johnson, David II-581
Johnson, John IV-188
Joneja, Ajay II-342
Jones, Anthony III-586
Jones, Edward L. IV-773
Joo, Bok-Gyu IV-260
José L. Valcarce II-470
Juhasz, Zoltan I-830
Julià, Carme I-555
June, Sun-Do II-1041
Jung, Bokrae IV-152
Jung, Eui Bong IV-244
Jung, Jason J. III-244
Jung, Myoung-Hee I-969
Jung, Seunho IV-293
Jung, Ssang-Bong II-1041
Jung, Sungwon II-1065
Jurczyk, Tomasz II-302
Juszczyszyn, Krzysztof III-224

Kaczmarek, Paweł L. I-904
Kaihuai, Qin I-864
Kaiser, Tim I-379
Kamci, A. Kerim II-996
Kaminski, Thomas IV-591
Kaminski, Wieslaw A. II-94
Kanaujia, A. III-554
Kang, Byung-Heon III-329
Kang, Byung-Su II-969, II-1058
Kang, Dazhou IV-95
Kang, Eui Chul II-371
Kang, Lishan I-340
Kang, Mikyung IV-962, IV-970
Kang, Minho IV-152
Kang, Minhyung II-977
Kar, T. I-372
Karaivanova, Aneta III-616
Karakaya, Ziya II-375
Karam, Noha Makhoul I-148
Karcanias, Nicos I-798, II-399
Karimabadi, Homa II-41

Karl, Wolfgang II-502
Karniadakis, G.E. III-538
Karpowicz, Michał III-791
Kasperska, Elżbieta I-24
Kasprzak, Andrzej I-100
Katarzyniak, Radosław III-224, III-891
Kawasaki, Yohei IV-436
Keim, Daniel IV-196
Keller, Gabriele II-920
Kempe, David III-514
Kennedy, Catriona III-562
Keriven, Renaud IV-212, IV-236
Kernan, Warnick I-179
Kharche, Rahul V. IV-558
Khattri, Sanjay Kumar I-860, II-102, II-239
Khonsari, A. I-744
Khrenov, Alexey A. III-879
Khrustaleva, Ekaterina Yu. I-117
Kim, ByungChul II-1033
Kim, Byunggi III-1040
Kim, Chang-Hun II-279
Kim, Choelmin IV-962
Kim, Dae Sun II-1016
Kim, Daehee IV-922
Kim, Dongryung III-807
Kim, Duck Bong II-371
Kim, Han-Doo I-1067
Kim, Hanil IV-962
Kim, H.-K. III-425
Kim, Hwa-sung IV-954
Kim, Hyeoncheol II-830
Kim, HyoJin II-1101
Kim, Hyung-Jun III-895, III-899
Kim, Hyunmyung IV-293
Kim, Hyunsook I-940
Kim, Hyun-Sung I-634
Kim, I.S. III-489
Kim, In-Young I-164
Kim, Jaegwan IV-152
Kim, JangSub I-956
Kim, Jinbae II-977
Kim, Jin Ho II-1016
Kim, Jinhwan IV-970
Kim, Jong G. IV-533
Kim, Jong Hwa IV-264
Kim, Juil IV-902
Kim, Jung-Hyun IV-886
Kim, Kee-Won III-329, IV-661
Kim, Ki-Hyung II-1073
Kim, LaeYoung I-1013
Kim, Mihui II-1024
Kim, Mingon IV-152
Kim, Moonseong I-30, I-38
Kim, Myungho I-1059
Kim, Sanghun II-961
Kim, Seki I-30, I-38
Kim, Seongbaeg IV-962
Kim, Soo-Kyun II-279
Kim, Sung-Ryul IV-289
Kim, Sun I. I-164
Kim, Sun-Jeong II-279, II-326
Kim, Sun Yong II-961
Kim, Taeil I-969
Kim, Tae-Sun IV-862
Kim, Tae-Wan IV-280
Kim, Ung Mo IV-356
Kim, Won-Sik III-648
Kirby, R.M. II-147
Kirley, Michael III-248
Kisiel-Dorohinicki, Marek III-831, III-839, III-908
Kitowski, Jacek IV-252
Kleijn, C.R. II-10
Klein, Dino III-608
Klie, Hector III-384
Knight, D. III-473
Knüpfer, Andreas II-526
Ko, Young-Bae II-1097
Kobayashi, Hidetsune I-924
Kolaczek, Grzegorz III-224
Kolberg, S. IV-607
Kong, Xiaohong I-514
Koo, Jahwan I-948
Korkhov, Vladimir V. I-530
Korpeoglu, Ibrahim II-996
Kosloff, Todd J. II-215
Köstler, H. II-185
Kotsokalis, C.A. III-956
Kotulski, Leszek III-887
Kou, Gang IV-476, IV-485
Kowalczyk, W. I-1071
Kowarz, Andreas IV-541
Kožaný, Jan III-711
Koźlak, Jarosław III-703
Kozlowski, Alex II-215
Kramer, Robin II-855
Kranzlmüller, Dieter II-557
Krawczyk, Henryk I-904
Krawczyk, M.J. I-665

Kreaseck, Barbara IV-582
Krefft, Bogdan I-904
Kriksciuniene, Dalia IV-316
Krishnamoorthy, Sriram I-267
Kroc, Jiří IV-847
Kryza, Bartosz IV-252
Krzhizhanovskaya, Valeria V. I-530
Krznaric, M. III-956
Kubota, Tetsuyuki II-34
Küçükosmanoğlu, Alp I-814
Kujawski, Bernard III-1024
Kułakowski, K. I-665
Kulikov, Gennady Yu. I-117, I-781
Kulikova, Maria V. I-473
Kulvietienė, Regina II-407
Kulvietis, Genadijus II-407
Kurc, Tahsin III-384
Kurz, Haymo II-86
Küster, Uwe I-196
Kwak, Jong Wook IV-979
Kwoh, Chee Keong II-718
Kwon, Jeong Ok I-977
Kwon, Tai-Gil II-1041
Kwon, Younggoo I-973

Labatut, Patrick IV-212
Laclavik, Michal III-980
Lai, Kin Keung I-790, IV-493
Lai, Poh-Chin I-884
Laidlaw, D.H. III-538
Lambert, T. I-641
Lambiotte, Renaud III-1114
Lane, Terran II-895
Lang, E.W. I-234, I-649
Langer, Malgorzata I-498
Larriba-Pey, Josep-L. I-156
Laskar, Jacques II-446
Lastovetsky, Alexey III-1008
Lau, L.C. II-758
Lavergne, Christian IV-372
Laws, Joseph IV-870
Lazarov, Raytcho III-393
Ledoux, Veerle IV-716
Lee, Bong-Keun I-908
Lee, Chilgee II-1089
Lee, Dongeun IV-922
Lee, Dong Hoon I-977
Lee, Doo-Soo I-164
Lee, Hakjoo II-1065
Lee, Hanku IV-272, IV-293
Lee, Hongseok I-960
Lee, Hyeon-Seok II-1041
Lee, Hyewon K. I-932
Lee, Hyongwoo IV-168
Lee, Hyukjoon IV-930
Lee, Hyungkeun IV-954
Lee, Hyung Su IV-910
Lee, Inbok IV-289
Lee, J. III-425
Lee, Jaemyoung I-443, I-761
Lee, Jae-Woo IV-264
Lee, Jae Yong II-1033
Lee, Joo-Haeng II-362
Lee, Joowan II-977
Lee, Ju-Hong I-63
Lee, Junghoon I-985, IV-962, IV-970
Lee, Junguck II-1065
Lee, Jysoo I-1059
Lee, Kang Woong I-108
Lee, Keon-Myung I-908
Lee, Kwangyong IV-902
Lee, Kwan H. II-371
Lee, Kyu Min III-895, III-899
Lee, Sang-Ho I-908
Lee, Sangkeon I-1059, III-972
Lee, SeoungYoung I-1018
Lee, Seung-Heon III-200
Lee, Seung-Jun I-952
Lee, Seung-Que IV-168
Lee, Sooyoung II-1065
Lee, SuKyoung I-1013, II-985
Lee, Sung-Hee II-1097
Lee, Sung-Woon I-634
Lee, Tae-Jin I-989, II-1041
Lee, Tong-Yee II-295
Lee, William III-964
Lee, Woojin IV-902
Lee, Yeung-Hak IV-862
Lee, Yuqin IV-805
Lees, Janet M. I-834
Lefeuve-Mesgouez, Gaëlle II-50
Leonard II, J. III-425
Lepp, Dmitri I-928
Leshchinskiy, Roman II-920
Leupi, Célestin II-58
Levrat, B. I-603
Lew, A.J. III-456
Lewis, Andrew I-720
Li, Bigang III-656, III-687
Li, Deng III-522

Li, Dongdong I-435
Li, Donghai IV-645
Li, Guoqing I-880, III-17
Li, Hong IV-918
Li, Hongxing IV-404
Li, Jianping IV-501
Li, Jianyu I-997
Li, Jianzhong II-662
Li, Jingtao I-896
Li, Kuan-Ching I-900
Li, Liang III-988
Li, Minglu III-948
Li, Tao I-250
Li, Wei III-522
Li, Wen-hui II-223
Li, Xiaowei II-1069
Li, Xiaowen III-9
Li, Xing IV-176
Li, Yanhui IV-95
Li, Yiming IV-599
Li, Yin I-818
Li, Ying I-1055
Li, Yong IV-73, IV-87
Li, Yuan III-440
Li, Z. III-554
Li, Zhenhua IV-404
Li, Zhong II-358
Liao, Sheng-hui IV-839
Liao, Xiaofei IV-380
Liatsis, P. III-767
Lie, Knut-Andreas IV-220
Liljeros, Fredrik III-1048
Lim, Azman Osman IV-9
Lin, Chao-Hung II-295
Lin, Chuang IV-41
Lin, Po-Feng IV-184
Lin, Woei IV-49
Lin, Yongmin III-216
Lin, Yu IV-128
Ling, Yun III-184
Linkens, D.A. I-993
Lipscomb, William H. IV-533
Lisik, Zbigniew I-498
Little, L. II-169
Liu, Bo III-593
Liu, Chia-Lung IV-49
Liu, Dingsheng I-880, III-17
Liu, Fei I-818, III-1004
Liu, Feng II-686
Liu, Jia I-449
Liu, Jing I-514
Liu, Kun III-695
Liu, Ming I-1030
Liu, Wei II-646
Liu, Weijiang I-1022, IV-120, IV-144
Liu, Xiaojian I-769
Liu, Yang IV-188
Liu, Zhaodong IV-781
Liu, Zhiyu IV-404
Lloret, I. I-316
Lluch, Javier II-287, II-310
Lo, Shih-Ching I-1038
Lodder, Robert J. III-393
Loh, Woong-Kee III-648
Loitière, Yannick III-570
López, Antonio I-555
López-Ruiz, R. III-353
Lu, Dongming IV-57
Lu, Feng I-884
Lu, Huimei IV-81
Lu, Jianjiang IV-95
Lu, Qingda I-267
Lu, Ssu-Hsuan I-900
Lu, Yijuan II-686
Lucas, Philipp IV-200
Lumbreras, Felipe I-555
Luo, Fei IV-380
Luo, Ying I-292, I-876, III-1
Luque, Emilio I-539
Lursinsap, Chidchanok II-838
Lv, Rui I-997
Lv, Song IV-396

Ma, Fanyuan I-818, III-1004
Ma, Guangsheng IV-645
Ma, Jixin II-775, IV-781
Ma, Lizhuang II-358
Ma, Min I-769
Ma, Xi-min I-826
Ma, Yan I-880
Madey, Gregory R. III-417
Mahfouf, M. I-993
Mahinthakumar, Kumar III-401
Maik, Vivek IV-922
Makowiec, Danuta III-256
Malaschonok, Gennadi II-486
Maliekal, J. II-169
Malinowski, Krzysztof III-791
Malkowski, Konrad I-242
Mamat, Ali IV-1071

Manceny, Matthieu III-1056
Mandel, Alan K. III-522
Mandel, Jan III-522
Măndoiu, I.I. II-758
Măndoiu, Ion I. II-742
Manfroi, Fairus I-68
Manzoni, Sara III-289
Mao, Zhihong II-358
Marchese, Fabio M. III-264
Marchiori, E. I-1071
Marcjan, Robert III-775
Margalef, Tomàs I-539
Marín, Mauricio I-611
Marroquín-Alonso, Olga II-207
Martínez, A. IV-685
Martins, Ana Mafalda II-255
Martyniak, J. III-956
Marzo, Jose L. IV-136
Maskell, Douglas L. I-522
Masteika, Saulius IV-332
Mastroianni, Carlo IV-1047
Mat Deris, M. IV-1071
Mateja-Losa, Elwira I-24
Matéo-Vélez, Jean-Charles II-1
Matossian, Vincent III-384
Matsukubo, Jun III-1106
Matsumoto, Noriko IV-436
Mauch, Sean P. II-122
Maurizio, Marchese I-547
Mavridis, Pavlos II-271
McCalley, James D. III-440
McCourt, Frederick R.W. II-193
McCrindle, Rachel I-868
McGough, A.S. III-956
McGough, A. Stephen III-964
McInnes, Lois Curfman I-242
Meeker, William Q. III-440
Mei, Jian IV-669
Meiron, Daniel I. II-122
Mellema, Angela III-433
Melnik, Roderick V.N. II-114
Memon, Ashraf III-920
Meng, Yu II-223
Merelli, Emanuela III-1012
Merkevicius, Egidijus IV-364
Merkulov, Arkadi I. I-117
Merschmann, Luiz II-863
Merzky, Andre III-97
Mesgouez, Arnaud II-50
Metaxas, D. III-554
Michener, William K. III-912
Michopoulos, John G. II-131, III-456
Miehe, Philipp III-120
Mihaylova, Lyudmila III-624
Miklaszewski, Wiesław III-256
Milhous, Wilbur K. I-387
Milledge, Tom II-694, II-702
Miller, Gavin S.P. IV-228
Milthorpe, Josh I-218
Min, Jun-Ki I-364
Min, Yong I-140, III-1032
Mingarelli, Angelo B. III-360
Mishra, Bud II-638, II-654
Mitrouli, Marilena II-399
Mix, Hartmut II-526
Mo, Hongwei I-997
Mochena, M.D. III-128
Möller, Kim II-565
Monfroy, E. I-641
Monnier, J. II-26
Moon, Jongbae I-1059
Moon, Sanghoon I-276, I-284
Moreira, José E. I-2
Moreno, A. I-316
Moreno-Vozmediano, Rafael IV-1031
Morimoto, Shoichi IV-797
Morisse, Karsten II-565
Morley, Chris T. I-834
Morvan, Michel III-321
Mou, Tai-yong IV-452
Mould, David II-318
Mu, Fei III-687
Mukherjee, Joy I-46
Muldoon, C. III-727
Mun, Sung-Gon I-960
Munagala, Kamesh III-409
Muñoz Masqué, J. II-438
Muñoz, Miguel A. III-1075
Munoz, Roberto III-1091
Muntés-Mulero, Victor I-156
Murugesan, K. I-457
Murugesh, V. I-457

Nagar, Atulya I-802
Nagel, Wolfgang E. II-526
Najim, M. I-697
Namachchivaya, N.S. III-448
Namiki, Takefumi II-34
Nandigam, Viswanath III-920
Narasimhan, Giri II-694, II-702, II-807

Narayanan, Babu I-16
Nassif, Nabil R. I-148
Natvig, Jostein R. IV-220
Naumov, Maxim I-258
Navarro, Gonzalo I-611
Navas-Delgado, Ismael III-936
Nawarecki, Edward III-839
Nedjalkov, Mihail III-616
Nenortaitė, Jovita I-1034
Neves, José III-240
Ng, Kam-Wing IV-1007
Nguyen, Ngoc Thanh III-208, III-224
Nichols, Daniel A. I-387
Nicole, Denis A. III-928
Nilsson, Patric II-879
No, Jaechun IV-1063
Noël, Alfred G. II-422
Nooijen, Marcel I-267
Norris, Boyana I-242
Nou, Ramon I-84
Novák, Stanislav I-806
Nowak, Leszek I-300
Nutaro, James J. IV-814
Nygaard, Jens Olav IV-204

Oberg, Carl III-514
Oden, J.T. III-530
O'Grady, M.J. III-727
O'Hare, G.M.P. III-727
Oh, Donsung II-977
Oh, Hyukjun IV-991
Oh, Jai-Boo IV-661
Oh, Seungtak II-1089
Oladunni, Olutayo O. I-188
Olanda, R. III-13
Oliveira, Rafael Sachetto I-68, I-76
Oliveira, S. II-726
Oliver, Timothy F. I-522
Olman, Victor II-855
Olsson, Björn II-879
Osguthorpe, David I-308
Ospina, Juan I-920
Ould-Khaoua, M. I-744
Overbye, T.J. III-448
Ozaki, T. I-132

Pace, Brigida IV-724
Pachter, R. I-372
Pai, M.A. III-448
Paik, Joonki IV-922
Paik, Juryon IV-356
Paik, Woojin IV-894
Pajarola, Renato B. II-371
Pak, Jinsuk IV-180
Palekar, M. III-425
Palkow, Mark II-1050
Pan, Gang I-435
Pan, Xuezeng IV-156
Pan, Yi II-646, II-710
Pang, Wei II-223
Papaioannou, Georgios II-271
Papavassiliou, Dimitrios V. I-188
Papini, Alessandra IV-677
Paprzycki, Marcin III-208
Parashar, Manish III-384
Parasuk, Vudhichai III-136
Parasuk, Waraporn III-136
Park, Chang Won IV-1063
Park, DongGook III-232
Park, Geunyoung IV-946
Park, Gwitae IV-987
Park, Gyung Leen I-985, II-587, IV-962
Park, Hong-Shik I-1018
Park, Hyungjun II-362
Park, Jaehyung I-969
Park, Kisoeb I-30, I-38
Park, Kiyong I-936, II-1081
Park, Namhun IV-168
Park, Neungsoo IV-244
Park, Sangjoon III-1040
Park, Sang Soon I-944
Park, SeongHoon I-736
Park, Sung Soon IV-1063
Park, Taehyung II-992
Park, Taesoon III-807
Park, Tae-Su I-63
Parker, Steven G. III-393
Passalis, Georgios II-271
Pasztor, Egon II-215
Paszynski, Maciej III-751
Paternoster, Beatrice IV-700
Patist, J.P. I-1071
Paul, Samit I-16
Paventhan, A. III-928
Pazo-Robles, M.E. III-337
Peachey, Tom I-720
Pecheanu, Emilia II-199
Pei, Pengjun II-734
Peláez, Ignacio I-872
Peng, Bo I-140, III-1032

Peng, Yanbing I-1022, IV-120
Peng, Yi IV-476, IV-485
Pennington, Deana D. III-912
Pereira, António II-454
Perelman, Alex II-215
Pérez, Mariano III-13
Perumalla, Kalyan II-41
Peterson, Janet II-177
Phipps, Eric T. IV-525
Pickin, Simon IV-765
Pieczykolan, Jan IV-252
Pieczynska, Agnieszka III-224, III-891
Ping, Lingdi IV-156
Pisa, Ivan T. I-1005
Pitera, Jed II-846
Pitman, Michael C. II-846
Pitzer, Russell M. I-267
Pivkin, I.V. III-538
Plastino, Alexandre II-863
Plaza, Antonio I-888, III-24
Plaza, Javier III-24
Pokrywka, Rafał III-855
Politi, T. IV-708, IV-732
Pons, Jean-Philippe IV-212
Popolizio, M. IV-708
Posse, C. II-871
Pota, Szabolcs I-830
Pouchard, Line C. IV-814
Prank, Rein I-928
Primavera, Leonardo I-465
Príncipe, José III-546
Prăjescu, Claudia II-742
Prudhomme, S. III-530
Przekwas, Andrzej IV-822
Przytycka, Teresa M. II-620
Pugliese, A. IV-732
Puglisi, Giovanni II-334
Puntonet, Carlos G. I-234, I-316, I-356, I-649
Pyle, David Leo I-403

Qian, Jixin III-593
Qian, Liang II-904
Qiao, Daji III-440
Qin, Guan III-393, III-522
Qiu, Shibin II-895
Qu, Youli III-216
Queiruga Dios, A. II-438
Quirós, Ricardo II-310

Rafe, Vahid III-578
Raghavan, Padma I-242
Rahmani, Adel Torkaman III-578
Rajasekaran, Sanguthevar II-822
Rajasethupathy, K. II-169
Ramakrishnan, Naren I-46
Ramalingam, M. III-143
Ramanujam, J. I-267
Ramasami, K. III-143
Ramasami, Ponnadurai III-153
Ramírez, J. I-234, I-356, I-395, I-649
Ramos, J.I. II-106
Ramos, Luis IV-582
Ramsamy, Priscilla II-595, II-603
Rangel-Kuoppa, Risto II-318
Ranjithan, Ranji III-401
Rasmussen, Craig E II-945
Rasúa, Rafael A. Trujillo I-324
Ravela, Sai III-497
Ravi, R. II-799
Raychaudhuri, Dipankar IV-930
Rayshubskiy, Aleksandr II-846
Redaelli, Stefano III-289
Regensburg, Henrik II-1050
Reis, Artur E. I-842
Rendell, Alistair P. I-218, II-155
Reynolds, Paul III-570
Rhee, Seung Hyong IV-930
Rhymend Uthariaraj, V. IV-388
Richard, Adrien II-887
Richardson, P.D. III-538
Ridge, Oak II-41
Ridley, A. III-489
Riensche, R. II-871
Rigopoulos, Stelios II-18
Rizzi, Romeo II-783
Roberts, Ronald A. III-440
Roch, Jean-Louis IV-999
Rodgers, G.J. III-1024
Rodrigues, Rosália II-454
Rodríguez, D. IV-789
Rogier, Francois II-1
Roh, Yong-Wan IV-886
Rojek, Gabriel III-823, III-855
Roman, Eric II-215
Romero, L.F. II-106
Romero, Sergio II-912
Ronald, Nicole III-248
Ronchieri, E. III-956
Rong, Chunming I-794, IV-693

Ros, E. II-518
Rossello, Damiano IV-324
Rossman, T. III-473
Rountev, Atanas I-267
Rouquier, Jean-Baptiste III-321
Roux, Olivier II-887
Roy Mahapatra, Debiprosad II-114
Rubio, Bartolomé II-912
Rüde, U. II-185
Ruiz, R. IV-789
Russell, A. II-758
Ryan, J. II-66
Ryan, Sarah M. III-440
Ryba, Przemyslaw I-100
Rylander, M.N. III-530
Ryu, Jeha II-610
Ryu, Seungwan II-977, II-1033, IV-168

Sadayappan, P. I-267
Safaei, F. I-744
Sagianos, E. I-798
Sahingoz, Ozgur Koray III-192, III-903
Saidane, Mohamed IV-372
Sakalauskas, Virgilijus IV-316
Sakurai, Kouichi I-977
Salinas, Luis I-856
Saltenis, Vydunas I-704
Saltz, Joel III-384
Sameh, Ahmed III-465
Sánchez, J.R. III-353
Sanchez, Justin C. III-546
Sanchez, Maribel II-686
Sancho Chust, S. II-350
Sandu, Adrian I-712, III-120, IV-550
Sanfilippo, A. II-871
Sankoff, David II-791
Santana, Miguel IV-999
Santini, Cindy I-379
Sappa, Angel I-555
Sappa, Angel D. I-563
Sarin, Vivek I-92
Saubion, F. I-603, I-641
Sauer, P.W. III-448
Sautois, B. II-391
Sbert, Mateu II-263
Schaefer, Robert III-783, III-799
Schaubschläger, Christian II-557
Schloissnig, Siegfried II-502
Schmidt, Bertil I-522
Schmidt, Thomas C. II-1050
Schneider, Adrian II-630
Schöbel, Rainer IV-340
Schwan, K. III-425
Schwartz, Jacob T. II-654
Schwartz, Russell II-799
Seber, Dogan I-379
Sedighian, Saeed III-578
Segal, Cristina I-172, II-199
Segura, J.C. I-356, I-395
Semoushin, Innokenti V. I-473
Seok, S.C. II-726
Sequeira, Adélia II-78
Serrat, Joan I-555
Seshasayee, B. III-425
Sethuraman, V. III-143
Sevinc, Bahadir IV-638
Sfarti, Dr. Adrian II-215
Shafi, Aamir II-953
Shahinpoor, Moshen II-131
Shakhov, Vladimir V. I-948
Shang, Wenqian III-216
Sharifi, Mohsen I-981
Sharma, Abhishek III-514
Shi, Hanxiao III-184
Shi, Yong IV-452, IV-476, IV-485, IV-509
Shi, Zhong-ke IV-878
Shi, Zhongzhi I-777
Shiffler, D.A. I-372
Shigezumi, Takeya II-815
Shih, Wen-Chung I-810
Shim, Choon-Bo III-232
Shin, Chang-Sun III-232
Shin, Dongchun II-977
Shin, DongRyeol I-956, III-895, III-899
Shin, HoJin I-956
Shin, In-Hye I-985
Shin, Jeongho IV-922
Shin, Jitae I-952, IV-25
Shindin, Sergey K. I-781
Shirayama, Susumu III-1063
Shu, Jiwu III-663, III-687, III-695
Sicilia, M.A. IV-789
Simha, Rahul III-679
Simutis, Rimvydas IV-332
Siwik, Leszek III-831, III-871
Słota, Damian I-24, I-786
Ślusarczyk, Grażyna III-883
Smetek, Marcin II-549
Smoliy, Elena F. III-879

Smołka, Maciej III-799
Śnieżyński, Bartłomiej III-703, III-759
So, Won-Ho III-232
Soler, Enrique II-912
Son, Hyung-Jin III-506
Son, Jeongho IV-180
Song, Byunghun IV-910
Song, Chang-Geun II-326
Song, Hyoung-Kyu II-969, II-1058
Song, In-Ho I-164
Song, JooSeok I-1013, II-1101
Song, Jungwook I-936, II-1081
Song, Mingli I-449
Song, Minseok I-1075
Song, Sung Keun II-587
Song, Yong Ho IV-244
Song, Young Seok III-105, III-113
Song, Zhanjie I-427, I-822
Sonmez, A. Coskun III-192, III-903
Soofi, M.A. III-48
Sosonkina, Masha I-54
Sottile, Matthew J. II-945
Soukiassian, Yeran I-148
Spezzano, Giandomenico IV-1047
Spiegel, Michael III-570
Sreepathi, Sarat III-401
Sridhar, Srinath II-799
Srinivasan, Kasthuri I-92
Srovnal, Vilém III-711
Stadlthanner, K. I-234
Stafford, R.J. III-530
Stagni, Rita IV-831
Stavrou, Pavlos II-271
Stauffer, Beth III-514
Stefanescu, Diana II-199
Strout, Michelle Mills IV-574, IV-582
Su, Fanjun IV-156
Su, Hui-Kai IV-184
Su, Sen IV-73, IV-87, IV-104, IV-164
Su, Xianchuang I-140, III-1032
Sudholt, Wibke III-69
Suh, Jonghyun II-1065
Suits, Frank II-846
Sukhatme, Gaurav III-514
Sun, Bing II-654
Sun, Jizhou I-419
Sun, Jun III-847
Sunderam, Vaidy I-1
Sung, Hocheol IV-260
Susitaival, Riikka IV-420
Sussman, A. III-448
Škvor, Jiří I-806
Šuvakov, Milovan III-1098
Švec, Martin I-806
Swiecicki, Mariusz I-300
Swierszcz, Pawel II-534
Swope, William II-846
Sygkouna, Irene I-892
Sykas, Efstathios I-892
Szabo, Gabor III-417
Szczerba, Dominik II-86
Székely, Gábor II-86
Szychowiak, Michał I-753

Tabakow, Iwan III-168
Tabik, S. II-106
Tadić, Bosiljka III-1016, III-1024, III-1098
Tahar, M. II-169
Taherkordi, Amirhosein I-981
Tai, Phang C. II-710
Takahashi, Tadashi I-924
Takaoka, Tadao I-595
Takeda, Kenji III-928
Taleghan, Majid Alkaee I-981
Tan, Feng II-678
Tan, X.G. IV-822
Tang, Kai II-342
Tao, Jie II-502
Tawfik, H. III-767
Tawfik, Hissam I-802, III-60
te Boekhorst, Rene III-367
Tembe, B.L. III-161
Theodoropoulos, Georgios III-562
Theoharis, Theoharis II-271
Therón, Roberto III-32
Thivet, Frédéric II-1
Thomas, Sunil G. III-384
Thurner, Stefan III-1016, III-1067
Tian, Jing IV-412, IV-428
Tian, Qi II-686
Tibiletti, Luisa IV-324
Tinnungwattana, Orawan II-838
Tiyyagura, Sunil R. I-196
Tong, Ruo-feng IV-839
Toporkiewicz, W. III-783
Torosantucci, L. IV-460
Torres, Jordi I-84
Tošić, Predrag T. III-272

Trafalis, Theodore B. I-188, III-506
Trappe, Wade IV-930
Trapp, John I-490
Trebacz, Lechoslaw II-549
Trelles, Oswaldo III-936
Triantafyllou, Dimitrios II-399
Trinh, Thanh Hai IV-1
Troya, José M. II-912
Tsai, Ming-Hui II-1008
Tsechpenakis, G. III-554
Tseng, Shian-Shyong I-810
Tsukerman, Igor I-54
Tu, Shiliang I-1043
Turek, Wojciech III-775
Turias, I. I-649
Tučník, Petr III-711
Tuzun, R. II-169

Uber, Jim III-401
Uchida, Makoto III-1063
Ufuktepe, Ünal I-916
Uhruski, P. III-783
Ülker, Erkan II-247
Uribe, Roberto I-611
Urra, Anna IV-136

Vaidya, Sheila IV-188
Vaiksaar, Vahur I-928
Valaitytė, Akvilina IV-348
Valakevičius, Eimutis IV-348
Valencia, David I-888, III-24
van Bloemen Waanders, Bart III-481
Van Daele, Marnix IV-716
Vanden Berghe, Guido IV-716
Van Hamont, John E. I-387
Vanneschi, Leonardo III-289
van Veldhuizen, S. II-10
Varadarajan, Srinidhi I-46
Vazirani, V.V. II-758
Venuvanalingam, P. III-143
Versteeg, Roelof III-384
Vialette, Stéphane II-622, II-783
Vianello, M. IV-685
Vianna, Gizelle Kupac I-842
Vidal, Antonio M. I-324, I-348
Virtamo, Jorma IV-420
Visscher, Lucas III-97
Viswanathan, M. III-200
Vivó, Roberto II-310
Vodacek, Anthony III-522
Volkert, Jens II-557
von Laszewski, Gregor III-401
Vuik, C. II-10

Wählisch, Matthias II-1050
Wais, Piotr I-300
Wajs, Wieslaw I-300
Walkowiak, Krzysztof I-618, I-626
Walther, Andrea IV-541
Wan, Wei I-292, I-876, III-1
Wan, Zheng IV-156
Wang, Chaokun II-662
Wang, Chien-Lung I-1026
Wang, Chunshan IV-468
Wang, Di III-695
Wang, Fang I-1063, III-671, IV-396
Wang, Feng III-82
Wang, Guoping I-411
Wang, Hanpin III-988
Wang, Hao II-678, IV-428
Wang, Heng I-411
Wang, Hongbo IV-128
Wang, Hsiao-Hsi I-900
Wang, Jian I-880
Wang, Jianqin III-1
Wang, Kuang-Jui I-900
Wang, Lin I-728, IV-308, IV-517
Wang, Pei-rong I-826
Wang, Qijia IV-938
Wang, Shaowei I-340
Wang, Shouyang I-790, IV-308, IV-444, IV-493, IV-517
Wang, Shuanghu I-851
Wang, Xin IV-9
Wang, Xueping I-896
Wang, Yan II-223
Wang, Yang III-663
Wang, Yueming I-435
Wang, Yufeng II-686
Wang, Zhengfang I-292
Ward, Richard C. IV-814
Ward, T.J. Christopher II-846
Wasson, Glenn I-681
Weber dos Santos, Rodrigo I-68, I-76
Wedemann, Roseli S. I-842
Wei, Anne IV-17
Wei, Guiyi III-184
Wei, Guozhi IV-17
Wei, Hu I-864
Wei, Huang I-790

Wei, Wei III-948, IV-57
Wei, Xue III-656
Weihrauch, Christian III-632, III-640
Welch, Samuel I-490
Wen, Wanzhi I-851
Wheeler, Mary F. III-384
Whitlock, P.A. III-608
Wieczorek, Damian IV-1039
Wierzbowska, Izabela III-719
Wilhelm, Reinhard IV-200
Wittevrongel, Sabine IV-65
Wojcik, Grzegorz M. II-94
Wojtowicz, Hubert I-300
Wong, Kim-Sing IV-260
Wozny, Janusz I-498
Wu, Chaolin I-292, I-876, III-1
Wu, Chin-Chi IV-49
Wu, Haiping III-656
Wu, Jianping IV-33
Wu, Kun IV-33
Wu, Meiping I-689
Wu, Song IV-1055
Wu, Yuanxin I-689
Wu, Zhaohui I-435, I-1055, IV-918, IV-938
Wu, Zhongqiang IV-748
Wu, Zhauhui IV-468

Xia, Peng III-671
Xia, Yu I-8, I-124
Xia, Yunni III-988
Xiang, Quanshuang IV-81
Xiao, Rui I-1030
Xie, Jiang IV-669
Xie, Wen IV-444
Xie, Xia I-1051
Xiong, Muzhou IV-1055
Xu, Anbang II-775
Xu, Baowen IV-95, IV-748
Xu, Chunxiang III-988
Xu, Dong II-855
Xu, Fuyin I-838
Xu, HaiGuo IV-293
Xu, Ke IV-17, IV-33
Xu, Shanying IV-444
Xu, Weixuan IV-501
Xu, Wenbo I-514, I-1043, III-847
Xu, Xian II-670
Xu, Ying II-855
Xu, Zhe I-826
Xue, Wei I-250, III-663, III-695
Xue, Xiangyang IV-9
Xue, Yong I-292, I-876, III-1, III-9

Yaşar, O. II-169
Yaghmaee, Mohammad Hossien I-588
Yaikhom, Gagarine II-929
Yan, Chung-Ren II-295
Yang, Chao-Tung I-810
Yang, Fangchun IV-73, IV-87, IV-164
Yang, Geng I-794, IV-693
Yang, Jun III-409
Yang, Kun II-662
Yang, Mao IV-412, IV-428
Yang, Mijeong I-969
Yang, Shouyuan I-427
Yang, Shuzhong I-997
Yang, Wang I-1022
Yang, Xin-She I-834
Yang, Xuejun II-904
Yang, Yingchun I-435
Yang, Young-Kyu III-200
Yang, Zhanxin I-997
Yanami, Hitoshi II-462
Yangchun, Liang I-547
Yantır, Ahmet I-916
Yao, Jialiang III-60
Yao, Yonglei IV-164
Yazici, Ali II-375
Ye, Chaoqun I-769
Yegül, Umut I-814
Yélamos, P. I-356
Yeo, So-Young II-969
Yeon, Eunja IV-894
Yi, Huizhan II-904
Yi, Sangho IV-946
Yildirim, Tulay IV-615
Yim, Soon-Bin II-1041
Yoo, Chae-Woo I-965, II-510
Yoo, Chuck IV-160, IV-172
Yoo, Kee-Young II-1000, III-329, IV-661
Yoon, Eun-Jun II-1000
Yoon, Jungwon II-610
Yoon, Seokho II-961
Yoon, Tae-Sun II-830
Yoon, Won-Sik II-1073
Yoshida, Norihiko IV-436
You, L.H. II-231
You, Mingyu I-449

You, Young-Hwan II-969, II-1058
Youn, Choonhan I-379
Youn, Hee Yong II-587, IV-1, IV-910
Youn, Jaehwan III-807
Yu, Chiu-Man IV-1007
Yu, Dongjin I-1055
Yu, Hongliang III-656
Yu, Jun III-184
Yu, Lean I-790, IV-308, IV-444, IV-493, IV-517
Yu, Mei II-367
Yu, Shao-Ming I-1038
Yu, Shaokai II-1073
Yu, Zhiping I-250
Yuan, Ding I-179
Yuan, Ruifeng IV-404
Yue, Wuyi IV-509
Yun, Hyunho IV-152

Żabińska, Małgorzata III-735
Zechman, Emily III-401
Zelikovsky, Alex II-750
Zelikovsky, Alexander II-767
Zeng, Lingfang I-1063, III-671, IV-396
Zeng, Yurong I-728
Zenger, Ch. I-673
Zhang, Aidong II-670, II-734
Zhang, Bin III-514
Zhang, Chengwen IV-104
Zhang, Shunda IV-396
Zhang, Gendu I-896
Zhang, Guang-Yan III-663
Zhang, Guiling I-419
Zhang, Hao IV-380
Zhang, Jian J. II-231
Zhang, Jianting III-912
Zhang, Jinlong I-728
Zhang, Liguo II-478
Zhang, Qianli IV-176
Zhang, Qin I-1051, IV-380
Zhang, Shensheng III-948
Zhang, Shijia I-411
Zhang, Shunda IV-396
Zhang, Wanjun III-17
Zhang, Weide I-681
Zhang, Wenju I-818, III-1004
Zhang, Wenyi III-17
Zhang, Wu I-1047, IV-669
Zhang, Y. III-530
Zhang, Yanqing II-678
Zhang, Yingzhou IV-748
Zhang, Yiying II-830
Zhao, Bo I-657
Zhao, Chen I-204
Zhao, Guiping I-851
Zhao, Guoxing II-775, IV-781
Zhao, Jun III-593
Zhao, Liqiang IV-9
Zhao, Mingxi II-358
Zhao, Peng III-1008
Zhao, Wenyun IV-805
Zheng, Bo IV-41
Zheng, Chunfang II-791
Zheng, Gaolin II-694, II-702
Zheng, Lei I-292, I-876, III-1
Zheng, Nenggan IV-938
Zheng, Weimin III-656, III-687
Zhestkov, Yuri II-846
Zhong, Shaobo I-292, I-876, III-1, III-9
Zhong, Wei II-710
Zhou, Deyu II-718
Zhou, Mingzhong IV-112
Zhou, Ruhong II-846
Zhou, Xingwei I-427, I-822
Zhou, Yanmiao IV-938
Zhou, Yi I-1055
Zhou, Yu II-367
Zhou, Zhongmei IV-468
Zhou, Zong-fang IV-452
Zhu, Haibin III-216
Zhu, Qiuping I-340
Zhu, Wei-yong I-826
Zhuang, Yueting I-204
Zieliński, Krzysztof III-940, IV-1023, IV-1039
Zory, J. IV-653
Zou, Deqing IV-1055
Zuo, Xin III-1008
Zuzarte, Calisto I-156

# Lecture Notes in Computer Science

For information about Vols. 1–3906

please contact your bookseller or Springer

Vol. 4011: Y. Sure, J. Domingue (Eds.), The Semantic Web: Research and Applications. XVIII, 723 pages. 2006.

Vol. 4007: C. Àlvarez, M. Serna (Eds.), Experimental and Efficient Algorithms. XI, 329 pages. 2006.

Vol. 4004: S. Vaudenay (Ed.), Advances in Cryptology - EUROCRYPT 2006. XIV, 613 pages. 2006.

Vol. 4003: Y. Koucheryavy, J. Harju, V.B. Iversen (Eds.), Next Generation Teletraffic and Wired/Wireless Advanced Networking. XVI, 582 pages. 2006.

Vol. 3998: T. Calamoneri, I. Finocchi, G.F. Italiano (Eds.), Algorithms and Complexity. XII, 394 pages. 2006.

Vol. 3994: V.N. Alexandrov, G.D. van Albada, P.M.A. Sloot, J. Dongarra (Eds.), Computational Science – ICCS 2006, Part IV. XXXV, 1096 pages. 2006.

Vol. 3993: V.N. Alexandrov, G.D. van Albada, P.M.A. Sloot, J. Dongarra (Eds.), Computational Science – ICCS 2006, Part III. XXXVI, 1136 pages. 2006.

Vol. 3992: V.N. Alexandrov, G.D. van Albada, P.M.A. Sloot, J. Dongarra (Eds.), Computational Science – ICCS 2006, Part II. XXXV, 1122 pages. 2006.

Vol. 3991: V.N. Alexandrov, G.D. van Albada, P.M.A. Sloot, J. Dongarra (Eds.), Computational Science – ICCS 2006, Part I. LXXXI, 1096 pages. 2006.

Vol. 3990: J. C. Beck, B.M. Smith (Eds.), Integration of AI and OR Techniques in Constraint Programming for Combinatorial Optimization Problems. X, 301 pages. 2006.

Vol. 3987: M. Hazas, J. Krumm, T. Strang (Eds.), Location- and Context-Awareness. X, 289 pages. 2006.

Vol. 3986: K. Stølen, W.H. Winsborough, F. Martinelli, F. Massacci (Eds.), Trust Management. XIV, 474 pages. 2006.

Vol. 3984: M. Gavrilova, O. Gervasi, V. Kumar, C.J. K. Tan, D. Taniar, A. Laganà, Y. Mun, H. Choo (Eds.), Computational Science and Its Applications - ICCSA 2006, Part V. XXV, 1045 pages. 2006.

Vol. 3983: M. Gavrilova, O. Gervasi, V. Kumar, C.J. K. Tan, D. Taniar, A. Laganà, Y. Mun, H. Choo (Eds.), Computational Science and Its Applications - ICCSA 2006, Part IV. XXVI, 1191 pages. 2006.

Vol. 3982: M. Gavrilova, O. Gervasi, V. Kumar, C.J. K. Tan, D. Taniar, A. Laganà, Y. Mun, H. Choo (Eds.), Computational Science and Its Applications - ICCSA 2006, Part III. XXV, 1243 pages. 2006.

Vol. 3981: M. Gavrilova, O. Gervasi, V. Kumar, C.J. K. Tan, D. Taniar, A. Laganà, Y. Mun, H. Choo (Eds.), Computational Science and Its Applications - ICCSA 2006, Part II. XXVI, 1255 pages. 2006.

Vol. 3980: M. Gavrilova, O. Gervasi, V. Kumar, C.J. K. Tan, D. Taniar, A. Laganà, Y. Mun, H. Choo (Eds.), Computational Science and Its Applications - ICCSA 2006, Part I. LXXV, 1199 pages. 2006.

Vol. 3979: T.S. Huang, N. Sebe, M.S. Lew, V. Pavlović, M. Kölsch, A. Galata, B. Kisačanin (Eds.), Computer Vision in Human-Computer Interaction. XII, 121 pages. 2006.

Vol. 3978: B. Hnich, M. Carlsson, F. Fages, F. Rossi (Eds.), Recent Advances in Constraints. VIII, 179 pages. 2006. (Sublibrary LNAI).

Vol. 3976: F. Boavida, T. Plagemann, B. Stiller, C. Westphal, E. Monteiro (Eds.), Networking 2006. Networking Technologies, Services, and Protocols; Performance of Computer and Communication Networks; Mobile and Wireless Communications Systems. XXVI, 1276 pages. 2006.

Vol. 3975: S. Mehrotra, D.D. Zeng, H. Chen, B. Thuraisingham, F.-Y. Wang (Eds.), Intelligence and Security Informatics. XXII, 772 pages. 2006.

Vol. 3973: J. Wang, Z. Yi, J.M. Zurada, B.-L. Lu, H. Yin (Eds.), Advances in Neural Networks - ISNN 2006, Part III. XXIX, 1402 pages. 2006.

Vol. 3972: J. Wang, Z. Yi, J.M. Zurada, B.-L. Lu, H. Yin (Eds.), Advances in Neural Networks - ISNN 2006, Part II. XXVII, 1444 pages. 2006.

Vol. 3971: J. Wang, Z. Yi, J.M. Zurada, B.-L. Lu, H. Yin (Eds.), Advances in Neural Networks - ISNN 2006, Part I. LXVII, 1442 pages. 2006.

Vol. 3970: T. Braun, G. Carle, S. Fahmy, Y. Koucheryavy (Eds.), Wired/Wireless Internet Communications. XIV, 350 pages. 2006.

Vol. 3968: K.P. Fishkin, B. Schiele, P. Nixon, A. Quigley (Eds.), Pervasive Computing. XV, 402 pages. 2006.

Vol. 3967: D. Grigoriev, J. Harrison, E.A. Hirsch (Eds.), Computer Science – Theory and Applications. XVI, 684 pages. 2006.

Vol. 3966: Q. Wang, D. Pfahl, D.M. Raffo, P. Wernick (Eds.), Software Process Change. XIV, 356 pages. 2006.

Vol. 3965: M. Bernardo, A. Cimatti (Eds.), Formal Methods for Hardware Verification. VII, 243 pages. 2006.

Vol. 3964: M. Ü. Uyar, A.Y. Duale, M.A. Fecko (Eds.), Testing of Communicating Systems. XI, 373 pages. 2006.

Vol. 3962: W. IJsselsteijn, Y. de Kort, C. Midden, B. Eggen, E. van den Hoven (Eds.), Persuasive Technology. XII, 216 pages. 2006.

Vol. 3960: R. Vieira, P. Quaresma, M.d.G.V. Nunes, N.J. Mamede, C. Oliveira, M.C. Dias (Eds.), Computational Processing of the Portuguese Language. XII, 274 pages. 2006. (Sublibrary LNAI).

Vol. 3959: J.-Y. Cai, S. B. Cooper, A. Li (Eds.), Theory and Applications of Models of Computation. XV, 794 pages. 2006.

Vol. 3958: M. Yung, Y. Dodis, A. Kiayias, T. Malkin (Eds.), Public Key Cryptography - PKC 2006. XIV, 543 pages. 2006.

Vol. 3956: G. Barthe, B. Gregoire, M. Huisman, J.-L. Lanet (Eds.), Construction and Analysis of Safe, Secure, and Interoperable Smart Devices. IX, 175 pages. 2006.

Vol. 3955: G. Antoniou, G. Potamias, C. Spyropoulos, D. Plexousakis (Eds.), Advances in Artificial Intelligence. XVII, 611 pages. 2006. (Sublibrary LNAI).

Vol. 3954: A. Leonardis, H. Bischof, A. Pinz (Eds.), Computer Vision – ECCV 2006, Part IV. XVII, 613 pages. 2006.

Vol. 3953: A. Leonardis, H. Bischof, A. Pinz (Eds.), Computer Vision – ECCV 2006, Part III. XVII, 649 pages. 2006.

Vol. 3952: A. Leonardis, H. Bischof, A. Pinz (Eds.), Computer Vision – ECCV 2006, Part II. XVII, 661 pages. 2006.

Vol. 3951: A. Leonardis, H. Bischof, A. Pinz (Eds.), Computer Vision – ECCV 2006, Part I. XXXV, 639 pages. 2006.

Vol. 3950: J.P. Müller, F. Zambonelli (Eds.), Agent-Oriented Software Engineering VI. XVI, 249 pages. 2006.

Vol. 3947: Y.-C. Chung, J.E. Moreira (Eds.), Advances in Grid and Pervasive Computing. XXI, 667 pages. 2006.

Vol. 3946: T.R. Roth-Berghofer, S. Schulz, D.B. Leake (Eds.), Modeling and Retrieval of Context. XI, 149 pages. 2006. (Sublibrary LNAI).

Vol. 3945: M. Hagiya, P. Wadler (Eds.), Functional and Logic Programming. X, 295 pages. 2006.

Vol. 3944: J. Quiñonero-Candela, I. Dagan, B. Magnini, F. d'Alché-Buc (Eds.), Machine Learning Challenges. XIII, 462 pages. 2006. (Sublibrary LNAI).

Vol. 3943: N. Guelfi, A. Savidis (Eds.), Rapid Integration of Software Engineering Techniques. X, 289 pages. 2006.

Vol. 3942: Z. Pan, R. Aylett, H. Diener, X. Jin, S. Göbel, L. Li (Eds.), Technologies for E-Learning and Digital Entertainment. XXV, 1396 pages. 2006.

Vol. 3941: S.W. Gilroy, M.D. Harrison (Eds.), Interactive Systems. XI, 267 pages. 2006.

Vol. 3940: C. Saunders, M. Grobelnik, S. Gunn, J. Shawe-Taylor (Eds.), Subspace, Latent Structure and Feature Selection. X, 209 pages. 2006.

Vol. 3939: C. Priami, L. Cardelli, S. Emmott (Eds.), Transactions on Computational Systems Biology IV. VII, 141 pages. 2006. (Sublibrary LNBI).

Vol. 3936: M. Lalmas, A. MacFarlane, S. Rüger, A. Tombros, T. Tsikrika, A. Yavlinsky (Eds.), Advances in Information Retrieval. XIX, 584 pages. 2006.

Vol. 3935: D. Won, S. Kim (Eds.), Information Security and Cryptology - ICISC 2005. XIV, 458 pages. 2006.

Vol. 3934: J.A. Clark, R.F. Paige, F.A. C. Polack, P.J. Brooke (Eds.), Security in Pervasive Computing. X, 243 pages. 2006.

Vol. 3933: F. Bonchi, J.-F. Boulicaut (Eds.), Knowledge Discovery in Inductive Databases. VIII, 251 pages. 2006.

Vol. 3931: B. Apolloni, M. Marinaro, G. Nicosia, R. Tagliaferri (Eds.), Neural Nets. XIII, 370 pages. 2006.

Vol. 3930: D.S. Yeung, Z.-Q. Liu, X.-Z. Wang, H. Yan (Eds.), Advances in Machine Learning and Cybernetics. XXI, 1110 pages. 2006. (Sublibrary LNAI).

Vol. 3929: W. MacCaull, M. Winter, I. Düntsch (Eds.), Relational Methods in Computer Science. VIII, 263 pages. 2006.

Vol. 3928: J. Domingo-Ferrer, J. Posegga, D. Schreckling (Eds.), Smart Card Research and Advanced Applications. XI, 359 pages. 2006.

Vol. 3927: J. Hespanha, A. Tiwari (Eds.), Hybrid Systems: Computation and Control. XII, 584 pages. 2006.

Vol. 3925: A. Valmari (Ed.), Model Checking Software. X, 307 pages. 2006.

Vol. 3924: P. Sestoft (Ed.), Programming Languages and Systems. XII, 343 pages. 2006.

Vol. 3923: A. Mycroft, A. Zeller (Eds.), Compiler Construction. XIII, 277 pages. 2006.

Vol. 3922: L. Baresi, R. Heckel (Eds.), Fundamental Approaches to Software Engineering. XIII, 427 pages. 2006.

Vol. 3921: L. Aceto, A. Ingólfsdóttir (Eds.), Foundations of Software Science and Computation Structures. XV, 447 pages. 2006.

Vol. 3920: H. Hermanns, J. Palsberg (Eds.), Tools and Algorithms for the Construction and Analysis of Systems. XIV, 506 pages. 2006.

Vol. 3918: W.K. Ng, M. Kitsuregawa, J. Li, K. Chang (Eds.), Advances in Knowledge Discovery and Data Mining. XXIV, 879 pages. 2006. (Sublibrary LNAI).

Vol. 3917: H. Chen, F.-Y. Wang, C.C. Yang, D. Zeng, M. Chau, K. Chang (Eds.), Intelligence and Security Informatics. XII, 186 pages. 2006.

Vol. 3916: J. Li, Q. Yang, A.-H. Tan (Eds.), Data Mining for Biomedical Applications. VIII, 155 pages. 2006. (Sublibrary LNBI).

Vol. 3915: R. Nayak, M.J. Zaki (Eds.), Knowledge Discovery from XML Documents. VIII, 105 pages. 2006.

Vol. 3914: A. Garcia, R. Choren, C. Lucena, P. Giorgini, T. Holvoet, A. Romanovsky (Eds.), Software Engineering for Multi-Agent Systems IV. XIV, 255 pages. 2006.

Vol. 3911: R. Wyrzykowski, J. Dongarra, N. Meyer, J. Waśniewski (Eds.), Parallel Processing and Applied Mathematics. XXIII, 1126 pages. 2006.

Vol. 3910: S.A. Brueckner, G.D.M. Serugendo, D. Hales, F. Zambonelli (Eds.), Engineering Self-Organising Systems. XII, 245 pages. 2006. (Sublibrary LNAI).

Vol. 3909: A. Apostolico, C. Guerra, S. Istrail, P. Pevzner, M. Waterman (Eds.), Research in Computational Molecular Biology. XVII, 612 pages. 2006. (Sublibrary LNBI).

Vol. 3908: A. Bui, M. Bui, T. Böhme, H. Unger (Eds.), Innovative Internet Community Systems. VIII, 207 pages. 2006.

Vol. 3907: F. Rothlauf, J. Branke, S. Cagnoni, E. Costa, C. Cotta, R. Drechsler, E. Lutton, P. Machado, J.H. Moore, J. Romero, G.D. Smith, G. Squillero, H. Takagi (Eds.), Applications of Evolutionary Computing. XXIV, 813 pages. 2006.